Air Sampling Instruments

for evaluation of atmospheric contaminants

9th Edition, 2001

Technical Editors

Beverly S. Cohen
Charles S. McCammon, Jr.

1330 Kemper Meadow Drive
Cincinnati, Ohio 45240-1634

www.acgih.org

ISBN: 1-882417-39-9

Published in the United States of America by

ACGIH
Kemper Woods Center
1330 Kemper Meadow Drive
Cincinnati, Ohio 45240-1634

Telephone: (513) 742-6163
Fax: (513) 742-3355
E-mail: publishing@acgih.org
http://www.acgih.org

Contents

PART I. THE MEASUREMENT PROCESS

Overview

Chapter 1   Sampling and Analysis of Air Contaminants: An Overview

Strategies

Chapter 2   Occupational Air Sampling Strategies

Chapter 3   Approaches for Conducting Air Sampling in the Community Environment

Procedures

Chapter 4   Particle and Gas Phase Interactions in Air Sampling

PART II. INSTRUMENTATION

General

Aerosols

Gases and Vapors

Special Topics and Applications

Dedication

In Memory of Robert Jeremy "Jerry" Sherwood

Robert Jeremy Sherwood, retired Senior Lecturer on industrial hygiene, died in a car accident on July 4, 2000, while in Queensland, Australia for the International Occupational Hygiene Association Conference, where he was to receive IOHA's highest award. Also killed were Mr. Sherwood's wife, Naomi, and another passenger, Patricia Beyger. Mr. Sherwood was 75.

Workers throughout the world are healthier as a result of Jerry Sherwood's accomplishments. He entered the field of occupational safety and health from mechanical engineering. After World War II service in the British Navy, he received his master's degree in industrial hygiene from the Harvard School of Public Health in 1950. He returned to Britain and began a career marked by pioneering research in a variety of areas, including aerosol measurements and respirator design. The concepts and practices of personal exposure measurement are largely founded upon equipment and techniques his research team developed in 1957. Among Mr. Sherwood's many contributions is the vast archive of exposure information on benzene that he collected over a period of nearly 50 years. As an industrial hygienist responsible for implementing programs at 22 refineries for Esso Europe, he developed new exposure monitoring methods and sampling equipment, as well as the first pharmacokinetic model for benzene. Later in his career, Jerry applied his experience and judgment to the formulation of exposure limit criteria as a member of the Chemical Substances TLV and BEI committees of ACGIH®.

In addition to his research, Mr. Sherwood was passionate about improving health and safety programs in developing nations. He served as an advisor and consultant to the World Health Organization, the International Labour Organization, and the United Nations Industrial Development Organization. He was an honorary member of the American Industrial Hygiene Association (AIHA), and a founding member of the British Occupational Hygiene Society (BOHS). Among his many awards and distinctions are the William P. Yant Award (1987) from AIHA, and the Meritorious Achievement Award (1994) from ACGIH. He was a former president of both BOHS (1966), and the Institute of Radiation Protection (1986). At the time of his visit to Australia, he was to be presented the Lifetime Achievement Award by the International Occupational Hygiene Association.

As he progressed on his remarkable career in occupational hygiene, Jerry Sherwood helped, challenged, and inspired countless others. Many in the United States became acquainted with him as Senior Lecturer on industrial hygiene at the Harvard School of Public Health, where he taught and directed the internship program for the industrial hygiene program. He is remembered as a gifted teacher, ready with advice and friendly guidance. Those who had the good fortune to spend time with him also learned of his utterly irreverent sense of humor. He recalled, for example, that he found it amusing that he had to fight for permission to publish a paper describing the development of the personal sampler, as it was considered ". . . demeaning to the standard of the health physics papers published by the Health Physics Division." He also noted that the payment he received of $1 for assigning the patent rights for the personal sampler to the Atomic Energy Authority was ". . . rather less rewarding than one might have hoped. . . ." Mr. Sherwood's legacy includes the hygienists, physicians, nurses, and other health professionals who are practicing throughout the world today using his approach. Jerry Sherwood mentored those who were his students, and all those who employ the method he pioneered in their own practice. Jerry Sherwood will be missed.

Acknowledgments

The first compendium of air sampling instruments was the *Encyclopedia of Instrumentation for Industrial Hygiene*, published in 1956 by the University of Michigan, Institute of Industrial Health, Ann Arbor, Michigan. The *Encyclopedia* contained descriptive information on instruments exhibited at a symposium on "Instrumentation for Industrial Hygiene," held at the Institute in May 1954.

Air Sampling Instruments was first published by the American Conference of Governmental Industrial Hygienists (ACGIH®) in 1960 as a successor to the *Encyclopedia*. Subsequent editions of *Air Sampling Instruments* appeared in 1962, 1967, 1972, 1978, 1983, 1989, and 1995. These volumes provided the basis for this edition, and the efforts of their authors are gratefully acknowledged.

This edition of *Air Sampling Instruments* was produced through the cooperative efforts of the members of the Air Sampling Instruments Committee of ACGIH, with the invaluable assistance of other industrial hygienists and air quality specialists in this country and abroad. This international pool of expertise and experience ensured the successful, responsive revision that is the 9th Edition.

The Committee wishes to recognize the significant contributions of authors from outside the Committee: Sidney C. Soderholm, Ph.D.; Bean T. Chen, Ph.D.; John G. Watson, Ph.D.; Harriet A. Burge, Ph.D.; Richard H. Brown, Ph.D.; Kenneth Rubow, Ph.D.; Constantinos Sioutas, Sc.D.; Barbara J. Turpin, Ph.D.; Judith C. Chow, Sc.D.; Lee Kenny, Ph.D.; Da-Ren Chen; Jeff Bryant, M.S., CIH; John N. Zey, M.S., CIH; Charles E. Billings, Ph.D.; and Maire S.A. Heikkinen, Ph.D.

Many manufacturers and distributors provided literature and photographs for the instruments described here. We appreciate their cooperation. Many of the illustrations were taken from technical journals and books, as noted in the figure captions. We thank the publishers for their generally prompt response in granting permission for their reproduction.

Finally, we thank the many unnamed institutions that have provided support, and the many unnamed colleagues of the Committee members who lent their services, suggestions, and encouragement in the production of this manual.

Air Sampling Instruments Committee Members

Yung-Sung Cheng, Ph.D.
Beverly S. Cohen, Ph.D.
Melvin W. First, Sc.D., CIH
Berenice Goelzer, MPH, CIH
Martin Harper, Ph.D., CIH
Susanne V. Hering, Ph.D.
Judson L. Kenoyer, MS, CIH, CSP
David K.N. Leong, Ph.D., CIH
Paul J. Lioy, Ph.D.
Morton Lippmann, Ph.D., CIH
Dale A. Lundgren, Ph.D.
Janet M. Macher, Sc.D., MPH
Charles M. McCammon, Jr., Ph.D., CIH
Lee E. Monteith, MS, CIH
Owen R. Moss, Ph.D.
John Palassis, CIH, CHMM, CSP
David Y.H. Pui, Ph.D.
James C. Rock, Ph.D., CIH, PE
Mary Lynn Woebkenberg, Ph.D.
Lori A. Todd, Ph.D. (consultant)

Foreword

Air Sampling Instruments is a comprehensive guide to the sampling of airborne contaminants. It addresses both occupational and environmental air sampling issues, and presents measurement methods for both gaseous and particulate air contaminants. In addition, this guide describes available air sampling instruments, and provides information for their use.

This 9th Edition is divided into two major areas: "Part I. The Measurement Process" and "Part II. Instrumentation." Each part has been further refined into specific topical sections, thus facilitating the learning/information process, and access to the appropriate data.

"Part I. The Measurement Process" encompasses four sections and 11 chapters. The first section presents an overview of the sampling and analysis of air contaminants, while the second discusses occupational and environmental air sampling strategies. Air sampling procedures are the focus of the third section, including particulate and gas phase interactions, size-selective health hazard samplers, and measurement and presentation of aerosol size distributions. The fourth section is devoted to instrument calibration and quality assurance. It covers airflow, gas and vapor samplers, and aerosol samplers. This section also examines precision, accuracy, and validity in measurements as well as performance testing criteria.

"Part II. Instrumentation" offers 12 chapters covering four main topic areas. In the first section, air movers and samplers are discussed. The second section is devoted to aerosols, and covers such topics as filters and filter holders; impactors, cyclones, and other inertial and gravitational collectors; electrostatic and thermal precipitators; and direct-reading instruments. Gases and vapors are the subject of the third section. Topics include sample collectors; detector tubes, direct-reading passive badges, and dosimeter tubes; and direct-reading instruments. The final section addresses special topics and applications. Areas of discussion are denuder systems and diffusion batteries, sampling from ducts and stacks, sampling airborne radioactivity, and sampling airborne microorganisms and aeroallergens, and direct-reading Fourier transform infrared spectroscopy for occupational and environmental air monitoring.

The instrument descriptions in Chapters 12–23 contain tables in addition to the individual instrument descriptions and photographs. These tables provide a concise list of available air sampling instruments and their major features. They are intended as a guide and supplement to the individual instrument descriptions. The descriptions are numbered and cross-referenced to the tables. Commercial vendors are listed in the concluding table of these chapters.

The instrument information in this manual was assembled based on literature submitted by the manufacturers. Every effort was made to assure that the data presented are factual and correct; however, the Air Sampling Instruments Committee does not assume responsibility for inaccuracies or false claims for the instruments described herein. Similarly, caution should be exercised with regard to calibration values that may be supplied. The Committee has not attempted to check the accuracy of instruments described in this manual. Furthermore, instrument calibrations can change due to use, handling, shipment, and use under conditions other than those assumed by the manufacturer. The user must take the responsibility for checking the calibrations over the range of concentrations and conditions for which the instrument is to be used.

The Committee also wishes readers to be aware that mention of company names or products does not constitute endorsement by any federal agency with which the authors may be affiliated. Among these entities are the Centers for Disease Control and Prevention and the Department of Energy.

The preparation of *Air Sampling Instruments* is a continuing activity of the ACGIH® Air Sampling Instruments Committee. The Committee asks for your comments to improve this volume. Information would be appreciated on new instruments or new uses for instruments, as well as corrections, omissions, or inaccuracies.

Chapter 1

Sampling and Analysis of Air Contaminants: An Overview

Melvin W. First, Sc.D.

Harvard University, School of Public Health, Boston, Massachusetts

CONTENTS

Introduction

The predecessor publication of *Air Sampling Instruments* is the *Encyclopedia of Instruments for Industrial Hygiene*, a compilation of papers presented at a conference entitled "Symposium on Instrumentation in Industrial Hygiene," which was held at the University of Michigan, Ann Arbor, May 24–27, 1954.[1] An interesting feature of the Conference and the encyclopedia (and a harbinger of what was to come) was a large section on "home made", i.e., non-commercially available air sampling instruments. The scope of *Air Sampling Instruments* has been enlarged with each succeeding edition to encompass additional environmental health aspects. Fortunately, the basic principles of air sampling, and much of the equipment that is used, are common to all aspects, making it possible to keep the size of this book within reasonable bounds.

Air sampling equipment intended for evaluation of airborne exposures has undergone marked evolution over the past several decades in the direction of miniaturization and automation. The trend has been especially conspicuous for equipment intended for long period personal sampling in the workplace and in the indoor environment, and it has been made possible by developments in diffusive sampling (often referred to as passive sampling) and in customized microelectronic circuitry. So universally has personal sampling been adopted as the most acceptable way to evaluate human exposures to airborne contaminants that it is a surprise to realize that the concept and equipment were first introduced in 1960 by Sherwood and Greenhalgh[2] who stated that,

"The personal air sampler has been developed to permit more precise assessment of the average air concentration to which individuals are exposed."

The trend toward miniaturization of sampling equipment has also been assisted greatly by enormous improvements in the smallest quantity measurable by modern analytical methods (permitting satisfactory analytical procedures on microgram, and, in many cases, nanogram quantities of air contaminants) and made desirable by the vast array of sampling devices required to cope with an ever-increasing number and variety of chemical and biological substances of health significance. For example, the first list of Threshold Limit Values (TLVs®) for Chemical Substances in the Workplace published by the American Conference of Governmental Industrial Hygienists (ACGIH®) in 1951 contained 162 substances; in 2000, the list encompasses more than 700 chemicals.[3]

By contrast with the shrinking size of personal sampling devices, stationary monitoring equipment used for radiation and air pollution measurements has grown. This has been most noteworthy with respect to instruments that incorporate size-selective inlets as well as the automatic analytical instruments that draw in samples continuously, perform many complex analytical steps automatically, instantly print the results, and store them for a permanent record. Many are capable of displaying cumulative average concentrations at any time this information is called for. These kinds of analytical instruments not only conserve highly skilled manpower, they may even be essential for monitoring critical exposures having a well-defined ceiling limit designation in the ACGIH list of TLVs, e.g., hydrogen cyanide and ethylene glycol.[3]

In addition, many automatic continuous air monitoring instruments perform functions that are impossible with older instruments and methods. For example, automatic particle sampling, counting, and sizing instruments, discussed in detail in Chapter 15, make it possible to examine airborne particles that have not been subjected to agglomeration or shattering in the sampling and analytical operations and to measure all the parameters that are needed to evaluate dust exposures. It is often exceedingly difficult to distinguish sampling from analytical operations in these complex automatic recording instruments, but, to the extent possible, it is desirable to do so to understand the exact nature of the results they display.

The sharp rise in the cost of energy during the 1970s resulted in structural tightening and drastic reductions in fresh air exchanges in residential and commercial buildings, whether ventilated by natural or

mechanical means. As a direct consequence, indoor air pollutants that were formerly diluted and flushed out as they evolved now become concentrated. Sometimes, they produce acute discomfort, and a number may induce chronic diseases having a chemical (formaldehyde), radioactive (radon), or biological (mold spores) source. Diffusive samplers have found wide use for estimating levels of formaldehyde and radon in residences. The householder exposes the unit and returns it to the vendor at the conclusion of the recommended exposure period for analysis and a report.

Active sampling by professionals is more usual in commercial buildings experiencing what is popularly referred to as "Sick building syndrome." Because such buildings are often equipped with heating, cooling, and humidifying facilities that represent a suitable habitat for a variety of molds and other microorganisms that release spores to the ventilation air when the systems cycle from wet to dry operation, microbiological sampling is likely to be undertaken in such facilities in addition to a search for irritants such as formaldehyde, systemic poisons such as mixed volatile organic compounds (VOCs), and indicators of deficient air exchange such as carbon dioxide. Air sampling for microbiological agents is an important activity in hospitals, microbiological laboratories, and research institutes, as well as in the vicinity of water cooling towers that may harbor Legionnaire's Disease bacteria. In response to these needs, a number of new microbiological sampling instruments have been developed and commercialized that impact airborne organisms directly onto culture media in a state ready for incubation. Details of instruments suitable for sampling and analyzing airborne microorganisms are described in Chapter 22.

Sampling and analysis of work atmospheres are simplified by two factors. First, industrial hygienists usually know the contaminants present in the workroom air from the nature of the process plus a knowledge of the raw materials, end products, and wastes. Therefore, identification of workroom contaminants is rarely necessary and, as a rule, only quantification is required. Second, usually, only a single contaminant of importance is present in the workroom atmosphere and, in the absence of obvious interfering substances, great simplification of procedures is possible. Nonetheless, one must be on guard continually to detect the presence of subtle and unsuspected interferences that may take the form of trace substances affecting color development or the shade of indicator dyes. By contrast, community air sampling, indoors and out, is made complex by the simultaneous presence of many substances of concern, plus the much lower concentrations that prevail compared to those observed in workroom atmospheres. A notable example is the need to identify and quantify specific components of total outdoor hydrocarbons for health assessment purposes.

Other reasons for sampling air include routine surveillance and evaluating the effectiveness of engineering control measures and process changes. The most frequent occupational health purpose is "to measure the dose of the hazardous agent absorbed by the worker at his place of work. This means that the assessment of the environment is not just an exercise in physical or chemical analysis but has its base in the biological characteristics of man, and the relevance of the results depends on the adequacy of the 'biological calibration' of the analytical procedures."[4] In addition, sampling is conducted to determine compliance with occupational and community air regulations or commonly accepted standards. Epidemiology of diseases of environmental origin and many other areas of research associated with environmental health are dependent on accurate evaluations of both occupational and nonoccupational exposures to toxic substances. Real-time sampling (with videotaping) may be used effectively to locate and evaluate sources and poor work practices, and to assist in the engineering design of work stations.[5]

Nature of Air Contaminants

Contaminants may be divided into a few broad categories depending on physical characteristics.

Gases and Vapors

Gases are fluids that occupy the entire space of their enclosure. They require increased pressure and decreased temperature for liquification, e.g., hydrogen sulfide and carbon monoxide, whereas vapors are the evaporation products of substances that are also liquid at normal temperatures and pressure, e.g., water and methanol. The sole reason for making the distinction is because in many instances they are collected by different devices, although thermodynamically they behave similarly.

Particulate Matter

Sampling considerations make it convenient to characterize particulate matter by size and phase (solid or liquid) as well as by chemical composition. Whether a particle is solid or liquid is important in determining its behavior in aerosol samplers and particle size is an important factor for evaluating deposition in the lung and transport in the environment.

Dusts are solid particles formed from inorganic or organic materials reduced in size by mechanical processes such as grinding, crushing, blasting, drilling, and pulverizing. These particles range in size from the visible to the submicroscopic, but the principal concern of industrial hygienists is with those below 5 μm because they reach the deepest parts of the lung and remain suspended in the atmosphere for a long period of time. Fibers are a special subcategory of dusts because of the effect of shape on their aerodynamic and toxic properties. Fiber exposures continue to be sampled by methods that permit evaluation by count, whereas other pneumoconiosis-producing dusts are sampled by methods that permit exposure evaluation by mass.

Fumes are fine particles formed from solid materials by evaporation, condensation, and gas phase molecular reactions. When heated, metals such as lead produce a vapor that condenses in the atmosphere to form metallic particles that oxidize, e.g., to lead oxide. These particles are usually less than 0.05 μm when formed. They may agglomerate and grow up to 1.0 μm in diameter. Solid organic materials, such as waxes (chlorinated naphthalenes), can form fumes by the same method. Current interest in atmospheric pollution studies has focused on the mass of particles in the size range below 2.5 μm, which includes the fraction formed by gas phase reactions in the atmosphere. Their prevalence can be correlated with specific health effects.

Smokes and *soot* are products of incomplete combustion of organic materials and are characterized by optical density. The size of smoke particles is usually less than 1.0 μm in diameter.

Liquid particles are produced by atomization or by condensation from the gaseous state. Droplets formed from atomization are generally greater than 5 μm in diameter. Industrial spray processes produce droplets up to several hundred μm in diameter. Condensation of low volatility organic and inorganic species usually produces submicrometer aerosols, e.g., concentrated H_2SO_4. High boiling organic liquids, such as refined petroleum oils, are sometimes found as fine particles in the workroom atmosphere. Photochemical reactions in smoggy atmospheres lead to the production of secondary aerosols containing droplets below 0.25 μm.

Odors

In some instances, the amount of material in the atmosphere is so small that it is only detectable by odor and the environmental health specialist should neglect no opportunity to exercise and sharpen the sense of smell. Because many substances of industrial hygiene importance have well-defined odor and irritation (nose, eyes) thresholds, the experienced hygienist is often able to distinguish nonacceptable air concentrations on this basis alone. Indeed, certain of the ACGIH TLVs are based on the criteria of eye and nose irritation or unpleasant odor. For example, the TLV and air quality standard for ozone is only a little above the odor threshold. Therefore, it may be concluded that, when this compound is detectable by odor, it would be desirable to test the atmosphere chemically to determine whether a standard is being exceeded.[6] Some substances, notably H_2S and ozone, rapidly anesthetize the odor receptors and for these substances, absence of odor is not a criterion of safety after the first few seconds of exposure.

Sampling Considerations

The volume of sample to be collected is dependent on an estimate of the amount of material to be found in the atmosphere, the sensitivity of the analytical method, and the hygienic or air quality standard. When dealing with occupational exposures, for example, sufficient sample must be collected for a reliable estimation of a concentration equal to one-half the hygienic standard, i.e., the proposed "action level"; however, a method capable of reliably estimating at least one-tenth the hygienic standard is preferable.

Personal and Area Sampling

Ideally, one wishes to characterize the environment in the breathing zone of individuals to evaluate their specific exposures. Passive dosimeters and compact, battery-operated personal sampling devices for particulate matter (filters) and gases (absorbers and adsorbers) are especially useful for monitoring those who move from place to place and engage in a variety of activities that involve interaction with different amounts of air contaminants. Sampling strategies for evaluating occupational exposures are contained in Chapter 2.

Formerly, the most frequent method of evaluating occupational exposure was to measure workroom contamination in the vicinity of workers at about the elevation of the breathing zone, but workroom sampling introduces uncertainty when evaluating the precise exposure of the most exposed worker and, therefore, fails to comply with Occupational Safety and Health Administration (OSHA) requirements.[7] Area sampling systems continue to have utility in workplaces to monitor the effectiveness of engineering controls. Short period personal samples are used to measure maximum exposures and, when excessive levels are found, to help determine which machine or part of a process is responsible so that corrective actions may be taken.

As discussed in Chapter 3, fixed station sampling has been the predominant mode for outdoor environmental measurements up to the present. Over the past several years, a number of important studies have been conducted using 24-hr personal samplers to combine exposures to indoor and outdoor air pollutants into an integrated personal daily exposure level. Being a natural time-weighted average of outdoor, work or school, commuting, and residential conditions, this sampling method is believed to yield a better estimate of total exposure for epidemiologic purposes than widely spaced, fixed-station outdoor air monitors.[8] When combined with single source differential sampling at home, at work, etc., it becomes possible to identify the major contributors for purposes of control.

Small personal sampling instruments have been developed that incorporate a particle size-selective inlet with a collection stage for respirable particulate matter plus a stage for trapping gases and vapors. They make it increasingly possible to measure round-the-clock exposures to a wide variety of indoor and outdoor air contaminants. For example, a study of the exposure of children of elementary school age to environmental tobacco smoke used 24-hr personal samples collected in a multistage instrument consisting of a 10-mm nylon cyclone preseparator and a tared 37-mm Fluoropore filter for the respirable particulate fraction, followed by a sodium-bisulfate-treated, all-glass fiber final filter for retention of nicotine vapor.[9]

Sampling Duration

Brief period samples are often referred to as "instantaneous" or "grab" samples, whereas longer period samples are termed "average" or "integrated" samples. Although there is no sharp dividing line between the two categories, grab samples are obtained over a period of less than 5 min, usually less than 1 min, whereas average samples are taken for longer periods.

ACGIH's 2000 TLVs and BEIs book[3] defines sampling periods in relation to specific physiological responses. Brief period samples include the "Threshold Limit Value–Ceiling (TLV-C)—the concentration that should not be exceeded during any part of the working exposure" and the "Threshold Limit Value–Short-Term Exposure Limit (TLV–STEL)—the concentration to which workers can be exposed continuously for a short period of time without suffering from 1) irritation, 2) chronic or irreversible tissue damage, or 3) narcosis of sufficient degree to increase the likelihood of accidental injury, impair self-rescue or materially reduce work efficiency, and provided that the daily TLV–TWA is not exceeded." The longest period sample defined by ACGIH is the "Threshold Limit Value–Time-Weighted Average (TLV–TWA)—the time-weighted average concentration for a normal 8-hour workday and a 40-hour workweek, to which nearly all workers may be repeatedly exposed, day after day, without adverse effect."

Closely similar short-term and long-term sampling periods are specified in the *Code of Federal Regulations* (29 CFR 1910.-1000, Tables Z-1 through Z-3) and are incorporated into current legal standards for evaluating the exposure of workers to airborne contaminants in the workplace. OSHA's legal, enforceable standards are referred to as permissible exposure limits (PELs). Recommended exposure limits (RELs), prepared by the National Institute for Occupational Safety and Health (NIOSH), usually refer to up to 10 hrs per day and 40 hrs per week. Workplace sampling strategies are covered in Chapter 2.

Environmental monitoring for community air pollution control is covered in Chapter 3. This sampling specialty has its own set of averaging times, dictated by custom and the physiological response of the body to specific pollutants. For example, ozone, an irritant, has a 1-hr averaging period; carbon monoxide has a 1-hr averaging time for high-level acute exposures plus an 8-hr averaging time for lower level exposures; suspended particulate matter has a 24-hr averaging period for maximum daily concentration and the daily results are averaged over a full calendar year for comparison with another air quality standard.

Brief sampling periods are best for following the phases of a cyclic process and for determining peak airborne concentrations of brief duration, but they usually require analytical methods capable of detecting and measuring very low concentrations of the collected airborne contaminants. Instantaneous sampling is a characteristic of most direct-reading instruments for gases and vapors (Chapter 18) as well as for aerosols (Chapter 15). Short period sampling may also be accomplished by trapping a small portion of the atmosphere in a suitable vessel such as a previously evacuated glass or stainless steel canister. Grab sampling methods can be used to investigate malodorous air pollution by organoleptic techniques. Short period sampling is suitable, physiologically, for primary irritants such as HCl, whereas continuous sampling methods are best for evaluating cumulative systemic poisons such as lead and mercury. Each method has special value and it is essential to develop a capability to do both. It is always important that the sample contain sufficient contaminant to be above the minimum quantity that can be measured reliably by the chosen analytical method.

Sampling Rate

Active gas sampling presents no special problems with respect to sampling rate and, more particularly, velocity of entry into a sampling device, because gas mixtures resist separation into components under the influence of usually applied centrifugal or inertial forces.

This is not the case for particulate matter and, more especially, for particles greater than 5 μm in aerodynamic equivalent diameter (AED), defined as the diameter of a hypothetical sphere of unit density having the same terminal settling velocity in air as the particle in question, regardless of its geometric size, shape, and true density. The need for isokinetic sampling rates in ducts and stacks, in which velocities are usually in excess of 5 m/s and often exceed 20 m/s, is unquestioned. The sampling errors introduced by anisokinetic sampling in rapidly moving air streams are detailed in Chapter 20.

On the other hand, studies by Davies[10] have shown that most sampling rates give representative results when the sample is drawn from still or nearly still air whenever certain criteria for inlet conditions are met. This was found to be especially correct for particle sizes in the range of hygienic concern, i.e., <10 μm. Subsequently, Bien and Corn[11] applied Davies' criteria to the inlet configuration of commonly used air sampling devices and found a number that did not measure up, including the 10-mm cyclone of the coal mine personal sampler. Nevertheless, by test, these cyclones were found to give accurate results for respirable particles.[12]

Additional empirical studies by Breslin and Stein[13] have shown "that published criteria for inlet conditions for correct sampling are overly restrictive and that respirable-size particles are sampled correctly in the normal range of operation of most dust sampling instruments." This observation was confirmed by Agarwal and Liu,[14] who concluded that there are no real inlet restrictions when sampling respirable particles in still air. There are still some with a contrary opinion, but "no restrictions" remains the mainstream view.

Although there is little evidence to show that interactions between typical indoor air velocity conditions, sample intake orientation, and inlet velocity are likely to affect the capture of particles in the respirable size range, this is not necessarily true for larger particles that can deposit in the nose and throat. When these larger particles are corrosive to tissues (e.g., chromic acid) or are systemic poisons (e.g., lead, arsenic), they may be absorbed where they deposit or be swallowed and exert their toxic action in that manner. For such substances, sampling errors associated with nonrespirable, but inhalable, particles can become a matter of concern (see

Chapter 5), but not enough is known about the systemic effects of exposures to particles larger than 10 μm to make firm judgments.

The effects of sampling rate and inlet configuration for outdoor air sampling have been principally concerned with the effect of anisokinetic sampling velocities on particles greater than 10-μm AED in moderate wind velocity fields. Investigations of the combined effects of anisokinetic sampling velocity and angle of yaw have shown that when both are seriously awry simultaneously, undesirable effects on sample recovery can also occur for particle sizes below 10-μm AED.[15] Such conditions occur during outdoor air sampling but are minimized by the use of geometrically symmetrical inlets and careful control of sampling rate.

Size-selective Sampling

The AED of aerosol particles is of special importance for evaluating toxicologic effects because certain particle sizes deposit preferentially in different parts of the respiratory system. Evaluation of toxic potential can be simplified by the use of special sampling devices that select out of an aerosol cloud only those particle sizes that would reach specific compartments of the human lung. The characteristics of this type of sampler were first specified in 1970[16] by the joint Aerosol Hazards Evaluation Committee of ACGIH and the American Industrial Hygiene Association (AIHA).

In the United States, miniature cyclones with carefully regulated characteristics have been used most frequently as sampling precollectors. They permit a predetermined fraction of each particle size to penetrate to a second sampling device that simulates the respiratory system and retains all particles passed by the size-selective cyclone. Multicompartmented gravitational settling chambers are also used as size-selective precollectors, e.g., the British MRE coal mine dust precollector,[17] but because they are relatively large devices and sensitive to orientation, a cyclone or impactor is preferred for use in conjunction with personal sampling devices.

Although the particle size retention characteristics of size-selective presampling devices have been chosen to simulate as closely as possible a standardized human lung particle size rejection curve, it is important to keep in mind that the range of human variation for lung retention is probably as great as for most physiological characteristics and that changes in breathing rate and volume per breath profoundly affect the size retention characteristics of the respiratory system. Nevertheless, the use of size-selective samplers has been recognized in the Mine

Safety Act of 1969 and the Occupational Safety and Health Act of 1970 as an important refinement in particle sampling for assessment of occupational risk. For similar reasons, the U.S. Environmental Protection Agency (U.S. EPA) modified its high-volume atmospheric sampling device for measuring total suspended particulate matter by adding a size-selective air intake with a cutpoint at 10 μm (referred to as a PM_{10} mass sampler inlet).[18]

Current research interest in sampling suspended particulate matter for health effects is focused on developing instruments with cut sizes of 5 and 2.5 μm, inasmuch as the smallest particles originate from gas phase reactions in the atmosphere that produce acid aerosols, whereas the largest ones tend to come from stack emissions of solid particles plus windblown mineral particles. The importance of the fine particle fraction lies in correlations between elevated levels and increased mortality.[19] Size-selective sampling is treated in Chapter 5.

Other Sampling Techniques

Many other types of sampling are needed by environmental health scientists from time to time. Those most frequently used are:

1. Microbiological and aeroallergen sampling to evaluate occupational exposures to pathogenic bacteria, viruses (hospitals and microbiological laboratories), fungi (indoor air pollution studies), and biological matter formed by recombinant DNA techniques. This type of sampling is the subject of Chapter 22. It has been enlarged to reflect the growing importance of bioaerosols in indoor air pollution studies and the rapidly growing knowledge of the nature of aeroallergens, their effects, and methods for collection, culture, and quantification.
2. Rafter samples to determine the long-time average size distribution and composition (e.g., percent quartz) of settled airborne dusts.
3. Product samples to estimate the hazard potential associated with handling specific materials.
4. Bulk air samples by high volume sampling to obtain sufficient material for in-depth qualitative and quantitative analysis. Appropriate statistical criteria must be applied as a guide in obtaining representative samples.[20]
5. Multiday diffusive sampling for radon inside buildings with activated carbon-filled canisters followed by measurement of alpha-ray-emitting radon daughter products. This is a widely used screening procedure recommended by the U.S. EPA. More quantitative

studies of radon in air are conducted with diffusion battery sampling to measure size distribution of the daughter products for evaluating differential lung deposition and by membrane filter sampling followed by prompt alpha activity counting with a bank of scintillation detectors for short period measurements. Sampling for this purpose and many other types of gaseous and particulate airborne radioactivity is covered in Chapter 21.

6. Air sampling to detect flammable and explosive concentrations. They are considerably higher than most hygienic standards (usually in the % by volume range for gases and g/m³ range for dusts). Some gas and vapor concentrations within the explosive range may also be anesthetic, asphyxial, or otherwise immediately dangerous to life and health (IDLH). Direct-reading instruments for detecting explosive atmospheres are described and illustrated in Chapter 18.

7. Analysis of a respirator pad or chemical cartridge worn by a worker. This type of analysis gives an integrated sample of the air that would have reached the respiratory system, although the exact air volume sampled can only be estimated.

8. An exposed worker without respiratory protection. Exposed workers have proven to be excellent, though sometimes involuntary, biological sampling devices. Analysis of appropriate body fluids or exhaled air gives an indication of absorbed dose and often reflects the average atmospheric concentrations of the exposure. For example, routine blood lead concentrations have been used for decades to supplement area and personal air samples and are particularly useful for locating individuals who may be exposed to the same toxic material during and outside working hours. Another example is the use of exhaled air samples at the conclusion of a work shift to measure the absorbed dose of substances such as toluene or carbon monoxide and estimate from this value a TWA exposure. Breath can be exhaled directly into a direct-reading instrument or collected in non-absorbing plastic gas sampling bags for later analysis.

Evaluating Sampling Results

Because it is impossible to examine the total air environment in which a person works, it is necessary to take small samples and generalize from them concerning the true nature of the entire environment. As might be expected, the larger the number of samples, the more faith can be put in the reliability of the derived information respecting the average concentration and the variability of the concentration from work station to work station and from time to time. Conversely, the degree of improvement in reliability obtainable by each additional sampling decreases as the total number of samples increases. Therefore, for economy, it is necessary to know the minimum number of samples required to characterize the environment to a degree of accuracy consistent with the maintenance of working comfort and safety. Information on sampling strategy is contained in Chapters 2 and 3 and statistical methods for assessing precision, accuracy, and validity of sampling data are contained in Chapter 10.

Industrial hygienists use field sampling data to compare the measured values to a TLV, PEL, or REL. When evaluating how well air samples represent the working environment, it is necessary to recognize the presence of instrument and analytical errors as well as normal variations in workroom concentrations over space and time. There are two kinds of errors: systematic errors that relate to imperfectly representative sampling and result from the use of a finite number of sampling points and a limited sampling time [21] and nonsystematic errors that result from random fluctuations in the process under study. Random fluctuations may be of long duration relative to the sampling time and produce marked variability from sample to sample at the same location, or they may be randomly distributed in space and produce extreme and uncontrollable variations in samples taken simultaneously at different locations.

Although a larger number of replicate samples gives greater faith in the estimate of the true average concentration derived from them, small sample numbers are most frequent. For this reason, an "interval estimate" is a more useful measure of how well sample averages correspond with the true average in the entire environment than is a "single value estimate." Confidence intervals bracket the true average value and are associated with a confidence coefficient that defines the probability that the true average will be included within that confidence interval; i.e., for a 95% confidence interval, it may be stated with 95% confidence that the true average is greater than a lower interval value and less than an upper interval value. Larger numbers of samples tend to give narrower confidence intervals for the same level of confidence and come closer to the true average. Methods for calculating confidence intervals are contained in Chapters 2 and 10.

Action Level

NIOSH has developed "predictive and analytical statistical methods"[22] for the evaluation of field sampling results and has recommended to OSHA that these statistical methods be used to "minimize the probability that even a very low percentage of actual daily employee exposure (8-hr TWA) averages exceed the standard" when "only one day's exposure measurement is used to draw conclusions regarding compliance on unmeasured days."[23] It was noted, on the basis of numerous studies, that "concentrations in random occupational environmental samples are log normally and independently distributed both within one eight hour period and over many daily exposure averages" and, therefore, sample results are not distributed symmetrically around the average.[22] This comes about because, although airborne concentrations tend to cover a wide range of values, most will lie closer to the zero concentration limit but a few will show very large values. Therefore, the distribution tends to peak toward the low concentration values with a long, flat "tail" on the high concentration side. This would be very difficult to handle mathematically were it not for the observation that a logarithmic transformation of the original data is often normally distributed and, by this transformation, a median and a geometric standard deviation can be easily determined.

Those familiar with particle size analysis will recognize the statistical methodology. It has been found applicable to air pollution exposure data as well. More details will be found in Chapter 2, "Occupational Air Sampling Strategies", Chapter 3, "Approaches for Conducting Air Sampling in the Community Environment", and Chapter 10, "The Measurement Process". The empirical observation that workroom measurements tend to follow a logarithmic probability distribution that has, on the average, a geometric standard deviation of 1.22 (i.e., the slope of the distribution curve) has been used by NIOSH to recommend an "action level" when only a small number of samples is used to estimate the true average air concentration.[23]

Figure 1-1 shows the effect that day-to-day variability in true daily exposure averages has on the probability that at least 5% of all unmeasured 8-hr TWA daily exposure averages will exceed the standard when a single day's measurement falls an identified fraction below the standard. This figure is the basis for the NIOSH recommendation that a measurement at or above one-half the standard should be an "action level" and call for remeasurement of exposures at least every 2 months. Two consecutive measurements (at least 1 week apart) showing employee exposures less than 50% of the federal limit are adequate to permit termination of the sampling program. Exposures above the federal limit call for more effective control measures and monthly remeasurements until the exposure is reduced to less than the federal

limit. This obviously puts a premium on obtaining a low value for each trial.[24]

A critique of the NIOSH action level and sampling strategy proposals was prepared by Rock and Cohoon.[25] They concluded that, "The OSHA implementation of a multiple sample decision strategy is arbitrarily stringent and by design ignores day-to-day variability. It is therefore not suitable for general use by industrial hygiene professionals other than OSHA compliance officers." This finding was based on their belief that "statistically sound strategies are elusive" and that "selection and proper use [of statistical strategies] requires disciplined professional judgment."[25] The authors emphasize their belief in the primacy of professional judgment over mechanistic evaluation schemes designed for subprofessionals and dictated by legal imperatives. The need for professional judgment in the design and conduct of all occupational-health-oriented sampling programs is emphasized in Chapter 2. Chapter 10 deals more fully with sampling statistics and should be consulted for guidance and details in conjunction with other literature citations.[22–25]

It seems reasonable to believe that epidemiologic studies involving exposure assessments should include a detailed analysis of the reliability of the exposure values that have been employed (as well as a similar analysis of the "effects" data) so that the derived TLV may be expressed in terms of statistical confidence limits instead of a single number that implies a false degree of certainty regarding the accuracy of the cited value. In the United Kingdom, it is usual to take into account day-to-day variability errors when setting limit values.[26]

Sampling Methods

Methods of Sampling Gases and Vapors

Gases and vapors offer less difficulty in air sampling than aerosols because they diffuse rapidly, mix freely with the general atmosphere, and can, in a short time, reach equilibrium with it. For air sampling purposes, contaminants can be grouped with regard to solubility and vapor pressure. Many gases and vapors of hygienic significance are water soluble and can be collected in aqueous media with or without a dissolved reacting chemical to suppress the vapor pressure of the solute. Gases and vapors that are not water soluble, but are soluble or reactive in other agents, can be absorbed in a suitable solvent. Gases and vapors that are neither soluble nor reactive may be collected on adsorbents, e.g., activated charcoal, silica gel, and molecular sieves, in both active or diffusive (passive) samplers. Adsorbents have tended to become

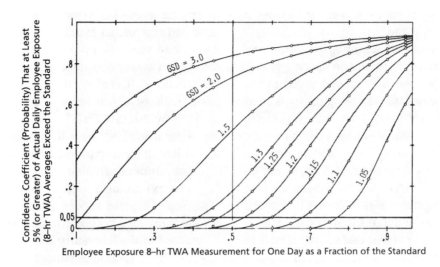

FIGURE 1-1. Probability that standard has been exceeded based on exposure measurements. *(Reprinted with permission from Leidel et al.[23])*

the sample collection medium of choice for all gases and vapors because of their convenience and generally high collection efficiency. (See Chapter 16 for details.)

Instantaneous gas and vapor samples may be collected in rigid glass or metal flasks or in soft plastic bags made of polyethylene, Saran®, Mylar®, Tedlar®, and combinations of these with aluminum foil in sizes up to 120 L.[27] For plastic bag sampling, "A sample is introduced into the bag by a hand or battery operated pump or a squeeze bulb. Bags can be re-used after purging with clean air and checking for any residual components. Certain contaminants cannot be sampled or stored in any type of plastic bag due to their reactivity with surrounding substances or with themselves, styrene being an example. Whether a substance can be sampled and stored in a plastic bag should be determined in the laboratory prior to field use."[28] (See discussion in Chapter 16.)

Pyrex® gas collecting tubes of 300-mL capacity with a capillary standard taper stopcock at each end may be used when the atmosphere sampled contains components incompatible with plastic bag materials. In practice, evacuation may be done in the laboratory and the flask opened in the environment to be sampled, or the flask may be evacuated in the field by a pump. In either case, it is necessary to know the volume of the sampling flask and the internal pressure prior to opening in order to calculate the volume of air sampled. This may be done easily when using evacuated flasks with stopcocks by connecting the flask to a mercury manometer or vacuum gauge and opening the stopcock for a reading just before sampling. Specially passivated stainless steel pressure vessels, known as "SUMMA canisters," (Moletrics, Inc., Cleveland, Ohio) are sometimes used for short period

sampling of complex mixtures of gases and vapors. The quantity of material trapped by these techniques is small and reactions between trapped chemicals, water vapor, and the walls of the container can occur during even brief storage.

Glass or metal sampling flasks not under vacuum are also used by purging them in the field with a pump or squeeze bulb. However, in this case, the amount trapped reflects the fact that concentration buildup is a semilogarithmic function. In addition, condensation can occur when sampling in saturated atmospheres and can result in continuous absorption in the condensate, thereby falsely increasing the apparent air concentration. Instantaneous gas and vapor sampling may also be done with direct-reading instruments that have a response time measured in seconds. Chapter 18 treats the subject of direct-reading instruments for gases and vapors and includes details of most of the commercially available items in this category.

Non-specific, direct-reading survey instruments such as photoionization and flame ionization meters that respond to broad classes of organic gases and vapors are widely used by industrial hygienists as survey instruments for initial hazard evaluations in environments that are likely to contain a number of poorly identified contaminants. When there is a single gas or vapor of interest present, the instruments can be "tuned" to respond strongly to that component and give reliable quantitative results. In addition they have found wide usage for investigations of leaking underground storage tanks and old chemical waste disposal sites. Detection of chemical vapors in gases found in shallow holes newly drilled in the ground in the vicinity of such facilities is a rapid and simple means of locating chemical leakage even when it is occurring deep underground.

For long-term, integrated sampling of gases and vapors, the sampling rate will depend on the type of collection device employed and the reaction speed of the contaminant. In some cases, the sampling rate will be as low as 50 cm³/min. However, with the analytical methods now available, this usually imposes no handicap. Integrated gas and vapor samples may be collected in a solvent with wash bottles, impingers, and absorbers; on adsorbents, e.g., activated charcoal; by condensation in freeze-out traps; and in large plastic bags filled at the rate of 50 mL/min with the aid of low-volume, battery-operated personal sampling pumps that contain a built-in accumulator counter to record total air volume. After sampling, the contents may be analyzed in the field or laboratory by nondispersive infrared (CO), gas chromatography (hydrocarbons, chlorinated solvents), etc.

Absorbers vary in characteristics depending on the gas or vapor to be collected. Simple bubbling devices (e.g., impingers and Drechsel bottles) are adequate for readily soluble gases such as HCl, HF, and SO_2. For less easily absorbable materials such as Cl_2 and NO_2, multiple contact washing is required (as with fritted glass absorbers). Sometimes, it is desirable to burn the gas or vapor in a furnace and sample the oxidation products, e.g., HC for chlorinated hydrocarbons. Gas absorbers of this type were discussed and illustrated in earlier editions of this manual and are summarized in Chapter 16.

Adsorption tubes are the method of choice for insoluble or nonreactive vapors. Commonly used adsorbents (in 6–20 mesh sizes) include activated charcoal, silica gel, and molecular sieves. Gas adsorption traps are sometimes preceded by one or two water vapor adsorption stages containing calcium chloride, calcium sulfate, or silica gel, all of which have excellent water vapor adsorption characteristics and poor adsorption capacity for most organic molecules. In the laboratory, the collected vapors may be desorbed thermally or stripped from the adsorbent with carbon disulfide and the recovered vapors quantified by gas or high pressure liquid chromatography using a suitable detector.

Adsorption tubes used for personal integrated sampling of many organic gases and vapors contain two interconnected chambers in series filled with gas adsorption charcoal. The first chamber, containing 100 mg of charcoal, is separated from a back-up section, containing 50 mg of carbon, by a plastic foam plug. Sampling can be conducted for as long as 8 hrs at 50 mL/min without saturating the first chamber when occupational exposures are at or below the TLV. The contents of the two chambers are analyzed separately to determine whether the first-stage adsorbent has become saturated and lost an excessive amount of the sample to the second stage. Sampling results are discarded when the second adsorption stage contains more than 20% of the amount collected on the first stage. Personal sampling with two-chambered charcoal tubes is recommended by NIOSH for vinyl chloride, benzene, CCl_4, etc. Similar methods have been used for environmental and indoor, nonoccupational, sampling.

Gas chromatographic column packing materials (e.g., Tenax) are also used in adsorbent traps for field sampling of organic vapors. They are particularly useful for sampling high boiling compounds. The resealed trap is returned to the laboratory for analysis. Absorption of inorganic atmospheric constituents by liquid coatings on solid supports has been used to collect NO_2 on triethanolamine-coated molecular sieves.

Chemically active compounds may react with each other or with oxygen in the air after adsorption and make it difficult or impossible to recover and quantify the original adsorbed gases and vapors. In such instances, the best collection method may be to react them chemically to a stable derivative or to condense the contaminants at low temperature using a mixture of dry ice and acetone ($-78°C$) or liquid nitrogen ($-196°C$) as the coolant.

Methods of Sampling Aerosols

Collection methods for liquid and solid aerosol particles are likely to differ, inasmuch as high liquid loadings call for special collection devices to handle the large liquid accumulations. In addition, liquid or solid aerosol particles that possess significant vapor pressure and are not in equilibrium with the conveying airstream experience evaporation losses during the sampling period when filtration or dry inertial collection methods are used. For these conditions, the particle collector must be followed by an appropriate vapor collector to account for the entire sampled quantity. Samplers of this type are covered in Chapter 19. When particle loadings are light and evaporation after collection is not a concern, solid and liquid particles can often be sampled successfully by identical methods.

Dusts may be grouped into 1) relatively insoluble mineral dusts such as silica, granite, asbestos, and insoluble metal oxides; 2) soluble mineral dusts such as limestone and dolomite that dissolve in weak acids; and 3) organic dusts such as trinitrotoluene, flour, soap, leather, wood, and plastics. Many of the last group are explosive when the concentration in air is high.

Instantaneous dust samples may be collected with a device that takes in a small measured volume of air (25–50 mL) and blasts it at high velocity against a glass plate on which the particles are deposited. After deposition, the particles are examined and enumerated by bright or dark field microscopy. Today, these instruments are seldom used; they are discussed and illustrated in older publications on air sampling.[1, 29]

For integrated or continuous sampling of particulate matter, several physical forces (gravity, inertia, electrophoresis, thermophoresis, and diffusion) are employed. Particle size determines the sampler to be used, and sample volume depends on the air concentration. Collectors for particulate matter can be divided into the following categories: 1) elutriators; 2) centrifugal devices (e.g., cyclones); 3) impingers and impactors; 4) electrostatic precipitators; 5) thermal precipitators; 6) filters; and 7) diffusion batteries. The first five are the principal subject matter of Chapter 14; the sixth of Chapter 13; the seventh of Chapter 19.

Elutriators use gravitational sedimentation to collect or reject large particles. They are sometimes used for size-selective sampling.

Centrifugal devices include small cyclones and curved surface traps. Cyclones can be used to provide a size-selective separation upstream of another particle collector, such as a filter, or they may be cascaded together to separate particles into several size classes. Aerosol centrifuges are not commonly used for field studies; they are capable of a high degree of precision in the size segregation of particles.

Impingers and impactors use inertial properties of particles to effect collection. The impinger consists of a glass nozzle submerged in water or other liquid. The velocity at the nozzle ranges from 60 to 113 m/s. Particles are impacted on the bottom of the flask and trapped in the liquid.[30] Cascade impactors[31] collect particles on a dry or sticky slide. They contain a number of impingement stages in series with graduated nozzle velocities to effect a progressive separation of smaller and smaller particles as the aerosol travels through the unit. Individual impactor stages may be of the single jet or multijet variety. Particles deposited on each stage may be examined microscopically, but when the impactor has been calibrated to define the AED 50% cutpoint of each stage, size distributions by mass can be measured more simply by using weighings, radioactivity, or chemical analysis to determine the amount of deposited material. Many types of cascade impactors are described in Chapter 14. The Anderson cascade impactor was developed for the U.S. military's biological warfare program during the 1950s (see Chapter 22). It consists of multijet stages that deposit airborne micoorganisms directly onto agar media-filled dishes ready for culturing, below each stage.[32]

Absorbers designed for gas collection in a liquid medium are seldom effective for particulate material of small size, but small-scale models of commercial high-efficiency fume collecting devices (e.g., a Venturi scrubber) are useful for obtaining large dust samples from very hot gases.

Filters are among the best methods of sampling solid particulate matter. Many kinds of filter media are available and their use requires a minimum of equipment. Many types of filters are described in Chapter 13. Cellulose, Teflon® and polyester membrane filters are often used for collecting metallic dusts for chemical analysis and mineral dusts for gravimetric or X-ray diffraction analysis. Absolute-type (HEPA), all-glass filter papers containing superfine glass fibers with diameters well below 0.25 µm are, as the name suggests, virtually 100% efficient for all particles of hygienic importance. All-glass HEPA filters resulted from an intensive joint development effort by the U.S. military and Atomic Energy Commission during the 1950s and 1960s to replace the standard crocodilite asbestos-esparta pulp gas mask filter that had been developed originally by the German army during World War II.[33] Particles of high boiling liquids, such as sulfuric acid mist, may be collected with equally good results on all-glass papers. Absolute-type, all-glass filters have low airflow resistance, adsorb little water vapor, and, because glass interferes with only a few analyses, have application for gravimetric, chemical, and physical analysis. They are widely used for air quality monitoring. Exceptions are noted in Chapter 13.

Membrane filters have been used for collecting sulfuric acid and similar mists but only for low concentrations. Membrane filters are widely used for collecting mineral dusts for mass respirable fraction evaluation. Membrane filters were an importation from Germany during World War II and were first evaluated intensively by the U.S. military for applications to biological warfare surveillance. When cleared for publication their use for airborne fiber counting and particle examination by microscopy was first reported by First and Silverman in 1953.[34]

Electrostatic precipitators, homemade or commercial, have been used since the 1920s for industrial air sampling in workrooms.[29] A specially designed point-to-plane electrostatic precipitator, consisting of an ionizing needle with the point mounted directly above an opposite-polarity, carbon-coated electron microscope grid, is used to collect dust specimens onto grids for direct examination in the electron microscope (see Chapter 14). Some direct-reading instruments for aerosols have used electrostatic precipitation as the dust precipating component.

Thermal precipitators were used in former times for particle enumeration and sizing to evaluate occupational dust exposures.[29] Collection efficiency is near 100% for small particles but the dust-free zone around the hot body is very limited and consequently sampling rate is only a few mL/min. At present, thermal precipitators are used as a research tool for collecting dust specimens directly onto grids for examination in the electron microscope. More information on thermal precipitators is contained in Chapter 14. Diffusion batteries are used for collecting and size-separating particles in the size range below the lower limit of optical airborne particle counters, i.e., those less than approximately 0.1 µm. Diffusion batteries are constructed in a variety of configurations. They are discussed and illustrated in Chapter 19.

Separating Volatile Aerosol Particles from Their Vapors

An air sampling task of considerable complexity arises when it is necessary to collect a vapor uncontaminated by its particulate phase in contact with it, or the reverse (see Chapter 4). It is not possible to remove the particulate phase first by filtration because unsaturated air passing through the filter will vaporize the liquid or sublimate the solid on the filter and contaminate the vapor phase collector that follows. It is equally unsatisfactory to pass the sampled air through a bed of gas-adsorbing granules as a first-stage vapor collector because the adsorbent will remove at least some of the particulate phase as well.

A requirement of this nature arises when one wishes to evaluate separately the vapor phase and liquid phase workplace exposure to, for example, middle distillate fractions of petroleum. The reason for doing this is to evaluate the precise nature of the exposure of workers who handle these products: the vapor-only exposure is limited by the vapor pressure of the many specific compounds encompassed by the designation "middle distillates," whereas the particulate exposure is not so limited and inhalation will have different physiological and toxicological effects.

A solution to this problem described by Bertoni et al.[35] is an "annular diffusive sampler" consisting of an annular inlet passage lined with activated charcoal to adsorb vapors by diffusion while the particulate phase penetrates the annular passage and collects on a filter. Samplers of this type are called diffusion denuders. When it is desirable to analyze both the vapor phase and the particulate phase of a volatile liquid, it is necessary to add a vapor adsorbing third stage to trap the vapors volatilized from the liquid droplets caught on

the second stage filter as the sampled air passes through it.[36]

Similar requirements occur when sampling outdoor air to separate nitrate and sulphate salts of ammonia from their precursor gases and from larger, inert particles, e.g., by the use of an impactor/honeycomb denuder/filter pack system.[37] Two honeycomb denuders in series, each containing over 200 hexagonal glass tubes sealed inside an outer glass tube of 4.7-cm diameter, were used to separate NH_3 from HNO_3 by coating the surfaces of the first denuder with a basic coating to collect acid gases and coating the second denuder with an acidic coating to collect basic gases. A diagram of the four-stage sampler is shown in Figure 1-2. The impactor has a 50% cut-off size of 2.1 µm. More information on "Denuder Systems and Diffusion Batteries" may be found in Chapter 19.

Collection Efficiency

Collection efficiency is an important factor in the selection of sampling devices. Efficiency need not be 100% as long as it is high, known, and constant over the range of concentrations being evaluated. It should, preferably, be above 90%. An important advantage of evacuated sampling flasks is that efficiency is normally considered to be 100%, provided wall adsorption is negligible and correction is made for completeness of evacuation of the container. For concentrating sampling devices, the collection efficiency of the concentrating device must be measured.

A widely used method for measuring sampling efficiency for gases and vapors is to place two or more identical samplers in series and analyze the catch in each. When efficiency is independent of gas concentra-

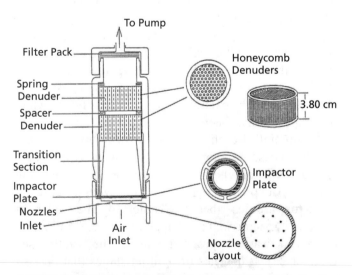

FIGURE 1-2. Honeycomb denuder sampler. (Source: Bertoni et al. Annali di Chimica 74:2497.[32])

tion, each sampler in the series will remove the same percentage of the concentration that reaches it, i.e., a log-decrement relationship.

This may not be the case when sampling trace concentrations of gases and vapors by absorption or adsorption because collection efficiency is sometimes proportional to the concentration difference driving force. Therefore, it is often necessary to resort to other methods for measuring collection efficiency. These methods include the use of permeation and diffusion tubes of known emission strength, comparison sampling in parallel with a known high efficiency device utilizing a different measurement principle, measurement of the loss of material from a concentration accurately known at the start and end while sampling in a closed circuit at a known flow rate, sampling from pressurized cylinders containing known concentrations of various gases, and following the sampler under test with a different sampling device known to have close to 100% efficiency for the air contaminant under study. An example of the latter method would be the use of a high efficiency particulate air (HEPA) filter following a particle collection device of unknown efficiency. Chapter 8 is concerned with instrument calibrations of gas and vapor instruments and Chapter 9 is concerned with aerosol instrument calibrations.

Direct-reading Instruments

Direct-reading instruments combine sampling and analytical functions and usually display results rapidly. Many in this category are also capable of storing continuous readings and displaying on command averages for selected time intervals. Devices that provide an immediate answer are useful for industrial hygiene workroom appraisal. The value of an immediate measurement for prevention of further injury and the ability to demonstrate worker exposures to management cannot be overemphasized.

An innovative teaching method that combines a direct-reading analytical instrument, such as a photoionization device for organic vapors or a light-scattering chamber for aerosols, with a videotape recorder makes it possible to instantly play back precisely how a production operation involving an exposure to a health hazard was conducted with a continuous record of the instrument readings superimposed.[5, 38] This technique makes it possible to involve the workers as well as management in a unique safety education experience. Fortunately, the trend in recent years has been toward the development of more direct-reading devices (Chapters 15, 17, and 18).

Direct-reading Instruments for Gases and Vapors

Numerous direct-reading instruments for gases and vapors have been in constant use by industrial hygienists for many decades. They include 1) a halide meter, 2) combustible gas detectors, and 3) thermal conductivity instruments. The halide meter produces an increase in the violet nitrogen spectrum from an electric arc in the presence of a halide vapor, e.g., methyl chloride. Recent versions of this instrument are capable of measuring concentrations below 1.0 ppm expressed as the halide. Combustible gas detectors measure the heat of combustion released when a gas or vapor is burned on a platinum wire. These detectors are not specific, but they can be calibrated for a single vapor or known mixtures. Thermal conductivity instruments form the basis for most of the less expensive gas-detecting devices used for analyzing higher concentrations of carbon dioxide, such as in engine exhausts or oil burner flue gases.

Small, hand-carried direct-reading instruments are available commercially for measuring hydrogen and mercury vapor (using ultraviolet radiation). Mercury vapor in air may be measured directly with instruments that detect the fogging of a gold leaf as it amalgamates with mercury. Hand-operated instruments based on Orsat analytical methods have been used for measuring CO_2 and oxygen in tanks, manholes, and underground excavations to determine the life supporting properties of the atmosphere prior to entry of workers, but electrochemical meters have largely displaced them.

Oxygen meters are often paired with combustible gas detectors and a toxic gas detector in a single compact instrument case for such gases as H_2S or CO. These instruments are portable and are designed to operate continuously. They contain microprocessor circuitry and software features that provide "automatic self-diagnostic check, automatic zeroing, and automatic calibration."[39] Oxygen measurements are needed not only to evaluate the presence of satisfactory concentrations for breathing, but also to evaluate the readings of combustible gas indicators correctly. Combustible gas indicators give a false indication of safety when the oxygen concentration is less than 8%. However, when oxygen is low, the inlet to the combustible gas meter can be equipped with a tee connection and two sampling hoses of equal length, one of which is placed inside the enclosure to be tested, the other in outside air. The meter reading must then be multiplied by two.

Indicator (or detector) tubes are outstanding direct-reading industrial hygiene air analysis instruments because they are small, light, hand operated, safe in all atmospheres, and give an immediate readout. In addition, an indicator tube is the simplest and most economical air analysis method available for many common air contaminants. During World War II, the National Bureau of Standards (NBS) produced an improved CO indicator tube based on a reaction between CO and a palladium silicomolybdate complex that produces molybdenum blue; the intensity of the color being proportional to the concentration of CO. The success of the NBS CO indicator tube stimulated the development and commercial production of a large variety of reliable detector tubes in the sensitivity ranges useful to industrial hygienists. Detector tubes for more than 160 chemicals are currently available. Some are available in more than one concentration range and the most popular types are offered by many commercial sources. Most of the chemicals measurable by indicator tube are included in the current tabulation of ACGIH TLVs.

Because they are so simple to operate, it is commonly assumed by those ignorant of industrial hygiene theory and practice that indicator tubes can be used by unskilled personnel for monitoring work environments. It has been repeatedly demonstrated in practice that serious errors in sampler operation, in selection of sampling locations and times, and in interpretation of results occur unless the tubes are in the hands of a trained operator who is closely supervised by a competent professional. This point is also made in Chapter 17 and cannot be overemphasized.

Some indicator tubes have a long shelf life, e.g., H_2S, but many deteriorate within a year or two. It is customary to extend the shelf life of tubes by storing them under refrigeration, but because the speed of most chemical reactions is sensitive to temperature, the tubes must be warmed to ambient conditions prior to use if the calibration charts accompanying the tubes are to be relied upon. A general certification recommendation for the accuracy of tubes in the United States is ± 25% of the true value when tested at 1–5 times the TLV and ± 35% at one-half the federal standard.[40] Although all tube manufacturers have improved their quality control, checking a suitable sample of each batch of tubes purchased is advisable and rechecking after a period equivalent to a large fraction of the normal shelf life is prudent.

NIOSH formerly certified detector tubes for a number of workroom exposures but they suspended their detector tube certification program in 1982. The Safety Equipment Institute (SEI), a nonprofit organization funded by safety equipment manufacturers, began a third-party certification program of their own in 1987. Tube testing is performed by contract laboratories holding AIHA laboratory accreditation status; the test protocols employed are those previously established by

NIOSH and Military Standard 414.[41] In addition to tube testing in the laboratory, the contract laboratories conduct periodic quality assurance audits of each participant's manufacturing facility. A listing of SEI certified tubes will be found in Chapter 17.

One of the most important developments in air sampling technology for measuring exposures to low concentrations of airborne substances has been the commercial appearance of passive dosimeters for a broad list of volatile substances. Many use the principle of diffusion to a nonspecific adsorbent with subsequent laboratory analysis. Direct-reading devices use permeation through a plastic film barrier to a compound-specific chemical bonding color-developing reagent.

The adsorbent or reactive layer represents an infinite sink for the diffusing or permeating compound, preventing back pressure of the captured material. For that reason, the rate at which volatile airborne substances reach the sensitive surface is proportional to the air concentration in the immediate vicinity of the dosimeter. A review of passive dosimeters[42] cited potential sources of inaccuracy in the measurement of airborne concentrations of gases and vapors but concluded that "passive systems appear to be as reliable as the now accepted active sampling systems." Diffusive dosimeters are dealt with in Chapter 17. Although only a few passive dosimeters are direct-reading devices, e.g., CO, formaldehyde, and Hg vapor (most must be returned to a laboratory for analysis), the rapid commercialization of all types of passive dosimeters gives promise of an early appearance of additional direct-reading types.

Direct-reading Instruments for Aerosols

Direct-reading field instruments for aerosols are combination sampling and analytical instruments. In addition to the economy of effort and instantaneous readout they provide, many permit measurement of the principal characteristics of liquid and solid particles in an unaltered airborne state.

Airborne particle mass may be measured by depositing the particles on a piezoelectric sensor by electrostatic precipitation.[43] The rate of change of the resonant frequency of the sensor is directly proportional to the mass of material deposited on it. From a continuous trace of resonant frequency with time on a strip chart, short period slopes can be analyzed to measure concentration fluctuations, or the electrical output of the instrument can be digitized and averaged electronically to produce dust concentrations over an averaging period of 24–120 seconds. The instrument weighs approximately 9 kg and is capable of measuring particle concentrations in air as low as a fraction of a mg/m³ at a sampling rate of 1.0 L/min over a 2-min sampling period. Another aerosol sampling and analyzing instrument, based on an oscillating rod's response to changing mass loading, collects particles continuously on a filter attached to a tapered rod and displays aerosol concentration in μg/m³ based on changes in oscillation frequency and air flow rate.[44] When fitted with a PM₁₀ sampling head (most of the incoming air is bypassed), the instrument gives results in ambient air that qualify for certification by U.S. EPA as an equivalent air monitoring method.[45]

An automatic sampling and analyzing instrument based on beta-ray attenuation was developed through a contract issued by the U.S. Bureau of Mines for use in coal mines.[46] This unit is no longer manufactured, but others that use beta-ray detectors are available. The advantage of beta-ray measurement is that attenuation is directly related to the mass of collected particles in the beam. Sampling may be accomplished by membrane filtration or by electrostatic deposition of particles on a thin film. Either method produces a clearly defined dust deposit that can be evaluated quantitatively by beta-ray attenuation.

Another means of assessing airborne particle concentrations is by light scattering. A British unit, called SIMSLIN II (Safety in Mines Scattered Light Instrument), uses a single plate horizontal elutriator to separate the respirable dust fraction from the sampled aerosol and measures the light attenuation of the aerosol stream that penetrates the elutriator. The results are displayed to the machine operator at 1 second intervals, with a 15-min and full-shift cumulative average available on demand. A German unit, called Hund TM Digital μP (Hund Corporation, Wetzlar, Germany), employs a scattering angle of 70° and monochromatic light of wave length 940 nm to optimize sensitivity to respirable dust. It also displays instantaneous and cumulative dust concentration data over full shift periods. When the instruments were evaluated using coal dust ranging in concentration from 1.0 to 9.9 mg/m³ and the results were compared with the British MRE dust sampler, it was found that on average, the SIMSLIN II measurement was 18% greater and the TM-Digital 25% less than the MRE.[47]

Joining the list of portable, direct-reading aerosol survey instruments is a small, hand-held, battery-operated, light-scattering instrument called the Miniram, which stands for "Miniature Real-Time Aerosol Monitor."[48] Because light scattering intensity is influenced by particle size, color, shape, and index of refraction as well as by particle numbers, it is not possible to use light-scattering instruments for quantitative evaluations of aerosol concentrations of unknown composition and size distribution. After calibration with a specific aerosol, it may,

however, be used as a semiquantitative survey instrument for repeat measurements of nonvarying operations to determine whether engineering controls have maintained their effectiveness.

Several automatic particle counting and sizing instruments capable of making measurements on flowing aerosols are available commercially. Most use optical systems and count light pulses scattered from particles that flow, one by one, through an intensely illuminated sensing zone. When instruments are equipped with an electronic pulse height analyzer, it is possible to obtain information on the size of each particle passing through the illuminated sensing zone. Sampling rate may be as high as 1 L/min. For conventional light particle counters, the smallest detectable size is about 0.3 μm. Portable models weighing about 10 kg are available that give a simultaneous readout of the entire size spectrum sensed by the instrument in 8–12 contiguous intervals. It is possible to make an airborne dust count and size analysis in 3–4 min with one of these instruments.

The use of laser illumination has improved the reliability of optical particle counters and sizers for the smallest particles because of the better collimation and greater light intensity that can be obtained with a laser beam. The most important advance in the use of lasers for counting and sizing airborne particles has been the development of intercavity lasers. The aerosol stream is introduced into the interior of the laser itself and only the scattered light is collected and reimaged at 10X or greater magnification against a dark field. In one commercial model,[49] the laser beam at the sensing volume is approximately 500 μm in diameter and produces a power density in excess of 500 Wcm⁻². It is capable of measuring particles as small as 0.08 μm in diameter and can cover the size range up to 20 μm (using two probes in series) in about 80 size intervals.

A notable development in automatic, real-time machine sampling and analysis of airborne dust is a battery-operated, portable (12 kg), Laser Fiber Monitor (FAM–7400).[50] In principle, it causes sampled airborne fibers to rotate rapidly in a rotating, high-intensity electric field and measures scattered light when the fibers are illuminated by a 2-mW He-Ne polarized laser operated at 632.8 nm. "Each fiber generates a pulse train as it rotates in a helical trajectory that results from the combined effects of the rectilinear airflow [through the sensing volume] and the perpendicular field-induced rotation. Because longer fibers produce narrower pulses than shorter ones, the monitor is able to discriminate between fibers of different length by sensing the sharpness of individual pulses."[50] Tests at the U.S. Bureau of Mines' laboratory "showed that the instrument response is linearly correlated to concentration data obtained using the optical membrane filter count technique" recommended by NIOSH.[51] A later study concluded that the instrument is "recommended as a screening method for monitoring airborne asbestos fibers [but] the device cannot be used as a substitute for the standard monitoring and analysis method."[52] For counting fibers when large numbers of nonfibrous minerals are present, a virtual impactor accessory is available.

Areas for Development

Many sampling and analytical instruments of excellent characteristics and small size have been developed for air measurements. When carefully calibrated and intelligently operated, they are capable of measuring atmospheric contaminants with an accuracy and reliability that is well within the requirements of most environmental health needs. Unfortunately, testing for function and calibration is difficult, time consuming, or costly, and is frequently all three. The introduction of permeation tubes and diffusion tubes has been a major step forward in the calibration of some instruments at low gas concentrations, but the tubes require a prolonged equilibration period prior to use and the equipment to house them in a constant temperature environment is bulky. Storage of calibration gases in compressed gas cylinders is satisfactory for unreactive gases such as CO, but is unsuitable for many substances of interest. Calibration of gas sampling instruments in covered in Chapter 8.

No standard aerosols are available for calibration purposes. This is a serious deficiency because all devices depend on calibration for reliability. Monodisperse polystyrene spheres are aerosolized for calibration of automatic particle sizing instruments, but this hardly constitutes a calibration aerosol in the fullest sense. Aerosol generation for testing purposes is covered in Chapter 9.

Even when correctly calibrated, it is important to remember that, in field usage, damage by vibration and impact from poor handling are factors that can alter the response of many components. In practical use, it is essential that recalibration be done at frequent intervals to ensure accuracy and reliability. Therefore, a continuing need exists for more simple, cheap, reliable, "off-the-shelf" field and laboratory calibration systems and devices to cover the wide range of gases, vapors, and aerosols of interest. A notable advance in the ability to test equipment in the field is the commercial appearance of small, battery-operated personal sampling pump flowrate calibrators.

Although a great deal of progress has been made in reducing the weight of field instruments, more must be done to reach the point where industrial hygienists can carry all the field instruments they need in their jacket pockets. In the area of environmental sampling instruments, considerably more miniaturization is needed. It seems reasonable to expect that solid state circuitry combined with greater instrument sensitivity will reduce electrical current needs and sample volume requirements (and hence, pumping power needs) to the point where increasing numbers of small, light, self-contained instruments of great versatility will become available soon. A step in the right direction is the "Micro Air Sampler," (Spectrex, Redwood City, California) a 114-g battery-operated pump and electronics package for charcoal adsorber tubes that fits in a shirt pocket and samples up to 200 mL/min.

Standard methods of sampling and analysis are essential for demonstrating compliance with OSHA and U.S. EPA exposure standards. A great deal has been accomplished in this area by publication of approved methods by NIOSH[53] and U.S. EPA and consensus methods by the Intersociety Committee[54] and the American Society for Testing and Materials (ASTM) D-22 Standards Committee.[55] Completion of the program to standardize sampling and analytical methods for every substance of interest to environmental health scientists is occurring rapidly in the United States and in united Europe (see Chapter 12).

New methods for capturing chemical species of special interest and rejecting all others are on the horizon. Instruments employing these methods are especially powerful when they can be miniaturized and equipped with sensors that give a real-time readout, thereby combining selectivity with identification and quantification. For example, man-made zeolites are porous crystalline minerals that can be constructed with uniform pore diameters in the nanometer size range that are big enough to admit large molecules or are so small that only individual atoms can get through. A chemical sensor has been described that consists of "a thin layer of tiny zeolite crystals affixed to an acoustic wave device, a larger crystal that, because of a feedback mechanism, constantly vibrates at a set frequency. If molecules come along that fit into the cavities of the zeolite, they will subtly change the weight of the whole device, altering its resonance frequency. That change can be detected electronically. By assembling a number of such zeolite sensors, each with pores with different sizes and shapes, it might be possible to create a sensor that responds only to one or a few closely related molecules."[56]

From this single example, it can be concluded that many new and useful sampling and analytical devices are on the way to make environmental and occupational health monitoring easier, as well as instantaneous, more accurate, more convenient, and more specific.

Air Sampling Equipment Selection

As noted in Chapter 12, "more than 1,000 instruments manufactured by 240 companies have been identified" for the purposes of this book. Even taking into account many of the instruments' duplicate functions, this is a rich menu from which to make choices. For industrial hygienists employed in industry the selection process may be simpler because of the unitary nature of the manufacturing operations. For example, hydrogen sulfide measuring instruments would be all-important at sour crude oil refineries as would carbon monoxide monitors and alarms at an iron blast furnace operation. On the other hand, some companies' operations are so huge and comprehensive that they require a huge inventory of instruments of great variety to satisfy all their needs; at major nuclear complexes, a full complement of radiation sampling instruments must be added. For regulatory agencies in large industrial states and countries, there will be a need for a large instrument inventory to satisfy major industrial diversity. Some needs can be met from instrumental rental agencies, but fees and shipping costs become high for routine use of these sources.

This is not intended to imply that well operated company and enforcement agency owned and operated instrument programs are cheap. Far from it. At a minimum, they require a well equipped mechanical and electronic repair shop with well qualified technicians; a calibration facility for each instrument must include verification sampling rate, zero, span, repeatability and accuracy each time an instrument is returned to the shop. This activity requires a large inventory of different certified compressed gases with many at more than one concentration. An instrument clerk whose duties are to check-out, check-in, and maintain surveillance on the whereabouts of every instrument in the collection is as essential as a competent director and crew of instrument technicians.

This magnitude of instrument inventory and degree of service and control are clearly beyond the budget and trained personnel resources of most industrial companies and small industrializing countries. What can we recommend for them? Versatility, ruggedness, and simplicity of design and function become critical when manufacturers that stock spare parts are far away and hard currency is in short supply. The solution, I believe, is to go back to the earliest editions of ASI, even including the original encyclopedia, and select only the essential number of multipurpose items that can be employed usefully and maintained by local resources. This may give the appearance of being an "old fashioned" operation when compared to the instrument shops of the large affluent agencies that are filled with the latest electronic gadgetry. But it should be remembered that a hand-held stopwatch and a liquid manometer provide all the accuracy needed for air sampling purposes, considering the confidence limits associated with even the best planned and executed sampling programs.

Careful operation and frequent instrument calibration are some of the keys to successful air sampling results as our pre-electronic age predecessors demonstrated with their simple manually-operated sampling devices. The message here is to select from the vast array of instruments those that serve the specific purposes of the users and that are within their capability to operate and maintain correctly with the resources at their disposal.

References

1. Yaffe, C.D.; Byers, D.H.; Hosey, A.D.; Eds.: Encyclopedia of Instrumentation for Industrial Hygiene. University of Michigan, Ann Arbor, MI (1956).
2. Sherwood, R.J.; Greenhalgh, D.M.S.: A Personal Air Sampler. Ann. Occup. Hyg. 2:127 (1960).
3. American Conference of Governmental Industrial Hygienists: 2000 TLVs® and BEIs®. ACGIH, Cincinnati, OH (2000).
4. World Health Organization: Environmental and Health Monitoring in Occupational Health. WHO Technical Report Series No. 535. WHO, Geneva (1973).
5. Rosén, G.: PIMEX. Combined Use of Air Sampling Instruments and Video Filming: Experience and Results During Six Years of Use. Appl. Occup. Environ. Hyg. 8:344 (1993).
6. American Industrial Hygiene Association: Odor Thresholds for Chemicals with Established Occupational Health Standards. AIHA, Fairfax, VA (1989).
7. Occupational Safety and Health Administration: OSHA Technical Manual, Chapter 1. CPL 2–2.208. OSHA, Washington, DC (Feb. 5, 1990).
8. Spengler, J.K.; Treitman, R.D.; et al.: Personal Exposures to Respirable Particulates and Implications for Air Pollution Epidemiology. Environ. Sci. Technol. 19:200 (1985).
9. McCarthy, J.: Physical and Biological Markers to Assess Exposure to Environmental Tobacco Smoke. Doctoral Dissertation, Harvard School of Public Health, Boston, MA (October 28, 1987).
10. Davies, C.N.: The Entry of Aerosols into Sampling Tubes and Heads. Brit. J. Appl. Phys. (J. Phys. D.) Sec. 2, 1:921 (1968).
11. Bien, D.T.; Corn, M.: Adherence of Inlet Conditions for Selected Aerosol Sampling Instruments to Suggested Criteria. Am. Ind. Hyg. Assoc. J. 32:453 (1971).
12. Pickett, W.E.; Sansone, E.B.: The Effect of Varying Inlet Geometry on the Collection Characteristics on a 10-mm Nylon Cyclone. Am. Ind. Hyg. Assoc. J. 34:421 (1973).

13. Breslin, J.A.; Stein, R.L.: Efficiency of Dust Sampling Inlets in Calm Air. Am. Ind. Hyg. Assoc. J. 36:576 (1975).

14. Agarwal, J.K.; Liu, B.Y.H.: A Criterion for Accurate Aerosol Sampling in Calm Air. Am. Ind. Hyg. Assoc. J. 41:191 (1980).

15. Tufto, P.A.; Willeke. K.: Dependence of Particulate Sampling Efficiency on Inlet Orientation and Flow Velocities. Am. Ind. Hyg. Assoc. J. 43:437 (1982).

16. American Industrial Hygiene Association Aerosol Technology Committee: Interim Guide for Respirable Mass Sampling. Am. Ind. Hyg. Assoc. J. 31:133 (1970).

17. Dunmore, J.H.; Hamilton, R.J.; Smith, D.S.G.: An Instrument for the Sampling of Respirable Dust for Subsequent Gravimetric Assessment. J. Sci. Inst. 41:669 (1964).

18. U.S. Environmental Protection Agency: Ambient Air Monitoring Reference and Equivalent Methods. Fed. Reg. 49(55):10454 (March 10, 1984).

19. Dockery, D.W.; Pope, A., III; et al.: An Association Between Air Pollution and Mortality in Six U.S. Cities. New Eng. J. Med. 329:1753 (1993).

20. Silverman, L.; Billings, C.E.; First, M.W.: Particle Size Analysis in Industrial Hygiene. Academic Press, New York (1971).

21. MacDonald, J.R.: Are the Data Worth Owning? Science 176 (4042) :1377 (June 1972).

22. Leidel, N.A.; Busch, K.A.: Statistical Methods for the Determination of Noncompliance with Occupational Health Standards. DHEW (NIOSH) Pub. No. 75-159. NIOSH, Cincinnati, OH (1975).

23. Leidel, N.A.; Busch, K.A.; Crouse, W.E.: Exposure Measurement Action Level and Occupational Environmental Variability. DHEW (NIOSH) Pub. No. 75-159. NIOSH, Cincinnati, OH (1975).

24. Leidel, N.A.; Busch, K.A.; Lynch, J.R.: Occupational Exposure Sampling Strategy Manual. Pub. No. PB-274-792. National Technical Information Service, Springfield, VA (1977).

25. Rock, J.C.; Cohoon, D.: Some Thoughts About Industrial Hygiene Sampling Strategies, Long Term Average Exposures, and Daily Exposures. Report OEHL 81–32. National Technical Information Service, Springfield, VA (July 1981).

26. Brown, R.H.: Personal Communication (1994).

27. VanderKolk, A.L.: Sampling and Analysis of Organic Solvent Emissions. Am. Ind. Hyg. Assoc. J. 28:588 (1967).

28. American Public Health Association Intersociety Committee: Methods of Air Sampling and Analysis, 2nd ed. M. Katz, Ed. APHA, Washington, DC (1977).

29. Drinker, P.; Hatch, T.: Industrial Dust, 2nd ed. McGraw-Hill, New York (1954).

30. Greenburg, L.; Smith, W.G.: A New Instrument for Sampling Aerial Dust. U.S. Bureau of Mines, Report. Invest. 2392 (1922).

31. Lodge, J.P.; Chan, T.L.; Eds.: Cascade Impactor Sampling and Data Analysis. American Industrial Hygiene Association, Akron, OH (1986).

32. Kruse, R.H., Barbeito, M.S.: A History of the American Biological Safety Association. Part I, Safety Conferences 1954–1965. J. Am. Biolog. Safety Assoc. 2(4), pp. 10–25 (1997).

33. First, M.W. HEPA Filters. J. Am. Biological Safety Assoc. 3(1) pp. 33–42 (1998).

34. First, M.W. ; Silverman, L.: Air Sampling with Membrane Filters. A. M. A. Arch. Of Ind. Hyg. And Occ. Med. 7:1, pp. 1–11 (1953).

35. Bertoni, G.; Febo, A.; Perrino, C.; Possanzini, M.: Annular Active Diffusive Sampler: A New Device for the Collection of Organic Vapors. Annali di Chimica 74:97 (1984).

36. Gottfried, G.; Yarko, J.; Olinger, C.; Lewis, R.D.: A Pilot Study to Develop a Method for Sampling and Analysis of Middle Distillate Fuels and Petroleum Solvents. Presented as Paper No. 148 at the American Industrial Hygiene Conference, San Francisco, CA, May 15–20, 1988.

37. Koutrakis, P.; Sioutis, C.; et al.: Development and Evaluation of a Glass Honeycomb Denuder/Filter Pack System to Collect Atmospheric Gases and Particles. Environ. Sci. Technol. 27:2497 (1993).

38. Gressel, M.G.; Heitbrink, W.A.; Eds.: Analyzing Workplace Exposures Using Direct Reading Instruments and Video Exposure Monitoring Techniques. DHHS(NIOSH) Pub. No. 92–104. NIOSH, Cincinnati, OH (August 1992).

39. Arenas, R.V.; Carney, K.R.; Overton, E.B.: Portable Multigas Monitors for Air Quality Evaluation, Part II: Survey of Current Models. Amer. Lab. p. 25 (July 1993).

40. National Institute for Occupational Safety and Health: Certification of Gas Detector Tube Units. Fed. Reg. 38:11458 (May 8, 1973).

41. National Institute for Occupational Safety and Health: Certified Equipment List. DHHS (NIOSH) Pub. No. 80–144. U.S. Government Printing Office, Washington DC (June 1980).

42. Rose, V.E.; Perkins, J.L.: Passive Dosimetry—State of the Art Review. Am. Ind. Hyg. Assoc. J. 43:605 (1982).

43. TSI, Inc.: Model 3500 Respirable Aerosol Mass Monitor, Piezobalance. TSI, Inc., St. Paul, MN.

44. Patashnick, H.; Rupport, E.G.: Continuous PM–10 Measurements Using the Tapered Element Oscillating Microbalance. J. Air Waste Mgmt. Assoc. 41:1079 (1991).

45. U.S. EPA: Ambient Air Monitoring Reference and Equivalent Methods. 40 CFR Part 53.

46. MIE Corp.: Respirable Dust Monitor Model RDM-101. MIE Corp., Bedford, MA.

47. Thompson, E.M.; et al.: Laboratory Evaluation of Instantaneous Reading Dust Monitors, USDL/MSHA. Presented at the American Industrial Hygiene Conference, Houston, TX, May 19–23, 1980.

48. Miniram, Model PDM–3: MIE, Bedford, MA 01730.

49. Particle Measuring System, Inc.: 1855 South 57th Court, Boulder, CO 80301.

50. Lilienfeld, P.: Selective Detection of Asbestos Fiber Aerosols by Electromagnetic Alignment and Oscillation. In: Proc. of the seminar, Trends in Aerosol Research II. Schmidt–Ott, A., Ed. University of Duisburg, Duisburg, Germany (June 17, 1991).

51. Page, S.J.: Correlation of the Fibrous Aerosol Monitor with the Optical Membrane Filter Count Technique. U.S. Bureau of Mines, Report of Investigations No. 8467. U.S. Dept. of the Interior, Washington, DC (1980).

52. Phanprasit, W.; Rose, V.E.; Oestenstad, R.K.: Comparison of the Fibrous Aerosol Monitor and the Optical Fiber Count Technique for Asbestos Measurement. Appl. Ind. Hyg. 3:28 (1988).

53. National Institute for Occupational Safety and Health: NIOSH Manual of Analytical Methods, 3rd ed. U.S. Government Printing Office, Washington, DC (February 15, 1984). First Supplement, May 15, 1985. Second Supplement, August 15, 1987.

54. Intersociety Committee: Methods of Air Sampling and Analysis, 3rd ed. J.P. Lodge, Jr., Ed. Lewis Publishers, Inc., Chelsea, MI (1988).

55. American Society for Testing and Materials, D-22 Committee: Book of ASTM Standards, Part 23, Industrial Water; Atmospheric Analysis. ASTM, Philadelphia, PA (annual issue).

56. Pool, R.: The Smallest Chemical Plants. Science 263:1698 (1994).

Chapter 2

Occupational Air Sampling Strategies

James C. Rock, Ph.D., CIH, PE

Occupational Health and Safety Institute, Texas A&M University, College Station, Texas

CONTENTS

Introduction—Why Sample Workplace Air?

Occupational air sampling is a key element of occupational exposure control programs. It is an effective tool for setting priorities for interventions to improve workplace health and safety. Air sampling also directly influences health and liability insurance eligibility and insurance rates, and it supports compliance decisions based on occupational exposure limits.

Key Definitions

An **Acceptable** (or Tolerable) occupational exposure is one that is acceptable in light of available evidence, including applicable Occupational Exposure Limits (OELs).

The **Action Level (AL)** is defined here as 50% of the OEL.

An **Area Sample** is an air sample taken at a fixed location in a workplace.

Area Monitoring is the process of using real-time air monitoring devices to chart airborne concentration as a function of time at fixed locations.

A **Breathing Zone Sample** is an air sample taken in such a way that the air sampled is within 30 cm (1 ft) of the nostrils of the person being sampled. It is often called a breathing zone sample even when the subject is wearing a respirator.

A **Confidence Interval** is a range of values bounded by two statistics, U and L. When θ is the true value of the parameter, $P(L \leq \theta \leq U) = (1 - \alpha)$. The **Confidence Level** is $\gamma = (1 - \alpha)$, sometimes referenced as $100 \, \gamma \%$ **Confidence**. For example, a 95% confidence interval about the mean will include the mean in about 19 of 20 trials. Contrast with Tolerance Interval, which bounds a proportion of population data values.

A useful **Goal** is an observable, achievable, and exceedable endpoint.

A **Homogeneous Exposure Group (HEG)** is a group of employees having similar exposures in the sense that monitoring airborne exposures for any one of them provides data useful for evaluating the exposure conditions for all members of the group. An individual may be a member of several homogeneous exposure groups. An HEG may be confirmed by examining data from several of its members.

A **Maximum Likelihood Estimator (MLE)** is a statistic characterized by minimum variance. The MLE statistics in this chapter are slightly biased, but they exhibit significantly smaller variance than the unbiased Sample Mean and Sample Variance. Any function of an MLE is itself an MLE statistic. This is not true, in general, of unbiased estimators.

A **Minimum Variance Unbiased Estimator (MVUE)** is an unbiased statistic with variance comparable to an MLE estimator. These are often computed from infinite series, and are impractical for the calculator-aided calculations discussed in this chapter.[1]

An **Occupational Air Sampling Strategy** is a plan to use available air sampling resources to quantify airborne concentrations in support of an Occupational Exposure Control Program. The strategy guides industrial hygiene actions and decisions toward the goal of accurate exposure estimates for each worker.[2]

An **Occupational Exposure Control Program** is that set of resources devoted by management to ensure that all airborne chemical concentrations are kept below appropriate OELs. Typically, it includes air sampling, biological monitoring, worker hazard communication training, mandatory work practices, material substitution, process isolation or enclosure, local exhaust ventilation, general ventilation, administrative work rules, and personal protective equipment (while better options are being installed or after all other exposure control options have been exhausted).

The **Occupational Exposure Limit (OEL)** is a generic term for a pair of numbers: 1) the criterion airborne concentration and 2) the time period over which workplace concentrations are averaged.[2] Some substances have more than one OEL: a Threshold Limit Value–time-weighted average (TLV®–TWA) for 8-hr exposures, a Threshold Limit Value–short-term exposure limit (TLV–STEL) for 15-min exposures, and a ceiling value (TLV–C). The OEL may be a consensus limit such as the American Conference of Governmental Industrial Hygienists (ACGIH®) TLV, a regulated limit such as the Occupational Safety and Health Administration Permissible Exposure Limit (OSHA PEL), or an internal exposure limit set by a chemical supplier or by local management. The goal of an occupational exposure program should be to keep all exposures below OELs and as low as reasonably achievable.

A **Parameter** is a number describing the population of exposures being sampled. The two parameters for the Gaussian distribution are μ and σ, and the two parameters for the lognormal distribution are $g\mu$, $g\sigma$.

The **Population** is the total number of exposures available for sampling. One data set containing n measured exposures is a statistical sample of the population.

Random Sampling occurs when samples are selected from the statistical population of all relevant air samples so that each sample has an equal probability of being selected. This is a three-step process: 1) define the population of relevant samples, 2) randomly sample that population, and 3) verify that the data support the HEG hypothesis.

A **Sample** in a statistical context is one data set (n measurements). In an air sampling context, it may refer to one measurement, as in collecting a breathing zone sample.

Data are statistically **Significant** at level α when the P-value $\leq \alpha$. The P-value is the probability that the test statistic will have a value as extreme as the one observed when the null hypothesis is true. α is the level of **Significance** (see Confidence).

A **Strategy** is a careful plan to use available resources effectively to achieve a goal. This chapter outlines a coherent strategy for occupational air sampling that allows reasonable estimates of worker exposure levels with reasonable numbers of samples.

A **Statistic** is a number that can be computed from data (n measurements). There is no need to assume a value for any unknown parameter. Key descriptive statistics for any occupational exposure sample include M, S, UCL, X_{95}, UTL, GM, GS, $GUCL$, GX_{95}, and $GUTL$.

The **Time Constant** of a system, tau τ, is the time it takes the output to fall to $1/e = 36.8\%$ of its initial value, or rise to $1 - 1/e = 63.2\%$ of its final value after all stimuli are removed. Many systems, including toxicokinetic and toxicodynamic pathways, are dominated by one slow subsystem. Thus, to a first approximation, their impulse response can be modeled by $y(t) = y_0 e^{-(t/\tau)}$, where τ is called the natural time constant of the system. When possible, the time constant of a direct-reading instrument should match the time-constant of the physiological response to a workplace stressor.

A $100\gamma\%$ **Tolerance Interval** for the $100p$ percentile is a range of values that with a probability, γ, contains a fraction, p, of the data; γ is the probability that $100p\%$ of data lie within the stated interval. For example, let UTL be the 90% one-sided upper tolerance limit for the 95th percentile of a distribution. When $UTL < OEL$, there is 90% confidence that no more than 5% of all data are greater than OEL. Contrast with Confidence Interval, which bounds a range of values for a parameter of a parametric distribution representing the data.

Mathematical Symbols

This chapter uses a common statistical convention. Uppercase letters represent population statistics or parameters estimated from data. Lowercase letters represent individual values of the data. Lowercase Greek letters describe the true, and usually unknown, parameters of the population being sampled. A symbol preceded by G or g represents a statistic or parameter for a lognormal distribution.

x = value of a data point

X_p = $(100)(p)$ percentile, a proportion p of a Gaussian population is $\leq X_p$: Eq. 7

GX_p = $(100)(p)$ percentile, a proportion p of a lognormal population is $\leq GX_p$: Eq. 7

M = sample mean, an estimate of population mean, μ: Eq. 2 and 5

Me = sample median, estimate of population median: $Me = X_{0.5}$ or $X_{50\%}$: Eq. 1 and 5

Mo = sample mode, the most frequent value of discrete data or the value at which a continuous probability density function reaches its peak: Eq. 5

S = sample standard deviation, an estimate of σ: Eq. 2

GM = sample geometric mean, an estimate of $g\mu$: Eq. 3

GS = sample geometric standard deviation, an estimate of $g\sigma$: Eq. 3

R = range of data, $[x_{max} - x_{min}]$

p = proportion of a population, $0 \leq p \leq 1$

$P(\bullet)$ = probability of the event in parentheses, $0 \leq P \leq 1$

α = probability of a type I error, rejecting the null hypothesis when it is true. The confidence level, $\gamma = (1 - \alpha)$.

β = probability of a type II error, accepting the null hypothesis when it is false or rejecting an alternate hypothesis when it is true.

μ = population mean

σ = population standard deviation

$g\mu$ = population geometric mean

$g\sigma$ = population geometric standard deviation

γ = confidence that a hypothesis is true, $0 \leq \gamma \leq 1$

n = number of data points in a population sample

v = degrees of freedom, $v = df = n - 1$, for S or GS.

UCL = upper confidence limit about M for Gaussian distribution: Eq. 6a

LCL = lower confidence limit about M for Gaussian distribution: Eq. 6a

UTL = upper tolerance limit for a proportion p of a Gaussian distribution: Eq. 8

$GUCL$ = upper confidence limit about GM for lognormal distribution: Eq. 6b

$GLCL$ = lower confidence limit about GM for lognormal distribution: Eq. 6b

$LLCL$ = Land's lower confidence limit about M for lognormal distribution: Eq 6c

$LUCL$ = Land's upper confidence limit about M for lognormal distribution: Eq 6c

$GUTL$ = upper tolerance limit for a proportion p of a lognormal distribution: Eq 8

$k_{p, n, \gamma}$ = coefficient for one-sided tolerance interval for proportion p, confidence γ, based on n data points from a Gaussian distribution: Table 2-7.

$t_{p, n-1}$ = Student's t coefficient, one-sided, with confidence p, for n data values: Table 2-11.

Zp = normalized z variate, $z_p = (X_p - \mu)/\sigma \approx (X_p - M)/S$: Table 2-10

F(·) = cumulative probability distribution function (CDF): Eq. 4

f(·) = probability density function (PDF): Eq. 4

The Cost of Air Sampling

The 20–80 rule is a great management principle to ensure proper investment of overall exposure control program resources. Devote about 20% of industrial hygiene resources to air sampling and priority setting. Devote 80% to training, engineering controls, process improvement, personal protective equipment, recordkeeping, and occupational medicine. This budget allows proper assessment of the relative risk of workplace exposures so that corrective action can be taken on a "worst first" basis.

It is reasonable to expect that over a period of time all airborne concentrations can be reduced below appropriate OELs and personal protective equipment can be eliminated in most workplaces. The half life for process changes is on the order of 3 years, meaning that exposure levels can be expected to drop by one-half about every 3 years once top management starts supporting such a program. As with all other aspects of quality engineering, resulting production efficiencies and product quality improvements generally more than pay for the cost of such efforts.

Purposes of Air Sampling

There are three purposes for occupational air sampling that merit attention in this publication:

- To characterize Air Quality for Occupational Exposure Control
- To characterize Process Emissions
- To characterize Air Quality for Regulatory Enforcement.

Sampling for Occupational Exposure Control—Breathing Zone Sampling

Breathing zone measurements are used 1) by compliance officers to enforce OSHA regulations, 2) by insurance carriers to assess underwriting risk, and 3) by industry to set priorities for process improvement and to establish locations and tasks requiring interim personal protective equipment. Direct-reading continuous monitors provide warning to workers with potential for exposure to conditions that are immediately dangerous to life and health (IDLH).

Sampling for Process Emissions—Area Sampling

Area measurements are used: 1) to continuously monitor for leaks to prevent fires, explosions, and IDLH airborne concentrations, 2) for process characterization for engineering control design and engineering control verification (such as laboratory safety hoods or biosafety cabinets), and 3) as basis for design and verification of breathing zone sampling strategies.

Area Monitoring for IDLH Conditions. The area measurements that are most critically important in an occupational exposure control program are those that warn of and prevent IDLH exposures. These are provided by continuous monitors connected to both alarms and process control devices. These monitors may be networked to provide coverage not only of workplace emissions but also of plant boundaries, providing community emergency responders with an early warning alarm as well.

Area Monitoring for Process Improvement and Verification of Installed Engineering Controls. Continuous area monitoring at fixed points in space can be used to correlate airborne contaminant concentrations and concentration gradients with process variables. This technique is widely used in nuclear industries.[3] There is a large and growing literature related to area monitoring as a means for confirming the satisfactory performance of ventilation and air cleaning systems. For principles of air movement, containment, evaporation rates, scrubbing, and filtration factors relevant to this task, see references such as Chapter 20 of this publication on stack and duct sampling, ACGIH's *Industrial Ventilation: A Manual of Recommended Practice,*[4] *ASHRAE Handbooks,*[5] Industrial Ventilation: Engineer-ing Principles,[6] and *The EPA Engineering Handbook.*[7]

Meteorological data and process data should be recorded simultaneously with area monitoring data. Ventilation systems that work as designed under prevailing wind conditions may fail to exhaust, or worse, may re-entrain contaminants through outside air intakes when winds approach from other directions. Careful area monitoring can reveal and quantify these problems, producing data directly useful to the process engineering team responsible for process improvements.

Sampling for Regulatory Compliance

Statistical tools for compliance sampling are well documented.[8–11] Compliance samples are seldom random samples. A compliance officer uses all available air sampling data, industrial hygiene reports, and his or her own observations of the workplace to identify locations of likely emissions, and selects a maximally exposed employee for a breathing zone measurement. If that measurement, corrected for all known bias and uncertainty in the sampling and analytical processes, exceeds the PEL, then a citation is possible. Although it is common to make noncompliance decisions on the basis of one measurement, it is rare to make a compliance decision on the basis of one measurement. A statistically significant number of representative measurements are required to build the case that a large exposure measurement is a rare event. The norm is that routine surveillance requires many measurements to demonstrate the distribution of exposures, whereas noncompliance may be declared on the basis of one measurement that exceeds the standard.

This reflects the rigorous reality of statistical hypothesis testing. The hypothesis of compliance can be rejected at a chosen level of significance, or it can be accepted, but it cannot be proven. In contrast, whenever the compliance hypothesis is rejected, noncompliance appears proven.

Reasons for an Air Sampling Strategy

A perfect axiomatic air sampling strategy exists.[12] Shannon's Sampling Theorem gives the number of measurements needed to obtain perfect information about airborne concentrations in the workplace as a function of space and time.[13] More than 250,000 measurements per cubic meter per hour are needed. Assuming a reasonable cost of $10 per data point, an axiomatic air monitoring program would cost $2.5 million per hour per cubic meter of breathing zone air.

The affordable alternative to perfect information is statistical sampling to estimate true exposure patterns from a small number of measurements. The uncertainty associated with statistical exposure estimates is inversely proportional to a fractional power of the number of data points. More measurements provide better information than fewer. The air sampling strategy of this chapter provides a framework for determining how many measurements are needed. It also provides guidelines for corrective actions to be taken in the workplace based on the distribution of exposures about the applicable OEL.

Occupational Exposure Limits and Averaging Times

An OEL has two parts: a concentration and an averaging time. Common examples include a long-term average exposure (LTA), an 8-hour time-weighted average exposure (TLV–TWA), a 15-minute short-term

exposure (TLV–STEL), or a 5-minute ceiling limit (TLV–C). It is inappropriate to pool, for statistical analysis, data collected with different averaging times. The statistical tools of this chapter work equally well for TLV–TWA, TLV–STEL, and TLV–C data sets. They do not work for a pooled data set containing a mixture of some LTA, some TWA, some STEL, and some C values.

Occupational Exposure Control Program Goals

Primary Goal—Know and Control All Exposures

A primary goal of an occupational exposure control program is to be sure that all hazardous exposures are known and controlled to a tolerable level. This means that employees, supervisors, and managers know their exposures, know and accept their possible health consequences, and participate in mitigating the risk associated with these exposures. A credible occupational air sampling strategy is a necessary foundation for this goal.

Secondary Goal—Continuous Improvement

Establish priorities for remedial action based on air sample data and OELs. Keep the priority list current by adding new risks as they are identified and deleting older risks as engineering controls or process changes achieve tolerable exposure levels. Use the priority list to invest available remedial resources to maximize marginal improvement of workplace health and safety. This results in continuous improvement of working conditions.

Components of a Sampling Strategy

The recommended air sampling strategy is a closed loop with six components:[2] 1) characterization, 2) risk assessment and sampling priorities, 3) air sampling and analysis, 4) data interpretation, 5) recommendations and reporting, and 6) the reevaluation schedule.

Characterization

The goal of this step is to completely characterize the workplace, the airborne stressors, and the workforce for the purpose of designing a suitable sampling strategy. If homogeneous exposure groups exist, they provide the considerable benefit of permitting stratified sampling plans with reasonable statistical power from a small number of workplace measurements. Characterization should proceed cautiously because a growing body of experience shows that each HEG must be confirmed by data analysis.[14–16]

Workplace characterization starts with a complete inventory of chemical, physical, and biological agents. Plant diagrams should be highlighted to show the location of all process steps and equipment likely to release contaminants into the workplace air. If prior air sampling data exist, those data should be used to confirm the inventory and improve the maps.

Stressor characterization uses OSHA-mandated material safety data sheet (MSDS) data and process models to compile estimates of airborne concentration levels under process conditions.[17] Concentration estimates should be compared with applicable OELs and estimated concentration contours overlayed on plant and process diagrams to identify locations and personnel where air sampling may be needed.

Workforce characterization involves obtaining task assignment data, demographic data, and health data for all members of the workforce. Evaluate these data to discover clusters of occupationally related symptoms. If present, task locations should be compared with the concentration contours from the workplace characterization. This often leads directly to high priority sampling requirements.

An HEG is a group of employees who have comparable exposures. Each group may be tentatively identified from data on the workplace, the stressors, and the workforce. The HEG promotes air sampling efficiencies because a breathing zone air measurement from any member of the group is considered representative of the exposures received by all members of the group. Each HEG must be confirmed by statistical analysis of breathing zone measurements representing that group (see "Descriptive Statistics," page 25).

Risk Assessment and Sampling Priorities

Risk assessment is a multidimensional problem involving factors such as exposure levels, health consequences of overexposure, whether the risk is voluntarily accepted, whether the perception is that alternate risks are higher, the immediacy of consequences if the risk is accepted, and who is at risk. Two parameters suffice to manage an occupational air sampling strategy: the exposure ranking and the health effect ranking.

Exposure ranking is a monotonic semisubjective scale based on all available air sampling data and on process models that predict both airborne concentrations and their frequency of occurrence. There are many ways to define the scale, and an appropriate one should be selected in each instance. Table 2-1 illustrates the concept.

The **very low exposure level** of OEL/10 is the level below which recurring sampling is not required. It is synonymous with the unexposed worker level in the 7th

TABLE 2-1. Exposure Ranking

Rank	Description	Comment
0	Very low exposure	Observed concentrations less than 10% of the OEL
1	Low exposure	Observed concentrations less than the AL (50% OEL)
2	Moderate exposure	Frequent exposure at concentrations below the AL or infrequent exposure at concentrations between the AL and OEL
3	High exposure	Frequent exposure at concentrations near the OEL or infrequent contact with the stressor at concentrations above the OEL
4	Very high exposure	Frequent contact with the stressor at concentrations above the OEL

edition of this text.[18] It was based on an American Society of Heating, Refrigerating and Air-Conditioning Engineers, Inc. (ASHRAE) guideline that concentrations in office spaces should be kept below 10% of the OEL. Additionally, the population standard for ionizing radiation is 10% of the occupational standard.[19] A U.S. government report recommends 10% of the annual intake limit for radioisotopes as the threshold below which occupational breathing zone monitoring is not required.[3] The low exposure level is the level below which minimal recurring breathing zone sampling is required. It is based on the NIOSH action level concept (50% of the OEL).[8]

Health effect ranking is based on the consequences of exposure to the stressor. It, too, is structured on a five-point semisubjective scale that can be tailored to fit various situations. Table 2-2 displays commonly used definitions.

Sampling Priority

Sampling priorities should be set on the basis of the exposure ranking (ER) and the health effects ranking (HER) scales. As Table 2-3 shows, there is a very low priority for sampling exposures with a score of (ER,HER) = (0,0) and a very high priority for sampling exposures with a score of (4,4). In fact, it is good policy to require immediate cessation of any process producing a (4,4) ranking until steps can be taken to reduce the exposures. It is appropriate for an industrial hygienist to use professional judgment for intermediate priority scores and to move the assigned priority up or down one level depending on the size of the exposed population or other factors not considered by the health effect and exposure rankings.

TABLE 2-2. Health Effect Ranking

Rank	Description	Comment
0	NONE	No known permanent health effects. No treatment needed; no sick leave is involved.
1	MILD	Reversible health effects with suspected consequences. Medical treatment usually not required for recovery. Sick leave is seldom involved.
2	SERIOUS	Severe reversible health effects. Medical treatment required for recovery. Lost time and sick leave are usually involved.
3	CRITICAL	Irreversible health effects; not treatable. New life style required to adapt to the disability.
4	IDLH	Life threatening or totally disabling injury or illness.

Air Sampling and Analysis

The resources available for air sampling should be allocated in a way that ensures priority for the highest ranked classes while reserving adequate resources for lower priority classes because unexpected results do occur during sampling campaigns. About 80% of the available effort should be devoted to the high and very high risk categories and about 20% to the other categories, as in Table 2-4.

Data Interpretation

Air sampling data must be interpreted in the context of all available information. There are three decisions possible with regard to each exposure scenario:

- The exposures are acceptable (H_A: μ < OEL)
- The exposures are unacceptably high (H_A: μ > OEL)
- There are insufficient data to make a decision (H_0: μ = OEL)

The sampling program plus professional judgment should provide data for making one of the first two

TABLE 2-3. Breathing Zone Sampling Priorities

Health Effect	Sampling Priority				
4	M	H	H	VH	VH
3	L	M	H	H	VH
2	L	M	M	H	H
1	T	L	M	M	H
0	T	T	L	L	M
	0	1	2	3	4	**Exposure**

TABLE 2-4. Budget Proportions for Air Sampling

Rank	Budget	20% for Sampling	80% for Intervention
Very High	55%	11.0%	44.0%
High	24%	4.5%	19.2%
Moderate	12%	2.4%	9.6%
Low	6%	1.2%	4.8%
Very Low	3%	0.6%	2.4%

decisions. The goal of an exposure control program is to make process and administrative changes until all exposures become tolerable. As emphasized by Mulhausen and Damiano,[16] exposure assessments are made in view of all available data: models, qualitative and quantitative. Thus, the insufficient data problem must be dealt with by either collecting more data or making the decision that exposures are unacceptable. Assuming there are sufficient data of sufficient quality, it must be determined whether the data support the decision of acceptable or unacceptable exposures. Statistical tools for making this decision are summarized below.

Descriptive Statistics

The first step in analysis of every data set is to compute descriptive statistics and plot the data on probit and log-probit scales. Methods for computing descriptive statistics, and confidence and tolerance intervals are given in "Statistical Tools for IH Decision-making," page 32. An example is provided in "Example of Data Interpretation," page 37. The descriptive statistics are used in setting priorities and at decision points in the decision flow diagram (Figure 2-1). The graphs are used to determine whether parametric hypothesis testing is appropriate. If the data do plot as a straight line on either graph, then the parametric tests described in the following sections can be used to make statistically sound decisions.

If the data do not plot as a straight line, or if the data have a GS > 4.5, it is likely that a bimodal or multimodal distribution exists.[3] Such data usually do not represent an HEG and an analysis of variance may be used to confirm the nonhomogeneity of the presumed HEG.[15] When data fail to support the hypothesis of an HEG, decisions cannot be based on parametric hypothesis testing. In these cases, a nonparametric tolerance interval test may be used to supplement a decision based on professional judgment (Table 2-5 at the end of the chapter).

Statistics for Compliance

The NIOSH sampling strategy is at the heart of compliance decisions.[8] It is based on a 95% confidence

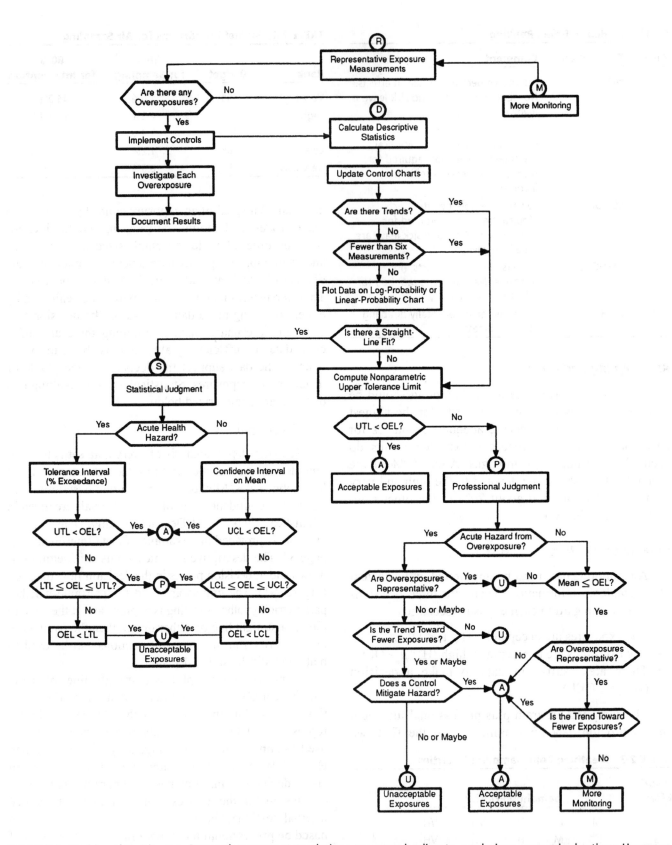

FIGURE 2-1. Decision Flow Diagram. Rectangles represent conclusions, processes leading to conclusions, or required actions. Hexagons represent decision points that branch through the decision flow diagram. Circles are connectors. Circles alone represent branching out of the decision flow diagram. Circles touching a rectangle represent the entry point for decisions that branched out elsewhere. Circles are identified by a letter loosely representing their functions in the decision-making process: A – the exposures are acceptable (tolerable); M – more monitoring is needed; P – use professional judgment to decide; R – restart the monitoring process; S – use parametric statistical hypothesis testing; U – the exposures are unacceptable.

interval about the mean of a group of measurements. If the 95% upper confidence limit (*UCL*) is below the OEL, the exposure observed is in compliance. If the 95% lower confidence limit (*LCL*) is above the OEL, the exposure observed is out of compliance. If the OEL is between the *LCL* and the *UCL*, no decision is possible at 95% confidence.

An extension of this concept is applied to a single measurement. Under the assumption that the relative standard deviation of a measurement is less than 25% (the variability allowed for an approved air sampling method), a confidence interval is assumed for a single measurement. This ignores spatial and temporal concentration gradients in the workplace air, and accounts only for random variation in air sampling and chemical analysis.

Finally, it is recommended that managers everywhere consider using an Action Level, *AL* = OEL/2 as the basis for internal decisions. Setting goals that keep exposures compliant with the *AL* is in the spirit of two important principles: first, it helps foster continuous improvement, and second, it supports the goal of keeping all exposures as low as reasonably achievable. Clearly, it also helps insure compliance during any regulatory inspection or audit.

Statistics for Exposures to Chronic Stressors

For stressors that pose a chronic risk, the confidence interval around the mean is a good estimator of long-term risk. Exposures for which the 95% *UCL* << OEL are usually deemed tolerable.

The data used should be representative of long-term exposures. This means that the sampling strategy should have an equal probability of sampling during all relevant conditions: day of the week, week of the year, under surge production or changeover production, and shift (if there is more than one). If your program is based on the *AL*, mentally substitute that value for the OEL in the discussion that follows.

Statistics for Exposures to Acute Stressors

For stressors that pose an acute risk, a one-sided tolerance interval (at 90% confidence) for the 95th percentile of the exposure distribution is recommended. A determination that the *UTL* << OEL establishes 90% confidence that no more than 5% of HEG exposures exceed the OEL.

Guidelines for the Number of Measurements Needed

The required number of measurements depends on the goals for a specific sampling campaign. The original NIOSH sampling strategy monograph provides guidance for estimating peak exposures, short-term exposures, and 8-hr TWA exposures.[8] The resultant rules of thumb are

that at least 6 measurements are required for a valid estimate of the confidence interval around the mean and more than 11 are required to estimate the variance.[2] At the upper end, one major corporation restricts to 18 the maximum number of measurements for characterizing a single exposure scenario.[21]

For **compliance sampling**, one or more measurements determine whether the exposure sampled is in or out of compliance. This technique does not predict overall conditions in the workplace; it is valid only for estimating the meaning of the exposure(s) observed at the time and place of the compliance sampling campaign, as specified in a regulation.

For stressors posing **chronic risk**, a minimum of six representative measurements is recommended to estimate the 95% *UCL* on the mean of all exposures in the workplace, for populations that converge rapidly under the Central Limit theorem. However the workplace will need a very low average exposure to show that *UCL* < OEL under these conditions.

There are better ways to estimate the number of samples required to test exposure measurements using the null hypothesis, H_0: M = OEL. An employer wants to reject the null with the alternate hypothesis of compliance, H_A: μ < OEL, while avoiding the alternate hypothesis of noncompliance, H_A: μ > OEL. Each sample should have sufficient measurements to achieve the desired confidence, $(1 - \alpha)$, of rejecting the null hypothesis (deciding either compliance or noncompliance), and sufficient power, $(1 - \beta)$, to identify a borderline workplace by accepting the null hypothesis. Table 2-6 is one such compilation, created for α = 0.05 and β = 0.10 and using a test statistic based on MLE estimators for the sample mean and sample standard deviation.[22] Table 2-6 was based on the robustness of the t-test for a sample drawn from a skewed distribution. Rappaport and Selvin tested their model with both simulated and real data. An alternative analysis appeared recently. Observing the incomplete convergence of the distribution for the sample mean to the t-distribution when *n* is small, Hewett shows higher precision with minimum variance unbiased estimators (MVUE) for the sample mean and sample standard deviation.[23] He also replaces the *t*-values in the equation for the upper and lower confidence limit with Land's *C*-values.[23]

Both articles emphasize the need for qualified professional judgement, and both dramatically illustrate the trade-off between number of measurements and cleanliness.[22, 23] The bottom line is that statistically significant decisions can be made with a few measurements in clean workplaces, but require many more measurements in workplaces characterized either by higher average exposures or higher variance.

TABLE 2-6. Sample Size for Testing Mean of a Sample from Lognormal Population with Two-Sided 95% Confidence Level and 90% Power[21] ($\alpha = 0.05$, $\beta = 0.10$, F = $g\mu$/OEL[A])

F	$g\sigma$ = 1.5	2.0	2.5	3.0	3.5
0.10	2	6	13	21	30
0.25	3	10	19	30	43
0.50	7	21	41	67	96
0.75	25	82	164	266	384
1.25	25	82	164	266	384
1.50	7	21	41	67	96
2.00	2	6	11	17	24
3.00	1	2	3	5	6

The population is assumed to be lognormal with geometric mean, $g\mu$, geometric standard deviation, $g\sigma$, arithmetic mean, μ, and arithmetic standard deviation, σ. The sample statistics for this test are arithmetic mean, M, sample standard deviation, S, and standard deviation of the mean, $S_M = S/\sqrt{n}$. The MLE estimators for these statistics can be used when n>10. The critical t-values are $t_{0.05, (n-1)}$ and $t_{0.95, (n-1)}$. The test statistic is T = (M – OEL)/S_M. The employer's test rejects H_0:μ = OEL, and accepts H_A:μ > OEL when T < $t_{0.05, (n-1)}$ ≤–1.645. The OSHA test rejects H_0:μ = OEL and accepts sH_A:μ > OEL when T > $t_{0.95, (n-1)}$ ≥+1.645.

[A]Modified from Rappaport and Selvin[22]; Table I.

For **estimating the variance** of any data set, a minimum of 11 measurements is recommended. For an HEG with *n* members, the larger of 11 measurements or $\sqrt{n}$ measurements is recommended.

For stressors posing **acute risk**, to test the hypothesis that no more than $100\alpha\%$ of exposures exceed the OEL, the one-sided *UTL* can be estimated heuristically from a minimum of $(1+1/\alpha)$ measurements. Thus, when $\alpha = 0.05$, at least 21 measurements are recommended to obtain a usable *UTL* to test that no more than $100\alpha = 5\%$ of all exposures exceed the OEL. Note that $p = 1 - \alpha$, and an equivalent statement for the hypothesis is that at least $100p = 95\%$ of all exposures are less than the OEL. (See Tables 2-7 and Equation 8 for the parametric test and Table 2-5 for the equivalent nonparametric test).

Nonparametric data sets usually require 30–100 measurements to achieve useful statistical power for decision-making. Nonparametric data sets are not suited to the hypothesis tests mentioned above; industrial hygiene data often fit either a normal or lognormal distribution. It is common to find industrial hygiene reports that declare an HEG acceptable on the basis that all observed measurements were below the OEL. When the data are nonparametric, the obtainable confidence level is surprisingly small. Two examples follow:

First, consider a data set where the largest observation of 25 nonparametric measurements is less than the OEL. Table 2-5 shows that there is only 36% confidence that at least 95% of the population lies below the OEL (see Table 2-5, at the end of the chapter, to find $\gamma = 0.36$ under $n = 25$, and $p = 0.95$).

Second, to assert that 95% of exposures are less than the OEL, with at least 99% confidence using nonparametric data, 90 to 120 measurements must all be smaller than the OEL. (See Table 2-8 at the end of the chapter to find $m = 1$ under $\gamma = 0.99$, $p = 0.95$, and $n = 90$ to 120).

For **IDLH situations**, higher confidence levels are required, and good practice dictates continuous monitoring. Statistical considerations confirm this. Ceiling values should be considered, and 99% confidence that no more than 1% of exposures exceed the OEL are recommended.

If the data are parametric, Table 2-7 at the end of the chapter shows that 50 measurements are required to obtain $k = 3.124$. Then, $UTL = M + k * S$ for normal data or $\ln(GUTL) = \ln(GM) + k*\ln(GS)$ for lognormal data. The desired confidence is obtained if UTL < OEL or if $GUTL$ < OEL (see "Tolerance Interval on the nth Percentile").

For nonparametric data, Table 2-5 shows that with 100 measurements, all below the OEL, there is only 26% confidence that 99% of all exposures are less than the OEL (read $\gamma = 0.26$ under $p = 0.99$ and $n = 100$). Table 2-8 shows that all of 500 measurements must be below the OEL to obtain 99% confidence that no more than 1% of all exposures exceed the OEL (read n = 500 under $\gamma = 0.99$, $p = 0.99$, $m = 1$).

In an 8-hr shift, there is little difference between 500 measurements and continuous monitoring. In fact, when using an instrument like a variable path infrared gas analyzer with a time constant greater than 20 seconds, there is no meaningful difference. To be statistically independent, samples should be collected three or more time constants apart. The analyzer mentioned here is only capable of one independent measurement per minute.

The Decision Flow Diagram

The decision flow diagram of Figure 2-1 systematizes the relationship between statistical data analysis and professional judgment. An earlier version was published in the AIHA Exposure Assessment Strategies Monograph.[2]

Figure 2-1 shows the overall flow diagram for deciding whether workplace exposures are tolerable. Starting at the top, in circle R, the first step is to collect representative samples. The small branch at the upper left emphasizes the importance of investigating and documenting every overexposure.[A] The main path emphasizes the importance of evaluating every data set, even

---

[A] OSHA treats exposure as the measured breathing zone concentration, irrespective of personal protective equipment. Thus, overexposures observed in air sampling do not always represent health threats.

when an observed overexposure requires stopping or altering the process. These two paths run parallel and both paths should be followed.

The second step is to compute the descriptive statistics. The third step is to select "Statistical Judgment" or "Professional Judgment." If Statistical Judgment is selected, the nature of the hazard determines whether a tolerance interval or a confidence interval test is to be applied. If Professional Judgment is selected, the decision is based on the data using rules of thumb established by a consensus of practicing industrial hygienists.[2] Such decisions should always be checked with nonparametric statistics so that the decision-maker understands the level of confidence associated with the decision.

The fourth step depends on the decision. If exposures are unacceptable, immediate action is required: the process must be stopped and changed before restarting. At a minimum, personal protective equipment should be issued until such time as engineering controls can be designed, budgeted, and installed. If more sampling is needed to make a decision, it should be done immediately. If the exposures are deemed acceptable, a time should be established for the next sampling campaign.

The flow diagram for descriptive statistics is entered through circle D touching the rectangle "Calculate Descriptive Statistics." Note that the first step is to update X-bar and R charts and check for trends (if prior data exist). If there are trends, prior data cannot be combined with present data for statistical decision-making; professional judgment is needed. Professional judgment is also used if there are fewer than six total measurements. Finally, the data should be checked for fit to normal or lognormal distributions by plotting on linear probability or log probability paper and checking for a straight line. Note that a straight-line fit is unlikely if either of the following is true: the identified homogeneous exposure group is not homogeneous (could also be tested by an analysis of variance,[15] given sufficient data); or the exposure conditions are changing between measurements. If a good straight-line fit is found, statistical decision making is entered through circle S. If the data are not well represented by either of these distributions, the nonparametric UTL is computed (Table 2-5). If UTL < OEL, the workplace is considered acceptable; otherwise, professional judgment is the most likely decision-making tool.

The Professional Judgment aspect of the decision-making process is entered through circle P touching the Professional Judgment rectangle. If all observed measurements are below the OEL and there are at least six of

them, the workplace is often reported as acceptable.[B] If there is at least one overexposure, the decision depends on whether there is an acute hazard and whether the measurements represent normal process conditions.

The Statistical Judgment aspect of the decision-making process is entered through circle S touching the Statistical Judgment rectangle. The first decision is whether the health effect is acute or chronic. If chronic, hypothesis testing about the confidence interval on the mean (six measurements minimum) is used. If acute, hypothesis testing about a chosen tolerance interval (90th or 95th percentile of the exposure distribution) is used. This usually requires at least 18 measurements. If a decision can be made, the appropriate branch leads to that decision point: A for acceptable and U for unacceptable. If a statistically significant decision cannot be made, the appropriate branch is P for Professional Judgment.

The "Unacceptable Exposure" branch is entered through circle U. Note that each overexposure must be investigated and documented by both the industrial hygiene and the occupational medicine teams. The industrial hygiene purpose is to determine whether the events were representative or resulted from process upset conditions. The occupational medicine purpose is to treat the patient if disease or injury resulted and to document the clinical status for future epidemiology. Corrective measures are implemented before the offending process is restarted. Air sampling to verify the corrective measures is restarted at circle R.

The "Acceptable Exposure" branch is entered through circle A. Note that results are documented. More monitoring or investigation is called for if the data are acceptable but showing an adverse trend. Otherwise, scheduling the next sampling campaign may be based on program policies using guidelines in the "Reevaluation" section.

Recommendations and Reporting

The industrial hygiene decision about each stressor associated with each task must be recorded and communicated. The purpose of this step of the exposure assessment strategy is to document the air quality in the workplace, inform employees of their exposures, set priorities for process improvements, and schedule follow-up evaluations of each process.

If breathing zone measurements represent environmental concentrations, but do not represent personnel

---

[B] When using this criterion in decision-making, be sure to check the table for the one-sided tolerance interval for nonparametric data. This will provide the level of confidence of the decision at hand.

exposures because of personal protective equipment, these conditions must be stated. Further guidance on written reports is available from a variety of sources.[2]

Reevaluation

An experienced industrial hygienist should walk through every process at least once a year and schedule air sampling of each process at least once every 3–5 years. Higher-risk processes need more frequent attention in the form of air samples taken on monthly to annual intervals.

The risk ranking is a good basis for assigning frequency of reevaluation (see Table 2-9). Because the risk ranking changes with process changes, this linkage ensures appropriate changes in industrial hygiene sampling frequency in response to material substitutions and process improvements.

Obviously, there are circumstances that will accelerate scheduled review of a process, task, or homogeneous exposure group: 1) a diagnosed occupational illness, 2) an employee concern, or 3) any significant changes in operations or facilities.

It is recommended that policy require an industrial hygiene sign-off on all plans for facility or process changes. Otherwise, such process changes may go unnoticed until the next scheduled industrial hygiene walk-through. The opportunity for such changes to occur unnoticed is the primary reason for an annual walk-through of all areas, including those not otherwise scheduled for air sampling.

Professional Judgment versus Statistical Requirements

Number of Measurements

An HEG with many workers should receive a higher proportion of the sampling budget than one with a few workers. It is reasonable to make the number of measurements proportional to the square root of the number of workers in the HEG.[18] For example, 16 workers need 4 measurements (but use 6, the minimum n for statistical analysis), 36 workers need 6 measurements, and 100

TABLE 2-9.- Reevaluating Frequency from Risk Ranking

Risk Ranking	Reevaluation Frequency
Very High	Continuous monitoring
High	Monthly to quarterly
Moderate	Quarterly to annually
Low	1–3 years
Very Low	3–5 years

workers need 10 measurements.[c] If there is need to check the homogeneity of the HEG, it is appropriate to double these numbers for each HEG.

If each full-shift exposure estimate exceeding an OSHA PEL is treated as a violation, there are incentives to reduce the number of measurements taken so that the number of such violations is minimized.[24] Under such conditions, too few measurements are taken to adequately determine the degree of risk from airborne contaminants. A better policy would encourage numerous voluntary measurements so long as corrective measures are taken when justified by the data.

Finally, the number of measurements needed depends on the concentration gradients over space and time. When this is the case, early papers discussing the concepts of correlation, covariance, and averaging time should be reviewed[25-27] and supplemented with current quality control and reliability references.[28, 29] If one seeks 95% confidence that not more than 5% of all exposures exceed the OEL, then the required number of samples is very much larger than indicated above. For a population with its 95th percentile at 2/3 of the OEL, the number of samples required to prove this with power = 0.80 ranges from 58 samples if the GSD = 2.03 to 249 measurements if the GSD = 4.86.[16, 30]

Whom to Sample

The usual procedure for obtaining representative samples is to randomly pick the set of exposures for air sampling so that every possible exposure has an equal chance of being selected. The distribution of exposures observed in these random samples is assumed to be similar to the distribution of exposures in the sampled population. In contrast to this conventional wisdom, a recent report shows that there are circumstances where a random sampling strategy requires more measurements than a properly selected deterministic sampling scheme.[31] For now, deterministic schemes for industrial hygiene sampling are rare.

Sometimes, intentionally biased samples are desired. In industrial hygiene, the worst-case exposures and the most exposed individual(s) are important.[32] If there exists a definable group of such exposures, then it is reasonable to sample them randomly and to apply statistical tests to the resulting data. Clearly, if such exposures are within the OEL, it can be safely concluded that ordinary exposures in the same homogeneous exposure group are acceptable. However, caution is in order if the

---

[c] Note from Tables 2-5 and 2-8 that n must be much larger than these numbers to obtain ($\gamma > 0.75$) reasonable confidence.

measurements are all drawn from the 2% of exposures that are truly the worst exposures, because it is unlikely that the data will fit either a normal or a lognormal distribution. The most likely distribution will be an extreme value distribution, such as the Fisher-Tippet distribution included in *Mathematica*[33] or the simpler exponential distribution.[34] The tools in "Example of Data Interpretation," page 37, will not help with analysis of such data.

To intentionally sample high exposure periods, it is necessary to identify those periods. Where there is visible dust, a powerful odor, or eye and upper respiratory irritation, the exposures are obvious. In these cases, the workers at the operation can furnish invaluable information, if asked. Do not overlook other operations, which by their nature, cause increased contaminant generation and higher concentrations; for example, confined space entry for maintenance, response to a local exhaust system failure while the protected process continues to run, continuing a normally enclosed operation with covers off, wiping or spraying a volatile solvent. These and similar operations have inadequate control and should be sampled separately from routine operations.

Limitations of Statistical Decision-making

The use of statistical criteria is based on the premise that the sample truly represents the population of interest. When some underlying feature of the sampled operation changes, previous data have limited utility for predicting the distribution of exposures in the "new" operation. Further, as a rule of thumb, a data set containing fewer than 20 data points is unable to make usable predictions about the upper 5% of the distribution, and a data set containing fewer than 100 data points is ill-suited for the upper 1% of the distribution (see "Guidelines for the Number of Measurements Needed," page 27).

For example, high concentrations of metals in urban areas occur significantly more frequently than is predicted by the lognormal distribution that adequately describes the central tendencies of the metal samples.[35] Predicting the concentrations associated with the upper 1%, 2%, or even 5% of exposures from statistical air sampling data is seldom (if ever) justified.

Averaging Period

An OEL is an ordered pair of numbers: an average allowable airborne concentration and an averaging period of time appropriate to the health hazard involved.[35] For pneumoconiosis-producing dusts, it is the cumulative exposure over a period of months and

years which determines the probability of a diagnosis of pneumoconiosis and the severity of the adverse health effects. For coal mine dust exposures, an average over 10 working shifts was used in the original legislation to determine the application of the mandatory respirable dust standard. For lead, in most cases, it is the integrated exposure over weeks which determines the probability of an excessive blood lead concentration. For carbon monoxide, averaging times measured in minutes are necessary to provide adequate protection at concentrations above 50 ppm.

There is a biological basis for single-shift averages for many chemicals. During non working hours, the body can metabolize and/or excrete the material inhaled or absorbed over the work period. For this reason, the ACGIH Chemical Substances TLV Committee has used a TLV–TWA over full work shifts since the first TLV list.[18] However, these values are modified by STEL or C requirements when toxicodynamic or toxicokinetic behavior requires it.

Other OEL pairs are used by various standard setting bodies. The ACGIH TLV Committee defines 8-hr TLV–TWA, 15-min TLV–STEL, TLV–C, and excursion limits.[36] The TLV–TWA is measured over an 8-hr shift in a 5-day, 40-hr work week. The TLV–STEL is measured over a 15-min period during a shift, and that concentration should not occur more than 4 times per day nor be repeated within a period of 60 mins.[17] The TLV–C should not be exceeded at any time and should be measured over as short a period as possible, not exceeding 15 minutes. When more than one TLV is assigned, each must be observed to protect against all possible pathways of injury and illness.D

The ACGIH guidance on styrene exposures is a useful means for illustrating how these values work together.[36] The average exposure over an 8-hr period is to be kept below the TWA of 213 mg/m³ and no 15-min average is to exceed 426 mg/m³. Further, the concentration of 426 mg/m³ should not be approached more than 4 times per day and should not be approached more than once in any hour. Finally, the semi-quantitative biological exposure index (BEI) limits the styrene in venous blood to 0.55 mg/L at the end of a work shift and to 0.02 mg/L at the beginning of a work shift. A properly designed workplace involving exposures near the STEL would need to keep other exposures well below the TWA to keep the 8-hr exposure below the TWA. If the full shift exposures approach the TWA, then 16 hrs of rest are likely to be needed to achieve the allowed beginning-of-shift BEI of 0.02 mg/L.

D See also 29 CFR 1910.1000, Tables Z–1, Z–2, and Z–3 for OSHA PELs using 8 hr, 15 min, and other specified averaging times.

Statistical Tools for IH Decision-making

Any set of data can be characterized by its descriptive statistics and by its probability distribution as plotted on probability paper. This should always be the first step in data interpretation. The sections that follow provide a concise review of statistical concepts and the section entitled "Example of Data Interpretation," page 37, provides an example of the application of these concepts to industrial hygiene data.

Descriptive Statistics

Descriptive statistics are easily computed for any set of numbers and provide an invaluable starting point for data interpretation. Common and useful descriptive statistics include measures of central tendency (mean, median, mode, and geometric mean) and measures of dispersion (range, standard deviation, and geometric standard deviation).

Most scientific calculators and spreadsheets have a function, "$\Sigma +$," in which values can be accumulated. One or two additional commands will calculate the mean, M, and the sample standard deviation, S, for any data set. The geometric mean, GM, and the geometric standard deviation, GS, are calculated nearly as easily. First, the logarithm of every data point, $\ln(x)$, is computed. Using this transformed set of data, the calculator or spreadsheet returns $\ln(GM)$ from its function for the mean and $\ln(GS)$ from its function for the sample standard deviation. The antilog of these two numbers is the geometric mean, GM, and the geometric standard deviation, GS. Finally, the mode (Mo), the median (Me), and the range (R) are estimated. The mode is the value that occurs most frequently. The median is the value at the midpoint of the distribution: 50% of all data are larger and 50% are smaller. The range is the difference between the largest and smallest data point. The descriptive statistics for a set of n data points are summarized below. In Equation 1 the symbol $x_{(m)}$ indicates that the data points are ordered from smallest to largest, and (m) is the running index on that ordered set.

$$R = x_{max} - x_{min} = x_{(n)} - x_{(1)}; \text{ for a sample of size } n,$$
$$x_{(i)} \le x_{(i+1)} \text{ for all } i < n$$

$$Me = x_{(m)} \text{ for } n \text{ odd and } m = \frac{n+1}{2} \qquad (1)$$

$$Me = \frac{x_{(m)} + x_{(m+1)}}{2} \text{ for } n \text{ even and } m = \frac{n}{2}$$

Equation 2a shows the sample mean, the sample standard deviation and the sample standard deviation of M, the estimate of the population mean. Equation 2b uses symbols with hats to indicate maximum likelihood estimators (MLE) for the same statistics. Sample statistics are unbiased and commonly used for small samples (n < 11). The MLE statistics are biased, especially for small samples, but have smaller variance. MLE statistics are preferred for large samples (n > 50).

$$M = \sum_{1}^{n} \frac{x_i}{n}; \ S = \sqrt{\sum_{1}^{n} \frac{(x_i - M)^2}{n-1}}; \text{ and } S_M = \frac{S}{\sqrt{n}} \quad (2a)$$

$$\widehat{M} = M; \ \hat{S} = S\sqrt{\frac{n-1}{n}}; \text{ and } \hat{S}_{\hat{M}} = \frac{\hat{S}}{\sqrt{n}} \quad (2b)$$

Equations 3a-c are introduced for ready reference. These expressions are equations for the probability density function of the standard Gaussian, the fitted Gaussian, and the fitted lognormal distributions.

$$N[0,1,z] = f(z) = \frac{dF(z)}{dz} = \frac{e^{-\left[\frac{z^2}{2}\right]}}{\sqrt{2\pi}}; \ z = \frac{x-\mu}{\sigma} \quad (3a)$$

$$N[\mu,\sigma,x] = f(x) = \frac{dF(x)}{dx} = \frac{e^{-\left[\frac{1}{2}\left(\frac{x-\mu}{\sigma}\right)^2\right]}}{\sigma\sqrt{2\pi}} \quad (3b)$$
$$\approx \frac{e^{-\left[\frac{1}{2}\left(\frac{x-M}{S}\right)^2\right]}}{S\sqrt{2\pi}}$$

$$N[\ln(GM),\ln(GS),\ln(x)] = \frac{dF(\ln(x))}{dx} \quad (3c)$$
$$= \frac{e^{-\left[\frac{1}{2}\left(\frac{\ln(x)-\ln(g\mu)}{\ln(g\sigma)}\right)\right]}}{x\ln(g\sigma)\sqrt{2\pi}} \approx \frac{e^{-\left[\frac{1}{2}\left(\frac{\ln(x)-\ln(GM)}{\ln(GS)}\right)^2\right]}}{x\ln(GS)\sqrt{2\pi}}$$

A remarkable amount of insight is available from descriptive statistics. If $M = Me = Mo$, then the data are symmetrically distributed around the mean and are likely to fit the Gaussian distribution. If $Me = GM$, and $Mo < GM < M$, then the data are skewed to the right and the lognormal distribution is likely to fit the data fairly well (see Equation 5). If neither of these rules fits, or if $R < S$, the data are not satisfactorily modeled by either the normal or the lognormal distribution and nonparametric statistical tests are advised.

Because one or the other of these two distributions will approximate almost any industrial hygiene data set, a brief review is presented in the following paragraphs.

33

When data are symmetrically distributed around the average value, the normal distribution provides considerable computational advantage. When data are skewed so that they cluster between zero and the mean on the low side and spread out from the mean on the high side, the lognormal distribution is the distribution of choice. Its computational advantage is that after taking the logarithm of each data point, the resulting transformed data set is approximately normally distributed. Industrial hygienists competent with the statistical techniques for normally distributed data are qualified to interpret their air sampling data. Important statistical tools include hypothesis testing, confidence intervals on the mean, tolerance intervals about the proportion of the distribution, and analysis of variance (ANOVA) testing.

Normal Distribution

The normal distribution is a symmetrical distribution for which all three measures of central tendency are equal: $M = Me = Mo$. It has the "nice" property that it is entirely described by two parameters: its population mean, μ, and its population standard deviation, σ. It is widely tabulated[2, 20, 37] in terms of the normalized variable, $z = (x - \mu)/\sigma$. In most industrial hygiene situations, the true values of μ and σ are unknown and unknowable. Instead, unbiased estimators for the sample mean, M, and the sample variance, S^2, are used. The normalized variable for industrial hygiene becomes $z \approx (x - M)/S$. The proportion, p, of a sample less than some value X_p can be read from a z-table entered at $Zp = (Xp - M)/S$.

The mathematical expression describing the probability density function (f($\bullet$) = PDF) for a Gaussian (or normal) distribution is expressed in terms of parameters of the population. Since those are unknown for an occupational exposure data set, the PDF is approximated by statistics estimated from the data (see Equation 3). Use the best statistics you have, whether sample mean and variance, MLE mean and variance, or mean and variance.[1] Minimum variance unbiased estimators (MVUE) are unbiased like the sample mean and variance and have the small variance associated with the MLE estimators. They are not widely used because they are complex expressions, and they work best when the PDF for the sampled population is truly known.

Lognormal Distribution

The lognormal distribution is a right-skewed distribution with the "nice" property that it is entirely described by two parameters: its population geometric mean, gμ, and its population standard deviation, gσ.

There is one word of caution. For a lognormal distribution, the "nice" measure of central tendency is the geometric mean. It is equal to the median value of the data, not the mean value. The geometric mean should not be compared to exposure limits; rather, the arithmetic mean is compared to exposure limits. To emphasize the point, consider a set of data adequately characterized by two numbers, GM and GS. Estimates for the mean, median, and mode for this distribution are derived from GM and GS:[37]

$$ln(M) = ln(GM) + \frac{1}{2}[ln(GS)]^2$$
$$ln(Me) = ln(GM) \qquad \textbf{(4)}$$
$$ln(Mo) = ln(GM) - [ln(GS)]^2$$

Further, the logarithmic transformation allows lognormal data to be analyzed using Gaussian tools in terms of the standard normal variate defined as in Equation 3c. The true values of the parameters gμ and gσ are seldom available for industrial hygiene air sample data so analysis depends on the approximate form for z. Equation 5a gives unbiased estimates of the statistics for the sample geometric mean, GM, and the sample geometric standard deviation, GS. Equation 5b gives the minimum variance or MLE statistics for GM and GS. Because occupational exposure standards are based on average exposure, it is necessary to estimate M and S for occupational exposure data sets. The variance for M and S is very large for small samples from lognormal populations. Equation 5c gives the MLE statistics for M and S based on a sample from a lognormal population. These are preferred over Sample statistics, whenever the data appear to arise from a lognormal Distribution.

$$ln(GM) = \sum_1^n \frac{ln(x_i)}{n}; \quad ln(GS) = \sqrt{\sum_1^n \frac{[(ln(x_i) - ln(GM)]^2}{n-1}};$$
$$ln(GS_{GM}) = \frac{ln(GS)}{\sqrt{n}} \qquad \textbf{5(a)}$$

$$ln(\widehat{GM}) = ln(GM); \quad ln(\widehat{GS}) = ln(GS)\sqrt{\frac{n-1}{n}};$$
$$ln(\widehat{GS}_{GM}) = \frac{ln(\widehat{GS})}{\sqrt{n}} \qquad \textbf{5(b)}$$

$$ln(\hat{M}) = ln(\widehat{GM}) + \frac{(ln(\widehat{GS}))^2}{2};$$
$$\hat{S}^2 = e^{+2(ln(\widehat{GM}) + (ln(\widehat{GS})^2)} - e^{+2(ln(\widehat{GM}) + (ln(\widehat{GS})^2)}; \qquad \textbf{5(c)}$$

$$\hat{S}_M = \frac{\hat{S}}{\sqrt{n}} \qquad \textbf{5(d)}$$

MLE statistics have a "nice" property not true of either simple unbiased or MVUE statistics: any function of an MLE statistic is, itself, an MLE statistic. The probability, p, that a sample is less than $\widehat{GX}_p$ can be read from the z-table (Table 2-10 at the end of the chapter) entered at $\hat{z}_p = (\ln(\widehat{GX}_p) - \ln(\widehat{GM}))/\ln(\widehat{GS})$, and p is the MLE estimator based on observed data.

The lognormal distribution is easily used for hypothesis testing about its median, and the principles can be extended to hypothesis testing about the mean of a lognormal population using Land's coefficients.[39] The procedure is more easily illustrated than described (see "Example of Data Interpretation," page 37).

Industrial hygiene standards are based on average exposures, and an unbiased estimate of an average may be computed as the sum of all data points divided by the number of those points, no matter the shape of the underlying probability distribution. Recently, a major reference recommended that analysts consider using complex estimators called MLE and MVUE because of two features.[16] The MLE has lower variance, but is a bit biased. The MVUE has the lowest possible variance for an unbiased estimator. Equations for the MLE estimators are included, but the MVUE estimators are beyond the scope of this chapter.

Finally, the reader is reminded that only like kinds of data may be averaged. For example it is appropriate to average all members of a set of 8-hr samples, or all members of a set of 15-min samples, but it is not appropriate to average some 8-hr with some 4-hr and some 15-min samples.

Probability Plotting

Probability paper is available in two forms: linear versus probability axes (probit) and logarithmic versus probability axes (log-probit). These are useful because Gaussian distributed data plot as a straight line on probit paper and lognormal data plot as a straight line on log-probit paper. From such plots, it is easy to accommodate missing data points, as from measurements that were below the analytical detection limit or where breakthrough showed that the concentration was above the range of the sampling method.

Data are prepared for plotting by arranging data points in order of increasing value and numbering them from 1 to n. In a second column, the logarithms of the data points are arranged in order of increasing value. The plotting position on the probability axis is computed by the rank order of the data point divided by $(n + 1)$. The data are then plotted and inspected for straight line behavior.

Standard scientific calculators or spreadsheets are sufficient to calculate the sample mean and the sample standard deviation from rank ordered data. The best fit straight line is then easily plotted by noting that $[M–S]$ plots at 16%, M at 50%, and $[M + S]$ at 84% for normal data on linear-probability paper, and that $[GM/GS]$ plots at 16%, GM at 50%, and $[GM * GS]$ at 84% for lognormal data on log-probability paper.

Truncated or Censored Data

In some exposure populations, several data points in each sample may be at or below the detection limit. In other samples, the true concentration is known to be above the reported concentration because breakthrough was detected at the time of analysis. In these cases, care is required to calculate an estimate for GS. It is not correct to assign some small value to samples below the detection limit and a large value to those with breakthrough and simply calculate descriptive statistics. If this is done, unusually large estimates of GS are often seen. A better technique is to assign a rank order to each data point, whether it has a usable value or not. The usable data points should be plotted in their assigned position and a line drawn through them. The intersection of that line with the plotting position for missing data is a usable estimate for the true value of those samples that were either below the detection limit or above the saturation limit.

Confidence Interval on the Mean

For normally distributed data, the confidence interval on the mean is computed using Student's t distribution (see Table 2-11 at the end of the chapter). Values for the coefficient $t_{p,n}$ are tabulated in terms of df $= n - 1$, and the probability, p, that the true mean, μ, lies below the UCL. Using the value of t appropriate to a sample from a Gaussian data set, the one-sided UCL and LCL are easily computed from Equation 6a. For most data sets with $n > 11$, the MLE statistics are preferred, in spite of their slight bias.[1] The hat over the symbol for a statistic identifies the MLE form of the calculation.

$$UCL = M + (t_{p,n-1})\left(\frac{S}{\sqrt{n}}\right)$$
and
$$LCL = M - (t_{p,n-1})\left(\frac{S}{\sqrt{n}}\right)$$

6(a)

$$\ln(GUCL) = \ln(GM) + (t_{p,n-1})\left(\frac{\ln(GS)}{\sqrt{n}}\right)$$
and
$$\ln(GLCL) = \ln(GM) - (t_{p,n-1})\left(\frac{\ln(GS)}{\sqrt{n}}\right)$$

6(b)

$$ln(LUCL)=ln(\hat{M})+(C_{up(GS,n)})\left(\frac{(ln(\widehat{GS}))}{\sqrt{n}}\right)$$

and   **6(c)**

$$ln(LUCL)=ln(\hat{M})-(C_{low(GS,n)})\left(\frac{(ln(\widehat{GS}))}{\sqrt{n}}\right)$$

For lognormally distributed data, three confidence intervals are useful in guiding professional judgment: 1) Equation 6a for the *CI* about the sample median, *GM*, given by *GLCL* and *GUCL*;[38] 2) Equation 6b, based on the Central Limit Theorem, for the Student-*t CI* about the MLE mean, given by *UCL* and *LCL*;[22] and 3) Equation 6c Land's *CI* about the MLE mean, given by *LLCL* and *LUCL*.[39]

First, the normal confidence interval about the sample mean computed from Equation 6a is very reliable when the population is Gaussian, and is surprisingly robust when the population being sampled is skewed.[22] As the number of samples of any population grows indefinitely, the other two confidence intervals (Equations 6b and 6c) converge to this one.

Second, the geometric confidence interval about the *GM*, is useful in understanding how tightly exposures are clustered about their median. That confidence interval is computed from Equation 6b using the *t* coefficient from Table 2–11 at the end of the chapter.

Third, Land's confidence interval on the mean is useful for comparing *M* with OEL standards. It is computed using Equation 6c on the basis of two coefficients, C_{up} and C_{low}. The C-coefficients, whose values may be read from Figure 2-2, represent a 90% two-sided confidence interval. Figure 2-2 a and b was plotted from approximate equations reported to have less than 5% error.[38] Coefficients read from Figure 2-2 a and b have two significant digits of accuracy. Although Land's method produces an exact estimate of the *confidence limits* for a sample drawn from a perfectly lognormal population, occupational exposure data are not drawn from exactly lognormal populations, and each measurement suffers analytical uncertainty of up to ± 25%. Thus, Land's *CI* remains an approximation.[40]

For further information, the recent AIHA monograph, edited by Mulhausen & Damiano, has a rather complete treatment that compares these three algorithms for estimating the *UCL*, and cautions that the *GUCL* underestimates the *LUCL* for most industrial hygiene data sets with small *n*.[16]

Percentiles of a Distribution

When peak exposures are of concern rather than long-term average exposures, it is important to know the percentiles of normally distributed data (see Table 2-10). These values are found by use of standard *z* tables. To find the value $X_{\hat{p}}$ such that the proportion of the PDF less than $X_{\hat{p}}$, is *p*, find $z_{\hat{p}}$ from the *z* table and then compute $X_{\hat{p}}$:

$$\hat{X}_p = \hat{M} + z_p(\hat{S});\text{ Gaussian data}$$
$$ln(\widehat{GX}_p) = ln(\widehat{GM}) + z_p(ln(\widehat{GS}));\text{ Lognormal data}\qquad\textbf{(7)}$$

Tolerance Interval on the nth Percentile

When peak exposures represent IDLH conditions, it is important to have higher confidence in exposure data than is given by the best estimate of the percentile. For that purpose, the Gaussian one-sided tolerance interval provides a quantitative level of confidence (Table 2-7).[41] It says that the 100*p* percentile is less than the *UTL* with a probability γ when it is estimated from a sample containing n data points. The *UTL* is easily computed from a table of *k* factors. The value of $k_{p,n,\gamma}$ depends on the desired value of *p*, the number of data points (*n*) and the desired confidence (γ). Rappaport and Selvin have pointed out that the power of tolerance interval tests is intrinsically poor.[21] Others have adopted a test for the size of the exceedance fraction to replace the UTL.[16]

$$UTL = M + k_{p,n,\gamma}(S);\text{ Gaussian data}$$
$$ln(GUTL) = ln(GM) + k_{p,n,\gamma}(ln(GS));\text{ Lognormal data}\qquad\textbf{(8)}$$

When the available data fail to fit either the Gaussian or the lognormal distribution, professional judgment is needed to make a decision about the quality of the workplace. Although the decision flow diagram provides some guidelines in the professional judgment portion of the diagram, it should not be used in isolation. When a set of data has no measured exposures above the OEL, those data should be considered in terms of the tabulated confidence levels in the table of nonparametric tolerance intervals. Only after the level of confidence is evident should a decision be made that a workplace is acceptable with no further improvements.

On most occasions when data are neither normally nor lognormally distributed, there is a problem with the data. Often the data are from two or more HEGs. This important possibility is illustrated in the "Example of Data Interpretation." A second common possibility is that there are undetected changes in the workplace occurring between samples, so that the samples are not from a single distribution.

Workplace exposures vary both systematically and randomly. Statistical analysis of data that contain

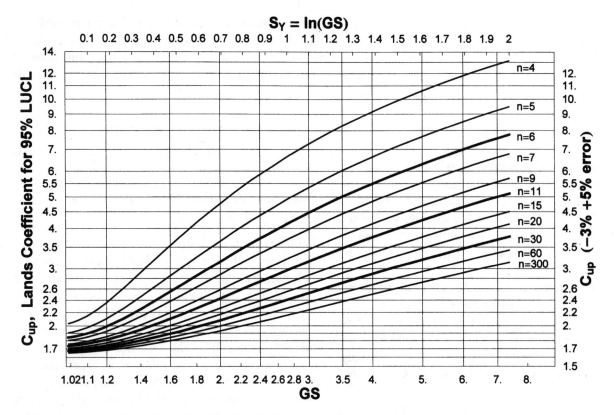

FIGURE 2-2a. Land's Coefficient for LUCL with one-sided γ = 0.95.

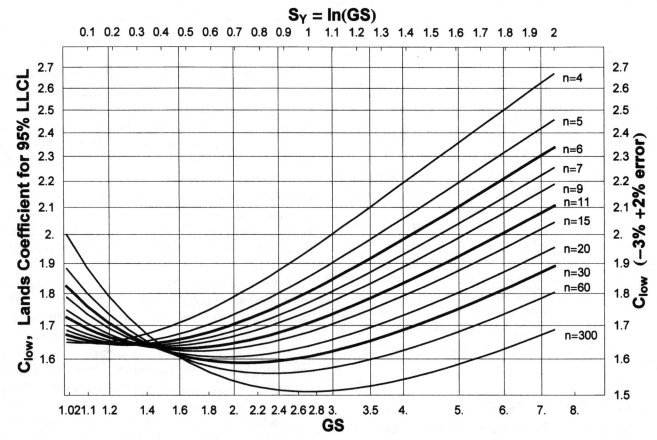

FIGURE 2-2b. Land's Coefficient for LLCL with one-sided γ = 0.95.

systematic variations can be misleading, sometimes leading to an overestimate of the variance of the data and an overprediction of the *UCL* and the *UTL* as well as an overstatement of the acute health risk represented by the data.

Software and Useful References

The calculations described in this section were initially computed with the now obsolete LOGAN software package for DOS-based computers.[21] LOGAN supported decision-making with either Gaussian or lognormal statistics for sample sizes from 6 to 18. When data did not fit parametric assumptions very well, the program defaulted to a nonparametric decision algorithm. Finally, it forced a decision with 18 samples instead of requiring the additional sampling called for by the NIOSH sampling strategy.[8] The author is aware of two newer packages released for the Windows 95/98 environment. LogNorm2 was released in 1997.[42] It retains all critical features of LOGAN and adds many new ones. It has user-friendliness; allows up to 16,384 measurements; uses Land's Confidence Interval when data are approximately lognormal; incorporates three goodness-of-fit statistics to test for normal, lognormal or nonparametric fit; provides decision support for nonparametric data sets; easily handles truncated or censored data sets; and supports Analysis of Variance (ANOVA).[42] A less comprehensive Excel Spreadsheet application, "IHSTAT.xls" is included with the AIHA monograph.[16] IHStat is well-suited for learning the contents of the monograph and is useful for many simple applications, but it is limited to 50 measurements and omits several important capabilities of LogNorm2. IHStat does not handle non-parametric statistics, censored or truncated data sets, or Analysis of Variance, and it does not allow user selection of confidence levels.

Example of Data Interpretation

The following sections present an example scenario incorporating the air sampling strategies discussed in this chapter. For this example, an HEG of 91 painters was surveyed. These workers sprayed paint two shifts per day. The data from the survey show that airborne solvent vapor concentrations were 49, 89, 24, 61, 287, 76, 72, 96, 83, 67, 54, 190, 50, 125, and 82 ppm. The question to be answered from the data is whether this HEG has acceptable solvent exposures. To determine the answer to this example scenario, the decision flow diagram in Figure 2-1 should be followed.

Calculate Descriptive Statistics

First, the descriptive statistics should be calculated and the data plotted on probability paper. Plotting is recommended even in cases like the present example where some exposures are observed to be above the TLV. This helps to develop the informed professional judgment necessary in all such cases.

Next, the data should be arranged for calculation by being placed in rank order (see Table 2-12). The data should be ranked from $r = 1$ to $r = n$ (see column 1 of Table 2-12). Then, the plotting position should be calculated, $p = 1/(n + 1)$ (column 2). Data are entered in order in column 3 and the logarithm of the ratio of the data values to some reference value (column 4) can be calculated. For this example, the reference value is 1 ppm. Statistical tables can be used to locate values for z_p, $t_{p,n-1}$, and $k_{p,n,\gamma}$.

The sample mean and the sample standard deviation for the numbers in columns 3 and 4 are easily calculated using Equation 2. For column 3, the calculation yields M and S. For column 4, the calculation yields $\ln(GM)$ and $\ln(GS)$. The median for both columns 3 and 4 should be estimated with Equation 1b.

From the calculated values for M and S and the relevant values of z_p, $t_{p,n-1}$, and $k_{p,n,\gamma}$ it is a simple matter to calculate *LCL*, *UCL*, percentile, and *UTL* for both columns 3 and 4. Finally, the antilog function should be applied to values listed in column 4 to obtain the decision thresholds in concentration units for column 5. Note that symbols in column 5 are preceded by *G* to signify that they are obtained from analysis assuming a lognormal distribution of data.

At this point, data are sufficient to make a preliminary judgment about whether a normal distribution is likely to fit either the data (column 3) or the log-transformed data (column 4). This is done by comparing the value of the calculated mean of the numbers in each column to the estimated median of each column. A useful test statistic is the ratio of the difference between M and Me to the sample standard deviation S. (This value is found in the last row of the data table.) A truly representative sample of a normal data set would return a value of zero because the mean equals the median for normal populations. Large values of this test statistic indicate significant asymmetry in the data. A poor fit to a normal distribution is indicated if the larger value is found in column 3; a poor fit to a lognormal distribution is indicated if the larger value is found in column 4.

For Table 2-12, the median test statistic on the last line is 6% of the standard deviation for the log-transformed data (column 4) and 27% of the standard deviation for the data (column 3). These results suggest that

TABLE 2-12. Descriptive Statistics Example: Painters' Solvent Exposure Date Analysis

		DATA PREPARATION		
Rank [r]	Plot Position [rf(n + 1]	Exposure (ppm) [x]	ln(data/1ppm) (unitless) [in(x)]	Tabulated Values [t, z, & k]
1	0.063	24.0	3.178	$p = 0.95$
2	0.125	49.0	3.892	$n = 15.$
3	0.188	50.0	3.912	$\gamma = 0.90$
4	0.250	54.0	3.989	$z(p) = 1.645$
5	0.313	61.0	4.111	$t(p, n-1) = 1.761$
6	0.375	67.0	4.205	$k(p, n, \gamma) = 2.329$
7	0.438	72.0	4.277	
8	0.500	76.0	4.331	
9	0.563	82.0	4.407	
10	0.625	83.0	4.419	
11	0.688	89.0	4.489	
12	0.750	96.0	4.564	
13	0.813	125.0	4.828	
14	0.875	190.0	5.247	
15	0.938	287.0	5.659	

	DESCRIPTIVE STATISTICS			
NORMAL		LOGNORMAL		
X		ln(GX)	GX = (1ppm)*exp[ln(GX)]	
$S = 65.9$		0.586	1.80	= GS unitless
$M = 93.7$		4.367	78.8	= GM ppm
$M - S = X_{16\%} = 27.8$		3.781	43.9	= GX_{16%} ppm
$M + S = X_{84\%} = 159.6$		4.953	142.	= GX_{84%} ppm
$OEL = 100.$		4.605	100.	= OEL ppm
$LCL = 63.7$		4.101	60.4	= GLCL ppm
$UCL = 124.$		4.634	103.	= GUCL ppm
$M + z_{95\%} * S = X_{95\%} = 202.$		5.331	207.	= GX_{95%} ppm
$UTL = 247.$		5.732	309.	= GUTL ppm
$ME = \text{median} = 76.$		4.331	61.4	= LLCL ppm
$(M - Me)/S = 0.268$ (closer to 0 is better).		**0.062**	109.	= LUCL ppm
M - Me)/S => lognormal fits better than normal distribution.				

the lognormal distribution will fit these data better than the normal distribution. That assumption is verified by the plotted data, as shown in the next section. As a point of interest, IHStat rejects the fit to the normal distribution using the W-test; LogNorm2 rejects the normal distribution with all three of its tests: W-test, skewness test, and the median test mentioned above; and when tested with the proprietary "omega" statistic from the LOGAN[21] software, the normal distribution was rejected. In sum, all readily available curve fitting test statistics agree, this data set is better modeled by a lognormal distribution.

Probability Plotting Exercise

Figure 2-3 shows the data plotted on normal probit paper and Figure 2-4 shows the data plotted on lognormal probit paper. The straight line is the line plotted by use of the estimated parameters M, S, GM, and GS for the normal and lognormal distributions. The mean, M, occurs at the 50% point, M–S occurs at the 15.87% point, M + S occurs at the 84.13% point, and M + 1.645 * S occurs at the 95% point for normal data. For lognormal data, GM occurs at the 50% point, GM/GS occurs at the 15.87% point, GM * GS occurs at the 84.13% point, and GM * GS^{1.645} occurs at the 95% point.

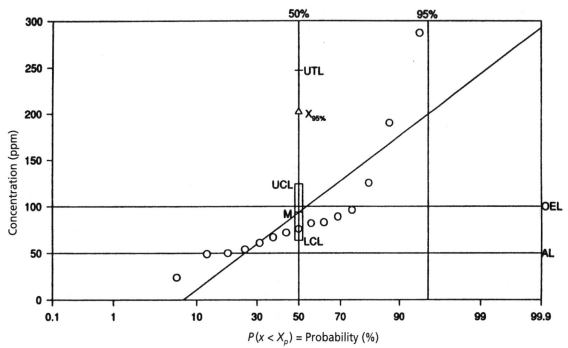

FIGURE 2-3. Normal Probability Plot–Painters' Exposures.

From Table 2-12, $M - S = 28$, $M = 94$, and $M + S = 160$. Thus, the best fit line in Figure 2-3 goes through the points (28 ppm, 16%), (94 ppm, 50%), and (160 ppm, 84%). A glance at the data shows that they are not a good fit to the line, confirming the indication in the "Calculate Descriptive Statistics" section that the normal distribution would not be a good fit.

For the data in column 4, the 50% point is given by $\ln(GM)$, the 16% point by $[\ln(GM) - \ln(GS)]$, and the 84% point by $[\ln(GM)+\ln(GS)]$. Thus, the plotting points in log-transformed notation are (3.78, 16%), (4.37, 50%), and (4.95, 84%). Although these can be plotted directly on linear-probability paper, it is usually more intuitive to take the antilog of these numbers and plot the line on logarithm-probability paper in preferred concentration units. This is accomplished by taking the antilog and multiplying it by the reference concentration (1 ppm). The plotting positions for Figure 2-4 are (44, 16%), (79, 50%), and (142, 84%). Inspection of this line and the data confirms that there is a better fit under the lognormal assumption than was true under the normal assumption. The fit looks somewhat irregular, but adequate for making some decisions. In fact, the Logan software calculates an "omega" of 0.103 (near the low end of the range recommended for decision-making);[21] and the LogNorm2 W-test statistic is 0.95, barely larger than the critical value of 0.76 for this data.

When evaluating the quality of fit visually, any substantial departure from linearity in the center 80% of the distribution should be cause for concern. A departure of this type indicates the distributional assumption is a poor approximation of the data. When predicting the probabilities that exposures exceed points in the upper end of the tail, predictions beyond the observed data should be made with caution. For this example, the highest observed data point was 287 ppm and it was estimated to represent the 93.8% point on the distribution. With 15 samples, any prediction above the 95% point is suspect. Likewise, with six samples, where the plotting position of the largest data point is 6/7 or 85.7%, it is wise to be cautious in estimating even the 90% point.

Decision-making with Descriptive Statistics

It is convenient to mark the AL, the OEL, the confidence interval about the mean (from LCL to UCL), and the range from the 95th percentile to the $\gamma = 90\%$ UTL, right on the probability plot. Decision-making with an explicit knowledge of the level of confidence becomes a very intuitive process. The decisions identified in the statistical judgment portion of Figure 2-1 are all available by inspection.

Further, the line of best fit can be used to read estimated exceedance levels directly. Two levels of interest are the proportion of exposures in the population[D] that exceed the OEL and the proportion that exceed the action level. For the data set in Table 2-12, Figure 2-4

[D] This is the population of all exposures in the HEG. Its distribution is estimated from GM and GS in this example.

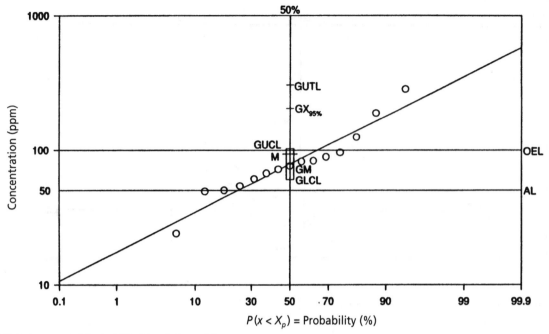

FIGURE 2-4. Lognormal Probability Plot–Painters' Exposures.

shows that the 95% *UCL* is above the OEL, that less than 20% of exposures are below the AL, and less than 70% are below the OEL. It appears from all evidence at hand that these data represent an unacceptable workplace. Immediate intervention is required.

Professional Judgment in Statistical Interpretation

Even though the statistical judgment says this workplace is unacceptable, an experienced industrial hygienist might have misgivings about the decision, particularly after personally inspecting the paint shop itself. A closer examination of data show them scattered around the line of best fit to a lognormal distribution. Also, looking at Figure 2-3, there are nine points that seem to lie in a very straight line. If the other six were discarded, these would be a good fit to a normal distribution.

Upon careful investigation of the data, it is discovered that among 91 painters assigned to this shop and working two shifts, six samples were taken on the night shift and nine were taken during the day shift. Table 2-13 shows the descriptive statistical analysis for the night shift and Table 2-14 for the day shift. Sure enough, the nine samples that looked so neat on Figure 2-3 were all taken from the day shift!

These stratified samples clearly show that the exposures of individual painters, who often rotate between day and night shifts, are not homogeneous. The probability of overexposure at night far exceeds the probability of overexposure during the day. This may be due to lack of

helpers or to lax supervision at night. Priority attention is needed to improve working conditions at night.

Further, it is apparent from Table 2-13 and Figure 2-5 that the nighttime exposures are approximately lognormally distributed and from Table 2-14 and Figure 2-6 that the daytime exposures are approximately normally distributed. Based on Figure 2-5, it is reasonable to classify the night shift exposures as unacceptable because the average of observed exposures is above the OEL. Based on Figure 2-6, it is reasonable to classify the day shift exposures as tolerable because the average of observed exposures, the 95% *UCL*, and the best estimate of the 95% of all exposures are all less than the OEL. The industrial hygienist making such a decision with the aid of Figure 2-6 would know that the level of confidence, γ, is less than 90% because the 90% *UTL* on the *X*95% is above the OEL. This ability to understand the confidence with which industrial hygiene decisions are made illustrates the benefit of plotting data in the form of Figures 2-3 to 2-6.

Confidence Intervals in the Example

Figure 2-7 shows the fitted lognormal distributions for the data in this example, and Table 2-15 summarizes the statistics computed from the data. Three of the panels display confidence and tolerance intervals as the length of a line. All confidence limits are based on $(1 - \alpha) = 0.95$ for one-sided intervals. This is a much higher confidence level than the author believes can be supported by such

TABLE 2-13. Descriptive Statistics Example: Solvent Exposure Data Analysis for Night Shift Painters

			DATA PREPARATION	
Rank [r]	Plot Position [rl(n+l)]	Exposure (ppm) [x]	ln(data/1ppm) (unitless) [ln(x)]	Tabulated Values [t, z, & k]
1	0.143	24.0	3.178	$p = 0.95$
2	0.286	50.0	3.912	$n = 6.$
3	0.429	82.0	4.407	$\gamma = 0.90$
4	0.571	125.0	4.828	$z(p) = 1.645$
5	0.714	190.0	5.247	$t(p, n-1) = 2.015$
6	0.857	287.0	5.659	$k(p, n, \gamma) = 3.091$

	DESCRIPTIVE STATISTICS		
NORMAL	LOGNORMAL		
X	ln(GX)	GX = (1 ppm)*exp[in(GX)]	
S = 98.1	0.906	2.47 = GS	unitless
M = 126.	4.539	93.6 = GM	ppm
$M - S = X_{16\%}$ = 28.2	3.633	37.8 = $GX_{16\%}$	ppm
$M + S = X_{84\%}$ = 224.	5.445	232. = $GX_{84\%}$	ppm
OEL = 100.	4.605	100. = OEL	ppm
LCL = 45.6	3.793	44.4 = GLCL	ppm
UCL = 207.	5.284	197. = GUCL	ppm
$X_{95\%}$ = 288.	6.029	415. = $GX_{95\%}$	ppm
UTL = 430.	7.339	1540. = GUTL	ppm
Me = median = 72.0	4.618	51.8 = LLCL	ppm
$(M - Me)/S$ = 0.554 (closer to 0 is better).	**-0.087**	384. = LUCL	ppm

$(M - Me)/S \Rightarrow$ lognormal fits better than normal distribution.

sparse data. Neverthless, the graphics illustrate key concepts about data interpretation.

The plot in the upper left corner compares the three fitted probability density functions with each other and with the OEL. Comparing the distribution of exposures during the day and night shifts clearly shows that the night shift is the proper place to intervene.

Moving clockwise to the upper right panel, find the composite PDF with its data points indicated by closed circles. At the top, the median, GM, is plotted as a circle and its 95% two-sided CI is indicated by a horizontal line drawn from the GLCL to the GUCL; its GX95 quantile is a rectangle and its one-sided GUTL is represented by a line to the right of the rectangle. Below that, the open triangles indicate the positions of the maximum likelihood estimate of the mean, M_{MLE}, and of the minimum variance unbiased estimate of the mean, M_{MVUE}, computed for an assumed lognormal population. Land's 95% two-sided confidence interval from LLCL to LUCL

is the horizontal line through M_{MVUE} and M_{MLE}. The circle in the last line indicates the unbiased sample mean, M, and in a line representing a normal 95% two-sided Student-t confidence interval on the mean, from LCL to UCL, based on an assumed Gaussian population. To the right is a rectangle marking the 95th percentile and a line extending to the 95% upper tolerance limit. Although, the point estimates of the mean and median are smaller than the OEL, the LUCL and UCL are above it. On that basis, the 15All data set does not pass any form of the employer's test at 95% confidence because none of the point estimates of critical statistics are less than the OEL: OEL < GUCL < LUCL < UCL < X95 < GX95 < UTL95 < GUTL95.

The PDF for the troublesome night shift is displayed in the lower right panel. Here, due to small $n = 6$ and large GS = 2.47, Land's estimate of the LUCL is significantly larger than either the GUCL or the UCL estimate. With 3 of 6 measurements over the OEL, this is pretty

TABLE 2-14. Descriptive Statistics Example: Solvent Exposure Data Analysis for Day Shift Painters

DATA PREPARATION

Rank [r]	Plot Position [r/(n+1)]	Exposure (ppm) [x]	ln(data/1 ppm) (unitless) [ln (x)]	Tabulated Values [t, z, & k]
1	0.100	49.0	3.892	$p = 0.95$
2	0.200	54.0	3.989	$n = 9.$
3	0.300	61.0	4.111	$\gamma = 0.90$
4	0.400	67.0	4.205	$z(p) = 1.645$
5	0.500	72.0	4.277	$t(p, n-1) = 1.860$
6	0.600	76.0	4.331	$k(p, n, \gamma) = 2.649$
7	0.700	83.0	4.419	
8	0.800	89.0	4.489	
9	0.900	96.0	4.564	

DESCRIPTIVE STATISTICS

NORMAL		LOGNORMAL		
X		ln(GX)	GX = (1 ppm)*exp[in(GX)]	
$S = 15.8$		0.226	1.25 = GS	unitless
$M = 71.9$		4.253	70.3 = GM	ppm
$M - S = X_{16\%} = 56.1$		4.027	56.1 = $GX_{16\%}$	ppm
$M + S = X_{84\%} = 87.7$		4.479	88.1 = $GX_{84\%}$	ppm
OEL = 100.		4.605	100. = OEL	ppm
$LCL = 62.1$		4.113	61.1 = GLCL	ppm
$UCL = 81.7$		4.393	80.9 = GUCL	ppm
$X_{95\%ile} = 97.9$		4.625	102. = $GX_{95\%}$	ppm
$UTL = 114.$		4.852	128. = GUTL	ppm
Me = median = 72.00		4.277	62.0 = LLCL	ppm
(M − Me)/S = −0.007	(closer to 0 is better)	−0.105	81.1 = LUCL	ppm

(M − Me)/S ⇒ normal fits better than lognormal distribution.

clearly an opportunity for creative industrial hygiene intervention.

The PDF for the careful day shift is displayed in the lower left panel. This is the best fit lognormal PDF, and it is nearly Gaussian in shape. This workplace passes the employer's *UCL* test at 95% confidence, as all estimates of the *UCL* are smaller than the OEL. Both the 95th percentile and the 95% *UTL* are above the OEL, so this workplace cannot pass *UTL* version of the employer's test. Professional judgment is clearly necessary. A plausible plan of action based on this data set is to implement immediate personal protective equipment for night shift workers. Investigate and correct the causes of those overexposures. Continue to sample both night and day shift operations, set priorities on improving the night shift working conditions, and use capital investment opportunities to reduce day shift exposures. The special problems occurring during the night shift may offer insight into means to reduce exposures on the day shift.

Figure 2-7 and Table 2-15 both show several interesting comparisons. For the data in this example, there is little difference between *M* and M_{MVUE}. For data that are approximately Gaussian, as in "15all" and "9day" data sets, there is little difference between Land's Confidence Interval and the *t* Confidence Interval. The estimate of the GUTL at 95% confidence is, in each of the three data sets, well above the largest observed measurement. In the "6night" data set, $GUTL_{95} = 1540$ ppm. There is little physical reality in such numbers when the largest observed measurement was 287 ppm. This author prefers to assess the *UTL* at $\gamma = 0.8$ to $\gamma = 0.9$, to stay closer to reasonable models of fugitive emissions.

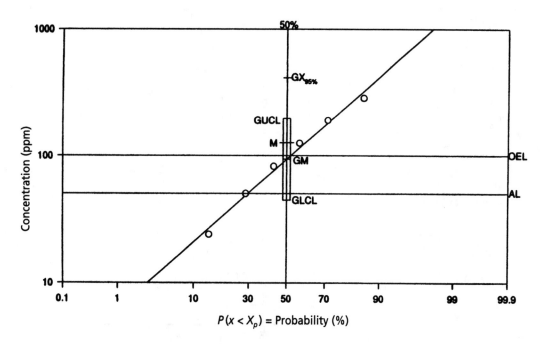

FIGURE 2-5. Lognormal probability plot, night shift. The *GUTL* = 1540 ppm plots above the axis on this figure because of the small number of samples (*n* = 6).

Summary

A formally defined air sampling strategy helps to provide maximum information for worker protection through good stewardship of available resources. A formal strategy ensures the resources needed to support a healthy industrial hygiene surveillance program. A successful strategy considers:

1. Whether an effect is acute or chronic;
2. Government regulations and mandatory PELs;
3. Consensus standards, recommended TLVs, and applicable OELs;
4. Patterns of exposure;
5. Presence of symptomatic workers;
6. Classification of workers.

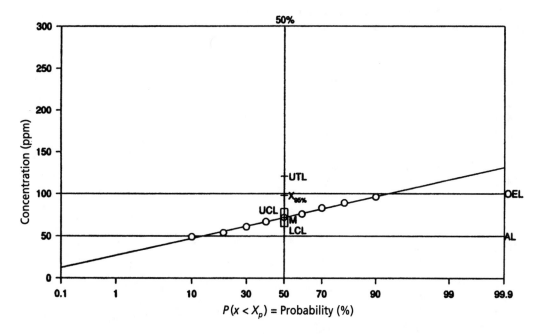

FIGURE 2-6. Normal probability plot, day shift.

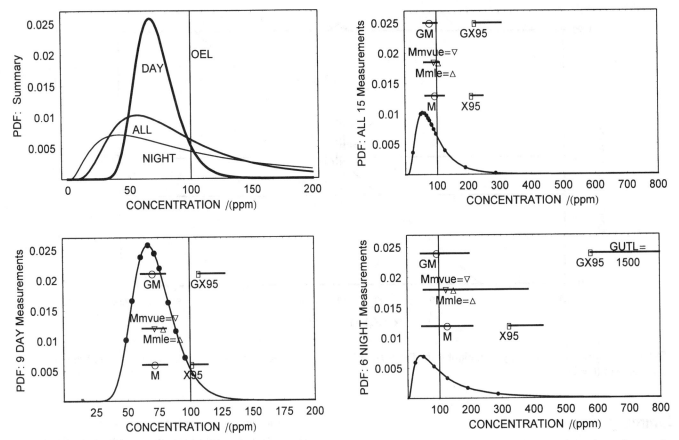

FIGURE 2-7. Best-fit lognormal PDFs for day, night, and total sample sets. The first panel compares the PDFs, others show three point estimates for central statistics and two for extreme statistics. In each panel, the top line shows the Student-T based estimate for the median = GM, and its 90% two-sided confidence interval plus a point estimate for the 95th percentile and its 90% upper tolerance limit under the lognormal assumption. The middle line shows the maximum likelihood estimate of the mean, Mhat, plus Land's confidence interval under the lognormal assumption. The bottom line shows estimates for the mean plus Student-T based 90% confidence interval about the mean, point estimate for the 95th percentile and the 90% upper tolerance limit for the 95th percentile under the normal assumption. For all data sets, all central estimates are within 90% confidence intervals. Parametric assumptions change extreme estimates, portrayed vividly by two widely spaced estimates for the 95th percentile for the night measurements (X95 & GX95).

TABLE 2-15. Descriptive Statistics for Exposure Assessment Example (see Figure 2-7). All entries in ppm except GS, which is unitless.

	$\gamma = 0.95$		
	15all	9day	6night
M	93.7	71.9	126.0
S	65.9	15.8	98.1
LCL	63.7	62.1	45.6
UCL	124.0	81.7	207.0
X95	202.0	97.9	288.0
UTL95	247.0	114.0	430.0
GM	78.8	70.3	93.6
GS	1.8	1.25	2.47
GLCL	60.4	61.1	44.4
GUCL	103.0	80.9	197.0
M_{MLE}	106.0	78.7	147.0
LLCL	61.4	62.0	51.8
LUCL	109.0	81.1	384.0
M_{MVUE}	91.4	71.7	123.0
GX95	207.0	102.0	415.0
GUTL95	309.0	128.0	1540.00

Statistical methods are helpful in designing the strategy and in interpreting the sampling data. The exact criteria to be used varies from one situation to another. There is no substitute for professional judgment. Air sampling is a major part of an exposure control program to protect workers against air contaminants. It is one of the tools used to ensure proper priorities for engineering controls, work practices, emergency response plans, and respiratory protection to reduce worker exposure.

A well-designed strategy includes criteria calling for additional control measures as necessary and specifies the speed with which they must be implemented. The careful reader will have noted that the exposure rating scheme proposed here interacts with the decision flow diagram to reduce surveillance frequency only after representative exposures fall below the action level (50% OEL). The highest intensity air sampling efforts occur where the average exposure is between 50% and 100% of the OEL and the *UCL* on the mean is above the OEL. The lowest intensity air sampling efforts occur for those activities where the average exposure is below 10% OEL and the *UCL* is below the *AL*.

Finally, the use of statistical principles does not change the general basis for the health protection strategy used by industrial hygienists for many years. Although statistics assist the industrial hygienist in determining the relative uncertainty associated with any decision that is made, the criteria for such decisions still must be determined by consensus standards, by applicable regulations, or by the employer when external guidelines are unavailable. This chapter has outlined a structure that allows an improved level of quantification for IH decisions while retaining the traditional flexibility required of all professionals. In the end, professional judgment remains the key to success for industrial hygienists who practice in a wide variety of occupational settings and provide safe and healthful working conditions in the face of an ever-growing quantity and diversity of workspace stressors.

Useful Statistical Tables

- Table 2-5. Confidence and Sample Size for a Two-Sided Distribution-Free Tolerance Limit
- Table 2-6. Sample Size for Testing Mean of a Sample from Lognormal Population with Two-sided 95% Confidence Level and 90% Power
- Table 2-7. Factors for One-Sided Tolerance Limits for Normal Distributions
- Table 2-8. Confidence and Sample Size for a One-Sided Distribution-Free Tolerance Limits

- Table 2-10. Normal Distribution z values—for Xp
- Table 2-11. Normal Distribution t values—for UCL
- Table 2-15. Descriptive Statistics for Exposure Assessment Example

References

1. Attfield, M.D., Hewett, P.: Exact Expressions for the Bias and Variance of Estimators of the Mean of a Lognormal Distribution. Am. Ind. Hyg. Assoc. J. 53(7):432-435 (1992).
2. Hawkins, N.C.; Norwood, S.K.; Rock, J.C.: A Strategy for Occupational Exposure Assessment. American Industrial Hygiene Association, Akron, OH (1991).
3. Hickey, E.E.; Stoetzel, G.A.; Strom, D.J.; *et al.*: Air Sampling in the Workplace, Final Report. NUREG-1400. U.S. Nuclear Regulatory Commission, Division of Regulatory Applications, Office of Nuclear Regulatory Research, Washington DC (September 1993).
4. American Conference of Governmental Industrial Hygienists, ACGIH, Committee on Industrial Ventilation: Industrial Ventilation—A Manual of Recommended Practice, 23rd ed. Cincinnati, OH (1998).
5. American Society of Heating, Refrigerating and Air-Conditioning Engineers: Fundamentals Volume, Systems Volume, Equipment Volume, Applications Volume. ASHRAE, 1791 Tullie Circle, N.E., Atlanta, GA (Various Copyright Dates).
6. Heinsohn, R.J.: Industrial Ventilation: Engineering Principles. John Wiley & Sons, Inc., New York (1991).
7. U.S. Environmental Protection Agency: Chemical Engineering Branch Manual for the Preparation of Engineering Assessments, Vol I: CEB Engineering Manual. U.S EPA, Office of Toxic Substances, Washington, DC (1991).
8. Leidel, N.A.; Busch, K.A.; Lynch, J.R.: Occupational Exposure Sampling Strategy Manual. NIOSH 77-173. National Institute for Occupational Safety and Health, Cincinnati, OH (1977).
9. Leidel, N.A.; Busch. K.A.: Statistical Design and Data Analysis Requirements. In: Patty's Industrial Hygiene and Toxicology, 2nd ed., Vol. 3A, Theory and Rationale of Industrial Hygiene Practice: The Work Environment. L.J. Cralley and L.V. Cralley, Eds. John Wiley & Sons, New York (1985).
10. Rock, J.C.: The NIOSH Action Level—A Closer Look. In: Measurement and Control of Chemical Hazards in the Workplace Environment, Chap. 29. American Chemical Society, New York (1981).
11. Tuggle, R.M.: The NIOSH Decision Scheme. Am. Ind. Hyg. Assoc. J. 42:493 (1981).
12. Shannon, C. E.: Communication in the Presence of Noise, Proc IRE, 37(10):10-21 (1949). Referenced in Hancock, J.C.: An Introduction to the Principles of Communication Theory. Chapter 1. McGraw-Hill (1961).
13. Rock, J.C.: Can Professional Judgment be Quantified? Am. Ind. Hyg. Assoc. J. 47(6):A-370 (1986).
14. Heederik, D.; Hurley, F.: Occupational Exposure Assessment: Investigating Why Exposure Measurements Vary. Appl. Occup. Environ. Hyg. 9:71B76 (1994).
15. Rappaport, S.M.: The Rules of the Game: An Analysis of OSHA's Enforcement Strategy. Am. J. Ind. Med. 6:291 (1984).
16. Mulhausen, J.R., Damiano, J.: A Strategy for Assessing and Managing Occupational Exposures. 2nd edition. AIHA Press, Fairfax, VA (1998).
17. Tuggle, R.M.: The Relationship Between TLV–TWA Compliance and TLV–STEL Compliance. Appl. Occup. and Envir. Hyg. 15(4):380–386 (2000).
18. Ayer, H.E.: Occupational Air Sampling Strategies. In: Air Sampling Instruments, 7th ed., Chap. B. Susanne V. Hering, Ed.

American Conference of Governmental Industrial Hygienists, Cincinnati, OH (1989).

19. National Council on Radiation Protection and Measurements: Recommendations on Limits for Exposure to Ionizing Radiation. NCRP Report No 91. NCRP, Bethesda, MD (June 1987).

20. Natrella, M.G.: Experimental Statistics. National Bureau of Standards Handbook 91. U.S Government Printing Office, Washington, DC (1966).

21. Dupont Statistics Group: LOGAN Workplace Exposure Evaluation System. American Industrial Hygiene Association, Fairfax, VA (1990).

22. Rappaport, S.M., Selvin, S.: A Method for Evaluating the Mean Exposure from a Lognormal Distribution. Am. Ind. Hyg. Assoc. J. 48(4):374-379 (1987).

23. Hewett, Paul: Sample Size Formulae for Estimating the True Arithmetic or Geometric Mean of Lognormal Exposure Distributions, Am. Ind. Hyg. Assoc. J. 56:219-225 (1995).

24. Rappaport, S.M.: The Rules of the Game: An Analysis of OSHA's Enforcement Strategy. Am. J. Ind. Med. 6:291 (1984).

25. Roach, S.A.: A More Rational Basis for Air Sampling Programs. Am. Ind. Hyg. Assoc. J. 27:1B12. (1966).

26. Roach, S.A.: A Most Rational Basis for Air Sampling Programmes. Ann. Occup. Hyg. 20:65B84 (1977).

27. Spear, R.C.; Selvin, S.; Francis, M.: The Influence of Averaging Time on the Distribution of Exposures. Am. Ind. Hyg. Assoc. J. 47(6):365B368 (1986).

28. The 'MEMORY JOGGER'™—A Pocket Guide of Tools for Continuous Improvement. GOAL/QPC, Methuen, MA (1988).

29. Grant, E.L.; Leavenworth, R.S.: Statistical Quality Control. 4th ed. McGraw-Hill Book Co., New York (1980).

30. Lyles, R.H., Kupper, L.L.: On Strategies for Comparing Occupational Exposure Data to Limits. Am. Ind. Hyg. Assoc. J. 57(1):6-15 (1996).

31. Traub, J.F.; Wozniakowski, H.: Breaking Intractability. Scientific American, pp. 102–114 (January 1994).

32. Rock, J.C.: A Comparison Between OSHA Compliance Criteria and Action-Level Decision Criteria. Am. Ind. Hyg. Assoc. J. 45:297 (1982).

33. Wolfram, S.: The Mathematica Book, 3rd Edition pg. 764, Cambridge University Press, New York, (1996).

34. Gumbel, E.J.: Statistics of Extremes. Columbia University Press, New York (1958).

35. Saltzman, B.E.; Cholak, J.; Shaefer, L.S.; et al.: Concentrations of Six Metals in the Air of Eight Cities. Environ. Sci Technol. 19:328 (1985).

36. American Conference of Governmental Industrial Hygienists: 2000 TLVs and BEIs. ACGIH, Cincinnati, OH (2000).

37. Box, G.E.P.; Hunter, W.G.; Hunter, J.S.: Statistics for Experimenters. John Wiley & Sons, New York (1978).

38. Aitchison, J.; Brown, J.A.C.: The Lognormal Distribution with Special Reference to its Uses in Economics. Cambridge University Press, Cambridge, UK (1957).

39. Hewett, P.; Ganser, G.H.: Simple Procedures for Calculating Confidence Intervals Around the Sample Mean and Exceedance Fraction Derived from Lognormally Distributed Data. Appl. Occup. Environ. Hyg. 12(2): 132-142 (1997).

40. Armstrong, Ben G.: Confidence Intervals for Arithmetic Means of Lognormally Distributed Exposures. Am. Ind. Hyg. Assoc. J. 53(8): 481-485 (1992).

41. Tuggle, R.M.: Assessment of Occupational Exposure Using One-sided Tolerance Limits. Am. Ind. Hyg. Assoc. J. 43:338 (1982).

42. Vos, Gordon: LogNorm2 Software. AIHA Press, American Industrial Hygiene Association, Fairfax, VA (1997).

TABLE 2-5. Confidence and Sample Size for a Two-sided Distribution-free Tolerance Limit (non-parametric)

Confidence γ with which we may assert that $100p$ percent of the population lies between the largest and smallest of a random sample of n from that population (continuous distribution assumed).

n	p =.75	p =.90	p =.95	p =.99	n	p =.75	p =.90	p =.95	p =.99
3	.16	.03	.01	.00	17	.95	.52	.21	.01
4	.26	.05	.01	.00	18	.96	.55	.23	.01
5	.37	.08	.02	.00	19	.97	.58	.25	.02
6	.47	.11	.03	.00	20	.98	.61	.26	.02
7	.56	.15	.04	.00	25	.99	.73	.36	.03
8	.63	.19	.06	.00	30	1.00−	.82	.45	.04
9	.70	.23	.07	.00	40	-	.92	.60	.06
10	.76	.26	.09	.00	50	-	.97	.72	.09
11	.80	.30	.10	.01	60	-	.99	.81	.12
12	.84	.34	.12	.01	70	-	.99	.87	.16
13	.87	.38	.14	.01	80	-	1.00−	.91	.19
14	.90	.42	.15	.01	90	-	—	.94	.23
15	.92	.45	.17	.01	100	-	—	.96	.26
16	.94	.49	.19	.01					

Source: Natrella.[20] Table A–32.

TABLE 2-7. Factors for One-Sided Tolerance Limits for Normal Distributions (k-table)

Factors $k_{p,n,\gamma}$ such that the probability is γ that at least a proportion p of the distribution will be less than $M + KS$ (greater than $M - KS$), where M and S are estimates of the mean and the standard deviation computed from a sample size of n.

n\p	γ=0.75 0.75	0.90	0.95	0.99	0.999	γ=0.90 0.75	0.90	0.95	0.99	0.999	n\p	γ=0.95 0.75	0.90	0.95	0.99	0.999	γ=0.99 0.75	0.90	0.95	0.99	0.999
3	1.464	2.501	3.152	4.396	5.805	2.602	4.258	5.310	7.340	9.651	3	3.804	6.158	7.655	10.552	13.857	—	—	—	—	—
4	1.256	2.134	2.680	3.726	4.910	1.972	3.187	3.957	5.437	7.128	4	2.619	4.163	5.145	7.042	9.215	—	—	—	—	—
5	1.152	1.961	2.463	3.421	4.507	1.698	2.742	3.400	4.666	6.112	5	2.149	3.407	4.202	5.741	7.501	—	—	—	—	—
6	1.087	1.860	2.336	3.243	4.273	1.540	2.494	3.091	4.242	5.556	6	1.895	3.006	3.707	5.062	6.612	2.849	4.408	5.409	7.334	9.550
7	1.043	1.791	2.250	3.126	4.118	1.435	2.333	2.894	3.972	5.201	7	1.732	2.755	3.399	4.641	6.061	2.490	3.856	4.730	6.411	8.348
8	1.010	1.740	2.190	3.042	4.008	1.360	2.219	2.755	3.783	4.955	8	1.617	2.582	3.188	4.353	5.686	2.252	3.496	4.287	5.811	7.566
9	0.984	1.702	2.141	2.977	3.924	1.302	2.133	2.649	3.641	4.772	9	1.532	2.454	3.031	4.143	5.414	2.085	3.242	3.971	5.389	7.014
10	0.964	1.671	2.103	2.927	3.858	1.257	2.065	2.568	3.532	4.629	10	1.465	2.355	2.911	3.981	5.203	1.954	3.048	3.739	5.075	6.603
11	0.947	1.646	2.073	2.885	3.804	1.219	2.012	2.503	3.444	4.515	11	1.411	2.275	2.815	3.852	5.036	1.854	2.897	3.557	4.828	6.284
12	0.933	1.624	2.048	2.851	3.760	1.188	1.966	2.448	3.371	4.420	12	1.366	2.210	2.736	3.747	4.900	1.771	2.776	3.410	4.633	6.032
13	0.919	1.606	2.026	2.822	3.722	1.162	1.928	2.403	3.310	4.341	13	1.329	2.155	2.670	3.659	4.787	1.702	2.677	3.290	4.472	5.826
14	0.909	1.591	2.007	2.796	3.690	1.139	1.895	2.363	3.257	4.274	14	1.296	2.108	2.614	3.585	4.690	1.645	2.592	3.189	4.336	5.651
15	0.899	1.577	1.991	2.776	3.661	1.119	1.866	2.329	3.212	4.215	15	1.268	2.068	2.566	3.520	4.607	1.596	2.521	3.102	4.224	5.507
16	0.891	1.566	1.977	2.756	3.637	1.101	1.842	2.299	3.172	4.164	16	1.242	2.032	2.523	3.463	4.534	1.553	2.458	3.028	4.124	5.374
17	0.883	1.554	1.964	2.739	3.615	1.085	1.820	2.272	3.136	4.118	17	1.220	2.001	2.486	3.415	4.471	1.514	2.405	2.962	4.038	5.268
18	0.876	1.544	1.951	2.723	3.595	1.071	1.800	2.249	3.106	4.078	18	1.200	1.974	2.453	3.370	4.415	1.481	2.357	2.906	3.961	5.167
19	0.870	1.536	1.942	2.710	3.577	1.058	1.781	2.228	3.078	4.041	19	1.183	1.949	2.423	3.331	4.364	1.450	2.315	2.855	3.893	5.078
20	0.865	1.528	1.933	2.697	3.561	1.046	1.765	2.208	3.052	4.009	20	1.167	1.926	2.396	3.295	4.319	1.424	2.275	2.807	3.832	5.003
21	0.859	1.520	1.923	2.686	3.545	1.035	1.750	2.190	3.028	3.979	21	1.152	1.905	2.371	3.262	4.276	1.397	2.241	2.768	3.776	4.932
22	0.854	1.514	1.916	2.675	3.532	1.025	1.736	2.174	3.007	3.952	22	1.138	1.887	2.350	3.233	4.283	1.376	2.208	2.729	3.727	4.866
23	0.849	1.508	1.907	2.665	3.520	1.016	1.724	2.159	2.987	3.927	23	1.126	1.869	2.329	3.206	4.204	1.355	2.179	2.693	3.680	4.806
24	0.845	1.502	1.901	2.656	3.509	1.007	1.712	2.145	2.969	3.904	24	1.114	1.853	2.309	3.181	4.171	1.336	2.154	2.663	3.638	4.755
25	0.842	1.496	1.895	2.647	3.497	0.999	1.702	2.132	2.952	3.882	25	1.103	1.838	2.292	3.158	4.143	1.319	2.129	2.632	3.601	4.706
30	0.825	1.475	1.869	2.613	3.454	0.966	1.657	2.080	2.884	3.794	30	1.059	1.778	2.220	3.064	4.022	1.249	2.029	2.516	3.446	4.508
35	0.812	1.458	1.849	2.588	3.421	0.942	1.623	2.041	2.833	3.730	35	1.025	1.732	2.166	2.994	3.934	1.195	1.957	2.431	3.334	4.364
40	0.803	1.445	1.834	2.568	3.395	0.923	1.598	2.010	2.793	3.679	40	0.999	1.697	2.126	2.961	3.866	1.154	1.902	2.365	3.250	4.255
45	0.795	1.435	1.821	2.552	3.375	0.908	1.577	1.986	2.762	3.638	45	0.978	1.669	2.092	2.897	3.811	1.122	1.857	2.313	3.181	4.168
50	0.788	1.426	1.811	2.538	3.358	0.894	1.560	1.965	2.735	3.604	50	0.961	1.646	2.065	2.863	3.766	1.096	1.821	2.269	3.124	4.096

Adapted by permission from Industrial Quality Control, Vol. XIV, No. 10, April 1958, from article entitled "Tables for One-Sided Statistical Tolerance Limits" by G. J. Lieberman.

TABLE 2-8. Confidence and Sample Size for a One-sided Distribution-free Tolerance Limit

Largest values of *m* such that we may assert with confidence at least γ that 100*P* percent of a population lies below the mth largest (or above the mth smallest) of a random sample of *n* from that population (no assumption of normality required)

n	γ = 0.75				γ = 0.90				γ = 0.95				γ = 0.99			
P	.75	.90	.95	.99	.75	.90	.95	.99	.75	.90	.95	.99	.75	.90	.95	.99
50	10	3	1	—	9	2	1	—	8	2	–	—	6	1	—	—
55	12	4	2	—	10	3	1	—	9	2	—	—	7	1	—	—
60	13	4	2	—	11	3	1	—	10	2	1	—	8	1	—	—
65	14	5	2	—	12	4	1	—	11	3	1	—	9	2	—	—
70	15	5	2	—	13	4	1	—	12	3	1	—	10	2	—	—
75	16	6	2	—	14	4	1	—	13	3	1	—	10	2	—	—
80	17	6	3	—	15	5	2	—	14	4	1	—	11	2	—	—
85	19	7	3	—	16	5	2	—	15	4	1	—	12	3	—	—
90	20	7	3	—	17	5	2	—	16	5	1	—	13	3	1	—
95	21	7	3	—	18	6	2	—	17	5	2	—	14	3	1	—
100	22	8	3	—	20	6	2	—	18	5	2	—	15	4	1	—
110	24	9	4	—	22	7	3	—	20	6	2	—	17	4	1	—
120	27	10	4	—	24	8	3	—	22	7	2	—	19	5	1	—
130	29	11	5	—	26	9	3	—	25	8	3	—	21	6	2	—
140	31	12	5	1	28	19	4	—	27	8	3	—	23	6	2	—
150	34	12	6	1	31	10	4	—	29	9	3	—	26	7	2	—
170	39	14	7	1	35	12	5	—	33	11	4	—	30	9	3	—
200	46	17	8	1	42	15	6	—	40	13	5	—	36	11	4	—
300	70	26	12	2	65	23	10	1	63	22	9	1	58	19	7	—
400	94	36	17	3	89	32	15	2	86	30	13	1	80	27	11	—
500	118	45	22	3	113	41	19	2	109	39	17	2	103	35	14	1
600	143	55	26	4	136	51	23	3	133	48	21	2	126	44	18	1
700	167	65	31	5	160	60	28	4	156	57	26	3	149	52	22	2
800	192	74	36	6	184	69	21	5	180	66	30	4	172	61	26	2
900	216	84	41	7	208	79	37	5	204	75	35	4	195	70	30	3
1000	241	94	45	8	233	88	41	6	228	85	39	5	219	79	35	3

Source: Natrella[20]; Table A–31.

TABLE 2-10. Cumulative Normal Distribution (z-table)

z is the standard normal variable. The value of P for $-z_p$ equals one minus the value of P for $+z_p$, e.g., the P for -1.62 equals $1 - .9474 = .0526$.

z_p	.00	.01	.02	.03	.04	.05	.06	.07	.08	.09
.0	.5000	.5040	.5080	.5120	.5160	.5199	.5239	.5279	.5319	.5359
.1	.5398	.5438	.5478	.5517	.5557	.5596	.5636	.5675	.5714	.5753
.2	.5793	.5832	.5871	.5910	.5948	.5987	.6026	.6064	.6103	.6141
.3	.6179	.6217	.6255	.6293	.6331	.6368	.6406	.6443	.6480	.6517
.4	.6554	.6591	.6628	.6664	.6700	.6736	.6772	.6808	.6844	.6879
.5	.6915	.6950	.6985	.7019	.7054	.7088	.7123	.7157	.7190	.7224
.6	.7257	.7291	.7324	.7357	.7389	.7422	.7454	.7486	.7517	.7549
.7	.7580	.7611	.7642	.7673	.7704	.7734	.7764	.7794	.7823	.7852
.8	.7881	.7910	.7939	.7967	.7995	.8023	.8051	.8078	.8106	.8133
.9	.8159	.8186	.8212	.8238	.8264	.8289	.8315	.8340	.8365	.8389
1.0	.8413	.8438	.8461	.8485	.8508	.8531	.8554	.8577	.8599	.8621
1.1	.8643	.8665	.8686	.8708	.8729	.8749	.8770	.8790	.8810	.8830
1.2	.8849	.8869	.8888	.8907	.8925	.8944	.8962	.8980	.8997	.9015
1.3	.9032	.9049	.9066	.9082	.9099	.9115	.9131	.9147	.9162	.9177
1.4	.9192	.9207	.9222	.9236	.9251	.9265	.9279	.9292	.9306	.9319
1.5	.9332	.9345	.9357	.9370	.9382	.9394	.9406	.9418	.9429	.9441
1.6	.9452	.9463	.9474	.9484	.9495	.9505	.9515	.9525	.9535	.9545
1.7	.9554	.9564	.9573	.9582	.9591	.9599	.9608	.9616	.9625	.9633
1.8	.9641	.9649	.9656	.9664	.9671	.9678	.9686	.9693	.9699	.9706
1.9	.9713	.9719	.9726	.9732	.9738	.9744	.9750	.9756	.9761	.9767
2.0	.9772	.9778	.9783	.9788	.9793	.9798	.9803	.9808	.9812	.9817
2.1	.9821	.9826	.9830	.9834	.9838	.9842	.9846	.9850	.9854	.9857
2.2	.9861	.9864	.9868	.9871	.9875	.9878	.9881	.9884	.9887	.9890
2.3	.9893	.9896	.9898	.9901	.9904	.9906	.9909	.9911	.9913	.9916
2.4	.9918	.9920	.9922	.9925	.9927	.9929	.9931	.9932	.9934	.9936
2.5	.9938	.9940	.9941	.9943	.9945	.9946	.9948	.9949	.9951	.9952
2.6	.9953	.9955	.9956	.9957	.9959	.9960	.9961	.9962	.9963	.9964
2.7	.9965	.9966	.9967	.9968	.9969	.9970	.9971	.9972	.9973	.9974
2.8	.9974	.9975	.9976	.9977	.9977	.9978	.9979	.9979	.9980	.9981
2.9	.9981	.9982	.9982	.9983	.9984	.9984	.9985	.9985	.9986	.9986
3.0	.9987	.9987	.9987	.9988	.9988	.9989	.9989	.9989	.9990	.9990
3.1	.9990	.9991	.9991	.9991	.9992	.9992	.9992	.9992	.9993	.9993
3.2	.9993	.9993	.9994	.9994	.9994	.9994	.9994	.9995	.9995	.9995
3.3	.9995	.9995	.9995	.9996	.9996	.9996	.9996	.9996	.9996	.9997
3.4	.9997	.9997	.9997	.9997	.9997	.9997	.9997	.9997	.9997	.9998

TABLE 2-11. Cumulative *t* Distribution (*t-table*)[A]

(df = n −1)	$t_{.60}$	$t_{.70}$	$t_{.80}$	$t_{.90}$	$t_{.95}$	$t_{.975}$	$t_{.99}$	$t_{.995}$
1	.325	.727	1.376	3.078	.6.314	12.706	31.821	63.657
2	.289	.617	1.061	1.886	2.920	4.303	6.965	9.925
3	.277	.584	.978	1.638	2.353	3.182	4.541	5.841
4	.271	.569	.941	1.533	2.132	2.776	3.747	4.604
5	.267	.559	.920	1.476	2.015	2.571	3.365	4.032
6	.265	.553	.906	1.440	1.943	2.447	3.143	3.707
7	.263	.549	.896	1.415	1.895	2.365	2.998	3.499
8	.262	.546	.889	1.397	1.860	2.306	2.896	3.355
9	.261	.543	.883	1.383	1.833	2.262	2.821	3.250
10	.260	.542	.879	1.372	1.812	2.228	2.764	3.169
11	.260	.540	.876	1.363	1.796	2.201	2.718	3.106
12	.259	.539	.873	1.356	1.782	2.179	2.681	3.055
13	.259	.538	.870	1.350	1.771	2.160	2.650	3.012
14	.258	.537	.868	1.345	1.761	2.145	2.624	2.977
15	.258	.536	.866	1.341	1.753	2.131	2.602	2.947
16	.258	.535	.865	1.337	1.746	2.120	2.583	2.921
17	.257	.534	.863	1.333	1.740	2.110	2.567	2.898
18	.257	.534	.862	1.330	1.734	2.101	2.552	2.878
19	.257	.533	.861	1.328	1.729	2.093	2.539	2.841
20	.257	.533	.860	1.325	1.725	2.086	2.528	2.845
21	.257	.532	.859	1.323	1.721	2.080	2.518	2.831
22	.256	.532	.858	1.321	1.717	2.074	2.508	2.819
23	.256	.532	.858	1.319	1.714	2.069	2.500	2.807
24	.256	.531	.857	1.318	1.711	2.064	2.492	2.797
25	.256	.531	.856	1.316	1.708	2.060	2.485	2.787
26	.256	.531	.856	1.315	1.706	2.056	2.479	2.779
27	.256	.531	.855	1.314	1.703	2.052	2.473	2.771
28	.256	.530	.855	1.313	1.701	2.048	2.467	2.763
29	.256	.530	.854	1.311	1.699	2.045	2.462	2.756
30	.256	.530	.854	1.310	1.697	2.042	2.457	2.750
40	.255	.529	.851	1.303	1.684	2.021	2.423	2.704
60	.254	.527	.848	1.296	1.671	2.000	2.390	2.660
120	.254	.526	.845	1.289	1.658	1.980	2.358	2.617
∞	.253	.524	.842	1.282	1.645	1.960	2.326	2.576

[A] *From National Bureau of Standards.*

Chapter 3

Approaches for Conducting Air Sampling in the Community Environment

Paul J. Lioy, Ph.D.

Division of Exposure Measurement and Assessment, Environmental and Occupational Health Sciences Institute, UMDNJ-Robert Wood Johnson Medical School and Rutgers University, Piscataway, New Jersey

CONTENTS

Introduction

Investigations of community air pollution can require one or more approaches to identify and examine problems and point the way toward a solution. The types of studies designed and completed on point sources, photochemical smog, long-range transport, and health effects have evolved substantially in scope since the passage of the 1970 Clean Air Act (CAA).[1] For instance, prior to that time, community air sampling programs used very simple monitoring tools, including the dustfall bucket, to measure pollutants, and they collected data in a limited number of specific geographic-population centers. The locations studied could have been a rural center, a small town or a city, a suburb of a large urban center, or a selected portion of a city. Usually, one or more fixed monitoring stations measuring a few pollutants (e.g., total particulate matter, sulfur dioxide) were placed at selected locations and these comprised a sampling program. Some of these sites eventually became part of well-established monitoring programs for examining attainment of National Ambient Air Quality Standards (NAAQS). The sampling programs have evolved from simple manual monitoring techniques, such as the high volume samplers and spot samplers, into networks that have multiple continuous and integrating monitors.[2]

For pollutants measured routinely with integrating samplers (e.g., volatile organic compounds, size-selected particulate matter), a statistically designed sampling schedule is needed for population-based sampling, and the data are available only as reports well after the day of sampling. Today, depending on the information needs, a number of pollutants can be measured continuously at a site or multiple sites, and much of the monitoring data are usually telemetered to a central data acquisition and validation center. In some cases, "on line" world wide web updates of pollution levels or indicators accessible to the general public. These are usually provided in the form of pollution indices that range from good to unhealthful for criteria pollutants.

Since the promulgation of the 1970 CAA,[1] one of the basic objectives of community air sampling has been to measure the concentrations of one or more pollutants originating from a number of different sizes and types of sources. This premise remains a cornerstone of the evaluation tools in the 1990 CAA amendments.[3] For most state and local agencies, the minimum set includes criteria pollutants, i.e., carbon monoxide, ozone, nitrogen dioxide, lead, suspended thoracic particulate matter (PM_{10}), and sulfur dioxide; however, other pollutants, especially air toxics (e.g., benzene), and pollutant indicators (e.g., visibility) are also measured at specific sites. In 1998, fine particulate matter less than 2.5 μm in diameter ($PM_{2.5}$) was added to the list of criteria pollutants.[4, 5] The actual techniques being employed to measure community air pollutants are outlined or illustrated in many chapters throughout this book. The concentration data obtained from these types of routine monitoring programs are normally reported in ppb or μg/m³ and used to evaluate compliance with the NAAQS, but the data can also be used as indicators or estimates of ambient exposure in epidemiological or long-term trend studies. When specific air pollution problems are addressed, the data from the routine monitoring programs are augmented by more detailed community air sampling programs.

Overall, community air sampling programs have been designed to examine the following: compliance with the NAAQS and National Emissions Standards for Hazardous Air Pollutants (NESHAPS); human exposures to pollutants or pollutant classes; pollutant formation, transport, and deposition; and the potential for human health effects, damage to vegetation and materials, etc. The primary U.S. NAAQS are based on the prevention of adverse human health effects, whereas the secondary NAAQS are based on prevention of ecological disturbances and resource degradation effects. The current 1998 NAAQS are shown in Table 3-1.[5] These are augmented internationally by the World Health Organization Air Pollution Guidelines, which are currently under revision.[6] Community air pollution studies can also be designed and conducted to investigate and understand basic chemical and physical processes in the atmosphere, to assist risk managers in attempts to reduce the intensity of air pollution episodes, and to develop more effective control strategies for primary and secondary pollutants.

TABLE 3-1. National Ambient Air Quality Standards[A]

Pollutant	Averaging Time	Primary Standard	Secondary Standard	Measurement Method
Carbon monoxide	8 hrs	10 mg/m³ (9 ppm)	Same	Nondispersive infrared spectroscopy
	I hr	40 mg/m³ (35 ppm)	Same	Nondispersive infrared spectroscopy
Nitrogen dioxide	Annual average	100 μg/m³ (0.05 ppm)	Same	Colorimetric using Saltzman method or equivalent continuous
Sulfur dioxide	Annual average	80 μg/m³ (0.03 ppm)		Pararosaniline method or equivalent continuous
	24 hrs	365 μg/m³ (0. 14 ppm)		
	3 hrs		1300 μg/m³ (0.5 ppm)	
$PM_{2.5}$	Annual average 24 hrs average	15 μg/m³ 65 μg/m³		Size-selective samplers
PM_{10} (≤10 μm)	Annual arithmetic mean 24-hrs	50 μg/m³ 150 μg/m³	50 μg/m³	Size-selective samplers
Ozone	8 hrs	80 ppb	Same	Chemiluminescent method or equivalent
Lead	3-month average	1.5 μg/m³	Same	Atomic absorption spectrophotometry

[A] Standards, other than those based on the annual average, are not to be exceeded more than once a year.

Types of Community Studies

Some of the air sampling programs developed to address the above issues are categorized as special short-term studies. Other special survey sites have evolved into the current National Air Monitoring Station (NAMS) network or the State and Local Air Monitoring Station (SLAMS) network for examining attainment of criteria pollutants monitored by regulatory agencies. Both of these networks are important and have either grown or have been augmented because they measure indicators of new or traditional emission sources, including fossil fuels, combustion, and industrial processes. In the 1980's, the Urban Air Toxic Monitoring Program (UATAP) for volatile organics was initiated, and in the 1990s the U.S. Environmental Protection Agency (U.S. EPA) deployed enhanced monitoring networks for identifying and tracking the individual volatile organic and nitrogen oxide constituents in the air that lead to the formation of ozone in many nonattainment areas.[7] These sites have assisted research in the Southern Oxidant Study and the Northeastern National Atmospheric Research Strategy for Tropospheric Ozone. Other monitoring networks include the IMPROVE and International Atmospheric Deposition Network (IADN).[8–10] The next major change to the National and State and Local air monitoring network will be the selection and use of over 1000 monitoring sites to begin compliance measurements for fine particulate matter, $PM_{2.5}$.

Research investigations on community air pollution can use short-term exposure studies to assess the acute exposures that produce health effects and the chemical characteristics of the atmosphere. In addition, long-term community studies are used to investigate the nature of acid rain, visibility or pollutant trends, and chronic health risks among the general population. Depending on the objectives of any particular study and the resources available, variations from program to program consider the size of the area studied, the specific site locations, the number of samplers at each site, the pollutants measured, the frequency and length of sampling, and the duration of the sampling intervals. The area associated with a community air sampling program can be defined by the meteorological influence on a microscale, mesoscale, and/or synoptic scale, which translate, for community air pollution, to radial distances of ~10 km, ~100 km, and ~3000 km.[11] The microscale investigations can have subcategories because problems may exist in a specific neighborhood, in a section of a city, around a group of small sources (<10 tons/year emission), or downwind of a single point source. The mesoscale influence can involve emissions from a number of point or line sources, which ultimately combine to produce the urban plume and its downwind impacts. Some of the major studies have been summarized in the U.S. EPA Criteria Document for Particulate Matter.[12] A study was also recently completed in Europe.[13] Two notable studies, the Midwest Interstate Sulfur Transport and Transformation Study (MISTT) and the Cross-Appalachian Tracer Experiment (CATEX), examined the relationship between mesoscale meteorology and pollution formation and transport.[14, 15] Synoptic scale events are associated with high or low pressure weather systems. High pressure systems, for example, are strongly associated with the formation and transport of secondary pollutants such as ozone, sulfate, oxidized hydrocarbons, and nitrates.

The preceding discussion has described a rather traditional approach to community air sampling. Since the later 1970s, concerns about community air pollution have also been extended to the indoor environment[16, 17] and, in some cases, the total exposure of an individual to specific pollutants.[18–21] These issues have evolved because major outdoor source controls have reduced emission of many pollutants and now pollutant concentrations can be higher in indoor environments relative to outdoor environments, e.g., particulate matter, toxic volatile organics etc.[22, 23] Increases in indoor pollution levels have been partly a result of activities designed to reduce energy costs by sealing up homes, construction of public and commercial buildings with windows that do not open, and use of synthetic materials for furniture and other personal products. In other cases, indoor air pollution has been due to the migration of hazardous wastes under a house and volatilization of organics within a basement or first floor.[22, 23] Because there are no indoor air standards at the present time, regulatory derived community-based studies are directed toward source identification and reduction, pollutant characterization, indoor-outdoor relationships, risk assessment, and health effects. However, indoor and personal exposures to many pollutants can be more important than exposure derived from outdoor pollutants. Therefore, future studies in the United States and other countries must factor contacts that may be derived from other than outdoor air in assessing human health effects and potential human risk to a variety of air pollutants. In this regard, major air pollution studies have been directed specifically toward the evaluation of indoor or personal air issues for criteria air pollutants, biological aerosols, and volatile organics.[21, 24–29]

The intent of this chapter is to examine the features of various types of community air sampling programs and the parameters that must be considered in the design of individual programs. A discussion of representative features of different studies is used to illustrate major design issues.

Community Air Sampling

Fixed Outdoor Sampling Sites

The selection of protocols and methods for measuring air pollution is mainly dependent upon the specific goals of the investigation. The earliest efforts in air monitoring focused on a central monitoring station; for instance, in New York City, it was the 121st Street Laboratory.[30] Various pollutants were measured, and originally, mechanized sampling techniques, such as bubblers and high-volume samplers, integrated ambient concentrations of pollutants for periods of 24 hrs or longer. This was done on a daily basis or on a statistically selected number of days (e.g., every sixth day) at regular intervals throughout the year. Air quality data from these types of monitoring networks provided valuable information on long-term trends and now provide a basis for assessing compliance with local and national standards. An example of the value of such a long-term network is shown in Figure 3-1, which illustrates the concentrations of sulfur dioxide found at the New York City site from the 1950s through 1993; the levels remain at approximately the 1993 levels today. It should be noted, however, that the measurement techniques have changed over time. This type of historical record provides a context that can place current concerns within an overall frame of reference for discussing reported health and environmental effects and the improvement of air quality due to source control. The original network was called the National Air Surveillance Network (NASN). It was a community-based voluntary effort conducted at selected locations throughout the United States.[31] It was superseded by the

NAMS and SLAMS networks, which select sites and pollutants to be measured using specific siting criteria and which utilize standard reference methods for sampling of criteria pollutants. Many of the standard methods or equivalent methods for sampling the criteria pollutants are described throughout the chapters in this manual and are available from the U.S. EPA (*http://www.epa.gov*). During 1999–2000, the U.S. EPA is attempting to establish "super" monitoring sites to conduct detailed ambient particle characterization and formation studies. This is to augment its plans for $PM_{2.5}$ mass measurement and mass speciation sites (SO_4^{-6}, carbon, NO_3^{-1}, total non-volatized organics, and trace elements.[32]

The ambient pollution siting criteria have been developed by the U.S. EPA for ambient air criteria pollutants such as photochemical oxidants.[5, 7] In addition, Ott recommended six types of outdoor sites, or monitoring categories, that could assist in identifying situations where a variety of human exposures could be measured in the community (Table 3-2).[33] The list still provides an excellent basis for establishing the initial criteria for site selection. Specific criteria will depend on the size of the area, density of the population, and the nature and accumulation patterns for individual primary or secondary air pollutants. (see also Chapter 10, page 204) These general concepts can be applied for both compliance and research studies. Population-based sites have been augmented by sites set up by the U.S. EPA and others to measure precursor pollutants for ozone Photochemical Assessment Monitoring Stations (PAMS) air toxins.

Although a standardized approach may be suitable for an air monitoring network, other approaches to sampling may be more appropriate for site-specific types of studies. For example, a state or local control agency may initially conduct a short-term, multiple station, intensive survey for particular ambient air pollutants. This has been done for specific pollutant classes such as polycyclic aromatic hydrocarbons (PAHs) and volatile organic compounds (VOCs) and for specific pollutants such as ozone, acid sulfates, and nitric acid. After assessing levels and distribution of the measured concentrations over time, the network may be reduced in number to a few strategically located stations. At a minimum, monitoring should be performed in an area that frequently receives the maximum ambient concentration, as well as in an area that periodically receives only background ambient concentrations. Any monitoring strategy can be modified as required to examine impacts from emissions increases (or decreases) as new control technology is applied to a source, process characteristics are altered, or fuel reformulations or conversions are implemented at the facility. Sometimes, monitoring strategies are designed to measure the amount of a

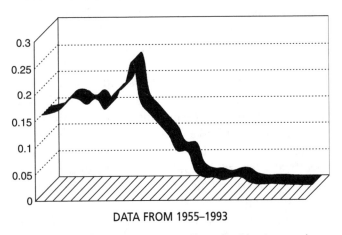

DATA FROM 1955–1993

FIGURE 3-1. Annual average sulfur dioxide in northern Manhattan, New York *(adapted from Eisenbud[14])*.

TABLE 3-2. Recommended Criteria for Siting Monitoring Stations

TYPE A	Downtown Pedestrian Exposure Station. Locate station in the central business district (CBD) of the urban area on a congested, downtown street surrounded by buildings (i.e., a "canyon" type street) and having many pedestrians. Average daily travel (ADT) on the street must exceed 10,000 vehicles/day, with average speeds less than 15 mph. Monitoring probe is to be located 0.5 m from the curb at a height of 3 ± 0.5 m.
TYPE B	Downtown Background Exposure Station. Locate station in the CBD of the urban area but not close to any major streets. Specifically, no street with ADT exceeding 500 vehicles/day can be less than 100 m from the monitoring station. Typical locations are parks, malls, or landscaped areas having no traffic. Probe height is to be 3 ± 0.5 m above the ground surface.
TYPE C	Residential Population Exposure Station. Locate station in the midst of a residential or suburban area but not in the CBD. Station must not be less than 100 m from any street having a traffic volume in excess of 500 vehicles/day. Station probe height must be 3 ± 0.5 m.
TYPE D	Mesoscale Meteorological Station. Locate station in the urban area at appropriate height to gather meteorological data and air quality data at upper elevations. The purpose of this station is not to monitor human exposure but to gather trend data and meteorological data at various heights. Typical locations are tall buildings and broadcasting towers. The height of the probe, along with the nature of the station location, must be carefully specified along with the data.
TYPE E	Nonurban Background Station. Locate station in a remote, nonurban area having no traffic and no industrial activity. The purpose of this station is to monitor for trend analyses, for nondegradation assessments, and large-scale geographical surveys, The location or height must not be changed during the period over which the trend is examined. The height of the probe must be specified.
TYPE F	Specialized Source Survey Station. Locate station very near a particular air pollution source under scrutiny. The purpose of the station is to determine the impact on air quality, at specified locations, of a particular emission source of interest. Station probe height should be of 3 ± 0.5 m unless special considerations of the survey require a nonuniform height.

Reprinted with permission from J. Air Pollut. Control Assoc., 27:543 (1977).[33]

particular pollutant transported into a state, country, or province from another jurisdiction. One type of monitoring network that looks at atmospheric deposition is the IADN which is designed to examine the deposition of persistent organic substances in the Great Lakes.[34] This type of monitoring, which includes PaHs, PCBs and pesticides, was significantly improved through the 1980s and 1990s; it is beginning to establish sources and receptor relationships for the Great Lakes.[34–37]

At times industry may take a different approach to ensure compliance with existing air quality standards. Managers of an industrial plant responsible for the control of a single pollutant may be most interested in averaging concentrations from their own source over specific sampling times when it is operating at different production levels. Therefore, the industry may design short-term studies. In fact, a plant's total monitoring effort may be focused on relating fence line pollution concentrations to production emissions as a basis for choosing the correct control devices. Consequently, these monitoring efforts could evolve into long-term studies that measure levels both before and after the implementation of a specific control strategy. The sites would be located primarily at the plant fence line and at particular locations either upwind or downwind of the facility. The 1990 revisions of the CAA require more emissions and fence

line monitoring efforts to quantify reductions of the approximately 189 air toxics targeted for reduction in Title III.[3] Emissions have been estimated in the Toxic Release Inventory (TRI) which, unfortunately, is not based on emissions tests, although they are used to estimate toxic pollutant exposure and impact. Clearly, this is a point of uncertainty that raises caution about its use by industry, government, and the general public to assess population exposure.

In the past 20 years, fixed outdoor sampling studies have been designed to investigate the nature of pollutants deposited in acid rain and their relationship to potential ecological effects in lakes and on forest vegetation.[38, 39] In addition, visibility degradation in the scenic areas of the western United States and in the urban-rural areas of the eastern United States, as well as compliance problems with prevention of significant deterioration of statutes, have been the subject of multipollutant, measurements at fixed sites, and mobile platform sampling studies.[12, 40, 41]

Epidemiological and field health effects studies require an understanding of the distribution of a pollutant and the potential locations of maximum impact. However, each health investigation requires sampling over time periods and in locations that are appropriate for relating species or pollutant exposures to a potential effect. At a minimum, the information derived for health purposes should be appropriate for estimating the incremental inhalation exposure, $c(t)dt$, of an individual in a particular environment; where $c(t)$ is the time varying concentration and dt is the differential interval of time associated with a biologically plausible effect used to define a sampling interval. The theoretical framework for conducting exposure measurement studies on environmental pollution, including air pollution, has been evolving since the late 1980s and is discussed in a number of articles.[42–46] Table 3-3 illustrates the general types of studies that can be considered to define exposures and develop exposure-response relationships. In the most recent revision of the particulate matter standard, the estimates of the association between daily mortality and particles were derived from measurements made at the routine monitoring sites. Clearly, the approach is reasonable, but more accurate analyses will require new and more sophisticated measurement tools.

Indoor and Personal Sampling

Many pollutants may have both outdoor and indoor sources, which will alter or augment a person's inhalation exposure to potentially toxic air pollutants. Situations involving exposures to a particular chemical that lead to health effects may require a thorough evaluation of a person's total inhalation exposure or, at a minimum, the quantification of the important non-ambient

microenvironments that contribute to exposure and confound analysis used in association with epidemiological investigation. This will ensure that the major sources of exposure can be accurately identified for risk assessment and/or health effects studies. Other confounding factors, such as occupation, weather, secondary products, etc., must also be explored to be sure that any potential effect is associated with air pollution.[43]

For pollutants with major indoor sources, the use of a fixed outdoor monitoring site in a population center may not accurately reflect the exposures for any given individual or for the population as a whole.[47] Prior to major initiatives on outdoor emission controls in the 1970s, this measurement design was probably more reasonable for compounds such as sulfur dioxide, benzo(a)pyrene, or particulate matter. However, in most areas of the United States, strict pollution control regulations have reduced the levels of a number of pollutants to the point where indoor concentrations (e.g., particulate matter, nitrogen dioxide, VOCs) are often equivalent to or higher than the outdoor concentrations.[6, 26, 27, 42, 48, 49] Further, there are a number of other studies which have demonstrated the need to measure personal or microenvironmental exposure in the cabins of the vehicles we drive.[50–52] In developing nations, the outdoor levels for some pollutants may still be well above indoor concentrations;[53] however, for situations in which unvented cooking occurs, the indoor concentrations of both criteria and non-criteria pollutants can be excessively high.[54] For pollutants like VOCs,

TABLE 3-3. Spatial Considerations: Summary of Sampling Designs and When They Are Most Useful

Sampling Design	Condition for Most Useful Applications
Haphazard sampling	Only valid when target population is homogeneous in space and time; hence, not generally recommended
Purposive sampling	Target population well defined and homogeneous so sample-selection bias is not a problem; or specific environmental samples selected for unique value and interest rather than for making inferences to wider populations
Probability sampling	
Simple random sampling	Homogeneous population
Stratified random sampling	Homogeneous population within strata (subregions); might consider strata as domains of study
Systematic sampling	Frequently most useful; trends over time and space must be quantified
Multistage sampling	Target population large and homogeneous; simple random sampling used to select contiguous groups of population units
Cluster sampling	Economical when population units cluster (e.g., schools of fish); ideally, cluster means are similar in value, but concentrations within clusters should vary widely
Double sampling	Must be strong linear relation between variable of interest and less-expensive or more easily measured variable

Source: NRC, 1991[39]

radon, nitrogen dioxide, and carbon monoxide, indoor and personal air sampling are required to identify situations where high exposures occur for such pollutants in many nations.[29, 55–58] For secondary pollutants such as ozone and acid aerosols and near street emissions, outdoor monitors may still provide adequate metrics of population exposures.[59–61] In the case of acid aerosols, most of the acid is neutralized by ammonia generated indoors.[62, 63] However, there is increasing evidence that dark phase chemical reactions occur in the indoor environment, which produce secondary gas phase and particle phase organics, including reactive intermediates, from such outdoor pollutants as ozone and nitrogen oxides etc.[64–66] This phenomenon will require new strategies for air indoor sampling in the years to come, which can be based, in part, on the recent work completed by Weschler *et al.*[67, 68]

Chronic health effects studies require identifying areas or sub-populations that would be subjected to conditions conducive to high, medium, and low pollution exposure. Acute effects studies may be designed in one location where significant temporal changes are possible and are on a scale comparable to the potential effect. Personal monitoring of individuals is very desirable in many situations, especially those identified at significant risk to a particular disease.[18, 20] For some pollutants, the monitors have become inexpensive to produce and inconspicuous to wear (e.g., passive diffusion monitors).[25, 69–71] This has been demonstrated in the Total Exposure Assessment Methodology (TEAM) studies of the 1980s and the National Human Exposure Assessment Survey (NHEXAS) studies which were initiated in the mid 1990's.[70–74] For the latter, both personal and microenvironmental sampling were conducted in a probability-based population study in the midwestern U.S., and a concurrent study conducted in Arizona.[74, 75] These types of investigations will provide a meaningful starting point for designing future studies of pollution that exist in multiple and single media and pathways of exposure.

Pollutant Characterization Sampling

Other common types of community air sampling studies are directed toward understanding the physical, chemical, and biologically active nature of the atmosphere (both indoor and outdoor). For an outdoor situation, the focus or foci can be: 1) the formation or decay processes of individual compounds; 2) the transport and transformation of pollutants in industrial or urban plumes; 3) the wet or dry deposition of pollutants; 4) the dynamics of pollution accumulation and removal in urban and rural locales; 5) the persistence of photochemical

smog and other types of episodes; 6) the long-range transport of pollutants; and 7) source tracer measurements. Such investigations can include fixed-site sampling, mobile vans and trailers, and airborne sampling platforms.

Indoor studies may emphasize characterization of sources, outdoor pollution penetration, adsorption, absorption, transformation products, and transformation rates. In these cases, samplers will be located in one or more rooms in a house or more commonly, a group of houses. An important example that is outside the realm of traditional air pollution is the infiltration of contaminated groundwater within the vadose zone (unsaturated soil) into the basement of a home.[22, 23] This will lead to the emission of volatile species into the basement and living quarters.

General Features of Community Studies

Using the previous section as a framework for identifying the types of community air pollution studies, it is immediately apparent that a number of factors must be considered when designing a community air pollution sampling program. Key articles or books can be useful in designing the details of a specific type of study and representative examples are found in the reference list. There are, however, some basic or fundamental steps that must be considered prior to any community air sampling study, and these are outlined in the following sections.

Sampler Location

Selection of an outdoor air monitoring site requires addressing a number of considerations that will affect pollution values recorded at any given time. These include: 1) the type(s) of point or area sources; 2) obstructions or changes to air flow caused by tall buildings, trees, etc.; 3) abrupt changes in terrain; 4) height above ground for a sampler or sampler probe; 5) topography; and 6) proximity to impacted or potentially impacted individuals or areas. The above implies, however, that most monitoring of air pollution for compliance or research requires the deployment of more than one monitoring site. The number of sites must be weighed for usefulness and providing enough information to address the questions to be answered by the study. In some instances, the monitoring site density may be low. This is frequently the case when a long-term trend network is established to determine changes in air quality across a community or rural area. In contrast, studies used to quantify the impact of a particular source may necessitate a high density network within close proximity to the source and the impacted area. A good example is the attempt to characterize the impact of a large stationary source. In all cases, background or non-impact sites are necessary.

Beyond considerations of the physical location of the monitor, each site must be representative in terms of the questions to be answered by the study. All too often, an agency or the investigator can select a location that seems practical in terms of availability, security, and electrical needs, yet its selection would severely compromise the intent of the study. At times, these practical problems cannot be easily resolved, but the suitability of the site for answering pollutant-related questions (e.g., regulatory or health) must be the prime consideration. If not, serious questions can be raised about the validity of the study and the answers to the questions and hypotheses posed prior to designing the study.

For example, if the objective is to monitor the outdoor air concentrations of carbon monoxide (or other tailpipe or evaporative compounds) derived from automobile emissions within the center of a city, the concentrations inhaled would be found in a breathing zone approximately 1–2 m above the street, and not on the roof of a 10-story building. In addition, some of the highest values would be found at street level between tall buildings that form a street canyon. Actual personal exposure to carbon monoxide is enhanced by the driving habits of an individual, time spent in parking garages, and time spent indoors with poorly ventilated combustion sources. Thus, the maximum carbon monoxide concentration for a specific sub-population may be due to exposure in a specific microenvironment such as the cabin area of a car or bus, or an office above a parking garage.[29, 76–80]

In contrast, ozone monitors are normally placed at some distance from the primary sources of its precursors, nitrogen oxides and hydrocarbons. In the early 1970s, Coffey and Stasiuk made a major observation in a rural area of New York State.[81] Their results showed the presence of ozone concentrations in a rural area to be at or above levels found in major urban areas. This finding required the scientific and regulatory community to reevaluate where and when high ozone would occur in the Eastern United States, where population exposures could be significant, and what is required to achieve attainment.[4, 82] Today, ambient ozone monitors are located in rural and suburban areas throughout the United States and other countries. During the summer, the 8-hr average ozone concentration can be above the new NAAQS of 80 ppb occur throughout the U.S. and Canada.[4, 83] Because high ozone concentrations occur

in both rural and suburban areas, a sampling study must be designed in such a way that it is not confounded by the presence of locally generated scavenger pollutants, e.g., nitrogen dioxide at high concentrations, > 10 ppb.

Studies of acid aerosol exposures display many design features characteristic of ozone, e.g., maximum during the summer; outdoor exposure during periods of activity.[63, 84] In addition, it is imperative that the monitors are located at a distance from local ammonia sources to obtain an accurate regional characterization of the potential exposure. However, a site near an ammonia source might also be useful because ammonia does neutralize the acidity. This site may identify the population or ecological area unaffected by the acid aerosol (a local control).

Sampler Location for Health Studies

Generally, air sampling conducted in support of health effects studies requires the data to be representative of population exposures. Ideally, the measurements should be completed coincidentally with any measurements of a health outcome. Once a population or populations are defined, the need for indoor sampling or personal monitoring must be assessed by an investigator.[41–43, 73] Further, the number of central sampling stations, personal monitors, and/or indoor air samplers must be positioned to obtain adequate representation of where the selected population (affected and control) lives.[85, 86] The size of the population to be studied, as well as the number of homes to be sampled, can be determined from epidemiologic principles using power calculations.[87, 88] The need for personal versus microenvironmental monitoring and the use of activity pattern or source questionnaires and biological markers of exposure must be considered by the investigator. He/she must also evaluate a variety of typical personal habits and lifestyles that could confound analysis. A typical confounder is the exposure attributable to environmental tobacco.[89]

The positioning of the outdoor monitors requires a full understanding of the nature of pollutant accumulation under ambient conditions. From our previous examples, adequate measurement of the distribution of ambient carbon monoxide requires the placement of many more samplers in an urban environment than are required to measure exposure to ozone. The basic reason is that carbon monoxide accumulates in confined spaces and produces a large range in concentrations over a short spatial range. Since short-term exposures in high concentration environments, e.g., garages, may lead to highest personal exposure, a personal carbon monoxide monitor provides data that are more closely coupled to the biological marker of internal dose: carboxyhemoglobin.[79] In contrast, ambient ozone concentrations will vary rather uniformly over a large area, although some local differences will be observed because of high local concentrations of nitrogen dioxide.[4] Thus, for the latter, a few well-placed samplers may provide data adequate for both exposure and health analyses. In the case of ozone, an adequate indicator of exposure appears to be increased visits to hospital emergency rooms.[82] Clearly, this is not a direct indicator of personal exposure, but due to the spatial distribution of ozone over large segments of a population, it is an adequate indicator of the acute effects experienced by a sensitive subgroup of the population.

For some pollutants, the mobility of the study population is important in defining the outdoor sampling situations. Consideration must be given to the variations in exposure to a pollutant emitted near a residence, a place of employment, transportation routes, school, and recreational activities (Figure 3-2).

Sampling Frequency and Duration

In a community air pollution study, the length of the sampling program is shaped by a number of factors, not the least of which is the resources available to conduct the study. However, this perennial problem aside, the purpose of a community air pollution research program will have a major influence on the duration of the sampling activities. For example, long-term trend studies should be conducted for multiple years to obtain data on the range of concentrations, overall influence of meteorology, and changes in emission strength.[55, 90] Examination of peak concentrations, diurnal variations, and episode conditions requires a study design that ensures a sufficient number of sampling days or hours to include a representative number and range of events. In many instances, only one or two major events will occur. For example, the study of photochemical smog exposures requires sampling to be conducted for an extensive period during the summer (approximately 45 days). This provides a

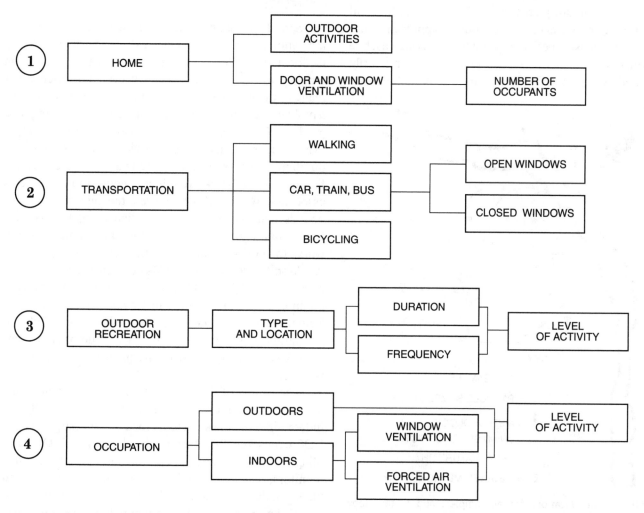

FIGURE 3-2. Personal activities with potential outdoor pollution exposure. 1: Home; 2: Transportation; 3: Recreation; 4: Occupation.

sufficient number of sampling days for measuring the impact of one or more episodes.[82, 91] If continuous samplers are able to be used for measuring reactive hydrocarbons, ozone, selected organic species, and nitrogen oxides measurements, the kinetic processes associated with the accumulation of secondary gaseous and particle species can be examined and subsequently modeled for the region.[4, 92]

Another example is the examination of local source impacts on the surrounding neighborhood. In this case, an intensive community air sampling program would have an array of sampling sites placed around the facility. The size of the emission source would dictate the extent of the array, e.g., an industrial source versus a gasoline station versus wood burning, but the number should be sufficient to detect concentration variability due to changes in wind direction and source strength (Figure 3-3). This type of spatial coverage provides an investigator with an opportunity to determine the contributions in background or upwind air. The duration of the study would have to be sufficient to identify the meteorological conditions conducive to maximum plume impact.[1, 93] The study could be at least 1 or 2 years in duration, but the approach would include an intensive period of study for gathering baseline information, and then a second phase that is primarily activated when specific meteorological conditions are

anticipated to occur. Determining this time frame is not an easy task. For individual pollutants present in a plume, it may also be important to consider the use of indoor and/or personal samplers within an impacted area. The true picture of the pollutant loading contributed by a plant to a community is better assessed by quantifying the extent of outdoor pollutant(s) penetration indoors. Further, for some pollutants, it may also be necessary to consider the amount deposited on the soil or in the water.

In cases such as lead, mercury or pesticide emissions, the deposition of semi-volatile or non-volatile particulate matter on the soil or in water may be potentially an important indirect source of air pollutant exposure to people, plants, or animals. Obviously, this situation will require determining a pollutant's concentration in the surface soil and/or the groundwater, and subsequently estimating exposure from ingestion and dermal contact in addition to inhalation.[23, 43, 46, 85]

Averaging Time

Sample averaging time is dependent upon the instrumentation available to conduct a study, the detection limits for a particular compound, and the time resolution required to discriminate particular events or effects. For many of the gaseous criteria pollutants, this will not pose a problem because most devices are continuous samplers.

For VOCs, the devices primarily integrate a sample over an interval that in the past was usually 24 hours in duration.[94] Unfortunately, there is a wide range of artifact formation, breakthrough, and equilibrium problems associated with VOCs that limit the number of compounds that can be measured reliably, see Chapter 18. Some state agencies set up routine air monitoring networks for VOCs. In the 1980s, U.S. EPA initiated a Toxic Air Monitoring system (TAMs) that measured approximately at least 26 VOCs using a stainless steel canister on an every sixth day sampling cycle.[95] Currently, the Summa[7] canister is the sampling methodology of choice for routine VOC investigations.[96] Volatile organic sampling has been expanded for frequent use at hazardous waste sites.

The results of the TEAM studies have indicated that the major route of population exposure to VOCs is indoor air, which has led to re-focusing investigations on VOC exposures.[18, 70] Further development of VOC sampling techniques, such as continuous monitors, passive monitors, canisters, and human breath analysis and blood analysis as biological markers of exposure, will occur as more researchers attempt to measure concentrations within specific or new volatile compounds present

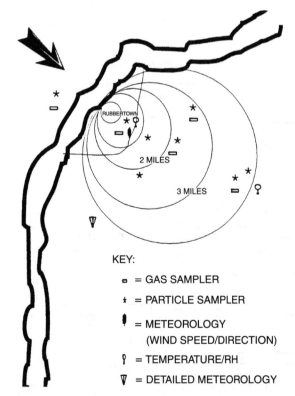

KEY:

    ◽ = GAS SAMPLER

    ✶ = PARTICLE SAMPLER

    ❙ = METEOROLOGY
          (WIND SPEED/DIRECTION)

    ♀ = TEMPERATURE/RH

    ▽ = DETAILED METEOROLOGY

FIGURE 3-3. Site locations for air samplers and meteorology around a major source *(adapted from Stern[2])*.

in the outdoor and indoor environment. Outdoor studies would include areas around specific toxic pollutant sources, landfills, hazardous waste sites, and near source fugitive emissions. It has become clear, however, that studies of VOC exposures must attempt to couple the measurement period with the biological half-time of metabolism or elimination to understand the significance of an exposure within a population.[85] Such concerns inevitably lead to designing longer duration studies with shorter sampling time when trying to establish source-exposure relationships. However, longer periods of sampling with active or passive monitors are still adequate for conducting risk assessments on compounds that cause long-term effects like cancer.

Particulate matter sampling has normally been conducted with devices that integrate samples, although continuous monitors exist that measure specific particle size ranges (Chapter 5). For most routine U.S. EPA and state regulatory monitoring efforts through the early 1980s, total suspended particulate (TSP) samples were collected on a statistically based, every sixth day schedule and were 24 hrs in duration. This approach was marginally adequately for the determination of mass and selected inorganic and organic constituents. Work is currently underway to develop continuous particle mass monitors; however, these will need validation. For specific contaminants, such as semivolatile organics and hydrogen ion (which represents aerosol acidity), other samplers with much shorter sampling periods (<6 hrs) and more frequent number of samples (four/day) are needed for a community air sampling program. For some compounds, gas and vapor denuders must be placed upstream of a filter; for others, additional samplers are required after the filter to reduce artifacts due to filter absorption or desorption of semi-volatile compounds during a specific sampling period.

The need for conducting size-selective particle sampling is discussed in Chapter 5, which also identifies instruments and methods for sampling. Presently, a number of size-selective devices are used in ambient, personal, and indoor studies. These are designed to collect particles presented to various regions of the lung (see Chapter 5, Figure 5-12). The most common are the thoracic samplers ($d_{50} = 10$ μm), respirable samplers ($d_{50} = 4.0$ μm), and fine particle samplers ($d_{50} = 2.5$ μm). D_{50} is the aerodynamic diameter of particles collected by a sampler with an efficiency of 50%. The current revision of the particulate matter standard now focuses attention on the development and application of integrated and continuous monitors for fine particulate matter. The attempt will be to obtain 24 hrs resolution for the integrated sampler with a desired sampling frequency of

every day. Devices have been developed that can be used to detect particle size fractionated mass and many inorganic and organic constituents in time intervals of 4 hrs or less. In addition, for the dichotomous filter sampler automated models provide the opportunity to obtain up to 36 consecutive samples over various time durations.[97] The minimum or maximum duration of any particular sample would depend on the detection limits for the compounds measured and the range of pollution levels present in a particular area. Currently, continuous methods for measuring particle mass and constituents are under development for ambient and other types of monitoring.[12, 32, 98] Excellent summaries on PM_{10} measurements throughout the United States and PM methods are found in Chow et al.[99] and Chow[100], respectively. The current status of information on $PM_{2.5}$ and PM_{10} measurements and measurement techniques have been presented in the PM criteria document.[12] An updated criteria document is currently in preparation.

Personal samplers with respirable particle inlets are presently used in indoor air pollution studies.[101] Personal samplers for PM_{10} and $PM_{2.5}$ have also become available (see Chapter 5), and have been used to study pollution exposures in a limited number of studies, including the Particle Total Exposure Assessment Methodology, (PTEAM) and Total Human Environmental Exposure Study, (THEES).[20, 44, 102, 103] New studies involving the next generation of personal monitoring studies for PM are being conducted at the present time.[32]

An important current activity for measurements of particulate matter is fugitive dust emissions at hazardous waste sites. This is much more complicated than routine monitoring of particles because high levels are usually sporadic and dependent upon local conditions, including wind speed and direction, unpaved road traffic density, and personal activities. Long-term data would probably not find the PM_{10} concentrations much above those typically found in ambient air, but sporadic events could lead to high concentrations of toxics redistributed through the air (inhalation exposure) and deposited at offsite surfaces (ingestion exposure).

Chemical Analyses

Presently, community air sampling surveys for particulate matter can require analyses beyond a traditional mass determination. Some fairly routine measurements include the concentrations of numerous trace elements (e.g., Pb, Cd, Cs, Fe, Se, As, and Br) and water-soluble ions (sulfate, ammonium, chloride, and nitrate).[104, 105] More sophisticated studies measure the acidity of sulfate particles,[106] the organic mass fractions,[98, 107] specific organic species,[12, 108, 109] and elemental carbon.[12, 109, 110] Such measurements have been instrumental in the development of receptor-based modeling approaches, which quantify the contributions made by particular sources to the ambient air.[12, 58, 111, 112]

Samplers have been developed for outdoor studies that can also collect particulate organics on a filter and have a vapor trap that contains polyurethane foam. The trap collects semivolatile species that evaporate from the filter during sampling. These devices have been housed in a Hi-volume sampler shell. One major concern with the new $PM_{2.5}$ standard is the vast number of chemical components that may contribute to the mass and which components are biologically active. (See Chapter 5 and 13 in this volume). Thus, an important component of the Particulate Matter (PM) research strategy will be the inclusion of sites that can measure the chemical constituents of PM and on a continuous basis. The latter need development. One major challenge will be the ability to select the right components to measure and establish the levels of potentially biologically active components of PM. In contrast, others will be measured as indicators or tracers of sources or the products of chemical reactions in the atmospheres.

Area particulate matter samplers for indoor air pollution measurements have been developed by a number of investigators. Included are the Marple-Spengler-Turner samplers, which collect mass on a Teflon filter and can have single or multiple aerodynamic cut sizes between 10 and 1 μm.[113] The samples are collected for various time intervals from <1 day to 4 days. Most analyses performed for ambient samples can be conducted on indoor impactor filters if the mass loading is sufficiently high and if a limited number of destructive analyses are completed on the samples.

A compendium of typical analytical techniques for air sampling can be found in Lodge.[104] Newer sampling techniques, including automatic and remote measurements of trace gases and aerosols, measurements of peroxy radicals and hydroxyl radicals in the atmosphere, and updates on advances in the measurements of acids, nitrates and non-methane hydrocarbons are the subject of a series of reviews edited by Newman.[105] Other useful information is found in the works of a number of authors listed in the reference section and from the many instrument chapters in this manual.

Biological Assay of Air Samples

The identification of potentially carcinogenic compounds in the atmosphere has led to the development and application of techniques for the measurement of the biological activity of particulate and gaseous samples. In principle, actual animal bioassays of carcinogenic air pollution are the most direct measure of potential effect;[114] however, these assays are difficult to conduct on ambient air. As an alternative to actual animal bioassays, short-term in vitro bioassays have been used to examine the mutagenic properties of organic materials. Of the known carcinogens, over 80% have been shown to be mutagenic in the Ames Assay,[115] a short-term in vitro technique that uses alterations in the Salmonella DNA to demonstrate the mutagenic properties of organic material in particulate matter.

The Ames Assay has been one method of choice for application to air pollution samples.[116] The concentration results of such assays have been reported as revertant colonies per cubic meter of air; for mutagenic potency, these have been reported as revertant colonies per microgram of sample. Many studies in the 1980s used the Ames Assay on air pollution samples, and the results with references were summarized by Louis et al.[117] At present, further research is being directed to the identification of the actual compounds contributing to the mutagenic activity of the air pollution samples.[118, 119] An excellent investigation called the Integrated Air Cancer Study was conducted to assess the contributions made by wood smoke and automobiles to the biologically active portions of particulate matter. The study employed a variety of chemical and biological assays to quantify the contributions to ambient air.[120] Some of the most significant advances in biological nuclear measurements for use in human populations in recent years have been made in the measurement of VOC, metals, blood and urine.[121, 122] Examples are found in Table 3-4 at the end of this chapter. Further information on Biological Markers can be found an article published by Henderson.[118] A Germany study of both adults and children has established baseline levels of metals in the blood and urine.[123]

The field application of biological measures as markers of exposure is beginning to evolve. It is anticipated that techniques will continue to be developed and validated which will permit the precise measurement of biologically effective exposure and dose to individuals in a community. This could include the use of cytotoxicity tests as well as the application of breath analyses.[124, 125]

Data Retrieval and Analysis

For the state monitoring networks required to assess attainment of the NAAQS, most data are sent by telemetry to a central station for computerized data recording, validation, and processing of many of the continuously measured pollutants. At a minimum, the NAAQS pollutants NO_x, O_3, SO_2, and CO are monitored continuously. These data are formatted, with identifying parameters, and sent to U.S. EPA for archival and use in the Aeromatric Information Retrieval System (AIRS).[126] (see *http://www.epa.gov/namti1/datamans.html/*). Manually collected particulate pollution data are eventually entered by tape or by hand into AIRS or local air monitoring databases after a number of chemical analyses have been performed on a series of samples.

Intensive field studies conducted at a given location have become much more sophisticated in both study design, monitoring equipment, and data gathering practices. The former will be discussed in a separate section; however, improvements in the use and application of sampling and data retrieval equipment have been significant. This should continue in the future with the advent of optical fiber transmission, high-speed microcomputers, and CD-ROM storage of data.

Beyond the construction of fixed monitoring sites with telemetry systems, a number of groups have developed and utilized fully equipped mobile vans or trailers. Sophisticated mobile units used in field studies can include complete computerized systems for continuous samplers, calibration facilities, and integrated samplers;[127] even wet chemical analysis laboratories are available. The interior a typical trailer is shown in Figure 3-4.

Air Pollution Modeling

When a community air pollution program or study is designed, one of the primary activities to be considered for use of the collected data is model development. Models play an important role in the examination of air pollution because they can be used to: 1) determine the effectiveness of control strategies; 2) predict pollutant concentrations downwind of sources; 3) examine chemical processes; 4) examine regional transport questions; 5) establish source-receptor relationships; and 6) estimate human exposure. In all cases, though, an important final step is testing the performance or the validation of a model using community sampling data. The acquisition

FIGURE 3-4. Atmospheric Research Laboratory of General Motors Research Laboratories (GMRL) *(supplied by G. Wolff, GMRL, Warren, Michigan).*

of these types of pollution data requires carefully selected protocols because an application of a model(s) poses certain constraints on the selection of variables, study duration, sampling frequency, number of samples, and sampling site selection.

The most common form of models applied to air pollution is based on the Gaussian dispersion model. It estimates the contributions at a downwind receptor site of a pollutant emitted by a source after dispersal in the atmosphere under a variety of meteorological conditions. The technique has been used in community air pollution for many years and is available for specific source and terrain applications as off-the-shelf models. Turner has published a text book on dispersion modeling that covers the general features of the technique and the source and meteorological information necessary to apply the various types of models.[93] In 1984, Hidy completed a review of air pollution modeling issues but focused on regional models and their application to the source-receptor relationships of acid deposition.[128] In 1988, Seinfeld examined the status of available models for predicting the behavior of photochemical smog.[129] The statistical aspects of environmental pollution data have been developed in detail by Ott.[130] It examines the probabilistic aspects of complex data sets and provides a detailed understanding of lognormal distributions, which are typical of most air pollutants. Another modeling approach that has been used extensively since the late 1970s is receptor-source apportionment. The general technique involves constructing a model that determines the sources that contribute to pollution levels observed at a receptor location. In contrast to dispersion modeling, it derives information from compositional data collected at the receptor (sampler). The significance of the source is obtained by measuring the concentration of a source tracer at the receptor and other pollutants emitted by the source by using the principle of conservation of mass.[131]

A number of different approaches are available that can be used to develop receptor models, including Chemical Element Balance[131] Factor-Analysis Multiple Regression,[132, 133] and Target Transformation.[112, 134] Applications of these models are published by Thurston and Lioy.[135] The results of the application of source apportionment models for particulate matter are summarized in the 1996 EPA Criteria Document.[12]

A more recent approach to modeling that examines the frequency and distribution of a pollutant among a population is *human exposure modeling*.[23, 44, 45] This type of modeling involves the acquisition of pollutant concentration data, identification of the activity patterns of an individual or a target population, and the estimation of time spent by an individual in a particular

microenvironment. From these data, a model can be constructed that links the presence of an individual at a location in a microenvironment to the concentration of a pollutant present at that location for a certain length of time.[44, 45, 136] The estimated exposures are usually the average or integrated values for each microenvironment and can be represented by probabilistic distributions using Monte Carlo Techniques.[85, 137, 138]

Conducting Community Air Pollution Studies

Since 1970, many studies have been conducted to examine a variety of air pollution issues. The complexity and scope of each are varied, but there are some fundamental concepts that are common to all and many issues about how to design a study are found in detail in books written by Hidy, et al.,[139] and Lioy, et al.[140] In most, if not all, investigations, there is a principal investigator and possibly one or more co-principal investigators. Depending upon the size of the study, there are usually six basic components to any field investigation: 1) The management team; 2) the field investigation team; 3) the logistics personnel; 4) the sample analysts; 5) the quality assurance personnel; and 6) the data compilation personnel. Within a given study, there can be a project manager, field coordinators, and specialized sampling personnel. After the field and sample analyses have been completed, a number of these individuals and the investigators analyze the compiled data. This includes statistical analyses, model development and or validation, and overall interpretation with respect to individual hypotheses and possible integration of the results. Within all these groups, there exist many possible combinations of personnel from a single institution, but more often from multiple organizations. Most of the time, major field studies are composed of the latter, and for a very simple reason: a single institution usually cannot employ all the specialized personnel or have all the sampling and or analysis techniques required to conduct a particular study.

Because of the many components that need to be tested and coordinated in an individual study, there are many trial runs or performance tests of individual components of the study or the entire sampling and analysis protocols. In recent years, the studies have incorporated standard operating procedures for many, if not all, sampling and analysis instruments employed within an individual study. U.S. EPA has called these documents "The Quality Systems and Implementation Plan (QSIP)." It is required for major field status to achieve high levels of accuracy and precision for all samples by reducing the opportunity for operator and analytical error. In any case all personnel must also track samples from their point of departure and from the logistics location to the point of quality assurance of the analytical results.

Studies can now include bar coding of field samples and placement of all field logging information using portable laptop computers. If there is one new aspect that has been added to field studies which, in the long term will help improve the amount and completeness of information on an individual sample, it is the computerization of sample collection data in the field.

Applications

It is difficult to describe all the major studies with any degree of completeness in a chapter such as this, and do justice to each. The following briefly identifies some important community air pollution studies designed to analyze outdoor, indoor, and personnel air. Each is identified by at least one major reference which the reader can use to obtain details on the scientific information derived from the study and instrumentation and sampling platforms used in each study. In some studies, only fixed ambient sampling sites were required to meet the goals and objectives of the study. Others required mobile sampling platforms, including trailers and/or airplanes. Others required sampling of buildings or personal residences, and finally we have begun to find it necessary to collect personal samples of the air we breathe and to take personal biological fluid samples to determine the internal dose received from an air pollutant.

One of the most comprehensive early investigations of outdoor air pollution, which is still a classic, was California Aerosol Characterization Experiment, (ACHEX).[141, 142] The major objectives of the study focused on describing the physical and chemical characteristics of photochemical smog aerosols and establishing relationships with the reduction of visibility in Southern California. As an adjunct to this study, the three-dimensional distribution and transport of a number of pollutants were examined in the region. The results of this study were augmented in the late 1980s with the more recent version, the Southern California Air Quality Study (SCAQS), which involved several government and industry laboratories and universities. It was conducted in 1987, and the results are found in a number of publications.[143-147] Before ACHEX, little was known about the ambient particulate pollutants, and that study provided a comprehensive picture of particulate pollutants. In contrast, SCAQS was designed to gather measurement inputs needed to improve basinwide modeling of the formation of particulate and gaseous pollutants. The SCAQS used intensive, simultaneous, ground-based

monitoring at nine sites, with 4- to 7-hr time resolution for particulate samples. It also included wind and temperature profiled aloft (by radio-equipped weather balloons called rawinsondes), and gaseous pollutant profiles aloft (measured by aircraft).

A number of extensive and intensive air pollution studies have been conducted; one in particular, the Regional Air Pollution Study (RAPS), was conducted in 1976, in the area surrounding St. Louis, Missouri. RAPS focused on identifying the sources of regional pollutants that covered a distance of 100 km. The spatial resolution was 10 km and the sampling focused on 1- to 3-day episodes. It identified the products of SO_2 oxidation downwind of major sources and included apportionment of major sources of both fine coarse particulate matter.[148, 149]

The Brown Cloud is a conceptual representation of the color of the greater Denver ambient atmosphere during the winter. It encompasses a mesoscale (<100 km) pollution phenomenon that is topographically induced and affects the metropolitan Denver area. There have been a number of "Brown Cloud" studies, including programs in 1973,[150] 1978,[151-154] 1982,[155] and 1987-1988.[156]

A major feature of the 1978 investigation, which was conducted in November and December, was the measurement of both the organic and elemental carbon content of size-fractionated particulate mass. Chemical analyses provided the opportunity to conduct source apportionment studies. In addition, the investigators estimated contributions of various species to visibility reduction in the Denver area.[154]

A very elaborate community sampling program was established for this intensive investigation (Figure 3-5). It included surface-based sites to measure: 1) the maximum impact of the pollution contained in the cloud; 2) the background concentrations; and 3) the characteristics of the cloud throughout various parts of the city. Aircraft measurements were made in the vertical and horizontal direction to examine cloud dynamics.

In 1976, the New York Metropolitan Area was studied in extensively in a collaborative effort called the New York Summer Aerosol Study.[157, 158] It was followed by a second investigation during 1977.[159] Many properties of the atmospheric aerosol were measured at multiple sites, and it was one of the first to systematically characterize the organic fraction and the acidity of the atmospheric aerosol around a major metropolitan area. The design included field locations in and on the perimeter of New York City, which provided an opportunity to examine the transport of particles into and out of the metropolitan area.

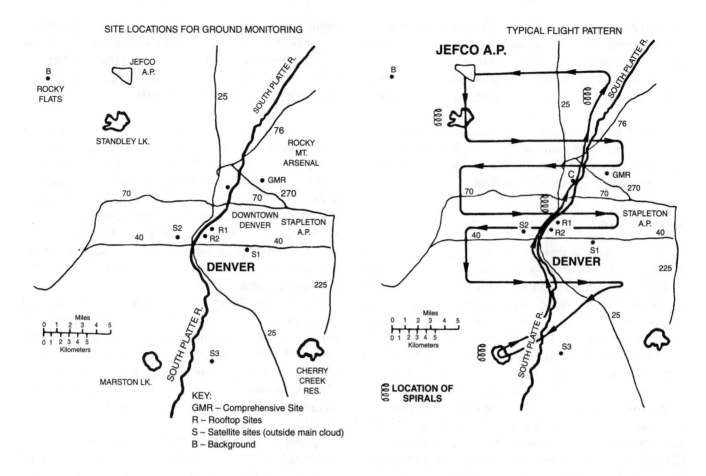

FIGURE 3-5. The Denver "Brown Cloud" Study (reproduced from Wolff et al.[151] with permission.

The Airborne Toxic Element and Organic Substance (ATEOS) project was an extensive community air pollution characterization study conducted at three urban and one rural site in New Jersey.[160] It was designed to investigate not only the atmospheric dynamics and distribution of pollutants, but also estimate potential risks to human health from biologically active materials present in the outdoor air. Each ambient measurement site was selected to reflect a different type of industrial–commercial–residential interface that was representative of a different type of exposure situation rather than the entire city. Only one sampler was placed in each city. More than 50 pollutants were measured simultaneously daily. This study was segmented into six week long summer and winter intensive sampling periods from 1981 through 1983. For example, the Newark site was within a residential area surrounded by small and moderately sized industrial facilities, e.g., body shops and chemical manufacturers (Figure 3-6).

The approaches used in ATEOS were the basis of a new form of source-related monitoring study, the Total Human Environmental Exposure Study (THEES).[20, 161] It focused on examining the influence of multimedia

pathways on an individual's exposure to benzo(a)pyrene (BaP) and determined the contribution of BaP emitted by a smelter to personal air exposure to PM_{10}. The results

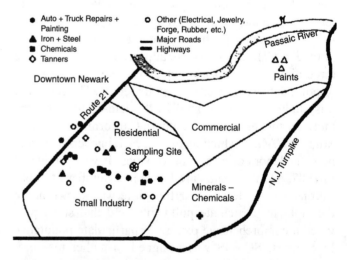

FIGURE 3-6. The distribution of sources around the subpopulation studied in ATEOS (from Lioy and Daisey;[160] adapted from Final Report to N.J. Dep. 1985).

showed that the smelter's contribution, after controls, was less than the BaP from indoor air sources.

The community air sampling was conducted during the Harvard Six Cities Study, and it included outdoor, indoor, and personal air monitoring.[12, 162] Most of these were in support of the 10-year prospective examination of respiratory symptoms and pulmonary function of children and adults living in the six communities of Topeka, Kansas; Portage, Wisconsin; Watertown, Massachusetts; Kingston, Tennessee; St. Louis, Missouri; and Steubenville, Ohio. Indices of acute and chronic respiratory effects and pulmonary function performances were examined in relation to any adverse effects of ambient and indoor air pollutants. In the Harvard six cities study, outdoor air sampling sites were located in each community. However, because exposure to a number of air pollutants, such as nitrogen dioxide, respirable particles, and formaldehyde, can be associated with a number of microenvironments, indoor and personal samples were required for individuals participating in the study. The use of each of the above types of samples provides microenvironment and personal pollution data for the development of exposure models as well as estimation of the influence of various activity

patterns. The most extensive indoor database developed in the Harvard study is for nitrogen dioxide and respirable particles. As can be seen in Figure 3-7, the respirable particles measured or monitored indoors can contribute the dominant proportion of personal exposure. This led to the 24 cities (U.S. and Canada) health study which examined the health effects related to ambient acid aerosols and ozone exposures in moderately sized cities in North America. Speizer published a paper in 1989 on the study design required to collect acid aerosol data that were relevant to the health measurements.[163] One major focus of the research was the city of Philadelphia, Pennsylvania, and the inner city and suburban levels of the acidic species which contributed to human exposure.[63]

The TEAM Studies provided a unique approach to the study of community pollution.[18] The investigation was actually a series of studies conducted in a number of cities in the United States.[18, 164–167] The series of studies concluded with a personal exposure to particulate matter in California called P-TEAM.[103] Because the TEAM was a statistically designed study, inferences could be drawn about the general population living in certain areas of Elizabeth/Bayonne, New Jersey; the South Bay

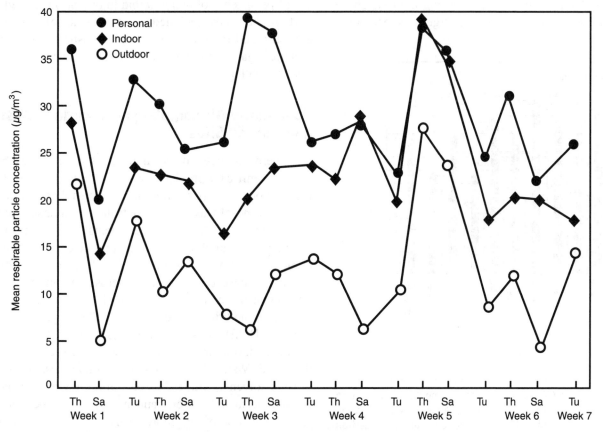

FIGURE 3-7. Personal, indoor, and outdoor levels of repirable particles in Topeka, Kansas, during the Harvard Health Study *(reprinted with permission. J. Spengler, Harvard School of Public Health, 1987).*

of Los Angeles, California; and Antioch/Pittsburgh, Pennsylvania.

The investigation primarily involved measuring the personal exposures of individuals to 20 VOCs and the corresponding body burden. Fixed-site, outdoor sampling was conducted next to the homes of the participants. Personal monitoring was completed on each individual, and the levels of the compounds were measured in exhaled breath as an indication of levels that could be found in an individual's blood. Drinking water and beverage concentrations of the VOCs were determined, and detailed questionnaires were administered concerning occupation, hobby, and home VOC sources. Sub-studies conducted in the TEAM included indoor microenvironmental sampling.[165] A second phase of TEAM studies was conducted in the late 1980s to follow up on the hypotheses generated in the initial experiments. Included was a much more extensive use of breath analysis as a biological marker of exposure.[166, 167]

A major result of the TEAM Study has been to demonstrate that the outdoor environment is not the primary contributor to personal VOC exposures (Figure 3-8). In many cases, the concentrations were 10 to 100 times higher indoors than those observed outdoors. Also, significant variations in the VOC concentrations can occur in a small geographic area during the day. These results have led to a number of the large scale personnel

and indoor air pollution studies that have been conducted in a variety of locales.[21–28, 49, 51, 61, 64–66, 74–79] One recently completed study investigated personal, indoor, and outdoor exposure to manganese in $PM_{2.5}$. The focus was on the distribution of manganese that could occur because of emission from gasoline with MMT as an octane additive.[49]

Recently, a European study called EXPOLIS has been conducted as a multi-country study of microenvironmental and personal exposure to traditional air pollutants.[168] The research includes examining exposure of adults during the work week for CO, $PM_{2.5}$ and VOC, and establishing personal population exposure distributions and microenvironmental (homes, shopping malls, streets) distributions for the cities of Athens, Greece; Basel, Switzerland; Grenoble, France; Helsinki, Finland; Milan, Italy; and Prague, Czech Republic. It is the first comprehensive examination of personal exposure to traditional air pollution in Europe. The design provides a new approach for the packing and transport of personal and microenvironmental routes. The design considerations for EXPOLIS are found in a recent article by Jantunen, et al.[153] At about the same time, the National Human Exposure Assessment Survey was conducted, and it included examination of total exposure to a number of persistent pollutants found in one or more media. Part of the study included measurements of personal exposure and microenvironmental exposure to airborne volatile organics and pesticides, and total inspirable particulate matter.[74]

Community Air Sampling at Hazardous Waste and Landfill Sites

Chemical waste disposal in the United States has been a significant problem for many years. Historically, the methods of disposal were not responsibly developed, and the disposal was usually controlled by industry, with insufficient regulation by governments.[169, 170] With the passage of a number of federal and state regulations, numerous National Priorities List (NPL) chemical disposal sites have been identified across he nation, and a number of these sites are associated with human health risk.[170] In addition, there are over 3,700 Department of Energy (DOE) sites that are associated with nuclear weapons development activities which accrued during the Cold War. These contain radionuclides, volatile organic solvents, and lead as part of the waste deposited in well defined locations and across vast expenses of lead.[171] A few of the more well-known examples are Rocky Mountain Arsenal, Colorado; the Savannah River Site, South Carolina; and Hanford Washington. Other

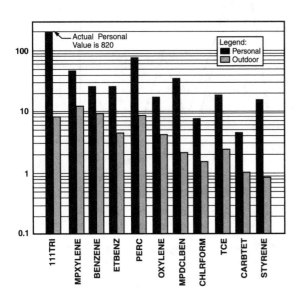

FIGURE 3-8. Estimated arithmetic means of 11 toxic compounds in daytime (6:00 AM to 6:00 PM) air samples for the target population (128,000) of Elizabeth and Bayonne, New Jersey, between September and November 1981. Personal air estimates based on 340 samples; outdoor air estimates based on 88 samples. *(From Wallace;[18] reprinted with permission.)*

waste sites have included Love Canal, New York; Hyde Park Landfill, Niagara Falls, New York; Rollins Landfill, Baton Rouge, Louisiana; Chemical Control, Elizabeth, New Jersey; and Jersey City Landfill, New Jersey. In addition to the NPL sites, there are thousands of sites controlled by the states and other federal agencies.[169–173]

Individuals can be exposed to any air emissions from the many types of disposal sites; included are persons living and working in communities adjacent to the area. Their air pollution exposures may result from the inhalation of particles (fugitive dusts), fumes, or vapors dispersed in the ambient air from a dump before remediation, or during the removal of the waste. Volatilization of VOCs may also occur as a result of contaminated leachate moving through groundwater to a well that provides water used for showering and tap water in a home, or the evaporation of VOCs in basements and their circulation throughout the house.[22, 23, 174]

In contrast to general community air sampling studies, the nature of the emissions may be quite variable because the waste will generally not be homogeneously distributed within a dump.[175] Also, any active or continuing deposition of wastes or any activities associated with waste removal could change the emission rates (e.g., active landfills). Thus, a series of emissions tests may be required throughout the period encompassed by the study. Other screening activities would involve the use of hydrogeologists and meteorologists to model the movement of contaminants through the water and air, respectively.

During the remedial investigation and the actual site remediation, the nature and extent of the offsite population exposure to the pollution should be thoroughly documented.[175] This effort would primarily involve microenvironmental sampling. A study should include the identification of the most probable downwind directions for outdoor air and indoor air contamination, the downstream direction of flow for groundwater contamination, and the identification of a control area. After a series of samples are taken, the most exposed area could be subject to follow-up investigation. The design issues were recently discussed by Georgopoulos and Lioy[175] and a summary of some the needs for addressing single and multimedia exposure issues at DOE hazardous waste sites was published by Daisey in 1998.[176]

Summary

There are multiple reasons for taking air pollution measurements within the community environment. A basic program is used to examine source compliance with regulations or the long-term progress toward attaining national or other ambient air quality standards for health or welfare effects. There are, however, a wide variety of research-based strategies needed and available to assess direct and indirect human or ecological exposure or effects caused by air pollution. These can become very sophisticated in terms of the types of measurements, the number of measurements, the number of sites, and duration of a study. Because of financial costs and the need to design a coherent sampling and analysis strategy, investigators must clearly establish the hypotheses and data quality objectives prior to the development of any monitoring study or research investigation. Today the investigator have many monitoring approaches and options available to him or her for addressing a community problem. Included are the typical outdoor monitoring site, airborne platforms, mobile vehicles, indoor air monitors, personal air monitors, and biological markers of human exposure. Thus, the measurements can be structured to provide a grid of information that covers thousands of kilometers down to a single focal point: the human receptor.

What is clear is that at this time community air sampling has progressed to a level where many pollutants can be measured simultaneously at a large number of sites and a large number of platforms, and the data and supporting information can be placed in large computer databases. However, major research studies still require teams from a variety of institutions and laboratories to conduct comprehensive investigations or at least multiple-disciplinary research teams within the same institution.

Acknowledgments

The author wishes to thank Ms. Roberta Salinger for processing the manuscript and Mr. Jason Lioy for the graphics. Dr. Lioy's work is in part supported by NIEHS Center Grant # ES 005022.

References

1. U.S. Federal Register 36:8186 (1971).
2. Stem, A.C., Ed.: Air Pollution, Vol. I-V. Academic Press, New York (1977).
3. Public Law 101-549, 104 Stat. 2399, Clean Air Act. U.S. Government Printing Office, Washington, DC (Nov. 15, 1990).
4. National Research Council: Rethinking the Ozone Problem in Regional Air Pollution. National Academy Press, Washington, DC (1991).
5. Proposing the new standard, National Ambient Air Quality Standards for Particulate Matter and Ozone. U.S. Federal Register 62:38761-38896 (July 18, 1997).
6. World Health Organization: Air Quality Guidelines for Europe, Regional Publications, European Series, No. 23. WHO: Copenhagen, Denmark (1987).
7. U.S. Environmental Protection Agency: Enhanced Ozone Monitoring Network Design and Siting Criteria Guidance Document. EPA-45/4-9/-033. OAQPS, Research Triangle Park, NC (November 1991).

8. Eldred, R.A.; Cahill, R.A.; Flocchini, R.G.: Comparison of PM_{10} and $PM_{2.5}$ aerosols in the IMPROVE network. In: Proceedings of an International Specialty Conference on Aerosol and Atmospheric Optics: Radiative Balance and Visual Air Quality, Volume A. Air and Waste Management Association. Pittsburgh, PA (1994).

9. Biegalski, S.R.; Landsberger, S.; Hoff, R.M.: Sour-receptor modeling using trace metals in aerosols collected at three Rural Canadian Great Lakes Sampling Stations, J. Air Waste Manage. Assoc. 48:227B237 (1998).

10. Eldred, R.A.; Cahill, T.A.; Feeney, P.J.: Regional Patterns of Selenium and Other Trace Elements in the Improve NETWORK. In: Proceedings of the International Specialty Conference on Aerosol and Atmospheric Optics: Radiative Balance and Visual Air Quality, Volume A. Air and Waste Management Association, Pittsburgh, PA (1994).

11. Wolff, G.T.: Mesoscale and Synoptic Scale Transport of Aerosols. In: Aerosols: Anthropogenic and Natural—Sources and Transport, pp. 338, 379–388. T.J. Kneip and P.J. Lioy, Eds. Ann. N.Y. Acad. Sci. (1980).

12. U.S. Environmental Protection Agency: Air Quality Criteria for Particulate Matter, Vols. 1–3, EPA/600/p095/001af. ORD, Washington, DC, (1996).

13. Van Dop, H.; Nodop, K., Eds.: ETEX, A European Tracer Experiment. Atmos. Environ. 32:4089–4378 (1998).

14. Wilson, W.E.: Sulfates in the Atmosphere: A Progress Report on Project MISTT. Atmos. Environ. 12: 537–547 (1978).

15. Goduwitch, J.M.: Evaluation and Sensitivity Analysis Results of the MESOPUFFII Model with the CAPTEX Measurements. EPA/600/3-89/056. NERL, Research Triangle Park, NC (1989).

16. National Research Council, Committee on Indoor Pollutants: Indoor Pollutants. National Academy Press, Washington, DC (1981).

17. Yocum, J.: Indoor–Outdoor Air Quality Relationships CA Critical Review. J. Air Pollut. Control Assoc. 32:500 (1982).

18. Wallace, L.: Total Exposure Assessment Methodology (TEAM) Study: Summary and Analysis, Vol. 1. Final Report. Contract #68-0-02-3679. U.S. Environmental Protection Agency, Washington, DC (1986).

19. Ott, W.: Exposure Estimates Based Upon Computer Generated Activity Patterns. J. Toxicol. Clin. Toxicol. 21:97 (1983-84).

20. Lioy, P.; Waldman, J.; Greenberg, A.; et al.: The Total Human Environmental Exposure Study (THEES) to Benzo(a)pyrene: Comparison of the Inhalation and Food Pathways. Arch. Environ. Health 43:304 (1988).

21. Quackenboss, J.L.; Spengler, J.D.; Kanarek, M.S.; et al.: Personal Exposure to Nitrogen Dioxide: Relationship to Indoor/Outdoor Air Quality and Activity Patterns. Environ. Sci. Technol. 20:775–783 (1986).

22. Little, J.C.; Daisey, J.M.; Nazaroff, W.: Transport of Subsurface Contaminants Into Buildings. Environ. Sci. Technol. 26:2058–2065 (1992).

23. Georgopoulos, P.G.; Walia, A.; Roy, A.; Lioy, P.J.: An Integrated Exposure and Dose Modeling and Analysis System: Part I—Formulation and Testing of Microenvironmental and Pharmacokinetic Components, Environ. Sci. Technol. 31:17–27(1997).

24. Daisey, J.M.; Hodgson, A.T.; Fisk, W.J.; et al.: Volatile Organic Compounds in Twelve California Office Buildings: Classes, Concentrations and Sources, Atmos. Environ. 28: 3557–3562 (1994).

25. Clayton, C.A.; Perritt, R.L.; Pellizzari, E.D.; et al.: Particle Total Exposure Assessment Methodology (PTEAM) Study: Distributions of Aerosol and Elemental Concentrations in Personal, Indoor, and Outdoor Air Samples in a Southern California Community. J. Exposure Anal. Environ. Epid. 3:227–250 (1993).

26. Lewis, C.W.: Sources of Air Pollutants Indoors: VOC and Fine Particulate Specie. J. Exposure Anal. Environ. Epid. 2:31–44 (1991).

27. Wallace, L.: Indoor Particles: a Review. J. Air Waste Manage. Assoc. 46: 98–126 (1996).

28. Platts-Mills, T.A.E.; Chapman, M.D.: Dust Mites: Immunology, Allergic Disease, and Environmental Control. J. Allergy Clin. Immunol. 80:755–775 (1987).

29. Jo, W-K; Moon, K.C.: Housewives' Exposure to Volatile Organic Compounds Relative to Proximity to Roadside Service Stations. Atmos. Environ. 33:2921–2928 (1999).

30. Eisenbud, M.: Levels of Exposure to Sulfur Oxides and Particulates in New York City and Their Sources. Bull. N.Y. Acad. Med. 54:99 (1978).

31. U.S. Environmental Protection Agency: Air Quality Criteria for Particulate Matter and Sulfur Oxides, Vol. I–IV. EPA-600/8-82-029a. ECAO, Research Triangle Park, NC (December 1982).

32. U.S. Environmental Protection Agency: Atmospheric Observations: Helping Build the Scientific Basis for Decisions Related to Particulate Matters, HEI/WOAA Report, pp. 1–48. Cambridge, MA (October 1998).

33. Ott, W.: Development of Criteria for Siting of Air Monitoring Stations. J. Air Pollut. Control Assoc. 27:543 (1977).

34. Baker, J.E.; Eisenreich, S.J.: Concentrations and Fluxes of Polycyclic Aromatic Hydrocarbons and Polychlorinated Biophenyls Across the Air–Water Interface of Lake Superior. Environ. Sci. Technol. 24: 342–352 (1990).

35. Cotham, W.E.; Bidleman, T.F.: Polycyclic Aromatic Hydrocarbons and Polychlorinated Biophenyl in an Urban and Rural Site Near Lake Michigan. Environ. Sci. Technol. 29: 2782–2789 (1995).

36. Simcik, M.F.: Zhang, H.; Eisenreich, S.J.; Franz, T.P.: Urban Contaminating of the Chicago/Coastal Lake Michigan Atmosphere by PCBs and PAI-Is during AEOLOS. Environ. Sci. Technol. 31:2141–2147 (1997).

37. McVeety, B.D.; Hites, R.A.: Atmospheric Deposition of Polycyclic Aromatic Hydrocarbons to Water Surfaces: A Mass Balance Approach., Atmos. Environ. 22:511–536 (1988).

38. U.S. Environmental Protection Agency: The Acid Deposition Phenomenon and Its Effects. EPA-600/8-83-016BF. OAQPS, Research Triangle Park, NC (July 1984).

39. National Deposition Study: U.S. National Acid Precipitation Assessment Program, Acid Deposition State of Science and Technology, Vol. I–IV. U.S. Government Printing Office, Washington, DC (December 1990).

40. White, W.H., Ed.: Plumes and Visibility: Measurements and Model Components. Atmos. Environ. 15: 785 (1981).

41. Malm, W.C.; Gebhart, K.A.; Molenar, J.; et al.: Examining the Relationship Between Atmospheric Aerosols, and Light Extinction at Mount Rainer and Northern Cascades, National Parks, Atmos. Environ, 28: 47–360 (1994).

42. National Research Council, Committee on Advances in Assessing Human Exposure to Airborne Pollutants: Human Exposure Assessment for Airborne Pollutants: Advances and Opportunities. National Academy Press, Washington, DC (1991).

43. Lioy, P.J.: Assessing Total Human Exposure to Contaminants. Environ. Sci. Technol. 24: 38–945 (1990).

44. Georgopoulos, P.G.; Lioy, P.J.: Conceptual and Theoretical Aspects of Exposure and Dose Assessment, J. Exp. Analysis & Environ. Epid. 4: 253–285 (1994).

45. Duan, N.; Mage, D.T.: Combination of direct and indirect approaches for exposure assessment. J. Expos. Anal. Environ. Epid. 8(4):439–470 (1997).

46. Zartarian, V.G.; Ott, W.R.; Duan, N.: Quantitative definition of exposure and related concepts. J. Expos. Anal. Environ. Epid. 7(4):411–437 (1997).

47. National Research Council, Committee on Air Pollution Epidemiology: Epidemiology and Air Pollution. National Academy Press, Washington, DC (1985).

48. Valerio, F.; Pala, M.; Lazarotto, A.; Balducci, D.: Preliminary Evaluation, Using Passive Tubes of Carbon Monoxide Concentrations in Outdoor and Indoor Air at Street Shops in Genoe. Atmos. Environ. 32:2871–2876 (1997).
49. Pellizzari, E.D.; Clayton, D.A.; Rodes, C.E.; et al.: Particulate Matter and Manganese Exposures in Toronto, Canada. Atmos. Environ. 33:721–734 (1999).
50. Weisel, C.; Lawryk, N.; Lioy, P. J.: Exposure to Emissions from Gasoline within Automobile Cabins., J. Exp. Anal. and Environ. Epid. 2: 79–96(1992).
51. Lioy, P.J.; Weisel, C.; Pellizzari, E.; Raymer, J.: Microenvironmental and Personal Measurements of Methyl-tertiary Butyl Ether Associated with Automobile Use Activities. J. Expo. Anal. Environ. Epid. 4: 427–441 (1994).
52. Lawryk, N.J.; Lioy, P.J.; Weisel, C.: Exposure to Volatile Organic Compounds in the Passenger Compartment of Automobiles During Periods of Normal and Malfunctioning Operation, J. Exp. Anal. Environ. Epid. 5: 511–531 (1995).
53. World Health Organization, Global Environmental Monitoring System, (GEMS) Geneva, Switzerland Annual Reports through 1996.
54. Zhang, J.; Smith, K.: Hydrocarbon Emissions and Health Risks from Cookstoves in Developing Countries. J Expo. Anal. Environ. Epid.6: 147–161 (1996).
55. Väkeva, M.; Hameri, K.; Kulmala, M.; et al.: Street Level Versus Roof Top Concentrations of Submicron Particles and Gaseous Pollutants in an Urban Street Canyon. Atmos. Environ. 33: 1385–1397 (1999).
56. Morawska, L.; Thomas, S.; Bofinger, N.; et al.: Comprehensive Characterization of Aerosols in a Subtropical Urban Atmosphere: Particle Size Distribution and Correlation with Gaseous Pollutants. Atmos. Environ. 32: 2467–2478 (1998).
57. Acker, K.; Möller, D.; Marquarett, W.; et al.: Atmospheric Research Program for Studying Emission Patterns After German Unification. Atmos. Environ. 32: 3435–3443 (1998).
58. Harrison, R.M.; Smith, D.J.T.; Pio, C.A.; Castro, L.M.: Coparative Receptor Modeling Study of Airborne Particulate Pollutants in Birmingham (UK), Coimbra (Portugal) and Labore, (Pakistan). Atmos. Environ. 31: 3309–3321 (1997).
59. Kirchstellar, T.W.; Harley, R.A.; Kreisberg, N.M.; et al.: On-Road Measurements of Fine Particles and Nitrogen Dioxide Emissions from Light and Heavy-Duty Motor Vehicles. Atmos. Environ. 33: 2955–2968 (1999).
60. Lioy, P.J.; Waldman, J.M.: Acidic Sulfate Aerosols: Characterization of Exposure. Environ. Health Persp. 79: 15–34 (1989).
61. Brauer, M.; Kontrakis, P.; Spengler, J.D.: Personal Exposure to Acidic Aerosols and Gases. Environ. Sci. Technol. 23: 1408–1412 (1989).
62. Norwood, D.M.; Waimnan, T.; Lioy, P.J.: Waldman, J.M.: Breath Ammonia Depletion and Its Relevance to Acidic Aerosol Exposure Studies, Archives of Environ. Health. 47: 309–313 (1992).
63. Suh, J.H.; Allen, G.A.; Koutrakis, P.; Burton, R.M.: Spatial variation in acidic sulfate and ammonia concentrations within metropolitan Philadelphia, J. Air Waste Manage. Assoc. 45: 442–452 (1995).
64. Zhang, J.; Lioy, P. J.: Ozone in Residential Air: Concentrations, I/O Ratios, Indoor Chemistry, and Exposures. Indoor Air 4: 95–105 (1994).
65. Zhang, J.; Lioy, P. J.: Characteristics of Aldehydes: Concentrations, Sources, and Exposure for Indoor Outdoor Residential Microenvrionments. Environ. Sci. Technol., 2: 146–152 (1994).
66. Zhang, J.; Wilson, W. E.; Lioy, P. J.: Sources of Organic Acids Indoors: A field Study. JEAEE 4(l) 25–47 (1994).
67. Weschler, C.; Shields, H.C.; Naik, D.V.: Indoor Chemistry Involving O3, NO, and NO2 as Evidenced by 14 Months of Measurements at a Site in California. Environ. Sci. Technol., 28: 2120–2132 (1994).
68. Weschler, C.; Hodgson, A.F.; Wooley, J.D.: Indoor Chemistry: Ozone Volatile Organic Compounds and Carpet, Environ. Sci. Technol. 26: 2331–3277 (1992).
69. Lioy, P. J.: Measurement of Personal Exposure to Air Pollution: Status and Needs. In: Measurement Challenges in Atmospheric Chemistry, L. Newman, Ed.ACS Series, 232:373–390, Washington, D. C. (1993).
70. Wallace, L.; Nelson W.; Ziegenfus, R.; et al.: The Los Angeles TEAM Study: Personal Exposures, Indoor Outdoor Air Concentrations, and Breath Concentrations of 25 Volatile Organic Compounds. J. Exposure Anal. Environ. Epidemiol. 1: 157–192 (1991).
71. Wallace, L.; Ott, W.R.: Personal Monitors: A State of the Art Survey. J. Air Pollut. Control Assoc. 32: 601 (1982).
72. Sexton, K.; Kleffman, D.E.; Callahan, M.A.: An Introduction to the National Human Exposure Assessment Survey (NHEXAS) and Related Phase I Field Studies, J. Expo. Anal. Environ. Epid. 5:229–232 (1995).
73. Lioy, P.J.; Pellizzari, E.D.: Conceptual Framework for Designing a National Survey of Human Exposure, J. Expo. Anal. Environ. Epid. 3: 425–444 (1995).
74. Pellizzari, E.; Lioy, P.; Quackenboss, J.; et al.: The Design and Implementation of Phase I National Human Exposure Assessment Study in EPA Region V. J. Expo. Anal. Environ. Epid. 3: 327–358 (1995).
75. Lebowitz, M.D.; O'Rourke, M.K.; Gordon, S.; et al.: Population-Based Exposure Measurements in Arizona: A Phase I Field Study in Support of the National Human Exposure Assessment Survey. J. Expo. Anal. Environ. Epid. 5:297–326 (1995).
76. Akland, G.G.; Hartwell, T.D.; Johnson, T.R.; Whitmore, R.W.: Measuring Human Exposure to Carbon Monoxide in Washington, D.C. and Denver, CO during the Winter of 1982–1983. Environ. Sci. Technol. 19:911 (1985).
77. Ott, W.R.; Mage, D.T.; Thomas, J.: Comparison of Microenvironmental CO Concentrations in Two Cities for Human Exposure Modeling. J. Expo. Anal. Environ. Epid. 2:249–267 (1992).
78. Cortese, A.M.; Spengler, J.D.: Ability of Fixed Monitoring Stations to Represent Personal Carbon Monoxide Exposure. J. Air Pollut. Control Assoc. 26: 1144–1150 (1976).
79. Wallace, L.; Thomas, J.; Mage, D.; Ott, W: Comparison of Breath CO, CO Exposure, and Coburn Model Predictions in the U.S. EPA Washington–Denver (CO) Study. Atmos. Environ. 22(10):2183–2193 (1988).
80. Law, P.L.; Zelenka, M.P.; Huber, A.H.; McCurdy, T.R.; Lioy, P.J.: Evaluation of a Probabilistic Exposure Model Applied to Carbon Monoxide (pNEM/CO) Using Denver Personal Exposure Monitoring Data. J Air Waste Manage. Assoc. 47: 491–500 (1997).
81. Coffey, P.E.; Stasiuk, W.N.: Evidence of Atmospheric Transport of Ozone into Urban Areas. Environ. Sci. Technol. 9:59 (1975).
82. U.S. Environmental Protection Agency: Review of the NAAQS for Ozone. Preliminary Assessment of Scientific and Technical Information. OAQPS, Research Triangle Park, NC (March 1986).
83. Rombout, P.J.A.; Lioy, P.J.; Goldstein, B.: Rationale for an Eight–Hour Ozone Standard. J. Air Pollut. Control Assoc. 36: 913 (1986).
84. Spengler, J.D.; Keeler, G.J.; Koutrakis, P.; et al.: Exposures to Acid Aerosols. Environ. Health Perspectives 79: 43–51 (1989).
85. Lioy, P.J.: The Analysis of Human Exposures to Contaminants in the Environment, Chapter 3. In: Oxford Text Book of Public Health, 3rd Ed., Vol 2, R. Detels, W.W. Holland, J. McEwen and

G.S Omenn, Eds. Oxford Medical Publications, New York (1997).

86. Lioy, P.J.: Exposure Analysis and Assessment for Low Risk Cancer Agents. Intern. J. Epid. 19 (Suppl. 1) 553–561 (1990).

87. Morris, J.: Uses of Epidemiology. Churchill Livingston, New York (1975).

88. Lilenfeld, A.M.; Lilenfeld, D.E.: Foundations of Epidemiology, 2nd ed. Oxford University Press, New York (1980).

89. Leaderer, B P.: Assessing Exposure to Environmental Tobacco Smoke. In: Risk Analysis 10:19–28 (1990).

90. Lioy, P.J.; Mallon, R.P.; Kneip, T.J.: Long Term Trends in Total Suspended Particulates, Vanadium, Manganese, and Lead at a Near Street Level and Elevated Sites in New York City. J. Air Pollut. Control Assoc. 30:153 (1980).

91. Lioy, P.J.; Samson, P.J.: Ozone Concentration Patterns Observed During the 1976–1977 Long Range Transport Study. Environ. Int. 2:77 (1979).

92. Seinfeld, J.H.: Atmospheric Chemistry and Physics of Air Pollution. Wiley Interscience New York (1986).

93. Turner, B.D.: Workbook of Atmospheric Dispersion Estimates, 2nd Ed. Lewis Publishers, Boca Raton, FL (1994).

94. Thompson, R.: Air Monitoring for Organic Constituents. In: Air Sampling Instruments for the Evaluation of Atmospheric Contaminants, 6th ed. P.J. Lioy and M.J. Lioy, Eds. American Conference of Governmental Industrial Hygienists, Cincinnati, OH (1983).

95. Walling, J.F.: The Utility of Distributed Air Volume Sets When Sampling Ambient Air Using Solid Adsorbents. Atmos. Environ. 18:855 (1984).

96. U.S. Environmental Protection Agency: Compendium Method TO–: The Determination of Volatile Organic Compounds (VOC's) in Ambient Air Using Summal Passivated Canister Sampling and Gas Chromatographic Analysis. U.S. EPA, Washington, DC (May, 1988).

97. Lou, B.W.; Jaklevic, J.M.; Goulding, F.S.: Dichotomous Virtual Impactors for Large Scale Monitoring of Airborne Particulate Matter. In: Fine Particles: Aerosol Generation Measurement, Sampling and Analysis, pp. 311–350. B.Y.H. Liu, Ed. Academic Press, New York (1976).

98. Eatough, D.T.; Obeidi, F.; Pang, Y.; Ding, Y.; Eatough, N.L.; Wilson, W.E.: Integrated and Real-Time Diffusion Sampler for $PM_{2.5}$. Atmos. Environ. 33: 2835–2844 (1999).

99. Chow, J.C.; Watson, J.G.; Ono, D.M.; Mathai, C.V.: PM_{10} Standards and Non–Traditional Particulate Source Controls: A Summary of the AWMA/EPA Conference, pp. 43, 74–84 (1993).

100. Chow, J.C.: Critical Review: Measurement Methods to Determine Compliance with Ambient Air Quality Standards for Suspended Particles. J. Air Waste Manage. Assoc. 45: 320–382 (1995).

101. Spengler, J.D.; Treitman, R.D.; Losteson, T.D.; et al.: Personal Exposures to Respirable Particulates and Implications for Air Pollution Epidemiology. Environ. Sci. Technol. 19:700 (1985).

102. Lioy, P. J.; Waldman, J. M.; Buckley, T.; et al.: The Personal, Indoor and Outdoor Concentrations of PM–10 Measured in an Industrial Community During the Winter. Atmos. Environ. 24B: 57–66 (1990).

103. Thomas, K.W.; Pellizzari, E.D.; Clayton, C.A.; et al.: Particle Team Exposure Assessment Methodology (PTEAM) Study: Method Performance and Data Quality for Personal, Indoor and Outdoor Aerosol Monitoring for 178 Homes in Southern California. J. Exp. Analy. Environ. Epid. 3:203–226 (1993).

104. Lodge, J.P.: Methods of Air Sampling and Analysis, 3rd Edition. Lewis Publishers, Chelsea, MI (1988).

105. Newman, L., Ed.: Measured Challenges in Atmospheric Chemistry, ACS–#232. American Chemical Society, Washington, DC (1993).

106. Koutrakis, P.; Wolfson, T.M.; Brauer, J.M; Spengler, T.D.: Design of a Glass Impactor for an Annular Denuder/Fillerpack System. Aerosol Sci. Technol. 12: 607–612 (1990).

107. Daisey, J.J.: Organic Compounds in Urban Aerosols. In: Aerosols: Anthropogenic and Natural Sources and Transport. T.J. Kneip and P.J. Lioy, Eds. Ann. N.Y. Acad. Sci. 338:50 (1980).

108. Lee, M.L.; Goates, S.R.; Markides, K.E.; Wise, S.A.: Frontiers in Analytical Techniques for Polycyclic Aromatic Compounds. In: Polynuclear Aromatic Hydrocarbons: Chemistry Characterization and Carcinogenesis, 9th International Symposium, pp. 13–40. M. Cooke and A.J. Dennis, Eds. Battelle Press, Columbus, OH (1986).

109. Puxbaum, H.; Novakov, T.; Hotzenberger, R., Eds.: Carbonaceous Particles in the Atmosphere. 33: 2601–2822 (1999).

110. Cadle, S.H.; Groblicki, P.J.: An Evaluation of Methods for the Determination of Organic and Elemental Carbon in Particulate Samples. In: Particulate Carbon: Atmospheric Life Cycle, pp. 89–110. G.T. Wolff and R.L. Klimesch, Eds. Plenum Press, New York (1982).

111. Watson, J.G.: Overview of Receptor Model Principles. J. Air Pollut. Control. Assoc. 34:619–623 (1984).

112. Hopke, P.K.: Receptor Modeling in Environmental Chemistry, John Wiley and Sons, New York, NY (1985).

113. Marple, A.; Rubow, K.L.; Spengler, J.D.: Low Flow Rate Sharp Cut Impactors for Indoor Air Sampling: Design and Calibration. Particle Technology Laboratory Pub. No. 623 (December 1986).

114. U.S. Environmental Protection Agency: Review and Evaluation of the Evidence for Cancer Associated with Air Pollution. EPA 450/5–83–006R. QAQOS, Research Triangle Park, NC (1984).

115. Maron, D.M.; Ames, B.N.: Revised Methods for the Salmonella Mutagenicity Test. Mutat. Res. 113:173 (1983).

116. Ames, B.N.; McCann, J.; Yamasaki, E.: Methods for Detecting Carcinogens and Mutagens with the Salmonella/Microsomal Mutagenicity Test. Mutat. Res. 113:347 (1974).

117. Louis, J.B.; McGeorge, L.J.; Atherholt, T.B.; et al.: Mutagenicity of Inhalable Particulate Matter at Four Sites in New Jersey. In: Toxic Air Pollution, P.J. Lioy and J. M. Daisey, Eds. Lewis Publishers, Chelsea, MI (1987).

118. Henderson, M.; Bechtold, W.E.; Maples, K.R.: Biological Markers of Exposure. J. Exp. Analysis Environ. Epid. 3:1–14 (1992).

119. Butler, J.P.; Kneip, T.P.; Daisey, J.M.: An Investigation of Interurban Variations in the Chemical Composition and Mutagenic Activity of Airborne Particulate Organic Matter Using an Integrated Chemical Class Bioassay System. Atmos. Environ. 21:883 (1987).

120. Cupitt, L.T.; Glen, W.G.; Lewtas, J.: Exposure and Risk from Ambient Particlebound Pollution in an Airshed Dominated by Residential Wood Combustion and Mobile Sources. In: Symposium of Risk Assessment of Urban Air: Emissions, Exposure, Risk Identification, and Risk Quantitation; May–June 1992; Stockholm, Sweden. Environ. Health Perspect. 102(suppl. 4):75–84 (1994).

121. Ashley, D.L.; Bonin, M.A.; Cardinali, F.L.; et al.: Determination of Volatile Organic Compounds in Human Blood From a Large Sample Population by Pulse and Trap Gas Chromatography in Mass Spectrometry. Anal. Chem. 64: 1021–1029 (1992).

122. Pirkle, J.L.; Needham, L.L.; Sexton, K.: Improving Exposure Assessment by Monitoring Human Tissue for Toxic Chemicals. J. Expos. Analysis Environ. Epid. 5:405–424 (1995).

123. Krause, C.; Babisch, W.; Becker, K.; et al.: Studien Beshrebiuns and Human Biomonitoring, Wabolu, Hefte, Umweld, Bundes Amt (1993).

124. Lioy, P.J.: Exposure Analysis and the Biological Response to a Contaminant: A Melding Necessary for Environmental Health Sciences. J. Exposure Anal. Environ. Epid. 2 (Suppl. 1) 19–24 (1992).

125. Schulte, P.A.; Perers, E.P.: Molecular Epidemiology, Acadmic Press, San Diego, CA (1993).

126. AIRS, Aerometric Information Retrieval System [database]. U.S. Environmental Protection Agency: Office of Air Quality Planning and Standards, Research Triangle Park, NC (1998).

127. Lioy, P.J.; Spektor, D.; Thurston, G.; et al.: The Design Consideration for Ozone and Acid Aerosol Exposure and Health Investigations: The Fairview Lake Summer Camp. Photochemical Smog Case Study. Environ. Int. 13:271 (1987).

128. Hidy, G.M.: Source–Receptor Relationships for Acid Depositing: Pure and Simple. J. Air Pollut. Control Assoc. 34:518 (1984).

129. Seinfeld, J.H.: Ozone Air Quality Models: A Critical Review. J. Air Pollut. Control Assoc. 38:616–645 (1988).

130. Ott, W.R.: Environmental Statistics and Data Analysis, Lewis Publishers (CRC Press) Boca Raton, FL (1998).

131. Miller, M.S.; Fiedlander, S.K.; Hidy, G.M.: A Chemical Balance for the Pasadena Aerosol. J. Coll. Interface Sci. 47:165 (1972).

132. Kleinman, M.T.; Pasternack, B.; Eisenbud, M.; Kneip, T.J.: Identifying and Estimating the Relative Importance of Sources of Airborne Particles. Environ. Sci. Technol. 14:62 (1980).

133. Henry, R.C.; Lewis, C.H.; Hopke, P.K.; Williamson, H.J.: Review of Receptor Model Fundamentals. Atmos. Environ. 18:1507–1515 (1984).

134. Hopke, P.E.; Alpert, D.J.; Roscoe, B.A.: Fantasia—A Program for Target Transformation Factor Analysis to Apportion Sources in Environmental Samples. Computers in Chem. 7:149 (1983).

135. Thurston, G.D.; Lioy, P.J.: Receptor Modeling and Aerosol Transport. Atmos. Environ. 21:687 (1987).

136. Nazaroff, W.W.; Cass, G.R.: Mathematical Modeling of Chemically Reactive Pollutants in Indoor Air. Environ. Sci. Technol. 20:924–934 (1986).

137. Price, P.S.; Sample, J.; Strieter, R.: Determination of Less than Lifetime Exposures to Point Source Emissions, 12: 3 67–3 82 (1992).

138. Thompson, K.M.; Burmaster, D.E.; Crouch, E.A.C.: Monte Carlo: Techniques for Quantitative Uncertainty Analysis in Public Health Risk Assessments. In: Risk Analysis 12:53–63 (1992).

139. Hidy, G.M.; Mueller, P.K.; Grosjean, D.; et al.: The Character and Origin of Smog Aerosols: A Digest of Results from the California Aerosol Characterization Experiment (ACHEX). John Wiley & Sons, New York (1980).

140. Lioy, P.J.; Daisey, J.M., Eds.: Toxic Air Pollutants: Study of Non–Criteria Pollutants. Lewis Publishers, Chelsea, MI (1986).

141. Hidy, G.M.; et al.: Summary of the California Aerosol Characterization Experiment J. Air Pollut. Control Assoc. 25:1106 (1975).

142. Appel, B.R.; Kothny, E.L.; Hoffer, E.M.; et al.: Sulfate and Nitrate Data from the California Aerosol Characterization Experiment (ACHEX). Environ. Sci. Technol. 12: 418–425 (1978).

143. Hering, S.V.; Blumenthal, D.L.: Southern California Air Quality Study (SCAQS), Description of Measurement Activities. Sonoma Technology, Final Report to California Air Resources Board, Sacramento, A5–151–32 (1992).

144. Lawson, D.R.: The Southern California Air Quality Study. J. Air Waste Manag. Assoc. 40:156–165 (1990).

145. Chow, J.C.; Watson, J–G.; Lowenthal, D.H.; et al.: PM$_{10}$ Source Apportionment in California's San Joaquin Valley. Atmos. Environ. Part A 26:3335–3354 (1992b).

146. Chow, J..C.; Watson, J.G.; Fujita, E.M.; et al.: Temporal and Spatial Variations of PM$_{2.5}$ and PM$_{10}$ Aerosol in the Southern California Air Quality Study. Atmos. Environ. 28: 2061–2080 (1994a).

147. Watson, J.G: Chow, J.C.; Lu, Z.; et al.: Chemical Mass Balance Source Apportionment of PM$_{10}$ During the Southern California Air Quality Study. Aerosol Sci. Technol. 21: 1–36 (1994a).

148. Burton, C.S.; Hidy, G.M.; Regional Air Pollution Study Program Objectives and Plans, Environmental Protection Agency Rep. EPA–650/3–75–009 (NTIS PB–247769), Rockwell International Science Center, Thousand Oaks, CA (1975).

149. Deubay, T.G.: Chemical Element Balance Method Applied to Dichotomons. Sample DA 17A. In: Aerosols: Anthpogenic and Natural Sources and Transport, T.J. Kneip and P.J. Lioy, Eds. 338:126–144 (1980).

150. Russell, P.A.: Denver Air Pollution Study–1973. Proceedings of a Symposium, Vol. 1: EPA Report No. 600/9–76–007a; Vol. 2: EPA Report No. 600/9–77–001.

151. Wolff, G.T.; Groblicki, P.J.; Countess, R.J.; Ferman, M.A: Design of the Denver Brown Cloud Study. GMR–3050. General Motors Research Laboratories, Larsen, MI (August 1979).

152. Countess, R.J.; Wolff, G.T.; Cadle, S.H.: The Denver Winter Aerosol: A Comprehensive Chemical Characterization. J. Air Pollut. Control Assoc. 30:1194 (1980).

153. Countess, R.J.; Cadle, S.H.; Groblicki, P.J.; Wolff, G.T.: Chemical Analysis of Size Segregated Samples of Denver's Ambient Particulate. J. Air Pollut. Control Assoc. 31:247 (1981).

154. Wolff, G.T.; Countess, R.J.; Groblicki, P.J.; et al.: Visibility Reducing Species in the Denver Brown Cloud, Part 11, Sources and Temporal Patterns. Atmos. Environ. 15:2485 (1981).

155. Lewis, C.W.; Baumgardner, R.E.; Stevens R.K.: Receptor Modeling Study of Denver Winter Haze. Environ. Sci. Technol. 20: 1126 (1986).

156. Watson, J.G.; Chow, J.C.; Richards, L.W.; et al.: The 1987–88 Metro Denver Brown Cloud Study, Vol. 1, Program Plan; Vol. 2, Measurements; Vol. 3, Data Interpretation. Final Report from the Desert Research Institute, DRI Document No. 8810–F (October 1988).

157. Lioy, P.; Wolff, T.; Leaderer, B.P.: A discussion of the New York Summer Aerosol Study, 1976. Ann. N.Y. Acad. Sci., 322: 153–162 (1979).

158. Leaderer, B. P.; Bernstein, D. M.; Daisey, J. M.; et al.: Summary of the New York Summer Aerosol Study (NYSAS). J. Air Pollut. Control Assoc., 28: 321–327 (1978).

159. Lioy, P. J; Samson, P. J.; Tanner, R.L.; et al.: The Distribution and Transport of Sulfate "Species" in the New York Metropolitan Area During the 1977 Summer Aerosol Study. Atmos. Environ. 14: 1391–1407 (1980).

160. Lioy, P.J.; Daisey, J.M.: Airborne Toxic Elements and Organic Substances. Environ. Sci. Technol. 20:8 (1986).

161. Waldman, J.; Lioy, P.J.; Greenberg, A.; Butler, J.: Analysis of Human Exposure to Benzo(a)pyrene Via Inhalation and Food Ingestion in the Total Human Environmental Exposure Study (THEES). J. Exposure Anal. Environ. Epid. 1:197–226 (1991).

162. Ferris, B.G.; Spengler, J.D.: Harvard Air Pollution Health Study in Six Cities in the USA. Tokai J. Exp. Clin. Med. 10:263 (1985).

163. Speizer, F.E.: Studies of Acid Aerosols in Six Cities and in a New Multi–city Investigation—Design Issues. Environ. Health Persp. 79:61–67 (1989).

164. Pellizzari, E.; Sheldon, L.; Sparcino, C.; et al.: Volatile Organic Levels in Indoor Air. In: Indoor Air, Vol. 4, Chemical Characterization and Personal Exposure. pp. 303–308. Swedish Council for Building Research, Stockholm, Sweden (1984).

165. Lioy, P.J.; Wallace, L.; Pellizzari, E.: Indoor/Outdoor and Personal Monitoring and Breath Analysis Relationships for Selected Volatile Organic Compounds Measured at Three Homes During New Jersey TEAM—1987. J. Expo. Anal. Environ. Epid. 1:45–61 (1991).

166. Raymer, J.H.; Pellizzari, E.D.; Thomas, K.W.; Cooper, S.D.: Elimination of Volatile Organic Compounds in Breath After Exposure to Occupational and Environmental Micro-environments. J. Expo. Anal. Environ. Epidem. 1:439–451 (1991).

167. Wallace, L.A.; Pellizzari, E.D.; Hartwell, T.D.; et al.: The California TEAM Study: Breath Concentrations and Personal Exposures to 26 Volatile Compounds in Air and Drinking Water of 188 Residents of Los Angeles, Antioch and Pittsburgh. Atmos. Environ. 22:2141–2163 (1988).

168. Jantunen, M.J.; Hanninen, O.; Katsouyanni, K.; et al.: Air Pollution Exposure in European Cities, The EXPOLIS Study. J. Exp. Analy. Env. Epid. In Press, (1998).

169. Landrigan, P.J.: Epidemiologic Approaches to Persons with Exposures to Waste Chemicals. Environ. Health Persp. 48:93 (1983).
170. National Research Council: Environmental Epidemiology: Public Health and Hazardous Wastes. National Academy of Sciences, Washington, DC (1991).
171. National Research Council: Improving the Environment: An Evaluation of DOE's Environmental Management Program. National Academy of Sciences, Washington, DC (1996).
172. Campbell, D.L.; Quintrell, W.N.: Cleanup Strategy for Rocky Mountain Arsenal. In: Proceedings of the 6th National Conference on Management of Uncontrolled Hazardous Waste Sites, pp. 36–42 (November 4–6, 1985).
173. Melius, J.M.; Costello, R.J.; Kominsky, J.R.: Facility Siting and Health Questions. The Burden of Health Risk Uncertainty. Natural Resources Lawyer 17:467 (1985).
174. Hawley, J.K.: Assessment of Health Risk from Exposure to Contaminated Soil. Risk Analysis 5:289–302 (1985).
175. Georgopoulos, P.; Lioy, P.J.: Exposure Measurement Needs for Hazardous Waste Sites: Two Case Studies. J. Toxicology and Ind. Health. 12(5) 651–665 (1996).
176. Daisey, J.M.: Feature Article: A Report on the Workshop on Improving Exposure Analysis for DOE Sites. J. Expo. Anal. Environ. Epid. 8:3–9 (1998).

TABLE 3-4. Human Tissue Measurements Currently Performed at the National Center for Environmental Health (NCEH) of the Centers for Disease Control and Prevention (CDC)

Metals (typical urine or blood sample—3 mL, typical limit of detection—low parts-per-billion)

Lead	Arsenic	Chromium
Mercury	Vanadium	Nickel
Cadmium	Beryllium	Thallium

Polychlorinated dibenzo-dioxins, polychlorinated dibenzo-furans, coplanar polychlorinated biophenyls (PCBs) (all analytes measured in serum from one 25-mL blood sample if exposure is near background levels—smaller samples are adequate for higher exposures; typical limit of detection—low parts-per-trillion on a lipid-weight basis, low parts-per-quadrillion on a whole weight basis)

2,3,7,8-Tetrachlorodibenzo-p-dioxin (TCDD)	2,3,4,7,8-Pentaclilorodibenzofuran (PnCDF)
1,2,3,7,8-Pentachlorodibenzo-p-dioxin (PnCDD)	1,2,3,4,7,8-Hexachlorodibenzofuran (HxCDF)
1,2,3,4,7,8-Hexachlorodibenzo-p-dioixin (HxCDD)	1,2,3,6,7,8-Hexachlorodibenzofuran (HxCDF)
1,2,3,6,7,8-Hexachlorodibenzo-p-dioxin (HxCDD)	1,2,3,7,8,9-Hexachlorodibenzofuran (HxCDF)
1,2,3,7,8,9-Hexaclilorodibenzo-p-dioxin (HxCDD)	2,3,4,6,7,8-Hexachlorodibenzofuran (HxCDF)
1,2,3,4,6,7,8-Heptachlorodibenzo-p-dioxin (HpCDD)	1,2,3,4,6,7,8-Heptachlorodibenzofuran (HpCDF)
(HpCDD)	1,2,3,4,7,8,9-Heptachlorodibenzofluran (HpCDF)
1,2,3,4,6,7,9-Heptachlorodibenzo-dioxin	1,2,3,4,6,7,8,9-Octachlorodibenzofuran (OCDF)
	3,3',4,4'-Tetrachlorobiphenyl (TCB)
1,2,3,4,6,7,8,9-Octachlorodibenzo-p-dioxin (OCD)	3,4,4',5-Tetrachlorobiphenyl (TCB)
2,3,7,8-Tetrachlorodibenzofuran(TCDF)	3,3',4,4',5-Pentachlorobiphenyl (PnCB)
1,2,3,7,8-Pentachlorodibenzofuran (PnCDF)	3,3',4,4',5,5'-Hexachlorobiphenyl (HxCB)

Volatile organic compounds (VOCs) (all analytes measured in one 10-mL blood sample: typical limit of detection—low parts-per-trillion)

1,1,1-Trichloroethane	Acetone	Hexachloroethane
1,1,2,2-Tetrachloroetilane	Benzene	m-/pXylene
1,1,2-Trichloroethane	Bromdichloromethane	Methylene chloride
1,1-Dichloroethane	Bromolform	o-Xylene
1,1-Dichloroethene	Carbon Tetrachloride	Styrene
1,2-Dichlorobenzene	Chlorobenzene	Tetrachlorethene
1,2-Dichloroethane	Chloroform	Toluene
1,2-Dichloropropane	cis-1,2-Dichloroethene	trans-1,2-Dichloroethene
1,3-Dichlorobenzene	Dibromochloromethane	Trichloroethene
1,4-Dichlorobenzene	Dibromomethane Ethylbenzene	
2-Butanone		

Chlordinated pesticides and noncoplanar polychlorinated biphenyls (all analytes measures in serum from one 5-mL blood sample; typical limits of detection—low parts-per-billion)

Aldrin	DDE
Chlordane, alpha	DDT
Chlordane, gamma	Dieldrin
beta-Hexachlorocyclohexane	Endrin
gamma-Hexachlorocyclohexane	Heptachlor
Biphenyls, Polychlorinated (total)	Heptachlor epoxide
Biphenyls, Polychlorinated (individual congeners)	Hexachlorobenzene
DDD	Mirex
Trans-nonachlor	Oxychlordane

Nonpersistent pesticides (all analytes measured in one 10-mL urine sample; typical limits of detection—low parts-per-billion)

Urine metabolites	Parent pesticide(s)
2-isopropoxyphenol (IPP)	Propoxur
2,5-Dichlorophenol (25DCP)	1,4-Dichlorobenzene
2,4-Dichlomphenol (24DCP)	1,3-Dichlorobenzene, dichlolenthion, prothiofos, phosdiphen
Carbofuranphenol	Carbofuran, benfuracarb. Carbosulfan. Furathiocarb
2,4,6-Trichlorophenol (246TCP)	1,3,5-Trichlorobenzene, hexachlorobenzene, Lindane
3,5,6-Trichloro-2-pyndinol (TCPY)	Chlorpyrifos, chlorpyrifos-methyl
4-Nitrophenol (NP)	Parathion, methylparathion, nitrobenzene, EPN
2,4,5-Trichlorophenol (245TCP)	1,2,4-Trichlorobenzene, fenchlorphos, inchloronate
1-Naphthol (1NAP)	Napthalene, carbaryl
2-Naphthol (2NAP)	Napthalene
2,4-Dichlorophenoxy acetic acid (24D)	2,4-D
Pentachlorophenol (PCP)	Pentachlorophenol

Chapter 4

Particle and Gas Phase Interactions in Air Sampling

Sidney C. Soderholm, Ph.D.

National Institute for Occupational Safety and Health, Morgantown, West Virginia

---

CONTENTS

Introduction

An aerosol consists of airborne particles and surrounding gases. The particle phase may include solids and liquids. The gas phase normally includes air and water vapor and may include vapors of organic and inorganic compounds as well as contaminant gases, such as sulfur dioxide. The distinction between a vapor and a gas is somewhat arbitrary. Following common practice, the term "vapor" refers to the gas phase portion of a material which can exist as a liquid or solid at room temperature and atmospheric pressure. A "gas" cannot exist as a liquid or solid at normal conditions.

Molecules of each species in the gas phase continually bombard the surface of each particle, giving ample opportunity for chemical reactions, i.e., formation of new molecules, and physical interactions, e.g., transfer of mass between the particle and gas phases. This chapter does not emphasize the most general case in which both chemical reactions and physical interactions occur simultaneously in the atmosphere and during sampling. This is important in particular situations, e.g., when sampling ammonia, acids, and ammonium salts in the ambient atmosphere, and has been discussed elsewhere.[1-4] This chapter emphasizes information needed to make decisions about which phase(s) to sample and techniques for avoiding erroneous sampling results when chemical reactions or condensation/evaporation may occur during sampling. The words "condensation" and "evaporation" are used to indicate the net transport of mass from the gas phase to the particle phase or from the particle phase to the gas phase, respectively.

Air sampling errors being considered include allowing unwanted chemical reactions to occur during sampling, measuring the concentration of a contaminant only in the particle or gas phase when there is a significant fraction in the unsampled phase, and using inappropriate techniques to measure the distribution of a contaminant between the two phases. Avoiding unwanted chemical reactions with reactive gases during sampling is discussed briefly in the next section. This issue has received substantial attention in the literature of environmental air sampling. In the remaining sections, the emphasis is on effects of interphase mass transfer in sampling. The term "semivolatile" is sometimes associated with compounds for which such effects are important. First, the equilibrium distribution of a material between the particle and gas phases will be discussed. Several factors that tend to disturb equilibrium and the time-scale of the approach to equilibrium will then be presented. Finally, techniques for measuring a contaminant's total airborne concentration (concentration in both the particle and gas phases) and approaches to measuring a contaminant's distribution between the two phases will be offered. A draft prestandard has been prepared by a working group of the European Committee on Standardization (CEN) to offer guidance on such issues.[5]

Much of this chapter contains detailed discussions intended to aid the reader who needs a deeper understanding of how particle-gas interactions can influence air sampling. Readers who wish to focus on the practical implications may be most interested in the following: 1) the "Chemical Reactions" section for a brief discussion of how to avoid chemical reactions between collected material and reactive gases; 2) the "General Guidance" subsection of the "Physical Equilibrium" section for rules-of-thumb to guide decisions about whether to sample the particle, gas, or both phases of an aerosol; and 3) the "Sampling Approaches" section for a review of the applicability of different types of sampling instrumentation.

Chemical Reactions

Gases, such as HNO_2, HNO_3, NH_3, NO_2, SO_2, HCl, and HF, in an atmosphere may react with sampled particulate or the filter material used to collect the particulate and lead to erroneous results. This is a more serious problem when sampling times are long as is often the case when sampling outdoors.

A common approach to avoiding these problems is to remove the reactive gas from the air stream with a denuder before the air reaches the filter (see Chapter 19). The sampled air passes through one or more channels in the denuder before reaching the filter. The channel walls are coated with a chemical that will adsorb or react with the unwanted gas and retain it at the wall (Figure 4-1). A properly designed denuder

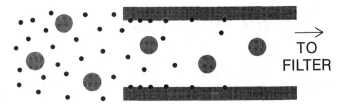

FIGURE 4-1. Schematic diagram of a denuder showing the collection of a reactive gas (small black circles) at the wall by adsorption or chemical reaction. Particles (large gray circles) penetrate the denuder and are transmitted to the filter.

removes the reactive gas from the air stream without removing a significant fraction of the particles being sampled.

Ideally, a denuder reduces artifacts caused by a gas reacting with collected particles on the filter. However, calculations illustrate the potential for a denuder to introduce artifacts. A denuder section of a sampler collecting particulate from an atmosphere containing HNO_3 and NH_3 gases and NH_4NO_3 particulate in equilibrium with the gases would remove the NH_3, disturb the equilibrium, and tend to cause the chemical species in the particulate to change, possibly leading to an incorrect assessment of the aerosol mass concentration and chemical species.[6, 7] Denuders can be very helpful in avoiding sampling artifacts due to chemical reactions between a reactive gas and collected particles or the filter, but all the possible influences of a denuder should be considered and tested before deciding to include one in a sampling system.[8]

Physical Equilibrium

Gas molecules collide with the surface of suspended particles. At the same time, molecules are ejected from particle surfaces due to thermal energy. At equilibrium, there is no net flux of any species across any solid/gas, liquid/gas, or solid/liquid interface. A solid/liquid interface occurs within particles containing a liquid and undissolved material.

Experimental data relevant to the equilibrium between identical particles and the gas phase can be written in the form:

$$C_{G,i} = f(d_p, C_{P,1}, C_{P,2}, ..) \qquad (1)$$

The symbol C will always imply an airborne mass concentration in this chapter (mass of contaminant per volume of air). The subscript G or P refers to the gas or particle phase, respectively. The mass concentration of the ith contaminant in the gas phase just outside the particle surface $C_{G,i}$ is some function of the composition of the particle surface, which is described here by the mass concentration $C_{P,j}$ of each species in the particle phase of the aerosol. A relation like Equation 1 holds for each species in the gas phase. All the relations are linked because each depends on all constituents in the particle phase. Additional relations exist between the concentration in the liquid phase and the surface composition of solid constituents, if there are any solid/liquid interfaces within particles.

The particle diameter d_p appears in the function because of the Kelvin effect, which increases the equilibrium concentration above a convex surface compared to a planar surface.[9] This factor can be significant for particles that are submicrometer in diameter. For example, a pure submicrometer droplet in an atmosphere which is saturated with the vapor is expected to evaporate completely because it requires a supersaturated atmosphere to reach equilibrium.

When an aerosol consists of particles that differ markedly in composition or size, the equilibrium of each particle class should be considered separately. Each particle size and composition class may influence the composition of the other classes through the gas phase that they share.

In principle, it is possible to predict the distribution of a material between the particle and gas phases if the total airborne mass concentration of each constituent is known and if each of the possible liquid/gas, solid/gas, and solid/liquid interfaces in the system has been sufficiently characterized. Typically, this information is obtained by measuring the concentrations of each constituent on both sides of the interface in a laboratory system. It is often not clear whether available experimental data are relevant because the equilibrium distribution is sensitive to temperature and pressure and may be sensitive to the addition or removal of small quantities of other constituents. Because real atmospheres are often complex and contain many constituents, rigorous prediction of equilibrium conditions is a difficult task. The approach taken here is to present the general form of the relations needed to predict equilibrium and to make some general observations for guidance when making sampling decisions.

Absorption

For aerosols, absorption refers to the dissolution of gas phase materials in droplets. The relation between the gas phase concentration of one of the components adjacent to a droplet surface and the composition of the droplet surface can be expressed in terms of an activity coefficient γ_i[10, 11]

$$C_{G,i} = \gamma_i x_i\, C_{G,i}^{S} \qquad (2)$$

where x_i is the mole fraction of the ith material in the droplet surface and $C_{G,i}^{S}$ is its saturated vapor concentration at the droplet temperature. The superscript S indicates saturation. The saturated vapor concentration is the mass concentration of the vapor above a planar surface of the pure material and can be calculated from the saturated vapor pressure using the Ideal Gas Law. In general, γ is a function of the droplet composition, but it is constrained to assume the value of unity when x_i is unity (pure droplet). The Kelvin effect is not included explicitly in Equation 2, although it should be included for submicrometer particles.[10]

Saturated vapor pressure is a property of the pure material. Caution should be exercised if using saturated vapor pressure data from the literature. The value is temperature sensitive, often changing by an order of magnitude for a 10 to $20^{\circ}C$ change in temperature, so application of predictive equations for temperature correction, e.g., the Clausius–Clapeyron equation, may be necessary.[11] In many cases, a literature search may be worthwhile; early measurements of small saturated vapor pressures may be in substantial error and yet may be widely cited. Some saturated vapor pressure information may be available in material safety data sheets, handbooks, and exposure limit documentation.

Equation 2 can be expressed in terms of mass concentrations which are commonly measured when describing an atmosphere

$$C_{G,i} = \gamma_i\, \frac{M_P}{M_i}\, \frac{C_{P,i}}{C_P}\, C_{G,i}^{S} \qquad \text{droplets} \qquad (3)$$

where a substitution has been made for the mole fraction

$$x_i = \frac{M_P C_{P,i}}{M_i C_P} \qquad (4)$$

The molecular weight of the ith material is M_i and the average molecular weight of the droplet M_P is

$$M_P^{-1} = \frac{M_1^{-1} C_{P,1} + M_2^{-1} C_{P,2} + \ldots}{C_P} \qquad (5)$$

The concentration of all materials in the particle phase C_P is

$$C_P = C_{P,1} + C_{P,2} + \ldots \qquad (6)$$

Equation 3 is very general. It can be used to summarize experimental data on liquid/gas equilibria for most systems across the whole range of possible mole fractions for each component, even when the liquids are not completely miscible.[10, 11]

Specifying the activity coefficient as a function of droplet composition can be quite complex. Simpler and less general relations can successfully characterize liquid/gas equilibria in special cases that commonly occur. Three such special cases will be discussed: ideal liquids, Raoult's Law for solvents, and Henry's Law for dilute solutions.

For ideal liquids, the activity coefficient for each component is unity for all compositions. This is likely to occur for similar liquids, for example, propylene glycol and water.[10,12] The equilibrium of an airborne system of ideal liquids in uniform droplets can be described by

$$C_{G,i} = \frac{M_P}{M_i} \frac{C_{P,i}}{C_P} C_{G,i}^S \quad \text{ideal liquids} \quad (7)$$

Very few aerosols are likely to consist of ideal liquids.

Raoult's Law has been found to apply to the solvent in many liquid systems even if the solution is not ideal. It can be derived from Equation 2 or 3 by setting the activity coefficient to unity (it should be nearly equal to unity for any material that is nearly pure in the liquid) and applying the resulting equation only to the solvent in a dilute solution[12]

$$C_{G,solv} = \frac{M_P}{M_{solv}} \frac{C_{P,solv}}{C_P} C_{G,solv}^S \quad \text{solvents} \quad (8)$$

The ratio of the molecular weights and the ratio of the particle phase mass concentrations are both near unity for the solvent in a dilute solution. The Kelvin effect should be included for submicrometer droplets. A common application of Raoult's Law is to describe the relationship between the composition of aqueous droplets and relative humidity RH (in percent) resulting in

$$RH = \frac{C_{G,water}}{C_{G,water}^S} 100 \quad (9)$$

Henry's Law often applies to a solute in a dilute solution. It states that the partial pressure of the material is proportional to its mole fraction in the solution[12]

$$P_i = x_i H_i \quad (10)$$

The proportionality constant H_i is the Henry's Law coefficient.

Henry's Law is compatible with Equation 2, the general equation for liquids. As the mole fraction of a material approaches zero, the activity coefficient approaches

a constant γ_i^0.[11] Thus, for a material in a dilute solution, Equation 2 becomes

$$C_{G,i} = \gamma_i^0 x_i C_{G,i}^S \quad (11)$$

Converting the gas phase mass concentration to partial pressure P_i by the Ideal Gas Law

$$P_i = C_{G,i} \frac{RT}{M_i} \quad (12)$$

(R is the universal gas constant and T is the absolute temperature) gives for a dilute solution

$$P_i = \gamma_i^0 x_i C_{G,i}^S \frac{RT}{M_i} \quad (13)$$

The last result is in the form of Henry's Law. Comparing Equations 10 and 13 shows that the Henry's Law coefficient H_i is related to the activity coefficient for a material in a dilute solution γ_i^0 by

$$H_i = \gamma_i^0 C_{G,i}^S \frac{RT}{M_i} = \gamma_i^0 P_i^S \quad (14)$$

The saturated vapor pressure P_i^S has been related to the saturated vapor concentration C_{Gi}^S by the Ideal Gas Law, Equation 12, applied to the saturated state. Solving Equation 14 for γ_i^0 and substituting for γ_i in Equation 3 gives the relation for materials in droplets as dilute solutions in terms of the Henry's Law coefficient

$$C_{G,i} = \frac{H_i}{P_i^S} \frac{M_P}{M_i} \frac{C_{Pi}}{C_P} C_{Gi}^S \quad \text{dilute solutions} \quad (15)$$

The form has been chosen to be similar to that for ideal solutions and solvents.

Many liquid/gas systems, especially those consisting of water and an organic liquid, have been characterized in terms of Henry's Law coefficients either by direct measurement or by extrapolation from measurements in similar systems. All Henry's Law coefficients that have been tabulated have something in common: each is a proportionality constant between the amount in the gas phase and the amount in a dilute (usually aqueous) solution at equilibrium. Unfortunately, different authors define the proportionality constant differently, so Henry's Law coefficients have a variety of units and some Henry's Law coefficients are related to the inverse of another. For example, one list is limited to aqueous solutions and tabulates a Henry's Law coefficient that is related to the inverse of that defined here.[12] For that tabulation, the units of the quantity specifying the amount of the material in the dilute aqueous solution are moles per liter and the units of the quantity specifying the amount in the gas phase (partial pressure) are atmospheres, so the Henry's Law coefficient is tabulated with units of moles per liter per atmosphere. The relationship between that Henry's Law coefficient and the one defined in Equation 10 can be derived by relating the concentration of a contaminant in a dilute aqueous solution to its mole fraction. Great care must be taken in applying Henry's Law coefficients obtained from the literature because of the confusing number of definitions and units. The Henry's Law coefficient used in Equations 10, 14, and 15 has units of pressure.

Adsorption

Gas phase materials may adsorb onto the surface of solid particles.[13] Adsorption theories have been discussed in the environmental literature, but there seems to be no definitive compilation of experimental results. One problem has been the difficulty in obtaining valid data on the distribution of an airborne substance between the particle and gas phases due to the artifacts discussed later in this chapter. Some data have been summarized by a relation of the form

$$\log\left(\frac{C_{G,i} C_P}{C_{P,i}}\right) = \frac{m_Y}{T} + b_Y \tag{16}$$

where m_Y and b_Y are constants determined by fitting the data.[14] This relation has been shown to be consistent with the Junge equation which had been proposed earlier.[15] The Junge equation relates the fraction of the material of interest in the particle phase to the saturated vapor pressure of the material at that temperature, P_i^S; the surface concentration of particles, C_S (the total surface area of particles per volume of air); and the Junge constant, c_J, which is determined experimentally,

$$\frac{C_{P,i}}{C_{T,i}} = \frac{c_J C_S}{P_i^S + c_J C_S} \tag{17}$$

where the total airborne concentration of the contaminant $C_{T,i}$ is

$$C_{T,i} = C_{P,i} + C_{G,i} \tag{18}$$

Pankow[15] has derived the theoretical temperature dependence of the Junge constant assuming linear Langmuir adsorption and has shown that the resulting equation describing equilibrium conditions is equivalent

to Equation 16. Solving Equations 17 and 18 for $C_{G,i}$ results in the equilibrium relation

$$C_{G,i} = \frac{P_i^S C_{P,i}}{c_J C_S} \tag{19}$$

For materials with a melting point higher than the temperature of the atmosphere being sampled, it has often been found preferable to use the saturated vapor pressure of the liquid extrapolated down to the atmospheric temperature (called the subcooled liquid saturated vapor pressure) rather than the saturated vapor pressure above the pure crystalline solid.[16] Application of the Ideal Gas Law and straightforward algebraic manipulation lead to a form similar to Equations 3, 7, 8, and 15 for the equilibrium relation for adsorption

$$C_{G,i} = \frac{RT}{M_i c_J \sigma_p} \frac{C_{P,i}}{C_P} C_{G,i}^S \quad \text{adsorption} \tag{20}$$

where the specific surface area (area per mass) of the particle phase σ_P is

$$\sigma_p = \frac{C_S}{C_P} \tag{21}$$

Sorption

The association of gas phase material with particles by either absorption or adsorption is called sorption. It has been suggested that it may be adequate to describe the gas phase mass concentration of an air contaminant in equilibrium with either solid or liquid particles by a fairly simple relation:[17]

$$C_{G,i} = B_i \frac{C_{P,i}}{C_P} C_{G,i}^S \tag{22}$$

$C_{P,i}$ is the mass concentration of the ith contaminant in the particle phase. C_P is the mass concentration of all materials in the particle phase. $C_{G,i}^S$ is the saturated vapor concentration of the ith material or the extrapolated saturated vapor concentration for the subcooled liquid, if the melting point of the pure material is above the temperature of the atmosphere. The Kelvin correction may be needed for submicrometer particles. The proportionality "constant" B_i is temperature dependent and also depends on the materials involved. One utility of this approximation is that although B_i depends on a variety of characteristics of the particles and contaminant gases, it may be slowly varying. For example, it may be nearly constant for similar solid/gas and liquid/gas systems.[17]

Equation 22 emphasizes that the distribution of a material between the particle and gas phases depends strongly on its saturated vapor concentration and the amount of particulate available. One advantage of this equation for summarizing air sampling data on the distribution of a particular contaminant of interest between the particle and gas phases is that it asserts a relationship between the equilibrium concentration of the contaminant in the gas phase and the particle phase mass concentrations that might be measured in a field study, as well as the contaminant's saturated vapor concentration. Generally, B_i and its variability (including its temperature dependence) will be determined experimentally for the particular situation.

Formulating the particle/gas distribution relationship in terms of a new proportionality constant B_i does not mean that all the previous data on such systems must be discarded. Available data on adsorption of material onto solid particles and absorption into droplets can be converted to information about B_i. Comparing Equation 22 with Equations 3, 7, 8, 15, and 20 shows that the form of each is similar and gives relationships which can be expected to hold between the proportionality constant B_i in Equation 22 and other experimental quantities that are sometimes tabulated for solid/gas and liquid/gas systems

$$B = \gamma_i \frac{M_P}{M_i} \quad \text{droplets} \tag{23}$$

$$B = \frac{M_P}{M_i} \quad \text{ideal liquids} \tag{24}$$

$$B = \frac{M_P}{M_{solv}} \quad \text{solvents} \tag{25}$$

$$B = \frac{H_i}{P_i^S} \frac{M_P}{M_i} \quad \text{dilute solutions} \tag{26}$$

$$B = \frac{RT}{M_i c_J \sigma_p} \quad \text{adsorption} \tag{27}$$

Additional field studies will be needed to indicate whether the relatively simple form of Equation 22 is adequate to describe particle-gas equilibria in real systems. However, it seems to have a solid theoretical basis for both adsorption and absorption and is related to previously tabulated quantities, as shown in Equations 23–27.[17] The form of Equation 22 suggests quantities that may be important to measure when dealing with the distribution of materials between the particle and gas phases and indicates a relationship which should be tried when attempting to summarize experimental data.

In the special case of the sorption of semi-volatile organic compounds on atmospheric particles, observations suggest that an even better description of the equilibrium conditions than Equation 22 is

$$C_{G,i} = \text{fn}(K_{OA}) \, C_{P,i} \tag{28}$$

where K_{OA} is the octanol-air partition coefficient for the compound and fn is some function determined from experimental data.[18-20] The utility of this relation is that the function appears to be the same for all semi-volatile organic compounds in several classes. There is no experimental evidence that this relation will apply to most workplace aerosols, although it is plausible that it would apply in workplaces having a substantial organic component in the airborne particles.

Single Component System

Consider the case of a single component aerosol. This occurs when the airborne particles consist of only one constituent because none of the other constituents of the gas phase dissolve into or adsorb onto the particles. Calculation of the distribution of the contaminant between the gas and particle phases at equilibrium is straightforward. If the total airborne concentration, i.e., the concentration in the particle phase plus that in the gas phase, is larger than the saturated vapor concentration, the difference between the two is the mass concentration in the particle phase. The saturated vapor concentration is the concentration in the gas phase. Otherwise, if the total airborne mass concentration is less than the saturated vapor concentration, any particles evaporate completely and all the contaminant is in the gas phase. For small particles, the gas phase concentration that is in equilibrium with the pure droplet may be significantly higher than the concentration over a planar surface (the saturated vapor concentration), so a Kelvin effect correction may be needed.

Multiple Component System

In most cases, particles are not pure and two or more materials must be considered. For example, water vapor is ubiquitous on Earth and dissolves into or adsorbs onto most particles to some extent. Some general discussions of the equilibrium of simplified or actual airborne systems have been presented in the literature of environmental and occupational air sampling.[21-26] Calculation of the Kelvin effect is more complicated for a two-component droplet than for a single component.[27] The correction for the equilibrium vapor concentration above a curved surface is often small for droplet diameters larger than a micrometer and will not be explicitly included here.

Ignoring the Kelvin effect and assuming identical particles, there are four unknowns in describing the equilibrium of a two-component system: a particle phase and a gas phase mass concentration for each of the components. If the total mass concentrations of both components are specified and the two interdependent relations between the mass concentrations of the two components in the gas phase and the composition of the particles similar to Equations 1 or 22 are known, then the four equations can be solved for the four unknowns.[21] In principle, the equilibrium of a system consisting of more than two components can be predicted using analogous information for each component.

General Guidance

Consideration of two-component systems emphasizes the importance of four basic facts that also hold for more complex systems:

A. The mass concentration of a contaminant in the gas phase is less than or equal to its saturated vapor concentration at that temperature.

B. If the particle phase has significant excess capacity to adsorb or dissolve the contaminant, the mass concentration of the contaminant in the gas phase is significantly less than its saturated vapor concentration.

C. The mass concentration of a contaminant in the particle phase is the difference between the mass concentration in the gas phase and its total mass concentration in the atmosphere.

D. The mass concentration of each contaminant in the particle phase is less than or equal to the total mass concentration of all airborne particles.

Applying these somewhat elementary facts leads to three useful generalizations (rules-of-thumb):

1. Sample only the particle phase when the saturated vapor concentration of a substance $C^S_{G i}$ is much less than its total airborne mass concentration $C_{T,i}$

$$C^S_{G,i} << C_{T,i} \qquad (29)$$

2. Sample only the gas phase when the total mass concentration of all constituents in the particle phase C_P is much smaller than the total airborne mass concentration of a contaminant $C_{T,i}$

$$C_p << C_{T,i} \qquad (30)$$

3. Sample both the particle and gas phases when neither of the above two conditions is met.

When the saturated vapor concentration of a substance is of the same order as or is larger than its total airborne mass concentration, much of the mass might be expected to be in the gas phase. However, if the material occurs as a dilute solution in the particle phase or is strongly bound to particle components, a significant fraction of the mass may be in the particle phase. For example, both observations and calculations suggest that some pesticides and other organic compounds can have a significant fraction of the total airborne material in the particle phase at high humidities.[21,22,28]

The three rules-of-thumb can be applied once some information is known about the system under consideration. The most helpful pieces of information are the identities of the major constituents and their saturated vapor concentrations and some estimate of their total airborne mass concentrations. These ideas are consistent with previous publications that emphasized considering the ratio of the saturated vapor concentration to the exposure limit for the material under consideration when deciding whether to sample the particle, gas, or both phases of an atmosphere.[21, 22] This ratio has also been called the "vapor/hazard ratio number."[29]

Substances for which the American Conference of Governmental Industrial Hygienists (ACGIH®) had established Threshold Limit Values (TLVs®) were reviewed to provide examples of materials for which too little might be known to judge whether the particle, gas, or both phases should be sampled as well as examples of substances for which the available information might seem to suggest erroneously that only one phase should be sampled.[22]

Table 4-1 illustrates the results from applying the rules-of-thumb to a number of assumed atmospheres when deciding whether to sample the particle, gas, or both phases. Each atmosphere contains a contaminant with a known saturated vapor concentration of 0.001, 0.01, 1, or 100 mg/m³. For an assumed molecular weight of 100, the Ideal Gas Law gives the corresponding saturated vapor pressures: 3×10^{-6}, 3×10^{-4}, 3×10^{-2}, and 3 Pa (2×10^{-8}, 2×10^{-6}, 2×10^{-4}, and 2×10^{-2} mm Hg). The mass concentration of all airborne particles, C_P, is assumed to be either 0.05 mg/m³, which may be typical of the general outdoor environment and many workplaces, or 5 mg/m³, which may be typical of dusty workplaces or light fog. Table 4-1 illustrates the large range of conditions for which both the particle and gas phases should be sampled. Whenever it is not clear that only one phase should be sampled, both phases should be sampled.

The discussion in this section applies if an airborne system is in or near equilibrium. Sometimes a system is not near equilibrium. The following two sections consider

TABLE 4-1. Examples of Sampling Decisions in a Range of Assumed Atmospheres

Mass concentration of all airborne particulate, C_p (mg/m³)	Total airborne mass concentration of contaminant, $C_{T,i}$ (mg/m³)	Saturated vapor concentration of contaminant at 25°C, $C_{G,i}^s$ (mg/m3)			
		0.0001	0.01	1	100
0.05	0.0001	PG	PG	PG	PG
0.05	0.01	P	PG	PG	PG
0.05	1	*	*	G	G
0.05	100	*	*	*	G
5	0.0001	PG	PG	PG	PG
5	0.01	P	PG	PG	PG
5	1	P	P	PG	PG
5	100	*	*	*	G

P = Sample particulate phase only, because $C_{G,i}^s \ll C_{T,i}$

G = Sample gas phase only, because $C_P \ll C_{T,i}$

PG = Sample both phases because neither condition is met.

* = Unphysical, because no atmosphere can meet both conditions.

conditions that disturb the equilibrium of a system and the time a system needs to reach equilibrium.

Disturbance of Equilibrium

Some kinds of changes will disturb the equilibrium of an airborne system and require mass to be transferred between the phases to attain a new equilibrium consistent with the new conditions. For example, an atmosphere may not be in equilibrium immediately after a source injects a gaseous or particulate contaminant into it or just after it is pulled into a sampler. Detailed calculations of the system dynamics can be performed using similar experimental information to that needed to predict equilibrium (equilibrium concentrations across interfaces), plus information about transport through a phase, e.g., the diffusion of gas molecules through air. The equations are complicated by the transport of latent heat which occurs during condensation and evaporation. The formulation and results of detailed computational models are beyond the scope of this discussion.

Atmospheric Changes

Conditions that may disturb an airborne system's equilibrium include a change in temperature or atmospheric pressure, addition or removal of contaminant mass, or addition of clean air (dilution).[30] The direction in which each of these changes tends to move the equilibrium will be discussed briefly.

An increase in temperature tends to increase the fraction of a substance in the gas phase due to higher thermal energy promoting the ejection of molecules from surfaces. A decrease in temperature tends to decrease the fraction in the gas phase.

A decrease in atmospheric pressure with no change in temperature, for example, due to the slow expansion of a volume, decreases the gas phase concentration at a particle surface and tends to lead to evaporation of material from the particle. The decrease in the mass concentration of the particle phase due to the expansion does not influence the new equilibrium, because equilibrium depends most fundamentally on gas phase concentrations and particle phase composition, not particle phase concentration. Increases in atmospheric pressure have an opposite effect.

Sources or sinks of vapor disturb the equilibrium and may lead to significant condensation onto or evaporation from particles. For example, an increase in humidity is likely to lead to increased water in the particle phase. The loss of particles, for example, by sedimentation, or the addition of particles which have the same composition as the particles that are already equilibriated with the atmosphere does not disturb the equilibrium of the system. However, the addition of particles of a different composition will disturb the equilibrium in general.

Dilution with clean air decreases the vapor concentration at the particle surface and tends to promote evaporation of volatile contaminants from the particles.

Sampling

The process of sampling an atmosphere has the potential to disturb the particle-gas equilibrium in the sampled air. This can lead to biased sampling results. Characteristics of four types of sampling instruments, filter samplers, cascade impactors, denuders, and electronic monitors, will be discussed briefly. Some effects of inhaling an aerosol will also be mentioned.

After a filter sampler collects particles (see Chapter 13), they remain on the filter surface and are exposed to the air that is sampled subsequently. Even if there are no chemical changes due to reactive gases, physical changes in the particles can occur. During extended sampling, the atmosphere is likely to change in temperature or in the concentration of gas phase species, including water vapor. Such changes will require volatile components to evaporate from or absorb/adsorb onto the particles collected on the filter, possibly leading to substantial decreases or increases in the overall mass or the mass of volatile components.[31] For example, ammonium nitrate evaporates from filters, if the air sampled subsequently is relatively clean.[32, 33] Much less evaporation into clean air occurs from ammonium nitrate collected in an impactor (see Chapter 14) than in a filter, apparently because of the relatively thick boundary layer of air and the decreased surface area of the collected material.[34] Another situation that can occur with extended sampling times is clogging of the filter with particulate. The decreased pressure experienced by some of the collected particulate may lead to evaporation of some materials.

Another type of error can occur in filter sampling. Some filter materials adsorb some vapors from the air stream.[35–38] This contributes to an increase in mass and may allow the materials to be extracted from the filter for chemical analysis. Such errors lead to an overestimate of the mass of material in the particle phase. This problem can be reduced by choosing a filter material that is not prone to adsorbing the types of vapors which are in the atmosphere to be sampled. Filter adsorption can be detected in the field by placing a second filter downstream of the first and checking for an increase in mass, assuming the first filter is not such an efficient adsorber that all the vapor is scrubbed from the air stream. Adsorption on the back-up filter may not parallel that on the front filter, so correction of the mass increase on the front filter by that on the back-up filter may not be accurate.[35] Inefficient particle collection by the first filter would confound detection of vapor adsorption by this method, because it would also lead to an increase of mass of the second filter.

In a cascade impactor (see Chapter 14), the sampled air is drawn through a nozzle to accelerate the particles, so the larger ones will impact on a collection surface and be separated from the smaller particles and the air stream. Then the process is repeated with smaller nozzles that achieve higher air velocities and allow smaller

particles to be collected on other surfaces. During its passage through each nozzle, the air experiences a pressure drop. The decrease in pressure tends to promote evaporation of volatile components from particles.[39] In smaller nozzles and higher speed jets, the air may be cooled somewhat. This tends to promote condensation of volatile materials onto particles.[40] Because the two effects act in opposite directions, predicting the net effect of passage through a cascade impactor depends on the details of the situation, but the potential for significant particle-gas interactions should not be overlooked.

If a sampler incorporates a denuder as the first stage to remove reactive gases (see Chapter 19), any particle components that are in equilibrium with that gas will tend to adjust to the removal of the gas from the sampled air stream.[6] Some have suggested that measuring the distribution of a material between the particle and gas phases of an atmosphere could be accomplished using a denuder to collect material from the gas phase followed by a filter to collect particulate. One difficulty with this approach is that there will be a tendency for the particles to evaporate in the denuder. This situation will be considered in more detail later in this chapter.

Electronic sampling instruments, e.g., real-time monitors, may have high speed inlets and contain pumps, motors, and lights that generate heat. Drawing air into warm instruments will promote evaporation of volatile components from particles; decreasing the pressure will also promote evaporation; and cooling in the jet will promote condensation.[40-43]

Inhaling air typically warms and humidifies it. Warming the air tends to cause volatile materials to evaporate from particles. High humidity tends to cause water-soluble particles to absorb water vapor. This typically dilutes other constituents in the particles and causes other volatile components to be absorbed also.[10] However, more complex interactions can occur. For example, water might drive off less polar compounds adsorbed on some materials. If the particles consist entirely of substances that are immiscible with water, high humidity may have little effect. It is difficult to generalize about the effects of inhaling an atmosphere on the distribution of material between the gas and particle phases. It is clear that the distribution of a material between the particle and gas phases that is measured in the atmosphere may not be the same distribution that occurs in the respiratory tract.

Time-scale of Evaporation and Condensation

When considering the effects of evaporation and condensation on sampling results, it is often useful to know the time-scale, an estimate of the time required for a significant fraction of the evaporation or condensation to occur as the system approaches its new equilibrium. This can be helpful in judging whether a recently generated system, e.g., one consisting of droplets that were ejected into relatively clean air, has had sufficient time to approach equilibrium. One approach for estimating the time-scale for evaporation of a contaminant from the particle phase is to divide the decrease in contaminant mass in the particle phase by the initial evaporation rate.[10] The resulting time-scale for evaporation τ_e is

$$\tau_e = \frac{C_{P,i}^0 - C_{P,i}^f}{C_P} \frac{\rho_p d_p^2}{12 D_i C_{G,i}^S \left(B_i \dfrac{C_{P,i}^0}{C_P} - S_i \right)} \quad \textbf{(31)}$$

The initial and final mass concentrations of the contaminant in the particle phase are $C^o{}_{Pi}$ and $C^f{}_{Pi}$, respectively. The particle density is ρ_p and the diffusion coefficient of the vapor in air is D_i. The saturation S_i is the gas phase concentration of the contaminant (far from a particle surface) divided by the saturated vapor concentration. It has been assumed that the mass concentration of vapor at the particle surface can be described successfully by Equation 22. Equation 31 neglects the slower evaporation that occurs due to the latent heat of the evaporating material cooling the particles as well as the Kelvin effect and the correction for gas transport from small particles.[10]

Substituting plausible values for the density (1 g/cm³) and the diffusion coefficient (0.1 cm²/s) and applying the equation to the special case of complete evaporation of the contaminant ($C^f{}_{Pi} = 0$) into air that is substantially depleted of vapor (S_i negligible) results in the order-of-magnitude estimate:

$$\tau_e \approx 8\,\text{sec}\,\frac{\left(\dfrac{d_p}{1\mu m}\right)^2}{\left(\dfrac{B_i\ C_{G,i}^S}{1\,\text{mg/m}^3}\right)} \quad \textbf{(32)}$$

If the value of B_i is unknown, a first estimate of the time-scale of evaporation can be obtained by assuming B_i is unity. For pure droplets ($B_i = 1$) near room temperature, the time-scale for evaporation of a 1-μm water droplet $C^S{}_{G,i} = 17,000$ mg/m³) into dry air is about a half millisecond; that of a 10-μm glycerol droplet ($C^S{}_{G,i} = 0.5$ mg/m³) is about 27 min. Equation 32 also provides a reasonable estimate for the time-scale of growth by condensation for a particle that contains no contaminant initially and grows into one with diameter d_p.[10]

Sampling Approaches

There has been little discussion of the issue in the literature, but it could be argued that in most health-related air sampling, the quantity that is the most important to measure accurately is the total airborne mass concentration without regard to whether the material is in the gas or particle phase. Volatile materials may move between the gas and particle phases because of a change in conditions, including after inhalation, but the total airborne mass concentration is unchanged.[10] This makes it a practical first choice for measurement.

In some cases, it may be desirable to measure the distribution of a compound between the two phases instead of only the total airborne mass concentration. For example, there may be compounds for which the toxic effects are thought to differ significantly depending on whether the material existed in the particle or gas phase just prior to inhalation.[44] Also, the choice of control technology may depend on which phase dominates. Both sampling the total airborne concentration and sampling the distribution between the two phases are discussed in the following sections.

Total Airborne Concentration

Measuring the total airborne concentration requires an efficient particle collector and an efficient vapor collector. Two approaches have been found suitable. One is to place the vapor collector, such as a tube containing an efficient sorbent, downstream of an efficient particle collector, such as a filter.[44-51] One problem with the reverse configuration, placing the vapor collector upstream of the particle collector, is that material might evaporate from the particles which are collected on the filter and be lost. The second approach that has been used successfully is to apply a coating of a material to the filter that either adsorbs or reacts chemically with the air contaminant being sampled.[52, 53] A variation of this approach is to preload the filter with a particulate sorbent.[54] If done properly, the approach of using a coated or preloaded filter results in a compact and convenient sampler with all the collected material on one substrate for analysis.

Other approaches might provide satisfactory results, but care must be taken to ensure that the collection efficiency is sufficiently high for both phases in each sampling situation. Impingers are versatile in allowing the liquid to be changed to match the compound being collected, but the collection efficiency for submicrometer particles is low.[55] Impingers collect solid isocyanates more effectively than coated filters, since the required derivatization occurs more quickly.[56]

The particle collection efficiency of some sorbent beds may be sufficient to allow measurements of the total airborne concentration of some compounds.[49,57,58] These devices should not be used without experimental validation of their suitability for the specific application.

Techniques for measuring the total airborne concentration have been much less widely discussed in the literature of workplace air sampling than in the literature of air sampling in the general environment. However, such techniques have been proposed for organophosphorus pesticides, fluoride, formaldehyde, and isocyanates.[5, 59–64]

Particle/Gas Distribution

In cases where high precision is not required and some experimental phase distribution data are available to validate a relation like Equation 22, it would be possible to estimate the distribution between the particle and gas phases after measuring only the total airborne concentration of the contaminant and the concentration of all airborne particulate. In order to use a relation like Equation 22, it would be necessary to assume that the atmosphere is near equilibrium, i.e., sufficient time has passed since the atmosphere's equilibrium was last disturbed, as discussed in the previous section. In most cases, reliable information about the distribution of a material between the particle and gas phases of an atmosphere can only be obtained by measurements in that or similar atmospheres.

There does not appear to be any universally accepted method for measuring the mass concentrations of an air contaminant in the particle and gas phases separately. Many approaches have been described in the literature, but questions arise about the accuracy of the results. Three approaches are presented here that seem suitable for a range of sampling conditions, but each has characteristics that might lead to erroneous results in some sampling situations. Because of the ease with which the particle/gas equilibrium is disturbed and mass is transferred between the two phases, measurement of the distribution of an air contaminant between the particle and gas phases is much more difficult than measurement of the total airborne mass concentration.

A sampling system consisting of an efficient particle collector followed by an efficient vapor collector was one of the two configurations recommended for measuring the total airborne mass concentration of a contaminant. It has also been used in attempts to measure the distribution of a contaminant between the gas and particle phases. The vapor collector must be removed from the particle collector at the end of the sampling period and stored separately to avoid evaporation of the volatile contaminant from the particles and transport to the vapor collector.

The most common configuration, called a "filter pack," consists of a filter as the particle collector and a sorbent as the vapor collector. The particle phase concentration is calculated from the mass collected on the filter and the gas phase concentration from that collected in the sorbent (Figure 4-2).[16, 65] A filter pack might be suitable for measuring the average concentration in each phase over the sampling period if two conditions are met: 1) the filter does not adsorb the vapor and 2) the chemical composition and physical characteristics of the sampled atmosphere are constant during the entire sampling period.[66] If the filter adsorbs vapor, the particle phase mass concentration will be overestimated and the gas phase underestimated.[35–38, 67] As mentioned previously, adsorption onto a filter might be detected by placing a second filter downstream of the first, but correcting for such adsorption is problematic.[35, 68] If the sampled atmosphere does not stay constant, mass collected from the particle phase onto the filter will tend to evaporate and be transferred into the vapor collector whenever the mass concentration of the contaminant in the gas phase decreases. Also, mass in the gas phase of the air being sampled will be transferred to collected particles on the filter whenever the mass concentration in the gas phase increases. As a result, filter packs may not be valid for health-related sampling from the occupational or general environment, although they may be valid for sampling from a controlled atmosphere, e.g., inhalation exposure chambers or a process line.

Replacing the filter by a cascade impactor appears to reduce the interaction between particles that have deposited on the collection surfaces and the air drawn

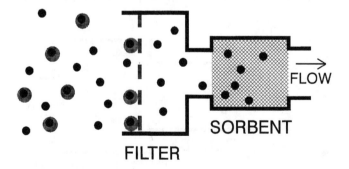

FIGURE 4-2. Schematic diagram of a common approach to measuring the distribution of a contaminant between the particle and gas phases of an atmosphere. Particles (large gray circles) with associated molecules of a contaminant in the gas phase (small black circles) are collected on the filter. Vapor penetrating the filter is collected in the sorbent. Particle and gas phase concentrations are calculated from the masses collected on the filter and in the sorbent, respectively. Potential errors include vapor adsorption by the filter material and transfer of contaminants between the gas phase and particles already collected on the filter if the atmosphere changes temperature, humidity, or composition.

through the device subsequently, reducing the severity of some of the artifacts.[69] The problem remains that the air which enters the vapor collector has passed over all the previously collected particles and significant mass transfer between the gas and collected particles may have occurred.

Replacing the filter by a virtual impactor reduces this problem, because the larger particles are not collected on a surface over which air subsequently passes on the way to the vapor collector.[70-72]

A second approach uses a denuder to remove vapor from the sampled air stream (see Chapter 19), followed by an efficient collector of both particles and vapor (Figure 4-3). The particle and vapor collector may be a treated filter or a filter followed by an adsorbent. The gas phase concentration is calculated from the mass collected in the denuder and the particle phase concentration from that collected in the particle and vapor collector.[73–79]

If the material collected in the denuder cannot be removed quantitatively for analysis, the "denuder difference method" may be used.[80] Two samplers are operated in parallel. One is a particle and vapor collector and the other consists of a denuder to collect the material from the gas phase followed by a particle and vapor collector. The difference between the masses of material collected in the two particle and vapor collectors is taken to be the mass in the gas phase.

In another variation, the vapor may be transported preferentially into a parallel air stream because it has a higher diffusion coefficient than the particles.[81]

These denuder-related approaches might be expected to give valid results if: 1) there is insignificant particle deposition in the denuder;[82, 83] 2) the denuder collects all or a known fraction of the contaminant mass which was in the gas phase when the air entered the sampler without breakthrough;[84] and 3) there is insignificant particle evaporation and subsequent transfer of that contaminant mass to the denuder walls. Item 2 is easier to attain using denuders with narrow channels because the small distance between the bulk of the sampled air and the coated walls allows efficient collection of vapor at higher flow rates. The third item is the most troublesome and may be impossible to attain for some materials. During the time the particles reside in an atmosphere that has been depleted of vapor, some material will evaporate from the particles. In order to have this mass transfer be negligible, the amount that evaporates must be a negligible fraction of the amount in the particle phase and must be a negligible fraction of the amount that was in the gas phase before the aerosol entered the denuder.

Consideration of a simple case, particles with a single volatile component and the vapor in equilibrium with them, is instructive. The sampled air's residence time in the denuder τ_r can be estimated as the denuder's volume divided by the sampling flow rate. The characteristic time for complete evaporation/desorption into vapor-depleted air is given in Equation 32. Only a negligibly small fraction of the volatile component's mass in the particles will evaporate during the transit time if

$$\tau_r << \tau_e = 8\,\text{sec}\,\frac{\left(\dfrac{d_p}{1\mu m}\right)^2}{\left(\dfrac{B_i\,C_{G,i}^S}{1\,\text{mg/m}^3}\right)} \tag{33}$$

Another condition is necessary to ensure that the mass of contaminant that does evaporate from the particles is negligibly small compared to the mass of material which was initially in the gas phase. Otherwise, the mass of contaminant collected at the denuder walls will be in error. Recalling that the characteristic time of evaporation is defined here as the volatile contaminant's mass in the particle phase divided by the initial evaporation rate, an order of magnitude estimate of the mass of material that evaporates from the particles per volume of air can be written as the particle mass divided by the characteristic time of evaporation times the residence time. A relationship sufficient to ensure that the amount of evaporated contaminant is negligibly small compared to the amount originally in the gas phase, expressed in terms of saturation, is:

$$\frac{C_{P,i}}{\tau_e}\tau_r << S_i C_{G,i}^S \tag{34}$$

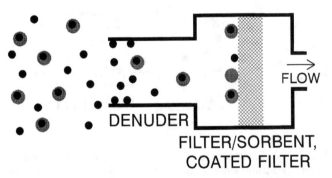

FIGURE 4-3. Schematic diagram of a denuder-based approach to measuring the distribution of a contaminant between the particle and gas phases of an atmosphere. Vapor molecules (small black circles) are collected at the wall of the denuder. Particles (large gray circles) with associated volatile contaminants are collected by a filter that is either treated or followed by a suitable sorbent to trap vapor molecules. Potential errors include particle deposition in the denuder and significant evaporation of the particles while surrounded by vapor-depleted air.

Substituting for τ_e using Equation 32 and substituting $C_{G,i}$ using Equation 22 for $S_i\,C^S_{G,i}$ under the assumption that the particles were in equilibrium with the vapor before entering the denuder gives:

$$\tau_{r\,<<}\quad 8\,\sec\;\frac{\left(\dfrac{d_p}{1\mu m}\right)^2}{\left(\dfrac{C_p}{1\,mg/m^3}\right)} \qquad (35)$$

The residence time cannot be made arbitrarily small because all or a known fraction of the vapor molecules must have time to diffuse to the wall of the denuder and be collected (see Chapter 19). Evaporation of particles in a denuder will lead to negligible errors if both the contaminant's saturated vapor concentration and the concentration of all materials in the particle phase are sufficiently small according to Equations 33 and 35, given the assumptions stated in their derivation. Negligible evaporation of particles in a denuder is more likely for outdoor environmental sampling than for sampling in dusty/misty workplaces where C_P is likely to be higher. These generalizations should hold even for atmospheres containing several volatile and nonvolatile components.

A third approach to measuring the distribution of a contaminant between the gas and particle phases has been described.[85–87] The approach is to measure the total concentration in both phases by methods outlined already and place a suitable passive sampler nearby to measure the vapor concentration (Figure 4-4). The concentration in the particle phase would be estimated by subtraction. If only a small fraction of the total airborne contaminant were in the particle phase, this estimate might be quite imprecise because it is determined by subtracting two measurements. Another source of significant error might be the deposition of particles on the front surface of the diffusive sampler. If such particles were allowed to remain in that position sufficiently long and the surrounding vapor concentration fluctuated, material in the particles might evaporate during periods of lower vapor concentration and erroneously contribute a significant amount to the mass collected on the sorbent of the diffusion sampler. This would lead to an overestimate of the fraction of contaminant in the gas phase.

Summary and Conclusions

Particle and gas phase interactions can lead to errors in air sampling results. Some errors include allowing unwanted chemical reactions to occur during sampling,

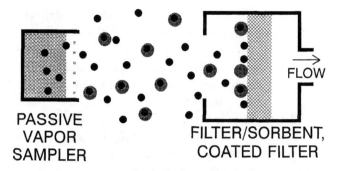

FIGURE 4-4. Schematic diagram of an approach to measuring the distribution of a contaminant between the particle and gas phases of an atmosphere. Vapor molecules (small black circles) and particles (large gray circles) are collected in an active sampler containing a treated filter or filter followed by a sorbent to measure the total airborne concentration of the contaminant. Vapor only is collected in the passive sampler. The particle concentration is estimated by subtraction. Potential errors include deposition of particles on the passive sampler and high uncertainty in estimating the particle concentration when it is much smaller than the vapor concentration.

measuring the concentration of a contaminant only in the particle or gas phase when there is a significant fraction in the unsampled phase, and using inappropriate techniques to measure the distribution of a contaminant between the two phases.

Errors due to chemical reactions between a reactive gas and particles that have been collected on a filter can be reduced by passing the air through a suitable denuder before it reaches the filter.

It is important to understand the distribution of an air contaminant between the particle and gas phases and the potential for changes in that distribution during sampling. Three pieces of information are helpful in judging whether it is necessary to sample the gas, particle, or both phases of an atmosphere: the saturated vapor concentration of the contaminant, a rough estimate of the total airborne concentration of the contaminant in the atmosphere being sampled, and a rough estimate of the concentration and composition of the particles in the atmosphere. The distribution of a contaminant between the particle and gas phases depends on temperature, pressure, particle composition, and the gas phase concentration of the contaminant. The processes of sampling or inhalation may change the distribution between the phases significantly.

It is arguably more important to measure the total airborne concentration of a contaminant, i.e., the concentration in the particle phase plus that in the gas phase, than to measure the distribution between the phases. The total airborne concentration can be measured by a sampler consisting of an efficient filter followed by an efficient sorbent or can be measured by a filter that is treated with a material which adsorbs or reacts with the vapor.

When it is necessary to measure the distribution of a contaminant between the particle and gas phases, no universally acceptable technique appears to be available. The common approach of placing a sorbent downstream of a filter is prone to errors if the filter adsorbs vapor or if the atmosphere's composition changes. Replacing the filter with a virtual impactor reduces many of these errors. The approach of placing a denuder capable of collecting the vapor upstream of a collector of the particles and vapor that penetrate the denuder is prone to errors if the denuder is not optimized for the application. In some cases, it may be difficult to avoid significant errors due to evaporation of the contaminant from the particles while they reside in vapor–depleted air in the denuder. A third approach to measuring the distribution of a contaminant between the gas and particle phases is available. The total airborne concentration is measured by an active sampler and the vapor concentration is measured by a passive sampler. This approach has not been widely used, yet. Other approaches are likely to be suggested in the future, but measuring the distribution will always be much more difficult than measuring the total airborne concentration.

References

1. Pio, C.A.; Nunes, T.V.; Leal, R.M.: Kinetic and Thermodynamic Behavior of Volatile Ammonium-Compounds In Industrial and Marine Atmospheres. Atmos. Environ. 26A:505–512 (1992).
2. Pandis, S.N.; Wexler, A.S.; Seinfeld, J.H.: Dynamics of Tropospheric Aerosols. J. Phys. Chem. 99:9646–9659 (1995).
3. Pankow, J.F.; Mader, B.T.; Isabelle, L.M.; et al.: Conversion of Nicotine in Tobacco Smoke to its Volatile and Available Free-Base Form Through the Action of Gaseous Ammonia. Environ. Sci. Technol. 31:2428–2433 (1997).
4. Bowman, F.M.; Odum, J.R.; Seinfeld, J.H.; Pandis, S.N.: Mathematical Model for Gas-Particle Partitioning of Secondary Organic Aerosols. Atmos. Environ. 31:3921–3931 (1997).
5. European Standardization Committee (CEN): Workplace Atmospheres—Measurement of Chemical Agents Present as Mixtures of Airborne Particles and Vapour—Requirements and Test Method. Document CEN/TC137/WG3/N217. CEN, Brussels (1998).
6. Pratsinis, S.E.; Xu, M.; Biswas, P.; Willeke, K.: Theory for Aerosol Sampling through Annular Diffusion Denuders. J. Aerosol Sci. 20:1597–1600 (1989).
7. Lu, C.Y.; Bai, H.L.; Lin, Y.M.: A Model for Predicting Performance of an Annular Denuder System. J. Aerosol Sci. 26:1117–1129 (1995).
8. Perrino, C.; DeSantis, F.; Febo, A.: Criteria for the Choice of a Denuder Sampling Technique Devoted to the Measurement of Atmospheric Nitric and Nitrous Acids. Atmos. Environ. 24A:617–626 (1990).
9. Hinds, W.C.: Aerosol Technology: Properties, Behavior, and Measurement of Airborne Particles, 2nd ed., John Wiley & Sons, New York (1999).
10. Soderholm, S.C.; Ferron, G.A.: Estimating Effects of Evaporation and Condensation on Volatile Aerosols During Inhalation Exposures. J. Aerosol Sci. 23:257–277 (1992).
11. Reid, R.C.; Prausnitz, J.M.; Poling, B.E.: The Properties of Gases & Liquids, 4th ed., Chapter 8. McGraw–Hill, New York (1987).
12. Betterton, E.A.: Henry's Law Constants of Soluble and Moderately Soluble Organic Gases: Effects on Aqueous Phase Chemistry, Chapter 1, Gaseous Pollutants: Characterization and Cycling, pp. 1–50. J.O. Nriagu, Ed. John Wiley & Sons, New York (1992).
13. Gordieyeff, V.A.: Adsorption of Gases and Vapors on Aerosol Particulates. American Industrial Hygiene Association Quarterly. 17:411–417 (1956).
14. Yamasaki, H.; Kuwata, K.; Miyamoto, H.: Effects of Ambient Temperature on Aspects of Airborne Polycyclic Aromatic Hydrocarbons. Environ. Sci. Technol. 16:189–194 (1982).
15. Pankow, J.F.: Review and Comparative Analysis of the Theories on Partitioning Between the Gas and Aerosol Particulate Phases in the Atmosphere. Atmos. Environ. 21:2275–2283 (1987).
16. Bidleman, T.F.: Atmospheric Processes: Wet and Dry Deposition of Organic Compounds are Controlled by their Vapor-Particle Partitioning. Environ. Sci. Technol. 22:361–367 (1988).
17. Pankow, J.F.: An Absorption Model of Gas/Particulate Partitioning in the Atmosphere. Atmos. Environ. 28:185–188 (1994).
18. Finizio, A.; Mackay, D.; Bidleman, T.; Harner, T.: Octanol-Air Partition Coefficient as a Predictor of Partitioning of Semi-Volatile Organic Chemicals to Aerosols. Atmos. Environ. 31:2289–2296 (1997).
19. Pankow, J.F.: Further Discussion of the Octanol/Air Partition Coefficient K-oa as a Correlating Parameter for Gas/Particle Partitioning Coefficients. Atmos. Environ. 32:1493–1497 (1998).
20. Harner, T.; Bidleman, T.F.: Octanol-Air Partition Coefficient for Describing Particle/Gas Partitioning of Aromatic Compounds in Urban Air. Environ. Sci. Technol. 32:1494–1502 (1998).
21. Soderholm, S.C.: Aerosol Instabilities. Appl. Ind. Hyg. 3:35–40 (1988).
22. Perez, C.; Soderholm, S.C.: Some Chemicals Requiring Special Consideration when Deciding Whether to Sample the Particle, Vapor, or Both Phases of an Atmosphere. Appl. Occup. Environ. Hyg. 6:859–864 (1991).
23. Pankow, J.F.; Storey, J.M.E.; Yamasaki, H.: Effects of Relative-Humidity On Gas-Particle Partitioning of Semivolatile Organic-Compounds to Urban Particulate Matter. Environ. Sci. Technol. 27:2220–2226 (1993).
24. Saxena, P.; Hildemann, L.M.; McMurry, P.H.; Seinfeld, J.H.: Organics Alter Hygroscopic Behavior of Atmospheric Particles. J. Geophys. Res. 100D:18755–18770 (1995).
25. Latimer, H.K.; Kamens, R.M.; Chandra, G.: The Atmospheric Partitioning of Decamethylcyclopentasiloxane (D5) and 1-Hydroxynonamethylcyclopentasiloxane (D4TOH) on Different Types of Atmospheric Particles. Chemosphere. 36:2401–2414 (1998).
26. Jang, M.; Kamens, R.M.: A Thermodynamic Approach for Modeling Partitioning of Semivolatile Organic Compounds on Atmospheric Particulate Matter: Humidity Effects. Environ. Sci. Technol. 32:1237–1243 (1998).
27. Nair, P.V.N.; Vohra, K.G.: Growth of Aqueous Sulphuric Acid Droplets as a Function of Relative Humidity. J. Aerosol Sci. 6:265–271 (1975).
28. Gotfelty, D.E.; Seiber, J.N.; Liljedahl, L.A.: Pesticides in Fog. Nature 325:602–605 (1987).
29. McFee, D.R.; Zavon, P.: Solvents. In: Fundamentals of Industrial Hygiene, 3rd ed., p. 103. B.A. Plog, G.S. Benjamin, and M.A. Kerwin, Eds. National Safety Council, Chicago (1988).
30. Kamens, R.; Odum, J.: Fan, Z.H.: Some Observations on Times to Equilibrium for Semivolatile Polycyclic Aromatic-Hydrocarbons. Environ. Sci. Technol. 29:43–50 (1995).

31. Raynor, P.C.; Leith, D.: Evaporation of Accumulated Multicomponent Liquids from Fibrous Filters. Ann. Occup. Hyg. 43:181–192 (1999).

32. Cheng, Y.H.; Tsai, C.J.: Evaporation Loss of Ammonium Nitrate Particles During Filter Sampling. J. Aerosol Sci. 28:1553–1567 (1997).

33. Hering, S.; Cass, G.: The Magnitude of Bias in the Measurement of $PM_{2.5}$ Arising from Volatilization of Particulate Nitrate from Teflon Filters. J. Air & Waste Manage. Assoc. 49:725–733 (1999).

34. Zhang, X.Q.; McMurry, P.H.: Evaporative Losses of Fine Particulate Nitrates During Sampling. Atmos. Environ. 26A:3305–3312 (1992).

35. Cotham, W.E.; Bidleman, T.F.: Laboratory Investigations of the Partitioning of Organochlorine Compounds between the Gas Phase and Atmospheric Aerosols on Glass Fiber Filters. Environ. Sci. Technol. 26:469–478 (1992).

36. McDow, S.R.; Huntzicker, J.J.: Vapor Adsorption Artifact in the Sampling of Organic Aerosol. in: Sampling and Analysis of Airborne Pollutants Chap. 12, pp. 191–208. E.D. Winegar and L.H. Keith, Eds. Lewis Publishers, Inc., Boca Raton, FL (1993).

37. Lofroth, G.: Phase Distribution of Nicotine in Real Environments as Determined by 2 Sampling Methods. Environ. Sci. Technol. 29:975–978 (1995).

38. Batterman, S.; Osak, I.; Gelman, C.: SO_2 Sorption Characteristics of Air Sampling Filter Media Using a New Laboratory Test. Atmos. Environ. 31:1041–1047 (1997).

39. Zhang, X.Q.; McMurry, P.H.: Theoretical Analysis of Evaporative Losses from Impactor and Filter Deposits. Atmos. Environ. 21:1779–1789 (1987).

40. Biswas, P.; Jones, C.L.; Flagan, R.C.: Distortion of Size Distributions by Condensation and Evaporation in Aerosol Instruments. Aerosol Sci. Tech. 7:231–246 (1987).

41. Mallina, R.V.; Wexler, A.S.; Johnston, M.V.: Particle Growth in High-Speed Particle-Beam Inlets. J. Aerosol Sci. 28:223–238 (1997).

42. Bergin, M.H.; Ogren, J.A.; Schwartz, S.E.; McInnes, L.M.: Evaporation of Ammonium Nitrate Aerosol in a Heated Nephelometer: Implications for Field Measurements. Environ. Sci. Technol. 31:2878–2883 (1997).

43. Finlay, W.H.; Stapleton, K.W.: Undersizing of Droplets from a Vented Nebulizer Caused by Aerosol Heating During Transit Through an Anderson Impactor. J. Aerosol Sci. 30:105–109 (1999).

44. Soderholm, S.C.; Anderson, D.A.; Utell, M.J.; Ferron, G.A.: Method of Measuring the Total Deposition Efficiency of Volatile Aerosols in Humans. J. Aerosol Sci. 22:917–926 (1991).

45. Hill, Jr., R.H.; Arnold, J.E.: A Personal Air Sampler for Pesticides. Arch. Environ. Contam. Toxicol. 8:621–628 (1979).

46. König, J.; Funke, W.; Balfanz, E.: Testing a High Volume Air Sampler for Quantitative Collection of Polycyclic Aromatic Hydrocarbons. Atmos. Environ. 14:609–613 (1980).

47. Kirton, P.J.; Ellis, J.; Crisp, P.T.: Investigation of Adsorbents for Sampling Compounds Found in Coke Oven Emissions. Fuel 70:4–8 (1991).

48. Hawthorne, S.B.; Miller, D.J.; Langenfeld, J.J.; Krieger, M.S.: PM-10 High-Volume Collection and Quantitation of Semivolatile and Nonvolatile Phenols, Methoxylated Phenols, Alkanes, and Polycyclic Aromatic-Hydrocarbons From Winter Urban Air and Their Relationship to Wood Smoke Emissions. Environ. Sci. Technol. 26:2251–2262 (1992).

49. Sturaro, A.; Parvoli, G.; Doretti, L.; et al.: Determination of 2-4 Condensed Ring Aromatic-Hydrocarbons in Air Using 2 Specific Sampling Methods. Annali Di Chimica. 86:319–328 (1996).

50. McConnell, L.L.; Bidleman, T.F.: Collection of Two-Ring Aromatic Hydrocarbons, Chlorinated Phenols, Guaiacols, and Benzenes from Ambient Air Using Polyurethane Foam Tenax-GC Cartridges. Chemosphere. 37:885–898 (1998).

51. Scobbie, E.; Dabill, D.W.; Groves, J.A.: The Development of an Improved Method for the Determination of Coal Tar Pitch Volatiles (CTPV) in Air. Ann. Occup. Hyg. 42:45–59 (1998).

52. Levin, J.-O.; Fängmark, I.: High-Performance Liquid Chromatographic Determination of Hexamethylene-tramine in Air. Analyst 113:511–513 (1988).

53. Wang, L.: Air Sampling Methods for Diisocyanates: Dynamic Evaluation of SUPELCO ORBO(TM)-80 Coated Filters. Am. Ind. Hyg. Assoc. J. 59:490–494 (1998).

54. Markell, C.; Hagen, D.F.; Bunnelle, V.A.: New Technologies in Solid-Phase Extraction. LC•GC 9:332–337 (1991).

55. Spanne, M.; Grzybowski, P.; Bohgard, M.: Collection Efficiency for Submicron Particles of a Commonly Used Impinger. Am. Ind. Hyg. Assoc. J. 60:540–544 (1999).

56. Key-Schwartz, R.J.; Tucker, S.P.: An Approach to Area Sampling and Analysis for Total Isocyanates in Workplace Air. Am. Ind. Hyg. Assoc. J. 60:200–207 (1999).

57. Kogan, V.; Kuhlman, M.R.; Coutant, R.W.; Lewis, R.G.: Aerosol Filtration by Sorbent Beds. J. Air Waste Manag. Assoc. 43:1367–1373 (1993).

58. Brouwer, D.H.; Ravensberg, J.C.; de Kort, W.L.A.M.; van Hemmen, J.J.: A Personal Sampler for Inhalable Mixed-Phase Aerosols: Modification to an Existing Sampler and Validation Test with Three Pesticides. Chemosphere. 28:1135–1146 (1994).

59. Occupational Safety and Health Administration (OSHA): Diisocyanates: Method 42. OSHA Methods Manual. U.S. Dept. of Labor, OSHA, Salt Lake City, UT (1994).

60. American Society for Testing and Materials (ASTM): D5836-95 Standard Test Method for Determination of 2,4-Toluene Diisocyanate (2,4-TDI) and 2,6-Toluene Diisocyanate (2,6-TDI) in Workplace Atmospheres (1-2 PP Method). American Society for Testing and Materials, West Conshohocken, PA (1995).

61. Occupational Safety and Health Administration (OSHA): Methylene Bisphenyl Isocyanate (MDI): Method 47. In: OSHA Methods Manual. U.S. Dept. Of Labor, OSHA, Salt Lake City, UT (1985).

62. National Institute for Occupational Safety and Health (NIOSH): Determination of Airborne Isocyanate Exposure. In: NIOSH Manual of Analytical Methods, 4th ed., 2nd supplement. M.E. Cassinelli and P.F. O'Connor, eds. DHHS (NIOSH) Publication No. 98-119. U.S. Dept. Health and Human Services, Public Health Service, Centers for Disease Control and Prevention, NIOSH, Cincinnati, OH (1998).

63. National Institute for Occupational Safety and Health (NIOSH): Isocyantes: Method 5522 (supplement issued 5/15/96). In: NIOSH Manual of Analytical Methods, 4th ed., 1st supplement. P.M. Eller, ed. DHHS (NIOSH) Publication No. 96-135. U.S. Dept. Health and Human Services, Public Health Service, Centers for Disease Control and Prevention, NIOSH, Cincinnati, OH (1996).

64. Streicher, R.P.; Kennedy, E.R.; Lorberau, C.D.: Strategies for the Simultaneous Collection of Vapours and Aerosols with Emphasis on Isocyanate Sampling. Analyst 119:89–97 (1994).

65. Pankow, J.F.; Bidleman, T.F.: Effects of Temperature, TSP and Per Cent Nonexchangeable Material in Determining the Gas-Particle Partitioning of Organic Compounds. Atmos. Environ. 25A:2241–2249 (1991).

66. Hart, K.M.; Isabelle, L.M.; Pankow, J.F.: High-Volume Air Sampler for Particle and Gas Sampling. 1. Design and Gas Sampling Performance. Environ. Sci. Technol. 26:1048–1052 (1992).

67. Ogden, M.W.; Maiolo, K.C.; Nelson, P.R.; Heavner, D.L.; Green, C.R.: Artifacts in Determining the Vapor-Particulate Phase Distribution of Environmental Tobacco-Smoke Nicotine. Environ. Technol. 14:779–785 (1993).

68. Hart, K.M.; Pankow, J.F.: High-Volume Air Sampler for Particle and Gas Sampling. 2. Use of Backup Filters to Correct for the Adsorption of Gas-Phase Polycyclic Aromatic-Hydrocarbons to the Front Filter. Environ. Sci. Technol. 28:655–661 (1994).

69. Kaupp, H.; Umlauf, G.: Atmospheric Gas-Particle Partitioning of Organic Compounds: Comparison of Sampling Methods. Atmos. Environ. 26A:2259–2267 (1992).

70. Sioutas, C.; Koutrakis, P.: Development of a Low Cutpoint Size Slit Virtual Impactor for Sampling Ambient Fine Particles. J. Aerosol Sci. 25:1321–1330 (1994).

71. Sioutas, C.; Koutrakis, P.; Burton, R.M.: A High-Volume Small Cutpoint Virtual Impactor for Separation of Atmospheric Particulate from Gaseous-Pollutants. Particulate Sci. and Technol. 12:207–221 (1994).

72. Xiong, J.Q.; Fang, C.; Cohen, B.S.: A Portable Vapor/Particle Sampler. Am. Ind. Hyg. Assoc. J. 59:614–621 (1998).

73. Gunderson, E.C.; Anderson, C.C.: Collection Device for Separating Airborne Vapor and Particulates. Am. Ind. Hyg. Assoc. J. 48:634–638 (1987).

74. Caka, F.M.; Eatough, D.J.; Lewis, E.A.; et al.: An Intercomparison of Sampling Techniques for Nicotine in Indoor Environments. Environ. Sci. Technol. 24:1196–1203 (1990).

75. Krieger, M.S.; Hites, R.A.: Diffusion Denuder for the Collection of Semivolatile Organic Compounds. Environ. Sci. Technol. 26:1551–1555 (1992).

76. Gundel, L.A.; Lee, V.C.; Mahanama, K.R.R.; et al.: Direct Determination of the Phase Distributions of Semi-Volatile Polycyclic Aromatic Hydrocarbons Using Annular Denuders. Atmos. Environ. 29:1719–1733 (1995).

77. Lane, D.A.; Gundel, L.: Gas and Particle Sampling of Airborne Polycyclic Aromatic Compounds. Polycyclic Aromatic Compounds. 9:67–73 (1996).

78. Eatough, D.J.; Obeidi, F.; Pang, Y.; et al.: Integrated and Real-Time Diffusion Denuder Samplers for $PM_{2.5}$. Atmos. Environ. 33:2835–2844 (1999).

79. Rando, R.J.; Poovey, H.G.: Development and Application of a Dichotomous Vapor/Aerosol Sampler for HDI-Derived Total Reactive Isocyanate Group). Am. Ind. Hyg. Assoc. J. 60:737–746 (1999).

80. Coutant, R.W.; Brown, L.; Chuang, J.; Lewis, R.G.: Field Evaluation of Phase Distribution of PAH. Proceedings of the 1986 EPA/APCA Symposium on Measurement of Toxic Air Pollutants, pp. 146–155. Report No. 600/9-86-013. U.S. Environmental Protection Agency, Washington, DC (1986).

81. Turpin, B.J.; Liu, S.-P.; Podolski, K.S.; et al.: Design and Evaluation of a Novel Diffusion Separator for Measuring Gas/Particle Distributions of Semivolatile Organic Compounds. Environ. Sci. Technol. 27:2441–2449 (1993).

82. Ye, Y.; Tsai, C.-J.; Pui, D.Y.H.; Lewis, C.W.: Particle Transmission Characteristics of an Annular Denuder Ambient Sampling System. Aerosol Sci. Tech. 14:102–111 (1991).

83. Sioutas, C.; Koutrakis, P.; Wolfson, J.M.: Particle Losses in Glass Honeycomb Denuder Samplers. Aerosol Sci. Technol. 21:137:148 (1994).

84. Bemgard, A.; Colmsjo, A.; Melin, J.: Assessing Breakthrough Times for Denuder Samplers with Emphasis on Volatile Organic-Compounds. J. Chromatog. 723:301–311 (1996).

85. Malek, R.F.; Daisey, J.M.; Cohen, B.S.: The Effect of Aerosol on Estimates of Inhalation Exposure to Airborne Styrene. Am. Ind. Hyg. Assoc. J. 47:524–529 (1986).

86. Cohen, B.S.; Brosseau, L.M.; Fang, C.-P.; et al.: Measurement of Air Concentrations of Volatile Aerosols in Paint Spray Applications. Appl. Occup. Environ. Hyg. 7:514–521 (1992).

87. Brosseau, L.M.; Fang, C.-P.; Snyder, C.; Cohen, B.S.: Particle Size Distribution of Automobile Paint Sprays. Appl. Occup. Environ. Hyg. 7:607–612 (1992).

Chapter 5

Size-Selective Health Hazard Sampling

Morton Lippmann, Ph.D.

Nelson Institute of Environmental Medicine, New York University School of Medicine, Tuxedo, New York

CONTENTS

Introduction

Sampling for Respiratory Hazard Evaluation

Air sampling techniques have been used to obtain information for a variety of purposes. The following discussion concerns the specific purpose of sampling for the evaluation of the toxicological insult arising from the inhalation of airborne particles and compliance with particle size-selective Threshold Limit Values (PSS–TLVs®) recommended by the American Conference of Governmental Industrial Hygienists (ACGIH®). Air sampling techniques used to obtain information for other purposes (e.g., performance testing of ventilation systems and air cleaners, contamination monitoring in so-called "white room" or "cleanroom" operations, and basic scientific studies of atmospheric reactions, composition, and capacity for pollutant dispersion) may differ and are beyond the scope of this discussion.

If the objective is to obtain information on the nature and magnitude of the potential health hazard resulting from the inhalation of airborne particles, the techniques should provide an index of the contaminant concentration within the size range that reaches the critical organ for toxic action. In other words, the choice of methods must recognize the size-selecting characteristics of the human respiratory tract in addition to the usual factors affecting the selection of methods, e.g., the physical limitations of the collection process, and the sensitivity and specificity of the analytical procedures.

There has been an increasing recognition of the importance of size-selective sampling of airborne particles. Before 1952, the size-selecting characteristics of the human respiratory tract were largely ignored. The only standard method that had provided a means for discriminating against nonrespirable particles was the impinger sampling–light field counting technique for pneumoconiosis-producing dusts. The Greenburg–Smith impinger, developed in 1922–1925 through the cooperative efforts of the U.S. Bureau of Mines, the U.S. Public Health Service, and the American Society of Heating and Ventilating Engineers,[1] and the midget impinger, developed in 1928 by the Bureau of Mines,[2] collect particles larger than about 0.75 μm in a liquid medium. Such samples were analyzed by counting the particles that settled to the bottom of an aqueous counting cell and were visible when viewed through a 10X objective lens. Particles larger than 10 μm that were observed during the count were rejected as "nonrespirable." The alternative approach was gravimetric analysis of the total airborne particulate sample, in which there was neither any practical way to discriminate against oversized particles

nor any upper limit on the particle size that would penetrate through the sampler's inlet.

The use of the terms "respirable" and "nonrespirable" were first applied to those mineral dusts known to produce pneumoconioses, i.e., dust diseases of the nonciliated gas-exchange region of the lungs (also referred to as the alveolar or pulmonary region). The particles that deposit in the oral or nasal airways of the upper respiratory tract (head airways region) or in the conductive airways of the tracheobronchial region are cleared from the deposition sites by mechanical processes such as mucociliary transport and cough; they do not contribute to the pathogenesis of the pneumoconioses. Such particles, which are generally considered "nonrespirable," can, however, contribute to the development of other diseases such as bronchitis and cancers of the nasal and bronchial airways. The section that follows outlines the factors affecting the deposition of particles within the major functional regions of the human respiratory tract and the quantitative data available on deposition in these regions as a function of aerodynamic particle size. This will be followed by a discussion of the criteria that have been proposed and/or used for sampling "respirable" dusts and for sampling the larger particles that can deposit in the head airways and tracheobronchial airways.

Regional Deposition, Clearance, and Dose

The hazard from airborne particles varies with their physical, chemical, and/or biological properties. These properties determine the fate of the particles and their interactions with the host after they are deposited. A basic consideration is that this fate, in any given individual, varies greatly with the site of deposition within the respiratory tract.

There are a number of major subdivisions within the respiratory tract that differ markedly in structure, size, and function, and they have different mechanisms for particle elimination. Thus, a complete determination of dose from an inhaled toxicant depends on the regional deposition and the retention times at the deposition sites and along the elimination pathways, in addition to the chemical and surface properties of the particles.

Anatomical and Physiological Factors in Respiratory Tract Particle Deposition and Clearance

The succeeding paragraphs present a brief summary of the factors controlling particle deposition and clearance. More complete descriptions of the anatomy of the respiratory tract and of some of the factors controlling particle deposition and clearance are presented elsewhere.[3–5]

Head Airways Region

Nasal Passages. Air enters through the nares or nostrils, passes through a web of nasal hairs, and flows posteriorly toward the nasopharynx while passing through a series of narrow passages winding around and through shelflike projections called turbinates. The air is warmed and moistened in its passage and partially depleted of particles. Some particles are removed by impaction on the nasal hairs and at bends in the air path; others are removed by sedimentation and diffusion. Except for the anterior nares, the surfaces are covered by a mucous membrane composed of ciliated and goblet cells. The mucus produced by the goblet cells is propelled toward the pharynx by the beating of the cilia, carrying deposited particles along with it. Particles deposited on the anterior unciliated portion of the nares and at least some of the particles deposited on the nasal hairs usually are not carried posteriorly to be swallowed, but rather are removed mechanically by nose wiping, blowing, sneezing, etc.

Oral Passages, Pharynx, Larynx. In mouth breathing, some particles are deposited, primarily by impaction, in the oral cavity and at the back of the throat. Diffusion may also be important for ultrafine particles. These particles are rapidly eliminated to the esophagus by swallowing.

Tracheobronchial Region

The conductive airways in the tracheobronchial region have the appearance of an inverted tree, with the trachea analogous to the trunk and the subdividing bronchi to the limbs. The branching pattern is normally asymmetric in a regular pattern, as described by Horsfield et al.[6] However, for purposes of discussion, it will be clearer if Weibel's simplified anatomic model,[7] in which there are 16 generations of bifurcating airways, is adopted. As illustrated by Table 5-1, the diameter decreases from generation to generation, but because of the increasing number of tubes, the total cross section for flow increases and the air velocity decreases toward the ends of the tree. In the larger airways, particles too large to follow the bends in the air path are deposited by impaction. At the low velocities in the smaller airways, particles deposit by sedimentation and, if small enough, by diffusion.

Ciliated and mucus-secreting cells are found at all levels of the tracheobronchial tree. Most of the inert, nonsoluble particles deposited in this region are thus carried within hours toward the larynx on the moving mucous sheath that is propelled proximally by the beating of the cilia. Beyond the larynx, the particles enter the esophagus and pass through the gastrointestinal tract.

Cigarette smoke and air contaminants can affect mucociliary transport along the tracheobronchial tree. As demonstrated by Lippmann et al.,[8] brief exposures at low doses of irritants such as cigarette smoke and submicrometer H_2SO_4 can accelerate mucus transport, while higher doses of the same pollutants can slow or temporarily halt mucus transport. Chronic exposures to these pollutants can result in more variable rates of clearance and persistent changes in clearance rates which may predispose the individual to, or initiate a sequence of changes leading to, the development of chronic bronchitis.

Persistent defects in clearance of particles from the bronchial tree would also lead to increased residence times for particles containing toxic and carcinogenic chemicals, thereby increasing the dose to the underlying tissues from those chemicals and resulting in increased systemic uptake. In this manner, defective clearance may contribute to a variety of disease conditions.

Gas-exchange Region

The region beyond the terminal bronchioles is the region in which gas exchange takes place. The epithelium is nonciliated and, therefore, insoluble particles deposited in this region by sedimentation and diffusion are removed at a very slow rate, with clearance half-times on the order of a month or more. The mechanisms for particle clearance from this region are only partly understood, and their relative importance remains a matter of some debate. Some particles are engulfed by phagocytic cells which are transported onto the ciliary "escalator" of the bronchial tree in an undefined manner. Others penetrate the alveolar wall and enter the lymphatic system. Still others dissolve slowly in situ. "Insoluble" dusts all have some finite solubility, which is greatly enhanced by the large surface-to-volume ratio characteristic of particles small enough to penetrate to the alveolar region of the lung. Morrow et al.[9] demonstrated that the late-phase clearance half-times of many "insoluble" dusts in the lung are proportional to their solubilities in simulated lung fluids. Alveolar clearance rates may differ for different dusts. Jammet et al.[10] studied the clearance of hematite, silica, and coal in cats, rats, and hamsters. Three clearance phases were observed. The first phase, representing bronchial clearance, had a half life of less than 1 day. An intermediate phase, with a half life of 10–12 days was seen in all species for hematite and in the cat for coal dust. When silica dust was inhaled, this phase was not seen. The slow third clearance phase, with a half

TABLE 5-1. Architecture of the Lung Based on Weibel's[A] Model A: Regular Dichotomy Average Adult Lung With Volume 4800 cm³ at About Three-Fourths Maximal Inflation

Name of Airway	Generation[A]	Number/Generation[A]	Diameter (mm)[A]	Length (mm)[A]	Cumulative Length (mm)[A]	Total Cross Section (cm²)[A]	Volume (cm³)[A]	Cumulative Volume (cm³)[A]	Velocity (cm/s)[B,C]	Residence Time (ms)[B,C]	Cumulative Time (ms)[B,C]	Pressure Difference (μmH₂)[B,C,D]	Cum. Press. Diff. (μmH₂)[B,C,D]	Reynolds No.[B,E]
Trachea	0	1	18.0	120.0	120.0	2.54	30.5	30.5	393	30.5	31	87	87	4350
Main bronchus	1	2	12.2	47.6	167.4	2.33	11.3	41.8	427	11.1	41	82	169	3210
Lobar bronchus	2	4	8.3	19.0	186.6	2.13	4.0	45.8	462	4.11	45	76	246	2390
	3	8	5.6	7.6	194.2	2.00	1.5	47.2	507	1.50	47	73	320	1720
Segmental	4	16	4.5	12.7	206.9	2.48	3.5	50.7	392	3.23	50	147	467	1110
bronchus	5	32	3.5	10.7	217.6	3.11	3.3	54.0	325	3.29	53	170	638	690
Bronchi with	6	64	2.8	9.0	226.6	3.96	3.5	57.5	254	3.55	57	174	812	434
cartilage in wall	7	128	2.3	7.6	234.2	5.10	3.9	61.4	188	4.04	61	162	974	277
	8	256	1.86	6.4	240.6	6.95	4.5	65.9	144	4.45	65	160	1134	164
	9	512	1.54	5.4	246.0	9.56	5.2	71.0	105	5.15	80	143	1277	99
	10	1.02K	1.30	4.6	250.6	13.4	6.2	77.2	73.6	6.25	77	120	1397	60
Terminal bronchus	11	2.05K	1.09	3.9	254.5	19.6	7.6	84.8	52.3	7.45	85	103	1500	34
	12	4.10K	0.95	3.3	257.8	28.8	9.8	94.6	34.4	9.58	94	75	1576	20
Bronchioles with	13	8.19K	0.82	2.7	260.5	44.5	12.5	107	23.1	11.7	106	55	1632	11
muscle in wall	14	16.4K	0.74	2.3	262.8	69.4	16.4	123	14.1	16.2	122	35	1667	6.5
	15	32.8K	0.66	2.0	264.8	113	21.7	145	8.92	22.4	144	24	1692	3.6
Terminal bronchiole	16	65.5K	0.60	1.65	266.5	180	29.7	175[F]	5.40	30.6	175	14	1707	2.0
Respiratory bronchiole	17	131K	0.54	1.41	267.9	300	41.8	217	3.33	42.3	217	10	1716	1.1
Respiratory bronchiole	18	262K	0.50	1.17	269.0	534	61.1	278	1.94	60.2	277	5	1722	0.57
Respiratory bronchiole	19	524K	0.47	0.99	270.0	944	93.2	371	1.10	90.0	368	3	1725	0.31
Alveolar duct	20	1.05M	0.45	0.83	270.9	1.60K	140	510	0.60	138	506	1.4	1726	0.17
Alveolar duct	21	2.10M	0.43	0.70	271.6	3.22K	224	735	0.32	213	719	0.74	1727	0.08
Alveolar duct	22	4.19M	0.41	0.59	272.1	5.88K	350	1085	0.18	326	1047	0.37	1727	0.04
Alveolar sac	23	8.39M	0.41	0.50	272.6	11.8K	591	1675	0.09	553	1602	0.16	1728	—
Alveoli, 21 per duct		300M[C]	0.28[C]	0.23[C]	272.9[C]		3200[C]	4875[C]						

[A] From Weibel. (7)

[B] At flow rate = 1.0 L/sec = 60 L/min.

[C] Added by W. Briscoe (personal communication).

[D] Pressure difference from mouth if flow were laminar.

[E] Added by B. Altshuler (personal communication).

[F] Dead space from larynx.

life of >100 days, was unaffected, except that it accounted for more of the clearance. Further tests on cats and rats with carbon, quartz, titanium dioxide, and hematite were reported by LeBouffant.[11] The alveolar clearance was found to be a function of the species used, the pulmonary dust load, the time since exposure, and the nature of the particles. Coal, even in small quantities, slowed clearance in the rat. With heavy exposures, the clearance rate did not recover appreciably.

For asbestos and man-made mineral fibers, long-term fiber retention in the lungs depends on fiber length, fiber diameter, and leaching rates of various elements from the fibers. Bellmann et al.[12] showed that crocidolite fibers longer than 5 μm did not clear from rat lungs in 1 year, whereas chrysotile fibers longer than 5 μm increased in numbers, presumably due to longitudinal splitting. The glass fibers longer than 5 μm were lost, with a half-time of 55 days, primarily by dissolution. Short fibers of all types were cleared rapidly by comparison.

Gaseous air contaminants can also affect the clearance of particles from the alveolar region. Brief periods of exposure to irritant gases such as SO_2[13] and O_3[14] have been shown to stimulate the early alveolar clearance of rats, while prolonged exposure to SO_2 slowed clearance.[13] McFadden et al.[15] showed that cigarette smoke reduced the more rapid phase of alveolar clearance of asbestos fibers in guinea pigs.

Considering the recognized importance of the alveolar retention of relatively insoluble particles in the pathogenesis of chronic lung disease, it is somewhat surprising that examination of the literature yields so little useful data on the rates or routes of alveolar particle clearance in people.

In a study reported by Albert and Arnett,[16] eight normal human males inhaled neutron-activated metallic iron particles. For three subjects, there was sufficient residual activity after the completion of the bronchial clearance for continued measurement of retention. For a 32-year-old nonsmoking male and a 27-year-old male who was a moderate smoker, the postbronchial clearance occurred in two phases, a fast phase lasting about 1 month and a much slower terminal phase. The faster phase was missing in a 38-year-old, two-pack-a-day cigarette smoking male with chronic cough. Although it is not possible to draw firm conclusions from these limited data, they are consistent with the findings of Cohen et al.,[17] who studied the alveolar clearance rates of magnetite particles in nine nonsmokers and three smokers using an external magnetometer for the particle retention measurements. The clearance rates in all three smokers were much lower than in any of the nine nonsmokers.

Thus, it appears that the fast alveolar phase can be detected in man, and that cigarette smoking may increase dust retention beyond the retention of the smoke particles themselves. Low doses of cigarette smoke have been shown to inhibit macrophage phagocytosis.[18]

Another study that provides confirmation for the hypothesis that cigarette smoking can severely retard the clearance of particles from the alveolar region was performed by Bohning et al.[19] They exposed five healthy nonsmokers, six healthy ex-smokers, eight smokers, and six persons with chronic obstructive lung disease to 3.6-μm-diameter polystyrene latex particles tagged with ^{85}Sr. The nonsmokers and ex-smokers essentially had the same clearance patterns. There were two clearance phases: one with a $T_{1/2}$ of 30 ± 23 days, which accounted for 27% ± 13% of the total alveolar clearance. The $T_{1/2}$ of the slower phase was 296 ± 98 days. Only three smokers had a measurable fast phase, accounting for 6% to 13% of the clearance, with $T_{1/2}$ of 4, 18, and 20 days. The average $T_{1/2}$ for the slower phase for the eight smokers was 534 days. The slow phase $T_{1/2}$ was linearly correlated ($r = 0.99$) with the amount of smoking, increasing 14.7 ± 3.0 days per pack-year. The obstructive lung disease subjects had an average $T_{1/2}$ for the faster phase of 26.6 days and an average $T_{1/2}$ for the slower phase of 660 days.

It is difficult to imagine that prolonged retention of particles in the alveolar regions of the lungs is beneficial. Therefore, it is important to develop a better understanding of the normal patterns and rates of particle clearance from the alveoli, as well as the dose-related influences of air contaminants on that clearance. Prolonged retention of inhaled particles in the alveolar region increases both the doses of those particles to the underlying tissues and the potential for systemic uptake. If the particles are fibrogenic, they could contribute to the development of pneumoconiosis and emphysema. Cigarette smoke from either passive or active smoking contains a variety of carcinogens, and greater retention in the alveoli could cause an increased risk from both lung cancer and cancer in other organs that accumulate these chemicals after their dissolution in the lungs.

Much of the preceding remains speculative. It is unfortunate that our current knowledge of the quantitative aspects of the normal rates of clearance and of the effects of inhaled pollutants on clearance rates and pathways is too meager to permit a more definitive assessment.

Whereas variations in clearance dynamics for particles deposited in the alveoli may be critical determinants of toxicity and should be considered in the establishment of TLVs, these variables cannot be simulated by size-selective samplers that can only subdivide the airborne suspension on the basis of where the particles are expected to deposit. Thus, the establishment of size-selective sampling criteria has been dependent primarily on regional deposition data in healthy adults.

Regional Deposition and Clearance Dynamics

To estimate toxic dose from inhaled particles, the respiratory tract can be divided into five functional regions that differ grossly from one another in retention time at the deposition site, the elimination pathway, or both. These regions are:

1. Gas-exchange region (for both nose and mouth breathing).
2. Tracheobronchial region (for both nose and mouth breathing).
3a. Oral cavity, pharynx, and larynx (for mouth breathing).
3b. Nasopharynx, pharynx, and larynx (for nose breathing).
4. Ciliated nasal passages (for nose breathing).
5. Anterior unciliated nares (for nose breathing).

The fractional deposition in each of these regions is dependent on the aerodynamic particle size and the subject's airway dimensions and respiratory characteristics (e.g., flow rate, breathing frequency, tidal volume). Ideally, air sampling data should provide data on the deposition to be expected in each functional region or at least in regions 1, 2, and 3–5 inclusive.

Experimental Deposition Data

Total Deposition

Total deposition as a function of particle size and respiratory parameters has been measured experimentally by numerous investigators. Many previous reviews on deposition have called attention to the very large difference in the reported results.[3–5, 20–23] Figure 5-1 shows data from studies done with mouth breathing. Tidal volumes varied from 0.5 to 1.5 L. All appear to show the same trend with a minimum of deposition at diameters between 0.1 and 1.0 μm.

It is also apparent that in most studies involving more than one subject, there was considerable individual variation among the subjects. Davies et al.[24] showed that some of this variation could be eliminated by standardizing the expiratory reserve volume (ERV) and thereby the size of the air spaces. They found that deposition decreases as ERV increases. This was confirmed by Heyder et al.,[25] who reported that there was little intrasubject variation among six subjects when

their deposition tests were performed at their normal ERVs. Some of the variability was also due to the variations in breathing frequency and flow rate among the various subjects, and Heyder et al.[26] showed how these variable factors affect total respiratory tract deposition. However, when all of the controllable factors are taken into consideration, there is still variability in deposition due to the intrinsic variability of airway and air space sizes among individuals in a population. The extent and significance of such variability has been discussed.[27–30] Using aerosol deposition data to estimate bronchial airway sizes, Chan and Lippmann[29] reported a coefficient of variation of 0.23 among healthy young nonsmokers. For alveolar air space dimensions, Lapp et al.[31] found a coefficient of variation of 0.21 using an aerosol deposition technique, while Matsuba and Thurlbeck[32] reported a coefficient of variation of 0.25 based on measurements of lung sections taken at autopsy.

The data of Heyder et al.,[25,33] Muir and Davies,[34] and Davies et al.[24] appear to represent deposition minima for normal men. Their test protocols were precisely controlled. Their aerosols were charge neutralized. With more natural aerosol and respiratory parameters, higher deposition efficiencies would be expected. The deposition data in Figure 5-1 were based on the difference between inhaled and exhaled particle concentrations, except for the data of Lippmann,[21] Foord et al.,[35] and Stahlhofen et al.,[36] which are based on external in vivo measurements of γ-tagged particle retention. The large amount of scatter among the individual data points for the larger particles is due to a quite variable deposition in the head and tracheobronchial tree. Cigarette smokers have a similar median behavior for head deposition, but even more scatter. The median and upper limits of tracheobronchial deposition are higher for cigarette smokers than for nonsmokers, but the lower limit is about the same.[37]

Regional Deposition

Some inhaled particles deposit within the air passages between the point of entry at the lips or nares and the larynx. The fraction depositing can be highly variable, being dependent on the route of entry, the particles' sizes, and the flow rates. In most cases, the nasal route is a more efficient particle filter than the oral, especially at low and moderate flow rates. Thus, those people who normally breathe part or all of the time through the mouth may be expected to deposit more particles in their lungs than those who breathe entirely through the nose. During exertion, the flow resistance

SOURCE	TIDAL VOL, ml	RES. RATE, breaths/min	SOURCE	TIDAL VOL, ml	RES. RATE, breaths/min
○ LANDAHL et al. (1951)	500	15	▼ LEVER (1974)	600	16
□ LANDAHL et al. (1952)	1500	15	◆ MUIR & DAVIES (1967)	500	15
△ ALTSHULER et al. (1957)	500	15	◑ DAVIES et al. (1972)	600	16
▽ GEORGE AND BRESLIN (1967)	760	11	◩ HEYDER et al. (1975)	1000	15
◇ GIACOMELLI-MALTONI et al. (1972)	1000	12	▲ SHANTY (1974)	1140	18
● CHAN & LIPPMANN (1980)*	1000	14	▼ STAHLHOFEN et al. (1980)	1500	15
■ FOORD et al. (1976)	1000	15	◆ STAHLHOFEN et al. (1980)	1000	7.5
▲ MARTENS & JACOBI (1973)	1000	14	◇ SWIFT et al. (1977)	500	15
*USED MMD FOR D <0.5 μm			▨ HEYDER et al. (1973b)	500	15

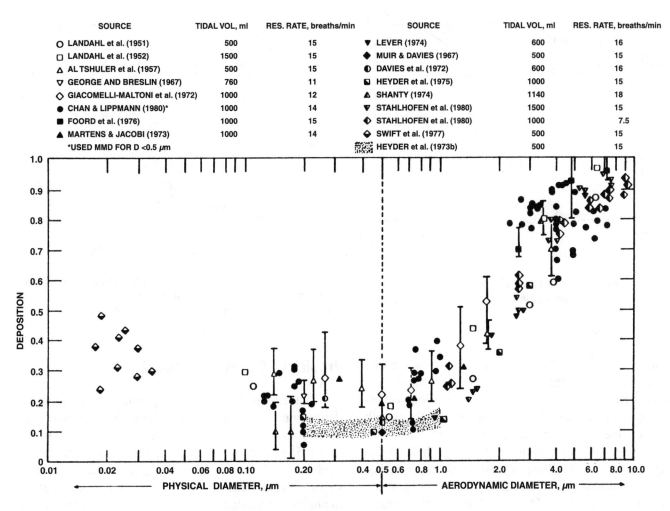

FIGURE 5-1. Deposition of monodisperse aerosols in the total respiratory tract for mouth breathing in humans as a function of aerodynamic diameter, except below 0.5μm, where deposition is plotted versus physical diameter. The data are individual observations, averages, and ranges as cited by various investigators.

of the nasal passages causes a shift to mouth breathing in almost all people.

Available data on the regional deposition of inhaled particles in the human respiratory tract were recently summarized by the U.S. Environmental Protection Agency (U.S. EPA) in their criteria document for particulate matter.[38] The data considered reliable for deposition in the head (extrathoracic), tracheobronchial (TB) tree, and nonciliated pulmonary (alveolar) regions of healthy humans are summarized in Figures 5-2 through 5-5. There is a great amount of intersubject variability in deposition in all regions, due both to their inherent variability in airway and air space dimensions, and the variability in breathing rates and patterns. Deposition in the head is primarily by impaction, and particle collections in the nasal passages are much more efficient than those in the oral passages. In the TB airways, impaction is the dominant removal mechanism for particles larger than about 2.0

μm under most conditions, while sedimentation is the major collection mechanism for particles between about 0.5 and 2.0-μm. As an impactor, the TB region is much more efficient than the oral airways, but somewhat less efficient than the nasal airways. Thus, particle deposition within the lungs for 1.0- to 10-μm particles is very much dependent on whether the individual breathes through the nose or mouth. Deposition in the pulmonary region is primarily by sedimentation for particles larger than 0.5 μm and by diffusion for smaller particles. Particles of approximately 0.5 μm, having a minimal intrinsic mobility, have a minimum in deposition probability. For particles larger than approximately 3 μm, there is less pulmonary deposition with increasing size because these larger particles have a diminishing penetration through the conductive airways.

Figure 5-5 also shows an estimate of the alveolar deposition that could be expected when aerosol is

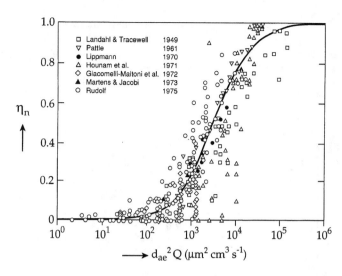

FIGURE 5-2. Inspiratory deposition of the human nose as a function of particle aerodynamic diameter and flow rate ($d^2_{ae}Q$).[38]

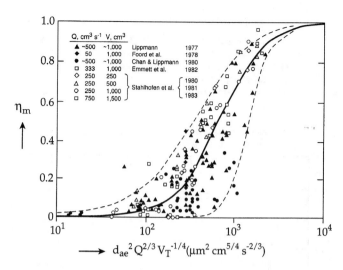

FIGURE 5-3. Inspiratory extrathoracic deposition data in humans during mouth breathing as a function of particle aerodynamic diameter, flow rate, and tidal volume ($d^2_{ae}Q^{2/3}V_T^{-1/4}$).[38]

inhaled via the nose. This estimate is based on the difference in head retention during nose breathing and mouth breathing from the straight line relations developed by Lippmann.[21] It can be seen that for mouth breathing, the size for maximum deposition is approximately 3 μm and that approximately one-half of the inhaled aerosol at this size deposits in this region. For nose breathing, there is a much less pronounced maximum of approximately 25% at 2.5 μm, with a nearly constant alveolar deposition averaging about 20% for all sizes between 0.1 and 4 μm.

Predictive Deposition Models

Mathematical models for predicting the regional deposition of aerosols were developed by Findeisen[39] in 1935, Landahl in 1950[40] and 1963,[41] by Beeckmans[42] in 1965, and by Yu[43] in 1978. Findeisen's simplified anatomy, with nine sequential regions from the trachea to the alveoli, and his impaction and sedimentation deposition equations were used in the International Commission on Radiological Protection (ICRP) Task Group's 1966 model.[23] For diffusional deposition, the

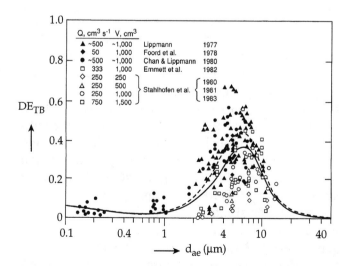

FIGURE 5-4. Tracheobronchial deposition data in humans for mouth breathing as a function of particle aerodynamic diameter (d_{ae}). The solid curve represents the approximate mean for all of the experimental data; the broken curve represents the data excluding the data of Stahlhofeneld[38].

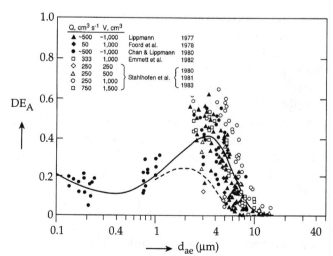

FIGURE 5-5. Alveolar deposition data for humans as a function of particle aerodynamic diameter (d_{ae}). The solid curve represents the mean of all of the data; the broken curve is an estimate of deposition for nose breathing[38].

Task Group used the Gormley-Kennedy[43] equations, and for head deposition, they assumed entry through the nose with a deposition efficiency given by the empirical equation of Pattle.[44]

The ICRP Task Group's 1966 model was adopted by ICRP Committee II in 1973, with numerical changes in some clearance constants. The Task Group report has been widely quoted and used within the health physics field. One of the significant conclusions of the Task Group study was that the regional deposition within the respiratory tract can be estimated using a single aerosol parameter, the activity median aerodynamic diameter (AMAD). For a tidal volume of 1450 cm^3, there are relatively small differences in estimated deposition over a very wide range of geometric standard deviations ($1.2 < \sigma g < 4.5$).

None of these earlier models provide reliable estimates of aerosol deposition in healthy normal adults. Their predictions for total and alveolar deposition efficiencies differ from the best available experimental data for adult normals illustrated in Figures 5-1 through 5-5. Furthermore, they do not give any measure of the very large variability in deposition efficiencies among normals, nor of the changes produced by cigarette smoking and lung disease. However, there have been significant advances in the measurement of deposition in recent years, and considerable effort was devoted to improve theoretical understanding and predictive models.

Latest Predictive Deposition Models of ICRP and NCRP

In 1984, both the ICRP and the National Council on Radiation Protection (NCRP) appointed task groups to review the dosimetric model of the respiratory tract to propose revisions or a new model. Although intergroup liaison members were appointed, each task group produced its own model with significant differences between them.

ICRP Deposition Model[46]

The 1994 ICRP Task Group directed its efforts toward improving the model adopted in 1973 rather than developing a completely new model. The objective was a model that would: 1) facilitate calculation of biologically meaningful doses; 2) be consistent with the morphological, physiological, and radiobiological characteristics of the respiratory tract; 3) incorporate current knowledge; 4) meet all radiation protection needs; 5) be no more sophisticated than necessary to meet dosimetric objectives; 6) be adaptable to development of computer software for calculation of relevant radiation doses from knowledge of a few readily measured exposure parameters; 7) be equally

useful for assessment purposes as for calculating recommended values for limits on intake; 8) be applicable to all members of the world population; 9) allow for use of information on the deposition and clearance of specific materials; and 10) consider the influence of smoking, air pollutants, and diseases on the inhalation, deposition, and clearance of radioactive particles from the respiratory tract. Although it was intended that this new or revised model be applicable to all members of the world's population, i.e., to both sexes for all ages, to smokers and nonsmokers, and to those with healthy and diseased respiratory tracts, the lack of data prevented the full achievement of this goal.

The 1994 ICRP Task Group's clearance model identifies the principal clearance pathways within the respiratory tract that are important in determining the retention of various radioactive materials, and thus doses received, by respiratory tissues and/or other organs. The deposition model is required to estimate the amount of inhaled material that enters each clearance pathway. These discrete pathways are represented by the compartment model shown in Table 5-2 and Figure 5-6.

Extrathoracic Airways

The extrathoracic airways are partitioned into two distinct clearance and dosimetric regions: the anterior nasal passages (ET_1) and all other extrathoracic airways (ET_2), i.e., the posterior nasal passages, the naso- and oropharynx, and the larynx. Particles deposited on the surface of the skin lining the anterior nasal passages

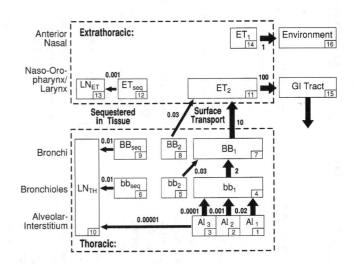

FIGURE 5-6. Compartment model to represent time-dependent particle transport form each region in 1994 ICRP model. Particle transport rate constants shown beside the arrows are reference values in d^{-1}. Compartment numbers (shown in the lower right-hand corner of each compartment box) are used to define clearance pathways. Thus, the particle transport rate from bb$_1$ to BB$_1$ is denoted m$_{4,7}$ and has the value 2 d^{-1}.

TABLE 5-2. Morphometry, Cytology, Histology, Function, and Structure of the Respiratory Tract and Regions Used in the 1992 ICRP Dosimetry Model

Functions	Cytology (Epithelium)	Histology (Walls)	Generation Number	Anatomy	Regions used in Model — New	Regions used in Model — Old*	Zones (Air)	Location (Thoracic/Extrathoracic)	Location (Pulmonary/Extrapulmonary)	Airway Surface	Number of Airways
Air Conditioning; Temperature and Humidity, and Cleaning; Fast Particle Clearance; Air Conduction	Respiratory Epithelium with Goblet Cells: Cell Types: – Ciliated Cells – Nonciliated Cells: • Goblet Cells • Mucous (Secretory) Cells • Serous Cells • Brush Cells • Endocrine Cells • Basal Cells • Intermediate Cells	Mucous Membrane, Respiratory Epithelium (Pseudostratified, Ciliated, Mucous), Glands		Anterior Nasal Passages	ET_1		Conditioning — 0.175 × 10⁻³m³ (Anatomical Dead Space)	Extrathoracic	Extrapulmonary	$2 \times 10^{-3}\,m^2$	—
		Mucous Membrane, Respiratory or Stratified Epithelium, Glands		Nose / Mouth, Pharynx Posterior, Larynx, Esophagus	ET_2 LN_{ET}	(N–P)				$4.5 \times 10^{-2}\,m^2$	—
		Mucous Membrane, Respiratory Epithelium, Cartilage Rings, Glands	0 / 1	Trachea / Main Bronchi				Thoracic			
		Mucous Membrane, Respiratory Epithelium, Cartilage Plates, Smooth Muscle Layer, Glands	2–8	Bronchi	BB		Conduction — 0.2 × 10⁻³m³		Pulmonary	$3 \times 10^{-2}\,m^2$	511
	Respiratory Epithelium with Clara Cells (No Goblet Cells) Cell Types: – Ciliated Cells – Nonciliated Cells: • Clara (Secretory Cells)	Mucous Membrane, Respiratory Epithelium, No Cartilage, No Glands, Smooth Muscle Layer	9–14	Bronchioles	bb $LN_{TH}^{\dagger}$	(T–B)				$2.6 \times 10^{-1}\,m^2$	6.5×10^4
		Mucous Membrane, Single-Layer Respiratory Epithelium, Less Ciliated, Smooth Muscle Layer	15	Terminal Bronchioles							
Air Conduction; Gas Exchange; Slow Particle Clearance	Respiratory Epithelium Consisting Mainly of Clara Cells (Secretory) and Few Ciliated Cells	Mucous Membrane, Single-Layer Respiratory Epithelium of Cuboidal Cells, Smooth Muscle Layers	16–18	Respiratory Bronchioles			Gas-Exchange Transitory — 4.5 × 10⁻³m³			$7.5\,m^2$	4.6×10^5
Gas Exchange; Very Slow Particle Clearance	Squamous Alveolar Epithelium Cells (Type I), Covering 93% of Alveolar Surface Areas	Wall Consists of Alveolar Entrance Rings, Squamous Epithelial Layer, Surfactant	**	Alveolar Ducts	AI	P					
	Cuboidal Alveolar Epithelial Cells (Type II. Surfactant-Producing), Covering 7% of Alveolar Surface Area	Interalveolar Septa Covered by Squamous Epithelium, Containing Capillaries, Surfactant	**	Alveolar Sacs						$140\,m^2$	4.5×10^7
	Alveolar Macrophages			Lymphatics	L						

* Previous ICRP Model

** Unnumbered because of imprecise information

† Lymph nodes are located only in BB region but drain the bronchial and alveolar interstitial regions as well as the bronchial region.

are assumed to be subject only to removal by extrinsic means (e.g., nose blowing, wiping). The bulk of material deposited in the naso-oropharynx or larynx (ET$_2$) is subject to fast clearance in the layer of fluid that covers these airways. The new model recognizes that diffusional deposition of ultrafine particles in the extrathoracic airways can be substantial, whereas the earlier models did not.

Thoracic Airways

Activity deposited in the thorax is divided between bronchial (BB) and bronchiolar (bb) regions, which are subject to relatively fast ciliary clearance, and the alveolar-interstitial (AI) region, which is not subject to fast ciliary clearance.

For dosimetry purposes, deposition of inhaled material is divided between the trachea and bronchi (BB), and in the more distal, small airways, the bronchioles (bb). However, the subsequent efficiency with which cilia in either type of airway are able to clear deposited particles is controversial. In order to be certain that doses to bronchial and bronchiolar epithelia will not be underestimated, the Task Group assumes that as much as half the number of particles deposited in these airways is subject to "slow" clearance. The likelihood that a particle is cleared slowly by the mucociliary system appears to depend on its physical size.

Material deposited in the AI region is subdivided among three compartments (AI$_1$, AI$_2$, and AI$_3$) that are each cleared slowly but at different characteristic rates.

NCRP Deposition Model

The 1997 NCRP deposition model[47] made fewer changes to the old ICRP Task Group model format than did the new ICRP model. The original name of the head airways region was changed from nasopharynx (NP) to naso-oro-pharyngo-laryngeal (NOPL) to be more fully descriptive of the nature of the region. The names of the other regions remained tracheobronchial (TB) for the trachea and the conductive airways within the thorax, and pulmonary (P) for the gas exchange and interstitial lung areas. NCRP also retained the lymph nodes as a separate compartment for clearance and dose calculations. Their model permits assumptions about different proportions depositing in each region, and different rate constants for clearance from subcompartments within each region, but does not designate them as separate compartments. Thus, NCRP's NOPL corresponds to ICRP's ET$_1$ + ET$_2$, and their TB region corresponds to ICRP's BB + bb regions. The new NCRP model also recognizes that substantial deposition of ultrafine particles in the NP region can take place.

Comparison of ICRP and NCRP Predictive Models

Both NCRP and ICRP used the same databases in formulating their mathematical predictive regional deposition models, but they made some different decisions about the appropriate formulations and mathematical forms to fit the available data. Figure 5-7 shows the size dependence of the deposition fractions for the light exercise category of ICRP. Figure 5-8 shows the corresponding curves for the NCRP model, along with those of ICRP when ET$_1$ is combined with ET$_2$, and when BB is combined with bb. It can be seen that the inclusion of the inspirability factor, i.e., the aspiration efficiency for entry of ambient particles into the human nose or mouth, strongly affects the predicted deposition in the head airways. Notable differences are the modal particle size and maximal efficiencies for AI versus P and for TB versus BB + bb.

Some of the other differences between the two deposition models are:

- The NCRP model's code permits a wider range of user flexibility.
- The NCRP model uses generation-by-generation data for calculating deposition and clearance; the ICRP model uses broad classes of airways.
- The ICRP model provides for early sequestration in bronchial airways; the NCRP model does not.
- The NCRP model provides more guidance for evaluation of novel materials and situations.

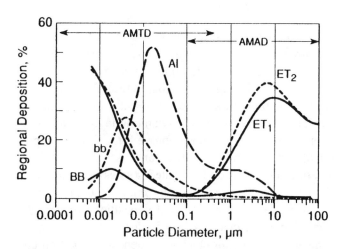

FIGURE 5-7. Fractional deposition in each region of respiratory tract for reference light worker (normal nose breather) in 1994 ICRP model.[46] Deposition is expressed as a fraction of activity present in volume of ambient air that is inspired, and activity is assumed to be lognormally distributed as a function of particle size (for particles of density 2.25 g/cm³ and shape factor 1.5). The activity median particle diameter (AMAD) applies to larger particles, while the activity median thermal diameter (AMTD) applies to smaller particles where deposition is by diffusion.

Respirable Dust Versus Lung Dust

For the pneumoconiosis-producing dusts, the aerosol of direct interest is that fraction retained in the alveoli for long periods of time. All of the dust that penetrates the ciliated airways and reaches the alveolar region is not retained. Some is exhaled without deposition, and some of the dust that does deposit in the alveoli is cleared out relatively rapidly. Thus, it might seem that the best and most direct way to determine how much dust has a long retention time is to compare the size-mass distribution of inhaled dust with dust actually retained in the lungs. Unfortunately, such studies are difficult to perform in animals and cannot be performed in humans. Human lungs obtained after accidental deaths could be analyzed, but the amount of dust inhaled would not be accurately known.

Cartwright and Skidmore[57] exposed rats to well-characterized clouds of coal dust and glass microspheres. The comparison of airborne, respirable coal dust levels (determined by Hexhlet thimble samples) with lung dust was limited in accuracy because the thimble samples included aggregate particles, whereas lung dust can be sized only after complete redispersion. The glass sphere aerosols lacked this complication. They found that the lung dusts from animals with high dust retentions had the same size distributions as the lung dust from animals with low dust retention. Also, comparisons of the recoveries of glass spheres from animals killed 6 months after the exposure with those killed 5 days after the exposure showed that one-half of the dust retained at 5 days was eliminated 6 months later, without any change in size distribution. Thus, they concluded that a sampler following the British Medical Research Council (BMRC) criteria for "respirable" dust, as defined in the next part of this review, had retention characteristics corresponding reasonably well with rat-lung retention.

Carlberg et al.[58] measured total dust, free silica, and trace metal concentrations in 65 West Virginia bituminous coal miners' lungs and compared the values obtained with those reported by others for English, Welsh, and German miners. Although coal and total dust had nearly equal concentrations in the lungs and hilar lymph nodes, the silica was more concentrated in the lymph nodes by a factor of about 3.6.

Total or Gross Air Concentration Measurements

Aerosol sampling is most commonly performed using single-stage collectors, and the collected samples are analyzed to determine the mass concentration of the

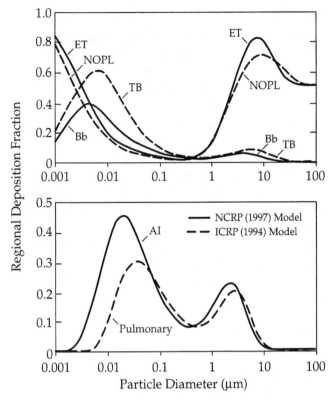

FIGURE 5-8. Comparison of regional deposition fractions predicted by the proposed National Council on Radiation Protection (NCRP) model with those of the International Commission on Radiological Protection (ICRP) Publication 66 (1994) model. Predictions are for unit density, spherical particle inhaled through the nose of an adult male with a tidal volume of 750mL, respiratory frequency of 12 min⁻¹, and functional residual capacity (FRC) of 3300mL. See text for an explanation of abbreviations for respiratory tract compartments.

Perhaps the message from this intercomparison of the careful work of two distinguished expert committees is that considerable uncertainty remains about regional deposition, reflecting our incomplete understanding of airway anatomy, flow factors, and pathways for particle clearance, and their variability among individuals in the population. As techniques improve for controlled experimental studies of deposition, and as more data from carefully designed and executed studies accumulate, it should be possible to reduce current uncertainties.

Because total and regional deposition are particle-size dependent, changes in size due to droplet growth can cause significant changes in deposition pattern and efficiency. If a hygroscopic aerosol with a dry size of 1.0 μm or larger is released, its growth in the atmosphere[48–52] or the respiratory tract[53–56] could have a major effect on its regional deposition. For the larger droplets, the fractional deposition in the head and tracheobronchial zones increases rapidly with increasing droplet size (see Figures 5-3, 5-4, and 5-5).

overall sample or constituents thereof. It is also often used for determining number concentration, as for fiber counting and viable aerosol counting after incubation and growth of colonies (see Chapter 22).

In the past, reports of air concentration measurements often implied that there was something called "total airborne dust" or "total suspended particulate" that could be measured simply by drawing air through a collector, without regard to the design of the inlet. Because most aerosols are polydisperse, with a geometric standard deviation (σg) > 2, the mass median size approaches the diameter of the largest particles in the sample. However, particles above a certain size may not be aspirated into the sampler inlet. When the aerosol being sampled contains very large particles, the gross air concentrations determined using various samplers may differ from one another and from the true total concentration. The aerosol could include large particles that dominate the measured mass concentration and yet have little biological significance. Alternatively, when large particles are important, representative samples may not be collected. Very large particles may be important; for example, wood dusts that cause nasal cancers, or highly soluble materials that deposit in the nasal or oral passages and are taken up systemically.

There has been some progress in recent years toward defining "total" dust for workplace and community air sampling purposes. One approach has been to define the biologically important fraction, i.e., the "inspirable" or "inhalable" fraction, defined as the fraction of the total aerosol that enters the nose and mouth. This principle has been adopted by the International Standards Organization (ISO)[59] and by ACGIH.[60] They both proposed that future exposure limits be based on the "inspirable" fraction. An alternate approach to the biological one is to define a standard sampling method without prejudging what is thereby collected. A separate proposal to ISO was that "total" dust should be defined as that collected by a sampling device into which air enters at a velocity of between 1.1 and 3 m/s and in which the volumetric flow rate is between 0.5 and 4 L/min. Ogden[61] examined the results of three computational studies in order to estimate what particle size-range such a device would collect provided that it is sharp-edged and operating in calm air. He found that a particle of aerodynamic diameter d_a (cm) would be collected with better than 90% efficiency by a sharp-edged sampler of diameter D (cm) and entry velocity V (cm/s) in an external wind W (cm/s) provided that

and

$$d_a < 0.003D^{0.2}V^{0.09}$$

$$w < 0.002(D^2V / d_a{}^4)^{1/3}$$

Thus, a sharp-edged sampler with this proposed ISO specification would efficiently collect particles up to about 40-μm aerodynamic diameter, but this would be limited to winds less than about 10 cm/s. For blunt samplers, the diameter limit may be about half as much. The theory for moving air is less well-developed, and sampler shape would affect efficiency. One cannot, therefore, say what the ISO "total" dust proposals correspond to in moving air. However, experience gained from efficiency measurements on practical samplers should make it possible to make static and personal samplers that meet the "inspirable" specification, and the indications are that such a sampler would, under most conditions, collect more than a sampler meeting the proposed "total" specification.

In practice, there is a broad range of inlet efficiencies among the samplers used in the field. Buchan *et al.*[62] described the inlet penetration of open- and closed-face 37-mm filter cassettes. Chung *et al.*[63] reported on the inlet penetration of 12 different personal samplers mounted on a tailor's dummy in a wind tunnel. Vincent and Mark[64] did additional wind tunnel tests on nine personal samplers and five static samplers, and summarized their results along with those of Chung *et al.*[63] The most recent study of personal inhalable particulate matter samplers was reported by Kenny *et al.*[65] There is clearly an enormous range of inlet efficiencies for large particles among the commonly used samplers.

Measurement of Mass Concentrations Within Size-graded Aerosol Fractions

Because the dose from inhaled toxicants is dependent on the regional deposition, which is dependent on particle size, the best dose estimates for a material whose toxicity is proportional to absorbed mass can be derived from a knowledge of the mass concentrations within various size ranges. Such information can be obtained in several ways: 1) by separating the aerosol into size fractions corresponding to anticipated regional deposition during the process of collection; 2) by making a size distribution analysis of the airborne aerosol, e.g., with an aerosol centrifuge, cascade impactor, or light-scattering aerosol spectrometer; and 3) by making a size distribution analysis of a collected sample.

The most reliable information can be obtained using methods in which the aerosol is fractionated on the basis of aerodynamic diameters in much the same manner as it is fractionated within the respiratory tract. Thus, differences in particle shape and density are compensated for automatically.

Light-scattering instruments that sort the pulses resulting from the scattered light from individual particles can provide information on the distribution of airborne particle diameters. In converting this information to a size-mass distribution, an average particle density must be assumed. Furthermore, the accuracy of the diameter distribution is dependent on the particle shape, index of refraction, and surface roughness. For example, Whitby and Vomela[66] reported that for India ink particles, which absorb light and have a rough surface, the indicated size was one-half to one-fifth of the true size for the three different instrument designs tested.

Further opportunities for error arise when the size distribution analysis is performed on collected samples. It is almost impossible to examine the sample in the original state of dispersion. Thus, particles that were unitary in the air may be analyzed as aggregates and vice versa. Furthermore, particles analyzed by microscopy will be graded by a linear dimension or by projected area diameter, and these are normally larger than the true average diameter. Also, there is no way to distinguish between toxic and nontoxic particles.

Standards and Criteria for Respirable Dust Samplers—Historical Review (1952–1980)

British Medical Research Council

In 1952, the BMRC adopted a definition of "respirable dust" applicable to pneumoconiosis-producing dusts. It defined respirable dust as that reaching the alveolar region. The BMRC selected the horizontal elutriator as a practical size selector, defined respirable dust as that passing an ideal horizontal elutriator, and selected the elutriator cutoff to provide the best match to experimental lung deposition data. The same standard was adopted by the Johannesburg International Conference on Pneumoconiosis in 1959.[67]

To implement these recommendations, it was specified that

1. For purposes of estimating airborne dust in its relation to pneumoconiosis, samples for compositional analysis or for assessment of concentration by a bulk measurement, such as that of mass or surface area, should represent only the "respirable" fraction of the cloud.

2. The "respirable" sample should be separated from the cloud while the particles are airborne and in their original state of dispersion.

3. The "respirable fraction" is to be defined in terms of the free falling speed of the particles, by the equation $C/Co = 1-(f/fc)$, where C and Co are the concentrations of particles of falling speed, f, in the "respirable" fraction and in the whole cloud, respectively, and fc is a constant equal to twice the falling speed in air of a sphere of unit density 5 μm in diameter.

A sampling device meeting these requirements would have a sampling efficiency versus size curve suggested by Davies,[68] as illustrated in Figure 5-9.

U.S. Atomic Energy Commission

A second standard, established in January 1961 at a meeting sponsored by the U.S. Atomic Energy Commission (AEC), Office of Health and Safety,[3] defined "respirable dust" as that portion of the inhaled dust penetrating to the nonciliated portions of the lung. This application of the concepts of respirable dust and concomitant selective sampling was intended only for "insoluble" particles that exhibit prolonged retention in

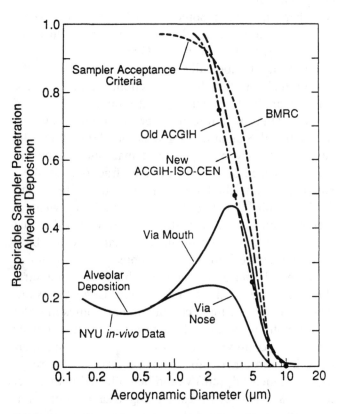

FIGURE 5-9. Comparison of respirable sampler acceptance curves of BMRC and ACGIH with new ACGIH–ISO–CEN criteria, and with median human *in vivo* alveolar deposition data.[21]

the lung. This was not intended to include dusts having appreciable solubility in body fluids and those that are primarily chemical toxicants. Within these restrictions, "respirable dust" was defined as being 0% at 10 μm, 25% at 5 μm, 50% at 3.5 μm, 75% at 2.5 μm, and 100% at 2.0 μm, all sizes being aerodynamic diameters.

American Conference of Governmental Industrial Hygienists

The application of respirable dust sampling concepts to other toxic dusts and the relationships between respirable dust concentrations and accepted standards such as the ACGIH TLVs are more complicated. Unlike the exposure limits for radioisotopes, which are based on calculation, most TLVs are based on animal and human exposure experience. Thus, even if the data on which these standards were based could be related to the particle size of the dust involved, which unfortunately is unlikely, there probably would be a different correction factor for each TLV rather than a uniform factor.

ACGIH initiated the process of adopting "respirable" dust limits at its 1968 annual meeting by including, in their "Notice of Intended Changes," alternate mass concentration TLVs for quartz, cristobalite, and tridymite (three forms of crystalline free silica) to supplement the TLVs based on particle count concentrations. For quartz, the alternative mass values proposed were:[69]

1. For respirable dust in mg/m³,

$$\frac{10 \text{ mg/m}^3}{\% \text{ Respirable Quartz} + 2}$$

Note: Both concentration and % quartz for the application of this limit are to be determined from the fraction passing a size-selector with the following characteristics:

(Aerodynamic Diameter) – μm ≤	2.0	2.5	3.5	5.0	10
% Passing Selector –		90	75	50	25	0

2. For "total dust" (respirable and nonrespirable),

$$\frac{30 \text{ mg/m}^3}{\% \text{ Quartz} + 2}$$

For both cristobalite and tridymite, use one-half the value calculated from the count or mass formula for quartz.

The size-selector characteristic specified by ACGIH was almost identical to that of the AEC, differing only at

2 μm, where it allowed for 90% passing the first stage collector instead of 100%. The difference reflected a recognition of the characteristics of real particle separators. For practical purposes, the two standards may be considered equivalent.

The proposed mass concentration limits were obtained by a comparison of simultaneous impinger and size-selective samples collected in the Vermont granite sheds.[70] Because the original impinger sampling and microscopic particle counting standards were based on epidemiological investigations, which had been performed 3–4 decades earlier in some of the same granite cutting sheds, it was possible to make a valid comparison of "respirable" mass and particle count.

In 1969, the U.S. Department of Labor adopted the ACGIH size-selector criteria for respirable dust and extended its application to coal dust and inert or nuisance dust. In the revised Safety and Health Standards for Federal Supply Contracts published in the Federal Register,[71] the ACGIH quartz, tridymite, and cristobalite TLVs were adopted along with the following respirable dust limits:

Coal Dust—2.4 mg/m³ or

$$\frac{10 \text{ mg/m}^3}{\% \text{SiO}_2 + 2} \text{ (Respirable fraction} < 5\% \text{ SiO}_2)$$

Inert or Nuisance Dust—15 million particles per cubic foot (mppcf) or

5 mg/m³ (Respirable fraction)

The Federal Coal Mine Health and Safety Act of 1969[72] specified that:

"References to concentrations of respirable dust in this title means the average concentration of respirable dust if measured with an MRE instrument or such equivalent concentrations if measured with another device approved by the Secretary (of Interior) and the Secretary of Health, Education and Welfare. As used in this title, the term 'MRE instrument' means the gravimetric dust sampler with four channel horizontal elutriator developed by the Mining Research Establishment of the National Coal Board, London, England."

Although the 1969 Act specified the MRE instrument, which closely follows the BMRC sampling criteria, the Federal Mine Safety and Health Act of 1977,[73] which superceded it, did not. The National Research Council Committee on Measurement and Control of

Respirable Dust in Mines[74] noted that it may be more appropriate to use the definition of respirable dust adopted by ACGIH because human deposition data demonstrate that the ACGIH curve is a better representation of respirable dust than the BMRC curve.

The Occupational Safety and Health Act of 1970[75] has led to the adoption of only a few permanent standards, and none of them addressed the issue of "respirable" dust. As a result, the Occupational Safety and Health Administration (OSHA) enforced numerous interim permissible exposure limits (PELs), including 22 maximum allowable concentrations (MACs) of the American National Standards Institute (ANSI) and approximately 280 of the ACGIH 1968 TLVs,[69] including the silica TLVs which specify either dust counts or respirable mass concentrations. The Mine Safety and Health Administration (MSHA) of the U.S. Department of Labor operates under different enabling legislation and uses the 1973 TLVs as PELs which, for silica, are the same as the 1968 values. Most of the transitional OSHA PELs were superceded by the adoption of permanent PELs in 1989,[76] but this rulemaking was overturned by the U.S. Eleventh Circuit Court of Appeals in 1992, thus restoring the transitional PELs.

Comparison of Standards for Respirability

Basically, there are two sampler acceptance curves described in the preceding discussion, and they have similar, but not identical characteristics (see Figure 5-9). The shapes of the curves differ because they are based on different collector types. The BMRC curve was chosen to give the best fit between the calculated characteristics of an ideal horizontal elutriator and lung deposition data, whereas the AEC curve was patterned more directly after the Brown et al.[77] upper respiratory tract deposition data and is simulated by the separation characteristics of cyclone-type collectors. In most field situations, where the geometric standard deviation of the particle size distribution is 2, samples collected with instruments meeting either criterion will be comparable. For example, Mercer[78] calculated the predicted pulmonary (alveolar) deposition according to the ICRP Task Group deposition model[23] for a tidal volume of 1450 cm$_3$ and aerosols with $1.5 < \sigma g < 4$. He found that a sampler meeting the BMRC acceptance curve would have about 10% more penetration than a sampler meeting the AEC curve.[78]

Maguire and Barker[79] made eight coal mine tests with SIMPEDS cyclones adjacent to MRE elutriators. The average respirable dust ratio was 0.97, with a standard deviation of 0.11; i.e., there was no statistical difference.

Comparative "respirable" mass sampling in the Vermont granite sheds for granite cutters operating their equipment without exhaust ventilation produced the following concentrations: a 10-L/min NIOSH elutriator—11.6 mg/m^3; the MRE (Isleworth) 2.5-L/min sampler —10.7 mg/m^3; the HASL 1/2-in cyclone at 10 L/min—10.9 mg/m^3; and the 10-mm nylon cyclone at 1.7 L/min—10.7 mg/m^3. Thus, for practical purposes, they were equivalent in performance.[80]

It is apparent from the preceding discussions that the various definitions of respirable dust are somewhat arbitrary. The BMRC and AEC definitions are based on the aerosol that reaches the alveolar region. Thus, they do not predict alveolar deposition because part of the aerosol that penetrates to the alveoli remains suspended in the exhaled air. The portion that does not deposit is a variable that depends on particle size.

Standards and Criteria for Health-based, Inhalable, Thoracic and Respirable Particulate Matter Samplers—Recent Decades (after 1980)

Comprehensive definitions are clearly needed for particles that deposit in the head and tracheobronchial regions, causing diseases such as nasal and bronchial cancers and chronic bronchitis. Three groups have addressed this need. The first was the U.S. EPA on the basis of its responsibility to protect the public health from diseases associated with the inhalation of airborne particles. The second was the ISO on the basis of their desire to have better sampling specifications for test methods used to determine potential inhalation hazards in both the workplace and general community atmospheres. Similar criteria were subsequently adopted by ACGIH for use with PSS–TLVs.

U.S. Environmental Protection Agency

In developing a revised primary ambient air quality standard for particulate matter (PM) to protect the public health, U.S. EPA concluded that the diseases which could be related to the inhalation of ambient aerosols were associated with particles that penetrated through the upper respiratory tract and were available for deposition in the tracheobronchial and/or alveolar regions. They initially called this fraction "inhalable" dust.[81] Because they were only concerned about the particles entering the trachea, they took a conservative position on the selection of the appropriate cut size for a precollector, proposing a D_{50} (50% cut size) at an aerodynamic diameter of 15 μm on the basis of published data indicating that about 10% of the particles of this size would enter the trachea of a mouth-breathing person.

The use of the word "inhalable" to designate particles penetrating though the upper respiratory airways and entering the thorax was in conflict with the usage of the word in Europe, where it was defined as the particles that entered the nasal or oral air passages.[82, 83]

On the basis of public comment, the subsequent recommendations of an ISO Task Group (discussed in the next paragraph), and the recommendation of the U.S. EPA Clean Air Scientific Advisory Committee, the U.S. EPA Office of Air Quality Planning and Standards recommended to the U.S. EPA Administrator that U.S. EPA revise the criterion for the particulate matter primary standard for ambient air to include a D_{50} of 10 μm. The fraction below the 10-μm cut, designated by ISO as thoracic particulate (TP) or by U.S. EPA as PM_{10} (particulate matter below a 10-μm cut size), replaced total suspended particulate (TSP) as the basic ambient air particulate pollution parameter in 1987. The adoption of the PM_{10} standard provided a basis for the collection of ambient air concentration data of better relevance to potential inhalation hazards.[84] The actual specification for inlets matching the PM_{10} criteria is shown in Table 5-3. The ideal sampler as defined by U.S. EPA, is one that matches particle penetration to the thorax as illustrated in Figure 5-10.

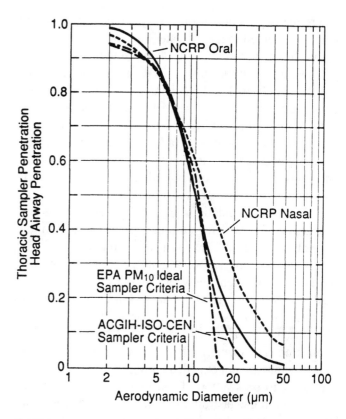

FIGURE 5-10. Comparison of thoracic sampler acceptance curves of U.S. EPA (PM_{10}) and ACGIH-ISO-CEN with NCRP thoracic penetration models for nasal and oral inhalation based on curves fitted to *in vivo* experimental data for aerosol penetration through the airways in the human head.[47]

TABLE 5-3. U.S. EPA's Performance Specifications for PM_{10} Samplers

Performance Parameter	Units	Specification
1. Sampling effectiveness A. Liquid particles	%	Such that the expected mass concentration is within ± 10% of that predicted for the ideal sampler.
B. Solid particles	%	Sampling effectiveness is no more than 5% above that obtained for liquid particles of same size.
2. 50% cutpoint	μm	10 ± 0.5-μm aerodynamic diameter.
3. Precision	μg/m³ or %	5 μg/m³ or 7% for three collocated samplers.
4. Flow rate stability	%	Average flow rate over 24 hours within ± 5% of initial flow rate; all measured flow rates over 24 hours within ± 10% of initial flow rate.

International Standards Organization (ISO)

Technical Committee 146—Air Quality of the ISO appointed an *ad hoc* working group to prepare recommendations on size definitions for particle sampling to be used in preparing standard methods for the sampling and analysis of air contaminants in both occupational and general environmental settings. The working group used the available human regional deposition data to define a series of aerosol fractions related to particle deposition within specific regions of the human respiratory tract.[57] To avoid conflict with the proposed U.S. EPA definition of "inhalable," the fraction drawn in by the nose or mouth was called "inspirable," that part collected in the head was called "extrathoracic," and that part penetrating through the larynx was called "thoracic" and was further subdivided into "tracheo-bronchial" and "alveolar." The ISO adopted a thoracic D_{50} cut of 10 μm. They also provided two options for the "respirable" size cut. Their recommendations accommodated both the BMRC and ACGIH criteria, according to national preference. They also endorsed an alternate alveolar convention for ambient air sampling where the target population is very young or infirm and may be expected to have greater tracheobronchial deposition. It was very similar in shape to the ACGIH "respirable" dust criteria, but with all the diameter values reduced by 29%. The ISO thoracic cut was essentially consistent with U.S. EPA's PM_{10} standard that specifies a 10-μm D_{50}, but differed in that the 10-μm cut was applied to the aerosol penetrating a precollector following the "inspirable" cut convention. The ISO standard was revised in 1995 and is now identical to the latest criteria adopted by ACGIH and CEN.[85]

The recommendations of the ISO working group also provided a basis for a thorough reexamination of air concentration limits for occupational exposures. For some, such as droplets or soluble components of solid particles, deposition anywhere in the respiratory tract leads to absorption by the tissues, and the current total concentration limits may be appropriate. For other particles, the biological effect may depend on the region of deposition. For example, particles depositing extrathoracically that are not expelled through the nose or mouth are likely to be swallowed and may cause a hazard by absorption in the gastrointestinal tract. Particles depositing in the tracheobronchial region and cleared by the mucociliary escalator are also likely to be swallowed, so that gastrointestinal absorption is a possible route for these particles also. Particles depositing in the alveolar region may also be cleared by this route, or through the lymphatic system, or may cause a reaction in the alveolar region itself.

American Conference of Governmental Industrial Hygienists

In 1982, the ACGIH Board appointed an *ad hoc* Committee on Air Sampling Procedures (ASP) to prepare general recommendations for size-selective sampling appropriate to size-selective TLVs for particulate materials.

The ASP Committee of the ACGIH had as its primary charge

". . . to recommend size-selective aerosol sampling procedures which will permit reliable collection of aerosol fractions which can be expected to be available for deposition in the various major subregions of the human respiratory tract, e.g., the head, tracheobronchial region, and the alveolar (pulmonary) region."

It was anticipated from the outset that the work of this committee would lead to an approach for establishing PSS–TLVs for many airborne agents. The ASP Committee reviewed the relevant literature and the recommendations of other groups on size-selective aerosol sampling; its report and recommendations were presented to the Board of Directors and the ACGIH membership at the 1984 Annual Membership Meeting. The report of the ASP Committee and its background documentation were published in the 1984 *Transactions of the American Conference of Governmental Industrial Hygienists*[86] and in 1985 as a separate document entitled *Particle Size-Selective Sampling in the Workplace*.[60] The latter publication was revised and expanded and released by ACGIH in 1999 under the title *Particle Size-selective Sampling for Health-related Aerosols*.

The following paragraphs summarize the recommendations of the ASP Committee in 1984.

The ASP Committee report was a background document summarizing the available data on 1) airway anatomy and physiology that influence the deposition and retention of inhaled particles, 2) penetration of inhaled particles into the major functional regions of the respiratory tract, 3) the particle size collection characteristics of available size-selective aerosol samplers, and 4) evaluation of the performance of samplers. The ASP Committee also reviewed the basis for its particular recommendations on size-selective sampling criteria and how and why they differed from the recommendations of others.

The major functional regions of the human respiratory tract were given different names and/or abbreviations than those used by others, but were anatomically equivalent, as indicated in Figure 5-11 and Table 5-4. The designations chosen were, in the ASP Committee's view, more anatomically correct and unambiguous.

Deposition within the head airways region (HAR) was associated with an increased incidence of nasal cancer in wood and leather workers and in ulceration of the nasal septum in chrome refinery workers. Within the tracheobronchial region (TBR), deposited particles can contribute to the pathogenesis of bronchitis and bronchial cancer. Particles depositing within the gas-exchange region (GER) can cause emphysema and fibrosis. On the other hand, the hazards from inhaled materials that exert their toxic effects on critical sites outside the respiratory tract, after dissolution into circulating fluids, depend on total respiratory tract deposition rather than on deposition within one region.

The ASP Committee considered several options for size-selective sampling of fractions of the aerosol that represent hazards for specific health endpoints. The major options were 1) samplers that would mimic deposition in the specific regions of interest and 2) samplers that would collect those particles that would penetrate to,

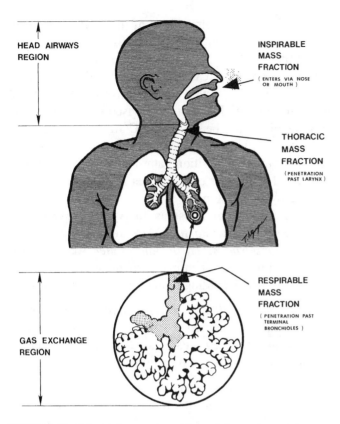

FIGURE 5-11. Schematic representation of the major respiratory tract regions of ACGIH, i.e., head airways region (HAR), tracheobronchial region (TBR), and gas-exchange region (GER).

practical because this approach had proven to be effective in "respirable" dust sampling. It was recognized that "respirable dust" concentrations may be as much as 5–10 times greater than the fraction actually depositing in the lungs because 80% to 90% of particles in the 0.1-μm- to 1.0-μm-diameter range may be exhaled. However, because the fraction deposited in the GER is relatively constant over the whole "respirable dust" size range, the "respirable dust" concentration is an adequate index of the hazard. A sampler that would mimic GER deposition would be much more difficult to design and operate, and it would not give a materially better index of hazard.

The aerosol that enters the HAR is called the inspirable particulate mass (IPM) fraction. The aerosol that penetrates the HAR and enters the TBR is called the thoracic particulate mass (TPM) fraction. Here, the ASP Committee chose to define the thoracic mass fraction on the basis of data for HAR deposition during mouth breathing. The difference between the IPM and the TPM fractions approximates the deposition fraction in the HAR occurring during mouth breathing. Therefore, because nasal inhalation would almost always produce more HAR deposition than oral inhalation, actual HAR deposition during nasal breathing would be greater than that calculated. Similarly, the TPM fraction would overestimate the hazard to the TBR region for nose-breathing workers.

The ASP Committee's selection of a mouth-breathing model rather than a nose-breathing model was made in order to be conservative. Occupational diseases of the lung airways are much more common than are diseases of the head airways. Also, heavy work in industry is

but not necessarily deposit in, the specific region of interest. The ASP Committee opted for the latter approach as the one requiring simpler and less expensive samplers. They also concluded that it would be more

TABLE 5-4. Respiratory Tract Regions as Defined in Particle Deposition Models

ACGIH Region	Anatomic Structure Included	ISO Region	1966 ICRP Task Group Region	1994 ICRP Task Group Region	1994 NCRP Task Group Region
1. Head airways (HAR)	Nose Mouth Nasopharynx Oropharynx Laryngopharynx	Extrathoracic (E)	Nasopharynx (NP)	Anterior nasal passages (ET$_1$) All other extrathoracic (ET$_2$)	Naso-oro-pharyngo-laryngeal (NOPL)
2. Tracheobronchial (TBR)	Trachea Bronchi Bronchioles (to terminal bronchioles)	Tracheobronchial (B)	Tracheobronchial (TB)	Trachea and large bronchi (BB) Bronchioles (bb)	Tracheobronchial (TB)
3. Gas-exchange (GER)	Respiratory bronchioles Alveolar ducts Alveolar sacs Alveoli	Alveolar (A)	Pulmonary (P)	Alveolar-interstitial (AI)	Pulmonary (P)

believed to cause a significant fraction of workers to engage in mouth breathing during periods of maximal activity, which may coincide with maximal levels of airborne dust. The algebraic difference between TPM and respirable particulate mass (RPM) approximates tracheobronchial region deposition during oral breathing. For nasal breathing individuals, the difference between TPM and RPM is a poor estimate of tracheobronchial deposition.

In general, mass concentrations tend to be dominated by the largest size-fraction collected. In consideration of all these factors, the ASP Committee recommended samplers that follow its IPM criteria be used for sampling those materials which are hazardous when deposited in the HAR or when systemic toxicity can follow from deposition anywhere in the respiratory tract. For those materials that represent a hazard when deposited on the conductive airways of the lungs, the ASP Committee recommended using a sampler that follows its criteria for TPM. Finally, for those materials, such as silica, which are hazardous only after deposition in the GER, the ASP Committee recommended using a sampler that follows its RPM sampling criteria.

The ASP Committee's recommendations for the performance specifications of samplers that would mimic aerosol penetration into these regions are similar, but not identical, to those of ISO. The most notable differences are in the IPM criteria and the RPM criteria. In terms of the former, the ACGIH ASP Committee had the advantage over ISO of access to deposition data in the head for particles larger than 40 μm in aerodynamic diameter that were not available to the ISO Working Group. The ISO Group made the reasonable, but inadequate, assumption that the < 40-μm data could be extrapolated to zero deposition at 185 μm. For RPM, the major difference was in not having the alternate criteria based on the BMRC recommendations.

The ASP Committee's recommendations for sampling TPM were quite similar to those of ISO[59] and U.S. EPA.[84] The recommendations also contained sampler acceptance envelopes about the recommended curves.

Following acceptance of the ASP Committee's recommendations by the ACGIH Board of Directors in 1984, activities to implement the recommendations proceeded in two ACGIH Technical Committees. The Chemical Substances TLV Committee addressed the use of the Particle Size-Selective Criteria for Airborne Particulate Matter by listing the criteria as an issue under study in the TLV/BEI booklet for 1986–1987. In the following year, ACGIH adopted these criteria as a separate appendix in the booklet.

In 1989, Soderholm, Chair of the ASP Committee, with the endorsement of the full ASP Committee, proposed modified particle size-selective sampling criteria for adoption by ACGIH, ISO, and the European Community (CEN), with the objective of international harmonization.[87] In effect, his revised sampling criteria split the difference between the ACGIH and ISO criteria consistent with matching the best available total and regional human deposition data. His initiative was well received by the interested parties and is being implemented by all concerned.[85, 88, 89]

The Soderholm revisions to Appendix D of the TLV/BEI booklet, *Particle Size-selective Sampling Criteria for Airborne Particulate Matter,* were adopted by ACGIH in 1993. The three particulate mass fractions were redefined according to the following equations:

A. *Inhalable Particulate Mass* consists of those particles that are captured according to the following collection efficiency regardless of sampler orientation with respect to wind direction:

$$SI(d) = 50\% \times (1 + e^{-0.06d})$$
$$\text{for } 0 < d \leq 100 \ \mu m$$

where: $SI(d)$ = the collection efficiency for particles with aerodynamic diameter d in μm

B. *Thoracic Particulate Mass* consists of those particles that are captured according to the following collection efficiency:

$$ST(d) = SI(d) \ [1 - F(x)]$$

where:

$$x = \frac{ln(d/\Gamma)}{ln(\Sigma)}$$
$$\Gamma = 11.64 \text{ mm}$$
$$\Sigma = 1.5$$

$F(x)$ = the cumulative probability function
of a standardized normal variable, x

C. *Respirable Particulate Mass* consists of those particles that are captured according to the following collection efficiency:

$$SR(d) = SI(d) \ [1 - F(x)]$$

The most significant difference from previous definitions was the increase in the median cut point for a respirable dust sampler from 3.5 μm to 4.0 μm; this was in accord with the International Standards Organization/European Standardization Committee (ISO/CEN) protocol.[90] No change was recommended for the measurement of respirable dust using a 10-mm nylon cyclone at a flow rate of 1.7 L/min. Two analyses of available data indicated that the flow rate of 1.7 L/min allows the 10-mm nylon cyclone to approximate the dust concentration which would be measured by an ideal respirable dust sampler as defined herein.[91]

Collection efficiencies representative of several sizes of particles in each of the respective mass fractions are shown in Table 5-5. References 85 and 87 provide documentation for the respective algorithms representative of the three mass fractions. The respirable and thoracic sampling criteria are illustrated in Figures 5-9 and 5-10, respectively.

An issue initially raised in Appendix F of the *1986–87 TLV/BEI* booklet concerned the changing of all TLVs that were explicitly defined in terms of "total dust" to "inspirable particulate mass" without changing the numerical values. At its 1993 Annual Meeting, ACGIH endorsed the need to examine each TLV for airborne particles for conversion to PSS–TLVs. Since then, a number of substances have been specified in new or proposed TLVs in terms of the inhalable fraction.

The ASP Committee undertook several additional activities related to the development of size-selective TLVs. By extension of its initial activities, it responded to comments received and made some modifications to its recommendations.

Bartley and Doemeny[92] of the National Institute for Occupational Safety and Health (NIOSH) objected to the original ACGIH sampler acceptance criteria. The ASP Committee had recommended that calibrations be performed at equal size intervals between 2- and 10-μm aerodynamic diameter, and that the correlation coefficient (r^2) of the linear fit be larger than 0.90.

The Bartley and Doemeny critique had two primary concerns: 1) that the statistical tests required for determining satisfactory sampler performance were too difficult to meet for real samplers; and 2) that the criteria would, for certain aerosol size distributions, permit instruments that differed too greatly in their measured RPM concentration. In part, these criticisms arose because of differences in perspective. Although NIOSH and ACGIH both provide technical information and professional guidance to facilitate the protection of worker health, NIOSH has the further role of certifying samplers for compliance purposes. In this regard, NIOSH develops performance criteria for size-selective samplers.

Bartley and Doemeny's concern about the sampler performance criteria was well founded and was addressed by changes to the criteria for acceptable test performance adopted by the ASP Committee. The simplified performance criteria for an RPM sampler required only that tests be performed at 10 monodisperse particle sizes between 2 and 10 μm, and that 9 out of the 10 points fall within the acceptance bands given. A point is considered to be within the band if more than 50% of replications at that particle size lie within the acceptance band.

The simplification of the criteria greatly reduced the range of RPM concentration that could be measured with instruments meeting the criteria. When an RPM sampler performing according to the lower bound is compared to a sampler performing according to the upper bound (a worst-case difference in performance which is extremely unlikely in practice), there is only about a factor of two difference in RPM for the worst-case size distribution given by Bartley and Doemeny, a

TABLE 5-5. Inhalable, Thoracic, and Respirable Dust Criteria of ACGIH-ISO-CEN

Inhalable		Thoracic		Respirable	
Particle Aerodynamic Diam. (μm)	Inhalable Particulate Mass (IPM) (%)	Particle Aerodynamic Diam. (μm)	Thoracic Particulate Mass (TPM)(%)	Particle Aerodynamic Diam (μm)	Respirable Particulate Mass (RPM) (%)
0	100	0	100	0	100
1	97	2	94	1	97
2	94	4	89	2	91
5	87	6	80.5	3	74
10	77	8	67	4	50
20	65	10	50	5	30
30	58	12	35	6	17
40	54.5	14	23	7	9
50	52.5	16	15	8	5
100	50	18	9.5	10	1
		20	6		
		25	2		

size distribution for coal mine dust with mass median diameter (MMD) of 18.5 μm and geometric standard deviation (σg) of 2.3. For this condition, the RPM represents only 2.3% to 4.7% of the total mass. The nuisance dust TLV of 10 mg/m³, if enforced, would limit dust concentration such that RPM would be well below the coal mine dust TLV of 2 mg/m³, and the twofold difference in measured RPM concentration would therefore be relatively insignificant.

Bartley and Doemeny[92] and Liden and Kenny[93] recommended that sampler equivalence be measured by simulated performance with various test particle size distributions for known types of dust exposure such as coal mine dust. However, the ASP Committee had reservations about the sampler equivalence approach as being appropriate for professional practice recommendations that would be applied to an extremely wide range of size distributions.

McCawley[94] argued that "respirable" dust is not a good index for inhalation hazard for dust which deposits in the gas-exchange region of the lung, he suggested it would be better to design a sampler that collects a dust fraction more closely related to what actually deposits in that region. This approach to hazard evaluation has a long history of application in Germany.[95]

Knight[96] presented a rationale for an approach to the specification of size-selective sampling criteria, which had been considered and rejected by the ASP Committee. He suggested that "The mean anatomical regional depositions can be obtained with a reasonable degree of accuracy" from linear combinations (adding, subtracting, and scaling) of the three sampler results. Although the regional dose approach is conceptually sound, the ASP Committee maintained that its regional exposure approach has several advantages and no serious disadvantages. It is simpler, more operationally reliable, and, in many cases, more conservative (i.e., protective).

The development of PSS–TLVs for specific substances other than mineral dusts has advanced slowly. Technical papers have documented the basis for particle size-selective sampling for beryllium,[97] wood dust,[98] and sulfuric acid aerosol.[99] The first step in deriving a PSS–TLV is the identification of the chemical substance that constitutes a potential air pollutant, including examination of all available physicochemical properties related to its airborne and biological behavior. Concomitantly, the literatures of epidemiology, industrial hygiene, and toxicology should be searched to identify diseases that may be associated with the chemical substance affecting specific regions of the respiratory tract or systemic organ systems.

New data gathered from these searches, including experimental animal studies, especially on recently developed substances, should be incorporated with existing TLV documentation for insight into possible disease mechanisms.

If no potential diseases related to the chemical substance are found, then the evaluation can be terminated. If a disease potential exists, but the physicochemical nature of the chemical substance is such that no airborne particle phase can be produced, the procedure can revert to the traditional procedure for establishing a TLV.

However, if the physicochemical properties of the chemical substance suggest that it may become airborne as an aerosol, the analysis proceeds. At this stage, the physical and chemical properties of the substance are evaluated under conditions likely to be encountered by workers.

The aerodynamic particle size distribution will determine the mass fraction of the workplace aerosol that will enter the head airways, tracheobronchial, or gas-exchange regions of the respiratory tract. Particle size-selective sampling is then used to estimate the actual quantity of chemical substance that will be presented to the three principal regions of the respiratory tract during the course of each working day. Thus, the mass of the substance presented to each region will be established as the critical value in airborne hazard evaluation. Once the chemical substance is deposited in a particular region or regions of the respiratory tract, the critical factor in selecting the appropriate particle mass fraction (respirable, thoracic, or inspirable) is the extent of dissolution of the substance within each region.

Concurrent examination of the clinical diseases that may affect any systemic organ will identify extrapulmonary sites of action. Subsequently, it will be determined whether the incorporated dose of the substance is a critical dose that is likely to cause acute or chronic injury. Once the particle size and particulate mass fraction are determined and the hazard analyses are completed, a critical mass concentration will be determined for an appropriate size fraction. This review will result in a recommendation for a PSS–TLV.

If the inhaled chemical material is likely to dissolve only slowly or is essentially insoluble after deposition in any of the three principal regions of the respiratory tract, selection of the appropriate particle size-selective sample should be based on the specific site of action within the respiratory tract that is associated with the most restrictive PSS–TLV, as based on comparing each potential disease.

Other Size-selective Criteria for Inhalation Hazards

Size-selective criteria adopted for sampling cotton dust, asbestos, diesel exhaust particles in coal mines, and fine particulate matter in ambient community air are different from those previously discussed. A brief review of the rationale and practice used for each of these special cases follows.

Cotton Dust Sampling

Because byssinosis or "brown lung" is characterized by an allergic response producing airway constriction, it was recognized that particles depositing in the tracheo-bronchial airways should not be excluded. Thus, conventional "respirable" dust criteria were judged to be inappropriate. On the other hand, the mass of the dust in cotton ginning and textile operations tends to be dominated by very large cotton fibers which are too large to be inspirable. These considerations led to the recommendation of a vertical elutriator with a nominal 50% cut size at 15 μm as the first stage of a standard sampler.[100] The second stage filter is analyzed for the mass concentration of the particles judged most likely to be related to the health effects.

Asbestos Sampling

In asbestos and other mineral fiber analyses, the size-selectivity is applied after the sampling. There is no sampling selectivity specified in the NIOSH[101] or ACGIH–American Industrial Hygiene Association (AIHA)[102] sampling recommendations, although the specified inlet configurations to the filter holders will, of course, impose some. In the analyses by phase contrast optical microscopy, there is an effective lower limit for fiber diameter imposed by the resolving power of the objective lens. There are also other limits specified by the methods, whereby particles with an aspect ratio (length to diameter) of less than 3 or a length of less than 5 μm, are not counted. The rationale for these exclusions is based on toxicological and epidemiological studies which showed that the toxic effects were primarily associated with long thin fibers. Asbestosis, a pneumoconiosis, and mesothelioma, a cancer of the pleural or peritoneal surfaces, are presumably related to long fibers depositing in the alveolar regions, while bronchial cancer may be related to the long fibers depositing on bronchial airways.

In a critical review of the literature on asbestos toxicity and human disease in relation to the dimensions of

TABLE 5-6. Summary of Recommendations on Asbestos Exposure Indices

Disease	Relevant Exposure Index
Asbestosis	Surface area of fibers with: Length >2 μm, diameter >0. 15 μm
Mesothelioma	Number of fibers with: Length >5 μm, diameter <0.1 μm
Lung cancer	Number of fibers with: Length >10 μm, diameter >0.15 μm

the fibers, Lippmann[103] identified the critical dimensions for each of the asbestos-associated diseases. These are summarized in Table 5-6.

Diesel Exhaust Particles in Coal Mines

It has been suggested by McCawley and Cocalis[104] that sampling for diesel particulate in underground coal mines in the presence of coal mine dust should be carried out using an impactor with a cut-point at $d_{ae} = 1$ μm. In this specific case, the rationale was that separation of the diesel particulate from coal mine dust could be achieved on a particle size-selective basis, since the coal mine dust particle size distribution had been well characterized as mechanical-mode size dust with 90% of the mass greater than 1 μm.[105, 106] This approach assumed that diesel particulate would then account for any carbon found in the fraction of dust below 1 μm. The intent of this proposal was very different from that for established particle size-selective criteria. The particle size selection was performed in order to differentiate between two distinct classes of particles, for which there did not exist at the time an established chemical separation technique.

Sampling for the Fine Particulate Mass Fraction of Ambient Air

One major responsibility of the Administrator of the U.S. EPA under The Clean Air Act is to periodically review the National Ambient Air Quality Standards (NAAQS) and, if necessary, to propose and promulgate new or revised standards.[107] The primary NAAQS are concentration limits for specified averaging times that are intended to prevent adverse acute and chronic health effects associated with exposures to pollutants having numerous and widespread sources. Furthermore, they are intended to protect sensitive or susceptible segments of the general population with an adequate margin of safety. One of the NAAQS is for ambient particulate matter (PM), without reference to chemical composition. In this sense, it is analogous to historic occupational exposure limits (OELs) for nuisance dusts, such as the ACGIH TLVs for particulates not otherwise classified (PNOCs). However, the toxicity of mixtures of chemicals in ambient air PM is generally greater than that of nuisance dust in the occupational setting.

Ambient air PM has long been associated with excess mortality and morbidity, as well as the nuisances of cinders in the eye and soiling of materials and surfaces. Such associations have generally been stronger than those for the ubiquitous pollutant gases that are also associated with combustion sources. This strength of association is remarkable, considering the historical reliance, by the epidemiological studies, on crude indices of ambient PM, especially black smoke (BS) and total suspended particulate matter (TSP). More recent epidemiological studies, many of them using monitoring data for thoracic particulate matter (measured as particles passing a pre-collector with a 50% cut-size at d_{ae} = 10 μm, i.e., PM_{10}) have generally produced even stronger associations between ambient PM and mortality and morbidity. For those relatively few studies for which fine particulate indices of exposure (particles passing a pre-collector with a d_{ae} = 2.5 μm, i.e., $PM_{2.5}$, and/or sulfate ion) have been available, the associations are generally the strongest of all.[108]

Sulfate ion ($SO_4^=$), most of which is formed in the atmosphere from the oxidation of the SO_2 emitted from fossil fuel combustion sources, is a major mass constituent of $PM_{2.5}$ in central and eastern North America. Some of it is associated with hydrogen ion (H^+), while most of it is associated with ammonium ion (NH_4^+) as a result of neutralization by ambient ammonia gas (NH_3) arising from anaerobic decay processes at ground level. Other major constituents of $PM_{2.5}$ include ammonium nitrate (especially in western North America) and fixed and organic carbon. Much of the fixed carbon is attributable to diesel engine exhaust, and most of the organic carbon is formed in the atmosphere in the photochemical reaction sequence that also leads to ambient air O_3.

The sulfate, nitrate, and fixed carbon components of $PM_{2.5}$ are unlikely causal factors for excess mortality and morbidity and we cannot, with any confidence, identify causal constituents at the time. Some of the prime suspect agents are listed in Table 5-7. Sulfate ion, which is by itself a very unlikely causal agent, may be a good surrogate index because of association in the ambient air with H^+ and peroxides. In any case, EPA staff selected $PM_{2.5}$ mass concentration as the exposure index for a new fine particle standard for ambient air.[109] The background and rationale for this selection is presented in detail in the 1996 PM Criteria Document (CD)[110] and Staff Paper[109] prepared by EPA. Additional background is available in a recent Workshop report by Lippmann et al.[111] The PM CD concluded that:

TABLE 5-7. Components of Ambient Air Particulate Matter (PM) that May Account for Some or All of the Effects Associated with PM Exposures

Component	Evidence for Role in Effects	Doubts
Strong Acid (H^+)	• Statistical associations with health effects in most recent studies for which ambient H^+ concentrations were measured	• Similar PM-associated effects observed in locations with low ambient H^+ levels
	• Coherent responses for <u>some</u> health endpoints in human and animal inhalation and *in vitro* studies at environmentally relevant doses	• Very limited data base on ambient concentrations
Ultrafine Particles (D ≤ 0.1 μm)	• Much greater potency per unit mass in animal inhalation studies (H^+ and TiO_2 aerosols) than for same materials in larger diameter fine particle aerosols	• Absence of relevant data on responses in humans
	• Concept of "irritation signalling" in terms of number of particles per unit airway surface	• Absence of relevant data base on ambient concentrations.
Peroxides	• Close association in ambient air with $SO_4^=$	• Absence of relevant data on responses in humans or animals
	• Strong oxidizing properties	• Very limited data base on ambient concentrations

- Airborne PM is not a single pollutant, but rather is a mixture of many subclasses of pollutants with each subclass containing many different chemical species. Atmospheric PM occurs naturally as fine-mode and coarse-mode particles that, in addition to falling into different size ranges, differ in formation mechanisms, chemical composition, sources, and exposure relationships (See Table 5-8).

- Fine-mode PM is derived from combustion material that has volatilized and then condensed to form primary PM or from precursor gases reacting in the atmosphere to form secondary PM. New fine-mode particles are formed by the nucleation of gas phase species and grow by coagulation (existing particles combining) or condensation (gases condensing on existing particles). Fine particles are present in two separate modes, i.e., (a) freshly generated particles, in a transient ultrafine or nuclei mode whose particles coagulate and move into; (b) an accumulation mode, so called because its particles remain in that mode until removed by precipitation.

- Coarse-mode PM, present by contrast in a single mode, is formed by crushing, grinding, and abrasion of surfaces, which breaks large pieces of material into smaller pieces. They are then suspended by the wind or by anthropogenic activity. Energy considerations limit the break-up of large particles and small particle aggregates generally to a minimum size of about 1 μm in diameter. Mining and agricultural activities are examples of anthropogenic sources of coarse-mode particles. Fungal spores, pollen, and plant and insect fragments are examples of natural bioaerosols also suspended as coarse-mode particles.

- Within atmospheric particle modes, the distribution of particle number, surfaces, volume, and mass by diameter is frequently approximated by lognormal distributions (as illustrated in Figure 5-12). Aerodynamic diameter, d_{ae}, which depends on particle density and shape, is defined as the diameter of a particle with the same settling velocity as a spherical particle with unit density (1 g/cm^3). Typical values of the mass median aerodynamic diameter (MMAD) and geometric standard deviation (σg) of each size mode of an aerosol, as reported by the U.S. EPA[110] are:

Nuclei mode: MMAD=0.05 to 0.07 μm; σg=1.8
Accumulation mode: MMAD=0.3 to 0.7 μm; σg=1.8
Coarse mode: MMAD=6 to 20 μm; σg=2.4

- Research studies indicate an atmospheric bimodal distribution of fine and coarse particle mass with a minimum in the distribution between 1 and 3 μm d_{ae}.

- The terms "fine" and "coarse" were originally intended to apply to the two major atmospheric particle distributions which overlap in the size range between 1 and 3 μm diameter. Now, fine has been defined by U.S. EPA as $PM_{2.5}$ and coarse as $PM_{10-2.5}$. However, $PM_{2.5}$ may also contain, in addition to the

TABLE 5-8. Comparison of Ambient Fine and Coarse Mode Particles

	Fine Mode	Coarse Mode
Formed from:	Gases	Large solids/droplets
Formed by:	Chemical reaction; nucleation; condensation; coagulation; evaporation of fog and cloud droplets in which gases have dissolved and reacted.	Mechanical disruption (e.g., crushing, grinding, abrasion of surfaces); evaporation of sprays; suspension of dusts.
Composed of:	Sulfate, $SO_4^=$; nitrate, NO_3^-; ammonium, NH_4^+; hydrogen ion, H^+; elemental carbon; organic compounds (e.g., PAHs, PNAs); metals (e.g., Pb, Cd, V, Ni, Cu, Zn, Mn, Fe); particle-bound water.	Resuspended dusts (e.g., soil dust, street dust); coal and oil fly ash; metal oxides of crustal elements (Si, Al, Ti, Fe); $CaCO_3$, NaCl, sea salt; pollen, mold spores; plant/animal fragments; tire wear debris.
Solubility:	Largely soluble, hygroscopic and deliquescent.	Largely insoluble and nonhygroscopic.
Sources:	Combustion of coal, oil, gasoline, diesel, wood; atmospheric transformation products of NO_x, SO_2, and organic compounds including biogenic species (e.g., terpenes); high temperature processes, smelters, steel mills etc.	Resuspension of industrial dust and soil tracked onto roads; suspension from disturbed soil (e.g., farming, mining, unpaved roads); biological sources; construction and demolition; coal and oil combustion; ocean spray.
Lifetimes:	Days to weeks	Minutes to hours
Travel Distance:	100s to 1000s of kilometers	< 1 to 10s of kilometers

Source: EPA[109]

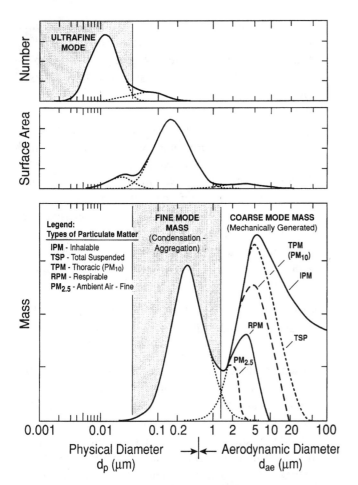

FIGURE 5-12. A hypothetical distribution of particles by number, surface area, and mass. Truncations in lower right indicate cuts imposed by aerodynamic size-selective inlets.

fine-particle mode, some of the smaller sized particles in the tail of the coarse particle mode between about 1 and 2.5 μm d_{ae}. Conversely, under high relative humidity conditions, the larger particles in the accumulation mode extend into the 1 to 3 μm d_{ae} range.

• Receptor modeling has proven to be a useful method for identifying contributions of different types of sources especially for the primary components of ambient PM. Apportionment of secondary PM is more difficult because it requires consideration of atmospheric reaction processes and rates. Results from western U.S. sites indicate that fugitive dust, motor vehicles, and wood smoke are the major contributors to ambient PM samples there, while results from eastern U.S. sites indicate that stationary combustion and fugitive dust are major contributors to ambient PM samples in the East. Sulfate and organic carbon are the major secondary components in the East, while nitrates and organic carbon are the major secondary components in the West.

• Fine and coarse particles have distinctly different sources, both natural and anthropogenic. Therefore different control strategies are likely to be needed.

• Dry deposition of fine particles is slow. Nuclei-mode (ultrafine) particles are rapidly removed by coagulation into accumulation-mode particles. Accumulation-mode particles are removed from the atmosphere primarily by forming cloud droplets and falling out in raindrops. Coarse particles are removed mainly by gravitational settling and inertial impaction.

• Primary and secondary fine particles have long lifetimes in the atmosphere (days to weeks) and travel long distances (hundreds to thousands of kilometers). They tend to be uniformly distributed over urban areas and larger regions, especially in the eastern United States. As a result, they are not easily traced back to their individual sources.

• Coarse particles normally have shorter lifetimes (minutes to hours) and only travel short distances (< 10's of km). Therefore, coarse particles tend to be unevenly distributed across urban areas and tend to have more localized effects than fine particles. (Dust storms occasionally cause long-range transport of the smaller coarse-mode particles.)

While mortality, morbidity and lung function decrements have been more closely associated with fine particle mass concentrations than with coarse particle mass, it is not clear that fine particle NAAQS alone would provide adequate public health protection. The coarse mode particles within the thoracic fraction (PM$_{10}$) deposit preferentially in the lung conductive airways where they may contribute to asthma exacerbations and to the development and/or progression of chronic bronchitis and bronchial cancer. Furthermore, even if the effects are due to the particles that deposit in the gas-exchange region of the lungs, there may be high concentrations of respirable coarse mode particles, i.e., particles with aerodynamic diameters below the RPM cut-size of 4 μm and the fine mode cut-size of 2.5 μm, in the arid portions of the western U.S., on windy days.

In consideration of current knowledge of the health effects of ambient air PM, as summarized above, the EPA adopted new daily and annual fine particle (PM$_{2.5}$) NAAQS and retained daily and annual thoracic (PM$_{10}$) NAAQS in July 1997. It rejected the alternate option of specific thoracic coarse mode NAAQS based on the mass concentration of PM$_{10}$ in particles larger than 2.5 μm in aerodynamic diameter as being too complex to be implemented. Their rationale is that with the existance of PM$_{2.5}$ NAAQS, the PM$_{10}$ NAAQS will serve adequately as limits against excessive exposures to coarse mode thoracic particles. With this approach, regulated communities could continue to use their existing PM_{10} monitors to supplement their newly installed $PM_{2.5}$ monitors. They could also monitor $PM_{2.5}$ and PM_{10} simultaneously using one instrument, such as a virtual dichotomous sampler with a 10 µm inlet and a 2.5 µm separator, provided that such an instrument could be demonstrated to provide equivalent performance to the established Federal Reference Methods for PM_{10} and $PM_{2.5}$.

The primary PM standards are designed to protect sensitive subgroups of the overall population from adverse health effects. The relevant subgroups include infants and the elderly, especially those with pre-existing respiratory diseases such as asthma, pneumonia, chronic obstructive pulmonary disease (COPD), and cardiovascular disease. Some of these people may be older workers or retired workers who have had long-term exposures to dusts and irritant vapors. Such workers are likely to constitute a susceptible subpopulation. It is well established that long-term exposure to mineral dusts can lead to chronic bronchitis and accelerated loss of lung function, and that these conditions can progress even after the occupational exposures end. Miners and other workers in dusty trades may retire with lung function in the normal or near normal range and then become progressively disabled. This is because of dust that accumulates around small airways, and the loss of lung recoil, which is a normal part of lung aging, results in airflow obstruction. They would be a sensitive population group. Registries of such workers, if enrolled in prospective cohort studies, may be ideal populations for further studies of both acute and chronic air pollution mortality and morbidity.

In designing future studies of the health effects of occupational exposures to PM or to mixtures of PM and irritant vapors, consideration should be given to the hypotheses generated by the recent epidemiological research on ambient air PM. Attention should be paid to cardiovascular and respiratory symptoms and functions in occupationally exposed workers, as well as to daily variations in their responses that may be related to variations in their exposures to environmental pollutants.

Instruments for Size-selective Sampling

The goal of obtaining air concentration data related to health hazards can be approached in several ways. For "insoluble" dusts, multistage samplers, consisting of one or more collectors with cutoff characteristics like those of the upper respiratory and tracheobronchial airways, followed by an efficient final stage, can provide the desired information with minimal sampling and analytical effort.

For other toxic materials or for aerosols where the contaminant of interest is a minor mass constituent, it may be necessary to obtain the overall size-mass concentration within different size ranges appropriate to the sites of toxic action.

Because the size-selective particle criteria outlined in the preceding sections were intended for "insoluble" dusts, most of the samplers developed to satisfy them have been relatively simple two-stage devices. In recent years, multistage samplers designed to simulate deposition within more restricted subdivisions of the respiratory tract have been developed. These and other multistage samplers will be discussed in the sections to follow.

Two-stage, "Respirable" Particulate Mass Samplers

A two-stage respirable dust sampler consists of a first stage, whose collection efficiency falls from very high to very low as the aerodynamic particle size decreases from approximately 10 to 2 µm, and a second stage with a high collection efficiency for all particle sizes. Horizontal elutriators and cyclones have been most widely used as first-stage collectors, whereas filters have been used as the second stage in most two-stage samplers.

Other first-stage collectors that have been used for "respirable" dust sampling include the pre-impinger of May and Druett;[112] multiple-nozzle, single-stage impactors of Marple[113] and Willeke;[114] and an inefficient filter. Roessler[115] and Gibson and Vincent[116] proposed using plastic foam filters to simulate the respirable cut, while Parker et al.[117] and Cahill et al.[118] proposed using large pore Nuclepore filters. The use of Nuclepore filters for this purpose is inadvisable for two reasons. One, their primary collection mechanism in the size range and face velocity of interest is interception; therefore, they collect particles according to their linear dimensions more than on the basis of their aerodynamic diameters. Second, solid particles tend to bounce or be re-entrained off Nuclepore filter surfaces, as demonstrated by Spurny,[119] Buzzard and Bell,[120] and Heidam.[121] On the other hand, the porous foam filter sampler[116, 122, 123] offers some interesting advantages for "respirable" dust sampling. It has a collection efficiency close to that of the MRE elutriator over a broad range of filter face velocities. At the lower velocities, particle collection by sedimentation increases while collection by impaction decreases. At the higher flow rates, the reverse is true, and the overall efficiency is about the same.

In addition to filters, other second-stage collectors that have been used include impingers, impactors, cyclones, bubblers, thermal precipitators, and electrostatic precipitators.

Cyclones are the most commonly used respirable mass samplers. They are available in a wide range of flow rates, including miniature sizes for personal sampling. The sampling efficiency can be closely matched to that of a respirable curve (Figure 5-13). Cyclones have important practical advantages, such as minimal particle bounce and re-entrainment, large capacity for loading, and insensitivity to orientation. A disadvantage of the cyclone is the lack of a fundamental theory that can predict performance. However, empirical theories are available to assist the designer.[122–124]

Considerable data are available on the performance of the widely used 10-mm nylon cyclone.[125–129] Studies using an aerodynamic particle sizer to measure cyclone penetration, has been reported by Bartley et al.[125] and by Maynard and Kenny. Their results for the 10-mm nylon cyclone and the Higgins and Dewell design made by Casella and BGI, Inc. are shown in Figure 5-13.

Based on their analyses, they have concluded that the 10-mm nylon cyclone will most closely match the revised ACGIH–ISO–CEN criteria for respirable dust sampling at 1.7 L/min, while the best match for the Casella and BGI cyclones is 2.2 L/min.

Small cyclones were designed to approximate the separation characteristics specified by the BMRC. The Casella SIMPEDS, widely used in Europe, is a modified version of a design by Higgins and Dewell[130] that was fabricated for the British Cast Iron Association (BCIRA). In Sweden, SKC has produced a series of similar cyclones, i.e., SKCα, SKCβ, and SKCγ. These cyclones have been evaluated by Liden.[131] Figure 5-14 and Table 5-9 show the critical dimensions of these cyclones. Figure 5-15 shows performance data reported by Liden for the SIMPEDS and SKCα in relation to the BMRC and new ACGIH–ISO–CEN criteria. He also reported that the SKCγ is closest in performance to the SIMPEDS; however, the SKCα and SKCβ gave the closest performance to the new criteria of ACGIH–ISO–CEN.

Other miniature cyclones with nominal diameters between 1/2 and 3 in. (1.2 to 7.6 cm), have also been used for respirable and thoracic dust sampling.[122–131] Kenny and Gussman[132] have characterized and modeled the performance of two families of cyclone samplers that can be used for respirable and thoracic aerosol sampling.

An inertial spectrometer personal sampler that sorts the sampled aerosol onto a 47-mm membrane filter according to aerodynamic diameter has been developed by Prodi et al.[133, 134] The filter can be cut into strips with cut sizes ranging from >10 μm to <2.5 μm. It was

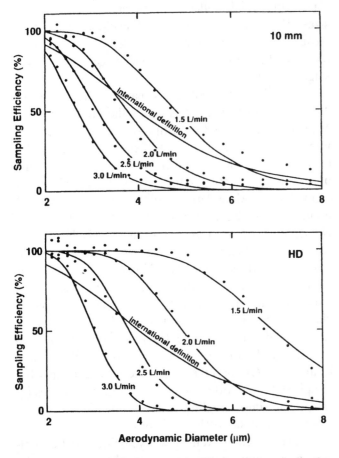

FIGURE 5-13. Aerodynamic particle diameter versus collection efficiency for the 10-mm nylon cyclone (top) and the Higgins and Dewell (HD) type of cyclone made by Casella and BGI, Inc. (bottom), in relation to the recently adopted ACGIH-ISO-CEN respirable mass sampling criteria (from Bartley et al.[125]).

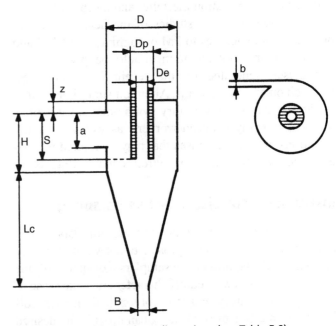

FIGURE 5-14. Critical cyclone dimensions (see Table 5-9).

TABLE 5-9. Dimensions (mm) of Three SKC Cyclone Generations, Single-hole Vortex-outlet SIMPEDS Cyclone, and Original Drawing by Higgins and Dewell (HD) (*from Liden*[131])

	HD	SIMPEDS	SKCα	SKCβ	SKCγ
D	9.5	9.5	9.3	9.3	9.5
D_e	3.2	3.2	3.9	3.1	3.2
D_p	4.7	5.0	4.9	4.5	4.7
b	1.6	1.5	0.9	1.6	1.7
a	9.5	10.3	9.6	9.1	9.5
S	9.5	10.3	9.3	19.5	9.5
H	12.1	12.8	12.2	9.3	12.1
L_c	25.4	26.2	36.8	23.7	25.4
z	0.0	0.0	0.0	0.0	0.0
B	2.4	2.3	2.2	2.4	2.4

reported that its performance for sampling TPM and RPM matches the ACGIH criteria and suggests that, when worn on the lapel, its inlet efficiency matches the IPM criteria.

Horizontal elutriators have been widely used outside of the United States. Their main advantage is the predictable performance based on gravitational settling of the particles during passage between horizontal collecting plates. Disadvantages include the restriction to a

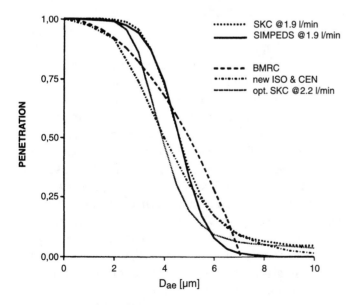

FIGURE 5-15. Average penetration for the SKCα and SIMPEDS cyclones at 1.9L/min, together with the BMRC and the new ACGIH-ISO-CEN-proposed respirable sampling conventions, and optimum SKCα flow rate for proposed ACGIH-ISO-CEN respirable sampling convention (*from Liden* [131]).

fixed orientation, the possible re-entrainment of particle deposits, and the difficulty of miniaturization.

The collection efficiency of an impactor can be accurately predicted by theory.[135] On the other hand, important details such as wall losses cannot be reliably predicted. Also, impactors suffer from the problems associated with particle bounce and re-entrainment. Impactors can be designed over a wide range of flow rates and can be operated in any orientation. Particle bounce and re-entrainment can be minimized by using virtual impactors. It should be emphasized, however, that the cutoff curves of most existing impactors are sharp, and hence do not conform to the human respirable curve. The multiple nozzles of varying diameters in the Marple[113] and Willeke[114] designs coarsen the cut in order to match the respirable sampling criteria.

The deposition models proposed by the ICRP[23, 46] and NCRP[47] Task Groups, which were previously discussed, define deposition in the head airways, tracheo-bronchial airways, and gas-exchange regions on the basis of a single parameter that they call the AMAD. This parameter is a type of mass median diameter, specifically for radioactive aerosols, and it can be determined in industrial environments using cascade impactor samplers under some circumstances. However, cascade impactors cannot be used effectively to monitor materials with very low concentration limits because background dust will overload the collection plates, resulting in unacceptable re-entrainment and wall losses long before detectable levels of the radioisotope of interest are collected. Two-stage samplers cannot provide an estimate of the aerosol AMAD and might appear to be inapplicable to the estimation of "respirable" concentrations as defined by the original Task Group model. However, Mercer[78] demonstrated that the second-stage collection of a two-stage sampler, whose first stage conforms to the BMRC or AEC criteria, could be related to the original Task Group deposition prediction by a simple fraction. For a variety of lognormal aerosol distributions with σg between 1.5 and 4, the predicted pulmonary (alveolar) deposition was very close to 30% of the fraction collected on the second-stage collector for both elutriator and cyclone.

Inspirable Particulate Mass Samplers

It is desirable that IPM sampling eventually replace the present method of total dust sampling using open-face filter holders. So-called total dust samplers, such as open-face filter cassettes, do not measure total dust and are unsuitable for most monitoring of airborne particles larger than a few micrometers because their sampling efficiency for large particles is sensitive to wind velocity and direction. Implementation of IPM sampling will require the development and testing of suitable sampling instruments.

Inhalable aerosol sampling in calm air has been summarized by several investigators.[136–138] However, in the workplace environment, it is rare for the air to be sufficiently calm for this still air analysis to hold. Studies of blunt samplers led to the conclusion that for calm air the particle aerodynamic diameter that results in a 90% sampling efficiency is roughly one-half that predicted for thin-walled tube samplers.[61]

The inlet characteristics of currently used inhalable personal and static samplers under various conditions were discussed in detail by Kenny et al.[139]

Mark et al.[140] developed an area sampler, the IOM/STD1, that comes close to matching the recommended IPM criteria over the range of 0- to 100-mm aerodynamic diameter. The sampling head of their device is a vertical axis cylinder about 5 cm in diameter and 6 cm high. A horizontal axis, oval-shaped inlet slot (about 3 mm high H 16 mm wide) is located midway up the side of the cylinder. The device samples at 3 L/min through a 37-mm filter mounted in a weighable cassette inside the cylinder. The sampling head is mounted on a larger vertical axis cylinder about 15 cm in diameter and about 18 cm high that houses batteries, pump, and flow control. The sampling head rotates continuously at about 2 rpm. Test results indicate reasonable agreement with the ACGIH IPM criteria. Mark and Vincent[141] described a 2 L/min personal lapel sampler whose collection characteristics closely match the ACGIH IPM criteria (Figure 5-16). These samplers are now available commercially with both plastic and metal cassettes. However, plastic cassettes can take up moisture, interfer-

ing with gravimetric analyses. Vaughan et al.[142] described a field comparison of personal samplers for inhalable dust.

Thoracic Particulate Mass Samplers

Probably the simplest approach to sampling for TPM is to use a sampler whose collection efficiency as a function of particle aerodynamic diameter falls within the acceptance envelope. Such a TPM sampler consists of an inlet, a size-fractionating stage, which is sometimes integral with the inlet, and a particle collector, which is usually a filter.

One of the principal criteria used in the selection of samplers is the flow rate. TPM samplers can be classified into low volume ($Q < 20$ L/min), medium volume (20 L/min $< Q < 150$ L/min), and high volume ($Q > 150$ L/min) samplers. In the low volume category, the dichotomous sampler[143, 144] is a virtual impactor having a flow rate of 16.7 L/min. The TPM fraction is selectively passed through the inlet; the virtual impactor further fractionates the aerosol into coarse and fine fractions with a d_{50} of 2.5 mm.

Several inlet designs are available. The UMLBL inlet[145] is a single-stage impactor with a grooved impaction surface and an internal flow pattern designed to suppress particle bounce. Independence of wind direction is assured by cylindrical symmetry about the vertical axis. For TPM sampling alone, the virtual impaction stage of the dichotomous sampler is unnecessary. The fractionating inlet can be coupled directly to a filter to form a sampler which has been called the PASS (Particulate Automatic Sampling System [Graseby Andersen, Atlanta, GA]). Such a sampler, using the earlier U.S. EPA 15-mm cutpoint dichotomous sampler inlet, performed well.[146]

Medium volume samplers have been developed.[147, 148] One version employs a sampler geometry that fractionates particles by a combination of impaction and sedimentation. The tortuous air path also suppresses particle bounce. A high volume sampler based on a similar geometry, called the Size-Selective Inlet (SSI), converts a standard hi-vol into a thoracic mass sampler.[149] The SSI can be used only with quartz or glass fiber filters.

A small, portable sampler has been developed within a thoracic cut provided by the inlet, which contains a single-stage impactor with an oil-soaked porous plate to suppress particle bounce.[150] Kenny and Gussman[132] described a cyclone that can be used for TPM sampling.

The foregoing samplers are all area samplers. Personal impactor samplers designed for the collection of the TPM fraction have been developed.[151–153]

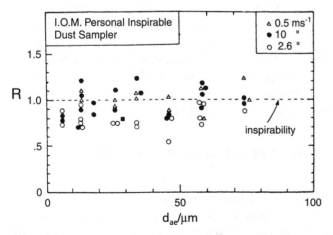

FIGURE 5-16. Collection ratio (R) to the IPM personal "inspirable" dust sampler as a function of particle aerodynamic diameter (d_{ae}).[132] *Reprinted with permission from the British Occupational Hygiene Society.*

Miniature cyclones can also be used to determine the TPM fraction.[127,132]

The measured sampling efficiencies of two of the samplers discussed above are compared to the TPM sampling criteria in Figure 5-17. The data points lie within the tolerance band established by EPA for PM_{10} samplers. These particular samplers were chosen for illustrative purposes only; a number of other samplers also satisfy the criteria.

Filter Pack Aerosol Samplers

A filter pack sampler that provides estimates of the deposition in each of the ICRP Subcommittee II Task Group subdivisions was described by Shleien et al.,[154] who calculated their regional deposition estimates using a linear programming approach in which they combined their calibration data on the relative collection efficiencies of the stages in the filter pack and the characteristics of the Task Group's deposition model. The filter stage collections themselves do not correspond to particular regions of the respiratory tract.

Reiter and Potzl[155] developed a filter pack in which the particle penetration characteristics of each filter were selected so that the particles retained on

each filter represent deposition in a particular region of the respiratory tract. Their calibration data indicate that the stages of their filter pack are similar to the regional deposition estimates from the calculations of Findeisen[41] and Landahl.[42] Thus, the first filter collection represents the deposition in the trachea and bronchi of the first order, the second filter represents the remainder of the bronchial deposition, the third filter represents alveolar deposit, and the fourth or final filter represents the exhaled aerosol. Air enters the filter pack through a moistening vessel which serves to maintain a selected water vapor partial pressure. The filter pack is housed within a Faraday cage and the electric charge deposited on each filter by the collected particles can be measured. This model does not consider nasopharyngeal deposition and thus appears to represent deposition during mouth breathing.

Ambient Fine Particle Samplers

Samplers that match the relatively sharp 2.5 μm cut criteria established by U.S. EPA in 1997 for fine particle collection have been developed for fixed site monitors and, as of this writing, are undergoing validation testing. Personal $PM_{2.5}$ samplers and fixed site $PM_{2.5}$ monitors that make size-cuts that approximate the 1997 U.S. EPA criteria have been used in various research studies over the past 20 years. These include the two-stage virtual impactor for fixed site monitoring and the Marple personal impactor-filter sampler.

Summary of Two-stage Size Selective Aerosol Samplers

The characteristics of two-stage samplers developed for health related size-selective sampling are summarized in Table 5-10.

Other Airborne Particle Classifiers

There are a number of inertial particle classifier samplers whose cutoff characteristics were not designed to match human respiratory tract deposition or penetration. These include two-stage samplers designed to make some other specific size cut and multiple-stage instruments which make a series of specific cuts and/or an estimate of the overall size-mass distribution.

Samplers with a single impaction stage include the single-stage impactors. These impactors deposit the large particle fraction on adhesive-coated plates or agar-filled dishes.

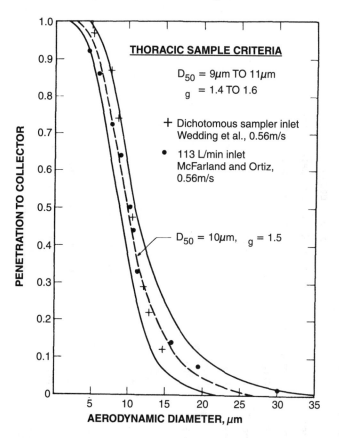

FIGURE 5-17. TPM sampline criteria with data for two thoracic particulate mass samplers.[151,156]

TABLE 5-10. Size-selective Samplers for Respirable, Thoracic (PM$_{10}$), Inhalable, and Fine Particle Sampling

A. Two-Stage "Respirable" Dust Samplers and Monitors

Type of First-stage Collector	Type of Second-stage Collector	Instrument or Precollector Name	Sampling Rate (L/min)	Suction Source	Reference	Commercial[A] Sources	Descriptions in Other Sections
Elutriator	Filter thimble	Hexhlet	50[B]	Air ejector	Wright [117]	CAS	14-38
Elutriator	Thermal ppt	Long period thermal ppt	0.002	Piston pump	Hamilton[118]		
Elutriator	Filter	SMRE semiautomatic hand pump	80 ml/stroke	Hand-pump	Dawes & Winder[119]		
Elutriator	Filter	High-volume elutriator	1250	Turbine blower	Shanty & Hemeon[120]		
Elutriator	Filter	MRE gravimetric dust sampler (Isleworth)	2.5	Diaphragm pump	Dunmore et al.[121]	CAS	14-39
Elutriator	Photometer[C]	Simslin	0.625	Vane pump	Blackford & Harris[122]	RML	
Spiral tube	Midget impinger		2.8	Impinger pump	Hatch & Hemeon[104]		
Pre-impinger	Porton impinger	Pre-impinger	11	Various	May & Druett[105]		
Centripeter	Filter	Personal centripeter	2	Diaphragm pump	Langmead & O'Conner[106]	BGI	14-16
Filter	Filter	Polyurethane foam prefilter	1130	Turbine blower	Roesler[109]		
Impactor	Filter		28.3	Pump	Marple[107]	MSP	14-15
Impactor	Filter	Personal environmental monitoring impactor	4	Pump			
Impactor			10				
Impactor	Electrostatic[C]	Respirable aerosol mass monitor	1	Pump	Sem et al.[123]	TSI	
Impactor	Photometer[C]	Respirable aerosol photometer	4	Pump		TSI	
Cyclone	Filter	Aerotec 3/4	25	Various	Lippmann & Chan[124]	AA	14-34, 14-35,
Cyclone	Filter	Aerotec 2	430	Turbine blower	Lippmann & Chan[124]	GMW, BGI	
Cyclone	Filter	10-mm Dorr-Oliver Cyclone	1.7	Various	Bartley et al.[125]	MSA, SEN, SKC	14-36
Cyclone	Filter	1/2" HASL cyclone	9	Various	AIHA Aerosol Technology Committee[126]	SEN	14-35
Cyclone	Filter	1" HASL cyclone	75	Turbine blower	Lippmann & Chan[171]	SEN	14-35
Cyclone	Filter	Personal dust sampler, SIMPEDS 70 MK2	2.2	Diaphragm pump	Higgins & Dewell[127]	BGI, CAS; RML	14-31
Cyclone	Filter	Gravimetric dust sampler vT/BF	15.4	Pump	John & Reischl[128]		
Cyclone	Cyclone	Respirable dust mass monitor	50	Air ejector	Breuer[94]		
Cyclone	Impactor[C]	RAM	2	Pump		MIE	
Cyclone	Photometer[C]	RAM	2	Pump		MIE	

TABLE 5-10 (con't). Size-selective Samplers for Respirable, Thoracic (PM$_{10}$), Inhalable, and Fine Particle Sampling

Type	Type of Size Selector	Nominal Cut-Size	Downstream Collector(s)	Sampling Rate (L/min)	Reference	Commerical[A] Source(s)	Descriptions in Other Sections
B. Samplers and Monitors with Inlet Cut-Sizes at 10-15 µm							
Cotton dust samplers	Vertical elutriator	100% @ 15 µm	Filter—37mm	7.4	NIOSH[100]	GMW	14-40
PCAM	Vertical elutriator	100% @ 15 µm	Photometer	7.4	Shofner et al.[129]	PPM	
Wedding PM$_{10}$ inlets:							
Dichotomous sampler	Cyclone	50 % @ 10 µm	37-mm virtual impactor filters collecting 10-15 µm, < 2.5 um	16.7		N/A	
High vol. PM$_{10}$	Cyclone	50 % @ 10 µm	Filter—8 x 10 in	1130		WED	14-29
PM$_{10}$ ambient samplers:							
Dichotomous samplers	Impaction baffles	50% @ 10 µm	37-mm virtual impactor filters 10-2.5 µm, < 12.5 µm	16.7		GRA, GMW	14-28
Medium flow samplers	Impaction baffles	50% @ 10 µm	Filter—102 mm	113		GRA, GMW	14-27
Size-selective hi-vols.	Impaction baffles	50% @ 10 µm	Filter—8 x 10 in	1130		GRA, GMW	14-26
PM$_{10}$ personal sampler	Impactor	50% @ 10 µm	Filter	4	Buckley et al.[130]	MSP	14-15
C. Samplers with Inhalable Inlet Cut-Sizes							
IOM/STD 1	Rotating slit	ACGIH IPM	Filter capsule	3	Mark et al.[131]	RML	
IOM personal sampler	15-mm inlet tube	ACGIH IPM	Filter capsule	2	Mark & Vincent[132]	SKC	

[A]See Table 5-11 for explanation of manufacturer's codes.
[B]For revised design—original unit had smaller plate spacing and operated at 100 L/min.
[C]Provides for direct readout of "respirable" mass concentration.

To overcome the limited collection capability of agar and adhesive-coated collection plates, Conner[156] directed the impaction jet into a still air region. Most of the flow must pass through an annular slit in the sampling tube and on to the back-up filter, carrying with it the smaller particles. The large particles continue down the tube beyond the slit along with a small volume of bleed air and are retained on a filter at the exit of the still air chamber. With a sampling rate of 39.7 L/min and a bleed rate of 0.5 L/min, the 50% cutoff is at 1.4 µm for polystyrene latex spheres. With other jet dimensions and flow rates, this type of sampler could make other particle size cuts, e.g., a "respirable" size cut.

Multiple-stage samplers that project large particles into still air have also been described. The cascade centripeter of Hounam and Sherwood[157] consists of three such inertial separation stages followed by a final filter stage. At the back of each still air chamber is a filter which collects the oversize particles and limits the amount of bleed air flow.

A similar approach is followed in the so-called "virtual impactors" recently developed for two-stage air pollution sampling applications.[158, 159] Samplers used in aerosol characterization studies have had a 2.5-µm design cutoff, whereas others, used in health-effects studies, were operated with a 3.5-µm cutoff.

A three-stage virtual impactor that can be used as a personal sampler at 3.1 L/min has been described by Keady.[160] Its dimensions are 82 × 70 × 50 mm; it weighs 290 grams; and its nominal aerodynamic diameter cut sizes are 100 µm (inlet), 10 µm, and either 4 or 2.5 µm, depending on whether it is used for occupational exposure assessment (with a 4 µm respirable cut) or air pollution exposure assessment (with a 2.5 µm fine particle cut). The particles are collected on 37 mm diameter filters. The filters for the two smaller fractions ($PM_{2.5}$ and $PM_{10-2.5}$) have a small center hole for the nozzles carrying the oversized particles to the next collection stage.

The most widely used type of multistage sampler is the cascade impactor (see Chapter 14), which is available commercially in a variety of designs. One major limitation to its application is that only limited sample masses can be collected without re-entrainment. Further limitations, which are shared by other multistage instruments such as the cascade centripeter, are wall losses between collection stages and the increased number of analyses per sample which are required. Their major advantage is that the full particle size-mass distribution can be determined. Recent developments include cascade impactors small enough for use as personal samplers.[161, 162]

The first development of a portable sampler using a parallel array of cyclone-filter series samplers was described by Lippmann and Kydonieus.[163] Each cyclone had a different cut size and the overall size-mass distribution of the total aerosol or any of its chemical constituents could be determined from analyses of the collection on the filter following each cyclone and a parallel filter sampler operated without a cyclone precollector. The performance characteristics of an improved version of this sampler were described by Blachman and Lippmann.[164] A much larger version of this sampler for fixed-station, ambient air pollution sampling was described by Bernstein et al.[165] A parallel multicyclone sampling train for stack sampling applications was described by Chang;[166] in addition, Smith et al.[167] developed a five-stage series cyclone sampler for the same purpose.

The major advantage in using cyclones instead of impactors for such applications is that their performance is not significantly affected by the amount of sample collected.[164] They share with impactors the advantage of relatively low fabrication cost. Air centrifuges such as the Stöber spiral centrifuge,[168] Conifuge,[169] and Goetz Aerosol Spectrometer,[170] and the horizontal elutriators of Timbrell[171] and Walkenhorst[172] deposit the aerosol sample in a continuous trace that can be subdivided into particle size subgroups. These instruments have very low sampling rates, and their use in the field is usually restricted to research studies rather than routine monitoring.

Other Techniques for Size Classification

All of the preceding methods separated the airborne particles according to their aerodynamic diameters. This parameter is of primary interest when considering the deposition probabilities of aerosols. However, useful data can be obtained by measuring other size parameters, such as light scatter,[173, 174] provided that a basis can be established for reliable conversion of the diameter measured to aerodynamic diameter.

Limitations of Selective Sampling and Selective Samplers

The effective application of selective sampling concepts to respiratory hazard evaluation requires 1) adequate knowledge of the regional deposition and clearance of particles in man and 2) reliable, reproducible, and accurately calibrated selective samplers. In both areas, the current state-of-the-art leaves much to be desired.

It is apparent from the review of the human deposition and clearance data and models presented earlier that the data are far from consistent and that the available models are, at best, crude approximations. Furthermore, as recent studies have demonstrated, there are very large variations in both regional deposition efficiencies and clearance rates among normal populations. Thus, even if the data were highly precise and reproducible, and population average figures met with general acceptance, the potential toxicity from the inhalation of a given aerosol would vary over a wide range.

The technology for designing a size-selective sampler and characterizing its collection characteristics is relatively more advanced than that for determining regional deposition and clearance dynamics in the human respiratory tract. Yet, even here, there are many conflicting data in the literature, and many instrument designs have required modifications to meet their original specifications. Also, some laboratory calibration data have been found to be erroneous, due to differences between laboratory test and field conditions as well as errors in measurement and/or data conversion errors.

For example, the original standard British elutriator sampler, the Hexhlet,[175] when operated at its design flow rate of 100 L/min, was found to be passing oversize particles onto the filter thimble. One major cause was re-entrainment of settled dust from the plates. To minimize this problem, the plate spacing was increased and the sampling rate reduced to 50 L/min.

The chief advantage of the horizontal elutriator over the cyclone as a precollector is that its performance can be predicted on the basis of gravitational sedimentation theory and the physical dimensions of the device. Thus, it was sometimes claimed that laboratory calibrations with carefully characterized test aerosols were not needed. This generalization was true, at least in a relative sense, in comparison to cyclone collectors where no adequate predictive relations exist for collection efficiency. However, in an absolute sense, the prediction of performance of actual elutriator samplers is not completely reliable, as discussed in the preceding paragraphs. Thurmer[176] presented a theoretical basis for describing some of the discrepancies observed in elutriator performance on the basis of technical shortcomings in the manufacture of the elutriators, especially in the nonuniformity of the plate spacings.

One of the major limitations of all elutriator precollectors is that it is difficult if not impossible to recover the collected material for analysis. In many cases, it is even difficult to periodically clean out the collected dust to minimize contamination of the second-stage collection by re-entrained dust. In most cases, the use of elutriators

has been restricted to sampling for pneumoconiosis-producing dusts where the hazards and concentration standards are based entirely on the "respirable" fraction. For research studies and other situations where the concentrations of both fractions are to be determined, cyclone precollectors are generally used.

Another limitation of the elutriator type of sampler is that it must be operated in a fixed horizontal position. The same limitation also applies to the water-filled preimpinger. Cyclones, on the other hand, can be operated in any orientation without significant change in their collection characteristics.[164, 177] The only precaution necessary is to avoid turning them upside down during or after sampling, which could cause dust from the cyclone to fall onto the filter or out through the inlet. This independence of orientation, combined with their smaller physical size at comparable flow rates, is one of the reasons that most of the recent two-stage, personal sampler designs have been built around miniature cyclones as the precollectors. These samplers, combined with filter collectors as the second stage and miniature battery-powered air pumps, are small and light enough to be worn throughout a work shift.

Most of the miniature battery-powered pumps are diaphragm or piston-type air movers and therefore produce a pulsating flow. This could render them unsuitable for pulling air through precollectors whose collection characteristics are flowrate dependent. Initially, personal gravimetric samplers distributed in the United States did not have pulsation dampers. Caplan et al.[178] demonstrated that the instantaneous flow was as much as 4 times the average for some units. There was a greater increase in collection efficiency at flows above the average than there was a decrease for flows below it. The net effect was to produce reduced cyclone penetration in pulsating flow as compared to a constant flow at the average rate. As a result, field samplers with pulsating flows underestimated the respirable mass. In recent years, all of the manufacturers have been producing samplers with built-in pulsation dampers. The performance of pulsation dampers used in personal sampling pumps and the effects of the residual pulsations on cyclone penetration were reviewed by Berry.[179]

For small variations in flow rate, changes in cyclone collection efficiency are not a severe problem, at least for those applications when the parameter of interest is the "respirable" mass measured on the second stage. Variations in air flow are corrected to some extent by changes in cyclone collection efficiency. As the flow rate increases, the aerosol mass entering the cyclone increases proportionally, but so does the collection efficiency. In an elutriator, the effect would be the opposite; an increase in sampling rate would result in an increase in penetration to the filter.

One potential problem with the 10-mm nylon cyclone is that, being an insulator, it can accumulate a static charge. When sampling aerosols with very high charge levels, this can significantly affect collection efficiency. Blachman and Lippmann[164] showed that highly charged aerosols with aerodynamic diameters below approximately 4 μm were collected with higher efficiencies than charge-neutralized aerosols of the same aerodynamic diameter. Almich and Carson[180] reported that the average collection efficiency for 4- to 5-μm charged particles was not significantly increased, but the variability in collection efficiency in replicate runs was increased. This variability was absent when using 10-mm cyclones of the same design which were constructed of stainless steel.

Lippmann and Chan[181] calibrated three larger, commercially available cyclones. They found that the 1/2-in HASL, the Aerotec 3/4, the 1-in HASL, and Aerotec 2 cyclones matched the original ACGIH criteria at 9, 25, 75, and 430 L/min, respectively. Other calibrations of these cyclones at the flow rates recommended by Lippmann and Chan have been inconsistent. Yablonsky et al.[182] reported that the concentration of a clay dust cloud passing the Unico 240 (HASL 1-in) was 1.3 times the concentration passing the Aerotec 3/4 when both were operated at the flow rates recommended by Lippmann and Chan. Thompson et al.,[183] in using the Unico 240 and the Aerotec 3/4 at recommended flow rates, noted that the Unico 240 passed 1.24 times as much dust as the Aerotec 3/4. Using an Andersen Impactor, Yablonsky determined that the Aerotec 3/4 was passing essentially that fraction of the total dust which would be expected for a cyclone with characteristics that corresponded to the retention criteria specified by ACGIH. Thompson et al. used a Coulter counter to determine the efficiency of the cyclone for various size fractions and determined that the Aerotec 3/4, operating at its recommended flow rate, performed very close to the ACGIH criteria. Thus, independent analyses by differing methods confirmed the calibration of Lippmann and Chan for the Aerotec 3/4, but produced rather different results for the Unico 240.

It is suggested that many of the inconsistent results which have been observed in "respirable" dust cyclone calibration are not the result of experimental error or because of minor variations in construction between the experimental cyclones. The way Lippmann and Chan connected the cyclone outlet to the filter holder was used by Thompson and Yablonsky for the Aerotec 3/4 cyclones, and comparable results were observed. However, whereas Lippmann and Chan used a close-coupled filter for both the Unico 240 and the Unico 18, this configuration was not used by Thompson et al. in their experiments and the results were not comparable. Likewise, Yablonsky did not use a close-coupled filter for the Unico 240 and again the penetration was some 25% greater. These findings are consistent with those of Knight,[184] who observed different penetration through the 10-mm nylon cyclone in the two MSA configurations, i.e., the coal mine dust cassette and the silica dust apparatus which uses the Millipore field monitor cassette.

Although the relationship of the outlet configuration of miniature cyclones to efficiency has not yet been investigated in detail, it is already quite clear that cyclone outlet configurations are important and that standardizations of pump and cyclone for a personal sampler will not characterize the performance of the sampler unless the outlet configuration through the filter is also standardized.

The remaining uncertainty about the collection characteristics of such simple devices is unfortunate because if the correct cut is not made during the process of collection, it cannot be made with much assurance later. A rough approximation could be made on the basis that an equivalent mass would have been collected on the filter at another flow rate, as previously discussed. In this respect, multistage collectors, such as the cascade impactor, have an advantage over the two-stage collector, even though there are conflicting data in the literature on the collection efficiencies of the various collection stages. In this case, even if the stage constants and the resulting size-mass distribution are incorrect, the basic stage collection data are valid, and a corrected size-mass plot can be made at a later time using more reliable stage calibrations. Also, as discussed previously, a major advantage of multistage sampling is that overall size-mass distributions can be used to estimate deposition at all levels of the respirable tract, not just the nonciliated level. Also, the MMD, as determined from multistage sampling data, can serve as an indicator of the percent "respirable" dust.

It is unfortunate that there are few multistage samplers which can provide useful data. Cascade impactors can make reasonably sharp particle size cuts, but those with collection plates cannot collect large sample masses without overloading them.

It appears that the best prospects for successful multistage sampling lie in the further development and application of parallel cyclone-filter systems and virtual impactor-filter systems.

Applications

The wide variety of equipment available for two-stage respirable, thoracic, and inhalable dust sampling is amply documented in Table 5-10, with reference to commercial suppliers (Table 5-11). All of these instruments have similar purposes; however, they differ in cutoff characteristics and sampling rates, which often imposes restrictions on the types of analyses that can be performed.

Conclusions

The potential health hazards arising from the inhalation of insoluble toxic aerosols can be related to the concentration of "inspirable" particles. The term "inspirable" in this context can refer to those particles which are sufficiently small to be aspirated into the human respiratory tract. More commonly, especially when considering pneumoconiosis-producing dusts and other insoluble dusts whose site of toxic action is the alveolar region of the lung, the term "respirable" dust is used to refer to a more restricted size spectrum, i.e., the particles small enough to penetrate the tracheobronchial region of the lung.

In either case, the commonly measured parameter of air concentration, i.e., the gross air concentration, provides a crude and sometimes misleading indication of inhalation hazard. Samplers designed to separate particles during the process of collection into "respirable" and "nonrespirable" fractions are available and have the potential of providing more realistic measurements of pneumoconiosis hazard.

In practice, there are several factors that limit the precision of "respirable dust" samples as hazard indicators. One is the questionable accuracy of the standardized deposition curves whose cutoff characteristics the samplers attempt to simulate. Human deposition studies using monodisperse spherical test aerosols indicate large individual differences in regional deposition among normal, nonsmoking males, as well as indications that tracheobronchial deposition is increased among some cigarette smokers. Thus, the "respirable" fraction of a given aerosol will differ with the individual.

Finally, there are instrumental uncertainties arising from several sources. One is the basic design and

TABLE 5-11. List of Instrument Manufacturers

BGI	BGI Incorporated 58 Guinan Street Waltham, MA 02451 (781)891-9380 FAX (781)891-8151 www.bgiusa.com	MIE	MIE, Inc. 7 Oak Park Bedford, MA 01730 (781)275-1919 or (888)643-4968 FAX (781)275-2121 www.mieinc.com	SEN	Sensidyne, Inc. 16333 Bay Vista Drive Clearwater, FL 33760 (727)530-3602 or (800)451-9444 www.sensidyne.com
CAS	Casella London, Ltd. Regent House, Brittania Walk London NI 7ND, England www.casella.co.uk	MSP	MSP Corporation 1313 Fifth Street SE, Suite 206 Minneapolis, MN 55414 (612)379-3963 FAX (612)379-3965 msp.sales@mspcorp.com www.mspcorp.com	SKC	SKC Inc. 863 Valley View Road Eighty Four, PA 15330-9614 (412)941-9701 or (800)752-8472 FAX (412)941-1369 www.skcinc.com
GRA	Graseby Andersen 500 Technology Court Smyrna, GA 30082-5211 (770)319-9999 or (800)241-6898 FAX (770)319-0336 andersen@graseby.com www.graseby.com	PPM	ppm Enterprises, Inc. 11428 Kingston Pike Knoxville, TN 37922 (615)966-8796 FAX (615)675-4795	TSI	TSI Incorporated 500 Cardigan Road P.O. Box 64394 St. Paul, MN 55164 (651)483-0900 Fax (651)490-2748 info@tsi.com www.tsi.com
GMW	Graseby GMW General Metal Works, Inc. 145 S. Miami Ave. Cleves, OH 45002 (513)941-2229 FAX (513)941-1977	RML	Rotheroe and Mitchell Ltd. Victoria Road Ruislip, Middlesex HA4 OYL England	WED	Wedding & Associates, Inc. 209 Christman Drive, #2 Fort Collins, CO 80524 (303)221-0678 or (800)367-7610 FAX (303)221-0400

calibration of size-selective samplers. There is still dis-agreement about the collection efficiency characteristics of some of the instruments used in the field. Other uncertainties arise from field applications. Because the efficiency of these devices is flow-rate dependent, oper-ation at nonstandard flows will cause erroneous results in both total concentration and percent inspirable, tho-racic, or "respirable." With elutriators, errors can arise from re-entrainment, departures from the desired orien-tation, and high velocity pressures at the inlet. With cyclones, very high dust concentrations may cause parti-cle agglomeration and an increased collection efficiency within the cyclone.

Multistage aerosol sampler data can provide esti-mates of the fractions depositing in several functional regions, as well as an overall particle size distribution curve. However, field sampling application of these devices has been limited for several reasons. One is the increased number and cost of sample analyses. More important perhaps is the lack of suitable instrumentation. Most of the commercially available instruments of this type are cascade impactors, which as a group suffer from a very limited mass collection capability, especially for aerosols of respirable size.

One of the major factors limiting the application of selective sampling concepts in the United States has been the absence of recognized criteria for size-selective mass concentrations. With the introduction of an alter-nate respirable mass concentration limit for quartz in 1968 by ACGIH, and the specification of respirable mass limits for coal dust by the U.S. Department of Labor in 1969, larger-scale applications of "respirable" sampling began. Now, with the harmonization of defini-tions and guidelines by ISO, CEN, and ACGIH,[85, 88, 89] much more activity and development of size-selective exposure criteria can, and should be, expected.

Despite all of the uncertainties and instrumental problems, inhalation hazard evaluations based on inspirable, thoracic, and/or "respirable" mass are clearly superior to estimates based on gross air concentrations for insoluble dusts whose site of toxic action is the deep lung. Gross concentration sampling protocols should be redefined as "inspirable" particulate mass sampling and be limited to situations where the entire aerosol is absorbed, e.g., for some highly soluble aerosols, or where the particle size distribution is relatively con-stant, and there is a known fixed ratio between the "inspirable" concentration and the concentration in the size range of interest.

References

1. Katz, S.H.; Smith, G.W.; Myers, W.M.; *et al.*: Comparative Tests of Instruments for Determining Atmospheric Dust. Public Health Bull. No. 144. DHEW, Public Health Service, Washington, DC (1925).

2. Littlefield, J.B.; Schrenk, H.H.: Bureau of Mines Midget Impinger for Dust Sampling. Bureau of Mines RI 3360. U.S. Department of the Interior, Washington, DC (1937).

3. Hatch, T.F.; Gross, P.: Pulmonary Deposition and Retention of Inhaled Aerosols. Academic Press, New York (1964).

4. Brain, J.D.; Valberg, P.A.: Deposition of Aerosol in the Respiratory Tract. Am. Rev. Resp. Dis. 120:1325 (1979).

5. Lippmann, M.; Yeates, D.B.; Albert, R.E.: Deposition, Retention, and Clearance of Inhaled Particles. Br. J. Ind. Med. 37:337 (1980).

6. Horsfield, K.; Dart, G.; Olson, D.E.; *et al.*: Models of the Human Bronchial Tree. J. Appl. Physiol. 31:207 (1971).

7. Weibel, E.R.: Morphometry of the Human Lung. Academic Press, NewYork (1963).

8. Lippmann, M.; Schlesinger, R.B.; Leikauf, G.; *et al.*: Effects of Sulphuric Acid Aerosols on Respiratory Tract Airways. In: Inhaled Particles V, pp. 677–690. W.H. Walton, Ed. Pergamon Press, London (1982).

9. Morrow, P.E.; Gibb, F.R.; Johnson, L.: Clearance of Insoluble Dust from the Lower Respiratory Tract. Health Phys. 10:543 (1964).

10. Jammet, H.; Lafuma, J.; Nenot, J.C.; *et al.*: Lung Clearance: Silicosis and Anthracosis. In: Pneumoconiosis—Proceedings of the International Conference, Johannesburg, 1969, pp. 435–437. H.A. Shapiro, Ed. Oxford University Press, Capetown (1970).

11. LeBouffant, L.: Influence de la Nature des Poussieres et de la Charge Pulmonaire sur l'Epuration. In: Inhaled Particles III, pp. 227–237. W.H. Walton, Ed. Unwin Bros., London (1971).

12. Bellmann, B.; Konig, H.; Muhle, H.; Pott, F.: Chemical Durability of Asbestos and of Man-Made Mineral Fibers *In Vivo*. J. Aerosol Sci. 17:341 (1986).

13. Ferin, J.; Leach, L.J.: The Effect of SO_2 on Lung Clearance of TiO_2 Particles in Rats. Am. Ind. Hyg. Assoc. J. 34:260 (1973).

14. Phalen, R.F.; Kenoyer, J.L.; Crocker, T.T.; McClure, T.R.: Effects of Sulfate Aerosols in Combination with Ozone on Elimination of Tracer Particles by Rats. J. Toxicol. Environ. Health 6:797 (1980).

15. McFadden, D.; Wright, J.L.; Wiggs, B.; Chung, A.: Smoking Inhibits Asbestos Clearance. Am. Rev. Resp. Dis. 133:372 (1986).

16. Albert, R.E.; Arnett, L.C.: Clearance of Radioactive Dust from the Lung. Arch. Ind. Health 12:99 (1955).

17. Cohen, D.; Arai, S.F.; Brain, J.D.: Smoking Impairs Long Term Dust Clearance from the Lung. Science 204:514 (1979).

18. Haroz, R.K.; Mattenberger-Kreber, L.: Effects of Cigarette Smoke on Macrophage Phagocytosis. In: Pulmonary Macrophages and Epithelial Cells, pp. 36–57. C.L. Sanders *et al.*, Eds. CONF-76092. National Technical Information Service, Springfield, VA (1977).

19. Bohning, D.E.; Atkins, H.L.; Cohn, S.H.: Long Term Particle Clearance in Man: Normal and Impaired. Ann. Occup. Hyg. 26:259 (1982).

20. Davies, C.N.: Deposition and Retention of Dust in the Human Respiratory Tract. Ann. Occup. Hyg. 7:169 (1964).

21. Lippmann, M.: Regional Deposition of Particles in the Human Respiratory Tract. In: Handbook of Physiology, Section 9. D.H.K. Lee, H.L. Falk, and S.D. Murphy, Eds. The American Physiological Society, Bethesda, MD (1977).

22. Stuart, B.O.: Deposition of Inhaled Aerosols. Arch. Intern. Med. 131:60 (1973).

23. International Commission on Radiological Protection, Task Group on Lung Dynamics Committee II: Deposition and Retention Models for Internal Dosimetry of the Human Respiratory Tract. Health Phys. 12:173 (1966).

24. Davies, C.N.; Heyder, J.; Subba Ramu, M.C.: Breathing of Half Micron-Aerosols; 1: Experimental. J. Appl. Physiol. 32:592 (1972).

25. Heyder, J.; Armbruster, L.; Stahlhofen, W.: Deposition of Aerosol Particles in the Human Respiratory Tract. In: Aerosole in Physik, Medizin und Technik, pp. 122–125. Gesellschaft fur Aerosolforschung, Bad Soden, Federal Republic of Germany (1973).

26. Heyder, J.; Gebhart, J.; Rudolf, G.; Stahlfofen, W.: Physical Factors Determining Particle Deposition in the Human Respiratory Tract. J. Aerosol Sci. 11:505 (1980).

27. Tarroni, G.; Melandri, C.; Prodi, V.; et al.: An Indicator on the Biological Variability of Aerosol Total Deposition in Humans. Am. Ind. Hyg. Assoc. J. 41:826 (1980).

28. Yu, C.P.; Nicolaides, P.; Soong, T.T.: Effect of Random Airway Sizes on Aerosol Deposition. Am. Ind. Hyg. Assoc. J. 40:999 (1979).

29. Chan, T.L.; Lippmann, M.: Experimental Measurements and Empirical Modelling of the Regional Deposition of Inhaled Particles in Humans. Am. Ind. Hyg. Assoc. J. 41:399 (1980).

30. Stahlhofen, W.; Gebhart, J.; Heyder, J.: Biological Variability of Regional Deposition of Aerosol Particles in the Human Respiratory Tract. Am. Ind. Hyg. Assoc. J. 42:348 (1981).

31. Lapp, N.L.; Hankinson, J.L; Amandus, H.; Palmes, E.D.: Variability in the Size of Airspaces in Normal Human Lungs as Estimated by Aerosols. Thorax 30:293 (1975).

32. Matsuba, K.; Thurlbeck, W.M.: The Number and Dimensions of Small Airways in Non-emphysematous Lungs. Am. Rev. Resp. Dis. 104:516 (1971).

33. Heyder, J.; Gebhart, J.; Heigwer, G.; et al.: Experimental Studies of Total Deposition of Aerosol Particles in the Human Respiratory Tract. Aerosol Sci. 44:191 (1973).

34. Muir, D.C.F.; Davies, C.N.: The Deposition of 0.5-mm Diameter Aerosols in the Lungs of Man. Ann. Occup. Hyg. 10:161 (1967).

35. Foord, N.; Black, A.; Walsh, M.: Regional Deposition of 2.5–7.5 mm Diameter Particles in Healthy Male Nonsmokers. J. Aerosol Sci. 9:343 (1978).

36. Stahlhofen, W.; Gebhart, J.; Heyder, J.: Experimental Determination of the Regional Deposition of Aerosol Particles in the Human Respiratory Tract. Am. Ind. Hyg. Assoc. J. 41:385 (1980).

37. Palmes, E.D.; Lippmann, M.: Influence of Respiratory Air Space Dimensions on Aerosol Deposition. In: Inhaled Particles IV, W.H. Walton, Ed. Pergamon Press, Oxford (1977).

38. U.S. Environmental Protection Agency: Air Quality Criteria for Particulate Matter. EPA-600/P-95/001F, U.S. EPA, Research Triangle Park, NC (1996).

39. Findeisen W.: Uber das Absetzen Kleiner, in der Luft Suspendierten Teilchen in der Menschlichen Lunge bei der Atmung. Pfluger Arch. fd. ges. Physiol. 236:367 (1935).

40. Landahl, H.D.: On the Removal of Airborne Droplets by the Human Respiratory Tract; 1: The Lung. Bull. Math. Biophys. 12:43 (1950).

41. Landahl, H.D.: Particle Removal by the Respiratory System. Bull. Math. Biophys. 25:29 (1963).

42. Beeckmans, J.M.: The Deposition of Aerosols in the Respiratory Tract; 1: Mathematical Analysis and Comparison with Experimental Data. Can. J. Physiol. Pharmacol. 43:157 (1965).

43. Yu, C.P.: A Two-Component Theory of Aerosol Deposition in Lung Airways. Bull. Math. Biol. 40:693 (1978).

44. Pattle, R.E.: The Retention of Gases and Particles in the Human Nose. In: Inhaled Particles and Vapors, C.N. Davies, Ed. Pergamon Press, Oxford (1961).

45. Gormley, P.G.; Kennedy, M.: Diffusion from a Stream Flowing through a Cylinder. Proc. Roy. Irish Acad. A52:163 (1949).

46. International Commission on Radiological Protection: Human Respiratory Tract Model for Radiological Protection. Report of Committee II. ICRP, Washington, DC (1994).

47. National Council on Radiation Protection: Deposition, Retention and Dosimetry of Inhaled Radioactive Substances. Report S.C. 57–2. NCRP, Bethesda, MD (1994).

48. Charlson, R.J.; Vanderpol, A.H.; Covert, D.S.; et al.: $H_2SO_4(NH_4)2SO_4$ Background Aerosol: Optical Detection in St. Louis Region. Atmos. Environ. 8:1257 (1974).

49. Cooper, D.W.; Byers, R.L.; Davis, J.W.: Measurements of Laser Light Backscattering vs. Humidity for Salt Aerosols. Environ. Sci. Technol. 7:142 (1973).

50. Orr, C.; Hurd, F.K.; Corbett, W.J.: Aerosol Size and Relative Humidity. J. Coll. Sci. 13:472 (1958).

51. Winkler, P.; Junge, C.: The Growth of Atmospheric Aerosol Particles as a Function of the Relative Humidity. Part I: Method and Measurements at Different Locations. J. Recherches Atmospheriques 6:617 (1972).

52. Winkler, P.: The Growth of Atmospheric Aerosol Particles as a Function of the Relative Humidity. Part II: An Improved Concept of Mixed Nuclei. Aerosol Sci. 4:373 (1973).

53. Milburn, R.H.; Crider, W.C.; Morton, S.D.: The Retention of Hygroscopic Dusts in the Human Lungs. AMA Arch. Ind. Health 15:59 (1957).

54. Porstendorfer, J.: Untersuchungen zur Frage des Wachstums von Inhalierten Aerosolteilchen im Atemtrak. Aerosol Sci. 2:73 (1971).

55. Held, J.L.; Cooper, D.W.: Theoretical Investigation of the Effects of Relative Humidity on Aerosol Respirable Fraction. Atmos. Environ. 13:1419 (1979).

56. Austin, E.; Brock, J.; Wissler, E.: A Model for Deposition of Stable and Unstable Aerosols in the Human Respiratory Tract. Am. Ind. Hyg. Assoc. J. 40:1055 (1979).

57. Cartwright, J.; Skidmore, J.W.: The Size Distribution of Dust Retained in the Lungs of Rats and Dust Collected by Size-Selective Samplers. Ann. Occup. Hyg. 7:151 (1964).

58. Carlberg, J.R.; Crable, J.V.; Limtiaca, L.P.; et al.: Total Dust, Coal, Free Silica, and Trace Metal Concentrations in Bituminous Coal Miners' Lungs. Am. Ind. Hyg. Assoc. J. 32:432 (1971).

59. International Standards Organization: Air Quality-Particle Size Fraction Definitions for Health Related Sampling. ISO/TR 7708-1983 (E). ISO (1983).

60. American Conference of Governmental Industrial Hygienists, Air Sampling Procedures Committee: Particle Size-Selective Sampling in the Workplace, 80 pp. ACGIH, Cincinnati, OH (1985).

61. Ogden, T.L: Inhalable, Inspirable, and Total Dust. In: Aerosols in the Mining and Industrial Work Environment, Vol. 1, pp. 185–205. V.A. Marple and B.Y.H. Liu, Eds. Ann Arbor Science Publishers, Ann Arbor, MI (1983).

62. Buchan, R.M.; Soderholm, S.C.; Tillery, M.I.: Aerosol Sampling Efficiency of 37-mm Filter Cassettes. Am. Ind. Hyg. Assoc. J. 47:825 (1986).

63. Chung, K.Y.K.; Ogden, T.L.; Vaughan, N.P.: Wind Effects on Personal Dust Samplers. J. Aerosol Sci. 18:159 (1987).

64. Vincent, J.H.; Mark, D.: Entry Characteristics of Practical Workplace Aerosol Samplers in Relation to the ISO Recommendations. Ann. Occup. Hyg. 34:249 (1990).

65. Kenny, L.C.; Aitken, R.; Chalmers, C.; et al.: Outcome of a Collaborative European Study of Personal Inhalable Sampler Performance. Ann. Occup. Hyg. 41(2):135–154 (1997).

66. Whitby, K.T.; Vomela, R.A.: Response of Single Particle Optical Counters to Nonideal Particles. Environ. Sci. Technol. 1:801 (1967).

67. Orenstein, A.J., Ed.: Proceedings of the Pneumoconiosis Conference, Johannesburg, 1959, A.J. Orenstein, Ed. J. and A. Churchill, Ltd., London (1960).

68. Davies, C.N.: Dust Sampling and Lung Disease. Br. J. Ind. Med. 9:120 (1952).

69. American Conference of Governmental Industrial Hygienists. Threshold Limit Values for Airborne Contaminants for 1968, p. 17. ACGIH, Cincinnati, OH (1968).

70. Sutton, G.W.; Reno, SJ.: Respirable Mass Concentrations Equivalent to Impinger Count Data. Presented at American Industrial Hygiene Conference, St. Louis, MO (May 1968).

71. U.S. Dept. of Labor: Public Contracts and Property Management. Fed. Reg. 34(96):7946 (May 20, 1969).

72. Public Law 91-173, Federal Coal Mine Health and Safety Act of 1969, 91st Congress (December 10, 1969).

73. Public Law 95-164, Federal Mine Safety and Health Act of 1977, 95th Congress (November 9, 1977).

74. National Research Council: Measurement and Control of Respirable Dust in Mines. NMAB-363. National Academy of Sciences, Washington, DC (1980).

75. Public Law 91-596, Occupational Safety and Health Act of 1970, 91st Congress (December 29, 1970).

76. 29CFR Part 1910. Air Contaminants; Final Rule. Federal Register 54(12):2329–2984 (January 19, 1989).

77. Brown, J.H.; Cook, K.M.; Ney, F.G.; Hatch, T.: Influence of Particle Size Upon the Retention of Particulate Matter in the Human Lung. Am. J. Pub. Health 40:450 (1950).

78. Mercer, T.T.: Air Sampling Problems Associated with the Proposed Lung Model. Presented at the 12th Annual Bioassay and Analytical Chemistry Meeting, Gatlinburg, TN (October 13, 1966).

79. Maguire, B.A.; Barker, D.: A Gravimetric Dust Sampling Instrument (SIMPEDS): Preliminary Underground Trials. Ann. Occup. Hyg. 12:197 (1969).

80. Ayer, H.E.; Dement, J.M.; Busch, K.A.; et al.: A Monumental Study—Reconstruction of a 1920 Granite Shed. Am. Ind. Hyg. Assoc. J. 34:206 (1973).

81. Miller, F.J.; Gardner, D.E.; Graham, J.A.; et al.: Size Considerations for Establishing a Standard for Inhalable Particles. J. Air Pollut. Control Assoc. 29:610 (1979).

82. Vincent, J.H.; Armbruster, L.: On the Quantitative Definition of the Inhalability of Airborne Dust. Ann. Occup. Hyg. 24:245 (1981).

83. Vincent, J.H.; Mark, D.: The Basis of Dust Sampling in Occupational Hygiene: A Critical Review. Ann. Occup. Hyg. 24:375 (1981).

84. Ambient Air Monitoring Reference and Equivalent Methods. Fed. Reg. 52(126):24727 (July 1,1987).

85. International Organization for Standardization (ISO): Air Quality-Particle Size Fraction Definitions for Health-related Sampling. ISO Standard 7708. ISO, Geneva (1995).

86. American Conference of Governmental Industrial Hygienists, Technical Committee on Air Sampling Procedures: Particle Size-Selective Sampling in the Workplace. Ann. Am. Conf. Govt. Ind. Hyg. 11:21 (1984).

87. Soderholm, S.C.: Proposed International Conventions for Particle Size-Selective Sampling. Ann. Occup. Hyg. 33:301–320 (1989).

88. European Standardization Committee (CEN): Size Fraction Definitions for Measurement of Airborne Particles in the Workplace. Approved for publication as prEN 481. CEN, Brussels (1991).

89. American Conference of Governmental Industrial Hygienists, 1994–1995 Threshold Limit Values and Biological Exposure Indices. ACGIH, Cincinnati, OH (1993).

90. International Organization for Standardization (ISO): Air Quality—Particle Size Fraction Definitions for Health-Related Sampling. Approved for publication as CD 7708. ISO, Geneva (1991).

91. Lidén, G.; Kenny, L.C.: Optimization of the Performance of Existing Respirable Dust Samplers. Appl. Occup. Environ. Hyg. 8(4):386–391 (1993).

92. Bartley D.L.; Doemeny, L.J.: Critique of 1985 ACGIH Report on Particle Size- Selective Sampling in the Workplace. Am. Ind. Hyg. Assoc. J. 47:443 (1986).

93. Lidén, G.; Kenny, L.C.: The Performance of Respirable Dust Samplers: Sampler Bias, Precision, and Inaccuracy. Ann. Occup. Hyg. 36:1 (1991).

94. McCawley, M.A.: Should Dust Samplers Mimic Human Lung Deposition. Appl. Occup. Environ. Hyg. 5:829 (1990).

95. Breuer, H.: Problems of Gravimetric Dust Sampling. In: Inhaled Particles III, pp. 1031–1042. W.H. Walton, Ed. Unwin Bros., London (1971).

96. Knight G.: Definitions of Alveolar Dust Deposition and Respirable Dust Sampling. Ann. Occup. Hyg. 29:526 (1985).

97. Raabe, O.G.: Basis for Particle Size-Selective Sampling for Beryllium. In: Advances in Air Sampling, pp. 39–51. ACGIH, Lewis Publishers, Inc., Chelsea, MI (1988).

98. Hinds, W.C.: Basis for Particle Size-Selective Sampling for Wood Dust. Appl. Ind. Hyg. 3:67 (1988).

99. Lippmann, M.; Gearhart, G.M.; Schlesinger, R.B.: Basis for a Particle Size- Selective TLV for Sulfuric Acid Aerosols. Appl. Ind. Hyg. 2:188 (1987).

100. National Institute for Occupational Safety and Health: Criteria for a Recommended Standard—Occupational Exposure to Cotton Dust. DHEW (NIOSH) Pub. No. 75-118. U.S. Government Printing Office, Washington, DC (1975).

101. Leidel, N.A.; Bayer, S.G.; Zumwalde, R.D.; Busch, K.A.: USPHS/ NIOSH Membrane Filter Method for Evaluating Airborne Asbestos Fibers. DHEW (NIOSH) Pub. No. 79-127. NIOSH, Rockville, MD (February 1979).

102. ACGIH-AIHA Aerosol Hazards Evaluation Committee: Recommended Procedures for Sampling and Counting Asbestos Fibers. Am. Ind. Hyg. Assoc. J. 36:83 (1975).

103. Lippmann, M.: Asbestos Exposure Indices. Environ. Res. 46:86 (1988).

104. McCawley, M.A.; Cocalis, J.C.: Diesel Particulate Measurement Techniques for Use with Ventilation Control Strategies in Underground Coal Mines. Ann. Am. Conf. Govt. Ind. Hyg. 14:271 (1986).

105. Burkhart, J.E.; McCawley, M.A.; Wheeler, R.W.: Particle Size Distributions in Underground Coal Mines. Am. Ind. Hyg. Assoc. J. 48:122–126 (1987).

106. Seixas, N.S.; Hewett, P.; Robins, T.G.; et al.: Variability of Particle Size-specific Fractions of Personal Coal Mine Dust Exposures. Am. Ind. Hyg. Assoc. J. 56:243–250 (1995).

107. Lippmann, M.: Role of Science Advisory Groups in Establishing Standards for Ambient Air Pollutants. Aerosol Sci. Tech. 6:93–114 (1987).

108. Lippmann, M.; Thurston, G.D.: Sulfate Concentrations as an Indicator of Ambient Particulate Matter Air Pollution for Health Risk Evaluations. J. Exposure Anal. Environ. Epidemiol. 6:123–146 (1996).

109. U.S. Environmental Protection Agency, Office of Air Quality Planning and Standards (OAQPS). Review of the National Ambient Air Quality Standards for Particulate Matter: OAQPS Staff Paper, EPA-452/R-96-013. U.S. EPA, Research Triangle Park, NC (July 1996).

110. United States Environmental Protection Agency: Air Quality Criteria for Particulate Matter. EPA/600/P-95/001. U.S. EPA, Washington, DC (1996).

111. Lippmann, M.; Bachmann, J.D.; Bates, D.V.; et al.: Report of the Particulate Matter (PM) Research Strategies Workshop, Park City, UT. Appl. Occup. Environ. Hyg. 13:485–493 (1998).

112. May, K.R.; Druett, HA.: The Pre-Impinger. Br. J. Ind. Med. 10:142 (1953).

113. Marple, V.A.: Simulation of Respirable Penetration Characteristics by Inertial Impaction. J. Aerosol Sci. 9:125 (1978).

114. Willeke, K.: Selection and Design of an Aerosol Sampler Simulating Respiratory Penetration. Am. Ind. Hyg. Assoc. J. 39:317 (1978).

115. Roessler, J.F.: Application of Polyurethane Foam Filters for Respirable Dust Separation. J. Air Pollut. Control Assoc. 16:30 (1966).

116. Gibson, H.; Vincent, J.H.: The Penetration of Dust Through Porous Foam Filter Media Ann. Occup. Hyg. 24:205 (1981).

117. Parker, R.D.; Buzzard, G.H.; Dzubay, T.G.; Bell, J.P.: A Two-Stage Respirable Aerosol Sampler Using Nuclepore Filters in Series. Atmos. Environ. 11:617 (1977).

118. Cahill, T.A.; Ashbaugh, L.L; Barone, J.B.; et al.: Analysis of Respirable Fractions in Atmospheric Particulate via Sequential Filtration. J. Air Pollut. Control Assoc. 27:675 (1977).

119. Spurny, K.: Discussion: A Two-Stage Respirable Aerosol Sampler Using Nuclepore Filters in Series. Atmos. Environ. 11:1246 (1977).

120. Buzzard, G.H.; Bell, J.P.: Experimental Filtration Efficiencies of Large Pore Nuclepore Filters. J. Aerosol Sci. 435 (1980).

121. Heidam, N.Z.: Review: Aerosol Fractionation by Sequential Filtration with Nuclepore Filters. Atmos. Environ. 15:891 (1981).

122. John, W.; Reischl, G.: A Cyclone for Size-Selective Sampling of Ambient Air. J. Air Pollut. Control Assoc. 30:872 (1980).

123. Chan, T.; Lippmann, M.: Particle Collection Efficiencies of Air Sampling Cyclones: An Empirical Theory. Environ. Sci. Technol. 11:377 (1977).

124. Saltzman, B.: Generalized Performance Characteristics of Miniature Cyclones for Atmospheric Particulate Sampling. Am. Ind. Hyg. Assoc. J. 45:671 (1984).

125. Bartley, D.; Chen, C-C.; Song, R.; Fischbach, T.J.: Respirable Aerosol Performance Testing. Am. Ind. Hyg. Assoc. J. 55:1036 (1994).

126. Blachman, M.W.; Lippmann, M.: Performance Characteristics of the Multicyclone Aerosol Sampler. Am. Ind. Hyg. Assoc. J. 35:311 (1974).

127. Maynard, A.D.: Characterizatiion of Six Thoracic Aerosol Samplers Using Spherical Particles. Report 1R/A/97/13. Health and Safety Laboratory, Sheffield, UK.

128. John, W.: Thoracic and Respirable Particulate Mass Samplers: Current Status and Future Needs. In: Advances in Air Sampling, pp. 25–38. ACGIH, Lewis Publishers, Inc., Chelsea, MI (1988).

129. Maynard, A.D.; Kenny, L.C.: Sampling Efficiency Determina-tion for Three Models of Personal Cyclone, Using an Aero-dynamic Particle Sizer. J. Aerosol Sci. 26(4):671–684 (1995).

130. Higgins, R.I.; Dewell, P.: A Gravimetric Size-Selecting Personal Dust Sampler. In: Inhaled Particles and Vapours II, pp. 575–585. C.N. Davies, Ed. Pergamon Press, London (1967).

131. Liden, G.: Evaluation of the SKC Personal Respirable Dust Sampling Cyclone. Appl. Occup. Environ. Hyg. 8:178 (1993).

132. Kenny, L.C.; Gussman, R.A.: Characterization and Modelling of a Family of Cyclone Aerosol Preseparators. J. Aerosol Sci. 28:677 (1997).

133. Prodi, V.; Belosi, F.; Mularoni, A.: A Personal Sampler Following ISO Recommendations on Particle Size Definitions. J. Aerosol Sci. 17:576 (1986).

134. Prodi, V.; Sala, C.; Belosi, F.: PERSPEC, Personal Size Separating Sampler: Operational Experience and Comparison with Other Field Devices. Appl. Occup. Environ. Hyg. 7:368 (1992).

135. Marple, V.A.; Willeke, K.: Impactor Design. Atmos. Environ. 10:891 (1976).

136. Davies, C.N.: The Entry of Aerosols into Sampling Tubes and Heads. Br. J. Appl. Phys. D. 2s(1):921 (1968).

137. Yoshida, H.; Uragami, M.; Masuda, H.; Linoya, K.: Particle Sampling Efficiency in Still Air. Kagaku Kagalcu Robunshu 4:123 (1978).

138. Agarwal, J.K.; Liu, B.Y.H.: A Criterion for Accurate Sampling in Calm Air. Am. Ind. Hyg. Assoc. J. 41:191 (1980).

139. Kenny, L.C.; Aitken, R.J.; Baldwin, P.E.J.; et al.: The Sampling Efficiency of Personal Inhalable Samplers in Low Air Movement Environments. J. Aerosol Sci. 30:627–638 (1999).

140. Mark, D.; Vincent, J.H.; Gibson, H.; Lynch, G.: A New Static Sampler for Airborne Total Dust in Workplaces. Am. Ind. Hyg. Assoc. J. 46:127 (1985).

141. Mark, D.; Vincent, J.H.: A New Personal Sampler for Airborne Total Dust in Workplaces. Ann. Occup. Hyg. 30:89 (1986).

142. Vaughn, N.P.; Chalmers, C.P.; Botham, R.A.: Field Comparison of Personal Samplers for Inhalable Dust. Ann. Occup. Hyg. 34:553–573 (1990).

143. Loo, B.W.; Adachi, R.S.; Cork, C.P.: A Second Generation Dichotomous Sampler for Large-Scale Monitoring of Airborne Particulate Matter. LBL-8725. Lawrence Berkeley Laboratory, Berkeley, CA (January 1979).

144. Liu, B.Y.H.; Pui, D.Y.H.: Aerosol Sampling Inlets and Inhalable Particles. Atmos. Environ. 15:589 (1981).

145. Shaw, Jr., R.W.; Stevens, R.K.; Lewis, C.W.; Chance, J.H.: Comparison of Aerosol Sampling Inlets. Aerosol Sci. Technol. 2:53 (1983).

146. John, W.; Wall, S.M.; Wesolowski, J.J.: Validation of Samplers for Inhaled Particulate Matter. EPA-600/4-83-010. National Technical Information Service Report No. PB 83-191395. Springfield, VA (March 1983).

147. McFarland, A.R.; Ortiz, C.A.: A 10-mm Cutpoint Ambient Aerosol Sampling Inlet. Atmos. Environ. 16:2959 (1982).

148. Wedding, J.B.; Weigand, M.A.; Carney, T.C.: A 10-mm Cutpoint Inlet for the Dichotomous Sampler. Environ. Sci. Technol. 16:602 (1982).

149. McFarland, A.R.; Ortiz, C.A.; Bertch, Jr., R.W.: A High Capacity Preseparator for Collecting Large Particles. Atmos. Environ. 13:761 (1979).

150. Bright, D.S.; Fletcher, R.A.: New Portable Ambient Aerosol Samplers. Ind. Hyg. Assoc. J. 44:528 (1983).

151. Buckley, T.J.; Waldman, J.M.; Freeman, N.C.G.; Lioy, P.J.: Calibration, Intersampler Comparison, and Field Application of a New PM10 Personal Air Sampling Impactor. Aerosol Sci. Tech. 14: 380 (1991).

152. Lioy, P.J.; Waldman, J.M.; Buckley, T.; et al.: The Personal Indoor and Outdoor Concentrations of PM10 Measured in an Industrial Community During the Winter. Atmos. Environ. 24B: 57 (1990).

153. Marple, V.A.; McCormack, J.E.: Personal Sampling Impactor with Respirable Aerosol Penetration Characteristics. Am. Ind. Hyg. Assoc. J. 44:916 (1983).

154. Shleien, B.; Friend, A.G.; Thomas, H.A.: A Method for the Estimation of the Respiratory Deposition of Airborne Materials. Health Phys. 13:513 (1967).

155. Reiter, R.; Potzl, K.: The Design and Operation of a Respiratory Tract Model. Staub 27:19 (English Translation) (1967).

156. Conner, W.D.: An Inertial-type Particle Separator for Collecting Large Samples. J. Air Pollut. Control Assoc. 16:35 (1956).

157. Hounam, R.F.; Sherwood, R.J.: The Cascade Centripeter: A Device for Determining the Concentration and Size Distribution of Aerosols. Am. Ind. Hyg. Assoc. J. 26:122 (1965).

158. Dzubay, T.G.; Stevens, R.K.: Ambient Air Analysis with Dichotomous Sampler and X-ray Fluorescence Spectrometer. Environ. Sci. Technol. 9:663 (1975).

159. Loo, B.W.; Jacklevic, J.M.; Goulding, F.S.: Dichotomous Virtual Impactors for Large Scale Monitoring of Airborne Particulate Matter. In: Fine Particles, pp. 311–350. B.Y.H. Liu, Ed. Academic Press, New York (1976).

160. Keady, P.: A New Size-Selective Personal Particle Sampler: Description and Performance. Presented at ACGIH 1998 Applied Workshop, Chapel Hill, NC, (February, 1998).

161. Gibson, H.; Vincent, J.H.; Mark, D.: A Personal Inspirable Aerosol Spectrometer for Applications in Occupational Hygiene

Research. Ann. Occup. Hyg. 31:463 (1987).

162. Rubow, K.L.; Marple, V.A.; Olin, J.; McCawley, M.A.: A Personal Cascade Impactor: Design, Evaluation, and Calibration. Am. Ind. Hyg. Assoc. J. 48:532 (1987).

163. Lippmann, M.; Kydonieus, A.: A Multi-Stage Aerosol Sampler for Extended Sampling Intervals. Am. Ind. Hyg. Assoc. J. 31:730 (1970).

164. Blachman, M.W.; Lippmann, M.: Performance Characteristics of the Multicyclone Aerosol Sampler. Am. Ind. Hyg. Assoc. J. 35:311 (1974).

165. Bernstein, D.; Kleinman, M.T.; Kneip, T.J.; et al.: A High-Volume Sampler for the Determination of Particle Size Distributions in Ambient Air. J. Air Pollut. Control Assoc. 26:1069 (1976).

166. Chang, H-c.: A Parallel Multicyclone Size-Selective Particulate Sampling Train. Am. Ind. Hyg. Assoc. J. 35:538 (1975).

167. Smith, W.B.; Wilson, Jr., R.R.; Harris, D.B.: A Five-Stage Cyclone System for In-Situ Sampling. Environ. Sci. Technol. 13:1387 (1979).

168. Stöber, W.; Flachsbart, H.: Size Separating Precipitation of Aerosols in a Spinning Spiral Duct. Environ. Sci. Tech. 3:1280 (1969).

169. Sawyer, K.F.; Walton, W.H.: The Conifuge—A Size Separating Sampling Device for Airborne Particles. J. Sci. Instr. 27:272 (1950).

170. Goetz, A.; Stevenson, H.J.R.; Preining, O.: The Design and Performance of the Aerosol Spectrometer. J. Air Poll. Control Assoc. 10:378 (1960).

171. Timbrell, V.: The Terminal Velocity and Size of Airborne Dust Particles. Br. J. Appl. Phys. Suppl. No. 3:86 (1954).

172. Walkenhorst, W.; Bruckmann, E.: Mineral Analysis of Suspended Dusts Classified According to Particle Sizes. Staub 26:45 (English Translation) (1966).

173. Armbruster, L.: A New Generation of Light-Scattering Instruments for Respirable Dust Measurement. Ann. Occup. Hyg. 31:181 (1987).

174. Roebuck, B.; Vaughn, N.P.; Chung, K.Y.K.: Performance Testing of the Osiris Dust Monitoring System. Ann. Occup. Hyg. 34:263 (1990).

175. Wright, B.M.: A Size-Selecting Sampler for Airborne Dust. Br. J. Ind. Med. 11:284 (1954).

176. Thurmer, H.: Investigations with the Horizontal Plate Precipitator. Staub 29:35 (English Translation) (1969).

177. Watson, H.H.: Dust Sampling to Simulate the Human Lung. Br. J. Ind. Med. 10:93 (1953).

178. Caplan, K.J.; Doemeny, L.J.; Sorenson, S.D.: Performance Characteristics of the 10 mm Cyclone Respirable Mass Sampler; Part I: Monodisperse Studies; Part II: Coal Dust Studies. Am. Ind. Hyg. Assoc. J. 38:83-162 (1977).

179. Berry, R.D.: The Effect of Flow Pulsations on the Performance of Cyclone Personal Respirable Dust Samplers. J. Aerosol Sci. 22:887 (1991).

180. Almich, B.P.; Carson, G.A.: Some Effects of Charging on 10-mm Nylon Cyclone Performance. Am. Ind. Hyg. Assoc. J. 35:603 (1974).

181. Lippmann, M.; Chan, T.: Calibration of Dual-Inlet Cyclones for "respirable" Mass Sampling. Am. Ind. Hyg. Assoc. J. 35:189 (1974).

182. Yablonsky, J.; Ayer, H.E.; Svetlik, J.; Horstman, S.W.: Calibration System for Dust Sampling. Final Report, Contract No. DAMD 17-74-c-4024. University of Cincinnati, Cincinnati, OH.

183. Thompson, E.M.; Treaftis, H.N.; Tomb, T.F.: Comparison of Recommended Respirable Mass Dust Sampling Devices. Presented at American Industrial Hygiene Conference, Atlanta, GA (May 1976).

184. Knight, G.: Personal communication (1978).

Chapter 6

Measurement and Presentation of Aerosol Size Distributions

Constantinos Sioutas, Sc.D.

Department of Civil and Environmental Engineering, School of Engineering, University of Southern
California, 3620 South Vermont Avenue, Los Angeles, CA 90089-2531

CONTENTS

Introduction

The two most important properties of particles are their size and chemical composition. Particle size is the most important parameter in describing particle behavior and the origin, formation, removal, and residence time in any environment. This chapter discusses types of methods for measuring particle size, as well as the methods that have been found useful in communicating particle size information. The chemical composition of particles is beyond the scope of this chapter.

Measuring the size distribution of particles is a complicated task and over the past 20 years one of the main goals of environmental scientists has been to improve instrumentation for determining the size distribution of particles. In general, there are two main strategies for particle sampling and measuring, namely direct and integrated sampling. In the first method, aerosol size distributions are determined continuously (or near-continuously), i.e., in time intervals ranging from a few seconds to a few minutes. The second category consists of instruments, which determine aerosol size distribution over time periods ranging from hours to days.

In the following paragraphs, different definitions of particle size will be presented first. Subsequently, a general survey of instruments and methods for aerosol measurement will be given, and finally methods for presenting particle size distributions will be discussed. This chapter has the same structure and outline as that in the previous edition, authored by Drs. Lioy and Knutson, whose contributions are gratefully acknowledged.

Particle Size

In almost every case, particle size data are presented in a x-y plot or in the form of a table. What follows is a description of different definitions of particle "size" depending on the method used for particle measurement as well as the quantitative parameters (or "amounts") typically used to obtain the size distribution.

Liquid aerosol particles are always spherical, whereas solid particles usually have complex shapes.[1] For spherical particles, the most common definition of particle size is their diameter as measured by a microscope. Microscopy is a direct sizing method (as opposed to methods that will be discussed in subsequent paragraphs); hence linear measurements made with a microscope can be calibrated accurately and serve as the primary measurement upon which all other aerosol methods are based. In measuring particle size with a microscope, it is essential to assign to each particle a size based on a two-dimensional projected image. In the case of irregularly shaped particles, diameters, which are equivalent to that of a perfect sphere, are used to describe particle size. Martin's diameter, Feret's diameter, and projected area diameter are typically used in microscopy to size irregularly shaped particles. Martin's diameter is equal to the length of the line parallel to a given reference line that divides the projected area of the particles in two equal areas. Feret's diameter is defined as the length of a projection of a particle along a given reference line. Projected area diameter is the diameter of a circle that has the same projected area as the particle silhouette. The three different diameters are shown schematically in Figure 6-1. Particle size resolution achieved by means of conventional light microscopes is limited to about 1 µm. However, the resolving power of microscopes has been enhanced significantly by using very short wavelength radiation from high-speed electron beams (electron microscopy). Particle size resolution down to about 1–5 nm is quite common using electron microscopes.[2]

In addition to these diameters measured directly, the equivalent volume diameter is also a size related to particle geometry. The equivalent volume diameter is defined as the diameter of a sphere with the same volume as the particle in question.

Along with direct methods of measuring particles (such as microscopy), technologies have been developed for indirect measurement of particle size. These technologies include inertial, electrical, diffusional, and optical methods to determine particle size.

One of the most common methods is the use of particle inertia to measure particle size. Inertial devices such as conventional or virtual impactors [3–5] or time-of-flight spectrometers[6–8] are based on this principle. The specific property measured by inertial particle devices is called aerodynamic diameter. It is defined as the diameter of a unit density sphere that has the same settling velocity as the particle in question.[9] For particles smaller than about 20 µm, the "speed of sedimentation," or terminal settling is given by Stokes' law. Stokes described the competition between two forces acting on the particles; the resistance of the air to the particles' motion, which is proportional to the particle diameter; and the force of gravity, which is proportional to the particle mass (or the cube of particle diameter). The diameter of a sphere with the same density and settling velocity with the particle in question is called Stokes diameter. The formula for the settling velocity, V, is given as follows:

$$V = \frac{\rho d_p^2 g C_c}{18\mu} = \frac{d_a^2 g C_c}{18\mu} \quad \textbf{(1)}$$

where (in cgs units):
  V = terminal settling velocity
  ρ = particle density (g/cm³)
  g = gravitational acceleration (981 cm²/s)
  d_p = particle Stokes diameter (cm)
  d_a = aerodynamic particle diameter (cm)
  μ = kinematic viscosity of the air (1.810⁻⁴ poise)
  C_c: slip correction factor
The slip correction factor is given by the equation:[1]

$$C_c = 1 + \frac{2}{Pd_p} \left[6.32 + 2.01 \exp\left(-0.1095 Pd_p\right) \right] \quad \textbf{(2)}$$

where P is the absolute pressure in cm Hg and d_p is the particle diameter in µm.

Thus the definition of this diameter is based on particle physics rather than on geometry and takes into account intrinsic density of particle material. One of the most important aspects of the aerodynamic particle diameter is that it is the most useful parameter for predicting the probability and the site of deposition in the human respiratory tract (discussed in greater detail in Chapter 5).

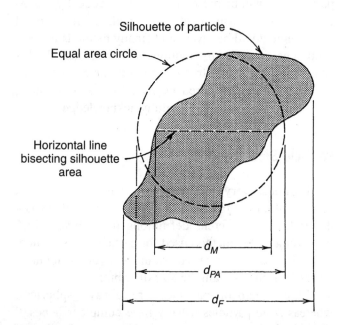

FIGURE 6-1. Martin's (d_m), projected area (d_{PA}), and Feret's (d_f) diameters for an irregular particle *(from Hinds, 1982. Reproduced with permission. Copyright John Wiley & Sons).*

When dealing with particles smaller than 0.1 µm, particle inertia is not a particularly useful property to characterize particle size. Instead, the thermodynamic particle diameter is often used. This is defined as the diameter of a sphere having the same diffusion coefficient as the particle in question.[10]

Particle diameter is often determined using optical methods. Optical instruments use light absorption or scattering properties to measure particle mass.[11, 12] Those methods are commonly dependent on the chemical composition and size of the particles. Hence, light scattering equivalent diameter is the diameter of a polystyrene latex sphere that scatters the same amount of light as the particle in question. This definition is rather weak because it depends considerably on the design of the particular optical counter used for particle measurement.

Electrical mobility sizing instruments are based on charging the sampled aerosols and measuring the ability of particles to traverse an electrical field. The most widely used instrument of this category is the Differential Mobility Analyzer manufactured by TSI Inc. (St. Paul, MN). This technology has been proven to be very useful in generating monodisperse aerosol in the size range 0.01–1.0 µm. The particle size measured by electrical counters is known as mobility equivalent diameter (d_{me}). This diameter is related to the equivalent volume diameter by the following equation:[13]

$$d_{me} = \frac{d_{ve} \chi \, C_m}{C_v} \qquad (3)$$

where d_{ve} is the volume equivalent diameter; χ is the particle dynamic shape factor, which accounts for the fact that irregularly shaped particles experience higher resistance when moving in the air and thus have a lower settling velocity compared to a spherical particles; and C_v and C_m are the slip correction factors corresponding to the equivalent volume and mobility diameters, respectively. The Scanning Mobility Particle Sizer (SMPS, Model 3934, TSI Inc.) is an improved version of the DMA, as it provides a more rapid readout of the aerosol size distribution by continuously changing the intensity of the electrical field and hence the particle size range that is counted.

The different particle sizes are listed in Table 6-1. Particles are found in a very wide range of sizes. For practical purposes, however, it is customary to separate particles into discrete size ranges. The size increments in these ranges may progress arithmetically or logarithmically (the latter being more frequent than the former). Although there is no particular methodology in dividing particles in sub-ranges or classes, classes of equal width

TABLE 6-1. Parameters Used to Represent Size of Individual Particles

Direct (geometrical) parameters	Indirect (physical) parameters
Feret's diameter	Stokes diameter
Martin's diameter	Aerodynamic diameter
Equivalent projected area diameter	Electrical mobility diameter
Fiber length, fiber diameter	Thermodynamic diameter
Equivalent volume diameter	Optical equivalent diameter

are usually preferred because they make data analysis easier. For example, cascade impactors typically are designed so that the increment of the logarithm of particle diameter would be 0.301. Equivalently, the ratio of the upper to the lower sizes of a specific size range is 2.[14, 15]

Basic Categories of Aerosol Instruments

Measurement of aerosol size distribution can be carried out by means of different instruments based on entirely different physical principles. Regardless of the type of the instrument used, there are some general criteria that need to be considered prior to making a decision on the sampling strategy. Friedlander[15] classified the range of aerosol instrumentation in terms of resolution of particle size, time, and chemical composition. The ideal instrument should be one that measures each individual particle in real-time and identifies its chemical composition. However, actual instruments represent compromises of this ideal. The different types of currently available instrumentation for particle measurement and their size-, time-, and composition-based responses are shown in Table 6-2.

Direct-reading instruments provide instantaneous information on the concentration and size distribution of aerosols. (These instruments are discussed extensively in Chapter 15, and only a brief description is given in the following paragraphs). In these instruments, the aerosol is drawn into a "sensing" region. The presence of particles gives rise to a change in some property of the sensing region, which is a function of a property of the sampled aerosol. It must be noted that these instruments provide information on particle size, as well as number and mass concentration.

Optical counters make use of the interaction between light and particles. The theory of optical aerosol behavior and its application to particle measurement is discussed by Willeke and Liu.[12] Most of the optical systems count light pulses scattered from

TABLE 6-2. Classification of Aerosol Instruments in Terms of their Size, Time and Composition Resolution[A]

Instrument	Resolution Size (μm)	Time	Chemical Composition
Ideal Single Particle Analyzer	0.001–100.0	Continuous	✓
Optical Counter	0.1–50.0	Continuous	
Electrical Aerosol Analyzer	0.01–0.5	Continuous	
Condensation Nuclei Counter	Non-specific[B]	Continuous	
Aerodynamic Particle Sizer	0.3–20.0	Continuous	
Cascade Impactor	0.03–100.0	Time-integrated	✓
Particle Time-Of-Flight Mass Spectrometer	0.3–10.0	Continuous	✓
Filter sampler	Non-specific	Time-integrated	✓

[A]*Adapted from Friedlander 1971.*[15]
[B]These instruments cannot provide information on particle size.

particles that flow, one by one, through an intensely illuminated zone. Sampling flow rate is low, and the smallest detectable particle size is about 0.1 μm when laser beams are used.[11] High particle concentration may result in coincidence errors. Coincidence errors arise when two or more particles are in the sensitive volume at the same time, causing a spurious signal that leads to an underestimation of the true particle concentration and an overestimation of the true particle size.[16] Typically, a concentration of 100 particles/cm³ or lower has been recommended to ensure that coincidence errors are below 5%.

Electrical mobility counters are based on charging the sampled aerosols and measuring the ability of particles to traverse an electrical field. Most of these counters draw particles through a cloud of either unipolar or bipolar ions, and each of the particles acquires a quantity of charge that is simply related to its size. Subsequently, the particles are drawn into a radially symmetric electrical field where particles smaller than a certain size, which depends on the intensity of the field, are collected onto the walls of the collecting device. By changing the field voltage, the particle size distribution can be obtained.[17, 18]

The Aerodynamic Particle Sizer (APS 3310, TSI Inc.) sizes and counts particles by measuring their time-of-flight in an accelerating flow field, thereby determining the aerodynamic particle diameter.[16, 19] This measurement technique, often referred to as Time-Of-Flight (TOF), is based on the principle that the magnitude of a particle's lag in an accelerating flow field is directly proportional to the particle's aerodynamic diameter. This time lag is measured as the particles' transit

time between two laser beams perpendicular to the aerosol flow.[16] A timer measures the time difference between light pulses produced as the particle passes through each beam. The main shortcoming of the APS is that it cannot determine size for particles smaller than about 0.5 μm. In addition, coincidence errors could also lead to erroneous readings of the APS concentrations.

Condensation nuclei counters[20–22] have been used extensively over the past 20 years to measure particle counts. Although particle number concentration is determined continuously, they cannot provide any information on either the size or the chemical composition of the particles. These instruments are typically used in conjunction with a particle-sizing instrument (such as the DMA or SMPS) to yield continuous data on the aerosol size distribution.

Particle collection on a filter is the simplest method for determining particle concentration and chemical composition (particle collection using filters is discussed extensively in Chapters 5 and 13 of this edition). Filters, however, do not provide any information on particle size, and concentration measurement is based on sampling over several hours to several days, depending on the sampling flow rate and the type of chemical compound to be measured. The simplicity of the filter techniques is a major advantage for particle sampling, but there are still potential artifacts (i.e., sampling errors) associated with these methods. These artifacts, discussed in great detail in Chapter 4, can result in an underestimation (negative artifact) or overestimation (positive artifact) of the concentration of particles. Artifacts are due to phenomena such as the following:[5] 1) Acid gas phase compounds, such as SO_2, HNO_3, and a variety of organic compounds, can be adsorbed on filter media; 2) Particle interactions on the filter media during and after sampling, e.g., reactions of fine particle (< 2.5 μm) acidic sulfates with coarse (> 2.5 μm) alkaline particles such as $CaCO_3$, and salts such as $NaCl$ or NH_4NO_3, can form neutral sulfate salts and acid gases such as CO_2, HCl, or HNO_3; 3) Volatilization of unstable ammonium particulate salts, such as NH_4NO_3, and NH_4Cl, forms NH_3 and HNO_3, or HCl, or volatilization of compounds such as semi-volatile organic matter from collected particles. Cascade impactors can overcome the problem of filter samplers by fractionating the aerosol into discrete size intervals prior to collecting the particles for further analysis. They cannot, however, provide the time-resolution that is needed in certain sampling cases, such as atmospheric studies aimed at understanding particle formation mechanisms, such as coagulation and gas-to-particle conversion.

The discussion in the previous paragraphs shows that, despite drastic improvements in the aerosol instrumentation area since Friedlander's original assessment, no technique yet approaches the ideal instrument. Thus, the choice of the instrument largely depends on the intended use of the data to be collected.

Additional Criteria for Aerosol Instrumentation

Along with the physical principles and the criteria for different types of resolution that aerosol instruments need to meet, there are also practical criteria that may have to be considered prior to selecting a particular method for aerosol measurement.

Convenience of use is a major issue especially in field studies. Current trends tend to minimize both the size and the power requirements of aerosol instruments. Of particular interest lately have been instruments that can be programmed to operate unattended for periods as long as a week. Data acquisition can be completely automated in many instruments currently used in the field.

Finally, ruggedness and reliability are also important criteria to consider when selecting an appropriate method for particle measurement. Often field studies require the use of instrumentation in extreme environments; examples include sampling in the arctic or stratosphere sampling, with temperatures typically below –50°C, or stack sampling, in which temperatures can be as high as 500°C. Many aerosol instruments are too complicated or too sensitive to be used in the field, and their use is inevitably restricted to laboratory settings where variations in parameters such as temperature, humidity, vibration, and other challenging factors are kept to a minimum.

Presentations of Particle Size Distribution

One graphical representation of particle size distribution data is a histogram, in which the width of the rectangle represents the size range, whereas the height represents the amount (i.e., number, mass or any other quantity of interest) in each size interval. The following example illustrates how data from a size-classifying instrument can be converted to a size distribution plot through a histogram. The data have been obtained with the Micro-Orifice Uniform Deposit Impactor (MOUDI) and represent particulate sulfate levels in the city of Philadelphia, Pennsylvania, collected over 12 hours.[23] The MOUDI operates at 30 L/min and classifies particles in the following size intervals: 0.094, 0.094–0.17, 0.17–0.3, 0.3–0.56, 0.56–1.0, 1.0–2.5, and 2.5–5.0 μm.[4] Each impaction stage removes particles larger than a certain size (known as the cutpoint), while particles smaller than the cutpoint follow the deflected air stream and are removed in a subsequent stage. Upon completion of sampling, the substrates of each impaction stage (which in the case of the MOUDI are 2 μm-pore Teflon filters) are removed for further analysis. This analysis could be: 1) gravimetric, by simply weighing the filters to determine the mass of the collected particles; 2) some type of chemical analysis (ion chromatography, thermal desorption, X-ray fluorescence, etc); or 3) microscopy to determine the number and morphometry of the particles. The type of analysis is determined by the type of information of a certain particle property that is desired. For the specific data shown in Table 6-3, sulfate concentrations have been determined by means of ion chromatography.

Table 6-3 shows results from this study. The first column shows the size ranges of the cascade impactor. Values represent aerodynamic diameters because this is the particle property measured by impactors. The second column shows the sulfate concentrations (in μg/m³), whereas the logarithm of the ratio of the upper to lower size in each range is shown in the third column. This value essentially represents the "width" of each size range when the horizontal scale of the histogram is logarithmic. When this scale is linear, the difference rather than the ratio of the upper to lower sizes is equal to the width of the specific size range.

The widths of particle size ranges are often not equal; thus in order to normalize the concentrations in a histogram, the amount on each stage is divided by the width of that range. This normalization allows comparisons of results obtained using different methods by simply overlaying the plots. This parameter is shown in the fourth column. The fraction of sulfate concentration in each stage, shown in the fifth column, is determined by

TABLE 6-3. Example of Size Distribution Data; Particulate Sulfate in Philadelphia[A]

Particle Size Range (µm)	Sulfate Concentration (µg/m³)	Δlog dₚ	ΔM/Δlog dₚ (µg/m³)	Mass Fraction	(%) Cumulative Fraction
0.094–0.16	2.2	0.23	9.56	0.08	0.08
0.16–0.27	13.8	0.23	60.0	0.50	0.58
0.27–0.56	14.9	0.31	48.1	0.39	0.97
0.56–1.0	0.45	0.25	1.7	0.015	0.985
1.0–2.5	0.22	0.39	0.564	0.006	0.991
2.5–5.0	0.21	0.30	0.70	0.009	1.000
Total	31.78		120.624		1.0

[A] *Source: Koutrakis et al.*[23]

dividing the amount in the fourth column by the total amount collected. Finally, the cumulative fraction, shown in the sixth column, is determined by adding the mass fractions of all previous size ranges to the next size range (the cumulative fraction of the last size range is always 1.0). Figure 6-2 shows a histogram of the concentrations for each size range.

Figure 6-3 shows a similar histogram plot of atmospheric aerosols.[24] The horizontal axis is logarithmic and represents particle size, whereas the vertical axis shows the volume of particles associated with a specific size range. The size distributions were obtained with electrical aerosol analyzers; hence the shown diameters are the electrical mobility diameters. This figure shows a classic size distribution of atmospheric aerosols, displaying

three distinct modes: one that peaks at about 0.2–0.04 µm; another that peaks at about 0.3–0.6 µm; and a third that peaks at about 5–10 µm. The particle size range from 0.01 to 0.1 µm is known as the "ultrafine mode," or "Aitken nuclei mode." These particles have relatively short residence times in the atmosphere because they are physically mobile due to their Brownian diffusive motion. They are products of homogeneous nucleation of supersaturated vapors (SO_2, NH_3, NO_x, and combustion products).

The size range 0.1 to 2.5 µm is known as the "accumulation mode" or "fine mode." These particles are formed by coagulation of ultrafine mode particles, through gas-to-particle conversion processes known as heterogeneous nucleation, and by condensation of gases onto pre-existing particles in the accumulation mode. The major constituents of these particles are sulfate (SO_4^{2-}), nitrate (NO_3^-), ammonium (NH_4^+), elemental (EC) and organic (OC) carbon. In addition they contain a variety of trace metals formed in combustion processes. Fine

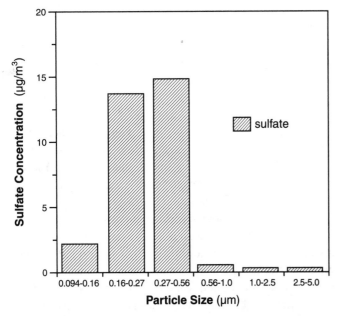

FIGURE 6-2. Histogram of particulate sulfate concentrations in the city of Philadelphia. 12-hour averages (*data from Koutrakis et al.* [23]).

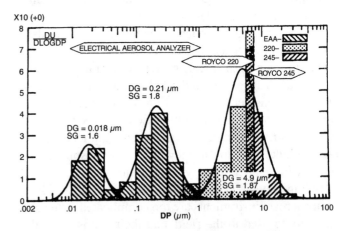

FIGURE 6-3. Generalized histogram plot of atmospheric aerosol size distributions (*from Whitby, 1978. Reproduced with permission. Copyright® Pergamon Press, Ltd.*).

particles are, in general, too small to settle out (by gravity) and too large to coagulate into larger particles; thus they have lifetimes in the atmosphere on the order of days and can be transported over long distances.

Particles in the size range 2.5 to 100 µm are referred to as coarse particles. These particles primarily contain soil and sea salt elements, such as silica, aluminum, calcium, iron, magnesium, sodium, and potassium. They are produced by mechanical processes (such as grinding, erosion, or resuspension by the wind). They are relatively large, settle out of the atmosphere by gravity within hours, and are only found near the source (a few tens of km with stack gas height of 100 meters).

By properly choosing a large number of size ranges, a smooth curve could be drawn through the rectangle tops, thereby providing a curve that is a graphical representation of the frequency function, or probability density function. This is shown in Figure 6-3, for the three different modes of ambient particles.

An alternative way of depicting particle size data is cumulative plots. Figure 6-4 shows a cumulative plot of the sulfate data shown in Table 6-3. While the particle size is plotted at the geometric midpoint of the interval on a logarithmic scale (similar to the histogram), the particle cumulative amount is plotted on a special scale called probit. The combination of the two scales is also known as a log-probability graph. Data from a lognormally distributed aerosol are depicted in such a graph as a straight line. For example, the data shown in Figure 6-4 could be represented by two lognormal distributions: one covering the range 0.1–0.6 µm and another lognormal distribution covering the size range 0.56–5.0 µm.

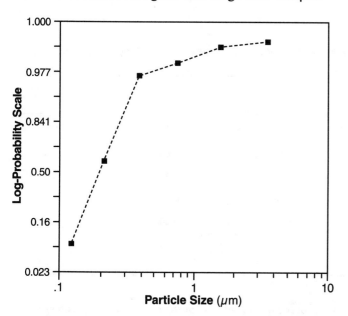

FIGURE 6-4. Log-probability of particulate sulfate size distribution in Philadelphia (*Source: Kourtrakis et al., 1989*).

When plotting data in a cumulative plot, one needs to account for the total amount of particles. For example, optical counters are only capable of measuring particle counts down to about 0.1 µm. As the particle counts in the atmosphere are dominated by particles smaller than 0.1 µm, dividing the number of large particles, which is often very small, by the total number of particles is misleading and could lead to erroneous conclusions about the actual size distribution.

To summarize, particle size distribution can be depicted in a histogram as an x-y plot. The x scale, which could be either linear or logarithmic, typically represents particle size, whereas the y-scale (usually linear) is used for the quantity of particles in a specific range.

Size Distribution Statistics

In order to better present particle distribution data, it is often desirable to summarize the size distribution by employing two parameters, one that identifies the location of the center of the distribution and another that identifies the scatter or spread of the distribution.

The most commonly used quantities defining the center of a distribution are the mean, mode, median, and geometric mean. The median is simply defined as the particle size for which one-half of the particles are smaller and one-half are bigger. It is also the diameter that divides the frequency distribution curve into two equal parts and it corresponds to a cumulative fraction of 0.5. For example, in the sulfate distribution data shown in Table 6-3, the median diameter is about 0.2 µm. Since sulfate concentrations have been expressed in terms of mass, 0.2 µm is the mass median diameter. If a different particle property was measured, the median diameter could be expressed analogously. For example, in data obtained with an optical counter, the particle size corresponding to a cumulative fraction of 0.5 would be the count median diameter. The mode is the most frequent size, or the size associated with the highest point in the frequency distribution curve. For example, for the sulfate data shown in Figure 6-2, the mode would be the middle size of the 0.27–0.56 µm range (i.e., about 0.38 µm).

Mean sizes, along with median, are often employed to describe a particle size distribution. In general, the mean diameter of a particle distribution is defined as:

$$x_m = \frac{\sum A_i x_i}{\sum A_i} \qquad \text{(4)}$$

where x_m is the arithmetic mean (or average) size, A_i is the amount of particles in the ith size interval, and x_i is the midpoint in the ith size interval. Similarly to the median sizes, the arithmetic means are dependent on the quantity (i.e., counts, mass, surface, radioactivity) used to express the size distribution. Thus, if instead of the mass concentrations of the data shown in Table 6-3, the particle counts were used, the mean diameter would be the mean diameter of the count-weighted particle distribution. Along with the arithmetic mean, the geometric mean size is very frequently used to express aerosol data. This is defined as follows:

$$x_g = \exp\left[\frac{\sum A_i \log x_i}{\sum A_i}\right] \qquad (5)$$

For symmetrical distributions, such as the normal distribution, the mean, median, and mode sizes will have the same value, which will be that associated with the size with the highest point in the frequency function. For an asymmetrical or skewed distribution, such as the lognormal distribution, the values of these quantities will differ. Most aerosol size distributions have a long tail to the right, which is typical of lognormal distributions. In this case, the mode is smaller than the median, which is smaller than the mean, i.e., mode< median< mean.[25]

An example of how these equations can be applied to actual particle data is shown in Table 6-4. Again, the data are borrowed from Koutrakis et al.[23] The arithmetic diameter of this aerosol can be then calculated as follows:

$$x_m = \frac{\sum A_i x_i}{\sum A_i} = \frac{10.4}{31.78} = 0.33 \mu m \qquad (6)$$

Similarly the geometric mean of the sulfate distribution, x_g, can be calculated as follows:

$$x_g = \exp\left[\frac{\sum A_i \log x_i}{\sum A_i}\right] = \exp\frac{-17.42}{31.78} = 0.28 \mu m \qquad (7)$$

Thus, the arithmetic mean is larger than the median aerodynamic diameter (0.20 μm) and the geometric mean diameter (0.28 μm). As mentioned earlier in the text, these diameters do not have to be identical. In the example shown in Tables 6-3 and 6-4, particulate sulfate concentrations were expressed in terms of mass concentrations for different size intervals. Thus, mean diameters (arithmetic or geometric) were weighted on mass contributions. If the amounts expressed in Table 6-4 as A_i were surface areas or particle counts, the mean size would have been, by analogy, called the surface area or particle count mean diameter, respectively. Again, the measuring method determines the property on which the size distribution is weighted and in terms of which the mean diameter is expressed.

In general, every weighted mean diameter of an aerosol distribution can be expressed in terms of two moment averages. The general form for a diameter of average property proportional to the particle diameter raised to the power of p (i.e., x^p) can be expressed as follows:

$$d_p = \left[\frac{\sum A_i x_i^{p}}{\sum A_i}\right]^{1/p} \qquad (8)$$

where A_i is the number of particles in the ith size range, and x_i is the midpoint of the ith size range. If instead of

TABLE 6-4. Example Calculation of Arithmetic and Geometric Mean Particle Sizes; Particulate Sulfate Concentrations in Philadelphia[A]

Particle Size Range (μm)	Sulfate Concentration, A_i (μg/m³)	Range Midpoint, x_i(μm)[B]	$A_i x_i$	$\log(x_i)$	$A_i\log(x_i)$
0.094–0.16	2.2	0.12	0.26	−0.92	−2.03
0.16–0.27	13.8	0.21	2.9	−0.68	−9.38
0.27–0.56	14.9	0.39	5.81	−0.41	−6.11
0.56–1.0	0.45	0.75	0.34	−0.125	−0.06
1.0–2.5	0.22	1.58	0.35	0.20	0.044
2.5–5.0	0.21	3.54	0.74	0.55	0.116
Total	31.78		10.4		−17.42

[A]Data obtained from Koutrakis et al.[23]

[B]The midpoint has been calculated as the square root of the product of the upper and lower sizes of the range (thus the geometric mean of the two sizes) as the distribution appears to be lognormal. For a normal distribution plot, it could also be calculated as the arithmetic average of the upper and lower sizes of the range.

weighting the property x^p according to the number (or counts) of particles, we choose to weight it according to a different parameter (e.g., surface area or volume), which depends on the particle size raised to the power of q (i.e., x^q), the resulting average is called the p-moment average of the q-weighted distribution. The general form of the p-moment average of the q-weighed size distribution (d_{qp}) is given by the following equation:

$$d_{qp} = \left[\frac{\sum A_i x_i^q x_i^p}{\sum a_i x_i^q} \right]^{1/p} \quad (9)$$

For example, the mass mean diameter (d_{mm}) of an aerosol can be calculated by the following equation:

$$d_{mm} = \frac{\sum m_i x_i}{\sum m_i} = \frac{(\pi \rho_p / 6) \sum n_i x_i^3 x_i}{(\pi \rho_p / 6) \sum n_i x_i^3} = \frac{\sum n_i x_i^4}{\sum n_i x_i^3} \quad (10)$$

where n_i is the number concentration of particles in the size range with midpoint x_i, and m_i is the mass concentration of particles in the size range with midpoint x_i. Thus $p = 1$ and $q = 3$ for the mass mean diameter. When $q = 0$ weighting is done by the number and when $p = 1$ and $q = 0$, the average property is the arithmetic mean. Number and surface mean diameters can be defined in an analogous manner.

Representation of an aerosol size distribution by means of just one average number would be incomplete without providing a measure of the spread of particles around this mean size. This spread is expressed in terms of the arithmetic or geometric standard deviation, expressed as follows:

$$\sigma = \left[\frac{\sum A (x_i - x_m)^2}{\sum A_i} \right]^{1/2} \quad (11)$$

and:

$$\sigma_g = \exp \left[\frac{\sum A_i \log^2 \left(\frac{x_i}{x_g} \right)}{\sum A_i} \right]^{1/2} \quad (12)$$

where σ and σ_g are the arithmetic and geometric standard deviations, and x_m and x_g the arithmetic and geometric mean sizes. It should be noted that a particle size distribution does not have to be normally or lognormally distributed to have a standard deviation. These two quantities are defined strictly by the formulas shown in equations 11 and 12.

It is very common sampling practice to derive a size distribution based on a certain property from measurements of particle distribution based on a different property. For example, ambient particle size data have been continuously measured by monitoring particle counts in discrete size intervals using the Differential Mobility Analyzer (DMA) or a combination of instruments such as a DMA with an Aerodynamic Particle Sizer or an optical counter.[26–28] The surface and volume-weighted distributions can then be determined by assuming that: 1) particles have a spherical shape; and 2) the midpoint of each size interval is representative of that size interval.

While the validity of the first assumption is in most cases true (with the exception of cases such as sampling of fibers), the validity of the second assumption is questionable. This is because the size resolution of most aerosol instruments is limited to a certain finite number of size ranges. The higher this number, of course, the more representative the midpoint becomes of the size range. If, however, particles within a size range are not uniformly distributed across the sizes of this range, but are somewhat skewed toward the upper or lower size of the range, converting particle counts to surface area (proportional to the square of the particle size) or volume (proportional to the cube of the particle size) may result in large errors. Thus, this transformation of the size distribution must be carried out with great caution and estimates of uncertainty must be always provided along with the converted data.

Table 6-5 illustrates how the aforementioned transformations from particle counts are performed. The data have been taken from Abt et al.[29] and represent 1-hour average indoor particle number concentrations, determined with the TSI Scanning Mobility Analyzer (TSI SMPS 3934, TSI Inc., St. Paul MN) for particles below 0.7 μm and with the TSI Aerodynamic Particle Sizer (TSI APS 3310) for particles larger than 0.7 μm. The sixth column of Table 6-5 is the surface area, computed for each size range by multiplying the quantity $\pi x_i^2 / 4$ with the number concentration. Here, it is assumed that particles are perfect spheres and that the midpoint of each range (i.e., the geometric mean of the upper and lower sizes) is representative of the entire range. As noted above, this is an assumption that should be applied with caution, particularly when the widths of the size ranges are quite large. Similarly to the surface area, the volume of each size class has been calculated by multiplying the quantity $\pi x_i^3 / 6$ with the number concentration of that size class. If the particle density of the indoor aerosol shown in Table 6-5 were known, it would be straightforward to calculate the mass-weighted size distribution. Multiplication of the ninth column in Table 6-5 by the particle density yields particle mass in each size interval.

TABLE 6-5. Example of Converting Particle Number Distributions to Surface and Volume Distributions

Particle size Range (μm)	Midpoint (μm)	Number Concentr. (p/cm³)	$\Delta N/\Delta\log d_p$ (p/cm³)	Surface Concentr. (μm²/cm³)[A]	$\Delta S/\Delta\log d_p$ (μm²/cm³)	Volume Concentr. (μm²/cm³)[B]	$\Delta V/\Delta\log d_p$ (μm³/cm³)	$\Delta\log d_p$
0.02–0.05	0.031	17,852	44,860	14.0	35.23	0.29	0.75	0.398
0.05–0.1	0.071	54,998	182,699	215.98	717.46	10.2	33.8	0.301
0.1–0.2	0.14	19,572	65,017	307.44	1,021.3	29.0	96.3	0.301
0.2–0.5	0.316	707	1,777	55.54	139.56	11.7	29.4	0.398
0.5–1.0	0.71	4.5	14.93	1.77	5.86	0.8	2.8	0.301
1.0–2.0	1.412	1.67	5.56	2.63	8.75	2.5	8.23	0.301
2.0–5.0	3.162	0.29	0.74	2.3	5.77	4.9	12.2	0.398
5.0–10.0	7.07	0.02	0.067	0.8	2.66	3.8	12.5	0.301

[A]The surface area is calculated by multiplying the number concentration with the product $\pi x_i^2/4$, where x_i is the midpoint diameter.
[B]The volume of a given range is calculated by multiplying the number concentration with the product $\pi x_i^3/6$, where x_i is the midpoint.

Methods for Particle Size Distribution Analysis and Presentation

After obtaining raw data from instruments measuring the size distribution of aerosols, the data need to be inverted to obtain the true size distribution of the aerosol, $f(x)$. Inversion of size distribution data to find the "true" particle size distribution is not a straightforward problem, and different methods have been proposed in the literature for solving this problem.[30–32] Mathematically, the relationship between the measured data and the size distribution is given by the following formula:

$$y_i = \int k_i(x)f(x)dx + \varepsilon_i, \qquad i = 1, \ldots, p \qquad (13)$$

where p is the number of measurements, y_i is the data in the ith size interval as measured by the instrument, $k_i(x)$ is commonly referred to as the kernel function of the instruments, and is proportional to the known response of the instrument to a monodisperse aerosol of size x, and $f(x)$ is the aerosol true size distribution. The quantity ε_i is the error in the measurement by the instrument for particles in the ith size channel.

Finding the true size distribution from measured data is an "ill-posed" problem. One aspect of the problem arises from the fact that the particle collection characteristics of different instruments are not sharp. Ideally, aerosol samplers are designed to minimize the range of particles contributing to each datum, so that the kernel function $k_i(x)$ depends monotonically on x. The data shown in Figure 6-5,[32] however, indicate that this may or may not be true for different instruments. The data on the left side have been collected by the Differential Mobility Analyzer (DMA), whose typical kernel functions are also shown in this figure. Particle classification by means of the DMA is based on charging the particles and measuring their ability to traverse an electrical field, which is a function of the electrostatic migration velocity of each particle in the field. A detailed description of this system is given in Chapter 15 of this edition. Briefly, the migration velocity is proportional to the number of charges acquired by each particle and is inversely proportional to the particle physical diameter. By fixing the DMA voltage to measure, for example, 0.1 µm particles having acquired one charge, the DMA also measures doubly-charged 0.2 µm particles, triply-charged 0.3 µm particles, etc. This is best illustrated in Figure 6-5, where the primary peak at 720 volts is associated with 0.04 µm particles, whereas the small secondary peak at about 0.08 µm is due to the doubly-charged particles. Similarly, the primary peak at a DMA voltage of 5760 volts is associated with singly-charged 0.11 µm particles, whereas the secondary peaks at approximately 0.22 µm, 0.33 µm, etc., are associated with particles having multiple charges.

If the variance in the concentrations of the true size distribution, (shown by the dotted line in the left hand side of Figure 6-5) is large compared to the variance of

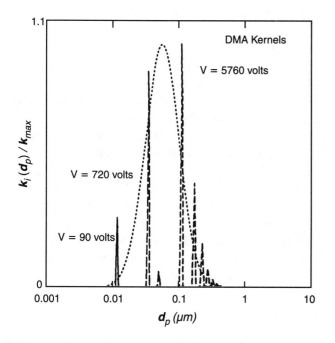

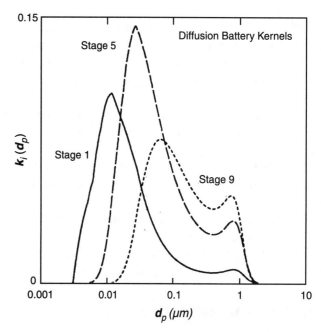

FIGURE 6-5. Kernel functions for diffusion battery and differential mobility analyzer. The DMA kernels (dashed lines) are plotted for three different inner rod voltages. The dotted line represents a distribution that the DMA can easily resolve *(from Wolfenbarger and Seinfeld, 1990; reproduced with permission. Copyright® Pergamon Press, Ltd.).*

the DMA peaks, the size distribution can then be approximated as a constant in the size interval corresponding to a given DMA peak. This is not true in the case of data obtained by means of a diffusion battery.[33] Particle classification by means of diffusion batteries is described in Chapter 19. The kernel functions for each size channel are broad, thus there is considerable overlap between particles of different size ranges (i.e., the chance of particle misclassification increases as this overlap increases) and obtaining the true size distribution becomes a difficult task. In addition, the problem is undetermined because, in principle, there are an infinite number of curves that may be fitted to these discrete data points (i.e., a finite number of measurements is used to match the infinity of the continuous size distribution).

Many different approaches have been pursued to solve the data inversion problem. Authors have approached the inversion problem by assuming an *a priori* knowledge of the form of the kernels. Specifically, the true size distribution *f(x)* is determined by superposition of kernel functions that are lognormal,[34] or normal, [35] or have a known functional form.[31] This type of approach is called fitting, in which a function with free parameters is chosen as *f(x)* and a mathematical algorithm is applied to fit the data. An example of this approach is shown in Figure 6-3; the atmospheric multimodal aerosol size distributions can be well represented by three overlapping lognormal distributions.[24] Dzubay and Hasan [36] have fitted lognormal distributions to data obtained with cascade impactors. Raabe[37] describes the fitting process in more detail.

Other approaches have attempted to overcome the "ill-posedness" of the inversion problem by assuming that the size distribution function, *f(x)* must be smooth. This technique is called regularization and is described in more detail by Wolfenbarger and Seinfeld.[38] A similar process has been described recently by Ramachadran et al.[39] Two different inversion techniques are used for data obtained with a personal cascade impactor. Both methods attempt to match the solution to type measured data with an *a priori* expectation on the smoothness of the size distribution.

Inversion methods can sometimes be rigorous and may require considerable expertise in this field. Nonspecialists are encouraged to use any of the fitting methods described in the following paragraph.

Available Software for Data Inversion

DISTFIT. The Distfit Aerosol Data Fitting Program is available from TSI Inc., St. Paul, MN. The program is used to generate normal, lognormal, or Rosin-Rammler fits for partial distributions, as well as linear, logarithmic, power and exponential fit for particle size data. It provides multi-modal curve fitting with optional constraint fits. This allows individual parameters to be constrained to predetermined values so that only unconstrained parameters are determined by the fitting procedure. It should be noted that the program does not do data inversion as this is dependent on the specific aerosol sampler to be used. It is a particularly powerful program in cases where size distribution data are obtained in settings with different and varying sources, which tend to change the characteristics of the size distribution. In determining size distributions in more stable cases, for example, the output of an aerosol generator, such as an atomizer, computer graphics or spreadsheets can be used as successfully.

STWOM. This is a FORTRAN code, developed by Markowski[40] to convert data from impactors and/or electrical aerosol analyzers. The user supplies the measurement uncertainty and the kernel functions (i.e., the instrument response functions) in an array. The code then provides an inversion that is meant to yield optimum "smoothness" of the data.

MICRON. This inversion technique was developed by Wolfenbarger and Seinfeld.[38] This technique is based on regularization of the ill-posed problem of equation 13. Solutions to equation 13 are sought that are smooth (i.e., generate no discontinuity in the size distribution function) and at the same time are faithful to the data generated by one, or a combination of, instruments. A novelty of the method is that it considers the dependent nature of the data. Particle data generated by a single instrument are generally subjected to the same noise sources affecting all readings. The need to account for dependent errors becomes more apparent when more than one instrument is used to obtain size distributions; data from a single instrument are generally self-consistent, whereas data from different instruments generally disagree due to the differing error sources of the instruments.

EVE. This program, developed by Paatero[41] inverts data by means of a method called extreme value estimation. A family of acceptable solutions, rather than a single solution, is given to a specific data inversion problem. According to Knutson and Lioy,[27] EVE is a rather rigorous program and requires expertise that often exceeds that available to the occasional user.

Summary

Particle size and amount are the two variables used to describe data on the size distribution of an aerosol. There are many definitions of these two parameters, depending on the instrument(s) used to obtain the data as well as the particle property of interest. Measuring the size distribution of particles is a complicated task and over the past 20 years one of the main goals of environmental scientists has been to improve instrumentation for determining the size distribution of particles. In general, there are two main strategies for particle sampling and measuring, namely direct and integrated sampling. In the first method, aerosol size distributions are determined continuously (or near-continuously), i.e., in time intervals ranging from a few seconds to a few minutes. The second category consists of instruments, which determine aerosol size distribution by collecting samples over time periods ranging from hours to days.

An "average" size is often employed to define a particle distribution. This may be a median diameter based on particle counts, surface, mass or any other property of interest. A standard deviation (arithmetic or geometric) is used to characterize particle "spread" around this average value. It should be emphasized that these two parameters can describe the size distribution of aerosols that are not normally or lognormally distributed.

After obtaining raw data from instruments measuring the size distribution of aerosols, the data need to be inverted to obtain the true size distribution of the aerosol. Inversion of size distribution data to find the "true" particle size distribution is not a straightforward problem, but several methods have been developed for solving this problem and are currently available to the scientific community.

References

1. Hinds, W.C.: Aerosol Technology: Properties, Behavior and Measurement of Airborne Particles. John Wiley & Sons, New York (1982).
2. Hidy, G.M.: Measurement of Aerosol Properties. In: *Aerosols: An Industrial and Environmental Science*, pp.167–254. Academic Press Inc., New York (1984).
3. Wang, H.C.; John, W.: Characteristics of the Berner Impactor for Sampling Inorganic Ions. Aerosol Sci. Technol. 8 (2): 157–172 (1988).
4. Marple V.A.; Rubow K.L.; Behm, S.M.: A Micro-Orifice Uniform Deposit Impactor (MOUDI). Aerosol Sci. Technol. 14:434–446 (1991).
5. Koutrakis P.; Thompson K.M.; Wolfson J.M.; *et al.*: Determination of Aerosol Strong Acidity Losses due to Interactions of Collected Particles: Results from Laboratory and Field Studies. Atmos. Environ., 26 A:987–995 (1992).
6. Nordmeyer, T.; Prather, K.A.: "Real-Time Measurement Capabilities Using Aerosol Time-of-Flight Mass Spectrometry." Anal. Chem., 66:3540 (1994).
7. Noble, C.A.; Prather, K.A.: Real-time Measurement of Correlated Size and Composition Profiles of Individual Atmospheric Aerosol Particles. Environ. Sci. Technol., 30:2667–2680 (1996).
8. Weiss, M.; Verheijen, P.J.T.; Marijnissen, J.C.M; and Scarlett, B.: On the Performance of an On-line Time-of-Flight Mass Spectrometer for Aerosols. J. Aerosol Sci. 28:159–171 (1997).
9. Mercer, T.; Stafford, R.G.: Impaction from Round Jets. Ann. Occup. Hyg., 12:41–48 (1969).
10. Davies, C.N. Deposition from Moving Aerosols. In: *Aerosol Science*, C.N. Davies, Ed. Academic Press, London (1966).
11. Knollenberg, R.G.; Luehr, R.: Open Cavity Laser Active Scattering Particle Spectrometry from 0.05 to 5 µm. In: *Fine Particles*, pp. 669–696, B.Y.H. Liu, Ed. Academic Press, New York (1976).
12. Willeke, K.; Liu, B.Y.H.: Single Particle Optical Counters: Principle and Applications. In: *Fine Particles*, pp. 697–730, B.Y.H. Liu, Ed. Academic Press, New York (1976).
13. Kasper, G. Aerosol Sci. Technol. 1:187–199 (1982).
14. Mitchell, R.; Pilcher, J.: Improved Cascade Impactor for Measuring Aerosol Particle Sizes in Air Pollutants, Commercial Aerosols, Cigarette Smokes. Ind. Eng. Chem. 47:1039 (1952).
15. Friedlander, S.K.: *Smoke, Dust, and Haze,* John Wiley, New York (1976).
16. Heitbrink, W.A.; Baron, P.A.; Willeke, K.: Coincidence in Time-of-Flight Aerosol Spectrometers: Phantom Particle Creation. Aerosol Sci. Technol. 14:112–126 (1991).
17. Liu, B.Y.H.; Pui, D.Y.H. Unipolar Charging of Aerosol Particles in the Continuum Regime. J. Aerosol Sci. 6:249 (1975).
18. Whitby, K.T.: Electrical Measurement of Aerosols. In: *Fine Particles*, pp. 581–624, B.Y.H. Liu, Ed. Academic Press, New York (1976).
19. Wilson, J.C.; Liu, B.Y.H.: Aerodynamic Particle Sizes Measurement by Laser-Doppler Velocimetry. J. Aerosol Sci. 11: 139 (1980).
20. Agarwal, J.K.; Sem, G.J.: Continuous Flow, Single-Particle-Counting Condensation Nucleus Counter. J. Aerosol Sci., 11:343–357 (1980).
21. Kousaka, Y.; Niida, T.; Okyuama, K.; Tanaka, H.: Development of a Mixing Type Condensation Nucleus Counter. J. Aerosol Sci. 13:231 (1982).
22. Sinclair, D.; Hoopes, G.S.: Continuous Flow Condensation Nucleus Counter. J. Aerosol Sci. 6:1 (1975).
23. Koutrakis, P.; Wolfson, J.M.; Spengler, J.D.; *et al.*: Equilibrium Size of Atmospheric Sulfates as a Function of Relative Humidity. J. Geophys. Res. 94:6442–6448 (1989).
24. Whitby, K.T.; Svendrup, G.M.: California Aerosols: Their Physical and Chemical Characteristics. Adv. Environ. Sci. Technol. 10:477 (1980).
25. Herdan, G.: Small Particle Statistics. Butterworth and Co., London (1960).
26. Whitby, K.T.: The Physical Characteristics of Sulfur Aerosols. Atmos. Environ. 12:135–159 (1978).
27. Knutson, E.O; Lioy, P.J.: Measurement and Presentation of Aerosol Size Distributions. In: Air Sampling Instruments for Evaluation of Atmospheric Contaminants, pp.121–137, B.S. Cohen. Ed. American Conference of Governmental Industrial Hygienists, Cincinnati, OH (1993).
28. Allen, G.A.; Sioutas, C.; Koutrakis, P.; *et al.*: Evaluation of the TEOM Method for Measurement of Ambient Particulate Mass in Urban Areas. J. Air Waste Manag. Assoc., 47:682–689, (1997).
29. Abt, E.; Sioutas, C.; Suh, H.H.; Koutrakis, P.: Characterization of Indoor Particulate Sources in Homes in the Boston Metropolitan Area. J. Air Waste Manag. Assoc., (in press).
30. Twomey, S.: Information Content in Remote-Sensing. Appl. Opti. 13:942–945 (1974).
31. Pandis, S.N.; Baltensperger, U.; Wolfenbarger, J.K.; Seinfeld, J.H. Inversion of Aerosol Data from the Epiphaniometer. J. Aerosol Sci. 22:417–428 (1991).

32. Wolfenbarger, J.K.; Seinfeld, J.H.: Inversion of Aerosol Size Distribution Data. J. Aerosol Sci. 21:227–247 (1990).

33. Cheng, Y.S.; Keating, J.A.; Kanapilly, G.M.: Theory and Calibration of a Screen-Type Diffusion Battery. J. Aerosol Sci., 11:549-556 (1980).

34. Heitzenberg, J.: Determination *in situ* of the Size Distribution of the Atmospheric Aerosol. J. Aerosol Sci. 6:231–303 (1975).

35. Lloyd, J.J.; Taylor, C.J.; Shields, R.A.: Analysis of Diffusion Battery Data Using Simulated Annealing. Proc. 8th Annual conference of the U.K. Aerosol Society, York, England (1994).

36. Dzubay, T.G.; Hasan, H.: Fitting Multimodal Lognormal Size Distributions to Cascade Impactor Data. Aerosol Sci. Technol. 13:144–160 (1990).

37. Raabe, O.G.: A General Method for Fitting Size Distributions to Multicomponent Data Using Weighted Least Squares. Environ. Sci. Technol. 12:1162–1167 (1978).

38. Wolfenbarger, J.K.; Seinfeld, J.H.: Regularized Solutions to the Aerosol Data Inversion Problem. SIAM J. Sci. Stat. Comput. 12:342–361 (1991).

39. Ramachadran, G.; Johnson, E.W.; Vincent, J.H.: Inversion Techniques for Personal Cascade Impactor Data. J. Aerosol Sci. 27:1083–1097 (1997).

40. Markowski, G.: Improving Twomey's Algorithm for Inversion of Aerosol Measurement Data. Aerosol Sci. Technol. 7:127–142 (1987).

41. Paatero, P.: EVE Reference Manual 04.03.1991; University of Helsinki, Helsinki, Finland (1991).

Chapter 7

Airflow Calibration

Morton Lippmann, Ph.D.

Nelson Institute of Environmental Medicine, New York University School of Medicine, Tuxedo, New York

CONTENTS

Flow Rate and Volume Metering Instruments

Accurate measurement of air flow rate and volume is an integral part of the calibration of most air sampling instruments. The various instruments and techniques involved in the measurement of flow rate and volume are discussed in this chapter and elsewhere.[1, 2] These can be divided into two general categories: primary and secondary standards. Primary measurements generally involve a direct measurement of volume on the basis of the physical dimensions of a defined enclosed space. Secondary standards are reference instruments or meters that trace their calibration to primary standards and which have been shown to be capable of maintaining their accuracy with reasonable handling and care in operation (Table 7-1).

Primary Standards

Spirometers

The spirometer (Figure 7-1) is a cylindrical bell with its open end under a liquid seal. The bell is supported by a chain or cord and is balanced by a counterweight. The volume of air entering the spirometer is determined by calculating the change in height times the cross section. With the gas valve open, the cylindri-

cal bell should remain stationary. If it does not, the counterweight should be adjusted accordingly. Some spirometers do not have a cycloid counterpoise and, in these units, the bell will not remain stationary with the valve open to ambient air, but rather will slowly move toward the geometric center of the bell. Spirometers are often calibrated by the manufacturer; however, it is prudent to check the calibration after proper alignment by measuring the spirometer bell's inside dimensions.

The Mariotte bottle (Figure 7-2) is an instrument similar to the spirometer that measures displaced water instead of air. When the valve at the bottom of the bottle is opened, water drains out of the bottle by gravity, and air is drawn into the bottle via a sample collector to replace it. The volume of air drawn in is equal to the change in water level multiplied by the cross section at the water surface.

"Frictionless" Piston Meters

Cylindrical air displacement meters with nearly frictionless pistons are frequently used for primary flow calibrations at flow rates of 0.001–10 L/min. The simplest of these is the soap bubble meter illustrated in Figure 7-3. A soap bubble is created in a wetted wall graduated tube (i.e., volumetric laboratory burette) by squeezing a rubber bulb and raising the soap solution above the gas

TABLE 7-1. Apparatus for Air Sampling Flow Rate Calibration

Type of Meter	Quantity Measured	Range	Commercial Sources*
Spirometer	Integrated volume	0.2-20 ft³ (6-600 L)	AMC, BRO, GRA, WEC
Soap film flowmeter	Integrated volume	2-10,000 mL	BIO, BUC, GIC, SEN, SKC, SPE
Mercury sealed piston	Integrated volume	1-12,000 mL	BRO
Wet test meter	Integrated volume	Unlimited volumes; max. flow rates from 1-480 ft³/hr (0.5-230 L/min)	AMC, PSC
Dry test meter	Integrated volume	Unlimited volumes; max. flow rates from 20-325 ft³/hr (10-150 L/min)	AMC, GRA
Electronic mass flow rate	Mass flow rate	0-10 mL/min up to 0-3000 L/min	BRO, KRZ, MGE, SIE, SKC, THR
Laminar flowmeter	Volumetric flow rate	0.00005-2000 ft³/hr (0.02 mL/min-1 m³/min)	AFP, CME, MIC, RAD
Venturi meter	Volumetric flow rate	Depends on pipe and orifice diameters	HIQ, RAD, TSI
Orifice meter	Volumetric flow rate	Depends on pipe and orifice diameters	BGI, MIC
Rotameter	Volumetric flow rate	From 1.0 mL/min up	AFP, BRO, FPC, GIC, GIL, KEY, MGE, SKC, SKD
Thermo-anemometer	Velocity	From 10 fpm (0.3 m/min) up	ALN, KRZ, SIE, TSI
Pitot tube	Velocity	From 1000 fpm (300 m/min) up	DWY, GRA, MIC

*Refer to Table 7-2 at the end of this chapter for the full company names and addresses for these sources.

inlet level. As the gas passes through the soap solution, it creates bubbles that are then timed as they traverse through a known volume within the tube. In this case, the bubbles act as the frictionless pistons.

Soap film flowmeters are generally accurate to within 1.0%, although greater accuracy can be achieved under select conditions. However, a correction may be needed for the humidification of the air during contact with the soap solution. Also, at high flow rates, the accuracy of soap bubble meters declines because of gas permeation through the soap film.

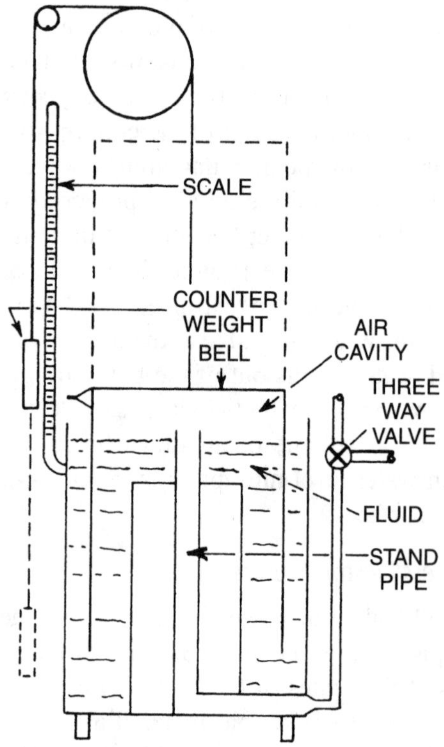

Pitot Tubes

The spirometer and frictionless piston are considered the primary standard for measuring volume and flow rate. The Pitot static tube (commonly referred to as Pitot tube) is the primary standard for measuring gas velocities, and it needs no calibration. It consists of a tube (the Pitot tube) whose opening faces directly into the flow and a static tube formed by a concentric tube with holes placed equally around it in a plane that is 8 diameters from the impact opening. The pressure in the

FIGURE 7-1. Schematic of a spirometer or gasometer (*reprinted from The Industrial Environment—Its Evaluation and Control*).[1]

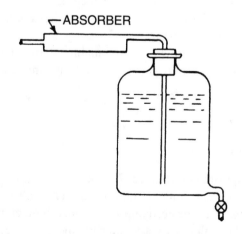

FIGURE 7-2. Mariotte bottle (*reprinted from The Industrial Environment—Its Evaluation and Control*).[1]

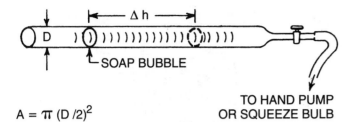

A = π (D /2)²

TO HAND PUMP
OR SQUEEZE BULB

FIGURE 7-3. Soap bubble meter (*reprinted from The Industrial Environment—Its Evaluation and Control*).[1]

Pitot tube is the impact pressure, whereas that in the static tube is the static pressure. The difference between the static and impact (total) pressure is the velocity pressure. Bernoulli's theorem applied to a Pitot tube in an air stream simplifies to the formula:

$$V = \left(\frac{2g_c P_V}{\rho}\right)^{\frac{1}{2}} \quad (1)$$

where: V = linear velocity
P_V = velocity pressure = impact pressure − static pressure
g_c = gravitational constant (English units)
ρ = gas density

If the Pitot tube is to be used with air at 70°F and 1 atm, Equation 1 reduces to the convenient dimensional formula

$$V = 4005\sqrt{h_V} \quad (2)$$

where: h_V = velocity pressure in inches of water
V = velocity in feet per minute (fpm)

For air at 20°C and 1 atm, the equation in metric form reduces to

$$V = 12.8\sqrt{h_V} \quad (3)$$

where: h_V = velocity pressure in cm of water
V = velocity in m/s

The acceptable accuracy of the Pitot tube is limited by the ability to measure the velocity pressure. Above 2500 fpm (12.7 m/s), a U-tube manometer is satisfactory. However, for lower velocities, an inclined manometer or low-range Magnehelic® gauge is necessary. With such a manometer, velocities of 1000 fpm (5.1 m/s) can be measured accurately (at 1000 fpm [5.08 m/s], h_V = 0.1 in. H₂O [0.25 cm H₂O]). Electronic capacitance pressure gauges permit measurements of h_V

down to 0.001 in. H₂O, corresponding to a velocity of about 100 fpm (0.5 m/s or 50 cm/s).

Secondary Standards

Secondary standards are reference instruments that trace their calibration to primary standards. Among secondary standards, however, there are a number of instruments that provide an accuracy nearly comparable to that of primary standards but which, of themselves, cannot be calibrated by internal volume measurement. These instruments are sometimes referred to as intermediate standards and provide an accuracy of approximately 1.0%. These instruments include wet test meters and dry gas meters.

Wet Test Meter

A typical wet test meter is shown in Figure 7-4. It consists of a cylindrical container in which there is a partitioned drum half submerged in water with openings at the center and periphery of each radial chamber. Air or gas enters at the center and flows into an individual compartment with the buoyant force causing it to raise, thereby producing rotation. This rotation, and therefore the volume, is indicated by a dial on the face of the instrument. The volume measured will depend on the fluid level in the meter because the liquid is displaced by air. This liquid level must be maintained at a calibrated height that is indicated by a sight gauge. In addition, level screws and a sight bubble are provided to level the instrument horizontally. Once the instrument is filled with water, the water should be saturated with the gas in question by running the gas through the instrument for several hours. When calibrated against a spirometer, wet test meters should exhibit an accuracy of 0.5% or better. However, they can only be used within a narrow pressure range near ambient pressure. The manometer used with the instrument acts, in effect, as a pressure limiting valve.

Care has to be taken in the use of wet test meters. If they are used with gas that can produce a potentially corrosive solution upon contact with water, the internal drum and moving parts may corrode. In addition, it is necessary to overcome the inertia of the mechanical parts at low flow rates, and there is the possibility that the liquid might surge and break the water seal at the inlet or outlet at high flow rates.

Dry Gas Meter

The dry gas meter shown in Figure 7-5 is similar to that used for domestic natural gas metering. It consists of

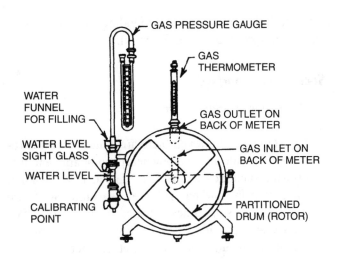

FIGURE 7-4. Wet test meter (*reprinted from The Industrial Environment—Its Evaluation and Control*).[1]

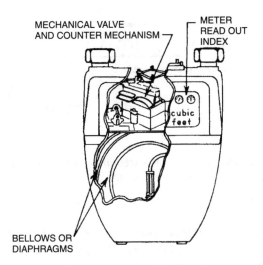

FIGURE 7-5. Dry gas meter (*reprinted from The Industrial Environment—Its Evaluation and Control*).[1]

two bags interconnected by mechanical valves and a cycle-counting device. The air or gas fills one bag while the other bag empties itself. When the cycle is completed, the valves are switched, and the second bag fills while the first one empties.

In using dry gas meters, operators should be cognizant of the mechanical drag of the instrument, especially at low flow rates, and of the resulting pressure drop and the possibility of leaks. Instruments of this type, however, can be used to measure flow rates from 5 to 5000 L/min. At pressures up to 250 lb/in.² (17 atm, 1717 kPa), an accuracy of approximately 1.0% can readily be obtained, and when calibrated against a spirometer, the accuracy can be improved. If calibration indicates an error in flow rate, the dry gas meter can be adjusted by means of tangential adjusting weights associated with the linkage to the volume dials.

Additional Secondary Standards

The remaining secondary standards have an accuracy that is usually less than that of the preceding instruments. Among these are a variety of positive displacement meters as well as air velocity meters and electromechanical devices.

Positive Displacement Meters

Positive displacement meters consist of a tight-fitting, moving element with individual volume compartments that fill at the inlet and discharge at the outlet ports. A lobed rotor design is illustrated in Figure 7-6. Another multicompartment continuous rotary meter uses interlocking gears. When the rotors of such meters

are motor driven, these units become positive displacement air movers.

Exchange of Potential and Kinetic Energy

The following secondary standards for flow rate operate on the principle of the conservation of energy. Specifically, they utilize Bernoulli's theorem for the exchange of potential energy for kinetic energy and/or frictional heat. Each consists of a flow restriction within a closed conduit. The restriction causes an increase in the fluid velocity and therefore an increase in kinetic energy, which requires a corresponding decrease in potential energy, i.e., static pressure. The flow rate can be calculated from a knowledge of the pressure drop, the flow cross section at the constriction, the density of the fluid, and the coefficient of discharge, which is the ratio of actual flow to theoretical flow and makes allowance for stream contraction and frictional effects.

Flowmeters that operate on this principle can be divided into two groups. The larger group includes

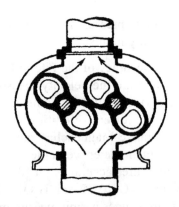

FIGURE 7-6. Cycloidal or roots-type gas meter.

orifice meters, Venturi meters, and flow nozzles; these have a fixed restriction and are known as variable-head meters because the differential pressure head varies with flow. Flowmeters in the other group, which includes rotameters, are known as variable-area meters because a constant pressure differential is maintained by varying the flow cross section.

Rotameters

A rotameter consists of a "float" that is free to move up and down within a vertical tapered tube that is larger at the top than the bottom. The fluid flows upward, causing the float to rise until the pressure drop across the annular area between the float and the tube wall is just sufficient to support the float. The floats achieve stability through their rotation within the tapered tube, providing the basis for the term "rotameter." The tapered tube is usually made of glass, metal, or clear plastic and has a flow rate scale etched directly on it. The height of the float indicates the flow rate. Floats of various configurations have been used, as indicated in Figure 7-7. Such shaped floats are conventionally read at the highest point of maximum diameter of the float, unless otherwise indicated. The float used in most rotameters now is spherical and is read at the center of the ball.

Most rotameters have a range of 10:1 between their maximum and minimum flows. The range of a given tube can be extended by using heavier or lighter floats. The tubes are made in sizes from about 1/8 to 6 in. (0.32–15 cm) in diameter, covering ranges from a few mL/min to over 1000 cfm (28.3 m³/min).

Both in the laboratory and in commercial air sampling devices, rotameters are the most commonly used devices for measuring flow rate. Depending on the accuracy required, they range in length from about 5 cm to approximately 50 cm. Whereas very small rotameters

may not have very good accuracy, most laboratory rotameters are supplied with a calibration curve by the manufacturer that is accurate to ∫5%. Accuracies of ∫1% to 2% are obtainable when the rotameters are calibrated in the system. Most rotameters are calibrated against a primary or an accurate secondary standard. These calibrations are usually performed with one port of the rotameter at standard "room" temperature (20°C) and pressure (760 torr). The accuracy is therefore limited by the reproducibility or correction to these conditions. If one side of the rotameter is not at standard temperature and pressure, the calibration supplied with the rotameter is no longer valid. In this case, the instrument has to be calibrated in the system in which it is used (temperature and pressure) against a known standard, or the supplied calibration must be corrected for these variations. It should be noted that not correcting for these factors is one of the most common errors encountered in the use of rotameters. At excessive pressure differences, inaccuracies of a factor of 5 to 10 can readily be encountered.

For rotameters with linear flow rate scales, the actual sampling flow will approximately equal the indicated flow rate times the square roots of the ratios of absolute temperatures and pressures of the calibration and field conditions.[3] The ratios change when the field pressures and temperatures differ from those in the calibration laboratory. Thus, if the flowmeter was accurate at standard conditions and the flow resistance of the sampling medium was relatively low (e.g., 30 torr), the flow rate that would be indicated on the rotameter for a standard flow rate of 11 L/min would be $11/(760/730)^{1/2} = 10.8$ L/min, a difference of only 1.8%. On the other hand, for a 25-mm diameter AA Millipore filter with a 3.9-cm² filtering area and a flow resistance of 190 torr, the indicated flow rate would be $11/(760/570)^{1/2} = 9.5$ L/min, 14% below the standard flow rate.

A further correction will be needed when the sampling is done at atmospheric pressures and/or temperatures that differ substantially from those used for the calibration. For example, at an elevation of 5000 ft above sea level, the atmospheric pressure is only 83% of that at sea level. Thus, the actual flow rate would be 9.6% greater than that of standard air, based on the altitude correction alone. If the temperature in the field was 35°C while the meter was calibrated at 20°C, the actual flow rate in the field would be $[(273 + 35)/(273 + 20)]^{1/2} \times 100 = 2.5\%$ greater than that of standard air.

These corrections can be summarized as follows:

$$Q_{ind} = Q_{std} \left(\frac{T_{amb}}{T_{std}} \cdot \frac{P_{std}}{P_{amb}} \cdot \frac{P_{amb}}{P_{rot}} \right)^{\frac{1}{2}}$$

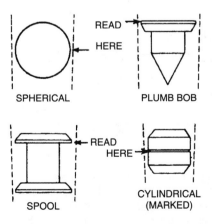

FIGURE 7-7. Types of rotameter floats (*reprinted from The Industrial Environment—Its Evaluation and Control*).[1]

where: Q_{ind} = rotameter reading on site
Q_{std} = flow rate at 760 torr and 20°C
T_{amb} = ambient temperature in °K
T_{std} = 20°C = 293°K
P_{std} = 760 torr
P_{amb} = ambient pressure in torr
P_{rot} = pressure at rotameter inlet in torr

In a situation where there were corrections needed for the pressure drop of the sampler, high altitude, and high temperature, the overall correction could be, for the examples cited, 1.14/1.096/1.025 = 1.28 or 28%.

Head Meters

For a closed channel with a stream of fluid flowing within it, an increase in velocity is experienced whenever the fluid passes through a restriction with a corresponding increase in kinetic energy at the point of constriction. The overall energy balance as determined by the first law of thermodynamics (Bernoulli's theorem) indicates that there must be a corresponding reduction in pressure as a result of the constriction. The mass rate of discharge from such a constriction can be determined by the following general working equation that is applied to both orifice and Venturi meters.

$$W = q_1\rho_1 = KYA_2\sqrt{2g_c(P_1 - P_2)}\,\rho_1 \qquad (4)$$

where: W = weight–rate of flow (lb/s) for English units
= mass–rate of flow (kg/s) for metric units
q_1 = volumetric flow at upstream pressure and temperature (ft³/s or m³/s)
ρ_1 = density at upstream pressure and temperature (lb/ft³ or kg/m³)
$K = C/(1-\beta^4)^{1/2}$
β = ratio of throat diameter to pipe diameter, dimensionless (see Figure 7-8)
Y = expansion factor (dimensionless, see Figure 7-8)
A_2 = cross-sectional area of throat (ft² or m²)
g_c = 32.17 ft/sec² for English units
= 1 for metric units
P_1 = upstream static pressure (lb/ft² or Pa)
P_2 = downstream static pressure (lb/ft² or Pa)

This equation should be used with caution because it is sometimes difficult to determine the actual coefficients for a given system.

Orifice Meters

The simplest form of variable-head meter is the square-edged or sharp-edged orifice illustrated in Figure 7-9. It is also the most widely used because of its ease of

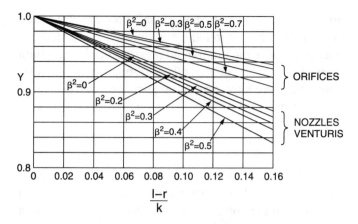

FIGURE 7-8. Expansion factor Y for variable head meters (*reprinted from The Industrial Environment—Its Evaluation and Control*).[1]

installation and low cost. If it is made with properly mounted pressure taps, its calibration can be determined from Equation 4 and Figures 7-8 and 7-10. However, even a nonstandard orifice meter can serve as a secondary standard, provided it is carefully calibrated against a reliable reference instrument.

The five most common tap locations for square-edged orifice meters are:

1. Flange taps: taps located 1.0 in. (2.54 cm) upstream and 1.0 in. (2.54 cm) downstream from the plate.
2. Radius taps: taps located 1.0 pipe diameter upstream and 0.5 pipe diameters downstream from the plate.
3. Vena Contracta taps: taps located upstream 0.5 to 2 pipe diameters from the plate. Downstream tap located at position of minimum pressure.
4. Corner taps: taps drilled one in the upstream and one in the downstream flange with openings as close as possible to the orifice plate.

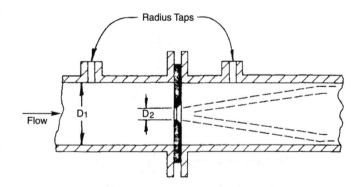

FIGURE 7-9. Square-edged or sharp-edged orifice meter. The plate at the orifice opening must not be thicker than 1/30 of pipe diameter, 1/8 of the orifice diameter, or 1/4 of the distance from the pipe wall to the edge of the opening. The orifice can be cylindrical or have a taper as shown above (*reprinted from Chemical Engineering Handbook*).[2]

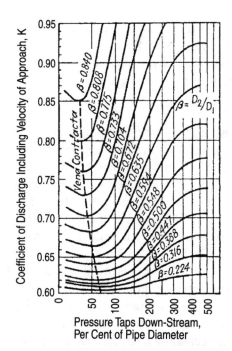

FIGURE 7-10. Coefficient of discharge for square-edged circular orifices (NRE 30,000), with upstream tap located between 1 and 2 pipe diameters from orifice plate (*reprinted from Chemical Engineering Handbook*).[2]

5. Pipe taps: taps located 2.5 pipe diameters upstream and 8 pipe diameters downstream from the plate.

The permanent pressure loss for a square-edged orifice meter with either radius or vena contracta taps is approximated by the following equation:

$$\frac{P_1 - P_4}{P_1 - P_2} = 1 - \beta^2 \qquad (5)$$

where: P_1 = upstream pressure
$\quad\quad\;\; P_2$ = downstream pressure
$\quad\quad\;\; P_4$ = fully recovered pressure (4–8 diameters downstream of orifice)
$\quad\quad\;\; \beta$ = diameter ratio (orifice to pipe)

If, for air, the downstream pressure P_2 is less than $0.53\,P_1$ (the upstream pressure) and the ratio of the upstream cross-sectional area to the orifice area is greater than 25, the orifice is said to be critical, producing a sonic velocity at the orifice gas exit. With these conditions, a constant flow is obtained. However, a critical orifice meter should be calibrated against a primary or secondary standard because it is difficult to take into account all the factors that affect the flow rate through such a system.

Venturi Meters

The large energy loss of an orifice is, for the most part, a result of the sudden increase of area after the air

has passed through the orifice restriction.[4] This pressure loss occurs because of the dead space in the corners between the pipe and orifice plate directly downstream of the orifice. This dead space causes large eddies that account for much of the energy loss. The Venturi meter minimizes this energy loss by essentially eliminating the dead space area by using a cone as shown in Figure 7-11. Venturi meters have optimal converging and diverging angles of 21 and 5 to 15, respectively. The potential energy that is converted to kinetic energy at the throat is reconverted to potential energy at the discharge, with an overall energy loss of only about 10%.

For air at 70°F and 1.0 atm and for $1/4 < \beta < 1/2$, a standard Venturi has a calibration described by

$$Q = 21.2\,\beta^2 D^2 \sqrt{\Delta h} \qquad (6)$$

where: Q = flow (cfm)
$\quad\quad\;\; \beta$ = ratio of throat to duct diameter, dimensionless
$\quad\quad\;\; D$ = duct diameter (inches)
$\quad\quad\;\; \Delta h$ = differential pressure (inches of water)

In metric units Equation 6 becomes

$$Q = 58.4\,\beta^2 D^2 \sqrt{\Delta h} \qquad (7)$$

where: units of Q are L/min
$\quad\quad\;\;$ units of D are cm
$\quad\quad\;\;$ units of Δh are cm H_2O

Laminar Flow Meters

In the laminar flow type of variable-head meter, the pressure drop is directly proportional to the flow rate. In orifice meters, Venturi meters, and related devices, the flow is turbulent and flow rate varies with the square root of the pressure differential.

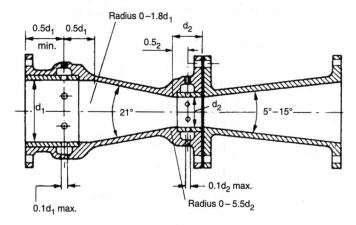

FIGURE 7-11. Standard Venturi (*reprinted from The Measurement of Air Flow*).[4]

Laminar flow restrictors used in commercial flowmeters consist of egg-crate or tube bundle arrays of parallel channels. Alternatively, a laminar flowmeter can be constructed in the laboratory using a tube packed with beads or fibers or a filter as the resistive element. Figure 7-12 illustrates this homemade kind of flowmeter. It consists of a "T" connection, pipet or glass tubing, cylinder, and packing material. The outlet arm of the "T" is packed with a porous plug and the leg is attached to a tube or pipet projecting down into the cylinder filled with water or oil. A calibration curve of the depth of the tube outlet below the water level versus the rate of flow should produce a linear curve. Saltzman[5] has used such tubes to regulate and measure flow rates as low as 0.01 cm³/min.

Pressure Transducers

All of the variable-head meters require a pressure sensor, sometimes referred to as the secondary element. Any type of pressure sensor can be used, with the three most common types being manometers, mechanical gauges, and electrical transducers.

Liquid-filled manometer tubes, when properly aligned and filled with a liquid whose density is accurately known, provide the most accurate measurement of differential pressure. In most cases, however, it is not feasible to use liquid-filled manometers in the field, and pressure differentials are measured with mechanical gauges with scale ranges in centimeters or inches of water. For the low pressure differentials most often encountered in air flow measurement, the most commonly used gauge is the Magnehelic (see DWY, Table 7-2). These gauges are accurate to ± 2% of full scale and are reliable pro-

vided they and their connecting hoses do not leak and their calibration is periodically rechecked. More sensitive measurements of pressure can be made with electronic capacitance pressure gauges.

Bypass Flow Indicators

In most high-volume samplers, the flow rate is strongly dependent on the flow resistance, and flowmeters with a sufficiently low flow resistance are usually bulky or expensive. A commonly used metering element for such samplers is the bypass rotameter, which actually meters only a small fraction of the total flow; a fraction, however, that is proportional to the total flow. As shown schematically in Figure 7-13, a bypass flowmeter contains both a variable-head element and a variable-area element. The pressure drop across the fixed orifice or flow restrictor creates a proportionate flow through the parallel path containing the small rotameter. The scale on the rotameter generally reads directly in cfm or L/min of total flow. In the versions used on portable high-volume samplers, there is usually an adjustable bleed valve at the top of the rotameter that should be set initially and periodically readjusted in laboratory calibrations so that the scale markings can indicate overall flow. If the rotameter tube accumulates dirt, or the bleed valve adjustment drifts, the scale readings can depart greatly from the true flows.

Heated Element Anemometers

Any instrument used to measure velocity can be referred to as an anemometer. In a heated element (hot

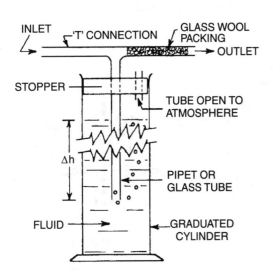

FIGURE 7-12. Packed plug flowmeter (*reprinted from The Industrial Environment—Its Evaluation and Control*).[1]

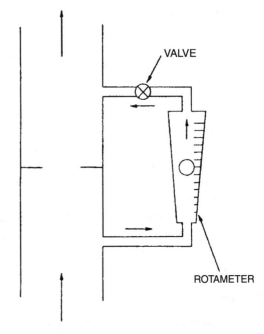

FIGURE 7-13. Bypass flow indicator.

TABLE 7-2. Sources for Calibration Instruments and Apparatus Code

AFP	AccuRa Flow Products, Inc. P.O. Drawer 100 Warminster, PA 18974-0100	GRA	Graseby Andersen 500 Technology Court Smyrna, GA 30082-5211 (770)319-9999 or (800)241-6898 FAX (770)319-0336 andersen@graseby.com www.graseby.com	SEN	Sensidyne, Inc. 16333 Bay Vista Drive Clearwater, FL 33760 (727)530-3602 or (800)451-9444 FAX(727)539-0550 www.sensidyne.com
ALN	Alnor Instrument Company 7555 N. Linder Avenue Skokie, IL 60077-3223 (800)424-7427 www.alnor.com	GIL	Gilmont Instruments Div. Barnant Company 28W092 Commercial Avenue Barrington, IL 60010 (847)381-7050 or (800)637-3739 FAX(847)381-7053	SIE	Sierra Instruments, Inc. 5 Harris Court, Bldg. L Monterey, CA 93940 (408)373-0200 or (800)866-0203 FAX (408)373-4402 www.sierrainstruments.com
AMC	American Meter Company 300 Welsh Road, Bldg. 1 Horsham, PA 19044-2234 (215)830-1800 Fax (215)830-1890 www.americanmeter.com	HIQ	Hi-Q Environmental Products Co. 7386 Trade Street San Diego, CA 92121 (619)549-2820 FAX (619)549-9657 info@HI-Q.net www.HI-Q.net	SKC	SKC Incorporated 863 Valley View Road Eighty Four, PA 15330-9614 (724)941-9701 or (800)752-8472 FAX (724)941-1369 www.skcinc.com
BGI	BGI Incorporated 58 Guinan Street Waltham, MA 02451 (781)891-9380 FAX (781)891-8151 www.bgiusa.com	KEY	Key Instruments 250 Andrews Road Trevose, PA 19053	SKD	Schutte and Koerting Div. Ametek 2233 State Road Bensalem, PA 19020 (215)639-0900 FAX(215)639-1597 www.s-k.com
BIO	BIOS International 230 West Parkway - Unit 1 Pompton Plains, NJ 07444-1029 (973)839-6960 FAX (973)839-7445 sales@biosint.com www.biosint.com	KRZ	Kurz Instruments, Incorporated 2411 Garden Road Monterey, CA 93940 (408)646-5911 or (800)424-7356 FAX (408)646-8901 www.Kurz-instruments.com	SPE	Spectrex Corporation 3580 Haven Avenue Redwood City, CA 94063 (650)365-6567 or (800)822-3940 FAX (650)365-5845 www.spectrex.com
BRO	Brooks Instrument Division 407 W. Vine Street Hatfield, PA 19440-0903 (215)362-3500 FAX(215)362-3745	MGE	Matheson Gas Equipment 166 Keystone Drive Montgomeryville, PA 18936 (215)641-2700 FAX(215)641-2714	THR	Teledyne Hastings Raydist P.O. Box 1275 Hampton, VA 23661 (757)723-6531 FAX(757)723-3925
BUC	A.P. Buck, Inc. 7101 President's Dr. #110 Orlando, FL 32809 (407)851-8602 or (800)330-BUCK FAX (407)851-8910	MIC	Meriam Instrument Company 10920 Madison Avenue Cleveland, OH 44102 (216)281-1100 FAX(216)281-0228 www.meriam.com	TSI	TSI Incorporated P.O. Box 64394 St. Paul, MN 55164 (651)490-2760 FAX (651)490-2704 info@tsi.com www.tsi.com
CME	CME, Inc. 1314 West 76th St. Davenport, IA 52806-1305	PSC	Precision Scientific 2777 West Washington Bellwood, IL 60104	WEC	Warren E. Collins, Inc. 220 Wood Road Braintree, MA 02184
DWY	Dwyer Instruments, Inc. P.O. Box 373 Michigan City IN 46361 (219)879-8000 FAX(219)872-9057 www.dwyer-inst.com	RAD	SAIC-RADeCO 1461 Campus Point Ct. San Diego, CA 92121-9416 (800)962-1632 FAX(619)646-9009		
FPC	Fischer & Porter Company 125 E. County Line Road Warminster, PA 18974				

wire) anemometer, the flow of air cools the sensor in proportion to the velocity of the air. Instruments are available with various kinds of heated elements, e.g., heated thermometers, thermocouples, films, and wires. They are all essentially nondirectional (i.e., with single element probes); they measure the airspeed but not its direction. They all can accurately measure steady-state airspeed, and those with low mass sensors and appropriate circuits can also accurately measure velocity fluctuations with frequencies above 100,000 Hz. Because the signals produced by the basic sensors are dependent on ambient temperature as well as air velocity, the probes are usually equipped with a reference element that provides an output which can be used to compensate or correct errors due to temperature variations. Some heated element anemometers can measure velocities as low as 10 fpm (0.05 m/s) and as high as 8000 fpm (40.6 m/s). Arrays of point sensors can be coupled electronically to provide indications of flow rate as well as velocity.

Other Velocity Meters

Vane Anemometers

There are several other ways to utilize the kinetic energy of a flowing fluid to measure velocity besides the Pitot tube. One way is to align a jeweled-bearing turbine wheel axially in the stream and count the number of rotations per unit time. Such devices are generally known as rotating vane anemometers. Some are very small and are used as velocity probes. Others are sized to fit the whole duct and become indicators of total flow rate. These are sometimes called turbine flowmeters.

Automated vane anemometers are currently available that permit measurement of air flow in circular tubes in the range from 7 to 2500 cfm (0.2–71 m³/min). The systems are generally included with a sensor and associated electronics which provide a digital readout of velocity or flow rate. The systems also have a linearity of between 0.5% and 1.0% with a pressure drop of a few inches of H_2O depending on pipe diameter and flow rate.

The velometer or swinging vane anemometer is widely used for measuring ventilation air flows, but it has few applications in sampler flow measurement or calibration. It consists of a spring-loaded vane whose displacement is indicative of velocity pressure. Its value is in its simplicity, lack of power requirement, and intrinsic safety in explosive atmospheres. These instruments are more fully described in *Industrial Ventilation: A Manual of Recommended Practice*.[6]

Mass Flow and Tracer Techniques

Thermal meters measure mass air or gas flow rate with negligible pressure loss. A unit consists of a heating element in a duct section between two points at which the temperature of the air or gas stream is measured. The temperature difference between the two points is dependent on the mass rate of flow and the heat input.

Mixture metering has a principle similar to that of thermal metering. A contaminant is added and its increase in concentration is measured, or clean air is added and the reduction in concentration is measured. This method is useful for metering corrosive gas streams. The measuring device may react to some physical property such as thermal conductivity or vapor pressure.

Ion flowmeters generate ions from a central disc which flow radially toward the collector surface. Air flow through the cylinder causes an axial displacement of the ion stream in direct proportion to the mass flow.

Procedures of Calibrating Flow and Volume Meters

In this limited space, it is not possible to provide a complete description of all of the techniques available or to go into great detail on those that are commonly used. This discussion will be limited to selected procedures which should serve to illustrate recommended approaches to some calibration procedures commonly encountered.

Comparison for Primary and Secondary Standards

Figure 7-14 shows the experimental setup for checking the calibration of a secondary standard (in this case a wet test meter) against a primary standard (in this case a spirometer). The first step should be to check out all of the system elements for integrity and proper functioning and to determine that there are no leaks within each system or in the interconnections between them. Both the spirometer and wet test meter require specific internal water levels and leveling. The operating manuals for each should be examined because they will usually outline simple procedures for leakage testing and operational procedures.

After all connections have been made, it is a good policy to recheck the level of all instruments and determine that all connections are clear and have minimum resistance. If compressed air is used in a calibration procedure, it should be purified and dried.

Actual calibration of the wet test meter shown in Figure 7-14 is accomplished by opening the bypass

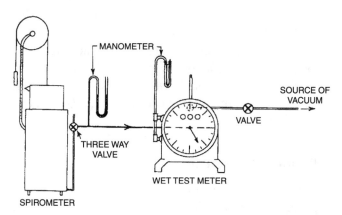

FIGURE 7-14. Calibration of wet test meter with a spirometer (*reprinted from The Industrial Environment—Its Evaluation and Control*).[1]

valve and adjusting the vacuum source to obtain the desired flow rate. The optimum range of operation is between one and three revolutions per minute. Before actual calibration is initiated, the wet test meter should be operated for at least a half hour to stabilize the meter fluid relative to temperature and absorbed gas, and to work in the bearings and mechanical linkage. After all elements of the system have been adjusted, zeroed, and stabilized, several trial runs should be made. During these runs, should any difference in pressure in the sampling system be indicated, the cause should be determined and corrected. The actual procedure would be to instantaneously divert the air to the spirometer for a predetermined volume indicated by the wet test meter (minimum of three revolutions) or to near the maximum capacity of the spirometer, then return to the bypass arrangement. Readings, both quantity and pressure of the wet test meter, must be taken and recorded while it is in motion, unless a more elaborate system is set up. In the case of a rate meter, the interval of time that the air is entering the spirometer must be accurately measured. The bell should then be allowed to come to equilibrium before displacement readings are made. A sufficient number of different flow rates are used to establish the shape or slope of the calibration curve with the procedure being repeated three or more times for each point. For an even more accurate calibration, the setup should be reversed so that air is withdrawn from the spirometer via a personal pump or vacuum source. In this way, any unbalance due to pressure differences would be canceled.

A permanent record should be made of a sketch of the setup, data, conditions, equipment, results, and personnel associated with the calibration. All readings (volume, temperature, pressures, displacements, etc.) should be legibly recorded, including trial runs or known faulty

data, with appropriate comments. The identifications of equipment, connections, and conditions should be so complete that the exact setup with the same equipment and connections could be reproduced by another person solely by use of the records.

After all of the data have been recorded, the calculations, such as corrections for variations in temperatures, pressure, and water vapor, are made using the ideal gas laws:

$$V_S = V_1 \times \frac{P_1}{760} \times \frac{273}{T_1} \qquad (8)$$

where: V_S = volume at standard conditions (in this case: 760 mm H_g at 0°C)

V_1 = volume measured at conditions P_1 and T_1

T_1 = absolute temperature of V_1 $^\circ$(K)

P_1 = pressure of V_1 (mm Hg)

In most cases, the water vapor portion of the ambient pressure is disregarded. Vapor pressure, however, can be a source of error when using such an instrument following a collection bubbler that contains a high vapor pressure liquid. In this case, an appropriate dryer should be placed between the bubbler and flowmeter. Also, the standard temperature of the gas in most industrial hygiene applications is normal room temperature, i.e., 25°C rather than 0°C. The manipulation of the instruments, data reading and recording, calculations, and resulting factors or curves should be done with extreme care. Should a calibration disagree with previous calibrations or the supplier's calibration, the entire procedure should be repeated and examined carefully to assure its validity. Upon completion of any calibration, the instrument should be tagged or marked in a semi-permanent manner to indicate the calibration factor and, where appropriate, the date and who performed the calibration.

Reciprocal Calibration by Balanced Flow System

In many commercial instruments, it is impractical to remove the flow-indicating device for calibration. This may be because of physical limitations, characteristics of the pump, unknown resistance in the system,[7] or other limiting factors. In such situations, it may be necessary to set up a reciprocal calibration procedure; that is, a procedure where a controlled flow of air or gas is compared first with the instrument flow, then with a calibration source. Often a further complication is introduced by the static pressure characteristics of the air mover in the instrument.[8] In such instances, supplemental pressure or vacuum must be applied to the system to offset the resistance of the calibrating device. An example of such a system is illustrated in Figure 7-15.

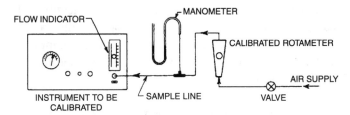

FIGURE 7-15. Schematic for balanced flow calibration (*reprinted from The Industrial Environment—Its Evaluation and Control*).[1]

The instrument is connected to a calibrated rotameter and a source of compressed air. Between the rotameter and the instrument, an open-end manometer is installed. The connections, as in any other calibration system, should be as short and as resistance-free as possible.

In the calibration procedure, the flow through the instrument and rotameter is adjusted by means of a valve or restriction at the pump until the manometer indicates zero pressure difference from the atmosphere. When this condition is achieved, the instrument and rotameter are both operating at atmospheric pressure. The indicated and calibrated rates of flow are then recorded and the procedure repeated for other rates of flow.

Dilution Calibration

Normally, gas-dilution techniques are employed for instrument response calibrations; however, several procedures[9, 10] have been developed whereby sampling rates of flow could be determined. The principle is essentially the same except that different unknowns are involved. In air flow calibration, a known concentration of the gas (i.e., carbon dioxide) or submicrometer-sized aerosol tracer is contained in a vessel. Uncontaminated air is introduced and mixed thoroughly in the chamber to replace the volume removed by the instrument to be calibrated. The resulting depletion of the agent in the vessel follows the theoretical dilution formula:

$$C = C_o e^{-bt} \qquad (9)$$

where: C = concentration of agent in vessel at time t
C_o = initial concentration at $t = 0$
e = base of natural logarithms
b = air changes in the vessel per unit time
t = time

The concentration of the gas or aerosol in the vessel is determined periodically by an independent method. A linear plot should result from plotting concentration of agent against elapsed time on semi-log paper. The slope of the line indicates the air changes per minute (b), which can be converted to the rate (Q) of air withdrawn

by the instrument from the following relationship: $Q = bV$, where V is the volume of the vessel.

This technique offers the advantage that virtually no resistance or obstruction is offered to the air flow through the instrument; however, it is limited by the accuracy of determining the concentration of the agents in the air mixture.

Summary and Conclusions

Because the accuracy of air sampling instruments is dependent on the precision of measurement of the sampled volume, extreme care should be exercised in performing all flow calibration procedures. The following comments summarize the key features of airflow calibration:

1. Use standard devices with care and attention to detail.

2. All calibration instruments and procedures should be checked periodically to determine their stability and/or operating condition.

3. Perform calibrations whenever a device has been changed, repaired, received from a manufacturer, subjected to use, mishandled or damaged, and at any time when there is a question as to its accuracy.

4. Understand the operation of an instrument before attempting to calibrate it, and use a procedure or setup that will not change the characteristics of the instrument or standard within the operating range required.

5. When in doubt about procedures or data, ensure their validity before proceeding to the next operation.

6. All calibration train connections should be as short and free of constrictions and resistance as possible.

7. Extreme care should be exercised in reading scales, timing, adjusting and leveling, and in all other operations involved.

8. Allow sufficient time for equilibrium to be established, inertia to be overcome, and conditions to stabilize.

9. Enough data should be obtained to give confidence in the calibration curve for a given parameter. Each calibration point should be made up of at least three readings to ensure statistical confidence in the measurement.

10. A complete permanent record of all procedures, data, and results should be maintained. This should include trial runs, known faulty data with appropriate comments, instrument identification, connection sizes, barometric pressure, temperature, etc.

11. When a calibration differs from previous records, the cause of change should be determined before accepting the new data or repeating the procedure.

12. Calibration curves and factors should be properly identified as to conditions of calibration, device calibrated and what it was calibrated against, units involved, range and precision of calibration, data, and who performed the actual procedure. Often, it is convenient to indicate where the original data are filed and attach a tag to the instrument indicating the above information.

Acknowledgment

The author wishes to thank D.M. Bernstein and R.T. Drew for their dedicated work and collaboration on this chapter for previous editions of this manual.

References

1. DiNardi, S., Ed.: The Industrial Environment—Its Evaluation and Control, 3rd ed.

2. Perry, J.H.; *et al.*, Eds.: Chemical Engineering Handbook, 6th ed. McGraw-Hill, New York (1984).

3. Leidel, N.A.; Busch, K.A.; Lynch, J.R.: Occupational Exposure Strategy Manual. USDHEW, PHS, CDC, NIOSH, Cincinnati, OH (January 1977).

4. Ower, E.; Pankhurst, R.C.: The Measurement of Air Flow, 5th ed. Pergamon Press, New York (1977).

5. Saltzman, B.E.: Preparation and Analysis of Calibrated Low Concentrations of Sixteen Toxic Gases. Anal. Chem. 33:1100 (1961).

6. American Conference of Governmental Industrial Hygienists: Industrial Ventilation: A Manual of Recommended Practice, 23rd ed. ACGIH, Cincinnati, OH (1998).

7. Tebbens, B.D.; Keagy, D.M.: Flow Calibration of High Volume Samplers. Am. Ind. Hyg. Assoc. Q. 17:327 (September 1956).

8. Morley, J.; Tebbens, B.D.: The Electrostatic Precipitator Dilution Method of Flow Measurement. Am. Ind. Hyg. Assoc. J. 14:303 (December 1953).

9. Setterlind, A.N.: Preparation of Known Concentrations of Gases to Vapors in Air. Am. Ind. Hyg. Assoc. J. 14:113 (June 1953).

10. Brief, R.S.; Church, F.W.: Multi-Operational Chamber for Calibration Purposes. Am. Ind. Hyg. Assoc. J. 21:239 (June 1960).

Chapter 8

Calibration of Gas and Vapor Samplers

Owen R. Moss, Ph.D.

Chemical Industry Institute of Toxicology, Research Triangle Park, North Carolina

CONTENTS

Introduction

Our understanding of the impact of an airborne contaminant is based on the accurate measure of concentration; the amount present in a unit volume of air. Calibration apparatus and techniques used to accurately measure a volume of air are discussed in Chapter 7, "Airflow Calibration." The following discussion, on the calibration of both sampling equipment and techniques used to measure the amount of contaminant present, will dwell on the generation of the gas and vapor contaminant and the operation of static or dynamic calibration atmosphere systems.

Units of Gas and Vapor Concentration

In the field of Industrial Hygiene, gas or vapor concentrations are usually discussed in terms of parts per million (ppm); a measure of the number of compound-specific molecules per million molecules of total atmosphere present. The ideal gas law can be used to show that a concentration given in "ppm" is equivalent to the volume-to-volume relationship of milliliters of pure gas or vapor per cubic meter (mL/m³) or microliters per liter (μL/L) of atmosphere. A concentration of 1.0 μL of SO_2 vapor per liter of air mixture thus becomes 1.0 ppm SO_2.

$$1.0 \left(\frac{\mu L\ SO_2}{L\ air} \right) = 1.0 \left(\frac{mL\ SO_2}{m^3\ air} \right) = 1.0\,(ppm\ SO_2) \quad \textbf{(1)}$$

Occasionally, with direct-reading instruments and, more frequently, with chemical analysis of a sample collected from the atmosphere, confusion and subsequent errors arise in converting to "ppm" from concentration measured as mass per unit volume. A two-step process can be used to make this conversion.

Step 1: Calculate $C(\text{milli-moles}_x/m^3)$ given: $C(mg_x/m^3\ air)$

x = trace contaminant
G_x = gram molecular weight of the trace gas or vapor (grams/mole)

$$C\left(\frac{\text{milli-moles}_x}{m^3}\right) = C\left(\frac{mg_x}{m^3 \text{ air}}\right) \cdot \frac{1}{G_x}\left(\frac{\text{milli-moles}_x}{m^3}\right) \quad (2)$$

Step 2: Calculate $C(mL/m^3)$ given:
$C(\text{milli-moles}_x/m^3)$
T = temperature (°K)
P = pressure (mm Hg)

$$C\left(\frac{ml_x}{m^3}\right) = C\left(\frac{\text{milli-moles}_x}{m^3}\right) \cdot 22.4 \left(\frac{ml_x \text{ at STP}}{\text{milli-moles}_x}\right) \quad (3)$$

$$\cdot \frac{T}{273}\left(\frac{°K}{°K}\right) \cdot \frac{760}{P}\left(\frac{mm\ Hg}{mm\ Hg}\right)$$

$$C\,(\text{ppm}_x) = C\left(\frac{ml_x}{m^3}\right)$$

When the two steps are combined, the concentration in ppm becomes:

$$C\,(\text{ppm}_x) = C\left(\frac{mg_x}{m^3 \text{ air}}\right) \cdot \frac{1}{G_x} \cdot 22.4 \cdot \frac{T}{273} \cdot \frac{760}{P} \quad (4)$$

Conversely, given the concentration of "x" in ppm (1 ppm = 1 mL/m³), the mass concentration is:

$$C\left(\frac{mg_x}{m^3 \text{ air}}\right) = C\left(\frac{ml_x}{m^3}\right) \cdot G_x \cdot \frac{1}{22.4} \cdot \frac{273}{T} \cdot \frac{P}{760} \quad (5)$$

Sample Calculation: A field sample of SO_2 was taken on a hot day in July (99°F, with barometric pressure of 586 mm Hg). The concentration obtained was 0.06 mg SO_2 per m³ of air. What is the SO_2 concentration in ppm?

$$99°F = 37.2°C = 37.2 + 273 \text{ or } 310°K$$

$$C\,(\text{ppm}) = 0.06\left(\frac{mgSO_2}{m^3 \text{ air}}\right) \cdot \frac{1}{64} \cdot 22.4 \cdot \frac{310}{273} \cdot \frac{760}{586} \quad (6)$$

$$C = 0.03\,(\text{ppm } SO_2)$$

Generation of Gas and Vapor Calibrations

Gases or vapors that comprise a pollutant must be generated specifically for the calibration of air sampling instruments. The accurate analysis of pollutant concentration, whether it be through direct-reading instrumentation, wet chemical techniques, or indirect methods, is only as

good as the calibration system. In order to test the collection efficiency of a sampler for a given contaminant, it is necessary either: 1) to make the measurements in the field using a proven reference instrument or technique such that expected concentrations are bracketed; or 2) to reproduce the expected field atmosphere in a calibration chamber or flow system and conduct the measurements in the laboratory. Techniques and equipment for producing such test atmospheres are discussed here and in detail in various other sources.

Accurate calibration standards of a given gas or vapor are needed either in temporary storage containers or generated and maintained in a continuous stream. Temporary storage containers, or "*static* calibration systems," may be composed of flexible or solid walled containers used to provide calibration atmospheres that can be "grabbed" by the monitoring instrument or sampling system. In the case of an instrument being operated in a closed loop configuration, the static calibration system is contained completely within the sensing device. The continuously generated stream of calibration atmosphere produced by a "*dynamic* calibration system" may be directly sampled by the sensing device, or the calibration atmosphere may completely surround the instrument as it sits in a chamber. The following description of both of these systems draws heavily from the recent publication by Nelson,[1] Wong,[2] and others.[3-7]

Test atmospheres generated for the purpose of calibrating instrument response and collection efficiency should be checked for accuracy by using mass balance relations or, where applicable, by using reference instruments or sampling and analytical procedures whose reliability and accuracy are well documented. Professional groups and governmental agencies that publish such recommendations, guidelines, and standards are listed in Table 8-1.

Static Calibration Systems

The term "static" is used to describe a gas and vapor calibration system where the calibration gas is stored in a container for periodic delivery of an aliquot to the sensing instrument or sampling train. Static calibration systems are operated with the calibration atmosphere stored either at or near atmospheric pressure or at elevated pressures.

Storage of Calibration Atmosphere at Atmospheric Pressure

Storage containers with flexible or rigid walls have been used to hold calibration atmospheres temporarily. When there is no adsorption or diffusion of material

TABLE 8-1. Organizations Publishing Recommended or Standard Methods and/or Test Procedures Applicable to Air Sampling Instrument Calibration

Abbreviation	Full Name and Address
ANSI	American National Standards Institute, Inc. 1430 Broadway New York, NY 10018
AWMA	Air and Waste Management Association (formerly the Air Pollution Control Association) P.O. Box 2861 Pittsburgh, PA 15230
ASTM	American Society for Testing and Materials D-22 Committee on Sampling and Analysis of Atmospheres and E-34 Committee on Occupational Health and Safety 1916 Race Street Philadelphia, PA 19103
EPA/EMSL	U.S. Environmental Protection Agency Environmental Monitoring Systems Quality Assurance Division (MD-77) Research Triangle Park, NC 27711
ISC	Intersociety Committee on Methods for Air Sampling and Analysis c/o Dr. James P. Lodge, Editor Intersociety Manual—3rd edition 385 Broadway Boulder, CO 80303
NIOSH	National Institute for Occupational Safety and Health NIOSH Manual Coordinator Division of Physical Sciences and Engineering 4676 Columbia Parkway Cincinnati, OH 45226

through the walls, these containers will hold a calibration standard when a known amount of gas, or liquid that can be vaporized, is injected into a known volume of dilution air.

Flexible Walled Storage Containers. Bags composed of a wide variety of material from Tedlar® and Teflon® to aluminized Mylar® have been used as flexible walled containers for temporary storage of calibration atmospheres.[1] Usually these bags come with a valved inlet port that can accept some type of gas chromatograph septum. The O-rings in the valve and the septum must have very little or no affinity for the components of the calibration atmosphere.

The bag should first be evacuated as thoroughly as possible prior to metering in the dilution gas and injection of calibration gas or liquid. Calibrated syringes provide a simple method for injection of these materials. The syringe should be flushed several times with the component of interest.

Side-port needles should be used to penetrate septa. This is especially important for septa that have an inert surface, like Teflon, coating a material, such as rubber, which may absorb the calibration gas or vapor. The side port will ensure that no piece of the latter material is cut out and injected into the bag along with the gas or liquid to be vaporized. The actual injection should be performed by gently depressing the syringe one time.

A rigid container such as a bottle can be modified to function as a collapsible bag by insertion of a balloon. Inflation and deflation of the balloon allows loading and unloading of the calibration atmosphere with no dilution. The fittings necessary for controlling the sampling can be attached directly to the exposure container, separate from the air source for the balloon.

Rigid Walled Storage Containers. Single and multiple rigid walled containers, for calibration gases held at atmospheric pressure, are used in the few cases where flexible walled bags are not available or are not appropriate due to loss of material on the walls or diffusion of the components of the calibration atmosphere through the walls of the bag. Even though the concentration will be diluted with replacement air whenever gas is transferred for analysis, the concentration can be accurately determined provided fan blades or stirring devices are present to ensure that instantaneous mixing occurs in each rigid walled container. The storage bottles are usually equipped with a valved inlet and a similar outlet (Figure 8-1). A third inlet or pass-through port for introduction of the contaminant may also be provided.

An alternative method of introducing a calibration atmosphere is to produce glass ampoules containing a

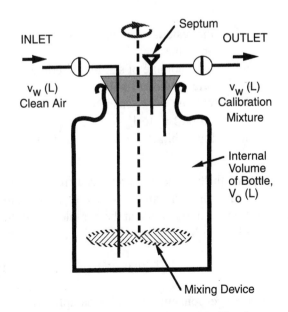

FIGURE 8-1. Single bottle gas and vapor static calibration system.

known amount of pure contaminant and then to break them within the fixed volume of the static system. Setterlind[8] has discussed the preparation of ampoules in detail. Other devices, such as gas burettes, displacement manometers, and small pressurized bombs, have all been used successfully.[9, 10] Gaseous concentrations can also be produced by adding stoichiometrically determined amounts of reacting chemicals to the bottle. Some instruments, such as gas chromatographs, may be calibrated by direct injection of standard solutions.

In practice, after the mixture inside the bottle has come to equilibrium, samples are drawn from the outlet while replacement air is allowed to enter through the inlet tube, diluting the contents of the bottle during the process of getting the sample. Under ideal conditions, the concentration remaining is a known fraction of the number of air samples removed from the bottle. If one assumes instantaneous and perfect mixing of the incoming air with the entire sample volume in the bottle, the concentration change, dC, as a small volume, dv_w, is withdrawn, is equal to the residual concentration, C, times the fraction, dv_w/V_o, of the total volume withdrawn:

$$dC = C\frac{dv_w}{V_o} \qquad (7)$$

The residual concentration, C, is:

$$C = C_o e^{-\left(\frac{v_w}{V_o}\right)} = C_o e^{-a}, \qquad (8)$$
$$\text{for } a = \frac{v_w}{V_o}$$

$$\ln\left(\frac{C_o}{C}\right) = a \qquad (9)$$

where: C = the residual concentration in the bottle at any time
v_w = total volume of sample withdrawn
C_o = original concentration
V_o = volume of the chamber
a = ratio of v_w to V_o
$\ln$ = logarithm to the base e ($\ln(x) = 2.3 \log 10(x)$)

If one-tenth the volume is removed from the bottle and replaced with clean air ($v_w = 0.1\ V_o$), the residual concentration, C, in the bottle at the completion of drawing a sample of volume v_w is:

$$C = C_o e^{-01} = 0.9047\,C_o \qquad (10)$$

The average concentration of a sample withdrawn from a single container is

$$\bar{C}_1 = C_o \frac{\left(1 - e^{-\left(\frac{v_w}{V_o}\right)}\right)}{\left(\frac{v_w}{V_o}\right)} = C_o\frac{(1 - e^{-a})}{a}, \text{ for } a = \frac{v_w}{V_o} \qquad (11)$$

For the case of $v_w = 0.1\ V_o$, $\bar{C}_1 = 0.953\ C_o$ and, as expected, is larger than the residual concentration, $C = 0.905\ C_o$. If instantaneous mixing does not occur, the average concentration, $\bar{C}_1$, for the time of sampling may be even higher.

This example presents a basic limitation encountered when drawing a calibration sample from a fixed-walled container. If the concentration of the sample must be held to within 5% of the initial concentration, only about 10% of the volume of the bottle can be sampled. Setterlind[8] has shown that this limitation can be overcome by using two or more bottles of equal volume (V_o) in series, with the initial concentration in each bottle being the same. When the mixture is withdrawn from the last bottle, it is not displaced by air but by the mixture from the preceding bottle. If, as above, a maximum of 5% drop in concentration can be tolerated, two identical bottles in series provide a usable sample of 0.6 V_o. With five bottles in series, the usable sample will increase to about 3 V_o. The general relation for the residual concentration in the last of n containers and the average concentration of the withdrawn sample can be extracted from the discussions by Setterlind,[8] Lodge[11] and Nelson:[1]

$$C_n = C_o\left\{1 + \left(\sum_{j=1}^{n-1}\frac{1}{j!}a^j\right)e^{-a}\right\} \qquad (12)$$

and

$$\bar{C}_n = C_o\left\{n + \left(\sum_{j=1}^{n-1}\frac{(n-j)}{j!}a^{j-1}\right)e^{-a}\right\} \qquad (13)$$

for, $n > 1$, $j < n - 1$, and $a = \frac{v_w}{V_o}$

This problem of calibration-concentration changing as samples are withdrawn from a fixed-walled container can be eliminated for some gases and vapors by recirculation. For example, when an analysis instrument such as the Miran infrared gas analysis unit is operated in a closed loop configuration, the body of the unit can be used as the rigid calibration atmosphere container. The unit is set to sample itself and a known amount of contaminant is injected into the internal volume.

Storage of Calibration Atmosphere at Elevated Pressure

Rigid storage containers for calibration atmospheres can be filled to pressures significantly greater than atmosphere pressure. During preparation, these standard cylinders are evacuated and then filled with 1) a measured mass, 2) a measured partial pressure, or 3) a known volume of gas or liquid, and then, as required, repressurized with a diluent gas such as air or nitrogen to produce the concentration required. Nelson[1] has recently reviewed these three main techniques (gravimetric, partial pressure, volumetric) for producing pressurized standard cylinders. These techniques have also been previously discussed by Cotabish et al.[11] and reviewed by Roccanova.[8]

Gravimetric. Measuring the weight of compound that is placed inside the standard cylinder is the most accurate method of producing a calibration atmosphere. In fact, when purchasing standard cylinders from a commercial source, the calibration is usually provided in terms of gravimetric analysis of the weight of the cylinder before and after the pure compound was introduced. The cylinder is repressurized with a known number of moles of diluent gas such as air or nitrogen. The actual cylinders should always be checked by an independent analysis procedure because the trace gas may not be adequately mixed or may be partially lost due to wall adsorption. The mixture of the gas produced in cylinders cannot be assumed to be uniform. In fact, the cylinders have to be rolled, or otherwise treated, in order to ensure that the injected gases are thoroughly mixed.

Partial Pressure. In this method of producing a calibration atmosphere, the relative concentrations of the different component gases that comprise the calibration atmosphere are determined by the partial pressure used to introduce each gas into the cylinder. The cylinder or cylinders are filled in sequence beginning with the component that will have the lowest partial pressure (concentration) in the gas mixture.

Volumetric. Standard cylinders can be filled with known volumes of contaminant gas and diluent gas by having known flows of each of these gases supplying the input line to the compressor. The concentration in the standard cylinder is determined by the ratio of these flows.

Dynamic Calibration Systems

The term "dynamic" is used to describe a gas and vapor calibration system where the calibration atmosphere is continuously produced and fed into a sampling train or calibration chamber following dilution and mixing. The air sampling instrument is placed to sample either the output of the sampling train or it is completely immersed in the calibration atmosphere within the chamber. In either case, the generation, dilution, and operation of the calibration system must be incorporated in the documentation along with the nature and frequency of calibration checks. This information must be present to support the scientific measurements and is often necessary to meet legal requirements. Any measurements made to document the presence or absence of excessively high or low concentrations will only be as reliable as the calibrations upon which they are based.

All instruments should be checked against such standard calibration atmospheres immediately upon receipt and periodically thereafter. Verification of the concentrations of test atmospheres should be performed whenever possible using analytical techniques that are referee-tested or otherwise known to be reliable. Procedures for establishing continuously generated atmospheres involve dilution of gas streams produced from "static" sources, generation and dilution of atmospheres produced from liquid sources through active vaporization, or controlled diffusion or permeation of gas molecules. The calibration atmosphere is introduced into dynamic calibration systems that may be as simple as a single delivery line into the inlet of the air sampling instrument or as complex as a calibration system that includes a chamber of sufficient size to allow the air sampling instrument to be completely immersed in the calibration atmosphere. Less frequently used approaches to calibration atmosphere generation, in addition to the ones presented here, are discussed by Nelson[1] and Wong.[2]

Gas Dilution

Dilution of the sample may be necessary when the concentration from the generator is above the range of the analytical instrument. In each stage of a dilution system, air and the contaminant gas are metered through restrictions and then mixed. The output is used as is, or it is diluted further by passing it through a similar system (Figure 8-2). Such dilution systems are subject to instabilities that make them difficult to control. The precision of the output concentration is very sensitive to the variations inherent in controlling each of the flows. This sensitivity, plus constraints on the acceptable variation in the calibration concentration, limits the number of "dilutions" that can be made in series. The schematic in Figure 8-2 includes the generalized equation for the diluted concentration at the nth stage. In practice, n is seldom larger than three.

Most dilution units operate on the assumption that the pressure drop across each restriction to flow is constant. Cotabish et al.[12] have described a system originally

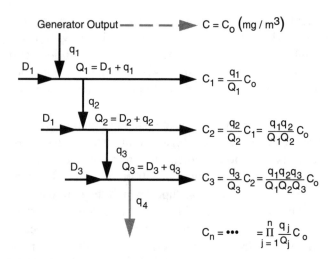

FIGURE 8-2. Basic schematic of a multi-stage dilution system. (D_j is the dilution air flow, in L/min, for the jth stage of the dilution system. Likewise, q_j is the flow of airborne mixture from the previous stage.)

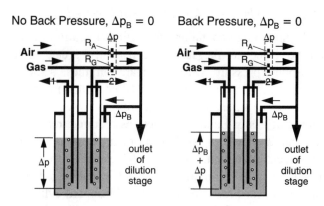

FIGURE 8-3. Automatic back-pressure compensator. (R_A and R_G are air and gas line restrictions with pressure drop Δp. Excess dilution air and gas vent at "1" and "2.")

patented by Mase for compensation of back pressure (Figure 8-3). In this system, both the air and contaminant gas flow are regulated by the height of a water column which, in turn, is controlled by the back pressure of the calibration system. Thus, an increase in back pressure causes an increase in the delivery pressure of both air and contaminant gas. This system is limited to those gases that are not significantly soluble in the water or oils used to regulate the pressure drop. This concept is also included, in part, in devices described by Saltzman.[13–15]

In general, commercially available dilution units (for example, from MEL and MGP, Table 8-2) use constant sources of flow that are relatively insensitive to back pressure (Figure 8-2). These dilution units control flow with the use of devices such as positive displacement pumps or mass flow controllers.

Generation of Dynamic Calibration Atmospheres

Static systems are limited by two factors: loss of vapor by surface adsorption and the finite volume of the mixture. Although the influences of the latter can be included in the calculations, the loss to the walls may be significant.

In dynamic systems, the rate of air flow and the addition of source material to the air stream are carefully controlled to produce a known dilution ratio. Dynamic systems offer a continuous supply of material, allow for rapid and predictable concentration changes, and minimize the effect of wall losses as the contaminant comes to equilibrium with the interior surfaces of the system. Both gases and liquids can be used with dynamic systems. With liquids, however, provision must be available for conversion to the vapor state.

Liquid Delivery: Vapor Generation. When the contaminant is a liquid at normal temperature, a vaporization step

must be included. One procedure is to use a motor-driven syringe [3, 10, 12] and meter the liquid onto a wick or a heated surface. Such systems consist of an air cleaner, a solvent injection device, a heating/vaporizing/dilution zone, and a zone for mixing and cooling. A large range of solvent concentrations can be produced (0.01 to 50,000 ppm). These devices permit rapid changes in the concentrations and can be accurate to better than 1.0%.

A second generation method is to saturate an air stream with vapor and then dilute the air stream to the desired concentration. The amount of vapor in the saturated air stream is directly proportional to the ratio of the vapor pressure, p_v, to the atmospheric pressure P_o;

$$C(\text{ppm}) = \frac{P_v}{P_o} 10^6 \qquad (14)$$

The vapor pressure is proportional to the temperature and can be estimated from equations such as the Antione relationship for those gases where the characteristic constants A, B, and C have been reported (Nelson[1] and Reid, et al.[16]).

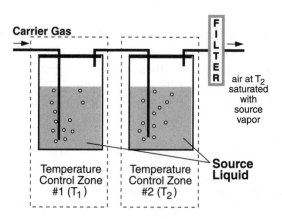

FIGURE 8-4. Schematic of a two-stage vapor saturator (T_1 is usually greater than T_2).

TABLE 8-2. Sources for Calibration Instruments and Apparatus

AID	Analytical Instrument Development, Inc. Division of Thermo Environmental Instruments, Inc. 8 West Forge Parkway Franklin, MA 02038	KTK	Kin-Tek Laboratories, Inc. 504 Lourel LaMarque, TX 77568	SKC	SKC Incorporated 863 Valley View Road Eighty Four, PA 15330-9614 (724)941-9701 or (800)752-8472 FAX (724)941-1396
APC	Air Products and Chemicals, Inc. Specialty Gas Department 7201 Hamilton Blvd. Allentown, PA 18195	LCC	Liquid Carbonic Corp. 901 Embarcadero Oakland, CA 94606	SKD	Schutte & Koerting 2233 State Rd. Bensalem, PA 19020
BGI	BGI Incorporated 58 Guinan Street Waltham, MA 02154 (781)891-9380 FAX (781)891-8151	MDC	Mast Development Company Suite 3, 736 Federal Street Davenport, IA 52803	SSG	Scott Specialty Gases 6141 Easton Rd. P.O. Box 310 Plumsteadville, PA 18949
CAL	Calibrated Instruments, Inc. 200 Saw Mill River Road Hawthorne, NY 10532 (914)741-5700 or (800)969-2254 FAX (914)741-5711	MEL	Meloy Laboratories, Inc. 6715 Electronic Drive Springfield, VA 22151	UNI	Univetrics Corp. 501 Earl Road Shorewood, IL 60436
		MGP	Matheson Gas Products 30 Seaview Drive P.O. Box 1587 Secaucus, NJ 07096 (201)867-4100	VCM	VICI Metronics 2991 Corvin Drive Santa Clara, CA 95051
CSI	Columbia Scientific P.O. Box 203190 Austin, TX 78720 (512)258-5191	MSA	Mine Safety Appliances P.O. Box 427 Pittsburgh, PA 15230 (412)776-8600 or (800)MSA-INST FAX (412)776-3280		
HAM	Hamilton Company P.O. Box 10030 Reno, NV 89520	NBS	National Bureau of Standards Office of Standard Reference Materials Room B311 Chemistry Bldg. National Institute of Standards and Technology Gaithersburg, MD 20899		
HOR	Horiba Instruments, Inc. 1021 Duryea Avenue Irvine, CA 92714				
KEC	KECO R & D Inc. 10034 Clay Road Houston, TX 77080	SEN	Sensidyne, Inc. 16333 Bay Vista Dr. Clearwater, FL 33760 (727)530-3602 or (800)451-9444		

$$\ln(p_v) = A - \frac{B}{C+T} \qquad (15)$$

A simple vapor saturator is shown in Figure 8-4. The carrier gas passes through two gas washing bottles in series which contain the liquid to be volatilized. The first bottle is kept at a higher temperature than the second one, which is immersed in a constant temperature bath. By using the two bottles in this fashion, saturation of the exit gas is assured. A filter is sometimes included to remove any droplets entrained in the air stream as well as any condensation particles. Such a system has even been used to generate a constant concentration of mercury vapor (Nelson[17]).

Diffusion of Vapors: Diffusion Cells. Diffusion cells have been used to produce known concentrations of gaseous vapors.[1, 2, 18] All such devices have the basic design shown in Figure 8-5. Liquid is placed in a reservoir that is connected to a mixing zone by a tube of known length and cross section. The concentration in the outlet of the diffusion cell is a function of the diffusion rate of the vapor, q_d, and the total flow, Q_T, in the system.

$$C(\text{ppm}) = \frac{q_d}{Q_T} 10^6 \qquad (16)$$

When designing a diffusion cell for a specific application or output rate, the diffusion rate of the vapor produced above the liquid in the reservoir can be calculated from the environmental conditions in the system, the diffusion coefficient of the vapor, and the length and cross sectional area of the diffusion tube. If the diffusion coefficient is unknown, it can be estimated from the molecular weights of the vapor and the diluent gas.[1] In general, these calculations for estimating the diffusion rate are subject to assumptions that cause a significant difference between the calculated and measured values.

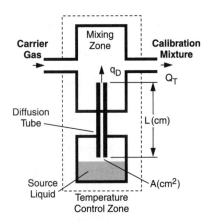

FIGURE 8-5. Basic design of all diffusion cells.

In practice, the diffusion rate of the vapor through the specific geometry of the diffusion tube can be estimated for a given temperature by measuring the change in weight of the reservoir or by measuring the output concentration with calibrated chemical detectors. In operating such devices, the accuracy and stability of the output concentrations are directly proportional to the ability to control temperature in the reservoir, diffusion tube, and mixing zone.

Diffusion of Vapors: Permeation Devices. Permeation methods for producing controlled atmospheres used in calibrating air sampling instruments have recently been reviewed by Nelson[1] in addition to previous discussions by O'Keefe and Ortman[19] and Lodge.[11] In general, the permeation of molecules of a source material through plastics can be used to reproducibly generate a controlled atmosphere provided the critical temperature of the source material is above 20 to 25°C. Plastics such as fluorinated ethylene propylene (FEP Teflon), tetrafluoroethylene (TFE Teflon), polyethylene, polyvinyl acetate, and polyethylene terephthalate (Mylar) are a few of the materials that have been used in permeation devices.[1] In the operation of these devices, the source material will usually dissolve in the plastic and permeate through it. The rate of permeation is primarily a function of the thickness of the plastic material, the total internal area exposed to the source material, and the temperature.

These devices are sensitive to the temperature to the extent that a 0.1°C change in temperature can result in a 1% change in the permeation rate through the plastic container.

All existing permeation devices have one of three basic formats (Figure 8-6): source liquid in a permeation tube; permeation tube in the source liquid; and source liquid reservoir with permeable and impermeable walls. In all cases, diluent air at flow rate Q_o (L/min) passes over or through the device and the source liquid held at constant temperature. The concentration in the output flow is proportional to the ratio of the permeation rate of the source material, q_p, to the output flow rate, Q_T.

Like diffusion devices, the output of a permeation device can be estimated from the thickness of the wall material, the type of the wall material, the source liquid, the magnitude of the permeable areas exposed to the source material, the pressure drop, and the temperature.[1] These approximations can be accurate to within 10% and are useful mainly in designing a device with specific output capabilities.

For permeation devices to be used in calibration of air sampling instruments, they must be either weighed before and after use, calibrated with chemical detectors just prior to use, or purchased as precalibrated units. Table 8-3 lists manufacturers who produce permeation devices for applicable source materials that may be of interest in calibration of air sampling instruments. The advantages of permeation devices are that they are extremely simple and, under adequate temperature control, can be highly precise. A wide range of concentrations in the high (hundreds of ppm) to very low (hundredths of ppm) range can be produced. However, the total output rate and, thus, the achievable concentrations from these permeation devices, is low.

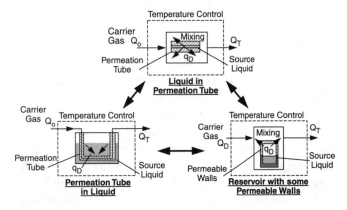

FIGURE 8-6. Three basic formats of permeation devices.

TABLE 8-3 Commercially Available Items for Gas and Vapor Calibration.

Types	Sources*
Microsyringes	HAM, UNI
Calibrated ampoules	MSA, KEC
Compressed gases	APC, LCC, MGP, SSG
Permeation devices and systems	AID, KEC, KTK, MDC, CSI, VCM
Gas blenders and diluters	CAL, CSI, HOR, MGP
Gas phase titration	CSI
Gas sampling bags	BGI, CAL

*See Table 8-2.

Dynamic Calibration

Calibration systems for air sampling instruments may be as simple as a single delivery line from the generator to the inlet of the air sampling instrument, or they may be as complex as an exposure system that includes a chamber of sufficient size to allow several air sampling instruments to be completely immersed in the calibration atmosphere. Sampling conditions encountered in the industrial environment can often be duplicated by placing instruments inside a calibration chamber or wind tunnel. When such facilities are not available, the instruments are tethered about a gas distribution system in a manner much like that used for nose-only exposure of animals to test atmospheres. Techniques, insights, and pitfalls common in whole body and nose-only exposure of animals to test atmospheres are directly applicable to completing accurate air sampling instrument calibrations.[20]

Regardless of whether a calibration duct or calibration chamber is used, the operator should have a basic understanding of the air flow through the calibration system, the assumptions contained in the equations used to predict concentration at different points in the calibration system as a function of time, and the influence of deviations from optimal operating conditions on the stability of the calibration atmosphere.[21-28]

Operation of an Ideal Calibration Chamber. Calibration systems in which the air sampling instrument will be immersed in the calibration atmosphere can have one of three basic configurations (push-only, push-pull, and pull-only) with respect to the total flow of calibration atmosphere through the chamber (Figure 8-7). The simplest and most commonly used configuration is the latter case where the output from the generator is mixed with dilution air whose flow is controlled by a system pulling air from the exhaust of the calibration chamber (Figure 8-8).

When there is instantaneous mixing of each incremental volume of calibration atmosphere entering the chamber, an exponential buildup of concentration is seen:

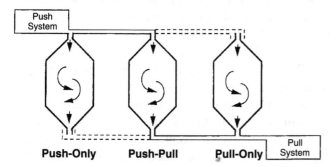

FIGURE 8-7. Calibration chamber systems for "immersion" of air sampling instruments (push-only, push-pull, pull-only.)

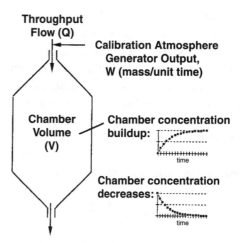

FIGURE 8-8. Air flow and concentration change in a dynamic calibration system.

$$C = \frac{W}{Q}\left[1 - e^{-\left(\frac{Q}{V}t\right)}\right] \quad (17)$$

where: C = concentration (mass/unit volume)
W = output of the test atmosphere generator (mass/unit time)
Q = total flow through the chamber (volume/unit time)
V = calibration chamber volume
t = time since the generator started

Silver[22] credited this equation to ventilation engineers; however, the relation is universal, occurring whenever the rate of change of a quantity is directly proportional to its current value:

$$\frac{dC}{dt} = \frac{(W - QC)}{V} \quad (18)$$

Dilution air mixes with and begins to dilute the test or calibration atmosphere when the generator in Figure 8-8 is shut off. In this case, the concentration within the chamber decreases at a rate, dC/dt, directly proportional to a multiple of its current value, $-(Q/V)C$. The concentration within the chamber decreases exponentially from an initial value, C_o:

$$C = C_o e^{-\left(\frac{Q}{V}t\right)} \quad (19)$$

where: t = time since the generator stopped

The time for concentration to approach 50% closer to its equilibrium value when the generator is on, or the time for concentration to reduce to one-half of its current value when the generator is off, will be the same for the system shown in Figure 8-8 provided that, in both cases, the total flow, Q, and volume, V, are constant. Operation of the system can be described by a characteristic response time because the exponential terms are the same in the equations for buildup (Equation 17) and clearance (Equation 19). When the test atmosphere generator is first turned on, the concentration in the test box increases from zero to an equilibrium concentration, C_e. The expected time or "half time" ($t_{1/2}$) for the concentration to become 50% closer to an equilibrium concentration is calculated from the ratio of V/Q:

$$t_{1/2} = \frac{V}{Q} \ln(2) \qquad (20)$$

where: $\ln(2)$ = the natural logarithm of 2 ($= 0.693$)

If it takes 6 min, $V/Q=6$, for one chamber volume of air to be sucked into the chamber inlet, then the half time for concentration change in this chamber will be 0.693×6 or 4.16 min. Every 4.16 min, the chamber concentration will come 50% closer to equilibrium concentration. Within two half times, or 8.32 min from the start, the concentration will be 75% of the equilibrium concentration [50% + (100%–50%)/2]. These "times" are independent of whether the test atmosphere generator is on, causing the concentration in the chamber to increase to an equilibrium level, or off, causing the equilibrium level in the chamber to decrease toward zero. The relations can be generalized—the time it takes the concentration to change some percentage, P, of the difference between the current (or initial) and equilibrium concentration is:

$$t_{P/100} = -\frac{V}{Q} \ln\left(\frac{100-P}{100}\right) \qquad (21)$$

where $P = 50\%$
$t_{0.50} = t_{1/2} = 0.693(V/Q)$

Likewise:

$t_{0.75} = 1.386\ (V/Q)$
$t_{0.90} = 2.30\ (V/Q)$
$t_{0.95} = 3.00\ (V/Q)$
$t_{0.99} = 4.61\ (V/Q)$

If the air exchange in the calibration chamber is three air changes an hour, then $V/Q = 20$ min, and it will take 92 min (4.61×20) to clear the air to 1% of the original concentration.

These relations (Equations 17–21) are valid only if the following two conditions are met:

1. The output of the source or calibration atmosphere generator is thoroughly mixed with the clean dilution air within the inlet line before entering the chamber.

2. Mixing of all air inside the worksite or chamber is rapid and thorough and there is little or no loss to surfaces.

Both conditions are difficult to achieve in practice.

Sampling from an Ideal Calibration Chamber. Ideally, a sample of test atmosphere should be obtained without changing the concentrations anywhere in the chamber.[26] When a percentage, P, of the flow through the chamber is removed for sampling purposes and replaced with clean dilution air, the equilibrium concentration in the system will shift to a new level:

$$C = C_o\left(1 - \frac{P}{100+P}\right) \qquad (22)$$

The half time for this shift will be shorter than the normal half time for the chamber by the same percentage $[P/(100 + P)]$. By keeping P small, the effect becomes negligible.

The problem of diluting the concentration within the chamber during sampling can be solved by returning the sample air to the exhaust line, if possible, or by operating the calibration system in the push-only mode (i.e., pushing the calibration atmosphere from the generator, past the sampling point, to an exit that is open to atmosphere; Figure 8-7). In either case, when this is done, the total flow into the chamber remains constant, as does the equilibrium concentration, ($C_o = W/Q$). Specific strategies[29] for sampling from a calibration chamber are discussed in the book by Willeke and Baron.[21]

When self-contained air sampling instruments are placed in a test chamber, their exhaust often must be directed into the chamber volume instead of into the exhaust line. The dilution effect should be undetected if the sample flow is less than 2% of the total flow through the chamber, Q, and if the instrument exhaust is directed away from sampling points.

Calibration Chamber Performance: Poor Mixing in the Inlet. When the output of the calibration atmosphere generator is not thoroughly mixed with the clean dilution air before entering the chamber, at least two air streams will enter: one relatively clean and the other having a concentration higher than the expected equilibrium concentration.

The two streams may actually travel separate and distinct paths in the chamber, especially if the energy provided for mixing is not large. Consequently, there will be zones in the chamber that are continually diluted by clean air and zones that are continually fed by air having a higher concentration of source material than desired. The mean concentration obtained from different sampling points will equal the target equilibrium concentration,

$C_o = W/Q$ (Figure 8-8), but the variation between sampling points would be large and would adversely affect any comparison between air sampling instruments being calibrated instantaneously.

An air leak into the chamber produces the same effect as little or no mixing of dilution air in the inlet line. Concentrations upstream from the leak should be stable and higher than concentrations below the leak. Initially, if the inlet air is several degrees centigrade higher than the chamber air, mixing within the chamber will be less because portions of the warm air will tend to remain above the colder air. Even with the temperature differences, the equilibrium concentration will be the same throughout the chamber, although the half time needed to reach them will be longer.

Calibration Chamber Performance: Poor Mixing in the Chamber.

Uniform mixing within a chamber is achieved with fans, baffles, cyclone injectors, or manifolds.[23] When this is not done, zones of little or no air movement, i.e., stagnant zones, are created in the chamber. The equilibrium concentration throughout the chamber will still be the same as expected, $C_o = W/Q$, and will be independent of sampling position, provided enough time has passed. The half time for buildup or decline of concentration at a given point will not be the same throughout the chamber. Concentration at sampling points within stagnant zones will change at a slower rate than the concentration at sampling points within the major streamlines of flow that exist, in such cases, between chamber inlet and exhaust.

A self-contained monitor placed within a stagnant zone in the chamber would tend to clean the air in that zone if it captures the contaminant. The instrument would suck in the calibration atmosphere and expel partially clean air back into the chamber. A new and lower equilibrium concentration would be reached in the stagnant zone around the monitor. Any alteration that provides for an improvement of mixing within the chamber, such as the addition of a fan, will resolve the problem by breaking up the stagnant zone and carrying the instrument exhaust away from the sampling port.

Measurement of Chamber Performance

Four areas should be tested in evaluating chambers before use in an instrument calibration program:

1. Uniform mixing of the test atmosphere before it reaches the chamber inlet.
2. Excessive loss of test material to chamber surfaces.
3. Possible leaks of air into the chamber.
4. Uniform mixing within the chamber.

Uniform mixing at the inlet is the most difficult to evaluate, and it is usually the area least considered. A process of elimination to evaluate mixing of dilution air with calibration atmosphere in the inlet line is recommended. If there are no leaks of clean air into the chamber, and if enough time has passed to reach some percent of the equilibrium concentration but a concentration distribution is seen within the chamber, then the cause must be poor mixing of test atmosphere in the inlet line. Poor mixing in the inlet line can be compensated by inserting simple deflectors or by inserting a surge chamber with a volume large enough to have 2–5 air changes per min.

Measurement of Chamber Performance: Loss of Material to Walls.

Excessive loss of test material to surfaces inside the chamber is specific for the source compound and the materials composing the surface. Silver[22] provides the most complete discussion to date, devoting over a third of his paper to this subject. He gives experimental results on the effects of walls, animal fur (in inhalation chambers), and even clothing (in walk-in chambers) on concentration levels. Concentration drops can be due to absorption, adsorption, or chemical reaction at the surfaces. Silver[22] reported that the concentration can be reduced over 80% if an absorbing surface is used on the walls. He also concluded that "excessive concentration lowering occurs when the volume of animals is more than 5% of the chamber volume." Actual deposition rates (μg/min/m²) cannot be estimated from his paper because only relative concentrations are given. Nonetheless, Silver clearly demonstrates the magnitude that surface effects can reach. Wall loss should be considered when setting up a calibration system for a source gas or vapor and should be estimated by measuring concentration at the inlet and exhaust of the chamber when empty and when the air sampling instruments being calibrated are in place.

Measurement of Chamber Performance: Leaks.

Chamber leaks can be evaluated in several ways. A messy but simple test is to pressurize the chamber slightly (about 1.0 cm of water) and squirt a soap solution along all the edges where leaks might occur. The bubbles produced by escaping air locate the problem area but do not quantitate the magnitude of the leak. This is appropriate for chambers operated at a slightly positive pressure but not for calibration chambers operated slightly below atmosphere pressure unless an observer enters the chamber during the test.

Another approach is to measure the rate at which air leaks into a chamber that is sealed under slight vacuum.[30] The change in vacuum or pressure difference, Δp, between chamber and room is measured over some preset time period. A slow change in pressure difference would indicate an inward leak, Q (L/min), given by:

$$f = \frac{V \Delta p \ln 2}{P_r T_{1/2}} \qquad (23)$$

where V = volume of the chamber

Δp = pressure difference between chamber pressure, P, and room pressure, P_r

$T_{1/2}$ = time for the pressure difference to change from Δp to $\Delta p/2$

In practice, a decision must be made on what leak rate is acceptable. In general, this should be less than 2% of the total flow through the chamber.

Measurement of Chamber Performance: Uniform Concentration. Measurements of the uniformity or degree of mixing within a chamber have been made by using dynamically similar models,[23] sampling from many different points within the chamber (point tests), and monitoring the entrance and exit of a bolus of gas (dynamic flow tests).[31, 32] In practice, it is only practical to conduct point tests and dynamic flow tests.

Point tests (sampling many different points within a chamber) should always indicate uniform concentration regardless of the type of the chamber, provided enough time has passed for equilibrium to be reached. A measure of degree of mixing within the chamber is obtained from the difference between buildup half times at each sampling point; the greater the difference, the poorer the mixing.

Dynamic flow tests[23, 31] consist of continuously monitoring a bolus of gas injected into the air stream as it passes the inlet and exhaust ports of the chamber. The mean time of the concentration versus time plot is calculated for the sampling point at the chamber inlet, t_1, and at the exhaust port, t_2. Chamber operation is evaluated from these numbers by calculating the percent dead space, %DS,

$$\%DS = 100 \left(\frac{T - T_m}{T} \right) \qquad (24)$$

where: T = the theoretical residence time, V/Q

T_m = the measured residence time, $t_2 - t_1$

The percent dead space is a direct measure of uniformity of mixing within the chamber. A high percent dead space indicates that stagnant zones exist and that a fraction of the inlet air is being directly shunted through the chamber.

Summary

Calibration of gas and vapor samplers involves the production of a stable calibration atmosphere. The calibration atmosphere may be stored in a container for

TABLE 8-4. Summary of Recommended and Standard Methods on Air Sampling and Instrument Calibration Related to Gases and Vapors

Organization	Type of Methods
ANSI	Sampling airborne radioactive materials
AWMA (APCA)	Recommended standard methods for continuous air monitoring of fine particulate matter
ASTM	Test methods for sampling and analysis of atmospheres
ASTM	Recommended practices for sampling and calibration procedures, nomenclature, guides, etc.
EPA/EMSL	Reference methods for air contaminants
ISC	Methods of air sampling and analysis
NIOSH	Analytic methods for air contaminants

periodic delivery of an aliquot to the air sampling instrument or it may be continuously generated through vaporization, diffusion, or permeation, and then diluted prior to delivery into the instrument being calibrated or into a calibration chamber containing several air sampling instruments that are being compared. Further discussions on the techniques for calibration of gas and vapor sampling instruments may be found in recommendations from organizations (Tables 8-1 and 8-4) and in reviews such as those by Nelson,[1] Willeke and Baron,[21] McClellan and Henderson,[20] and Lodge.[11]

References

1. Nelson, G.O.: Gas Mixtures Preparation and Control. Lewis Publishers, Chelsea, MI (1992).
2. Wong, B.A.: Generation and Characterization of Gases and Vapors. In: Concepts in Inhalation Toxicology, 2nd Ed., pp. 67–90. R.O. McClellan and R.F. Hendeson, Eds. Taylor & Francis, Washington DC (1995).
3. Barrow, C.S.: Generation and Characterization of Gases and Vapors. In: Concepts in Inhalation Toxicology, pp. 63–84. R.O. McClellan and R.F. Henderson, Eds. Hemisphere Publishing Corp., Washington, DC (1989).
4. Lippmann, M.: Calibration of Air Sampling Instruments. In: Air Sampling Instruments for Evaluation of Atmospheric Contaminants, 7th Ed., pp. 73–109. S.V. Hering, Ed. American Conference of Governmental Industrial Hygienists, Cincinnati, OH (1989).
5. Chapman, R.L.; Sheesley, D.C.: Calibration in Air Monitoring. ASTM Pub. 598. American Society for Testing and Materials, Philadelphia, PA (1976).
6. Hersch, P.A.: Controlled Addition of Experimental Pollutants to Air. J. Air Poll. Control Assoc. 19:164 (March 1969).
7. Raabe, O.G.: The Generation of Aerosols of Fine Particles. In: Fine Particles, pp. 50–110. B.Y.H. Liu, Ed. Academic Press, New York (1976).
8. Setterlind, A.N.: Preparation of Known Concentrations of Gases to Vapors in Air. Am. Ind. Hyg. Assoc. Q. 14:113 (June 1953).
9. Roccanova, G.: The Present State-of-the-Art of the Preparation of Gaseous Standards. Presented at the Pittsburgh Conference

on Analytical Chemistry and Spectroscopy; available from Scientific Gas Products, Inc. (1968).

10. Silverman, L.: Experimental Test Methods. In: Air Pollution Handbook, pp. 12:1–12:48. P.L. Magill, F.R. Holden, and C. Ackley, Eds. McGraw-Hill, New York (1956).

11. Lodge, J.P.: Methods of Air Sampling and Analysis, 3rd Ed. Lewis Publishers, Inc., Chelsea, MI (1989).

12. Cotabish, H.N.; McConnaughey, P.W.; Messer, H.C.: Making Known Concentrations for Instrument Calibration. Am. Ind. Hyg. Assoc. J. 22:392 (1961).

13. Saltzman, B.E.: Preparation and Analysis of Calibrated Low Concentrations of Sixteen Toxic Gases. Anal. Chem. 33:1100 (1961).

14. Saltzman, B.E.; Gilberg, N.: Microdetermination of Ozone in Smog Mixtures: Nitrogen Dioxide Equivalent Method. Am. Ind. Hyg. Assoc. J. 20:379 (1959).

15. Saltzman, B.E.; Wartburg, Jr., A.F.: Precision Flow Dilution System for Standard Low Concentrations of Nitrogen Dioxide. Anal. Chem. 37:1261 (1965).

16. Reid, R.C.; Prausnitz, J.M.; Sherwood, T.K.: The Properties of Gases and Liquids, 3rd Ed. McGraw–Hill, New York (1977).

17. Nelson, G.O.: Simplified Method for Generating Known Concentrations of Mercury Vapor in Air. Rev. Sci. Instr. 41:776(1970).

18. Altshuller, A.P.; Chohe, L.R.: Applications of Diffusion Cells to the Production of Known Concentrations of Gaseous Hydrocarbons. Anal. Chem. 32:802 (1960).

19. O'Keefe, A.E.; Ortman, G.O.: Primary Standards for Trace Gas Analysis. Anal. Chem. 38:760 (1966).

20. McClellan, R.O.; Henderson, R.F.: Concepts in Inhalation Toxicology, 2nd Ed. Taylor & Francis, Washington DC (1993).

21. Willeke, K.; Baron, P.Q.: Aerosol Measurement Principles, Techniques and Application. Van Nostrand Reinhold, New York (1993).

22. Silver, S.K.: Constant Gassing Chambers: Principles Influencing Design and Operation. J. Lab. Clin. Med. 31:1153 (1946).

23. Moss, O.R.: Comparison of Three Methods of Evaluating Inhalation Toxicology Chamber Performance. In: Proceedings of the Inhalation Toxicology and Technology Symposium, pp. 19–28. B.K.J. Leong, Ed. Ann Arbor Science Publishers, Inc., Ann Arbor, MI (1981).

24. Drew, R.T.; Laskin, S.: Environmental Inhalation Chambers. In: Methods of Animal Experimentation, Vol. IV, Environmental and Special Senses, pp. 1–42. W.I. Gray, Ed. Academic Press, New York (1973).

25. Griffis, L.C.; Wolff, R.K.; Beethe, R.L.; *et al.*: Evaluation of a Multitiered Inhalation Exposure Chamber. Fund. Appl. Toxicol. 1:8 (1981).

26. MacFarland, H.N.: Design and Operational Characteristics of Inhalation Exposure Equipment: A Review. Fund. Appl. Toxicol. 3:603 (1983).

27. Carpenter, R.L.; Beethe, R.L.: Airflow and Aerosol Distribution in Animal Exposure Facilities. In: Generation of Aerosols and Facilities for Exposure Experiments, pp. 459–474. K. Willeke, Ed. Ann Arbor Science Publishers, Inc., Ann Arbor, MI (1980).

28. Cheng, Y.S.; Moss, O.R.: Inhalation Exposure Systems. In: Concepts in Inhalation Toxicology. R.O. McClellan and R.F. Henderson, Eds. Hemisphere Publishing Corporation, Washington, DC (1989).

29. Moss, O.R.: Inhalation Toxicology: Sampling Techniques Related to Control of Exposure Atmospheres. In: Aerosol Measurement Principles, Techniques, and Applications, pp. 833–842. K. Willeke and P.A. Baron, Eds. Van Nostrand Reinhold, New York (1993).

30. Mokler, B.V.; White, R.K.: Quantitative Standard for Exposure Chamber Integrity. Am. Ind. Hyg. Assoc. J. 44(4):292 (1983).

31. Hemenway, D.R.; Carpenter, R.L.; Moss, R.R.: Inhalation Toxicology Chamber Performance: A Quantitative Model. Am. Ind. Hyg. Assoc. J. 43:120 (1982).

32. Whitaker, S.: Introduction to Fluid Mechanics. Prentice-Hall, Englewood Cliffs, NJ (1968).

Chapter 9

Aerosol Sampler Calibration

Yung-Sung Cheng and Bean T. Chen*

Lovelace Respiratory Research Institute, P.O. Box 5890, Albuquerque, NM 87185
*National Institute for Occupational Safety and Health, Morgantown, WV

CONTENTS

Introduction

All sampling instruments require calibration. Some instruments, such as photometers, have no means to theoretically predict response. These instruments rely on empirical calibrations using standard test aerosols. Other instruments, such as the optical counter, impactor, and condensation nucleus counter (CNC), can be described by theories that predict their instrument responses. Yet, the actual performance of these instruments may differ from the theory because not all factors are taken into account in these theoretical models. For example, impactor cutoff diameters can be calculated for ideal

sticky aerosols; however, such calculations do not take into account particle bounce, re-entrainment, electrostatic charge effects, and wall losses.[1, 2] Evaluation of the actual performance of the impactor requires experimental calibration. To obtain data that are of high quality, reproducible, and defensible for the intended purposes, the instrument must be calibrated. Furthermore, such calibration must be appropriate to the intended application.

Most instruments are calibrated and evaluated by the manufacturer or the inventor before being used by others. For an instrument intended to collect and analyze an aerosol, collection efficiency and wall losses are generally determined in the calibration. For a real-time, direct-reading instrument, calibration establishes the relationship between the instrument response (e.g., electronic signal or channel number) and the value of the property (e.g., particle size, number concentration, or mass concentration) being measured. However, the operating conditions and the parameters used during the original calibration can vary from those under which the eventual user operates. As a result, the original calibration data may not apply, and the user must recalibrate the instrument to operate it with confidence. In general, a reliable and accurate calibration process requires: 1) a sufficient knowledge of the capabilities and limitations of an instrument; 2) appropriate test facilities; 3) proper selection of a desired test aerosol; 4) a thorough investigation of relevant parameters; 5) sufficient knowledge about the conditions that can be encountered during operation; and 6) a quality assurance program that is followed throughout the test.

In the last two decades, developments in the generation and classification of monodisperse aerosol, along with the improvement of test facilities, have made instrument calibration easier and the results more reproducible. Calibration methods for air sampling instruments have been reviewed.[3, 4] Chapter 7 discusses the calibration and use of flow monitoring devices that play an integral role in aerosol sampling and instrument calibration. This chapter reviews the calibration techniques relevant to aerosol samplers, such as sizing instruments, devices for concentrations, and size-selective samplers. Test facilities, generation of test aerosols, and testing procedures will be emphasized.

General Considerations

Before embarking on a rigorous instrument testing/calibration program, decisions must be made on the frequency of calibration. General descriptions of instrument components and the measured parameter for each aerosol instrument are also discussed here. Finally, the sampling environments, which will influence the selection of appropriate test facility and procedures, must also be considered.

Components of Aerosol Samplers

An aerosol sampler usually consists of a sampling inlet, a detection or collection section, an air mover, and flow controllers. The sampling inlet, which is the entrance to the instrument, is connected to the detection section with a short transport line. The air mover (usually a vacuum pump) draws air into the sampler, and flow controllers control flow rates. Personal samplers and some area samplers, including impactors and filters, have separate pumps and flow meters. A few passive samplers, such as personal photometers, do not have an air mover and flow meter; they rely on air currents in the atmosphere to bring the aerosol into the detector. Most direct-reading aerosol instruments, such as optical counters, CNCs, and photometers, include the sampling inlet, transport line, detector, pump, and flow meter in a single unit. Each unit is indeed a complete system. For other systems such as filter samplers, individual components must be assembled.

It is important to know that there can be significant particle losses in the sampling inlet (aspiration efficiency) and transport lines (transport losses), especially for very large (> 5 µm) and small particles (< 0.01 µm). In addition, every instrument has finite detection limits and detection efficiency. The flow meter controls a constant volumetric flow rate so that the instrument can be operated properly and that accurate sampling volume and, therefore, aerosol concentration can be determined. Calibration of the flow meter is a part of the total instrument calibration. Recalibration of a flow meter is required when it is used at a high altitude or under ambient conditions (temperature, pressure, and gas composition) different from the original factory calibration. Calibration of the flow meter is usually the first step in the process of instrument calibration.

Measured Parameters

Depending on the function of the instrument, the measured parameter can be separated into particle concentration (number or mass) and particle size distribution. For size-selective samplers, mass concentrations for inspirable, thoracic, or respiration fraction will be determined. Often both parameters have to be considered in the instrument calibration. For example, in impactor calibration, the collection efficiency as a function of particle size taking into account the sampling and transport losses has to be determined. We recommend that the investigator measure the detection efficiency of the whole system including the aspiration efficiency of the inlet, losses in the transport lines, and efficiencies of the detector or sensor.

Each instrument has finite detection range and is, therefore, useful in that range. The applicable size range is based on the sampling principle, the detector efficiency, and the inlet design. For example, inertial-type instruments such as impactors usually collect particles between 0.5 and 15 µm; for diffusion batteries, the size ranges between 0.005 and 0.5 µm. For optical instruments, the lower detection limit is about 0.2 μm. Therefore, the instrument influences the selection of the test aerosol. In most calibrations, particles having a size in the applicable range of the instrument are required to establish the calibration curves. Also, most instruments have minimum and maximum detection limits in aerosol concentration. For example, instruments that are based on the detection of scattered light for single particles have very low maximum concentration limits (in the order of 100 particles cm^{-3}). At higher concentrations, increased coincidence errors are due to the presence of two or more particles in the sensing volume of the detector. For aerosol collectors, overloading the substrates causes sampling errors; therefore, appropriate sampling time and mass concentration have to be considered.

Sampling Environments

Depending on the wind speed in an environment, the sampling procedure can be classified as calm air sampling or sampling in the flow stream. Calm air sampling generally refers to a wind speed less than 50 cm/s and applies to indoor environments, including residential homes, offices, and factories. Flow stream sampling refers to environments with higher wind speed, such as ambient atmosphere or inside ventilation ducts and stacks. The airflow pattern in an environment affects particle movement and, therefore, is an important parameter for the inlet aspiration efficiency. Criteria for calm air and flow stream sampling have been discussed by Davies,[5] Hinds,[6] and Brockmann.[7] To simulate various flow conditions in the environment, test facilities with different capabilities should be considered. Instrument chambers with uniform, low air speeds are suitable for testing samplers under calm air conditions, and aerosol wind tunnels are required for testing samplers under flow stream conditions.

Test Programs

Aerosol instruments can be tested on several levels for performance. The decision on the appropriate test program is largely driven by regulatory and/or scientific needs. The three levels of test programs include:

1. Flow calibration and system integrity
2. Single point check
3. Full-scale calibration.

The simplest test procedure checks flow rates and system leaks. Scheduled and frequent checks are recommended for routine use of any instrument. However, pas-

sive samplers such as a personal mass monitor or personal DataRAM, do not require flow calibration.

The next level of testing involves instrument response for a single point. For an aerosol sizing instrument, a test aerosol (usually polystyrene latex [PSL] particles) of a defined size is used, and the response is compared to an existing calibration curve. This procedure assumes that a full-response calibration curve is available, and the user performs the test to make sure the instrument is functioning normally. A full-scale calibration requires testing the instrument response over its full operational range. Therefore, the calibration curve in terms of response as a function of particle size or concentration can be established. The efforts, equipment, and test facilities needed to perform these programs increase substantially from the simple flow calibration to the full-scale test. Only a standard flow meter and a pressure gauge are needed for a flow calibration and system check. The effort is minimum, and the benefit to the user in terms of improved quality of data is high. For a single point test, aerosol generation and monitoring systems are needed, but if they are available in the laboratory, the effort required is relatively small. A full-scale calibration requires more elaborate test facilities and extensive efforts in terms of time and labor for the exercise. Essentially, any user can perform a regular flow check, and any laboratory with experience in aerosol instruments can check a single point. However, only aerosol laboratories with appropriate facilities and equipment can perform a full-scale calibration.

When Should an Instrument Be Calibrated?

Full-scale calibration is needed to establish a calibration curve for each new instrument. Therefore, one would assume that the manufacturer or instrument developer should provide such data; however, there are some commercial instruments lacking calibration data. In many cases, independent investigators provide careful evaluation and calibration of such instruments, and their results are usually published in the open literature.

A user needs to obtain aerosol measurement data of high quality in order to meet scientific guidelines or regulatory standards established by government agencies. The user must ensure that the instrument performs according to its specifications. The flow and system integrity should be checked whenever possible, and a single point check should be considered for scientific validation. Decisions on full-scale calibration to a large extent depend on the scientific justifications and regulatory requirements for each study. Aerosols are usually measured for scientific research, regulatory compliance for health protection purposes, and toxicity testing.

For scientific research, calibration data provided by the manufacturer or published in a scientific journal can be utilized, when the instrument is used under normal conditions. If the instrument is used under different ambient pressures or flow rates, then it may need to be re-calibrated under the actual operating conditions. At a minimum, the flow meter must be calibrated and a single point check performed to see whether the instrument response differs from the original calibration.

Aerosols in work environments are measured by industrial hygienists because of mandates by government agencies including the U.S. Occupational Safety and Health Administration (OSHA) or Mine Safety and Health Administration (MSHA) regulations. Similarly, determination of release of aerosols to the ambient atmosphere from fixed-point and mobile sources is required by the U.S. Environmental Protection Agency (EPA) regulations. These regulations often specify standard sampling methods or equivalent methods that follow the same performance specifications. For example, the EPA performance specifications and test procedures applicable to a size-selective instrument, PM_{10} are contained in 40 CFR Part 53-Ambient Air Monitoring Reference and Equivalent Methods.[8] The PM_{10} samplers should be tested in a wind tunnel with liquid particles (10 sizes ranging from 3 to 25 μm in aerodynamic diameter) at wind speeds of 2, 8, and 24 km/h. The 50% cut-off determined for each speed must be 10 ± 0.5 μm. The precision for determination of concentration and flow stability is also specified.[9]

Finally, toxicity tests using inhalation exposures for new products or chemicals must be conducted under Good Laboratory Practice Standards established by the U.S. Food and Drug Administration (FDA)[10] and the Toxic Substances Control Act Test Guidelines promulgated by EPA.[11] Based on these guidelines, specific procedures for the calibration of mass monitors to determine aerosol concentrations in exposure chambers are required. For example, the laboratory must provide data on instrument responses of real-time aerosol monitors for the test material, with filter samples as a reference standard in the concentration range used for the study.

Calibration Standards

Calibration curves are generated by comparing responses from the test instrument to those of a calibration standard. Several standard methods are now available for calibrating aerosol instruments. The primary standard method for particle size and number concentration determination is the microscopic examination, whereas the gravimetric method is the primary standard for mass concentration determination. However, several secondary standard methods have been developed and frequently used because they are often easier to use than the primary method. Table 9-1 lists aerosol instruments and test standards that have been used for their calibration. The following describes each test standard in detail.

Direct Measurement with Microscopy

The primary standard method of instrument calibration is direct measurement of collected particles under an optical or electron microscope. This technique provides information on particle size distribution and particle number concentration. For example, a number concentration determined from an aerosol collected in a liquid impinger has long been used to determine dust levels in occupational environments. Also the response of a CNC was calibrated against a photographic CNC[12] in which the particles are photographed and the total number of the particles counted to yield the number concentration. However, this technique is labor-intensive, and the observations varied substantially among different operators. With advanced imaging techniques, the procedure can be automated, and the accuracy of the measurements in both size and number concentration determinations is greatly improved. The microscopic technique is still used routinely in determination of particle size and number concentration of fiber aerosols. It is also used to determine the number concentrations of bioaerosols. It is a good quality assurance practice to check the size of standard test aerosols with the primary method during the test.

One problem in using the microscopic method to determine the diameter of a liquid droplet is deformation of the droplet on a collection surface. Because of surface tension, a droplet loses its spherical shape and forms a shape similar to a segment of sphere with a height of h and diameter of D as shown in Figure 9-1. This diameter, D, is larger than the actual diameter, d, and the ratio of the projected and the actual diameter, $B = d/D$, is called the spread factor. This factor is a function of the surface tension of the droplet and the adhesion force between the particle and the surface; the spread factor is needed for accurate size measurement of droplets using the microscopic technique. For routine collection of known test aerosols such as dioctyl phthalate (DOP), oleic acid, and dibutyl phthalate (DBP) on glass slides

TABLE 9-1. Calibration Standards of Aerosol Instruments

Instrument	Measured Parameter	Particle Size Range (μm)	Calibration Standard
Size Measurement			
Cascade impactor	Flow rate, gas medium, physical dimension in and around the nozzle	0.05–30	Monodisperse spherical particles with a known size and density
Aerodynamic particle sizing instrument	Flow rate, pressure, gas medium	0.5–20	Monodisperse spherical particles with a known size, shape, and density
Optical particle counter	Wavelength of the light source, range of scattering angles, sensitivity of detector	0.3–15	Monodisperse, spherical particles with a known size and refractive index
Electrical mobility analyzer	Flow rate, charging mechanism, electric field strength	0.001–0.1	Monodisperse, spherical particles with a known size
Diffusion battery	Flow rate, temperature, deposition surface	0.001–0.1	Monodisperse, spherical particles with a known size
Number Concentration Measurement			
Condensation nuclei counter	Flow rate, saturation ratio, temperature gradient	0.001–0.5	Electrical classifer with electrometer
Mass Concentration Measurement			
Photometer	Wavelength of the light source, range of scattering angles, sensitivity of detector	0.3–1.5	Gravimetric measurement of filter samples
β-attenuation monitor	Uniformity of particle deposit	1–15	Gravimetric measurement of filter samples
Quartz crystal mass balance	Sensitivity of the sensor	0.02–10	Gravimetric measurement of filter samples

coated with a special agent, spread factors can be developed.[13–15] Using a test aerosol generated from a vibrating orifice monodisperse aerosol generator (VOMAG) as described in the section on "Test Aerosol Generation," the spread factor can be determined by comparing the measured diameter, D, and the theoretically calculated aerosol diameter, d. A more general but direct method has been described by Cheng *et al.*[16] This technique is used to determine the volume of a deformed droplet by examining the droplet under a microscope with an angle other than 90° (Figure 9-1). From the measured axes of the projected image, the height of the spherical segment is calculated, and therefore the volume equivalent diameter d is also calculated.

Particle Size

Monodisperse, spherical solid particles of PSL have been used widely as a secondary test standard to determine the responses of aerosol sizing instruments including optical counters, impactors, and other real-time monitors. These particles are uniform, spherical, and smooth with the aerosols having very small standard deviations. The particle sizes have been determined by the manufacturer

using microscopic techniques. They have well-defined physicochemical properties, including the refractive index, density, and chemical composition. They are commercially available in the size range of 0.03 to 200 μm and are relatively easy to use.

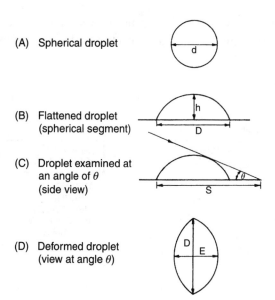

(A) Spherical droplet

(B) Flattened droplet (spherical segment)

(C) Droplet examined at an angle of θ (side view)

(D) Deformed droplet (view at angle θ)

FIGURE 9-1. Schematic of drop particles collected on a surface.

A VOMAG or an electrostatic classifier (EC) is also frequently used to produce monodisperse spherical test aerosols with well-defined characteristics. The VOMAG produces particles in the size range of 1.5 to 30 μm and the EC in the size range of 0.003 to 1 μm. The sizes of these test particles can be accurately predicted based on the operational principles, when the devices are operated appropriately. Theoretical sizes have been verified with the primary standard using microscopic techniques.[17, 18]

Number Concentration

The primary standard for calibration of number concentration is to collect aerosol particles using a high efficiency membrane filter and then to count the particles using a microscope. From the number of particles counted in a unit area, the total collection area, and the volume of air sample, the number concentrations (particles cm^{-3}) can be estimated. A secondary standard and more commonly used method is the electrometer technique, which measures the electrical current of mobility-classified aerosols produced by the EC.[19] A monodisperse test aerosol generated by an EC carries mostly a single positive charge. By measuring the electrical current with an electrometer, the aerosol concentration can be determined by dividing the measured current with the volumetric flow rate and the elementary electric charge.[18] For particles greater than 0.05 μm, a significant portion of the charged particles (depending on the aerosol size distribution) would include doubly-charged particles; therefore, corrections for the doubly-charged particles are needed to obtain an accurate number concentration from this method.[18] The electrometer technique is very useful for particles smaller than 1 μm.

For micrometer-size particles, a secondary method is to determine the mass concentration by the gravimet-ric or colorimetric method of monodisperse, spherical test particles, and then to convert the mass concentration (C_m, mass/volume) to the number concentration (C_n, particles/volume):

$$C_n = \frac{C_m}{\frac{\pi}{6}\rho_p d_{avg}^3} \qquad (1)$$

where ρ_p is the particle density, and d_{avg} is the particle diameter of the average volume.

Mass Concentration

The primary standard for aerosol mass concentration is to pass a known volume of an aerosol through a high-efficiency filter and determine the increase in mass of the substrate due to the collected aerosol particles. The mass can be determined using either an analytical balance (10 μg precision) or an electronic balance (1 μg precision). Mass concentration is obtained by dividing the increased mass by the gas volume sampled. Colorimetric or fluorescent techniques can be used to determine the mass concentration of test aerosols tagged with dye. Generation of tagged particles will be discussed in the section on "Test Aerosol Generation." Criteria for representative samples in either the flow stream or calm air sampling conditions should be followed to minimize sampling biases.

Calibration Systems and Test Facilities
Components of Calibration Systems

Figure 9-2 is a schematic diagram of a typical calibration apparatus for aerosol instruments. It includes an aerosol generator, conditioning devices, a flow mixer, a

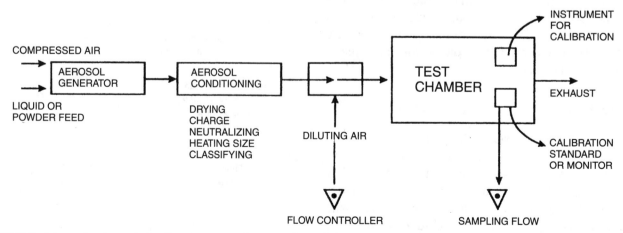

FIGURE 9-2. Schematic of a typical calibration system for aerosol samplers.

pump is sufficient to move air within the mixing chamber. For vertical chambers, air movers with capacities between 200 and 1000 L/min are required, whereas larger fans are part of the wind tunnel systems.

. The aerosol instrument is placed outside the chamber. Samples are taken via sampling probes inside the chamber and delivered to both the instrument to be calibrated and the calibration standard or monitor. Pressure in the chamber and flow rates through instruments are monitored. Multiple samples can be taken simultaneously to characterize the test aerosol, including aerosol concentration and size distribution. The key to successful calibration is to ensure that the same aerosol (size and concentration) is delivered to both the test standard/monitor and the instrument to be calibrated. The same sampling probes and transport line between the probe and the instrument inlet should be used. Also, the same flow rate through the test standard/monitor and the instrument is preferred. If the same flow rate can be used for the tested instrument and the standard, we recommend sampling a single aerosol flow from the chamber, then splitting the flow in two with a three-way valve. Alternate sampling by switching the valve between the tested instrument and the calibration standard minimizes experimental errors. In addition, an aerosol divider is a common sampling port for calibrating a mass monitor.[29] In the aerosol divider, the flow is split isokinetically into two streams: one passes directly into the instrument to be calibrated; the other flows through the calibration standard (Figure 9-3).

The mixing chamber is an inexpensive piece of equipment that is easy to setup and does not require a large working area. Therefore, many instrument user and factory calibrations have been performed in this fashion. The mixing chamber can be used to determine detector efficiency, such as the counting efficiency of a CNC; collection efficiency, as well as internal losses of impactors; and response of real-time monitors. It is especially useful for smaller particles; for particles larger than 3 to 5 μm, it is increasingly difficult to provide a stable aerosol stream in sufficiently high concentrations. Because the instrument is placed outside the chamber, it is not possible to test the aspiration efficiency of the instrument inlet under different wind conditions.

Test Chambers for Calm Air Conditions

Another way of calibrating an instrument is to introduce an aerosol into a test chamber that contains the

(A) MIXING CHAMBER

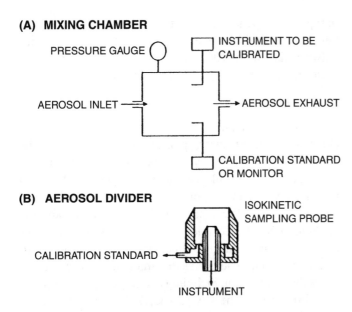

(B) AEROSOL DIVIDER

FIGURE 9-3. Schematic of a setup for instrument calibration using a mixing chamber (A) and an aerosol divider (B).

subject instrument and the test standard/monitor. This type of chamber usually has a large cross-sectional area in the test section, and therefore a low air velocity (<50 cm sec⁻¹), simulating quiescent atmosphere. For example, large exposure chambers for inhalation experiments[30] have been used for calibration of real-time aerosol monitors.[31] Also, dedicated instrument chambers have been designed for calibration of several instruments simultaneously.[32-34] A vertical flow instrument chamber with large test volume (1.8 m³) similar to the design of Marple and Rubow[34] is shown in Figure 9-4. The test aerosol enters vertically from the top of the chamber. The opposing air jets create turbulence to mix

the aerosol. To improve the uniformity of aerosol distribution, a 25-cm box fan is placed underneath the chamber entrance for mixing. A 10-cm-thick honeycomb structure is inserted above the test section of the chamber to reduce eddies created by flow turbulence and present the sampling area with a well-defined downward air flow. To ensure that spatial variations of concentration and size distribution are minimized, samplers are placed on a 75-cm platform rotating at 0.5 or 1 rpm. Variations of aerosol concentrations in the sampling platform based on filter samples are less than 5.5% for particles of 12 μm.

This type of instrument chamber provides uniform concentrations of aerosols in the test section for large particles. Several instruments could be placed inside the chamber for simultaneous calibration and comparison. The flow rate and turbulence intensity in the chamber are low, simulating calm air sampling conditions.

Aerosol Wind Tunnels

Aerosol wind tunnels have been used to test many aerosol samplers for wind speeds between 0.5 and 10 m sec⁻¹.[9, 35-37] Two types of wind tunnels are commonly used: an open circuit tunnel and a closed circuit tunnel. Figure 9-5 shows a diagram of an open circuit tunnel. Room air is taken into the system with the test aerosol and then it passes through honeycomb screens to reduce the turbulence and to provide uniform velocity. A section of contraction cone connects the test section and the screens. The test section usually has a rectangular shape and is large enough to accommodate the instrument and filter samplers with isokinetic inlets. The cross-sectional area of the test instrument should not occupy more than 10–15% of the test section so that the uniform velocity and aerosol profiles will not be affected. The aerosol then is filtered through a HEPA filter bank and clean air is exhausted to the room. A fan with sufficient capacity is used to move the air through the tunnel. Large open circuit wind tunnels that can accommodate a manikin are also used extensively to study the inspirability of large particles.[35, 38, 39]

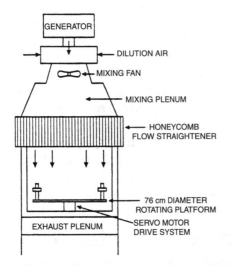

FIGURE 9-4. Schematic of a vertical aerosol instrument chamber.

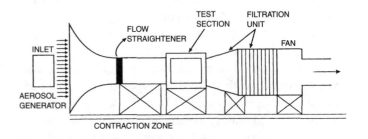

FIGURE 9-5. Schematic of an open circuit aerosol wind tunnel.

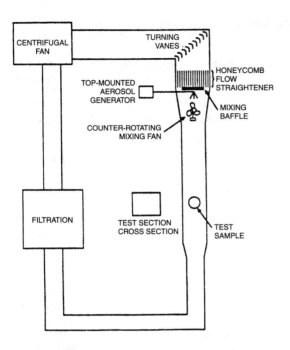

FIGURE 9-6. Schematic of a closed circuit aerosol wind tunnel.

Figure 9-6 shows a schematic of a closed-circuit wind tunnel which has a continuous path for the air. The HEPA filter removes aerosols after the test section, and the air is recirculated. Consistent flow profiles are easier to maintain in the recirculating tunnel, and less energy consumption is required for a given test section size and velocity as compared to an open tunnel. It is also less noisy than the open tunnel. However, it occupies at least twice the space and has a higher initial installation cost. To dissipate the heat produced, the closed tunnel may need to be cooled for continuous operation during the summer months.

Aerosol wind tunnels are similar to those used in aerodynamic tests, but they require provisions to introduce aerosols in the test section and remove test aerosols after the test section. In addition to a uniform velocity profile in the section, a uniform aerosol concentration profile is required. For example, EPA testing requirements specify that air velocity and aerosol concentration through the test section must be within ± 10% of the mean values in the test section.[9]

Test Aerosol Generation

Test aerosols contain either monodisperse or polydisperse, spherical or nonspherical, solid, or liquid particles.[3, 4, 6, 40, 41] The characteristics of an ideal generator are a constant and reproducible production of monodisperse and stable aerosol particles whose size and concentration can be easily controlled. For general instrument calibration, the test aerosol often contains monodisperse, spherical particles. To calibrate an instrument for a specific environ-

ment, the test aerosol should have similar physical and chemical properties to those of the aerosol of interest. For example, for sampling atmospheric particles, which can be either solid or liquid, impactor calibration should include tests with both solid and liquid particles. Solid particles are needed to assess the extent of particle bounce. Liquid particles are needed to assess wall losses. Another example is optical counters, which are most commonly calibrated with polystyrene latex particles. However, for measuring atmospheric aerosol size distributions, where the average refractive index is much lower, there can be significant errors in the particle size measurement. Calibration with atmospheric aerosols has been done.[42, 43]

In addition, the environment in which the instrument is to be operated must be considered when selecting the test aerosol. For example, if the instrument is to be operated in a high-temperature environment, the desired test aerosol could be a refractory metal oxide, such as cerium oxide, because of its thermal stability and chemical inertness. Generally, as long as the desired aerosol is determined, the appropriate method of generation can be identified. Table 9-2 lists the test aerosols frequently used for instrument calibration. Monodisperse aerosols containing spherical particles are the most widely used. Particles with nonspherical shapes are sometimes used in calibration to study the possible effect of shape on the instrument response. Polydisperse dust particles have also been used in calibrating dust monitors. This is important because most real aerosols contain nonspherical particles of different sizes and densities.

Size distribution and concentration of a test aerosol depend on the characteristics of the generator and the feed material. The information given in this section is intended to assist in the selection of appropriate generation techniques. The actual size distribution in each application should always be measured directly with the appropriate instruments.

Polydisperse Aerosols

Polydisperse aerosols are seldom used as test aerosols to calibrate sizing instruments; however, some polydisperse aerosols, such as coal dust and Arizona road dust, are frequently used in calibrating dust monitors, which provide information on the mass concentration of total and/or respirable dust. There are two ways to generate polydisperse aerosols: wet droplet dispersion and dry powder dispersion.

Wet Dispersion

The simplest way to disperse a droplet aerosol is by wet nebulization. Two types of nebulizers are often used

TABLE 9-2. List of Test Aerosols and Generation Methods Used for Instrument Calibration

Test Aerosol	Particle Morphology	Size Range[A,B] VMD (μm)	σ_g	Density (g/cm³)	Refractive Index[C]	Generation Method
PSL (PVT)	Spherical, solid	0.01–220	≤ 1.02	1.05 (1.027)	1.58	Nebulization
Glass	Spherical, solid	1.1–150	1.07-1.3	2.46	1.51	Dry powder dispersion
Fluorescent uranine	Irregular, solid	< 8	1.4-3	1.53	—	Nebulization
Dioctyl phthalate	Spherical, liquid	0.5–40	≤ 1.1	0.99	1.49	Vibrating atomization
Oleic acid	Spherical, liquid	0.5–40	≤ 1.1	0.89	1.46	Vibrating atomization
Ammonium fluorescein	Spherical, solid	0.5–50	≤ 1.1	1.35	—	Vibrating atomization
Fused ferric oxide	Spherical, solid	0.2–10	≤ 1.1	2.3	—	Spinning disc (top) atomization
Fused alummosilicate	Spherical, solid	0.2–10	≤ 1.1	3.5	—	Spinning disc (top) atomization
Fused cerium oxide	Spherical, solid	0.2–10	≤ 1.1	4.33	—	Spinning disc (top) atomization
Sodium chloride	Irregular, solid	0.002–0.3	≤ 1.2	2.17	1.54	Evaporation/ condensation
Silver	Irregular, solid	0.002–0.3	≤ 1.2	10.5	0.54	Evaporation/ condensation
Coal dust	Irregular, solid	~3.3	~3.2	1.45	1.54-0.5i	Dry powder dispersion
Arizona road dust	Irregular, solid	~3.8	~3.0	2.61	—	Dry powder dispersion

[A]Aerosol treatment of drying, charge neutralization, and size classification is generally used.

[B]VMD = volume median diameter; σ_g = geometric standard deviation

[C] — indicates refractive index (RI) unknown; i indicates imaginary RI for absorption coefficient

to produce droplet aerosols. Air-blast nebulizers[44] use compressed air (15–50 psig, 1 psig = 6.87 × 10⁴ dyne/cm²) to draw bulk liquid from a reservoir as a result of the Bernoulli effect. The high-velocity air breaks up the liquid into droplets, then suspends the droplets as part of the aerosol. Droplets produced from this method have a VMD of 1–10 μm and σ_g of 1.4–2.5 (Table 9-3). The aerosol size distribution can be modified by varying the pressure in the compressed air or the dilution ratio in the solution. One problem arises when the bulk liquid contains a volatile solvent that evaporates rapidly after formation of a droplet. The continuous loss of solvent increases the solute concentration in the reservoir and causes the particle size to increase gradually with

time. This problem can be circumvented by circulating the solution through a large reservoir,[45] by delivering the solution at constant rate[46] (Figure 9-7), and by cooling the nebulizer. Figure 9-7 shows a modified reservoir for the Retec nebulizers that maintain stable aerosol generation.

In the ultrasonic nebulizer, the mechanical energy necessary to atomize a liquid comes from a piezoelectric crystal vibrating under the influence of an alternating electric field produced by an electronic high-frequency oscillator. The vibrations are transmitted through a coupling fluid to a nebulizer cup containing the solution to be aerosolized. At a certain frequency (1.3–1.7 MHz), a heavy mist appears above the liquid surface of the cup. The diameter of the

TABLE 9-3. Operating Parameters of Air-Blast and Ultrasonic Nebulizers

| Nebulizer | Operating Conditions | | | Flow[A] Rate (L/min) | Aerosol Output (μL/L) | Droplet Size Distribution | | Commercial Source |
	Orifice Diameter (mm)	Air Pressure (psig)	Frequency (mHz)			VMD (μm)	σ_g	
Airblast type								
Collison	0.35	15		2.0	8.8	2.5–3	—	BGI
		25		2.7	7.7	1.9–2	—	
DeVilbiss[B] D–40	0.84	15		12.4	15.5	4.2	1.8	DEV
		30		20.9	12.1	2.8	1.9	
DeVilbiss D–45	0.76	15		9.4	23.2	4.0	—	DEV
		30		14.5	22.9	3.4	—	
Lovelace	0.26	20		1.5	40	5.8	1.8	INT
		50		2.3	27	2.6	2.3	
Retec X-70/N	0.46	20		5.0	46	5.7	1.8	INT
		50		9.7	47	3.2	2.2	
Ultrasonic Type								
DeVilbiss 880	(2)[C]		1.35	41.0	54	5.7	1.5	DEV
	(4)[C]		1.35	41.0	150	6.9	1.6	
Sono-Tek			0.025–0.12	10^{-6}–0.44	—	18–80	—	SON

[A]Output per orifice.
[B]Vent closed.
[C]Power settings.

droplets making up the mist is related to the wavelength of the capillary waves, which decreases with increasing frequency of the ultrasonic vibrations. Normally the VMD is 5–10 μm, with a σ_g of 1.4–2.0 (Table 9-2).

Aerosol particles with chemical properties different from those of the liquid feed material can be produced through wet dispersion by using suitable gas phase reactions, such as polymerization or oxidation. Production of spherical particles of insoluble oxides and aluminosilicate particles with entrapped radionuclides has been described by Kanapilly et al.[20] and Newton et al.[47]

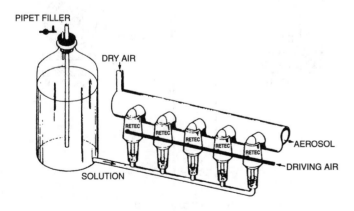

FIGURE 9-7. Schematic of a modified Retec Nebulizer for constant output.

Dry Dispersion

Aerosolization of dry powders and fibers requires different techniques than aerosolization of droplets or suspensions. The dry powder generation methods usually include a two-step operation: 1) feeding or delivery of dry powder at a constant rate to the disperser; and 2) dispersing the powder pneumatically in the aerosol form. The ease of dispersing a powder depends on the powder material, particle size, particle shape, electrostatic charge, and moisture content. Dry powder usually forms clumps or aggregates. A sticky powder is difficult to generate because the powder usually clogs the feeding system and requires high energy from the disperser to break apart the agglomerates into individual aerosol particles. Aerosols generated from dry powder dispersers are also highly charged and, therefore, require discharging to reduce the electrostatic charge.

Powder delivery systems[6, 48] can use hoppers, screw feeders, rotating disks, conveyor belts composed of chains, tubing, brushes, troughs, or compressed cylindrical packs, in which powder is delivered by scraping off the top layer (Figure 9-8). Gravity delivery systems are composed of simple hoppers designed to drop their contents into grooves cut in a plate or directly into a fluidized bed. The TSI fluidized-bed system uses a chain to deliver a

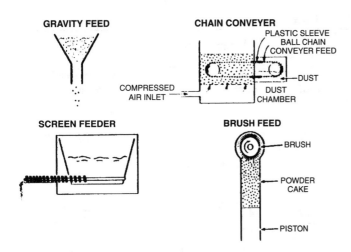

FIGURE 9-8. Schematic of dry powder delivery mechanisms.

powder from a hopper to the multi-component bed. Commercial screw feeders[49] and flexible-walled, brush-bristle augers[50, 51] have been used to successfully deliver powders to an airstream. In the instrument calibration and aerosol generation for inhalation exposures, the dry powder aerosol generator most often used is the Wright dust feed mechanism. Figure 9-9 is a view of the Wright Dust Feed scraper mechanism that consists of a cup in which the material is packed and a scraper blade moves through the packed material. As it moves, the blade removes material from the pack to the central tube. This generator is very dependable if the material can be uniformly compressed into the cup. If the cup is not well packed, air blowing across the blade of the scraper may cause large amounts of material to slough off and become airborne.

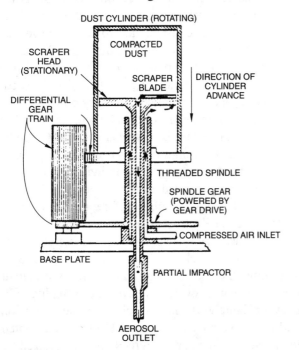

FIGURE 9-9. Schematic of the Wright Dust Feed.

The TSI small-scale powder generator uses a rotational table in which the powder is loaded. Aerosol concentration from any powder generator is to a large extent controlled and adjusted by the feeding rate. A device for dispersing fibers, similar to the Wright Dust Feed, has been designed by Timbrell[52] and is commercially available. The fibers are compressed into a cylindrical plug which is advanced by a threaded piston into a path of rotating blades. The motion of the blades disperses the fibers throughout the small chamber; air flowing through the chamber carries the dispersed fibers out of the device.

The powder is delivered to the disperser for aerosolization, dilution, and deagglomeration. Airstreams, Venturi tubes, air jet mills, and fluidized beds have been used to disperse powders as fine aerosols (Figure 9-10). The most common methods are to feed the dust into a high-velocity airstream or to blow air over the powder. The shear forces in the turbulent airstream disperse powder and break up agglomerates. In the Venturi design, a high-velocity air jet blows across a nozzle or restriction in the pipe to produce suction, which draws clumps of powders into the shear flow of air (Figure 9-11). Both the Venturi dry powder generator[53] and TSI small-scale powder generator use this principle. Fluid energy mills used for dispersion include the Trost jet mill,[51] Jet-O-Mizer,[49] and the microjet mill.[54] The fluid energy is delivered in high-velocity streams, which circulate around a grinding and classifying chamber, where turbulence and centrifugal forces deagglomerate particles. Fine particles carried by the fluid exit at the center of the chamber; coarse particles are recirculated for further size reduction. The Trost jet mill uses two opposing jets to grind and reduce particles with higher efficiency. In a one- or two-component fluidized bed, the minimal air flow required to cause the bed to fluidize is used so that only the smallest particles are released.

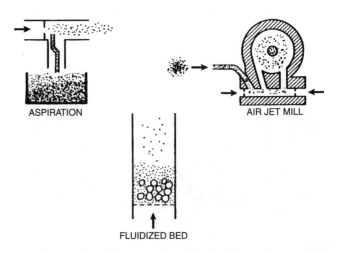

FIGURE 9-10. Schematic of dry powder dispersing mechanisms.

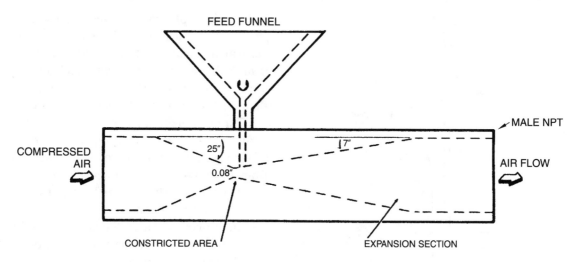

FIGURE 9-11. A Venturi powder disperser.

Commercial and laboratory powder generators listed in Table 9-4 consist of a combination of these delivery and dispersion mechanisms. Figure 9-12, on page 190, shows an example of a fluidized bed generator with gravity feed system. Other delivery systems, including a screw feed[55] and a chain belt,[56] have been used with the fluidized bed generator. The Jet-O-Mizer and Venturi with the screw feed are used to disperse sticky organic powders.[49, 53] Most dry powder generators are material-specific because delivery strongly depends on the bulk powder properties, including size, shape, compactness, and stickiness. The bulk material must be compact for delivery by a Wright Dust Feed, where the powder is packed under pressure to form a solid cylinder. Loose or uneven packing causes the packed powder to break up during the generating process.

Most dry dust generators work best for nonsticky, dry powders. When sticky materials are dispersed as powder,

they tend to form clumps that require more energy for dispersion, or they cannot be broken up, thus clogging the generator. The kinetic energy of an air dispersion system is proportional to the square of the air velocity. The fluidized bed has the least kinetic energy and therefore cannot be used to generate sticky powders. The Venturi and fluid energy mills have the highest velocity; therefore, they are more suitable for sticky powders.

One advantage of this dry dispersion method is that the aerosol generated has a similar size distribution as single powder particles when they are suspended in the air. In addition, because no solvent or heat treatment is required, the physical and chemical forms of the material are preserved. A problem common to dry-dispersion aerosols is the buildup of charge on particles as they touch and separate from the surface in the generator. This process reduces the output concentration due to

TABLE 9-4. Operating Parameters of Commercial Dry Powder Dispersers

	Wright Dust Feed	Fluidized Bed	Small Scale Powder Disperser	Jet-O Mizer Model 00
Type of Operation	Scraping the packed plug and dispersing it with air	Feeding the powder to the bed on a conveyor and air fluidizing it	Using rotating plate to deliver the powder and dispersing it with venturi suction	Using venturi suction to feed the powder into a fluid energy mill in which centrifugal force and air velocity are used to break up the agglomerate and to disperse the powder
Air flow rate, L/min	8.5–40	5–20	12–21	14–113
Feed flow rate, mm³/min	0.24–210	1.2–36	0.9–2.5	2000–30,000
Output mass concentration, g/m³ ($\rho=1$ g/cm³)	0.012–11.5	0.13–4.0	0.0003–0.04	10–1500
Source	BGI	TSI	TSI	FLU

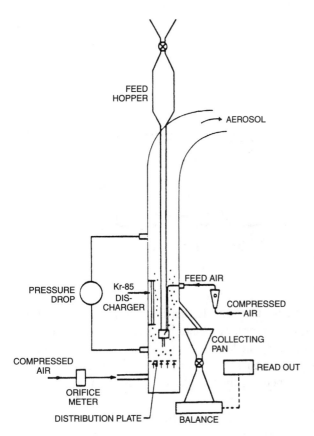

FIGURE 9-12. A fluidized bed powder generator with a gravity feed.

particle losses to the wall of the system. The problem can be solved by passing the aerosol through a chamber containing a bipolar ion source as previously described.

Monodisperse Aerosols with Spherical Particles

The methods of producing monodisperse aerosols with spherical particles have been reviewed by Fuchs and Sutugin,[57] Mercer,[40] and Raabe.[41] These methods include the atomization of a suspension of monodisperse particles, the formation of uniform droplets by dispersion of liquid jets with periodic vibration, electrical fields, or a spinning disc (top), and the growth of uniform particles or droplets by controlled condensation.

Atomization of Suspensions of Monodisperse Particles

The simplest way of producing monodisperse aerosols is by air-blast nebulizing a dilute liquid suspension containing monodisperse PSL or polyvinyltoluene (PVT) latex spheres. These spheres are commercially available in a size range from 0.01 to 30 μm (Duke Scientific, Palo Alto, CA; Dyno Industrier A. S., Lillestrom, Norway; Japan Synthetic Rubber, Tokyo, Japan; Polysciences, Warrington, PA; Seragen Diagnostics, Indianapolis, IN; 3M, Minneapolis, MN).

PSL particles of different sizes can also be concurrently produced in an aerosol to obtain more than one data point per experimental run. Monodisperse latex particles containing fluorescent dye or radiolabeled isotopes are also used in calibrations when quantitative measurements by fluorometric or radiometric techniques are needed.[47, 58]

Three problems arise in the generation of these latex particles: 1) measurement of particle size; 2) formation of aggregates; and 3) existence of a secondary aerosol from the non-latex components in the spray droplets. The diameters of latex particles reported by the manufacturer can be different from those measured by an electron microscope because the particles tend to evaporate and shrink due to electron irradiation, or to increase in size due to absorption of a contaminant under irradiation. Porstendörfer and Heyder[59] and Yamada et al.[60] recommend measurement of these particles by an electron microscope, but the particle size should be determined without particle shadowing and with minimal electron beam intensity and short exposure time.

The second problem in generating latex particles is the formation of aggregate latex particles in the aerosol. The percentage of aggregates can be reduced by diluting the suspension. Assuming that the probability of the number of particles in an atomized droplet can be described by Poisson statistics, and that the droplet-size distribution can be approximated by a log normal distribution, Raabe[61] derived the following equation to calculate the latex dilution factor, Y, necessary to give a desired singlet ratio, R, which is the number of droplets containing single particles relative to the total number of droplets containing particles:

$$Y = \frac{F(VMD)^3 \, exp(4.5 ln^2 \sigma_g) \left[1 - 0.5 \, exp(ln^2 \sigma_g) \right]}{(1 - R)d_p^3} \quad (2)$$

where F is the volumetric fraction of individual particles of diameter d_p in the original latex suspension, and VMD and σ_g are the volume median diameter and the geometric standard deviation of the droplet size distribution, respectively. The values of VMD and σ_g of commonly used air-blast atomizers are listed in Table 9-3. This equation is limited to values of $\sigma_g < 2.1$ and $R > 0.9$.

The third problem arises when nonlatex residual particles are present in the aerosol as a result of the impurities and surfactant in the liquid suspension. The nonlatex particles could either distort the size of an individual particle, if they attach to it, or become individual particles with a size different from that of the latex particles in the aerosol. To reduce the impurity in the diluting water, a system that provides deionized and double distilled water is normally used. To remove the surfactant from the suspension, a procedure of diluting, centrifuging, and discarding the supernate is generally followed.

Vibrating-Orifice and Spinning Disc (Top) Atomizers

The generators that produce monodisperse droplets can also produce monodisperse aerosols. Two methods are available. The VOMAG is an atomizer that creates a jet of liquid through an orifice and breaks the jet into droplets of a uniform diameter by applying certain disturbing frequencies[17, 62, 63] as shown in Figure 9-13. The stability, monodispersity, and diameter of the droplets depend on the diameter of the orifice, the density and surface tension of the liquid, the velocity of the jet, and the disturbing frequency. The advantage of this method is that the diameter of the droplet, d_d (μm), and the aerosol particle, d_p (μm), can be determined by the following equations:[17]

$$d_d = 10^4 \left(\frac{Q_l}{10\pi f} \right) \qquad (3)$$

$$d_p = C^{1/3} d_d \qquad (4)$$

where Q_l is the liquid feed rate in cm³/min, f is the vibrating frequency in Hz, and C is the volumetric concentration of the solute in the solution. In the frequency range for stable generation of droplets, the droplet size is about twice the orifice diameter. Therefore, with orifices of 5 to 30 μm, the droplet sizes range from 10 to 60 μm. The minimum aerosol particle size after evaporation of the solvent is about 1.5 μm, because the lowest concentration of aerosol solution, C, is limited by impurity in the solution,

which is on the order of 0.001 volume fraction. The output aerosol concentrations are low, about 30 to 400 cm⁻³, depending on the primary droplet size. In addition, liquid suspensions can clog the orifice and should not be used to feed the orifice. In the case of producing monodisperse solid aerosols (e.g., ammonium fluorescein), the drying process for suspended droplets is crucial to the surface smoothness of the final particles. Inclusion of small amounts of oil can improve sphericity by preventing crystallization, but density may be affected if the particle dries too fast. The aerosol generated has a narrow size distribution; however, the instrument is difficult to operate.

The second method of producing monodisperse droplets is by the spinning-disc (top) atomizer, in which a liquid jet is fed at a constant rate onto the center of a rotating disc (top). The liquid spreads over the disc (top) surface in a thin film, accumulating at the rim until the centrifugal force discharges it, and a droplet is thrown off. Droplet size d_d depends on disc (top) diameter, d_s (in μm), and rotating speed, ω_s (in rpm) as follows:

$$d_p = (W\gamma / \rho_t \varpi_s^2 d_s)^{1/2} \qquad (5)$$

where γ is the surface tension, ρ_t is the density of the liquid, and W is a constant. The application of this process has been investigated by Walton and Prewett[64] and May[65] using an air-driven, spinning top, and by Whitby et al.[66] and Lippmann and Albert[67] using a motor-driven, spinning disc. Unlike a vibrating-orifice atomizer, aqueous suspensions, as well as solutions, can be used. A disadvantage of this method is that undesired satellite droplets are frequently formed and must be removed from the useful aerosol produced by the primary droplets. In addition, the constant W (Equation 5) varies with the instrument and the feed material used, and the droplet size and the final particle size cannot be as easily calculated as with the vibrating-orifice atomizer.

Controlled Condensation Techniques

Condensation is also a method that produces monodisperse aerosols for calibration purposes. In this method, the heated vapor of a substance that is normally liquid or solid at room temperature is mixed with the nuclei on which it condenses when it passes in laminar flow through a cooling zone. If the condensation process is diffusion controlled, the surface area of the growing droplet will increase at a constant rate, producing a particle having a diameter, d_t, at time t related to the initial diameter, d_o, of the nucleus, by

$$d_t^2 = d_o^2 + bt \qquad (6)$$

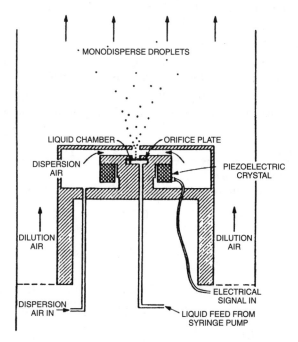

FIGURE 9-13. Schematic of a vibrating orifice aerosol generator.

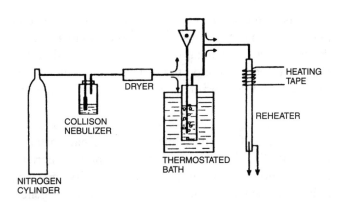

FIGURE 9-14. Schematic of a MAGE generator.

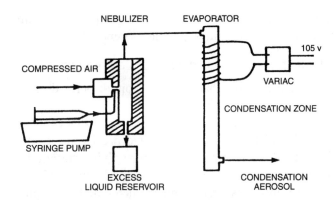

FIGURE 9-15. Schematic of a constant output controlled condensation aerosol generator.

where b is a constant, related to the concentration and diffusivity of the vapor and to the temperature. If bt is the same for all particles and much larger than d_o, the diameter of the nucleus has little effect on the final diameter of the particle, so that an aerosol containing monodisperse particles is produced. In practice, uniform temperature profile, sufficient vapor concentration, and sufficient residence time in the condensation region are the essential controls, and a constant nuclei concentration provides a stable aerosol concentration. Condensation generators are usually for high-boiling-point, low-vapor-pressure liquid droplets, e.g., DOP, triphenylphosphate, and di-octyl sebecate. Solid test aerosols, such as a carnauba paraffin aerosol, can also be generated. Aerosol concentrations are generally in the range of 10^5 to 10^7 particles/cm^3 with narrow size distribution (geometric standard deviation < 1.2).

Sinclair and LaMer[68] described the first condensation liquid aerosol generator based on this principle. Simplifications and modifications of the Sinclair and LaMer generator have been used in the laboratory[69, 70] Figure 9-14 shows a photograph of a modified Sinclair-LaMer generator (MAGE) based on Prodi's design.[70] The size range of particles generated is 0.2 to 8 μm for liquid droplets, and 0.2 to 2 μm for solid particles (waxes and paraffins). A somewhat different version of

the condensation generator as described by Liu and Lee[46] is illustrated in Figure 9-15.

A Collison-type nebulizer is used to atomize liquid into polydisperse droplets. An aerosol solution is delivered to the nebulizer by a syringe pump to control the feed rate and to prevent the evaporation of solvent and subsequently the increased concentration of the aerosol solution. The droplets are heated to vaporize in the heat section, and residue nuclei consisting of impurities in the solution are formed. The nuclei number concentration and the vapor pressure are constant, providing a stable environment for condensation in the down flow section of the glass tube. Both the Liu-Lee generator and MAGE are commercially available (Table 9-5).

Condensation generators for ultrafine solid test particles have been designed and used in instrument calibration[71–74] Figure 9-16 illustrates a basic solid aerosol generator using a tube furnace.[71, 74] The solid material is placed in a quartz boat inside the furnace, a stream of inert gas (e.g., nitrogen) carries the vapor out of the boat, and the vapor condenses to form particles in the condensation chamber. Test aerosols whose particle is carnauba wax, sodium chloride, or silver are commonly used. The aerosol has a narrow size distribution with a geometric

TABLE 9-5. Operating Parameters of Monodisperse Aerosol Generators

Generator	Principles	Aerosol Flow Rate (L/min)	Particle Size Range (μm)	Source
Electrostatic classifier (Model 3071)	Electrical mobility	2–4	0.005–1	TSI
Constant output generator (Model 3076)	Controlled condensation	2–4	0.04–1.3	TSI
CMAGE	Controlled condensation	1–3	0.9–2.5	TSI
Spinning-top aerosol generator	Atomization	120	1–100	BGI
Vibrating orifice aerosol generator (Model 3050)	Atomization	100	1–40	TSI

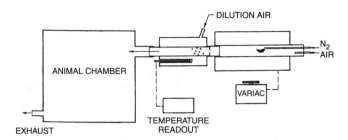

FIGURE 9-16. Schematic of a condensation aerosol generator for solid particles.

standard deviation between 1.1 to 1.3. More narrow size distributions (with geometric standard deviations smaller than 1.1) can be obtained by classifying the aerosol using an EC as described later in this section. The temperature control in the furnace is essential for the stable generation of aerosols. For generation of metal aerosols of lead, zinc cadmium, and antimony, a high-frequency induction furnace has been used.[75]

Electrospray Techniques

A method for producing monodisperse droplets using electrostatic spray of a semi-conductive fluid has recently been described.[76–79] Electrosprays refer to generation of liquid droplets through a capillary tube by feeding liquid with a finite electric conductivity and applying an electrical field. When the potential is sufficiently high, the liquid meniscus at the capillary outlet takes the shape of a cone from whose tip a very thin liquid jet emerges in the cone-jet mode.[80, 81] The microjet breaks by varicose wave instabilities into a stream of charged droplets, having diameters roughly twice as large as the jet diameter.[82, 83] The droplets have diameters an order of magnitude smaller than the capillary diameter. A stable spray can be generated if the applied electrical field exceeds a critical value. The droplet size is smaller than those produced in a vibrating orifice. The droplet size is a function of nozzle diameter, liquid feed rate, and surface tension, conductivity, and viscosity of the liquid.[84] Thus it can generate very small droplets without clogging problems in the liquid line such as are frequently encountered in the vibrating orifice aerosol generator. The mean droplet size is usually in the range of 0.3 to 50 μm. Tang and Gomez[83] have demonstrated a generation system producing monodisperse droplets in the size range from 2 to 12 μm with σ_g of 1.15 for small droplets and 1.05 for large droplets. Monodisperse droplets of 0.3 to 4 μm have been produced with σ_g of 1.1[82] More recently, monodisperse droplet sizes of 40 nm to 1.8 μm have been generated (σ_g of 1.1) by varying the liquid feed rate and electrical conductivity.[85] With evaporation of droplets, this technique can generate monodisperse particle of nanometer size (4 nm).

Monodisperse Aerosols with Nonspherical Particles

The effects of particle shape on instrument response are important, especially for the sizing instruments in which the measured properties are generally dependent on particle shape. Information concerning the effects of shape on instrument response can be obtained by using monodisperse aerosols of nonspherical particles during calibration. One way of generating these aerosols is to nebulize the liquid suspension containing the monodisperse, nonspherical particles. Various techniques have been used to produce monodisperse particles of highly uniform particle size and shape. Matijevic[86] produced inorganic and polymer colloid particles of cubic, spindle, and rhombohedral shapes by chemical reactions. Fiber-like particles of a narrow size range were also produced using different methods.[87–92] The vibrating-orifice and spinning-disc (top) aerosol generators described above can also be used to generate irregularly shaped particles, such as crystalline sodium chloride particles. Although the generators produce spherical droplets, the crystal form of the solid particles becomes the shape of the final aerosol after drying the liquid vapor.

In addition, naturally occurring materials, such as fungal spores, pollens, and bacteria or the fortuitous occurrence of multiplets of spheres, are also frequently used as test aerosols of nonspherical particles.[93, 94] The aerosols of fungal spores and pollens are commonly generated by using the dry powder dispersion technique described in the previous section.

Size Classification of Polydisperse Aerosols

Although polydisperse aerosols may be used for instrument calibration or to simulate the actual use of equipment under controlled laboratory conditions, they can also be classified according to size in order to provide an aerosol with narrow size range for instrument calibration. For particles smaller than 0.2 μm, Liu and Pui[24] developed a differential electrical mobility analyzer to classify aerosol particles of the same electrical mobility. Figure 9-17 shows that the schematic of the EC consists of a bipolar discharger to maintain the aerosol charge distribution and a concentric mobility analyzer. The polydisperse aerosol is charged and separated in the mobility analyzer, and only particles having the same mobility can exit at the same time. Because most classified particles of submicrometer size are singly charged,

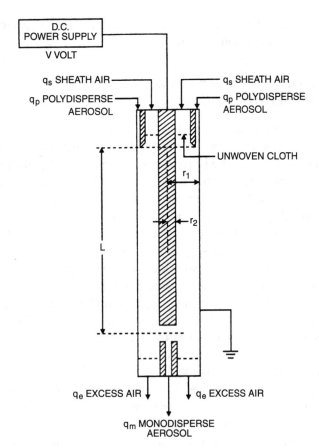

FIGURE 9-17. Schematic of an electrostatic classifier.

the aerosol produced is monodisperse. The particle size of the classified aerosol has a triangular size distribution with the median diameter d calculated from the following equation:

$$d = \frac{4ne\Lambda V\, C(d)\, l \times 10^7}{3\mu(q_s + q_e)} \qquad (7)$$

where n is the number of charges, e is the elementary charge (1.6×10^{-19} Coulomb), C is the Cunningham slip correction, μ is the gas viscosity (g/cm^{-1} sec^{-1}), q_s is the sheath flow rate (cm^3/sec^{-1}), q_e is the excess flow rate (cm^3/sec^{-1}), V is the applied voltage, $\Lambda = L/\ln(r_2/r_1)$, r_2 and r_1 are the outer and inner radii of the mobility analyzer, and L is the length of the analyzer. The geometric standard deviation of the classified aerosol is related to the flow rates:[18]

$$\sigma_g = exp\left[\left(\frac{1}{6}\right)^{1/2} \frac{(q_p + q_m)}{(q_p + q_m + q_s + q_e)}\right] \qquad (8)$$

where q_m and q_p are the monodisperse and polydisperse flow rates, respectively. This classification technique has been used to produce a submicrometer aerosol standard

in calibrating CNCs and diffusion batteries, and in determining particle deposition in human nasal, oral and tracheobronchial casts.[46, 95–97]

For particles greater than 1 μm, the size classifying technique based on aerodynamic properties is generally used. Two virtual impactors can be placed in a series to segregate the desired fraction of an input aerosol for use in instrument calibration.[25, 98] To classify aerosols in the 0.1–1.0 μm range, a technique that involves both the mobility analyzer and a single-stage micro-orifice impactor has been used.[26] One can also nebulize latex test spheres and use a mobility classifier to remove aggregates and other impurity. In fact, this method is recommended to generate PSL test particles in the size range from 0.03 to 0.1 μm. The above techniques are also used for removing undesired particles, such as PSL aggregates from an air nebulizer or satellite particles by the spinning-disc generator.

All devices and techniques described above classify aerosol particles when they are still airborne. Other instruments, such as elutriators, spectrometers, cascade impactors, and cascade cyclones, classify particles by means of collecting size-classified particles on a substrate, then resuspending the size-classified particles. For example, a spiral centrifuge collects aerodynamically classified particles on aluminum foil; resuspension of the particles caught on a narrow segment of the foil can be used to produce monodisperse aerosols.[99] The disadvantage of all size-classifying techniques is that only a small quantity of particles is produced.

Test Aerosols with Tagging Materials

For some applications, particle detection is often facilitated by incorporating dyes or radioisotope tags in the particles during their production. For example, test aerosols composed of fluorescent dyes can be analyzed in solutions containing as little as 10^{-10} g/m^{-3}. The radioactivity of the aerosols can be determined by counting the samples with much lower detection limits than can be obtained with electrobalances. Assuming tagged materials are evenly distributed in the aerosol material and therefore are proportional to the mass, the tagged material can be measured by colorimetric, fluorometric, or radioactive counting techniques to determine the particle mass with low detection limits. These techniques are especially useful for calibration of instruments having low flow rates smaller than 1 L min^{-1} such as Mercer impactors or low pressure impactors. Tagged particles deposited in the inlet section, transport lines, and internal wall of the detector can be washed and measured.[2, 100]

Commonly used dye materials include eosin,[2] methylene blue,[1] and ammonium fluorescein. Minute amounts of these chemicals can be dissolved in aqueous or alcohol solutions with oleic acid and DOP. Monodisperse-tagged liquid droplets can be generated by techniques described in the previous sections, including a VOMAG, a spinning-top generator, and condensation techniques. Commercial PSL particles that incorporate fluorescent dyes are also available as listed in Table 9-6. Radiolabeling techniques have been used in many forms and can usually be detected at extremely low concentrations.[47, 101] Radiolabels including ^{99}Tc, ^{51}Cr, and ^{137}Cs have been incorporated into PSL to produce monodisperse test particles. Similarly, several radiolabels have been incorporated into aluminosilicate clay by ion exchange; the clay solution is then nebulized and heat treated at $1150^{\circ}C$ to form insoluble spherical particles.[47, 102] Ultrafine particles (<0.2 µm) of metal oxides ($^{67}Ga_2O_3$, $^{144}CeO_2$, and $^{57}Co_3O_4$) have been produced by evaporation, subsequent thermal degradation, and condensation of organic ketone compounds.[71]

Calibration Procedures

This section describes the steps used in a typical calibration practice. It is always useful to describe detailed procedures in a test plan or protocol, and then follow through during the test.

Set up Test Facility

The criteria for selecting a proper test facility for a calibration program have been described in detail in previous sections. The selected test facility may already exist in the laboratory. In this case, one only needs to ensure the performance system for the specified procedure. If a new test facility or experimental apparatus is needed, then the system should be designed accordingly, and individual components of the test facility purchased and installed.

Select Test Aerosols

Proper selection of test aerosols is essential to instrument calibration. A variety of test aerosols and their generation methods have been described. For general purposes, standard test aerosols as listed in Table 9-2 provide sufficient choices. The selection is further narrowed by considering the particle size, monodisperse versus polydisperse aerosol, and liquid versus solid particles. Sometimes the choice is dictated by the standards specified in a regulation. In other cases, special test aerosols are required for certain classes of instruments. For example, fiber test aerosols should be used to calibrate the response of real-time fiber monitors. Once the test aerosol is selected, an appropriate aerosol generator and accessory parts are assembled and become part of the test system.

Test Calibration Systems

After the test system is assembled, the system should be checked for leaks. All meters used to regulate flows in the generator, dilution air, chamber supply, exhaust, and sampler should be calibrated and documented. The system should be tested without aerosol generation to determine the stability of the flow, the uniformity of flow in the test chamber, and the turbulence intensity in case a wind tunnel is used. Test runs with aerosols should be made to determine the stability of aerosol concentration at target levels, the particle size distribution, and distribution of aerosol concentration in the test section. The operating conditions can be adjusted and the test system modified during the test run to obtain the desired test conditions.

On-line detection of aerosol size and/or concentration in the calibration system is also recommended to assure the stable generation of aerosol. For example, an optical counter or an aerodynamic particle sizer can be used to monitor the satellite's generation of a VOMAG or spinning-disc generator.

TABLE 9-6. Commercial Sources of Monodisperse Latex Test Particles with Fluorescent or Radioactive Tags

Materials	Tags	Size Range (µm)	Source
Polystryene latex	Blue, green, red, yellow	0.44–7	BAN
Polystryene latex	Green, red, blue	0.025–3	DUK
Polystyrene latex with surface modification	Blue, orange, yellow-green, red, dark red	0.02–4	IDC
Polystyrene latex	Yellow-green	0.05–10	POL

Data Collection, Analysis, and Documentation

A calibration curve that contains the relationship between the instrument's responses and the values of a certain aerosol property is established after calibration. In the case of an instrument that directly indicates a value of the measured parameter, calibration provides an adjustment (or a correction factor) to the indicated value. In addition, resolution and sensitivity of the instrument should be examined and analyzed. The rule of thumb is to conduct the calibration based on all important parameters, to assemble all the data, and express those data in a generalized mathematical equation, relating the instrument response to a single parameter.

For data analysis, instrument manufacturers sometimes provide a built-in algorithm whose properties, accuracy, and limitations are often unknown to the user. Unless a user understands the algorithm, the analysis should be developed only based on the raw calibration data without any manipulation by the built-in algorithm.

All pertinent data including experimental conditions, aerosol data, and analysis should be documented carefully. This is especially important if the study is intended for regulatory or compliancy purposes. It is likely the information will be examined and scrutinized carefully. Examination of the documents is needed to remove any mistakes before submitting to the agency for approval.

Quality Assurance

To obtain high quality test data and to minimize experimental errors, steps should be taken to ensure that the test facility, instruments, and aerosol generator are functioning properly. Integrity of the test system and the instrument itself should be maintained by avoiding possible leakage or blockage of flow. Techniques and criteria of isokinetic sampling and sampling from still air should be followed, especially for aerosol particles greater than 5 µm in aerodynamic diameter. In order to generate reliable data, pre-calibration tests should be carried out to ensure that a stable aerosol can be maintained for a long period of time, that the aerosol concentration and size are constant, and that the aerosol is uniformly distributed in the test chamber. For each data point, at least three calibrations are required to provide statistically valid information. Also, it is a good practice to calibrate the test standard before the test is undertaken. For example, as noted above, PSL test particles may need to be examined in a microscope to assure the accuracy of particle size and the absence of impurities.

TABLE 9-7. Commercial Sources

Symbol	Source	Symbol	Source
BAN	Bangs Laboratories Inc. 979 Keystone Way Carmel, IN 46032 (317)844-7176 FAX (317)575-8801	IDC	Interfacial Dynamics Corp. 17300 SW Upper Boones Ferry Rd. Suite 120 Portland, OR 97224 (800)323-4810
BGI	BGI Inc. 58 Guinan St. Waltham, MA 02154 (781)891-9380 FAX (781)891-8151	INT	In-Tox Products 115 Quincy, NE Albuquerque, NM 87108 (505)286-2233
DEV	The DeVilbiss Co. P.O. Box 635 Somerset, PA 15501 (814)443-4881	POL	Polyscience Inc. 400 Valley Rd. Warrington, PA 18976 (800)523-2575
DUK	Duke Scientific Co. 2463 Faber Pl. Palo Alto, CA 94303 (800)334-3883	SON	Sono-Tek 313-A Main Mall Poughkeepsie, NY 12601 (914)471-6090
FLU	Fluid Energy Aljet 5136 Applebutter Rd. Plumsteadville, PA 18949 (215)766-0300.	TSI	TSI Inc. 500 Cardigan Rd. St Paul, MN 55164 (800)677-2708 FAX (651)490-3860

A quality assurance program should also include written documentation of the protocol, standard operating procedures for each procedure used in the protocol, and careful documentation of the results. After the procedures are finalized, the operator should adhere to the established procedures. If problems occur during the test, the cause should be identified, corrected, and documented before the test is continued. A quality assurance program is mandatory, if a study is to be conducted under the Good Laboratory Practices Guidelines or other compliance regulation.

Conclusions

Because the accuracy of measuring aerosols depends on the precision of the aerosol instrument, the instrument must be calibrated very carefully. The following comments summarize this chapter and the philosophy of aerosol instrument calibration:

1. The developer or manufacturer of an instrument has the responsibility for providing instrument performance data that include full-scale calibration data covering the whole applicable range (i.e., the particle size for sizing the instrument and the concentration for mass or number concentration monitors), the test aerosol used, and the conditions under which the instrument is tested (flow rates, ambient temperature, and pressure). This calibration should be performed for new instruments and after extensive repair.

2. The user should understand the principles of the instrument operation and be familiar with the calibration data. The decision on when to recalibrate the instrument is based on scientific and regulatory requirements. Frequent checks of the instrument and its flow meter are minimum requirements, and a single point check is recommended for routine use of any sampling instrument. Full-scale calibration requires substantial effort, appropriate equipment, and facilities that may not be available for every user.

3. Quality assurance is an important aspect of a successful test program. The calibration procedure should be carefully planned and documented. Test runs should be made to ensure that the system is functioning, that the correct procedures are used, and that the aerosol and flow stability are achieved and maintained. Standard operating procedures should be followed during the calibration, and any problems encountered should be resolved and documented. All records including test runs, data, problems with appropriate comments, instrument identification, barometric pressure, temperature, flow rate, properties of the test aerosol, and the name of the operator should be documented.

4. Proper test facilities should be used for an aerosol sampler intended for different environments. Instrument chambers with low air flow should be used for calm air sampling conditions, whereas aerosol wind tunnels should be used to simulate sampling in the flow streams or ambient environments.

5. Proper test aerosols should be used. Standard spherical test aerosols such as PSL and others provide data to compare the instrument response with other instruments using the same kind of aerosol. Whenever possible, test aerosols with physicochemical properties similar to the measured aerosol should be used.

References

1. Rao, A.K.; Whitby, K.T.: Non-Ideal Collection Characteristics of Inertial Impactors – II. Cascade Impactors. J. Aerosol Sci. 9:87 (1978).
2. Cheng, Y.S.; Yeh, H.C.: Particle Bounce in Cascade Impactors. Environ. Sci. Technol. 13:1392 (1979).
3. Lippmann, M.: Calibration of Air Sampling Instruments. In: Air Sampling Instruments, pp. 73–109. S.V. Hering, Ed. ACGIH, Inc., Cincinnati, OH (1989).
4. Chen, B.T.: Instrument Calibration. In: Aerosol Management: Principles, Techniques and Applications, pp. 493–520. K. Willeke; P.A. Baron, Eds. Van Nostrand Reinhold, New York (1993).
5. Davies, C.N.: The Aspiration of Heavy Airborne Particles into a Point Sink. Proc. Roy Soc. 279a:413 (1964).
6. Hinds, W.: Aerosol Tchnology. John Wiley and Sons, New York (1982).
7. Brockmann, J.E.: Sampling and Transport of Aerosols. In: Aerosol Measurement: Principles, Techniques and Applications, pp. 77–111. K. Willeke; P.A. Baron, Eds. Van Nostrand Reinhold, New York (1993).
8. U.S. Environmental Protection Agency: Ambient Air Monitoring Reference and Equivalent Methods. 40 CFR, Part 53. U.S. EPA, Washington, DC (1987).
9. Ranade, M.B.; Wood, M.C.; Chen, F.L.; et al.: Wind Tunnel Evaluation of PM-10 Samplers. Aerosol Sci. Technol. 13:54 (1990).
10. U.S. Food and Drug Administration: Good Laboratory Practice Regulations. 21 CFR, Part 58. U.S. FDA, Washington, DC (1984).
11. U.S. Environmental Protection Agency: Toxic Substances Control Act Test Guidelines. 40 CFR, Parts 796–798. U.S. EPA, Washington, DC (1985).
12. Jaenicke, R.; Kanter, H.J.: Direct Condensation Nuclei Counter with Automatic Photographic Recording, and General Problems of Absolute Counters. J. Appl. Meteology 15:620 (1976).
13. Schönauer, G.: Electron- and Light-Optical Determination of the Flattening Factor and Size of Oil Droplets. Staub-Reinhalt Luft (Eng.) 27(9):7 (1967).
14. Liu, B.Y.H.; Pui, D.Y.H.; Wang, X.Q.: Drop Size Measurement of Liquid Aerosols. Atmos. Environ. 16:563 (1982).
15. Olan-Figueroa, E.; McFarland, A.R.; Ortiz, C.A.: Flattening Coefficients for DOP and Oleic Acid Droplets Deposited on Treated Glass Slides. Am. Ind. Hyg. Assoc. J. 43:395 (1982).
16. Cheng, Y.S.; Chen, B.T.; Yeh, H.C.: Size Measurement of Liquid Aerosols. J. Aerosol Sci. 17:803 (1986).

17. Berglund, R.N.; Liu, B.Y.H.: Generation of Monodisperse Aerosol Standards. Environ. Sci. Technol. 7:147 (1973).

18. Cheng, Y.S.; DeNee, P.B.: Physical Properties of Electrical Mobility Classified Aerosols. J. Colloid Interface Sci. 80:284 (1981).

19. Liu, B.Y.H.; Kim, C.S.: On the Counting Efficiency of Condensation Nuclei Counters. Atmos. Environ. II:1097 (1977).

20. Kanapilly, G.M.; Raabe, O.G.; Newton, G.J.: A New Method for the Generation of Aerosols of Insoluble Particles. J. Aerosol Sci. 1:313 (1970).

21. Chen, B.T.; Cheng, Y.S.; Yeh, H.C.: A Study of Density Effect and Droplet Deformation in the TSI Aerodynamic Particle Sizer. Aerosol Sci. Technol. 12:278 (1990).

22. Yeh, H.C.; Teague, S.V.; Newton, G.J.: Fabrication and Use of Krypton-85 Aerosol Discharge Devices. Health Phys. 35:392–395 (1978).

23. Adachi, M.; Pui, D.Y.H.; Liu, B.Y.H.: Aerosol Charge Neutralization by a Corona Ionizer. Aerosol Sci. Tech. 18:45–58 (1993).

24. Liu, B.Y.H.; Pui, D.Y.H.: A Submicron Aerosol Standard and the Primary, Absolute Calibration of the Condensation Nuclei Counter. J. Colloid Interface Sci. 47:155 (1974).

25. Chen, B.T.; Yeh, H.C.; Rivero, M.A.: Use of Two Virtual Impactors in Series as an Aerosol Generator. J. Aerosol Sci. 19:137 (1988).

26. Romay-Novas, F.J.; Pui, D.Y.H.: Generation of Monodisperse Aerosols in the 0.1–1.0 μm Diameter Range Using a Mobility Classification-Inertial Impaction Technique. Aerosol Sci. Technol. 9:123 (1988).

27. Barr, E.B.; Hoover, M.D.; Kanapilly, G.M.; et al.: Aerosol Concentrator: Design, Construction, Calibration, and Use. Aerosol Sci. Technol. 2:437 (1983).

28. Yeh, H.C.; Cheng, Y.S.; Carpenter, R.L.: Evaluation of an In-Line Dilutor for Submicron Aerosols. Am. Ind. Hyg. Assoc. J. 44:358 (1983).

29. Marple, V.A.; Rubow, K.L.: An Evaluation of the GCA Respirable Dust Monitor 101-1. Am. Ind. Hyg. Assoc. J. 39:17 (1978).

30. Moss, O.R.: Sampling in Calibration and Exposure Chambers. In: Air Sampling Instruments: for Evaluation of Atmospheric Contaminants, 7th Ed., pp. 157–162. S.V. Hering, Ed. American Conference of Governmental Industrial Hygienists, Cincinnati, OH (1989).

31. Cheng, Y.S.; Barr, E.B.; Benson, J.M.; et al.: Evaluation of a Real-Time Aerosol Monitor (RAM-S) for Inhalation Studies. Fundam. Appl. Toxicol. 10:321 (1988).

32. Gibson, H.; Ogden, T.L.: Some Entry Efficiencies for Sharp-Edged Samplers in Calm Air. J. Aerosol Sci. 8:361 (1977).

33. Kuusisto, P.: Evaluation of the Direct Reading Instruments for the Measurement of Aerosols. Am. Ind. Hyg. Assoc. J. 44:863 (1983).

34. Marple, V.A.; Rubow, K.L.: An Aerosol Chamber for Instrument Evaluation and Calibration. Am. Ind. Hyg. Assoc. J. 44:361 (1983).

35. Vincent, J.H.; Mark, D.: Applications of Blunt Sampler Theory to the Definition and Measurement of Inhalable Dust. Am. Occup. Hyg. 26:3 (1982).

36. Fabries, J.F.; Carton, B.; Wrobel, R.: Equipment for the Study of Air Sampling Instruments with Real Time Measurement of the Aerosol Concentration. Saub Reinhalt. 44:405 (1984).

37. Blackford, D.B.; Heighington, K.: The Design of an Aerosol Test Tunnel for Occupational Hygiene Investigations. Atmos. Environ. 20:1605 (1986).

38. Armbruster, L.; Breuer, H.: Investigations into Defining Inhalable Dust. Ann. Occup. Hyg. 26:21 (1982).

39. Chung, I.P.; Dunn-Rankin, D.; Phalen, R.F.; Oldham, M.J.: Low-Cost Wind Tunnel for Aerosol Inhalation Studies. Am. Ind. Hyg. Assoc. J. 53:232 (1992).

40. Mercer, T.T.: Aerosol Technology in Hazard Evaluation. Academic Press, New York (1973).

41. Raabe, O.G.: The Generation of Fine Particles. In: Fine Particles:

Aerosol Generation, Measurement, Sampling and Analysis, pp. 57–110. B.Y.H. Liu, Ed. Academic Press, New York, NY (1976).

42. Hering, S.V.; McMurry, P.H.: Optical Counter Response to Monodisperse Atmospheric Aerosol. Atmos. Environ. 25A:463 (1991).

43. Stolzenburg, M.; Kreisberg, N.; Hering, S.V.: Atmospheric Size Distributions Measured by Differential Mobility Optical Particle Size Spectrometry. Aerosol Sci. Technol. 29:402 (1998).

44. Mercer, T.T.; Tillery, M.I.; Chow, Y.H.: Operating Characteristics of Some Compressed Nebulizers. Am. Ind. Hyg. Assoc. J. 29:66–78 (1968).

45. DeFord, H.S.; Clark, M.L.; Moss, O.R.: A Stabilized Aerosol Generator. Am. Ind. Hyg. Assoc. J. 42:602 (1981).

46. Liu, B.Y.H.; Lee, K.W.: An Aerosol Generator of High Stability. Am. Ind. Hyg. Assoc. J. 36:861 (1975).

47. Newton, G.J.; Kanapilly, G.M.; Boecker, B.B.; Raabe, O.G.: Radioactive Labeling of Aerosols: Generation Methods and Characteristics. In: Generation of Aerosols and Facilities for Exposure Experiments, pp. 399–425. K. Willeke, Ed. Ann Arbor Sci. Publ., Ann Arbor, MI (1980).

48. Moss, O.R.; Cheng, Y.S.: Generation and Characterization of Test Atmospheres: Particles. In: Concepts in Inhalation Toxicology, pp. 87–121. R.O. McClellan; R.F. Henderson, Eds. Hemisphere, New York, NY (1989).

49. Cheng, Y.S.; Marshall, T.C.; Henderson, R.F.; Newton, G.J.: Use of a Jet Mill for Dispensing Dry Powder for Inhalation Studies. Am. Ind. Hyg. Assoc. J. 46:449 (1985).

50. Milliman, E.M.; Chang, D.Y.P.; Moss, O.R.: A Dual Flexible-Brush Dust-Feed Mechanism. Am. Ind. Hyg. Assoc. J. 42:747 (1981).

51. Bernstein, D.M.; Moss, O.: Fleissner, H.; Bretz, R.: A Brush Feed Micronizing Jet Mill Powder Aerosol Generator for Producing a Wide Range of Concentrations of Respirable Particles. In: Aerosols, pp. 721–724. B.Y.H. Liu; D.Y.H. Pui; H.J. Fissan, Eds. Elsevier, New York (1984).

52. Timbrell, V.; Hyett, A.W.; Skidmore, J.W.: A Simple Dispenser for Generation Dust Clouds from Standard Reference Samples of Asbestos. Ann. Occup. Hyg. 11:273 (1968).

53. Cheng, Y.S.; Barr, E.B.; Yeh, H.C.: A Venturi Disperser as a Dry Powder Generator for Inhalation Studies. Inhal. Toxicol. 1:365 (1989).

54. Lee, K.P.; Kelly, D.P.; Kennedy, G.L.: Pulmonary Response to Inhaled Kevlar Synthetic Fibers in Rats. Toxicol. Appl. Pharmacol. 71:242 (1983).

55. Tanaka, I.; Akiyama, T.: A New Dust Generator for Inhalation Toxicity Studies. Ann. Occup. Hyg. 28:157 (1984).

56. Marple, V.A.; Liu, B.Y.H.; Rubow, K.L.: A Dust Generator for Laboratory Use. Am. Ind. Hyg. Assoc. J. 39:26 (1978).

57. Fuchs, N.A.; Sutugin, A.G.: Generation and Use of Monodisperse Aerosols. In: Aerosol Science, pp. 1–30. C.N. Davies, Ed. Academic Press, New York (1966).

58. Chen, B.T.; Cheng, Y.S.; Yeh, H.C.; et al.: Test of the Size Resolution and Sizing Accuracy of the Lovelace Parallel-Flow Diffusion Battery. Am. Ind. Hyg. Assoc. J. 52:75 (1991).

59. Porstendörfer, J.; Heyder, J.: Size Distribution of Latex Particles. J. Aerosol Sci. 3:141 (1972).

60. Yamada, Y.; Miyamoto, K.; Koizumi, A.: Size Determination of Latex Particles by Electron Microscopy. Aerosol Sci. Technol. 4:227 (1985).

61. Raabe, O.G.: The Dilution of Monodisperse Suspensions for Aerosolization. Am. Ind. Hyg. Assoc. J. 29:439 (1968).

62. Fulwyler, M.J.; Glascock, R.B.; Hiebert, R.D.: Device Which Separates Minute Particles According to Electronically Sensed Volume. Rev. Sci. Instrum. 40:42 (1969).

63. Raabe, O.G.; Newton, G.L.: Development of Techniques for Generating Monodisperse Aerosols with the Fulwyler Droplet Generator. In: Fission Product Inhalation Program Annual Report (LF-43), pp. 13–17 (1970).

64. Walton, W.H.; Prewett, W.C.: The Production of Sprays and Mists of Uniform Drop Size by Means of Spinning Disc Type Sprayers. Proc. Phys. Soc. B62:341 (1949).

65. May, K.R.: An Improved Spinning Top Homogeneous Spray Apparatus. J. Appl. Phys. 20:932 (1949).

66. Whitby, K.T.; Lundgren, D.A.; Peterson, C.M.: Homogeneous Aerosol Generator. Int. J. Air Wat. Poll. 9:263 (1965).

67. Lippmann, M.; Albert, R.E.: A Compact Electric-Motor Driven Spinning Disc Aerosol Generator. Am. Ind. Hyg. Assoc. J. 28:501 (1967).

68. Sinclair, D.; LaMer, K.: Light Scattering as a Measure of Particle Size in Aerosols. Chem. Rev. 44:245 (1949).

69. Rapaport, E.; Weinstock, S.E.: A Generator for Homogeneous Aerosols. Experientia 11:363 (1955).

70. Prodi, V.: A Condensation Aerosol Generator for Solid Monodisperse Particles. In: Assessment of Airborne Particles, pp. 169–181. T.T. Mercer, P.E. Morrow, W. Stöber, Eds. C.C. Thomas Publ., Springfield, IL (1972)

71. Kanapilly, G.M.; Tu, K.W.; Larsen, T.B.; Fagel, G.R.: Controlled Production of Ultrafine Metallic Aerosols by Vaporization of an Organic Chelate of the Metal. J. Colloid Interface Sci. 65:533 (1978).

72. Tu, K.W.: A Condensation Aerosol Generator System for Monodisperse Aerosols of Different Physicochemical Properties. J. Aerosol Sci. 13:363 (1982).

73. Scheibel, H.G.; Porstendörfer, J.: Generation of Monodisperse Ag and NaCl Aerosols with Particle Diameters Between 2 and 300 nm. J. Aerosol Sci. 14:113 (1983).

74. Cheng, Y.S.; Yamada, Y.; Yeh, H.C.; Su, Y.F.: Size Measurement of Ultrafine Particles (3 to 50 nm) Generated from Electrostatic Classifiers. J. Aerosol Res. 5:44 (1990).

75. Homma, K.; Kawai, K.; Nozaki, K.: Metal-Fume Generation and its Application to Inhalation Experiments. In: Generation of Aerosols, pp. 361–377. K. Willeke, Ed. Ann Arbor Science, Ann Arbor, MI (1980).

76. Hayati, I.; Bailey, A.; Tadros, T.F.: Investigations into the Mechanism of Electrohydrodynamic Spraying of Liquids, I. J. Colloid Interface Sci. 117:205 (1987).

77. Hayati, I.; Bailey, A.; Tadros, T.F.: Investigations into the Mechanism of Electrohydrodynamic Spraying of Liquids,, II. J. Colloid Interface Sci. 117:222 (1987).

78. Fernandez de la Mora, J.; Navascues, J.; Fernandez, F.; Rosell-Llompart, J.: Generation of Submicron Monodisperse Aerosols in Electrosprays. J. Aerosol Sci. 21:S673 (1990).

79. Meesters, G.M.H.; Versoulen, P.H.W.; Marijnissen, J.C.M.; Scarlett, B.: Generation of Micron-sized Droplets from the Tylor Cone. J. Aerosol Sci. 23:37 (1992).

80. Cloupeau, M.; Prunet-Foch, B.: Electrostatic Spraying of Liquid in Cone-jet Mode. J. Electrostatics 22:135–159 (1989).

81. Cloupeau, M.; Prunet-Foch, B.: Electrohydrodynamic Spraying Functioning Modes: A Critical Review. J. Aerosol Sci. 25:1021–1036 (1994).

82. Rosell-Llompart, J.; de la Mora, J.F.: Generation of Monodisperse Droplets 0.3 to 4 μm in Diameter from Electrified Cone-jets of Highly Conducting and Viscous Liquid. J. Aerosol Sci. 25:1093–1119 (1994).

83. Tang, K.; Gomez, A.: Generation by Electrospray of Monodisperse Water Droplets for Targeted Drug Delivery by Inhalation. J. Aerosol Sci. 25:1237–1249 (1994).

84. Smith, D.P.H.: The Electrohydrodynamic Atomization of Liquids. IEEE Trans. Ind. Appl. 1A-22:527 (1986).

85. Chen, D.R.; Pui, B.Y.H.; Kaufman, S.L.: Electrospraying of Conducting Liquids for Monodisperse Aerosol Generation in the 4 mm to 1.8 μm Diameter Range. J. Aerosol Sci. 26:963–977 (1995).

86. Matijevic, E.: Production of Monodisperse Colloidal Particles. Ann. Rev. Mater. Sci. 15:483 (1985).

87. Esmen, N.A.; Kahn, R.A.; LaPietra, D.; McGovern, E.D.: Generation of Monodisperse Fibrous Glass Aerosols. Am. Ind. Hyg. Assoc. J. 41:175 (1980).

88. Loo, B.W.; Cork, C.P.; Madden, N.W.: A Laser-Based Monodisperse Carbon Fiber Generator. J. Aerosol Sci. 13:241 (1982).

89. Vaughan, N.P.: The Generation of Monodisperse Fibers of Caffeine. J. Aerosol Sci. 21:453 (1990).

90. Hoover, M.D., Casalnuovo, S.A.; Lipowicz, P.J.; et al.: A Method for Producing Non-Spherical Monodisperse Particles Using Integrated Circuit Fabrication Techniques. J. Aerosol Sci. 21:569 (1990).

91. Chen, B.T.; Yeh, H.C.; Hobbs, C.H.: Size Classification of Carbon Fiber Aerosols. Aerosol Sci. Technol. 19:109 (1993).

92. Deye, G.J.; Gao, P.; Baron, P.A.; Fernback, J.E.: Performance Evaluation of a Fiber Length Classifier. Aerosol Sci. Technol. 30:420–437(1999).

93. Corn, M.; Esmen, N.A.: Aerosol Generation. In: Handbook on Aerosols, pp. 9–39. R. Dennis, Ed. Publ. TID-26608, National Technical Information Services, Springfield, VA (1976).

94. Adams, A.J.; Wennerstorm, D.E.; Mazunder, M.K.: Use of Bacteria as Model Nonspherical Aerosol Particles. J. Aerosol Sci. 16:163 (1985).

95. Scheibel, H.G.; Porstendörfer, J.: Penetration Measurements for Tube and Screen-Type Diffusion Batteries in the Ultrafine Particle Size Range. J. Aerosol Sci. 15:673 (1984).

96. Cheng, Y.S.; Yamada, Y.; Yeh, H.C.; Swift, D.L.: Deposition of Ultrafine Aerosols in a Human Oral Cast. Aerosol Sci. Technol. 12:1075 (1990).

97. Cohen, B.S.; Susman, R.G.; Lippmann, M.: Ultrafine Particle Deposition in a Human Tracheobronchial Cast. Aerosol Sci. Technol. 12:1082–1091 (1990).

98. Pilacinski, W.; Ruuskanen, J.; Chen, C.C.; et al.: Size-Fractionating Aerosol Generator. Aerosol Sci. Technol. 13:450 (1990).

99. Kotrappa, P.; Moss, O.R.: Production of Relatively Monodisperse Aerosols for Inhalation Experiments by Aerosol Centrifugation. Health Phys. 21:531 (1971).

100. Chen, B.T.; Yeh, H.C.; Cheng, Y.S.: A Novel Virtual Impactor: Calibration and Use. J. Aerosol Sci. 16:343 (1985).

101. Spurny, K.R.; Lodge, J.P.: Radioactivity Labelled Aerosols. Atom. Environ. 2:429 (1968).

102. Thomas, R.G.: Retention Kinetics of Inhaled Fused Aluminosilicate Particles. In: Inhaled Particles III, pp. 193–200. Unwin Bros., Surrey, UK (1971).

Chapter 10

The Measurement Process: Precision, Accuracy, and Validity

John G. Watson, Ph.D.[A]; Barbara J. Turpin, Ph.D[B]; and Judith C. Chow, Sc.D.[C]

[A]*Atmospheric Sciences Division, Desert Research Institute, Reno, Nevada;*

[B]*Department of Environmental Sciences, Rutgers University, New Brunswick, New Jersey;*

[C]*Atmospheric Sciences Division, Desert Research Institute, Reno, Nevada*

CONTENTS

Introduction

The primary product of any air quality study is the creation of a centralized database of measurements collected to address one or more specific objectives. The initiation of a workplace or an ambient air quality study begins with the definition of study goals. These goals drive the study design process, including the placement of monitors, selection of monitoring methods, sampling duration and frequency, and allowable measurement tolerances. Each *measurement* in such a database is comprised of a value and its associated uncertainty. This uncertainty is described as the precision and accuracy of the value. While there are several mathematical definitions of *precision* (see Table 10-1), each describes the expected variation of repeated measurements of the same observable collected during the same measurement period. The degree of correctness with which a measurement system yields the true value is termed the *accuracy*.

In addition to measurement uncertainties, measurement detection limits and expected values of observables (e.g., concentrations) need to be considered when selecting analytical instrumentation and designing sampling strategies. The *detection limit* is expressed in a number of ways (e.g., lower detectable limit, lower quantifiable limit; see Table 10-1), each describing the signal strength needed to detect or quantify an observable.

Prior to the initiation of an air quality study, the measurement strategy is designed to ensure that the resulting data will be adequate to address the specific objectives of the study. Likewise, *allowable tolerance levels* for measurement uncertainties are established based on the expected use of the data. For example, measurement uncertainties must be much smaller for data that will be used in modeling activities than for data to be used for descriptive analyses. Because the data quality tolerance levels depend on the application, it is useful to report data with validation codes that provide

TABLE 10-1. Definitions

Acceptance Test: Analysis of a representative number of sampling containers or substrates from a batch prior to use to assure that concentrations of observables on the unused substrates are acceptable (i.e., lower than allowable tolerance levels).

Accuracy [A]: The degree of correctness with which a measurement system yields the true value of an observable. Specifically, the percent difference between the measured and true value (the "true" value is determined by Standard Reference Materials or the use of two or more independent procedures to measure the same observable).

$$A = 100 \ (C_m - C_t)/C_t \qquad \textbf{(T1)}$$

where: A = accuracy (%)
C_m = measured value
C_t = true value

Allowable tolerance levels: The maximum deviation from ideal conditions that is acceptable given the application. For example, the maximum detection limits, precision, and accuracy, also the maximum change in flow rate during sampling, value of an unused substrate that would be considered to pass the acceptance test, and the lowest validity code that would yield data capable of meeting the study objectives. Allowable tolerance levels must be selected to provide a database of sufficient quality to test the stated hypotheses.

Bias [K]: The ratio of the measured value to the true value;

$$K = \frac{C_m}{C_t} \qquad \textbf{(T2)}$$

Coefficient of Variation: The ratio of precision of a measurement (s_m) to the value of the measurement (C_m). The coefficient of variation is also called the relative precision. It is expressed as a unitless number or, when multiplied by 100, as a percent.

Confidence Interval: A statistical estimate of probability that a values lies within a certain number of precision intervals. The 95% confidence interval is often used and is defined as:

$$C_m \pm t_{0.025, \, n-1} \ (s_m/n^{1/2}) \qquad \textbf{(T3)}$$

where $t_{0.025, \, n-1}$ is the t-distribution value for n-1 degrees of freedom where the upper and lower tails each make up 2.5% of the distribution (95% of the t-distribution falls within $\pm \, t_{0.025, \, n-1}$). The t-distribution depends on number of measurements taken, n. As more measurements are taken, there is a higher probability that C_m accurately reflects the true value of the observable, and the 95% confidence interval decreases.

Data Qualification Statement: A summary of data set quality that evaluates measurements with respect to allowable tolerance levels. It specifies data completeness, values above lower quantifiable limits, relative precision as a function of concentration, accuracy as determined by performance audits, and numbers of data validation flags. The data qualification statement defines the level of signal in an environmental cause that is needed to exceed the noise of the measurements system; it should be used by data analysts to evaluate the extent to which sought relationships are real or are an artifact of the measurement process.

Dynamic blank: The concentration in a sampling container or substrate that is involved in all aspects of the sampling process except for the deliberate collection of the measured observable. Clean samples are sent to field locations and returned to the laboratory without sampling air.

Interference: A positive or negative alteration of the measured observable concentration due to the presence of another species.

Lower detectable limit [LDL]: The smallest quantity or concentration of a chemical species for which an analytical method will show a recognizable positive response. Frequently also called *detection limit* and defined to be equal to three times the standard deviation of the dynamic blank. (Sometimes reported as one times the standard deviation of the dynamic blank or the quantity that yields a signal that is just perceptible above an instrument's baseline.)

Lower quantifiable limit [LQL]: The smallest quantity or concentration of a chemical species that can be quantified in an environmental sample. Defined as ten times the standard deviation of the dynamic blank. (Sometimes reported as one standard deviation of the dynamic blank or equal to the LDL of the analytical method, whichever is higher.)

Measurement: The amount of an observable quantified at a particular location and time by a measurement method with its associated precision, accuracy, and validity notes.

Measurement method: Combination of equipment, reagents, and procedures that provide the value of a measurement. A measurement method is defined by its Standard Operating Procedure.

Measurement method validity: Identification of measurement method assumptions, quantification of effects of deviations from those assumptions, ascertainment that deviations are within reasonable tolerances for the specific application, and creation of procedures to quantify and minimize those deviations during a specific application.

Observable: A substance or property that can be quantified by a measurement method.

Performance Audit: Independent verification of data quality characteristics of precision, accuracy, and detection limits through the blind presentation of independent standards and other reference samples and substrates.

Performance Test: Periodic testing procedures designed to identify measurement bias. Performance tests include analysis of independent standards, replicate samples, and laboratory blanks.

Pollutant: The material or contaminant under investigation.

Precision [s_m]: The standard deviation of repeated measurements of the same observable with the same measurement method. When n periodic measurements are made using a transfer standard of known value, precision is defined as:

$$s_m = \{[\textstyle\sum_i (C_i - \text{Avg}(C_i))^2]/(n - 1)\}^{1/2} \qquad \textbf{(T4)}$$

where C_i is the ith measurement of observable C in response to the same concentration, and $\text{Avg}(C_i)$ is the average concentration of the n measurements of C_i.

TABLE 10-1. Definitions (continued)

Primary Standard: Well characterized and protected gases, liquids, or solids, with stable concentrations to which all other standards are traceable. Primary standards are usually created and maintained by international organizations such as the National Institute of Standards and Technology.

Sample blank: The unused medium used to transfer the ambient sample to the measurement method.

Sample validity: Evaluation of the extent to which procedures were followed, application of internal and external consistency tests, assignment of validity flags, and removal of invalid measurements from a data set.

Standard Operating Procedures (SOP): Complete description of the measurement process. SOPs include: 1) summary of measurement method, principles, expected accuracy and precision, and the assumptions for validity; 2) materials, equipment, reagents, and suppliers; 3) individuals responsible for performing each part of the procedure; 4) traceability path, primary standards or reference

materials, tolerances for transfer standards, and schedule for transfer standard verification; 5) start-up, routine, and shut-down operating procedures and an abbreviated checklist; 6) data forms; 7) routine maintenance schedules, maintenance procedures, and troubleshooting tips; 8) internal calibration and performance testing procedures and schedules; 9) external performance auditing schedules; and 10) references to relevant literature and related SOPs.

Systems Audits: Independent examination of all phases of measurement and data processing to determine that SOPs define a valid measurement method and that these procedures are implemented in practice.

Transfer Standards: Easily produced or commonly available gases, liquids, or solids that are traceable to primary standards and used for calibration, performance testing, and auditing.

Uncertainty: The combination of the uncorrected biases and precision.

---

enough information to identify subsets of data that are appropriate for a particular use (e.g., modeling or descriptive analyses).

These issues are discussed in detail below, in the larger framework of quality control and quality assurance activities[1–4] that are performed to ensure the development of a centralized database with specified accuracy and precision for the purpose of air quality assessment. This chapter presents methods to estimate measurement attributes and provides examples of how they are applied. Examples focus on measurements of PM_{10} and $PM_{2.5}$ (particles with aerodynamic diameters less than 10 and 2.5 µm, respectively) because the recent promulgation of National Ambient Air Quality Standards (NAAQS) for these quantities[5] calls for the application of new measurement methodologies, community-representative siting criteria, and spatial and temporal averaging to better represent exposure and health effects than previous NAAQS. The methods demonstrated can also be applied to *in situ* and laboratory measurements of other pollutants such as sulfur dioxide (SO_2), oxides of nitrogen (NO_x), carbon monoxide (CO), ozone (O_3), chemical constituents of PM_{10} and $PM_{2.5}$, organic gases, and air toxics.

Study Design

Air quality monitoring is conducted for a variety of reasons, including notification of an immediate health threat (alert), monitoring for compliance with regulations or permits (compliance), development of effective air quality management plans (e.g., source evaluation), and

mechanistic research. Thus, the study design or measurement attributes (i.e., monitor siting, measurement duration and frequency, analyses to be conducted, and analytical methods) are tailored to best meet the specific objectives of the study. Common monitoring purposes are:

- **Alerts:** Notification of concentrations approaching or exceeding levels that result in immediate and adverse health consequences and require that rapid decisions be made. These types of monitoring programs are developed for confined workspaces where a process leak or lack of ventilation may cause hazardous pollutant levels to escalate to the extent that workers must evacuate or don breathing devices. In polluted urban areas, pollution control agencies may issue public warnings to curtail people's exposure or to reduce emissions when ambient concentrations exceed pre-set thresholds.
- **Compliance:** Identification of regions or facilities that are out of compliance with ambient air quality standards (e.g., NAAQS), permitted emission rates (e.g., New Source Performance Standards), or workplace air quality standards (i.e., established by the Occupational Safety and Health Administration [OSHA]). Periodic and continuous monitoring is conducted to determine where and when these standards are exceeded. When standards are exceeded, remedial actions are taken.
- **Management:** Characterization of emissions in order to develop effective emissions control strategies. Source sampling results are used to quantify contributions from excessive emitters and as inputs to models that elucidate the contributions of various

source types to ambient pollutant concentrations. The protocol for this type of source sampling can differ substantially from the source sampling protocols dictated for evaluation of permit compliance.

- **Research:** Examination of the physical and chemical mechanisms by which air quality is degraded (atmospheric science), behaviors that influence pollutant exposure (exposure assessment), and the association between pollutant indicators and health consequences (epidemiology). Ultimately, atmospheric science research is used to improve models that predict the effect of emissions changes on air quality. Exposure studies are conducted to identify changes in behavior that will impact public or workplace health and to address specific issues related to epidemiological measurement errors. Epidemiological studies are used to identify hazards and, together with toxicological studies, are used to set ambient and workplace air quality standards.

The intended use of measurements for one or more of these monitoring objectives dictates the allowable tolerance levels. Measurement attributes that are acceptable for one of the objectives are not necessarily adequate for the others, even though the same observables are quantified. For alerts, large measurement uncertainties and detection limits can often be tolerated as long as these methods immediately indicate exceedance of a high concentration in the vicinity of a susceptible population. Compliance determination often requires long-term monitoring over many years, depending on the form and values of the standards. Since public policy, financial or criminal penalties, or litigation result from these measurements, they must be well documented and be collected following procedures prescribed by regulation. State-of-the-art instrumentation is typically used in intensive (e.g., 6 week) field experiments to further the understanding of pollutant formation, transport, and transformation. In contrast, long-term monitoring is needed for epidemiological studies.

The siting of monitors is determined by the size of the spatial scale that the measurements are intended to represent. General guidance is given below.

Monitoring Locations

Monitoring sites are selected to represent several spatial scales as defined below.[6] Distances indicate the diameter of a circle, or the length and width of a grid square, with a monitor at its center.

- **Collocated or Confined Space Scale (1 to 10 m):** Collocated monitors are intended to provide replicate measures of the same air. This spatial scale is also used to characterize air quality in confined workspaces where the monitor is placed within 10 m of a worker to represent the maximum tolerable exposure prior to sounding an alert. Collocated measurements should not differ by more than the precision of the monitoring method. Monitors are operated on collocated scales to evaluate the equivalence of different measurement methods and procedures and to characterize measurement precision. The distance between collocated samplers should be large enough that the air sampled by any of the devices is not affected by any of the other devices, but small enough that all devices obtain air containing the same pollutant concentrations.

- **Micro-scale (10 to 100 m):** Micro-scale monitors show significant differences between locations separated by 10 to 100 m. Differences in air quality on this scale often occur near sources, such as a busy roadway, construction site, vent, or short stack. A micro-scale zone of representation is primarily useful for identifying excessive source emissions and for workplace compliance monitoring.

- **Middle Scale (100 to 500 m):** Middle-scale monitors show significant differences between locations that are ~0.1 to 0.5 km apart. Monitors with middle-scale zones of representation are often source-oriented, used to determine the contributions from emitting activities with multiple, individual sources to nearby community exposures.

- **Neighborhood Scale (500 m to 4 km):** Neighborhood-scale monitors do not show significant differences with spacing of a few kilometers. This dimension is often the size of emissions and modeling grids used in large urban areas for source assessment. Sources affecting neighborhood-scale sites typically consist of small individual emitters, such as clean, paved, curbed roads, uncongested traffic with a small number of heavy-duty vehicles, or neighborhood use of residential heating devices such as fireplaces and wood stoves.

- **Urban Scale (4 to 100 km):** Urban-scale monitors show consistency among measurements with monitor separations of tens of kilometers. Samples represent a mixture of pollutants from many sources within an urban complex, including those from the smaller scales. Urban-scale sites are often located at higher elevations and away from highly traveled roads, industries, and residential heating. Monitors on the roofs of two- to four-story buildings in the urban core area often represent the urban scale well.

- **Regional-Scale Background (100 to 1,000 km):** Regional-scale background monitors show consistency

CHAPTER 10: The Measurement Process: Precision, Accuracy, and Validity

among measurements with separations of a few hundred kilometers. Regional concentrations are often more consistent for secondary pollutants, such as ozone, sulfate, and nitrate, than they are for primary emissions such as sulfur dioxide, oxides of nitrogen, carbon monoxide, and directly emitted particles. Naturally occurring substances as well as pollutants generated in urban and industrial areas more than 100 km away both contribute to the concentrations observed by these monitors. Regional-scale sites are best located in rural areas away from local sources, and at higher elevations.

- **Continental-Scale Background (1,000 to 10,000 km):** Continental-scale background monitors show little variation even when they are separated by more than 1,000 km. They are hundreds of kilometers from the nearest significant emitters. Although these sites measure a mixture of natural and diluted manmade source contributions, the manmade component is at its minimum expected concentration (i.e., anthropogenic background).
- **Global-Scale Background (>10,000 km):** Global-scale background monitors are intended to quantify concentrations transported between different continents as well as concentrations emitted directly from natural sources (i.e., oceans, volcanoes, and windblown dust) and formed from precursor emissions.

The zone of representation for a monitoring site is often not evident in the absence of measurements from nearby locations. Figure 10-1 illustrates the variability in monthly-average PM_{10} measured on various scales in central California during winter, 1995–96.[6, 7] The larger concentrations within the urban area represent samplers located near roadways or in neighborhoods with woodburning. Most of the sites complying with the urban-scale criteria denoted above showed similar concentrations. Regional concentrations were achieved within distances outside the urban area that are comparable to the dimensions of the city and are fairly similar even over large distances. The zone of representation is different for different pollutants. Carbon monoxide and oxides of nitrogen, for example, would show much more variability than PM_{10}, depending on the proximity of the samplers to nearby roadways.

Even on the smallest spatial scale, the air volume actually measured is a small fraction of the volume being characterized. Typical sample volumes over a 24-hr period range from 0.5 to 2,000 m^3. A moderate-size (30 × 30 km) city with a 500 m mixing depth contains nearly 5 × 10^{11} m^3 of air. A factory or office building might contain 300 to 100,000 m^3 of air which is exchanged at a rate of approximately one air change per hour.

Nevertheless, the number of gas molecules or suspended particles is typically large enough within the sampled fraction that statistical counting errors are negligible compared to other uncertainties.

Sample Duration

Sequential, one-hour average concentrations for compliance are most commonly measured with *in situ* monitors. Workplace alert monitoring usually requires short duration monitoring, on the order of 5 minutes to 8 hours, while long-term health studies frequently use data averaged over days to weeks. Twenty-four hour integrated measurements are commonly used for PM_{10} and $PM_{2.5}$ monitoring to overcome detection limits associated with gravimetric analysis of filter-collected samples. *In situ* particle measurement technologies [8–10] report hourly particle concentrations in some networks and provide some insight into the effects of averaging time on different statistical indicators.

Figure 10-2 compares average, standard deviation, minimum, and maximum PM_{10} concentrations over a one-year period using 1 hour, 24 hour, 1 week, 1 month, and 1 year averaging times. Current NAAQS specify 24-hour and annual averaging times.[5] The annual average is identical in all cases, but the maximum concentration and the standard deviation of values decrease dramatically with increasing averaging time. The highest hourly-average concentration (1,453 $\mu g/m^3$), which occurred during a windblown dust episode, is not well represented by the highest 24-hour average (318 $\mu g/m^3$) because this short-duration event is averaged with low concentrations over the remainder of the day.

Chow[8] observed that new NAAQS for PM are based on associations between health endpoints and measurements acquired from networks designed to determine compliance with previous NAAQS for PM. This practice has perpetuated 24-hour filter sampling. Such a sampling strategy will not enable identification of short-term concentrations that might be associated with acute health effects, and it will not provide the data needed for epidemiological studies of acute exposures. Short-duration (or continuous) measurements collocated with long-duration measurements for a limited-time provide a practical method for evaluating the extent to which longer duration measurements are achieving the desired objectives. It should also be noted that 24-hour integrated filter measurements, which take several days to process, are not suitable for public alerts. Real-time continuous monitoring provides the opportunity to warn the public of unhealthy concentrations in a

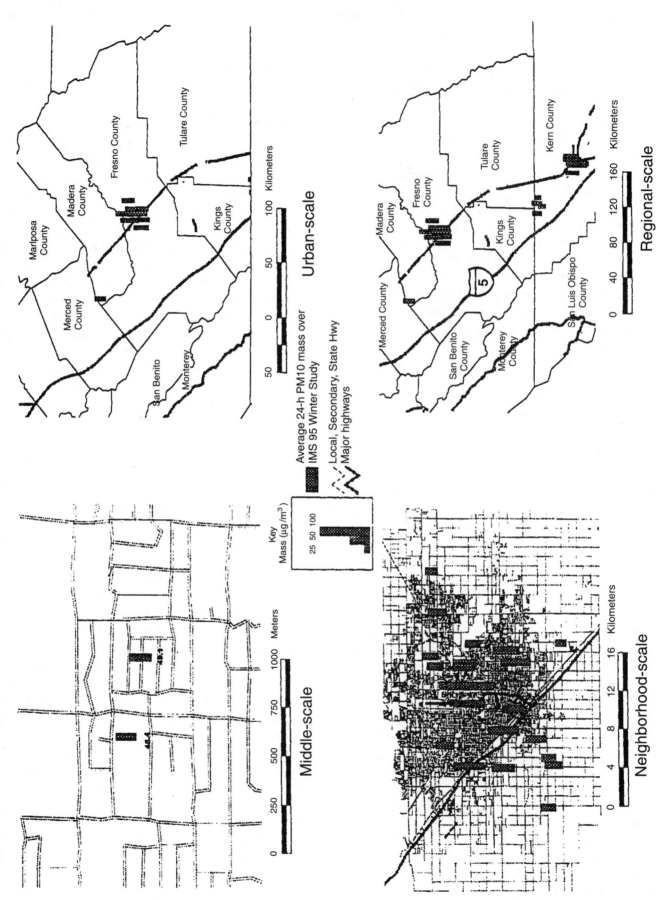

FIGURE 10-1. PM₁₀ concentrations at different nearby sites centered around Fresno, CA.

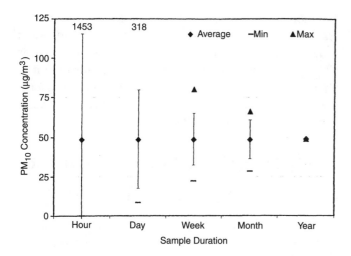

FIGURE 10-2. Changes in statistical indicators for PM$_{10}$ concentration with sample duration. (Data from Las Vegas, NV, 1995).

timely fashion so they can change their activities accordingly (i.e., to reduce emissions and exposure).

Sampling Frequency

Table 10-2 compares the annual average and standard deviation, highest, second highest, and several upper percentiles of 24-hour PM$_{10}$ concentrations for samples collected daily and at second, third, sixth, twelfth, and thirtieth day intervals. As can be seen in Table 10-2, the sampling frequency has little effect on the annual average (up to one in 12 days), but the sampling frequency has a substantial effect on measures of high concentration events (i.e., maximum, second maximum, 99th% and 98th%). Upper percentiles provide a more stable representation of high concentration events than the maximum measured concentration in cases when sampling is less frequent than every day, especially when more than 100 samples are collected. If maximum concentrations are desired, however, sampling frequencies must be continuous and complete. Extreme values

are rare events that do not obey standard statistical distributions.[11] These generalizations from Table 10-2 are consistent with those found from other locations and time periods.[12]

PM$_{10}$ compliance samples have typically been acquired every sixth day to minimize manpower. The sixth day schedule was originally selected so that sampling would regress through the week, preventing a bias due to weekly emissions variations. The PM$_{2.5}$ NAAQS requires sampling frequencies of every day to every sixth day, with more frequent sampling in more polluted communities. PM$_{10}$ and PM$_{2.5}$ NAAQS are expressed in terms of percentiles over 3-year intervals to take advantage of the added stability that this provides.

Measured Observables

Measurement technology and cost limit the specificity with which different observables can be quantified. For example, PM$_{10}$ mass concentrations are acquired because mass is an observable that can be practically measured. Unlike pure substances such as SO$_2$ or O$_3$, PM$_{10}$ consists of thousands of different compounds in an even greater variety of mixtures and sizes. Even the most advanced analytical techniques have characterized only ~10–20% of the organic compounds in PM$_{10}$.[13] These detailed methods are costly and require large sample volumes, necessitating long averaging times and limited sampling frequency. This illustrates some of the considerations that are made when sampling and analytical instrumentation is selected for a field campaign.

Light scattering, measured with a nephelometer,[14] is often used as a surrogate for particle measurements because it is correlated with PM$_{2.5}$ concentrations under certain conditions.[15] Nephelometers can be portable and inexpensive, making them useful for defining zones of representation (Figure 10-1) and for human exposure monitoring.[16] The lower precision and accuracy of

TABLE 10-2. Effects of sampling frequency on statistical indicators for 24-hour PM$_{10}$ concentrations (Data from Las Vegas, NV, 1995)

Sampling Frequency	Annual Average	Standard Deviation	Minimum	Maximum	Second Maximum	99th%	98th%	95th%	90th%
Every Day	49	31	8	318	219	176	128	92	76
Second Day	48	25	8	186	128	128	115	86	76
Third Day	48	29	11	219	136	128	104	92	77
Sixth Day	48	34	11	219	136	135	104	92	81
Twelfth Day	50	29	14	136	104	104	104	92	91
Thirtieth Day	58	56	11	219	91	91	91	91	91

$PM_{2.5}$ or PM_{10} mass concentrations estimated from nephelometer measurements may be acceptable, given the large spatial and temporal variability that can be characterized from their deployment.

A screening strategy is often used to identify a subset of samples for detailed analysis. For example, $PM_{2.5}$ mass concentration measurements can be used to identify a subset of samples for analyses of PM constituents. Frequently, this approach is used to select samples from high pollution episodes for further analysis. Understanding the composition and origin of pollutants during high pollution episodes assists in management of such episodes. However, as a result of this process, long-term averages of PM constituents constructed from analyses of selected samples tend to be positively biased and may not accurately reflect the composition, and thus the mixture of sources, that contributes to the annual average concentration.

Measurement Methods

Several choices of measurement methods often exist for the same observable. The accuracy, precision, detection limits, averaging time, labor, physical size, and cost all play a part in the method selection process. Different Standard Operating Procedures (SOP) for the same method and different applications of the same SOP may result in measurement discrepancies. Table 10-1 distinguishes between "measurement method validity" to evaluate the measurement methodology and "sample validity" to evaluate the data acquired by the method.

Interferences, or sampling artifacts, are the most common causes of measurement bias. Figure 10-3 shows how PM_{10} concentrations measured with a continuous Tapered Element Oscillating Microbalance (TEOM) compare with concentrations measured using a collocated filter sampler at two California locations. TEOM data in Figure 10-3 were collected with the sample compartment heated to 50 °C. During winter when ammonium nitrate was a substantial fraction of the mass and the ambient-sampler temperature differential was greatest, the TEOM consistently reported lower PM_{10} concentrations than the filter sampler. During summer when the temperature differential was small and stable particle species dominated, TEOM PM_{10} concentrations were in good agreement with filter-collected PM_{10}. In this example, volatilization of less stable PM constituents, specifically ammonium nitrate and more volatile particulate organic material, resulted in significant bias in wintertime TEOM measurements.[17] This bias is found in some locations and seasons and not in others, due to seasonal and spatial differences in particle composition. Other

common interferences that have been documented to result in biased measurements include: 1) inadvertent measurement of nitric acid and peroxyacetyl nitrate as NO_2 by chemiluminescent monitors;[18] 2) adsorption of organic gases on quartz-fiber filters;[19–21] and 3) adsorption of sulfur- and nitrogen-containing gases on excessively acidic or alkaline filters.[22–24]

Quality Control in Air Quality Measurement

Quality Control/Quality Assurance

The process by which the quality of the data is ensured is called *quality control*. Quality control is the responsibility of the laboratory conducting the work. Quality control tasks include: 1) efforts made to ensure the quality of the measurements are within established tolerance levels; 2) measurements and analytical performance checks needed to characterize data quality (i.e., detection limits, precision and accuracy); and 3) verification of comparability with related measurements and between operators/analysts. *Quality assurance* is the responsibility of an independent auditor. The tasks of the auditor are designed to address the following questions: Are the documented procedures complete and accurate? Are performance checks adequate so that a problem with the instrument will not go undetected? Are there laboratory/field practices that might adversely affect the quality of the data? Do the reported measures of data quality hold up to scrutiny? To address these questions, the auditor identifies deviations from standard operating procedures, evaluates laboratory quality control procedures, looks for gaps in sampling and analytical performance checks, observes sample handling, and looks for sources of contamination. In addition, the auditor often conducts a performance audit. Standards, blanks, spiked samples, and sometimes previously analyzed samples are presented to the laboratory for blind analysis. The auditor uses these results to independently assess measurement accuracy, precision, and detection limits. These values are compared to the laboratory-reported values.

Quality control activities begin with the design of the measurement strategy and continue through the final validation of the database. These activities include: 1) the development of standard operating procedures with comprehensive measurement performance checks; 2) the inclusion of measurements for the estimation of accuracy, precision, and detection limits (i.e., to characterize data quality); 3) documentation of field and analytical activities; and 4) data validation. *Standard operating procedures* (SOP) are detailed documents delineating

Collocated Comparison
SSI 1 and TEOM (Winter/Fall)

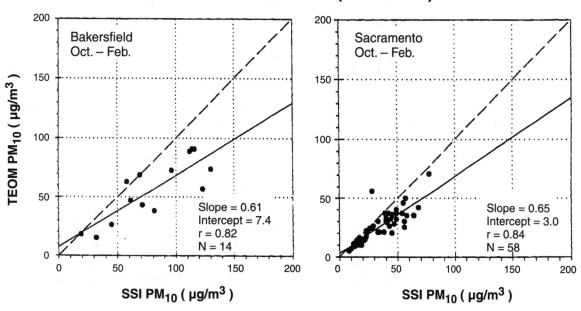

Collocated Comparison
SSI 1 and TEOM (Summer/Spring)

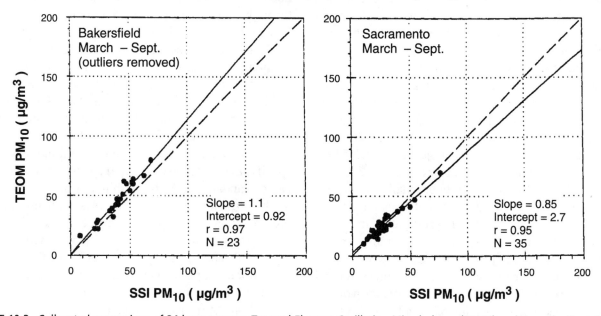

FIGURE 10-3. Collocated comparison of 24-hour-average Tapered Element Oscillating Microbalance (TEOM) and size-selective-inlet (SSI) PM_{10} during winter and summer at the Bakersfield and Sacramento monitoring sites in Central CA, 1993.

step-by-step the procedures to be followed for sample collection, analysis, substrate preparation, or other activity critical to the success of the study. The auditor expects that the SOP will accurately reflect laboratory practice and will be located where the measurement is being performed. For example, a SOP for thermal-optical carbon analysis should be found with the thermal-optical carbon analyzer.

PART I: The Measurement Process

Performance Checks for Standard Operating Procedures

During the development of a SOP, one must attempt to identify all possible sources of bias and incorporate adequate *performance checks* to ensure that, were measurements errors to occur, they would be identified. One critical performance check in air sampling is a leak check. A successful leak check of the sampling train before and after sampling ensures that the volume of air pulled through the sample collection device is the same as the volume of air pulled through the flow meter. To illustrate the importance of the leak check, consider a $PM_{2.5}$ filter sampler (inlet, filter holder, volume-flow meter, vacuum gage, and vacuum pump) that has a leak between the filter holder and flow meter downstream. The pump will pull air through the filter and through the leak. Thus, the volume of air measured by the flow meter will be greater than the volume pulled through the filter, and the $PM_{2.5}$ mass concentration (mass/volume) will be biased low. Flow calibrations and flow audits are also critical to the accurate collection of air samples. Ideally, air sampling equipment is calibrated (multi-point) in the configuration in which it will be used in the field, and the sampler flow rate is validated with an independent meter in the field.

For integrated samples, performance checks are also needed to ensure that the quality of substrates and samples is maintained and monitored during transit and storage. Optimal sample handling procedures depend on the specific analyses that will be conducted. For example, filter samples collected for analysis of organic particulate matter are typically collected on quartz-fiber filters. The filters must be baked prior to use to lower blank levels, and samples must be stored and archived cold to prevent volatile losses. Subjecting baked filters (some from each batch) to an *acceptance test* (i.e., requiring blank substrate values to be smaller than a certain value) is useful in ensuring that allowable tolerance levels for measurement uncertainties are met. Temperature monitors transported with samples can be used to determine whether or not samples were kept cold. This is an issue of importance to the measurement of organic particulate matter and other semi-volatile compounds because the heating of samples (e.g., on the hot express-mail airport runway) could result in substantial volatile losses. If appropriate *positive controls* can be produced (e.g., substrates spiked with a known amount of the species to be measured), they are extremely valuable in identifying volatile losses during sample transport and storage. This practice is frequently applied to the measurement of gaseous compounds.

Likewise, contamination of samples in transit and storage can be identified through the use of dynamic blanks. *Dynamic blanks* (or *field blanks*) are substrates that are prepared with the sample substrates and kept with them through analysis. To the greatest extent possible, these substrates are handled identically to sample substrates with the exception of actual sample collection. Dynamic blanks (and positive controls when available) should accompany each batch of substrates from substrate preparation, to the field and back, and through analysis. In addition to detecting sample contamination in transit or storage, these substrates are used to determine measurement detection limits.

Additional performance checks are designed for each specific analytical system to ensure that every component of the system is working correctly. One common performance check is the analysis of *independent standards,* standards that are independent of those used for calibration. Failure to include this independent check of accuracy can be devastating. For example, consider the laboratory that calibrates its carbon analyzer with an aqueous-sucrose solution which is being slowly degraded by microorganisms. Alternatively, consider the slow accumulation of mass to a 200 mg $PM_{2.5}$ calibration weight through soiling or loss of mass due to scratching. The bias that this introduces to the measurements would be easily identified through the analysis of independent standards. However, without this performance check, the problem could go undetected indefinitely, until new standards are prepared or until an audit. An additional tool used to address this issue is the setting of allowable tolerance levels on the response to standards. The SOP for carbon analysis might state that "the instrument response to the injection of a specified mass of carbon must be within 2% of the previous day's response," and the SOP might specify that "the instrument response to freshly prepared standards must be within 2% of previous standards." *Instrument blanks,* analyses with no substrate present, are also useful performance checks. Table 10-3 summarizes commonly measured atmospheric observables with appropriate primary and transfer standards for calibration, performance tests, and quality auditing. Methods for presenting these standards to instruments depend on the instrument audited.

Adequate standards are not available for every observable. For examples, different particulate carbon analysis methods often return comparable values for total carbon, but segregate total carbon differently into its "organic" and "elemental" fractions. [25–28] Since no absolute standards exist for these fractions, they are operationally defined. [28]

TABLE 10-3. Primary standards and transfer standards with recommended calibration, performance test, and performance audit frequencies for commonly measured observables in ambient air

Observable (Method)	Allowable Tolerance	Primary Standard	Calibration Standard	Calibration Frequency	Performance Test Standard	Performance Test Frequency	Performance Audit Standard	Performance Audit Frequency
NO/NO$_x$ (chemiluminescence)	±10%	NIST-traceable NO mixture	Certified NO mixture and dynamic dilution	Quarterly or when out of spec	Span with certified NO and zero with scrubbed air	Daily	Certified NO mixture and dynamic dilution	Yearly
O$_3$ (ultraviolet absorption)	±10%	Primary UV Photometer	Dasibi 1003H UV photometer	Quarterly or when out of spec	Span with internal ozone generator and zero with scrubbed air	Daily	Dasibi 1008 with temperature and pressure adjustments	Yearly
CO (infrared absorption)	±10%	NIST-traceable CO mixture	Certified CO mixture and dynamic dilution	Quarterly or when out of spec	Span with certified CO and zero with scrubbed air	Daily	Certified CO mixture and dynamic dilution	Yearly
SO$_2$ (pulsed fluorescence)	±10%	NIST-traceable HC mixture	Certified HC gas dilution	Quarterly or when out of spec	Span with certified HC and zero with scrubbed air	Daily	Certified HC gas dilution	Yearly
PM$_{10}$ and PM$_{2.5}$ flow rate (high volume filter sampler)	±5%	Spirometer (>1,000 L/min)	Calibrated orifice/rootsmeter	Quarterly	Calibrated orifice	Monthly	Calibrated orifice/rootsmeter	Yearly
PM$_{2.5}$ and PM$_{10}$ mass and chemistry (low volume filter sampler)	±5%	NIST-certified bubblemeter (1-25 L/min)	Mass flowmeter/bubblemeter	Quarterly	Calibrated bubblemeter	Monthly	Mass flowmeter	Yearly
Light scattering (nephelometer)	±10%	SUVA 134a refrigerant	Clean air/SUVA 134a refrigerant	Monthly	Clean air/SUVA 134a refrigerant	Weekly	Clean air/SUVA 134a refrigerant	Yearly
Light absorption (aethalometer)	±5%	Neutral density filter	Neutral density filter	Quarterly	Neutral Density filter	Weekly	Neutral Density filter	Yearly
Mass (electrobalance)	±10%	NIST Class 1.1 weights	NIST Class 1.1 weights	3 months	NIST weights	10 samples	NIST Class 1.1 weights	Yearly
Total elements (X-Ray Fluorescence)	±5%	EPA polymer films, NIST impregnated glass	Micromatter film deposits	6 months or when out of spec	Multi-element impregnated glass	15 samples	Micromatter film deposits	Yearly
Anions and cations (Ion Chromatography)	±5% µg/mL	Mineral salt solutions	Salt solution	100 samples	Mixed salt solution and distilled water	10 samples	Mixed salt solution	Yearly
Carbon (Thermal/Optical)	±0.2 µg/cm²	NIST CO$_2$ and CH$_4$	Pthalate and sucrose solutions	3 months or when out of spec	Methane	Every sample	Pthalate and sucrose solutions on filters CO$_2$ and CH$_4$	Yearly

Data Quality Characterization

Essential data quality measures are built into the measurement strategy in the form of dynamic blanks, independent standards, and collocated samplers (or replicate analyses). Measurement detection limits are proportional to the variability of the blank signal, and they are usually expressed as one to three times the standard deviation of the instrument baseline. Lower quantifiable limits for field study measurements are determined based on the variability of field blank values and that these are sometimes larger than the analytical detection limits. A number of other means of expressing detection limits are presented in Table 10-1. These include terms like *lower detectable limit* (LDL), which is defined as the smallest quantity of a chemical species for which the analytical method shows a positive response, and *lower quantifiable limit*, the smallest quantity that can be quantified by the measurement process. However, because a number of different definitions are in active use, it is important to indicate how detection limits were estimated when reporting data. Measurements below detection limits must be labeled appropriately (e.g., "nd" for non-detect).

The measurement accuracy (A) is usually expressed as the percent difference between the "true" (C_t) and measured (C_m) values of an observable as expressed by Equation T4 (see page 202). If the true value falls outside the range defined by the precision of the measurement, the measurement is considered *biased*. Multiple analyses of independent standards, not used for calibration, are used to determine analytical accuracy, with the equation above. The overall measurement accuracy includes all sources of measurement bias. For example, sampling artifacts (i.e., adsorption of gases on the sampling filter, volatilization of material from the filter, or chemical reactions altering collected materials) can result in substantial bias in the measurement of particulate organic carbon, nitrate, and other multiphase species. Care must be taken to minimize sampling artifacts, and their contribution to measurement bias should be acknowledged in the reporting of data at least qualitatively.

Precision describes the variation of repeated measurements of the same observable with the same measurement method. Precision is usually expressed as plus or minus one standard deviation ($\pm 1s$), or as a percentage using a coefficient of variation, or as a 95% confidence interval. For example, a measurement of $PM_{2.5}$ mass concentration might be written 63 ± 3 $\mu g/m^3$ where 3 $\mu g/m^3$ is one standard deviation, or 63 $\mu g/m^3 \pm 5\%$ where 5% is the coefficient of variation, or 63 ± 6 $\mu g/m^3$ where 6 $\mu g/m^3$ is the 95% confidence interval.

Again, it is important to note how precision was calculated when reporting data. In the rare case where many measurements were made of the same observable, C_m, the average of these measurements represents the value of the observable and the precision expressed as a standard deviation is computed according to Equation T4. The denominator is $(n-1)$ rather than n, since the calculation of the average concentration reduces the degrees of freedom by one. The precision expressed as a coefficient of variation (*CV*) is:

$$CV\,(\%) = 100 s_m/C_m \qquad (1)$$

It is also possible to define the precision based on the *95% confidence interval,* which defines a range of values within which the true value of the observable will fall with 95% confidence as expressed by Equation T2.

A t-table is provided in basic statistics books. Note that the student's t-distribution takes into consideration the number of measurements taken. As more measurements are taken, there is a higher probability that the average accurately reflects the true value of the observable, and therefore the 95% confidence intervals become narrower.

It is generally impractical in air quality studies to make many measurements of each observable. In some cases, overall measurement precision is obtained from side-by-side samplers operated concurrently. Otherwise, overall measurement precision can be estimated by propagation of error, which considers uncertainties in flow rates, analytical uncertainties, and other sources of error. Analytical precision is typically determined from replicate analysis of more than 10% of the samples. Using these data, the pooled standard deviation or pooled coefficient of variation is calculated. Given *k* pairs of replicates where each pair of replicates has a standard deviation of s_j, the pooled standard deviation is:

$$s_{pooled} = [(\Sigma_j s_j^2)/k]^{1/2} \qquad (2)$$

and the pooled coefficient of variation is:

$$CV_{pooled}\,(\%) = 100 s_{pooled}/C_{pairs}, \qquad (3)$$

where C_{pairs} is the average value of the replicate pairs. When the analytical precision and other major measurement uncertainties (i.e., flow volume) have been estimated, the overall measurement precision can be estimated by propagation of error. In the simplest case, where the quantity of interest, *C*, is the sum of several independent variables,

$$C = x_1 + x_2 + x_3 + \ldots + x_n, \qquad (4)$$

the precision of C is given by:

$$s_C = (s_1^2 + s_2^2 + s_3^2 + \ldots + s_n^2)^{1/2}, \qquad (5)$$

where s_i is the precision of variable x_i. For the general case, where $C = f(x_1, x_2, x_3, \ldots x_n)$,

$$s_C \approx \{s_1^2(dC/dx_1)^2 + s_2^2(dC/dx_2)^2 + \\ \ldots + s_n^2(dC/dx_n)^2\}^{1/2}. \qquad (6)$$

When propagation of error techniques are applied to precision estimates for filter-based $PM_{2.5}$ measurements, for example, the concentration, C, is given by:

$$C = m/v \qquad (7)$$

where m is the mass of particulate matter, v is sample volume, and s_m and s_v are the precision of m and v, respectively. The overall precision of C is:

$$s_C \approx \{(s_m/v)^2 + (ms_v/v^2)^2\}^{1/2}. \qquad (8)$$

Note, if a blank subtraction was performed, the equation for concentration becomes $C=(m_s-m_b)/v$ and the uncertainty calculated by propagation of error takes the uncertainty of the blank into consideration as well. When the blank value is less than one to three times the standard deviation of the blank, it can no longer be distinguished from zero. The inclusion of blank uncertainties in the propagation of error causes measurement uncertainty estimates (coefficient of variation) to increase as measured values approach detection limits. At measured concentrations, less than approximately five times the standard deviation of the blank, the precision of the blank dominates the measurement precision. At larger measured concentrations, the precision of the instrument calibration span dominates the measurement precision. The contribution of blank uncertainties to overall measurement precision is typically negligible when the instrument reading exceeds the blank value by a factor of ten or more.

Documentation

In addition to developing standard operating procedures with comprehensive performance checks, and inclusion of measurements to characterize data quality, the documentation of field and laboratory activities is important to the development of a high quality database. Documentation of field activities frequently takes place through the development and use of a standard field log. The field log prompts the field technician to record all critical sampling parameters and includes a column for comments. For integrated samples, the sample identification code, sampling start and stop times, start and stop flow rates, pressure gage readings, leak test results, and comments are recorded in the field log. A chain-of-custody form is another common tool used to track samples, avoid sample mix-ups, and identify affected samples should contamination occur. Chain-of-custody forms document the history of sample location and sample handling with a form that travels with the sample and is signed and dated each time a sample is handled. Documentation extends to sample analysis with the use of an analysis logbook. This documents, in ink, everything that is done with the analyzer in sequential order. Operator name, date, samples analyzed (by sample identification number), calibrations, leak tests, gas cylinder changes, instrument maintenance and repair, and other interactions with the instrument are recorded. This is an essential diagnostic tool. If an analytical problem is detected that affects the accuracy of the analyses (i.e., through the analytical performance checks), the analysis logbook facilitates the troubleshooting process to determine the cause of the problem, when the problem began, and which analyses were affected. This diagnostic process and conclusions are also recorded in the analysis log, the appropriate samples are re-analyzed, and the affected analyses are invalidated. It should be noted that all analyses, whether valid or invalid, are accounted for in this record. While it is necessary to note analyzer maintenance and repairs in the analysis logbook, many find it useful to also utilize a separate maintenance logbook to document in more detail maintenance and repairs to the instrument. This provides a forum for the documentation of systematic trouble-shooting and diagnostic performance testing of the instrument.

The database into which measurement values, uncertainties, and validity codes are entered must also contain documentation to indicate the definitions used for data quality characterization, sampling and analysis information, and codes used in the database tables (e.g., definitions of validation flags). The following types of information should be included in a project database:

- **Measurement locations:** Each measurement location is identified with a unique alphanumeric site ID accompanied by location (e.g., address, coordinates, elevation), its primary operator, and a summary of the types of measurements collected. Coordinates can be determined with the Geographical Positioning System using map basis NAD-83 (Federal Aviation Administration convention).

- **Variable definitions:** Each variable is assigned a unique code that is accompanied by its definition, units, averaging time, measurement method, applicable temperature and pressure adjustments, and data reporting format. Specifications are needed for: 1) daylight or standard time zones; 2) hour beginning or ending sample duration labels; 3) actual or standard temperatures and pressures for sample volumes; 4) missing value codes; and 5) different measurement methods used for the same variable.

- **Data validation flags:** Flags are defined that specify the validation level as well as specific deviations from SOPs.
- **Data:** Value, precision, accuracy, and validation level are provided for each sample. Missing or invalid measurements contain a "NULL" value or other missing data indicator. Separate tables are produced for different averaging times and for non-uniform data sets.
- **Validation tables:** Detailed information on specific samples indicating the nature of the data qualification.

Sample Validation

The final step of the quality control process is data validation. The first step is to examine field log, analysis log, and chain-of-custody forms to identify data that are invalid or suspect. The transcription of data from written log sheets to an electronic database also provides opportunities for database errors. It is useful to minimize data transcription, for example, by creating bar codes for sample identification numbers and direct electronic storage of analytical results. When results must be input manually by one individual, verification is usually more effective when done by another. Input strategies can also be developed that assist with verification. For example, thermal-optical carbon analysis yields organic carbon (OC), elemental carbon (EC), and total carbon (TC) loadings, where TC is the sum of OC and EC. Rather than verify that each OC and EC value was entered correctly by checking against a paper report, one can enter OC, EC, and TC, compute OC+EC in the column next to TC, and verify that TC is equal to OC+EC by checking each pair of adjacent numbers. Finally, graphical comparisons of data with measurements of the same observables made by other methods or measurements of related parameters add confidence in the validity of the database. Similarly, mass balance techniques can be used. For example, the sum of measured $PM_{2.5}$ species can be compared with measured particle mass. These methods can be used to identify outliers.

Data validity levels are designated in the validation tables for different stages of data acquisition and interpretation.

- **Level 0:** Data sets downloaded from a field instrument prior to examination. These measurements are used to evaluate instrument performance and to forecast conditions for special experiments. Level 0 data are not used for interpretive purposes.
- **Level 1:** Data evaluated by the measurement investigator prior to submission to the database. Values have been removed for instrument downtime and performance tests, adjustments for calibration deviations have been applied, extreme values were investigated, internal comparisons were made, blanks were subtracted, precision was estimated and propagated, and appropriate data qualification flags have been assigned. For sequential measurements, jump tests, standard deviation tests, and extreme value tests often identify values that need to be investigated. Jump tests evaluate the difference between sequential values against maximum expected changes. For example, ozone typically follows a smoothly varying diurnal pattern. A large increase or decrease between one hour and the next may indicate an instrument malfunction. Standard deviation tests compare the measured and expected standard deviation of a series of sequential measurements. When the standard deviation is zero, the instrument is probably not responding to environmental changes. Data loggers often carry more significant digits than those included in the final database so that this test can be performed.
- **Level 2:** Intercomparison tests between data sets have been completed. These tests often result in the investigation of several samples that do not follow the same pattern as other measurements (i.e., outliers). For example, typical tests for particle mass concentration data include: 1) comparison of collocated $PM_{2.5}$ and PM_{10} measurements to verify that $PM_{2.5} > PM_{10}$; 2) comparison of the sum of $PM_{2.5}$ or PM_{10} chemical components with mass concentrations (i.e., mass balance) to verify that the sum of the species is less than or equal to the gravimetric mass and quantify the unknown mass fraction; 3) balances of anions and cations; and 4) comparison of collocated measurements. These tests often identify outliers that warrant further investigation. Documentation pertaining to discrepancies are investigated, sometimes re-analyzed, and re-designated as valid, invalid, or suspect as a result of the investigation.

- **Level 3:** Values that are found to be contradictory to other values have been investigated. These measurements are used to test hypotheses. The quality of these measurements is especially important as they often indicate large deviations from conventional wisdom that should not be confused with measurement error. The first assumption upon finding a measurement inconsistent with physical expectations is that the unusual value is due to a measurement error. If, upon tracing the path of the measurement, nothing unusual is found, the value can be assumed to be a valid. Values that warrant further investigation before level 3 validation include: 1) extreme values; 2) values that do not track the values of other variables in a time series where covariance is expected; and 3) values deviating from a qualitatively predictable spatial or temporal pattern.

Summary and Conclusion

Quality control processes beginning with the study design phase and ending with the final database validation are discussed above. This chapter has presented definitions, procedures, and calculation methods for estimating the accuracy, precision, and validity of ambient and workplace air quality measurements. No environmental measurements can be considered complete without these attributes. A number of air quality models that relate ambient concentrations to their emissions sources now make use of uncertainty estimates.[29–31] There is a need to extend such error propagation to more predictive models of environmental concentrations and occupational exposure.

The assembly of an air quality database, with specified accuracy and precision, is hopefully only the first step toward an improved scientific understanding. However, it is essential to the success of the data interpretation activities that will follow. Because air quality studies frequently form the basis of regulatory actions and sometimes yield scientific knowledge that guides national and international policy, the importance of getting this first step right cannot be overlooked.

References

1. Hidy, G.M.: Jekyll Island Meeting Report: The Acquisition of Reliable Atmospheric Data. Environ. Sci. Technol. 19: 1032-1033 (1985).
2. Hook, L.A.; Cheng, M.D.; Boden, T.A.: NARSTO Quality Planning Handbook. Report No. 4786, prepared for U.S. Dept. of Energy, Office of Biological and Environmental Research, Environmental Sciences Division, North American Research Strategy for Tropospheric Ozone, Quality Systems Science Center. Oak Ridge National Laboratory, Oak Ridge, TN (1998).
3. Mueller, P.K.: Comments on the Advances in the Analysis of Air Contaminants. J. Air Pollut. Control Assoc. 30:988–990 (1980).
4. U.S. Environmental Protection Agency: Quality Assurance Handbook for Air Pollution Measurement Systems. Volume 2, Part I: Ambient Air Quality Monitoring Program Quality System Development. Report No. EPA-454/R-98-004, U.S. EPA, Research Triangle Park, NC (1998).
5. U.S. Environmental Protection Agency: National Ambient Air Quality Standards for Particulate Matter: Final Rule. Federal Register 62: 38651–38760 (1997).
6. Watson, J.G.; Chow, J.C.; DuBois, D.W.; *et al:* Guidance for Network Design and Optimal Site Exposure for $PM_{2.5}$ and PM_{10}. Prepared for U.S. EPA, Office of Air Quality Planning and Standards, Research Triangle Park, NC. Desert Research Institute, Reno, NV (1997).
7. Chow, J.C.; Egami, R.T.: San Joaquin Valley Integrated Monitoring Study: Documentation, Evaluation, and Descriptive Analysis of PM_{10} and $PM_{2.5}$, and Precursor Gas Measurements. Technical Support Studies No. 4 and No. 8 — Final report. Prepared for California Regional Particulate Air Quality Study, California Air Resources Board, Sacramento, CA. Desert Research Institute, Reno, NV (1997).
8. Chow, J.C.: Critical Review: Measurement Methods to Determine Compliance with Ambient Air Quality Standards for Suspended Particles. J. Air Waste Manag. Assoc. 45:320–382 (1995).
9. Klein, F.; Ranty, C.; Sowa, L.: New Examinations of the Validity of The Principle of Beta Radiation Absorption for Determinations of Ambient Air Dust Concentrations. J. Aerosol Sci. 15:391–394 (1984).
10. Patashnick, H.: On-line, Real-time Instrumentation for Diesel Particulate Testing. Diesel Prog. N. Amer. 53:43–44 (1987).
11. deNevers, N.; Lee, K.W.; Frank, N.H.: Extreme Values In TSP Distribution Functions. J. Air Pollution Control Assoc. 27:995–1000 (1977).
12. Watson, J.G.; Chow, J.C.; Shah, J.J.: Analysis of Inhalable and Fine Particulate Matter Measurements. Report No. EPA-450/4-81-035, prepared for Environmental Research and Technology, Inc. U.S. Environmental Protection Agency, Research Triangle Park, NC (1981).
13. Mazurek, M.A.; Simoneit, B.R.T.; Cass, G.R.; Gray, H.A.: Quantitative High-resolution Gas Chromatography and High-resolution Gas Chromatography\Mass Spectrometry Analyses of Carbonaceous Fine Aerosol Particles. Int. J. Environ. Anal. Chem. 29:119–139 (1987).
14. Heintzenberg, J.; Charlson, R.J.: Design and Application of the Integrating Nephelometer: A Review. J. Atmos. Oceanic Technol. 13:987–1000 (1996).
15. Zhang, X.Q.; Turpin, B.J.; McMurry, P.H.; *et al.*: Mie Theory Evaluation of Species Contributions to 1990 Wintertime Visibility Reduction in the Grand Canyon. J. Air Waste Manag. Assoc. 44:153–162 (1994).
16. Anuszewski, J.; Larson, T.V.; Koenig, J.Q.: Simultaneous Indoor and Outdoor Particle Light-Scattering Measurements at Nine Homes Using a Portable Nephelometer. J. Expo. Anal. Environ. Epidemiol. 8:483-494 (1998).
17. Allen, G.A.; Sioutas, C.; Koutrakis, P.; *et al.*: Evaluation of the TEOM Method for Measurement of Ambient Particulate Mass in Urban Areas. J. Air Waste Manag. Assoc. 47:682–689 (1997).
18. Winer, A.M.; Peters, J.W.; Smith, J.P.; Pitts, J.N., Jr.: Response of Commercial Chemiluminescence $NO-NO_2$ Analyzers to Other Nitrogen-Containing Compounds. Environ. Sci. Technol. 8:1118–1121 (1974).
19. McDow, S.R.; Huntzicker, J.J.: Vapor Adsorption Artifact in the Sampling of Organic Aerosol: Face Velocity Effects. Atmos. Environ. 24A:2563–2571 (1990).

20. Coutant, R.W.; Brown, L.L.; Chuang, J.C.; *et al.*: Phase Distribution and Artifact Formation in Ambient Air Sampling for Polynuclear Aromatic Hydrocarbons. Atmos. Environ. 22:403–409 (1988).

21. Turpin, B.J.; Huntzicker, J.J.; Hering, S.V.: Investigation of Organic Aerosol Sampling Artifacts in the Los Angeles Basin. Atmos. Environ. 28:3061–3071 (1994).

22. Coutant, R.W.: Effect of Environmental Variables on Collection of Atmospheric Sulfate. Environ. Sci. Technol. 11:873–878 (1977).

23. Spicer, C.W. and Schumacher, P.M.: Particulate Nitrate: Laboratory and Field Studies of Major Sampling Interferences. Atmos. Environ. 13:543–552 (1979).

24. Spicer, C.W.; Schumacher, P.M.: Interference in Sampling Atmospheric Particulate Nitrate. Atmos. Environ. 11:873–876 (1977).

25. Turpin, B.J.; Cary, R.A.; Huntzicker, J.J.: An *In-Situ*, Time-resolved Analyzer for Aerosol Organic and Elemental Carbon. Aerosol Sci. Technol. 12:161–171 (1990).

26. Turpin, B.J.; Huntzicker, J.J.; Adams, K.M.: Intercomparison of Photoacoustic and Thermal-Optical Methods for the Measurement of Atmospheric Elemental Carbon. Atmos. Environ. 24A:1831–1835 (1990).

27. Chow, J.C.; Watson, J.G.; Pritchett, L.C.; *et al.:* The DRI Thermal/Optical Reflectance Carbon Analysis System: Description, Evaluation and Applications in U.S. Air Quality Studies. Atmos. Environ. 27A:1185–1201 (1993).

28. Hering, S.V.; Appel, B.R.; Cheng, W.; *et al.:* Comparison of Sampling Methods for Carbonaceous Aerosols in Ambient Air. Aerosol Sci. Technol. 12:200–213 (1990).

29. Irwin, J.S.; Rao, S.T.; Petersen, W.B.; Turner, D.B.: Relating Error Bounds for Maximum Concentration Estimates to Diffusion Meteorology Uncertainty. Atmos. Environ. 21:1927–1937 (1987).

30. Freeman, D.L.; Egami, R.T.; Robinson, N.F.; Watson, J.G.: A Method for Propagating Measurement Uncertainties through Dispersion Models. J. Air Pollut. Control Assoc. 36:246–253 (1986).

31. Watson, J.G.; Cooper, J.A.; Huntzicker, J.J.: The Effective Variance Weighting for Least Squares Calculations Applied to the Mass Balance Receptor Model. Atmos. Environ. 18:1347–1355 (1984).

Chapter 11

Performance Criteria for Air Sampling and Monitoring Instruments

Judson Kenoyer, CIH, CHP[A] , David Leong, Ph.D., CIH[B], Richard H. Brown, Ph.D.[C], and Lee Kenny, Ph.D.[C]

[A]Battelle, P.O. Box 999, Richland, Washington; [B]Ontario Ministry of Labour, 400 University Ave., Toronto, Ontario, Canada; [C]Health and Safety Laboratory, Broad Lane, Sheffield, England

CONTENTS

Introduction

The objective of this chapter is to present a snapshot of current performance criteria and instrument testing, historical perspectives, and some projections for the future. The emphasis is on current performance testing standards.

Air sampling and monitoring instruments are used by industrial hygiene and safety personnel to establish safe working environments and to check the establishment and reliability of exposure control mechanisms in the workplace. The same types of instrumentation are often used for environmental measurements. This instrumentation includes but is not limited to air movers and pumps; particle size-selective sampling devices; passive dosimeters and adsorption tubes; radioactivity samplers; direct-reading instruments for aerosols, gases, and vapors; detector tubes for gases and vapors; and bioaerosol samplers.

Concern about performance criteria has grown substantially in recent years as more diverse advisory groups and regulatory bodies have become involved in establishing guidelines for air sampling instrument performance. Many different industrial hygiene instrument manufacturers have attempted to establish themselves as the "experts" on different air sampling needs, and it is very difficult for the potential users of the instrumentation to determine which instrument can best meet their needs while also meeting cost restrictions. The basic requirement is that the worker must be protected from exposure to toxic and hazardous materials by reducing such exposure to a level at which health and safety can be assured. The performance of the instruments used must be accurate, reproducible, and reliable.

Passage of the Occupational Safety and Health Act and the Clean Air Act Amendments in 1970 initiated activities related to the control of airborne contaminants in the workplace and the environment by United States government agencies. Reference and recommended methods for air sampling, monitoring, and analyses of specific chemicals, radionuclides, and other hazardous materials and conditions have been established by several different organizations including the U.S. Environmental Protection Agency (U.S. EPA), National Institute for Occupational Safety and Health (NIOSH), Occupational Safety and Health Administration (OSHA), and the American National Standards Institute (ANSI).

The nuclear industry has been thoroughly regulated for decades. Regulations were established to limit occupational exposures and set release limits for airborne radioactive effluents from nuclear facilities. These limits are adhered to by all licensees and facilities operated under the Nuclear Regulatory Commission (NRC) and the Department of Energy (DOE). Standards and guides on calibration of instruments,[1] performance testing of health physics instrumentation,[2] and sampling of airborne radioactive materials[3] have been written to establish minimum requirements to be met by both manufacturers and users of instruments.

In the industrial hygiene field, regulations and guidelines exist for compliance with the exposure limits for hazardous materials (i.e., chemicals) and conditions; however, guidance for the operation and performance criteria requirements of instrumentation used to measure the hazards does not exist at the level seen in the nuclear field. Threshold Limit Values (TLVs®) for chemical and physical agents are guidelines recommended by the American Conference of Governmental Industrial Hygienists (ACGIH®); they are used by industrial hygienists to establish control measures and personal protection equipment requirements.[4] A joint task group representing ACGIH and the American Industrial Hygiene Association (AIHA) developed a consensus set of standardized data elements to be captured with measurements and qualitative estimates of exposure to chemical hazards and noise.[5] AIHA has also provided some guidance on the use of sampling equipment (e.g., the Cascade Impactor Monograph[6]) and for the use of other equipment (e.g., noise), but, in general, a need exists for more cohesive guidance on the performance criteria for specific types of industrial hygiene instrumentation. Performance criteria and the testing results based on those criteria help users in making scientific as well as cost-effective selection of air sampling instruments, serving their needs for protecting the health of workers.

The First International Symposium on instrument performance testing criteria was held October 29–November 1, 1991, in Research Triangle Park, North Carolina. It was sponsored by the Air Sampling Instruments Committee of ACGIH. During that symposium, users, manufacturers, and regulators discussed air sampling instrument performance and criteria. Details of this symposium were published in *Applied Occupational and Environmental Hygiene*.[7] Some of

the information compiled from that symposium is included in this chapter.

A series of columns specifically addressing instrument performance criteria debuted in 1998 in *Applied Occupational and Environmental Hygiene*. These columns are written by experts in the field and have focussed on performance criteria for instruments such as inhalable aerosol samplers,[8] direct-reading monitoring devices for carbon monoxide,[9] diffusive sampling,[10] and portable gas chromatographs.[11]

The International Organization for Standardization (ISO),[12] ACGIH,[13] and the Comité Européen de Normalisation (CEN)[14] have all published quantitative criteria for particle size-selective sampling based on experimental data for the inhalation and regional deposition of airborne particles. Curves for the inhalable fraction, the thoracic fraction, and the respirable fraction have been developed by all three groups (Chapter 5).

In Europe, the task of standardization in the field of measurement is ultimately the responsibility of the European Commission, but it usually delegates this responsibility to Scientific Expert Groups, reporting to the Commission, or to Technical Committees of CEN. The development of standard measurement procedures, ideally meeting the CEN performance criteria, is also the responsibility of ISO.

Performance Criteria

The performance criteria discussed in this chapter have been divided into three general categories: occupational; environmental; and quality, environmental, and health and safety management systems. Standards for gas and vapor instrumentation and standards for aerosol instrumentation are discussed. Within each of these categories, the standards are discussed in subgroups of U.S. standards and international standards.

Occupational Standards

Gas and Vapor Standards

U.S. Standards. In the United States, performance criteria for gas detectors were established in a certification program sponsored by NIOSH in the 1970s. Procedures were defined for determining noncompliance that set the stage for performance criteria for sampling and analytical methods.[15, 16] The accuracy criterion was established as "The goal is to assure that a single measurement by a method will come within ± 25% of the corresponding 'true' air concentration at least 95% of the time."[17] This concept of accuracy encompasses both bias and precision and is currently being applied to several samplers. NIOSH published several monitoring methods for instruments after they were evaluated and also a number of technical reports on the performance of field portable instruments.[18–24] The evaluations included portability, reliability, calibration, interference, temperature and humidity effects, battery and sensor life, and warm-up and response time. The certification program has been discontinued. OSHA proposed regulations for butanone that stated a desired accuracy of ±25% for "above the Permissible Exposure Level (PEL)," ±30% "at or below the PEL and above the action level," and ±50% "at or below the action level."

Performance criteria and tests for instrumentation that measure airborne radioactive material are discussed in an ANSI standard originally published in 1990 and reaffirmed in 1994.[2] The scope of the standard includes both gas and particle monitors used in the workplace. The performance criteria are discussed in several categories; these include general, electronic, radiation response, interfering responses, environmental, and air circuit. General criteria include classification of radioactivity monitors, sampler design, units of readout, markings, alarm threshold, protection of switches and controls, power, and battery status indication. Electronic criteria include alarms, stability, response time, coefficient of variation, and line noise susceptibility. Radiation response criteria in this standard are minimum detectable activity, accuracy, beta-photon radiation overload, radiation type and energy, and response to unwanted radiations. Criteria and tests for interfering responses are for radiofrequency, microwave, electrostatic, and magnetic fields. Environmental criteria are for temperature, humidity, and ambient pressure. Air circuit criteria include flow or flowrate meter accuracy, air in-leakage, flowrate stability, filter pressure drop, power supply voltage effect on flowrate, power supply frequency effect on flowrate, particle collection, and gas collection.

ANSI/ISA Standards and Recommended Practices.

The Instrument Society of America (ISA) is an ANSI accredited organization. Several standards and recommended practices on specific gas detection instrumentation have been in the development process for the past several years. Some have been finalized and published;[25–30] some are still in the development and review process.[31–35] Finalized standards include those for instruments used to detect oxygen-deficient/oxygen-enriched atmospheres, hydrogen sulfide, and combustible gas. Draft standards being developed will address the detection of chlorine, carbon monoxide, and toxic gas. The recommended practices to be used with each of the standards are being developed concurrently and will discuss the installation, operation, and maintenance of each type of detection instrument.

Each of the standards follows the same format and includes approximately the same type of information. The specific performance requirements for the different instruments will vary to some extent because of the different types of detectors, different principles of operation, different ranges needed for occupational monitoring, and different environmental/workplace conditions during operation. Each standard includes sections on scope, purpose, definitions and terminology, general requirements, construction, instrument markings and instruction manuals, and performance tests. Construction requirements and guidance describe general items; meters, indicators, and outputs; alarm/output functions (where provided); trouble signals; controls and adjustments; and consumables. Performance requirements, tests, and guidance included in the standards are the following: general; sequence of tests; preparation of the instrument; conditions for test and test area; non-powered transportation; drop test; vibration; initial calibration and setup; accuracy; repeatability; step-change response and recovery; supply voltage variation; temperature variation; humidity variation; position sensitivity; air velocity variation; radiofrequency interference; long-term stability; battery and low-battery voltage alarm tests; and exposure to high level of gas.

Each of the recommended practice documents establishes user criteria for the installation, operation, and maintenance of the instruments. As is the case with the ANSI/ISA standards, each recommended practice follows essentially the same format and includes the same sections from one to the next. The sections of each document include scope, purpose, general requirements, unpacking, storage, user recordkeeping, maintenance, preparing instruments for use, installation of stationary instruments, equipment checkout procedures, general considerations, operational check, maintenance procedures, and external power supply systems. Examples of environmental and application checklists and instrument maintenance records are also provided. General considerations may include precautions, desensitizing agents, entering atmospheres potentially containing toxic gases, use of appropriate accessories, electromagnetic interference, and maintenance schedule. Maintenance procedures in the document may include general ones, preliminary checkout, detector head, flow system, readout devices, alarms, and shop calibration test. External power supply systems include requirements and guidance on AC and DC power supplies.

Degraded instrument performance due to electromagnetic fields has been shown to range from subtle deviations in readings to gross errors or even complete shutdown of the instrument.[7] The use of a human exposure standard (ANSI C95.1-1982)[36] as a basis for requirements to avoid electromagnetic susceptibility has been recommended. This is an example where a standard for combustible gas detectors exists,[29] but the performance requirements of the standard do not address the interference problem in adequate detail.

A standard exists for testing gas detector tube units. ANSI/ISEA 102-1990,[37] "American National Standard for Gas Detector Tube Units—Short Term Type for Toxic Gases and Vapors in Working Environments," establishes performance requirements for gas detector tube units and components that are used to determine the concentration of toxic gases and vapors in working environments. This standard includes the following major sections: required information, construction and performance requirements, sampling pump requirements, and quality assurance requirements. Required information includes instructions for the user to verify accurate flow rates, absence of leakage, and specific labeling requirements (e.g., contaminant for measurement, expiration dates of tubes). Construction and performance requirements include accuracy of the units, shelf life, and calibration specifics (e.g., correction factors for environmental effects, operating ranges, calibration concentrations). Sampling pump requirements such as limits on air in-leakage, flow rate accuracy, and prevention of backflow are discussed. Quality assurance details discussed include sampling plan information, acceptable quality levels, and inspection levels.

International Standards. Instruments are employed in a wide range of applications, most of which fall within five major categories: pollution monitoring and control, process monitoring and control, combustion monitoring and control, hazard monitoring, and occupational hygiene. During the past 20–30 years, a large number of instruments have been developed to meet these measurement tasks. More than 1000 instruments manufactured by 240 companies have been identified. For instruments for the measurement of flammable gases, initial standards were developed that related to the risk of the instrument becoming a source of ignition during operation. There are now minimum performance standards relating to parameters such as accuracy, repeatability, and drift. A five-part British Standard[38] was developed that formed the base document for the European Standard.[39] Both general and specific performance requirements are established for different applications.

The task of developing appropriate standards for workplace air quality measurements within the European Community has been carried forward by working groups (WGs) of CEN Technical Committee TC 137. This Committee took the view that air quality assessment standards should be in the form of Performance Requirements rather than Prescribed Methods. The advantage of this approach is that it does not stifle innovation and development since any method can be used which meets the requirements. The actual writing of standard measurement procedures was taken to be the role of ISO or member state regulatory bodies, such as the Health and Safety Executive (HSE) in the U.K., which publishes a series of "Methods for the Determination of Hazardous Substances." The approach taken by CEN has been to develop a hierarchy of Standards with a general performance requirements document at the top and a series of specialized standards under this umbrella.

The umbrella Standard[40] provides, among other things, definitions and minimum requirements for unambiguity (the uniqueness of the result), selectivity (which depends on whether detailed knowledge of the air composition is known in advance) and overall uncertainty (a combination of precision and bias closely analogous to the NIOSH concept of accuracy).

Second Tier Standards contain specific (minimum) performance requirements for measuring devices, together with the appropriate test methods. In this context, "measuring devices" means the same as "instruments" in the Introduction.

Standards for diffusive samplers,[41] pumped sorbent tubes,[42] detector tubes[43] and low volume sampling pumps[44] have been published. Standards are being prepared for high volume pumps, sampling and analysis of metal species (or, more generally, of chemical agents in airborne particles), samplers for mixed aerosols and vapors and, jointly with a CENELEC working group, direct-reading electrical apparatus. When published, this last standard will be complementary to the combustible gases standard[39] but have wider application. Guidance is also being prepared for chemically impregnated systems.

Aerosol Standards

U.S. Standards. Criteria and tests for instrumentation that measures airborne radioactive particulates are included in ANSI N42.17B; this standard was reaffirmed in 1994.[2] Most of the criteria and tests in this standard can be used for both gaseous and particulate monitors; some are specific for particulate monitors. This occupational U.S. standard is described in one of the paragraphs under Gas and Vapor Standards.

International Standards. Two CEN performance standards provide criteria associated with sampling airborne particles. The first is currently in voting and, therefore, has the reference prEN13205.[45] The second standard is a pre-standard on the sampling of particle-vapor mixtures.[46] The pre-standard corresponds to an ISO technical report.

prEN13205. The objective of prEN13205 is to provide test methods that determine whether, and under what conditions, aerosol samplers follow the aerosol sampling conventions presented in EN 481[14] (or ISO 7708[12]). This standard discusses performance requirements for sampling airborne particles. The complete title is "Workplace Atmospheres: Assessment of Performance of Instruments for Measurement of Airborne Particle Concentrations."[45] The specific criteria listed in this standard include accuracy, specimen variability, airflow stability (for samples with integral pumps), transportation and handling, sample identification, instructions for use, design safety, electrical safety, temperature stability, and time stability. For specimen variability, tests are unnecessary where manufacturers can demonstrate that dimensional tolerances are sufficiently stringent to reduce specimen variability below measurable levels. The criteria for temperature stability and time stability only apply to direct-reading instruments.

Influencing variables that need to be tested are also listed and discussed in this standard. These include the following variables: particle aerodynamic diameter, wind speed, wind direction, aerosol composition, aerosol agglomeration, sampled mass, aerosol charge, sampler specimen variability, flowrate variations, and particle collection surfaces.

CEN/TC137/WG3/N209. Performance requirements for sampling particle-vapor mixtures are discussed in the standard "Workplace Atmospheres-Measurement of Chemical Agents Present as Mixtures of Airborne Particles and Vapour-Requirements and Methods."[46] The intent is to publish this as a European pre-standard, not as a full standard at this time. Requirements (criteria) for airborne particle sampling include fraction to be sampled and accuracy (per prEN13205). Criteria for vapor sampling include recovery efficiency, concentration and loading, temperature and humidity, shelf life, and maximum capacity of sampler. Criteria for the combined particle-vapor system are overall uncertainty, storage, blank value, and dimension of result. Other requirements include airflow stability, flow resistance, sample identification, method description, design safety, and electrical safety.

Environmental Standards

Gas and Vapor Standards

U.S. Standards. The U.S. EPA is responsible for reducing the exposure of the United States population to the 189 hazardous air pollutants identified in the 1990 Clean Air Act Amendments.[47] Voluntary reduction in the release of these air pollutants is expected, followed by the establishment of engineering-based pollution control requirements. Five emerging issues exist: aerosol acidity, enhanced ozone monitoring, development of simple-to-operate and inexpensive air pollutant monitoring systems, indoor air pollution, and continuous measurement of stack gas flow rate.

Part 53 of 40 CFR[48] discusses ambient air monitoring reference and equivalent methods. Subpart B of this includes procedures for testing performance characteristics of automated methods for SO_2, CO, O_3, and NO_2. Performance criteria discussed in this subpart include range, noise, lower detectable level, interference equivalent, zero drift, span drift, lag time, rise time, fall time, and precision.

International Standards. An international document, published by the Organisation Internationale De Metrologie Legale (OIML), OIML D 22,[49] provides definitions and guidelines for selecting portable instruments to measure airborne pollutants at hazardous waste sites. It also provides background and literature references on the application of these instruments. Methods and requirements for testing and calibrating instruments are emphasized. Information is provided for the following types of instruments: compound-specific instruments; chemical detector tubes; total hydrocarbon analyzers; gas chromatographs; infrared analyzers; and dust monitors. Most of the information is guidance-only (i.e., "should" statements); however, there are a few basic requirements (i.e., "shall" statements). Metrological and technical characteristics for each type of instrument are listed and discussed in fairly general detail. These are specific to the type of instrument and may include the operational range (in ppm) the instrument is expected to perform, expected linearity over a specified range, repeatability, accuracy, and detection limit.

The performance of many of these instruments is affected by the presence of other air pollutants. Instrument manufacturers should provide information on the level of interference to be expected for specific pollutants. The OIML document does state that the manufacturer shall provide an accuracy statement for detection of specific chemical classes with each lot of detection tubes. The detector tubes shall be stored under environmental conditions specified by the manufacturer and the tubes shall be used on or before the date printed on the tube or its package, and discarded after that date. When using a simultaneous direct-reading detector tube system, the document states that the air flowrate through each tube shall be calibrated using a measuring device such as a soap bubble flow meter or one that has been calibrated against a soap bubble flow meter. With the more complicated devices such as the gas chromatographs and infrared spectrophotometers, it is stated that reference gas standards used shall be a certified reference standard or equivalent and that calibrations should be performed with several different compounds simultaneously.

The task of developing appropriate standards for ambient air quality measurements within the European Community has been undertaken by working groups (WGs) of CEN Technical Committee TC 264. Requirements for ambient air quality measurement in Europe are set by EC Directives, for example the "Directive on Ambient Air Quality Assessment and Management"[50] and its associated Daughter Directives. These Directives prescribe performance requirements as Data Quality Objectives (DQOs), including accuracy and precision, minimum data capture, and minimum time coverage. DQOs are set at different levels for different pollutants and also for different assessment methods (e.g., mandatory measurement, indicative measurement, modeling or objective estimation). Usually, a "reference" method is prescribed, but other methods, meeting the DQOs, may be used. The primary task of CEN TC 264 is to evaluate and recommend reference methods where these are not already prescribed in Directives. However, in the specific case of diffusive samplers, which are likely to be used for indicative measurement, performance requirements standards have been developed by CEN/TC 264/WG 11; these are analogous to EN 482[40] and EN 838[41] of CEN/TC 137/WG 2.

The International Electrotechnical Commission (IEC) has published a six-part standard on equipment for continuously monitoring radioactivity in gaseous effluents. Parts 1 through 6 cover the following topics: general requirements,[51] aerosol effluent monitors,[52] noble gas effluent monitors,[53] iodine monitors,[54] tritium effluent monitors,[55] and transuranic aerosol effluent monitors.[56] These are very thorough standards with each one having sections on general criteria, monitor design, test procedures, and documentation. Part 1 provides criteria relevant to all the different types of monitors while the other parts contain criteria specific

to the monitor type covered by the particular topical standard. The test procedures discussed in each standard describe standard test conditions, tests performed under these conditions (e.g., accuracy, overload, statistical fluctuations, stability, and alarm tests), and tests performed with variation of influence quantities (e.g., external radiation interference, warm-up time, voltage tests, frequency tests, ambient temperature, relative humidity, atmospheric pressure, and RF/MW interference). The test requirements and specific test procedures are provided.

Aerosol Standards

U.S. Standards.

ANSI. The update of ANSI N13.1[3] was completed in 1999. It discusses sampling and monitoring releases of airborne radioactive substances from the stacks and ducts of nuclear facilities. The majority of the standard is focussed on sampling and monitoring techniques and general requirements. The specific sections of the document discuss objectives and approaches for sampling programs, sampling locations, sampling system design, and quality assurance and quality control. Table 6 of the document provides a summary of performance criteria; this is referenced and included in the section on quality assurance and quality control. Specific performance criteria listed include the following: percent transport of 10 µm aerodynamic diameter particles from the free stream to the collector/analyzer; transmission ratio for the sampler nozzle inlet; aspiration ratio for the sampler nozzle; accuracy of the effluent and sample flowrates; and flow control limits. Also listed are acceptable characteristics of a suitable sampling location and conditions during which continuous measurement of effluent flowrate and continuous sample flowrate measurement and control are required.

U.S. EPA Regulations. PM_{10} and $PM_{2.5}$ refer to particulate matter with aerodynamic diameters less than or equal to a nominal 10 micrometers and 2.5 micrometers, respectively. Performance criteria for the environmental samplers for these respirable particles are discussed in detail in 40 CFR Parts 50, 53, and 58 which were updated recently to incorporate $PM_{2.5}$ criteria into the environmental air quality standards issued by U.S. EPA.[48, 57] Until this change, the regulations covered only PM_{10}.[58]

Appendices J and M of Part 50 of 40 CFR discuss the reference method for the determination of particulate matter as PM_{10} in the atmosphere. For now, EPA is retaining Appendix J in its current form. The new

Appendix M to 40 CFR Part 50 establishes the reference method for measuring PM_{10} in the ambient air. The only revision to the reference method for PM_{10} relates to the calculation of the volume of air supplied. Specifics on the acceptable range, precision, accuracy, and potential sources of error (volatile particles, artifacts, humidity, filter loading, flow rate determinations) are provided. Design characteristics of the sampling apparatus are also described; these include the sampler, filters, and flow rate transfer standard. Calibration requirements and procedures, operational procedures, sampler maintenance, and the calculations needed are also discussed.

Appendix L of Part 50 of 40 CFR discusses the reference method for the determination of fine particulate matter as $PM_{2.5}$ in the atmosphere. Specific performance criteria in this appendix include: accuracy; precision; sampler design specifications (including inlet, impactor, and filter characteristics); flow rate (cut off and measurement); leak test capability; range of operational conditions (temperature, humidity, and ambient pressure); temperature and barometric pressure sensors; and sampling control and measurement systems.

Subpart D of Part 53 of 40 CFR discusses procedures for testing performance characteristics of methods for PM_{10} and $PM_{2.5}$. Test procedures for specific performance specifications are discussed; these include ones for sampling effectiveness, 50% cut point, and flow rate stability.

Part 58 of 40 CFR discusses ambient air quality surveillance. Appendices A and B focus on quality assurance requirements for state and local monitoring status and for prevention of significant deterioration of air monitoring. Specific requirements include precision and accuracy for automated methods, precision and accuracy for manual methods, calculations, and reporting. Minimum data assessment requirements are stated for SO_2, NO_2, O_3, CO, TSP (Total Suspended Particulates), PM_{10}, $PM_{2.5}$, and lead.

International Standards.

ISO. ISO Standard 2889[59] discusses general principles for sampling airborne radioactive materials. It does not focus on performance criteria for instrumentation used to sample and monitor the environment. However, it does discuss the requirement that a sampler must not fractionate by particle size or in other ways distort the physical and chemical properties of the airborne radioactive constituents. It is recognized (and stated) that this requirement is difficult—often impossible—to accomplish perfectly. Any delivery line through which the sample is carried to the collection device will preferentially remove large particles, either through gravitational set-

tling when the air flowrate is too low or through turbulent impaction when the flowrate is too high. The standard also identifies density of the particles in the air as being of importance in the fractionation by particle size. An appendix presents some guidance on estimating particle loss in sampling lines due to gravitational settling and turbulent flow; however, it also states that characterization of the airborne constituents must be performed to be able to evaluate the potential error. Isokinetic sampling is recommended.

OIML. Operational criteria for dust monitors used at hazardous waste sites are discussed in OIML D 22,[49] specifically, quart piezoelectric oscillators, beta-ray attenuation instruments, and light-scattering instruments. General descriptions are provided with metrological and technical characteristics included. Specific operational characteristics are listed for each type of dust monitor. These include particle size ranges in which the instruments perform, use of pre-collectors when necessary, mass concentration ranges, and detection limits. Design requirements for pre-collectors are to limit progressively the collection of 50% and 100% of the specific sizes. In the U.S., the requirements are 50% of the particles of 3.5 μm aerodynamic diameter and 100% of the particles of 10.0 μm diameter; in the European community, the respective limits are 5.0 μm and 7.1 μm.[12]

IEC. Parts 2 and 6 of the IEC 761 Publication[52, 56] discuss specific requirements for aerosol effluent monitors. Part 6 is specific to transuranic aerosols. Each standard includes general criteria, monitor design requirements, test procedures, and documentation requirements. Reference conditions and standard test conditions are stated. Tests performed under the standard test conditions include accuracy, response to noble gases, overload, statistical fluctuations, stability, and alarm tests. Tests performed with variation of influence quantities include external radiation response, warm-up time, voltage tests, frequency tests, ambient temperature, relative humidity, ambient pressure, and RF/MW interference.

Quality, Environmental, and Health and Safety Management System Standards

Corporate ISO 9000 registration continues to gain international acceptance as a hallmark of quality system achievement; it brings business success and gains in both international and domestic trades. Worldwide ISO 9000 registrations came to a total of 127,389 in 99 countries as of December 1995, up by more than one-third from the previous survey conducted in March 1995.[60]

Quality has been defined by the ISO as "the totality of features and characteristics of a product or service that bear on its ability to satisfy stated or implied needs."[61]

The ISO 9000 series of quality management system (QMS) standards were developed in 1987 by the ISO Technical Committee 176 whose aim was to produce standards of a generic nature that could then be applied to a wide variety of applications and businesses. Quality standards were originally intended to act as a type of contractual agreement, written in very general terms, between customers and suppliers—a way for customers to be assured of the quality of a product. These standards were seen as an internationally acceptable system for rating quality management and quality assurance, applicable to all types of organizations, from manufacturing to service industries, such as hotels, restaurants and management consulting firms.[62]

Success with implementation of ISO 9000 series standards and the world's commitment to protection of the environment propelled the development of the ISO 14000 series of environmental management system (EMS) standards.[63, 64]

There are significant health and safety connections with ISO 9000 series standards,[65–67] and vast interests in the development of occupational health and safety management system (OHSMS) standard have been expressed.[68, 69] Several OHSMS standards emerged worldwide in 1996, including the BS 8800 Standard[70] and the AIHA Guidance Document.[71] The delegates attending the ISO workshop in September 1996 concluded there was no need for ISO to develop the OHSMS Standard, but the need for an international OHSMS standard will likely be re-visited in the future.

These quality, environmental, and occupational health and safety management system standards have common or similar system elements related to the following areas: Purchasing, Control of Customer-Supplied Product, Inspection and Testing, and Control of Nonconforming Product.

Testing and evaluation of the performance of air sampling instruments against instrument performance standards, as relevant to some of the above management system elements, is discussed in each of the following management system standards:

- ISO 9000 Series QMS Standards
- ISO 14000 Series EMS Standards
- British BS 8800 OHSMS Standard
- AIHA OHSMS and Guidance Document

ISO 9000 Series Quality Management System Standards

In 1979, ISO Technical Committee 176 was established to develop international quality system standards and to address consistent quality needs of growing international trade, especially within the European Union. In 1987, five international quality system standards (ISO 9000, 9001, 9002, 9003, and 9004) were published by ISO. In 1994, ISO 9000 series standards were revised and ISO 10000 series companion standards were produced.

ISO 9000 series standards are a set of generic and flexible systems with essential elements for quality management and quality assurance of any business operations, big or small, whether a manufacturer or service provider, or other type of organization, whether in public or private sector.

Within the series, the ISO 9000 standard is the guidance document on quality management and quality assurance, providing guidelines on the selection and use of the appropriate ISO 9001,[72] ISO 9002[73] or ISO 9003[74] standard for registration or certification based on a company's operations:

- ISO 9001 standard provides a quality systems model for quality assurance in design, development, production, installation, and servicing.
- ISO 9002 standard provides a quality systems model for production, installation and servicing.
- ISO 9003 standard provides a quality system model for quality assurance in final inspection testing.
- ISO 9004 standards[75–78] are guidance documents for quality management and quality system elements.

While both the ISO 9000 and ISO 9004 standards are intended for *internal* quality management, ISO 9001, 9002, and 9003 standards are quality conformance standards for providing *external* quality assurance to a purchaser via third-party, independent audits. The AIHA OHSMS and guidance document is structured after ISO 9001.

Table 11-1 shows the comparisons of quality elements in the ISO 9000 series standards. ISO 9000 series standards are based on concepts and quality elements familiar to health and safety management; these include management responsibility, commitment and involvement; inspection and testing; process control; corrective action; internal quality audits; and training.

ISO 9004 Standards explicitly address product safety and risk reduction;[79] they cover topics such as application of hazard analysis for risk assessment, identification of safety standards, design review and prototype testing, analysis of warnings, instruction manuals, promotion materials, and product traceability for recall

TABLE 11-1. Comparisons of ISO 9000 Series Standards

	9001	9002	9003	9004
Management responsibility	X	X	X	X
Quality system	X	X	X	X
Contract review	X	X	—	X
Design control	X	—	—	X
Document and data control	X	X	X	X
Purchasing	X	X	—	X
Control of customer-supplied product	X	X	—	—
Product identification and traceability	X	X	X	X
Process control	X	X	—	—
Inspection & testing	X	X	X	X
Control of inspection, measuring and test equipment	X	X	X	X
Inspection and test status	X	X	X	X
Control of non-conforming product	X	X	X	X
Corrective ad preventive action	X	X	—	X
Handling, storing, packaging, preservation & delivery	X	X	X	X
Control of quality records	X	X	X	X
Internal quality audits	X	X	—	X
Training	X	X	X	X
Servicing	X	—	—	X
Statistical techniques	X	X	X	X
Quality cost considerations	—	—	—	X
Product safety and liability	—	—	—	X

and follow-up. ISO 9004 Standards also state that "consideration has to be given to risks such as those pertaining to health and safety."

Substantial similarities between ISO 9000 Standards and the OSHA Process Safety Management Standard are such that the work done to achieve compliance with one can be used to achieve compliance with the other.[80]

While consensus has not been reached internationally on the development of an ISO OHSMS standard, it is widely recognized that workplace health and safety programs can be built into ISO 9000 quality programs through the quality elements in the ISO 9001 Standard.[81] For example, the quality system element under "inspection and testing" stresses the development and implementation of a quality assurance system for controlling the quality of incoming products. The quality assurance system can be extended to include health and safety monitoring equipment and personal exposure monitoring devices. Likewise, the quality system elements under "inspection and test status" and "control of non-conforming goods" ensure the performance of air sampling instruments is evaluated, verified, and tested to meet the needs for quality measurements to be made and the instrument perform-

ance standard as specified by the instrument manufacturer or the technical standard.

Registration of a company's quality management system conforming to ISO 9000 standards requires successful certification of the implemented system via third-party independent audit by the accredited registrar of the ISO quality system. However, independent audits of instrument performance tests and reports remain an unresolved issue when applying the ISO 9000 Standards to the inspection and testing of air sampling instruments.

ISO 14000 Series Environmental Management System (EMS) Standards

ISO 9000 series standards are based on concepts and quality elements also familiar to environmental management systems and provide a general framework for integrating environmental considerations into the quality system. With increased public concern over industry's impact on the world's environment, environmental advocacy groups, consumers, and the public have been urging businesses to take responsibility for their environmental effects.

The Rio Conference on the Environment in 1992 established the world's commitment to protection of the environment. Responsible companies recognize that environmental issues are critical to business success, and environmental activities must be integral to business planning and development, not afterthoughts resulting from incidents. The success of ISO in establishing international standards for quality management systems prompted the development of ISO 14000 series EMS standards. The ISO 14000 series EMS standards are voluntary consensus standards developed to provide business organizations with a structure for managing environmental impacts and to assist organizations to achieve environmental and economic gains through implementation of an effective environmental management system.

Included in the 14000 series are the following Standards:

ISO 14001	Environmental Management Systems—Specification with Guidance for Use
ISO 14004	Guide to Environmental Management—General Guidelines on Principles, Systems, and Supporting Techniques
ISO 14010	Guidelines for Environmental Auditing—General Principles of Environmental Auditing
ISO 14011	Guidelines for Environmental Auditing—Audit Procedures—Part 1; Auditing of Environmental Management Systems
ISO 14012	Guidelines for Environmental Auditing—Qualification Criteria for Environmental Auditors
ISO 14020	Environmental Labelling—General Principles
ISO 14021	Environmental Labelling—Self Declaration Environmental Claims—Terms and Definitions
ISO 14022	Environmental Labelling—Environmental Labels and Declaration Self-Declaration Environmental Claims—Symbols
ISO 14024	Environmental Labelling—Practitioner Programs, Guiding Principles, Practices, and Certification Procedures of Multiple Criteria
ISO 14031	Environmental Management—Environmental Performance Evaluation—Guidelines
ISO 14040	Life Cycle Assessment—General Principles and Practices
ISO 14041	Environmental Management—Life Cycle Assessment—Goal and Scope Definition and Inventory Analysis
ISO 14042	Environmental Management—Life Cycle Assessment—Life Cycle Impact Assessment
ISO 14043	Environmental Management—Life Cycle Assessment—Life Cycle Interpretation
ISO 14050	Environmental Management—Vocabulary
ISO GUIDE 64	Guide for the Inclusion of Environmental Aspects in Product Standards

ISO 14001 is a specification standard for management system elements that an organization must put in place to have its EMS successfully registered or certified.

Other ISO 14000 series standards are guidance documents, providing two types of tools to support EMS: 1) system-oriented tools, including the auditing and evaluation standards that help an organization assess its performance; and 2) product-oriented tools, including life cycle assessment and environmental labeling standards.

The EMS elements in the ISO 14001 Standard are: Environmental Policy; Planning; Implementation and Operation; Checking and Corrective Action; and Management Review.

Monitoring and measurement are stressed in checking and taking corrective action. The ISO 14001 Standard requires the organization to establish and maintain documented procedures to monitor and measure, on a regular basis, the key characteristics of its operations and activities that can have a significant impact on the environment. Guidance on monitoring and measurement is not provided in the annex of the Standard. It is, however, advisable to include testing, evaluation, and audit of air sampling instrument performance against the respective instrument performance standards in the planning and implementation of environmental air monitoring programs. The Standard does call for calibration and maintenance of monitoring equipment and for retaining the records according to the organization's procedures.

British Standard BS 8800: 1996 on Occupational Health and Safety Management Systems

The draft BS 8750 Standard on "Guide to Occupational Health and Safety Management Systems" published in December 1994 was finalized and released as

BS 8800 Standard on May 15, 1996. The Standard is based on the general principles of good management and is designed to: 1) enable the integration of occupational health and safety (OHS) management within an overall management system; and 2) improve the OHS performance of organizations in order to minimize risk to employees and others, to improve business performance, and to assist organizations to establish a responsible image within the marketplace.

The Standard contains guidance and recommendations and states that it should not be quoted as if it were a specification and should not be used for certification purposes. Various approaches could well be adopted for the OHS management system, and the Standard recommends that it may be based on HSE guidance "Successful Health and Safety Management" HS(G) 65, or BS EN ISO 14001, the environmental systems standard. The OHSMS elements based on the HS(G) 65 approach are the following: Initial Status Review; OHS Policy; Organizing; Planning and Implementing; Measuring Performance; Audit; and Periodic Status Review. Based on the BS EN ISO 14001 approach, the OHSMS elements are: Initial Status Review; OHS Policy; Planning; Implementation and Operation; Checking and Correction Action; and Management Review.

Whichever approach is used, implementation of the OHSMS for each system element will have to take both the feedback from performance measurement and the audit into consideration. This implies that the workplace air monitoring programs planned and implemented under the OHSMS will include testing, evaluation, and audit of air sampling instrument performance against the respective instrument performance standards.

AIHA OHSMS and Guidance Document

The AIHA Board of Directors approved in May 1996 the release of an OHSMS document submitted by the OHSMS task force formed in 1995.[71] The OHSMS is structured after ISO 9001 and may be used by practicing health and safety professionals for implementing and evaluating OHS management systems that may be established by organizations. Emphasis is placed on creation and continuous improvement of OHS systems, prevention of work-related illness and injury, employee involvement in health and safety matters, integration of health and safety into business practices, and prioritization of corrective measures

The OHSMS document is neither a U.S. national standard nor an international consensus standard. It specifies the basic elements of OHSMS for organizations to implement a deliberate, documented approach to anticipation, recognition, evaluation, prevention, and control of occupational health and safety hazards.

The OHSMS Guidance Document is the companion to the OHSMS document and provides illustrations, interpretations and explanations of the various elements found within the OHSMS.

The AIHA OHSMS is built on the following elements:

- Occupational Health and Safety Management Responsibility
- OHS Management System
- OHS Compliance and Conformance Review
- OHS Design Control
- OHS Document and Data Control
- Purchasing
- OHS Communication Systems
- OHS Hazard Identification and Traceability
- Process Control for OHS
- OHS Inspection and Evaluation
- Control of OHS Inspection, Measuring, and Test Equipment
- OHS Inspection and Evaluation Status
- Control of Nonconforming Process or Device
- OHS Corrective and Preventive Action
- Handling, Storage and Packaging of Hazardous Materials
- Control of OHS Records
- Internal OHS Management System Audits
- OHS Training
- Operations and Maintenance Services
- Statistical Techniques

The importance of instrument performance testing within the OHSMS is stressed in the following elements: OHS Inspection and Evaluation; Control of OHS Inspection, Measuring, and Test Equipment; OHS Inspection and Evaluation Status; and Control of OHS Nonconforming Process or Device.

OHS Inspection and Evaluation. The AIHA OHSMS requires an organization to establish procedures that prevent incoming and in-process personal protective devices, products, hazard abatement or control devices, contracted services, or data obtained from contract services from use until they have been inspected or otherwise verified as conforming to specified OHS requirements. The document recommends that the organizations should develop procedures for systematic review of personal protective equipment, engineering control devices, personal exposure sampling devices, calibration tools, and other equipment or consumables which may impact OHS—both by OHS staff and employees using them. The systematic review should be performed prior to use and be designed to verify materials are suitable for their intended purpose and in good working condition upon receipt.

This AIHA OHSMS element implies that a manufacturer will test its instruments against a performance standard, and such test report will be made available to its clients for review before purchasing. Upon receiving the purchased instruments, they will be inspected or further tested against the same performance standard by the clients.

Control of OHS Inspection, Measuring and Test Equipment.
According to the AIHA OHSMS, the organization shall establish and maintain documented procedures to control, calibrate, and maintain OHS inspection, measuring, and test equipment in conformance with the manufacturer's specifications. In addition, the OHS inspection, measuring, and test equipment shall be used only in a manner enabling the measurement uncertainty to be known and to be consistent with desired measurement reliability and validity parameters. The document recommends that where direct-reading devices, or the collection of atmospheric, environmental, biological, or medical samples are used to evaluate employee workplace exposures, the organization should ensure that the collection and analytical methodologies will meet specified federal, state, or other applicable good practice requirements.

This AIHA OHSMS element suggests the instrument performance and test reports will be reviewed by the organizations to select the appropriate instruments for the measurements to be made with the required specificity, sensitivity, accuracy, and precision.

OHS Inspection and Evaluation Status.
The AIHA OHSMS requires the identification of OHS inspection and evaluation status by suitable means be maintained, as defined in the OHS quality plan and/or documented procedures, throughout all stages of the process or device operation. Where continuous monitoring of working conditions is necessary to provide a safe and healthy place of employment (e.g., oxygen content in a confined space), or where ongoing periodic OHS evaluations are an essential aspect of operations (e.g., monitoring presence of explosive gases in underground mines), the document recommends that the sample data evaluation and reporting mechanisms should be explicitly documented in the OHS quality plan and strictly enforced.

Control of OHS Nonconforming Process or Device.
The AIHA OHSMS document stresses the needs for the organization to establish and maintain documented procedures for the review and disposition of nonconforming OHS processes or devices and to communicate findings to the affected parties. Perhaps the findings of nonconforming air sampling instruments may also be communicated to a central deposit such as the ACGIH Air Sampling Instruments Committee for sharing information and promoting best practices in instrument performance testing.

Other General or Specialized Performance Criteria

Other performance criteria and testing programs exist that have not been discussed in this chapter; these may be used by manufacturers and users of air sampling instruments nationally and internationally. Many instrument manufacturers will have their instruments "certified" in different countries by appropriate testing programs or services so their instruments can be available for purchase in those countries. The following examples are included to describe some of these programs.

One air sampling instrument manufacturer sent its instruments through no less than eight testing programs. These programs include: the Underwriters Laboratory Inc. (UL Listing), USA; Mine Safety and Health Administration (MSHA), USA; SIRA Certification Service, England; ISO 9002, International Standards Organization (ISO); Institute for Workers Safety (BIA), Germany; Government Mining Engineer (GME), South Africa; Work Cover Authority (WA), Australia; and National Association of Testing Authorities (NATA), Australia. The UL Listing certifies instruments of different classes of hazards (e.g., Class I [Group A, B, C, D]; Class II [Groups E, F, G]; and Class III). MSHA performs testing for intrinsic safety for use in specific gas-air mixtures. SIRA Certification Service, England, is the European equivalent to the American UL Listing for intrinsic safety approval. ISO 9002 is an internationally recognized quality standard that specifies management and production procedures. The other testing programs are used less frequently in the international market.

Many manufacturers have the resources needed to test their own units with company-owned equipment and with the use of procedures developed by their own personnel based on requirements found in standards or regulations (e.g., U.S. EPA, DOE, NRC, ANSI, ISO). Testing can be performed by outside services using similar procedures. One thorough testing program that is operated by a specific instrument manufacturer performs the following tests: calibration gas repeatability, long-term stability, gas-alarm, bounce/transportation, vibration, unpowered storage, linearity, step response time, gas velocity variation, gas saturation, supply voltage variation, temperature, humidity, ambient pressure, poisoning (hydrogen sulfide), cross sensitivity, radio frequency interference (RFI), time-weighted average (TWA), short-term exposure limit (STEL), battery lifetime, and air velocity for internal pumps.

Conclusions

Much work has been done in the past five years in the development of performance criteria for air sampling and monitoring instrumentation. Efforts in updating earlier standards and in developing new ones are continuing. This is in both the U.S. and international arenas.

The effort on the international side is still greater than that of the U.S.; however, with ANSI establishing ISA as an accredited organization, a number of U.S. standards are being developed and will eventually be finalized; some already have been. Other organizations such as ACGIH and the Health Physics Society (HPS) have expressed an interest in pursuing the sponsorship of ANSI or other accrediting organizations for the development of needed performance criteria.

In some areas such as passive dosimeters, NIOSH has developed validation criteria and manufacturers are validating their products in accordance with the NIOSH protocol; the testing results are then made available to the users. However, these testing results are not subject to third-person review. There is no agency responsible for coordinating the third-party review of testing reports and acting as a central depository for the audited reports. A partnership between ACGIH and ANSI within the framework of ISO 9000 or 14000 registrations may propel further development of performance criteria and coordination of third-party review of testing reports.

General procedures (standards) are currently available; however, there are many chemicals for which the general standards are not adequate. Several reactive species of volatile organics need specific criteria development. European working groups will also be examining isocyanates and aldehydes, metals, and methods for formaldehyde.

There are many inorganic species and particulates (dusts and fibers) for which there are not adequate NIOSH or ISO methods. Direct-reading instruments for gases and vapors and size-selective instruments for aerosols are poorly represented, although guidance on choices of the instrumentation is available.

Additional research funding will facilitate needed progress in further development of specific perform-

ance criteria for industrial hygiene instrumentation and in the verification of current sampling and analysis methods being used by the different types of instrumentation.

Useful Internet URLS

The URLs listed in Table 11-2 may be useful in finding resources related to instrument performance criteria available on the internet. Several refer to different standard development or governmental organizations that publish and/or develop specific criteria or guidance.

References

1. American National Standards Institute: American National Standard for Radiation Protection Instrumentation Test and Calibration. ANSI N323-1978. ANSI, New York (1978).
2. American National Standards Institute: American National Standard for Radiation Instrumentation, Performance Specifications for Health Physics Instrumentation–Occupational Airborne Radioactivity Monitoring Instrumentation. ANSI N42.17B-1989. ANSI, New York (1990, 1994).
3. American National Standards Institute: American National Standard for Sampling and Monitoring Releases of Airborne Radioactive Substances from the Stacks and Ducts of Nuclear Facilities. ANSI N13.1-1999. ANSI, New York (1999).
4. American Conference of Governmental Industrial Hygienists: 2000 TLVs® and BEIs®. ACGIH, Cincinnati, OH (2000; annual).
5. ACGIH–AIHA Task Group on Occupational Exposure Databases: Data Elements for Occupational Exposure Database–Guidelines and Recommendations for Airborne Hazards and Noise. Appl. Occup. Environ. Hyg. 11(11): 1244–1311 (1996).

Table 11-2. Useful Internet URLs

Group/Organization	URL
ACGIH	http://www.acgih.org
AIHA	http://www.aiha.org
ANSI	http://www.ansi.org
BSI	http://www.bsi.org.uk
CEN	http://www.cenorm.be
CENELEC	http://www.cenelec.be
EPA	http://www.epa.gov
HPS	http://www.hps.org
IEC	http://www.iec.ch
IEEE	http://www.ieee.org
ISA	http://www.isa.org
ISO	http://www.iso.ch
NIOSH	http://www.cdc.gov/niosh/homepage.html
OIML	http://www.oiml.org
OSHA	http://www.osha.gov

6. Lodge, J.P.; Chan, T.L., Eds.: Cascade Impactor Sampling and Data Analysis. American Industrial Hygiene Association, Akron, OH (1986).

7. Cohen, B.S.; McCammon, C.S.; Vincent, J.H., Eds.: Proceedings of The International Symposium on Air Sampling Instrument Performance. Appl. Occup. Environ. Hyg. 8(4):209–424 (1993).

8. Bartley, D.L.: Inhalable Aerosol Samplers. Appl. Occup. Environ. Hyg. 13(5):274–278 (1998).

9. Woebkenberg, M.L.: Direct-Reading Monitoring Devices for Carbon Monoxide. Appl. Occup. Environ. Hyg. 13(8):567–569 (1998).

10. Harper, M.: Diffusive Sampling. Appl. Occup. Environ. Hyg. 13(11):759–763 (1998).

11. Burroughs, G.E.; Tabor, M.E.: Portable Gas Chromatographs. Appl. Occup. Environ. Hyg. 14(3):159–162 (1999).

12. International Organization for Standardization (ISO): Air Quality—Particle Size Fraction Definitions for Health-Related Sampling. ISO Standard 7708. ISO, Geneva (1995).

13. American Conference of Governmental Industrial Hygienists: Particle Size Selective Sampling for Health-Related Aerosols. ACGIH, Cincinnati, OH (1999).

14. Comité Européen de Normalisation (CEN): Workplace Atmospheres—Size Fraction Definitions for Measurement of Airborne Particles in the Workplace. EN481, CEN, Brussels (1993).

15. Leidel, N.A.; Busch, K.A.: Statistical Methods for the Determination of Noncompliance with Occupational Health Standards. DHEW (NIOSH) Pub. No. 75-159; NTIS Pub. No. PB-83-180-414. National Technical Information Service, Springfield, VA (1975).

16. Leidel, N.A.; Busch, K.A.; Lynch, J.R.: Occupational Exposure Sampling Strategy Manual. DHEW (NIOSH) Pub. No. 77-173; NTIS Pub. No. PB-274-792. National Technical Information Service, Springfield, VA (1977).

17. Taylor, D.G.; Kupel, R.E.; Bryant, J.M.: Documentation of the NIOSH Validation Tests. DHEW (NIOSH) Pub. No. 77-185; NTIS Pub. No. PB-274-248. National Technical Information Service, Springfield, VA (1977).

18. Parker, C.D.; Lee, M.B.; Sharpe, J.C.: An Evaluation of Personal Sampling Pumps in Sub-Zero Temperatures. DHEW (NIOSH) Pub. No. 78-117; NTIS Pub. No. PB-279-615. National Technical Information Service, Springfield, VA (1977).

19. Bissette, L.W.; Parker, D.D.: Evaluation of Batteries Used in Sampling Pumps. Final Report. NIOSH Contract 210-75-0080. NTIS Pub. No. PB-83-109-694. National Technical Information Service, Springfield, VA (1975).

20. Tompkins, F.C., Jr.; Becker, J.H.: Evaluation of Portable Direct-Reading H_2S Meters. Report of NIOSH Contract HEW 210-75-0037. DHEW (NIOSH) Pub. No. 77-137. NIOSH, Cincinnati, OH (1976).

21. Parker, C.D.; Strong, R.B.: Evaluation of Portable, Direct-Reading Carbon Monoxide Meters, Part 1. DHEW (NIOSH) Pub. No. 75-106; NTIS Pub. No. PB-273-870. National Technical Information Service, Springfield, VA (1975).

22. Parker, C.D.; Strong, R.B.: Evaluation of Portable, Direct-Reading Sulfur Dioxide Meters, Part 2. DHEW (NIOSH) Pub. No. 75-137; NTIS Pub. No. PB-273-799. National Technical Information Service, Springfield, VA (1977).

23. Willey, M.A.; McCammon, C.S.: Evaluation of Portable, Direct-Reading Hydrocarbon Meters. DHEW (NIOSH) Pub. No. 76-166; NTIS Pub. No. PB-266-439. National Technical Information Service, Springfield, VA (1977).

24. Woodfin, W.J.; Woebkenberg, M.L.: An Evaluation of Portable, Direct-Reading Oxygen Deficiency Monitors. NTIS Pub. No. PB-85-196-442. National Technical Information Service, Springfield, VA (1985).

25. American National Standards Institute/Instrument Society of America: ANSI/ISA - S92.04.01, Part 1 - 1996, Performance Requirements for Instruments Used to Detect Oxygen-Deficient/Oxygen-Enriched Atmospheres (Standard). ISA, Research Triangle Park, North Carolina (1996).

26. Instrument Society of America: ISA-RP92.04.02, Part II-1996, Installation, Operation, and Maintenance of Instruments Used to Detect Oxygen-Deficient/Oxygen-Enriched Atmospheres (Recommended Practice). ISA, Research Triangle Park, North Carolina (1996).

27. American National Standards Institute/Instrument Society of America: ANSI/ISA - S12.15, Part I - 1990, Performance Requirements for Hydrogen Sulfide Detection Instruments (10-100 ppm) (Standard). ISA, Research Triangle Park, North Carolina (1990).

28. American National Standards Institute/Instrument Society of America: ANSI/ISA - RP12.15, Part II - 1990, Installation, Operation, and Maintenance of Hydrogen Sulfide Detection Instruments (10-100 ppm) (Recommended Practice). ISA, Research Triangle Park, North Carolina (1990).

29. American National Standards Institute/Instrument Society of America: ANSI/ISA - S12.13, Part I - 1995, Performance Requirements, Combustible Gas Detectors (Standard). ISA, Research Triangle Park, North Carolina (1995).

30. American National Standards Institute/Instrument Society of America: ANSI/ISA - RP12.13, Part II - 1995, Installation, Operation, and Maintenance of Combustible Gas Detectors (Recommended Practice). ISA, Research Triangle Park, North Carolina (1995).

31. Instrument Society of America: ISA-dS92.0.01, Part 1, Performance Requirements for Toxic Gas Detection Instruments (Draft Standard), Draft 4, March 1997. ISA, Research Triangle Park, North Carolina (1997).

32. Instrument Society of America: ISA-dRP92.0.02, Part II, Installation, Operation, and Maintenance of Toxic Gas Detection Instruments (Draft Recommended Practice), Draft 3, March 1997. ISA, Research Triangle Park, North Carolina (1997).

33. Instrument Society of America: ISA-dS92.02.01, Part I, Performance Requirements for Carbon Monoxide Detection Instruments (50-1000 ppm Full Scale) (Draft Standard), Draft 6, March 1997. ISA, Research Triangle Park, North Carolina (1997).

34. Instrument Society of America: ISA-dRP92.02.02, Part II, Installation, Operation, and Maintenance of Carbon Monoxide Detection Instruments (50-100 ppm Full Scale) (Draft Recommended Practice), Draft 5, March 1997. ISA, Research Triangle Park, North Carolina (1997).

35. Instrument Society of America: ISA-dS92.06.01, Performance Requirements for Chlorine Detection Instruments (0.5-30 ppm Full Scale) (Draft Standard), Draft 6, March 1997. ISA, Research Triangle Park, North Carolina (1997).

36. American National Standards Institute/The Institute of Electrical and Electronics Engineers: ANSI/IEEE C95.1-1982, American National Standard Safety Levels with Respect to Human Exposure to Radio Frequency Electromagnetic Fields, 300 kHz to 100 GHz, p. 10. IEEE, New York (1982).

37. American National Standards Institute: ANSI/ISEA 102-1990, American National Standard for Gas Detector Tube Units—Short Term Type for Toxic Gases and Vapors in Working Environments. ANSI, New York (1990).

38. British Standards Institution: Instruments for the Detection of Combustible Gases, Parts 1 to 5. Document BS6020. BSI, London (1981).

39. European Committee for Electrotechnical Standardisation: Electrical Apparatus for the Detection and Measurement of Combustible Gases. Document EN 50054-58. CENELEC, Brussels (1991).

40. Comité Européen de Normalisation: Workplace Atmospheres—General Requirements for the Performance of Procedures for the Measurement of Chemical Agents. EN 482. CEN, Brussels (1994).

41. Comité Européen de Normalisation: Workplace Atmospheres—Requirements and Test Methods for Diffusive Samplers for the Determination of Gases and Vapours. EN 838. CEN, Brussels (1995).

42. Comité Européen de Normalisation: Workplace Atmospheres—Requirements and Test Methods for Pumped Sorbent Tubes for the Determination of Gases and Vapours. EN 1076. CEN, Brussels (1997).

43. Comité Européen de Normalisation: Workplace Atmospheres—Requirements and Test Methods for Short Term Detector Tube Systems. EN 1231. CEN, Brussels (1996).

44. Comité Européen de Normalisation: Workplace Atmospheres—Requirements and Test Methods for Pumps for Personal Sampling of Chemical Agents. EN 1232. CEN, Brussels (1997).

45. Comité Européen de Normalisation: Workplace Atmospheres—Assessment of Performance of Instruments for Measurement of Airborne Particle Concentrations. prEN13205. CEN, Brussels (1999).

46. Comité Européen de Normalisation: Workplace Atmospheres—Measurement of Chemical Agents as Mixtures of Airborne Particles and Vapours—Requirements and Test Methods. CEN/TC137/WG3/N209. CEN, Brussels (2000).

47. U.S. Congress: Titles I, II, III, and IV of the Clean Air Act as Amended in 1990. Public Law 101-549. Washington, DC (November 15, 1990).

48. U.S. Environmental Protection Agency: Revised Requirements for Designation of Reference and Equivalent Methods for $PM_{2.5}$ and Ambient Quality Surveillance for Particulate Matter; Final Rule, 40 CFR Part 53. Fed. Reg. 62: 38763–38854 (1997).

49. Organisation Internationale De Metrologie Legale: OIML International Document D 22, Guide to Portable Instruments for Assessing Airborne Pollutants Arising from Hazardous Wastes. OIML, Paris (1991).

50. Comité Européen de Normalisation: Council Directive 96/62/EC on Ambient Air Quality Assessment and Management (1996).

51. International Electrotechnical Commission: IEC Publication 761-1: 1983, Equipment for Continuously Monitoring Radioactivity in Gaseous Effluents, Part 1: General Requirements. IEC Geneva (1983).

52. International Electrotechnical Commission: IEC Publication 761-2: 1983, Equipment for Continuously Monitoring Radioactivity in Gaseous Effluents, Part 2: Specific Requirements for Aerosol Effluent Monitors. IEC Geneva (1983).

53. International Electrotechnical Commission: IEC Publication 761-3: 1983, Equipment for Continuously Monitoring Radioactivity in Gaseous Effluents, Part 3: Specific Requirements for Noble Gas Effluent Monitors. IEC Geneva (1983).

54. International Electrotechnical Commission: IEC Publication 761-4: 1983, Equipment for Continuously Monitoring Radioactivity in Gaseous Effluents, Part 4: Specific Requirements for Iodine Monitors. IEC Geneva (1983).

55. International Electrotechnical Commission: IEC Publication 761-5: 1983, Equipment for Continuously Monitoring Radioactivity in Gaseous Effluents, Part 5: Specific Requirements for Tritium Monitors. IEC Geneva (1983).

56. International Electrotechnical Commission: IEC Publication 761-6: 1991, Equipment for Continuously Monitoring Radioactivity in Gaseous Effluents, Part 6: Specific Requirements for Transuranic Aerosol Effluent Monitors. IEC Geneva (1991).

57. U.S. Environmental Protection Agency: National Ambient Air Quality Standards for Particulate Matter; Final Rule. Fed. Reg. 62: 38651–38701 (1997).

58. U.S. Environmental Protection Agency: Revisions to the National Ambient Air Quality Standards for Particulate Matter. Fed. Reg. 52:24634–24750 (1987).

59. International Standards Organisation: ISO 2889: 1975, General Principles for Sampling Airborne Radioactive Materials. ISO, Geneva (1975).

60. Mobil Europe Limited: The Mobil Survey (Fifth Cycle) of ISO 9000 and Environmental Certificates Awarded Worldwide. London (August 6, 1996).

61. International Standards Organisation: ISO 8402:1994, Quality Management and Quality Assurance—Vocabulary. ISO, Geneva (1994).

62. Riswadkar, A.V.: ISO 9000, A Global Standard for Quality. Professional Safety, pp. 30–32 (April 1995).

63. Hansen, M.: ISO 9000: Effect on the Global Environmental Safety & Health Community. ASSE, pp. 44–47 (June 1994).

64. Sharifian, A.R.: Reap the Benefits of ISO 9000: Environmental Health and Safety Thrive under Global Standards. Safety and Health, pp. 68–71 (April 1995).

65. Blondin, R.: ISO 9000: The Health and Safety Connection. Accident Prevention, pp. 12–18 (September/October 1993).

66. Allen, H.: ISO 9000 Paves the Way to Global Safety. Safety and Health, pp. 33–37 (March 1994).

67. Mansdorf, Z.: The ISO Man Cometh: Moving the Global Standards. Occup. Hazards, pp. 43–46 (May 1995).

68. Dyjack, D.T.; Levine, S.P.: Development of an ISO 9000 Compatible Occupational Health Standard: Defining the Issues. Am. Ind. Hyg. Assoc. J. 56: 599–609 (1995).

69. Levine, S.P.; Dyjack, D.T.: Development of an ISO 9000-Compatible Occupational Health Standard–II: Defining the Potential Benefits and Open Issues. Am. Ind. Hyg. Assoc. J. 57: 387–391 (1996).

70. British Standards Institution: BS 8800:1996: Guide to Occupational Health and Safety Management Systems. London, United Kingdom (1996).

71. American Industrial Hygiene Association: AIHA Occupational Health and Safety Management System (OHSMS) and Guidance Document. AIHA Fairfax, VA. (May 1996).

72. International Standard Organisation: ISO 9001:1994, Quality Systems—Model for Quality Assurance in Design, Development, Production, Installation and Servicing. ISO, Geneva (1994).

73. International Standard Organisation: ISO 9002:1994, Quality Systems—Model for Quality Assurance in Production, Installation and Servicing. ISO, Geneva (1994).

74. International Standard Organisation: ISO 9003:1994, Quality Systems—Model for Quality Assurance in Final Inspection and Test. ISO, Geneva (1994).

75. International Standard Organisation: ISO 9004-1:1994, Quality Management and Quality System Elements—Part 1: Guidelines. ISO, Geneva (1994).

76. International Standard Organisation: ISO 9004-2:1991, Quality Management and Quality System Elements—Part 2: Guidelines for Services. ISO, Geneva (1991).

77. International Standard Organisation: ISO 9004-3:1993, Quality Management and Quality System Elements—Part 3: Guidelines for Processed Materials. ISO, Geneva (1993).

78. International Standard Organisation: ISO 9004-4:1993, Quality Management and Quality System Elements—Part 4: Guidelines for Quality Improvement. ISO, Geneva (1993).

79. Scott, G.C.; Whitehead, G.M.: Products Liability and Proper Implementation of ISO 9000. Hazard Prevention 29:24–26 (Second Quarter, 1993).

80. Kinder, D.: Process Safety Management: OSHA Standard Parallels Global Quality Efforts. Occup. Health Safety, pp. 47–52 (October 1994).

81. Kozak, B.; Clements, R.: Build Safety into Your ISO 9000 Program. ASSE, pp. 31–32 (September 1995).

Chapter 12

Air Movers and Samplers

Lee E. Monteith, CIH[A] and Kenneth L. Rubow, Ph.D.[B]

[A]*Department of Environmental Health, University of Washington, Seattle, WA*
[B]*Mott Corporation, 84 Spring Lane, Farmington, CT*

CONTENTS

Introduction

Active air sampling for airborne contaminants requires a system for moving air, a collection method, and a procedure to determine the quantity of contaminant collected. Because occupational exposure limits and air quality standards frequently are expressed in terms of concentrations, a method of determining the volume of air sampled is also needed.

The four principal components in a sampling train are shown in Figure 12-1. The inlet admits the air sample into the train; the collector(s) separates the gas, vapor, or aerosol from the air; the flowmeter measures the rate or total quantity of air sampled; and the pump (air mover) provides the suction required to draw an air sample through the train. Inlet sampling considerations are discussed in Chapter 20. Various types of collectors are described in subsequent chapters. Flowmeters are discussed in Chapter 7.

This chapter will describe the air mover portion of the sampling train and air sampling systems that contain three or four components in a convenient package. The chapter consists of two parts. The first part contains a brief description of the different types of air movers and air sampling systems. The descriptions of air movers have remained the same from previous editions but the air sampling systems have changed considerably over the last decade. New information on these types of samplers is included. The second part consists of tables and figures showing detailed information on a wide variety of commercially available air movers and sampling systems.

The experienced reader may want to proceed directly to the technical data tables that describe the types of air movers or samplers of interest. Students and others may find the basic descriptions of the different types of air movers informative. For the reader faced with the task of choosing or purchasing the most appropriate air sampling system, the Guide to the Selection of an Air Mover or Sampler beginning on page 9 will be helpful.

Air Movers

Air movers are classified according to the means by which an air flow is induced. These fall into three basic groups: volumetric displacement, centrifugal force (or acceleration), and momentum transfer. The following sections briefly describe the operating principles and salient features of many types of air movers within each of these three basic classifications. In particular, Tables 12-1 to 12-5 at the end of this chapter list characteristics for five different types of commercially available air movers. These are:

1. Diaphragm pumps
2. Piston pumps
3. Rotary vane pumps
4. Blowers
5. Ejectors

Information presented in these tables includes the source code, pump model number, dimensions, weight, maximum air flow rate, and motor power requirements. The source code can be used in conjunction with Table 12-11 to determine the company name, address, and telephone number.

Volumetric Displacement

One method of producing movement of a given volume of air is to displace it, either by mechanical means,

or by use of a second volume of gas. This principle is the basis of operation for a diverse group of air movers.

Air Displacement

One type of air displacement collector is an airtight flask or rigid-walled vessel in which a hard vacuum has been created. When the vessel is opened at the sampling location, a sample, which is often called a grab sample, that is equal in size to the free volume of the vessel is almost instantaneously collected. The most admirable feature of such a device is its simplicity. Its use is limited mainly by sample size restrictions. Also, care must be taken to prevent loss of vacuum before sampling. For practical purposes, sample size is limited by the portability of the flask, with a normal upper range of approximately 1 to 2 liters. Additional descriptions of this method, the equipment and a list of commercial sources are provided in Chapter 16.

This procedure is not limited to simple collection of air samples for subsequent gaseous analysis. The evacuated flask may contain an absorbing solution in which the desired component of the grab sample may be concentrated for analysis by wet chemical methods. For example, the U.S. Environmental Protection Agency (U.S. EPA) test method for "Determination of Nitrogen Oxide Emissions from Stationary Sources"[1] uses evacuated flasks containing a dilute sulfuric acid-hydrogen peroxide absorbing solution (U.S. EPA Appendix Method 7; see Table 20-7).

Air displacement samples can also be collected without prior evacuation of the flask by simply displacing the air in the flask with sample air. In this case, the flask should be flushed with at least 5-10 volumes of sample air before being sealed, so that the clean air initially inside is displaced completely.

A related method somewhat overcomes the portability problem and allows collection of much larger samples. The collecting vessel is a plastic bag. If it is mounted within a rigid outer container, it can be filled by creating a slight vacuum in the space around the bag. If it is used without an airtight outer container, it can be filled by directly pumping air into the bag, provided that the material being sampled is not absorbed or altered when passing through the pump. Commercial plastic bags are available in a variety of sizes up to 0.3 m^3 and come equipped with several types of leak-proof valves. A list of commercial sources can be found in Chapter 16.

The sample bag method is particularly useful for contaminants that can be analyzed by infrared spectroscopy or gas chromatography. Different types of plastics have been studied for use in this fashion, with the

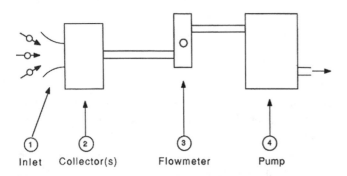

FIGURE 12-1. Principal components of sampling train.

relatively nonreactive fluorocarbons currently in favor. If the bag is to be used for aerosols, it should be foil-lined or have a conductive surface and be grounded. Unless the interaction of the contaminant of interest with the plastic bag, or the possible permeability of the contaminant through the bag, is known or is at least predictable, such bags should be used only for semiquantitative identification. For example, data on such wall losses have been discussed by Posner and Woodfin.[2]

Liquid Displacement

The evacuated flask normally is prepared in the laboratory and carried to the field. A similar method, which can be prepared in the field but does not require a pump, is liquid displacement. A vessel of any convenient size is filled with a liquid in which the suspected contaminant is insoluble. When the liquid is allowed to escape from the vessel, it is replaced by air from the atmosphere to be sampled. Such an air sample is an integrated sample rather than an instantaneous sample because the mass of fluid requires a finite time to empty. In fact, by employing a large vessel with controlled drainage, this method can be used to move air at low flow rates through a sample collector for extended intervals. Additional discussion of this method can be found in Chapter 16.

Diaphragm Pumps

A diaphragm pump, as shown in Figure 12-2, is a device in which a flexible diaphragm of metal or elastomeric material is moved back and forth. The diaphragm is clamped between the pump head and housing, forming a leak-tight seal between the pump chamber and the crankcase. Through the action of a rod or yoke, air in the chamber is displaced on one side of the diaphragm. By using a suitable arrangement of one-way check valves, a variable vacuum is produced in the chamber on the other side of the diaphragm. Mechanical damping is required for more uniform suction. Diaphragm pumps are fairly simple in construction and are used commonly in personal sampler pumps. Diaphragm pumps are oil-free and available in corrosion-resistant materials. They provide contamination-free pumping. Table 12-1 summarizes the characteristics of some commercially available diaphragm-type air movers.

Piston Pumps

Piston pumps are related to diaphragm pumps in that both use a mechanical reciprocating action to provide the motive force. In the piston pump, the piston oscillates in a cylinder equipped with inlet and outlet valves. Because the piston can displace a greater proportion of the air in the chamber above it, piston pumps can provide greater

differential pressure or vacuum. In either case, a surge chamber usually is required to smooth out irregularities in flow. In multiple-piston pumps, these irregularities are small and sometimes can be ignored. One new version of the piston pump is the linear-motor-driven free piston, as illustrated in Figure 12-3. This pump uses an electromagnet and return spring to alternately drive the reciprocating free piston. This design results in a compact structure, less vibration, and quieter operation than the conventional piston pump. It also has an oil-less construction. Table 12-2 summarizes the characteristics of some commercially available piston-type air movers.

Rotary Vane Pumps

Rotary vane pumps are used extensively as air movers for portable sampling instruments. There are two basic variants, both having the same operating principle. A rotor revolves eccentrically in a cylindrical housing, with multiple blades on the rotor providing the air-moving drive. Centrifugal force (or springs) keeps the outer edges of the blades in contact with the housing. However, because the rotary motion is eccentric, the vanes must be movable to retain constant contact. One common method is to place the vanes in slots in the rotor (see Figure 12-4). This guided or sliding vane type usually operates at fairly high speeds and it is subject to wear on the blade edge that contacts the casing. Such wear may lead to leakage and reduce capacity. Table 12-3 summarizes the characteristics of some commercially available rotary vane pumps.

Rotary vane pumps, as well as piston pumps, usually are available either in oil-less or lubricated models. In some air sampling procedures, concern may arise about generating contaminants with the sampling apparatus

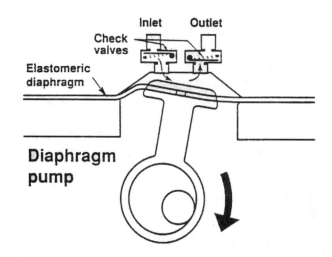

FIGURE 12-2. Schematic diagram of diaphragm pump *(KNF Neuberger)*.

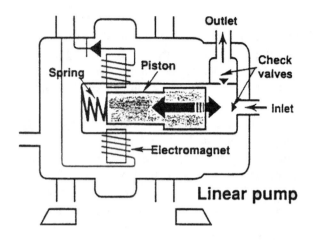

FIGURE 12-3. Schematic diagram of linear-motor-driven free piston pump *(KNF Neuberger)*.

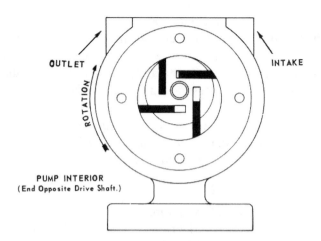

FIGURE 12-4. Schematic diagram showing principle of sliding vane pump operation.

itself. Lubricated pumps can introduce oil mists into the sample if the sample passes through the pump, or if the pump exhaust is resampled. Non-lubricated pumps, which frequently use graphite rings or vanes, can produce carbon dust that may also be an undesirable contaminant. The selection of pump must be based, in part, on the specific air sampling intended.

Gear Pumps

The gear pump is another type of positive displacement air mover. Like a vane pump, it usually is valveless and operates on a rotary principle. The typical gear pump has two shafts, each with a gear. The gears mesh on the interior side and contact the semi-cylindrical casing on the exterior side. The large number of teeth in contact with the outer surface reduces peripheral leakage.

Lobe Pumps

Lobe pumps are similar to gear pumps but use two counter-rotating impellers instead of meshing gears. The impellers can be either a two-lobed, figure-eight-shaped design (see Figure 12-5) or a three-lobed design. As each impeller passes the blower inlet, a volume of gas is trapped, carried through to the blower discharge, and expelled against the discharge pressure. As a result, the volumetric capacity varies little with changes in pressure. Lobe pumps with volumetric capacities from 0.3 to 100 m³/min. are available.

Hand-Operated Air Movers

Hand-operated air movers include manually actuated piston pumps, bellows pumps, and squeeze bulbs. The piston principle has been employed in several hand-operated air movers. The most familiar of these is the hypodermic syringe, which is used in several commer-

cial instruments and in countless homemade systems. A similar air mover is the hand-operated piston. For industrial hygiene purposes, it is usually calibrated carefully to provide an accurate air volume. This type of pump is often used with direct-reading colorimetric indicators as described in Chapter 17.

Squeeze bulbs or small bellows pumps have been used commonly as air movers on commercial air sampling devices, such as the direct-reading indicator tubes described in Chapter 17, and on several models of combustible gas meters, described in Chapter 18. Squeezing the bulb or bellows expels air through a one-way valve; the subsequent self-expansion of the bulb or bellows allows air to be drawn through the detector. With bulbs, the amount of air drawn by a single bulb compression varies according to the efficiency with which the air was expelled from the previous volumetric stroke. This variation may result in serious errors because the calibrations are based on sampling a constant volume of air. Possible sample size variation is less important in combustible gas meters, which indicate the concentration within the sensing zone (i.e., they are not integrating devices).

An example of a squeeze bulb application is found in the U.S. EPA test method for "Gas Analysis for Carbon Dioxide, Oxygen, Excess Air, and Dry Molecular Weight."[1] This method offers a choice of grab sampling with a one-way squeeze bulb, or integrated sampling with a leak-free pump (U.S. EPA Appendix Methods 3 and 3A; e.g., see Tables 20-7 and Figure 20-9A).

Centrifugal Force

A second basic method of inducing air movement is to produce kinetic energy by means of centrifugal force, with conversion of the resulting velocity pressure to

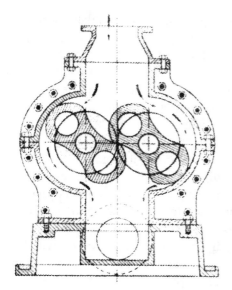

FIGURE 12-5. Schematic diagram showing principle of operation of twin-lubed positive displacement blower.

suction for moving the sampled air. Centrifugal fans that use this approach consist of an impeller and stationary casing, the impeller being a rotary device with vanes. There are two major types of centrifugal fans, either radial flow or axial flow, depending on the direction of air flow through the impeller. Table 12-4 summarizes the characteristics of commercially available centrifugal fans.

Radial-Flow Fans

The term centrifugal fan (or blower), while connoting all types of fans, usually is used in a restrictive sense to indicate just radial-flow fans. These fans are available in three basic types, differentiated by the direction of blade curvature at the delivery edge (see Figure 12-6).

The first of these types is the forward-curved blade fan which has its blade tips curved in the direction of fan rotation. This design usually is more compact, operates at low speeds with less noise than other types, and has a lower initial cost. It is relatively inefficient and not capable of producing high static pressures or high vacuums. These fans are found typically in comfort ventilation systems that have low static pressure or vacuum requirements and where low noise levels are desirable.

The second type is the radial blade fan in which the blades are straight and are aligned along the radii of the fan. This is similar to the old style straight (or paddle-wheel) blade, but the radial blade fan is more compact in design, capable of higher rotational velocity, and slightly more efficient. This type of fan is less prone to clogging than the other types and therefore finds application in systems that handle high particulate mass loads. Unlike the forward-curved fan, radial blade fans operate well in

parallel and can produce relatively high static pressures or vacuums.

The third type of centrifugal fan, the backward-curved blade fan, is characterized by blades that curve away from the direction of rotation. Such fans are highly efficient and, because of their high speed capability, are particularly well suited to use with electric motor drives. This type of fan has a distinct advantage over the other two types in that it is difficult to overload the backward-curved fan. Its power requirement peaks at its normal design loading, whereas the power requirements of each of the other types of centrifugal fans continue to increase with increasing flow volume requirements. The disadvantages of this fan design are relatively high noise levels and high susceptibility to clogging.

The backward-curved fan blade is used in several high-volume air samplers. The usual design is to use a 1/2- to 1-horsepower electric motor (AC or DC) to drive a two-stage turbine impeller. (Multiple-staging is required to increase suction.) The exhaust (sampled) air is often used to cool the motor, which imposes a lower flow-rate limit. Some units have separate cooling fans for the motor and therefore can be used with high-resistance filters at lower flow rates. Some representative types of this kind of blower are described in Table 12-4.

Axial-Flow Fans

Axial-flow fans also come in three types: propeller, tube-axial, and vane-axial. These types are named in increasing order of complexity, weight, cost, and static pressure.

The propeller fan is the most common. The fan blades are carried on a small hub in which the motor is often mounted. It can provide high flow volumes and usually operates at or near ambient static pressures. This type of fan is used on some electrostatic precipitator samplers (see Chapter 14), where the flow resistance is both very low and constant.

The tube-axial fan is a propeller fan enclosed in a short cylinder. The fan blades are mounted on a central ring slightly larger than the average propeller fan hub. This unit can operate at moderate static pressures.

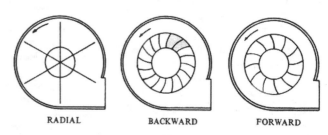

FIGURE 12-6. Curvature of centrifugal fan blades.

The vane-axial fan is a further modification of the tube-axial. It has the same parts, but the cylinder is slightly longer. The extra length accommodates a set of guide vanes that serve to convert the useless tangential velocity component of the discharge into useful static pressure. Vane-axial fans can be either single or multiple stage. In general, vane-axial fans can operate at much higher pressures than tube-axial fans. Because of the high fan speed, the vane-axial fan is sensitive to abrasion of the blades and thus should be used only to move clean air.

Momentum Transfer

A third basic method of inducing air movement is to transfer the momentum of one fluid to another. This process is relatively inefficient, but the equipment involved is very simple and reliable and, in certain cases, particularly well suited for portable sampling units because it does not need electrical power for operation. It has the added advantage of being useful in potentially explosive atmospheres.

A mechanism that utilizes this principle is called an ejector. An ejector consists of a source of high pressure primary fluid, a nozzle, a suction chamber containing a secondary fluid, and a diffuser tube (see Figure 12-7). The primary fluid enters the suction chamber at very

high velocity, entraining the available secondary fluid in the chamber and carrying it through the diffuser. It is discharged at a pressure higher than that of the suction chamber. The entrainment of suction chamber secondary fluid reduces the pressure in the chamber and subsequently causes entrainment of the remaining secondary fluid (sampled air).

The driving medium can be steam, high pressure air, water, or any compressible gas. For portable units, a small can of refrigerant can be used to power a compact, low-volume sampler. Mercury or oil diffusion pumps are sometimes used to obtain the optimum vacuum for the transfer of organic fractions from the sampling apparatus to the analytical instruments. Characteristics of some small, commercially available ejector air movers are summarized in Table 12-5.

Air Samplers

The previous discussion has been concerned solely with air movers, with minimum attention to the other three components of the sampling train illustrated in Figure 12-1. An air sampler combines all four functions in a convenient package and is widely used in industrial hygiene practice and indoor or ambient air sampling. Many commercial varieties of air samplers are available, employing many of the air moving devices previously

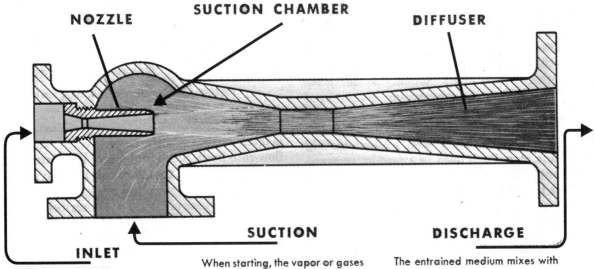

The motive fluid (steam, air, or liquid) enters the ejector through the inlet nozzle which converts the fluid pressure into a high-velocity jet stream.

When starting, the vapor or gases in suction chamber are continuously entrained by the jet stream emerging from the nozzle thereby lowering the pressure in suction chamber, causing the liquid, gases or vapor in the suction system to flow to the ejector.

The entrained medium mixes with the motive fluid and acquires part of its energy. In the diffuser the velocity of the mixture is reconverted to a pressure greater than the suction pressure but lower than the motive fluid pressure.

FIGURE 12-7. Schematic diagram showing principle of ejector operation.

discussed. Tables 12-6 through 12-10 list characteristics of several classes of commercial air samplers. Table 12-11 provides a list of the commercial sources for the air movers and samplers.

Over the last two decades personal sampling pumps have developed into one of the most used tools of the industrial hygienist and environmental evaluator. The pumps have become smaller, more reliable and more versatile. From large bulky pumps to the now small inconspicuous models, the newer pumps can be calibrated and controlled by internal programs. Feedback control by pressure sensors or by mass flow controllers has advanced their use and application. Multiple flow ranges for manufactured lines of pumps have broadened their use to cover practically all needs for low to medium high flow range: from a few milliliters per minute to 10-20 liters per minute. Pump manufacturers have responded to the requirements for indoor air sampling from the low flow over the long term and up to high flow for shorter term environmental sampling. One can now select the sampling pump for a specific purpose by balancing the features desired against the budgetary limitations: for example, the need for many low priced simple pumps vs. a few more expensive smart pumps.

Categories of Samplers

The listing of samplers presented in Tables 12-6 to 12-10 has been divided into five categories based on the sampling flow rate, source of power, sampler size, and primary application of the sampler. These categories are:

1. Personal sampler: battery powered
2. Low-volume area samplers: battery powered
3. Low-volume area samplers: portable
4. Medium and high volume samplers: portable
5. High volume samplers: with shelters

Flow Range for Each Category

Personal Samplers and Low Volume Area Samplers.
Personal samplers generally operate at a sampling air flow rate of up to 5 L/min. and are intended for personal exposure monitoring. Low-volume area samplers typically operate at flow rates in the 3-100 L/min. range. Medium and high volume samplers generally operate at flow rates from 0.1 to 0.3 m³/min. and 0.3 to 3 m³/min., respectively, and are most often used for ambient air sampling.

Types of Air Movers for Each Category

Information presented in Tables 12-6 to 12-10 includes the source code, model number, dimensions, weight, air flow-rate range, type of flowmeter, standard sampling attachments (inlets or collection devices), and motor power requirements. The source code can be used in conjunction with Table 12-11 to determine the company name, address, and telephone number.

Many of the air samplers listed in Tables 12-6 to 12-10 are designed to be used in conjunction with a variety of sampling inlets or sample collection devices. The air samplers are not sample collectors by themselves, but provide the air flow through the collectors. Several of the commonly used inlets and collection devices associated with these samplers are listed in these tables. For example, personal sampling pumps are used with charcoal and other sorbent tubes, filters, bubblers, cyclone-filter assemblies for respirable dust sampling, and gas collection bags. Medium- and high-flow area samplers are often used with PM_{10} or $PM_{2.5}$ inlets or collection devices. PM_{10} and $PM_{2.5}$ inlets or collection devices remove particulate matter greater than 10 or 2.5 micrometers, respectively. Detailed discussions of these inlets and devices are provided in later chapters.

A variety of air movers (pumps) are used in the samplers listed in Tables 12-6 to 12-10. Diaphragm pumps are generally used in personal sampler pumps. Low-volume area samplers usually use rotary vane, piston, or diaphragm pumps. Blowers are primarily used in medium- and high-volume samplers.

Flow Measurement in Personal Samplers

Flowmeters are frequently incorporated into the design of the samplers listed in Tables 12-6 to 12-10. The most commonly used integral flowmeters are either a rotameter or a calibrated orifice plate. Numerous other external flow meters such as the soap bubble meter and electronic flow meters can be used to set and calibrate the flow rate. To achieve reliable results from such instruments, a basic understanding of the limitations of these flow measuring devices is necessary. Procedures for calibrating air sampling flowmeters are discussed in Chapter 7.

Rotameters

A rotameter consists of a float inside a tapered vertical tube; the tube cross section increases in area from bottom to top. The position of the float in the tube is governed by establishing an equilibrium between the weight of the float and the force exerted on the float by the velocity pressure of the fluid or gas flowing through the annular space between the float and the tube wall. With increasing flow, the float rises and the annular area increases, permitting equilibrium to be established over a range of flow rates. Because the position of the float is related to fluid flow, flow rates can be etched directly on the tube after appropriate calibration. See Diagram in Chapter 7.

Rotameters are available for a wide range of gas flow rates. For a given unit, the range of accurate performance is usually about a factor of 10. For example, if the lowest accurate reading is 100 ml/min., the highest would be about 1000 ml/min. The accuracy of an individual rotameter depends mostly on the quality of construction. For a well-made unit with an individual calibration, accuracy to within ±2% of full scale is possible. However, the rotameters found as an integral part of "personal air samplers" are mass produced and are often far less accurate. The rotameters usually used with samplers are also imprecise because of their short length to diameter ratio. It is advisable to use their reading as a rough indication of flow rate. After an initial check of their calibration over their entire range against a secondary standard, rapidly check calibration at one or two points before and after each use. Because proper operation of a rotameter depends on clear annular space, these gauges are sensitive to accumulation of dirt or water vapor on either the float or tube walls. Periodic cleaning is recommended, usually in conjunction with a recalibration.

Because contamination of rotameter walls is a problem, it is general practice to place the meter downstream of the sampling device. This accomplishes the dual function of minimizing both wall buildup and sampling line losses. This practice, however, will usually introduce another problem. Most sampling mechanisms induce a pressure drop in the system, so the rotameter will operate in a partial vacuum. Because the manufacturer normally calibrates at atmospheric pressure, a significant error can be introduced. For a concise discussion of the problem, the reader is referred to the short note by Craig.[3] The air sampling system flowmeter needs to be calibrated each time it is used (see Chapter 7).

Calibrated Orifice Plate Assemblies

A calibrated orifice plate assembly is a second flow-controlling and flow-measuring device commonly used in air samplers, mainly because of simplicity and low cost. A thin metal plate with a carefully machined sharp-edged hole is placed in the air stream. The hole, or orifice, causes a convergence of streamlines downstream, with maximum contraction occurring at the *vena contracta*, the point of lowest pressure. A differential pressure device is then used to measure the pressure drop caused by the orifice and the flow rates can be calculated from the pressure differential using orifice equations.

However, with all physical parameters controlled and designated, the pressure gauge output can be empirically calibrated to directly read in flow units.

Limiting Orifices. Most problems with orifice meters can be associated with careless construction. The upstream edges of the orifice must be clean and sharp, with a 90° corner. The pressure taps must be positioned carefully with no roughness inside the nipple, or burrs or rough edges inside the hole. Three sets of pressure tap positions commonly are used: 1) flange taps are centered "from the nearest face of the orifice plate," 2) *vena contracta* taps have the upstream tap located one pipe diameter from the upstream face and downstream tap at the *vena contracta*, and 3) pipe taps are located 2.5 pipe diameters upstream and 5.0 pipe diameters downstream.

Critical Orifices. A special type of orifice known as a "critical flow orifice" is popular in air sampling work because it can maintain a moderately constant flow rate despite minor changes in inlet conditions. Flow through a critical orifice reaches a maximum value when sonic flow conditions are achieved in the throat of the orifice. Although increased sampler resistance during a run (e.g., filter buildup) could affect the critical flow rate, the pressure changes normally encountered do not produce major errors in flow rate measurement. A typical example of this type of device is found in the U.S. EPA reference method for the "Determination of Sulfur Dioxide in the Atmosphere,"[4] which offers a choice of several gauges (22, 23, or 27) of hypodermic needle to maintain a range of sample flow rates.

Laminar Flow Sensors

A recent development is a type of pump that incorporates an electronic laminar-flow sensor, which is insensitive to atmospheric pressure changes. The sensor is designed to have a linear relationship between the pressure drop and the volume flow rate. The intrinsic pressure sensor allows the pump to directly measure the flow rate. The control feedback acts as an internal secondary standard to control the pump flow.[5]

Constant Motor Speed

Several manufacturers now offer "constant flow" air samplers. These devices are designed to overcome flow rate variation problems that are inherent in many sampling situations (e.g., with a constant speed pump, an increase in sample media resistance will result in a decrease in flow rate). These samplers feature sophisticated flow rate sensors with feedback mechanisms that permit maintenance of a preset flow for the duration of the sample. A block diagram of this system is shown in Figure 12-8. Although these commercial devices are most commonly used for personal samplers with flow rates up to several liters per minute, the concept is readily applied to high volume samplers.[5]

Guide to the Selection of an Air Mover or Air Sampler

The following paragraphs discuss the many factors which should be considered in choosing the best air mover or sampler for the sampling task. If the sampler is to be used for a variety of applications, compromises must be evaluated to obtain an air sampler adequately versatile. The selection of an appropriate air mover for sampling purposes depends on a number of factors which can be divided into those which relate to the conditions of the sampling, to the requirements of the sample collector, and to the features of the air mover. The following discussion covers all three considerations.

Choice Factors Based on Sampling Considerations

The physical factors concerning the sampling mission are such things as the required sample volume, the necessary flows for the sampling device, the physical limitations of the characteristics of the air mover such as size and weight, and the particular type of collector such as particle size selectors vs. gas and vapor collectors. Governmental requirements or the official standard methods for the sampling may dictate the specifications that the air mover must meet.[6] The sampling location and the availability of power may limit the useful types of air movers. The requirement for operation in hazardous environments may limit the appropriate movers.

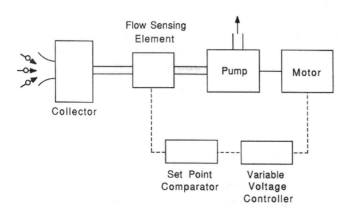

FIGURE 12-8. Block diagram of sampling system with feedback system for flow rate control.

Other desirable features that can influence the choice are serviceability, difficulty of calibration, and other features such as cost, security, noise, portability, and other attachments.

Sample Volume

The minimum or maximum amount of air to be sampled is usually dependent on the requirements of the analytical procedure, the anticipated concentration of the contaminant to be sampled, or the parameters set by applicable governmental standards. The minimum amount of sample is dictated by the lower detection limit of the analytical procedure. The maximum amount sampled is often a balance between the capacity of collection medium (e.g., charcoal, impinger solution) and sampling strategy (e.g., short-term versus time-weighted average [TWA] sampling). However, with some sampling systems, the collection period cannot be extended indefinitely because the accumulation of sample can cause changes in the operating characteristics of the system. This is particularly true of air movers for sampling with filters, which can cake up or crock off and the increasing resistance can affect flow rates and efficiency. Whereas, air movers for the cascade impactor, must limit the sample volume to prevent the sample buildup that encourages re-entrainment. Similarly air movers for the charcoal tube samplers must limit sample volume to prevent problems of sample breakthrough. Lastly, standards often require sampling at a certain flow rate, or within a set range (e.g., Asbestos Hazard Emergency Response Act [AHERA] standards for asbestos) and for a certain period of time (e.g., an 8-hour work shift) which may set the total sample volume and thus define the type of air mover. With adsorbent vapor collectors the problem of breakthrough and capacity place limits on the maximum sample volume.

Sampling Rate and Ranges of Flow Rates

Having defined the sample volume, the next choice is sampling rate. Two factors that can influence this decision are the total sampling time required and the dynamic characteristics of the collector.

Time Required

It is obvious that total sampling time is important in those cases where, because the required volume is large, only higher sampling rates will accomplish the procedure in a reasonable time. However, given a more moderate volume requirement, it must be decided whether to sample in a relatively instantaneous fashion, or obtain a longer-term, time-averaged sample. For a long-term sample, there is an additional choice of continuous versus intermittent sampling.

Requirements of the Sampler

Dynamic Character of the Sample Collector (Size Selection)

The sampling rate can affect both the overall efficiency and the size selectivity of a collecting device. The efficiency of a given collector, whether filter paper, liquid media, granular beds, or cyclone, will usually vary with sampling rate. This variation is seldom linear and must be determined experimentally. The cut-point of size selective aerosol collectors is very dependent on the controlled flow rate as detailed in Chapter 6. Furthermore, the sampling rate often determines what size of particulate material will be removed most efficiently. For those cases where the size distribution of the sample is of critical importance, desirable flow rates usually are established during instrument calibration and must not be altered. A flow controlling device will reduce this problem considerably. Lastly, the sampling flow rate may be specified by an applicable governmental standard. For example, a flow rate of 2 L/min. is required for sampling respirable coal mine dust,[6] whereas a flow rate of 1.13 m^3/min. (40 CFM) is required for PM_{10} sampling of ambient particulate matter.[7]

Flow Restriction/Back Pressure/Flow Resistance

A final factor related to sampling rate is the increase in sample media resistance as flow velocity increases. For small, battery-operated samplers, such a resistance increase places a practical upper limit on sampling rate. The collection efficiency of impingers and cold traps will be dependent on an optimum flow rate. Resistance increases with flow rate must be kept reasonable. The more restrictive the sample collector the more back pressure will increase with flow rate and eventually will effect the ability of the pump to control the flow.

Effects of Battery Life and Sample Loading

Proper attention must also be paid to the effect of battery life and sample loading on flow rate because both can lead to decreases in sampling air flow rate over a long sampling period. As the battery loses power or ages, the pump motor may slow down, reducing the flow rate. Also, as the collector loads the restriction to flow can cause the pump to move less air. Many samplers are now designed with feedback systems to maintain a constant air flow rate by continuously adjusting the pump or a flow control valve. These samplers are called constant

flow rate samplers. Menard et al.[8] described a computerized personal pump test system designed to test the ability of pumps to operate at a desired air flow rate. Early personal sampling pumps could be adjusted to a range of flows but could not hold the flow if conditions changed. The flow rate had to be determined at the start of sampling and at the end and an average flow rate applied to the sample.

Flow Control Devices

Constant flow devices using critical orifices or limiting orifices to maintain constant flow were previously described. The pump operating downstream of the orifice is not affected by minor changes in the back pressure of the sampler as long as the changes are less than the control range of the orifice. Newer pumps use an electronic feedback system to provide a constant flow. The system may have an internal pressure sensor which controls the speed of the pump or an internal mass flow device (laminar-flow sensor) such as that tested by Webber, et.al.[9] and was determined to provide consistent flow rate and maintain calibration under typical use.

Micro-Computerized Samplers

Most readers will know that the more modern pumps have programmable features such as timed sampling intervals and intermittent sampling. The pumps can be set for delayed start sampling time, specified on/off intervals or spaced sampling at constant intervals over an entire sampling period. The authors were not always provided with adequate descriptions to be able to include these features for all of the instruments in the technical tables.

Programmable Features. The pump's microcomputer will keep track of the overall sampling period and the total time sampled. Some pumps can also store and recall the total sampling time in case the pump shuts down because of excessive flow restriction caused by a kinked sampling line or a flow restricted beyond the pump control.

Controls. The controls of the pump are a feature which can play a roll in the selection of the most appropriate model. External switches have mostly been replaced with soft key buttons and the digital display of the controlled parameters. Security covers for the key pads prevent accidental changing of the set flow conditions.

Visualization of Flow

Internal Rotameter. Numerous models of pumps contain an integral rotameter to give a quick visual indication of the flow rate. As previously discussed, these rotameters are not as precision made as standard rotameters and should only be used as quick visual indications of flow rates and changes. Upstream flow restriction of sample collectors and other components of the sampling train will effect the indicated flow and must be compensated.

Digital Readout. Digital indication of flow is available in pumps with computerized flow measuring devices such as the laminar flow sensors.

Recorded/Logged. Micro-computerized pumps can record the flow rate at intervals and log the flow data for later use by downloading into an external computer.

Other Indicators. Other indicators are external to the sampling pump and must be used before and after the sampling periods. Their limitations are that no information is obtained about the changes in flow as they occur or the time scales of the changes.

Physical Characteristics of Personal Sampling Pumps

Obvious factors which can influence the choice of the personal sampling pump are as follows:

- The size, weight, and shape of personal sampling pumps have been approaching the convenient dimensions that allow them to be acceptable for all types of missions. Portability and unobtrusive operation is a key issue for many sampling applications.
- Controls, buttons, and displays are a matter of preference and as can be seen in the Tables, a wide variety are available.
- Clips and attachments are fairly standard, although some models have a holster for easy carrying.
- Depending on the variety of applications, both inlets and outlets may be desirable if the device will be used for both drawing samples through collectors and for filling bag samples.
- Security features may include covers for control buttons and switches and password mechanisms for microcomputer programs.

Applications

The application of the sampling pump may dictate which of the above features that will most accommodate the method. Most sampling pumps can handle all of the following types of collectors and tasks, but certain requirements might provide the overriding consideration.

Sampling Pumps for Particulate Sampling

In aerosol sampling the use of fine filters and impingers may dictate the type of pump that can handle greater back pressure. The use of size selective samplers, such as cyclones and impactors, will define the required flow range. High volume sampling of particulates requires the type of air movers described in Tables 12-8 to 12-10.

Sampling Pumps for Gases and Vapors

Sampling for gases and vapors can have different requirements than aerosols. The need to fill bag samplers requires the capability to exhaust the sample through the pump into the collection device. Fine fritted bubblers may require both lower flows and smooth operation against greater back pressure than for other vapor collectors. Sampling with adsorbent tubes for TWA over 8-hour periods may require constant low flow to prevent overloading the capacity of the collection tube. The alternative is to have a sampling pump that can do unattended intermittent sampling for set periods over the duration of the test period. For qualitative evaluation of hazardous situations the need for high volume sampling over a short time might specify that the sampling pump can handle medium to high flow rates.

Government Standards and Standard Sampling Methods

If the sampler is to be used primarily in following applicable governmental standards or other standard methods such as those found in the NIOSH Analytical Methods Manual,[10] the prescribed flow rates and sample collectors will dictate the most applicable types of sampling pumps, especially for low and medium flow rates.

Power Source and Location

The sampling location and power may be the deciding factor in the selection of the sampling pump. These factors tend to interact because the power source is often determined by the location of the sampling device. The choices of sampler location are fairly obvious, ranging from permanent fixed installations, to the movement of pollution clouds or of personnel through multiple locations requiring various degrees of portability and mobility, to the ultimate in lightweight, fully portable devices used for various forms of personal monitoring. For power, the choice usually is between line current, storage battery, hand power, or compressed gas cylinder or cans, such as refrigerant cans (momentum transfer ejectors). The use of portable, gasoline-powered generators is increasing for remote area monitoring, but their use in air sampling must be monitored carefully to ensure that their exhausts do not introduce contamination. The usefulness of air movers and samplers requiring batteries has been increased by the development of more compact, longer-life cells, many of which are rechargeable. Perkins[11] gives a good summary of the status of batteries for portable personal sampling pumps.

Batteries and Chargers—Trickle Charge, Automatic Discharge/Chargers

Because pumps, batteries and chargers are usually sold as a package, a brief discussion of the latter is included. The wide spread use of nickel-cadmium batteries has helped to reduce the size and improve the reliability of the samplers, but the problem of loss of capacity of the batteries has created a need for better chargers. A purchaser should select a charger that will keep the batteries operating at full capacity for as long as possible.[11] Such chargers should have the capability of cycling the batteries through charge and discharge periods, as well as providing a trickle charge between periods of use; otherwise, the user will need to manually sequence the batteries to make up for using a more economical charger. Multi-unit chargers are available to simultaneously charge a number of samplers.

Environments

Hazardous Environments

Hazardous environments may also be a factor when selecting an air mover or sampler. In flammable and explosive atmospheres, only explosion-proof electric devices or those instruments that are inherently explosion-proof (such as ejectors) may be used. The term "intrinsically safe" is often used to describe an electrically powered device that is capable of operating in an explosive atmosphere. Some of the air movers and samplers described in Tables 12-1 to 12-10 are suitable for such use and have been approved by the Mine Safety and Health Administration, Underwriters Laboratories, Inc. (for Class 1 Group D Flammable Atmospheres), National Fire Protection Association, and/or Factory Mutual Engineering Corporation.

EMI/RFI

Another type of hazardous environment involves the effect of electromagnetic interference/radio frequency interference (EMI/RFI) on the performance of samplers. In particular, this problem is of concern for sensitive, low power instruments such as personal sampler pumps where EMI/RFI can adversely affect the operating performance of the pump.[12] This degraded performance is called electromagnetic susceptibility. EMI/RFI effects on pumps can be minimized through proper shielding. Several of the newer personal sampler pumps are designed with EMI/RFI shielding.

Remote Operation

Other types of hazards may be present. For extremely toxic environments, remotely activated, automatically sequenced devices may be required. For corrosive or high-temperature environments, the materials used in constructing the device may be important. Also, although the original environment may be fairly innocuous, it is possible that the air mover itself could generate a hazardous or annoying condition, such as carbon monoxide from gasoline-powered samplers, or excessive noise. Noise generation can be extremely important when dealing with high volume samplers over extended sampling periods and samplers designed for indoor air sampling. Delayed start and timed shutoff features would allow remote operation of the sampler.

Servicing

Another consideration in sampler selection is the degree of difficulty involved with calibration and maintenance. The amount of effort expended in these areas varies with both the sampling apparatus and the degree of precision desired. Naturally, it is desirable that field instruments require a minimum of such care. Battery-powered personal air samplers require constant attention for maintenance, repair, and recharging. Flow calibration is now straightforward with automatic calibration devices. The newer samplers with laminar flow sensing have a limited self calibration capability. Ease of field maintenance, access to the inner workings and the availability and/or interchangeability of parts would be a positive factor in the choice of samplers.

Calibration

The Need for Calibration

With some of the newer micro-computer controlled pumps, the need for calibration may be less frequent than in the earlier models which depended on the reading of its rotameter and manually changing the controls. For each model of personal sampling pump the purchaser should at least follow the calibration frequency specified by the manufacturer, or even better, determine the actual appropriate calibration periods under the conditions of use. The latter can only be established by close observation and frequent calibration until a reliably stable period of use is determined. Some of the newer microcomputerized samplers log and retain a record of the calibration.

Desirable Features

Other features such as advanced electronics and programmability, timers, reduced size and weight, type of controls and readout, and safety provisions should be weighed against cost to determine their worth for the required applications.

Concluding Comment

A concluding comment addresses the current need for instrument performance criteria. A symposium sponsored by ACGIH was held in 1991 to specifically assess both the current state of performance of air sampling field instruments and instrument performance criteria.[13] In particular, one of the eight symposium workshops specifically dealt with the need for performance criteria for pumps and air movers in sampling systems.[14] The recommendations of this workshop covered the areas of standardized performance standards, including international cooperation, user technical education and training, use of generally acceptable terminology, and improvement needed in air samplers (see Chapter 11 for additional discussion).

References

1. Code of Federal Regulations, Title 40, Chapter 1, U.S. Environmental Protection Agency, Part 60. U.S. Govt. Printing Office, Washington, D.C. (1992).
2. Posner, J.C.; Woodfin, W.J: Sampling with Gas Bags. 1. Loss of Analyte with Time. Appl. Ind. Hyg. 1:163 (1986).
3. Craig, D.K.: The Interpretation of Rotameter Air Flow Readings. Health Physics 21:328–332 (1971).
4. Code of Federal Regulations, Title 40, Chapter 1, U.S. Environmental Protection Agency, Part 50.5, Appendix A. U.S. Govt. Printing Office, Washington, D.C. (1992).
5. Bernstein, D.M.: An Electronic Feedback Constant Flow Controller for High Volume Samplers and Air Movers. Am. Ind. Hyg. Assoc. J. 40:835–837 (1979).
6. Code of Federal Regulations, Title 30, Subchapter 0, Coal Mine Safety and Health Administration, Part 70. U.S. Govt. Printing Office, Washington, D.C. (1992).
7. Code of Federal Regulations, Title 40, Subchapter C, U.S. Environmental Protection Agency, Part 50, Appendix J. U.S. Govt. Printing Office, Washington, D.C. (1992).
8. Menard, L.; Caron, B.; Lariviere, P.: Computerized Personal Pump Tester and Cycler. Appl. Occup. Environ. Hyg. 8:327–333 (1993).
9. Webber, A.F. , et.al.: Field Testing a Personal Sampling Pump with Laminar-Flow Sensor As Secondary Calibration Standard. Appl. Occup. Environ. Hyg. 11:646–651 (1996).
10. National Institute for Occupational Safety and Health: NIOSH Manual of Analytical Methods, 4th Edition, DHEW (NIOSH) PUB. NO. 94-113 (1994).
11. Perkins, J.L.: Modern Industrial Hygiene, Volume 1, Van Nostrand Reinhold, New York, 1997.
12. Feldman, R.F.: Degraded Instrument Performance due to Radio Interference: Criteria and Standards. Appl. Occup. Environ. Hyg. 8:351–355 (1993).
13. Cohen, B.S.; McCammon, C.S.; Vincent, J.H.; Eds.: Proceedings of The International Symposium on Air Sampling Instrument Performance. Appl. Occup. Environ. Hyg. 8:223–411 (1993).
14. Rubow, K.L.: Report of Workshop #1 on Pumps and Air Movers. Appl. Occup. Environ. Hyg. 8:397–398 (1993).

TABLE 12-1. Diaphragm Pumps*

Source Code	Figure Number	Item, Model No., or Catalog No.	Dimensions(cm)			Wt. (kg)	Maximum Flow (L/min)	Remarks
			L	W	H			
ADI		Micro Dia-Vac	3.87	1.75	3.24	2	0–15	CE approved, non-contaminating oil-free diaphragm pumps. Corrosion-resistant materials available. Explosion-proof, air driven pumps, heated heads, or DC motors available.
		Mini Dia-Vac	7.8	4.2	5.5	8	38	
		Standard Dia-Vac	25.9	21.6	15.5	8.6	0–115	
		Micro-Mini- 19310V	12.2	6.2	11.0	1.1	7	
ACI	12-9	Dia-Pump, G3	27.2	14.2	19.1	7.0	22	Neoprene diaphragms standard, but corrosion-resistant materials are available. Model G4 requires separate power; 220 VAC/50 Hz motors available as option. Models G4 and G5A are rated as explosion-proof. See Figure 12-9 for vacuum flow characteristics.
		G4	11.9	15.2	15.2	2.5	28	
		G4A	39.9	16.5	19.8	16.3	22	
		G5/G5A	37.1	45.7	25.4	13.2	28–50	
ASF	12-10	ASF more than 10 models	—	—	—	—	0.9–4.3	Selected technical data on ASF diaphragm pumps in Figure 12-10; 6, 12, and 24 VDC and 115 VAC/60 Hz.
	12-11	WISA 113	11.6	6.2	8.4	0.8	4.2	Only data for selected models are listed. Corrosion-resistant diaphragms available for all pumps. Alternate power modes (220/240 VAC, 50 Hz, and 24 VDC/60 Hz) are also available. Model 115 has plastic case; Model 504 also has steel case.
		125	15.7	6.6	9.0	1.1	4.2	
		203	16.9	7.4	10.5	1.6	4.6	
		303	19.4	10.0	10.6	2.2	5.5	
		504	24.7	15.3	11.3	3.9	21	
ASF		ROMEGA 014	5.4	2.7	4.5	0.05	0.8	2.4, 6, and 12 VDC.
		020	5.8	2.7	4.8	0.08	1.9	
		041	7.4	3.5	4.8	—	2.9	12 VDC.
		044	7.4	3.5	4.8	0.12	2.9	
		074	13.0	5.9	9.2	1.0	9.2	220 VAC/50 Hz; 110 VAC/60 Hz; 12 VDC.
		084	9.7	6.0	6.0	0.22	4.0	12 VDC.
		094	9.7	6.0	6.0	0.22	3.9	12 VDC.
BAR		Series 400	17.8	10.0	12.7	2.5	4–23	15 models; 115 VAC/60 Hz, 230 VAC/50 Hz, 12 and 24 VDC; Fluorel®, Nitrile®, and Viton diaphragms; plastic pump head.
		Series 900	24.1	15.9	12.4	3.2	25–31	
BRA		TD-2A/1A	7.9	3.8	9.1	0.3	3.5	6, 9, 12, 16 and 24 VDC versions; all models have plastic housing with synthetic rubber diaphragm, optional adjustable stroke; all motors are brushless.
		TD-2L/2N/2S/1S	7.7	3.8	8.6	0.3	3.5	
		TD-4X2L/4X2N/4X2S/4X2	11.9	3.8	8.6	0.4	6.0	
		TD-3LL/3L/3LS	7.2	3.8	8.6	0.2	2.0	
		TD-4L/4N/4S/4	8.3	3.8	8.6	0.3	4.0	
		TD-4X23H	11.9	3.8	11.9	0.5	5.0	
		TD-45BD	8.8	3.8	8.6	0.4	3.0	

TABLE 12-1 (cont.). Diaphragm Pumps*

Source Code	Figure Number	Item, Model No., or Catalog No.	Dimensions(cm)			Wt. (kg)	Maximum Flow (L/min)	Remarks
			L	W	H			
GST	12-12	2D Series	3.2	1.6	2.6	0.02	.5	For technical data on Gast diaphragm pumps, see Figure 12-12. Also available with 110 V/50 Hz, 230 VAC/60 Hz, and 220 VAC/50 Hz and 12 and 24 VDC motors. (4, 6, and 9 VDC on the smaller miniature styles.)
		5D Series	4.2	2.3	3.6	0.04	.95	
		10D Series	7.6	3.2	5.3	.23	4.3	
		15D Series	9.8	4.6	7.5	.4-1.2	6.5	
		MOA Series	18.0	10.9	11.9	2.7	11-23	
		MAA Series	21.6	10.9	11.9	4.1	11-45	
		DOA Series	19.3	13.0	19.8	8.2	27-51	
		DOL Series	19.3	13.5	26.7	8.2	31	
		DAA Series	30.0	14.0	19.3	12.7	24-100	
KNF	12-13	Model OEM	4.2	10	7.5	—	6.5	2.3 to 24 VDC and 115 VAC/60 Hz; explosion-proof motor available. All KNF diaphragm pumps are oil-free and available in corrosion-resistant materials.
		More than 30 models	—	—	—	0.1 to 15.9	0.55 to 300	
MBC	12-14	MB-21/41	16.5	8.6	10.9	2.3	5-11	See Figure 12-15 for pressure volume characteristic; bellows made of 350 stainless; gaskets Teflon or Viton. Following available in high-temperature versions: MB-21/41/118/158/302/601. Following available as explosion proof: MB-21/41/118/158. All except MB-111/151 available in 230 VAC/50 Hz.
		MB-118/158	21.8	10.2	19.1	6.4	18-40	
		MB-111/151	27.9	14.2	22.4	10.9	28-40	
		MB-302	33.3	14.2	22.6	14.1	85	
		MB-601	33.8	33.3	22.9	21.8	140	
		MB-602	33.8	30.5	20.3	15.9	170	
MRM		Model EC500-LC	3.4	2.2	3.3	0.02	1.1	All models 1.5, 4.5, 6, and 9 VDC; chlorobutadine or Viton diaphragms.
		ACI series 100	8.9	5.7	5.7	0.19	3.0	
		ACI series 50	8.3	3.8	5.4	0.16	0.9-1.7	
		ACI series 25	4.9	2.5	3.5	0.07	0.8-1.07	
SEN	12-16	OEM sampling pumps	—	—	—	—	0.001-14	Available with 1-48 VDC and 120 VAC/60 Hz motors; silicon, neoprene, or Viton diaphragms.
SPE	12-17	AS-200	4.0	2.5	1.8	0.04	0.30	See Figure 12-17 for pressure-volume characteristic. Model AS-200 is 6 VDC; AS-300/350 series operate at 3-15 VDC, but are also available at 110 VAC/60 Hz. Pump bodies are polycarbonate resin.
		AS-300 series	7.9	6.4	5.8	0.2	—	
		AS-350 series	8.4	6.4	7.9	0.2	—	
THO	12-18	14 models available	—	—	—	—	17-175	See Figure 12-18 for specifications.

*Unless otherwise noted, all pumps listed above have aluminum cases and are rated for 115 VAC/60 Hz. Maximum flow is for free air. Where no maximum flow is listed, see cited figure.

TABLE 12-2. Piston Pumps*

Source Code	Item, Model No., or Catalog No.	Dimensions(cm)			Wt. (kg)	Maximum Flow (L/min)	Remarks
		L	W	H			
GST	1VAF-10	29.5	14.2	21.8	7.3	51	All pumps oil-less. Available with 110 VAC/50 Hz, 230 VAC/60 Hz, and 220 VAC/50 Hz motors.
	1VSF-10	31.8	14.8	11.4	8.6	85	
	1VBF-10	31.8	14.8	11.4	8.6	90	
	4VSF-10	38.1	30.3	22.6	15.0	120	
	4VCF-10	38.1	31.2	22.6	15.0	140	
	5VSF-10	52.5	30.3	22.6	24.0	180	
	5VDF-10	52.5	31.2	22.6	24.0	300	
	ROC-R LOA	18.9	10.9	14.8	3.0	24	Rocking piston type; all pumps oil-less. Available with 115 VAC/60 Hz, 230 VAC/50 Hz, 12 VDC, and 24 VDC motors.
	LAA	24.8	10.9	14.8	5.1	43	
	SOA	21.4	10.2	16.8	4.4	45	
	SAA	24.8	14.3	17.0	5.0	94	
	ROA	22.5	13.5	19.9	5.6	45	
	RAA	30.3	19.8	13.5	8.6	76	
MED	VP 0125	9.2	7.5	9.0	0.7	7	All pumps linear-motor-driven free piston system. 115 VAC/60 Hz and 230 VAC/50 Hz motors.
	VP 0140	9.2	7.5	9.0	0.7	3	
	VP 0435A	16.0	10.4	12.2	2.3	25	
	VP 0625	17.5	11.8	13.8	3.0	40	
	VP 0645	18.9	11.8	13.8	3.2	10	
	VP 0660	21.0	15.0	15.4	5.0	25	
	VP 0935A	19.3	15.0	15.4	4.4	60	
	VP 0945	20.6	15.0	15.4	4.9	12	
	VP 0550x2	28.6	10.6	13.5	4.7	30	
	VP 0660x2	39.6	15.0	15.4	10	50	
	VP 0950x2	37.0	15.0	15.4	9.5	80	
SCI	D-1000	5.1	8.9	10.2	0.4	1	All pump surfaces contacting gas are Teflon-reinforced with 15% glass fiber. Model D-1000 has 12 VDC motor. Model D-200 has 6, 12, or 24 VDC motor.
	A-1000	5.1	8.9	10.2	1.0	1	
	D-200	5.1	6.4	7.6	0.1	0.2	
	A-150	—	—	—	0.2	0.15	
SPE	ASB300 & -301	7.8	3.8	5.7	0.2	4	12 VDC and 115 VAC/60 Hz.
	AS-350 & -351	8.4	3.8	7.8	0.2	8	12 VDC and 115 VAC/60 Hz.

TABLE 12-2 (cont.). Piston Pumps*

Source Code	Item, Model No., or Catalog No.	Dimensions(cm) L	W	H	Wt. (kg)	Maximum Flow (L/min)	Remarks
THO	TASKAIR (11 models)	—	—	—	—	38–280	See Figure 12-19 for specifications.
	WOB-L (28 models)	—	—	—	—	10–200	See Figure 12-20 for specifications.
	Linear 5020V	11.1	12.8	14.8	2.2	33	All pumps are linear-motor-driven free piston system; 115 VAC/60 Hz motor.
	5030V	16.1	19.6	20.3	4.9	51	
	5040V	16.1	19.6	20.3	5.2	59	
	5060V	16.1	19.6	20.3	5.8	76	
	5070V	16.1	19.6	20.3	6.1	86	

*All pumps 115 VAC/60 Hz unless otherwise specified.

TABLE 12-3. Rotary Vane Pumps*

Source Code	Item, Model No., or Catalog No.	Dimensions(cm) L	W	H	Wt. (kg)	Maximum Flow (L/min)	Remarks
ALG	Allegro A100	—	—	—	—	3–20	Critical orifice control for 11, 13.5, or 15 LPM, Max. 16 LPM with 25mm cassette, 20 LPM with 37mm controlled flow regulation for constant flow.
ASF	Brey- (28 models available)	—	—	—	—	1–23	See Figure 12-21 for specifications.
BGI	ASB-II	13	9	7	6.8	15	
FJS	F&J Model LV-22	—	—	—	—	—	
GST	Gast 0532 series	10.4	9.0	7.6	2.0	18	All pumps oil-less. Available with 115 VAC 60/50 Hz, 220 VAC 60/50 Hz, and 24 VDC motors.
	0531 series	22.3	8.9	11.1	3.2	17	
	1031 series	22.9	8.9	12.4	3.2	31	
	1531 series	22.8	10.6	12.1	3.6	42	
	0211 series	29.6	16.8	17.9	7.7	31	
	0323 series	33.6	14.3	15.0	14.1	90	
	0523 series	33.6	14.3	15.0	13.6	130	
	0823 series	41.2	16.5	22.6	21.3	230	
	1023 series	41.2	16.5	22.6	23.6	283	
MRM	Model FZ 135	1.5	1.5	3.4	0.01	0.9	1.5, 3, and 4.5 VDC.
THO	TASKAIR (10 models)	—	—	—	—	52–600	See Figure 12-22 for specifications.

*All pumps 115 VAC/60 Hz unless otherwise specified.

TABLE 12-4. Centrifugal Blowers

Source Code	Model Number	Dimensions(cm) L[A]	W	H	Wt. (kg)	Power Requirements	Maximum Vacuum (cm water)	Maximum Flow (m³/min)	Description	Remarks
ALE	116642-01	14.5d	—	18.6	2.72	120VAC, 50/60 Hz	109	2.6	250W, bypass, 2 stage, high flow fan, electrical speed control	These blowers are sold through the AMETEK Rotron Technical Motor Division. All blowers are brushless. For immediate sampling, please go to our distributors Allied Electronics at (800)433-5700 or Grainger at (800)323-0620.
	117415-00	14.5d	—	20.2	2.72	240VAC, 50/60 Hz	302	2.4	1200W, bypass, 3 stage, standard fan, electrical speed control	
	117416-00	14.5d	—	20.2	2.72	240VAC, 50/60 Hz	302	2.4	1200W, bypass, 3 stage standard fan, mechanical speed control	
	117417-00	14.5d	—	20.2	2.72	120VAC, 50/60 Hz	223	1.9	800W, bypass, 3 stage, standard fan, electrical speed control	
	117418-00	14.5d	—	20.2	2.72	120VAC, 50/60 Hz	223	1.9	800W, bypass, 3 stage, standard fan, mechanical speed control	
	117642-00	14.5d	—	18.6	2.72	240VAC, 50/60 Hz	102	2.7	400W, bypass, 2 stage, high flow fan, electrical speed control	
BRA	TB1-1	6.6	5.3	5.8	0.2	6-24VDC	0.5	0.4	Tangential	All motors brushless.
	TB1-1.5	7.9	7.4	7.6	0.3	6-24VDC	0.9	0.3	Tangential	
	TBO-2.5	9.6	9.9	10.9	0.4	12-24VDC	1.1	1.0	Tangential	
	TBL-2.5	9.6	9.9	10.9	0.4	12-24VDC	2.0	1.4	Tangential	
	TB1-2F	—	7.9d	7.0	0.2	12-24VDC	—	3.4	Axial	
	TB1-3F	—	10.2d	7.3	0.2	12-24VDC	—	4.2	Axial	
CLE	Cadillac F-10	52.1	19.7	18.4	4.5	115VAC/60 Hz	—	3.1	Tangential	Self-contained blower.
	HP-33	54.6	23.5	24.8	7.6	115VAC/60 Hz	—	5.6	Tangential	Self-contained blower.
	G-12	—	—	—	5.9	115VAC/60 Hz	—	4.4	Tangential	Self-contained blower.
EGG		48 models available. See Figures 12-23 and 12-24 for specifications.								

[A]d indicates diameter.

TABLE 12-5. Ejectors

Source Code	Item, Model No., or Catalog No.	Dimensions(cm)		Wt. (kg)	Maximum Flow (L/min)	Operating Pressure (kPa [psig])	Remarks
		L	W				
AVE	TD series (25 models available)	15.5 to 50.3	7.6d to 22.6	001 to 7.3	65 to 3600	up to 540 [80]	Require compressed air; can handle solids; available in aluminum, brass, 316 stainless steel. See Figure 12-25.
	AV series	"	"	"	4 to 850	up to 540 [80]	Require compressed air; cannot handle solids; available in aluminum, brass, 316 stainless steel.
FOX	Series 250 AJV (more than 8 models available)	22.6 to 280	—	—	up to 28,000	—	Require compressed air; pipe thread or flanged ends; can handle solids; available in aluminum, brass, steel, stainless steel, ceramic, titanium.
	Mini-eductor No. 611210 (more than 5 models available)	—	—	—	5 to 90	up to 1,000 [150]	Require compressed air; available in brass, Teflon, CPCV, and stainless steel.
GST	VG series (27 models available)	7.1 to 26.9	1.6 to 8.2	0.09 to 6.4	5.7 to 3600	up to 610 [90]	Require compressed air; single and multistage units; available in anodized aluminum and Delrin®.

TABLE 12-6. Personal Samplers: Battery Powered

Source Code	Figure Number	Item, Model Number	Dimensions (cm) L	W	H	Wt. (g)	Rate (cm³/min)	Pump Type[A]	Flow Meter[B]	Standard Heads[C]	Remarks
ALT		GilAir 3R	10.3	10.0	5	638	0-750; 750-3000	d	r	1, 2, 3, 4, 6	
AVI		Pulse Pump 111	7.4	4.1	10.2	260	17-330; 1500	p	—	6	Pulsing or continuous flow.
AMG		Ametek ALPHA-2	7.5	3.8	13.3	450	2-200	d	—	1	Formerly manufactured by DuPont; Models ALPHA-1, ALPHA-2, ALPHA-LITE, P2500B, P4LC, AND P4L are intrinsically safe and electronically controlled constant flow.
		P2500B	10.1	3.8	15.8	740	300-2500	d	—	1	
		ALPHA-1	10.8	5.7	14.9	1060	5-5000	d	—	1	
		ALPHA-LITE	10.8	5.7	12.7	880	5-5000	d	—	1	
		P4LC	10.2	5.8	12.7	970	5-5000	d	—	1	
		P4L	10.2	5.8	12.7	970	500-4000	d	—	1	
		MG-4	10.5	5.1	10.8	630	5-4000	d	r	1	
	12-26	MG-5P	10.5	5.1	13.6	800	5-5000	d	r	1	Programmable.
ASC, ALT	12-27	Sipin SP-13	6.4	3.2	13.0	310	10-200	d	sc	1, 5, 6	
		SP-13P	6.4	3.2	13.0	310	10-200	d	sc	1, 5, 6	
		SP-15	6.4	3.2	13.0	310	2-200	d	sc	1, 5, 6	
		SP103	6.4	3.2	13.0	340	10-1000	d	sc	1, 2, 5, 6	
BIO	12-28	BIOS AirPro 6000D	9.1	5.6	11.7	940	5-6000	d	e	—	Electronically controlled constant flow; EMI resistant.
		AirPro Surveyor2	—	—	—	—	—	d	—	—	
		AirPro Sentry II	—	—	—	—	—	—	—	—	
BGI		BGI AFC123	7.4	4.4	11.7	450	1000-2500	d	—	4, 6	All models are electronically controlled constant flow; intrinsically safe.
		BF-1	7.4	4.4	11.8	590	1000-3000	d	—	4, 6	
		BF-1-A	7.4	4.4	11.8	590	200-1000	d	—	4, 6	
	12-29	ASBII	—	—	—	—	11-15	—	—	—	Critical orifice.
BUC, SUP	12-30	Buck GENIE	10.2	5.1	13.6	624	800-5000	d	e	1, 2, 3	All models are electronically controlled constant flow; intrinsically safe; EMI resistant, programmable.
		S.S.	10.2	5.1	13.6	624	600-5000	d	e	1, 2, 3	
		I.H.	10.2	5.1	13.6	624	5-5000	d	e	1, 2, 3	
		H.F.	10.2	5.1	18.4	1160	1000-10,000	d	e	1, 2, 3	
CAL		Pulse Pump III	7.6	5.1	12.7	—	17-1500	p	—	6	Pulsing or continuous flow.
CPI, ALT		Pulse Pump GB-7600-00	7.4	4.1	10.2	260	17-300; 1500	—	—	6	
MSA	12-31	Escort ELF	10.3	5.1	9.8	550	500-3000	d	—	1, 2, 3, 4, 6	Electronic flow control; intrinsically safe; optional heavy duty battery pack and programmable pump; EMI/RFI protected.

TABLE 12-6 (cont.). Personal Samplers: Battery Powered

Source Code	Figure Number	Item, Model Number	Dimensions(cm) L	W	H	Wt. (g)	Rate (cm³/min)	Pump Type[A]	Flow Meter[B]	Standard Heads[C]	Remarks
NUC		Model 08-430	8.9	5.7	10.5	540	5000-7000	—	—	2, 7	
SEN		Gilian LFS113	6.4	3.5	11.7	340	1-350	d	—	1, 2, 3, 4, 6	All models are electronically controlled constant flow; intrinsically safe;
		HFS513A	11.7	4.8	13.0	1020	1-5000	d	r	1, 2, 3, 6	optional EMI/RFI shield case;
		GilAir	9.1	5.1	10.0	600	1-3000	d	r	1, 2, 3, 4, 6	GilAir-3 and -5 are optionally
		GilAir-5	10.3	5.1	10.0	640	1-5000	d	r	1, 2, 3, 4, 6	programmable.
	12-32	GilAir II	10.3	5.1	10.0	640	1-5000	d	d	1, 2, 3, 4, 6	
		BDX-II	9.1	5.1	10.0	600	1-3000	d	r	1, 2, 3, 4	No constant flow control.
SKC	12-33	Model 224-PCXR8	11.9	4.9	13.0	970	1-5000	d	r	1, 2, 3, 4, 5, 6	All models constant flow, intrinsically safe, and
		224-PCXR4	11.9	4.9	13.0	970	1-5000	d	r	1, 2, 3, 4, 5, 6	EMI/RFI shielded case. Models PCXR7 and
		224-44XR	11.9	4.9	13.0	970	1-5000	d	r	1, 2, 3, 4, 5, 6	PCXR74 have elapsed time timed shut down.
		222-4	6.4	3.2	13.0	280	20-80	d	—	1	PCXR8 is programmable. 224-XR series have
		222-3	6.4	3.2	13.0	280	50-200	d	—	1	MSHA approved models for underground use.
		AirChek 52 series	7.6	4.4	12.7	560	5-3000	d	—	1, 2, 3, 4, 5, 6	Pocket Pump and AirChek 2000 are program-
		AirChek 2000	14.2	7.6	5.8	616	5-3000	d	—	1, 2, 3, 4, 5, 6	mable manually or with a personal computer.
		Pocket Pump	11.4	5.5	3.6	142	20-225	d	—	1	AirChek 2000 corrects for changes in pressure and temperature.
SPE		Spectrex PAS-1000	7.6	3.6	10.2	280	up to 2000	d	—	1, 2, 3,	All models electronically controlled constant
		PAS-2000	10.7	6.1	11.7	—	600-3000	d	—	1, 2, 3	flow.
	12-34	PAS-3000	10.7	6.1	11.7	910	5-3000	d	r	1, 2, 3	
		PAS-500	3.0	2.0	10.9	110	20-300	d	—	1, 5, 6, 7	Colorimetric, charcoal, and absorbent tubes mount directly onto pump.
STA		Staplex PST-5	13.1	7.1	18.4	1250	500-2000	—	—	2	Both models electronically controlled constant
		PST-2	10.8	6.3	11.7	910	500-2000	—	r	2	flow.
SUP	12-57	Q-MAX	8.5	5	10	550	5-3000*	d	—	1*, 2, 3, 4	Contains electronic laminar flow sensor to maintain constant flow and reduce calibration to monthly. *Twin Port Sampler required for low flow.

[A] Pump type: d = diaphragm   p = piston

[B] Flowmeter: sc = stroke counter   r = rotameter   e = electronic

[C] Standard sampling heads:
1 = adsorption tube
2 = filter
3 = bubbler
4 = cyclone-filter assembly
5 = colorimetric tube
6 = air bag
7 = activated-charcoal filter

TABLE 12-7. Low-Volume Area Samplers: Battery Powered

Source Code	Figure Number	Item, Model Number	Dimensions(cm) L^A	W	H	Wt. (kg)	Rate (L/min)	Flow Meter^B	Standard Heads^C	Remarks
AVI		Air Quality Sampler II	61d	—	117	10.9	0.033-0.17	—	3	Sample >500 hrs, multiple sample bags.
		Air Quality Sampler III	61	41	47	11.4	0.033-0.17	—	3	Sample up to 250 hrs, multiple sample bags.
BIO	12-35	BIOS AirPro Series	25	31	13	3.0	0.005-1	m	—	Sample >700 hrs; 1-4 parallel samples; constant flow control.
HIQ		Model LFRR	—	—	—	—	5-25	r	1	Three-position program timer.
MEI		Thunderbolt Series IDC	33	21	28	11.8	1-8.5	r	1, 2	Two models with constant flow control; optional shut-off timer.
SEN	12-36	AirCon-2	19	13	26	5.4	2-30	r	1, 2, 3, 4	Constant flow control; sampling time—maximum of 2 and 4 hrs; optional programmable.
STA	12-37	Model BN/BNA	22	11	18	4.5	5-17	r	1	Ni-Cad batteries, 1 hr.
		BS/BSA	22	11	18	4.5	5-17	r	1	Ni-Cad batteries, 2 hrs.
SUP	12-58	Model 1067	23	13	18	3.6	.005-0.5	—	Adsorption tube	Dual channel independently controlled

^{A}d indicates diameter

^{B}Flowmeter: r = rotometer
                          m = mass flowmeter

^{C}Standard sampling
heads: 1 = filter
                    2 = bubbler
                    3 = air bag
                    4 = Asbestos Hazard Emergency Response Act (AHERA) asbestos sampling

TABLE 12-8. Low-Volume Area Samplers: Portable[A]

Source Code	Figure Number	Item, Model Number	Dimensions (cm) L	W	H	Wt. (kg)	Rate (L/min)	Flow Meter[B]	Standard Heads[C]	Remarks
ASI		Model HV-108-5	58.4	50.8	22.9	17.2	0-15	r, o	1, 2	Up to five simultaneous samples; Model HV-108 EXP suitable for use in explosive atmospheres.
		HV-108-SP	45.7	38.1	15.2	9.5	5-25	r, o	—	
		HV-108-ES	33.0	30.5	15.2	5.0	0-27	r, o	—	
		HV-108 EXP	58.4	50.8	22.9	17.2	0-15	r, o	—	
ALG		Model A-100	—	—	152	4.3	3-20	—	1, 2	
ASC		Model SP-280	48.3	25.4	35.6	13.0	4-28	r	1, 2	Features constant flow controller.
BGI	12-38	Model ASB-11	33.0	20.3	17.8	6.8	11-15	o	1, 2	All models available with a choice of critical orifices; Model ASB-11-S designed with sound absorption material.
		ASB-11S	45.7	30.5	30.5	13.2	11-15	o	1, 2	
		ASB-111	45.7	30.5	30.5	12.7	15-24	o	1, 2	
DAW		High Volume Sampler	24.1	10.2	11.4	4.1	3-20	r, o	1, 2	Choice of critical orifices.
EIC	12-39	Model RAS-1	45.1	17.8	27.9	15.9	0-100	r	1, 2	Both models have constant flow controller.
		RAS-2	58.0	23.5	33.0	26.3	0-100	r	1, 2	
EPC		Model S0269	—	—	—	4.1	10, 12	o	1, 2	Choice of critical orifices.
		S0270	—	—	—	4.1	0-20	—	1,2	
FJS	12-40	Model LV-1	—	—	—	—	—	r, g, o	1	Constant pressure drop across an inline orifice by varying flow through a bypass valve. Electronic Flow Regulation Module.
	12-41	HV-1	—	—	—	—	—	—	—	
GRA		Series 110 (3 versions)	43.2	23.2	40.0	15.4	0-30	r	1	Constant flow controller.
	12-42	PM10 Median Flow Sampler Model SA254M	107	107	161	38.6	110	g	3	PM₁₀ inlet; constant flow control; weather-flow recorder.
	12-43	Universal Sampler Model 209087	34.3	31.8	54.6	27	1-50	o	—	Precalibrated orifices for 1, 2, 5, 14, 28, and 50 L/min; records time and volume; 113 L/min (4 cfm) free flow.
HIQ		Series CF971T	29.2	20.3	20.3	4.3	30-170	r	1	47mm 2 in., 4 in., and 8 in. x 10 in. fiber.
		Model LFRR	—	—	—	—	5-25	r	1, 2	Optional elapsed timer.
		CMP-0523CV	—	—	—	—	—	—	—	
		MRV-14C	—	—	—	22.2	110	—	1	Golf cart-type stand.
		MRV-0523CV	—	—	—	—	—	—	—	
MEI	12-44	Ultra Sampler series IAC	33.0	20.6	27.9	7.7	1-15	r	1, 2	Six models with constant flow control; optional shut-off timer; programmable.
		series 2AC	33.0	20.6	27.9	7.7	1-18	r	1, 2	
MSP		Micro-Environmental Monitor	24.0	23.0	30.0	3.6	10	g	1, 2, 4, 6	Constant flow control.
PMC		Tool Box Gas Sampler	50.8	24.8	26.7	—	0-6.5	r	5	Constant flow control.

TABLE 12-8 (cont.). Low-Volume Area Samplers: Portable[A]

Source Code	Figure Number	Item, Model Number	Dimensions(cm) L	W	H	Wt. (kg)	Rate (L/min)	Flow Meter[B]	Standard Heads[C]	Remarks
SCH		Model 3-AH	—	—	—	—	10-100	g	—	Flow regulator system for single or multiple sampling points.
SCI		Teflon Sampling Pump	—	—	—	3	0.15	—	—	Programmable up to 7 days for sampling gases; Teflon inner surfaces.
SEC		Asbestos Sampling Pump	26.7	15.2	24.1	7.4	4-22	r	1, 2	
SEN	12-36	AirCon-2	19.0	13.3	26.0	5.4	2-30	r	1, 2	Constant flow controller; optional programmable.
STA	12-45	Model VM3	11.4	14.6	23.4	5	3-25	r	1, 2	220 VAC version available.
	12-46	LV-1 and LV-2	10.2	10.2	17.8	2.7	15-35	r	1	LV-1 is 115/125 VAC, 50/60 Hz; LV-2 is 220/240 VAC, 50/60 Hz.
		EC-1	32.4	10.0	12.7	4.3	2-16	r	1, 2	

[A] Unless otherwise specified, all power is 115V/60 Hz.
[B] Flowmeter: r = rotometer
                o = orifice
                g = pressure gauge

[C] Standard sampling heads:
1 = filter
2 = AHERA asbestos sampling
3 = PM_{10}
4 = $PM_{2.5}$ or PM_1
5 = bag sampling
6 = diffusion denuder

TABLE 12-9. Medium- and High-Volume Samplers: Portable

Source Code	Figure Number	Item, Model Number	Dimensions (cm) L	W	H	Wt. (kg)	Standard Flow Rate (m³/min [cfm])	Speeds/Stages	Cool Air[A]	Flow Meter[B]	Sampling Head[C]	Notes[D]
GMW		Handi-Vol 2000	25.4	20.3	25.4	6	0.4-0.6 [15-20], 4" filter 0.6-1.7 [20-60], 8" x 10" filter	—	sam	o	1, 2	a
HIQ	12-47	Series CF-993B	—	—	—	—	—	—	—	—	—	—
		Series CF-1000BRL	—	—	—	—	—	—	—	—	—	—
		Series CF-920FT	—	—	—	—	—	—	—	—	—	—
		Series CF-900 (3 versions)	29.2	20.3	20.3	4	0.03-1.7 [1-60]	1/2	sep	r	1, 2	a, d
		Series CF-971T	29.2	20.3	20.3	—	—	—	—	r	—	—
		Series CF-972T	29.2	20.3	20.3	—	0.14-0.85 [5-30]	—	—	r	1, 2	a, d
		Series CF-973T	29.2	20.3	20.3	—	0.43-1.7 [15-60]	—	—	r	1, 2	a, d
		Model CF-18V	—	—	—	—	—	—	—	—	—	—
		Model CF-24B	29.2	20.3	20.3	5	0.08-0.22 [3-8]	—	—	r	1	b
		Model CF-1524-VBRL	—	—	—	—	—	—	—	—	—	—
		Model TFIA-4BC	—	—	—	—	—	—	—	—	—	—
	12-48	Model TFIA	—	—	—	—	—	—	—	—	—	—
MSP		Universal Air Sampler	152	76	142	43	0.3 [10.6]	—	—	g	3, 4, 5	a, c, e
NUC		Model 08-600ER	21.5	19.5	—	6	0.57 [20]	1/2	—	o	1	a, b, d, e
STA	12-49	Staplex, TFIA-Series	21.6	19.1	19.1	5	2.0 [70], free air	1/2	—	o	1, 2	a, b, c
GRA	12-50	Intermediate Flow	61	61	208	44	0.11 [4]	—	—	—	3, 4	a, d, e

[A]Cooling air: sam = through sample
sep = separate

[B]Flowmeter: g = pressure gauge
o = orifice
r = bypass rotometer

[C]Sampling Head: 1 = 4" diameter
2 = 8" x 10"
3 = PM_{10} inlet
4 = TSP inlet
5 = PUF sampler

[D]Notes:
a = 115VAC/60 Hz standard
b = 12V or 24VDC available
c = Shelter version available
d = 250VAC/50 Hz
e = Constant flow control

TABLE 12-10. High-Volume Samplers: With Shelters

Source Code	Figure Number	Item, Model No.	Dimensions(cm) L^A	W	H	Wt. (kg)	Flow Rate (m³/min [cfm])	Remarks
FJS	12-51	Model DH-50810	—	—	—	—	—	Microprocessor controlled system, flow volume correction to STP.
	12-52	Model TE-1000PUF	55	50	112	30	—	Pesticide particulate and vapor sampling.
		TE-5000TSP	55	50	112	20	[20-60]	TSP, mass flow controlled.
		TE-5100	—	—	—	—	—	
		TE-5200TSP	72	50	48	18	—	Tripod TSP monitor.
GRA	12-53	PM-10 High Volume Sampler	71d	—	155	45	1.1 [40]	Features constant flow control; timer, chart recorder, and other options available; PM₁₀ and TSP inlet; optional brushless motor.
GRA	12-53	TSP High Volume Sampler	61	46	132	27	1.1 [40]	
GMW	12-54	PS-1 PUF	61	46	132	—	0.2-0.3 [7-10]	Pesticide particulate and vapor collection system.
GMW	12-55	ACCU-Vol IP-10	71d	—	155	43	1.1 [40]	Features constant flow control; timer, chart recorder, and other options available, PM₁₀ inlet; optional brushless motor.
HIQ		Models HVP-2000	47	47	132	25	1.1 [40]	Optional features include constant flow control, elapsed timer, and brushless motor; PUF and IP₁₀ inlets.
		Models HVP-3500AFC	—	—	—	—	—	
		Models HVP-3000BRL	47	47	132	27	1.1 [40]	
STA		TSP	61	46	132	—	1.1 [40]	Metal or wooden shelters.
GRA	12-56	PM-10 Critical High Flow	47	47	220	47	1.1 [40]	Features constant flow control, timer; PM₁₀ TSP, or PUF inlets; optional brushless motor.
		TSP Sampler	61	46	132	29	1.1 [40]	
		PUF Sampler	61	46	132	40	0.2-0.3 [7-10]	

^A d indicates diameter.

Notes:

1. All meet requirements of U.S. EPA federal reference method (8" x 10" filter) and PM₁₀ sampling standard.

2. All 115VAC/60 Hz; other power options may be available.

3. See Chapter 14 for description of PM₁₀ inlets.

TABLE 12-11. Commercial Sources for Air Movers and Samplers

AVI	AeroVironment, Inc. 825 S. Myrtle Dr. Monrovia, CA 91016 (626) 357-9983 *www.aerovironment.com*	AMG	Ametek/Mansfield & Green Div. 8600 Somerset Dr. Largo, FL 34643 (813) 536-7831 FAX (813) 539-6882	CAL	Calibrated Instruments, Inc. 200 Saw Mill River Road Hawthorne, NY 10532 (914) 741-5700 or (800) 969-2254 FAX (914) 741-5711 *www.calibrated.com*
ACI	Air Control, Inc. 237 Raleigh Road, Box 1738 Henderson, NC 27536 (919) 492-2300 FAX (919) 492-9225	ASC	Anatole J. Sipin Co., Inc. 505 Eighth Ave. New York, NY 10018 (212) 695-5706 FAX (212) 695-5916 *ajsipinco@aol.com*	CLE	Clements National Company 6650 S. Narragansett Avenue Chicago, IL 60638 (708) 767-7900 or (800) 966-0016 *www.cadillacproducts.com*
ADI	Air Dimensions, Inc. 1015 West Newport Center Dr. Suite 101 Deerfield Beach, FL 33442 (954) 428-7333 or (800) 650-3267 FAX (954) 360-0987 *www.airdimensions.com*	ASF	ASF, Inc. 2100 Norcross Parkway Norcross, GA 30071 (404) 441-3611	CPI	Cole-Parmer Instrument Company 625 E. Bunker Court Vernon Hills, IL 60061 (800) 323-4340 FAX (847) 247-2929 *info@coleparmer.com* *www.coleparmer.com*
ASI	Air Systems International, Inc. 821 Juniper Crescent Chesapeake, VA 23320 (757) 424-3967 or (800) 866-8100 *www.airsystems.cc*	BAR	Barnant Co. 28W092 Commercial Ave. Barrington, IL 60010 (847) 381-7050 or (800) 637-3739 FAX (847) 381-3753 *www.barnant.com*	DAW	Dawson Associates P.O. Box 846 Lawrenceville, GA 30246 (404) 963-0207
AVE	AirVac Enginnering Co., Inc. Vacuum Pump Div. 100 Gulf Street Milford, Ct 06460 (203) 874-2541	BGI	BGI Incorporated 58 Guinan Street Waltham, MA 02451 (781) 891-9380 FAX (781) 891-8151 *www.bgiusa.com*	EIC	Eberline Instrument Corporation P.O. Box 2108 504 Airport Road Santa Fe, NM 87505 (505) 471-3232 FAX (505) 473-9221 *www.eberlineinst.com*
ALG	Allegro Industries 7221 Orangewood Ave. Garden Grove, CA 92841 (714) 899-9855 or (800) 622-3530 FAX (800) 362-7231 *custsvc@allegro.com* *www.allegrosafety.com*	BIO	BIOS International Corp. 10 Park Place Butler, NJ 07405 (973) 492-8400 FAX (973) 492-8270 *sales@biosint.com* *www.biosint.com*	EGG	EG&G Rotron Industrial Div. North Street Saugerties, NY 12477 (914) 246-3401
ALT	Alltech Associates, Inc. 2051 Waukegan Road Deerfield, IL 60015 (847) 948-8600 or (800) 255-8324 FAX (847) 948-1078 *alltech@alltechmail.com* *www.alltechweb.com*	BRA	Brailsford & Company, Inc. 670 Milton Road Rye, NY 10580 (914) 967-1820 FAX (914) 967-1820	EPC	Envirometrics, Inc. Envirometrics Product Co. 1019 Bankton Drive Charleston, SC 29406 (803) 740-1700 or (800) 255-8740
ALE	Ametek/Lamb Electric Div. 627 Lake Street Kent, OH 44240-2660 (330) 673-3452 FAX (330) 678-8227 *www.ametektmd.com*	BUC	A.P. Buck, Inc. 7101 President's Drive Suite 110 Orlando, FL 32809 (407) 851-8602 or (800) 330-BUCK FAX (407) 851-8910 *apbuck@apbuck.com* *www.apbuck.com*	FJS	F&J Specialty Products, Inc. P.O. Box 2888 Ocala, FL 34478-2888 (352) 680-1177 or (352) 680-1178 FAX (352) 680-1454 *www.fjspecialty.com*

TABLE 12-11 (cont.). Commercial Sources for Air Movers and Samplers

FMI	Fluid Metering, Inc. 5 Aerial Way Suite 500 Syosset, NY 11791 (516) 922-6050	MED	Medo U.S.A., Inc. 4525 Turnberry Dr. Hanover Park, IL 60103 (800) 843-6336 or (630) 924-8811 FAX (630) 924-0808	PMC	Pollution Measurements Corp. 1013 S. Lyman Ave. Oak Park, IL 60304 (708) 383-7794 FAX (708) 383-7877
FOX	Fox Valve Development Corp. Hamilton Business Park, Unit 6A Franklin Road Dover, NJ 07801 (973) 328-1011 FAX (973) 328-3651 www.foxvalve.com	MBC	Metal Bellows Corp. 1075 Providence Hwy. Sharon, MA 02067 (617) 784-1400	SCH	Schmidt Instrument Co. P.O. Box 111 San Carlos, CA 94070 (650) 591-5347
GST	Gast Manufacturing Corp. P.O. Box 97 2550 Meadowbrook Rd. Benton Harbor, MI 49023 (616) 926-6171 FAX (616) 925-8288 www.gastmfg.com	MEI	Midwest Environics, Inc. 10 Oak Glen Court Madison, WI 53717 (608) 833-0158	SCI	Science Pump Corporation 1431 Ferry Avenue Camden, NJ 08104 (609) 963-7700
GRA	Andersen Instruments, Inc. 500 Technology Court Smyrna, GA 30082-5211 (770) 319-9999 or (800) 241-6898 FAX (770) 319-0336 www.anderseninstruments.com	MIE	MIE, Inc. 7 Oak Park Bedford, MA 01730 (781) 275-1919 or (888) 643-4968 FAX (781) 275-2121 www.mieinc.com	SEN	Sensidyne, Inc. 16333 Bay Vista Drive Clearwater, FL 33760 (800) 451-9444 FAX (727) 539-0550 www.sensidyne.com
GMW	Graseby GMW General Metal Works, Inc. 145 S. Miami Ave. Village of Cleves, OH 45002 (513) 941-2229 FAX (513) 941-1977	MSA	Mine Safety Appliances Co. P.O. Box 427 Pittsburgh, PA 15230-0427 (800) 672-2222; Safety Products Division, portable units (800) 672-4678; permanently installed units www.msanet.com	SKC	SKC, Inc. 863 Valley View Road Eighty Four, PA 15330-9614 (412) 941-9701 or (800) 752-8472 FAX (412) 941-1369 www.skcinc.com
HIQ	HI-Q Environmental Products Co. 7386 Trade Street San Diego, CA 92121 (619) 549-2820 FAX (619) 549-9657 info@HI-Q.net www.HI-Q.net	MRM	MRM International 3905 Whitney Place Duluth, GA 30136 (770) 476-4040 FAX (770) 623-1465	SPE	Spectrex Corporation 3580 Haven Avenue Redwood City, CA 94063 (650) 365-6567 or (800) 822-3940 FAX (650) 365-5845 www.spectrex.com
KNF	KNF Neuberger, Inc. Two Black Forest Road Trenton, NJ 08691 (609) 890-8600 FAX (609) 890-8323 www.knf.com	MSP	MSP Corporation 1313 Fifth St., SE, Suite 206 Minneapolis, MN 55414 (612) 379-3963 FAX (612) 379-3965 sales@mspcorp.com www.mspcorp.com	STA	Staplex Company Air Sampler Div. 777 Fifth Avenue Brooklyn, NY 11232 (718) 768-3333 or (800) 221-0822 FAX (718) 965-0750 info@staplex.com
MDA	MDA Scientific, Inc. 405 Barclay Blvd. Lincolnshire, IL 60069 (708) 634-2800 or (800) 323-2000 FAX (708) 634-1371 www.zelana.com	NUC	Nuclear Associates P.O. Box 349 100 Voice Road Carle Place, NY 11514 (516) 741-6360 or (888) 466-8257 FAX (516) 741-5414 sales@nucl.com www.nucl.com	SUP	Supelco, Inc. Supelco Park Bellefonte, PA 16823-0048 (814) 359-3441 or (800) 247-6628 FAX (814) 359-3044

TABLE 12-11 (cont.). Commercial Sources for Air Movers and Samplers

THO   Thomas Industries, Inc.
         P.O. Box 29
         1419 Illinois Avenue
         Sheboygan, WI 53082
         (920) 457-4891
         FAX (920) 451-4276
         www.thomaspumps.com

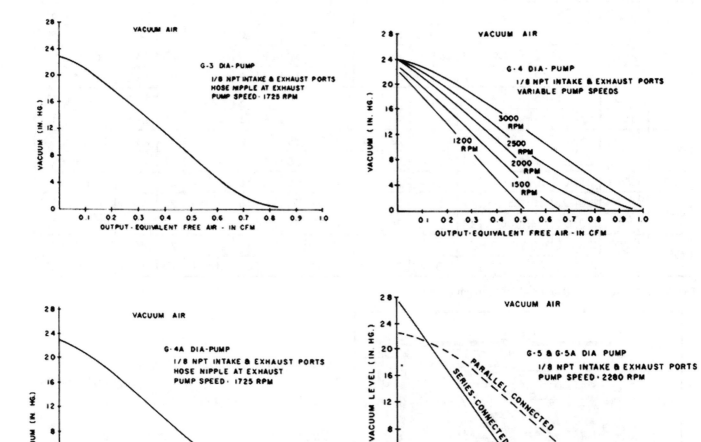

FIGURE 12-9. Vacuum-air flow characteristics for Air-Control Dia-pumps.

MODEL NO.	VOLTAGE	COMPRESSOR PERFORMANCE L/M VS PSIG								MAX PSIG		VACUUM PERFORMANCE L/M VS IN. HG.							MAX VAC IN. HG.	PHYSICAL SPECIFICATIONS	
		0	5	10	15	20	25	30	60	CONT	INT	0	5	10	15	20	25	28		WT.	HxWxL (in mm)
3003	6 VDC	0.9	0.4	0.1	—	—	—	—	—	3	10.5	0.9	0.4	0.1	—	—	—	—	10.5	1.5 oz.	36 x 23 x 42
	12 VDC	0.9	0.4	0.1	—	—	—	—	—	3	10.5	0.9	0.4	0.1	—	—	—	—	10.5	1.5 oz.	36 x 23 x 42
5002	6 VDC	2.2	1.5	0.8	0.1	—	—	—	—	5	16	2.2	1.5	0.6	0.1	—	—	—	15	6.5 oz.	53 x 30 x 82
	12 VDC	2.2	1.5	0.8	0.1	—	—	—	—	5	16	2.2	1.5	0.6	0.1	—	—	—	15	6.5 oz.	53 x 30 x 82
	24 VDC	2.2	1.5	0.8	0.1	—	—	—	—	5	16	2.2	1.5	0.6	0.1	—	—	—	15	6.5 oz.	53 x 30 x 82
	115/60	2.0	1.2	0.6	0.1	—	—	—	—	5	16	2.0	1.2	0.5	—	—	—	—	15	1.3 lb.	53 x 30 x 82
5010	6 VDC	3.7	1.9	0.9	0.1	—	—	—	—	5	16	3.7	2.1	1.1	0.3	—	—	—	17	10 oz.	68 x 40 x 85
	12 VDC	3.7	1.9	0.9	0.1	—	—	—	—	5	16	3.7	2.1	1.1	0.3	—	—	—	17	10 oz.	68 x 40 x 85
	24 VDC	3.7	1.9	0.9	0.1	—	—	—	—	5	16	3.7	2.1	1.1	0.3	—	—	—	17	10 oz.	68 x 40 x 85
7010	6 VDC	6.0	4.5	3.3	2.4	1.5	0.7	0.2	—	22	32	6.0	4.0	2.6	1.5	0.4	—	—	22	1.3 lb.	95 x 61 x 108
	12 VDC	6.0	4.5	3.3	2.4	1.5	0.7	0.2	—	22	32	6.0	4.0	2.6	1.5	0.4	—	—	22	1.3 lb.	95 x 61 x 108
	24 VDC	6.0	4.5	3.3	2.4	1.5	0.7	0.2	—	22	32	6.0	4.0	2.6	1.5	0.4	—	—	22	1.3 lb.	95 x 61 x 108
	115/60	6.0	4.5	3.3	2.4	1.5	0.7	0.2		22	32	6.0	4.0	2.6	1.5	0.4	—	—	22	2.1 lb.	95 x 61 x 108
7010Z	12 VDC PARALL	12.0	9.5	7.2	5.4	3.7	2.0	0.5	—	22	32	12.0	8.0	5.0	2.9	1.0	—	—	22	1.7 lb.	95 x 61 x 158
	12 VDC SERIES	6.0	5.3	4.8	4.2	3.6	3.2	2.7	0.3	22	72	6.0	4.1	2.8	1.8	1.0	0.2	—	27	1.7 lb.	95 x 61 x 158
	115/60 PARALL	12.0	9.5	7.2	5.4	3.7	2.0	0.5	—	22	32	12.0	8.0	5.0	2.9	1.0	—	—	22	3.3 lb.	95 x 61 x 158
	115/60 SERIES	6.0	5.3	4.8	4.2	3.6	3.2	2.7	0.3	22	72	6.0	4.1	2.8	1.8	1.0	0.2	—	27	3.3 lb.	95 x 61 x 158
7012	12 VDC	9.0	4.0	0.5	—	—	—	—	—	—	12	9.0	5.3	3.1	1.5	0.4	—	—	22	1.1 lb.	78 x 95 x 117
	24 VDC	9.0	4.0	0.5	—	—	—	—	—	—	12	9.0	5.3	3.1	1.5	0.4	—	—	22	1.1 lb.	78 x 95 x 117
7015	12 VDC	12.5	9.8	7.7	5.8	4.0	2.5	1.0	—	22	36	12.5	8.5	5.5	3.1	1.0	—	—	22	2.3 lb.	134 x 86 x 125
	115/60	12.5	9.8	7.7	5.8	4.0	2.5	1.0	—	22	36	12.5	8.5	5.5	3.1	1.0	—	—	22	3.5 lb.	134 x 86 x 125
8025	115/60	13.0	10.0	8.0	6.5	5.5	4.5	3.7	—	36	50	13.0	8.7	6.0	3.5	1.6	—	—	24	7.7 lb.	100 x 115 x 220
8050	115/60 PARALL	25.0	22.5	20.0	17.5	15	13	12	3.5	36	80	25.0	17.4	12.0	7.5	3.5	—	—	25	9 lb.	100 x 154 x 220
	115/60 SERIES	13.0	12.2	11.4	10.6	9.7	8.8	8.0	5.3	36	85	13.0	8.5	5.8	3.5	2.0	0.7	—	28	9 lb.	100 x 154 x 220
8050Z	115/60 PARALL	43.0	39.0	36.0	33.0	30.0	27.0	24.0	10.0	36	85	43.0	33.0	23.0	15.0	8.0	—	—	25	28.2 lb.	100 x 154 x 324
	115/60 SERIES	19.0	18.5	18.0	17.5	17.0	16.0	15.0	12.0	36	140	19.0	14.0	10.0	7.0	4.0	0.1	—	28	28.2 lb.	100 x 154 x 324

FIGURE 12-10. Technical data for ASF diaphragm pumps.

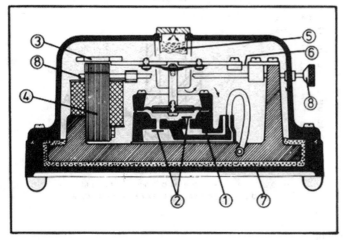

Section through type 3-1.000.010.0
Pressure pump with casing, output adjustable
by slider-type magnetic shunt.

1 *Diaphragm*   6 *Leaf spring*
2 *Valve plates*   7 *Foam bedding*
3 *Armature arm*   8 *Slider-type magnetic*
4 *Electro-magnet*   *shunt*
5 *Cotton wool filter*

FIGURE 12-11. Schematic diagram of ASF WISA pump.

Model	HP	RPM		Airflow				Maximum Vacuum	
				CFM @ 0" Hg		m³/h @ 1000 mbar			
		50 Hz	60 Hz	50 Hz	60 Hz	50 Hz	60 Hz	" Hg	mbar
MOA-V112-AE	1/16	1275	1575	0.40	0.49	0,68	0,83	24.0	200
MOA-V111-CD	1/16	1275	1575	0.40	0.49	0,68	0,83	24.0	200
MOA-V112-FB	1/16	2500	3000	0.60	0.72	1,02	1,22	24.0	200
MOA-V112-FD	1/16	2500	3000	0.60	0.72	1,02	1,22	24.0	200
MOA-V112-HB	1/16	1275	1575	0.40	0.49	0,68	0,83	24.0	200
MOA-V112-HD	1/16	1275	1575	0.40	0.49	0,68	0,83	24.0	200
MOA-V121-CG	1/16	1275	–	0.40	–	0,68	–	24.0	200
MOA-V111-JH	1/16	1800 (D.C.)		0.56	0.56	0,95	0,95	24.0	200
MOA-V111-KH	1/8	3200 (D.C.)		0.80	0.80	1,36	1,36	24.0	200
MOA-V111-JK	1/16	1800 (D.C.)		0.56	0.56	0,95	0,95	24.0	200
MAA-V103-HB	1/16	1275	1575	0.85	1.02	1,45	1,73	24.0	200
MAA-V109-HB	1/16	1275	1575	0.39	0.47	0,66	0,80	28.0	65
MAA-V103-HD	1/16	1275	1575	0.85	1.02	1,45	1,73	24.0	200
MAA-V109-HD	1/16	1275	1575	0.39	0.47	0,66	0,80	28.0	65
MAA-V103-MB	1/8	2500	3000	1.60	1.70	2,72	2,89	24.0	200
MAA-V109-MB	1/8	2500	3000	0.66	0.80	1,12	1,36	28.5	48
DOA-V191-AA	1/8	–	1575	–	1.10	–	1,87	25.5	150
DOA-V113-AC	1/8	–	1575	–	1.10	–	1,87	25.5	150
DOA-V111-AE	1/8	1275	–	0.95	–	1,62	–	25.5	150
DOA-V112-BN	1/8	1275	–	0.95	–	1,62	–	25.5	150
DOA-V110-BL	1/8	1275	1575	0.95	1.10	1,62	1,87	25.5	150
DOA-V113-DB	1/8	1275	1575	0.95	1.10	1,62	1,87	25.5	150
DOA-V119-DD	1/8	1275	1575	0.95	1.10	1,62	1,87	25.5	150
DOA-V112-FB	1/3	2500	3000	1.50	1.80	2,55	3,06	25.5	150
DOA-V114-FD	1/3	2500	3000	1.50	1.80	2,55	3,06	25.5	150
DOA-V113-FG	1/3	2500	–	1.50	–	2,55	–	25.5	150
DOA-V111-JH	1/8	2100 (D.C.)		1.26	1.26	2,14	2,14	25.5	150
DOA-V111-KH	1/6	3200 (D.C.)		1.70	1.70	2,89	2,89	25.5	150
DOA-V111-JK	1/8	2100 (D.C.)		1.26	1.26	2,14	2,14	25.5	150
DAA-V110-EB	1/4	1275	1575	1.80	2.20	3,06	3,74	25.5	150
DAA-V111-EB	1/4	1275	1575	0.85	1.16	1,45	1,97	29.0	31
DAA-V110-ED	1/4	1275	1575	1.80	2.20	3,06	3,74	25.5	150
DAA-V111-ED	1/4	1275	1575	0.85	1.16	1,45	1,97	29.0	31
DAA-V124-EG	1/4	1275	–	1.80	–	3,06	–	25.5	150
DAA-V134-EG	1/4	1275	–	0.85	–	1,45	–	29.0	31
DAA-V110-GB	1/2	2500	3000	3.00	3.60	5,10	6,12	25.5	150
DAA-V111-GB	1/2	2500	3000	1.40	1.95	2,38	3,32	29.0	31
DAA-V110-GD	1/2	2500	3000	3.00	3.60	5,10	6,12	25.5	150
DAA-V111-GD	1/2	2500	3000	1.40	1.95	2,38	3,32	29.0	31
DAA-V112-GG	1/2	2500	–	3.00	–	5,10	–	25.5	150
DAA-V111-GG	1/2	2500	–	1.40	–	2,38	–	29.0	31

FIGURE 12-12. Technical data for Gast diaphragm pumps.

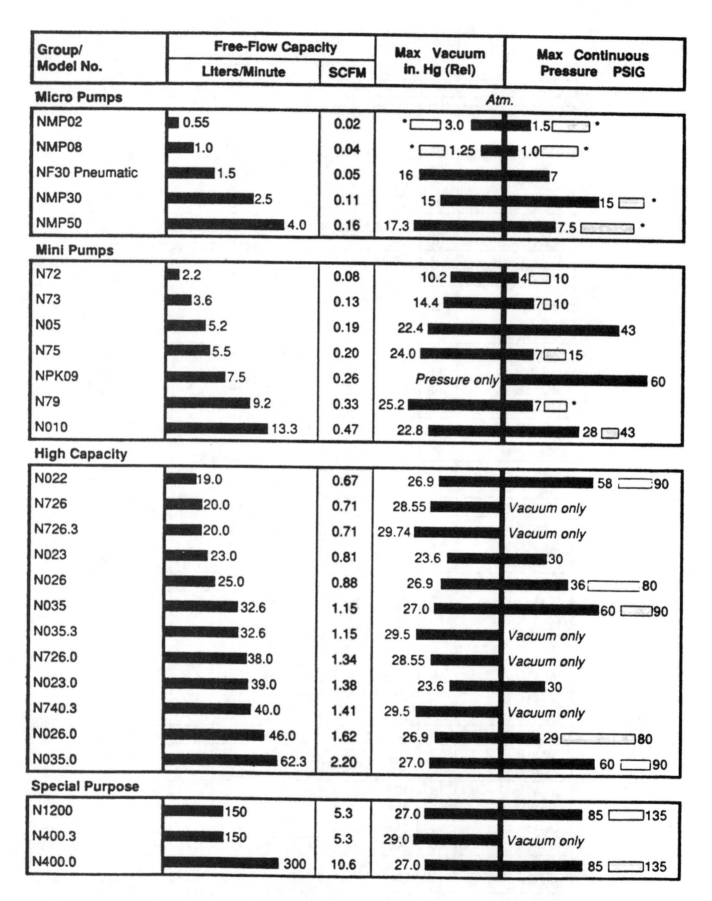

Group/ Model No.	Free-Flow Capacity		Max Vacuum in. Hg (Rel)	Max Continuous Pressure PSIG
	Liters/Minute	SCFM		
Micro Pumps				*Atm.*
NMP02	0.55	0.02	* 3.0	1.5 *
NMP08	1.0	0.04	* 1.25	1.0 *
NF30 Pneumatic	1.5	0.05	16	7
NMP30	2.5	0.11	15	15 *
NMP50	4.0	0.16	17.3	7.5 *
Mini Pumps				
N72	2.2	0.08	10.2	4 10
N73	3.6	0.13	14.4	7 10
N05	5.2	0.19	22.4	43
N75	5.5	0.20	24.0	7 15
NPK09	7.5	0.26	*Pressure only*	60
N79	9.2	0.33	25.2	7 *
N010	13.3	0.47	22.8	28 43
High Capacity				
N022	19.0	0.67	26.9	58 90
N726	20.0	0.71	28.55	*Vacuum only*
N726.3	20.0	0.71	29.74	*Vacuum only*
N023	23.0	0.81	23.6	30
N026	25.0	0.88	26.9	36 80
N035	32.6	1.15	27.0	60 90
N035.3	32.6	1.15	29.5	*Vacuum only*
N726.0	38.0	1.34	28.55	*Vacuum only*
N023.0	39.0	1.38	23.6	30
N740.3	40.0	1.41	29.5	*Vacuum only*
N026.0	46.0	1.62	26.9	29 80
N035.0	62.3	2.20	27.0	60 90
Special Purpose				
N1200	150	5.3	27.0	85 135
N400.3	150	5.3	29.0	*Vacuum only*
N400.0	300	10.6	27.0	85 135

FIGURE 12-13. Technical data KNF diaphragm pumps. Standard continuous performance ratings at sea level with an ambient temperature of 70°F (21°C) and nominal electrical supply.

MET. BEL.™ CONCEPT

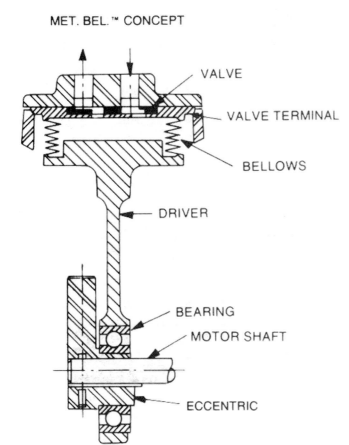

FIGURE 12-14. Schematic diagram of Metal Bellows pump.

FIGURE 12-16. OEM Micro-Air Pump.

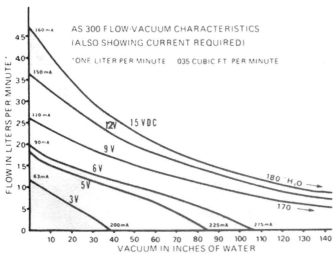

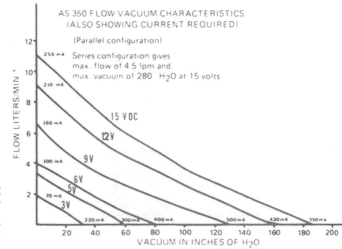

FIGURE 12-17. Vacuum-air flow characteristic for Spectrex Models AS-300 and AS-350.

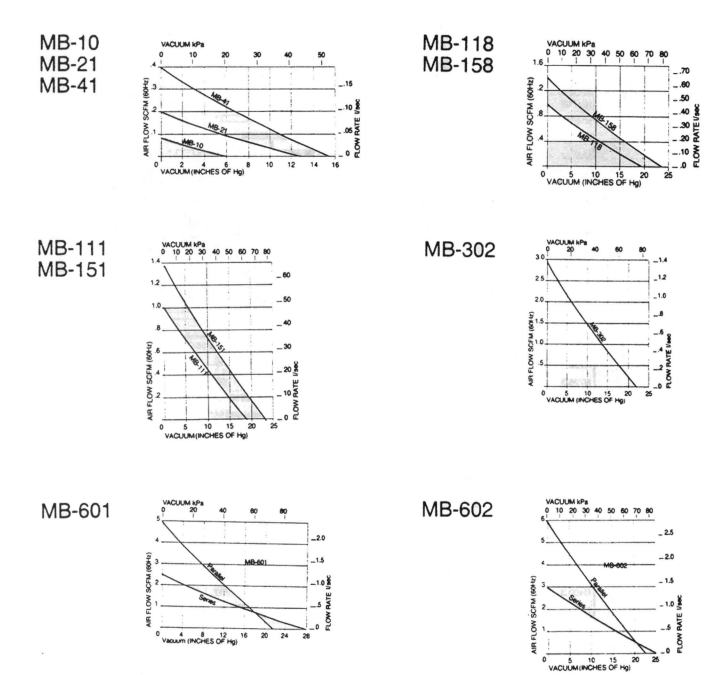

FIGURE 12-15. Vacuum-air flow characteristics for Metal Bellows pumps.

MODEL NUMBER	HP	kW	VOLTAGE	MOTOR TYPE	WT. LBS.	WT. KG.	H x W x L IN.	H x W x L CM	IN. HG. vs. CFM 0	5	10	15	20	25	MM HG. vs. LPM 0	127	254	381	508	635	MAX. VACUUM IN. HG.	MM HG.
007CDC19	1/32	.023	12VDC	Perm. Magnet	3.6	1.6	5.09 x 3.38 x 6.25	12.9 x 8.5 x 15.8	.65	.48	.34	.21			18.4	13.6	9.6	5.9			23.0	584.2
007BDC19	1/32	.023	12VDC	Perm. Magnet	3.6	1.6	5.04 x 3.35 x 6.16	12.8 x 8.5 x 15.6	1.26	.90	.58	.34			35.7	25.5	16.4	9.6			23.0	584.2
007CA13*	1/30	.024	115/60/1	Shaded Pole	2.5	1.1	5.12 x 3.50 x 6.00	13.0 x 8.8 x 15.2	.61	.38	.23	.11			17.3	10.8	6.5	3.1			20.3	515.6
107CAB18*	1/20	.037	115/60/1	Shaded Pole	5.1	2.3	4.70 x 4.25 x 6.84	11.9 x 10.7 x 17.3	.76	.58	.40	.23	.08		21.5	16.4	11.3	6.5	2.3		23.2	589.3
107CDC20*	1/10	.075	12VDC	Perm. Magnet	4.5	2.0	4.70 x 4.25 x 7.34	11.9 x 10.7 x 18.6	1.20	.86	.62	.41			34.0	24.4	17.6	11.6			22.9	581.7
2107CA20	1/20	.037	115/60/1	Shaded Pole	9.0	4.0	6.78 x 4.75 x 8.00	17.2 x 12.0 x 20.3	1.65	1.15	.77	.47	.17		46.7	32.6	21.8	13.3	4.8		22.4	569.0
2107CEF18	1/20	.037	115/60/1	PSC	7.0	3.1	5.31 x 5.25 x 8.84	13.4 x 13.3 x 22.4	1.50	1.10	.75	.41	.13		42.5	31.2	21.2	11.6	3.7		22.6	574.0
2107VA20	1/20	.037	115/60/1	Shaded Pole	6.3	2.8	4.70 x 4.25 x 8.18	11.9 x 10.7 x 20.7	.85	.65	.49	.33	.20		24.1	18.4	13.9	9.3	5.7		28.2	716.3
905CA18	1/15	.049	115/60/1	Shaded Pole	6.5	2.9	5.88 x 4.75 x 7.84	14.9 x 12.0 x 19.9	1.40	.87	.55	.32	.09		39.6	24.6	15.6	9.1	2.5		22.2	563.9
907CDC18*	1/10	.075	12VDC	Perm. Magnet	6.0	2.7	4.70 x 4.25 x 8.19	11.9 x 10.7 x 20.8	2.18	1.06	.64	.41	.16		61.7	30.0	18.1	11.6			23.2	589.3
917CA18*	1/8	.093	115/60/1	Shaded Pole	11.0	4.9	6.78 x 4.75 x 8.00	17.2 x 12.0 x 20.3	1.49	.89	.64	.36	.16		42.2	25.2	18.1	10.2	4.5		23.8	604.5
727CM39*	1/4	.186	115/60/1	Split Phase	21.0	9.5	8.81 x 5.81 x 11.50	22.3 x 14.7 x 29.2	3.10	1.72	1.27	.87	.48		87.8	48.7	36.0	24.6	13.6		25.7	652.8
2737CM39*	1/2	.372	115/60/1	Split Phase	25.0	11.3	6.56 x 11.12 x 15.34	16.6 x 28.2 x 38.9	6.20	4.30	2.90	1.80	.97		175.6	121.8	82.1	51.0	27.5		25.3	642.6
2737VM39	1/2	.372	115/60/1	Split Phase	25.8	11.7	6.56 x 11.12 x 15.34	16.6 x 28.2 x 38.9	3.10	2.40	1.90	1.40	.89	.34	87.8	68.0	53.8	39.6	25.2	9.6	29.0	736.6

FIGURE 12-18. Technical data for Thomas diaphragm pumps.

MODEL NUMBER	NOM. HP	# OF CYL.	VOLTAGE/ HERTZ/ PHASE	MOTOR TYPE	FRAME	WT. (LBS.)	H" x W" x L"	IN. HG vs. CFM 0	5	10	15	20	25	MAX. VACUUM IN. HG
LGH-1V	1/12	1	115/50/60/1	Split Phase	48	16	8.06x5.65x10.29	1.35	1.09	.83	.57	.31	.05	26.0
TA-1V1	1/12	1	115/50/60/1	Split Phase	48	16	8.06x5.65x10.29	1.35	1.09	.83	.57	.31	.05	26.0
LGH-2V	1/6	1	115/50/60/1	Split Phase	48	20	8.00x5.65x10.88	2.15	1.76	1.37	.98	.59	.20	27.5
TA-2V1	1/6	1	115/50/60/1	Split Phase	48	20	8.00x5.65x10.88	2.15	1.76	1.37	.98	.59	.20	27.5
TA-3V2	1/4	2	115/230/60/1	Capacitor Start	56	33	6.88x10.03x13.80	3.60	2.95	2.29	1.64	.98	.33	27.5
GH-3V1B	1/4	1	115/230/60/1	Capacitor Start	56	33	8.75x10.20x15.31	1.70	1.39	1.08	.77	.46	.15	27.5
GH-3V2B	1/4	2	115/230/50/1	Capacitor Start	56	34	8.75x11.50x15.31	3.10	2.54	1.97	1.41	.85	.28	27.5
GH-4VB	1/3	2	115/230/60/1	Capacitor Start	56	38	8.75x11.50x16.00	3.60	2.95	2.29	1.64	.98	.33	27.5
TA-4V2	1/3	2	115/230/60/1	Capacitor Start	56	37	6.88x10.03x14.03	4.20	3.44	2.67	1.91	1.15	.38	27.5
GH-5VB	1/2	2	115/230/60/1	Split Capacitor	56	41	8.75x12.30x16.19	4.20	3.44	2.67	1.91	1.15	.38	27.5
HP-100V	1	2	115/230/60/1	Capacitor Start, Capacitor Run	56	70	9.22x18.00x19.81	8.00	6.55	5.09	3.64	2.18	.73	27.5

FIGURE 12-19. Technical data for Thomas piston pumps.

MODEL NUMBER	HP	kW	VOLTAGE	MOTOR TYPE	WT LBS.	WT KG.	PHYS. SPEC. H×W×L IN.	PHYS. SPEC. H×W×L CM	IN.HG vs.CFM 0	5	10	15	20	25	MM HG. vs.LPM 0	127	254	381	508	635	MAX.VAC. IN.HG.	MAX.VAC. MM.HG.
014CDC20	1/30	.024	12VDC	Perm. Magnet	2.3	1.0	4.13×1.89×5.48	10.4×4.8×13.9	.39	.24	.18	.12	.06		11.0	6.8	5.1	3.4	1.7		25.3	642.6
014CA28	1/16	.046	115/60/1	Shaded Pole	4.7	2.1	5.06×4.50×5.89	12.8×11.4×14.9	.46	.37	.28	.19	.09		13.0	10.5	7.9	5.4	2.5		26.0	660.4
010CA26*	1/25	.029	115/60/1	Shaded Pole	3.3	1.4	4.25×3.00×5.17	10.8×7.6×13.1	.37	.27	.18	.10	.02		10.5	7.6	5.1	2.8	.6		22.3	566.4
010CDC26	1/25	.029	12VDC	Perm. Magnet	1.7	.7	4.01×2.36×4.82	10.1×5.9×12.2	.37	.27	.18	.10	.02		10.5	7.6	5.1	2.8	.6		22.3	566.4
405AE38	1/12	.061	115/60/1	PSC	6.5	2.9	6.00×5.25×8.75	15.2×13.3×22.2														
405ADC38*	1/10	.075	12VDC	Perm. Magnet	4.3	1.9	6.50×4.00×7.27	16.5×10.1×18.4														
415CDC30	1/10	.075	12VDC	Perm. Magnet	4.8	2.1	5.25×4.00×7.12	13.3×10.1×18.0	.92	.75	.54	.34	.20		26.1	21.2	15.3	9.6	5.7		24.6	624.8
215ADC38	1/6	.124	12VDC	Perm. Magnet	3.0	1.3	4.46×2.10×6.20	11.03×5.3×15.7														
315CDC50	1/5	.149	12VDC	Perm. Magnet	4.3	1.9	6.30×4.00×9.31	16.5×10.1×23.6	1.18	.86	.58	.29			33.4	24.4	16.4	8.2			21.0	533.4
317CDC56	1/5	.149	12VDC	Perm. Magnet	6.5	2.9	6.30×4.00×8.60	16.0×10.1×21.8	1.12	.86	.61	.36			31.7	24.4	17.3	10.2			22.0	558.8
607CA22*	1/8	.093	115/60/1	Shaded Pole	11.0	4.9	6.78×5.00×8.00	17.2×12.7×20.3	.84	.62	.46	.31	.15		23.8	17.6	13.0	8.8			25.9	657.9
607FA22	1/8	.093	115/60/1	Shaded Pole	25.0	11.3	14.82×7.00×16.50	37.6×17.1×41.9	.84	.62	.46	.31	.15		23.8	17.6	13.0	8.8			25.9	657.9
607CA32*	1/7	.105	115/60/1	Shaded Pole	11.0	4.9	6.78×5.00×8.00	17.2×12.7×20.3	1.22	.87	.65	.45	.23		34.6	24.6	18.4	12.7	6.5		26.8	680.7
607CE44*	1/3	.248	115/60/1	PSC	14.5	6.5	6.78×5.00×9.17	17.2×12.7×23.2	1.60	1.15	.90	.62	.33		45.3	32.6	25.5	17.6	9.3		27.6	701.0
619CE44*	1/5	.149	115/60/1	PSC	10.0	4.5	6.75×5.00×6.44	17.1×12.7×16.3	2.00	1.20	.96	.64	.36		56.6	34.0	27.2	18.1	10.2		26.0	660.4
2607CE22	1/4	.186	115/60/1	PSC	14.8	6.7	6.76×6.03×11.00	17.1×15.3×27.9	1.68	1.20	.93	.63	.33	.11	47.6	34.0	26.3	17.8	9.3	3.1	26.2	665.5
2608TE22/18	1/3	.248	115/60/1	PSC	17.8	8.0	6.94×5.00×11.04	17.6×12.7×28.0	1.60	1.20	.98	.69	.48	.16	45.3	34.0	27.8	19.5	13.6	4.5	29.0	736.6
2508VE44	1/3	.248	115/60/1	PSC	16.6	7.5	6.76×6.03×11.00	17.1×15.3×27.9	3.15	2.30	1.60	1.10	.64	.18	89.2	65.1	45.3	31.2	18.1	5.1	27.6	701.0
2618CE44*	1/3	.248	115/60/1	PSC	14.3	6.4	6.68×5.00×9.71	16.9×12.7×24.6	3.75	2.75	2.15	1.49	.86	.20	106.2	77.9	60.9	42.2	24.4		27.3	693.4
2619CE44*	1/3	.248	115/60/1	PSC	15.9	7.2	6.75×5.00×9.38	17.1×12.7×23.8	4.80	4.10	3.20	2.30	1.30	.30	135.9	116.1	90.6	65.1	36.8	8.5	27.5	698.5
2621CE564	1/3	.248	115/60/1	PSC	16.0	7.2	7.87×5.00×9.38	19.9×12.7×23.8	5.35	4.58	4.36	4.16	3.94	3.74	151.5	129.7	123.5	117.8	116.6	105.9	27.0	558.8
2750CE50	1/3	.248	115/60/1	PSC	20.0	9.0	9.29×5.37×10.09	23.5×13.6×25.6	7.05	6.27	4.69	3.17	1.77	.49	199.7	177.6	132.8	89.8	50.1	13.9	27.0	658.8
2750BE75	1/3	.248	115/60/1	PSC	20.0	23.5	9.29×5.37×10.09	23.5×13.6×25.6													27.0	658.8
707CK50	1/4	.186	115/60/1	Capacitor Start	26.0	11.7	10.00×6.38×10.72	25.4×16.2×27.2	2.70	1.99	1.54	1.12	.65	.20	76.5	56.4	43.6	31.7	18.4	5.7	27.7	703.6
807CK60	1/3	.248	115/60/1	Capacitor Start	26.0	11.7	10.00×6.38×10.72	25.4×16.2×27.2	3.25	2.07	1.60	1.14	.67	.21	92.0	58.6	45.3	32.3	19.0	5.9	27.7	703.6
1007CK72	1/2	.372	115/60/1	Capacitor Start	26.0	11.7	10.00×6.38×10.72	25.4×16.2×27.2	4.05	2.49	1.93	1.39	.83	.28	114.7	70.5	54.7	39.4	23.5	7.9	27.9	708.7
1107CK75	3/4	.559	115/60/1	Capacitor Start	26.0	11.7	10.00×6.38×10.72	25.4×16.2×27.2	4.25	2.70	2.10	1.40	.86	.29	120.4	76.5	59.5	39.6	24.4	8.2	27.4	696.0
2807CE72*	1	.745	115/60/1	PSC	39.0	17.6	10.03×6.54×15.71	25.4×16.6×39.9	6.60	3.60	2.66	1.80	.88		186.9	102.0	75.3	51.0	24.9		25.0	635.0

FIGURE 12-20. Technical data for Thomas WOB-L piston pumps.

Model	Available Voltages		Max. Flow L/M	Maximum Pressure PSI		Maximum Vacuum in. HG		Weight		Dimensions HxWxL (mm)
				INT.	CONT.	INT.	CONT.			
G01-K	6	VDC	1.0	0.3	0.3	0.6	0.6	1	oz	21x21x44
G01	3, 6,12,16	VDC	1.1	1	1	3	1.8	3	oz	26x26x55
G01-4	6,12,16	VDC	1.0	2	1	4.7	1.8	3	oz	26x26x55
G02	3, 6,12,16	VDC	2.6	2	1	3.5	1.8	3	oz	26x26x55
G02-LC	6,12	VDC	3.0	3	1	6	2	4	oz	26x26x71
G02-4	6,12,16	VDC	2.5	3	1	7	1.8	3	oz	26x26x55
G02-8	6,12	VDC	2.4	7	4	13	9.0	5	oz	26x26x71
G02-8-LC	6,12	VDC	2.5	6	4	12	9.0	4	oz	26x26x71
G04	3, 6,12,16	VDC	4.5	1	1	2.4	1.5	3	oz	26x26x61
G04-LC	6,12	VDC	5.0	2	1	3.0	1.5	4	oz	26x26x77
G04-4	6,12,16	VDC	4.5	3	1	6	1.5	3	oz	26x26x61
G045	6,12,24	VDC	6	7	4	15	9	5	oz	42x42x84
G045-LC	12	VDC	7	12	4	20	9	8	oz	42x42x89
G045-TP	12,24	VDC	10	7	4	15	9	11	oz	42x42x130
G045-TS	12,24	VDC	5	13	9	27	18	11	oz	42x42x130
W045-U	110V/60*	Hz	9	9		18		12	oz	59x42x93
W045-S	110V/60*	Hz	3.5	5	5	10	10	1.2	lbs	74x60x90
W05	110V/60*	Hz	6	9	6	18	18	3.2	lbs	63x90x119
W08	110V/60*	Hz	15	10	7	21	21	4.6	lbs	105x65x135
G07	12,24	VDC	14	10		21		1.3	lbs	51x51x135
G08	12,24	VDC	14	10	4	21	9	2.0	lbs	62x62x165
G08-TP	12,24	VDC	28	10	4	21	9	3.4	lbs	62x62x210
G08-TS	12,24	VDC	14	12	6	27	18	3.4	lbs	62x62x210
G09	12,24	VDC	20	10	4	21	9	2.0	lbs	50x50x165
TF1	110V/60*	Hz	20	10	10	22	22	9	lbs	156x95x175
TF1E	12,24	VDC	25.6	11	11	23	23	6.5	lbs	101x90x193
TF1.5	110V/60*	Hz	20	10	10	22	22	8.8	lbs	100x95x175
TF1.5-E	110V/60*	Hz	23	12	12	23	23	7	lbs	116x105x182
TF1.5-E-LC	110V/60*	Hz	23	12	12	23	23	7.3	lbs	114x110x196

FIGURE 12-21. Technical data for ASF Brey vane pumps.

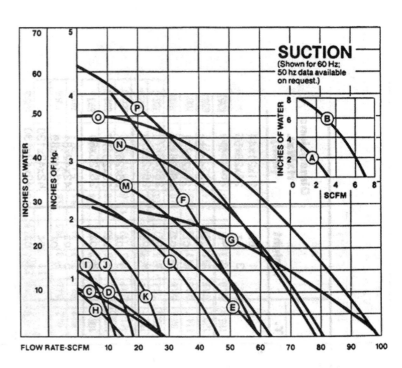

FIGURE 12-24. Vacuum-air flow data for EG&G Rotron blowers. A = Model SE; B = SE-DC; C = SL1P; D = SL1S; E = SL2P; F = SL4P; G = SL5P; H = RDC; I = DR068; J = DR083; K = DR101; L = DR202; M = DR303; N = DR353; O = DR404; P = DR513.

MODEL NUMBER	NOMINAL HP	VOLTAGE/ HERTZ/ PHASE	MOTOR TYPE	FRAME	PHYSICAL SPECIFICATIONS		VACUUM PERFORMANCE IN. HG VS. CFM						MAX. VACUUM IN. HG
					WT. (LBS.)	H" x W" x L"	0	5	10	15	20	25	
SR-0015-VP	1/10	115/50/60/1	Split Phase		8.5	4.94x4.19x7.62	1.50	1.19	.88	.56	.25	0	24.0
SR-0015-VP	1/10	115/50/60/1	PSC		8.0	4.94x4.19x6.32	1.50	1.14	.88	.56	.25	0	24.0
SR-0030	1/8	115/60/1	PSC		9.3	3.95x3.86x6.85	2.85	2.26	1.66	1.07	.48	0	24.0
TA-0015-V	1/8	115/50/60/1	Split Phase	48	17.0	6.00x5.75x9.00	1.85	1.45	1.05	.64	.24	0	23.0
TA-0030-V	1/6	115/50/60/1	Split Phase	48	20.0	5.75x5.75x10.96	3.00	2.38	1.75	1.13	.50	0	24.0
TA-0040-V	1/4	115/50/60/1	Split Phase	48	24.0	5.81x5.75x9.56	4.00	3.23	2.46	1.69	.92	.15	26.0
TA-0075-V	1/2	115/230/60/1	Capacitor Start	56	42.0	6.75x6.50x13.70	7.20	5.82	4.43	3.05	1.66	.28	26.0
TA-0100-V	3/4	115/230/60/1	Capacitor Start	56	44.0	6.75x6.50x13.95	10.00	8.08	6.15	4.73	2.31	.38	26.0
TA-0170-V	1	115/230/60/1	Capacitor Start	Close Coupled Rotary	71.0	7.50x8.50x21.25	17.00	13.85	10.70	7.56	4.41	1.26	27.0
TA-0210-V	1½	115/230/60/1	Capacitor Start, Capacitor Run	Close Coupled Rotary	84.0	7.50x8.50x21.25	21.00	17.11	13.22	9.33	5.41	1.56	27.0

FIGURE 12-22. Technical data for Thomas TASKAIR vane pumps.

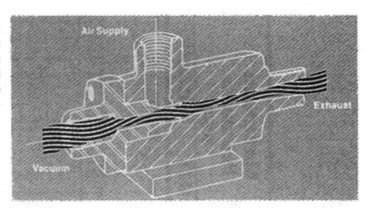

TD DESIGN
Straight-through vacuum passage allows material to pass directly through Vacuum Transducer with no reduction of vacuum flow. Compressed air enters through annular orifice.

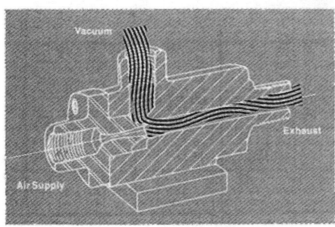

AV DESIGN
Vacuum passage has 90° change of direction. Compressed air flows directly through circular orifice into venturi section. AV design converts compressed air to vacuum more efficiently than TD design. No solid material should pass through Vacuum Transducer.

FIGURE 12-25. Schematic diagram of Air-Vac transducers.

FIGURE 12-26. Ametek MG-5P personal air sampler.

Technical data for EG&G Rotron blowers

Models	Motor Type[1]	Phase	Voltages	Weight (lbs.)	Part Number "S" Unit	Part Number "A" Unit	Max Flow (SCFM)	Max Press (IWG) "S" Units	Max Vac (Hg) "S" Units	Performance Curve Index
MINISPIRAL BLOWERS										
SE-B21	1 / A / F					036258	3.2	3.0	22	A
SPIRAL BLOWERS										
SL1P__	TE²	1	115V 50Hz S	22	036005		29	12	73	B
SL1S__	TE	3	220-230V 60Hz F	22	036006	036007	29	17	1.0	C
SL2P__	TE	3	A	22	036000	036008 / 036013	58	35	2.1	D
SL4P__	TE	3	A	27/23	036009	036027 / 036020	61	62	3.7	E
SL5P__	TE	3	A	43/37	036010	036261 / 036021	100	33	2.1	F
SL6P__	TE² / TE	1 / 3	F	43/37	036011	036023 / 036022	100	62	3.7	E/F
DR XOX BLOWERS										
DR068__	TEFC	1	115/230	14	037143		12	17	17	G
	TEFC	3	230/460	14	037144					
DR083__	TEFC	3	115/230	15	036862		18	24	23	H
	TEFC	3	230/460	15	037164					
DR101__	TEFC	1	230/460	27	036244	036245 / 036672	28	27	1.8	I
	TEFC	3	115/230	25						
			575							
DR202__	TEFC	1	115/230	32	037066	037067 / 036373	48	33	2.3	J
	TEFC	3	230/460	29						
			575	29						
DR303__	TEFC	1	115/230	36	036233	036234 / 036372	63	40	3.0	K
	TEFC	3	208-230/460	31						
			575	31						
DR353__	TEFC	1	115/230	56	037147	037149	88	48	4.3	L
	XP	1	115/230	56	037148		88	50	4.5	
	XP	3	230/460	56						
	XP	3	230/460	56						
			575							
DR404__	TEFC	1	115/230	75	037062	037058	98	56	3.6	M
	XP	1	115/230	89	037063	037150 / 037146				
	TEFC	3	208-230/460	61						
	XP³	3	230/460	61						
			575	76						
DR513__	TEFC	1	115/230	78	037209	036267 / 037059	80	75	60	N
	TEFC	3	230/460	78	037217					
DR BLOWERS										
DR312__	XP	1	115⁴	36		037048	48	26	1.7	O
DR313__	XP	1	115⁴	45		037047	53	50	3.1	P
DR4__	TEFC	1	115/230	68	036104	036108	100	74	5.9	Q
	XP	1	115/230	103	036103	036109 / 036106				
	TEFC	3	208-230/460	56						
	TEFC	3	230/460	56						
	XP³	3	230/460	79						

1. All 3 ph motors are factory tested and certified to operate on 200-230/460 VAC-3 ph-60 Hz and 220-240/380-415 VAC-3 ph-50 Hz. All 1 ph motors are factory tested and certified to operate on 115/230 VAC-1 ph-60 Hz and 220-240 VAC-1 ph-50 Hz.
2. Spiral motors are Rotron manufactured, totally enclosed within the blower body but open to the gas stream
3. Three phase explosion proof motors are shown as 230/460 volt. They are also available in 575 volt.
4. DR3__ – are shown in 115V, 1 phase. They are available by special order in many other voltages.
5. Performance shown for "A" units when no "S" unit is listed.
S. Cataloged, stocked by distributors
A. Cataloged, available from but not stocked by distributors
F. Non-cataloged, available from factory
X. Denotes Spiral blower horsepower. There are no optional horsepowers in Spiral models.

FIGURE 12-23. Technical data for EG&G Rotron blowers.

FIGURE 12-27. Sipin SP-15 personal air sampler.

FIGURE 12-29. ASB-II High Volume Air Sampler.

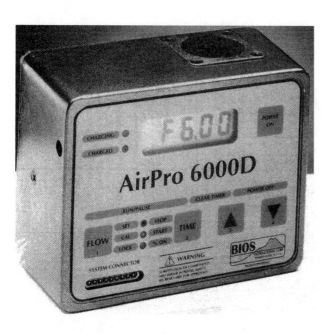

FIGURE 12-28. BIOS AirPro 6000D personal air sampler.

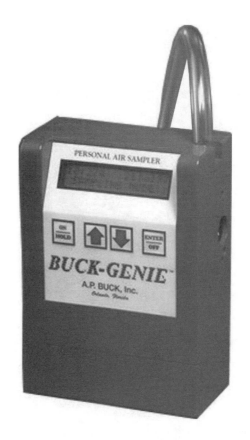

FIGURE 12-30. BUCK Personal Sampler SS.

FIGURE 12-31. Escort® ELF Sampling Pump.

FIGURE 12-33. SKC Personal Sampler.

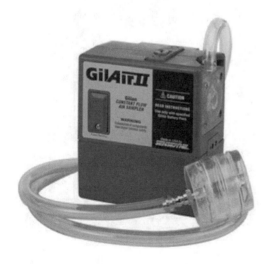

FIGURE 12-32. GilAir II Air Sampling Pump.

FIGURE 12-34. Spectrex PAS-3000, Model II personal air sampler.

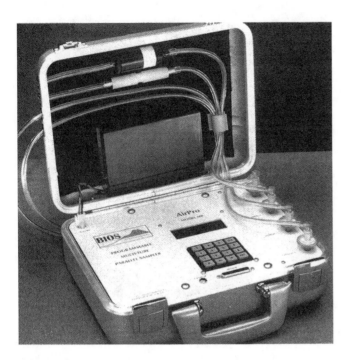

FIGURE 12-35. BIOS AirPro series area sampler.

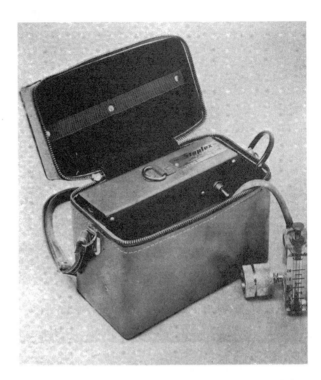

FIGURE 12-37. Staplex Model BN/BS battery-powered, low-volume air sampler.

FIGURE 12-36. Gilian AirCon-2 Air Sampling System.

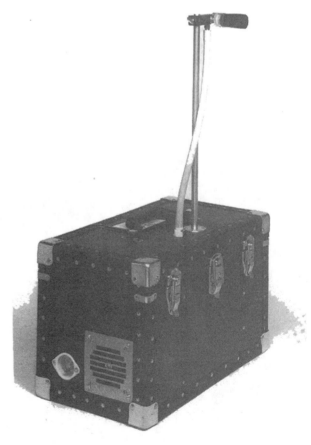

FIGURE 12-38. BGI Model ASB-II-S low-volume air sampler.

FIGURE 12-39. Eberline Model RAS-2 Environmental Air Sampler.

FIGURE 12-41. HV-I Standard High Volume Air Sampler.

FIGURE 12-40. LV-I Standard Low Volume Air Sampler.

FIGURE 12-42. Graseby Andersen PM10 Medium Flow Sampler.

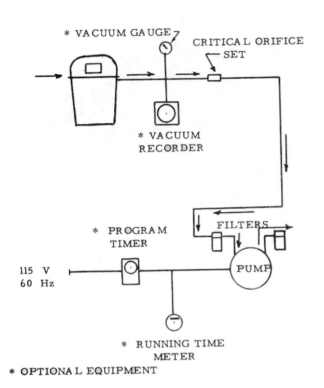

FIGURE 12-43. Flow diagram for Graseby Andersen Universal Sampler.

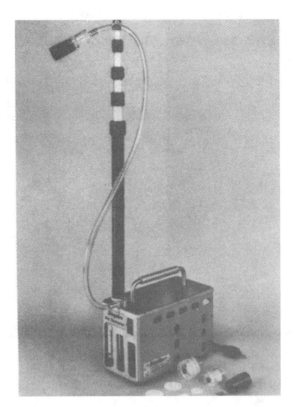

FIGURE 12-45. Staplex Model VM-3 low-volume air sampler.

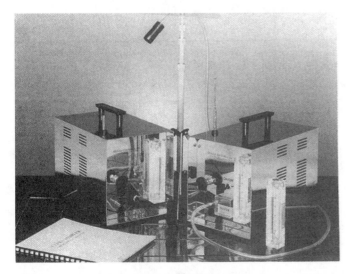

FIGURE 12-44. Midwest Environics Ultra Sampler.

FIGURE 12-46. Staplex Model VM-2 low-volume air sampler.

FIGURE 12-47. Hi-Q Model CF-993B sampler.

FIGURE 12-49. Staplex TFIA sampler.

FIGURE 12-48. Hi-Q TFIA High Volume Air Sampler.

FIGURE 12-50. Andersen Intermediate-Volume Sampler.

FIGURE 12-51. DH-50810 Digital High Volume Air Sampler.

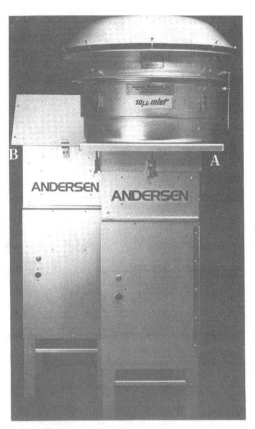

FIGURE 12-53. Graseby Andersen (A) PM-10 High Volume Sampler and (B) TSP High Volume Sampler.

FIGURE 12-52. TE-1000 PUF (Poly-Urethane Foam) Monitor in Ambient Air.

FIGURE 12-54. Graseby Andersen Model PS-1 PUF sampler.

FIGURE 12-55. General Metal Works Model ACCU-Vol IP-10 high
volume sampler.

FIGURE 12-56. Andersen Critical Flow High-Volume Sampler.

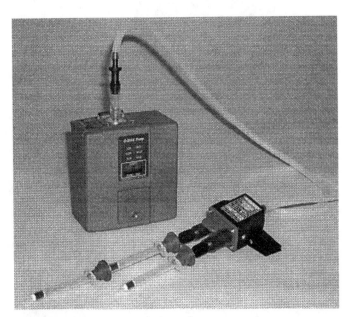

FIGURE 12-57.

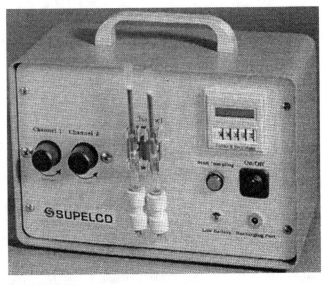

FIGURE 12-58.

Chapter 13

Filters and Filter Holders

Morton Lippmann, Ph.D.

Nelson Institute of Environmental Medicine, New York University School of Medicine, Tuxedo, NY

CONTENTS

Introduction

Filtration is the most widely used technique for aerosol sampling, primarily because of its low cost and simplicity. The samples obtained usually occupy a relatively small volume and may often be stored for subsequent analysis without deterioration. By appropriate choice of air mover (see Chapter 12), filter medium, and filter size, almost any sample quantity desired can be collected in a given sampling interval.

Figure 13-1 is a schematic representation of the elements of a filter sampling system. The component parts may include all or some of the following: sampling probe, filter holder, filter, pressure sensor, flowmeter, air mover, and a means of regulating the flow. A probe is needed only when sampling large particles from a mov-

ing stream, e.g., a duct or stack. For these applications, careful attention must be given to the probe's shape, size, and orientation with respect to the flowing stream in order to obtain representative samples. The factors affecting the entry of large particles into a sampling tube, i.e., particle inertia, gravity, flow convergence, and the inequality of ambient wind and suction velocity, have been critically evaluated by Davies.[1] They are also discussed in detail in Chapter 20. Errors can also arise from particle deposition onto the surfaces of plastic inlet probes due to electrostatic deposition, or to deposition between the probe inlet and the filter due to impaction at the bends and turbulent diffusion.[2]

The filter should be upstream of everything in the system but the nozzle, so that any dirt in the system, manometer liquid, or pump oil will not be carried

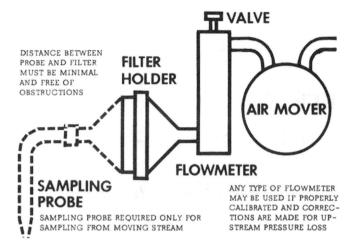

FIGURE 13-1. Elements of a filter sampling system.

accidentally onto it. The filter should be as close as possible to the sampling point, and all sampling lines must be free of contamination and obstructions.

The filter holder, designed for the specific filter size used, must provide a positive seal at the edge. A screen or other mechanical support may be required to prevent rupture or displacement of the filter in service. An in-line filter holder should also include a gradual expansion from the inlet to the filter. With a properly designed holder, the air velocity will be uniform across the cross section of the filter holder. Uniform flow distribution is especially desirable when analyses are to be performed directly on the filter, or where only a portion of the filter will be analyzed, so that the remainder can be archived or used for replicate analyses or analyses by other techniques.

The accurate measurement of either flow rate and sampling time or sample volume is as important as the measurement of sample quantity because aerosol concentration is determined by the ratio of sampled quantity to sampled volume. Unfortunately, air volume measurements are often inaccurate (see Chapter 7). When the volumetric capacity of the air mover is highly pressure dependent, as it is for turbine blowers, ejectors, and some other types of air movers, the flow cannot be metered by any technique that introduces a significant pressure drop itself. This precludes the use of most meters that require the passage of the full volume through them and limits the choice to low resistance flowmeters. Low resistance flowmeters include bypass meters, which measure the flow rate of a small volume fraction of the sampled air, and meters utilizing very sensitive measurements of vane displacement or pressure drop. These types of meters can provide sufficiently accurate measurements, but they often require more careful maintenance and more frequent calibration

and adjustment than they are likely to receive in field use.

Most flowmeters are calibrated at atmospheric pressure, and many require pressure corrections when used at other pressures. Such corrections must be based on the static pressure measured at the inlet of the flowmeter. The flowmeter should be downstream of the filter to preclude the possibility of sample losses within it. It will, therefore, be metering air at a pressure below atmospheric, due to the pressure drop across the filter. Furthermore, if the filter resistance increases with loading, as is often the case, the pressure correction will not be a constant factor. Chapter 7 provides a comprehensive discussion of air flow calibration.

If the sampling flow rate is to be controlled by a throttling valve, this valve should be downstream of the flowmeter to avoid adding to the pressure correction for the flowmeter. Flow rate adjustments can be made either with a throttling valve or by speed control of the air mover's motor, and they can be either manual or automatic. Automatic control requires pressure or flow transducers and appropriate feedback and control circuitry.

The discussion that follows is designed to provide the background necessary for the proper selection of filters for particular applications. Filtration theory is outlined, the various kinds of commercial filter media used for air sampling are described, and the criteria that limit the selection for various sampling situations are discussed.

Filtration Theory

Types and Structures of Filters

All filters are porous structures with definable external dimensions such as thickness and cross-section normal to fluid flow. They differ considerably in terms of flow pathways, flow rates, and residence times, and these factors are strongly influenced by their structure. One of the oldest and most common filter types for air sampling is the fibrous filter, which is comprised of a mat of cellulose, glass, quartz, asbestos, or plastic fibers in random orientation within the plane of the filter sheet. Another type of filter is the granular bed, in which solid granules are packed into a definable sheet or bed. In granular bed filters used in air sampling, the granules are usually sintered to the point where they form a relatively rigid mechanical structure. Granules of glass and aluminum oxide are frequently sintered in the form of a thimble for high-temperature stack sampling. Thin sintered beds of silver granules are used in disc

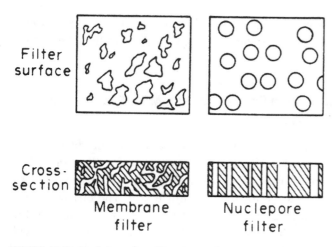

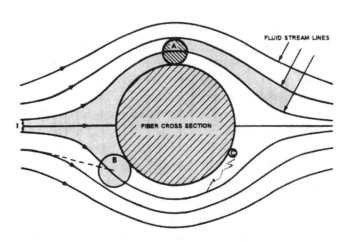

FIGURE 13-2. Surface and section views of porous gel-type membrane filter and Nuclepore (polycarbonate pore) filter.

FIGURE 13-3. Flow pattern around a filter fiber and particle capture mechanisms: Particle A–direction interception; Particle B–inertial impaction; Particle C–diffusional deposition.

form for a variety of applications and are generally known as silver membranes.

The term membrane filter was originally applied to discs of a cellulose ester gel having interconnected pores of uniform size. First and Silverman[3] described various applications of such filters for air sampling in 1953. Gel-type membrane filters are now also available in polyvinyl chloride (PVC), nylon, and other plastics. Whereas the method of production is quite different from those used to make fibrous filters or granular beds, the flow pathways of all three types of structures are quite similar in terms of the tortuosity of the air flow pathways. The Nuclepore® filter, a polycarbonate membrane filter, has a radically different structure, i.e., a series of nearly parallel straight-through holes. It is made by exposing a thin sheet (approx. 10 μm) of polycarbonate plastic to a flux of neutrons from a nuclear reactor and then chemically etching the neutron tracks. Simplified versions of the various filter structures are illustrated in Figure 13-2.

Flow Fields and Collection Mechanisms

Theoretical models of particle filtration have been developed using simplified flow field and particle motion in the vicinity of a single isolated cylindrical fiber. Extension of the theory to a filter mat depends upon taking proper account of the influence of adjacent fibers on the flow field.[4-6] The fluid motion and particle motion in the vicinity of a cylindrical fiber are illustrated in Figure 13-3. The corresponding flow fields around the pores in a polycarbonate pore filter are illustrated in Figure 13-4.

Filters remove particles from a gas stream by a number of mechanisms. These mechanisms include

direct interception, inertial deposition, diffusional deposition, electrical attraction, and sedimentation. The mechanisms that predominate in a given case will depend on the flow rate, the structure of the filter, and the nature of the aerosol.

Interception occurs when the radius of a particle moving along a gas streamline is greater than the distance from the streamline to the surface. This mechanism is important only when the ratio of the particle size to the void or pore size of the filter is relatively large.

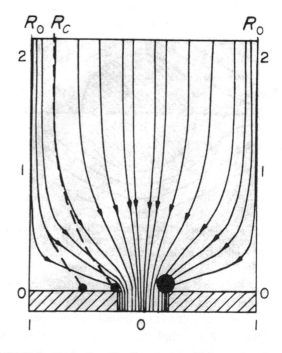

FIGURE 13-4. Streamlines for flow approaching a polycarbonate pore filter with porosity 0.05. The interception, impaction, and diffusion mechanisms are shown.

Inertial collection results from a change in direction of the gas flow. The particles, due to their relatively greater inertia, tend to remain on their original course and strike a surface. Capture is favored by high gas velocities and dense fiber packing. The factors affecting inertial deposition in a jet impactor are discussed at greater length in Chapter 14. The operation of the inertial mechanism in a variety of commercially available fibrous filters was demonstrated experimentally by Ramskill and Anderson.[7]

Diffusion is most effective for small particles at low flow rates. It depends on the existence of a concentration gradient. Particles diffuse to the surfaces of the fibers, where the concentration is zero. Diffusion is favored by low gas velocities and high concentration gradients. The root-mean-square displacement of the particles, and hence the collection efficiency, increases with decreasing particle size down to about 10 nm. Below this size, there is particle rebound from the filter surface, as discussed by Wang and Kaspar.[8]

Kirsch and Zhulanov[9] tested the performance of high efficiency fibrous filters made of glass and polymeric polydisperse fibers. They found good agreement with the theory proposed earlier by Kirsch, Stechkina, and Fuchs.[6]

Gentry et al.[10] studied the diffusional deposition of ultrafine aerosols on polycarbonate pore filters. They found that particles <0.03 μm were collected by diffusion on the upstream surface of the filter and that the efficiency was only slightly higher than the values predicted by theory. On the other hand, for particles of 0.04 to 0.10 μm, the particles were collected primarily around the rims of the pores, and the efficiencies were much higher than those predicted by theory.

Electrical forces may contribute greatly to particle collection efficiency if the filter or the aerosol has a static charge. Lundgren and Whitby[11] showed that image forces, i.e., the forces between a charged particle and its electrical image in a neutral fiber, can strongly influence particle collection. The factors controlling particle deposition on a filter suspended in a uniform electric field and the influence of such a field on the deposition of both charged and uncharged particles were described by Zebel.[12] Unfortunately, the data needed to predict the effect of electrostatic charges on the collection efficiency of sampling filters are seldom available.

Gravitational forces may usually be neglected when considering filter sampling. The settling velocities of airborne particles of hygienic significance are too low, and the horizontal components of the surface areas in the filters are too small, for gravitational attraction to

have any significant effect on particle collection efficiency unless the face velocity through the filter is very low, e.g., <5 cm/s.

Minimum Efficiencies and Most Penetrating Particle Sizes

Because a variety of collection mechanisms are involved in filtration, it is not surprising that, for a given aerosol and a given filter, the collection efficiency varies with face velocity and particle size. The efficiency of a given filter for a given particle size could be high at low flows, due primarily to the effects of diffusion. With increasing velocity, it could first fall off and then, with still higher velocities, begin to rise because of increased inertial deposition. This pattern has been observed in several experimental penetration tests[13, 14] and is illustrated in Figures 13-5 and 13-6. At very high velocities, the retention could decrease because of re-entrainment. Additional data showing these effects are presented in Table 13-1.

Filter retention by the interception and diffusion mechanisms is also strongly influenced by particle size,

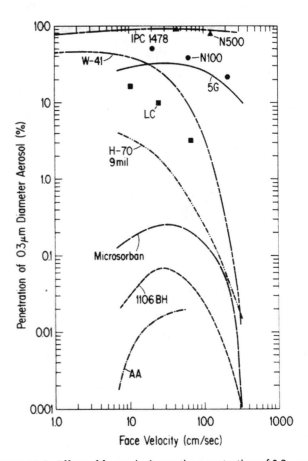

FIGURE 13-5. Effect of face velocity on the penetration of 0.3-μm-diameter particles through various air sampling filter media—based on data reported by Lockhart et al.,[14] Rimberg,[13] and Liu and Lee.[16]

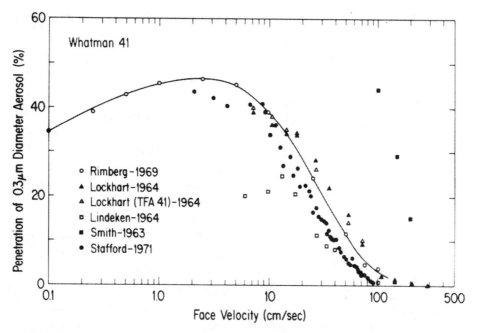

FIGURE 13-6. Effect of face velocity on the penetration of 0.3-µm-diameter particles through Whatman 41 filters—based on data reported by various investigators.

as illustrated in Figure 13-7. This figure presents experimental data from Spurny et al.[15] for polycarbonate pore (Nuclepore) filters with 5 µm-diameter pores at a face velocity of 5 cm/s. Figure 13-8, from Liu and Lee,[16] shows collection efficiency as a function of particle size for 1 µm Nuclepore filters at three different face velocities. Polycarbonate pore filters have a very different structure than other types of filters, as has been discussed, and exhibit a more extreme size dependence. Rimberg[13] has demonstrated experimentally that for fibrous filters, the size at which maximum penetration occurs increases with decreasing face velocity.

The theoretical basis for predicting the minimum collection efficiency and most penetrating particle sizes for fibrous filters was addressed by Lee and Liu.[17] They developed equations for such predictions, which compared favorably with experimental filter efficiency data. Lee[18] extended his analysis of minimum efficiency and most penetrating particle size to granular bed filters. Predictive theories for deposition in such filters were also developed by Schmidt et al.[19] and Fichman et al.[20]

Spurny[21] investigated the collection efficiencies of membrane and polycarbonate pore filters for aerosols of

TABLE 13-1. Flow Rate and Collection Efficiency Characteristics of Selected Air Filter Media[A]

Filter Type	Filter	mm Hg Pressure Drop 53	106	211	Percent Penetration of 0.3 µM DOP[B] 26.7	53	106	211	Flow Reduction Due to Loading[C] %/m³/cm²
Cellulose	Whatman 1	86	175	350	7	0.95	0.061	0.001	17.9
	41	36	72	146	28	16	2	0.30	5.0
	541	30	61	123	56	40	22	9	10.4
Glass	MSA 1106BH	30	61	120	0.068	0.048	0.022	0.005	0.43
	Pall–Gelman A	33	65	129	0.019	0.018	0.011	0.001	0.50
	E	28	57	114	0.036	0.030	0.014	0.004	0.53
	Whatman 934AH	37	74	150	0.010	0.006	0.003	0.001	0.47
	Whatman GF/A	29	60	118	0.018	0.015	0.008	0.001	0.37
Membrane	Millipore AA (0.8 µ)	142	285	570	0.015	0.020	—	—	1.6

[A]Data extracted from NRL Report No. 6054.[14]
[B]DOP = Di(2-ethylhexyl) phthalate.
[C]Normalized to the dust loading in the atmosphere on an "average" summer day (Washington, DC, 1964).

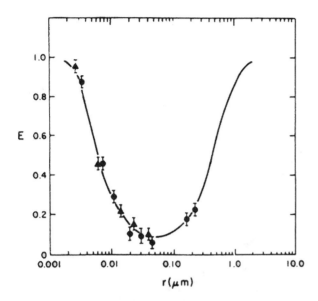

FIGURE 13-7. Effect of particle radius (r) on fraction collected (E) for a Nuclepore filter with 5-μm-diameter pores at a face velocity of 5 cm/s. Experimental data are shown for a selenium aerosol (•) and a pyrophosphoric acid aerosol (s). The line is computed from filtration theory. *(Reprinted from Environ. Sci. Technol. 3:463, 1969; courtesy American Chemical Society.)*

chrysotile asbestos. For Millipore membrane filters with 8 μm pores, the collection efficiency at a face velocity of 3.5 cm/s fell from 100% for fibers >5 μm in length to 75% for fibers of 2 μm in length and to 25% for fibers approximately 0.5 μm in length. For polycarbonate pore filters with pore diameters of 0.2, 0.4, and 0.8 μm, collection efficiencies began to drop for fiber lengths <3 μm and fiber diameters <0.2 μm. For 0.2 μm pores, the efficiencies did not drop below approximately 80%, whereas for 0.8 μm pores, the efficiencies dropped to near zero for fiber lengths below 0.5 μm and diameters below 0.05 μm.

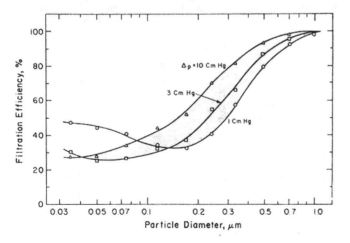

FIGURE 13-8. Efficiency of 1.0-μm polycarbonate pore filter.

Forces of Adhesion and Re-entrainment

The collection mechanisms discussed above act to arrest the motion of the particles in a gas stream as the gas flows through the voids of a filter. The particles removed from the gas stream are then subject to forces of adhesion. If the forces of adhesion on a particle are greater than the forces that tend to push the particle free, then that particle is "collected" and will be available for analysis. However, the forces exerted on the particle by the flowing gas stream may be greater than the forces of adhesion, resulting in re-entrainment of the particle. At present, it is at least as difficult to predict forces of adhesion from theoretical considerations as it is to predict the effectiveness of the collection mechanisms. One reason is that it is usually not possible to determine whether particles that penetrate a filter were blown off after collection because of inadequate adhesion, or whether they underwent elastic rebound upon initial contact with the filter surface. This question was theoretically and experimentally investigated by Loffler.[22] He concluded that the measured forces of adhesion were in good agreement with the Van der Waals forces calculated theoretically and that the flow velocity required for blowing collected particles off fibers is much higher than that normally used in air filtration. An increase in particle penetration with increasing velocity will usually be due to increased rebound or to the resuspension of particle flocs.

Commercial Filter Media

Filter media of many different types and with many different properties have been designed for, or adapted to, air sampling requirements. For purposes of discussion, they have been divided into groups determined by their composition. Air flow resistance and collection efficiency characteristics of some commonly used filters are tabulated in Table 13-1. (Also see Table 13-I-1 at the end of this chapter which summarizes the physical characteristics of commercially available filter media, based on vendor supplied or approved data.) These media have been subdivided on the basis of their composition and/or structure.

Cellulose Fiber Filters

Most cellulose fiber filters are used widely by analytical chemists for liquid-solid separations. They are made of purified cellulose pulp, are low in ash content, and are usually less than 0.25 mm thick. These filters are relatively inexpensive, are obtainable in an almost unlimited range of sizes, have excellent tensile strength, and show little tendency to fray during handling. Their disadvantages include nonuniformity, resulting in variable flow resistance and collection efficiency, and hygroscopicity, which makes accurate gravimetric determinations either difficult or impossible.

Whatman No. 41 is used for industrial hygiene air sampling. It has the advantages of low cost, high mechanical strength, and high purity typical of these papers and, in addition, has a moderate flow resistance. However, as shown in Figure 13-6, unless used at high face velocity, particle penetration can be significant. Cellulose filter papers also include hardened papers, such as Whatman No. 50, from which collected particles can be removed by washing.

Glass and Quartz Fiber Filters

Glass and quartz fiber filters are, in most cases, more expensive and have poorer mechanical properties than cellulose filters. They also have many advantages, i.e., reduced hygroscopicity, ability to withstand higher temperatures, and higher collection efficiencies at a comparable pressure drop. These properties, combined with the ability to make benzene, water, and nitric acid extracts from particles collected on them, led to the selection of a high-efficiency glass fiber filter as the standard collection medium for high-volume samplers in early air sampling networks. In recent years, quartz fiber filters have replaced glass fiber filters for many applications.

As described by Pate and Tabor,[23] a large number of tests were routinely performed on glass fiber filters. Nondestructive tests, e.g., weighing, gross-activity, and reflectance, were performed prior to the chemical extractions. Also, portions of the filters were stored untreated for possible use at later times to obtain background data on air concentrations whose need was not anticipated at the time of sample collection.

The types of chemical analyses that can be performed on extracts from the filters are determined by the sensitivity of the analyses and by the magnitude and variability of the extractable filter blank for the particular ion or molecule involved. The filter characteristics are determined by the process variables in at least four production stages, i.e., the production of the glass, the production of glass fiber from the bulk glass, the production of the fiber mat from the glass fiber, and the packaging of the individual filters. For the most widely used filter, for the National Air Sampling Network (NASN), Pate and Tabor[23] described four different types produced sequentially between 1956 and 1962, which differed in softening temperature, chemical composition, and extractability.

One of the determinations, made by NASN, was gross mass of particulate matter by gravimetric analyses. Many other investigators have used the same types of high-volume samplers and filters for routine monitoring and analyzed only for gross mass concentration. The validity of these determinations is suspect. The potential errors arising from inaccurate sample volume determinations, from inadequate temperature and humidity conditioning prior to weighing, and from the limited precision of the weighing procedure are well known; they were discussed by Kramer and Mitchel.[24] An additional potentially serious source of error is the loss of filter fibers drawn through the support screen into the air mover during sample collection. Flash-fired binderfree filters are soft and friable, and the loose fiber content is variable. Some NASN filters returned from the field had lower than tare weights, despite the presence of visible deposit on the filter face. If gross mass concentration analyses are to be performed, other nonhygroscopic filter media, which are both mechanically strong and efficient, should be used, or the filter should include a backing layer to prevent the loss of filter fibers.

All of the preceding discussion applied to glass fiber filters that are virtually 100% efficient for all particle sizes. For some applications, e.g., a filter-pack sampler designed to provide data on particle size distribution, less efficient glass fiber filters may be desirable. Shleien et al.[25] described the physical and collection efficiency characteristics of four less-efficient glass fiber filters, produced for gas cleaning and air conditioning applications, which they selected for their filter pack.

Membrane Filters

Filters consisting of porous membranes can be used for many applications where fibrous filters cannot. Organic membranes are produced by the formation of a gel from an organic colloid, with the gel in the form of a thin (approx. 150 μm) sheet with uniform pores. Membrane filters made from cellulose nitrate achieved widespread use for air sampling in the early 1950s.[3] In subsequent years, membrane filters made of cellulose triacetate, regenerated cellulose, polyvinyl chloride, nylon, polypropylene, polyimide, polysulfone, a copolymer of vinyl chloride and acrylonitrile, Teflon®, and silver also become available. Silver membranes are produced by a different technique and will be discussed separately at the end of this section.

Cellulose triacetate membranes are the most widely used and, as indicated in Table 13-I-1, are available in the widest range of pore sizes. The mass of these filters is very low, and their ash content is usually negligible.

Some are completely soluble in organic solvents. Cellulose nitrate filters dissolve in methanol, acetone, and many other organic solvents. Cellulose triacetate, nylon, and PVC filters dissolve in fewer solvents, while filters composed of Teflon and regenerated cellulose do not dissolve in common solvents. The ability to completely dissolve a filter in a solvent permits the concentration of the collected material within a small volume for subsequent chemical and/or physical analyses.

Nylon membrane filters are efficient collectors of nitric acid vapor and are used as back-up collectors in ambient air sampling filter packs to capture nitric acid generated by reactions, on a Teflon primary filter, between strong acid aerosols, i.e., sulfuric acid and ammonium bisulfate, and ammonium nitrate.[26] Such systems are used for accurate determinations of the inorganic ion content of ambient air.

For asbestos and for other mineral and vitreous fibers, fiber count measurements and fiber size analyses are performed by microscopic assays.[27] For occupational health applications, the standard NIOSH 7400 method specifies a 0.8 μm mixed cellulose ester membrane filter and analysis using a phase-contrast optical microscope. For schools and general occupancy buildings, the U.S. Environmental Protection Agency (U.S. EPA) has recommended use of a 0.45 μm mixed cellulose ester membrane filter and analysis of a filter surface segment transferred directly onto an electron microscope grid. Some investigators use an indirect transfer technique in which a segment of the sampling filter is dissolved and an aliquot is transferred to a new filter. The rationale is to obtain an optimal density of fibers per unit surface on the grid in terms of both sufficient fiber density for efficient surface scanning and avoidance of overlap of fibers and their associated coincidence errors. The disadvantage of the indirect transfer technique is that fiber clumps and bundles are disaggregated, increasing the fiber concentration in an unpredictable way.

Collection efficiency increases with decreasing pore size, but even the large pore size filters have relatively high collection efficiencies for airborne particles much smaller than their pores. Membrane filters do not behave at all like sieves when used for air filtration. As in fibrous filters, particles are removed primarily by impaction and diffusion. Early investigators believed that electrostatic forces played a major role in particle deposition in membrane filters, but experimental studies by Spurny and Pich[28] and Megaw and Wiffen[29] demonstrated that diffusional and inertial deposition account for most of the observed collection and that the contribution of direct interception and electrostatic deposition, if present, is less important.

Membrane filters differ from fibrous filters in that a much greater proportion of the deposit is concentrated at or close to the front surface. Lindeken et al.[30] and Lossner[31] measured the penetration depth using test aerosols tagged with alpha emitters. Lindeken et al. were interested primarily in the use of the filters for measuring the concentration of α-emitters in air. If the deposit was truly at the surface, there would be no need for correcting for differences in distance from the detector face or for absorption of α-energy in the filter. They found that, on a microscopic scale, the filter surfaces were not smooth. The surface roughness varied among different brands and, for Millipore Company filters, from the front surface to the back. They concluded that the smooth face of an SM Millipore was suitable for their application. Lossner[31] demonstrated the effect of pore size and face velocity on penetration depth for 0.55 μm SiO_2 particles.

The fact that particle collection takes place at or near the surface of the filter accounts for most of the advantages of membrane filters and also some of their disadvantages. The advantages arising from this property are:

1. It is possible to examine solid particles microscopically without going through a transfer step that might change the state or form of the particles. Examination can be by optical microscopy using immersion oil having the same index of refraction as the filter. The oil renders the filter transparent to light rays. Transmission electron microscopy can be performed on a replica of the filter surface produced by vacuum evaporation techniques, whereas scanning electron microscopy can be performed directly on a segment of the filter.

2. Direct measurements of the deposit can be made on the surface without interference caused by absorption in the filter itself. This is advantageous in radiometric counting of dust particles and in soiling index measurements made by reflectance.

3. Autoradiographs of radioactive particles can be produced by a technique whereby photographic emulsion is placed in contact with the membrane filter sample.[32]

The disadvantage arising from surface collection is that the amount of sample that can be collected is limited. When more than a single layer of particles is collected on a membrane filter, the resistance rapidly increases, and there is a tendency for the deposit to slough off the filter.

Silver membranes for air sampling applications are made by sintering uniform metallic silver particles.

These membranes possess a structure basically similar to that of the organic membranes previously described. They have a uniform pore size and, for a given pore size, about the same flow characteristics. For filters up to 47 mm in diameter, they are 50 μm thick. The membrane is an integral structure of permanently interconnected particles of pure silver, contains no binding agent or fibers, and is resistant to chemical attack by all fluids that do not attack pure silver. Thermal stability extends from −130° to +370°C (−200° to +700°F).

Richards *et al.*[33] described the use of silver membranes for sampling coal tar pitch volatiles. Other filter media evaluated were not suitable because of the high weight losses of blank filters in the benzene extraction step in the analysis, including 1106BH glass, cellulose acetate membrane, and Whatman 41 cellulose. The weight loss for the silver membrane was negligible. Another application of silver membrane filters is for sampling airborne quartz for X-ray diffraction analysis, as described by Knauber and VonderHeiden.[34] Most instruments satisfying the American Conference of Governmental Industrial Hygienists (ACGIH®) criteria for respirable dust samplers operate at low flow rates, and the sample masses on the backup filters are too small for conventional analyses. Using silver membranes, the X-ray diffraction background is very consistent, and quartz determinations can have a lower limit of sensitivity as low as 0.02 mg.

Polycarbonate Pore Filters

Polycarbonate pore filters are similar to membrane filters in that both contain uniform-sized pores in a solid matrix. However, they differ in structure and method of manufacture. They are made by placing polycarbonate sheets approximately 10 μm thick in contact with sheets of uranium into a nuclear reactor. The neutron flux causes ^{235}U fission, and the fission fragments bore holes in the plastic. Subsequent treatment in a caustic etch solution enlarges the holes to a size determined by the temperature and strength of the bath and the time within it. Commercial filters are available with pore diameters between 0.015 and 14 μm.

Polycarbonate pore filters possess many of the attributes erroneously attributed to membrane filters in earlier days. They have a smooth filtering surface, the pores are cylindrical, almost all uniform in diameter, and essentially perpendicular to the filter surface. The filters also are transparent, even without immersion oil.

The structure and air paths through polycarbonate pore filters are so simple that, as demonstrated by Spurny *et al.*,[15] it is possible to predict their particle collection efficiency on the basis of measured dimensions and basic particle collection theory.

Although their pore volume is much lower, polycarbonate pore filters have about the same flow rate–pressure drop relations as membrane filters of comparable pore diameter. However, as shown in Figure 13-7, the filter penetrations at 5 cm/s, reported by Spurny and Lodge, are much greater than those of membrane filters with the same pore sizes. Pore filters have a lower and more uniform weight, and because they are nonhygroscopic, they can be used for sensitive gravimetric analyses. The polycarbonate base is very strong and filter tapes do not require extra mechanical backing. They can be analyzed by light transmittance, or filter segments can be cut from discs or tapes for microscopy.

The very smooth surface makes polycarbonate pore filters good collectors for particles to be analyzed by electron microscopy and X-ray fluorescence analyses. Spurny *et al.*[15] show high resolution electron micrographs made from silicon monoxide replicas of the filter surface. The very smooth surface also permits good resolution of the collected particles by scanning electron microscopy. The very low collection efficiencies of polycarbonate pore filters under certain conditions, as illustrated in Figure 13-7, permit their use in particle classifications that separate aerosols into size-graded fractions. Cahill *et al.*[35] and Parker *et al.*[36] proposed using two Nuclepore filters in series, with the first having a cut-characteristic approximating the ACGIH "respirable" dust criterion (see Chapter 5). Heidam[37] reviewed the use of series polycarbonate pore filters for a variety of applications but cautioned that particle bounce may be a significant source of error. Particle bounce as a means of penetration of such filters was also noted by John *et al.*,[38] Buzzard and Bell,[39] and Spurny.[40] Figure 13-9 from John *et al.*[38] shows that the collection of solid particles was lower than that for liquid droplets of the same aerodynamic size, and this was attributed to the bouncing of solid particles off the collection surface.

Plastic Foam Filters

Gibson and Vincent[41] described the use of porous filter media to simulate the collection characteristics of the MRE elutriator (see Chapter 5) under a wide range of face velocities. They found that, for particles close to the respirable size, deposition by inertial impaction and gravitational sedimentation compete. As a result, the efficiency remains relatively constant over a substantial range of face velocities.

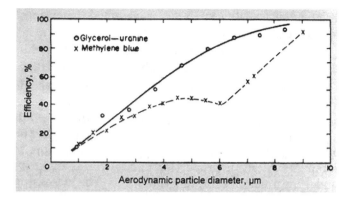

FIGURE 13-9. Measured filtration efficiency of an 8-μm pore size polycarbonate pore filter for methylene blue particles compared to that for glycerol-uranine particles. The flow rate was 5 L/min (face velocity 6 cm/s, based on an exposed area 42 mm in diameter).

Filters Occasionally Used for Air Sampling

Respirator Filters

Respirator filters of felt and/or cellulose fiber can be, and have been, used for air sampling. In many of them, the filter is manufactured in a pleated form, which increases the surface area without increasing the overall diameter. Filters of this type have the same advantages and disadvantages as the mixed fiber filters previously discussed. Wake *et al.*[42] described the collection characteristics of 18 respirator filters for radon daughter aerosols.

Thimbles

Filter thimbles, available in glass fiber, paper, and cloth, are sometimes filled with loose cotton packing to reduce clogging. Their advantage is that large samples can be collected. Other thimbles are made from aluminum oxide (Alundum) and sintered glass. These rigid filter thimbles are manufactured with a variety of porosities. They have considerably higher resistance to air flow than comparable paper cloth or glass fiber filters, but they can be used for higher temperature sampling.

Filter Selection Criteria

General Considerations

The selection of a particular filter type for a specific application is invariably the result of a compromise among many factors. These factors include cost, availability, collection efficiency, the requirements of the analytical procedures, and the ability of the filter to retain its filtering properties and physical integrity under the ambient sampling conditions. The increasing variety of commercially available filter media sometimes makes the choice seem somewhat difficult, but more importantly, it increases the possibility of a selection that satisfies all important criteria.

Efficiency of Collection

Before discussing experimental efficiency data, it is important that a distinction be made between particle collection efficiency and mass collection efficiency. The former refers to fractions of the total number of particles, while the latter refers to fractions of the total mass of the particles. These efficiencies will be numerically equivalent only when all of the particles are the same size, as in some laboratory investigations of filter efficiency. In almost all other cases, the mass collection efficiency will be significantly larger than the corresponding particle collection efficiency. When sampling for total mass concentration of particulate matter, or for the mass concentration of a component of an aerosol, the efficiency of interest is mass efficiency. Submicrometer particles often contribute only a small fraction of the total mass of an industrial dust, even when they represent the majority of the particles. Therefore, it is not always essential that an air sampling filter have a high efficiency for the smallest particles. Insistence on high efficiency for all size particles may restrict the selection to media with other limitations, such as high flow resistance, high cost, and fragility.

Collection efficiency data for a variety of filter media are given in Table 13-1 for 0.3 μm-diameter di(2-ethylhexyl)phthalate (DOP)(=octyl) droplets at various face velocities.[14] This is a commonly used particle size for a test aerosol because it is close to the size for maximum filter penetration for many commonly used sampling media operating at representative flow rates. On this basis, it is reasonable to assume that penetration of both smaller and larger particles would be lower, i.e., the collection efficiency would be higher. This assumption was confirmed by Stafford and Ettinger,[43] who showed that the collection efficiency of Whatman 41 is lowest for 0.264 μm particles at a face velocity of approximately 15 cm/s. It increases for both larger and smaller particles and is approximately 95% or greater for all sizes at face velocities above 100 cm/s. This information is also shown in Figure 13-5.

Liu and Lee[16] measured the collection efficiencies of Nuclepore and Teflon membrane filters for particles in the 0.03 μm to 1 μm-diameter range. For Teflon filters with 10 μm pores (Type LC), the collection efficiencies for 0.003 to 0.1 μm particles at low face velocities were in the 60%–65% range; for Type LS filters with 5 μm pores, they were in the 80%–85% range. For higher velocities and/or larger particles, the efficiencies were >99.99% under all conditions tested. For Nuclepore filters, the penetrations were much higher at comparable pore sizes, and reached 100% for small particles with 5 and 8 μm pore filters. The results were consistent with predictions based on interception, impaction, and diffusion collection.

Liu et al.[44] summarized the results of collection efficiency measurements at four particle sizes and four face velocities for 76 different air sampling filters. Key results of this extensive body of calibration data are summarized in Table 13-2.

The effect of particle shape on filter penetration was explored by Spurny[21] using aerosols of chrysotile asbestos, as discussed earlier. Collection efficiencies decreased substantially with fiber length for both membrane and polycarbonate pore filters of larger pore size. The orientation of the airborne fibers as they approach the filter pore entrances has an important effect on their ability to penetrate the filter.

Skocypec[45] measured the penetration of condensation nuclei in the 0.002 to 0.007 μm range through most of the commercially available membrane filters at a face velocity of 10 cm/s. Less than 1.0% of the particles penetrated through most of the filters. However, much higher penetrations were observed for some of them. Penetrations of 3% or more were only found for some of the large pore (3 μm) filters, i.e., nuclepore filters with <0.08 μm or 0.6 to 11 μm pores, silver membranes with 0.8 μm pores, and Type FG-0.2 μm polytetrafluoroethylene (PTFE) Fluoropore. Some of the large pore membranes, e.g., the cellulose ester filters of Millipore, cellulose triacetate filters of Gelman, and S & S nitrocellulose filters, retained very high efficiencies for these very small particles.

John and Reischl[46] also determined the collection efficiency of various air sampling filters for condensation nuclei. Efficiencies of >99% were found for a variety of Teflon membranes, including the Gelman Teflo filters with 1 to 3 μm and 2 to 4 μm pores, Gelman cellulose acetate with 5 μm pores (GA-1), and glass fiber filters (Gelman A and Spectrograde, MSA 1106BH, and EPA Microquartz). The Gelman Teflo membranes with 3 to 5 μm pores were almost as good, with efficiencies >98%. The Costar Nuclepore (0.8 μm pore) filters had efficiencies of 72%, 72%, and 89% at face velocities of approximately 25, 50, and 150 cm/s, respectively, whereas the efficiencies for Whatman 41 were 64% and 83% at approximately 50 and 150 cm/s, respectively.

Hoover and Newton[47] described the relative radon progeny collection efficiencies for 11 cellulose fiber, glass fiber, and membrane filters used in continuous air monitors for α-emitting radionuclides. Efficiencies above 90% were found for Millipore filters SMWP, AABP, AW19, Fluoropore (3 and 5 μm pores), Gelman A/E, and Whatman EPM 2000.

Lundgren and Gunderson[48] tested the effects of temperature, face velocity, and loading on the particle collection efficiency of glass fiber filters. At room temperatures, they found similar collection efficiencies for Gelman Type A, Gelman Type E, Gelman Spectrograde Type A, MSA 1106B, and the EPA Microquartz filters made of Johns-Manville "Microquartz" fibers by A.D. Little. The EPA filters had a low extractable background and are used for stack gas sampling at temperatures in excess of 500°C. All of the filters had similar pressure drop versus flow rate characteristics and filter masses per unit area, and the high temperature comparisons were limited to the Gelman Type A and "Microquartz" filters.

In all tests, aerosol penetrations of nonvolatile particles were less than about 0.10%. The highest penetrations were for particles approximately 0.1 μm in diameter at the highest face velocity tested, i.e., 51 cm/s. Penetrations dropped significantly with aerosol loadings of only several μg/cm^2. The effect of pinholes on filter efficiency was examined by punching two 0.75 mm pinholes though the filter mat. Although this action produced higher initial penetrations by up to 30 times, the penetrations were never more than a few percent and fell rapidly with loading. Thus, their effect on sample collection would be essentially negligible. Particle penetrations decreased with increasing temperature, except when the temperature was sufficient to volatilize the particles or to contribute to mechanical leakage of the filter holder.

Figure 13-5 shows additional data on the penetration of 0.3 to 11 μm-diameter DOP and clearly demonstrates that, for many filters, there is also a face velocity for maximum penetration. These curves were all plotted from the data of Lockhart et al.,[14] except for the Whatman 41 and IPC-1478 curves, which were extended to lower flow rates on the basis of the data of Rimberg.[13] The Nuclepore (polycarbonate pore) filter data points at 5 cm/s are from the data of Liu and Lee.[16]

On the basis of the data plotted in Figure 13-5, it can also be seen that the same filter can be inefficient at some face velocities and highly efficient at others. For example, Whatman 41 penetration below 10 cm/s exceeds 40%, while at 100 cm/s, it is only about 4%; at higher flow rates, it is much less than that. This filter is often used in industrial hygiene surveys with both low and high volume samplers. When sampling with a

TABLE 13-2. List of Filters Tested and Principal Results (from Liu et al.[44])

Filter	Material	Pore Size, μm	Filter Permeability Velocity, cm/sec (ΔP=1 cm Hg)	Filter Efficiency Range, %*
A. Cellulose Fiber Filter				
Whatman				
No. 1	Cellulose fiber	——	6.1	49–99.96
No. 2		——	3.8	63–99.97
No. 3		——	2.9	89.3–99.98
No. 4		——	20.6	33–99.5
No. 5		——	0.86	93.1–99.99
No. 40		——	3.7	77–99.99
No. 41		——	16.9	43–99.5
No. 42		——	0.83	92.0–99.992
B. Glass Fiber Filter				
Pall–Gelman				
Type A	Glass fiber	——	11.2	99.92–>99.99
Type E		——	15.5	99.6–>99.99
Spectrograde		——	15.8	99.5–>99.99
Microquartz		——	14.1	98.5–>99.99
MSA 1106B		——	15.8	99.5–>99.99
Pallflex				
2500 QAO	Quartz fiber	——	41	84–99.9
E70/2075W		——	36.5	84–99.95
T60A20	Teflon-coated glass fiber	——	49.3	55–98.8
(another lot)		——	40.6	52–99.5
T60A25		——	36.5	65–99.3
TX40H120		——	15.1	92.6–99.96
(another lot)		——	9.0	98.9–>99.99
Reeve Angel 934AH	Glass fiber	——	12.5	98.9–>99.99
(acid treated)		——	20	95.0
Whatman				
GF/A	Glass fiber	——	14.5	
GF/B		——	5.5	
GF/C		——	12.8	
EPM 1000		——	13.9	
C. Membrane Filter				
Millipore				
MF-VS	Cellulose acetate/nitrate	0.025	0.028	99.999–>99.999
MF-VC		0.1	0.16	99.999–>99.999
MF-PH		0.3	0.86	99.999–>99.999
MF-HA		0.45	1.3	99.999–>99.999
MF-AA		0.8	4.2	99.999–>99.999
MF-RA		1.2	6.2	99.9–>99.999
MF-SS		3.0	7.5	98.5–>99.999
MF-SM		5.0	10.0	98.1–>99.999
MF-SC		8.0	14.1	92.0–>99.9
Mitex-LS	Teflon	5.0	4.94	->99.99
Mitex-LC		10.0	7.4	->99.99
Fluoropore	PTFE-polyethylene reinforced			
FG		0.2	1.31	>99.90–>99.99
FH		0.5	2.32	>99.99–>99.99
FA		0.1	7.3	>99.99–>99.99
Gelman Metricel				
GM-6	Cellulose acetate/nitrate	0.45	1.45	>99.8–>99.99
VM-1	Polyvinyl chloride	0.5	51.0	49–98.8
DM-800	PVC/Acrylonitrile	0.8	2.7	99.96–>99.99
Gelman Teflon	Teflon	5.0	56.8	85–99.90

TABLE 13-2 (cont.). List of Filters Tested and Principal Results (from Liu et al.[44])

Filter	Material	Pore Size, μm	Filter Permeability Velocity, cm/sec (ΔP=1 cm Hg)	Filter Efficiency Range, %*
Ghia (Gelman)				
S2 37PL 02	Teflon	1.0	12.9	>99.97–>99.99
S2 37PJ 02		2.0	23.4	99.89–>99.99
S2 37PK 02		3.0	24.2	92–98.98
S2 37PF 02		10.0		95.4–>99.9
Selas Flotronics (Poretics)				
FM0.45	Silver	0.45	1.8	93.6–99.98
FM0.8		0.8	6.2	90–99.96
FM1.2		1.2	9.2	73–99.7
FM5.0		5.0	19.0	25–99.2
D. Polycarbonate Pore Filter				
Nuclepore (Costar)				
N010	Polycarbonate	0.1	0.602	>99.9–>99.9
N030		0.3	3.6	93.9–99.99
N040		0.4	2.9	78–>99.99
N060		0.6	2.1	53–99.5
N100		1.0	8.8	28–98.1
N200		2.0	7.63	9–94.1
N300		3.0	12	9–90.4
N500		5.0	30.7	6–90.7
N800		8.0	21.2	1–90.5
N1000		12.0	95	1–46
N1200		10.0	161.1	1–66
E. Miscellaneous Filter				
MSA Personal Air Sampler		——	12	89–99.97

*The range of filter efficiency values given generally corresponds to a particular diameter range of 0.035 to 1 μm, a pressure drop range of 1 to 30 cm Hg, and a face velocity of 1 to 100 cm/s.

25-mm filter at 25 L/min, the face velocity (based on an effective filtration area of 3.68 cm²) is 113 cm/s. When sampling with a 102 mm (4 in.) filter at 500 L/min (17.7 cfm), the face velocity (based on effective filtration area of 60 cm²) is 139 cm/s. On the other hand, when sampling at lower flow rates, as in personal air samplers, Whatman 41 would not be a good choice. With a 25-mm filter and a flow rate of 2.5 L/min, the face velocity would only be 11.3 cm/s. For such an application, other filters more efficient at this flow rate would be preferred.

The necessity for caution in interpreting filter efficiency data in the literature is illustrated in Figure 13-6, which shows the data of various authors for the penetration of Whatman 41 by 0.3 μm-diameter particles. The most reliable data appear to be those of Rimberg,[13] Stafford and Ettinger,[43] and Lockhart et al.,[14] which are in reasonably good agreement with one another. Lockhart et al.'s data are plotted for both Whatman 41 and TFA-41, which is Whatman 41 packaged and sold by the Staplex Company. The differences between the two sets of data are presumably the differences to be expected from randomly selected batches. The Smith and Surprenant[49] data were based on the same techniques as the data of Stafford and Ettinger[43] and of Lockhart et al.,[14] i.e., light-scattering measurements of 0.3 μm DOP droplets, and the large discrepancy is inexplicable.

Rimberg[13] measured the penetration of charge-neutralized polystyrene latex spheres using a light scattering photometer. The 0.3 μm points are actually interpolated from the corresponding data for 0.365 and 0.264 μm particles. Lindeken et al.[50] used a similar technique except that they did not neutralize the electrical charge on their polystyrene test aerosols. Thus, their data appear to reflect the influence of particle charge on filter penetration.

Stafford and Ettinger[51] also compared the collection efficiencies of Whatman 41 filters for 0.3 μm DOP and latex spheres of similar sizes. The efficiencies were higher for the solid particles, especially at face velocities below 20 cm/s. They also showed that efficiency increased with loading of solid particles but not for liquid DOP droplets. Thus, some of the differences in their efficiency test results could have been due to the increase with loading during the test with the latex.

In interpreting filter efficiency data, it is also important to consider that the test data are usually based on the efficiency of a "clean" filter. For most filters, collection efficiency increases with the accumulation of solid particles on the filter surfaces. The resistance to flow also increases with increasing loading but usually at a much slower rate. A theoretical basis for these phenomena was developed by Davies.[52] A practical implication is that even with reliable published filter efficiency and aerosol size distribution data, it is not possible to know precisely what the collection efficiency of a filter will be for a given sampling interval. The filter efficiency data can only provide an estimate of the minimum collection efficiency. The actual collection efficiency will usually be higher.

Biles and Ellison[53] reported on the increase in collection efficiency for three types of cellulose fiber filters, i.e., Whatman 1, 4, and 451, for collecting lead and "black smoke" from the air of London, England. At a face velocity of 6.5 cm/s, the clean paper efficiencies for lead were 50, 30, and 15%, respectively, and 70, 40, and 30% in terms of the light reflectance measurement for black smoke. As the percent soiling index approached 40%, the collection efficiencies of all three papers approached 100% for both lead and black smoke.

There have been reports in the literature that low concentrations of small atmospheric particles could have large penetration rates through filters such as the Millipore HA or glass fiber filters.[54, 55] Because numerous careful investigations have shown such filters to have almost complete collection for all particle sizes and flow rates, as discussed earlier and illustrated in Figure 13-5 and Table 13-2, it appears that such reports are most likely due to background or contamination problems associated with the analysis of the charcoal traps used by the investigators as back-up collectors. Kneip et al.[56] investigated the efficiency of Millipore AA and SC membrane filters and Gelman AE glass fiber filters for ambient air lead particles and laboratory-generated dye aerosol particles <0.07 μm in diameter at very low loadings and face velocities as low as 1.0 cm/s and found that all efficiencies were >99%.

Requirements of Analytical Procedures

Sample Quantity

In many instances, the limited sensitivity of an analytical method, when combined with a low aerosol concentration, makes it necessary for large volumes of air to be sampled in order to collect sufficient material for an accurate analysis. In addition to the material being stud-

ied, background dust and co-contaminants must, unavoidably, also be collected. Therefore, it is highly desirable that the filter medium selected have the capacity to collect and retain large sample masses. Furthermore, it is usually desirable to have the sampling rate nearly uniform over the length of the sampling period. The flow resistance of all filters increases with increased loading, but some do so at much lower rates than others. Table 13-1 shows the rates of resistance increase for a variety of filters when sampling the ambient air outside the Naval Research Laboratory. The loading rate would certainly differ for other aerosols, and these data generally would not be applicable. However, they do indicate the relative loading characteristics of these filters, i.e., those with low values load much more slowly than those with high values. Those filters with the lowest resistance build-up rate are most useful for collecting high-volume samples, especially when using pressure-sensitive, turbine-type blowers as air movers. In general, deep-bed fibrous filters have the lowest rates of resistance pressure increase. The relative rates of loading of some commonly used air sampling filters are illustrated in Figure 13-10.

Sample Configuration

Some analyses require that the sample be collected or mounted in a particular form. For example, microscopic particle size analysis can be performed only when the particles are on a flat surface. This is due to the limited depth of focus of the objective lens. In order to use fibrous filters for collecting samples for size analysis, it must be possible to remove the sample quantitatively and transfer it to a microscope stage without altering it. For such applications, the membrane and polycarbonate pore filters offer significant advantages over other filters.

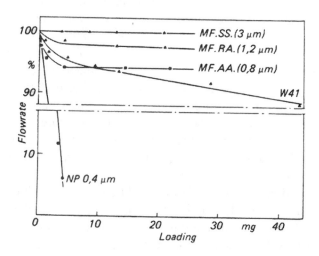

FIGURE 13-10. The effect of dust loading on the % of the initial flowrate for various filter media.

First, the samples can be analyzed directly on the filter surface. Second, because the sample does not have to be transferred, there is a greater likelihood that the sample observed is in the same form as when airborne.

Another situation in which the sample configuration may be important to the analysis is the determination of airborne radioactivity. Many radiation detectors such as Geiger-Mueller tubes and scintillation detectors are designed to view a limited surface area, usually a 2.5 cm-diameter circle. Thus, to make efficient use of the detector, the effective filtering area should be limited to a similar size. An additional consideration in radiometric analysis is the depth of penetration of the particles into the filter, especially for alpha and beta emitters.[47] The activity observed by the detector will be affected by the distance of the particles from the detector and by absorption of radiation by intervening filter fibers.

Other characteristics influence the choice of filters when quantitative particulate analysis by X-ray diffraction is desired. Davis and Johnson[57] examined seven filter substrates for both fiber and membrane construction and found that the degree to which the filters were suitable for X-ray diffraction analysis was primarily dependent on: 1) interfering background scatter; and 2) the mass per unit area of the particulate load collected. They found that Teflon filters were superior when mass loadings were <200 μg/cm². On the other hand, when mass loadings were >300 μg/cm², quartz and glass fiber filters were more suitable because of their particle retention qualities and their lack of a substrate spectrum in the diffraction pattern.

Sample Recovery from Filter

High collection efficiency is valueless if all of the sample is not available for analysis. For most chemical analyses, it is necessary to either remove the sample from the filter or to destroy the filter. Inorganic particles usually are recovered from cellulose paper filters by low temperature (plasma) ashing, wet ashing (digesting in concentrated acid), or muffling (incinerating) the filter. Samples collected on glass fiber and cellulose-asbestos filters can be recovered only by leaching or dissolving the sample from the filter. Samples can be recovered from membrane filters, polystyrene filters, and soluble granular beds by dissolving the filter in a suitable solvent.

Some of the membrane filters have a limited loading capacity in terms of the ability of the filter to retain the dust after it is collected. The dust retained on the surface may have very poor adhesion to the surface or to the dust

layer and slough off the surface. The problem is especially severe for polycarbonate pore filters.

Interferences Introduced by Filters

Before selecting a filter for a particular application, the filter's blank count or background level of the material to be analyzed must be determined. All filters contain various elements as major, minor, and trace constituents, and the filter medium of choice for analyzing particular elements must be one with little or no background level for the elements being analyzed. The components of the filter medium itself may introduce undesirable or unacceptable background to the subsequent analyses. If the filter is dissolved or digested, then all of the material in the filter will be mixed with the sample. If it is oxidized, then the residual ash content of the filter will be mixed with the sample. On the other hand, if the sample is extracted from the filter by a solvent, the sample will contain only those components of the filter matrix that are soluble. Finally, if a nondestructive analysis, such as X-ray diffraction, is performed, the contribution of the components of the filter will depend on both the content of the filter, its distribution in space, and the amount of X-ray absorption by the matrix and sample.

Data on the composition and interference levels of some commonly used sampling filters have been presented by Zhang et al.,[58] Gelman et al.,[59] and Mark.[60] There have also been problems with polycarbonate pore filters used in asbestos sampling because of the presence of such fibers on blank filters.

Polycarbonate pore filters build up an electrical charge that can cause a serious weighing error when they are used in gravimetric analysis. Figure 13-11 shows the change in weight over time due to the decay in charge on

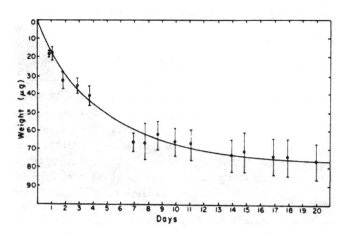

FIGURE 13-11. Microbalance error versus time. Data points represent the mean accumulative mass loss for eight filters. Data are seen to fit a logarithmic curve.

the filter observed by Engelbrecht et al.[61] The charge effect was attributed to electrostatic force between the charged filter on the weighing pan and the metal case of the electrobalance. Their 30-second exposure to a ^{210}Po source prior to the weighing was not sufficient to fully neutralize the charge on the filter.

Another type of interference is inaccessibility of the sample to a measurement or sensing device. For instance, in determining reflectance of filtered particulate matter, the more the particles penetrate the surface the less they will be visible. In such an application, the sensitivity of measurement on a membrane filter surface would be greater than on a fibrous filter.

Size or Mass of Filter

The mass of the filter itself may be important in gravimetric determinations. In determining the mass of collected aerosol, the mass of the filter should be as small as possible, relative to the mass of the sample. Also, other things being equal, the less the filter weighs and/or the smaller it is, the simpler the sample handling and processing. Collecting the sample on a smaller filter may save a concentrating step in the analysis and make it possible to use smaller analytical equipment and/or glassware.

Limitations Introduced by Ambient Conditions

Temperature

The temperature stability of a filter must be considered when sampling hot gases such as stack effluents. For such applications, combustible materials cannot be used, and a selection must be made from the several types of mineral, glass, or other refractory media. In order to select the appropriate medium, the peak temperature and duration of sampling must be known. Glass fiber filters are widely used for temperatures up to about 500°C.

Moisture Content

For sampling under conditions of high humidity, filter media that are relatively nonhygroscopic must be chosen. Some filters pick up moisture and this may affect their filtering properties. If their efficiency is partially dependent on electrostatic effects, moisture may reduce it. Also, when a filter picks up moisture, it may become mechanically weaker and rupture more easily.

For some airborne dusts, such as suspended particulate matter in the ambient air and coal mine dust, the standards are based on gravimetric analyses without regard to dust composition. Mass concentrations are determined from the gain in weight of the filter during the sampling interval, divided by the sampled volume.

Because the filter weighs much more than the sample collected on it, the accuracy of the analysis depends on the stability of the filter's weight. Serious errors can arise if some of the filter's mass is lost due to abrasion during handling between the tare and final filter weighings, or if there is a significant difference in atmospheric water vapor content at the time of analysis.

The highly variable water vapor retention characteristics of cellulose fiber filters usually rule out their selection for use when gravimetric analyses are to be performed.[62] However, even glass fiber and membrane filters, while much less affected by water vapor, may still have enough adsorption to cause problems in gravimetric analyses. Charell and Hawley[63] examined the weight changes at various humidities for cellulose ester, PVC, and polycarbonate membrane filters. They found that all changed their weights reversibly in proportion to the water vapor concentration, that the minimum uptake was seen with polycarbonate and some PVC filters, that other PVC filters took up 6.6 times as much water, and that cellulose ester membranes took up 40–50 times more water vapor. Thus, pre- and post-sampling weighings should be done at the same humidity conditions. Mark[60] examined the weight changes associated with changes in humidity for a variety of PVC membranes, some cellulose ester membranes, and a glass fiber filter. The results of his tests are illustrated in Figure 13-12. He also reported that a PVC-type filter developed an electrical charge that repelled particles onto the filter holder during sampling, reducing the apparent collection efficiency. He was able to overcome this source of error by pretreating the filters with a detergent solution.

Artifact Formation

Air sampling filters can collect gases and vapors as well as particles. The intentional collection of vapor phase chemicals by filters is discussed in Chapter 16. When they are collected unintentionally by adsorption or absorption onto filter surfaces, or onto particles collected on those surfaces, their presence in the sample can constitute an artifact. For example, ordinary glass fiber filters are slightly alkaline and collect SO_2 while sampling ambient air. This led to overestimation of the ambient aerosol sulfate concentrations for many years.

As shown by Coutant,[64] Spicer and Schumacher,[65] and Appel et al.,[66] artifact particulate matter can be formed by oxidation of acidic gases (e.g., SO_2, NO_2) or by retention of gaseous nitric acid on the surface of alkaline (e.g., glass fiber) filters and other filter types. The effect is a surface-limited reaction and, depending on the concentration of the acidic gas, should be especially

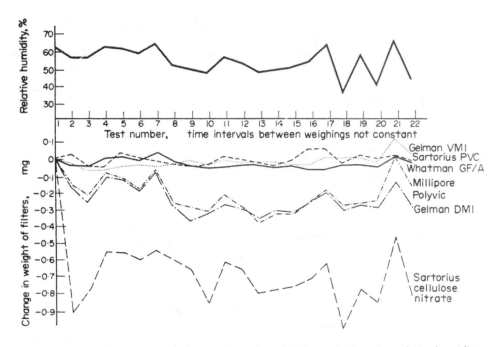

FIGURE 13-12. Variation in weight of filter materials due to absorption of moisture in changing relative humidity.

significant early in the sampling period. The magnitude of the resulting error depends on such factors as the sampling period, filter composition and pH, and the relative humidity. The magnitude and the significance of artifact mass errors are variable and dependent on local conditions. Excluding the uncertainty associated with the collection and retention of organic particulate matter with appreciable vapor pressure, artifact mass primarily reflects the sum of the sulfates and nitrates formed by filter surface reactions with sulfur dioxide and nitric acid vapors, respectively.

The study by Coutant[64] reported artifact sulfate for 24-hour samples from 0.3 to 3 $\mu g/m^3$. Stevens et al.[67] found 2.5 $\mu g/m^3$ average artifact sulfate sampling at eight sites around St. Louis, Missouri, and Rodes and Evans[68] noted 0.5 $\mu g/m^3$ artifact sulfate in West Los Angeles, California.

Artifact sulfate formation can also occur on nylon filters. Chan et al.[69] examined the extent of conversion of SO_2 to sulfate on Nylasorb nylon filters used as nitric acid vapor collectors. The percent conversion was found to depend on both the concentration of SO_2 and the relative humidity.

Appel and Tokiwa[70] reported that artifact particulate nitrate on glass fiber filters is limited only by the gaseous nitric acid concentration. Such filters approximated total inorganic nitrate samplers, retaining both particulate nitrate and nitric acid even when the latter was present at very high atmospheric concentrations, e.g., 20 ppb. Nitric acid was found to represent from approximately 25% to 50% of the total inorganic nitrate at Pittsburgh, Pennsylvania, and Lennox and Claremont, California. Based on an estimate of the most probable 24-hour artifact sulfate error, 3.0 $\mu g/m^3$, and of the most probable artifact particulate nitrate, 8.2 $\mu g/m^3$ in the Los Angeles Basin and 3.8 $\mu g/m^3$ elsewhere, typical errors in mass due to sulfate plus nitrate artifacts are estimated at 11.2 $\mu g/m^3$ in the Los Angeles Basin and 6.8 $\mu g/m^3$ elsewhere.

Nitrate salts can be rapidly lost from inert filters (e.g., Teflon, quartz) by volatilization[68] and by reactions with acidic materials.[71] Sampling artifacts are also of serious concern for organic contaminants in air. Schwartz et al.[72] showed that the apparent concentration of extractable organics collected on glass fiber filters varied with the duration of the sampling period. They found that moderately polar organics extracted by dichloromethane were increasingly difficult to recover as the sampling period became progressively longer. This could have been due to volatilization of sampled material during continued sampling or to their oxidation to a form not extracted by the solvent. For more polar organics extracted with cyclohexane, the apparent concentration increased with increasing sampling time, suggesting that the sampled material was behaving as a vapor adsorbent. Similar observations have been made by Appel et al.[73] Much more work is needed on the volatility of sampled material during further sampling, on chemical conversions which take place on filter substrates, and on adsorption of vapors by sampled materials before the extent and significance of those factors can be fully established. Increasingly, quartz fiber filters are being used to minimize such problems.

Limitations Introduced by Filter Holder

Filter Size

In order to use any filter, it must be held securely and without leakage in an appropriate filter holder. This limits the diameter of a filter disc to a particular size unless the filter holder is fabricated especially for the filter. Most filter media can be obtained in any desired size, but some, such as respirator filters, are preformed on molds and are available in only one size.

Mechanical Properties of Filters

Some filter holders can only be used with filters of high mechanical strength. A strong filter paper (e.g., Whatman 41) can be used in a simple head without a back-up screen, whereas softer filters (e.g., glass and quartz fiber) or brittle filters (e.g., the membrane filter) require a more elaborate holder with a firm back-up screen or mesh support to prevent rupture.

Materials Used in Filter Holders and Inlets

Experience has shown that nonconductive plastic inlet cowls remove asbestos fibers before they can reach the sampling filter. The use of conductive cowls has reduced, but has not eliminated, this problem.

Availability and Cost

There are great variations in the unit cost of filter media. For example, although cellulose-asbestos and glass cost about twice as much as cellulose filters, membrane filters may cost 10 times as much. For large-scale sampling programs, such price differentials can add up to significant annual cost increments. The less expensive filter should be chosen when the differences in performance are marginal. Ready availability is another factor to be considered. The cellulose and glass filters can be obtained from any chemical supply house, whereas other types may only be available from a limited number of suppliers. Information on the availability of filter holders is presented in Tables 13-I-2 and 13-I-3.

Filter holder characteristics are listed in Table 13-I-2. This table also contains cross references to filter holder illustrations that follow. Both tables provide code letters for instrument manufacturers; complete names and addresses are given in Table 13-I-3.

Summary and Conclusions

The advantages of sampling by filtration have been discussed, filtration theory has been outlined, commercial filter media have been described, and the criteria for selecting appropriate filters for particular applications have been reviewed.

Of all the particle collection techniques, filter sampling is the most versatile. With appropriate filter media, samples can be collected in almost any form, quantity, and state. Sample handling problems are usually minimal, and many analyses can be performed directly on the filter. No single filter medium is appropriate to all problems, but a filter appropriate to any immediate problem can usually be found.

References

1. Davies, C.N.: The Entry of Aerosols into Sampling Tubes and Heads. Br. J. Appl. Phys. Ser. 2, 1:921 (1968).
2. Davies, C.N.: Deposition from Moving Aerosols. In: Aerosol Science, pp. 393–446. C.N. Davies, Ed. Academic Press, London (1966).
3. First, M.W.; Silverman, L.: Air Sampling with Membrane Filters. Arch. Ind. Hyg. Occup. Med. 7:1 (1953).
4. Stenhouse, J.L.T.; Harrop, J.A.; Freshwater, D.C.: The Mechanisms of Particle Capture in Gas Filters. J. Aerosol Sci. 1:41 (1970).
5. Emi, H.; Okuyama, K.; Adachi, M.: The Effect of Neighboring Fibers on the Single Fiber Inertia-Interception Efficiency of Aerosols. J. Chem. Eng. Japan 10: 148 (1977).
6. Kirsch, A.A.; Stechkina, I.B.; Fuchs, N.A.: Efficiency of Aerosol Filters Made of Ultrafine Polydisperse Fibers. J. Aerosol Sci. 6:119 (1975).
7. Ramskill, E.A.; Anderson, W.L.: The Inertial Mechanism in the Mechanical Filtration of Aerosols. J. Coll. Sci. 6:416 (1951).
8. Wang, H.C.; Kasper, G.: Filtration Efficiency of Nanometer-Size Aerosol Particles. J. Aerosol Sci. 22:31 (1991).
9. Kirsch, A.A.; Zhulanov, U.V.: Measurement of Aerosol Penetration Through High Efficiency Filters. J. Aerosol Sci. 9:291 (1978).
10. Gentry, J.W.; Spurny, K.R.; Schoermann, J.: Diffusional Deposition of Ultrafine Aerosols on Nuclepore Filters. Atmos. Environ. 16:25 (1982).
11. Lundgren, D.A.; Whitby, K.T.: Effect of Particle Electrostatic Charge on Filtration by Fibrous Filters. I & EC Process Res. Develop. 4:345 (1965).
12. Zebel, G.: Deposition of Aerosol Flowing Past a Cylindrical Fiber in a Uniform Electric Field. J. Coll. Sci. 20:522 (1965).
13. Rimberg, D.: Penetration of IPC 1478, Whatman 41, and Type 5G Filter Paper as a Function of Particle Size and Velocity. Am. Ind. Hyg. Assoc. J. 30:394 (1969).
14. Lockhart, Jr., L.B.; Patterson, Jr., R.L; Anderson, W.L.: Characteristics of Air Filter Used for Monitoring Airborne Radioactivity. NRL Report No. 6054. U.S. Naval Research Laboratory, Washington, DC (March 20, 1964).
15. Spurny, K.R.; Lodge, Jr., J.P.; Frank, E.R.; Sheesley, D.C.: Aerosol Filtration by Means of Nuclepore Filters: Structural and Filtration Properties. Environ. Sci. Technol. 3:453 (1969).
16. Liu, B.Y.H.; Lee, K.W.: Efficiency of Membrane and Nuclepore Filters for Submicrometer Aerosols. Environ. Sci. Technol. 10:345 (1976).

17. Lee, K.W.; Liu, B.Y.H.: On the Minimum Efficiency and the Most Penetrating Particle Size for Fibrous Filters. J. Air Pollut. Control Assoc. 30:377 (1980).

18. Lee, K.W.: Maximum Penetration of Aerosol Particles in Granular Bed Filters. J. Aerosol Sci. 12:79 (1981).

19. Schmidt, E.W.; Gieseke, J.A.; Gelfand, P.; et al.: Filtration Theory for Granular Beds. J. Air Pollut. Control Assoc. 28:143 (1978).

20. Fichman, M.C.; Gutfinger, C.; Pnueli, D.: A Modified Model for the Deposition of Dust in a Granular Bed Filter. Atmos. Environ. 15:1669 (1981).

21. Spurny, K.: On the Filtration of Fibrous Aerosols. J. Aerosol. Sci. 17450 (1986).

22. Loffler, F.: The Adhesion of Dust Particles to Fibrous and Particulate Surfaces. Staub 28:29 (English trans.) (November 1968).

23. Pate, T.B.; Tabor, E.C.: Analytical Aspects of the Use of Glass Fiber Filters for the Collection and Analysis of Atmospheric Particle Matter. Am. Ind. Hyg. Assoc. J. 23:145 (1962).

24. Kramer, D.N.; Mitchel, P.W.: Evaluation of Filters for High Volume Sampling of Atmospheric Particulates. Am. Ind. Hyg. Assoc. J. 28:224 (1967).

25. Shleien, B.; Cochran, J.A.; Friend, A.G.: Calibration of Glass Fiber Filters for Particle Size Studies. Am. Ind. Hyg. Assoc. J. 27:253 (1966).

26. Koutrakis, P.; Wolfson, J.M.; Spengler, J.D.: An Improved Method for Measuring Aerosol Strong Acidity. Atmos. Environ. 22:157 (1988).

27. Lippmann, M.: Asbestos Exposure Indices. Environ. Res. 46:86 (1988).

28. Spurny, K.; Pich, J.: The Separation of Aerosol Particles by Means of Membrane Filters by Diffusion and Inertial Impaction. Int. J. Air Wat. Pollut. 8:193 (1964).

29. Megaw, W.J.; Wiffen, R.D.: The Efficiency of Membrane Filters. Int. J. Air Wat. Pollut. 7:501 (1963).

30. Lindeken, C.L.; Petrock, F.K.; Phillips, W.A.; Taylor, R.D.: Surface Collection Efficiency of Large-Pore Membrane Filters. Health Phys. 10:495 (1964).

31. Lossner, V.: Die Bestimmung der Eindringtiefe von Aerosolen in Filtern. Staub 24:217 (1964).

32. George, II, L.A.: Electron Microscopy and Autoradiography. Science 133:1423 (May 5, 1961).

33. Richards, R.T.; Donovan, D.T.; Hall, J.R.: A Preliminary Report on the Use of Silver Metal Membrane Filters in Sampling Coal Tar Pitch Volatiles. Am. Ind. Hyg. Assoc. J. 28:590 (1967).

34. Knauber, J.W.; VonderHeiden, F.H.: A Silver Membrane X-Ray Diffraction Technique for Quartz Samples. Presented at American Industrial Hygiene Conference, Denver, CO (May 14, 1969).

35. Cahill, T.A.; Ashbauch, L.L.; Barone, J.B.; et al.: Analysis of Respirable Fractions in Atmospheric Particulates via Sequential Filtration. J. Air Pollut. Control Assoc. 27:675 (1977).

36. Parker, R.D.; Buzzard, G.H.; Dzubay, T.G.; Bell, J.P.: A Two-Stage Respirable Aerosol Sampler Using Nuclepore Filters in Series. Atmos. Environ. 11:617 (1977).

37. Heidam, N.Z.: Review: Aerosol Fractionation by Sequential Filtration with Nuclepore Filters. Atmos. Environ. 15:891 (1981).

38. John, W.; Reischl, G.; Goren, S.; Plotkin, D.: Anomalous Filtration of Solid Particles by Nuclepore Filters. Atmos. Environ. 12: 1555 (1978).

39. Buzzard, G.H.; Bell, J.P.: Experimental Filtration Efficiencies of Large Pore Nuclepore Filters. J. Aerosol Sci. 11:435 (1980).

40. Spurny, K.: Discussion: A Two-Stage Respirable Aerosol Sampler Using Nuclepore Filters in Series. Atmos. Environ. 11:1246 (1977).

41. Gibson, H.; Vincent, J.H.: The Penetration of Dust Through Porous Foam Filter Media. Ann. Occup. Hyg. 24:205 (1981).

42. Wake, D.; Brown, R.C.; Trottier, R.A.; Liu, Y.: Measurements of the Efficiency of Respirator Filters and Filtering Facepieces Against Radon Daughter Aerosols. Ann. Occup. Hyg. 36:629 (1992).

43. Stafford, R.G.; Ettinger, H.J.: Filter Efficiency as a Function of Particle Size and Velocity. Atmos. Environ. 6:353 (1972).

44. Liu, B.Y.H.; Pui, D.Y.H.; Rubow, K.L.: Characteristics of Air Sampling Filter Media. In: Aerosols in the Mining and Industrial Work Environment, Vol. 3; Instrumentation, pp. 989–1038, V.A. Marple and B.Y.H. Liu, Eds. Ann Arbor Science, Ann Arbor, MI (1981).

45. Skocypec, W.J.: The Efficiency of Membrane Filters for the Collection of Condensation Nuclei. M.S. Thesis. University of North Carolina, School of Public Health, Chapel Hill, NC (1974).

46. John, W.; Reischl, G.: Measurements of the Filtration Efficiencies of Selected Filter Types. Atmos. Environ. 12:2015 (1978).

47. Hoover, M.D.; Newton, G.J.: Update on Selection and Use of Filter Media in Continuous Air Monitors for Alpha-Emitting Radionuclides. In: 1991–1992 Inhalation Toxicology Research Institute Annual Report, pp. 5–7, LMF-138, ITRI, Albuquerque, NM 87185 (December 1992).

48. Lundgren, D.A.; Gunderson, T.C.: Efficiency and Loading Characteristics of EPA's High-Temperature Quartz Fiber Filter Media. Am. Ind. Hyg. Assoc. J. 36:806 (1975).

49. Smith, W.J.; Surprenant, N.F.: Properties of Various Filtering Media for Atmospheric Dust Sampling. Presented at the American Society for Testing and Materials, Philadelphia, PA (July 1, 1963).

50. Lindeken, C.L.; Morgan, R.L.; Petrock, K.F.: Collection Efficiency of Whatman 41 Filter Paper for Submicron Aerosols. Health Phys. 9:305 (1963).

51. Stafford, R.G.; Ettinger, H.J.: Comparison of Filter Media Against Liquid and Solid Aerosols. Am. Ind. Hyg. Assoc. J. 32:319 (1971).

52. Davies, C.N.: The Clogging of Fibrous Aerosol Filters. Aerosol Sci. 1:35 (1970).

53. Biles, B.; Ellison, J. McK.: The Efficiency of Cellulose Fiber Filters with Respect to Lead and Black Smoke in Urban Aerosol. Atmos. Environ. 9:1030 (1975).

54. Robinson, J.W.; Wolcott, D.K.: Simultaneous Determination of Particulate and Molecular Lead in the Atmosphere. Environ. Lett. 6:321 (1974).

55. Skogerboe, R.K.; Dick, D.L; Lamothe, P.J.: Evaluation of Filter Inefficiencies for Particulate Collection Under Low Loading Conditions. Atmos. Environ. 11:243 (1977).

56. Kneip, T.J.; Kleinman, M.T.; Gorczynski, J.; Lippmann, M.: A Study of Filter Penetration by Lead in New York City Air. In: Environmental Lead, pp. 291–308. D.R. Lynam, L.G. Piantanida, and J.F. Cole, Eds. Academic Press, New York (1981).

57. Davis, B.L.; Johnson, L.R.: On the Use of Various Filter Substrates for Quantitative Particulate Analysis by X-ray Diffraction. Atmos. Environ. 16:273 (1982).

58. Zhang, J.; Billiet, S. J.; Dams, R.: Stationary Sampling and Chemical Analysis of Suspended Particulate Matter in a Workplace. Staub-Reinhalt. Luft. 41:381 (1981).

59. Gelman, C.; Mehta, D.V.; Meltzer, T.H.: New Filter Compositions for the Analysis of Airborne Particulate and Trace Metals. Am. Ind. Hyg. Assoc. J. 40:926 (1979).

60. Mark, D.: Problems Associated with the Use of Membrane Filters for Dust Sampling When Compositional Analysis is Required. Ann. Occup. Hyg. 17:35 (1974).

61. Engelbrecht, D.R.; Cahill, T.; Feeney, P.J.: Electrostatic Effects on Gravimetric Analysis of Membrane Filters. J. Air Pollut. Control Assoc. 30:391 (1980).

62. Demuynck, M.: Determination of Irreversible Absorption of Water by Cellulose Filters. Atmos. Environ. 9:623 (1975).

63. Charell, P.R.; Hawley, R.E.: Characteristics of Water Adsorption on Air Sampling Filters. Am. Ind. Hyg. Assoc. J. 42:353 (1981).

64. Coutant, R.W.: Effect of Environmental Variables on Collection of Atmospheric Sulfate. Environ. Sci. Technol. 11:873 (1977).

65. Spicer, C.W.; Schumacher, P.M.: Particulate Nitrate: Laboratory and Field Studies of Major Sampling Interferences. Atmos. Environ. 13:543 (1979).

66. Appel, B.R.; Wall, S.M.; Tokiwa, Y.; Haik, M.: Interference Effects in Sampling Particulate Nitrate in Ambient Air. Atmos. Environ. 13:319 (1979).

67. Stevens, R.F.; Dzubay, T.G.; Russwurm, G.; Rickel, D.: Sampling and Analysis of Atmospheric Sulfates and Related Species. In: Sulfur in the Atmosphere, Proceedings of the International Symposium, United Nations, Dubrovnik, Yugoslavia, September 7–14, 1977. Atmos. Environ. 12:55 (1978).

68. Rodes, C.E.; Evans, G.F.: Summary of LACS Integrated Measurements. EPA-600/4-77-034. U.S. Environmental Protection Agency, Research Triangle Park, NC (June 1977).

69. Chan, W.H.; Orr, D.B.; Chung, D.H.S.: An Evaluation of Artifact SO_4 Formation on Nylon Filters Under Field Conditions. Atmos. Environ. 20:2397 (1986).

70. Appel, B.R.; Tokiwa, Y.: Atmospheric Particulate Nitrate Sampling Errors Due to Reactions with Particulate and Gaseous Strong Acids. Atmos. Environ. 15:1087 (1981).

71. Harker, A.; Richards, L.; Clark, W.: Effect of Atmospheric SO_2 Photochemistry Upon Observed Nitrate Concentrations. Atmos. Environ. 11:87 (1977).

72. Schwartz, G.P.; Daisey, J.M.; Lioy, P.J.: Effect of Sampling Duration on the Concentration of Particulate Organics Collected on Glass Fiber Filters. Am. Ind. Hyg. Assoc. J. 42:258 (1981).

73. Appel, B.R.; Hoffer, E.M.; Haik, M.; et al.: Characterization of Organic Particulate Matter. Environ. Sci. Technol. 13:98 (1979).

TABLE 13-I-1. Summary of Air Sampling Filter Characteristics

Filter	Void Size (µm)	Fiber Diam. (µm)	Thickness (µm)	Weight/Area (mg/cm²)	Ash Content (%)	Max. Oper. Temp. (°C)	Tensile Strength (psi)	ΔP100[A] in H$_2$O	Source
A. Cellulose Fiber Filter Characteristics									
Whatman 1	2+	NA	180	8.7	0.06	150	4700	40.5	WLS
4	4+	NA	210	9.2	0.06	150	NA	11.5	WLS
40	2	NA	210	9.5	0.01	150	4600	54	WLS
41	4+	NA	220	8.5	0.01	150	4600	8.1	WLS
42	>1	NA	200	10.0	0.01	150	NA	NA	WLS
44	>1	NA	180	8.0	0.01	150	NA	NA	WLS
50	1	NA	120	9.7	0.025	150	NA	NA	WLS
541	4+	NA	160	7.8	0.008	150	NA	NA	WLS
S&S 589BR	NA	NA	210	7.6	NA	NA	NA	NA	SAS
B. Glass and Quartz Fiber Filter Characteristics									
MSA									
11064[B]	NA	NA	180–270	6.1	~ 95	540	625	19.8	MSA
1106BH[C]	NA	NA	180–460	5.8	~100	540	270	19.8	MSA
Pall–Gelman									
Type A/B[C]	1	NA	660	NA	NA	500	NA	NA	PGL
A/C[C]	1	NA	279	NA	NA	500	NA	NA	PGL
A/D[C]	3.1	NA	686	NA	NA	500	NA	NA	PGL
A/E[C]	1	NA	457	NA	NA	500	NA	NA	PGL
Extra Thick[B]	1	NA	1270	NA	NA	135	NA	NA	PGL
Pallflex									
800A	<0.4	0.4–0.7	230	3	~ 95	315	1000	6	PGL
2500A	<0.4	0.4–0.7	500	6.5	~ 96	315	1500	15	PGL
2500QAO-UP[E]	<0.2	0.4–0.7	530	6.5	100	1000	500–900	10–15	PGL
TX40H120WW[F]	<0.3	<0.5	175	5	~ 85	315	3500	20	PGL
T60A20[F]	<0.4	0.4–0.7	240	4	~ 80	315	1200	8	PGL
E70[G]	<0.4	0.4–0.8	175	3.5	~ 35	120–160	650	8	PGL
Millipore									
AP 15[B]	NA	<1	380	8.0	95	500	625	70	MIL
AP 20[B]	NA	<1	330	7.3	95	500	625	16	MIL
AP 25	NA	<1	NA	NA	NA	NA	NA	NA	MIL
AP 40[C]	NA	<1	410	6.9	100	500	450	18	MIL
Whatman									
GF/A[C]	<1	0.5–0.75	250	5.3	NA	540	500	NA	WLS
GF/B[C]	<1	0.5–0.75	680	14.3	NA	540	1000	NA	WLS
GF/-C[C]	<1	0.2–0.5	260	5.3	NA	540	500	NA	WLS
934AH[C]	<1	NA	330	6.4	NA	540	180	24.4	WLS
EPM-2000[C]	NA	NA	430	8.0	NA	540	700	NA	WLS
QM-A Quartz	NA	NA	450	8.5	NA	540	250–v300	15.3	WLS
H&V									
HD-2021	NA	1.6	500	9.4	95	500	1150	6.6	H&V
HB-5055	NA	0.6	475	9.0	95	500	1200	12.4	H&V
HB-5211	NA	0.6	400	6.0	95	500	980	13.7	H&V
LB-5211 A-O[D]	NA	0.6	400	8.8	65	175	2000	14.0	H&V
HA-8021[C]	NA	0.45	380	7.3	100	500	700	18.8	H&V
HA-8141	NA	0.45	410	7.8	95	500	650	19.4	H&V
HA-8071	NA	0.25	350	6.7	95	500	1000	65.0	H&V
Omega									
Fiber Glass	NA	<1	350	NA	NA	540	500	NA	OSI
Quartz	NA	<1	400	NA	NA	540	325	NA	OSI
S+S									
25[C]	NA	NA	430	7.3	NA	NA	NA	NA	SAS
30[C]	NA	NA	360	6.4	NA	NA	NA	NA	SAS

TABLE 13-I-1 (cont.). Summary of Air Sampling Filter Characteristics

Filter	Void Size (µm)	Fiber Diam. (µm)	Thickness (µm)	Weight/Area (mg/cm²)	Ash Content (%)	Max. Oper. Temp. (°C)	Tensile Strength (psi)	ΔP100[A] in H_2O	Source
31[C]	NA	NA	280	5.6	NA	NA	NA	NA	SAS
Osmonics GB-100R	2.0	NA	380	NA	NA	500	NA	NA	OLP

Filter	Composition	Pore Size (µm)	Thickness (µm)	Weight/Area (mg/cm²)	Ash Content (%)	Max. Oper. Temp. (°C)	Tensile Strength (psi)	Refractive Index	ΔP100[A] in H_2O	Source
C. Membrane Filter Characteristics										
Millipore										
SC	Mixed	8.0	135	NA	<0.001	125	175	1.515	20	MIL
SM	Cellulose esters	5.0	135	NA	<0.001	125	160	1.495	32	MIL
SS		3.0	150	3.0	<0.001	125	150	1.495	56	MIL
RA[H]		1.2	150	4.2	<0.001	125	300	1.512	75	MIL
AA[H,I]		0.80	150	4.7	<0.001	125	350	1.510	102	MIL
DA[H]		0.65	150	4.8	<0.001	125	400	1.510	112	MIL
HA[H]		0.45	150	4.9	<0.001	125	450	1.510	250	MIL
PH		0.30	150	5.3	<0.001	125	500	1.510	300	MIL
GS		0.22	150	NA	<0.001	125	700	1.510	450	MIL
VC		0.10	105	NA	<0.001	125	800	1.500	2290	MIL
VM		0.05	105	NA	<0.001	125	1000	1.500	3610	MIL
VS		0.025	105	NA	<0.001	125	1500	1.500	5100	MIL
LC	Teflon	10.0	125	8.0	NA	260	250	NA	125	MIL
LS		5.0	125	8.0	NA	260	150	NA	187	MIL
FS		3.0	200	NA	NA	130	NA	NA	NA	MIL
FA	PTFE-polyethylene	1.0	145	NA	NA	130	NA	NA	NA	MIL
FH	reinforced	0.5	175	NA	NA	130	NA	NA	NA	MIL
FG		0.2	175	NA	NA	130	NA	NA	NA	MIL
SLVP	polyvinylidene fluoride	5.0	125	NA	NA	NA		1.42	NA	MIL
DVPP		0.65	125	NA	NA	NA		1.42	NA	MIL
HVHP		0.45	125	NA	NA	NA		1.42	NA	MIL
HVLP		0.45	125	NA	NA	NA		1.42	NA	MIL
GVHP		0.22	125	NA	NA	NA		1.42	NA	MIL
GVWP		0.22	125	NA	NA	NA		1.42	NA	MIL
VVLP		0.10	125	NA	NA	NA		1.42	NA	MIL
PVCS	polyvinyl chloride	5.0	NA	NA	NA	NA		NA	NA	MIL
PVCO		0.8	NA	NA	NA	NA		NA	NA	MIL
Metricel										
GN6	Mixed	0.45	153	4.0	NA	74	NA	1.51	NA	PGL
GN4	Cellulose esters	0.8	153	4.0	NA	74	NA	1.51	NA	PGL
DM-800	PVC/Acrylo-nitrile	0.8	145	3.0	NA	66	NA	1.51	NA	PGL
DM-450		0.45	140	3.0	NA	66	NA	1.51	NA	PGL
GLA-5000	polyvinyl chloride	5.0	152	1.3	NA	52	NA	1.55	NA	PGL
Nylasorb	Nylon	1.0	90	NA	NA	NA	180	NA	NA	PGL
Zylon	PTFE	5.0	127	NA	NA	NA	NA	NA	NA	PGL
TF	PTFE w/poly-propylene support	1.0	178	NA	NA	NA	NA	NA	NA	PGL

TABLE 13-I-1 (cont.). Summary of Air Sampling Filter Characteristics

Filter	Composition	Pore Size (µm)	Thickness (µm)	Weight/Area (mg/cm²)	Ash Content (%)	Max. Oper. Temp. (°C)	Tensile Strength (psi)	Refractive Index	ΔP100[A] in H_2O	Source
		0.45	178	NA	NA	NA	NA	NA	NA	PGL
		0.20	178	NA	NA	NA	NA	NA	NA	PGL
Teflo	PTFE with polymethyl- pentene support ring	3.0	25	NA	NA	NA	NA	NA	NA	PGL
		2.0	25	NA	NA	NA	NA	NA	NA	PGL
		1.0	76	NA	NA	NA	NA	NA	NA	PGL
Zefluor	PTFE	3.0	152	NA	NA	NA	NA	NA	NA	PGL
		2.0	152	NA	NA	NA	NA	NA	NA	PGL
		1.0	165	NA	NA	NA	NA	NA	NA	PGL
		0.5	178	NA	NA	NA	NA	NA	NA	PGL
Omega										
MEC	Mixed cellulous esters	1.2	150	4.2	<0.0001	125	300	NA	75	OSI
		0.8	150	4.7	<0.0001	125	350	NA	102	OSI
		0.45	150	4.8	<0.0001	125	450	NA	250	OSI
TEF	PTFE	0.5	175	NA	NA	500	NA	NA	NA	OSI
		1.0	125	NA	NA	500	NA	NA	NA	OSI
		2.0	175	NA	NA	500	NA	NA	NA	OSI
		5.0	125	NA	NA	500	NA	NA	NA	OSI
Silicol®	PVC Homopolymer	5.0	70	NA	0.3	55	NA	NA	27	OSI
		2.0	70	NA	0.3	55	NA	NA	31	OSI
		0.8	70	NA	0.3	55	NA	NA	43	OSI
		0.5	70	NA	0.3	55	NA	NA	60	OSI
		0.2	70	NA	0.3	55	NA	NA	111	OSI
Ag	Silver	0.8	50	NA	NA	400	NA	NA	NA	OSI
		0.45	50	NA	NA	400	NA	NA	NA	OSI
S&S	Nitrocellulouse									
AE 100		12	133	NA	NA	NA	NA	NA	NA	SAS
AE 99		8	122	NA	NA	NA	NA	NA	NA	SAS
AE 98		5	119	NA	NA	NA	NA	NA	NA	SAS
BA 90		0.6	131	NA	NA	NA	NA	NA	NA	SAS
BA 85		0.45[H]	131	NA	NA	NA	NA	NA	NA	SAS
BA 83		0.2	133	NA	NA	NA	NA	NA	NA	SAS
BA 79		0.1	106	NA	NA	NA	NA	NA	NA	SAS
BA 75		0.05	105	NA	NA	NA	NA	NA	NA	SAS
	PTFE									
		0.45	161	NA	NA	NA	NA	NA	NA	SAS
		0.2	165	NA	NA	NA	NA	NA	NA	SAS
Poretics										
MCE	Mixed cellulose esters	8.0	NA	NA	NA	NA	NA	NA	NA	OLP
		5.0	NA	NA	NA	NA	NA	NA	NA	OLP
		1.2	NA	NA	NA	NA	NA	NA	NA	OLP
		0.8	NA	NA	NA	NA	NA	NA	NA	OLP
		0.65	NA	NA	NA	NA	NA	NA	NA	OLP
		0.45	NA	NA	NA	NA	NA	NA	NA	OLP
		0.22	NA	NA	NA	NA	NA	NA	NA	OLP
		0.10	NA	NA	NA	NA	NA	NA	NA	OLP
Nylon	nylon	5.0	110	NA	NA	NA	NA	NA	NA	OLP
		1.2	110	NA	NA	NA	NA	NA	NA	OLP
		0.8	110	NA	NA	NA	NA	NA	NA	OLP
		0.65	110	NA	NA	NA	NA	NA	NA	OLP
		0.45	110	NA	NA	NA	NA	NA	NA	OLP
		0.22	110	NA	NA	NA	NA	NA	NA	OLP

TABLE 13-I-1 (cont.). Summary of Air Sampling Filter Characteristics

Filter	Comp-osition	Pore Size (μm)	Thickness (μm)	Weight/Area (mg/cm²)	Ash Content (%)	Max. Oper. Temp. (°C)	Tensile Strength (psi)	Refrac-tive Index	ΔP100[A] in H₂O	Source
PVC	polyvinyl	5.0	NA	1.4	NA	NA	NA	NA	NA	OLP
	chloride	0.8	NA	2.1	NA	NA	NA	NA	NA	OLP
PTFE	PTFE with	1.0	110	6.5	NA	130	NA	NA	NA	OLP
	polypropylene	0.45	NA	NA	NA	130	NA	NA	NA	OLP
	support	0.2	160	4.5	NA	130	NA	NA	NA	OLP
Silver	Silver	5.0	50	NA	NA	550	NA	NA	NA	OLP
		3.0	50	NA	NA	400	NA	NA	NA	OLP
		1.2	50	NA	NA	350	NA	NA	NA	OLP
		0.8	50	NA	NA	300	NA	NA	NA	OLP
		0.45	50	NA	NA	300	NA	NA	NA	OLP
		0.2	50	NA	NA	250	NA	NA	NA	OLP
Corning MF	Mixed cellulose	0.22	NA	NA	NA	NA	NA	NA	NA	COR
		0.45	NA	NA	NA	NA	NA	NA	NA	COR
		0.65	NA	NA	NA	NA	NA	NA	NA	COR
		0.8	NA	NA	NA	NA	NA	NA	NA	COR
		1.2	NA	NA	NA	NA	NA	NA	NA	COR
		3.0	NA	NA	NA	NA	NA	NA	NA	COR
		5.0	NA	NA	NA	NA	NA	NA	NA	COR
		0.2	NA	NA	NA	NA	NA	NA	NA	COR
		0.45	NA	NA	NA	NA	NA	NA	NA	COR
		1.00	NA	NA	NA	NA	NA	NA	NA	COR

D. Polycarbonate Pore Filter Characteristics

Filter	Comp-osition	Pore Size (μm)	Thickness (μm)	Weight/Area (mg/cm²)	Ash Content (%)	Max. Oper. Temp. (°C)	Tensile Strength (psi)	Refrac-tive Index	ΔP100[A] in H₂O	Source
Nuclepore®		14	NA	NA	NA	NA	NA	NA	NA	COR
		12	NA	NA	NA	NA	NA	NA	NA	COR
		10	NA	NA	NA	NA	NA	NA	NA	COR
		8	9.0	1.0	0.04	140	>3,000	1.58+1.614	NA	COR
		5	NA	NA	NA	NA	NA	NA	NA	COR
		3.0	NA	NA	NA	NA	NA	NA	NA	COR
		2.0	NA	NA	NA	NA	NA	NA	NA	COR
		1.0	NA	NA	NA	NA	NA	NA	NA	COR
		0.8	NA	NA	NA	NA	NA	NA	NA	COR
		0.6	NA	NA	NA	NA	NA	NA	NA	COR
		0.4	10.0	0.8	0.04	140	>3,000	1.58+1.614	83	COR
		0.2	10.0	0.9	0.04	140	>3,000	1.58+1.614	208	COR
		0.1	NA	NA	NA	NA	NA	NA	NA	COR
		0.08	NA	NA	NA	NA	NA	NA	NA	COR
		0.05	NA	NA	NA	NA	NA	NA	NA	COR
		0.03	NA	NA	NA	NA	NA	NA	NA	COR
		0.015	NA	NA	NA	NA	NA	NA	NA	COR
Isopore	polycarbonate									
TM		5.0	10	NA	NA	NA	NA	1.6	NA	MIL
TS		3.0	9	NA	NA	NA	NA	1.6	NA	MIL
TT		2.0	10	NA	NA	NA	NA	1.6	NA	MIL
RT		1.2	11	NA	NA	NA	NA	1.6	NA	MIL
AT		0.8	9	NA	NA	NA	NA	1.6	NA	MIL
DT		0.6	10	NA	NA	NA	NA	1.6	NA	MIL
HT		0.4	10	NA	NA	NA	NA	1.6	NA	MIL
GT		0.2	10	NA	NA	NA	NA	1.6	NA	MIL
Ortho-pore	polycarbonate	0.8	10	1.0	0.01	140	>3,000	NA	NA	OSI
		0.4	10	1.0	0.01	140	>3,000	NA	NA	OSI

TABLE 13-I-1 (cont.). Summary of Air Sampling Filter Characteristics

Filter	Compo-sition	Pore Size (µm)	Thickness (µm)	Weight/Area (mg/cm²)	Ash Content (%)	Max. Oper. Temp. (°C)	Tensile Strength (psi)	Refrac-tive Index	ΔP100[A] in H₂O	Source
Poretics										
	Polycarbonate	20	3	1.1	NA	140	>3,000	1.584+1.625	NA	OLP
		14	6	0.6	NA	140	>3,000	1.584+1.625	NA	OLP
		12	8	0.9	NA	140	>3,000	1.584+1.625	NA	OLP
		10	10	1.1	NA	140	>3,000	1.584+1.625	NA	OLP
		8.0	7	0.8	0.01	140	>3,000	1.584+1.625	NA	OLP
		5.0	10	1.1	0.01	140	>3,000	1.584+1.625	NA	OLP
		3.0	9	0.9	0.01	140	>3,000	1.584+1.625	NA	OLP
		2.0	10	1.1	0.01	140	>3,000	1.584+1.625	NA	OLP
		1.0	11	1.1	0.01	140	>3,000	1.584+1.625	NA	OLP
		0.8	9	0.9	0.01	140	>3,000	1.584+1.625	NA	OLP
		0.6	9	1.0	0.01	140	>3,000	1.584+1.625	NA	OLP
		0.4	10	1.0	0.01	140	>3,000	1.584+1.625	NA	OLP
		0.2	10	1.1	0.01	140	>3,000	1.584+1.625	NA	OLP
		0.1	6	0.7	0.01	140	>3,000	1.584+1.625	NA	OLP
	Polyester	10	10	1.1	NA	NA	NA	NA	NA	OLP
		8	7	0.8	NA	NA	NA	NA	NA	OLP
		5	10	1.1	NA	NA	NA	NA	NA	OLP
		3	9	0.9	NA	NA	NA	NA	NA	OLP
		2	10	1.1	NA	NA	NA	NA	NA	OLP
		1	11	1.1	NA	NA	NA	NA	NA	OLP
		0.8	9	0.9	NA	NA	NA	NA	NA	OLP
		0.6	9	1.0	NA	NA	NA	NA	NA	OLP
		0.4	10	1.0	NA	NA	NA	NA	NA	OLP
		0.2	10	1.1	NA	NA	NA	NA	NA	OLP
		0.1	6	0.7	NA	NA	NA	NA	NA	OLP

Designation	Composition	Size	Void Size (µm)	Max. Oper. Temp. (°C)	Source	Remarks
E. Filter Thimble Characteristics						
D1013	Cellulose	43 x 123 mm	NA	120	GRA	Use with D1012 Paper Thimble Holder
D1016	Glass cloth	2-3/16 x 14"	NA	400	GRA	Use with D1015 Glass Cloth Thimble Holder
RA-98	Alundum	NA	Standard	High	GRA	Use with D1021 Alundum Thimble Holder
RA-360	Alundum	NA	Fine	High	GRA	Use with D1021 Alundum Thimble Holder
RA-84	Alundum	NA	Extra Fine	High	GRA	Use with D1021 Alundum Thimble Holder
S&S 603 GV	Glass fiber Heat treated	from 19 x 90 mm to 90 x 200 mm	NA	510	SAS	
Whatman	Cellulouse	from 10 x 50 mm to 90 x 200 mm	NA	120	WLS	
	Glass fiber	from 19 x 90 mm to 43 x 123 mm	NA	550	WLS	

[A]Pressure drop at face velocity of 100 ft/min (50cm/s).
[B]With organic binder.
[C]Without organic binder.
[D]Laminated on one side with polyester.
[E]Quartz fiber.
[F]Contains Teflon.
[G]Contains cellulose.
[H]Available with or without imprinted grid lines.
[I]Available with black color.
NA = Information not available or not applicable.
AP = Apiezon coated.

TABLE 13-I-2. Summary of Filter Holder Characteristics

Figure No.	Catalog No.	Type	Filter Size (mm)	Effective Area (cm²)	Fittings Supplied with Holder	Weight (g)	Body	Gasket	Filter Support	Fittings	Overall Size (cm) (w/o fittings)	Type of Closure	Max Temp (°C)	Source
13-13	1107	Open	25	3.7	1/4" ID Hose	NA	Delrin	Viton	Stainless screen	Nylon	3.5D x 2.0	Threaded	85	PGL
	1109	Inline	25	3.7	1/8" NPT to 1/4" ID hose	NA	Delrin	Viton	Stainless screen	Nylon	3.5D x 2.0	Threaded	85	PGL
	1209	Inline	25	NA	1/8" NPT to 1/4" ID hose	NA	Stainless	Viton	Stainless screen	Nylon	NA	Threaded	204	PGL
	4376	Open or inline	25	3.9	Female luer	NA	C-filled polypropylene	—	Pad	—	3.0D x 5	Press fit	NA	PGL
	1219	Open	37	4.9	1/8" NPT to hose barb	NA	Aluminum	NA	Stainless screen	Nylon	4.4D x 2.4	Threaded	NA	PGL
	4339	Open or inline	37	9.1	Female luer	NA	Styrene	–	Pad	–	4.2D x 3.8	Press fit	NA	PGL
	1119	Inline	47	9.6	1/4" NPT	NA	Polycarbonate	Silicone	Polysulfonate	Nylon	6.4D x 5.6	Threaded	NA	PGL
	1220	Open	47	9.6	1/8" NPT to 1/4" ID hose	NA	Aluminum	–	Stainless screen	Nylon	5.4D x 2.2	Threaded	NA	PGL
	1235	Inline	47	9.6	3/8" NPT to 1/4" ID hose	NA	Aluminum	Viton	Stainless screen	Nylon	5.9D x 2.2	Threaded	NA	PGL
13-14	2220	Inline	47	9.6	3/8" NPT to	NA	Stainless	Viton	Stainless screen	Nylon	5.9D x 5.7	Threaded	204	PGL
13-15	26800	Open	47 or 50	NA	Built-in hose barb	NA	Polysulfone	O-Ring	Polysulfone	Polysulfone	NA	Threaded	NA	SAS
		Open	47	NA	NA	NA	NA	NA	NA	NA	NA	Threaded	NA	SRD
		Open	50.0	NA	NA	NA	NA	NA	NA	NA	NA	Threaded	NA	SRD
		Open	102	NA	Built-in 1"D nipple	NA	NA	NA	NA	NA	NA	Threaded	NA	SRD
		Open	Charcoal cartridge	NA	NA	NA	PVC	NA	NA	NA	NA	Threaded	NA	SRD
		Open	47 or 50 & charcoal	NA	NA	NA	PVC	NA	NA	NA	NA	Threaded	NA	SRD
		Inline	47 or 50	NA	NA	NA	Aluminum or stainless	NA	NA	NA	NA	Threaded	NA	SRD

TABLE 13-I-2 (cont.). Summary of Filter Holder Characteristics

Figure No.	Catalog No.	Type	Filter Size (mm)	Effective Area (cm²)	Fittings Supplied with Holder	Weight (g)	Materials of Construction				Overall Size (cm) (w/o fittings)	Type of Closure	Max Temp (°C)	Source
							Body	Gasket	Filter Support	Fittings				
		Inline	Charcoal	NA	NA	NA	Aluminum or stainless	NA	NA	NA	NA	Threaded	NA	SRD
		Inline	47 or 50	NA	NA	NA	Aluminum or stainless	NA	NA	NA	NA	Threaded	NA	SRD
	SH-18	Open	28	4.9	Luer adaptor	42	Aluminum	–	–	–	NA	Threaded	NA	STA
	SH-20	Open	49	18.5	Luer adaptor	43	Aluminum	–	–	–	NA	Threaded	NA	STA
	SH-4	Open	110	69.3	Flanged w/4" threaded locking ring	NA	Aluminum	Neoprene	Stainless cross bar	–	NA	Threaded	NA	STA
	SH-69	Open	152 × 228	NA	Flanged w/4" threaded locking ring	NA	Stainless	Neoprene	Stainless screen	Aluminum	NA	Lock nuts (4)	NA	STA
	SH-810	Open	203 × 254	NA	Flanged w/4" threaded locking ring	NA	Stainless	Neoprene	Stainless screen	Aluminum	NA	Lock nuts (4)	NA	STA
13-16	SX0001300	Inline	13	0.7	Female Luer inlet, male Luer outlet	7.1	Polypropylene	Silicone	Polypropylene	–	1.7 D × 3.5	Threaded	NA	MIL
	XX0001200	Inline	13	0.81	Female Luer inlet, male Luer outlet	NA	Stainless	Teflon	Stainless screen	–	1.6D × .3	Threaded	NA	MIL
	SX0002500	Inline	25	3.34	Female Luer inlet, male Luer outlet	14.2	Polypropylene	Silicone	Polypropylene	–	3.2 D × 1.6	Threaded	NA	MIL
	XX3002500	Inline	25	NA	Female Luer inlet, male Luer outlet	NA	Stainless	Teflon	Stainless screen	–	3.2 D × 3.2	Threaded	NA	MIL
13-16	XX3002514	Inline	25	NA	1/4" NPT female inlet, male Luer outlet	NA	Stainless	Teflon	Stainless	–	3.2 D × 3.2	Threaded	NA	MIL
	M000025AC	Open or inline	25	3.9	Female Luer ports	NA	C-filled Polypropylene	–	Pad	–	3.0 D × 5.0	Press fit	NA	MIL
	M000025A0	Open or inline	25	3.9	Female Luer ports	NA	Styrene	–	Cellulose pad	–	2.8 D × 3.8	Press fit	NA	MIL

TABLE 13-I-2 (cont.). Summary of Filter Holder Characteristics

Figure No.	Catalog No.	Type	Filter Size (mm)	Effective Area (cm²)	Fittings Supplied with Holder	Weight (g)	Body	Gasket	Filter Support	Fittings	Overall Size (cm) (w/o fittings)	Type of Closure	Max Temp (°C)	Source
13-17	MO00037A0	Open or inline	37	9.0	Female Luer ports	21	Polystyrene	–	Cellulose pad	–	4.3 D x 3.5	Press fit	NA	MIL
13-18	XX5004700	Inline	47	9.6	7/16" OD hose	906	Aluminum & stainless	Teflon	Stainless steel	–	7.0 D x 17.8	Bayonet lock	NA	MIL
	XX5004720	Open	47	9.6	7/16" OD hose	1160	Aluminum & stainless	Teflon	Stainless screen	–	7.0 D x 17.8 or 10.2	Bayonet lock	NA	MIL
	XX4304700	Inline	47	9.6	1/4" NPT female inlet & outlets w/hose connectors for 1/4" to 3/8" ID tubing	NA	Glass-filled polystyrene	Silicone O-Ring	Polystyrene	–	7.6 D x 12.0	Bayonet lock	NA	MIL
	SX0004700	Inline	47	13.8	Female Luer & 1/4" NPT male inlet, Female Luer & hose connector output	43	Polystyrene	Silicone O-Ring	Polystyrene	–	5.72 D x 5.4	Threaded	NA	MIL
	420100	Inline	13	0.8	Female Luer inlet, male Luer outlet	NA	NA	NA	NA	NA	NA	NA	NA	COR
	420200	Inline	25	2.4	Polycarbonate	NA	NA	NA	NA	NA	NA	NA	NA	COR
	420300	Open	25	2.4	Polycarbonate	NA	NA	NA	NA	NA	NA	NA	NA	COR
	425610	Inline	25	3.9	Stainless	NA	NA	NA	NA	NA	NA	NA	NA	COR
	420400	Inline	47	11.3	Polycarbonate	NA	NA	NA	NA	NA	NA	NA	NA	COR
	430500	Open	47	11.3	Polycarbonate	NA	NA	NA	NA	NA	NA	NA	NA	COR
	272	Inline	37	7.1	1/4" FPT	500	Stainless	Viton	Stainless screen	–	5.6 D x 4.5	Threaded	NA	GRA
	272-LI	Inline	37	7.1	1/4" FPT	500	Stainless	Viton	Stainless screen	–	5.6 D x 8.4	Threaded	NA	GRA
	272-AL	Inline	37	7.1	1/4" FPT	200	Aluminum	Viton	Stainless screen	–	5.6 D x 4.5	Threaded	NA	GRA

TABLE 13-I-2 (cont.). Summary of Filter Holder Characteristics

Figure No.	Catalog No.	Type	Filter Size (mm)	Effective Area (cm²)	Fittings Supplied with Holder	Weight (g)	Materials of Construction Body	Gasket	Filter Support	Fittings	Overall Size (cm) (w/o fittings)	Type of Closure	Max Temp (°C)	Source
	272-O	Open	37	7.1	1.18"D x 0.75"L inlet tube, 1/4" FPT outlet	500	Stainless	Viton	Stainless screen	–	5.6 D x 5.5	Threaded	NA	GRA
	272ALO	Open	37	7.1	1.18"D x 0.75"L inlet tube, 1/4" FPT outlet	200	Aluminum	Viton	Stainless screen	–	6.5 D x 5.5	Threaded	NA	GRA
	273	Inline	47	9.6	3/8" FPT	600	Stainless	Viton	Stainless screen	–	6.5 D x 5.2	Threaded	NA	GRA
	273-LI	Inline	47	9.6	3/8" FPT	700	Stainless	Viton	Stainless screen	–	6.5 D x 8.9	Threaded	NA	GRA
	273-AL	Inline	47	9.6	3/8" FPT	200	Aluminum	Viton	Stainless screen	–	6.5 D x 5.2	Threaded	NA	GRA
	273-O	Open	47	9.6	1.38ID x 0.75" inlet tube, 3/8" FPT outlet	600	Stainless	Viton	Stainless screen	–	6.5 D x 6.1	Threaded	NA	GRA
	273AL-O	Open	47	9.6	1.38 ID x 0.75" inlet tube, 3/8" FPT outlet	200	Aluminum	Viton	Stainless screen	–	6.5 D x 6.1	Threaded	NA	GRA
	274	Inline	63.5	23	1/2" FPT	700	Stainless	Viton	Stainless screen	–	7.5 D x 5.9	Threaded	NA	GRA
	275AL	Inline	100	55	1/2" FPT	1500	Aluminum	Viton	Stainless screen	–	13.3 D x 6.7	Thumb nuts	NA	GRA
	275AL-O	Open	100	55	1/2" FPT	1300	Aluminum	Viton	Stainless screen	–	13.3 D x 4.6	Thumb nuts	NA	GRA
	GMW3000	Cartridge	203 x 254	406	Fits into Model FH2100 8 x 10 filter holder	NA	Aluminum	Viton	Stainless screen	–	31.1 x 23.8 x 2.5	Thumb nuts	NA	TEI

TABLE 13-I-2 (cont). Summary of Filter Holder Characteristics

Figure No.	Catalog No.	Type	Filter Size (mm)	Effective Area (cm²)	Fittings Supplied with Holder	Materials of Construction					Overall Size (cm) (w/o fittings)	Type of Closure	Max Temp (°C)	Source
						Weight (g)	Body	Gasket	Filter Support	Fittings				
13-19	FH-2100	Open	203 x 254	406	4" locking cap	NA	Aluminum	Rubber	Stainless	–	31.1 x 23.8 x 15.2	Wing Nuts	NA	TEI
	23505-1	Inline	203 x 254	406	4" locking cap	NA	Aluminum	Rubber	Stainless	–	31.1 x 23.8 x 15.2	Wing Nuts	NA	TEI
	RVPH-20[A]	Open	50	NA	3/8" FPT	NA	Aluminum	–	–	–	NA	Threaded	NA	HIQ
	RVPH-25[A]	Open	47	NA	3/8" FPT	NA	Aluminum	–	–	–	NA	Threaded	NA	HIQ
	RVPH-102	Open	102	NA	3/8" FPT	NA	Aluminum	–	–	–	NA	Bayonet	NA	HIQ
13-20	CFPH-20[A]	Open	50	NA	1.75" D x 11.5 TPI	NA	Aluminum	–	–	–	NA	Threaded	NA	HIQ
	CFPH-25[A]	Open	47	NA	1.75" D x 22.5 TPI	NA	Aluminum	–	–	–	NA	Threaded	NA	HIQ
	CFPH-40	Open	108	NA	1.75" D x 11.5	NA	Plastic	–	–	–	NA	Bayonet	NA	HIQ
	ILPH-20	Inline	50	NA	3/8" FPT	NA	Aluminum	–	–	–	NA	Threaded	NA	HIQ
	ILPH-47	Open	47	NA	3/8" FPT	NA	Aluminum	NA	NA	–	NA	Threaded	NA	HIQ
	ILPH-5	Open	47	NA	2" FPT x 3/8" FPT	NA	Aluminum	NA	NA	–	NA	Threaded	NA	HIQ
	ILPH-13	Open	47	NA	3/4" FPT x 3/8" FPT	NA	Aluminum	NA	NA	–	NA	Threaded	NA	HIQ
	ILPH-17	Open	47	NA	1" FPT x 3/8" FPT	NA	Aluminum	NA	NA	–	NA	Threaded	NA	HIQ
	91135	Open	25	3.8	1/8" D hose or Luer slip	NA	Polypropylene	Silicone	Screen	–	NA	Threaded	128	OLP
	91275	Open	47	13.5	1/4" NPTM	NA	Polypropylene	PTFE	Screen	–	6.4 x 5.1	Threaded	128	OLP
13-21	F-1	Inline	478	12.97	1/4" NPT x 3/8" hose barb	306	Aluminum	Silicone	Stainless screen	–	7.0	Threaded	260	BGI
	F-2	Open	47	12.97	1/4" NPT x 3/8" hose barb	243	Aluminum	Silicone	Stainless screen	–	5.0	Threaded	260	BGI
	A002550-2	Inline	25	NA	Female Luer	NA	Various[B]	–	Pad	–	–	Press fit	NA	OSI
	A002550-3	3-piece	25	NA	Female Luer	NA	Styrene	–	Pad	–	–	Press fit	NA	OSI
	A003750-2	Inline	37	NA	Female Luer	NA	Various[B]	–	Pad	–	–	Press fit	NA	OSI
	A003750-3	3-piece	37	NA	Female Luer	NA	Various[B]	–	Pad	–	–	Press fit	NA	OSI
	APP-47	Inline	47	NA	Female Luer	NA	Polypropylene	–	Pad	–	–	Press fit	NA	OSI

TABLE 13-I-2 (cont.). Summary of Filter Holder Characteristics (Continued)

Figure No.	Catalog No.	Type	Filter Size (mm)	Effective Area (cm²)	Fittings Supplied with Holder	Weight (g)	Materials of Construction				Overall Size (cm) (w/o fittings)	Type of Closure	Max Temp (°C)	Source
							Body	Gasket	Filter Support	Fittings				
13-22	SS-1	Thimble	30 × 100	70.7	5/8" compression × 5/8" tube	595	Stainless	Silicone or fiber	–	–	20.0 L × 5.0 D	Threaded	900	BGI
	SS-3	Thimble	30 × 100	70.7	5/8" compression × 5/8" tube	726	Stainless	Silicone or fiber	–	–	21.3 L × 5.0 D	Threaded	1500	BGI
	D1021	Thimble	Uses Ra-98, Ra360 & Ra-84 Alundum thimbles (see Table 13-I-1-E)	NA	1/2" NPT female sockets	NA	Stainless	Fiber	–	–	NA	Threaded	NA	GRA
	D1015	Thimble	Uses D1016 glass, cloth thimbles (see Table 13-I-1-E)	NA	1/2" NPT female sockets	NA	Stainless	Special	–	–	7.5 D X 48 Lock nuts (4)		400	GRA

AAlso available as combination holder for filter paper plus metal cartridge for vapor sampling.
BCassette available in styrene (clear and opaque) and polypropylene (transparent, opaque, white, and conductive black (carbon filled)).

TABLE 13-I-3. Commercial Sources for Filters and Filter Holders

BGI	BGI Incorporated 58 Guinam Street Waltham, MA 02451 (781) 891-9380 FAX: (781) 891-8151 *www.bgiusa.com*	MIL	Millipore Corporation 80 Ashby Road Bedford, MA 01730 (617) 533-2125 or (800) 645-5476 *www.millipore.com*	SRD	SAI/RADECO 4161 Campus Point San Diego, CA 92121-9416
COR	Corning, Inc. Science Prod. Div.-Separations 45 Nagog Park Acton, MA 01720-3413 (978) 635-2200 or (800) 882-7711 FAX: (978) 635-2490 Email: separation@corning.com	OLP	Osmonics Laboratory Products Micron Separation Products P.O. Box 1046 135 Flanders Road Westborough, MA 01581 (508) 366-8212 FAX: (508) 366-5840 *www.osmonics.com*	STA	Staplex Company 777 Fifth Avenue Brooklyn, NY 11232 (718) 768-3333 or (800) 221-0822 FAX: (718) 965-0750 Email: *info@staplex.com*
GRA	Graseby Andersen 500 Technology Court Smyrna, GA 30082-5211 (770) 319-9999 or (800) 241-6898 FAX: (770) 319-0336 E-mail: *andersen@graseby.com* *www.graseby.com*	OSI	Omega Specialty Instrument Co. P.O. Box 160 Chelmsford, MA 01824 (978) 256-5450 or (800) 346-8253 FAX: (978) 256-8015 Email: *omega@omegaspec.com* *www.omegaspec.com*	TEI	Tisch Environmental, Inc. 145 S. Miami Avenue Village of Cleves, OH 45002 (513) 467-9000 or (877) 263-7610 FAX: (513) 467-9009 Email: *Tisch@cinci.infi.net*
HIQ	Hi-Q Environmental Products Co. 7386 Trade Street San Diego, CA 92121 (619) 549-2820 FAX: (619) 549-9657 Email: *info@HI-Q.net* *www.HI-Q.net*	PGL	Pall–Gelman Laboratory 600 South Wagner Road Ann Arbor, MI 48103-9019 (734) 665-0651 or (800) 521-1520 FAX: (734) 913-6114 *www.pall.com/gelman*	WLS	Whatman Lab. Sales P.O. Box 1359 Hillsboro, OR 97123-9981
H&V	Hollingsworth & Vose Co. 112 Washington Street East Walpole, MA 02032 (508) 668-0295 FAX: (508) 668-6526 *www.hollingsworth-vosc.com*	SAS	Schleicher and Schuell, Inc. 10 Optical Avenue P.O. Box 2012 Keene, NH 03431-2062 (603) 352-3810 FAX: (603) 357-3627 Email: *custserv@s-and-s.com*		

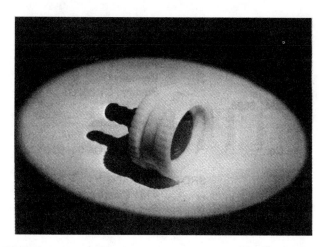

FIGURE 13-13. Pall–Gelman #1107 25-mm Delrin open filter holder.

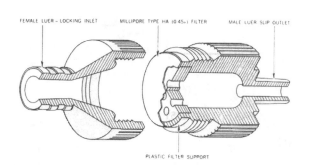

FIGURE 13-16. Schematic of Millipore Corp. Swinnex-13 polypropylene Swinny-type in-line filter unit.

FIGURE 13-14. Pall–Gelman #2220 47-mm stainless in-line filter holder.

FIGURE 13-17. Millipore 37-mm aerosol monitor. At left is a sealed unit, as supplied. The top can be removed with a coin for use as an open filter holder.

FIGURE 13-15. S&S Aerosol holder, 47 or 50 mm.

FIGURE 13-18. Millipore aerosol universal filter holder (#XX50 047-20) with a set of limiting orifices. It consists of the front ends of both the open (#XX50 047 10) and standard (#XX50 047 00) holders and an interchangeable base.

FIGURE 13-19. General Metal Works #23505-1 8-in. x 10-in. in-line filter holder.

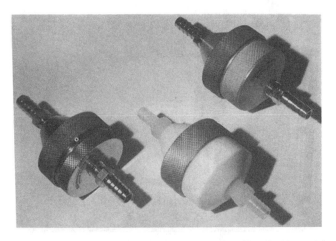

FIGURE 13-21. BGI, Incorporated 47-mm in-line filter holders with stainless steel, Teflon, and aluminum bodies.

FIGURE 13-20. HI-Q Paper filter open-face holders for high volume samplers for 47-mm-, 2-in.-, and 4-in.-diameter filters.

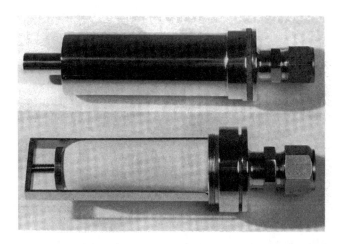

FIGURE 13-22. BGI, Incorporated 30 x 100 Alundum Thimble Adaptor; closed, and with cover sleeve removed.

Chapter 14

Impactors, Cyclones, and Other Particle Collectors

Susanne V. Hering, Ph.D.

President, Aerosol Dynamics Inc., 2329 Fourth Street, Berkeley, CA

CONTENTS

Overview

This chapter presents four types of particle collectors: inertial collectors, gravitational collectors, electrostatic precipitators, and thermal precipitators. Impactors, cyclones, and aerosol centrifuges are all types of inertial collectors. They collect particles when the air flow changes direction, and the larger particles cannot follow the air flow streamlines. Elutriators are a type of gravitational collector and collect particles using gravitational settling. Electrostatic precipitators collect charged particles in an electric field. Thermal precipitators collect particles on surfaces that are colder than the surrounding air.

In contrast to filters that generally collect particles of all sizes, inertial and gravitational samplers collect particles in characteristic size ranges. Some of these samplers are used for size-selective sampling, defined as the collection of particles from one specific particle size fraction. Others of these samplers are used for size-segregated particle sampling, which refers to the physical separation of airborne particles into several size fractions. The particle sizing by these instruments depends on particle "aerodynamic diameter," defined below.

For size-selective sampling, cyclones, elutriators, and single-stage impactors are used to remove larger particles from the air stream and are followed by a filter for collection of the smaller particles. Some cyclones can be operated to approximate the respirable collection efficiency curve, as discussed in Chapter 5. Other cyclones mimic the thoracic collection efficiency curve. Similarly, elutriators have been used in size-selective sampling to measure respirable or thoracic particle mass. Single-stage impaction heads are used to adapt Hi-Volume samplers for PM_{10} sampling.

For size-segregated sampling, cascade impactors or cascaded cyclones can be used. In a cascade impactor, the different stages collect different sizes of particles. The stages can be analyzed to determine aerosol mass in each size fraction to yield the mass size distribution. They may also be analyzed to assess chemical composition as a function of particle size. Cascaded cyclones have been designed for stack gas sampling in which a robust system is needed to handle the elevated temperatures. Aerosol centrifuges, inertial spectrometers, and some horizontal elutriators deposit particles by size along a single substrate, providing a continuous size spectrum. They can be used to determine aerodynamic shape factors for irregularly shaped particles.

Electrostatic and thermal precipitators can be an effective means of collecting small particles at low pressure drops. Although they are not as widely used for particle sampling as are inertial collectors, they are appropriate for applications such as concentrating particles for microscopic analysis. Unlike gravitational and inertial collectors, electrostatic or thermal precipitation is not dependent on aerodynamic diameter.

Aerodynamic and Stokes Diameters

Particle Stokes and aerodynamic diameters are defined here because they are important to the characteristics of the collectors discussed in this chapter. The parti-

cle separation characteristics of inertial and gravitational collectors depend on particle "aerodynamic diameter." For particles greater than about 0.5 μm, the aerodynamic diameter is the parameter that enters into the equations for particle transport and respiratory tract deposition. Respirable, thoracic, and inhalable particle sampling, as described in Chapter 5, are based on particle aerodynamic diameter. For electrostatic deposition, the precipitation step depends on the particle Stokes diameter.

The Stokes diameter d_p of a particle is the diameter of a sphere having the same density and settling velocity as the actual particle. When a particle settles, it moves with respect to the air around it. The particle encounters resistance from the surrounding air proportional to its relative velocity. This resistance is called a drag force, F_{drag}, and is given by the relation:

$$F_{drag} = \frac{3\pi\mu}{C} d_p (V_p - V_{air}) \qquad (1)$$

where μ is the air viscosity, V_p is the velocity of the particle, and V_{air} is the velocity of the surrounding air. The particle diameter that enters into this equation, d_p, is called the particle Stokes diameter. For a smooth, spherically shaped particle, d_p exactly equals the physical diameter of the particle. For irregularly shaped particles, d_p is the diameter that characterizes the drag force on the particle. In other words, it is the value that must be used in Equation 1 to obtain the correct value for the drag force.

The factor C that enters Equation 1 is a called the Cunningham slip factor. It is an empirical factor that accounts for the reduction in the drag force when the particle diameter is small compared to the mean free path length in air. The reduction in the drag force for these small particles arises from the "slip" of the gas molecules at the particle surface. It is important for small particles, less than 1 μm in diameter, for which the surrounding air cannot be modeled by a continuous fluid. The slip factor is a function of the ratio between particle diameter and mean free path of the suspending gas; it is given by the following expression:[1]

$$C = 1 + \frac{\lambda}{d_p}\left[2.514 + 0.800\, exp\left(-0.55\frac{d_p}{\lambda}\right)\right] \qquad (2)$$

where: d_p = particle size diameter as defined in
              Equation 1
       λ = mean free path in air

At normal atmospheric conditions (i.e., temperature = 20°C, pressure = 1 atmosphere), λ = 0.066 μm. For large particles ($d_p > 5$ μm), $C = 1$; for smaller particles, $C>1$.

Aerodynamic diameter is defined as the diameter of a smooth, unit density ($\rho_o = 1$ g/cm³) sphere that has the same settling velocity as the particle. It is dependent on the particle density and particle shape, as well as the particle size. The general expression for the particle aerodynamic diameter, d_a, is:

$$d_a = \left(\frac{\rho C}{\rho_o C_a} \right)^{1/2} d_p \qquad (3)$$

where d_p is the Stokes diameter as defined in Equation 1, ρ is the particle density, ρ_o is a constant equal to 1 g/cm³, C_a and C are the Cunningham slip factors, as defined in Equation 2, evaluated for diameters d_a and d_p respectively.

The relationship between the physical particle size, Stokes, and aerodynamic diameters is illustrated in Figure 14-1.[1] Particles with the same physical size and shape, but different densities, will have the same Stokes diameter but different aerodynamic diameters. For two particles of the same physical size but differing densities, the particle with the larger density will have the larger aerodynamic diameter. If the density of a particle is greater than 1 g/cm³, then its aerodynamic diameter is larger than its Stokes diameter. Conversely, for particles of densities less than 1 g/cm³, the aerodynamic diameter is smaller than the Stokes diameter.

For particles with diameters much greater than the mean free path, the aerodynamic diameter given by Equation 3 can be approximated by:

$$d_a = \sqrt{\frac{\rho}{\rho_o}} \, d_p \qquad (\text{for } d_p \gg \lambda) \qquad (4)$$

In this approximation, the aerodynamic diameter is directly proportional to the square root of the particle density. This expression holds for large particles for which the slip factor equals one. It is often used for particles as small as 0.5 μm, which is acceptable if the particle density is at all close to 1 g/cm³. For example, a density of 2 g/cm³ and a Stokes diameter of 0.5 μm gives an aerodynamic diameter calculated from Equation 3 of 0.68 μm. The approximation of Equation 4 gives 0.71 μm, an error of only 4%.

For particles with diameters much smaller than the mean free path, the slip factor C is inversely proportional to particle diameter, which makes the aerodynamic diameter directly proportional to the particle density:

$$d_a = \frac{\rho}{\rho_o} \, d_p \qquad (\text{for } d_p \ll \lambda) \qquad (5)$$

This small particle limit is applicable for low-pressure systems, such as low-pressure or inertial devices used in stratospheric sampling.

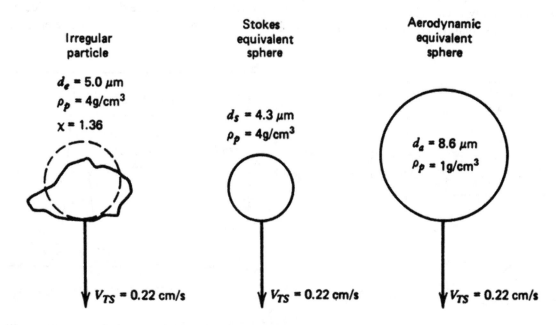

FIGURE 14-1. An irregularly shaped particle and its equivalent Stokes and aerodynamic spheres.[1] Reprinted with permission of John Wiley & Sons, Inc.

Impactors

Description and Operational Principle

The term "impactor" encompasses a large category of aerosol collection instruments in which particle impaction in a nonrotating flow is the primary mechanism of particle capture. Particle impaction refers to the collection of particles that by virtue of their inertia deviate from the air flow streamlines. Impaction occurs when streamlines bend as the air flow bypasses a solid object.

Conventional flat-plate impactors employ a collection surface located internal to the device, as illustrated in Figure 14-2. Particle-laden air passes through the nozzle and impinges on a collection plate oriented perpendicular to the nozzle axis. The air flow is laminar, and particles within the nozzle are accelerated to a nearly uniform velocity. At the nozzle exit, the streamlines of the gas are deflected sharply by the collection plate. Larger particles are propelled across the air streamlines and deposit on the plate. Smaller particles follow the streamlines more closely and remain suspended in the air.

The minimum size collected by an individual stage depends on the jet diameter and the air stream velocity in the jet. In low-pressure impactors, it also depends on the pressure at which the stage operates. Typically, the collection of smaller particles is achieved by using smaller diameter jets with higher jet velocities. Within limits, particle impaction is insensitive to the spacing between the collection plate and the jet exit, and to the geometry of the stage.

Cascade impactors were introduced by May in 1945[3] and are widely used. These impactors are multistage instruments that fractionate particle samples into several size fractions. Cascade impactors have several impactor stages mounted in series. Each stage consists of one or more jets followed by a collection plate. Air enters at the top, passes through each of the impactor stages, and is exhausted through a back-up filter. The largest particles are collected in the first stage, successively smaller particles are collected by the subsequent stages. Those particles that penetrate the last impaction stage are collected by the back-up filter. Air flow is generated by means of a vacuum pump and controlled by a valve or critical orifice downstream of the back-up filter.

Size-fractionated samples from cascade impactors are used to determine the distribution of aerosol mass or chemical species with respect to particle size. When cascade impactor samples are analyzed chemically, they yield species size distributions as shown in Figure 14-3.[4] Alternatively, the impactor samples can be assayed gravimetrically to provide an aerosol mass distribution. Simultaneous data on particle size and mass or chemical composition are important for assessing health effects and particle transport in the atmosphere or in a room. Discussion of these instruments, their use, and data analysis procedures is given by Lodge and Chan.[5]

The overall size range covered by an impactor depends on its design. Conventional cascade impactors can be designed to collect particles as small as 0.4 µm. Low-pressure and micro-orifice impactors can collect particles as small as 0.05 µm. Others extend to the nanometer size range.[6-8] Some impactors, such as the Andersen microbial sampler, are designed to collect very large particles, as much as 30 µm in diameter. Rotary impactors

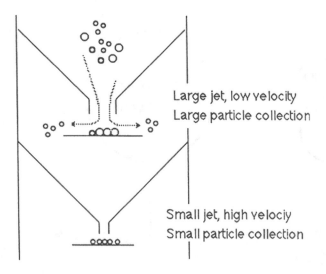

FIGURE 14-2. Schematic of two impactor stages showing large and small particle trajectories. (original drawn by S. Hering)

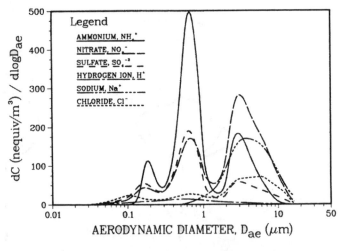

FIGURE 14-3. Inorganic ion particle size distributions collected with the Berner Impactor in Claremont, CA.[4] *Reprinted with permission of American Association for Aerosol Research.*

have been used with high efficiency to sample ambient air particles as large as 250 μm. In these, a rod moving through the ambient air impacts and collects particles larger than the characteristic cut-size for the sampler.

Impactors differ in the nature of the particle collection surface. Most collect particles on a solid plate located immediately downstream of the accelerating jet, as shown in Figure 14-2. However, unless the collection plate is greased, particles may bounce and be re-entrained in the flow. To avoid this problem, the virtual impactor uses a nearly stagnant air flow to transport the size-fractionated sample to a filter. Although it does not have an impaction surface, the air flow streamlines are similar to those in conventional impactors. Virtual impactors are also used to concentrate particles in an air flow. Applications for concentrators include particle counting[9] and particle exposure studies.[10]

Commercially available impactors are listed in Table 14-I-1. The impactor jets may be round or rectangular in cross section. Many use multiple jets per stage to permit larger collection of particles at larger flow rates. Large flow rate cascade impactors have been designed for use with Hi-Volume samplers.[11, 12] Low flow rate impactors are used for personal and ambient sampling.[13–17] Some impactors are designed for specific purposes such as cloud water sampling or measurement of the pharmaceutical aerosols.[18, 19] Impactors are also used for stack sampling[20, 21] and viable particle sampling,[22] as discussed in

Chapters 20 and 22. Some impactors, such as the electrical low-pressure impactor[23] are used in real-time monitors for particle concentration, as discussed in Chapter 15.

Impactor Theory

Impactor performance is characterized by a set of collection efficiency curves. Those from two of the more commonly used cascade impactors, the Andersen and MOUDI, are shown in Figure 14-4[24] and Figure 14-5.[25] Each curve shows the efficiency with which a particle entering the stage is collected. The point corresponding to a collection efficiency of 50% is referred to as the cut-point, or cutoff diameter, d_{50}. The curve shape indicates the sharpness of the size-segregation. For an infinitely steep collection efficiency curve, all particles above the cutoff diameter would be collected, and all below that size would pass onto the next stage. In practice, efficiency curves have a finite slope, which gives rise to crossover in particle size between neighboring stages.

Generally, impactors are designed such that the efficiency curves are as steep as possible. This is especially desirable because the most common data reduction methods use only the cutoff diameter to characterize stage performance. More sophisticated inversion methods for data reduction take into account the actual shape of the efficiency curves and produce a smoothed size distribution (e.g., see Figure 14-3).

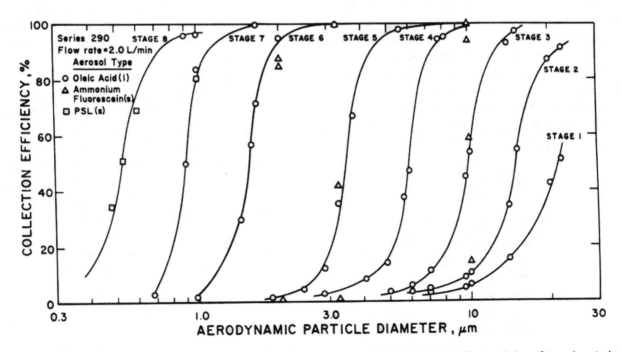

FIGURE 14-4. Collection efficiency curves for the Sierra/Anderson personal sampler.[24] *Reprinted with permission of American Industrial Hygiene Association Journal.*

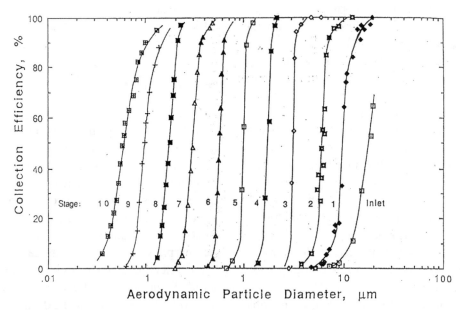

FIGURE 14-5. Particle collection efficiency curves for the MOUDI. *Reprinted with permission of American Association for Aerosol Research.*

Impactor theory can be used to predict the cutoff diameter and the shape of the collection efficiency curves. Theory does not account for nonideal effects such as particle rebound from the surface, but it is applicable for sticky particles. The first impactor theories were advanced by Ranz and Wong[26] and Davies and Aylward.[27] Currently used models include those of Marple and Liu[28, 29] and Rader and Marple,[30] which use numerical solutions to the fluid dynamics and particle trajectory equations in impactors. Other models of note are those of Mercer and coworkers[31, 32] and Ravenhall and Forney.[33] Results from these models are used as guidelines in the design of impactors.[34, 35]

In an impactor, whether a particle impacts depends on the drag force on the particle, the particle momentum, and the effective transit time across the plate. Impactor theory combines these parameters into a dimensionless parameter called the Stokes number, given by:

$$St = \frac{\rho \, d_p^{\,2} \, CV}{9 \, \mu \, W} \qquad (6)$$

where: ρ = particle density
  d_p = particle Stokes diameter
  C = Cunningham slip factor, as defined in Equation 2
  V = mean velocity in the jet
  μ = air viscosity
  W = jet diameter or width

Physically, the Stokes number is proportional to the ratio of the particle-stopping distance to half the jet diameter. (The stopping distance is the distance traveled by a particle before stopping when injected into still air.) Alternatively, it may be viewed as the ratio of particle relaxation time to the transit time of the air flow through the impaction region. (The relaxation time is the time for a particle initially at rest to accelerate within $1/e$ of the velocity of the air stream, which is 63%, where e is the base of natural logarithms.) The larger the Stokes number, the greater the impaction efficiency.

One of the most important uses of the Stokes number is to predict the cutoff diameter, d_{50}. Impactor stages with similar geometry but varying jet diameters or flow rates will have collection efficiencies that tend to fall on a common curve when plotted as a function of St. The cutoff diameter d_{50} corresponds to a single Stokes number, referred to as the critical Stokes number, St_{50}. The value of St_{50} is approximately the same for different impaction stages, and even for different impactors of similar geometry, and it can be used to predict impactor performance.

It is useful to express the cutoff diameter in terms of the critical Stokes number; the sampler volumetric flow rate, Q; and the number of jets per stage, n. This is accomplished by writing the jet velocity in Equation 6 as the ratio of the flow rate to the jet cross-sectional area. For round jet impactors, the expression is:

$$d_{50}^{\,2} \, C = \frac{9 \, \mu \, \pi \, n \, W^3 \, (St_{50, \, round})}{4 \, \rho \, Q} \qquad (7)$$

The Cunningham slip factor, C, depends on the particle diameter and thus has been placed on the left-hand side of the equation. For rectangular jet impactors, with jets of width W and length L, and with n jets (or slots) per stage, the cutoff diameter is given by:

$$d_{50}^2\, C = \frac{9\,\mu\,\pi\, n\, L\, W^2\, (St_{50,\,rect.})}{\rho\, Q} \quad (8)$$

For most slotted impactors, the value of $St_{50,\,rect.}$ is close to 0.59. For round jet impactors, $St_{50,\,round}$ is about 0.24.

Equations 7 and 8 are used to calculate stage cutoff diameters for impactors operated at different flow rates, or at temperatures and pressures other than the design conditions. Changes in temperature affect μ; changes in pressure affect C. Impactor cutoff diameters decrease with increasing flow rate per jet and decrease with decreasing jet diameter. Because the cutoff diameter is relatively insensitive to the distance between the jet exit and the collection plate, this parameter does not appear in the Stokes number.

The shape of an impactor collection efficiency curve depends on the jet Reynolds number, defined as:

$$Re = \frac{\rho_{air}\, VW}{\mu} \quad (9)$$

Impactor collection efficiency curves tend to be steeper at higher Reynolds numbers, as shown in the model calculations of Figure 14-6.[28] The performance at $Re = 500$ is much better than at $Re = 100$. For very low Reynolds numbers, below 100, impactors are not very effective, and collection efficiencies may never reach 100%. Once the Reynolds number is above about 200, the impactor will perform well, and the effect of Re on the efficiency curves is relatively small.

The effect of the jet-to-plate spacing on impactor cutoff diameters is shown in Figure 14-7.[28] The jet-to-plate spacing is the distance between the outlet of the impactor nozzle and the impaction plate. Figure 14-7 plots the nondimensional cutoff diameter, expressed in terms of the critical Stokes number, against the ratio of the jet-to-plate spacing, S, to the jet diameter or width, W. For values of S/W between 1 and 5, the impactor stage d_{50} is almost unaffected. At much smaller jet-to-plate spacings, the cut-sizes are smaller and strongly affected by the spacing. At large S, greater than $5W$ or $10W$, depending on the jet Reynolds number, the cut-sizes increase because of the expansion of the jet. The recommended jet-to-plate spacing corresponds to S/W values near 1 for round jet impactors and 1.5 for rectangular jets.

The foregoing discussion applies to spherical particles. For sampling fibers or other nonspherical particles, the collection efficiency depends on the aspect ratio as well as the particle Stokes number. The aspect ratio for a cylindrical particle is its length divided by its diameter. The collection of these particles has been modeled numerically.[36] For fibers of the same diameter, the longer fibers will be collected more efficiently than will the shorter ones.

Theoretical Considerations for Multijet Impactors

Some impactors, such as the Andersen or the MOUDI, use many orifices per stage, some times as many as one hundred or more. This is done to maintain small cutpoints at higher flow rates. But experience with multijet impactors has shown that the flow characteristics of each orifice may not be the same as for other equally sized orifices on the same stage. May[37] observed that the deposit spots for orifices located at the center of the some of the stages of the Andersen multijet impactor contained fewer particles than those at the edge of the stage. He found that the pressure drop for the flow

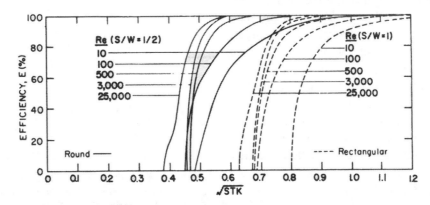

FIGURE 14-6. Model calculations of impactor collection efficiency curves for round and rectangular jet impactors at various jet Reynolds numbers. *Reprinted with permission from* Marple, V.A.; Liu B.Y.H.: Characteristics of Laminar Jet Impactors, Environ. Sci. Technol. 8:648–654 (1974). *Copyright 2001 American Chemical Society.*

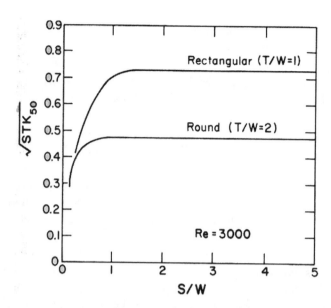

FIGURE 14-7. Impactor 50% cutoff size as a function of the jet-to-plate spacing, S, expressed as a fraction of the jet diameter, W. Curves are shown for round and rectangular jets with a throat length T and diameter or width W.[28] *Reprinted with permission of American Chemical Society.*

to move from the center of the impaction plate to the edge was comparable to the pressure drop across a single orifice. Thus the flow velocities, and hence the deposits, were less in the center than at the edge. This effect was most pronounced on the upper impaction stages for which the orifice pressure drops were small.

Another concern with multijet impactors is the influence of cross-flow of spent air from adjacent nozzles. Spent air refers to the air flow after it has impinged on the impaction plate and is traveling across the top of the plate to reach the edge. When the flows through the individual orifices are approximately equal, the mass flow of spent air increases toward the edge of the orifice plate, as it must include the flow from all of the enclosed orifices. This spent air flow can deflect the air jets at the edges, thereby changing the particle collection characteristics. The spent air cross-flow interferes mostly with the outer jets, leading to lighter deposits at the edges. This is very different from the situation of the Andersen impactor, where the deposits were least for orifices located in the center of the collection plate because of reduced flow through the center orifices.

Fang *et al.*[38] derived a cross flow parameter to characterize this effect. The cross flow is largest at the periphery of the orifice cluster. The cross-flow parameter for N uniformly distributed nozzles of diameter D_n contained within a cluster of diameter D_c is:

$$\frac{1}{m}\frac{S}{D_n} = \frac{D_n N}{4 D_c} = \frac{\pi D_n n D_c}{16} \qquad (10)$$

where: m = ratio of jet mass flow rate to cross mass flow rate
S = jet-to-plate separation distance
n = $4N/\pi D_c^2$ = number of jets per unit area

The effect of the cross flow parameter on impactor performance is shown in Figure 14-8.[38] In this study, the cross-flow parameter was varied by masking the jets, and the flow rate adjusted to maintain constant velocity through the nozzles. Their results show that 100% efficiency is only reached for cross flow parameter below 1.2.

Particle Bounce

Impactor theory assumes that all particles striking the collection surface adhere to it. In practice, this criterion is not always met. Dry, solid particles may bounce from the surface on impaction and be re-entrained in the air stream. If collected on a subsequent stage, the size distribution will be further distorted. This problem is perhaps the greatest limitation in the use of impactors. It was recognized in 1945 by May[3] in the initial development of the impactor and has been raised by many others since.[39-53]

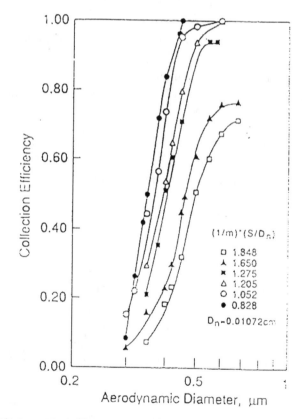

FIGURE 14-8. Influence of cross-flow on particle collection efficiency. The cross-flow parameter is $(1/m)(S/D_n)$ as defined in Equation 10. *Reprinted from Fang, C.P.; Marple, V.A.; Rubow, K.L.: Influence of Cross-Flow on Particle Collection Characteristics of Multi-Nozzle Impactors. J. Aerosol Sci. 22(4): 403–415 (1991) with permission of Elsevier Science.*

TABLE 14-1. Adhesive Coatings Used for Impaction Surfaces

	Source*	Author (Ref. No.)
A. Recommended Coatings:		
Apiezon L grease	1.	Wesolowski et al., [31] Lawson,[37] Vanderpool et al.,[49] and Pak et al.[46]
Dow Corning Antifoam A silicone adhesive	2.	Mercer and Chow[22]
Dow Corning oil (200 & 600 cst)	2.	Rao and Whitby,[32, 33] Mercer and Stafford,[23] Vanderpool et al.,[49] and Pak et al.[46]
Dow Corning silicone grease	2.	Cushing et al.,[29] Wesolowski et al.,[31] and Vanderpool et al.[49]
Flypaper mixture: one part rosin to three parts castor oil	—	May[4]
Halocarbon	3.	Wang and John[3]
One part methylated starch to three parts tricresyl phosphate	—	May[4]
Polyisobutene	—	May[4]
Petroleum jelly (Vaseline)	—	May[4], Hering et al.,[58] Rao and Whitby,[32, 33] Lawson,[37] Cushing et al.,[29] and Vanderpool et al.[49]
Oil-sintered metal	—	Reischl and John[41]
Oiled membrane filters	—	Turner and Hering[47]
B. Ineffective Coatings:		
Sticky tape		Wesolowski et al.[31]
Paraffin		Lawson[37] and Wesolowski et al.[31]
C. For Microscopy:		
Sticky carbon tape		Kromidas and Leifer[1996]

* Source: 1. Apiezon Products Ltd., England; available through most scientific supply houses.
2. Dow Corning Co., Midland, Michigan 48686; available through most scientific supply houses.
3. Halocarbon Products Co., 82 Burlews Court, Hackensack, New Jersey 07601.

The theory of particle interactions with surfaces shows there is a critical approach velocity below which the particle will stick on a clean surface, and above which it will bounce.[54–56] This velocity depends on the coefficient of restitution, which is a measure of the particle's tendency to rebound. Cheng and Yeh[57] proposed a criterion for impactor design that would maintain jet velocities below typical critical approach velocities, thereby minimizing particle bounce. However, obtaining the desired cut-sizes at low jet velocities requires small orifices, which often is not a practical option. Generally, substrate coatings must be used.

Submicrometer as well as supermicrometer particles are subject to particle bounce. Particles as small as 0.2 μm have been observed to bounce from uncoated surfaces. Particles that bounce are often lost to the walls of the impactor.[47, 57] Thus, bounce can underestimate mass loadings as well as distort size distributions. Bounce-off errors do not affect all types of particles equally. For the same operating conditions, liquid particles adhere, whereas solid particles may not.

To obtain reliable data, it is necessary to take precautions to avoid particle bounce. In some cases, the sampled aerosol itself will be sticky and no coating will be needed, but this must be evaluated on a case-by-case basis. In practice, only collection surfaces that show good retention of solid particles are considered "bounce-free."

An adhesive coating can be used on the impaction stage to ensure that all sampled particles will stick. The effectiveness of different impaction surfaces has been evaluated in several laboratory[40, 43–45, 47–49] and field[39, 42, 46, 49] studies, as given in Table 14-1. In general, greases and oils are quite effective in reducing particle bounce. They can be used when analyzing for inorganic compounds or specific organic compounds (such as polycyclic aromatic hydrocarbons) which are not present in the grease or oil coating. Sticky tapes and paraffin have been found to be much less effective. However, for microscopic analysis, a sticky carbon tape has been used that provides for good imaging of the particles, even though it was found to be only about 60% effective in eliminating bounce of solid polystyrene latex test particles.[58]

It is important that the coating be sufficiently thick and that substrates not be overloaded with particles. A laboratory study of the effect of coating thickness showed that Apiezon L (Apiezon Products Ltd., England; available through most scientific supply houses) coatings less than 0.7 μm thick were not as effective as 9-μm-thick coatings of the same grease for capturing 0.56- and 1-μm latex particles.[59] These same

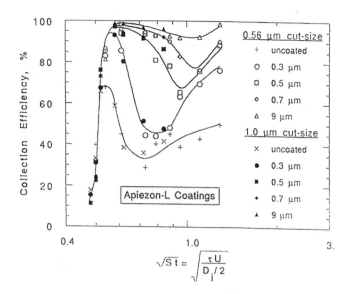

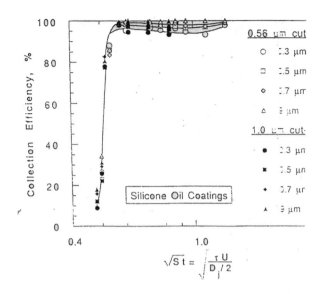

FIGURE 14-9. Particle collection efficiency of Apiezon-L coatings for coating thicknesses of 0.3 to 9 μm.[59] *Reprinted with permission of American Association for Aerosol Research.*

FIGURE 14-10. Particle collection efficiency of silicone coatings for coating thicknesses of 0.3 to 9 μm.[59] *Reprinted with permission of American Association for Aerosol Research.*

investigators found that for silicone oil, the coating thickness did not have as large an effect on capture efficiencies. Graphs of the dependence of collection efficiency curves on the coating thickness are shown in Figure 14-9 for Apiezon-L, and in Figure 14-10 for silicone oil. Both of these figures correspond to particle loadings of much less than one monolayer.

When sampling solid particles on greased surfaces, it is important not to overload the substrate. John and coworkers[52, 53] found that greased surfaces become ineffective after becoming partially coated with particles. The effect is noticeable at submonolayer loadings. With half of the grease coating covered with particles, the incoming particle is equally likely to impact on top of a deposited particle as on the greased surface. Data showing the decrease in collection efficiency of solid particles on a silicone-greased substrate are shown in Figure 14-11.

To eliminate the effect of substrate loading, Reischl and John[53] used an oil-soaked sintered metal disk. This surface is bounce-free even for high substrate loadings. The oil is drawn up onto the depositing particles by capillary action; thus, incoming particles are always presented with an oily surface. The porous metal serves to hold the oil in place under the impactor jet. Sticking efficiencies do not drop, even for large accumulations of deposited aerosol, as shown by the open symbols in Figure 14-11. This concept has been used in some commercial devices such as the Wedding PM_{10} particle inlet for Hi-Volume samplers.[60] The disadvantage for some applications is that the surface is not amenable to chemical analysis.

An alternative approach to reducing particle bounce is to humidify particles prior to collection. At higher relative humidity, the hygroscopic particles become wet and do not bounce upon impact. Stein *et al.*[61] measured collection efficiencies of ambient particles as a function of relative humidity, as shown in Figure 14-12. These data show that at low relative humidity, below about 20%, almost all particles bounced upon impact. In contrast, when the relative humidity was above about 80%, the collection efficiency was 90% or greater.

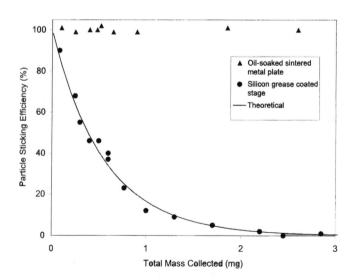

FIGURE 14-11. Sticking efficiency of 20 μm potassium biphthalate particles impacting at 8 m/s onto an oil-soaked sintered metal disk (triangles) and onto a silicone grease coated plate (circles). *Adapted from data of Reischl and John (1978).*

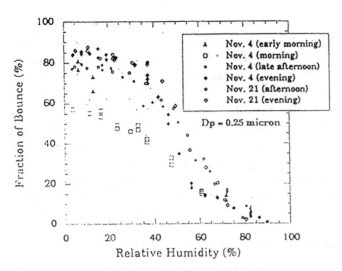

FIGURE 14-12. Measurements of particle bounce vs. relative humidity for atmospheric particles in Minneapolis, MN. *Reprinted from Stein, S.W.; Turpin, B.J.; Cai, X.P.: Measurements of Relative Humidity-Dependent Bounce and Density for Atmospheric Particles Using the Dma-Impactor Technique. Atmos. Environ. 28(10): 1739–1746 (1994) with permission of Elsevier Science.*

Various impactor plate geometries have been investigated as a means of trapping bouncy particles upon impact. Examples include the cylindrical,[62] wide-mouthed traps,[63] and a grooved surface[64]. Grease and oil coatings minimize particle bounce by absorbing the kinetic energy of the incoming particle. The trap designs rely on multiple collisions to dissipate the particle kinetic energy. The performance of these traps appears to be variable. Some investigators report that the traps do not eliminate particle bounce, whereas other investiga-

tors have found them to be effective. Collection efficiencies for various trap designs investigated by Tsai and Cheng[63] are shown in Figure 14-13. Collection efficiencies for the traditional flat plate (design 1) are initially low on uncoated surfaces. As the surface becomes coated with particles, the final collection efficiency is between 50–60%. Similarly, the trap designs show a low initial collection efficiency which increases to near 80% as the particle loading on the surface increases.

Fiber filters are not effective impactor collection substrates for solid particles. Rao and Whitby[45] found that while fiber filters reduce particle bounce, they do not eliminate it (Figure 14-14). Furthermore, the filter has the effect of shifting and flattening the efficiency curve because a fraction of the air stream penetrates the filter mat and is, in effect, filtered. These curves no longer follow impactor theory. The ineffectiveness of filters has been confirmed by several investigators including Dzubay *et al.*,[42] Walsh *et al.*,[46] Willeke,[65] Vanderpool *et al.*,[66] and Newton *et al.*[47]

Pressure and Flow Induced Size Changes

Large pressure drops in an impactor can lead to a change in the size of particles during sampling. These large pressure drops are found in impactors with high jet

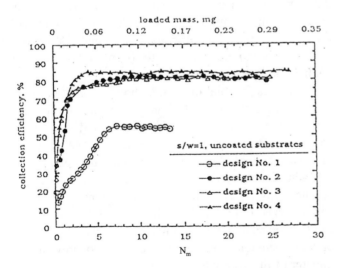

FIGURE 14-13. Collection efficiency versus the cross-sectional area of deposited particles relative to the jet-cross-sectional area, N_m and loaded particle mass for uncoated impaction surfaces. D_{pa} = 3.2 ± 0.04 µm; total running time = 90 minutes.[63] *Reprinted with permission of American Association for Aerosol Research.*

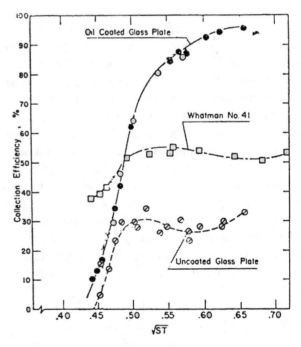

FIGURE 14-14. Collection characteristics of a single jet impactor for solid, polystyrene latex particles with uncoated, coated, and fiber filter impaction surfaces (adapted from Reference 45). *Reprinted from Rao, A.K.; Whitby, K.T.: Non-Ideal Collection Characteristics of Inertial Impactors I: Single-Stage Impactors and Solid Particles. J. Aerosol Sci. 9:77–86 (1978) with permission of Elsevier Science.*

velocities, as are needed to capture very small particles. The magnitude of the change depends on several factors: the chemical nature of the aerosol (its volatility and hygroscopicity); the ambient relative humidity; and the operating pressures and geometry of the impactor. For sufficiently high velocity jets, the flow expands as it accelerates, resulting in a decrease in temperature and corresponding increase in relative humidity. For hygroscopic particles, this can cause the particles to grow. High jet velocities also lead to reduced operating pressures of the collection stages, and this decreases the relative humidity. Correspondingly the size of a hygroscopic particle passing through this jet will first increase, and then decrease. Size changes are most pronounced for sampling of hygroscopic particles at relative humidities above about 80%.[67, 68]

Calibration

Several means have been used to calibrate impactor cutpoints. One can generate a monodisperse aerosol, which contains an easily analyzed compound such as fluorescein. After sampling these particle of a known size, the deposits on each collection stage are analyzed. Additionally, the walls of the impactor can be washed and analyzed to determine wall losses. The test is repeated for particles of many sizes to generate a complete calibration curve.

An alternative on-line calibration method is to measure particle concentrations immediately upstream and downstream of the test impaction stage. For impactors with cutpoints above 1 μm, it is possible to detect and precisely size the test particles with an on-line aerodynamic particle sizing instrument. The test stage is challenged with a polydisperse aerosol, and the efficiency curve is derived by comparing the upstream and downstream particle concentrations as a function of particle size. For impactors with cutpoints below 1 μm, the monodisperse particles are generated by electrical mobility selection. The concentrations above and below the test stage can be measured by an aerosol electrometer.[25] By electrically isolating each collection stage, the current corresponding to the depositing aerosol can be measured directly.[69]

Literature references to calibrations of commercially available impactors are listed in Table 14-I-1.

Impactor Operation: Guidelines for Use

The mechanical and theoretical simplicity of impactors has made them popular instruments for particle sampling; however, they are easily misused, leading

to the generation of erroneous data. Correct operation requires: 1) proper preparation and loading of the collection substrates; 2) leak-tight assembly of the instrument; 3) measurement and regulation of the flow rate; 4) appropriate choice of sample time; 5) a suitable inlet system; and 6) a precut device, where appropriate. Yet when properly deployed, direct field comparisons show that the data are consistent among impactors of similar design operated by different groups[70] and among impactors of different designs.[71] With proper calibration, impactor sizing is also consistent with that measured by physical sizing instruments.[72, 73]

Substrate Coatings and Preparation

One of the most critical factors in impactor operation is the preparation of the collection surface. Except for virtual impactors, sampling of solid aerosols requires an adhesive coating to prevent errors from particle bounce. Although many manufacturers supply fibrous filter substrates, these substrates degrade impactor performance and do not eliminate bounce-off, as discussed above.

The choice of the adhesive surface depends on the application. Greases work well for chemical or elemental analyses of nonorganic species. They have also been used for determining the size distributions of specific organic species. The size distributions of polycyclic aromatic hydrocarbon compounds have been measured using Vaseline-coated substrates.[74] Other commonly used adhesive greases are Apiezon L and M, Halocarbon (Halocarbon Products Co., 82 Burlews Court, Hackensack, NJ 07601), and Dow silicone (Dow Corning Co., Midland, MI 48686). Although these vacuum greases have the advantage that they do not volatilize during sampling, Vaseline has the advantage of lower blank values for sulfur and trace metals.[49] Various types of oils and greases that have been used in impactor applications are listed in Table 14-1.

Analyses for total organic carbon remain a problem because there are no noncarbon greases or oils. To date, these samples are collected on uncoated substrates, which are suspect except in cases such as the sampling of cigarette smoke, where the aerosol particles may be self-adhesive. In some cases, investigators have operated parallel, single-stage impactors with different cut-points and analyzed only the after-filters. The impactor stages can then be coated without interfering with the analyses, provided suitable precautions are taken to prevent any transfer of the grease to the after-filter.

Installation of Collection Substrates

In using any impactor, the operator must be careful to ensure that the impaction stages are installed correctly.

Jet-to-plate spacings often equal the diameter of the jet and can be quite small. An improperly installed collection surface that is too close to the jet or, worse yet, one that partially blocks the flow, can sharply affect the cutoff diameter. Impactor cutoffs are significantly affected when jet-to-plate spacings are less than 0.4 jet diameters. It is generally a good idea to inspect each stage before sampling to ensure proper installation.

Flow Rate Regulation

As with most inertial samplers, the particle cut-sizes depend on the sampler flow rate. Thus, proper operation requires a steady flow at a known rate. Simply knowing the sampled volume is not sufficient. Pumps that produce pulsating flows, such as some of the small diaphragm pumps, should not be used for impactor sampling because the cut-sizes will fluctuate. Likewise, a large drop in the flow rate during the course of sampling will affect the sharpness of the size fractionation. It is recommended that the operator measure the flow rate at the beginning and end of sampling. This can be done using a low pressure-drop device such as a rotometer or a dry-test meter placed at the inlet. Rotameter response depends on pressure, and readings will be different if the rotometer is placed downstream of the flow-regulating valve or orifice. If a rotometer is used on the low-pressure side, its pressure must be noted and the appropriate calibration applied.

Sample Duration

With impactors, it is possible to sample for too long a period as well as too short. Minimum sample durations are chosen on the basis of expected particle concentrations, analytical requirements, and substrate blanks. Maximum sample times are limited by the buildup of particle deposits on the collection surface. For sampling solid particles, greases can become ineffective at substrate loadings of a fraction of a monolayer, and particle bounce errors can reappear if sampling times are excessive. If a porous oiled substrate is used, the particle deposit can grow to be quite high, and jet-to-plate distances can decrease enough to lower the particle size cutoff. In some applications, the first impactor stage may become overloaded prior to collection of enough sample on subsequent stages. This problem can be avoided by use of a precutter, as discussed below.

Inlets

If impactor size distributions are to be representative, the sampler placement and inlet configuration must not exclude particles of the size range of interest. This can be quite significant for sampling large particles (greater than about 5 μm). Long stretches of tubing on the impactor inlet can cause unaccounted-for losses. If the impactor inlet is a small tube oriented perpendicular to the air currents in the room or atmosphere being sampled, larger particles will not follow the streamlines into the impactor. The problem is lessened by using a wider inlet with a lower intake velocity, or pointing the probe inlet into the flow. The accurate collection of coarse particles requires isokinetic sampling, as discussed in Chapter 20.

Precutters

In many applications, it is necessary to prevent very large particles from entering the impactor. This is accomplished by means of a precut device such as a cyclone or size-selective inlet. Precutters exclude large particles that would otherwise bounce or overload the first impactor stage and thereby distort the impactor size distribution measurement. They are appropriate for applications that call for size distributions below a specified particle diameter, such as in respirable or thoracic sampling.

Data Reporting

Impactors provide data on aerosol mass or chemical composition in one or more size ranges. To obtain mass or chemical species size distributions, one of several data reduction procedures can be employed. The approaches include: 1) histogram and cumulative plots based on stage cutoff diameters; 2) data inversion methods that take into account the shape of the collection efficiency curves; and 3) extraction of mass median diameters and distribution widths.

Histograms, such as those shown in Chapter 6, are a straightforward means of presenting impactor data. In this approach, each impactor stage is characterized by its cutoff diameter, and crossover between neighboring stages is neglected. This is the same as assuming infinitely steep collection efficiency curves. These graphs plot the quantity $\Delta M_i / \Delta \log d_a$ against $\log d_a$, where ΔM_i is the mass collected on the ith stage, d_a is aerodynamic diameter, and $\Delta \log d_a = \log(d_{50, i-1} / d_{50, i})$ is the difference between logarithms of the aerodynamic cutoff diameters, $d_{50, i-1}$ and $d_{50, i}$ for the stage immediately preceding stage i (labeled $i-1$) and stage i itself. The denominator $\Delta \log d_a$ is a normalizing factor, such that the area under the histogram is proportional to the mass collected. It also accounts for whatever nonuniformity may exist in the spacing of the impactor cut-sizes, so that the shape of the histogram reflects the mass distribution. The data reduction procedures are described in Chapter 6 and will not be repeated here.

The histogram presentations do not account for cross-sensitivity in the impactor calibration curves. They assume infinitely sharp collection efficiencies such that the impactor stage collects all particles at or above the cutoff and no particles below that size. Actual efficiency curves have a finite slope, and particles of equal size will collect on several stages. When the impactor calibration efficiency curves are known, it may be desirable to take them into account in the data reduction. These procedures are known as data inversion methods.

Data inversion techniques have been applied to a variety of problems for which instrument responses are multivalued. There is no unique solution to the inversion problem. Mathematically, it is possible to have several mass distributions yielding the same loadings on the impactor stages. The inversion methods that have been developed are constrained to produce physically reasonable solutions. Inversion results are to be considered "best estimates" and will vary somewhat depending on the algorithm used.

One of the more widely used inversion methods for aerosol instruments is that of Twomey.[75] The data of Figure 14-3 were reduced using this method, and they show a smooth curve for the chemical species size distributions. Another method with similar output is that of Wolfenbarger and Seinfeld.[76] Hasan and Dzubay[77] developed an inversion method that assumes a lognormal form for the aerosol size distribution. The accuracy of these methods depends on how well the efficiency curves are known.

Sometimes the investigator is not interested in the details of the aerosol size distribution but simply wishes to extract certain parameters, such as the mass median diameter or the fraction of aerosol in the respirable size range. These calculations are facilitated by presenting the data in terms of a cumulative distribution, as shown in Chapter 6. Cumulative distributions display the percentage of the aerosol in particles with diameters equal to or smaller than the diameter indicated.

Aerosol size distributions can often be approximated by lognormal distributions, which have a Gaussian shape when displayed against the logarithm of the particle diameter. When the cumulative distribution is plotted on a log-probability graph, the result is a straight line. These plots are useful for evaluating whether a distribution is lognormal and for extracting the median diameter and the geometric standard deviation, which is the measure of the width of the distribution. For lognormal distributions, the mass median and count median diameters are related through the geometric standard deviation. For a detailed treatment of this approach to the analysis of impactor data, refer to Hinds[78] and Chapter 6.

For measurement of bioaerosols, different data reduction procedures are used, as discussed in Chapter 22.

For example, when sampling with multijet impactors, one can count the number of "colony forming units," i.e., the number of impactor deposit spots that result in the formation of a colony on a multijet impaction stage. If the airborne concentration of microbial particles is small, only a small fraction of the holes on a impactor stage will result in a colony-forming unit. Methods for relating the number of "colony forming units" to the airborne concentrations are discussed by May,[37] Macher,[79] and Somerville and Rivers.[80]

Special Types of Impactors

Micro-orifice and Low-Pressure Cascade Impactors

Traditional impactors do not offer much size resolution for submicrometer particles; typically, their finest cut-size size is around 0.4 μm. Yet for many aerosol applications, it is useful to be able to size-segregate smaller particles. Diesel emissions, welding fumes, cigarette smokes, and photochemically generated smog aerosols typically exhibit mass median diameters between 0.1 and 0.6 μm. When sampling these aerosols with a conventional impactor, 50% or more of the aerosol mass can penetrate the final impactor stage. Although the material can still be collected on a back-up filter, the filter gives no size resolution; as a result, the investigator has no size information on a substantial portion of the sample. Often a lower cutoff diameter of 0.1 μm or less is needed to size-segregate the majority of the aerosol mass.

Two types of impactors have been developed to obtain smaller cutoff diameters: low-pressure impactors and micro-orifice impactors. Both instruments can provide size cut-points as small as 0.05 μm. Low-pressure impactors were introduced more than 30 years ago,[81] and a variety of these samplers are in use today. Examples are the ambient impactors of Berner[4, 82, 83] and Hering et al.,[84, 85] and the in-stack sampler of Vanderpool et al.[86] Micro-orifice impactors have been developed at the University of Minnesota[87, 88] and have been further investigated by Gudmundsson et al.[89]

Low-pressure and micro-orifice impactors use two different approaches to achieve their small particle-size cutpoints. Low-pressure impactors resemble ordinary impactors but operate at reduced pressures of 5–40 kPa (0.05–0.4 atm). They take advantage of the decreased aerodynamic drag on particles that occurs when the mean free path in the air is as large or larger than the particle diameter. Micro-orifice impactors operate closer to atmospheric pressure (0.8–0.9 atm), but employ very small orifices (40–200 μm in diameter). The streamlines of the air impinging on the impaction plate have correspondingly smaller radii of curvature; the air is accelerated more quickly, making it more difficult for the particles to follow. The basic operating principles for both types of impactors are evident from the particle Stokes number, defined in Equation 6 above. To collect small particles, the quantity (W/CV) must be small. Low-pressure impactors operate at large values of the slip factor C; micro-orifice impactors operate at small jet diameters W. Both types of impactors use relatively high jet velocities ($V > 100$ m/s) for the particle cutpoints of 0.1 µm and lower. Micro-orifice impactors have multiple jets per stage to accommodate the flow.

The pressure drops in an impactor can lead to a change in the size of particles during sampling. The magnitude of the change depends on the chemical nature of the aerosol, namely its volatility and hygroscopicity, on the ambient relative humidity, and on the operating pressures and geometry of the impactor as discussed above. For micro-orifice multijet impactors, the collection efficiency curves are dependent on the jet Reynolds number and the ratio of the distance between the orifice plate and the impaction plates to the orifice diameter.

Virtual Impactors

Virtual impactors do not have a collection plate. Instead, an axial probe is placed below the impactor jet. Only a small fraction of the flow passes through the probe; the majority of the flow bends around the tip of the probe to pass onto the next stage. The streamlines above the probe tip resemble those of a conventional impactor, and the particles are separated by size into the two air streams. One is the minor flow, which passes through the probe; the other is the major flow, which bypasses the probe. The minor flow through the probe carries with it all of the large particles from the total sample flow plus the small particles from the minor flow. The major air flow that bypasses the probe contains smaller particles only. Particles are collected by filtration of the two air streams.

A major advantage of virtual impactors is that they are not subject to errors resulting from particle bounce or re-entrainment, and grease coatings are not required. Aerosols may be collected on whatever filter medium is best suited for the analyses to be performed. A limitation is that unless they are carefully designed and constructed, they are subject to significant wall losses for liquid particles near the cutpoint size.[90, 91] Experimentally determined criteria for minimizing these losses are given by Loo and Cork.[92]

The first type of virtual impactor was the "aerosol centripeter" introduced by Hounam and Sherwood.[93] The most widely used virtual impactor is the dichoto-

mous sampler shown in Figure 14-15.[94] This instrument was introduced by Conner[95] and developed by Dzubay and Stevens[96] and Loo et al.[97] It operates at a sample rate of 16.7 L/min, or 1 m³/hr with aerosol collection onto two 37-mm filters. The commercially available instrument provides a fine particle cut at 2.5 µm, although earlier versions had a 3.5-µm cut. The unit is generally operated with a PM_{10} inlet as described in Chapter 5, and it is most frequently used for ambient air monitoring. Calibration curves are given by McFarland et al.[94] and John and Wall.[90]

Collection efficiency curves for the virtual impaction stage of the dichotomous sampler are shown in Figure 14-16.[94] For particle sizes below the cutoff diameter, collection efficiencies reach a minimum value equal to the fraction of the total flow passing through the receiving probe. The cutoff diameter decreases as the fraction of the flow through the receiving probe is increased. Wall losses are most significant at the cutoff diameter. A critical factor in minimizing wall losses is the radius of curvature at the inlet of the receiving probe. John and Wall[90] found that alignment of the jet and receiving probe is critical and that deviations of more than 0.05 mm in concentricity can increase wall losses and affect the cutpoint.

Several other virtual impactors have been developed. Solomon et al.[98] and Marple et al.[99] report a high-volume virtual dichotomous sampler that operates at 500 L/min and employs 100-mm-diameter filters. This sampler has the advantage of providing larger sample volumes, permitting analyses of trace species, or facilitat-

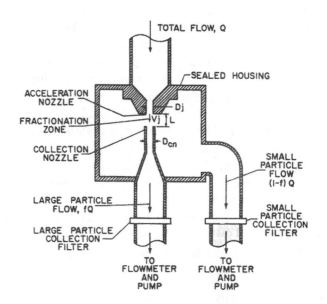

FIGURE 14-15. Dichotomous sampler, showing the fine particle (2.5-m cutpoint) virtual impaction stage.[94] *Reprinted with permission of American Chemical Society.*

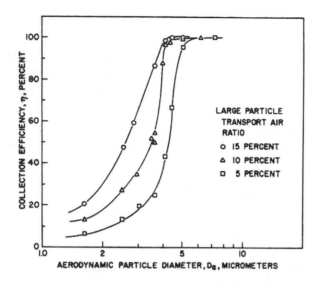

FIGURE 14-16. Virtual impaction efficiencies for large particle transport air rations of 5%, 10%, and 15%. *Reprinted from McFarland, A.R.; Ortiz, C.A.; Bertch, R.W.: Particle Collection Characteristics of a Single-Stage Dichotomous Sampler. Environ. Sci. Technol. 12:679–682 (1978) with permission of Elsevier Science.*

ing collection in cases of low airborne concentrations. Marple *et al.*[100] report a virtual impactor system for separating coal dust from diesel exhaust in mines. Novick and Alvarez[101] have designed a three-stage virtual impactor. Sioutas *et al.*[10, 102] developed a small cutpoint virtual impactor that is used for concentrating submicrometer particles for exposure studies to assess health effects associated with the inhalation of fine, ambient particles.

Several investigators[103, 104] have developed virtual impactors that use a particle-free core air stream to eliminate the fine particle collection in the minor (coarse) particle flow. These have been used for particle size classification, that is selecting particles in a specific size fraction,[105] and for reducing the phantom particle counts in aerodynamic sizing instruments.[106] Experimental optimization of the position and diameter of the clean air flow core has been investigated by Li and Lundgren.[107]

Noone *et al.*[108] developed a counterflow virtual impactor for separately sampling cloud droplets and interstitial aerosol. Like the clean-air core virtual impactors, the counter flow impactor has no fine particle in its coarse particle collection.

Theoretical analyses of virtual impactors are given by Forney[109, 110] and Marple and Chien.[111] and Asgharian *et al.*[36] Practical guidelines for virtual impactor design are reviewed by Loo and Cork.[92]

Impactors for Coarse Particle Sampling

Several types of impactors have been designed for collection of particles in the 10- to 100-μm size range.

They are used for collecting fogs, cloudwater, and coarse airborne particles. Collett *et al.*[112] report a three-stage cloud impactor for size-resolved measurement of cloud drop chemistry. One of the first large particle collectors is the WRAC.[66] It uses four parallel impactors with rectangular slots. Each samples at 55 scfm, using a standard Hi-Volume blower. Particles are collected on greased metal foils. Two independent calibrations give equal cutpoints for liquid and solid particles of 9, 18, 34, and 48 ± 1 μm.[66] Collection efficiencies on ungreased plates or fiber filters are 7% because of bounce.

Rotary impactors are another type of coarse particle collector. These samplers collect particles on a rapidly rotating rod or tube that moves through the air much like a large propellor. There are no jets and no accelerated air streams. Instead, the relative motion is achieved by rotation of the collection surfaces. Air streamlines bend around the collection surface, and particles too large to follow are intercepted.

The Noll rotary impactor,[113, 114] designed for atmospheric coarse particle sampling, has four stages with size cuts from 6 to 29 μm. Collection surfaces are external and greased to prevent particle bounce. Deposits are analyzed gravimetrically and microscopically. Air sampling rates are inferred from the rotational speed and collection surface cross-sectional area. Regtuit *et al.*[115] developed a tunnel impactor that uses a tube pointing into the wind with four parallel impactors mounted inside. The tube is suspended to give a well-defined sampling rate. Calibration data show cutpoints between 10 and 60 μm.

A rotating coarse particle sampler developed at the University of Minnesota[116] uses internal collection surfaces and a rotating L-shaped sampling probe that is aspirated at a speed equal to the speed of the probe tip. Particles are deposited at the elbow inside the probe. Comparative tests with open-faced filters and sedimentation plates indicate greater than 90% collection efficiencies for particle diameters between 40 and 250 μm.

As with conventional impactors, sampler collection efficiencies can be calculated from the particle Stokes number and flow Reynolds number. The size fraction collected is dependent on particle aerodynamic diameter, instrument geometry, and rotational speed. For a rotating rod, narrower widths and higher rod velocities give smaller cutoff diameters. Some collectors use a slot in the leading edge of the rotating rod. Examples are the fog water collectors of Mack and Pillie,[117] Kramer and Schultz,[118] and Jacob *et al.*,[119] and the large-particle sampler of Tomic and Lilienfeld as reported by McFarland *et al.*[120] The performance of these collectors also depends on Stokes numbers, as has been modeled by Lesnic and coworkers.[121]

Pre-cutters

Single-stage impactors are often used as inlets for size-selective sampling. Examples include PM_{10} inlets used in Hi-Volume and dichotomous samplers[122–124] and the indoor air sampling impactor.[125] Some personal samplers for PM_{10} and $PM_{2.5}$ measurements also use a single-stage impactor.

Diffusion denuder systems, as discussed in Chapter 19, also use either a cyclone or a single-stage impactor that is precut to remove particles above 2 or 3 µm. One example is the glass impactor[126] used in the Harvard annular denuder system.[127] This impactor was especially designed to pass nitric acid.

Real-Time Impactor Sensors

Some impactors have been equipped with real-time sensors on each collection plate for providing direct-reading size distribution. The QCM uses quartz crystal impaction surfaces, which detect the mass of deposited aerosol.[128–130] Tropp et al.[131] and Keskinen et al.[132] have designed "electrical" impactors that use a charger and impactor in series. Deposited particles are detected by the electrical current. These impactors are described in more detail in Chapter 15.

Inertial Spectrometer

The inertial spectrometer illustrated in Figure 14-17 was developed by Prodi and coworkers.[133, 134] With this device, particles are separated by size in a laminar air flow and then collected by filtration. Aerosol is injected into a clean air flow in a rectangular channel immediately upstream of a 90° bend. Particles are separated aerodynamically in the bend and then collected on a membrane filter. The position of particle deposition on the filter corresponds to aerodynamic size. Aerosol sample rates are < 0.1 L/min. Total flow rates, including the sheath air, are 3–10 L/min.

The instrument provides aerodynamic separation for particles in the 1- to 10-µm size range. Particles outside this range are collected, but not separated by size. Unlike conventional impactors, the inertial spectrometer uses filtration for particle collection and is not susceptible to particle bounce sampling errors. The basic theory of operation for this instrument is given by Prodi et al.[133] and Belosi and Prodi[134] and has been modified by Aharonson and Dinar[135] to include gravitational effects. As with other inertial instruments, the sizing by the inertial spectrometer is dependent on the aerodynamic diameter of the particle. Applications include the measurement of fibrous particles.[136] Calibration data under various operating conditions are given by Mitchell and Nichols.[137]

Impingers

Impingers operate much like an impactor, except that the sampled air stream jet is immersed in water at the bottom of a flask. The sampled air stream is accelerated in the impinger orifice to velocities of 60 m/s or greater. The air stream exits underneath the liquid surface immediately above an impaction plate or at a specified distance above the bottom of the collection flask. Particles impinge on the plate or flask bottom, stop, and are subsequently retained by the liquid.

Impingers were developed in 1922[138] and until 1984 were recommended by ACGIH for dust counting. "Dust counting" is the determination of the particle number concentration (i.e., millions of particles per cubic foot of air) for particles such as graphite, mica, and mineral wool fibers. The actual number concentration of insoluble particles collected by an impinger is determined by microscopic examination of an aliquot of the sample using a dust-counting cell to immobilize the liquid in a 0.1-mm layer between two glass surfaces. Dust concentrations measured in this way were correlated with the incidence and severity of respiratory disease in trades such as mining, quarrying, smelting, and the manufacture of metallic and mineral (stone or clay) products. ACGIH also used such dust concentration measurements to set more than a dozen threshold limit values (TLVs®) for occupational exposure. These TLVs

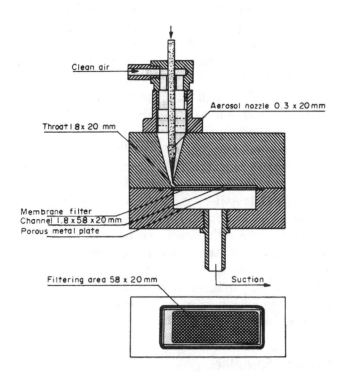

FIGURE 14-17. Cross section of the inertial spectrometer showing aerosol and clean air flows, and the membrane filter collection surface.[136] *Reprinted with kind permission of Elsevier Science.*

have since been converted from particle number concentrations to respirable mass concentrations, and impinger sampling for particles has been largely replaced by respirable mass sampling, as described in Chapter 5.

Although developed for dust counting, impingers are now also used for the collection of gases, vapors, acid mists, and viable aerosols. They are used for sampling toxic organic vapors such as formaldehyde, as described in Chapter 16. They are also used to collect moisture and condensible materials in the U.S. EPA Method 5 particulate stack sampling train, as described in Chapter 20. The use of impingers for viable particles, such as bacteria, is described in Chapter 22. For vapor sampling, modifications of the impinger include a spill-proof design and/or use of a fritted glass in place of the orifice.

Dust counting impingers include the Greenburg–Smith impinger,[138–140] which used a 2.3-mm-diameter jet located 5 mm above an attached impinging plate. It was designed to operate at a flow rate of 28 L/min (1 cfm) with a jet velocity of 100 m/s and collected particles greater than 1 μm in diameter. The Hatch modification[141, 142] of this impinger used the flat bottom of the collection bottle for the impinging surface; however, it had the same size jet, flow rate, and performance characteristics as the standard Greenburg–Smith impinger.[143] The midget impinger[144, 145] was developed as a more portable instrument. It used a smaller jet (1 mm in diameter) and operated at 2.8 L/min. It had a lower jet velocity of 60 m/s and was operated with a smaller pump. The midget impinger is also used for vapor sampling, and it is available in spill-proof designs or with fritted glass in place of the orifice (see Chapter 16).

The all-glass impingers, such as the AGI-4 (All Glass Impinger, 4 mm) and AGI-30, are used for sampling microbial aerosols. Both operate at 8.5–12.5 L/min, corresponding to 70% to 100% of sonic velocity at the jet exit. The AGI-4 uses a submerged jet located 4 mm above the bottom of the collection bottle. The exit of the AGI-30 is 30 mm above the bottom of the collection bottle and is generally operated such that the level of the collection liquid is a few millimeters below the jet exit. Although not as efficient a collector as the AGI-4, the AGI-30 is gentler and bacterial cells are not as likely to be shattered or damaged during collection. Impingers for bioaerosol collection are described in Chapter 22.

Cyclone Samplers

Cyclone samplers use a vortical flow inside a cylindrical or conical chamber. A typical "reverse flow"

cyclone is illustrated in Figure 14-18. Air is introduced tangentially near the top, creating a double vortex flow within the cyclone body. The flow spirals down the outer portion of the chamber and then reverses and spirals up the inner core to the exit tube. Particles having sufficient inertia are unable to follow the air streamlines, and they impact onto the cyclone walls. The particles are either retained on the cyclone walls, or they migrate to the bottom of the cyclone cone. The work of Ranz[147] shows that a wall flow in the boundary layer plays an important role in transporting the particles along the walls to the collection cup at the bottom.

Usually cyclones are used to provide a particle precut to another aerosol collector. One of the most common applications is respirable particle sampling, wherein the cyclone is operated upstream of a filter. The cyclone is used to remove the larger, nonrespirable particles such that the material collected on the filter is representative of that penetrating into the nonciliated deep lung spaces of humans. The filter can be assayed for particle mass or specific chemical compounds. Usually the material collected in the cyclone itself is not assayed. There has been considerable research on the development and calibration of cyclones to mimic the ACGIH respirable curve, as described in Chapter 5.

Another application of cyclones is for ambient monitoring of $PM_{2.5}$, which refers to the measurement of particles below 2.5 μm in aerodynamic diameter. Cyclones have been used for many years as the precutter for $PM_{2.5}$

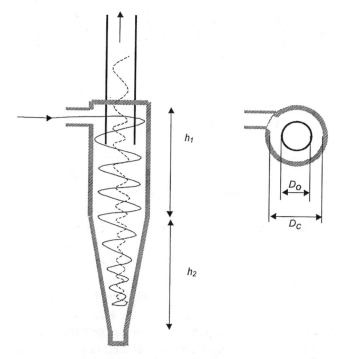

FIGURE 14-18. Flow patterns and characteristic dimensions in a cyclone collector. (drawn by S. Hering)

in the IMPROVE network, which monitors visibility-reducing particles throughout the National Parks in the United States.[148] It is also used in many special studies in urban areas, such as those done in the Los Angeles,[149] the western United States,[150] and California's central valley.[151] The separation of $PM_{2.5}$ from larger particles is of interest because these smaller particles are often derived from anthropogenic sources, and are the primary constituent in urban haze. Measurement of $PM_{2.5}$ has taken on greater meaning with the EPA promulgation of an ambient air standard for the mass of $PM_{2.5}$. Although the Federal Reference Method for $PM_{2.5}$ does not use a cyclone, many of the proposed speciation samplers, for assaying chemical composition of $PM_{2.5}$ use cyclone precutters.

Another application of cyclones is for stack sampling. Smith et al.[152] developed a cascaded cyclone system consisting of five cyclones arranged in series much like a cascade impactor. Each cyclone collects a smaller size fraction. This system was designed for collection in high-temperature process streams, as described in Chapter 20.

Cyclones have several advantages in air sampling, including their relatively low cost of construction and ease of operation. They have no moving parts, and they are easily maintained. Unlike impactors, they are not as subject to errors caused by particle bounce or re-entrainment. They are especially suited to applications requiring the removal of coarse aerosols prior to sample collection.

Theory of Operation

Cyclones are characterized by a collection efficiency curve much like that described for impactors in the preceding section. Again, the particle size collected with a 50% efficiency is referred to as the cutpoint of the cyclone, or d_{50}. As with impactors, the collection efficiency depends on the particle aerodynamic diameter. The two points of interest in cyclone design are: 1) how the cyclone cutpoint depends on the cyclone dimensions and flow rate; and 2) the shape of the collection efficiency versus particle size curve.

Over the years, several theories have been proposed to predict cyclone performance. Because the flow pattern inside cyclones is complex, the particle collection characteristics are not easily modeled. The various cyclone theories present somewhat different expressions for the dependence of cyclone cutpoints on the cyclone dimensions, flow rate, gas viscosity, and temperature. Some conventional theories include as a parameter the effective number of turns the air flow makes within the cyclone, which is largely unknown. Summaries and

comparison with experimental data are given by Chan and Lippmann,[153] Leith and coworkers,[154–157] DeOtte,[158] and Moore and McFarland.[159]

At present, there is no generally accepted fundamental relationship to describe cyclone performance. One must rely on empirical correlations of the dependence of cutpoint on flow rate and cyclone geometry.

Types of Cyclones and Calibration Data

Many of the types of cyclones used for airborne particle sampling are listed in Table 14-I-2. The 10 mm, or "Dorr–Oliver" cyclone, is often used for personal sampling of respirable particles.[160] The HASL cyclone has a similar reverse-flow configuration but is somewhat larger, and has been used in ambient air sampling networks.[150] The SRI cyclones were designed as a five-stage cascade cyclone system for stack-sampling.[152, 161] The AIHL cyclone[162] follows the design of the midsize SRI cyclone and has been used in several ambient sampling studies. A large Stairmand cyclone has been designed for sampling cotton dust, as a substitute for the cotton dust elutriator.[163]

The cutpoints of many of these cyclones have been measured with laboratory-generated particles. Calibration techniques are similar to those described above for impactors. One method is to generate a monodisperse aerosol of an easily assayed material such as fluorescein, and assay the penetration, (e.g., Moore and McFarland,[159] Tsai and Shih[164]). Another is to count particle penetration as a function of size for a polydisperse challenge aerosol (e.g., Bartley, et al.[165]; and Maynard and Kenny[166]).

There have been many calibrations of the 10-mm, "Dorr-Oliver" cyclone.[160, 167–170] Recent calibration curves for two of the more commonly used respirable cyclones, the 10-mm cyclone and the SKC cyclone, are shown in Figures 14-19a and 14-19b.[164] Data for the 10-mm cyclone are given for a flow rate of 1.7 L/min, used to match the original 1984 ACGIH respirable criteria; and for a flow rate of 1.3 L/min, chosen to match the new 1993 ISO/ACGIH criteria. Data for the SKC cyclone are shown for a range of flow rates. Additional literature references to laboratory calibrations for other cyclones[171–173] are listed in Table 14-I-2.

Empirical Correlations for Predicting Cyclone Performance

The characteristics of cyclone performance are perhaps best illustrated by experimental data. The effect of flow rate is illustrated in Figure 14-20, which shows

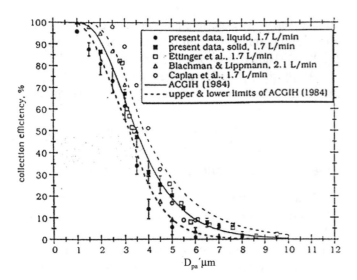

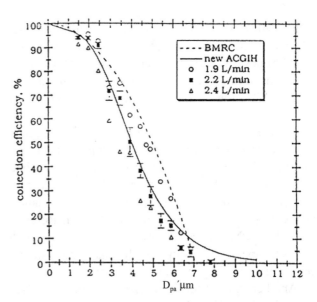

FIGURE 14-19a. Comparison of present particle collection efficiency data with previous work, 10-mm nylon cyclone sampler. *Reprinted from* Tsai, C.J.; Shih, T.S.: Particle Collection Efficiency of Two Personal Respirable Dust Samplers. Am. Ind. Hyg. Assoc. J. 56(9): 911–918 (1995) *with permission of American Industrial Hygiene Association Journal.*

FIGURE 14-19b. Comparison of solid particle collection efficiency at different flow rates. SKC cyclone sampler. *Reprinted from* Tsai, C.J.; Shih, T.S.: Particle Collection Efficiency of Two Personal Respirable Dust Samplers. Am. Ind. Hyg. Assoc. J. 56(9): 911–918 (1995) *with permission of American Industrial Hygiene Association Journal.*

collection efficiency curves for the AIHL cyclone for six different flows ranging from 8.4 to 26.6 L/min. These data, taken from John and Reischl,[162] are for a short cone cyclone with a body diameter of 3.66 cm. As is readily apparent, the cyclone cutpoint decreases with increasing flow rate. The shape of the collection efficiency curve appears steeper at the higher flow rates. However, when plotted as a function of the normalized particle diameter, $(d_p - d_{50})/d_{50}$, a common collection efficiency curve describes the behavior at all flow rates. This is shown in Figure 14-21.

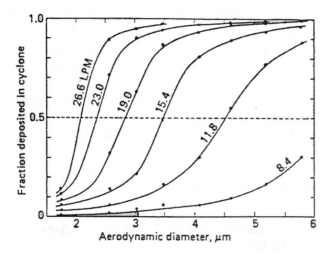

FIGURE 14-20. Fraction of solid particles deposited in the AIHL cyclone as a function of aerodynamic diameter. The curves are labeled with the flow rate.[162] *Reprinted with permission of Air & Waste Management Association.*

The flow rate dependence of a cyclone of fixed geometry is often described by the relation:

$$d_{50} = K Q^n \qquad (11)$$

where: d_{50} = cutpoint
Q = flow rate
n and K = empirically determined constants

The values of K and n vary for different cyclones. Several investigators[152, 153, 160, 162, 174, 175] have correlated cyclone cutpoints with flow rate according to this relation. Resulting values of K and n are given in Table 14-2.

An exception to this simple dependence of cutpoint on flow rate is the 10-mm cyclone. Table 14-2 lists several values of K and n corresponding to different ranges in flow rate. This is because the 10-mm cyclone performance is not described by a single curve. As described below, the outlet flow loses its vorticity and becomes laminar at low flow rates.

Several correlations have been advanced to predict cyclone performance as a function of cyclone body size and flow rate. Generally, smaller cyclones of the same proportions yield smaller cutpoints at the same flow rate, as shown by Smith *et al.*[152] Saltzman[176] correlated cutpoints of many different cyclones to body diameter and the Reynolds number in the outlet flow (see Equation 9 for definition of Reynolds number). Leith and coworkers[156, 157] used flow visualization methods to better understand and model cyclone performance. Kim and

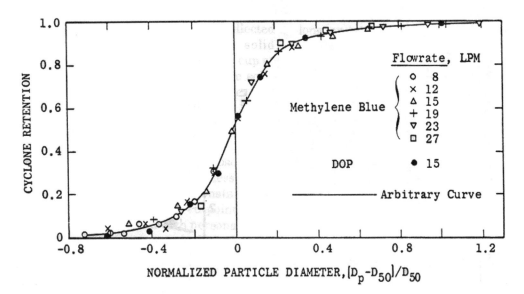

FIGURE 14-21. Particle deposition in the AIHL cyclone taken at various flow rates versus the normalized particle diameter. Methylene blue data points refer to solid particle collection, DOP points are for liquid particle collection.[162] *Reprinted with permission of Air & Waste Management Association.*

Lee's systematic experimental evaluation of nine different cyclones at three flow rates provides qualitative guidelines for understanding cyclone performance.[177] These studies are of special interest in the design of new cyclones.

Moore and McFarland[159, 178, 179] developed a correlation for predicting the performance of Stairmand-type cyclones of varying size at different flow rates. Stairmand cyclones have the simple geometry shown in Figure 14-18. They have a cylindrical body section of diameter D_c and height $h_1=2D_c$, and a conical section of height $h_2=2D_c$. Moore and McFarland investigated the effect of changing the body diameter D_c and the outlet tube diameter D_o. They defined a flow Reynolds number

$$Re_f = \frac{\rho (D_c - D_o)U_i}{2\mu} \qquad (12)$$

where U_i is the velocity of the air at the inlet, D_c is the radius of the cyclone body and D_o is the diameter of the outlet tube. The parameters ρ and μ are the air density and viscosity, respectively. The inlet velocity U_i is important because it persists into the body of the cyclone. The dimension $(D_c - D_o)/2$ is the distance inside the cyclone between the outlet tube and the cyclone wall. This dimension characterizes the radial constraint of the flow inside the cyclone. The dependence on flow rate is reflected through the value of U_i.

For six different cyclones, with three different body diameters D_c and three different outlet tube dimensions D_o, Moore and McFarland found the dependence of the

aerodynamic cutoff diameter D_p on flow rate, D_c and D_o is described by the relation:

$$\ln\left(\frac{D_{0.5}}{D_c}\right) = \ln(a) + b\ln(Re_f) \qquad (13)$$

The constants a and b are derived from an empirical fit to the data. Results for single-inlet and multiple inlet cyclones are shown in Figure 14-22.[178, 179] For a long-body, single inlet cyclone (with $h_1=h_2=2D_c$,) the fit to Equation 13 gives $a=0.0517$ and $b=.812$. They found the same form of the correlation also holds for multiple inlet cyclones with long bodies ($h_1=h_2=2D_c$) or short bodies ($h_1=h_2=D_c$), although the values of a and b are somewhat different, as shown in Figure 14-22.[179]

Kenny and Gussman[180] evaluated the performance of two families of cyclones, those of the SRI cyclone design, and those of their own design called GK cyclones. Their GK cyclones resemble one of the smaller SRI cyclones except that the collection cup at the bottom of the cyclone is smaller. They tested cyclones of several sizes for each of these designs and fit the aerodynamic cutpoint diameter D_{50} by the following empirical relationship:

$$\ln(D_{50}) = a + b\ln(D_c) - c\ln(Q) \qquad (14)$$

where D_c is the body diameter and Q is the volumetric flow rate. This relationship is similar to that of Equation 11 only it takes into account changes in the overall size of the cyclone, as indicated by the body diameter D_c. Kenny and Gussman also compared their data to the formulation

TABLE 14-2. Cyclones and Their Performance Characteristics

Description	Manufacturer	Cyclone Name	Flow Rate Range (L/min)	d_{50} Range (µm)	Internal Dimensions		Coefficients[1] $d_{50} = KQ^n$		Reference Author (Ref. No.)
					Body (cm)	Outlet (cm)	K	n	
14-31	BGI	Respirable	2.3	4.0	—	—	—	—	Bartley (1994) Maynard & Kenney (1995)
	BGI	GK-2.05	4	2.5	2.05	0.4	12.2	-1.14	Kenney & Gussman (1998)
14-32, 14-33	GRA, ITP	SRI V	7–28	0.3–2.0	1.52	0.36	14.0	-1.11	Smith et al.[152]
14-32, 14-33	GRA, ITP	SRI IV	7–28	0.5–3.0	2.54	0.59	17.6	-0.98	Smith et al.[152]
14-32, 14-33	GRA, ITP	SRI III	14–28	1.4–2.4	3.66	0.83	22.7	-0.84	Smith et al.[152]
—	—	AIHL	8–27	2.0–7.0	3.66	1.05	52.48	-0.99	John and Reischl[162]
14-32, 14-33	GRA, ITP	SRI II	14–28	2.1–3.5	4.13	1.05	22.2	-0.70	Smith et al.[152]
	—	Aerotec 3/4	22–55	1.0–5.0	4.47	0.75	214.17	-1.29	Chan and Lippmann[153]
14-32, 14-33	GRA, ITP	SRI I	14–28	5.4–8.4	5.08	1.50	44.6	-0.63	Smith et al.[152]
14-34	MSA, SEN	10-mm cyclone	0.9–5	1.8–7.0	1	0.25	6.17	-0.75	Blachman & Lippmann[160]
14-35	SEN	1/2″ HASL	8–10	2–5	3.11	0.50	—	—	—
14-35	SEN	1″ HASL	65–350	1.0–5.0	7.6	1.09	123.68	-0.83	Chan and Lippmann[153]
	—	BK 76	400–1100	1.0–3.0		3.8	221.48	-0.77	Beeckmans and Kim[131]
		Casella	0.9–1.9	3.5–6			11.32	-.715	Ogden, 1983, Maynard, 1995

[1]Q is flow rate in L/min, K and n are constants. Units of K are $\mu m\,(L/min)^{-n}$, see text, equation 11.

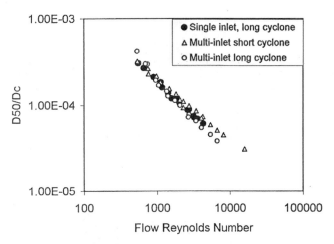

FIGURE 14-22. Ratio of cyclone cutpoint diameter to body diameter for three types of cyclones as a function of the flow Reynolds number defined in Equation 12. Data from Moore and McFarland, 1983, 1996.[178, 179]

of Moore and McFarland and found that those cyclones of the GK family fell on a common curve when plotted as a function of the flow Reynolds numbers. However, for otherwise similar cyclones, those with longer cones or larger outlet diameters had larger cutpoints.

The effect of temperature on cyclone performance is of interest for sampling stack gases and high temperature streams. Air viscosity increases with temperature, which in turn increases the cyclone cutpoint. The data of Smith et al.[152] on cyclones used for stack gas sampling show that cutpoint increases in direct proportion to the increase in the gas viscosity. It becomes evident that even for a fixed geometry, cyclone performance is not solely dependent on particle Stokes number.

Comparison of Solid and Liquid Particle Collection Efficiencies

Unlike impactors, cyclones are not easily subject to errors due to particle bounce. John and Reischl[162] calibrated their AIHL cyclone with both liquid and solid particles, and they found very little difference in the overall cyclone collection efficiency as shown in Figure 14-21. However, they observed that the region of the cyclone in which the particles deposited was quite different for the two aerosol types. Liquid particles tended to remain on the walls of the cyclone and were collected within the cyclone body. In contrast, the solid methylene blue particles tended to collect in the cup at the bottom of the cyclone cone. The heavier the solid particle-loading within the cyclone, the greater the proportion deposited in the cup. Moore and McFarland[179] also investigated collection of large, solid particles in their cyclones and found that it was generally less than 1%. In

contrast, Tsai and Shih[164] observed some differences in liquid and solid particle collection curves for the 10-mm cyclone, as shown in Figure 14-19.

Sources of Sampling Errors for Cyclones

Cyclones are one of the easier types of aerosol sampling instruments to use properly; nonetheless, some words of warning are appropriate. First, a constant flow rate is needed to ensure a constant cutpoint. Diaphragm pumps used in conjunction with personal samplers produce a fluctuating flow, which degrades the cyclone cutpoint. These oscillations are not necessarily eliminated by the commonly used pulsation dampers. The pulsations degrade the cutpoint[164, 165] and can change the penetration through a personal cyclone.[181]

For nylon and other nonconducting plastic cyclones, the particle collection efficiency can be influenced by electrostatic effects,[182] leading to retention of small particles. If the cyclone carries a net charge, particles of the same charge will be repelled. Briant and Moss[183] demonstrated that particles of like charge can be repelled by the electric field surrounding the cyclone and thus are not sampled efficiently. Obviously, this effect is more pronounced when sampling charged aerosols. Electrostatic effects can bias the collection of net neutral aerosols because "neutral aerosols" contain many charged particles. This artifact could be eliminated by using metal or electrically conducting plastic for the construction of cyclones.

Although the empirical correlations presented above offer a means to predict cutpoints at flow rates other than those previously measured, these relationships are not always applicable. For the 10-mm Dorr–Oliver, Lippmann and Chan[184] noted that the plot of cutpoint versus flow rate does not give the simple power law dependence observed for other cyclones. Saltzman and Hochstrasser[185] also noted that the shape of the collection efficiency curves for the 10-mm cyclone differs significantly for flows above and below 5 L/min, with the sharpness of the cutpoint greatly reduced at the higher flows. This has been attributed to a flow instability in the cyclone that leads to laminar flow in the outlet tube at low flow rates, as observed by Saltzman and Hochstrasser.[185] Thus caution must be used such that empirical correlations are not used beyond the flow rates for which they were derived.

Although cyclones are designed to collect large particles, they can also be a sink for reactive gases. This is of concern in gaseous sampling systems that employ a cyclone upstream of the gaseous collection. Often this arrangement is used for diffusional collection systems such as the annular denuder, the transition flow reactor, or the denuder difference method described in Chapter 19.

Appel *et al.*[186] studied the penetration of nitric acid through several types of Teflon and Teflon-coated cyclones and found that losses were as high as 40%–70% for freshly cleaned cyclones, but were small for those preconditioned by operation in ambient air.

Aerosol Centrifuges

Description

Aerosol centrifuges refer to a class of aerosol samplers that spin at high rotational speeds in order to subject particles to large centrifugal forces. One example is the spiral centrifuge[187–189] illustrated in Figure 14-23.[190] This sampler has a long, narrow duct that starts near the center of a flat plate, and spirals towards the outside edge. Aerosol and particle-free sheath air are introduced into the duct at the center, with the aerosol flow confined along the inner wall of the spiral. The air flow travels along the spiral exits at the outer edge. During operation, the plate spins at a typical rate of 3000 revolutions per minute. The centrifugal forces drive the particles through the particle-free sheath air toward the outside wall of the spiral. The position of the deposit along the spiral is a function of the particle aerodynamic diameter.

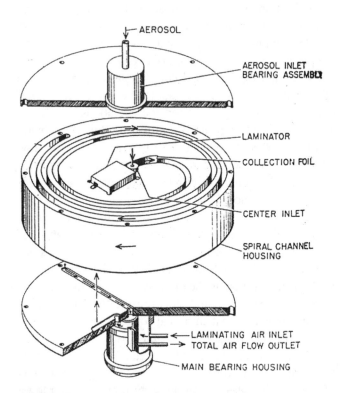

FIGURE 14-23. Schematic illustration of the Los Alamos Scientific Laboratory (LASL) centrifuge. *Reprinted from Martonen, T.B.: Theory and Verification of Formulae for Successful Operation of Aerosol Centrifuges. Am. Ind. Hyg. Assoc. J. 43(3): 154–159 (1982) with permission of Elsevier Science.*

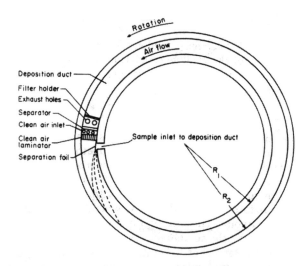

FIGURE 14-24. Top view of a cylindrical duct aerosol centrifuge, with particle trajectories shown by dashed lines.[198] *Reprinted with permission of American Industrial Hygiene Association Journal.*

Typical particle trajectories in the rotating, curved duct are illustrated in Figure 14-24.[198] As with the spiral centrifuge described above, the particles are introduced along the inner radius of the duct. As the duct spins, the particles are driven across the sheath air flow to the outer wall of the duct. Particles with large aerodynamic diameters deposit first, near the point opposite the flow entrance. Smaller particles require more time to traverse the sheath air and are collected further down the length of the duct. Generally, particles are collected on a foil that lines the outer wall of the channel. The foil is removed after collection. The linear distance along the foil at which the particle deposits is directly related to its aerodynamic diameter.

The reason for spinning the sampler is to subject the particles to a much greater centrifugal force than can be accomplished by air flow alone. Because of viscosity, the air in the duct rotates with the sampler. Consider a rotating duct like that in Figure 14-24, with a cross-sectional area A and a volumetric flow rate Q. From the rotating frame of reference of the sampler, the mean air stream velocity is Q/A, but from the surrounding inertial reference frame (i.e., as viewed from the table on which the sampler sits), the mean air stream velocity is $Q/A + 2\pi Rf$, where R is the mean radius of curvature of the duct and f is the rotational frequency. In most cases, $2\pi Rf \gg Q/A$, and the centrifugal force on the particles is dominated by the spinning of the duct.

Calibration curves for the spiral duct centrifuge are shown in Figure 14-25, which gives the particle aerodynamic diameter as a function of the deposition distance from the aerosol entrance. The instrument calibration depends on the speed of rotation, the total flow rate, and

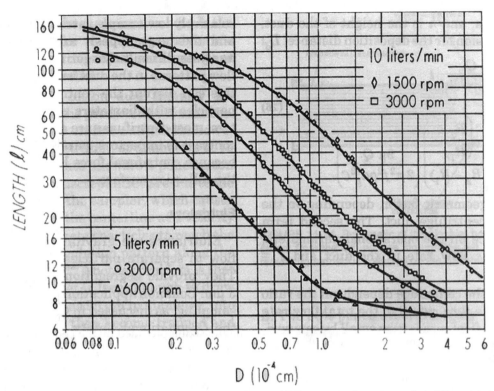

FIGURE 14-25. Calibration curves for the Stoeber spiral duct centrifuge at different rotation rates and at different total flow rates. Total flow refers to both aerosol sample flow plus the particle-free sheath air flow.[187] *Reprinted with permission of American Chemical Society.*

the geometry of the duct. The resolution of the instrument depends on the ratio of sample to sheath air flow rates. The data of Figure 14-25[187] were collected for aerosol flows ranging between 0.6% and 15% of the total flow.

Different types of aerosol centrifuges are summarized in Table 14-I-3. At one time, both the Lovelace[192] and Stoeber[187–189] spiral centrifuges were manufactured, but currently none are available commercially. Some of the first aerosol centrifuges, such as the Goetz spectrometer,[193, 194] were designed to operate without sheath air; but for these instruments, the deposition pattern represents a cumulative distribution, which is generally not as desirable. Another geometry once used for aerosol centrifuges is the conifuge.[195, 196] In this device, the aerosol is introduced into the annular space between two coaxial cones that spin together. The net flow is a descending spiral of increasing radius. Clean sheath air makes up the outer portion of the flow, so that the aerosol is initially confined next to the inner cone.

Most of the centrifuges have low air sampling rates, most are less than 1L/min. An exception is the high-volume drum centrifuge,[197] which employs particle deposition on the inner surface of a porous, rotating drum. This sampler is a unique design that has been used to collect large amounts of aerosol for trace analyses.

Principle of Operation

An annular centrifuge duct with a constant radius of curvature is used to illustrate the theory of operation of the aerosol centrifuge (Figure 14-24). Centrifuges of this design are reported by Tillery[191, 198] and Hochrainer.[199, 200] Although centrifuges such as that shown in Figure 14-24 use ducts of varying curvature, the principle is the same.

In the radial direction, particles are subject to a centrifugal force:

$$F_{r,centr} = \frac{m}{R}(2\pi R f + U)^2 \qquad (15)$$

and to an aerodynamic resistance

$$F_{r,aero} = \frac{3\pi\mu}{C}\frac{dR}{dt}d_p \qquad (16)$$

(As before, C = Cunningham slip factor and μ = gas viscosity.) Note that the radial component of the fluid velocity is zero.

In Equation 15 for centrifugal force, the fluid velocity U is usually neglected because it is small by comparison to the tangential velocity due to the spinning of the duct. With this approximation, the equation for the radial particle velocity becomes:

$$\frac{dR}{dt} = \frac{2\pi^2 R f^2 \rho d_p^2 C}{9\mu} \qquad (17)$$

where $|F_{r,centr}| = |F_{r,aero}|$ (i.e., the magnitude of the centrifugal and aerodynamic resistance forces is equal).

The quantity of interest in the aerosol centrifuge is the farthest distance along the outer channel wall at which particles of a specified diameter will deposit, L_d. This is found by evaluating the transit time, t, for the particle to travel across the entire duct from the inner radius, R_1, to the outer radius, R_2. Integration of Equation 17 gives:

$$t = \ln\left(\frac{R_2}{R_1}\right)\frac{9\mu}{2\pi^2 f^2 \rho d_p^2 C} \qquad (18)$$

Because particles must traverse the entire width of the channel, the distance they travel down the channel prior to capture depends only on the average duct velocity. The average velocity is just the ratio of the total volumetric flow to the duct cross-sectional area, $Q/[h(R_2-R_1)]$, where Q is the sum of the aerosol sample and sheath air flows, and h is the height of the duct. This gives an expression for the deposition distance, L_d:

$$\begin{aligned} L_d &= \frac{Q}{h(R_2-R_1)}t \\ &= \left(\frac{\ln\left(\frac{R_2}{R_1}\right)}{h(R_2-R_1)}\right)\left(\frac{9\mu Q}{2\pi^2 f^2 \rho d_p^2 C}\right) \end{aligned} \qquad (19)$$

The first term is a geometric factor, depending on the dimensions of the centrifuge duct. The second term gives the dependence on the operational parameters, $Q =$ total volumetric flow rate in the duct and $f =$ rotational frequency.

The resolution of the centrifuge depends on the ratio of the aerosol sample flow rate S to the total flow rate Q. When the aerosol flow rate is very small by comparison to the total flow Q, then all of the particles will have to cross the entire duct radius, R_2-R_1, to be collected, and the deposition distance is given by Equation 19. However, when the aerosol flow is a significant portion of the total flow, then some of the particles will start at a position closer to the outer duct wall, and they will not have to travel as far to be collected. As a result, there will be a range of deposition distances.

In the case where the duct width (R_2-R_1) is small compared to the radius, R_1, the radial velocity of the particle across the duct is essentially constant, and the dif-

ference between the maximum and minimum deposition distances for a particles of uniform aerodynamic diameter is given by:

$$\frac{\Delta L_d}{L_d} = \frac{S}{Q} \qquad (20)$$

where: S = aerosol sample flow
         Q = sum of the sample and sheath flows

Applications

Aerosol centrifuges can provide exceptionally high particle-size resolution, but they generally operate at relatively low sample flow rates. One of their major applications has been to measure aerodynamic shape factors for particle clusters. They have also been used to measure densities for spherical particles by comparing the aerodynamic sizing provided by the centrifuge with microscopically determined geometric diameters. In some inhalation exposure applications, centrifuges have been used to preselect a narrow particle size range for subsequent resuspension as a nearly monodisperse aerosol.

In principle, centrifuge deposits can be analyzed chemically to provide species size distributions similar to those obtained with impactors. However, their application in this field has been limited. Limitations are the relatively low sample rates, usually less than 1 L/min, and the large deposit area, which makes chemical analyses difficult. Larger flow rates lead to secondary flow patterns in the duct because of Coriolis effects, and this can disrupt the centrifuge sizing capability. For particles with diameters greater than a few micrometers, most centrifuges are subject to inlet losses. But in contrast to impactors, particle bounce is not a problem because centrifugal force holds the particles tightly to the collection surface.

Elutriators

Elutriators use gravitational settling in a laminar flow to separate particles by aerodynamic diameter. They provide segregation for particles greater than 3 μm. Common applications are respirable and thoracic sampling, as discussed in Chapter 5. Examples include the Occupational Safety and Health Administration (OSHA)-recommended cotton dust sampler,[201, 202] which is a vertical elutriator, and the Hexhlet[203] and MRE[204] dust samplers, which are horizontal elutriators used for respirable sampling. Horizontal elutriators can also be operated as spectrometers to measure distributions of particle size.

In still air, particle sedimentation is characterized by a terminal settling velocity, which is reached when the aerodynamic resistance exactly balances the gravitational force. The terminal settling velocity depends on aerodynamic diameter, and is given by:

$$V_{TS} = \frac{\rho d_p^2 gC}{18\mu} = \frac{\rho_o d_a^2 gC_a}{18\mu} \qquad (21)$$

Settling velocities for aerosols are relatively small. A 10-μm aerodynamic diameter particle has a settling velocity of 0.305 cm/s; for 1 μm, it is only 0.0035 cm/s.

Vertical Elutriators

The vertical elutriator consists of a vertical duct through which air flows slowly upward. Particles whose sedimentation velocity is greater than the duct velocity cannot follow the air flow and settle out. In laminar flow with a known velocity profile, particle penetration characteristics can be calculated. The sharpness of the cutpoint is reduced by the parabolic distribution of velocities in the duct; nonetheless, it is an effective method for removing large particles.

The OSHA-required sampler for cotton dust sampling is a vertical elutriator. It has been used in epidemiological studies to establish correlation between dust and the prevalence of byssinosis among cotton mill workers. The device is 15 cm in diameter and 70 cm high. Air enters a 2.7-cm-diameter conical inlet at the bottom, and a 37-mm filter is mounted at the top. At the recommended flow rate of 7.4 L/min, the average upward velocity in the main section equals the terminal settling velocity for a 15-μm aerodynamic diameter particle. However, the actual performance of this sampler is more complicated.

Calibration data for the cotton dust sampler are shown in Figure 14-26, as is the predicted performance.

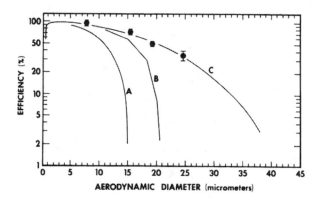

FIGURE 14-26. Comparison of theoretical and experimental collection efficiencies for the cotton dust elutriator. A — theory for laminar plug flow; B — theory for a parabolic velocity profile; C — theory for separated flow.[202] *Reprinted with permission of American Industrial Hygiene Association Journal.*

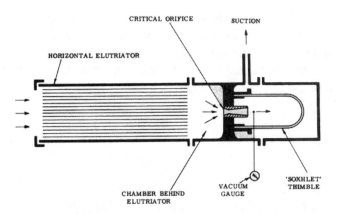

FIGURE 14-27. Schematic of the Hexhlet horizontal elutriator. Air flow enters from the left, passes through the set of parallel plates, and the remaining aerosol is collected on the "thimble" filter.

Although calculations based on flat and parabolic velocity profiles show 50% penetration at diameters of 10 and 15 μm, the calibration data show 50% penetration for 20-μm particles.[202] The discrepancies are due to the conical inlet at the bottom of the sampler, which causes a jet of air to travel up along the centerline at a velocity sufficient for the 30-μm-diameter particles to reach the filter. It also induces a recirculation pattern along the walls within the main duct.[201] Another problem is that the velocity at the 2.7-cm inlet is large enough to draw very large particles, as much as 95 μm in diameter, into the sampler. These particles then become trapped and can act as a floating filter for the upward moving air stream.

Horizontal Elutriators

The principle of the horizontal elutriators is illustrated by the MRE Gravimetric Dust Sampler[204] shown in Figure 14-27. Air travels slowly through a set of closely spaced parallel plates oriented horizontally. All particles whose settling velocity is greater than the ratio of the plate separation to transit time will be trapped. Smaller particles will be trapped at less than 100% efficiency. Particles that penetrate are collected by filtration.

One of the advantages of the horizontal elutriator is that its performance is predicted easily from basic principles. Consider an elutriator containing a total of n horizontal channels, each with a rectangular cross-sectional area, A, a separation distance, h, and length, L. The 50% penetration efficiency will occur at

$$\rho\, Cd_p^2 = \frac{hQ}{2nAL}\left(\frac{18\mu}{g}\right) \qquad (22)$$

where: Q = volumetric flow rate
g = gravitational constant

This corresponds to the settling velocity of a particle that enters along the centerline between the plates and just reaches the bottom plate at the exit of the elutriation section.

More generally, the penetration, P, for particles with a settling velocity V_{TS} is given by:

$$P = 1 - \frac{V_{TS}F}{Q} \qquad for \ V_{TS} < Q/F$$
$$= 0 \qquad\qquad for \ V_{TS} > Q/F \qquad (23)$$

where: F = horizontal area for collection.

This relation also holds for elutriators of variable cross-sectional area and plate separation, provided the flow is laminar and the particle trajectories are not affected by inertia.

In practice, there can be discrepancies between the theory and the performance. If the air stream velocity is too high, or if the device has not been cleaned, dust that has settled can be re-entrained and transported to the collection filter. This difficulty was observed in the original Hexhlet elutriator and corrected by reducing the flow rate and increasing the plate spacing. When flow velocities are very low, elutriator performance will be sensitive to thermal convection. If not operated in a fixed, horizontal orientation, its effective cutoff diameter will increase. In some older commercial units, discrepancies in elutriator performance have been traced to nonuniformity in plate spacings.

Horizontal elutriators can also be used as aerosol spectrometers by employing particle-free sheath air as the main carrier gas. Only a single channel consisting of two parallel plates is used. Aerosol is introduced as a thin stream along the upper plate. Particles of different sizes settle at different speeds, and they deposit at different positions along the bottom plate. The resolution with respect to particle size depends on the ratio of aerosol flow to sheath air flow. The principle is very much the same as for the aerosol centrifuge spectrometers, except that particles are drawn across the air flow by gravitational force rather than centrifugal force.

Electrostatic Precipitators

Electrostatic precipitation is the collection of charged particles in an electrostatic field. Electrostatic precipitation is often used for removal of particles from process streams or power plant stacks. It is less often used for airborne particle sampling, but it offers distinct advantages. Because electrostatic force is exerted directly on the particles instead of on the whole gas volume, relatively less energy is required to precipitate the particles. The sample flow rate is not affected by mass

loading, and the sample is in a readily recoverable form.

In a single-stage or Cottrell-type precipitator, both charging and precipitation take place within the same region. Some precipitators have two-stages, with physically distinct regions for charging and precipitating the particles. The precipitator collection surface can be large, as in many axially symmetric electrostatic precipitator samplers. Alternatively, it can be small, such as the "point-to-plane" precipitators used for collection of particles onto electron microscope grids.

A schematic of the point-to-plane electrostatic precipitator is shown in Figure 14-28. There are two electrodes; a high voltage needle electrode and plane electrode at ground voltage. A "corona" of ionized air molecules forms near the needle electrode, and those of the same sign are driven towards the opposite electrode. Much of the region between the two electrodes contains unipolar ions, that is, the ions present carry the same sign, they are either all positively charged or all negatively charged. The aerosol flowing through this region becomes charged by attachment of the unipolar ions. The charged particles then deposit on the plate electrode.

Principles of Electrostatic Precipitation

The collection of a particle by electrostatic precipitation involves two separate and distinct operations. First, the particle must acquire electric charges, and second, the charged particle must be accelerated toward an electrode of opposite polarity by an electric field. Most commonly, precipitators charge particles by using a corona discharge to produce large numbers of unipolar ions which then attach to the particles. The charged particles are subsequently deposited on an electrode of the opposite polarity.

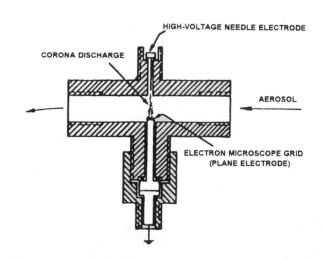

FIGURE 14-28. Diagram of point-to-plane electrostatic precipitator from Cheng, Yeh, and Kanapilly.[225]

Characteristics of the Corona Discharge

A schematic diagram of corona discharge is shown in Figure 14-29. A wire at high potential is located within a coaxial cylinder or above a flat plate held at ground potential. Although air and other gases are normally good electrical insulators, at high electric fields they become conductive. With the coaxial geometry shown in Figure 14-29, the electric field is much stronger near the surface of the wire, and the electrical breakdown of the air is confined to a small region near the wire surface. In this region, the electric field accelerates free electrons to a velocity that is large enough to knock an electron from the gas molecule upon collision. The result is a avalanche of ion production forming a cloud of free electrons and ionized gas molecules and that produce a characteristic corona glow.

Within the glow region along the wire, equal numbers of both negative and positive charge carriers are found. However, beyond the narrow confines of the glow region, the space between the electrodes is occupied almost entirely by ions of the same polarity as the wire. A corona can also be maintained between a point electrode at high voltage and a grounded plane surface. Corona formation requires a spatially nonuniform electric field whereby the ionization of the air is confined to a portion of the region. If the field is uniform, as with parallel plate electrodes, then a high potential will ionize the air across the entire gap, producing an arc.

The appearance of positive and negative coronas is very different. Positive coronas (where the wire is positive) are manifested by a smooth uniform glow. Negative coronas appears as a series of localized glow points or brushes and which, on a clean wire, appear to dance along the wire surface. The glow points are spread more or less uniformly along the wire and increase in number with increasing voltage and current.

In positive corona, free primary electrons are drawn to the positive electrode, creating electron-positive ion pairs by impact ionization. In a wire-cylinder configuration, the glow region occupies much less than 1.0% of the cross section. The remainder of the cross section is occupied by positive ions that are moving toward the receiving electrode. They interact with the particles, charging them positively, so that they can then be accelerated toward the receiving electrode.

The corona current is the electrical current between the wire and the receiving electrode. No current flows until a minimum voltage level is reached, which is that required to begin ionizing air molecules. Beyond this point, current increases rapidly with increasing voltage. As the voltage is increased still further, either of two limiting conditions will be reached. At normal atmospheric pressures, the field concentration at a localized point on one of the electrodes can become too great causing a spark discharge at that point. During the duration of such a spark discharge, the corona current disappears. When the breakdown is caused by a temporary occurrence, such as the passage of a large conducting particle or a fluctuation in the ambient humidity, the normal corona can return. On the other hand, when the breakdown at a given voltage is inherent in the electrode design, the breakdown would become continuous and normal corona could only be obtained by lowering the applied voltage. If the electrode design is conservative and there is no spark discharge, the second limitation on corona current will be reached. This occurs when the potential gradient is high enough to cause a generalized ionization of the air between the electrodes. Here, the glow region, containing both negative and positive air ions, is no longer confined to the vicinity of the corona wire, but rather fills the entire air gap. There is no longer a large region filled with unipolar ions and thus particles can no longer be given unipolar charges.

One practical consideration is that a small change in applied voltage makes a large difference in the magnitude of the corona current. For submicrometer particles, where the efficiency is strongly dependent on the ion density in the charging zone, it is important to maximize the corona current in order to obtain maximum collection efficiency.

Particle Charging Mechanisms

Two charging mechanisms important in electrostatic precipitators are field charging and diffusion charging. Field charging is the attachment of ions in a strong electric field. Ions follow the electric field lines to the particle surface, until the particle acquires enough charge to repel further incoming ions. In diffusion charging, the attachment of ions occurs due to random collisions

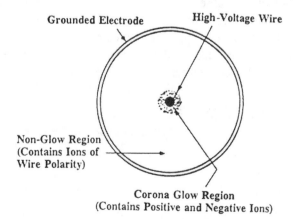

FIGURE 14-29. Axial view of high voltage corona discharge (adapted from White[205]).

between the ions and particles. It does not require an electric field, but is important for small particles even when an electric field is present. Both mechanisms require exposure of the aerosol to unipolar ions. A third mechanism of particle charging, static electrofication, is generally not important to electrostatic precipitators. Static electrofication refers to the charge a particle obtains when it is separated from bulk material or from contact with a surface, such as often happens when particles are formed by atomization.

Field charging depends on the interaction of the ions moving with the electric field and the particles passing through the field. For this mechanism, the maximum or saturation number of electron charges, n_s, that can be acquired by a particle is given by White[205] in the expression:

$$n_s = \left(1 + 2\frac{\varepsilon - 1}{\varepsilon + 2}\right)\frac{E_o a^2}{e} \tag{24}$$

where: n_s = maximum number of charges/particle
ε = dielectric constant of particle
E_o = electric field, volt/cm
a = particle radius, cm
e = charge per electron, 1.6×10^{-19} coulombs

The actual number of charges, $n(t)$, acquired in time t will be:

$$n(t) = n_s\left(\frac{\pi N_o eKt}{\pi N_o eKt + 1}\right) \tag{25}$$

where: $n(t)$ = number of charges/particle in time t
N_o = ion density in charging zone, ions/cm^3
k = ion mobility, cm^2/volt$\cdot$sec
t = time, seconds

According to White,[205] a typical value for N_o is 5×10^8 ions/cm^3, whereas an appropriate value of K for negative gas ions is 2.2 cm^2/volt$\cdot$sec.

The second mechanism is known as diffusion or thermal charging. In this mechanism, the ions come in contact with the particles by virtue of the Brownian movement of the particles. White's[205] equation for the number of charges acquired by diffusion charging in time t, $n(t)$, of an initially uncharged particle is:

$$n(t) = \frac{akT}{e^2}\ln\left(1 + \frac{\pi a v_{rms} N_o e^2 t}{kT}\right) \tag{26}$$

where: $n(t)$ = number of charges/particle in time t
a = particle radius
e = charge per electron
k = Boltzmann constant, 1.38×10^{-16} erg/°K
T = absolute temperature, °K
v_{rms} = root-mean-square velocity of ions, cm/sec

Table 14-3 shows a comparison of the charges for various particle sizes and charging periods calculated according to Equations 25 and 26 at identical space charges (according to Lowe and Lucas[206]). For particles <1 μm, charging by diffusion is seen to predominate; above 1 μm, ion bombardment is the predominant method. In the latter case, 80% of the maximum charge is already reached within 0.1 seconds. Neither the diffusion charging nor the field charging equations adequately account for the observed charging of particles on the order of 0.1-μm diameter and smaller where the predicted number of charges per particle approaches unity. They fail because they do not take into account the mechanisms whereby the particle acquires its initial charges. Assumptions made in the development of the diffusion charging equation limit its validity to relatively low charging rates and do not take into account the effect of the electric field.

It has been demonstrated that an enhanced rate of diffusion charging of up to 10 fold can be realized with free electrons rather than ions.[207] Enhanced precipitation in sampling instruments can, in principle, be realized by providing high free electron concentrations (such as by an electron beam precharger) and in precipitator devices either at high temperature or low

TABLE 14-3. Number of Charges on a Particle after Time t for Field and Diffusion Charging*

Particle Diameter μm	Fielding Charging Time, sec				Diffusion Charging Time, sec			
	0.01	0.1	1.0	∞	0.01	0.1	1.0	10
0.2	0.7	2.0	2.4	2.5	3	7	11	15
2.0	72	200	244	250	70	110	150	190
20.0	7200	2×10^4	2.5×10^4	2.5×10^4	1100	1500	1900	2300

*From Lowe and Lucas[206]

concentration of electronegative gases (e.g., CO_2, O_2, or H_2O).

Equation 26 does not hold strictly for the transition or free molecular particle size regime which, for normal atmospheric pressure, is below particle diameters of 0.2 μm. Diffusion charging of particles in the range of 0.004–0.050 μm was measured by Pui et al.[208] and found to fit the theoretical predictions of Fuchs[209] for particles above 0.010 μm and that of Marlow and Brock[210] for particles in the range of 0.004–0.010 μm. The review of Yeh[211] contains a more complete discussion of both field and diffusion charging.

Collection of Charged Particles

The separation force acting on a charged particle is given by Coulomb's Law, which states that the force is proportional to the product of the particle charge and the strength of the collecting field. The Coulomb force is opposed by inertial and viscous forces. For small particles, the inertial forces are usually negligible, and the viscous or retarding force can be approximated from Stokes' Law (Equation 1). The migration velocity of the particle can be calculated by balancing the Stokes and Coulomb forces. For streamline flow, relatively simple calculations could be made of collection efficiency. However, purely streamline flow is seldom achieved in electrostatic precipitators. By assuming completely turbulent flow, collection efficiency can be calculated by probability theory. This leads to an exponential-type formula for the probability of capturing a given charged particle and, by extension to the case of a large number of particles which do not interact, it leads to precipitator efficiency. It follows that 100% collection efficiency is approached only as an asymptotic limit.

Factors Influencing Collection Efficiency and Deposition Patterns

The collection efficiency of any precipitator sampler is dependent on operating parameters, e.g., current, voltage, and flow rate; the particle parameters, e.g., particle size, shape, and dielectric properties; and the carrier gas parameters, e.g., humidity, ambient pressure, temperature, and composition. Collection efficiency is aided by high charging currents, high voltage gradients, and low flow rates. For particles >0.5-μm diameter, the charging, and hence the collection efficiency, is strongly dependent on the potential gradient in the charging field; whereas for smaller particles, the charging current, i.e., the number of air ion charge carriers, is more significant. The flow rate affects performance in several ways. In most instruments, it exerts a drag force vector normal to

the electrical force vector. In addition, the tendency of collected particles to be re-entrained or eroded from the collection surface by the air stream is strongly dependent on the linear air velocity.

For effective precipitation, the particles should have some electrical conductivity. Nonconductive particles precipitated on the collection electrode within the corona zone can form an insulating barrier that will reduce the corona current. However, the electrical conductivity of particles is not necessarily the same as that of the parent material. For most dusts and fumes of mineral origin in the temperature range below 200°F, the humidity of the air influences the particle conductivity. Water vapor is absorbed on the surface, and the resultant surface conductivity aids precipitation. On the other hand, very high humidity can have an adverse effect on precipitator performance because electrical breakdown takes place at lower voltages in humid atmospheres.

In many electrostatic precipitators, the deposition pattern is dependent upon the particle size and dielectric properties. Variations in particle size of deposited dust as a function of length were described by Drinker et al.[212] and Fraser,[213] for a coaxial precipitator design. An exception is the two-stage precipitator of Liu et al.[214] which uses a pulsed voltage to precipitation particles of all sizes uniformly over the collection surface.

Considerations in the Use of Electrostatic Precipitators

Maximum collection efficiency is obtained at the higher voltages, provided the sparking is minimized. Within the limits discussed above, the higher voltages increase the unipolar ion concentration, which increases the charging efficiency for small particles. The higher voltage also increases the rate at which charged particles move toward the collector surface. However, operation at higher voltages also leads to an increase in the chance of sparking in the collector. High voltage sparking in precipitators results in localized re-entrainment. The spark creates a "crater" in the dust layer on the ground electrode, resuspending the displaced dust. When collecting for microscopy, such high-voltage discharges can also destroy the surface film of an electron microscope grid. Thus, while high collection voltages are generally desired, the rate of sparking must be held to a minimum.

One disadvantage of corona charging is that ozone is produced by the corona discharge. Ozone production is much larger for negative coronas. Ohkubo et al.[215] measured ozone production rates and typical room concentrations for positive and negative corona discharge in a wire and plate geometry at similar high voltages

(10–18 kV). They reported that the ozone production rate was 10 times greater for negative corona. In a typical indoor room, positive corona precipitation did not raise the ozone level above background (outdoor) levels (0.02 ppm), whereas a steady-state ozone concentration of 0.10 ppm was measured with negative corona. Modest wire heating was found to decrease the ozone production rate for the negative corona significantly. Similarly, for a point-to-plane geometry, Brandvold et al.[216] reported that negative corona produced 7.2 times more ozone than positive corona and that ozone production rates for both cases increased linearly with current.

Applications of Electrostatic Precipitators

The precipitator sampler described by Barnes and Penney[217] used a negative corona central electrode in an axially symmetric arrangement and operated at 85 L/min with high voltage adjustable from 8000–15,000 volts. Area samples for gravimetric or particulate analysis could be obtained with the instrument. Steen[218] described an "isokinetic sampler" employing electrostatic collection. This instrument has a specially designed nozzle and Venturi screen to provide for isokinetic flow through the sampler at the air velocity outside the sampler. The collection tube is 15 cm long and has a 2.5-cm i.d.; and it has an axially mounted wire electrode that is maintained at 12 kV AC. The front 1.0 cm of this electrode is 1.0 mm in diameter for corona discharge, whereas the remainder is 2.5 mm. Particles drift outward radially and are collected on the outer tube. Decker et al.[219] reported a high-volume collector that could concentrate particulate matter from 10,000 L into 10 ml of a collecting fluid. Instruments employing electrostatic collection for radioactive aerosols were described by Wilkening[220] and Bergstedt;[221] and Thomas.[222]

A number of electron microscope grid samplers based on electrostatic precipitation have been designed in which a single electron microscope grid of 3 mm diameter is the entire grounded collection surface. Most of these devices utilize a point-to-plane electrode configuration, with a needle point as the corona-emitting electrode and the electron microscope grid, backed and supported by a metal bar, as the grounded collection electrode. Samplers of this type are described by Billings and Silverman[223] and Morrow and Mercer.[224] Although the samples collected by these instruments are very small in terms of numbers and mass of particles, they are very dense in terms of numbers of particles per unit area of collection surface. They are not expected to collect all of the aerosol passing through them, but rather are designed to collect a representative sample for size analysis.[224–226]

An electrostatic precipitator, electron microscope grid sampler without an ionizing corona field was described by Mercer, et al.[227] In this instrument, the source of unipolar air ions for charging the particles is a tritium source. This permits the use of lower applied voltages (+2100 volts in this case) and avoids the possibility of undesirable high-voltage discharges that can destroy the collection surface film of an electron microscope grid.

In the two-stage sampler described by Liu et al.[214] the separate charging and precipitation zones allow for optimization of each. Particles of all sizes are uniformly distributed over the collection surface by the periodic application of 4200 volts to the precipitating region, with the overall collection efficiency varying from 60% for 0.28-μm-diameter particles to 80% for 3.2-μm particles. Furthermore, because the precipitation region does not have to carry current, it can utilize nonconducting particle collection surfaces such as glass slides and polymer-coated electron microscope grids.

Thermal Precipitators

A thermal precipitator removes particles from an aerosol by passing it through a relatively narrow channel having a significant temperature gradient perpendicular to the direction of flow. The movement of a particle in the direction of decreasing temperature, called its thermophoretic velocity, causes the particle to deposit on a collecting surface appropriate to the type of subsequent evaluation.

Figure 14-30 shows a cutaway view of a thermal precipitator.[228] In this device, air is drawn through the slit at 6 cm³/min. A nichrome wire, 0.254 mm in diameter, is centered in the 0.5-mm gap between the glass cover slips and is heated to approximately 120°C. The glass slips are held in place and kept at ambient temperature by contact with brass cylinders. As the aerosol passes through the slit, the particles are deposited as two strips on the cover slips opposite the heated wire. Examination of the cover slips with an optical microscope yields information about the size distribution and/or particle concentration of the aerosol.

Theory of Thermophoresis

The theory of thermophoretic motion of aerosol particles is discussed in detail by Waldmann and Schmitt.[229] For a particle whose diameter, d_p, is small with respect to the gas mean free path, 0.066 μm at 1 atm, 20°C), a free molecular theory has been developed, and experiments performed in this regime are in good

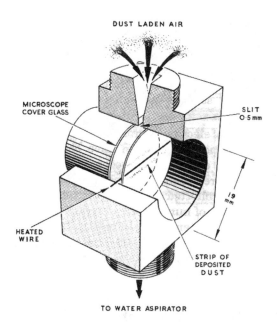

DUST LADEN AIR

MICROSCOPE
COVER GLASS

SLIT
0·5 mm

19
mm

HEATED
WIRE

STRIP OF
DEPOSITED
DUST

TO WATER ASPIRATOR

FIGURE 14-30. Sampling head of thermal precipitator from Watson[228] *(Crown Copyright with permission of Controller of H.M. Stationery Office, London).*

agreement with the theory. The thermophoretic velocity for a spherical particle in this regime is:

$$V_{th} = \frac{-0.55\eta\nabla T}{\rho_g T} \qquad (27)$$

where: η = gas viscosity (1.81×10^{-4} poise for air, 20°C)

ρ_g = gas density (1.2×10^{-3} g/cm³ for air, 1 atm, 20°C)

T = gas temperature, °K

∇T = temperature gradient, dT/dx, for direction x

V is proportional to the temperature gradient and is independent of the particle size. For air at 1 atm and 20°C, this condition holds for a particle diameter <<0.01 μm.

For particles that are large with respect to the mean free path ($d_p > 1$ μm), a simplified theoretical treatment was initially given by Epstein,[230] whose solution included the temperature gradient within the particle:

$$V_{th} = \frac{2Kg}{5P(2K_g + K_p)}\nabla T \qquad (28)$$

where: K_p = particle thermal conductivity, cal/cm•sec•°K

K_g = air thermal conductivity, 5.6×10^{-5} cal/cm•sec•°K

P = gas pressure, dyne/cm²

Thermal conductivity for particles varies widely, from a value of about 0.2 cal/cm•sec•°K for metallic iron to 2 × 10⁻⁴ for asbestos. The expression above was found to be in good agreement with experimentally measured velocities for particles of low thermal conductivity, but the predicted values for more conductive aerosol particles were low by more than an order of magnitude when compared to experiments of Schadt and Cadle.[231]

A more rigorous treatment of thermophoretic motion was given by Brock.[232] Brock's equation for the transition regime (0.01 μm < d_p < 1 μm) demonstrates particle diameter dependence. The predicted values have been compared to experimentally measured values by Waldmann and Schmitt[229] and Springer[233] and were found to be in good agreement. For particles in the transition regime, typical values of the thermal velocity per unit temperature gradient range from 1 × 10⁻⁴ to 2 × 10⁻⁴ cm2/°K. For d_p < 0.01 μm, the value increases to 2.5 × g 10⁻⁴. It can thus be seen that to achieve a reasonable thermal velocity, a rather steep temperature gradient is required.

Because the thermophoretic velocity of particles of diameter <0.01 μm is somewhat greater than that of larger particles, aerosols composed of a range of sizes will be differentially deposited in collection devices that employ temperature gradients. In a collector in which the aerosols flow through the temperature gradient, the smaller particles will be collected first and larger particles further downstream. This was demonstrated experimentally by Fuchs.[234]

Thermophoretic theory usually has dealt with a steady-state temperature gradient, whereas in a thermal precipitator, a particle's temporal experience corresponds to the application of a nonsteady gradient. The implication of this for collection has been considered by Reed and Morrison,[235] who showed that for particles less than 10 μm, the relaxation time was short enough to use the steady-state velocity approximation.

Precipitation Efficiency and Deposition Pattern

Provided that a sufficient thermal gradient is established in the sampling region (typically 10⁴ °K/cm), thermal precipitators collect essentially all particles from 5 to 0.005 μm and probably smaller. The lower limit of collection has not been determined experimentally, but the theory suggests that the collection efficiency should remain high down to sizes approaching molecular dimensions. For particles larger than 5 μm, the thermal force is adequate for collection, but upstream sampling difficulties due to gravitational and inertial effects in sampling devices may interfere.[236, 237]

The deposition pattern of a submicron platinum oxide aerosol in a thermal precipitator was described by Polydorova;[238] it was shown that the spatial distribution of the particles on the collection surface in a direction parallel to flow was approximately Gaussian, the deviations being less than 4% at any location. Therefore, if the total volume of aerosol sampled is known accurately, the aerosol concentration can be determined by extrapolation of the spatial distribution curve.

Advantages and Disadvantages of Thermal Precipitators

The very high efficiency of collection of submicron particles is one of the great advantages of the thermal precipitators over other collectors, such as liquid impingers or cascade impactors. The degree of charge on the particle appears to have little effect on the collection efficiency in a thermal precipitator. The low velocity of precipitation ensures that shattering or breakup of agglomerated particles does not occur during sampling.

Particles may be collected on a number of different surfaces according to the type of analysis desired; the sample may be evaluated by optical microscopy, electron microscopy, photometry, microscopic spot scanning, colony counting of viable airborne microbes, or radioactivity.

For some applications, the low sampling rate of thermal precipitators (ranging from 7 to 1000 cm³/min) is unsuitable. Sample evaluation may be very laborious compared to some of the direct-reading instruments for aerosol size or concentration determination. Many relatively volatile aerosols could not be collected in a thermal precipitator. By itself, the standard thermal precipitator has rather poor size selection characteristics; it should not be used when several distinct size fractions of an aerosol are to be separated. However, a sizing instrument using thermophoresis for the collection of transition regime particles has been proposed by Matteson and Keng.[239]

Precautions in the Use of Thermal Precipitators

Volatile aerosol particles should not be sampled in a thermal precipitator because of the likelihood of evaporation in the vicinity of the heated surface. If nonvolatile liquid aerosols are being collected, it is usually necessary to treat the collecting surface with a nonwetting agent or a fluorocarbon to prevent the drops from spreading. Even with these precautions, it is necessary to know the drop diameter to lens diameter ratio (which is a function of the liquid surface tension) for size evaluation.[240]

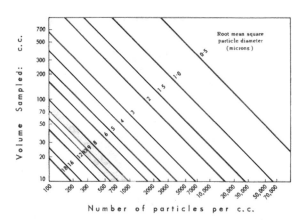

FIGURE 14-31. Maximum air volume sample of Watson thermal precipitator to keep particle overlap error <5% *(from Dust is Dangerous,[240] with permission of Farber & Farber, London).*

If too large a sample is taken, there will be significant particle overlap. This cannot be tolerated if the sample is to be analyzed by microscopy for particle size or concentration measurements. For the thermal precipitator, Davies[241] has established the conditions, given the particle diameter and concentration, for limiting this overlap error to 5% (Figure 14-31). If the aerosol size and concentration cannot be estimated beforehand, several samples of the same aerosol with volumes in a geometric progression should be taken to determine the true concentration or size distribution.[242]

Summary

This chapter has described several types of particle samplers, including inertial and gravitational collectors, and electrostatic and thermal precipitators. Each have their own advantages and limitations, and the instrument of choice will depend upon the application.

Impactors, inertial spectrometers, cyclones, and aerosol centrifuges collect particles in an accelerating air flow wherein the particles by virtue of their inertia cannot follow the flow stream lines, and deposit on a collecting surface. Elutriators collect particles by gravitational settling in a flow. Some of these instruments are used to provide precuts to other aerosol collection devices; others are used to provide size-segregated aerosol samples. All distinguish particles according to aerodynamic diameter, which is the parameter characterizing respiratory deposition for supermicrometer-sized particles. Without the physical separation of particles by size provided by these aerosol samplers, it would be difficult to determine the distribution of mass or chemical species with respect to particle diameter. Because adverse effects from airborne particles depend on both parameters, the simultaneous size and composition information obtained with these samplers is exceptionally valuable.

Electrostatic precipitator samplers for aerosols are based on the well-established principles of charged particle drift in electric fields. A number of devices that employ electrostatic collection have been designed and used in the past, some of which are still used in specific instances for particle collection. A major advantage of such devices is the absence of high impact collection velocities which may fracture or deform particles.

Thermal precipitators are useful aerosol sampling devices, particularly when high collection efficiency of submicron particles is required. Their rather low sampling rate compared to other samplers is a disadvantage for many situations, such as rapid sequential sampling of aerosols. There are many applications for thermal precipitators provided that the user understands the limitations of the particular instrument.

References

1. Hinds, W.C.: Aerosol Technology: Properties, Behavior and Measurement of Airborne Particles, p. 50. John Wiley, New York, NY (1982).

2. Mitchell, R.I.; Pilcher, J.M.: Improved Cascade Impactor for Measuring Aerosol Particles Sizes in Air Pollutants, Commercial Aerosols and Cigarette Smoke. Industrial and Engineering Chemistry 51:1039–1042 (1959).

3. May, K.R.: The Cascade Impactor: An Instrument for Sampling Coarse Aerosols. J. Scient. Instrum. 22:187–195 (1945).

4. Wang, H.C.; John, W.: Characteristics of the Berner Impactor for Sampling Inorganic Ions. Aerosol Sci. Technol. 8:157–172 (1988).

5. Lodge, J.P.; Chan,T.L.; Eds.: Cascade Impactor Sampling and Data Analysis. Am. Ind. Hyg. Assoc., Akron, OH (1986).

6. De la Mora, J.F.; Hering, S.V.; Rao, N.; McMurry, P. H.: Hypersonic Impaction of Ultrafine Particles. J. Aerosol Sci. 21(2): 169–187 (1990).

7. De la Mora, J.F.; Schmidt-Ott, A.: Performance of a Hypersonic Impactor with Silver Particles in the 2 nm Range. J. Aerosol Sci. 24(3): 409–415 (1993).

8. deJuan, L.; De la Mora, J.F.: Sizing Nanoparticles with a Focusing Impactor: Effect of the Collector Size. J. Aerosol Sci. 29(5/6): 589–599 (1998).

9. Keskinen, J.; Janka, K.; Lehtimaki, M.: Virtual Impactor as an Accessory to Optical Particle Counters. Aerosol Sci. Technol. 6(1): 79–83 (1987).

10. Sioutas, C.; Koutrakis, P.; Godleski, J.J.; et al.: Fine Particle Concentrators for Inhalation Exposures - Effect of Particle Size and Composition. J. Aerosol Sci. 28(6): 1057–1071 (1997).

11. Knuth, R.H.: Calibration of a Modified Sierra Model 235 Cascade Impactor, Report EML-360. Environmental Measurements Laboratory, New York (1979). Available from National Technical Information Service, Springfield, VA.

12. Burton, R.; Howard, J.N.; Penley, R.L.; et al.: Field Evaluation of the High-Volume Particle Fractionating Cascade Impactor. J. Air Pollut. Contr. Assoc. 23:277–281 (1973).

13. Mercer, T.T.; Tillery, M.I.; Newton, G.J.: A Multi-Stage, Low Flow-Rate Cascade Impactor. J. Aerosol Sci. 1:9–15 (1970).

14. Lippmann, M.: Review of Cascade Impactor for Particle Size Analysis and a New Calibration for the Casella Cascade Impactor. Am. Ind. Hyg. Assoc. J. 20:406–416 (1959).

15. Soole, B.W.: Concerning the Calibration Constants of Cascade Impactors, with Special Reference to the Casella MK-2. J. Aerosol Sci. 2:1–14 (1971).

16. Marple, V.A.; McCormack, J.E.: Personal Sampling Impactors with Respirable Aerosol Penetration Characteristics. Am. Ind. Hyg. Assoc. J. 44:916–922 (1983).

17. Rader, D.J.; Mondy, L.A.; Brockmann, J.E.; et al.: Stage Response Calibration of the Mark III and Marple Personal Cascade Impactors. Aerosol Sci. and Technol. 14:365–379 (1991).

18. Marple, V.A.; Olson, B.A.; Miller, N.C.: A Low-Loss Cascade Impactor with Stage Collection Cups: Calibration and Pharmaceutical Inhaler Applications. Aerosol Sci. Technol. 22(1): 124–134 (1995).

19. Olson, B.A.; Marple, V.A.; Mitchell, J.P.; Nagel, M.W.: Development and Calibration of a Low-Flow Version of the Marple-Miller Impactor (MMI). Aerosol Sci. Technol. 29(4): 307–314 (1998).

20. Pilat, M.J.; Ensor, D.S.; Bosch, J.C.: Source Test Cascade Impactor. Atmos. Environ. 4:671–679 (1970).

21. Pilat, M.J.; Ensor, D.S.; Bosch, J.C.: Cascade Impactor for Sizing Particles in Emission Sources. Am. Ind. Hyg. Assoc. J. 32:508–511 (1971).

22. Andersen, A.A.: A New Sampler for the Collection, Sizing and Enumeration of Viable Airborne Particles. J. Bacteriol. 76:471–484 (1958).

23. Keskinen, J.; Pieratinen, K.; Lehtimaeki, M.: Electrical low-pressure impactor. J. Aerosol Sci 23: 353 (1992).

24. Rubow, K.L.; Marple, V.A.; Olin, J.; McCawley, M.A.: A Personal Cascade Impactor: Design, Evaluation and Calibration. Am. Ind. Hyg. Assoc. J. 48:532–538 (1987).

25. Marple, V.A.; Rubow, K.L.; Behm, S.M.: A Microorifice Uniform Deposit Impactor (MOUDI): Description, Calibration and Use. Aerosol Sci. Technol. 14(4): 434–446 (1991).

26. Ranz, W.E.; Wong, J.B.: Impaction of Dust and Smoke Particles on Surface and Body Collectors. Industrial Engineering Chemistry, 44:1371–1381 (1952).

27. Davies, C.N.; Aylward, M.: The Trajectories of Heavy Solid Particles in a Two-Dimensional Jet of Ideal Fluid Impinging Normally Upon a Plate. Proc. Phys. Soc. B 64:889–911 (1951).

28. Marple, V.A.; Liu, B.Y.H.: Characteristics of Laminar Jet Impactors. Environ. Sci. Technol. 8:648–654 (1974).

29. Marple, V.A.; Liu, B.Y.H.: On Fluid Flow and Aerosol Impaction in Inertial Impactors. J. Colloid and Interface Sci. 53:31–34 (1975).

30. Rader, D.J.; Marple, V.A.: Effect of Ultra-Stokesian Drag and Particle Interception on Impaction Characteristics. Aerosol Sci. Technol. 4:141–156 (1985).

31. Mercer, T.T.; Chow, H.Y.: Impaction from Rectangular Jets. J. Colloid and Interface Sci. 27:75–83 (1968).

32. Mercer, T.T.; Stafford, R.G.: Impaction from Round Jets. Ann. Occup. Hyg. 12:41–48 (1969).

33. Ravenhall, D.G.; Forney, L.J.: Aerosol Impactors: Calculation of Optimum Geometries. J. Phys. E: Sci. Instrum. 13:87–91 (1980).

34. Marple, V.A.; Willeke, K.: Impactor Design. Atmos. Environ. 10:891–896 (1976).

35. Marple, V.A.; Rubow, K.L.: Theory and Design Guidelines. In: Cascade Impactor Sampling and Data Analysis, pp. 79–102. J.P. Lodge, Jr. and T.L. Chan, Eds. Am. Ind. Hyg. Assoc. Akron, OH (1986).

36. Asgharian, B.; Zhang, L.; Fang, C.P.: Theoretical Calculations of the Collection Efficiency of Spherical Particles and Fibers in an Impactor. J. Aerosol Sci. 28(2): 277–287 (1997).

37. May, K.R.: Calibration of Modified Andersen Bacterial Aerosol Sampler. Appl. Microbiol 12: 37–43 (1964).

38. Fang, C.P.; Marple, V.A.; Rubow, K.L.: Influence of Cross-Flow on

Particle Collection Characteristics of Multi-Nozzle Impactors. J. Aerosol Sci. 22(4): 403–415 (1991).

39. Hinds, W.C.; Liu, W.V.; Froines, J.R.: Particle Bounce in a Personal Cascade Impactor: A Field Evaluation. Am. Ind. Hyg. Assoc. J. 46:517–523 (1985).

40. Rao, A.K.; Whitby, K.T.: Non-Ideal Collection Characteristics of Inertial Impactors II: Cascade Impactors. J. Aerosol Sci. 9:87–100 (1978).

41. Cushing, K.M.; McCain, J.D.; Smith, W.B.: Experimental Determination of Sizing Parameters and Wall Losses of Five Source-Test Cascade Impactors. Environ. Sci. Technol. 13:726–731 (1979).

42. Dzubay, T.H.; Hines, L.E.; Stevens, R.K.: Particle Bounce in Cascade Impactors. Environ. Sci. Technol. 13:1392–1395 (1976).

43. Wesolowski, J.J.; John, W.; Devor, W.; et al.: Collection Surfaces of Cascade Impactors. In: X-Ray Fluorescence Analysis of Environmental Samples, p. 121. T. Dzubay, Ed. Ann Arbor Science Publishers, Ann Arbor, MI (1977).

44. Rao, A.K.; Whitby, K.R.: Nonideal Collection Characteristics of Single-Stage and Cascade Impactors. Am. Ind. Hyg. Assoc. J. 38:174–179 (1977).

45. Rao, A.K.; Whitby, K.R.: Nonideal Collection Characteristics of Inertial Impactors I: Single-Stage Impactors and Solid Particles. J. Aerosol Sci. 9:77–86 (1978).

46. Walsh, P.R.; Rahn, K.A.; Duce, R.A.: Erroneous Elemental Mass-Size Functions from a High-Volume Cascade Impactor. Atmos. Environ. 12:1793–1795 (1978).

47. Newton, G.J.; Cheng, Y.S.; Barr, E.B.; Yeh, H.C.: Effects of Collection Substrates on Performance and Wall Losses in Cascade Impactors. J. Aerosol Sci. 21:467–470 (1990).

48. Ellenbecker, M.J.; Leith, D.; Price, J.M.: J. Air Pollut. Control Assoc. 30:1244–1227 (1980).

49. Lawson, D.R.: Impaction Surface Coatings Intercomparison and Measurements with Cascade Impactors. Atmos. Environ. 14:195–199 (1980).

50. Boesch, P.: Practical Comparison of Three Cascade Impactors. J. Aerosol Sci. 14:325–330 (1983).

51. Aylor, D.E.; Ferrandino, F.J.: Rebound of Pollen and Spores During Deposition on Cylinders by Inertial Impaction. Atmos. Environ. 19:803–806 (1985).

52. Wang, H.-C.; John, W.: Comparative Bounce Properties of Particle Materials. Aerosol Sci. Technol. 7:285–299 (1987).

53. Reischl, G.P.; John, W.: The Collection Efficiency of Impaction Surfaces. Staub Reinhalt. Luft 38:55 (1978).

54. Wall, S.; John, W.; Wang, H.-C.: Measurements of Kinetic Energy Loss for Particles Impacting Surfaces. Aerosol Sci. Technol. 12:926–946 (1990).

55. Xu, M.; Willeke, K.: Right-Angle Impaction and Rebound of Particles. J. Aerosol Sci. 24:19–30 (1993).

56. Tsai, C.-J.; Pui, D.Y.H.; Liu, B.Y.H.: Capture and Rebound of Small Particles upon Impact with Solid Surfaces. Aerosol Sci. Technol. 12:497–507 (1990).

57. Cheng, Y.S.; Yeh, H.C.: Particle Bounce in Cascade Impactors. Environ. Sci. Technol. 13:1392–1395 (1979).

58. Kromidas, L.; Leifer, R.: An Innovative Application of a Commercially Available Double-Sided Adhesive For the Collection of Aerosols By Impaction. Atmos. Environ. 30(7): 1177–1180 (1996).

59. Pak, S.S.; Liu, B.Y.H.; Rubow, K.L.: Effect of Coating Thickness on Particle Bounce in Inertial Impactors. Aerosol Sci. Technol. 16:141–150 (1992).

60. Turner, J.R.; Hering, S.V.: Greased and Oiled Substrates as Bounce-free Impaction Surfaces. J. Aerosol Sci. 18:215–224 (1987).

61. Stein, S.W.; Turpin, B.J.; Cai, X.P.: Measurements of Relative Humidity-Dependent Bounce and Density For Atmospheric Particles Using the Dma-Impactor Technique. Atmos. Environ. 28(10): 1739–1746 (1994).

62. Biswas, P.; Flagan, R.C.: The Particle Trap Impactor. J. Aerosol Sci. 19(1): 113–121 (1988).

63. Tsai, C.J.; Cheng, Y.H.: Solid Particle Collection Characteristics on Impaction Surfaces of Different Designs. Aerosol Sci. Technol. 23(1): 96–106 (1995).

64. Liu, B.Y.H.; Pui, D.Y.H.; Wang, X.Q.; Lewis, C.W.: Sampling Carbon Fiber Aerosols. Aerosol Sci. Technol. 2:499–511 (1983).

65. Willeke, K.: Performance of the Slotted Impactor. Am. Ind. Hyg. Assoc. J. 36:6883–691 (1975).

66. Vanderpool, R.W.; Lundgren, D.A.; Marple, V.A.; Rubow, K.L.: Cocalibration of Four Large-Particle Impactors. Aerosol Sci. Technol. 7:177–185 (1987).

67. Biswas, P.; Flagan, R.C.: High Velocity Inertial Impactors. Environ. Sci. Technol. 18(8): 611–616 (1984).

68. Fang, C.P.; McMurry, P.H.; Marple, V.A.; Rubow, K.L.: Effect of Flow-Induced Relative Humidity Changes on Size Cuts for Sulfuric Acid Droplets in the Microorifice Uniform Deposit Impactor (MOUDI). Aerosol Sci. Technol. 14:266–277 (1991).

69. Keskinen, J.; Marjamaki, M.; Virtanen, A.; et al.: Electrical calibration method for cascade impactors. J. Aerosol Sci. 30(1): 111–116 (1999).

70. Francois, F.; Maenhaut, W.; Colin, J.L.; et al.: Intercomparison of Elemental Concentrations in Total and Size-Fractionated Aerosol Samples Collected During the Mace Head Experiment, April 1991. Atmos. Environ. 29(7): 837–849 (1995).

71. Howell, S.; Pszenny, A.A.P.; Quinn, P.; Huerber, B.: A Field Intercomparison of Three Cascade Impactors. Aerosol Sci. Technol. 29(6): 475–492 (1998).

72. Peters, T.M.; Chein, H.M.; Lundgren, D.A.; Keady, P.B.: Comparison and Combination of Aerosol Size Distributions Measured with a Low Pressure Impactor, Differential Mobility Sizer, Electrical Aerosol Analyzer, and Aerodynamic Particle Sizer. Aerosol Sci. Technol. 19(3): 396–405 (1993).

73. Hering, S.; Eldering, A.; Seinfeld, J.H.: Bimodal Character of Accumulation Mode Aerosol Mass Distributions in Southern California. Atmos. Environ. 31(1): 1–11 (1997).

74. Miguel, A.H.; Kirchstetter, T.W.; Harley, R.A.; Hering, S.V.: On-road Emissions of Particulate Polycyclic Aromatic Hydrocarbons and Black Carbon from Gasoline and Diesel Vehicles, Environ. Sci. Technol., 32: 450–455 (1998).

75. Twomey, S.J.: Comparison of Constrained Linear Inversion and Iterative Nonlinear Algorithm Applied to Indirect Estimation of Particle Size Distributions. Comput. Phys. 18:188–200 (1975).

76. Wolfenbarger, J.K.; Seinfeld, J.H.: Estimating the Variance in Solutions to the Aerosol Data Inversion Problem. Aerosol Sci. Technol. 14:348–357 (1991).

77. Hasan, H.; Dzubay, T.G.: Size Distributions of Species in Fine Particles in Denver Using a Micro-Orifice Impactor. Aerosol Sci. Technol. 6:29–40 (1987).

78. Hinds, W.C.: Data Analysis. In: Cascade Impactor Sampling and Data Analysis, Chapter 3. J.P. Lodge, Jr. and T.L. Chan, Eds. Am. Ind. Hyg. Assoc., Akron, OH (1986).

79. Macher, J.M.: Positive-Hole Correction of Multiple-Jet Impactors for Collecting Viable Organisms. Am. Ind. Hyg. Assoc. 50: 561–568 (1989).

80. Somerville, M.C.; Rivers, J.C.: An Alternative Approach for the Correction of Bioaerosol Data Collected With Multiple Jet Impactors. Am. Ind. Hyg. Assoc. J. 55(2): 127–131 (1994).

81. Stern, S.C.; Zeller, H.W.; Schekman, A.I.: Collection Efficiency of Jet Impactors at Reduced Pressures. I&EC Fundamentals 1:273–277 (1962).

82. Berner, A.; Luerzer, C.; Pohl, F.; et al.: The Size Distribution of the Urban Aerosol in Vienna. Sci. Tot. Environ. 13:245–261 (1979).

83. Hillamo, R.E.; Kauppinen, E.I.: On the Performance of the Berner Low-Pressure Impactor. Aerosol Sci. Technol. 14:33–47 (1991).

84. Hering, S.V.; Flagan, R.C.; Friedlander, S.K.: Design and Evaluation of a New Low-Pressure Impactor 1. Environ. Sci.

Technol. 12:667–673 (1978).

85. Hering, S.V.; Friedlander, S.K.; Collins, J.J.; Richards, L.W.: Design and Evaluation of a New Low-Pressure Impactor 2. Environ. Sci. Technol. 13:184–188 (1979).

86. Vanderpool, R.W.; Lundgren, D.A.; Kerch, P.E.: Design and Calibration of an In-Stack Low-Pressure Impactor. Aerosol Sci. Technol. 12:215–224 (1990).

87. Kuhlmey, G.A.; Liu, B.Y.H.; Marple, V.A.: Micro-orifice Impactor for Submicron Aerosol Size Classification. Am. Ind. Hyg. Assoc. J. 42:790–795 (1981).

88. Marple, V.A.; Rubow, K.L.; Behm, S.M.: A Microorifice Uniform Deposit Impactor (MOUDI): Description, Calibration, and Use. Aerosol Sci. Technol. 14:434–446 (1991).

89. Gudmundsson, A.; Bohgard, M.; Hansson, H.C.: Characteristics of Multi-Nozzle Impactors with 50 æm Laser-Drilled Nozzles. J. Aerosol Sci. 26(6): 915-931 (1995).

90. John, W.; Wall, S.M.: Aerosol Testing Techniques for Size-Selective Samplers. J. Aerosol Sci. 14:713–727 (1983).

91. Chen, B.T.; Yeh, H.C.; Cheng, Y.S.: A Novel Virtual Impactor: Calibration and Use. J. Aerosol Sci. 16:343–354 (1985).

92. Loo, B.W.; Cork, C.P.: Development of High-Efficiency Virtual Impactors. Aerosol Sci. Technol. 9:167–176 (1988).

93. Hounam, R.F.; Sherwood, R.J.: The Cascade Centripeter: A Device for Determining the Concentration and Size Distribution of Aerosols. Am. Ind. Hyg. Assoc. J. 26:122–131 (1965).

94. McFarland, A.R.; Ortiz, C.A.; Bertch, R.W.: Particle Collection Characteristics of a Single-Stage Dichotomous Sampler. Environ. Sci. Technol. 12:679–682 (1978).

95. Conner, W.D.: An Inertial-Type Particle Separator for Collecting Large Samples. J. Air Pollut. Control Assoc. 16: 35–38 (1966).

96. Dzubay, T.G.; Stevens, R.D.: Ambient Air Analysis with Dichotomous Sampler and X-ray Fluorescence Spectrometer. Environ. Sci. Technol. 9:663–668 (1975).

97. Loo, B.W.; Jaklevic, J.M.; Goulding, F.S.: Dichotomous Virtual Impactors for Large Scale Monitoring of Airborne Particulate Matter. In: Fine Particles, Aerosol Generation, Measurement, Sampling and Analysis, pp. 311–350. B.Y.H. Liu, Ed. Academic Press, New York (1976).

98. Solomon, P.A.; Moyers, J.L.; Fletcher, R.A.: High-Volume Dichotomous Virtual Impactor for the Fractionation and Collection of Particles According to Aerodynamic Size. Aerosol Sci. Technol. 2:455–465 (1983).

99. Marple, V.A.; Liu, B.Y.H.; Burton, R.M.: High-Volume Impactor for Sampling Fine and Coarse Particles. J. Air Waste Manag. Assoc. 40(5): 762–767 (1990).

100. Marple, V.A.; Rubow, K.L.; Olson, B.A.: Diesel Exhaust/Mine Dust Virtual Impactor Personal Aerosol Sampler: Design, Calibration, and Field Evaluation. Aerosol Sci. Technol. 22(2): 140–150 (1995).

101. Novick, V.J.; Alvarez, J.L.: Design of a Multistage Virtual Impactor. Aerosol Sci. Technol. 6:63–70 (1987).

102. Sioutas, C.; Koutrakis, P.; Burton, R.M.: Development of a Low Cutpoint Slit Virtual Impactor for Sampling Ambient Fine Particles. J. Aerosol Sci. 25(7): 1321–1330 (1994).

103. Chen, B.T.; Yeh, H.C.; Cheng, Y.S.: Performance of a Modified Virtual Impactor. Aerosol Sci. Technol. 5:369–376 (1986).

104. Chen, B.T.; Yeh, H.C.: An Improved Virtual Impactor: Design and Performance. J. Aerosol Sci. 18:203–214 (1987).

105. Chen, B.T.; Yeh, H.C.; Rivero, M.A.: Use of Two Virtual Impactors in Series as an Aerosol Generator. J. Aerosol Sci. 19(1): 137–146 (1988).

106. Chein, H.M.; Lundgren, D.A.: A High-Output, Size-Selective Aerosol Generator. Aerosol Sci. Technol. 23(4): 510–520 (1995).

107. Li, S.N.; Lundgren, D.A.: Effect of Clean Air Core Geometry on Fine Particle Contamination and Calibration of a Virtual Impactor. Aerosol Sci. Technol. 27(5): 625–635 (1997).

108. Noone, K.J.; Ogren, J.A.; Heintzenberg, J.; et al.: Design and Calibration of a Counterflow Virtual Impactor for Sampling of

Atmospheric Fog and Cloud Droplets. Aerosol Sci. Technol. 8:235–244 (1988).

109. Forney, L.J.: Aerosol Fractionator for Large-Scale Sampling. Rev. Sci. Instrum. 47:1264–1269 (1976).

110. Forney, L.J.; Ravenhall, D.G.; Lee, S.S.: Experimental and Theoretical Study of a Two-Dimensional Virtual Impactor. Environ. Sci. Technol. 16:492–497 (1982).

111. Marple, V.A.; Chien, C.M.: Virtual Impactors: A Theoretical Study. Environ. Sci. Technol. 14:976–985 (1980).

112. Collett, J.; Iovinelli, R.; Demoz, B.: A 3-Stage Cloud Impactor for Size-Resolved Measurement of Cloud Drop Chemistry. Atmos. Environ. 29(10): 1145–1154 (1995).

113. Noll, K.E.; Pontius, A.; Frey, R.; Gould, M.: Comparison of the Coarse Particles at an Urban and Non-Urban Site. Atmos. Environ. 19:1931–1943 (1985).

114. Noll, K.E.: A Rotary Inertial Impactor for Sampling Giant Particles in the Atmosphere. Atmos. Environ. 4:9–19 (1970).

115. Regtuit, H.E.; de Ruiter, C.J.; Vrins, E.L.M.; et al.: The Tunnel Impactor: A Multiple Inertial Impactor for Coarse Aerosol. J. Aerosol Sci. 21:919–933 (1990).

116. Hameed, R.; McMurry, P.H.; Whitby, K.T.: A New Rotating Coarse Particle Sampler. Aerosol Sci. Technol. 2:69–78 (1983).

117. Mack, E.; Pillie, R.: Fog Water Collector. U.S. Patent 3889532.

118. Kramer, M.; Schultz, L.: Collection Efficiency of the Mainz-Rotating-Arm-Collector. J. Aerosol Sci. 21(Suppl. 1):653–656 (1990).

119. Jacob, D.; Wang, R.-F.T.; Flagan, R.C.: Fogwater Collector Design and Characterization. Environ. Sci. Technol. 18:827–833 (1984).

120. McFarland, A.R.; Ortiz, C.A.; Cermak, J.E.; et al.: Wind Tunnel Evaluation of a Rotating-Element Large-Particle Sampler. Aerosol Sci. Technol. 12:422–430 (1990).

121. Lesnic, D.; Elliott, L.; Ingham, D.B.: A Mathematical Model for Predicting the Collection Efficiency of the Rotating Arm Collector. J. Aerosol Sci. 24:163–180 (1993).

122. McFarland, A.R.; Ortiz, C.A.; Bertch, Jr., R.W.: A 10 μm Cutpoint Size-Selective Inlet for Hi-Vol Samplers. J. Air Pollut. Contr. Assoc. 34:544 (1984).

123. McFarland, A.R.; Ortiz, C.A.; Bertch, Jr., R.W.: Particle Collection Characteristics of a Single-Stage Dichotomous Sampler. Environ. Sci. Technol. 12:679 (1978).

124. Wedding, J.B.; Weigand, M.A.: The Wedding Ambient Aerosol Sampling Inlet (D50 = 10 μm) for the High-Volume Samplers. Atmos. Environ. 19:535 (1985).

125. Georghlou, P.; Blagden, E.; Snow, D.E.; et al.: Mutagenicity of Indoor Air Containing Environmental Tobacco Smoke: Evaluation of a Portable PM$_{10}$ Impactor Sampler. Environ. Sci. Technol. 25:1496–1500 (1991).

126. Koutrakis, P.; Wolfson, J.M.; Brauer, M.; Spengler, J.D.: Design of a Glass Impactor for an Annular Denuder/Filter Pack System. Aerosol Sci. Technol. 12:607–612 (1990).

127. Koutrakis, P.; Fasano, A.M.; Slater, J.L.; et al.: Design of a Personal Annular Denuder Sampler to Measure Atmospheric Aerosols and Gases. Atmos. Environ. 23:2767–2773 (1989).

128. Fairchild, C.I.; Wheat, L.D.: Calibration and Evaluation of a Real-Time Cascade Impactor. Am. Ind. Hyg. Assoc. J. 45:205–211 (1984).

129.

130. Horton, K.D.; Ball, M.H.E.; Mitchell, J.P.: The Calibration of a California Measurements PC-2 Quartz Crystal Cascade Impactor (QCM). J. Aerosol Sci. 23:505–524 (1992).

131. Tropp, R.J.; Kuhn, P.J.; Brock, J.R.: A New Method for Measuring Particle Size Distributions of Aerosols. Rev. Sci. Instrum. 51:516–520 (1980).

132. Keskinen, J.; Pietarinen, K.; Lehtimaki, M.: Electrical Low-Pressure Impactor. J. Aerosol Sci. 23:353–360 (1992).

133. Prodi, V.; Melandri, C.; Tarroni, G.; et al.: An Inertial Spectrometer for Aerosol Particles. J. Aerosol Sci. 10:411–419 (1979).

134. Belosi, F.; Prodi, V.: Particle Deposition Within the Inertial Spectrometer. J. Aerosol Sci. 18:37–42 (1987).

135. Aharonson, E.F.; Dinar, N.: The Effect of Gravity on Deposition Distances in an Inertial Particle Spectrometer. J. Aerosol Sci. 18:193–202 (1987).

136. Prodi, V.; De Zaiacomo, T.: Fibre Collection and Measurement with the Inertial Spectrometer. J. Aerosol Sci. 13:49–58 (1982).

137. Mitchell, J.P.; Nichols, A.L.: Experimental Assessment and Calibration of an Inertial Spectrometer. Aerosol Sci. Technol. 9:15–28 (1988).

138. Greenburg, L.; Smith, G.W.: A New Instrument for Sampling Aerial Dust. Bureau of Mines R.I. 2392. U.S. Dept. of the Interior, Washington, DC (1922).

139. Katz, S.H.; Smith, G.W.; Meyers, W.M.; et al.: Cooperative Tests of Instruments for Determining Atmospheric Dusts, pp. 41–55. Public Health Bulletin #44, Washington, DC (1925).

140. Greenburg, L.; Bloomfield, J.J.: The Impinger Dust Sampling Apparatus as Used by the United States Public Health Service. Pub. Health Reports 47:654 (1932).

141. Hatch, T.; Warren, H.; Drinker, P.: Modified Form of the Greenburg–Smith Impinger for Field Use With a Study of its Operating Characteristics. J. Ind. Hyg. 14:301 (1932).

142. Hatch, T.; Pool, C.L.: Quantitation of Impinger Samples by Dark Field Microscopy. J. Ind. Hyg. 16:177 (1934).

143. DallaValle, J.M.: Note on Comparative Tests Made With the Hatch and Greenburg-Smith Impingers. Pub. Health Reports 42:1114 (1937).

144. Schrenk, H.H.; Feicht, F.L.: Bureau of Mines Midget Impinger. Bureau of Mines I.C. 7076. U.S. Dept. of the Interior, Washington, DC (1939).

145. Littlefield, J.B.; Schrenk, H.H.: Bureau of Mines Midget Impinger for Dust Sampling. Bureau of Mines R.I. 3360. U.S. Dept. of the Interior, Washington, DC (1937).

146. Ayer, H.E.; Hochstrasser, J.M.: Cyclone Discussion. In: Aerosol Measurement, pp. 70–79. D.A. Lundgren, F.S. Harris, W.H. Marlow, et al., Eds. University Presses of Florida, Gainesville, FL (1979).

147. Ranz, W.E.: Wall Flows in a Cyclone Separator: A Description of Internal Phenomena. Aerosol Sci. Technol. 4:417–432 (1985).

148. Malm, W. C.; Sisler, J. F.; Huffman, D.; et al.: J. Geophys. Rese, 99: D1: 1347–1370 (1994).

149. Solomon, P.A.; Salmon, L.G.; et al.: Spatial and Temporal Distribution of Atmospheric Nitric Acid and Particulate Nitrate Concentrations in the Los Angeles Area. Environ. Sci. Technol. 26(8): 1594–1601 (1992).

150. Chow, J.C.; Watson, J.G.; Bowen, J.L.; et al.: A Sampling System for Reactive Species in the Western United States. In: Sampling and Analysis of Airborne Pollutants. E.D. Winegar and L.H. Kei, Eds. Lewis Publishers, Boca Raton, FL (1993).

151. Chow, J.C.: PM$_{10}$ and PM$_{25}$ Compositions in California's San Joaquin Valley. Aerosol Sci. Tech. 18(2): 105–128.

152. Smith, W.B.; Wilson, Jr., R.R.; Harris, D.B.: A Five-Stage Cyclone System for in situ Sampling. Environ. Sci. Technol. 13: 1387–1392 (1979).

153. Chan, T.L.; Lippmann, M.: Particle Collection Efficiencies of Air Sampling Cyclones: An Empirical Theory. Environ. Sci. Technol. 11:377–382 (1977).

154. Leith, D.; Mehta, D.: Cyclone Performance and Design. Atmos. Environ. 7:527–549 (1973).

155. Dirgo, J.; Leith, D.: Cyclone Collection Efficiency: Comparison of Experimental Results with Theoretical Predictions. Aerosol Sci. Technol. 4:401–415 (1985).

156. Iozia, D.L.; Leith, D.: Effect of Cyclone Dimensions on Gas Flow Pattern and Collection Efficiency. Aerosol Sci. Technol. 10:491–500 (1989).

157. Kessler, M.; Leith, D.: Flow Measurement and Efficiency Modeling of Cyclones for Particle Collection. Aerosol Sci. Technol. 15:8–18 (1991).

158. DeOtte, Jr., R.E.: A Model for the Prediction of the Collection Efficiency Characteristics of a Small, Cylindrical Aerosol Sampling Cyclone. Aerosol Sci. Technol. 12:1055–1066 (1990).

159. Moore, M.E.; McFarland, A.R.: Design of Stairmand-Type Sampling Cyclones. Am. Ind. Hyg. Assoc. J. 51(3):151–159 (1990).

160. Blachman, M.W.; Lippmann, M.: Performance Characteristics of the Multicyclone Aerosol Sampler. Am. Ind. Hyg. Assoc. J. 35:311–326 (1974).

161. Dekeyser, E.; Dams, R.: In situ Comparison of a Multistage Series Cyclone System and a Cascade Impactor for In-Stack Dust Sampling. Environ. Sci. Technol. 22:1034–1037 (1988).

162. John, W.; Reischl, G.: A Cyclone for Size-Selective Sampling of Ambient Air. J. Air Pollut. Control Assoc. 30:872–876 (1980).

163. McFarland, A.R.; Hickman, P.D.; Parnell, Jr., C.B.: A New Cotton Dust Sampler for PM$_{10}$ Aerosol. Am. Ind. Hyg. Assoc. J. 48(3):293–297 (1987).

164. Tsai, C.J.; Shih, T.S.: Particle Collection Efficiency of Two Personal Respirable Dust Samplers. Am. Ind. Hyg. Assoc. J. 56(9): 911–918 (1995).

165. Bartley, D.L.; Chen, C.C.; Song, R.; Fischbach, T.J.: Respirable Aerosol Sampler Performance Testing. Am. Ind. Hyg. Assoc. J. 55: 1036–1046 (1994).

166. Maynard, A.D.; Kenny, L.C.: Performance Assessment of 3 Personal Cyclone Models, Using an Aerodynamic Particle Sizer. J. Aerosol Sci. 26(4): 671–684 (1995).

167. Ettinger, H.J.; Partridge, J.E.; Royer, G.W.: Calibration of Two-Stage Air Samplers. Am. Ind. Hyg. Assoc. J. 31: 537–545 (1970).

168. Seltzer, D.F.; Bernanski, W.J.; Lynch, J.R.: Evaluation of Size-Selective Presamplers II. Efficiency of the 10-mm Nylon Cyclone. Am Ind. Hyg. Assoc. J. 32: 441–446 (1971).

169. Caplan, K.J.; Doemeny, L.J.; Sorenson, S.D.: Performance Characteristics of the 10-mm Cyclone Respirable Mass Sampler: Part I- Monodisperse Studies. Am. Ind. Hyg. Assoc. J. 38: 83–95 (1977).

170. Bartley, D.L.; Breuer, G.M.: Analysis and Optimization of the Performance of the 10-mm Cyclone. Am. Ind. Hyg. Assoc. J. 43: 520–528 (1982).

171. Ogden, T.L.; Barker, D.; et al.: Flow-Dependence of the Casella Respirable-Dust Cyclone. Ann. Occup. Hyg. 27(3): 261–267 (1983).

172. Kenny, L.C.; Liden, G.: A Technique for Assessing Size-Selective Dust Samplers Using the APS and Polydisperse Test Aerosols. J. Aerosol Sci. 22(1): 91–100 (1991).

173. Harper, M.; Fang, C.P.; et al.: Calibration of the SKC, INC Aluminum Cyclone for Operation in Accordance with ISO/ACGIH Respirable Aerosol Sampling Criteria. J. Aerosol Sci. 29 (supp. 1): S347–S348 (1998).

174. Baxter, T.E.; Lane, D.D.; Asce, A.M.; et al.: J. Environ. Eng. 112:468–478 (1986).

175. Beeckmans, J.M.; Kim, C.J.: Analysis of the Efficiency of Reverse Flow Cyclones. Canadian J. Chem. Eng. 55:640–643 (1977).

176. Saltzman, B.E.: Generalized Performance Characteristics of Miniature Cyclones for Atmospheric Particulate Sampling. Am. Ind. Hyg. Assoc. J. 45:671–680 (1984).

177. Kim, J.C.; Lee, K.W.: Experimental Study of Particle Collection by Small Cyclones. Aerosol Sci. Technol. 12:1003–1015 (1990).

178. Moore, M.E.; McFarland, A.R.: Performance Modeling of Single-inlet Aerosol Sampling Cyclones. Environ. Sci. Technol. 27(9): 1842–1848 (1993).

179. Moore, M.E.; McFarland, A.R.: Design Methodology for Multiple Inlet Cyclones. Environ. Sci. Technol. 30(1): 271–276 (1996).

180. Kenny, L.C.; Gussman, R.A.: Characterization and Modeling of a Family of Cyclone Aerosol Preseparators, J. Aerosol Sci. 28: 677–688 (1997).

181. Berry, R.D.: The Effect of Flow Pulsations on the Performance of Cyclone Personal Respirable Dust Samplers. J. Aerosol Sci. 22:887–899 (1991).

182. Almich, B.P.; Carson, G.A.: Some Effects of Charging on 10-mm

Nylon Cyclone Performance. Am. Ind. Hyg. Assoc. J. 35:603–612 (1974).

183. Briant, J.K.; Moss, O.R.: The Influence of Electrostatic Charge on the Performance of 10-mm Nylon Cyclones. Am. Ind. Hyg. Assoc. J. 45:440–445 (1984).

184. Lippmann, M.; Chan, T.L.: Cyclone Sampler Performance. Staub-Reinhalt. Luft 39:7–11 (1979).

185. Saltzman, B.E.; Hochstrasser, J.M.: Design and Performance of Miniature Cyclones for Respirable Aerosol Sampling. Environ. Sci. Technol. 7:418–424 (1983).

186. Appel, B.R.; Povard, V.; Kothny, E.L.: Loss of Nitric Acid Within Inlet Devices for Atmospheric Sampling. Atmos. Environ. 22:2535–2540 (1988).

187. Stoeber, W.; Flachsbart, H.: Size-Separating Precipitation in a Spinning Spiral Duct. Environ. Sci. Technol. 3:1280–1296 (1969).

188. Stoeber, W.: Design Performance and Applications of Spiral Duct Aerosol Centrifuges. In: Fine Particles, Aerosol Generation, Measurement, Sampling and Analysis, pp. 351–398. B.Y.H. Liu, Ed. Academic Press, New York (1976).

189. Hoover, M.D.; Morawietz, G.; Stoeber, W.: Optimizing Resolution and Sampling Rate in Spinning Duct Aerosol Centrifuges. Am. Ind. Hyg. Assoc. J. 44:131–134 (1983).

190. Martonen, T.B.: Theory and Verification of Formulae for Successful Operation of Aerosol Centrifuges. Am. Ind. Hyg. Assoc. J. 43(3): 154–159 (1982).

191. Tillery, M.I.: Aerosol Centrifuges. In: Aerosol Measurement, pp. 3–23. D.A. Lundgren, F.S. Harris, W.H. Marlow, et al., Eds. University Presses of Florida, Gainesville, FL (1979).

192. Kotrappa, P.; Light, M.E.: Design and Performance of the Lovelace Aerosol Separator. Rev. Sci. Instrum. 43:1106–1112 (1972).

193. Goetz, A.; Stevenson, H.J.R.; Preining, O.: The Design and Performance of the Aerosol Spectrometer. J. Air Pollution Control Assoc. 10:378 (1960).

194. Gerber, H.E.: The Goetz Aerosol Spectrometer. In: Aerosol Measurement, pp. 36–55. D.A. Lundgren, F.S. Harris, W.H. Marlow, et al., Eds. University Presses of Florida, Gainesville, FL (1979).

195. Stoeber, W.; Flachsbart, H.: Aerosol Size Spectrometry with a Ring Slit Conifuge. Environ. Sci. Technol. 3:641–651 (1969).

196. Hochrainer, D.; Brown, P.M.: Sizing of Aerosol Particles by Centrifugation. Environ. Sci. Technol. 3:830–835 (1969).

197. Holländer, W.; Morawietz, G.; Pohlmann, G.; et al.: Very High-Volume Aerosols Sampling with a Novel Drum Centrifuge. Aerosol Sci. Technol. 7:67–77 (1987).

198. Tillery, M.I.: A Concentric Aerosol Spectrometer. Am. Ind. Hyg. Assoc. J. 35:62–74 (1974).

199. Hochrainer, D.: A New Centrifuge to Measure the Aerodynamic Diameter of Aerosol Particles in Submicron Range. J. Coll. Interface Sci. 36:191–194 (1971).

200. Hochrainer, D.; Stoeber, W.: A Stoeber-rotor with Recirculation of Particle-Free Air. Am. Ind. Hyg. Assoc. J. 39:754–757 (1978).

201. Claassen, B.J.: Effects of Separated Flow on Cotton Dust Sampling with a Vertical Elutriator. Am. Ind. Hyg. Assoc. J. 40:933–941 (1979).

202. Robert, K.Q.: Cotton Dust Sampling Efficiency of the Vertical Elutriator. Am. Ind. Hyg. Assoc. J. 40:535–541 (1979).

203. Wright, B.M.: A Size-Selecting Sampler for Airborne Dust. Brit. J. Industr. Med. 11:284–288 (1954).

204. Dunmore, J.H.; Hamilton, R.J.; Smith, D.S.G.: An Instrument for the Sampling of Respirable Dust for Subsequent Gravimetric Assessment. J. Sci. Instrum. 41:669 (1964).

205. White, H.J.: Industrial Electrostatic Precipitation. Addison-Wesley Publishing Co., Reading, MA (1963).

206. Lowe, H.T.; Lucas, D.H.: The Physics of Electrostatic Precipitation. Br. J. Appl. Phys. 24, Supp. 2:40 (1953).

207. O'Hara, D.B.; Clements, J.S.; Finney, W.C.; Davis, R.H.: Aerosol Particle Charging by Free Electrons. J. Aerosol Sci. 20(3):313 (1989).

208. Pui, D.Y.H.; Fruin, S.; McMurry, P.H.: Unipolar Diffusion Charging of Ultrafine Aerosols. Aerosol Sci. & Tech. 8(2):173 (1988).

209. Fuchs, N.A.: On the Stationary Charge Distribution on Aerosol Particles in a Bipolar Ionic Atmosphere. Geofis. Pura Appl. 56:185 (1963).

210. Marlow, W.H.; Brock, J.R.: Unipolar Charging of Small Aerosol Particles. J. Coll. Int. Sci. 50(1):32 (1975).

211. Yeh, H-C.: Electrical Techniques. In: Aerosol Measurement: Principles, Techniques, and Applications. K. Willeke and P.A. Baron, Eds. Van Nostrand Reinhold, New York (1993).

212. Drinker, P.; Thomson, R.M.; Fitchet, S.M.: Atmospheric Particulate Matter, II. The Use of Electric Precipitation for Quantitative Determinations and Microscopy. J. Ind. Hyg. 5(5):162 (1923).

213. Fraser, D.A.: The Collection of Submicron Particles by Electrostatic Precipitation. Am. Ind. Hyg. Assoc. J. 17(1):73 (1956).

214. Liu, B.Y.H.; Whitby, K.T.; Yu, H.S.: Electrostatic Aerosol Sampler for Light and Electron Microscopy. Rev. Sci. Inst. 38:100 (January 1967).

215. Ohkubo, T.; Hamasaki, S.; Nomoto, Y.; et al.: The Effect of Corona Wire Heating on the Downstream Ozone Concentration Profiles in an Air Cleaning Wire Duct Electrostatic Precipitator. IEEE Trans. on Ind. Appl. 26: 542 (1990).

216. Brandvold, D.K.; Martinez, P.; Dogruel, D.: Polarity Dependence of N_2O Formation from Corona Discharge. Atmos. Env. 23(9):1881 (1989).

217. Barnes, E.C.; Penny, G.W.: An Electrostatic Dust Weight Sampler. J. Ind. Hyg. Toxicol. 20(3):259 (1938).

218. Steen, B.: A New, Simple Isokinetic Sampler for the Determination of Particle Flux. Atmos. Environ. 11:623 (1977).

219. Decker, H.M.; Buchanan, L.M.; Frisque, D.E.: Advances in Large-Volume Air Sampling. Contamination Cont. 8:13 (1969).

220. Wilkening, M.H.: A Monitor for Natural Atmospheric Radioactivity. Nucleonics 10(6):36 (1962).

221. Bergstedt, B.A.: Application of the Electrostatic Precipitator to the Measurement of Radioactive Aerosols. J. Sci. Instr. 33:142 (1956).

222. Thomas, R.D.: Simplified Air-Sampling Method. Nucleonics. 17:134 (1959).

223. Billings, C.E.; Silverman, L.: Aerosol Sampling for Electron Microscopy. J. Air. Pollut. Control Assoc. 12(12):586 (1962).

224. Morrow, P.E.; Mercer, T.T.: A Point-to-Plane Electrostatic Precipitator for Particle Size Sampling. Am. Ind. Hyg. Assoc. J. 25(1):8 (1964).

225. Cheng, Y-S.; Yeh, H-C.; Kanapilly, G.M.: Collection Efficiencies of a Point-to-Plane Electrostatic Precipitator. Amer. Ind. Hyg. Assoc. J. 42(8):605 (1981).

226. Arnold, M.; Morrow, P.E.; Stöber, W.: Vergleichende Untersuchung über die Bestimmung der Korngrosssenverteilung fester Stauber mit Hilfe eines Hochspannungs-abscheiders und des Elektronenmikroskops. Koll. Z. Polymere 181(1):59 (1962).

227. Mercer, T.T.; Tillery, M.L.; Flores, M.A.: An Electrostatic Precipitator for the Collection of Aerosol Samples for Particle Size Analysis. LF-7. Lovelace Foundation for Med. Res. and Ed., Albuquerque, NM (1963).

228. Watson, H.H.: The Thermal Precipitator. Trans. Ins. Mining Metallurgy 46:176 (1936).

229. Waldmann, L.; Schmitt, K.H.: Thermophoresis and Diffusiophoresis of Aerosols. In: Aerosol Science, Chap. VI. C.N. Davis, Ed. Academic Press, London (1966).

230. Epstein, P.: On the Theory of Radiometer. Z. Phys. 54:537 (1929).

231. Schadt, C.F.; Cadle, R.D.: Thermal Forces on Aerosol Particles. J. Phys. Chem. 65:1689 (1961).

232. Brock, J.: Theory of Thermal Forces Acting on Aerosol Particles. J. Coll. Sci. 17:768 (1962).

233. Springer, G.S.: Thermal Forces on Particles in the Transition Regime. J. Coll. Inte. Sci. 34:215 (1970).

234. Fuchs, N.A.: The Mechanics of Aerosols, p. 66. Pergamon Press, Oxford (1964).
235. Reed, L.D.; Morrison, F.A.: Motion of an Aerosol Particle in an Unsteady Temperature Gradient. J. Coll. Inter. Sci. 42:358 (1973).
236. Prewett, W.G.; Walton, W.H.: The Efficiency of the Thermal Precipitator for Sampling Large Particles of Unit Density. Tech. Paper 63. Chemical Defense Experimental Establishment, Porton, England (1948).
237. Watson, H.H.: The Sampling Efficiency of the Thermal Precipitator. Br. J. Appl. Physics 2:78 (1958).
238. Polydorova, M.: Determining the Concentration of Ultrafine Aerosol Particles by Means of the Thermal Precipitator. Staub 27:448 (1967).
239. Matteson, M.J.; Keng, E.Y.H.: Aerosol Size Determination in the Submicron Range by Thermophoresis. J. Aerosol Sci. 3:45 (1972).
240. Bexon, R.; Ogden, T.L.: The Focal Length Method of Measuring Deposited Liquid Droplets. J. Aerosol Sci. 5:509 (1974).
241. Davies, C.N.: Dust is Dangerous. Farber and Farber, London (1954).
242. Roach, S.A.: Counting Errors Due to Overlapping Particles in Thermal Precipitator Samples. Br. J. Ind. Med. 15:250 (1958).

Instrument Descriptions

Cascade Impactors for Ambient Particle Sampling

14-1. Sierra/Marple Series 210 Ambient Cascade Impactor

Graseby Andersen Samplers, Inc.

This is a radial slot cascade impactor for ambient sampling. It is equipped with a cyclone preseparator and uses 47-mm-diameter, slotted impaction substrates. A built-in, 47-mm-diameter after-filter follows the impactor stages. Nominal cutpoints at the designed flow rate of 7 L/min are 0.16, 0.32, 0.53, 0.95, 1.7, 2.65, 4.4, 11, and 18 µm. Cutpoints are listed by the manufacturer at six additional flow rates of 0.3, 1, 3, 10, 14, and 21 L/min. Ten stages are available; however, stages with smaller nozzles cannot be operated at the higher flow rates. For example, at 7 L/min, only nine stages are operable. The impactor and impactor stages are made of 316 stainless steel. Dimensions: 6.4 cm diameter × 28 cm high (2.5 in. × 11 in.). Weight: 2 kg (4.5 lb).

14-2. Sierra/Marple Series 260 Ambient Cascade Impactor

Graseby Andersen Samplers, Inc.

This impactor has interchangeable nozzles that screw into the impaction stages. A set of six single, round, jet impactor nozzles and four rectangular nozzles give size cuts between 0.5 and 20 µm. Impactor flow rates may be varied from 0.3 to 20 L/min. Size cuts depend on flow rate and which impactor nozzles are used. Impactor nozzles are located off axis so that the location of the deposit can be varied by rotating

INSTRUMENT 14-1. Sierra/Marple Series 210 ambient cascade impactor.

stages relative to each other during collection. Impaction substrates are 18-mm diameter disks. Construction: aluminum. Size: 5 cm diameter × 40 cm high. Weight: 1.4 kg.

14-3. Andersen Low Pressure Impactor

Graseby Andersen Samplers, Inc.

This is a 13-stage, multijet impactor that operates at a fixed flow rate of 3 std L/min. The first eight stages are the same as those from the Andersen One ACFM ambient cascade impactor and provide size cuts at 35, 21.7,

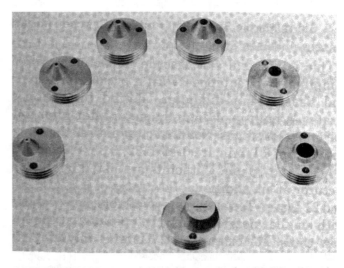

INSTRUMENT 14-2. Interchangeable nozzles for the Sierra/Marple cascade impactor.

INSTRUMENT 14-3. Anderson low pressure impactor.

14.7, 10.5, 6.6, 3.3, 2.0, and 1.4 μm. The last five stages operate at low pressure (≤ 0.14 atm) to provide size cuts at 0.90, 0.52, 0.23, 0.11, and 0.08 μm. The low pressure enables the capture of these smaller particles (see text). A critical orifice separates the low pressure and atmospheric pressure stages and controls the flow rate. An adapter kit is available for modifying Andersen One ACFM impactors. The complete kit includes a high-pressure vacuum pump and absolute pressure gauge.

14-4. Andersen One ACFM Ambient Cascade Impactor

Graseby Andersen Samplers, Inc.

This multijet cascade impactor has eight aluminum stages and a back-up filter holder, held together by three spring clamps and gasketed with O-ring seals. The first two stages contain 96 circular orifices each arranged in a radial pattern. The next five stages have 400 orifices

each; the last stage has 201 orifices. Cutpoints at a sampling rate of 28 L/min (1 cfm) range from 10 to 0.4 μm. Each stage has a removable, 8.2-cm- (3.25 in.-) diameter stainless steel or glass collection plate. An impactor preseparator is optional. The sampler is furnished as a complete system with a vacuum pump and carrying case. Dimensions: 11 cm diameter × 20 cm high (4.25 in. × 7.75 in.). Weight: 1.7 kg (3.75 lb).

14-5. Flow Sensor Ambient Cascade Impactor

Graseby Andersen Samplers, Inc.

This is a seven-stage, multijet cascade impactor with a preimpactor stage and a back-up filter holder. It is based on the design of Andersen (A.A., Sampler for Respiratory Health Hazard Assessment. Am. Ind. Hyg. Assoc. J. 27; 1966). Cutpoints at a sampling rate of 28 L/min (1 cfm) are 6, 4.6, 3.3, 2.2, 1.1, 0.7, and 0.4 μm. The sampler is furnished as a complete system with a flow controller, vacuum pump, and carrying case. Construction is of aluminum with O-ring seals. Dimensions of the impactor case: 14 cm × 14 cm × 30 cm high (6 in. × 6 in. × 12 in.). Impactor weight: 5.3 kg (7.5 lb); pump weight: 14 kg (31 lb).

14-6. Berner Impactor

Hauke KG

Hauke Cascade impactors follow the design of A. Berner. They are designed for a particle size analysis of ambient and industrial aerosols. Particle mass size distributions and size distributions of specific elements or chemical compounds can be measured in accordance with the type of analysis. These impactors have a wide

INSTRUMENT 14-4. Andersen One ACFM ambient cascade impactor.

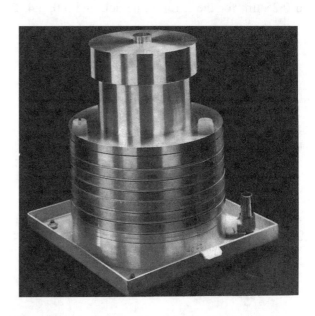

INSTRUMENT 14-5. Flow sensor ambient cascade impactor.

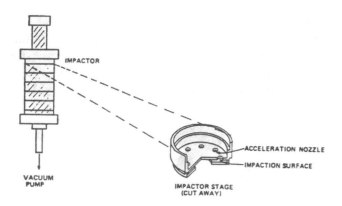

INSTRUMENT 14-6. Berner impactor.

measuring range, from 0.008 µm to 20 µm. The smallest particle sizes are collected by applying low pressure. They offer the advantage that the entire size distribution of atmospheric and industrial aerosols can be measured by one apparatus. Foils of any material can be used for particle collection. Flow rates may be varied according to requirements—Standard configurations are 25 Lmin^{-1}, 30 Lmin^{-1} 80 Lmin^{-1}; and temperature range from –30 °C up to +180 °C.

14-7. Mercer Seven-Stage Cascade Impactors

In-Tox Products

These are seven-stage, round-jet cascade impactors based on the design of Mercer et al.[8] Four models are available with flow rates from 0.1 to 5 L/min using one to four jets per stage. Effective cutpoint diameters are as follows: 3.1, 2.1, 1.6, 1.0, 0.85, 0.58, and 0.33 µm for the 0.1 L/min impactor; 4.5, 3.0, 2.1, 1.5, 1.0, 0.71, and 0.32 µm for the 1 L/min impactor; 5.0, 4.0, 3.0, 1.8, 1.0, 0.4, and 0.25 µm for the 2 L/min model; and 5.0, 3.4, 2.3,

1.5, 1.0, 0.7, and 0.5 µm for the 5 L/min model. Collection substrates are 22 mm in diameter, and the stages are sealed in O-rings. The impactors are made of brass (stainless steel versions are available upon request) and are 4.5 cm in diameter, 10 cm high, and weigh 0.9 kg (2 lb).

14-8. Multijet Cascade Impactors

In-Tox Products

These seven-stage cascade impactors are available in four models with flow rates of 10, 14, 20, and 28 L/min. They are similar in construction to the Mercer seven-stage impactor described above. The stages have round jets, 37-mm-diameter collection substrates, and are sealed with O-rings. The impactors are made of brass (stainless steel available upon request) and are 7 cm in diameter, 14 cm high, and weigh 2 kg (4.4 lb). The effective cutpoint diameters are as follows: 8.0, 5.0, 3.2, 2.0, 1.3, 0.8, and 0.5 for the 10 L/min, 14 L/min, and 20 L/min models; and 9.25, 5.7, 3.6, 2.2, 1.4, 0.8, and 0.5 for the 28 L/min model.

14-9. MOUDI (Model 100)

MSP Corporation

The Model 100/110 MOUDI (*M*icro-*o*rifice *u*niform *d*eposit *i*mpactor) with rotating impaction plates provides near-uniform particle deposits over circular impaction areas. It is an 8-stage cascade impactor with round jets, and constructed with hard-coated aluminum with stainless steel micro-orifice plates. The 37 mm or 47 mm removable impaction plates and final filter allow quick and easy change in field or lab. The flow rate is 30 liters per minute and the cutpoint diameters range from 10 µm to 0.056 µm. Together with the final filter, the impactor provides a

INSTRUMENT 14-7. Mercer seven-stage cascade impactor.

INSTRUMENT 14-8. In-Tox multijet cascade impactor.

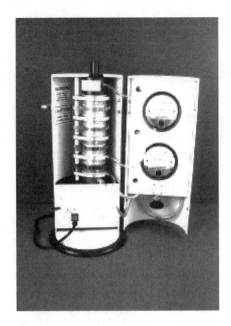

INSTRUMENT 14-9. Model 100 MOUDI.

complete size distribution measurement system. Dimensions: 13 cm diameter × 28 cm high (5 in. × 11 in.). Weight 5 kg (11 lb), plus 6 kg (13 1b) for the rotating unit.

14-10. Hi-Volume Fractionating Sampler (Model 65-800)

Graseby Andersen Samplers, Inc.

This is a multijet cascade impactor designed to mount on a Hi-Volume sampler. The impactor segregates particles by aerodynamic diameter on each of the stages, with the smallest particles collected by the Hi-Volume filter. The impaction stages use round impaction jets arranged in a circular pattern. Two versions of the impactor, with four or two stages, respectively, are designed for operation at 566 L/min (20 cfm). A third version has one stage for operation at 1132 L/min (40 cfm). For operation at 566 L/min, the size cuts for the four-stage impactor are 1.1, 2.0, 3.3, and 7 µm; for the two-stage version, the size cuts are 1.1 and 7.0 µm. At 1132 L/min, the single-stage version of the impactor has a cutpoint at 3.5 µm. Dimensions: 30 cm diameter × 13 cm high (12 in. × 5 in.). Weight: 8.6 kg (19 lb).

14-11. High-Volume Cascade Impactors (Series 230)

Graseby Anderson Samplers, Inc.

This is a rectangular-jet cascade impactor designed to mount on a Hi-Volume sampler. Collection substrates are 14.3 cm × 13.7 cm and must be slotted to allow air flow to the next stage. The Model 235 is designed for a nominal flow rate of 1.13 m³/min (40 cfm) and has five stages with cutpoints at 7.2, 3.0, 1.5, 0.95, and 0.49 µm. The Model 236 is designed for a flow rate of 0.566 m³/min (20 cfm) and has six stages with particle cutpoints at 10.2, 4.2, 2.1, 1.4, 0.73, and 0.41 µm in aerodynamic diameter. Single-stage versions of the impactors with 1.13 m³/min (40 cfm) cutpoints at 3.5 µm (respirable) and 2.5 µm (fine) are available. Stages are made of aluminum. Dimensions: 23 cm × 30 cm × 5 cm high (9.25 in. × 12 in. × 2 in.). Weight: 2.5 kg (5.5 lb).

Personal and Microenvironmental Sampling Impactors

14-12. Marple Personal Samplers (Series 290)

Graseby Andersen Samplers, Inc.

This impactor is designed to be worn on the lapel of a worker for personal monitoring in the workplace. It has a radial slot jet design, 34-mm-diameter collection substrates, and a 34-mm-diameter PVC back-up filter. The

INSTRUMENT 14-10. Andersen Hi-Volume fractionator.

INSTRUMENT 14-11. General Metal Works Series 230 high-volume cascade impactor.

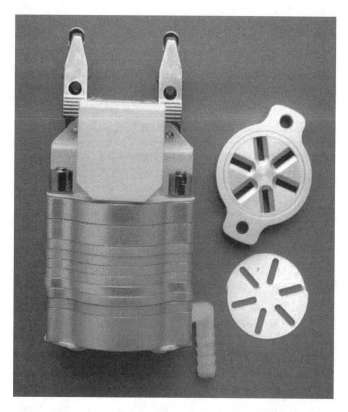

INSTRUMENT 14-12. Marple 290 Personal Cascade Impactor.

Dimensions for both: 5.7 cm wide and 7.2–8.6 cm high. Weight: 170 to 200 g (6–7 oz), depending on the model.

14-13. Personal Environmental Monitoring Impactor
MSP Corporation

The Model 200 PEM (Personal Environmental Monitor) is a lightweight personal impactor for collecting airborne particles in the PM$_{2.5}$ or PM$_{10}$ size ranges. PM$_{2.5}$ and PM$_{10}$ particles are airborne particles less than 2.5 μm and 10 μm, respectively. These classes of particles are widely used in indoor/outdoor pollution studies to assess the potential health effects of airborne particles. The sample air enters the Model 200 through multi-nozzle, single-stage aluminum impactor to remove particles above 2.5 μm or 10 μm, respectively. The impaction surface is an oil-soaked, porous, stainless steel plate. Particles smaller than the cut size are then collected on a 37 mm filter of choice. The collected particles can be analyzed gravimetrically to determine the particle mass or for specific chemical compounds. Dimensions: 2.5 cm high × 6 to 9 cm diameter (1 in. × 2.5-3.5 in). Weight: 55 g (2 oz.).

14-14. Model 400 Micro-Environmental Monitor (MEM)
MSP

The Model 400 MEM (Micro-Environmental Monitor) is a compact aerosol sampler for determining the concentration of airborne particles in the PM$_{2.5}$ and PM$_{10}$ size ranges. In the Model 400, aerosol particle are sampled through a single stage impactor to remove particles above 2.5 or 10 μm in aerodynamic diameter. Smaller particles are then collecting on a 37 mm diameter filter. The 10 L/min flow is maintained by a ultra-

design sample flow rate is 2 L/min. An inlet cowl excludes extraneous debris. The three models available are of four, six, and eight stages, respectively. For the eight-stage model, cutpoints are 0.6, 1, 2, 3.5, 6, 10, 14, and 20 μm; for six stages, the cutpoints are 0.6, 1, 2, 3.5, 6, and 10 μm; and for four stages, the cutpoints are at 3.5, 10, 14, and 20 μm. The impactors are machined from aluminum, and the impactor stages are nickel plated. The Andersen and GMW instruments were identical.

INSTRUMENT 14-13. Model 200 PEM™ Personal Environmental Monitor.

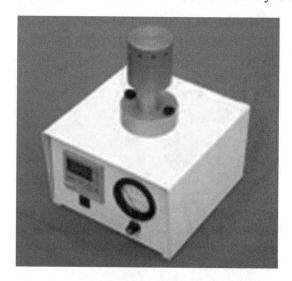

INSTRUMENT 14-14. Model 400 MEM™ Micro-Environmental Monitor.

INSTRUMENT 14-15. URG-2000-25A Personal Monitor.

quiet pump, to allow the sampler to be used indoors without disturbing the occupants.

14-15. Personal Monitor for Pesticides, PAHs and SVOCs with Size-Selective Inlet

URG Corporation

The URG-2000-25A is a personal monitor for ambient, indoor air and personal sampling. The 25A consists of a size-selective impactor inlet, which separates coarse particles from fine particles by impaction, a filter pack,

and PUF cartridge. The unit is housed in rugged, high-density polyethylene. This personal monitor operates at 4 liters per minute with a cut point of 1 µm, 2.5 µm, or 10 µm. Particles, on appropriate filter media, are analyzed for organic and elemental carbon, anions, cations, and trace metals. The PUF is analyzed for PAH's, pesticides, SVOC's and PCB's.

An optional URG-2000-25F consists of a size-selective inlet (1 µm, 2.5 µm, or 10 µm) followed by a filter pack in a rugged, high-density polyethylene housing. Both units can be used with tissuquartz or teflon filter media. The units require a 4 liters per minute portable sampling pump to operate.

Virtual Impactors

14-16. Cascade Centripeter

BGI Incorporated

This instrument is a type of multistage virtual impactor based on the design of Hounam and Sherwood.[68] The air stream passes through a series of orifices of diminishing diameter. Successively finer fractions of the aerosol are collected by sharp-edged nozzles located immediately downstream of each orifice. Particles are deposited on filters located behind the receiver nozzles. A final filter collects particles that

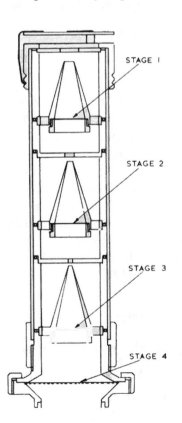

INSTRUMENT 14-16. Schematic diagram of cascade centripeter.

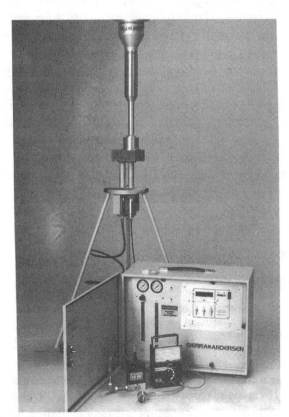

INSTRUMENT 14-17. Series 241, PM₁₀ manual dichotomous sampler.

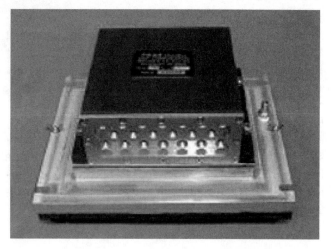

INSTRUMENT 14-18. Model 340 HVVI™ High Volume Virtual Impactors.

INSTRUMENT 14-19. Microcontaminant particle sampler.

escape removal by the three centripeter stages. The sampler flow rate is 30 L/min and corresponding cutpoints are at 1.2, 4, and 14 µm. Dimensions: 3.8 cm diameter × 18 cm high (1.5 in. × 7 in.).

14-17. PM-10 Manual Dichotomous Sampler (Series 241)

Graseby Andersen Samplers, Inc.

The dichotomous sampler has a PM_{10} inlet to provide a precut at 10 µm, followed by a virtual impaction stage that provides a second particle size cut at 2.5 µm. The inlet is based on the design of McFarland; the virtual impactor is based on the design of Loo.[71] It samples at 16.7 L/min (1 m³/hr) and provides samples in two particle size fractions: coarse (2.5 to 10 µm) and fine (<2.5 µm). Samples are collected on 37-mm filters. Air flow is regulated by a flow controller. The sampling module is made of aluminum, measures 162 cm high × 76 cm diameter (64 in. × 30 in.), and weighs 9 kg (20 lb). The control module measures 41 cm high × 56 cm wide × 28 cm diameter (16 in. × 22 in. × 11 in.) and weighs 27 kg (60 lb).

14-18. High Volume Virtual Impactor (Model 340)

MSP Corporation

The Model 340 PM 2.5 Particle Classifier is a high-flow virtual impactor classifier with multiple parallel nozzles to achieve a high volumetric flow rate at a low pressure drop. The classifier can be supplied with a 2.5 µm or 1.0 µm cut point in equivalent aerodynamic diameter and can be used for the inertial separation of coarse particles from fine particles for a variety of applications, such as air pollution and air quality studies and coarse

particle concentration. Air flow is drawn through two sets of parallel nozzles. Particles larger than the cut point diameter of the virtual impactor pass through the receiving tube with 5% of the inlet flow and then through the central chamber to the larger particle filter. Particles smaller than the cut size are carried by the major flow to the smaller particle filter. The HVVI may be used at flows less than 40 CFM by plugging opposing pairs of nozzle/receiving tube sets. Dimensions: 14 cm high × 25 cm × 30 cm (5 in. × 9.5 in. × 12 in.). Weight: 5 kg (11 lb).

14-19. Microcontaminant Particle Sampler

MSP Corporation

The microcontaminant particle sampler uses a 1-µm cut virtual impactor at 30 L/min to concentrate supermicrometer particles into a 1.5-L/min stream that passes though a two-stage impactor followed by a final 25-mm filter. The first impactor stage collects particles on a scanning electron microscope (SEM) stud. Particles that bounce from this ungreased SEM stud are collected on a greased SEM stud in the second impaction stage. The fine fraction from the virtual impaction stage is collected on a 37-mm filter. Construction is of aluminum. Dimensions: 14 cm high × 11 cm diameter (5.8 in. × 4.2 in.). Weight: 1.4 kg (3 lb).

Source Test Impactors

14-20. Series 220 In-Stack Cascade Impactor

Graseby Andersen Samplers, Inc.

This is a multijet, radial slot cascade impactor with six, eight, or ten impaction stages, a built-in 47-mm holder for the after-filter, and an optional cyclone

INSTRUMENT 14-20. Andersen in-stack cascade impactor, Model 226.

preseparator. At the nominal flow rate of 7 actual L/min, the impactor stage cutpoints are 0.16, 0.32, 0.53, 0.95, 1.7, 2.65, 4.4, 11, and 18 μm aerodynamic diameter. Iso-kinetic sampling nozzles are available. Impactor construction: nickel-plated aluminum or 316 stainless steel. Dimensions: 6.3 cm diameter × 28 cm high (25 in. × 11 in.). Weight: 2 to 4 kg (4 to 9 lb).

14-21. Andersen Mark III and Mark IV Stack Sampling Heads

Graseby Andersen Samplers, Inc.

These are nine-stage, multijet cascade impactors designed to adapt to stack sampling trains. Jets are round and arranged in concentric circles. Nominal flow rates are 2.8 to 21 actual L/min (0.1 to 0.75 acfm). At the 21-L/min flow rate, aerodynamic diameter cutpoints are 10.9, 6.8, 4.6, 3.2, 2.0, 1.0, 0.61, and 0.41 μm. The impactor is made of stainless steel and can be operated at 800°C. The Mark III is available with a stainless steel cyclone preseparator. The Mark IV uses an external right

INSTRUMENT 14-22. Andersen high capacity stack sampler, Model 70-900.

angle inlet nozzle preseparator. The entire assembly will fit through a 7.6-cm- (3-in.-) diameter port. Dimensions: 7 cm diameter × 25 cm long (2.8 in. × 10 in.).

14-22. High Capacity Stack Sampler (Model 70-900)

Graseby Andersen Samplers, Inc.

The high-capacity stack sampler has two impaction stages followed by a cyclone and back-up filter thimble. At the recommended flow rate of 14 actual L/min (0.5 acfm) and 25°C, the cutpoints are 10.8, 5.8, and 1.5 μm. A preimpactor with a 12-μm cutpoint is available. The assembled unit fits through a 7.6-cm- (3-in.-) diameter sampling port. Units are made of stainless steel.

14-23. Impactor Preseparator (Model 50-160)

Graseby Andersen Samplers, Inc.

The impactor preseparator is designed for stack sampling under conditions of high particulate loadings. It is a single-stage, high capacity impactor that screws directly into the inlet of the high capacity or Mark III stack samplers. It has a 10-μm cutpoint at 25°C and 21 L/min, is made of stainless steel, and fits through a 7.6-cm- (3 in.-) diameter sampling port.

14-24. High Temperature, High Pressure Cascade Impactor

In-Tox Products

This seven-stage cascade impactor is designed for process stream sampling and has been tested at pressures

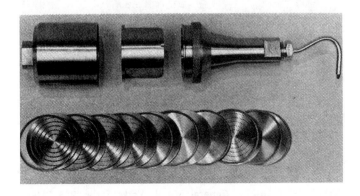

INSTRUMENT 14-21. Andersen stack head sampler.

INSTRUMENT 14-23. Andersen impactor preseparator, Model 50-160.

INSTRUMENT 14-25. Pilat Mark III Cascade Impactor.

of 10 atmospheres (140 psig) and 540°C (1000°F). It is made of stainless steel and uses gold wire seals. Collection substrates are constructed of 313 stainless steel shim stock and are 0.13 mm (0.005 in.) thick and 47 mm in diameter. At room temperature and a flow rate of 16 L/min, the 50% efficiency cutpoint diameters are 8.8, 6.4, 4.5, 2.5, 1.9, 1.3, and 0.62 µm. The impactor is 11 cm in diameter, 20 cm long, and weighs approximately 4 kg.

14-25. Pilat UW Source Test Cascade Impactor
Pollution Control Systems Corporation

The Pilat (UW) Mark 3 Cascade Impactor includes a single jet inlet stage in the sampling nozzle, 6 multi-jet stages, 7 particle collection plates, and an outlet section with built-in filter holder (47mm diameter filters). Gas

volumetric sampling flowrate at the Mark 3 Cascade Impactor temperature and pressure is in the 0.5 to 1.0 ft³/min. (or 14 to 28 liters/min) range. Various substrates (stainless steel foil, glass fiber filter, ultrapure quartz microfiber, Teflon, etc.) can be used on the particle collection plates. Pollution Control Systems also offers a PCSC precutter that connects to the inlet of the UW Source Test Cascade Impactor, and provides a 14 µm precut when sampling at 14 L/min.

PM₁₀ Inlets

14-26. PM-10 Size Selective Hi-Volume Inlet
Graseby Andersen Samplers, Inc.

The Hi-Volume sampler PM_{10} inlet removes particles greater than 10 µm at sampling rates of 1.1 m³/min (40 cfm); it can be mounted on a high-volume sampler to provide PM_{10} sampling. This inlet was designed by McFarland[90] to give a consistent size precut, independent of wind speed and coarse particle loading. The inlet is made of aluminum, weighs 16 kg (35 lb), and measures 70 cm (28 in.) in diameter.

14-27. PM-10 Medium Flow Samplers (Series 254)
Graseby Andersen Samplers, Inc.

The medium flow samplers operate at 112 L/min (4 cfm) and are equipped with a PM_{10} inlet to remove particles greater than 10 µm. Particles are collected onto 102-mm filters and flow rates are regulated by a flow controller. The inlet is based on a design of

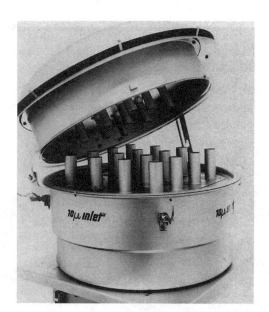

INSTRUMENT 14-26. General Metal Works PM-10 size-selective hi-volume inlet.

INSTRUMENT 14-27. General Metal Works Series 245 PM-10 medium flow fair sampler.

INSTRUMENT 14-28. Dichot inlet by General Metal Works.

McFarland.[90] The sampling module is made of aluminum, measures 134 cm high × 110 cm diameter (53 in. × 40 in.), and weighs 11 kg (25 lb). The control module measures 51 cm high × 74 cm wide × 46 cm diameter (20 in. × 29 in. × 18 in.) and weighs 44 kg (96 lb).

14-28. Andersen Dichot Inlet
Graseby Andersen Samplers, Inc.

The Andersen Dichot inlet is based on the design of McFarland.[90] It removes particles greater than 10 μm at

sampling rates of 16.7 L/min (1 m³/hr). It is designed to give a consistent size precut, independent of wind speed and coarse particle loading. The inlet is made of aluminum.

14-29. INSPEC Aerosol Spectrometer
BGI Incorporated

The aerosol spectrometer aerodynamically separates particles from 1 to 10 μm, as described in this chapter. Particles are collected on a single membrane filter wherein the position of deposition depends on particle diameter. The filter may be sectioned for analysis of mass or radioactivity, or it may be examined microscopically. The maximum aerosol sample rates are 0.1 L/min. No substrate coatings are needed.

Cyclones

14-30. Cyclade (Series 280)
Graseby Andersen Samplers, Inc.

The cyclade consists of a train of two to six cyclones (depending on model) followed by a 64-mm back-up filter. The cyclones are designed for stack sampling and are based on the design of Smith *et al.*[116] For stack temperatures of 140°C and a flow rate of 28 actual L/min, cutpoints for the model containing five cyclones are 0.57, 1.1, 2.7, 3.5, and 7.5 μm. Isokinetic sampling nozzles are available. Cyclones and the filter holder are

INSTRUMENT 14-29. INSPEC inertial particle spectrometer.

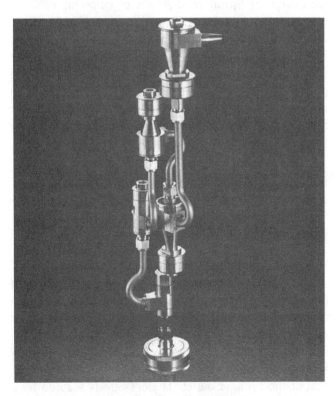

INSTRUMENT 14-30. Andersen Series 280 Cyclade.

INSTRUMENT 14-31. In-Tox cyclone system.

INSTRUMENT 14-32. BGI-4 respirable dust cyclone.

made of 316 stainless steel with C-ring seals. All units will fit through a 20-cm- (4-in.-) diameter sampling port. Length: 36 to 74 cm (14–29 in.), depending on the model. Weight: 3 to 6 kg (7–12 lb).

14-31. Cyclone Sampling Train

In-Tox Products

The cyclone train consists of five cyclones based on the design of Smith *et al.*,[116] with cutpoints of 0.32, 0.65, 1.4, 2.1, and 5.4 at a sampling rate of 28 L/min (1 cfm). The laboratory model cyclones are made of brass. The stack sampling train uses a folded configuration that can pass through a 10-cm- (4-in.-) diameter sampling port; it is constructed of stainless steel with Neoprene O-ring seals. Cyclones may be purchased individually. Sizes of individual cyclones range from 4 cm × 8 cm for the smallest to 5 cm × 13 cm for the largest.

Personal and Microenvironmental Sampling Cyclones

14-32. Respirable Dust Cyclone

BGI Incorporated

The BGI-4 respirable dust cyclone is a new unit designed to match a respirable curve with a 50% cutpoint at 4 μm at a sample flow rate of 2.3 L/min. The cyclone is all-metal construction with a black oxide finish. Aerosol samples are collected by a 37-mm disposable plastic filter cassette that presses over an O-ring seal at the cyclone outlet. The cassette is secured by a spring steel lapel clip.

14-33. "KTL" and "GK" Cyclones

BGI Incorporated

The KTL cyclone is a small cyclone designed to provide a 2.5 μm cutpoint at a flow rate of 4 L/min. It is based on the GK family of cyclones described by Kenny and Gussman (J Aerosol Sci, 1997). It was developed in response to a request from the Finish Institute of Public Health to provide cyclones for a European indoor air quality study. It is manufactured from nickel-plated aluminum, and is designed to hold a 37 mm diameter filter cassette. BGI has announced that it will be offering other GK designed cyclones that provide 2.5 μm or 4 μm cutpoints at other flow rates.

14-34. Cyclone for Personal Filter Cassette

SKC, Inc.

The SKC Cyclone is a lightweight aluminum respirable dust sampler that is used with a filter loaded into a three-piece filter cassette. The cyclone separates

INSTRUMENT 14-34. SKC cyclone for personal filter cassette.

INSTRUMENT 14-35. GS Cyclone.

dust particles according to size with respirable particles collecting on the filter and the larger particles falling into the grit pot to be discarded. The SKC Cyclone gives sharp size selection between respirable fractions at 4 or 5 μm meeting the ACGIH/ISO/CEN convention, at 2.5 L/min, specified in NIOSH Method 0600 and the BMRC (British Medical Research Council) curve that stipulates a 5 μm cut-point at 1.9 L/min. Its aluminum construction eliminates static charge build-up. Models are available in 25 or 37 mm for use with 3-piece cassettes. An aluminum calibration chamber that fits over the stem of the cyclone is available.

14-35. GS Cyclone

SKC, Inc.

The GS is a 10 mm cyclone made from conductive plastic, and holds a filter cassette for the collection of respirable dust particles. Its removable cassette adapter securely holds the filter in place. The GS Cyclone has a 50% cutpoint of 4.0 μm at a flow rate of 2.75 L/min,

and is designed to meet the ACGIH-CEN-ISO curve. The 2.75 L/min flow rate is higher than many other cyclones that provide this cutpoint. The conductive plastic prevents static collection. It also has a tangential inlet to lessen sampling errors from particle impaction on the wall opposite the inlet.

14-36. Plastic Cyclone

SKC, Inc.

The SKC Plastic Cyclone is a small lightweight conductive plastic unit constructed to collect respirable dust onto a 25 mm or 37 mm diameter filter. It is designed for a 50% cutpoint of 5.0 μm at 1.9 L/min and a 4.0 μm cutpoint at 2.2 L/min. The filter cassettes are reusable, and are made of lightweight conductive plastic with a stainless steel support grid to hold the filter securely in place during sampling. Cassettes are available in 25 or 37 mm diameters, and include a transport clip. The cyclone is equipped with an outlet n the cap that provides an easy means of calibrating without the use of an adapter.

14-37. Spiral Sampler

SKC, Inc.

The spiral sampler is a compact unit that uses a spiral inlet to provide a precut for respirable or $PM_{2.5}$ sampling. It is made of conductive plastic and accommodates a 25 mm diameter filter. Two designs are available, a $PM_{2.5}$ spiral that provides a 2.5 μm cutpoint at a flow rate of 2 L/min, and a respirable spiral that provides a 4.0 μm cutpoint at 2.5 L/min.

14-38. Cyclone Respirable Dust Sampler

Mine Safety Appliances Company

The MSA gravimetric dust sampler uses a 10-mm cyclone followed by a 37-mm filter. The cyclone

INSTRUMENT 14-36. SKC Plastic Cyclone.

INSTRUMENT 14-37. SKC Spiral Sampler.

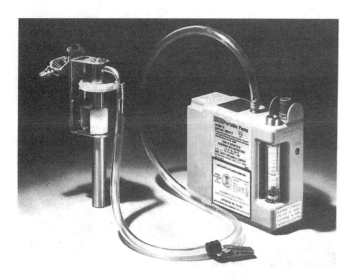

INSTRUMENT 14-38. MSA cyclone respirable dust sampler.

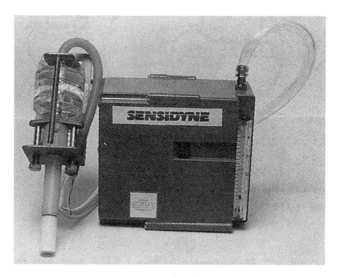

INSTRUMENT 14-39. Sensidyne Model BDX99R personal cyclone.

provides a respirable 3.5-μm precut at a flow rate of 2 L/min. The cyclone is used with a battery-powered pump capable of 8 hours of continuous operation. At the end of a sampling period, the weight of dust on the filter is established. From this measurement, the respirable dust concentration per cubic meter of air is determined.

14-39. Respirable Cyclones
Sensidyne, Inc.

The Sensidyne cyclone is designed to separate the respirable fraction of airborne dust from the nonrespirable fraction. It utilizes a 10 mm nylon cyclone, and is followed by a filter cassette. The cyclone approximates the ACGIH respirable sampling curve when operated at a flow rate of 1.7 Lmin⁻¹. The respirable fraction is captured by the filter membrane while the larger, nonrespirable particle fraction falls into the lower section of the cyclone.

Aerosol Centrifuges

14-40. LAPS Aerosol Centrifuge
In-Tox Products

The LAPS (*Lovelace Aerosol Particle Separator*) is an aerosol centrifuge with an expanding spiral duct. It is based on the design of Kotrappa and Light.[148] Particles are size segregated by aerodynamic diameter and deposited along a 3-cm × 46-cm foil that is mounted along the outside wall of the flow channel. For operation at 4500 rpm, with an aerosol flow rate of 0.4 L/min and a total flow rate of 5 L/min, particles between 0.4 and 4 μm are collected. The LAPS is 18 cm in diameter and weighs 14 kg (including motor).

Elutriators

14-41. Hexhlet
Casella London, Ltd.

The Hexhlet is a horizontal elutriator followed by a filter thimble for the collection of respirable particles (less than 3.5 μm in aerodynamic diameter). A schematic of the Hexhlet horizontal elutriator is found in Figure 14-27. The

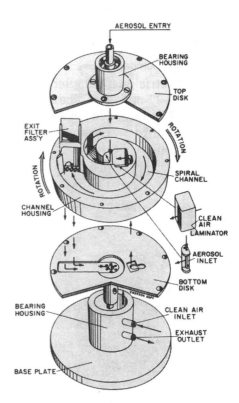

INSTRUMENT 14-40. Schematic of the LAPS aerosol centrifuge. Reprinted from *Aerosol Measurement* with permission, University Presses of Florida.

INSTRUMENT 14-41. Casella Hexhlet. The casing that fits over the soxhlet filter thimble is removed and stands at the right of the picture.

INSTRUMENT 14-42. Gravimetric Dust Sampler, Type 113A.

sampling rate is 50 L/min. A 42.5-mm-diameter filter can be used in place of the filter thimble. Construction is of aluminum. Dimensions: 17 cm × 17 cm × 50 cm (6.5 in. × 6.5 in. × 20 in.). Weight: 5 kg (11 lb).

14-42. Gravimetric Dust Sampler
Casella London, Ltd.

This is a horizontal elutriator followed by a filter. At a sampling rate of 2.5 L/min, the elutriator provides a respirable precut at 3.5 μm in aerodynamic diameter. The unit contains a diaphragm pump, a flowmeter, and an elapsed time counter; it is housed in a stainless steel case. Dimensions: 17 cm × 23 cm × 11 cm (7 in. × 9 in. × 4.5 in.). Weight: 4 kg (9 lb).

14-43. Cotton Dust Sampler
Graseby Andersen Samplers, Inc.

The cotton dust sampler is a vertical elutriator designed for the sampling of particles below 14 μm. The air flow moves upward through the elutriator and particles that penetrate are collected on a 37-mm filter

mounted at the elutriator exit. Construction is of aluminum. Weight: 10 kg (22 lb).

Impingers for Particle and Vapor Collection

14-44. Greenburg–Smith Impinger
Ace Glass

This impinger follows the design of the original Greenburg–Smith impinger for the collection of dusts. It is an all-glass impinger with a ground glass joint and a 500-ml capacity. One version of the impinger uses an attached impingement disk, while the other uses the bottom of the flask as the impingement plate. Nominal sampling rate is 28 L/min (1 cfm).

INSTRUMENT 14-45. MSA Monitaire Sampler, Model S, with MSA midget impinger.

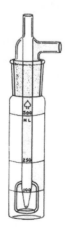

INSTRUMENT 14-44. Greenburg–Smith impinger.

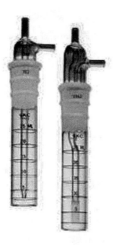

INSTRUMENT 14-46. SKC Midget Impingers.

14-45. Midget Impinger

Mine Safety Appliances Company

The MSA midget impinger can be used for both dust and vapor collection. It is an all-glass impinger with a ground glass joint. The collection volume is 25 ml. It can be operated at 2.8 L/min with a battery-powered pump.

14-46. Midget Impingers

SKC, Inc.

SKC laboratory-quality Midget Impingers (bubblers) accurately collect airborne hazards into a liquid medium. Impingers can be worn either in a holster near the breathing zone for personal sampling, or mounted on an air sample pump for area sampling. SKC offers three types of impingers: the Standard Midget with precisely-placed tip; the Special Midget with fritted tip to

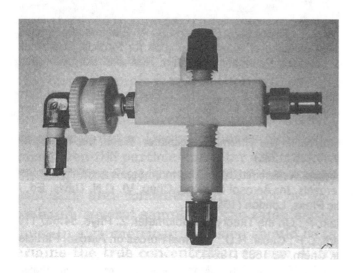

INSTRUMENT 14-47. Point-to-Plane Electrostatic Precipitator.

increase contact between the air sample and the liquid (many NIOSH and OSHA procedures specify fritted tip impingers); and the Spill-resistant Midget with a special outlet to prevent spills if the impinger is tipped. Each SKC Impinger contains a printed serial number on both sections to assist with sample identification and ensure impinger part matching. SKC recommends an impinger trap when using impingers to prevent impinger liquids from being drawn into the sample pump.

Electrostatic Precipitators

14-47. Point-to-Plane Electrostatic Precipitator

In-Tox Products

This instrument is the commercial version of the point-to-plane electrostatic precipitator described by Morrow and Mercer.[27] This device is useful for collecting aerosol samples for electromicroscopic examination. Samples are drawn into a 3/8-in. diameter cylindrical channel at a chosen volumetric rate between 50 cm³/min and 1.0 L/min for a period of 1–5 min. A sharp needle near one side of the channel serves as a high voltage electrode, producing a corona discharge with an electrical potential near 7000 volts DC. In opposition to this needle, on the other side of the channel, a carbon-substrated electron-microscope grid is mounted on a metal post which serves as the other electrode. Aerosols drawn through this device are unipolarly charged and collected at random on the grid by action of the electric field forces.

This point-to-plane electrostatic precipitator is constructed of Delrin plastic with a channel threaded at each end so that one end can be used to connect to the sample probe and the other end can be connected to a back-up filter holder and vacuum line. The body of the unit is 12 cm long. High-voltage electrodes are reversible so that the corona discharge can be either positive or negative as desired by the user.

The precipitator is available separately or with a solid-state power supply to provide the necessary 7000 DC high voltage at the normal operational current of 5 microamperes. No flowmetering equipment or vacuum pump is included; provision for these must be made by the user.

14-48. Combination Point-to-Plane Electrostatic Precipitator

In-Tox Products

This instrument is similar to the point-to-plane electrostatic precipitator (ESP), except that two particulate samples (one for transmission, one for scanning electron

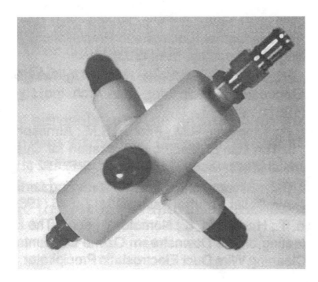

INSTRUMENT 14-48. Combination Point-to-Plane Electrostatic Precipitator.

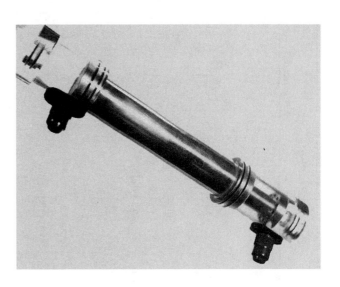

INSTRUMENT 14-49. Concentric Electrostatic Precipitator.

microscopy) can be collected simultaneously. A single needle provides corona discharge for both collectors. Flow and current characteristics are similar to the point-to-plane ESP. The power supply is portable and operates on 100 VAC, 60 cycle; high-voltage DC up to 5 kV is produced.

14-49. Concentric Electrostatic Precipitator

In-Tox Products

This particle collector is cylindrical in design with an axially mounted needle at the inlet end for charging.

Particles drift to a cylindrical foil collector within a 3/4-in.-diameter brass inner cylinder. The total length of the precipitator is 12 in., with a working length of 9 in. The outer cylinder is 2 in. in diameter, constructed of methacrylate plastic. The power supply is the same as for the combination point-to-plane electrostatic precipitator.

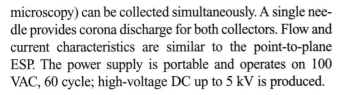

TABLE 14-I-1. Commercially Available Impactors

Description	Manufacturer[A]	Sampler Name	Flow Rate (L/min)	No. of Stages	Cutpoints (Range, μm)	Reference Author (Ref. No.)	Comments[B]
Cascade Impactors for Ambient Air Sampling							
14-1	GRA	Sierra/Marple Model 210	7	10	0.16-18		1
14-2	GRA	Sierra/Marple Model 260	0.3-20	6	0.5-20		2
14-3	GRA	Low pressure impactor	3	12	0.08-35	Vanderpool et al.[60]	3
14-4	GRA, GMW	One ACFM ambient impactor	28	8	0.4-10	Rao and Whitby[45]	
14-5	GRA	Flow sensor ambient impactor	28	7	0.4-6		
14-6	HAU	Berner impactor	30	9	0.063-16.7	Wang and John[4]	4
14-7	ITP	Mercer 7-stage impactor (02-100)	0.1	7	0.33-3.1	Mercer et al.[13]	
14-7	ITP	Mercer 7-stage impactor (02-130)	1	7	0.32-4.5	Mercer et al.[13]	
14-7	ITP	Mercer 7-stage impactor (02-150)	2	7	0.25-5.0	Mercer et al.[13]	
14-7	ITP	Mercer 7-stage impactor (02-170)	5	7	0.5-5.0	Mercer et al.[13]	
14-8	ITP	Multijet CI (02-200)	10	7	0.5-8	Newton et al.[239]	
14-8	ITP	Multijet CI (02-220)	15	7	0.5-8	Newton et al.[239]	
14-8	ITP	Multijet CI (02-240)	20	7	0.5-8	Newton et al.[239]	
14-8	ITP	Multijet CI (02-260)	28	7	0.5-9	Newton et al.[239]	
14-9	MSP	MOUDI (micro-orifice impactor)	30	10	0.56-10	Kuhlmey et al.[87] Marple et al.[243]	5
Impactors for Ambient HiVol Samplers							
14-10	GRA	HiVol impactor, Series 65-800	1130	1	3.5	Burton et al.[12]	8
14-10	GRA	Hivol impactor, Series 65-800	565	4	1.1-7.0	Burton et al.[12]	8
14-11	GRA, GMW	HiVol impactor, Series 230	1130	4	0.49-7.2	Willeke,[65] Knuth[11]	9

TABLE 14-I-1 (cont.). Commercially Available Impactors

Description	Manufacturer[A]	Sampler Name	Flow Rate (L/min)	No. of Stages	Cutpoints (Range, μm)	Reference Author (Ref. No.)	Comments[B]
14-11	GRA, GMW	HiVol impactor, Series 230	565	6	0.41–10		9
Personal Samplers							
14-12	GRA, GMW	Marple personal sampler (Model 290)	2	8	0.5–20	Rubow et al.,[24] Hinds[39]	1
14-12	SKC	Marple personal sampler	2	8	0.5–20	Rubow et al.,[24] Hinds[39]	1
14-13	MSP	Personal environmental monitor	4 or 10	1	2.5		
14-13	MSP	Personal environmental monitor	4 or 10	1	10		
14-14	MSP	Micro environmental monitor	10	—	—		
14-15	URG	Portable size selective impactor	4	1	2.5	Marple et al.[243]	
Virtual Impactors							
14-16	BGI	Cascade centripeter	30	3	1.2, 4, 14	Hounam and Sherwood[93]	
14-17	GRA, GMW	Dichotomous sampler	16.7	1	2.5	Loo,[97] McFarland et al.[94]	
14-18	MSP	High volume virtual impactor	1130	1	2.5	Marple et al.[243]	
14-19	MSP	Microcontaminant particle sampler	30	1	1		
Source Test Impactors							
14-20	GRA	In-stack air sampler, Series 220	7	9	0.16–18		
14-21	GRA	Stack sampling head (Mark III, IV)	3–21	8	0.4–11		1
14-22	GRA	High capacity stack sampler	14	3	1.5–11		
14-23	GRA	Impactor preseparator	1	21	10		
14-24	ITP	High temp., high pressure impactor	16	7	0.62–8.8		
14-25	PCS	UW source test cascade impactor	28	10	0.2–20	Pilat et al.[20]	

TABLE 14-I-1 (cont.). Commercially Available Impactors

Description	Manufacturer[A]	Sampler Name	Flow Rate (L/min)	No. of Stages	Cutpoints (Range, μm)	Reference Author (Ref. No.)	Comments[B]
14-25	PCS	UW high capacity source test impactor	28	3	1.5–11	Pilat et al.[20]	
14-25	PCS	UW low pressure source test impactor	28	14	0.05–20	Pilat et al.[20]	
PM$_{10}$ Inlets							
14-26	GRA, GMW	Hi-Volume PM-10 inlet	1130	1	10	McFarland et al.[122]	
14-27	GRA, GMW	Medium flow PM-10 inlet	112	1	10	McFarland et al.[122]	
14-28	GRA, GMW	Dichotomous sampler inlet	16.7	1	10	McFarland et al.[122]	
14-29	BGI	Aerosol Spectrometer	–	–	–	–	

[A]See Table 14-I-5 for explanation of manufacturers' codes.
[B]Comments:

1. Radial slot design.
2. Circular jets, interchangeable nozzles.
3. Four low-pressure stages.
4. One round jet per stage.
5. Micro-orifice plates of 2000 jets on bottom stages.

Comments (con't.)

6. Two low-pressure stages added to 1 CI.
7. Uses quartz crystal collection surfaces for continuous mass measurement.
8. Fits on Hi-Volume, round jets.
9. Fits on Hi-Volume, rectangular jets.
10. Collection directly onto agar plates.
11. Slot impactor with rotating turntable for agar plates.

TABLE 14-I-2. Cyclones and Their Performance Characteristics

Description	Manufacturer	Cyclone Name	Flow Rate Range (L/min)	d_{50}Range (μm)	Internal Dimensions		Reference Author (Ref. No.)
					Body (cm)	Outlet (cm)	
14-30, 14-31	GRA, ITP	SRI V	7–28	0.3–2.0	1.52	0.36	Smith et al.[152]
14-30, 14-31	GRA, ITP	SRI IV	7–28	0.5–3.0	2.54	0.59	Smith et al.[152]
14-30, 14-31	GRA, ITP	SRI III	14–28	1.4–2.4	3.66	0.83	Smith et al.[152]
	—	AIHL	8–27	2.0–7.0	3.66	1.05	John and Reischl[162]
14-30, 14-31	GRA, ITP	SRI II	14–28	2.1–3.5	4.13	1.05	Smith et al.[152]
	—	Aerotec	22–55	1.0–5.0	4.47	0.75	Chan and Lippmann[153]
14-30, 14-31	GRA, ITP	SRI I	14–28	5.4–8.4	5.08	1.50	Smith et al.[152]
Personal and Microenvironmental Sampling Cyclones							
14-32	BGI	Respirable	2.3	4.0	—	—	
14-33	BGI	KTL/GK	4	—	—	—	
14-34	SKC		18.5–29.6	0.1–1.0	—	—	Blachman & Lippmann[160]
14-35	SKC	GS Cyclone	2.75	—	—	—	
14-36	SKC	Plastic Cyclone	1.9–2.2	—	—	—	
14-37	SKC	Spiral Sampler	2–2.5	—	—	—	
14-38	MSA	10-mm cyclone	0.9–5	1.8–7.0	1	0.25	Blachman & Lippmann[160]
14-39	SEN	(also called Dorr-Oliver)	5.8–9.2	1.0–1.8			Blachman & Lippmann[160]
14-39	SEN	1/2" HASL	8–10	2–5	3.11	0.50	—
14-39	SEN	1" HASL	65–350	1.0–5.0	7.6	1.09	Chan and Lippmann[153]
		BK 76	400–1100	1.0–3.0		3.8	Beeckmans and Kim[131]

TABLE 14-I-3. Aerosol Centrifuges

Description	Manufacturer	Sampler	Duct Length (cm)	Aerosol Flow (L/min)	Total Flow (L/min)	Rotational Speed (rpm)	Particle Size Rar. (mm)	Reference Author (Ref. No.)	Comments
Spiral Duct Centrifuges									
		Stöber spiral centrifuge	180	0.05-2	5-19	1500-6000	0.08-6	Stöber and Flachsbart[187]	
		Stöber small rotor	60	0.05-	1-2	3000	0.15-2	Hochrainer and Stöber[200]	
14-40	ITP	LAPS Centrifuge	46	0.2-0.5	5-10	1500-4500	0.3-4	Kotrappa and Light[192]	
		Goetz spectrometer	30	1-5	—	6000-18000	0.05-1	Goetz et al.[193] and Gerber[194]	1
Cylindrical Duct Centrifuges									
—		Concentric spectrometer	40	0.05-0.5	1-5	2000-6000	0.3-4	Tillery[198]	
		Constant radius centrifuge	30	0.03	0.5	10000	0.2-1	Hochrainer[199]	
High Flow Rate Centrifuges									
—		Drum centrifuge	—	5000-20000	—	1000-3000	>0.5	Holländer et al.[197]	2
Conifuges and Annular Duct Centrifuges									
—		Ring slit conifuge	19	0.1-1	5-14	1500-9000	0.1-4	Stöber and Flachsbart[187]	
		Cylindrical centrifuge	3	0.01	0.6	3600-10000	0.5-2	Hochrainer and Brown[196]	

TABLE 14-I-4. Impingers for Particle Collection

Description	Manufacturer	Sampler Name	Material	Sample Rate (L/min)	Capacity (mL)	Impingement Distance (mm)
14-44	AGI	Greenburg-Smith	Glass	28	500	5
14-45	MSA	Midget	Glass	2.8	25	5
14-46	SKC	Midget	Glass	2.8	25	5

Impingers for vapor collection are described in Chapter 16.

TABLE 14-I-5. List of Instrument Manufacturers

AGI	Ace Glass Incorporated P.O. Box 688 1430 Northwest Blvd. Vineland, NJ 08362 (800) 223-4524 FAX: (800) 543-6752 *www.aceglass.com*	GRA	(Graseby) Andersen Instruments 500 Technology Ct. Smyrna, GA 30082-5211 (770) 319-9999 or (800) 241-6898 FAX: (770) 319-0336 *www.anderseninstruments.com*	PCS	Pollution Control Systems Corp. P.O. Box 15770 Seattle, WA 98115 (206) 523-7220 FAX: (206) 523-7221 *www.cascadeimpactor.com*
BGI	BGI Incorporated 58 Guinan Street Waltham, MA 02451 (781) 891-9380 FAX: (781) 891-8151 *www.bgiusa.com*	HAU	Hauke GmbH & Co. KG P.O. Box 103 A-4810 Gmunden, Austria 43-7612-63758 FAX: 43-7612-641338	SKC	SKC Inc. 863 Valley View Road Eighty Four, PA 15330 (724) 941-9701 or (800) 752-8472 FAX: (724) 941-1369 *www.skcinc.com*
CAI	California Measurements, Inc. 150 E. Montecito Avenue Sierra Madre, CA 91024-1934 (626) 355-3361 FAX: (626) 355-5320 *info@californiameasurements.com* *www.californiameasurements.com*	ITP	In-Tox Products P.O. Box 2070 Moriarty, NM 87035 (505) 832-5107 FAX: (505) 832-5092	SEN	Sensidyne, Inc. 16333 Bay Vista Drive Clearwater, FL 33760 (727) 530-3602 or (800) 451-9444 *www.sensidyne.com*
CLL	Casella London Limited Regent House Wolseley Rd. Kempston, Bedford, MK42 7JY 44 (0)1234 841441 FAX: 44 (0)1234 841490 *www.casella.co.uk*	MSP	MSP Corporation 1313 Fifth Street, SE Minneapolis, MN 55414 (612) 379-3963 FAX: (612) 379-3965 *www.mspcorp.com*	URG	URG Corporation 116 Merritt Mill Rd. P.O. Box 368 Carrboro, NC 27510 (919) 942-2753 FAX: (919) 942-3522
		MSA	Mine Safety Appliances Co. 121 Gamma Drive Pittsburgh, PA 15238-2937 or P.O. Box 426 Pittsburgh, PA 15230-0426 (412) 967-3000 or (800) 672-2222 FAX: (412) 967-3451 *www.msanet.com*		

Chapter 15

Direct-Reading Instruments for Analyzing Airborne Particles

David Y. H. Pui, Ph.D. and Da-Ren Chen, Ph.D.

Mechanical Engineering Department, University of Minnesota, 111 Church Street, S.E., Minneapolis, MN 55455

CONTENTS

Introduction

Aerosol sampling instruments described in previous chapters are used to collect particles for subsequent microscopic, gravimetric, or chemical analyses. Instruments considered in this section are more complex. Sampling and analysis are carried out within the instrument and the property of interest can be obtained immediately. These are called direct-reading instruments.

Recent development of direct-reading instruments capable of real-time measurement is largely the result of the availability of modern electronic components such as laser illumination sources, high sensitivity photo- or electrometer detectors, operational amplifiers, miniature power supplies and microprocessors. Direct-reading instruments for real-time aerosol measurements are

available to cover particles in the size range of 0.002 to 100 µm. These instruments have fast time response and can follow rapid changes in both particle size and concentration. Good counting statistics can also be obtained because repeated measurements can be performed in a short time. However, these instruments usually rely on indirect sensing techniques and more calibration efforts are usually required. Figure 15-1 summarizes the principal direct-reading instruments used in aerosol studies, their measuring ranges, and the monodisperse aerosol generators used for their calibration. A comprehensive review of the topic has been given in two papers by Pui[1] and Liu and Pui.[2] Detailed review of some direct-reading instruments is contained in the book by Willeke and Baron.[3] The monodisperse aerosol generators for the calibration of these direct-reading instruments have been

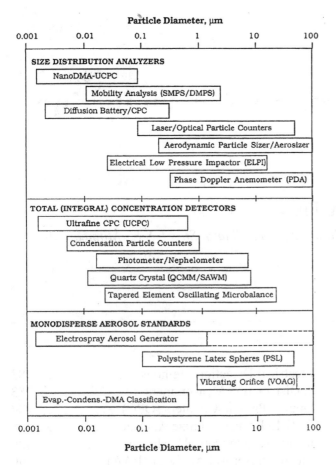

FIGURE 15-1. Measuring range of principal aerosol instruments and their calibration standards.

described in a previous chapter (Chapter 9) by Cheng and Chen.

Different direct-reading instruments measure different aerosol properties. Although many instruments provide data in terms of particle "size," this "size" is actually derived from one of many possible particle properties such as its gravimetric, optical, aerodynamic, mechanical, or force field mobility behaviors. Thus, the particle sizes measured by different principles may not be directly compared without some correction of the data to account for these differences. Other aerosol properties determined by direct-reading instruments include aerosol number concentration, aerosol mass concentration, size distribution, opacity, and chemical composition.

The sensitivity of these instruments is generally limited by one of two factors: 1) the random property fluctuations of the accompanying gas molecules or 2) the noise level of the electronic circuit which converts the property measured to an electronic signal. Accuracy is dependent upon the relationship between signals in a sensing zone and the aerosol property. Although this relationship is often based on first principles, it is more common to establish an empirical relationship using a

"well-calibrated" aerosol system. The danger of this approach is that the real aerosol measured may have a different, unknown relationship between the resulting signal and the aerosol property so that an inaccurate "particle size" may be obtained.

The users of direct-reading instruments must also beware of comparing properties of the same aerosol determined by several direct-reading instruments, particularly those using different principles because this comparison is likely to give contradictory information. It is important to know what property has been changed in the sensing zone and how this is assumed to be related to an aerosol property.

Atmospheric and Workplace Size Distributions

Aerosol particles may consist of microscopic bits of materials in the form of molecular clusters of only a few nanometers in diameter to particles of a near macroscopic size of a fraction of a millimeter in diameter. The nominal size range of interest in aerosol science is 0.002 to 100 μm. The lower size limit, 2 nm, corresponds to the smallest particles that can be detected by a direct-reading instrument, e.g., the ultrafine condensation particle counter. The largest particles are those that would begin to fall out from the gas medium too quickly to be considered an aerosol, e.g., rain drops. Particle size determination is important because many aerosol properties are a function of particle size. Particle deposition on wafers in semiconductor manufacturing and deposition in different regions of the human respiratory tract in health studies are determined by particle size.

Particle size distribution is usually presented as particle concentration in number, surface or volume (mass) weightings as a function of particle size. Chapter 6 gives a more detailed discussion of size distribution presentation. Vast databases have been collected on atmospheric size distributions from many air pollution studies. As a result, there is now a general understanding of the aerosol formation and removal mechanisms in the atmosphere. Figure 15-2 shows a volume-concentration size distribution of atmospheric aerosols measured near a combustion source, i.e., automobile traffic.[4] Atmospheric aerosol is seen to consist of three modes and cover a size range of 0.002–100 μm. The three principal modes are referred to as the nuclei mode (0.002 –0.05 μm), the accumulation mode (0.05–2 μm) and the coarse particle mode (2–50 μm). The accumulation mode and the coarse particle mode can be observed in nearly every atmospheric measurement, whereas the nuclei mode can only be measured near an aerosol source involving high temperature

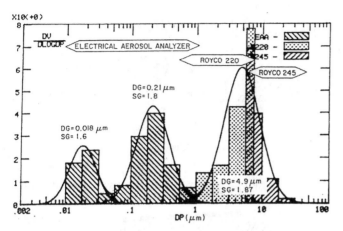

FIGURE 15-2. Trimodal size distribution of atmospheric aerosols according to Whitby.[4]

and other gas-to-particle conversion processes, e.g., combustion and photochemical aerosols. The relative amounts of material in the different modes depend on the locality and weather and on the pollution by local industry and motor vehicle traffic.

The chemical compositions of aerosol particles are different for the three modes. Particles in the coarse particle mode generally contain various crustal elements such as silicon, iron, aluminum, calcium, sea salt and plant particles. Particles in the nuclei and accumulation modes contain sulfates, nitrates, elemental and organic carbon, ammonium, lead and other trace constituents. This evidence suggests that the formation mechanisms

are different for the fine and coarse particle modes. It is expected that the removal mechanisms will also be different owing to the wide particle size range covered by the three modes. A comprehensive review of the formation and removal mechanisms of atmospheric aerosols can be found in the paper by Whitby.[4]

In the workplace, the particle size distribution is expected to contain the accumulation mode, which can come from the infiltrated atmospheric aerosol or from the aged indoor aerosol. The nuclei mode and the coarse particle mode are expected to vary greatly depending on the work processes involved. The particle size distribution can also be modified significantly by the filtration devices installed at the workplace.

Table 15-1 shows the modal parameters of some typical atmospheric and workplace aerosols.[1] Because the nuclei mode consists of a very small mass fraction in most cases, it has been incorporated into the accumulation mode. The combined nuclei and accumulation modes are referred to as the fine particle mode. As shown in Table 15-1, the workplace size distributions and mass concentrations vary a great deal depending on the processes involved. The mass mean diameters of the coarse particle mode in the workplace aerosols can be much larger than those in the atmospheric aerosols. In general, most mechanical processes, e.g., mining, sanding, machining, and harvesting, produce high concentrations of large particles in the coarse particle mode. For processes involving high temperature and gas-to-particle

TABLE 15-1. Modal Parameters of Some Atmospheric and Workplace Aerosols[1]

Aerosol Source	Fine Particle Mode			Coarse Particle Mode		
	Mass Mean Dia., μm	Geometrical Std. Dev.	Conc., μg/m³ or (% of Total)	Mass Mean Dia., μm	Geometrical Std. Dev.	Conc., μg/m³ or (% of Total)
Atmospheric Background[4]	0.32	2.0	4.45	6.04	2.16	25.6
Atmospheric Urban[4]	0.32	2.16	38.4	5.7	2.21	30.8
Power Plant Plume[4]	0.18	1.96	37.5	5.5	2.5	24.0
Coal Mine, Diesel[5]	0.17	2.0	880.	6.8	2.3	1800.
Coal Mine, Electric[5]	0.46	2.6	60.	7.2	2.0	1200.
Metal Mine, Diesel[6]	0.18	2.5	600.	5.1	3.1	700.
Soda Ash Mine, Diesel/Scrubber[6]	0.16	2.4	200.	6.1	1.8	7500.
Wood Dust, Sanding[7]	0.42	1.7	(3.1%)	35.	2.3	(97%)
Wood Dust, Machining[7]	5.5	3.1	(27%)	65.	2.1	(73%)
Spray Can[7]	1.2	1.5	(12%)	5.6	1.8	(88%)
Commercial Kitchen[8]	0.1	1.8	(50%)	3.0	2.5	(50%)
Rice Harvest[9]	Respirable dust		1180.	(Total - Resp.) dust		10700.

conversions, e.g., diesel exhaust and cooking emissions, a significant increase in mass concentrations is found in the fine particle mode. This observation is clearly demonstrated in the underground coal mines equipped with diesel engines. Most of the diesel exhaust appears in the fine particle mode while the coal dust appears in the coarse particle mode. This has led to the design of personal aerosol samplers that incorporate an inertial impactor to separate the diesel and coal masses in their respective collection substrates.[10, 11]

Integral Concentration Measurement

Integral concentration detectors are those that can be used to measure some integral parameters of an aerosol over its entire size distribution, such as the total number or mass concentration or total light-scattering or extinction coefficients. Several of the more widely used instruments are described below. These include the condensation particle counter (CPC) for number concentration determinations, the beta gauge, quartz-crystal microbalance and tapered-element oscillating microbalance (TEOM) used for mass concentration measurement, and other miscellaneous techniques based on light scattering, contact or diffusion charging.

Condensation Particle Counter (CPC)

The condensation particle counter (CPC), or condensation nucleus counter (CNC), is widely used to measure particles in the diameter range from approximately 0.005 to 1.0 µm. The instrument operates by passing the aerosol stream through a vapor-supersaturated region produced either by adiabatic expansion or direct contact cooling to cause vapor condensation on the particles. The particles are then grown to a size where they can be detected optically by light scattering. Recent development includes a continuous flow, direct contact type CPC[12, 13] and the mixing of a hot vapor stream and a cool aerosol stream to achieve a super-saturation condition.[14]

A schematic diagram of a commercially available, continuous flow CPC is shown in Figure 15-3.[15] In this instrument, butyl alcohol is used as the working fluid. An air stream is saturated with this vapor in a saturator kept at 35°C. The subsequent cooling of this alcohol-vapor laden air stream in a thermoelectrically cooled condenser tube kept at 10°C produces the required super-saturation for the vapor condensation on the particles. Particles emerging from he condenser tube at a size of approximately 12 µm are then detected optically by light scattering. For low particle concentrations, the individual particles are counted. Above a particle concentra-

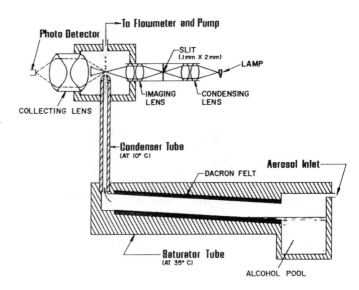

FIGURE 15-3. Schematic diagram of the condensation nucleus (particle) counter.[15]

tion of 10⁴ particles/cm³, the total light scattering from the droplet cloud is detected in a "photometric mode" to measure the total particle concentration. The concentration range of the instrument is from less than 0.01 particles/cm³ to more than 106 particles/cm³.

Detailed calibration studies of the CPC have shown that below a particle size of 0.005 µm the response of the instrument begins to drop off as a function of particle size.[16–22] The counting efficiency decrease can be attributed to particle loss in the flow passages in the instrument due to diffusion and the lack of 100% activation due to inhomogeneous vapor concentration distribution in the condenser.[23, 24] By introducing a clean sheath air around the aerosol stream in the CPC, Stolzenburg and McMurry[13] were able to increase the counting efficiency of the instrument to over 70% at a particle size of 0.003 µm. Figure 15-4 shows the details of such a sheath air CPC (Ultrafine Condensation Particle Counter, UCPC). Wilson et al.[25] using a similar design has developed a low pressure CPC for stratospheric aerosol measurements. Niessner et al.[26] have shown that by changing the supersaturation ratios in steps, the dependence of particle size on critical supersaturation for vapor condensation can be used for size distribution measurement.

Quartz-crystal Microbalance (QCM) and Surface Acoustic Wave Microbalance (SAWM)

Several sensors for near real-time mass concentration measurements have been developed. By depositing the particles on a quartz crystal, the natural vibrating frequency of the crystal can be affected and used as a measure of the deposited particle mass. The deposited

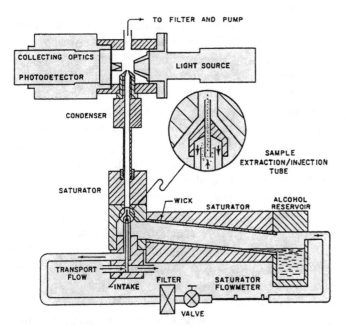

FIGURE 15-4. Schematic diagram of the sheath-air ultrafine aerosol condensation particle counter.[18]

particle mass is proportional to the frequency shift. The particle deposition can be achieved either by electrostatic precipitation[27, 28] or by inertial impaction.[29] The sensitivity of the quartz crystal microbalance is approximately about 109 Hz/g corresponding to a frequency shift of 1 Hz for a 10 MHz AT-cut quartz crystal. Figure 15-5 shows the schematic diagram of a battery operable piezoelectric microbalance for respirable aerosol detection.[30] The instrument incorporates a respirable impactor at the inlet to remove the non-respirable particles allowing the respirable particles to be deposited by electrostatic precipitation on the quartz crystal for measurement. The instrument can measure particle concentrations of approximately 0.05–5.5 mg/ml range.

A new development in QCMs enables the mass sensitivity to be significantly increased over the ones using AT-cut crystals. By modifying the quartz crystal and applying an electric field between two electrodes on the same surface (instead of through the crystal thickness) at several micrometers apart, the mode of excitation results in a natural frequency of up to 300 MHz.[31] The vibration on the same surface is called the surface acoustic wave (SAW) mode. The mass sensitivity of the SAW devices can achieve up to 10^{10}–10^{11} Hz/g.

Tapered-element Oscillating Microbalance (TEOM)

Recent advances in the vibrating mass sensing technique include the use of a low frequency vibrating mass in the form of a hollow tapered element[32] coupled to a filter collector or impactor. The operation principle of

TEOM is the same as the QCMs. Instead of relying on the natural frequency of a quartz crystal, the vibration of the hollow tapered element is initiated and maintained by an electronic feedback system. The oscillation of the tapered element is then monitored by a light-emitting diode and phototransistor aligned perpendicularly to the oscillation plane of the tapered element. Figure 15-6 shows a typical arrangement for the TEOM. The aerosol stream is drawn from the ambient air passed through the filter installed on the top of the tapered element. The collected particle mass is then inferred from the frequency difference before and after each sampling interval. Unlike QCMs, the collected particle mass is not directly proportional to the frequency shift. It is proportional to the difference of the inverse square of frequencies before and after sampling. This design extends the measurement range or the technique to mass concentration levels in the g/m³ range. The filter can usually hold the particle mass up to 2–6 mg. The application of the technique to particle measurement in high temperature and high-pressure gas streams of the pressurized, fluidized bed combustor has been reported by Wang.[33]

Beta-attenuation Mass Sensor

Instruments based on the attenuation of beta radiation through collected particle mass on a surface have been developed for respirable dust measurement in mining applications and for atmospheric studies.[34, 35] The particles can be deposited either by impaction or filtration. A two-stage continuous atmospheric mass monitor

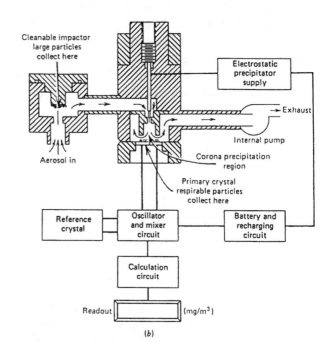

FIGURE 15-5. Schematic diagram of the respirable piezoelectric micro-balance (piezobalance).[30]

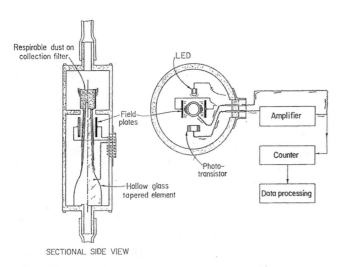

FIGURE 15-6. Schematic diagram of a Tapered Element Oscillating Microbalance (TEOM).

based on the beta attenuation principle has been constructed.[36] The specific instrument uses a ¹⁴C beta source and a solid-state silicon surface-barrier detector to measure the attenuated beta radiation. The particles are separated into two size fractions by impaction and filtration using a rotating tape.

Light-scattering Photometers and Nephelometers

For atmospheric studies, the total light-scattering coefficient of the airborne particles is important as it is related to atmospheric visibility or visual range. Measurement of the total light-scattering coefficient is usually made with a photometer or integrating nephelometer. For aerosols that differ only in concentration and with the same size distribution, the integral light-scattering measurement can be converted to mass concentration. Examples of such a correlation between total scattering and atmospheric mass concentration are given by Waggoner and Charlson.[37]

In the integrating nephelometer, shown schematically in Figure 15-7, the particles are illuminated in a sensing volume of approximately 1.0 L and scattered light from the particles reaches the photoreceptor at angles from 8° to 170° off axis. This simplifies the complex angular scattering relationship by summing the scattering over nearly the entire range of angles. Although the instrument was originally used to measure visual range, it has found its applications in the studies of the urban and rural atmospheric aerosol. In some cases, the scattering shows to be well correlated with the atmospheric mass concentration.[37, 38] The instrument is simple in construction and has been used in automobiles and aircraft for mapping the concentration of particles in the 0.1 to 1.0 μm range. These particles are chiefly responsible for degraded urban visibility. Some caution must be exercised when using the nephelometer in an environment with sooty particles since the scattering will be attenuated because of light absorption. In this case, the apparent concentration will be lower than expected.

Forward-scattering photometers, which employ a laser or incandescent light source and optics similar to dark field microscopy, have been commercially produced. A narrow cone of light converges on the aerosol cloud, but it is prevented from falling directly on the photoreceptor by a dark stop; only light scattered in the near forward direction falls on the receptor. The readout of these instruments is in mass or number concentration, but the calibration may change with composition and size distribution of the particles. Based on the solutions to Maxwell's equations, forward-scattering photometers are, however, less sensitive to the change in the refractive index of particles than are photometers at other commonly used sensing angles such as 30°, 45°, or 90°.

A forward-scattering photometer (45°–95°) which is a passive personal monitor for airborne particles has

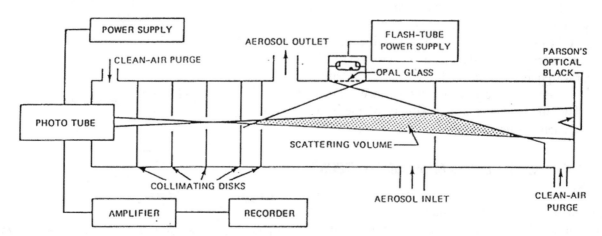

FIGURE 15-7. Schematic diagram of an integrating nephelometer.

been designed using microelectronics.[39] This instrument displays the current particle mass concentration for time intervals as small as 10 seconds and calculates time-weighted averages (TWAs) for up to a full shift for display or readout.

A multi-particle, light-scattering instrument that employs a long path and particle backward scattering is LIDAR.[40] A powerful pulsed laser is used, and the temporal analysis of back-scattering light intensity indicates the spatial distribution of particles. This type of instrument has been used to map smoke plume opacity in the vicinity of the stack. Unless the size distribution and composition of the particles are known, only a qualitative comparison of aerosol concentration at different locations can be made.

For aerosols composed of specific cations, such as Na^+, the detection of aerosol mass can be achieved by thermal excitation in an H_2 flame. One such instrument[41] has been used for laboratory filter testing. The number and size of particles is determined in a similar fashion to the conventional single particle counter, using a photomultiplier and a multi-channel pulse height analyzer.

Light-attenuating Photometers

Transmissometers and other light-attenuating photometers are based on the simple extinction of light by particles. In order to get a measurable change in extinction (>5%) , the sensing volume must contain a large number of particles. This means that there must be either a high concentration of particles or a long path length. Smoke stack transmissometers are used because of the high particle concentration within the stack. If the mass or number concentration is to be derived directly from theory, it is necessary to design such an instrument to exclude scattered light in the near forward direction, particularly for particles > 1.0 μm diameter. In practice, this is not done, and the calibrations of these transmissometers are empirical, either based on gravimetric or opacity (Ringelmann) comparisons. This procedure is acceptable if the stack particles consist of known and reproducible characteristics (refractive index, chemical composition, and absorption of light), but if these properties are different from the calibration aerosol, the results can only be qualitatively correct. Conner and Hodkinson[42] showed that oil and carbon plumes of similar mass concentration and particle size gave significantly different in-stack transmittance. In transmissometry, the source of light and the photoelectric receptor (usually a photomultiplier or photodiode) are coaxial, and the presence of particles attenuates the light reaching the receptor.

Direct-reading instruments that measure "soiling index" or "coefficient of haze" (COHs) detect changes in the reflectivity or transmission of a filter paper after a fixed volume of air has passed through the filter. This index is highly dependent upon particle size, opacity, and composition; thus, it is not considered as a scientifically established analytical technique for particle mass concentration.

Electrical Detection Methods

Concentration of aerosols can also be measured by imparting a charge to particles and measuring the resulting charge with an electrometer. Liu and Lee[43] uses unipolar diffusion charging and electrometer detection to measure particle charge for aerosol concentration measurement, and John[44] used contact electrification, or impact charging, for the same purpose. In general, these techniques can be used for precise concentration measurements only if the charging characteristics of the particles are constant and the size distributions of the aerosol being measured are similar in shape.

Ion interception by particles has been used by laboratory investigators to determine the number concentration and mean diameter of aerosol systems.[45] In this type of instrument, bipolar ions are produced by a Co source on an axial wire in a cylindrical chamber. As particles pass through the chamber, the ions are intercepted, and the current is attenuated and compared to a parallel chamber from which all particles are excluded by filtration. No commercial instrument using this principle is presently available.

Similar to the behavior of certain gases, aerosol particles passing through a H_2 flame alter the dielectric properties of the flame region. This alteration is the basis for a laboratory instrument described by Altpeter et al.[46] as an aerosol flame ionization detector. With appropriate dilution, the aerosol particles pass through the flame one-by-one, and for a given substance, the integrated response is simply related to the particle diameter. Response is significantly dependent upon the particle composition. Particle sizes suitable for detection in this device are similar to optical counters, i.e., 0.5 to 10.0 μm.

Table 15-2 summarizes some of the principal integral concentration detectors and their characteristics.

Size Distribution Measurement

Any of the integral concentration measuring techniques described above can be used with an appropriate particle size classification device to measure the size distribution of aerosols. Examples include the use of the CPC with a diffusion battery and an impactor with a

TABLE 15-2. Major Automatic Integral Concentration Detectors For Aerosol Measurement

Detector	Flow Rate(LPM)	Size Range(μm)	Lower Concentration Limit	References
1. Number Concentration Measurement				
Condensation Particle Counter (Single particle)	0.3–1.5	0.005–0.5	0.01 cm^{-3}	15–22
Ultrafine Condensation Particle Counter	0.3–1.5	0.003–0.5	0.01 cm^{-3}	18
Condensation Particle Counter (photometric mode)	3–10	0.005–0.5	10 cm^{-3}	15–22
Electrical aerosol detector	0.5–20	0.01–2	5/cm^{-3} at 1 μm	43, 44
Optical particle counter	0.1–28	0.065–20	0.001 cm^{-3}	47–63
Light-attenuating photometer	3–10	0.1–20	—	42
2. Mass Concentration Measurement				
Quartz crystal microbalance	1–5	0.01–20 electrostatic* 0.3–20 impaction*	10 μg m^{-3}	27–29
SAWM	1–5	0.01–20 electrostatic*	0.1 μg m^{-3}	31
TEOM	1–5	0.01–20 filtration*	10^{-3} μg/m^3	32
Beta-attenuation sensor	1–12	0.01–20 filtration* 0.3–20 impaction*	10^{-3} μg/m^3	34–36
Photometer, nephelometer (Light scattering)	1–100	0.1–2.0	10^{-3} μg/m^3	37–40

* particle collection method

quartz crystal microbalance for size distribution measurement. In this section, these and other approaches to size distribution measurement are described.

Optical Particle Counters (OPC)

The optical particle counter (OPC) is widely used for size distribution measurement both in the indoor and outdoor environments. Figure 15-8 shows the operating principle of the optical particle counter.[47] Single, individual particles are carried by an air stream through an illuminated viewing volume in the instrument and cause light to be scattered to a photo-detector. The photo-detector generates a voltage pulse in response to each particle passing through the viewing volume. The pulse amplitude is then taken as a measure of particle size. The pulse is then counted and processed electronically to yield a pulse-height histogram, which is then converted to a histogram for particle size distribution using an appropriate calibration curve. Many commercial counters using an incandescent light source have been developed for particle size distribution measurement in the range 0.3 μm to approximately 10 μm. Recent advances include the use of laser illumination to achieve lower detection limits down to 0.05 μm.

Optical particle counters (OPC) differ widely in their design and performance characteristics. Figure 15-9 shows the optical systems used in three commercial OPCs. The PMS counter uses the "active scattering" principle, in which the particles are passed through the resonant cavity of a helium-neon laser. The Hiac/Royco instrument uses a helium-neon laser and an external scattering volume. In contrast, the TSI instrument uses a

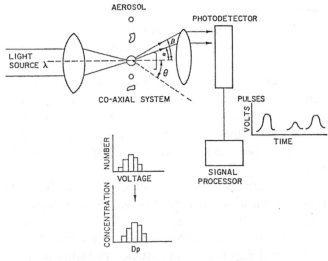

FIGURE 15-8. Operating principle of the optical particle counter.[47]

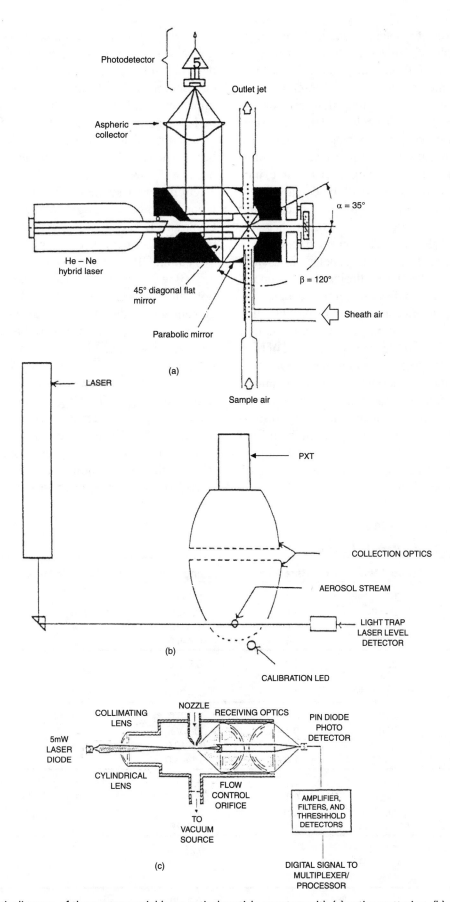

FIGURE 15-9. Schematic diagram of three commercial laser optical particle counters with (a) active scattering, (b) external scattering, (c) solid-state laser.

solid-state laser diode to obtain a small, light-weight portable sensor. Table 15-3 gives a selected list of optical particle counters available commercially.[47] They can be differentiated by the light source (laser, laser diode, or white light) they use, the sampling flow rate of the instrument, the number of channels of data the instrument provides, and other distinguishing characteristics, such as portability and ability to be interfaced with computers. In general, instruments using a laser source, particularly for the "active scattering" type, can detect smaller particles than a corresponding instrument using an incandescent light source because of the higher illuminating intensity of the laser. The lower detection limit of the white-light and laser diode counters is usually around 0.3 µm whereas "active scattering" laser counters can detect particles as small as 0.1 µm and below. A higher flow rate instrument can count more particles in a given time period than an instrument of a lower sampling flow rate. However, the particle coincidence level of high flow rate instruments is usually much lower. OPCs of this type are particularly important for particle counting in a low concentration environment, such as in cleanrooms. A lower flow rate instrument has higher resolution and can detect smaller particles than the high flow counter. A sampling flow rate of 1 cubic foot per minute (cfm) is usually considered high and a flow rate of 0.01 cfm is usually considered low.

The OPC response, which gives a functional relationship between the pulse height and the particle size, depends on both the instrument properties and the particle properties. The former includes optical design, illumination source, and electronics gain, and the latter includes particle size and shape, refractive index, and orientation of non-spherical particles with the incident beam. The relative response of the OPC as a function of particle size can be calculated by means of the theory of electromagnetic scattering developed by Mie. The calculation for some white light counters has been reported by Cook and Kerker.[48] More recent studies have concentrated on laser particle counters.[49]

For an OPC with an axisymmetric scattering geometry ($\theta = 0°$, see Figure 15-8) and near forward direction and narrow angle ($\alpha < \beta < 30°$), the simple geometry provides strong signals but with higher background noises. It is also susceptible to strong multi-valued response, i.e., different particle sizes giving the same pulse height. The geometry is relatively insensitive to variations in real and imaginary part of refractive index. Figure 15-10 gives the response of a near forward light scattering instrument (PMS-ASAS-300X; 4–22°).[49] For the wide angle counter, e.g., Climet CI-7300 (15–150°), the response is more sensitive to changes in both real and imaginary part of refractive index, but much less prone to multi-valued

TABLE 15-3. Selected Optical Particle Counters[47]

Manufacturer	Model No.	Flow Rate (cfm)	Size Range (µm)	No. of Size Channel	Illumination Source
Climet Instrument Co.	CI-7600	0.1	0.1–5.0	4	Multimode
P.O. Box 151	CI-500	1.0	0.3–25	6	Laser
Redland, CA 92373	CI-4224	1.0	0.3–10	4	Laser Diode
	CI-4200	0.1	0.3–5.0	2	Laser Diode
Hiac/Royco	2250	1.0	0.5–5.0	2	Laser Diode
11801 Tech Road	5190	1.0	0.09–1.0	6	Active Laser
Silver Spring, MD 20904	MicroAir 5230	1.0	0.3–25.0	8	Laser Diode
Met One	237B	0.1	0.3–5.0	6	Laser Diode
481 California Avenue	227B	0.1	>0.3/Var.	–	Laser Diode
Grants Pass, OR 97526	A2320	1.0	0.3–10.0	6	Laser Diode
	A2200	1.0	0.14–3.0	6	Active Laser
Particle Measuring	LAS-X	0.1	0.1–7.5	15	Active Laser
System, Inc.	HS-LAS	0.1	0.065–1.0	32	Active Laser
5475 Airport Boulevard	LASAIR 1001	0.01	0.1–2.0	8	Multimode
Boulder, CO 80301	LASAIR 2500	1.0	0.5–25	8	Multimode
Palas GmbH	PC2000	–	0.25–25	256	Xenon bulb
D-76229 Karlsruhe					
FRG					
TOPAS GmbH	LAPS	0.1	0.3–20	128	Laser
Dresden, FRG					

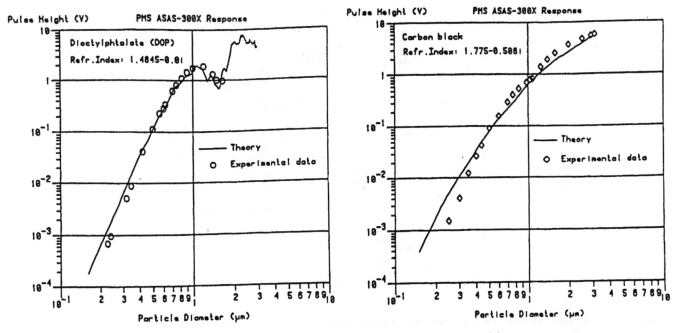

FIGURE 15-10. Theoretical and experimental responses for a near-forward, light-scattering sensor.[49]

response, particularly for white light illumination. Figure 15-11 shows the calculated response of the Climet CI-7300 counter.[48]

To determine the absolute voltage-size response of the OPCs, as well as other instrument characteristics, such as resolution, count coincidence, response to irregular particles, and inlet efficiency, experimental studies are generally required. Liu et al.[50] reported on the evaluation of

several commercially available white-light counters using monodisperse spherical particles. Wen and Kasper[51] and Liu and Szymanski[52] evaluated the counting efficiencies of several commercial OPCs. A novel technique to determine the OPC response to irregular coal dust particles has been developed by Liu, et al.[53] Marple and Rubow[54] made use of inertial impactors to obtain aerodynamic particle size calibration of the OPCs. Recent works mostly

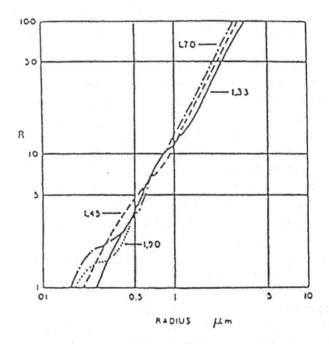

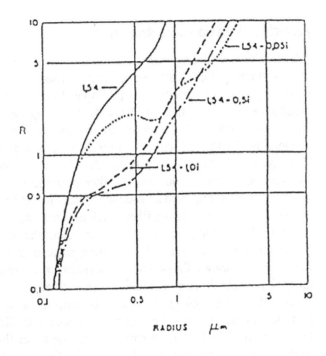

FIGURE 15-11. Theoretical response curves for a wide angle sensor.[48]

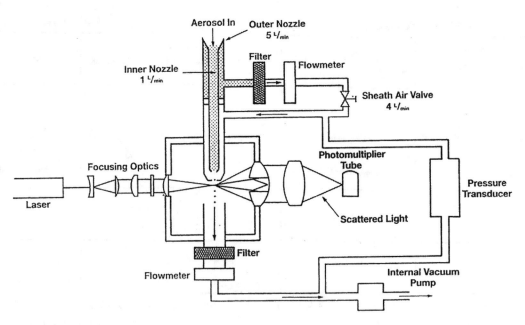

FIGURE 15-12. Schematic diagram of the aerodynamic particle sizer.[70]

involve the evaluation of laser OPCs.[55–60] Comprehensive discussions on the principle and application of the OPC may be found in the papers by Willeke and Liu,[61] Knollenberg and Luehr,[62] and by Gebhart *et al.*[63]

Recent developments on OPCs include increasing the count coincidence level while keeping high sampling flow rate measuring particle refraction indices,[64] and determining the particle shape using scattering intensity from multi-angle light scattering.[65–69] These latest advances had been presented at the 5th International Congress on Optical Particle Sizing.[68]

Particle Relaxation Size Analyzers

In addition to direct light-scattering measurement described above, light scattering can also be used in combination with other measurement principles to extend the measurement capabilities of the light-scattering technique. An example of this is the combination of vapor condensation with light scattering for sub-0.1 μm particle measurement[15] discussed earlier under the section on condensation particle counters. Another example is the use of light scattering with an oscillating electric field for the measurement of airborne fibers, including asbestos.[69]

A further example is the use of an accelerating nozzle in combination with light-scattering measurement. Figure 15-12 shows the schematic diagram of a commercially available Aerodynamic Particle Sizer (APS) described by Agarwal *et al.*[70] In this instrument, the particles are accelerated through a small nozzle to different speeds. The larger the particle size, the lower the speed of the particle due to particle inertia. The particle velocity at the nozzle exit is then measured by detecting the time required to pass through two laser beams with a fixed separate distance to provide a measure of particle size. This principle enables the "aerodynamic size" of the particles to be measured in the size range 0.5 to 30 μm. The aerodynamic size is related to the settling speed of the aerosol and to the particle deposition in the lung. Calibration studies on the APS have been reported by Chen *et al.*[71] and Baron.[72]

Recent APS developments include the incorporation of an UV pulse laser to detect the viability of bioaerosols and of a high energy laser,[73] such as Nd:Yag laser, and mass spectrometry for in-situ particle composition measurement.[74–77]

A second commercial aerodynamic sizing instrument, the Aerosizer, operates under the same time-of-flight principle as the APS. One significant difference is that particles are accelerated at sonic flow through a critical nozzle in the Aerosizer while they are subjected to a moderate acceleration at subsonic flow in the APS. The instrument is capable of measuring particles in a wider size range, 0.5 to 2000 μm and higher concentration up to 1100 particles/cm³ than the APS. However, the calibration curve is strongly dependent on the particle density. Cheng *et. al.*[78] calibrated two Aerosizers using uniform-sized spherical polystyrene latex particles and glass beads and with nonspherical natrojarosite particles.

An instrument which employs the principle of laser Doppler velocimetry (LDV) for size determination is the "SPART" analyzer (Single Particle Aerosol Relaxation Time).[79] In this instrument, the particle is subjected to sinusoidal force by an acoustic transducer at 27 kHz. The motion of the particle, detected by the LDV optics, lags behind the force sine wave by an amount which depends primarily on the particle aerodynamic diameter. The range of aerodynamic diameter measurable with acceptable sensitivity and resolution is stated to be 0.2 to 10.0 μm.

In-situ Sensing Optical Techniques

The optical techniques described above involve extracting aerosol from the environment and transporting it to a sensing zone for measurement. In situ techniques, often optically based, are noninvasive and measure aerosol in its natural state without extractive sampling. The noninvasive measurement is accomplished by locating the sensing volume external to the instrument, thereby eliminating the need for extractive sampling. The techniques are most suitable for measuring aerosols in hostile environments of extreme pressure and temperature ranges and in reactive or corrosive environments. The techniques have received significant interest over the last decade and were summarized in a review paper by Rader and O'Hern.[80] They classified the techniques into the following major categories: single-particle counters of intensity based, phase-based, or imaging; and ensemble techniques of particle field imaging, Fraunhofer diffraction, or dynamic light scattering. To determine the particle size, single-particle counters measure the scattering behavior of an individual particle as it passes through a well-defined sensing zone formed by two crossing laser beams, while the ensemble techniques analyze the collective scattering of a large number of particles. They can be used to measure individual particle sizes from about 0.25 to above 1000 μm, concentrations as high as 10^6 particles/cm^3, and speeds in the km/s range. With the ensemble techniques, particle mean diameters as low as 0.01 μm can be measured.

For a single-particle counter based on the scattering intensity principle, the scattered light intensity may be collected over some solid angle by reception optics and focused onto a photo-detector as individual particles pass through a sensing zone Similar to the OPCs, the particle size can be derived from the intensity peak collected. The forward-scattering spectrometer probe (Particle Measuring Systems, Inc, Boulder, CO), Polytec optical aerosol analyzers (Polytec GmbH, Waldbronn, Germany; Polytec Optronics, Costa Mesa, CA), particle counter sizer velocimeter (Insitec Measurement Systems, San Ramon, CA) are examples of this instrument type.

Polarization intensity differential scattering (PIDS) is another recent development for particle sizing. The technique is based on the observation that the scattered intensity in the horizontal direction of the electric vector

in an unpolarized incident light has a minimum at around 90 degrees for small particles. This minimum will shift to larger angles for larger particles. Thus, although the scattered intensities in the horizontal and vertical direction of the electric vector in an incident light have small contrast for small particles, their difference shows more distinguishable fine structures that make sizing small particles possible. Combining the polarization effect of light scattering with the wavelength dependence at high angles, one can extend the lower sizing limit to as small as 50 nm.[81, 82] In general, the PIDS signal can be easily detected for particles smaller than 0.2 μm (based on a laser wavelength of 450 nm). Therefore, this technique is often combined with other conventional light-scattering methods to extend the detectable particle size range.

An example of a single-particle counter based on the phase Doppler principle is shown in Figure 15-13.[80] The principles of a phase Doppler particle analyzer (PDPA) were described in the paper by Bachalo and Rouser.[83] The system consists of a laser and transmitting optics, and a receiver optics package with multiple photo-detectors to measure the spatial and temporal frequency of the Doppler-shifted light scattered by individual particles passing through the measuring volume. The spatial frequency gives a measure of the particle diameter, refractive index and receiver geometry, while the temporal frequency gives a measure of the particle velocity. A commercial instrument gives the specifications of either one or two velocity component measurement in addition to particle size, particle diameter range of 1–8000 μm and a dynamic range of 35:1, and a velocity range from 1 to 200 m/s. The maximum measurable number concentration is 10^6/cm^3, which is based on the number of particles passing through a calculated size-

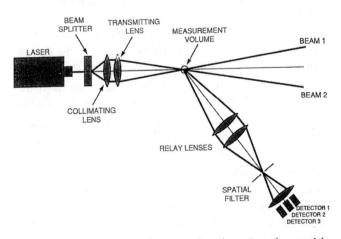

FIGURE 15-13. Schematic diagram of a phase Doppler particle analyzer.[80]

dependent measurement volume. More information on the recent development of phase Doppler particle-sizing velocimeters can be found in the Special Issue of Particle and Particle Systems Characterization.[84]

Single-particle counters based on the imaging technique are one of the early devices used in particle measurement. The technique avoids the problems associated with the particle shape and refraction index by directly imaging the particle shape on an array of photodiodes. The image can be in 1-D, 2-D, or grayscale. Particles in the size range from 200-1240 μm can be measured accurately.

Ensemble techniques based on particle field imaging are used to record the motion of an aerosol stream on films for subsequent analysis. With the increasing power of computers, more efficient image analysis software and advanced image digitizing techniques are now available. The analysis can be completed in several seconds, making direct particle sizing using this imaging technique a reality.

Another ensemble technique for particle sizing is based on the patterns resulting from the Fraunhofer diffraction when a cloud of particles is illuminated using lasers. The optical systems based on this technique are the same as others except for the reception optics. A Fourier lens and ring diode photo-detector are often seen on the reception optics in these systems. A pre-assumed particle size distribution is needed in order to invert the collected signals to the actual size distribution.[85, 86] A typical size range for which this technique is used is 0.5–2000 μm.

Dynamic light scattering (DLS) technique (also known as photon correlation spectroscopy, PCS or Quasi Elastic Light Scattering, QELS) becomes a unique optical tool to study particles with diameters much smaller than 0.1 μm, although the technique can also be used to study larger particles.[87–91] The technique measures the degree of spectral broadening of an incident laser beam caused by the Brownian motion of particles. The mean diffusion coefficient of particles can then be derived from the degree of this broadening effect. Particle size distributions can be inferred from the mean diffusion coefficient by pre-assumed particle size distribution forms. DSL technique usually requires high concentration of particles in the sensing zone for an ensemble measurement. The technique can be applied to both batch samples and flowing systems.[91] The typical size ranges for these systems are 0.003–5 μm. However, very limited particle concentration and shape information can be deducted from the measurements. Recent developments include the application of the technique to low particle concentration environments to reducing error

induced by multiple scattering by a two-color cross-correlation system, and to the improvement of correlators and optics for better signal processing.[92–94] A recent review on the technique can be found in Filella et al.[95]

Electrical Aerosol Analyzer (EAA)

The high electric mobility of submicrometer particles in an electric field makes it possible to separate and classify electrically charged aerosol particles. If the electric mobility is a monotonic function of the particle size, a size classification or size distribution measurement can be made on the basis of the particle electrical mobility.

A widely used instrument for size distribution measurement using the electrical mobility technique is the electrical aerosol analyzer (EAA) originally developed by Whitby[96] and further improved by Liu and Pui.[97] Figure 15-14 is a schematic diagram of the Liu-Pui version of the EAA. The aerosol to be measured is first sampled into the aerosol charger, where it is exposed to unipolar positive ions and becomes electrically charged. The charged aerosol then enters the mobility analyzer which functions as a low-pass filter to precipitate the high mobility particles while allowing the low mobility particles to pass through. The "cut-off" mobility of the analyzer is determined by the applied voltage on the analyzer. By varying this voltage and measuring the corresponding current carried by the charged particles with the electrometer current sensor. a voltage-current curve is generated which can be further analyzed to yield the desired particle size distribution curve. With the availability of a microcomputer-based data acquisition system, sophisticated data reduction softwares (e.g., Kapadia,[98] Helsper et al.[99]) can be used in near real time to obtain the size distribution of aerosols. This type of instrument is no longer manufactured commercially because of its low resolution on particle sizing.

Differential Mobility Analyzer (DMA)

The current electrical measurement technique has replaced the integral type of mobility analyzer with a differential type. The differential mobility analyzer (DMA) described by Liu and Pui[100] and Knutson and Whitby[101] allows high-resolution particle sizing using the mobility classification technique. The resolution is further increased using a bipolar charger rather than a unipolar charger. The bipolar charging-differential mobility analysis method for size distribution measurement was first proposed by Knutson.[102] While the technique is capable of high resolution sizing, the measurement sensitivity decreases due to the smaller

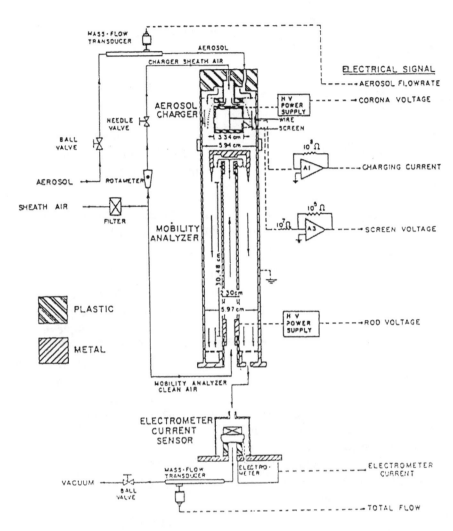

FIGURE 15-14. Schematic diagram of the electrical aerosol analyzer.[97]

current flow to the electrometer. With the introduction of high sensitivity, single particle counting CPC (UCPC), accurate size distribution measurements with the DMA becomes possible. A commercial differential mobility particle sizer (DMPS) is shown schematically in Figure 15-15 consisting of the neutralizer (bipolar charger), the differential mobility analyzer, and the condensation nucleus counter.[103] A microcomputer is used to control the instrument for automatic data acquisition and reduction. The general acceptance of the DMPS is in part due to the data reduction procedure developed by Fissan et al.[104] who also proposed the use of an impactor to remove the coarse particles which tend to interfere with particle measurement in the sub-0.1 μm range. Work by Hoppel,[105] Reischl,[106] and Winklmayr et al.[107] further enhanced the DMPS technique. A significant improvement of the accuracy of the technique has been provided by Kousaka et al.[108] who reported the particle loss data within the DMA and the applicable theory for bipolar charging.

A significant advance on the DMA technique was developed by Wang and Flagan.[109] They made use of a scanning electric field, in place of changing the electric field in discrete steps as in the traditional DMPS operation, to speed up the cycle time of the mobility analyzer considerably. Using an exponential ramp in the field strength, the particles are classified in the time-varying electric field while maintaining a one-to-one correspondence between the time a particle enters the classifier and the time it leaves. The scanning mobility particle sizer (SMPS) is the commercial instrument incorporating this scanning technique. Chen et al.[110] optimized a new DMA inlet for aerosol/sheath air using computer simulation and subsequently evaluated the DMA performance experimentally. This work demonstrates that sizing resolution as high as 1% is achievable with the new inlet design.

By arranging two DMAs in series and inserting a particle conditioner in between, it is possible to make use of the high resolution capability of the DMA to

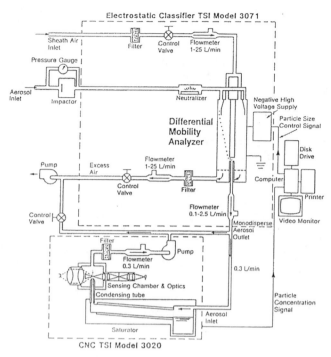

FIGURE 15-15. Schematic diagram of the differential mobility particle sizer.[103]

detect small changes in particle size caused by the conditioner. Liu et al.[111] used a humidifier to effect a particle size change for sulfuric acid aerosol detection. Rader and McMurry[112] made further refinement in the technique and applied the technique for precise droplet growth and evaporation studies.

Recently, nanoparticles have received significant attention because they possess special electrical, optical and/or magnetic properties. They may be used in many "high-tech" applications, e.g., industrial ceramics, coating to improve hardness and thermal properties, fuel cells and tunable lasers. They are also of increasing concern to the industrial hygiene community as nanoparticles are suspected carcinogens. To study nanometer aerosols, the DMA technique is a valuable tool because of its capability of high-resolution measurement and classification of nanoparticles. Although the DMA described above works well in the size range 20–500 nm, it becomes increasingly more difficult to perform accurate measurement and classification for nanoparticles smaller than 10 nm. It is because of the deposition loss and diffusion broadening of the nanometer aerosol in the DMA due to Brownian diffusion.

Fissan et al.[113] compared the performances of four DMAs, namely the TSI-long,[100, 102] the TSI-short,[114] the Hauke 3/150,[106] and the Spectromete de Mobilite Electrique Circulaire (SMEC).[115] The three principal types of DMAs are shown in Figure 15-16. Chen and Pui[116] developed numerical models for these DMAs.

Based on these validated models, a new DMA has been developed that is optimized for the nanoparticle size range.[117] The Nano-DMA provides a threefold increase in the resolution and a threefold decrease in the diffusional loss than the best available DMA in the work of Fissan et al.[113] Zhang et al.[118] has developed their own version of a radial differential mobility analyzer (RDMA) for nanometer particle measurements, which is similar to the SMEC design.

QCM Cascade Impactor and Electrical Low Pressure Impactor (ELPI)

By combining two or more principles of aerosol measurement and detection, novel instruments can be developed which can improve the instrument response speed or can increase the dynamic range of concentration measurement. One example is to combine the quartz crystal mass detection technique with cascade impaction technique for rapid mass distribution measurement.[29] Two drawbacks of this technique are that the crystals need frequent cleaning, and the concentration range is limited in order to avoid overloading of the crystal stages.

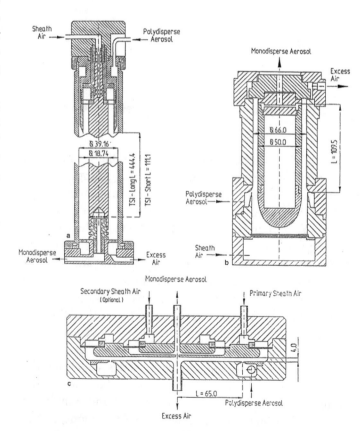

FIGURE 15-16. Schematic diagram of the three principal types of DMAs: (a) the TSI-long (444.4 mm) and the TSI-short (111.1 mm); (b) the Hauke 3/150; and (c) the Spectrometre de Mobilite Electrique Circulaire (SMEC).[113]

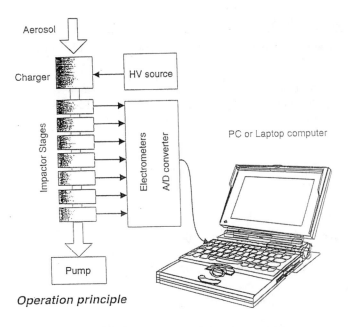

Operation principle

FIGURE 15-17. Schematic diagram of the electrical low pressure impactor (ELPI).[119]

The concentration range can be increased by the recent development of an electrical low-pressure impactor (ELPI), shown schematically in Figure 15-17.[119] The ELPI make use of a diffusion charger to charge the aerosol electrically. The charged aerosol will achieve a charge distribution similar to that given by Liu and Pui.[97] It can then be impacted onto collection stages equipped with current-detecting electrometers. The impactor provides the cut-size information and the current provides the concentration information. The sensitivity of the instrument is comparable to that of the electrical aerosol analyzer (EAA). The parallel current measurements allow a fast response compared with the EAA, and reduced maintenance compared with the QCM cascade impactors.

Diffusion Batteries

Since the rate of diffusion of aerosol particles to a solid surface is a function of particle size, particle loss in a diffusion collector can be used for size distribution measurement. The technique has been used for many years for size distribution measurements of small particles below 0.1 μm.[120] A simple diffusion battery design consists of a single capillary tube, or a capillary tube bundle through which the aerosol is passed. A condensation nucleus counter is then used to measure the upstream and downstream aerosol concentration. The aerosol penetration through the diffusion collector can then be taken as a measure of particle size. By arranging a number of these diffusion collectors in series, the size distribution of the aerosol can be measured. More

detailed information on diffusion devices is given in Chapter 19.

Several novel diffusion batteries have been developed by Sinclair (see the review paper by Sinclair [121]), who used metal disks containing uniform parallel holes of a finite length—the so-called collimated hole structure, and layers of fine stainless steel screens as diffusion collectors. Detailed studies of the wire screen as diffusion collectors have been reported.[122–124] Various sophisticated data reduction techniques based on Simplex and other minimization techniques for the diffusion battery have also been reported. The CNC/DB technique of aerosol size distribution measurement is now widely accepted.

The diffusion battery has also been used as a particle separator for size selective particle sampling. Lundgren and Rangaraj[125] use it for in-stack particle sampling. George[126] and Sinclair et al.[127] made use of the screen diffusion batteries to measure the submicron radioactive aerosols. One of the requirements for accurate diffusion battery measurement is that the counting efficiency of the condensation nucleus counter should be well characterized. Recent development in this area (see above section on number concentration measurement) has further improved the accuracy of the technique.

In addition to the principal techniques described above, other combinations of integral sensor with particle classifier are possible. One notable example is the use of the quartz crystal microbalance as the collecting surface of a cascade impactor to obtain a near-real-time measurement of aerosol mass distribution measurement. Two commercial quartz crystal sensors equipped cascade impactors have been developed.[29,128] Stober et al.[129] also mounted the sensors in a spiral aerosol centrifuge to measure aerosol mass distribution. Marple et al.[130] incorporate the use of quartz crystal sensor in the micro-orifice impactor for mass distribution measurement down to 0.05 μm.

Table 15-4 summarizes the major size distribution measuring techniques for direct-reading, near real-time measurement of aerosols.

Summary

Considerable advances have been made in recent years in the development of direct-reading instruments for analyzing airborne particles. Instruments are now available to measure aerosol number concentration up to 10^6 particles/cm^3, mass concentration up to 1000 mg/m^3, and size distribution over a particle size range of 0.003 μm to over 100 μm. Advances in instrumentation in the field are such that many of the measurement problems that were

TABLE 15-4. Major Techniques for Automatic Size Distribution Aerosol Measurement

Method of Instrument	Flow Rate (LPM)	Size Range (μm)	No. Channels	References
1. Light Scattering				
Optical particle counter	0.3–28	0.065–20	2–32	47–63
Fiber monitor	2–5	0.2 diameter	2–6	69
		2-20 length		
PIDS	—	0.02–0.2	—	81, 82
2. Particle Relaxation Sizing				
Aerodynamic particle sizer	1–5	0.5–30	up to 60	70–72
Aerosizer	2–5	0.2–200		78
3. Diff. Mobility Analyzers				
Differential mobility				
Particle Sizer	0.5–2.0	0.008–0.7	up to 32	103
Scanning mobility				
Particle Sizer	0.5–2.0	0.003–0.7	up to 32	104
4. Sedementation				
Laser setting velocimeter	—	10–20	2–6	
Ultramicroscope				
sedimentation	—	0.2–20	2–20	
Quartz crystal centrifuge	5–10	0.06–10	2–20	
5. Inertia Impaction				
Quartz crystal impactor	2–6	0.3–20	2–8	29
ELPI	10–30	0.03–10	11	119
6. Diffusion Batteries	4–6	0.005–0.2	2–11	121–124
7. Other Techniques				
Phase Doppler particle sizer	—	0.3–200	—	80, 83, 84
Light diffraction	—	0.5–200	—	85, 86
Particle imaging	—	200–12400	—	
Dynamic light scattering	—	0.003–5	—	87–95
Acoustic sizer	—	5–80	2–8	

considered too difficult only ten years ago can now be performed routinely with good experimental accuracy.

With a wide array of available commercial instruments, it is necessary for the practitioners to understand well the operating principles of the instruments and the aerosol system under study. Some of the criteria for selecting an appropriate instrument include the particle size range of interest; the system parameters to be studied, e.g., mass or number concentration vs. particle size distribution. aerodynamic property vs. light scattering property, etc.; and the cost and compactness of the instrument.

Figure 15-18 shows a flow chart for selecting a direct-reading instrument for analyzing airborne particles. The first step is to decide whether to perform an integral concentration measurement or a complete size distribution measurement. In principle, the integral concentration, i.e., mass concentration or number concentration, can be obtained by integrating the size distribution over the appropriate size range. However, there are many compact, inexpensive instruments available for integral concentration measurement compared to the more sophisticated, expensive instruments used for size distribution measurement. Considerable cost savings can be realized if an integral concentration detector is deemed appropriate for the application.

For mass concentration measurement, as required in many industrial hygiene applications, the particles may be directly captured on a surface for measurement by the quartz-crystal sensing technique or by the Beta-attenuation sensing technique. These techniques generally require longer sampling time and frequent cleaning of the surface. A quicker but indirect method would be to measure the light-scattering intensity of the aerosol and to infer its mass concentration through calibration. The accuracy of the technique depends on the measured aerosols having the same surface properties, nearly the same size distribution, and differing only in concentration. For number concentration measurement, as required in many cleanroom applications, two classes of instruments may be used depending on the size range of interest. For particle diameters less than 2 μm, several condensation particle counters may be used. For particle

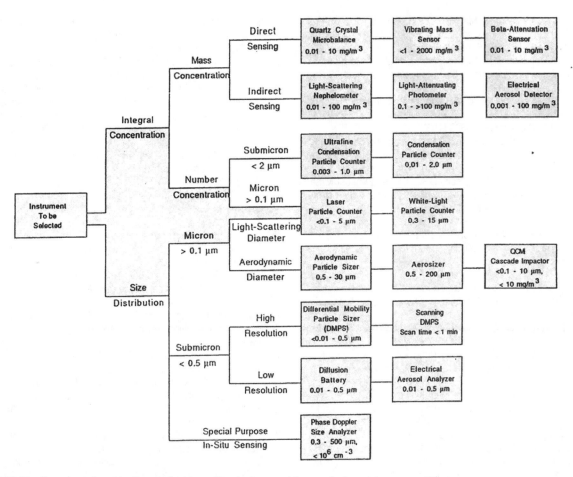

FIGURE 15-18. Flowchart for selecting a direct-reading instrument for analyzing airborne particles.

diameters larger than 0.1 μm, a large array of white-light or laser optical particle counters may be used.

For size distribution measurement, the instruments are divided into two major classes depending on their measuring size ranges. For particles larger than 0.1 μm, the optical particle counters or the particle relaxation size analyzers may be used. The latter are capable of measuring the aerodynamic particle size, which is important for studies of deposition in the respiratory system. For submicron particles the high-resolution differential mobility particle sizer or the low-resolution diffusion battery and the electrical aerosol analyzer may be used. A number of special purpose instruments are also available for difficult applications. For example, the phase Doppler size analyzer measures aerosol in its natural state without extractive sampling. The noninvasive measurement is accomplished by locating the sensing volume external to the instrument, thereby eliminating the need for extractive sampling. The technique is most suitable for measuring aerosols in hostile environments of extreme pressure and temperature ranges, and in reactive or corrosive environments. The cost of such a system, however, will be many times that of an integral concentration detector.

Instruments are also developed to meet specific applications. Examples include the development of a portable condensation particle counter for respirator fit testing, adaptation of the aerodynamic particle sizer for bioaerosol component measurement, and the development of high flow rate laser particle counters for cleanroom applications. Compact direct-reading instruments have been developed with modern electronics and transducer technologies. Significant efforts have also been devoted to improving the software for on-line data reduction. These hardware and software developments will make the next generation of aerosol instruments more sensitive, accurate, compact, and user friendly.

References

1. Pui, D.Y.H.: Direct-reading Instrumentation for Workplace Aerosol Measurements—A Review. Analyst 121:1215 (1996).
2. Liu, B.Y.H.; Pui, D.Y.H.: Aerosols. In: Encyclopedia of Applied Physics 1:415–441 (1991).
3. Willeke, K. ; Baron, P.A. : Aerosol Measurement—Principles, Techniques and Applications. Van Nostrand Reinhold, New York (1993).
4. Whitby, K.T.: The Physical Characteristics of Sulfur Aerosols. Atmos. Environ. 12:135 (1978).

5. Rubow, K.L.; Cantrell, B.K.; Marple, V.A.: In: Proceedings of VIIth International Pneumoconioses Conference, Pittsburgh, Penn., August 23-26, 1988, pp.645–650 (1988).

6. Cantrell, B.K.; Rubow, K.L.; In: Proceedings of VIIth International Pneumoconioses Conference, Pittsburgh, Penn., August 23-26, 1988, pp.651–655 (1988).

7. Whitby, K.T.: In: Aerosols in the Mining and Industrial Work Environment, V.A. Marple and B.Y.H. Liu, Ann Arbor Science, Ann Arbor, MI, Vol. 2, pp.363–380 (1982).

8. Annis, J. C.; Annis, P. J.: ASHRAE Transactions 95:735 (1989).

9. Scales, D., John, W., Lawson, R.; Schmidt, J.: Appl. Occup. Environ. Hyg. 10:685 (1995).

10. Marple, V.A.; Rubow, K.L.; Olson, B.A.: Diesel Exhaust/Mine Dust Virtual Impactor Personal Aerosol Sampler: Design, Calibration and Field Evaluation. Aerosol Sci. Technol. 22:140 (1995).

11. Marple, V.A.; Rubow, K.L.; Olson, B.A.: Diesel Exhaust/Mine Dust Personal Aerosol Sampler Utilizing Virtual Impactor Technology. Appl. Occup. Environ. Hyg. 11:721 (1996).

12. Bricard, J.; Delattre, P.; Madelaine, G.; Pourprix, M.: Detection of Ultra-fine Particles by Means of a Continuous Flux Condensation Nuclei Counter. In: Fine Particles, B.Y.H. Liu, Ed., Academic Press, New York (1976).

13. Sinclair, D.; Yue, P. C.: Continuous Flow Condensation Nucleus Counter, II. Aerosol Sci. Technol. 1:217 (1982).

14. Kousaka, Y.; Nuda, T.; Okyuama, K.; Tanaka, H.: Development of a Mixing Type Condensation Nucleus Counter. J. Aerosol Sci. 13:231 (1982).

15. Agarwal, J.K.; Sem, G.J.: Continuous Flow, Single-Particle-Counting Condensation Nucleus Counter. J. Aerosol Sci. 11:343 (1980).

16. Liu, B.Y.H.; Kim, C.S., Atmos. Environ. 11:1097 (1977).

17. Zhang, Z.Q.; Liu, B.Y.H.: Dependence of the Performance of TSI 3020 Condensation Nucleus Counter on Pressure, Flow Rate and Temperature. Aerosol Sci. Technol. 13:493 (1990).

18. Stolzenburg, M.R.; McMurry, P.H.: An Ultrafine Aerosol Condensation Nucleus Counter. Aerosol Sci. Technol. 14:48 (1991).

19. McDermott, W.T.; Ockovic, R.C.; Stolzenburg, M.R.: Counting Efficiency of an Improved 30A Condensation Nucleus Counter. Aerosol Sci. Technol. 14:278 (1991).

20. Keston, J.; Reineking, A.; Porstendorfer, I.: Calibration of a TSI Model 3025 Ultrafine Condensation Nucleus Counter. Aerosol Sci. Technol. 15:107 (1991).

21. Sinclair, D., Atmos. Environ. 16:955 (1982).

22. Noone, K.J.; Hansson, H.C.: Calibration of the TSI 3760 Condensation Nucleus Counter for Nonstandard Operating Conditions. Aerosol Sci. Technol. 13:478 (1990).

23. Engilmez, N.; Davies, C. N., J. Aerosol Sci. 15:177 (1984).

24. Brockmann, J.E., Ph.D: Thesis, Department of Mechanical Engineering, University of Minnesota, Minneapolis, Minnesota, (1981).

25. Wilson, J.C.; Hyun, J.H.; Blackshear, E.D.: The Function and Response of an Improved Stratospheric Condensation Nucleus Counter. J. Geophys. Res.88:6781 (1983).

26. Niessner, R.; Helsper, C.; Roenicke, G.: Application of a Multistep Condensation Nuclei Counter as a Detector for Particle Surface Composition. Aerosols. B.Y.H. Liu, D.Y.H. Pui and H. Fissan, Eds., Elsevier Sci. Publ., New York (1984).

27. Lundgren, D.A.; Carter, L.D.; Daley, P.S.: Aerosol Mass Measurement using Piezoelectric Crystal Sensors. In: Fine Particles B.Y.H. Liu, Ed., Academic Press, New York (1976).

28. Sem, G.J.; Tsurubayashi, K.: A New Mass Sensor for Respirable Dust Measurement. J. Am. Ind. Hyg. Assoc. 36:791 (1975).

29. Chuan, R.L.: Rapid Measurement of Particulate Size Distribution in the Atmosphere. In: Fine Particles, B.Y.H. Liu, Ed. Academic Press, New York (1976).

30. Sem, G.J.; Tsurubayashi, K.; Homma, K.: Performance of the Piezoelectric Microbalance Respirable Aerosol Sensor. J. Am. Ind. Hyg. Assoc. 38:580 (1977).

31. Bowers, W.D.; Chuan, R.L.: Surface Acoustic Wave Piezoelectric Crystal Aerosol Mass Monitor. Rev. Sci. Instru. 60:1297 (1989).

32. Patashnick, H.; Rupprechl, G.: A New Real-Time Aerosol Mass Monitoring Instrument: The TEOM. Proceedings: Advances in Particle Sampling and Measurement (Edited by W. B. Smith), p. 264, EPA-600/9-80-004, U.S. Environmental Protection Agency (1980).

33. Wang, J.C.F.: A Real-Time Particle Monitor for Mass and Size Fraction Measurements in a Pressurized Fluidized-Bed Combustor Exhaust. Aerosol Sci. Technol. 4:301 (1985).

34. Lilienfeld, P.: Design and Operation of Dust Measuring Instrumentation Based on the Beta-Radiation Method. Staub, 35:458 (1975).

35. Jaklevic, J.M.; Ga1li, R.C.; Goulding, F.S.; Loo, B.W.: A Beta-Gauge Method Applied to Aerosol Samples. Environ. Sci. Technol. 15:680 (1981).

36. Macias, E.S., Husar, R.B.: Atmospheric Particulate Mass Measurement with Beta Attenuation Mass Monitor. Environ. Sci. Technol. 60:904 (1976).

37. Waggoner, A.P.; Charlson, R.J.: Measurement of Aerosol Optical Parameters. In: Fine Particles. B.Y.H. Liu, Ed. Academic Press, New York (1976).

38. Butcher, S.S. ; Charlson, R.J. : An Introduction to Air Chemistry. Academic Press, New York (1972).

39. Lilienfeld, P.: Current Mine Dust Monitoring Instrument Development. In: Aerosols in the Mining and Industrial Work Environment. V.A. Marple and B.Y.H. Liu, Eds. Ann Arbor Science, Ann Arbor, MI (1982).

40. Cook, C.S.; et al.: Remote Measurement of Smoke Plume Transmittance using LIDAR. Appl. Optics 11 (1972).

41. Binek, B.; et al.: Using the Scintillation Spectrometer for Aerosols in Research and Industry. Staub 27:1 (in English) (September 1967).

42. Conner, W.O.; Hodkinson, J.R.: Optical Properties and Visual Effects of Smoke Stack Plumes. EPA AP-30. U .S. Environmental Protection Agency, Washington DC (1972).

43. Liu, B.Y.H.; Lee, K.W.: An Aerosol Generator of High Stability. Am. Ind. Hyg. Assoc. J. 36:861 (1975).

44. John, W.: Contact Electrification Applied to Particulate Matter Monitoring. In: Fine Particles. B.Y.H. Liu, Ed. Academic Press, New York (1976).

45. Mohnen, V.A.; Holtz, P.: The SUNY-ASRC Aerosol Detector. J. Air Pollut. Control Assoc. 18:667 (1968).

46. Altpeter, L.L.; Pilney, J.P.; Rust, L.W.; Senechal, A.J.; et al.: Recent Developments Regarding the Use of Flame Ionization Detector as an Aerosol Monitor. In: Fine Particles. B.Y.H. Liu, Ed. Academic Press, New York (1976).

47. Szymanski, W.W.: Lecture Notes on Optical Particle Counter, Minnesota Shortcourse on Aerosol and Particle Measurement, Minneapolis, MN (1998).

48. Cooke, D.D.; Kerker, M.: Response Calculations for Light Scattering Aerosol Particle Counters. Applied Optics 14:734 (1975).

49. Liu, B.Y.H.; Szymanski, W.W.; Pui, D.Y.H.: Response of Laser Optical Particle Counter to Transparent and Light Absorbing Particles. ASHRAE Transactions V.92, Pt. 1 (1986).

50. Liu, B.Y.H.; Berglund, R.N.; Agarwal, J.K.: Experimental Studies of Optical Particle Counters. Atmos. Environ. 8:717 (1974).

51. Wen, H.Y.; Kasper, G.: Counting Efficiencies of Six Commercial Particle Counters. J. Aerosol Sci. 17:947-961.

52. Liu, B.Y.H.; Szymanski, W.W.: Counting Efficiency, Lower Detection Limit and Noise Levels of Optical Particle Counters. In: Proceedings of the 33rd Annual Technical Meeting, Institute of Environmental Sciences, CA (1987).

53. Liu, B.Y.H.; Marple, V.A.; Whitby, K.T.; Barsic, N.J.: Size Distribution Measurement of Airborne Coal Dust by Optical Particle Counters. Am. Ind. Hyg. Assoc. J. 8:443 (1974).

54. Marple, V.A.; Rubow, K.L.: Aerodynamic Particle Size Calibration of Optical Particle Counters. J. Aerosol Sci. 7:425 (1976).

55. Hinds, W.C.; Kraske, G.: Performance of PMS Model LAS-X Optical Particle Counter. J. Aerosol Sci. 17:67(1986).

56. Kim, Y.J.; Boatman, J.F.: Size Calibration Corrections for the Active Scattering Aerosol Spectrometer Probe (ASASP-100X). Aerosol Sci. Technol. 12:665 (1990).

57. van der Meulen, A.; van Elzakker, B.G.: Size Resolution of Laser Optical Particle Counters. Aerosol Sci. Technol. 5:313 (1986).

58. Yamada, Y.; Miyamoto, K.; Koizumi A.: Size Measurements of Latex Particles by Laser Aerosol Spectrometer. Aerosol Sci. Technol. 5:377 (1986).

59. Chen, B.T.; Cheng, Y.S.; Yeh, H.C.: Experimental Responses of Two Optical Particle Counters. J. Aerosol Sci. 15:457 (1984).

60. Szymanski, W.W.; Liu, B.Y.H.: On the Sizing Accuracy of Laser Optical Particle Counters. Part. Charact. 3:1 (1986).

61. Willeke, K.; Liu, B.Y.H.: Single Particle Optical Counter: Principle and Application. In: Fine Particles, B.Y.H. Liu, Ed., Academic Press, New York (1976).

62. Knollenberg, R.C.; Luehr, R.: Open Cavity Laser "Active" Scattering Particle Spectrometry from 0.05 to 5 microns. ibid., p. 669 (1976).

63. Gebhan, J.; Heyder, J.; Rolh, C.; Stahlhofen, W.: Optical Aerosol Size Spectrometry below and above the Wavelength of Light—A Comparison. ibid., p. 793 (1976).

64. Dick, W.D.; McMurry, P.H.; Bottiger, J.R.: Size- and Composition-Dependent Response of DAWN-A Multiangle Single-Particle Optical Detector. Aerosol Sci. Technol. 20:345 (1994).

65. Sachweh, B.; Umhauer, H.; Ebert, F.; Buttner, H.; et al.: In Situ Optical Particle Counter with Improved Coincidence Error Correction for Number Concentrations up to 10^7 Particle cm^{-3}. J Aerosol Sci. 29:1075 (1998).

66. Sachweh, B.A.; Dick, W.D.; McMurry, P.H.: Distinguishing between Spherical and Nonspherical Particles by Measuring the Variability in Azimuthal Light Scattering. Aerosol Sci. Technol. 23:373 (1995).

67. Szymanski, W.W.; Schindler, C.: Response Characteristics of A New Dual Optics Particle Spectrometer. In: Proceedings 5th International Congress on Optical Particle Sizing, Minneapolis, Minnesota, August 10-14, pp. 219–220, 1998.

68. The Proceedings of 5th International Congress on Optical Particle Sizing, Minneapolis, Minnesota, August 10-14 (1998).

69. Lilienfeld, P.: Light Scattering from Oscillating Fibers at Normal Incidence. J. Aerosol Sci. 18:389 (1987).

70. Agarwal, J.K.; Remiaz, R.J.; Quant, F.J.; Sem, G.J.: Real-Time Aerodynamic Particle Size Analyzer. J. Aerosol Sci. 13:222 (1982).

71. Chen, B.T.; Cheng, Y.S.; Yeh, H.C.: Performance of a TSI Aerodynamic Particle Sizer. Aerosol Sci. Technol. 4:89 (1985).

72. Baron, P.A.: Calibration and Use of the Aerodynamic Particle Sizer (APS-3300). Aerosol Sci. Technol. 5:55 (1986).

73. Hairstone, P.P .; Ho, J.; Quant, F.R.: Design of an Instrument for Real-time Detection of Bioaerosols using Simultaneous Measurement of Particle Aerodynamic Size and Intrinsic Fluorescence. J. Aerosol Sci., 28:471 (1997).

74. Noble, C.A.; Prather, K.: A Real-time Measurement of Correlated Size and Composition Profiles of Individual Atmospheric Aerosol Particles. Environ. Sci. Technol. 30:2667 (1996).

75. Sylvia, H.W.; Prather, K.A.: Time-of-flight Mass Spectrometry Methods for Real Time Analysis of Individual Aerosol Particles. Trend in Analytical Chemistry 17:346 (1998).

76. Salt, K.; Noble, C.; Prather, K.A.: Aerodynamic Particle Sizing Versus Light Scattering Intensity Measurement as Methods for Real-time Particle Sizing Coupled with Time-of-flight Mass Spectrometry. Analytical Chemistry 68:230 (1996).

77. Silva, P.J.; Prather, K.A.: On-line Characterization of Individual Particles from Automobile Emissions. Environmental Science and Technology 31:3074 (1997).

78. Cheng, Y.S.; Barr, E.B.; Marshall, I.A.; Mitchell, J.P.: Calibration and Performance of an API Aerosizer. J. Aerosol Sci. 24:501 (1993).

79. Mazumder, M.K.; Kirsch, K.J.: Single Particle Aerodynamic Relaxation Time Analyzer. Rev. Sci. Instrum. 48:622 (1977).

80. Rader, D.J.; O'Hern, T.J.: Optical Direct-Reading Techniques: In Situ Sensing. In: Aerosol Measurement Principles, Techniques and Applications. K. Willeke and P.A. Baron, Eds. Van Nostrand Reinhold, New York (1993).

81. Bott, S.E.; Hart, W.H.: Particle Size Distribution II. In: ACS Symposium Series 472. T. Provder, Ed., American Chemical Society, Washington, DC, p. 106 (1991).

82. S.E. Bott and W.H. Hart, US Patents 4953978, (1990); 5104221 (1992).

83. Bachalo, W.D.; Rouser, M.J.: Phase Doppler Spray Analyzer for Simultaneous Measurements of Drop Size and Velocity Distribution. Opt. Eng. 23:583–90 (1984).

84. Special Issue of Particle and Particle Systems Characterization Volume 13 (2), page 57–176 (1996).

85. Felton, P.G.: A Review of the Fraunhofer Diffraction Particle Sizing Technique. In: Liquid Particle Size Measurment Techniques, ASTM STP 1083. E.D. Hirleman, W.D. Bachalo, and P.G. Felton, Eds., pp. 47–59. Philadelphia: American Society for Testing and Materials (1990).

86. Hirleman, E.D.: Optimal Scaling of the Inverse Fraunhofer Diffraction Particle Sizing Problem: the Linear System Produced by Quadrature. Part. Charact. 4: 128–133 (1987).

87. Berne, B.J.; Pecora R.: Dynamic Light Scattering. Wiley, New York (1976).

88. Photon Correlation and Light Beating Spectroscopy. Cummins, H.Z.; Pike, E.R., Eds., Plenum, New York (1974); Photon Correlation Spectroscopy and Velocimetry. Plenum, New York (1976).

89. Schmitz, K.S.: An Introduction to Dynamic Light Scattering by Macromolecules. Academic, San Diego (1990).

90. Appl. Optics 36 (20 Oct. 1997).

91. Taylor, T.W.; Sorensen, C.M.: Gaussian Beam Effects on the Photon Correlation Spectrum from a Flowing Brownian Motion System. Applied Optics 25:242 (1986).

92. Willemse, A.W.; Marijnissen, J.C.M.; van Wuyckhuyse, A.L.; Roos, R.; et al.: Low-Concentration Photon Correlation Spectroscopy. Part. Part. Sys. Character. 14: 157 (1997).

93. Drewel, M.; Ahrens, J.; Schatzel, K.: Suppression of Multiple Scattering Errors in Particle Sizing by Dynamic Light Scattering. In: Proceeding Second International Congress on Optical Particle Sizing. E.D. Hirleman, Ed., pp. 130–138. Arizona State University Printing Services (1990).

94. Williamse, A.W.; Merkus, H.G.; Scarlett, B.: Development of a Hterodune PCS Measuring Probe For Highly Concentrated Dispersions. In: Proceeding 5th International Congress on Optical Particle Sizing. P.H. McMuny; A.A. Naqwi, Eds., pp. II–12 (1998).

95. Filella, M.; Zhang. J.; Newman, M.E.; Buffle, J.: Analytical Applications of Photon Correlation Spectroscopy for Size Distribution Measurements of Natural Colloidal Suspensions: Capabilities and Limitations. Colloids Surfaces A: Physico-chemical and Engineering Aspects 120:27 (1997).

96. Whitby, K.T.; Clark, W.E.: Electrical Aerosol Particle Counting and Size Distribution Measuring System for the 0.015 to 1 μ Size Range. Tellus 18:573 (1966).

97. Liu, B.Y.H.; Pui, D.Y.H.: On the Performance of the Electrical Aerosol Analyzer. J. Aerosol Sci. 6:249 (1975).

98. Kapadia, A., Ph.D. Thesis, University of Minnesota, Minneapolis, Minnesota, (1980).

99. Helsper, C.; Fissan, H.; Kapadia, A; Liu, B.Y.H.: Data Inversion by Simplex Minimization for the Electrical Aerosol Analyzer. Aerosol Sci. Technol. 1:135 (1982).

100. Liu, B.Y.H.; Pui, D.Y.H.: A Submicron Aerosol Standard and the Primary, Absolute Calibration of the Condensation Nucleus Counter. J. Colloid Interface Sci. 47:155 (1974).

101. Knutson, E.O.; Whitby, K.T.: Aerosol Classification by Electrical Mobility: Apparatus, Theory and Applications. J. Aerosol Sci. 6:443 (1975).

102. Knutson, E.O.: Extended Electric Mobility Method for Measuring Aerosol Particle Size and Concentration. In: Fine Particles, B.Y.H. Liu, Ed., Academic Press, New York (1976).

103. Keady, P.B., Quant, F.R.; Sem, G.J.: TSI Quarterly 9, 3 (1983).

104. Fissan, H.J.; Helsper, C; Thielen, J.H.: Determination of Particle Size Distributions by Means of an Electrostatic Classifier. J. Aerosol Sci. 14:354 (1983).

105. Hoppel, W.A.: Determination of the Aerosol Size Distribution from the Mobility Distribution of the Charged Fraction of Aerosols. J. Aerosol Sci. 9:41 (1978).

106. Reischl, G.P.: Measurement of Ambient Aerosols by the Differential Mobility Analyzer Method: Concepts and Realization Criteria for the Size Range between 2 and 500 μm. Aerosol Sci. Technol. 14:5 (1991).

107. Winklmayr, W; Reischl, G.P.; Lindner, A.O.; Bemer, A.: A New Electromobility Analyzer for the Measurement of Aerosol Size Distributions in the Size Range from 1 to 1000 μm. J. Aerosol Sci. 22:289 (1991).

108. Kousaka, Y.; Okuyama, K.; Adachi, M.: Determination of Particle Size Distributions of Ultrafine Aerosols using a Differential Mobility Analyzer. Aerosol Sci. Technol. 4:209 (1985).

109. Wang, S.C.; Flagan, R.C.: Scanning Electrical Mobility Spectrometer. Aerosol Sci. Technol. 13:230 (1990).

110. Chen, D.R.; Pui, D.Y.H.; Mulholland, G.W.; Femandez, M.: Design and Testing of an Aerosol/Sheath Inlet For High Resolution Measurements With a DMA, J. Aerosol Sci., Vol. 30, No. 8, pp. 983–999 (1999).

111. Liu, B.Y.H.; Pui, D.Y.H.; Whitby, K.T.; Kittelson, D.B.; et al.: The Aerosol Mobility Chromatograph: A New Detector for Sulfuric Acid Aerosols. Atmos. Environ. 12, 99 (1978).

112. Rader, D.J.; McMurry, P.H.: Application of the Tandem Differential Mobility Analyzer to Studies of Droplet Growth or Evaporation. J. Aerosol Sci. 17:771 (1986).

113. Fissan, H.; Hummes, D.; Stratman, F.; Buscher, P.; et al.: Experimental Comparison of Four Differential Mobility Analyzers for Nanometer Aerosol Measurements. Aerosol Sci. Technol. 24:1 (1996).

114. Kousaka, Y.; Okuyama, K.; Adachi, M.; Mimura, T.: Effect of Brownian Diffusion on Electrical Classification of Ultrafine Aerosol Particles in Differential Mobility Analyzer. J. Chemical Engineering Japan, 19:401 (1986}.

115. Pourprix, M.: Selecteur de particules chargees, a haute sensibilite, Brevet francais No.94 06273, 24 Mal. (1994).

116. Chen, D.R.; Pui, D.Y.H.: Numerical Modeling of the Performance of Differential Mobility Analyzers for Nanometer Aerosol Measurements. J. Aerosol Sci. 28:985 (1997).

117. Chen, D.R.; Pui, D.Y.H.; Hummes, D.; Fissan, H.; et al.: Design and Evaluation of a Nanometer Aerosol Differential Mobility Analyzer (Nano-DMA), J Aerosol Sci., 29:497 (1998).

118. Zhang, S.H.; Akutsu, Y.; Russell, L.M.; Flagan, R.C.; et al.: Radial Differential Mobility Analyzer. Aerosol Sci. Technol. 23: 357 (1995).

119. Keskinen, J.; Pietarinen, K.; and Lehtimaki, M.: Electrical Low Pressure Impactor. J Aerosol Sci. 23:353 (1992)

120. Pollak, L.W.; Metnieks, A.L., Geofisica Pura e Applicata 37, 183 (1957).

121. Sinclair, D.: Measurement of Nanometer Aerosols. Aerosol Sci. Technol. 5:187 (1986).

122. Yeh, H.C., Cheng, Y.S.; Orman, M.M.: Evaluation of Various Types of Wire Screens as Diffusion Battery Cells. J. Colloid Interface Sci. 86, 12 (1982).

123. Scheibel, H.G.; Porstendorfer, J.: Penetration Measurements for Tube and Screen-Type Diffusion Batteries in the Ultrafme Particle Size Range. J. Aerosol Sci. 15:673 (1984).

124. Cheng, Y.S., Keating, J.A.; Kanapilly, G.M.: Theory and Calibration of a Screen-type Diffusion Battery. J. Aerosol Sci. 11:549 (1980).

125. Lundgren, D.A.; Rangaraj, C.N., ESL-TR-81-04, Defense Technical Information Center, Virginia (1981).

126. George, A.C.: Health Phys. 23, 390 (1972).

127. Sinclair, D.; George, A.C.; Knutson, E.O.: In: Airborne Radioactivity, p. 103, American Nuclear Society, La Grange Park, Illinois (1978).

128. Wallace, D.; Chuan, R.: A Cascade Impaction Instrument using Quartz Crystal Microbalance Sensing Elements for "Real- Time" Particle Size Distribution Studies: Special Publication 464, p. 199, U.S. National Bureau of Standards, Washington DC (1977).

129. Stober, W.; Monig, F.J.; Flachsbart, H.; Schwarzer, N.: Mass Distribution Measurements with an Aerosol Centrifuge with Quartz Sensors as Mass Detectors. J. Aerosol Sci. 10, 194 (1980).

130. Marple, V.A.; Rubow, K.L.; Ananth, G.; Fissan, H.: Microorifice Uniform Deposit Impactor. J. Aerosol Sci. 17:489 (1986).

Instrument Descriptions

Commercial sources for instruments presented in this section are listed in Table 15-I-1, which is located at the end of this chapter.

1. Light-scattering Photometers

15-1-1. Particulate Detection Apparatus

Air Techniques, Div. of Hamilton Associates, Inc.

The ATI, 2E Series particulate detectors are portable, forward-scattering photometers primarily intended for on-site integrity testing (leakage detection) of HEPA filters and other similar test applications. They can be used where the high sensitivity of forward-scattering optics is advantageous for detection and measurement of particles. Filter testing and other applications can be performed with di-2-ethylhexyl-phthalate (DEHP, formerly DOP) aerosols or other liquid or solid particle aerosols. The instrument yields a percent penetration as the ratio of the downstream to upstream aerosol concentration, as indicated by the degree of scattered light.

Sampling at 28.3 L min^{-1} (1 ft^3 min^{-1}) is provided by a vacuum pump, the only moving part of the apparatus, with either a handlemeter probe or a short length of tubing. The instrument is contained in an aluminum case 51 × 43 × 20 cm and weighs 16 kg; it operates on 115 VAC, 50/60 Hz, but can be adapted to other electrical service.

INSTRUMENT 15-I-1. TDA-2G Portable Photometer.

In 1994, ATI introduced the TDA-2G model which has similar operating parameters but is one-third the size and one-half the weight of the 2E series; this reduction is accomplished by microprocessor control and signal handling.

15-1-2. Aerosol Photometer

The Virtis Co., Inc.

The microcompressor-controlled JM-9000 is "user-friendly" and enables the operator to quickly set operating functions and parameters. The digital display has a Help menu and diagnostic codes for operator convenience. The unit is auto-ranging over 6 decades which eliminates the manual range selection switch. The digital display indicates % leakage with resolution from 100% down to .0001%. The aerosol concentration values for the 100% upstream and 0% downstream baselines are automatically set with the 100% and 0% push buttons which eliminate the gain and straylight controls. The set point for the audible and visible alarms can be easily and quickly programmed or disabled. The display panel and scanning probe "flash" when the alarm set point is exceeded. The new JM-9000 has 3 pre-programmed reference values for DOP, PAO or other aerosol agents.

15-1-3. Integrating Nephelometer

Belfort Instrument Co.

The Model 1590 series integrating nephelometer is designed to measure total light scattering of airborne particles. This measurement is presented as the light scattering coefficient, b_{scat}, and the visual range. Air is sampled at 100 L min^{-1} through a 10-cm diameter, 112-cm-long cylindrical tube. The particles are illuminated by a quartz halogen lamp and scattered light at angles ranging from 8-179° is detected by a photomultiplier (PM) tube. Zero, span, and linearity are established by particle-free measurement of several gases.

The instrument can be used in automobiles or aircraft for visibility traverses and can operate from -10 to 50°C at relative humidities ranging from 0% to 95%. The instrument consists of an optical assembly that measures 100 × 25 × 15 cm, and a blower box measur-

ing 24 × 21 × 30 cm. The total weight of the two units is 20 kg. An optional air sampling heater is available. The instrument operates on 105–125 VAC, 50 Hz. The minimum visual range detectable is 0.4 km.

15-1-4. Integrating Nephelometers

TSI Incorporated

The Integrating Nephelometer measures the scattering coefficient of aerosols in pristine environments, a key parameter when studying climate, visibility, and air quality. TSI Model 3563 provides sensitivity to aerosol-scattering coefficients lower than 2.0×10^{-7} meter^{-1}. It features three color detectors (blue, green, and red), total and backscatter measurements, a selectable averaging time, high vacuum integrity, and software for data collection and instrument control. TSI also offers a single-wavelength version (Model 3551) without the backscatter feature.

15-1-5. Respirable Dust Measuring Instruments

15-1-5a. Respirable Dust Measuring Photometers

Hund GmbH

The TM Digital μP and TM Data are hand-held, under 70° forward scattering working photometers which have a measuring range of 0.01–100mg/m^3. Factory calibration uses DEHS particles. The instruments' optical response curve provides close correlations to the Johannesburg convention and to the ACGIH7 curve. Intrinsic safety code Eexial and Eexiall CT4 is used for TM Digital μP and Eexial for TM Data.

Both instruments measure and display single values and averages over any time interval up to 24 hours. The memory of TM Data stores for more than 12 hours on fully charged internal accu. TM Data has analog and digital output. The instruments have many approvals worldwide. A separate gravimetric calibration kit is available. These instruments can be used even under rough conditions. Weight: 1 kg.

INSTRUMENT 15-I-4. TSI Integrating Nephelometer.

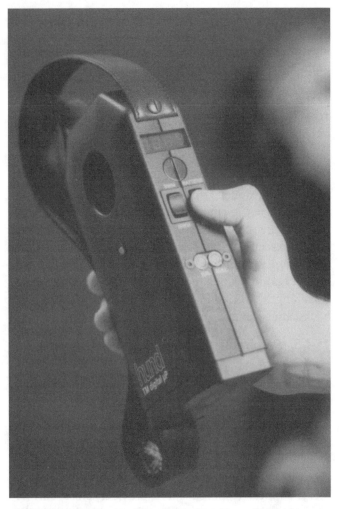

INSTRUMENT 15-I-5a. Hund Respirable Dust Measuring Photometer.

15-1-5b. Stationary Respirable Dust Measuring System FMA-TMS1

Hund GmbH

Model FMA-TMS 1 is a stationary, forward-scattering aerosol-photometer under 70°. Measuring range is 0.0–100 mg/m³. Work calibration uses DEHS particles. The instrument's optical response curve provides close correlations to the Johannesburg convention and to the ACGIH curve. Averaging between 1 min. and 199 days and continuous outputs of current measurement at display at intervals of 2 seconds. Automatic internal monitoring of work calibration and zero point. RS232 interface built in, analog (current-outputs, frequent outputs). FMA-TMS 1 has intrinsic code Eex ibl. The system encloses an operating and data processing unit, measuring head and 5m shielded cable. FMA-TMS 1 is mainly designed for use in hazardous areas as underground in coal mines or in the chemical industry. Total weight: approx. 28 kg.

15-1-6. Fine Dust Measuring Instruments

15-1-6a. Respirable Dust Measuring Instrument TM-M

Hund GmbH

Model TM-M is a hand-held, 70° forward-scattering and passive working photometer. Measuring ranges are 0-1 mg/m³ and 0-10 mg/m³. The factory calibration uses DEHS particles. Model TM-M displays single value measurement and averaging over any time interval up to 24 hours. An internal memory stores 4,095 single values or 1,023 independently determined aerosol concentration average values. Preselection of measuring time is available. Current output 4 to 20 mA; digital output. Lower practical detection limit is approximately 5microgram/m³. Model TM-M is mainly designed for measurement at low level concentrations at indoor and outdoor applications. Relative readings can be converted into gravimetric unit via gravimetric methods. Weight: approx. 1.2 kg.

15-1-6b. Stationary Fine Dust Measuring Sensor TM-SE

Hund GmbH

Model TM-SE is a stationary, passive working, forward-scattering aerosol sensor under 70°. Measuring ranges are 0-2 and 0-20 mg/m³, other ranges are possible. Factory calibration uses DEHS particles. Lower detection limit is approximately 5 μg/m³. Switchable

INSTRUMENT 15-I-5b. FMA-TMS1 Respirable Dust Measuring Instrument.

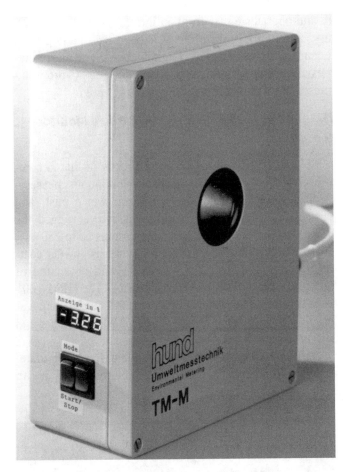

INSTRUMENT 15-I-6a. TM-M Respirable Dust Measuring Instrument.

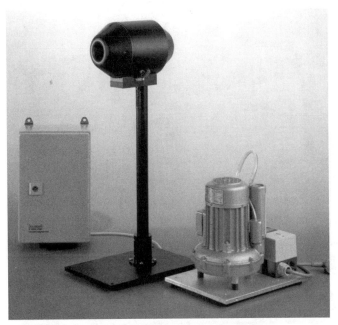

INSTRUMENT 15-I-6b. TM-SE Dust Measuring System.

self-test functions. Current output 4-20 mA. Built-in RS232C interface. Model TM-E consists of 2 units, 19 inch rack mount casing. The aerosol to be measured is sucked through the measuring chamber by a built pump. A built-in sheath air pump prevents the measuring chamber from soiling. Model TM-E is designed to monitor traffic junctions, filter-test-stands, and climate ventilation systems. Weight: approx. 45 kg.

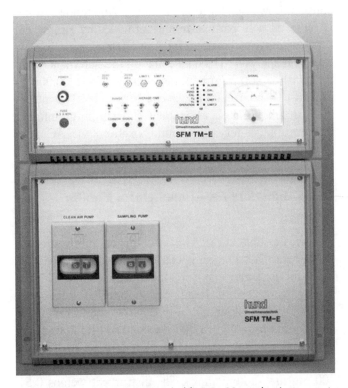

INSTRUMENT 15-I-6c. TM-E Respirable Dust Measuring Instrument.

integration time at 0-1-30-120 min. Two adjustable limit switches. Automatic internal monitoring of work calibration, zero point and malfunctions. Current output 4-20 mA. Model TM-SE is modular and designed for various requirements and applications. It is mainly designed for in situ measurement, filterplants, or filter-test-stands. Model TM-SE uses a pump assembly for sheath air supply to prevent the measuring chamber from soiling. It includes a central control and data processing unit and a separate sensorhead. Weight: approx. 13 kg.

15-1-6c. Stationary Fine Dust Measuring Instrument TM-E

Hund GmbH

Model TM-E is a stationary, forward-scattering aerosol-photometer system under 25-50°. The instrument can be fully remote controlled or manually operated. Measuring ranges are 0-100-250 microgram and 0-1-5 mg/m³. Factory calibration uses DEHS particles. Lowest practical detection limit is approximately 5 microgram/m³. Switchable integration time at 0-1-30-120 min. Two adjustable independent limit switches. Automatic internal calibration and zero point check and

INSTRUMENT 15-I-7. DataRAM Real-Time Aerosol Monitor.

15-1-7. Real-Time Aerosol Monitor

MIE, Inc.

The DataRAM Real-Time Aerosol Monitor is a portable, active sampling, light scattering photometer/ data logger that measures mass concentrations of airborne dust, smoke, mists, haze, and fumes and provides continuous real-time readouts. It is designed for walk-through surveying, fixed point monitoring and, with optional inlet accessories, for ambient air monitoring. Sampling is at a constant flow rate (1.7 to 2.3 L min^{-1}). Automatic zeroing is achieved with particle-free air provided by an internal filter which can be used for gravimetric calibration or chemical analysis. The concentration measurement range is 0.0001 to 400 mg m^{-3} (autoranging). Up to 20,000 1-second to 4-hour data blocks can be stored. The instrument weighs 5.3 kg and operates either from an internal rechargeable battery (24 hours), or from the AC line. Digital, analog and alarm outputs, and PC software for data tabulation and graphing are provided.

15-1-8. Miniature Real-Time Aerosol Monitor

MIE, Inc.

The Personal DataRAM is a compact, light scattering photometer/data logger. It can be worn on the waist or lapel for personal exposure monitoring, hand-held for surveying, or used for fixed point monitoring. It incorporates an LCD readout of real-time and average particulate concentrations. Up to 13,000 1-second to 4-hour averages with real-time tagging can be stored in up to 99 data groups. The concentration measurement range is 0.001 to 400 mg m^{-3} (autoranging). The instrument weighs 0.57

kg and operates either from an internal 9-V alkaline battery (24 hours), or from the AC line, or from an optional rechargeable battery pack. Digital, analog and alarm outputs are provided. Software is included for PC programming, data transfer, tabulation, and graphing.

15-1-9. HPM-1000 High-Pressure Particle Monitor

MIE, Inc.

Model HPM-1000 is installed directly in-line in compressed air/gas streams and provides continuous measurement of oil mist carryover, entrained water mist, and particulate contamination. Highly sensitive yet rugged, the HPM-1000 monitors air quality at pressures up to 250 psig (17 bar) and is ideal for applications such as powder coating deposition, critical parts cleaning, and protection of high-precision flow control components. The HPM-1000 has a resolution and repeatability to 0.001 mg/m^3 (equivalent to 0.0001 ppm in air at 120 psig.) Measurement and calibration are independent of flow rate.

While the HPM-1000 is designed for particulate/ aerosol monitoring in pressurized environments, it can

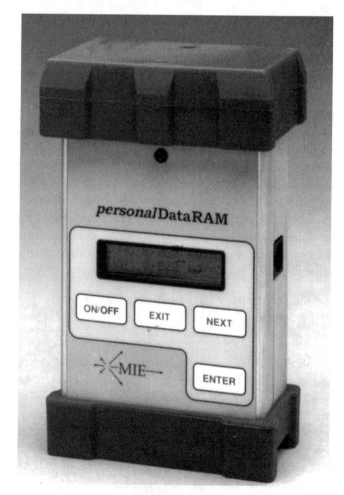

INSTRUMENT 15-I-8. Personal DataRAM.

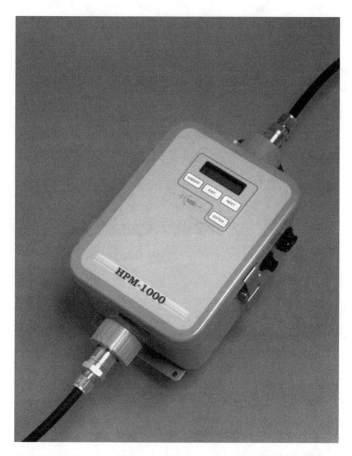

INSTRUMENT 15-I-9. HPM-1000 High-Pressure Particulate Monitor.

also be used at ambient pressures, such as in duct monitoring, as well as in vacuum applications.

15-1-10. Portable Dust Monitor

Negretti Ltd.

OSIRIS is a portable instrument for measuring respirable dust by forward light scattering and subsequent collection of the sampled dust on a filter. It has a horizontal elutriator preseparator that conforms to the BMRC respirability convention. Air is sampled through the elutriator at 0.62 L min^{-1} and passes through a sensing zone illuminated by a laser source. Scattered light in the near-forward direction is detected by a photodiode. The light-scattering intensity expressed as an aerosol concentration is displayed on an LCD panel. Additionally, averages of dust levels up to a full 8-hr workshift can be computed and stored in memory for subsequent analysis.

The instrument is powered by rechargeable Ni-Cd batteries that provide up to 30 hrs of continuous operation. The instrument can be hand carried or mounted for area measurement; it measures 41 × 11 × 15 cm and weighs 7 kg.

15-1-11. Aerosol Measuring Photometer

Casella London, Ltd.

The Microdust is a hand-held, forward-scattering aerosol photometer employing an infrared source and photodiode. It has two measuring ranges, 0.01–20 mg m^{-3} and 0.1–200 mg m dust concentration; each instrument is factory calibrated using AC fine test dust. Particle concentration can be displayed for 1- or 10-sec averaging times and stored for up to 10 hrs on fully charged Ni-Cd rechargeable batteries. The control unit measures 10 × 20 × 4 cm, the cylindrical probe measures 3.5 cm (diameter) × 36 cm (length) and the entire instrument weighs 1 kg. No air mover is needed to move particles through the cylindrical sensing zone in the probe.

The instrument can be used with an accessory gravimetric filter dust collector and a respirable dust cyclone preseparator if desired.

15-1-12. Hand-Held Aerosol Monitor

ppm Enterprises, Inc.

Model 1005 Aerosol Monitor is a portable, battery-powered instrument that measures dust concentration by forward scattering of LED light in the sensing zone. The instrument has no air mover, relying on convective air flow or movement of the instrument through still aft. The measurement sensitivity is stated to be 1 µg m^{-3} with three concentration ranges of 2, 20, and 200 mg m^{-3} maximum. The hand-held version operates on rechargeable batteries for 10 hrs and weighs 1 kg. An alternative

INSTRUMENT 15-I-11. Microdust Aerosol Photometer.

Model 1060 has the sampling head separate from the control/data logger unit. A field gravimetric calibration kit is available to provide calibration for a specific aerosol composition.

15-1-13. Particulate Monitor
Environmental Systems Corp.

Model P-5A is a back-scattering stack monitor that employs an LED source to illuminate a cylindrical sensing zone, 1 cm in diameter and 12 cm long. The range of particle concentrations detectable by the instrument is $1–10^4$ mg m^{-3} (measurement sensitivity 1 mg m^{-3}). The instrument mounts onto a standard 4-in. pipe. The instrument measures $170 \times 32 \times 46$ cm and weighs 41 kg. It operates on 115 VAC, 60 Hz, with option for 220 VAC, 50 Hz.

15-1-14. DUSTCHECK
Grimm Labortechnik GmbH & Co.

This portable fine dust spectrometer allows the real-time measurement of airborne particles up to a maximum concentration of 500,000 P/l in size ranges from 3–20 µm. The instrument presents the counts every 6 seconds, 1 minute, or longer on LCD or RS232 output in 15 different size channels. In addition, the instrument has a data logger with a memory card up to 1 Mb.

15-1-15. Laser Photometer
TSI Incorporated

Model 8587 laser photometer is intended for filter testing, respirator fit testing, and general dust monitoring. The instrument detects particles >0.l-µm diameter and measures aerosol concentrations from 0.001-1000 mg m^{-3}. A 30-m W laser diode illuminates particles in the sensing zone and a photodiode detector measures scattered light 45° off axis. Two separate sample ports allow mass concentration comparisons. Sheath air flow is provided to prevent dust degradation of the optics. The

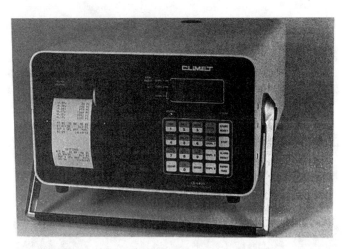

INSTRUMENT 15-2-1. Climet CI-7200 Particle Counter.

photometer has microprocessor-driven, automatic gain selection and a fast purge mode; operation and data storage can be PC linked. The instrument measures $42 \times 32 \times 13$ cm and weighs 7 kg.

2. Light-scattering Single Particle Counters

15-2-1. Laser Diode Particle Counter
Climet Instruments Co.

Series 7000 particle counters employ an elliptical mirror to collect scattered light from a single particle illuminated by a laser diode over the angular range of 15-150°, A particle concentration up to 14 p cm^{-3} can be measured without coincidence errors for particles in the size range 0.3–10.0 µm. Model CI-7350 counts particles in six size ranges from 0.3-10.0 µm, whereas Model CI-7200 counts in six ranges from 0.5–25.0 µm in diameter. Both instruments sample air at 28.3 L min^{-1}, measure $22 \times 31 \times 49$ cm, weigh 18.2 kg, and operate on 120/240 VAC, 50/60 Hz.

15-2-2. Portable Laser Diode Particle Counter
Climet Instruments Co.

Models CI-4102 and CI-4202 employ the same optical detection arrangement as the 7000 series, but are small, lightweight, and battery powered for portable use. They can be powered by 110/240 VAC, 50/60 Hz, as well as by a rechargeable battery for up to 4 hrs continuous use. The self-contained instruments sample air at 2.8 L min^{-1} and can store up to 200 sets of particle size information. These models measure $10 \times 15 \times 28$ cm and weigh 5 kg.

Model GI-500 is a battery-operated instrument employing the same particle detection system and provides

INSTRUMENT 15-I-13. P-5A Mass Concentration Monitor.

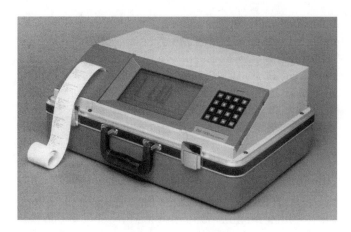

INSTRUMENT 15-2-5. MIE FM-7400 Laser Fiber Monitor.

for six channels of size information from 0.3- to 25.0-μm particle diameter. Its air sampling rate is 28.3 L min⁻¹. It measures 20 × 20 × 35 cm and weighs 10.7 kg.

15-2-3. Facility Monitoring Systems

Met One, Inc.

Met One's remote particle counters, the 4800 and 5800 series, combined with their customized software, provide a complete facility monitoring system. Minimum sensitivities range from 0.1 to 0.5 microns at 0.1 and 1.0 cfm flow rates. RS 232, RS 485, and 4 to 20 milliamp outputs are available for ease of integration. The 4800 counters are about the size of a stick of butter, making remote location mounting easy.

15-2-4. Portable Airborne Particle Counter

Met One, Inc.

Met One manufactures particle counters for a variety of applications. Models A237, 227 and 3113 are used for clean room spot-checking and monitoring. Sensitivities as low as 0.1 micron are available in flow rates from 0.1 to 1.0 cfm. Two of the most popular models are the Model 227, a two-channel hand-held counter with minimum sensitivity of 0.3 micron at 0.1 cfm flow rate. The other, Model 3113 is a recently introduced 0.3 micron 1.0 cfm six channel instrument weighing less than nineteen pounds.

15-2-5. Laser Fiber Monitor

MIE, Inc.

Model FM-7400 is a field instrument for real-time counting and sizing of airborne fibers. Air is sampled at 2 L min⁻¹ into a cylindrical sensing cell illuminated by a 3-mW He-Ne laser. Particles are subjected to an oscillating electric field (3400 V cm⁻¹) at 450 Hz. Fibers are detected by their variable scattering synchronous with

the field. The concentration range is 10^{-4} to 25 fibers cm⁻³; the minimum detectable fiber diameter is 0.2 μm. Fiber length distribution is measured from 2 to 30 μm. The instrument operates from either standard AC power or from a 12-VDC battery pack. With an optional virtual impactor, the instrument can be restricted to counting respirable fibers.

15-2-6. Particle Sizing Aircraft-Mountable Probes

Particle Measuring Systems, Inc.

This group of 15 aerosol measuring probes is designed to collect atmospheric particles of various diameters and shapes and measure their size characteristics one by one as they pass through a laser-illuminated sensing zone. Particles in the diameter range from 0.10–95 μm can be measured by forward light scattering, whereas particles of diameter 10–9,300 μm can be measured by optical array imaging. No single instrument in the group covers the entire size range; the sizing ranges include an instrument with a diameter range of 0.10–3.0 μm for fine particles and an instrument with a diameter range of 150–9,300 μm for precipitation particles. The former instrument has 15 size channels plus oversize; the latter has 62 size channels. Probes for measuring cloud droplets and precipitation droplets have optical arrays capable of two dimensional particle size analysis in accordance with their non-spherical shape.

All probes are contained in an 18-cm-diameter, 79-cm-long cylinder with hemispherical end caps; the particle sensing probe or sampler protrudes from the forward hemisphere into undisturbed air. The probe requires 120 watts of 115 VAC power for the electronics, with additional heating and deicing provided at 28 VDC; it weighs 20 kg.

15-2-7. Particle Sizing Ground-Based Probes

Particle Measuring Systems, Inc.

This series of 10 ground-based probes employs similar optical means for particle detection and sizing to the aircraft probes; because the probes are not moving

INSTRUMENT 15-2-6. PMS Aircraft-mountable Probe.

through the particle cloud, a sample of particles must be pumped through the sensing zone where laser illumination produces forward scattering or (for optical array instruments) the particles move by natural convection through the laser-illuminated beam.

Seven of the probes employ light scattering in which the particles interact either actively in the laser beam or passively with the emitted laser light. The most sensitive instrument detects particles down to a diameter of 0.045 µm and counts particles in 31 diameter channels up to 0.90 µm (plus oversize). Four of these instruments are housed in a sampling and sizing module which measures $18 \times 36 \times 64$ cm and weighs 20 kg. They operate on 115/220 VAC, 60/50 Hz and have a carry handle for transport.

The other three light-scattering instruments are laboratory *in situ* instruments in which particles are drawn by an aspirator fan into an external sampling horn to produce a small wind tunnel through the sensing zone. These instruments weigh 23 kg and measure $18 \times 24 \times 97$ cm. These instruments cover similar diameter ranges as the aircraft probes and count particles in either 15 or 32 diameter channels.

Three instruments are optical array probes for larger particles, ranging from 70 to 4,300 µm in diameter and from 200 to 12,400 µm in 62 size channels. Both one- and two-dimensional particle size instruments are included in this group. These instruments are similar in size, weight, and power requirement to the laboratory *in situ* scattering instruments.

15-2-8. Particle Size Analyzer HC

Polytec PI, GmbH

Model HC is a 900 scattering spectrometer for particle diameters from 0.5 to 40 µm. The sensing volume detects single particles for particle concentrations up to 10^5 p cm^{-3} without significant coincidence error. The instrument's optical head employs a halogen light source and a photomultiplier detector counts and sizes particle pulses into 256 channels. The instrument consists of a control-display module and a measuring optical head. The measuring head measures $19 \times 56 \times 10$ cm, weighs 16 kg, and operates on 110/220 VAC, 60/50 Hz. The control module incorporates a portable PC with internal data acquisition hardware and an external compact power supply.

15-2-9. Particle Size and Shape Analyzers

Galai Instruments, Inc.

Both Model CIS-1 and CIS-100 are *in situ* particle measuring instruments which are capable of determining particle dimensionality and shape by the principle of single particle shadowing. As particles pass through a rotat-

INSTRUMENT 15-2-9. Galai CIS-1 Particle Analyzer.

ing 1.0-µm focused beam, a particle shadows the beam; the resulting illumination falls on a PIN photodiode that provides a signal which is analyzed for size and shape by signal processing software. A 2-mW He-Ne laser provides the illumination for particles whose size can range from 0.5 to 6000 µm at concentrations (for 1.0-µm diameter particles) up to 109 p cm^{-3}.

The Model CIS-100 detector optical module measures $66 \times 28 \times 18$ cm, weighs 14 kg, and is powered by 115/230 VAC, 60/50 Hz, requiring 100 watts. The output of the unit is PC compatible where size and shape analysis is performed with provided software.

15-2-10. Aerosol Particle Analyzer

Wyatt Technology Corp.

Model DAWN-A is a multiangle instrument for particle analysis in a moving aerosol. A particle is illuminated by a laser beam within a spherical scattering chamber, 4-cm i.d. The spherical chamber contains 72 small and 2 large apertures for light-scattering detection. The large apertures are permanently fitted with optical fiber bundles at 25° and 155° with respect to the forward direction. Fourteen optical fibers with collimating optics may be placed into any selected apertures, the remaining apertures being sealed. The collimating optics on each fiber may be fitted with polarizing analyzers. Scattered light from the 14 collimated fibers and the 2 fiber bundles is conducted to PM detectors for rapid light measurement of scattering at the selected positions.

Signals from the PMs are analyzed to derive particle size and refractive index (for spherical particles) or for size and shape (for nonspherical particles). The instrument minimum data rate is 200 p sec^{-1}. Typical particles measured range in diameter from 0.2 to 4.0 µm. The detection limit is claimed to be 0.1 µm or less, depending on the laser source used.

15-2-11. Laser Air Particle Counter

Malvern Instruments, Inc.

The Model 300A AutoCounter is intended for use in cleanroom monitoring and other applications where high

INSTRUMENT 15-2-11. Malvern 300A Air Particle Counter.

particle concentrations do not result in coincidence errors (<40 p cm^{-3}). A particle in the sensing zone is illuminated by a 45-m W solid state laser; the light scattered over a wide angle is collected and processed to yield a particle count which is placed in one of eight size channels. Two particle diameter ranges, 0.3-10.0 µm and 0.5-25.0 µm, are available as options. Air is sampled at 28.3 L min^{-1}. Software for size analysis provides an interface between the instrument and a PC or other data logger. The instrument measures 42 × 56 × 19 cm, weighs 15 kg, and is powered by 115/230 VAC, 60/50 Hz.

15-2-12. Laser Diffraction Particle Analyzer

Malvern Instruments, Inc.

Mastersizer X is a series of instruments in which a He-Ne laser beam is expanded to illuminate a field several mm in diameter containing a number of particles. Scattered light from these particles is collected by a circular array of sensors which are scanned to obtain angular intensity information used to derive size distribution information.

15-2-13. Particle Counter Sizer Velocimeter

Process Metrix

Model PCSV is a group of single particle, forward-scattering instruments designed for a variety of applications including aerosol process streams. A 5-mW He-Ne laser beam is split and focused to form a non-uniformly illuminated sensing zone. Particle sizing is by near-forward scattering, whereas particle velocity is indicated by the width of the particle scattering intensity as it passes through the sensing zone. A deconvolution process is required to obtain these two particle properties. Particle diameters in the size range 0.2–200 µm are measurable at concentrations <10^7 p cm^{-3} (for submicronic particles). Particle velocities from 0.1 to 400 m sec^{-1} are measurable. The instrument comes in several forms, each of which includes a sensing module, a signal pro-

cessing module, a PM tube power supply, and motor and flow controls.

3. Light-attenuating Photometers

15-3-1. Opacity Monitor

ABB Automation

Type UC opacity monitor is a single pass, stack particulate monitor in which the transmitter and receiver units are mounted in opposition on the duct/stack walls but are maintained in independent alignment by a standard 2 1/2-in. metal pipe which extends across the duct and supports the two units. The optics are maintained dust free by an air pump which provides clean air for both transmitter and receiver units.

Opacity measurements employ solid-state electronics that provide for electronic chopping and autocalibration.

15-3-2. Visible Emission Monitor

Datatest

Model 1000 monitor comes in two versions, a single pass transmissometer, 1000 MPS; and a double pass instrument, 1000 MPD. Both instruments employ a tungsten/halogen light source. The light is collimiated to pass through 5-cm apertures in the stack containing the particulates. A stack path length <57 m is acceptable for the instrument optics. The double pass version is supplied with a retroreflector and a beam splitter to measure the light attenuation. Air purge is supplied to minimize optical dust degradation. A single silicon cell detector measures both the illuminating beam intensity and that of the attenuated light with the aid of a mechanical chopper.

The single pass model splits the source light into two beams, one of which passes through the stack, while the

INSTRUMENT 15-2-12. Malvern Mastersizer X Particle Analyzer.

INSTRUMENT 15-3-2. Datatest Model 1000 Opacity System.

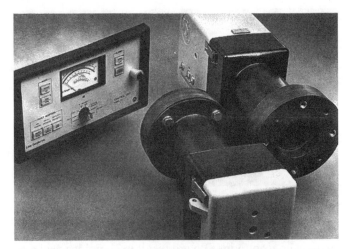

INSTRUMENT 15-3-3. Lear Siegler LS541 Opacity Monitor.

other is transmitted via a fiber optic cable to the receiver, the silicon cell detector. The instrument employs a microprocessor control unit that runs periodic checks of system parameters and controls printer function and averaging time.

The transmitter/receiver units measure 50 × 20 × 15 cm and weigh 10.5 kg. The retroreflector measures 30 × 20 × 15 cm and weighs 7 kg. The control unit and lamp power supply weigh 5.9 and 11.4 kg respectively. The system is powered by 115 VAC, 60 Hz supply.

15-3-3. Opacity Monitor

Lear Siegler Measurement Controls Corp.

Model LS541 is a double pass stack transmissometer in which a collimated light beam passes through a beam splitter to form a measurement and reference beam. The measurement beam can pass through a stack of up to 22 m to a retroreflector for return to the receiver detector in the main module. Light chopping provides for simultaneous measurement of the reference and attenuated beam intensity from which the opacity is determined.

The instrument consists of a transmitter/receiver unit, a reflector, a control unit, and air purge blowers integral to the two operating units. The transmitter/receiver is housed in a weather cover which measures 76 × 57 × 58 cm and weighs 45 kg. The control unit measures 17 × 48 × 25 cm; reflector units vary in size depending on the separation distance through the stack. The system can be operated on 115/230 VAC, 50/60 Hz.

15-3-4. Transmissometer/Opacity Monitor

Rosemount Analytical, Inc.

Model OPM2000 is a double pass stack transmissometer consisting of a transceiver, reflector, and control

unit. It provides for both instantaneous and averaged stack opacity measurements and, in common with most stack transmissometers, provides an internal heater to allow operation down to –40°C. Electronic chopping of the beam source is provided to separate the incident from the reflected beam intensity, from which opacity is determined. The control unit records and controls the operation of the instrument through simple keyboard menu choices.

15-3-5. Opacity Monitor

United Sciences, Inc.

Model 500C is a double pass transmissometer employing an LED source which is electronically modulated. Beam splitting is used to control fluctuations in the source and transmitted light intensity and a value of the opacity, zero and span is provided each second. The transceiver and reflector units are of similar size and weight to other transmissometers, and a signal processing card is located in the tranceiver. A remote microprocessor records data and controls instrument function. The instrument may be configured to serve as 1) a U.S. EPA compliance opacity monitor, 2) a noncompliance stack opacity monitor, 3) a backscatter dust analyzer, or 4) a long path extinction correlation device for monitoring dust levels in the work place.

4. Condensation Nucleus Counters

15-4-1. Condensation Nucleus Counter

Met One, Inc.

Applications include facility monitoring and ultraclean gas certification. Models range from a 10 nanometer, 0.1 cfm remote to the only 1.0 cfm product available.

INSTRUMENT 15-3-5. Model 500C Opacity Monitor.

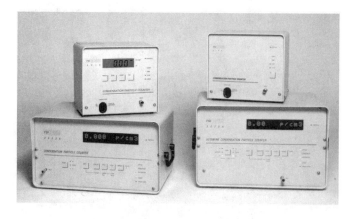

INSTRUMENT 15-4-2. TSI Condensation Particle Counters.

The product uses environmentally friendly glycerol as the working fluid. In addition to inert gases such as argon, helium, and nitrogen, the 1107 model safely sample oxygen and hydrogen. All units can operate for several days on their internal fluid reservoirs, allowing ease of operation in continuous sampling applications.

15-4-2. Condensation Particle Counter

TSI Incorporated

Condensation Particle Counters (CPCs) detect particles as small as 3 nanometers in diameter by condensing vapor onto particles in the sample flow, creating aerosol droplets large enough to be detected efficiently using a light-scattering technique. The sample passes through a heated saturator block, where alcohol evaporates into the sample stream and saturates the flow. The sample then moves into a cooled condensor tube, in which the alcohol vapor supersatuates and condenses onto virtually all particles in the sample, regardless of chemical composition. Droplets leaving the condenser pass through a single-particle counting optical detector.

TSI offers a family of scientific Condensation Particle Counters to meet a variety of research needs. Specific lower detection limits and features vary by model.

15-4-3. PortaCount Plus Respirator Fit Tester

TSI Incorporated

The PortaCount Plus is a quantitative respirator fit tester. It makes a direct measurement of respirator fit factors through the use of condensation nuclei counting (CNC) technology. CNC technology allows the PortaCount to measure the concentration of sub-micron particles in the ambient air as well as the concentration of those particles that leak into the respirator. The ratio

of the two concentrations is calculated and displayed as a fit factor. Measurements are made while the respirator wearer concurrently performs a series of dynamic moving and breathing exercises designed to stress the face seal in ways that simulate actual workplace activities. Once initiated, the fit test proceeds automatically without operator intervention. FitPlus™ for Windows software further automates the fit test and maintains records.

The N95-Companion™ is an accessory for the PortaCount Plus that allows the instrument to quantitatively fit test lower efficiency filtering-facepiece (disposable) particulate respirators that cannot be accurately fit testes using the PortaCount alone (due to filter penetration issues). This includes respirators with filtering efficiencies below 99 percent such as NIOSH N95, R95, and P95 disposable respirators as well as many international P1 and P2 disposables. The instrument uses electrostatic separation technology to limit the particle size used for the measurements to about 0.04 micrometer. Since these filter classes are very efficient at that particle size, this has the effect of making a lower efficiency filter look like a very high efficiency filter to the PortaCount.

5. Resonant Oscillation Aerosol Mass Instruments

15-5-1. Air Particle Analyzer

California Measurements, Inc.

Model PC-2 is a 10-stage cascade impactor in which each stage impaction plate contains a piezoelectric quartz crystal microbalance (QCM). Each QCM acts as a mass sensor; as particulate mass is added, the resonant frequency changes in a predictable manner, making the instrument a real-time indicator of aerodynamic mass

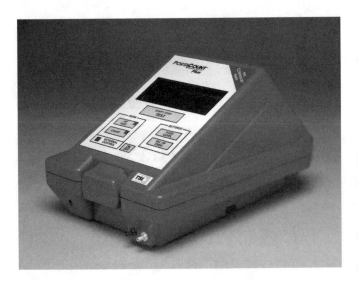

INSTRUMENT 15-4-3. PortaCount Plus.

INSTRUMENT 15-5-1. California Measurements PC-2 Ar Particle Analyzer.

distribution. The stages of the impactor have 50% aerodynamic diameter cutoffs in a geometric progression ranging from 25 to 0.05 μm. The stated mass concentration range of the instrument is from 5 to 50,000 μg m^{-3}. Correction for ambient conditions is provided by mounting a reference crystal beneath each sensing crystal where no particle can reach the crystal surface. The sensing crystals may be removed after an experimental measurement and mounted directly in an SEM or other type of particle characterizing instrument for particle shape and chemistry analyses.

The instrument flow rate is 240 cm^3 min^{-1}. Sampling times range from 1 to 2 sec (for high concentration) to 15 min for the minimum concentration. The instrument output is in the form of a tape containing the mass collected for each stage during the sampling period. The instrument consists of an impactor stack and a control unit. The stack measures 41 × 29 × 11.5 cm and weighs 5 kg, whereas the control unit measures 43 × 30.5 × 19 cm and weighs 10 kg. The instrument requires 50 watts and operates on 120 VAC, 50/60 Hz (240 VAC, 50 Hz version available).

15-5-2. Ambient Particle Monitor

Rupprecht & Patashnick Co., Inc.

The TEOM Series 1400a Ambient Particulate Monitor has received USEPA (EQPM-1090-079) and German EPA equivalency designation for particulate mass concentration measurements, as well as regulatory approvals in other countries. The instrument incorporates a true microweighing technology developed and patented by Rupprecht & Patashnick named the *tapered element oscillating microbalance*. This operating principal enables a direct measurement of the particle mass collected on a filter in real time and allows for unparal-

leled precision and quality. The Series 1400a monitor is used at locations ranging from highly polluted urban areas to background stations with low particulate concentrations. It can be fitted with inlets for PM-10, PM-2.5, PM-1 and TSP measurements.

6. Laser Doppler Particle Analyzers

15-6-1. Phase Doppler Particle Analyzer

Aerometrics, Inc.

The phase doppler particle analyzer (PDPA) consists of a laser-based optical transmitter, an optical receiver,

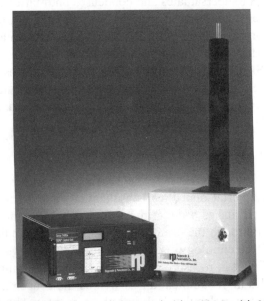

INSTRUMENT 15-5-2. Rupprecht & Patashnick 1400A Particle Monitor.

an electronic signal processor, and a PC system with software for data acquisition, analysis, and display. The principle used is that of light-scattering interferometry; a single particle passing through the fringes of the sensing zone yields velocity information while the phase relationship between different detectors yields particle size information. Model PDPA-100 uses a He-Ne laser and measures a single velocity component. Model PDPA-200 uses an argon-ion laser to simultaneously measure two velocity components.

The maximum particle concentration is stated to be 10^6 p cm^{-3} for submicronic particles. Particles of diameter 0.5–10,000 µm are measurable, with particle distributions up to 50 size classes. Particle velocity up to 200 m sec^{-1} is measurable and is also distributed into 50 velocity classes. The maximum data rate is 120,000 samples sec^{-1}. Because the instrument is an *in situ* device, no sample pumping is required; the particles move through the sensing zone by their own convective velocity.

The optical transmitter measures 62 × 22 × 22 cm and weighs 14 kg. The signal processor measures 43 × 32 × 10 cm and weighs 2.3 kg. The system operates on 115 VAC, 60 Hz.

15-6-2. Particle Dynamics Analyzer

Dantec Measurement Technology, Inc.

The Dantec Particle Dynamics Analyzer (PDA) provides simultaneous measurements of the size and velocity of spherical particles or aerosols on-line and in real-time. The instrument provides non-intrusive, non-contact diagnostics utilizing fiber optics transmission and reception of laser light for single particle measurements. The PDA can also be used for velocity only measurements up to 3-Dimensions. Advanced software provides complete analysis of the particle distributions, concentration, volume flux, size/velocity correlation, traverse control for 3-D surveys, turbulence and spectral analysis. The PDA is normally used in laboratory environments, but has been used in field applications.

15-6-3. Airborne Particle Sensor

Titan Spectron

This instrument measures the diameter, velocity, and concentration of particles *in situ* as they pass through a sensing zone illuminated by solid-state laser diodes. These diodes produce light at 800 nm. Avalanche photodetectors are the optical sensors; their cylindrical housing measures 10 cm in diameter and 30 cm long. The stated range of measurable particle diameters is 0.3-10 µm.

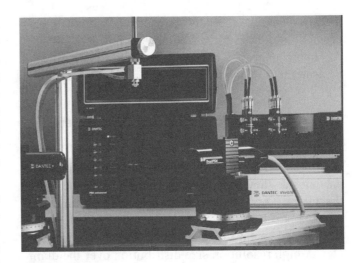

INSTRUMENT 15-6-2. Dantec Particle Dynamics Analyzer.

15-6-4. Adaptive Phase/Doppler Velocimeter

TSI Incorporated

Model APV provides *in situ* simultaneous measurement of particle diameter and velocity. These two measurements provide an estimate of local particle concentration and volume flux of the particulate medium.

Illumination of the sensing zone is achieved by a laser beam through a fiberoptic probe. Optical sensors are aligned on the sensing zone to permit the measurement of particle characteristics; the signals from these sensors are processed digitally and analyzed by a software package which displays particle diameter, velocity, and other flow parameters in real time. Geometric and Mie-scattering techniques are employed in a simulation package, SIMAP, which is included with the instrument system. Spherical particles of diameters from 0.5 to 1,000 µm are measured accurately. Particle velocities up to 200 m sec^{-1} are likewise accurately measured.

7. Particle Relaxation Size Analyzers

15-7-1. Aerosizer

Amherst Process Instruments, Inc.

Building on the proven Aerodynamic Time-of-Flight technology used in the original API Aerosizer®, the Aerosizer LD utilizes state-of-the-art laser diode optics for high resolution (2048 channels) particle analysis in the 0.2 to 700 micron range.

The addition of a CDRH/FDA approved access port allows quick and easy cleaning of the measurement zone optics. When combined with the revolutionary

AeroDispenser™ the new Aerosizer LD allows measurement of cohesive and free flowing powders DRY.

Interchangeable accessories are available for the measurement of inhalation devices, such as MDI's and DPI's as well as aerosols, smokes, and sprays.

15-7-2. Aerodynamic Particle Sizer Spectrometer

TSI Incorporated

Model APS measures the aerodynamic diameter of particles as they pass through a pair of laser beams in the vicinity of a high velocity jet. Each particle produces a pair of pulses from which particle velocity in the defined jet flow determines the aerodynamic diameter. A high resolution size distribution over the diameter range of 0.5–30 μm is obtained after sampling a suitable number of particles. The aerosol flow rate through the jet is controlled at 1.0 L min⁻¹, whereas the total flow rate (augmented by the sheath flow) is 5.0 L min⁻¹.

The entire instrument, including sampling inlet, jet nozzle, laser, controlling electronics, and pumps, is enclosed in a single module measuring 46 × 44 × 23 cm and weighing 22.7 kg. The signal processing, storage, and display of size distributions is performed in an accompanying PC. The instrument requires 200 watts and is powered by 115/220 VAC, 50/60 Hz.

8. Electric Mobility Aerosol Analyzers

15-8-1. Electrical Mobility Spectrometer

Hauke GmbH & Co. KG

The Hauke EMS VIE-08 opens new ways for differential mobility size analysis. It is used for fast and precise automatic measurements of particle size distributions and has already proven its capabilities in a

INSTRUMENT 15-8-1. Hauke EMS VIE-08.

wide field of application. Hauke offers two types of Differential Mobility Analyzers (DMA), each one designed for a specific particle size range (measuring ranges 1–40 n m; 3–150 n m; 10–1000 n m). The instruments can be combined with various particle sensors to achieve the best performance for an application. The EMS VIE-08 control system allows a synchronous and asynchronous operation of two DMAs—sensor systems to support advanced measurement techniques. Software: MS-Windows user interface to facilitate the operation and data reduction.

15-8-2. Scanning Mobility Particle Sizer

TSI Incorporated

TSI Scanning Mobility Particle Sizer (SMPS) systems provide high-resolution particle sizing of submicrometer aerosols in 60 seconds or less. A choice of three Condensation Particle Counters (CPCs) and three Differential Mobility Analyzers (DMAs) allows greater flexibility when configuring a system. Collectively, SMPS systems size particles from 3 to 1000 nanometers, in concentrations from 1 to 108 particles per cubic centimeter. (Actual specifications vary by configuration.) Scans are made continuously to

INSTRUMENT 15-7-2. Model APS.

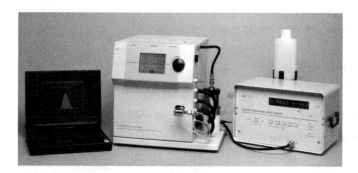

INSTRUMENT 15-8-2. TSI Scanning Mobility Particle Sizer (SMPS).

eliminate gaps in the data distributions. An innovative Electrostatic Classifier platform (Model 3080-series) allows you to switch between one of three DMAs and calculates required flow settings automatically. A control knob, built-in display, and state-of-the-art software make the SMPS quite easy to use.

9. Beta Gauge Particle Sampler

15-9-1. Beta Gauge Automated Particle Sampler

Andersen Instruments, Inc.

This instrument is similar to ambient outdoor particulate samplers that use a PM_{10} or TSP inlet configu-

ration (see Chapter 5), except that the particulate collected on the filter is continuously measured by the attenuation of beta radiation. Thus, the temporal accumulation of particulate mass can be recorded to detect spikes and other time trends of concentration. Although the ^{14}C source of beta radiation (<100 µCi) is located beneath the filter collector, the measuring counter is mounted above the filter, requiring the radiation to pass through the particulate matter where it is attenuated. The attenuation of beta radiation at this energy level is relatively independent of the atomic composition of the particles, and is thus an appropriate measure of total particulate mass that reaches the filter.

The sample flow rate of the instrument is :18.9 L min^{-1} and the range of detectable mass concentration is from 2 to 50,000 µm^{-3}. The collector and control module are housed in a weatherproof cabinet for most applications; the cabinet measures 44 × 51 × 31 cm and weighs 16 kg.

TABLE 15-I-1. Commercial Sources

ABB Automation 501 Merritt 7 Norwalk, CT 06856 (203) 750-2200 FAX (203) 961-7898 *www.abb.com/automation*	California Measurements 150 E. Montecito Ave. Sierra Madre, CA 91024 (626) 355-3361 FAX (626) 355-5320 *info@californiameasurements.com* *www.californiameasurments.com*	Environmental Systems Corp. 200 Tech Center Dr. Knoxville, TN 37912 (865) 688-7900 FAX (865) 687-8977 *www.envirosys.com*
Aerometrics 755 N. Mary Avenue Sunnyvale, CA 94086 (408) 738-6688 FAX (408) 738-6871	Casella Group Ltd. Regent House, Wolsely Road Kempston Bedford MK42 7JY 44(0)1234841441 FAX 44(0) 1234841490 *info@casella.co.uk* *www.casella.co.uk*	Galai Instruments, Inc. 577 Main St. Islip, NY 11751 (516) 581-8500 FAX (516) 581-8573 *www.galai.co.il*
Andersen Instruments, Inc. 500 Technology Court Smyrna, GA 30082 (800) 241-6898 or (770) 319-9999 FAX (770) 319-0336 *andersen@anderseninstruments.com* *www.anderseninstruments.com*		

TABLE 15-I-1 (cont.). Commercial Sources

Air Techniques
11403 Cronridge Drive
Owings Mill, MD 21117
(410) 363-9696
FAX (410) 363-9695
www.atitest.com

Amherst Process Instruments
7 Pomeroy Lane
Amherst, MA 01002
(413) 253-6966
FAX (413) 253-6960

Belfort Instrument Co.
727 S. Wolfe St.
Baltimore, MD 21231
(410) 342-2626
FAX (410) 342-7028
www.belfort-inst.com

Lear Siegler Corp.
74 Inverness Dr. East
Englewood, CO 80112
(303) 792-3300
FAX (303) 799-4853

Malvern Instruments, Inc.
10 Southville Rd.
Southborough, MA 01772
(508) 480-0200
FAX (508) 460-9692
www.malvern.de

Met One Instruments, Inc.
1600 Washington Blvd.
Grants Pass, OR 97526
(541) 471-7111
FAX (541) 471-7116
www.metone.com

MIE, Inc.
7 Oak Park
Bedford, MA 01730
(781) 275-1919 or (888) 643-4968
FAX (781) 275-2121
www.mieinc.com

Particle Measuring Systems, Inc.
5475 Airport Blvd.
Boulder, CO 80301
(303) 443-7100 or (800) 238-1801
FAX (303) 449-6870
www.pmeasuring.com

Climet Instruments Co.
1320 W. Colton Ave.
Redlands, CA 92374
(909) 793-2788
FAX (909) 793-1738
www.climet.com

Dantec Measuring Tech., Inc.
777 Corporate Drive
Mahway, NJ 07430
(201) 512-0037
FAX (201) 512-0120
www.dantecmt.com

Datatest
6850 Hibbs Lane
Levitton, PA 19057
(215) 943-0668
FAX (215) 547-7973
www.datatest-inc.com

Polytec PI, GmbH
Polytec Platz 5-7
P.O. Box 1140
W-7517 Waldbronn, Germany
0-72-43-604-0
FAX 0-72-43-699-44
www.polytec.com

ppm Enterprises, Inc.
11428 Kingston Pike
Knoxville, TN 37922
(615) 966-8796
FAX (615) 675-4795

Process Metrix
2110 Omega Rd., Suite D
San Ramon, CA 94583
(925) 837-1330
FAX (925) 837-3864
www.processmetrix.com

Rosemount Analytical, Inc.
1201 N. Main St.
Orrville, OH 44667
(330) 682-9010 or (800) 433-6076
FAX (330) 682-4434
www.frco.com/proanalytic

Rupprecht & Patashnik Co.
25 Corporate Circle
Albany, NY 12203
(518) 452-0065
FAX (518) 452-0067
info@rpco.com
www.rpco.com

Grimm Lobortechnik
GmbH
Dorfstrasse 9
D-83404 Ainring, Germany
49-86-545780
FAX 49-86-5457810

Hauke GmbH & Co. KG
Cumberlundstrasse 46-50
A-4810 Gmunden, Austria
43-7612-4133
FAX 43-7612-413385

Hund GmbH
Willhelm-Will-Str. 7
35580 Wetzlar, Germany
49-6441-20040
FAX 49-6441-200444
hundwetzlar@hund.de
www.hund.de

Titan-Spectron
1582 Parkway Loop, Suite B
Tustin, CA 92680
(714) 566-9060
FAX (714) 566-9055

TSI Inc.
P.O. Box 64394
St. Paul, MN 55164
(651) 490-2833 or
(800) 677-2708
FAX (651) 490-3860
info@tsi.com
www.tsi.com

United Sciences, Inc.
5310 N. Pioneer Rd.
Gibsonia, PA 15044
(724) 443-8610
FAX (724) 443-1025

The Virtis Company, Inc.
815 Route 208
Gardiner, NY 12525
(914) 255-5000 or (800) 765-6198
FAX (914) 255-5338
info@virtis.com
www.virtis.com

Wyatt Technology Corp.
30 S. LaPatera Ln., B-7
Santa Barbara, CA 93117
(805) 681-9009
FAX (805) 681-0123
info@wyatt.com
www.wyatt.com

Chapter 16

Gas and Vapor Sample Collectors

Richard H. Brown, Ph.D.[A] and Lee E. Monteith[B]

[A]*Health and Safety Laboratory, Broad Lane, Sheffield, S3 7HQ, UK ;*

[B] *Department of Environmental Health, University of Washington, Seattle, WA*

CONTENTS

List of Symbols

A = cross-sectional area of diffusion path (cm^2)

C = measured mass concentration of analyte in air (mg/m^3)

C_{corr} = C, corrected to standard temperature and pressure

C' = measured volume fraction of the analyte in air (ppm, v/v); ppm is the volume fraction, $(\varphi) = 10^{-6}$

D = coefficient of diffusion (cm^2/s)

E_d = desorption efficiency corresponding to m_1

E_s = sampling efficiency

M = molecular mass of the analyte of interest (g/mol)

P = actual pressure of air sampled (kPa)
T = absolute temperature of air sampled (°K)
U = sampling rate (cm³/min)
U' = sampling rate (ng/ppm-min)
V = volume of air sample (ml)
k = correction factor for non-ideal behavior
l = length of diffusion path (cm)
m = mass of analyte in sample (µg)
m_1 = mass of analyte on first tube section (µg)
m_2 = mass of analyte on backup tube section (if used) (µg)
m_{blank} = mass of analyte in blank (µg)
m_s = mass of analyte sorbed by diffusion (µg)
t = sampling time (s)
t' = sampling time (min)
ρ = actual mass concentration of analyte in air (mg/m³)
ρ_1 = actual mass concentration at the beginning of the diffusion layer ($l = 0$) (mg/m³)
ρ_2 = actual mass concentration at the end of the diffusion layer (mg/m³)
r_2 = actual mass concentration at the end of the diffusion layer (mg/m³)
f = actual mass concentration of analyte in air (ppm, v/v)
τ = time constant of diffusive sampler (sec)

Introduction

This chapter discusses the collection and analysis of gases and vapors commonly found in the ambient, indoor or workplace environment. It is limited to descriptions of sampling methods for subsequent laboratory analysis. It does not, therefore, include any discussions of direct-reading instruments, colorimetric indicators, tape samplers, or other "on-the-spot" testing devices.

Selection of Sampling Devices and Measurement Methods

There are two basic methods for collecting gas and vapor samples. In one, called grab sampling, an actual sample of contaminated air is taken in a flask, bottle, bag, or other suitable container; in the other, called continuous or integrated sampling, gases or vapors are removed from the air and concentrated by passage through a sorbing medium.

The first method usually involves the collection of instantaneous or short-term samples, usually within a few seconds or a minute, but similar methods can be used for sampling over longer periods.

Grab sampling is of questionable value when 1) the contaminant or contaminant concentration varies with time, 2) the concentration of atmospheric contaminants is low (unless a highly sensitive detector is used so that the mass of analyte collected is above the limit of detection), or 3) a time-weighted average exposure is desired. In such circumstances, continuous or integrated sampling is used instead. The gas or vapor in these cases is extracted from air and concentrated by 1) solution in a sorbing liquid, 2) reaction with a sorbing liquid (or reagent therein), or 3) collection onto a solid sorbent. Efficiency percentages must be determined for each case. Later in this chapter (and in Chapters 17 and 19), another technique, passive (or diffusive) sampling, is discussed.

There are many alternative approaches to gas and vapor sampling, and the best one will depend on the circumstances.

Factors which will need to be taken into account include:

- the measurement task;
- the concentration to be determined;
- the time resolution required;
- selectivity to the target gas or vapor and sensitivity to interfering gases and vapors;
- bias, precision, and overall uncertainty required;
- susceptibility of the sampler to environmental factors;
- fitness for purpose, e.g. weight, size, durability;
- training requirements for the reliable operation, maintenance, and calibration;
- the total cost of purchase and operation, including calibration and maintenance;
- compliance with the performance requirements of appropriate national or local governmental regulations;
- conformity to the user's quality system.

Having established the requirements, the next step in the selection of a sampling method or device and analytical procedure is to search the available literature. Primary sources are the compendia of methods recommended by the regulatory authorities or governmental agencies in the United States, i.e., the *NIOSH Manual of Analytical Methods*[1] and the *OSHA Analytical Methods Manual*.[2] Recommended methods from other countries, such as the United Kingdom,[3] Germany,[4] or Sweden,[5] might also be consulted. Other standards are available from the American Society for Testing of Materials (ASTM). Secondary sources are published literature references in, for example, the *American Industrial Hygiene Association Journal, Applied Occupational and Environmental Hygiene*, or books such as the *Intersociety Committee's Methods for Air Sampling and Analysis*.[6]

If a published procedure is not available, one can be devised from theoretical considerations. However, its suitability must be established experimentally before application.

The final stage is to review the performance characteristics of the available methods against the selection criteria already established. Important information on the performance characteristics of devices or procedures can be obtained from various sources. These include:

- the manufacturer's instructions for use;
- published commercial technical information;
- technical and research publications;
- national and international standards (see Chapter 11);
- user groups, e.g., HSE/CAR/WG5, which issues *The Diffusive Monitor*, a newsletter produced since 1988 and available from the Health and Safety Laboratory, Broad Lane, Sheffield S3 7HQ, UK.

Legislation and Directives

Details of relevant U.S. legislation and some international standardization, particularly with regard to performance requirements, are given in Chapter 11.

Grab Samplers
(Short-term Samplers—Seconds or Minutes)

Evacuated Flasks

These are containers of varying capacity and configurations. In each case, the internal pressure of the container is reduced, either to near zero (<1 millibar) or to a known absolute pressure. These containers are generally removed to a laboratory for analysis, although it is possible to achieve field readability if the proper equipment and direct-reading instrument are available. Some examples of evacuated flasks are heavy-walled containers, separation flasks, and various commercial devices.

"Passivated" Canisters

Stainless steel containers that have been specially treated to reduce sorption effects have been used for collecting trace organic gases, especially the less reactive hydrocarbons and halocarbons.[7]

Flexible Plastic Containers

Bags are used to collect air samples and prepare known concentrations that can range from parts per billion to more than 10% by volume in air. The bags are commercially available in sizes up to 250 L. However, 5- to 15-L bags are the most useful to industrial hygienists.

These bags are constructed from a number of materials, including polyester, polyvinylidene chloride, Teflon®, aluminized Mylar®, or other fluorocarbons. Bags have the advantages of being light, non-breakable, inexpensive to ship, and simple to use. However, they should be used with caution because storage stabilities for gases, memory effects from previous samples, permeability, precision, and accuracy of sampling systems vary considerably.

Plastic bags should be tested before they are used. Such testing should be done under ambient conditions that approximate those of the sampling environment. Some general recommendations are available in the published literature[6, 8–10] for the use of such bags for air sampling. A good review of specific applications up to 1967 is Schuette;[11] other specific applications are given in References 12 to 18. Posner and Woodfin[18] made a useful systematic study of five bag types for sampling of six organic vapors; they concluded that Tedlar® bags are best for short-term sampling, whereas aluminized Mylar bags are better for long-term storage prior to analysis. Storage properties, decay curves, and other factors, however, will vary considerably from those reported for a given gas or vapor because sampling conditions are rarely identical. Each bag, therefore, should be evaluated for the specific gas or gas mixture for which it will be used.

Continuous Active Samplers
(Long-term Samplers—Hours or Days)

Sorbers

The sorption theory of gases and vapors from air by solution, as developed by Elkins *et al.*,[19] assumes that gases and vapors behave like perfect gases and dissolve to give a perfect solution. The concentration of the vapor in solution is increased during air sampling until an equilibrium is established with the concentration of vapor in the air. Sorption is never complete, however, because the vapor pressure of the material is not reduced to zero but is only lowered by the solvent effect of the sorbing liquid. Some vapor will escape with continued sampling, but it is replaced. Continued sampling will not increase the concentration of vapor in solution once equilibrium is established.

According to formulas developed by Elkins *et al.*,[19] and verified by Gage[20] in his experiments with ethylene oxide, the efficiency of vapor collection depends on 1) the volume of air sampled, 2) the volume of the sorbing liquid, and 3) the volatility of the contaminant being collected. Efficiency of collection, therefore, can be increased by cooling the sampling solution (reducing the volatility of the contaminant), increasing the solution volume by adding two or more bubblers in series, or altering the design of the sampling device. Sampling rate and concentration of the vapor in air are not primary factors that determine collection efficiency.

Sorption of gases and vapors by chemical reaction depends on the size of the air bubbles produced in the bubbler, the interaction of contaminant with reagent molecules, the rate of the reaction, and a sufficient excess of reagent solution. If the reaction is rapid and a sufficient excess of reagent is maintained in the liquid, complete retention of the contaminant is achieved regardless of the volume of air sampled. If the reaction is slow and the sampling rate is not low enough, collection efficiency will decrease.

Four basic sorbers used for the collection of gases and vapors are 1) simple gas washing bottles, 2) spiral and helical sorbers, 3) fritted bubblers, and 4) glass bead columns. Sampling and sorbent capacities of these sorbers are found in Reference 6. Their function is to provide sufficient contact between the contaminant in the air and the sorbing liquid.

Petri, Dreschsel, and midget impingers are examples of simple gas washing bottles. They function by applying a suction to an outlet tube, which causes sample air to be drawn through an inlet tube into the lower portion of the liquids contained in these sorbers. They are suitable for collecting nonreactive gases and vapors that are highly soluble in the sorbing liquid. The sorption of methanol and butanol in water, esters in alcohol, and organic chlorides in butyl alcohol are examples. They are also used for collecting gases and vapors that react rapidly with a reagent in the sampling media. High collection efficiency is achieved, for example, when toluene diisocyanate is hydrolyzed to toluene diamine in Marcali[21] solution. Hydrogen sulfide reaction with cadmium sulfate and ammonia neutralization by dilute sulfuric acid are other examples.

Several methods for testing the efficiency of a sorbing device are available: 1) by series testing where enough samplers are arranged in series so that the last sampler does not recover any of the sampled gas or vapor; 2) by sampling from a dynamic standard atmosphere or from a gas-tight chamber or tank containing a known gas or vapor concentration; 3) by comparing results obtained with a device known to be accurate; and 4) by introducing a known amount of gas or vapor into a sampling train containing the sorber being tested.

Cold Traps

Cold traps are used for collecting materials in liquid or solid form primarily for identification purposes. Vapor is separated from air by passing it through a coiled tube immersed in a cooling system, i.e., dry ice and acetone, liquid air, or liquid nitrogen. These devices are employed when it is difficult to collect samples efficiently by other techniques. Water is extracted along with organic materials and two-phase systems result.

Sampling Bags

Bags (as used for grab sampling) can also be used for collecting integrated air samples. Samples can be collected for 8 hours, at specific times during the day, or over a period of several days. The bags may be mounted on workers for personal sampling or may be located in designated areas.

Solid Sorbents

Activated Charcoal

Charcoal is an amorphous form of carbon formed by partially burning wood, nutshells, animal bones, and other carbonaceous materials. A wide variety of charcoals are available; some are more suitable for liquid purification, some for decolorization, and others for air purification and air sampling.

Ordinary charcoal becomes activated charcoal by heating it with steam to 800–900°C. During this treatment, a porous, submicroscopic internal structure is formed that gives it an extensive internal surface area, as large as 1000 m² per gram of charcoal. This greatly enhances its sorption capacity.

Activated charcoal is an excellent sorbent for most organic vapors. During the 1930s and 1940s, it was used in the then well-known activated charcoal apparatus[22] for the collection and analysis of solvent vapor. The quantity of vapor in the air sample was determined by a gain in weight of the charcoal tube. Lack of specificity, accuracy, and sensitivity of the analysis and the difficult task of equilibrating the charcoal tube, however, discouraged further use.

Renewed interest in activated charcoal as a sorbent for sampling organic vapors appeared in the 1960s.[23–25] The ease with which carbon disulfide extracts organic vapors from activated charcoal and the capability of microanalysis by gas chromatography are the reasons for its current popularity.

Sample Collection

The sorption capacity of a sampler, i.e., the volume of air that can be collected without loss of contaminant, depends on the sampling rate, the quantity of sorbent, the sorbent surface area, the density of active sites and bulk density, the volatility of the contaminant, the ambient humidity, and the concentration of contaminant in the workroom air. For many organic vapors, a sample volume of 10 L can be collected without significant loss in NIOSH-recommended tubes. A breakthrough of more than 20% in the back-up section indicates that some of the sample was lost. Optimum sample volumes are found in NIOSH procedures.[1]

It is always best to refer to an established procedure for proper sampling rates and air sample volumes. In the absence of such information, breakthrough experiments must be performed before field sampling is attempted. Normally, these experiments are conducted using dynamic standard atmospheres prepared at twice the exposure limit (normally the ACGIH threshold limit value, TLV®) and 80% relative humidity to give a suitable margin of safety to the measured breakthrough volume. See Chapter 8 for the preparation of known concentrations.

After the procedure has been validated, field sampling may be performed. Immediately before sampling, the ends of the charcoal tube are broken, rubber or Tygon® tubing is connected to the back-up end of the charcoal tube, and air is drawn through the sampling train with a calibrated battery or electrically driven suction pump. A personal or area sample may be collected. The duration of the sampling is normally 8 hours, but may be as short as 15 minutes or up to 24 hours, depending on the information desired. When sampling is completed, plastic (but not rubber) caps are placed on the ends of the tube.

For each new batch of charcoal tubes, an analytical blank must be prepared to determine the aging, collection efficiency, and recovery characteristics for a given contaminant. This may be achieved by introducing a known amount of the contaminant into a freshly opened charcoal tube, passing clean air through it to simulate sampling conditions, and carrying through its analysis with the field samples. Another charcoal tube, not used to sample, is opened in the field and used as a field blank.

The first step in the analytical procedure is to desorb the contaminant from the charcoal. An early drawback to using charcoal for air sampling was the difficulty in recovering samples for analysis. Steam distillation was only partially effective. Extraction with carbon disulfide has been found, in many instances, to be quite satisfactory, although for the more volatile vapors, thermal desorption may also be used (see "Thermal Desorption" below).

The most frequently used liquid desorbant is carbon disulfide. Unfortunately, carbon disulfide does not always completely desorb the sample from charcoal. Recovery varies for each contaminant and batch of charcoal used. The extent of individual recoveries must be determined experimentally and a correction for desorption efficiency applied to the analytical result.[23] Over a narrow range of analyte concentrations, as used in the NIOSH validations,[26] this desorption efficiency is essentially constant, but it may vary widely over larger concentration ranges, particularly for polar compounds.[27, 28] Desorption efficiency can also be affected by the presence of water vapor and other contaminants.[29, 30] NIOSH[1] recommends that methods be used only where the desorption efficiency is greater than 75%; ideally, it should be greater than 90%.

The practical desorption step in charcoal analysis is also critical because, upon the addition of carbon disulfide to charcoal, the initial heat of reaction may drive off the more volatile components of the sample. This can be minimized by adding charcoal slowly to pre-cooled carbon disulfide. Another technique is to transfer the charcoal sample to vials lined with Teflon septum caps and to introduce the carbon disulfide with an injection needle. The sealed vial will prevent the loss of any volatilized sample. Headspace analysis is also possible.

It should be emphasized that carbon disulfide is a highly toxic solvent that produces severe health effects on the cardiovascular and nervous systems. Care should be exercised in handling the solvent, and the analytical procedure should be performed in a well-ventilated area.

Silica Gel

Silica gel is an amorphous form of silica derived from the interaction of sodium silicate and sulfuric acid. It has several advantages over activated charcoal for sampling gases and vapors: 1) polar contaminants are more easily desorbed by a variety of common solvents; 2) the extractant does not usually interfere with wet chemical or instrumental analyses; 3) amines and some inorganic substances for which charcoal is unsuitable can be collected; and 4) the use of highly toxic carbon disulfide is avoided.

One disadvantage of silica gel is that it will sorb water. Silica gel is electrically polar, and polar substances are preferentially attracted to active sites on its surface. Water is highly polar and is tenaciously held. If enough moisture is present in the air or if sampling is continued long enough, water will displace organic solvents (which are relatively non-polar in comparison) from the silica gel surface. With water vapor at the head of the list, compounds in descending order of polarizability are alcohols, aldehydes, ketones, esters, aromatic hydrocarbons, olefins, and paraffins. It is obvious, therefore, that the volume of moisturized air which can be effectively passed over silica gel is limited.

In spite of this limitation, silica gel has proven to be an effective sorbent for collecting many gases and vapors. Even under conditions of 90% humidity, relatively high concentrations of benzene, toluene, and trichloroethylene are quantitatively sorbed on 10 g of silica gel from air samples collected at the rate of 2.5 L/min for periods of at least 20 minutes or longer.[31, 32] Under normal conditions, hydrocarbon mixtures of 2–5 carbon paraffins, low molecular weight sulfur compounds (H_2S, SO_2, mercaptans), and olefins concentrate on silica gel at dry ice-acetone temperature if the sample volume does not exceed 10 L.[33] Significant losses of ethylene, methane, ethane, and other light hydrocarbons occur if the sampling volume is extended to 30 L.

Many of the same considerations apply to silica gel tubes as to the charcoal tubes; the sampling capacity and desorption efficiency for the compound of interest should be determined before use, or a reliable, officially established method should be used. A variety of desorption solvents will be needed for desorbing specific compounds with high efficiency; polar desorption solvents, such as water or methanol, are commonly applied.

Thermal Desorption

Because of the high toxicity and flammability of carbon disulfide and the labor-intensive nature of the solvent desorption procedure, a useful alternative is to desorb the collected analyte thermally.[34–37] Except in a few cases, this is not practical with charcoal as sorbent because the temperature needed for desorption (e.g., 300°C) would result in some decomposition of the analytes. Carbon molecular sieves or, more frequently, porous polymer sorbents (in particular Tenax®, Porapak Q® and Chromosorb 106®) are used instead.[38–40] Of these, Tenax has the lowest thermal desorption blank (typically less than a few nanograms per gram of sorbent, when properly conditioned) but only modest sorption capacity compared with carbon. The advantages of thermal desorption over solvent extraction have been recognized for ambient,[34, 35] workplace,[36] and indoor air applications.[37]

The thermal desorption procedure typically uses larger tubes than the NIOSH method; usually 200–500 mg

of sorbent are used, depending on type. Desorption can be made fully automatic, and analysis is usually carried out by gas chromatography. Some desorbers also allow automatic selection of sample tubes from a multiple-sample carousel. The whole sample can be transferred to the gas chromatograph, resulting in greatly increased sensitivity compared with the solvent desorption method. Alternatively, some desorbers allow the desorbed sample to be held in a reservoir from which aliquots are withdrawn for analysis, but then the concentrating advantage is reduced.

The main disadvantage of thermal desorption directly with an analyzer is that it is essentially a "one-shot" technique; normally, the whole sample is analyzed. This is why many such methods are linked to mass spectrometry. However, with capillary chromatography, it is usually possible to split the desorbed sample before analysis and, if desired, the vented split can be collected and reanalyzed.[41] Alternatively, the desorbate can be split between two capillary columns of differing polarity.[42]

Coated Sorbents

Many highly reactive compounds (e.g., isocyanates and lower molecular weight aldehydes) are unsuitable for sampling directly onto sorbents, either because they are unstable or cannot be recovered efficiently. In addition, some compounds may be analyzed more easily, or with greater sensitivity, by derivatizing them first, which can sometimes be achieved during the sampling stage.

Wet Chemistry and Spectrophotometric Methods

Several gases and vapors may be analyzed by wet chemical methods or by ultraviolet spectrophotometry. Spectrophotometric methods have now been replaced largely by direct-reading instruments or detector tubes (Chapter 18) or by high-performance liquid chromatography (HPLC) or other instrumental techniques.

Sampling Train

Except for grab samplers (described above) and diffusive samplers (described below), sampling devices are used in conjunction with a sampling pump and air metering device. To avoid contaminating the metering device and pump, these are usually placed downstream of the sampler during the sampling period. However, because many samplers introduce back-pressure, the sampling train should be pre-calibrated by the use of an external flowmeter upstream of the sampling head. The sampling train should also be calibrated after sampling and preferably should be calibrated periodically during sampling. Air movers are described in Chapter 12 and calibration in Chapter 8.

Calculations

The collected sample is analyzed, either directly if it is a gas phase or impinger sample, or after desorption if it is collected on a solid sorbent, using appropriate gas or liquid standard solutions to calibrate the analytical instrument. Gas phase samples give a result directly in ppm (v/v), but other types of samples will give a mass of analyte per collected sample, or a concentration, which can be converted to a mass by multiplying by the sample volume.

The mass concentration of the analyte in the air sample is then calculated using the following equations:

Impinger

$$C = (m - m_{blank}) / (E_s \times V) \qquad (1)$$

where: C = mass concentration of analyte in air (mg/m³)
  m = mass of analyte in sample (μg)
  E_s = sampling efficiency
  m_{blank} = mass of analyte in blank (μg)
  V = volume of air sample (litres)

Sorbent Tube

$$C = (m_1 + m_2 - m_{blank}) / (E_d \times V) \qquad (2)$$

where: m_1 = mass of analyte on first tube section (μg)
  m_2 = mass of analyte on backup tube section (if used) (μg)
  E_d = desorption efficiency corresponding to m_1

Note: If it is desired to express concentrations reduced to specified conditions, e.g., 25° C and 101 kPa, then,

$$C_{corr} = C \times (101/P) \times (T/298) \qquad (3)$$

where: P = actual pressure of air sampled (kPa)
  T = absolute temperature of air sampled (°K).

Volume Fraction

The volume fraction of the analyte in air, in ppm (v/v), is

$$C' = C_{corr} \times (24.5/M) \qquad (4)$$

where: M = molecular mass of the analyte of interest (g/mol⁻¹).

Diffusive Samplers

Overview

A diffusive sampler is a device that is capable of taking samples of gas or vapor pollutants from the atmosphere at a rate controlled by a physical process, such as diffusion through a static air layer or permeation through a membrane, but does not involve the active movement of the air through the sampler.[42] It should be noted that in the United States, the adjective "passive" is preferred in describing these samplers and should be regarded as synonymous with "diffusive."

This type of diffusive sampler should not be confused with the annular or aerosol denuders, which not only rely on diffusion to collect the gas or vapors but also upon the air in question being simultaneously drawn through the annular inlet into the sampler. Aerosol particles have diffusion coefficients too low to be collected on the annular inlet and are trapped on a back-up filter. More information on denuders can be found in Chapter 19.

Principles of Diffusive Sampling

A general overview is given in Berlin *et al.*[43] A specific review with environmental applications is given in Brown.[44]

The mass of the analyte which can diffuse to a suitable sorbent within a certain time is determined by the equation which is derived from Fick's first law of diffusion:

$$m_s = \frac{A \times D \times (\rho_1 - \rho_2) \times t}{l} \qquad (5)$$

where: A = cross-sectional area of diffusion path (cm²)
  D = coefficient of diffusion (cm²/sec)
  l = length of diffusion path (cm)
  m_s = mass of analyte sorbed by diffusion (ng)
  t = sampling time (seconds)
  ρ_1 = actual mass concentration at the beginning of the diffusion layer ($l = 0$) (mg/m³)
  ρ_2 = actual mass concentration at the end of the diffusion layer (mg/m³)

Ideally ρ_1 is equal to the concentration of the given analyte in the air outside the diffusive sampler (ρ), and ρ_2 equals zero ("zero sink"-condition). In that case, the magnitude of the diffusive uptake rate, $A \cdot D/l$, is dependent only on the diffusion coefficient of the given analyte and on the geometry of the diffusive sampler used.

In practice, there are a number of factors that can give rise to non-ideal behavior, so that:

$$m_s = \frac{A \times D \times \rho \times t \times k}{l} \qquad (6)$$

where: k = correction factor for non-ideal behavior

ρ = actual mass concentration of analyte in air (mg/m³)

Dimensions of Diffusive Uptake Rate

For a given concentration ρ in milligrams per cubic meter of gas or vapor, the diffusive uptake rate is given by:

$$U = \frac{m_s}{\rho \times t'} \qquad (7a)$$

where: U = sampling rate (cm³/min)

t' = sampling time (min)

Although the uptake rate, U, has dimensions of cubic centimeter per minute this is really a reduction of nanograms per milligrams per cubic meter per minute, ng. (mg/m³)⁻¹ .min⁻¹, and does not indicate a real volumetric flow of (analyte in) air.

Diffusive uptake rates are very often quoted in units of ng.ppm⁻¹.min⁻¹. These are practical units, since most environmental analysts use ppm for concentrations of gases and vapors. The dependency of uptake rates on temperature and pressure is explained later. Thus for a given concentration (ppm) of gas or vapor, the sampling rate is given by:

$$U' = \frac{m_s}{\phi \times t'} \qquad (7b)$$

where: U' = sampling rate (ng/ppm-min)

ϕ = actual mass concentration of analyte in air (ppm, v/v)

Ideal and practical diffusive uptake rates are related by:

$$U' = \frac{U \times M \times 298 \times P}{24.5 \times T \times 101} \qquad (8)$$

Bias Due to the Selection of Non-ideal Sorbents

The performance of a diffusive sampler depends critically on the selection and use of a sorbent or collection medium which has high sorption efficiency. The residual vapor pressure of the sampled compound at the sorbent surface (ρ_2) will then be very small in comparison to the ambient concentration, and the observed uptake rate will be close to its ideal steady-state value, which can usually be calculated from the geometry of the sampler and the diffusion coefficient of the analyte in air.

In the case where a weak sorbent is used, then ρ_2 in Equation 5 is non-zero and m_s/t will decrease with the time of sampling. Hence U in Equation 6 will also decrease with the time of sampling. The magnitude of this effect is dependent on the sorption isotherm of the analyte and sorbent concerned and may be calculated with the aid of computer models. [45, 46]

Another manifestation of the same effect is back diffusion, sometimes called reverse diffusion. This can happen where, some time after sampling has started, the vapor pressure of the analyte at the sorbent surface, ρ_2, is greater than the external concentration, ρ_1; for example, if a sampler is first exposed to a high concentration and then to a much lower or even zero concentration. This type of exposure profile can occur in certain applications, and the magnitude of any error introduced will depend on whether the period of high concentration occurs at the beginning, middle, or end of the sampling period. The phenomenon has been discussed in detail by Bartley and others[47–49] and a simple test proposed[50] to give an estimate of the maximum bias to be expected between a pulsed exposure and an exposure to a constant concentration, which normally provides the basis for the sampler calibration. The extent of back-diffusion can also be modelled theoretically.[45, 51]

It is therefore desirable to choose a sorbent with high sorption capacity and low vapor pressure of the sorbed material or of the reaction product formed by a reactive sorbent.

Environmental Factors Affecting Sampler Performance

Temperature and Pressure

For an ideal diffusive sampler, the dependence of U on absolute temperature and pressure is governed by that of the diffusion coefficient of the analyte. The latter dependence is given by:

$$D = f(T^{n+1}, P^{-1}) \qquad (9)$$

with $0.5 < n < 1.0$.

Hence, the dependence of U, expressed in units of $cm^3 \cdot min^{-1}$ or equivalent is:

$$U = f(T^{n+1}, P^{-1}) \qquad (10)$$

When U' is expressed in units of $ng \cdot ppm^{-1} \cdot min^{-1}$ or equivalent by application of Equation 8, then the dependence is given by:

$$U' = f(T^n) \qquad (11)$$

In the latter case, the dependence will be of the order of 0.2 to $0.4\% \cdot K^{-1}$. In the case of a non-ideal sampler, the temperature dependence of U' may be compensated by the temperature dependence of the sorption coefficient of the analyte.[52] In any case, accurate knowledge of the average temperature and pressure during the sampling period is important for a correct application of Equations 7a and 7b.

Humidity

High humidity can affect the sorption capacity of hydrophilic sorbents, such as charcoal and Molecular Sieve. This will normally reduce the sampling time (at a given concentration) before saturation of the sorbent occurs, when sampling becomes non-linear because of a significant ρ_2 term in Equation 5. High humidity can also alter the sorption behavior of the exposed inner wall of tube-type samplers or draft screen, particularly if condensation occurs.

Transients

Simple derivations of Fick's Law assume steady-state conditions, but in the practical use of diffusive samplers, the ambient level of pollutants is likely to vary widely. The question then arises whether a sampler will give a truly integrated response (ignoring sorbent effects) or will "miss" short-lived transients before they have had a chance to be trapped by the sorbent. The issue has been discussed theoretically [47, 53–55] and practically [53, 56, 57] and shown not to be a problem, provided the total sampling time is well in excess of (say 10 times) the time constant of the diffusive sampler, i.e. the time a molecule takes to diffuse into the sampler under steady-state conditions. The time constant, t, for most commercial samplers is between about 1 and 10 seconds. τ is given by:

$$\tau = l^2/D \qquad (12)$$

where: τ = time constant of diffusive sampler (sec)

l = length of diffusion path (cm)
D = coefficient of diffusion (cm^2/s)

The Influence of Air Velocity

Effect of Low and High Wind Speeds. Ambient air face velocity and orientation can affect the performance of a diffusive sampler because they may influence the effective diffusion path length.[58–61] The diffusive mass uptake of a sampler (Equation 6) is a function of the length, l, and the cross-sectional area, A, of the diffusion gap within the sampler. The nominal diffusion path length is defined by the geometry of the sampler and is the distance between the sorbent surface and the external face of the sampler. The cross-sectional area is also defined by the geometry of the sampler and if the cross-section of the diffusion gap is not constant along its length, is defined by the narrowest portion. The effective length, l, is not necessarily the same as the nominal length, and may be greater or less, depending on circumstances.

Under conditions of low external wind speeds, the effective diffusion path length may be increased.[60, 61] This is because a 'boundary layer'[58, 59] exists between the stagnant air within the sampler and the turbulent air outside and contributes to the effective diffusion path length, l. In reality, there is an area outside the sampler where there is a transition between static air and turbulent air, but this is equivalent to an extra length (δl) of static air which must be included in the value of l. The magnitude of δl depends on the external geometry of the sampler, being roughly proportional to the linear cross-section of the sampler collection surface, where this surface is flat. It also decreases with increasing air velocity. Its significance depends on the value of the nominal path length of the diffusive sampler. Thus, a sampler with a small cross-section and long internal air gap will be relatively unaffected by air velocity, while a short, fat sampler will be significantly affected. This is borne out in practice, as has been demonstrated with samplers of varying length.[60, 61] Low sampling rates are observed at low air velocities, but increase to a plateau value as the boundary layer effect becomes insignificant.

Under conditions of high external wind speeds, the effective diffusion path length may be decreased.[62–68] This is because turbulent air disturbs the static air layer within the sampler, which reduces the effective air gap by a factor δl. The magnitude of δl is small, provided the length-to-diameter ratio of the sampler air gap is greater than 2.5 to 3,[62] or it can be avoided, or greatly reduced, by incorporating a draft shield (e.g., a stainless steel screen or plastic membrane).

The overall effect is therefore sinusoidal.

Consequence for Different Sampler Geometries. Tube-type samplers are typically unaffected by low air velocities[53, 69, 70] but those without a draft shield may be affected by high speeds.

Badge-type samplers generally have a large surface area and small air gap, so that they may be more affected by air velocity than tube designs and typically require a minimum face velocity of between 0.5 and 0.2 m.s^{-1} [71–74] Some badges with an inadequate draft shield are also affected at high air velocities.[70, 72, 75]

Radial diffusive samplers[76] require a minimum face velocity of about 0.25 m.s^{-1}.

Calculations

The method of calculation of atmospheric concentrations is essentially the same as for pumped samplers, i.e., the collected sample is analyzed and the total weight of analyte on the sampler is determined. Then, as before,

$$C = (m_1 + m_2 - m_{blank}) / (E_d \times V) \qquad (2)$$

Notes: m_2 is relevant only to samplers with a back-up section, and an additional multiplication factor may be needed to account for differing diffusion path lengths to primary and back-up sections. m_2 and E_d are ignored for liquid sorbent badges.

V, the total sample volume, is calculated from the effective sampling rate (L/min) and the time of exposure (min).

This calculation gives C in mg/m^3; strictly speaking, an appropriate sampling rate for the ambient temperature and pressure should be made.

Alternatively, sampling rates can be expressed in units such as ng/ppm-min (dimensionally equivalent to cm^3/min), when C' is calculated directly in ppm:

$$\frac{C = (m_1 + m_2 - m_{blank}) \times 1000}{E_d \times U \times t'} \qquad (13)$$

Practical Applications

Canisters

The U.S. Environmental Protection Agency (U.S. EPA) has used passivated canisters for ambient air analysis alongside sorbent tubes.[77] A more recent study[78] has shown significant under- and over-estimations of some compounds by the canister method compared to a continuous cycling gas chromatograph.

Charcoal Tubes

Air sampling procedures using activated charcoal are widely used by industrial hygienists[79–81] and form the basis of the majority of the official analytical methods for volatile organic compounds recommended by the National Institute for Occupational Safety and Health (NIOSH) and the Occupational Safety and Health Administration (OSHA).[1, 2] An International Standards Organization (ISO) draft standard is in preparation (ISO/DIS 16200-1: VOCs by pumped tube/solvent desorption).

Analytical information on selected NIOSH procedures is given in Table 16-1. In general, the NIOSH procedures use a 100-mg charcoal tube (with 50-mg back-up), but very volatile analytes may require a larger tube. The NIOSH study showed that the charcoal tube method is generally adequate for hydrocarbons, halogenated hydrocarbons, esters, ethers, alcohols, ketones, and glycol ethers that are commonly used as industrial solvents. Compounds with low vapor pressure and reactive compounds (e.g., amines, phenols, nitro-compounds, aldehydes, and anhydrides) generally have low desorption efficiencies from charcoal and require alternative sorbents such as silica gel or porous polymers for collection, or alternative reagent systems for recovery.

Inorganic compounds, such as ozone, nitrogen dioxide, chlorine, hydrogen sulfide, and sulfur dioxide, react chemically with activated charcoal and cannot be collected for analysis by this method.

Even for substances recommended for sampling on charcoal, this sorbent may not always be ideal. Reference to Table 16-1 will indicate that carbon disulfide is the recommended desorption solvent for non-polar compounds, whereas a variety of desorption cocktails are required for the more polar compounds. Difficulties arise, therefore, when sampling mixtures of polar and non-polar compounds because each will give poor recoveries with the other's desorption solvent. Several more universal solvents have been investigated,[82–84] but none of these has achieved wide recognition. In such circumstances, it may be necessary to take two or more samples at the same time and desorb each one with a different solvent.

Silica Gel

In some cases, silica gel tubes (in similar sizes to the NIOSH range of charcoal tubes) are used instead of charcoal tubes. NIOSH recommends such tubes for a variety of more polar chemicals such as amines, phenols, amides, and inorganic acids (Table 16-1).

TABLE 16-1. Collection and Analysis of Gases and Vapors (solvent desorption)

Method Name	Test Compounds	Sorbent[A]	Desorption Solvent	NIOSH Method No.
Alcohols I	t-Butyl alcohol Isopropyl alcohol Ethanol	C	99:1 CS_2:2-butanol	1400
Alcohols II	n-Butyl alcohol Isobutyl alcohol sec-Butyl alcohol n-Propyl alcohol	C	99:1 CS_2:2-propanol	1401
Alcohols III	Allyl alcohol Isoamyl alcohol Methyl isobutyl carbinol Cyclohexanol Diacetone alcohol	C	99:5 CS_2:2-propanol	1402
Alcohols IV	2-Butoxyethanol 2-Ethoxyethanol 2-Methoxyethanol	C	99:5 CH_2Cl_2:methanol	1403
Amines	Aniline	S	95% ethanol	2002
Aromatic	o-Toluidine 2, 4-Xylidine N, N,-Dimethyl-p-toluidine N, N,-Dimethylaniline	S	95% ethanol	2002
Aminoethanol Compounds	2-Aminoethanol 2-Dibutylaminoethanol 2-Diethylaminoethanol	S	80% ethanol	2007
Esters I	n-Amyl acetate n-Butyl acetate 2-Ethoxyethyl acetate Ethyl acrylate Methyl isoamyl acetate n-Propyl acetate, etc.	C	CS_2	1450
Hydrocarbons BP 36-126° C	Benzene, Toluene Pentane thro' Octane Cyclohexane Cyclohexane	C	CS_2	1500
Hydrocarbons Aromatic	Benzene Cumene Naphthalene, etc.	C	CS_2	1501
Hydrocarbons Halogenated	Chloroform Tetrachloroethylene p-Dichlorobenzene Bromoform, etc.	C	CS_2	1003
Ketones I	Acetone Cyclohexanone Diisobutyl ketone 2-Hexanone Methul isobutyl ketone 2-Pentanone	C	CS_2	1300

TABLE 16-1 (cont.). Collection and Analysis of Gases and Vapors (solvent desorption)

Method Name	Test Compounds	SorbentA	Desorption Solvent	NIOSH Method No.
Ketones II	Camphor Ethyl butyl ketone Mesityl oxide 5-Methyl-3-heptanone Methyl n-amyl ketone	C	99:1 CS_2:methanol	1301
Naphthas	Kerosine Petroleum ether Rubber solvent Stoddard solvent, etc.	C	CS_2	1550
Nitro-benzenes	Nitrobenzene Nitrotoluene 4-Chloronitrotoluene	S	Methanol	2005
Nitrogylcerin and Ethylene glycol dinitrate		T	Ethanol	2507
Pentachloroethane		R	Hexane	2517
Tetrabromoethane		S	Tetrahydrofuran	2003
Vinyl chloride		C	CS_2	1007

AC = charcoal
 S = silica gel
 T = Tenax
 R = Porapak R

Thermal Desorption

Thermal desorption has been adopted as a (nonexclusive) recommended method for the determination of volatile organic compounds in the United Kingdom,[3] Germany,[4] and the Netherlands,[85] but it is less widely accepted in the United States. NIOSH[1] has relatively few methods based on thermal desorption (compared to those which use solvent desorption). U.S. EPA[86] has a number of methods based on thermal desorption and mass spectrometry.

Desorption efficiency is usually 100% for the majority of common solvents and similar compounds in a boiling range of approximately 50 to 250°C. Thus, the analysis of complex mixtures is easier than for charcoal or silica gel solvent desorption methods. However, if a wide boiling range is to be covered, more than one sorbent may be required. Thus, gasoline may be monitored by a Chromosorb 106 tube and carbon tube in series.[87] Extensive lists of recommended sampling volumes and minimum desorption temperatures for Tenax and other sorbents are given in Brown and Purnell[36] and the United Kingdom Health and Safety Executive Method MDHS 72.[88] An ISO draft standard is in preparation (ISO/DIS 16017-1: VOCs by pumped tube/thermal desorption).

Coated Sorbents

Methods have been developed that use coated sorbents, either sorbent tubes or coated filters. Table 16-2 lists a number of such methods. An ISO draft standard is in preparation (ISO/CD 16000-3: Formaldehyde and other carbonyl compounds—Active sampling method), which used the DNPH reagent and HPLC.

Wet Chemistry and Spectrophotometric Methods

There are two primary compendia of methods.[6, 89] Other useful sources are Hansen,[90] Jacobs,[91, 92] the *Methods for the Detection of Toxic Substances in Air* series,[93] Ruch,[94, 95] Thomas,[96] and Feigl.[97]

Diffusive Samplers

A variety of diffusive samplers have been described[98] and only a selection of the major types manufactured can be described here. Diffusive equivalents to the more familiar pumped methods exist for nearly all types; the main exception being the direct collection of gas samples, where the nearest equivalent is an evacuated canister. Thus, the diffusive equivalent of an impinger is a liquid-filled badge

TABLE 16-2. Collection and Analysis of Gases and Vapors (coated sorbents)

Test Compounds	Sorbent	Matrix[A]	Method No.
Acetaldehyde	2-(Hydroxymethyl)piperidine on Supelpak 20N	T	NIOSH 2538
Acrolein	2-(Hydroxymethyl)piperidine on Supelpak 20N	T	NIOSH 2501
Arsenic trioxide	Sodium carbonate	F	NIOSH 7901
Butylamine	Sulfuric acid	T	NIOSH S138
Diisocyanates	1-(2-Pyridyl)piperzine	F	OSHA 42
Formaldehyde	N-benzylethanolamine on Supelpak 20F	T	NIOSH 2502
Methylene dianiline		F	NIOSH 5029

[A] T = sorbent tube
  F = filter

such as the Pro-Tek™ inorganic monitor or the SKC badge; the diffusive equivalent of the charcoal tube is the charcoal badge such as the 3M OVM or the SKC Passive Sampler; and the diffusive equivalent of the thermal desorption method is the Perkin-Elmer tube or the SKC thermal desorption badge. There are also diffusive devices based on reagent-impregnated solid supports, but these are for specific analytes (see Instrument Section below) or are direct-reading and are dealt with in Chapter 18.

In general, the regulatory authorities have been reluctant to accept diffusive monitoring methods, except in the United Kingdom and the Netherlands where several such methods have been adopted as nonexclusive recommended methods.[3, 85] Extensive lists of recommended sampling rates for solvent desorption methods are given in *The Diffusive Monitor*[99] and the United Kingdom Health and Safety Executive Method MDHS 88.[100] An ISO draft standard is in preparation (ISO/DIS16200-2: VOCs by diffusive sampler/solvent desorption).

Extensive lists of recommended sampling rates for thermal desorption methods are given in *The Diffusive Monitor*[101] and the United Kingdom *Health and Safety Executive Method MDHS 80*.[102] A generic ISO draft standard for VOCs is in preparation (ISO/CD 16017-2: VOCs by diffusive tube/thermal desorption).

An ISO draft standard for aldehydes is in preparation (ISO/CD 16000-4: Formaldehyde-Passive/Diffusive sampling method).

Quality Systems and Quality Control

Canister sampling should be checked by using certified reference gas standards where available.

Several interlaboratory quality assurance schemes have been developed that apply to the charcoal tube method. One of these is the Proficiency Analytical Testing (PAT) Program[103] and the Laboratory Accreditation Program of the American Industrial Hygiene Association (AIHA).[104] Another is the Health and Safety Executive (HSE) Workplace Analysis Scheme for Proficiency (WASP). Details of these programs may be obtained from The Laboratory Accreditation Coordinator, AIHA, 2700 Prosperity Ave., Suite 250, Fairfax, VA 22031, and the WASP Coordinator, Health and Safety Laboratory, Broad Lane, Sheffield, S3 7HQ, United Kingdom.

The WASP scheme also includes test samples appropriate to the thermal desorption technique, at both occupational and ambient concentration levels. Certified Reference Materials are available from the EC BCR (Community Bureau of Reference) for aromatic hydrocarbons (CRM 112)[105] and chlorinated hydrocarbons (CRM 555).[106]

References

1. National Institute for Occupational Safety and Health: NIOSH Manual of Analytical Methods, 2nd ed. DHEW (NIOSH) Pub. No. 75-121 (1975); 3rd ed. DHEW (NIOSH) Pub. No. 84-100 (1984, revised 1990); 4th ed. DHHS (NIOSH) Pub. No. 94-113 (1994); (updated regularly).

2. Occupational Safety and Health Administration: OSHA Analytical Methods Manual. OSHA Analytical Laboratories, Salt Lake City, UT. Available from ACGIH, Cincinnati, OH (various methods, 1979–1995, updated regularly).

3. Health and Safety Executive: Methods for the Determination of Hazardous Substances. HSE Occupational Medicine and Hygiene Laboratory, Sheffield, UK (in series, 1981–2000).

4. Deutsche Forschungsgemeinschaft: Analytische Methoden zur Prufung Gesundheitsschadlicher Arbeitsstoffe. DFG. Verlag Chemie, Weinheim, FRG (1985).

5. Arbetarskyddsverket: Principer och Methoder for Provtagning och Analys av Amnen Upptagna pa Listan over Hygieniska Gransvarden. Arbete och Halsa. Vetenskaplig Skriftserie 1987:17. Solna, Sweden (1987).

6. Intersociety Committee: Methods of Air Sampling and Analysis, 3rd ed. Lewis Publishers, Inc., Chelsea, MI (1988).

7. Denyszyn, R.B.; Harden, J.M.; Hardison, D.L.; et al.: Analytical Facilities for the Analysis of Trace Organic Volatiles in Ambient Air. National Bureau of Standards Special Publication 519 (issued April 1979).

8. Nelson, G.O.: Controlled Test Atmospheres, Principles and Techniques. Ann Arbor Science Publishers, Ann Arbor, MI (1971).

9. Pellizzari, E.D.; Gutknecht, W.F.; Cooper, S.; Hardison, D.: Evaluation of Sampling Methods for Gaseous Atmospheric Samples. EPA 600/3-84-062. U.S. Environmental Protection Agency, Research Triangle Park, NC (1984).

10. Apol, A.G.; Cook, W.A.; Lawrence, E.F.: Plastic Bags for Calibration of Air Sampling Devices—Determination of Precision of the Method. Am. Ind. Hyg. Assoc. J. 27:149 (1966).

11. Schuette, F.J.: Plastic Bags for Collection of Gas Samples. Atmos. Environ. 1:515–519 (1967).

12. Levine, S.P.; Hebel, K.G.; Bolton, Jr., J.; Kupel, R.E.: Industrial Analytical Chemists and OSHA Regulations for Vinyl Chloride. Anal. Chem. 47:1075A (1975).

13. Seila, R.L.; Lonneman, W.A.; Meeks, S.A.: Evaluation of Polyvinyl Fluoride as a Container Material for Air Pollution Studies. J. Environ. Sci. Health All. 11:121 (1976).

14. Scheil, G.W.: Standardization of Stationary Source Method for Vinyl Chloride. EPA 600/4-77-026. U.S. Environmental Protection Agency, Research Triangle Park, NC (1977).

15. Rothwell, R.; Mitchell, A.D.: Plastic Bags for Sampling of C2–C6 Hydrocarbons. Clean Air 7:35–36 (1977).

16. Knoll, J.E.; Penney, W.H.; Midgett, M.R.: The Use of Tedlar Bags to Contain Gaseous Benzene Samples at Source Level Concentrations. EPA 600/4-78-057. U.S. Environmental Protection Agency, Research Triangle Park, NC (1978).

17. Knoll, J.E.; Smith, M.A.; Midgett, M.R.: Evaluation of Emission Test Methods for Halogenated Hydrocarbons. EPA 600/4-79-025. U.S. Environmental Protection Agency, Research Triangle Park, NC (1979).

18. Posner, J.C.; Woodfin, W.J.: Sampling with Gas Bags; 1: Losses of Analyte with Time. Appl. Ind. Hyg. 1:163B168 (1986).

19. Elkins, H.B.; Hobby, A.K.; Fuller, J.E.: The Determination of Atmospheric Contaminants; I: Organic Halogen Compounds. J. Ind. Hyg. 19:474–485 (1937).

20. Gage, J.C.: The Efficiency of Absorbers in Industrial Hygiene Air Analysis. Analyst 85:196–203 (1960).

21. Marcali, K.: Microdetermination of Toluene Diisocyanates in the Atmosphere. Anal. Chem. 29:552 (1957).

22. Elkins, H.B.: The Chemistry of Industrial Toxicology, 2nd ed. John Wiley & Sons, Inc., New York (1959).

23. Otterson, E.J.; Guy, C.U.: A Method of Atmospheric Solvent Vapor Sampling on Activated Charcoal in Connection with Gas Chromatography. In: Transactions of the 26th Annual Meeting, Philadelphia, PA, p. 37. American Conference of Governmental Industrial Hygienists, Cincinnati, OH (1964).

24. Fraust, C.L.; Hermann, E.R.: The Adsorption of Aliphatic Acetate Vapors onto Activated Carbon. Am. Ind. Hyg. Assoc. J. 30:494–499 (1969).

25. Reid, F.H.; Halpin, W.R.: Determination of Halogenated and Aromatic Hydrocarbons in Air by Charcoal Tube and Gas Chromatography. Am. Ind. Hyg. Assoc. J. 29:390–396 (1968).

26. National Institute for Occupational Safety and Health: Documentation of the NIOSH Validation Tests. DHEW (NIOSH) Publ. No. 77-185 (1977).

27. Posner, J.C.; Okenfuss, J.R.: Desorption of Organic Analytes from Activated Carbon; I: Factors Affecting the Process. Am. Ind. Hyg. Assoc J. 42:643–646 (1981).

28. Posner, J.C.; Okenfuss, J.R.: Desorption of Organic Analytes from Activated Carbon; II: Dealing with the Problem. Am. Ind. Hyg. Assoc J. 42:647–652 (1981).

29. Rudling, J.: Organic Solvent Vapor Analysis of Workplace Air. Arbete och Halsa. Vetenskaplig Skriftserie 1987:11. Arbetarskyddsverket, Solna, Sweden (1987).

30. Callan, B.; Walsh, K.; Dowding, P.: Industrial Hygiene VOC Measurement Interference. Chem. and Ind. X:250–252 (1991).

31. Elkins, H.B.; Pagnotto, L.D.; Comproni, E.M.: The Ultraviolet Spectrophotometric Determination of Benzene in Air Samples Adsorbed on Silica Gel. Anal. Chem. 34:1797 (1962).

32. Van Mourik, J.H.C.: Experiences with Silica Gel as Absorbent. Am. Ind. Hyg. Assoc. J. 26:498 (1965).

33. Altshuller, A.P.; Bellar, T.A.; Clemons, C.A.: Concentration of Hydrocarbons on Silica Gel Prior to Gas Chromatographic Analysis. Am. Ind. Hyg. Assoc. J. 23:164 (1962).

34. Zlatkis, A.; Lichtenstein, H.A.; Tishbee, A.: Concentration and Analysis of Volatile Organics in Gases and Biological Fluids with a New Solid Absorbent. Chromatographia 6:67–70 (1973).

35. Pellizzari, E.D.; Bunch, J.E.; Carpenter, B.H.; Sawicki, E.: Collection and Analysis of Trace Organic Vapor Pollutants in Ambient Atmospheres. Environ. Sci. Technol. 9:552–560 (1975).

36. Brown, R.H.; Purnell, C.J.: Collection and Analysis of Trace Organic Vapour Pollutants in Ambient Atmospheres. The Performance of a Tenax-GC Adsorbent Tube. J. Chromatogr. 178:79–90 (1979).

37. Wolkoff, P.: Volatile Organic Compounds—Sources, Measurements, Emissions, and the Impact on Indoor Air Quality. Indoor Air, Suppl. 3/95. (1995).

38. Bertoni, G.; Bruner, F.; Liberti, A.; Perrino, C.: Some Critical Parameters in Collection Recovery and Gas Chromatographic Analysis of Organic Pollutants in Ambient Air Using Light Adsorbents. J. Chromatogr. 203:263–270 (1981).

39. Ciccioli, P.; Cecinato, A.; Brancaleoni, E.; Frattoni, M.: Use of Carbon Adsorption Traps Combined with High Resolution Gas Chromatography—Mass Spectrometry for the Analysis of Polar and Non-polar C4–C14 Hydrocarbons Involved in Photochemical Smog Formation. J. High Res. Chromatogr. 15:75 (1992).

40. Rothweiler, H.: Active Sampling of VOC in Non-industrial Buildings. In: Clean Air at Work. R.H. Brown, M. Curtis, K.J. Saunders and S. Vandendriesche, Eds. CEC Pub. No. EUR 14214, Brussels–Luxembourg (1992).

41. Kristensson, J.: Diffusive Sampling and Gas Chromatographic Analysis of Volatile Compounds. Ph.D. Thesis. University of Stockholm, Sweden (1987).

42. Wright, M.D.: A Dual-capillary Column System for Automated Analysis of Workplace Contaminants by Thermal Desorption–Gas Chromatography. Anal. Proc. 24:309–311 (1987).

43. Berlin, A.; Brown, R.H.; Saunders, K.J., Eds.: Diffusive Sampling: An Alternative Approach to Workplace Air Monitoring. CEC Pub. No. 10555EN. Commission of the European Communities, Brussels–Luxembourg (1987).

44. Brown, R.H.: Diffusive Sampling. In Clean Air at Work. R.H. Brown, M. Curtis, K.J. Saunders,.S. Vandendriessche, S. Eds. pp 141–148. EC Publ. No EUR 14214, Brussels–Luxembourg (1992).

45. Van den Hoed, N.; van Asselen, O.L.J.: A Computer Model for Calculating Effective Uptake Rates of Tube-type Diffusive Air Samplers. Ann. Occup. Hyg. 35, 273–285 (1991).

46. Nordstrand, E.; Kristensson, J.: A Computer Model for Simulating the Performance of Thick-bed Diffusive Samplers. Am. Ind. Hyg. Assoc. J. 55:935–941 (1994).

47. Bartley, D.L.; Doemeny, L.J.; Taylor, D.g.: Diffusive Monitoring of Fluctuating Concentrations. Am. Ind. Hyg. Assoc. J. 44:241–247 (1983).

48. Bartley, D.L.: Diffusive Monitoring of Fluctuating Concentrations Using Weak Sorbents. Am. Ind. Hyg. Assoc. J. 44:879–885 (1983).

49. Bartley, D.L.; Woebkenberg, M.L.; Posner, J.C.: Performance of Thick-sorbent Diffusive Samplers. Ann. Occup. Hyg. 32:333–343 (1988).

50. Bartley, D.L.; Deye, G.J.; Woebkenberg, M.L.: Diffusive Monitor Test: Performance Under Transient Conditions. Appl. Ind. Hyg. 2(3):119–122 (1987).

51. Posner, J.S.; Moore, G.: a Thermodynamic Treatment of Passive Monitors. Am. Ind. Hyg. Assoc. J. 46:277–285 (1985).

52. Pfeffer, H.-U.; Breuer, L.; Ellermann, K.: Validierung von Passivsammlern für Immissionsmessungen von Kohlenwasserstoffen. Materialien Nr. 46, Landesumweltamt NRW, Essen (1998).

53. Brown, R.H.; Charlton, J.; Saunders, K.J.: the Development of an Improved Diffusive Sampler. Am. Ind. Hyg. Assoc. J. 42:865–869 (1981).

54. Hearl, F.J.; Manning, M.P.: Transient Response of Diffusion Dosimeters. Am. Ind. Hyg. Assoc. J. 41:778–783 (1980).

55. Underhill, D.W.: Unbiased Passive Sampling. Am Ind. Hyg. Assoc. J. 44:237–239 (1983).

56. Hori, H.; Tanaka, I.: Response Characteristics of the Diffusive Sampler at Fluctuating Vapor Concentrations. Am. Ind. Hyg. Assoc. J. 54:95–101 (1993).

57. Compton, J.R.; Dwiggins, G.A.; Feigley, C.E. et.al.: the Effect of Square-wave Exposure Profiles upon the Performance of Passive Organic Vapor Monitoring Badges. Am. Ind. Hyg. Assoc. J. 45:446–450 (1984).

58. Tompkins, F.C.; Goldsmith, R.L.: A New Personal Dosimeter for the Monitoring of Industrial Pollutants. Am. Ind. Hyg. Assoc. J. 38:371–377 (1977).

59. Underhill, D.W. and Feigley, C.E. Boundary Layer Effect in Diffusive Monitoring. Anal. Chem. 63:1011–1013 (1991).

60. Pozzoli, L.; Cottica, D.: An Overview of the Effects of Temperature, Pressure, Humidity, Storage and Face Velocity. In: Diffusive Sampling: An Alternative Approach to Workplace Air Monitoring. A. Berlin, R.H. Brown, and K.J. Saunders, Eds. CEC Pub. No. 10555EN. Commission of European Communities, Brussels–Luxembourg (1987).

61. Zurlo, N.; Andreoletti, F.:Effect of Air Turbulence on Diffusive Sampling. In Diffusive Sampling, an Alternative Approach to Workplace Air Monitoring. A. Berlin, R.H. Brown, K.J. Saunders, Eds. pp. 174–176, EC Publ. No. 10555 EN, Brussels–Luxembourg (1987).

62. Coleman, S.R.: A Tube-type Diffusive Monitor for Sulphur Dioxide. Am. Ind. Hyg. Assoc. J. 44:929–936 (1983).

63. Palmes, E.D.; Gunnison, A.f.; Dimattio, J. et. al.: Personal Sampler for Nitrogen Dioxide. Am. Ind. Hyg. Assoc. J. 37:570–577 (1976).

64. Atkins, D.H.F.; Sandalls, J.; Law. D.V. et. al.: The Measurement of Nitrogen Dioxide in the Outdoor Environment Using Passive Diffusion Tube Samplers. AERE-R12133, Harwell, UK (1986).

65. Gair, A.J.; Penkett, S.A.: The Effects of Wind Speed and Turbulence on the Performance of Diffusion Tube Samplers. Atmos. Env. 29:2529–2533 (1995).

66. Downing, C.E.C.; Campbell, G.W.; Bailey, J.C.: A Survey of Sulphur Dioxide, Ammonia and Hydrocarbon Concentrations in the United Kingdom Using Diffusion Tubes: July to December 1992. Warren Spring Laboratory. Report LR 964. Stevenage, UK (1994).

67. Ferm, M.: A Sensitive Diffusional Sampler. Institutet for Vatten och Luftvardsforskning, Sweden Report B1020 (1991).

68. Frenzel, W.; Grimm, E.; Gruetzmacher, G.: Evaluation of a Diffusive Sampling Method for the Determination of Atmospheric Ammonia. Fresenius J. Anal. Chem. 351:19–26 (1995).

69. Pannwitz, K.H.: Influence of Air Currents on the Sampling of Organic Solvent Vapours with Diffusive Samplers. In Diffusive Sampling, an Alternative Approach to Workplace Air Monitoring. A. Berlin, R.H. Brown, K.J. Saunders, Eds. pp. 157–160. EC Publ no 10555 EN, Brussels–Luxembourg (1987).

70. Mark, D.; Robertson, A.; Gibson, H. et.al.: Evaluation of Diffusive Samplers for Monitoring Toxic Gases and Vapours in Coalmines. Institute of Occupational Medicine, Report TM/90/11 Edinburgh (1990).

71. Harper, M.; Purnell, C.J.: Diffusive Sampling—a Review. Am. Ind. Hyg. Assoc. J. 48:214–218 (1987).

72. Hori, H.; Tanaka, I.: Effect of Face Velocity on Performance of Diffusive Samplers. Ann. Occup. Hyg. 40:467–475 (1996).

73. Lewis, R.G.; Mulik, J.D.; Coutant, R.W. et. al.: Thermally Desorbable Passive Sampling Device for Volatile Organic Chemicals in Ambient Air. Analytical Chemistry 57:214–219 (1985).

74. Kasper, A.; Puxbaum, H.: A Badge-type Passive Sampler for Monitoring Ambient Ammonia Concentrations. Fresenius J. of Anal. Chem. 350:448–453 (1994).

75. Samini, B.S.: The Effect of Face Air Velocity on the Rate of Sampling of Air Contaminants by a Diffusive Sampler. In Diffusive Sampling, an Alternative Approach to Workplace Air Monitoring. A. Berlin, R.H. Brown, K.J. Saunders, Eds. pp. 166–169. EC Publ. No. 10555 EN, Brussels–Luxembourg (1987).

76. Cocheo, V.; Boaretto. C.; Sacco, P.: High Uptake Rate Radial Diffusive Sampler Suitable for Both Solvent and Thermal Desorption. Am. Ind. Hyg. Assoc. J. 57:897–904 (1996).

77. Varns, J.L.; Mulik, J.D.; Williams, D.: Passive Sampling Devices and Canisters: Their Comparison in Measuring Air Toxics During a Field Study. Proceedings of the 1990 EPA/A&WMA International Symposium, Measurement of Toxic and Related Air Pollutants, pp. 219–223.

78. Ballesta, P.P.; Field, R.A.; De Saeger, E.: Field Intercomparison of VOC Measurements. EC Report EUR 18085 EN. ERLAP, JRC, Ispra, Italy (1998).

79. White, L.D.; Taylor, D.G.; Mauer, P.A.; Kupel, R.E: A Convenient Optimized Method for the Analysis of Selected Solvent Vapors in the Industrial Atmosphere. Am. Ind. Hyg. Assoc. J. 31:225–232 (1970).

80. Fraust, C.L.: The Use of Activated Carbon for Sampling Industrial Environs. Am. Ind. Hyg. Assoc. J. 36:278 (1975).

81. Reckner, L.R.; Sacher, J.: Charcoal Sampling Tubes for Several Organic Solvents. DHEW (NIOSH) Pub. No. 75-184 (June 1975).

82. Langvardt, P.W.; Melcher, R.G.: Simultaneous Determination of Polar and Non-polar Solvents in Air Using a Two-phase Desorption from Charcoal. Am. Ind. Hyg. Assoc. J. 40:1006–1012 (1979).

83. Johansen, I.; Wendelboe, F.: Dimethylformamide and Carbon Disulphide Desorption Efficiencies for Organic Vapours on Gas-sampling Charcoal Tube: Analyses with a Gas Chromatographic Backflush Technique. J. Chromatogr. 217:317–326 (1981).

84. Posner, J.C.: Comments on "Phase Equilibrium Method for Determination of Desorption Efficiencies" and Some Extensions for Use in Methods Development. Am. Ind. Hyg. Assoc J. 41:63–66 (1980).

85. Nederlands Normalissatie-Instituut: Methods in NVN Series (Luchtkwaliteit; Werkplekatmosfeer). NNI, Delft, The Netherlands (in series, 1986–92).

86. U.S. Environmental Protection Agency: EPA Compendium of Methods for the Determination of Toxic Organic Compounds in Ambient Air. U.S. EPA, Washington, DC (1984: updated regularly).

87. CONCAWE: Method for Monitoring Gasoline Vapour in Air. CONCAWE Report 8/86. Den Haag, The Netherlands (1986).

88. Health and Safety Executive: Methods for the Determination of Hazardous Substances. Volatile Organic Compounds in Air. MDHS 72. HSE Books, Sudbury, UK (1993).

89. American Industrial Hygiene Association, AIHA, Analytical Chemistry Committee: Analytical Abstracts. Akron, OH (1965).

90. Hanson, N.W.; Reilly, D.A.; Stagg, H.E.: The Determination of Toxic Substances in Air. Heffer, Cambridge, UK (1965).

91. Jacobs, M.B.: The Analytical Chemistry of Industrial Poisons, Hazards and Solvents, 2nd ed. Interscience Publishers, Inc., New York (1949).

92. Jacobs, M.B.: The Analytical Toxicology of Industrial Inorganic Poisons. Interscience Publishers, Inc., New York (1967).

93. Health and Safety Executive: Methods for the Detection of Toxic Substances in Air. HM Factory Inspectorate. HMSO, London, UK (in series, 1943–77).

94. Ruch, W.E.: Chemical Detection of Gaseous Pollutants. Ann Arbor Science Publishers, Ann Arbor, MI (1966).

95. Ruch, W.E.: Quantitative Analysis of Gaseous Pollutants. Ann Arbor Science Publishers, Ann Arbor, MI (1970).

96. Thomas, L.C.; Chamberlain, G.J.: Colorimetric Chemical Analytical Methods, 8th Ed. The Tintometer Ltd., Salisbury, UK (1974).

97. Feigl, F.: Spot Tests, 4th Ed. Elsevier Publ. Co., London, UK (1954).

98. Squirrell, D.C.M.: Diffusive Sampling—An Overview. In: Diffusive Sampling: An Alternative Approach to Workplace Air Monitoring. A. Berlin, R.H. Brown, and K.J. Saunders, Eds. CEC Pub. No. 10555EN. Commission of European Communities, Brussels-Luxembourg (1987).

99. Health and Safety Executive: The Diffusive Monitor, 9: 14–19 (1997).

100. Health and Safety Executive: Methods for the Determination of Hazardous Substances. Volatile Organic Compounds in Air. MDHS 88. HSE Books, Sudbury, UK (1997).

101. Health and Safety Executive: The Diffusive Monitor, 8: 14–16 (1996).

102. Health and Safety Executive: Methods for the Determination of Hazardous Substances. Volatile Organic Compounds in Air. MDHS 80. HSE Books, Sudbury, UK (1995).

103. National Institute for Occupational Safety and Health: NIOSH Proficiency Analytical Testing (PAT) Program. DHEW (NIOSH) Pub. No. 77-173 (1977).

104. American Industrial Hygiene Association: Laboratory Accreditation. AIHA, Akron, Ohio (1985).

105. Vandendriessche, S.; Griepink, B.; Hollander, J.C.Th. et. al.: Certification of a Reference Material for Aromatic Hydrocarbons in Tenax Samplers. Analyst 116:437–441 (1991).

106. Hafkenscheid, Th. L.: CRM 555: A Reference Material for the Determination of Chlorinated Hydrocarbons in Air Using Thermal Desorption-Gas Chromatography. The Diffusive Monitor, 9: 4–5 (1997).

Instrument Descriptions

After reviewing the fundamentals of the gas- and vapor-collecting devices discussed in the preceding section, readers will want to be able to find out what information is available on current models and to become aware of the changes that have occurred since the 8th edition of the ASI Manual. New models of bubblers and impingers, sorbent tubes, and sorbent badges have appeared. Some manufacturers' lines have changed and more has been published about some of the collecting devices. Efforts were made to contact all known producers of each type of device; however, unintentional omissions may have occurred. In addition, some producers elected not to respond. Therefore, no claim is made that the chapter is all-inclusive. The authors would appreciate notification of any omissions so that future editions may include appropriate information.

The instrument descriptions in this chapter are grouped into three main categories: A) Grab Samplers, in Sections 16-1 through 16-8, B) Active Samplers in Sections 16-9 through 16-11, and C) Diffusive Samplers and Monitors in Sections 16-12 through 16-17. In each section, the manufacturers have been asked to briefly describe the features of the instrument that will be most helpful to the prospective users. Please note that some manufacturers refer to their sampling devices as passive; whereas, the devices are referred to as diffusive in the preceding section. The instrument manufacturers' names, addresses, email, and telephone numbers are grouped in Tables 16-I-5 and 16-I-6. All tables appear at the end of the section.

Gas and Vapor Collectors

A. Grab Samplers

16-1. Evacuated Flasks

Ace Glass; Alltech Associates, Inc; Cole-Parmer Instrument Co.; Entech Instruments; Andersen Instruments; Restek Corp; Supelco, Inc.; XonTech, Inc.; Whitey Co.

The evacuated flasks are usually heavy-walled containers of 200 to 1000 ml capacity. By means of a heavy-duty vacuum pump, the internal pressure is reduced (nominally) to zero. Instrument 16-1a illustrates one such container. The neck of the container is drawn to a tip and sealed by heating during the final stages of evacuation. The sample is taken by breaking the sealed end. The barometric pressure and air temperature at the sampling site are noted. After sampling, the flask is resealed with a ball of wax, masking tape, rubber septum cap, or other suitable sealant and sent to the laboratory for analysis.

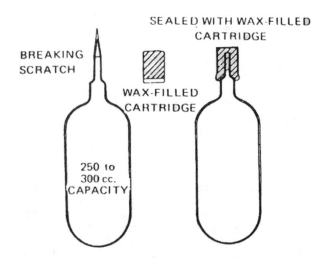

INSTRUMENT 16-1a. Glass Sampling Vacuum Flasks.

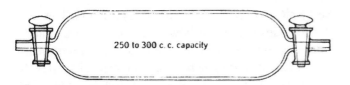

INSTRUMENT 16-1b. Gas or Liquid Displacement Type Sampling Bottle.

Instrument 16-1b illustrates a separatory flask fitted with glass-stoppered cocks on each end. Alltech Associates supplies 125, 250, and 500 ml gas sampling bulbs with septum ports. These tubes are suitable for partial evacuation. Evacuation is achieved by drawing a vacuum through one stem while the other is kept closed, then closing the open stem before the vacuum is turned off. These containers are available in glass, plastic, or metal. The samples are then taken by opening the valve and drawing in the sample. If previous evacuation was not possible, the sample is drawn through the flask by a sampling pump on the other stem until the flask is completely flushed with the sample.

Cole-Parmer distributes polypropylene bulbs that store 250 ml of gas samples at 25 psig. maximum pressure. They have a 4-mm bore stopcock and Teflon plug on one end. The stopcock accepts 1/2 to 3/8 in. i.d. tubing. Bulbs are 3 1/4 in. long and are 2 in. in diameter.

Cole-Parmer also distributes a gas washing bottle that consists of a 250 ml polypropylene graduated cylinder with extended base, a vinyl stopper with 1/4 in. o.d. tubes, and a removable gas dispersion fitting with 70 μm porosity polyethylene disc. Pressure capacity is 0.5 psig.

Alltech Associates also provides gas sampling bulbs with a choice of stopcocks and septum ports. The Alltech Vacuum Sampler is an evacuable 280 ml aerosol container that provides a unique and convenient way of sampling gases. The Vacuum Sampler is fitted with a unique septum port and needle guide through which the container is filled and samples transferred. A 4 in. long needle is supplied with each vacuum sampler. Because the container is supplied evacuated, no external pumps are required to obtain your sample. Vacuum samplers are reusable.

Entech Instruments features time integrated sampling into 6 L stainless steel canister which can have fill times from 1 hour to 1 week. Cleanup times can be reduced by the use of a high speed flush port. Electropolished surfaces reduce chemical adsorption. A glass filter frit has improved inertness compared to stainless steel.

Andersen Instruments and SKC have heavy-metal gas sampling spheres, which have been called Summa Canisters. They are pictured in Instrument 16-1c and are used as pre-evacuated stainless steel vessels that have been specially passivated for minimum interaction with sample analytes. They were designed to meet U.S. EPA TO-14 specifications for environmental sampling of volatile organic compounds; however, they find many applications for industrial hygiene sampling.

The Whitey Co. distributes a line of stainless steel and alloy sample cylinders, from 10 ml to 4000 ml in size, fitted with Nupro/Whitey™ valves and Swagelok™ connections.

Except for these and the heavy-walled glass containers illustrated in Instrument 16-1a, no attempt has been made to reduce the pressure to zero in other containers. Therefore, the degree of evacuation must be known and is determined from the manometer pressure or vacuum gauge. The sampling information, along with the barometric pressure and temperature at the sampling site, would be used to calculate the actual volume of air or gas collected.

16-1-1. MiniCan VOC Sampling Canister

Entech Instruments, Inc.

The 0.4L MiniCan VOC Sampling Canister is used to collect volatile organic compounds for analysis in a stationary or mobile laboratory. A vacuum tight, quick connect fitting eliminates the need for tools while simplifying vacuum/pressure checking before and after sampling. An internal fused silica lining improves gas phase stability, allowing reactive compounds such as hydrogen sulphide and mercaptans to be recovered. The MiniCan can be used for integrated sampling from 0.1 to

INSTRUMENT 16-1c. Passivated Canister Sampler (Graesby Andersen).

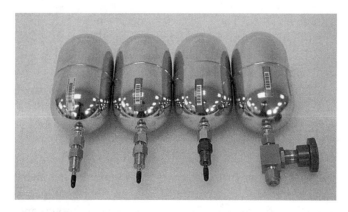

INSTRUMENT 16-1-1. MiniCan VOC Sampling Canister.

8 hours using only the vacuum in the canister to draw in the sample. Concentrations from 1 ppb to >1000 ppm can be stored from days to weeks without sample loss. Analysis in the laboratory is performed without the need for solvent extraction, making it compatible with both routine GC analysis or GC/MS investigation of unknowns. Applications include personal monitoring, indoor air analysis, and source emissions testing.

16-1-2. IH1200 Personal Sampler

Entech Instruments, Inc.

The IH1200 Personal Sampler has a controlled flow system used to collect VOCs into evacuated 0.4 L MiniCan canisters at flow rates from 0.3 to 100 ml/min. The sampling rate is controlled by an electropolished, stainless steel flow controller that uses a sapphire orifice to maintain a constant flow rate despite a changing vacuum level in the receiving canister. A built-in vacuum gauge validates the proper introduction of sample into

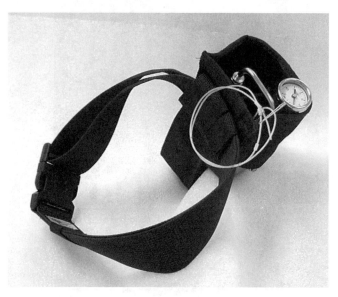

INSTRUMENT 16-1-2. IH1200 Personal Sampler.

the MiniCan over periods as long as 8 hours. The sampler includes a belt that is worn around the waist and a sampling line that is positioned to draw in a representative air sample during exposure monitoring. The IH1200 is utilized to verify compliance with TWA and STEL limits, allowing a wide range of analytes to be collected simultaneously. Fused silica coated samplers are also available for monitoring of sulfur compounds.

16-1-3. "Passive" Air Sampling Kits

Restek, Inc.

Restek offers a complete 'Passive' Air Sampling Kit that incorporates all of the necessary hardware for successful passive air sample collection using their SilcoCan™ canisters. (Ed. Note: 'Passive' in this case is not to be confused with "diffusive" samplers.) All cans, valves, filters, gauges and tubing are treated by Restek's Silcosteel™ process which adheres a thin layer of silica on the stainless steel, resulting in a very inert surface. A Veriflo flow controller is used to control the sampling flow rate, maintaining a constant mass flow as the change in pressure occurs from 30" Hg to 5" Hg.

Restek developed the Silcosteel™ process, a unique metal coating technique, in 1987. The Restek ambient air 'passive' sampling kit uses Silcosteel™-treated, wetted parts to provide a non-adsorptive pathway to the sampling canister. The 'passive' sampling kit includes the following parts: the low-flow-controlling device SC423XL made by Veriflo Corporation, an interchangeable sapphire critical orifice, a 5μm sintered filter, a vacuum gauge, and 2 m. x 1/8" stainless steel tubing. All the parts are Silcosteel™ treated, with the exception of the vacuum gauge. The interchangeable orifice on the inlet of the Veriflo flow controller operates in a specific flow range. There is a large stainless steel diaphragm in the body of the flow controller that uses a piston to control flow as the vacuum level changes in the sampling canister. The 5μm filter prevents clogging upstream of the orifice, during sampling. The vacuum gauge allows monitoring of the vacuum in the canister over the sampling period. The 2 m of Silcosteel™-treated stainless steel tubing features a small, 1/8" nut on the inlet, which, when bent down will act as a rain cap. These kits are sold though Restek Corporation and their distributors.

16-1-4. Volatile Organic Samplers, Models: 910A, 910PC, 911A, 912, 915

XonTech, Inc.

The samplers are designed to collect volatile organic compounds from whole air samples into electropolished canisters by USEPA Method TO-14 and they meet

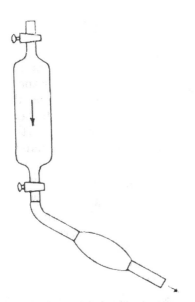

INSTRUMENT 16-2a. Filling container with rubber bulb hand aspirator.

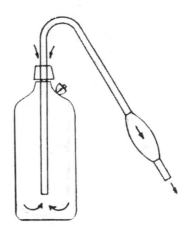

INSTRUMENT 16-2b. Filling bottle with rubber bulb hand aspirator.

PAMS (photochemical assessment monitoring stations) site requirements. Features include portability, rack mountable, rechargeable or batter/AC operation, all wetted components of non-reactive materials, remote control via modem, and adapter to collect up to 16 canisters.

16-1-5. Canister Cleaning System, Model 960

XonTech, Inc.

The Model 960 canister cleaning system is a top-loading, oil-free system which utilizes repeated cycles of evacuation, fill and bake at 120° C. The cycles are performed using humidified air, followed by final dry air cycles. Only the spheres of the canisters are heated. The valve protrudes out of the oven and is not heated to protect it from overheating. Two systems are available: the low-vacuum system (pumped down to –20 inches of Hg.) and the high-vacuum system (can be pumped down to 50 microns). Each system can clean up to four 6-liter canisters or eight 3.2 liter canisters. Each system can also operate two additional auxiliary systems of the same capacity. Each system requires a 115V, 15 amp AC circuit or 220V, 60 Hz. The control unit weighs approximately 300 lbs. and the auxiliary unit weighs about 150 lbs.

16-2. Gas/Liquid Displacement Flasks

Cole-Parmer Instrument Co.; Fisher Scientific Company; General Supply Houses (see Table 16-I-5)

Many kinds of ordinary, sealable containers can be used as gas/liquid displacement flasks. Numerous commercial firms sell suitable displacement flasks. A user's selection criteria should include the required volume of sample to be taken, reactivity of the analyte of interest, and

the practicality of using such a device (i.e., no pump is necessary with liquid displacement, but something must be done with the drained fluid). The major limitation is the solubility of the analyte in the displacement liquid. With bulb displacement (see Instruments 16-2a through 16-2c), the equivalent of three or more collector volumes must be flushed through the container to ensure 99% of the contents is pure sample. See Instruments 16-2a through 16-2d.

16-3. Flexible Plastic Containers (Sampling Bags)

Alltech Associates, Inc.; Anspec Company, Inc.; BGI Incorporated; Calibrated Instruments, Inc.; Cole-Parmer Instrument Co.; Edlon Products Inc.; Plastic Film Enterprises; PMC, Inc.; Supelco, Inc.

Sampling bags are widely used, are available from a number of manufacturers and distributors, made of a variety of materials, and come in a selection of sizes (from less than l to 250 L). The materials include polyester (e.g., aluminized Scotch Pak™, Scotch Pak™, and Mylar™); polyvinylidene chloride (e.g., Saran™); five-layer, high-density polyethylene (e.g., Cali-5-Bond™); and the fluorocarbons (e.g., Chemtron™, Kel F™, Aclor™, Kynar™, Tedlar™, Crinkle Tedlar™, and Teflon™). Instrument 16-3a illustrates a Teflon bag from Edlon Products, Inc. In addition to the less reactive materials, bags that are black or opaque are now avail-

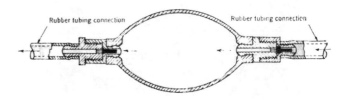

INSTRUMENT 16-2c. Rubber bulb hand aspirator.

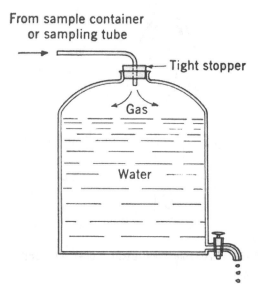

INSTRUMENT 16-2d. Aspirator bottle.

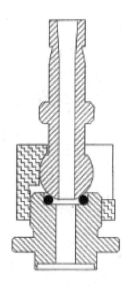

INSTRUMENT 16-3b. Valves for Plastic Sampling Bags (BGI).

able for the collection of light-sensitive gases and vapors. A crinkle surface bag was developed by Plastic Film Enterprises to facilitate filling, flushing, and complete evacuation of the sample bags. An array of valve types has been employed and is available from the different manufacturers. In addition to the familiar rectangular shape, sampling bags are designed in other shapes to improve sampling techniques, minimize dead volume problems, and relieve stresses around fittings (see Instrument 16-3b).

All bags should be leak tested, cleaned with pure compressed air, and conditioned before use. Three pump and cleaning cycles should be sufficient, unless condensable or reactive materials have deposited in the bag. The conditioning is first performed in the laboratory using test atmospheres; it is then repeated in the field before use by filling and emptying a bag several times at the sampling rate that will be used for taking the sample. Gas/vapor storage stability and decay curves for these devices must be determined. Advantages are their low

cost, convenient handling, and ease of use. Bags can be selected with appropriate valves, as well as with septum ports for injecting measured gas or liquid volumes for dilution to standard concentrations. Calibrated Instruments makes syringe systems for filling bags.

Alltech offers Teflon™, Tedlar™, Saran™ and 5-layer bags in a variety of sizes equipped with a variety of fittings to meet the application needs. Their selection has expanded to include Crinkle Tedlar™, black Tedlar™, Snap and Seal, and custom gas sampling bags. Saran™ Gas sampling bags are available with all of their standard fittings to meet sampling requirements.

Cole-Parmer markets Teflon™ and Tedlar™ bags from 0.5 to 85.7 L with on/off and septum valves for tubing connections and hypodermic needles, respectively. Tedlar™ bags come in dust-proof and light-free types for carbon monoxide, sulfur dioxide, hydrogen sulfide, radon, and mercaptans. Bags are 2 mils thick and can be used in cleanroom applications. Tedlar™ bags are double seamed and Teflon™ bags are single seamed. The Teflon™ bags

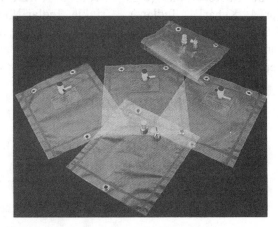

INSTRUMENT 16-3a. Plastic Gas Sampling Bags.

INSTRUMENT 16-3c. Bag Filling Pump (BGI).

meet National Aeronautics and Space Administration Specification Level 1 and are 5 mils thick.

BGI also markets Tedlar™ and Teflon™ sampling bags of 2 and 5 mil thickness, respectively, in a variety of sizes fitted with patented valve and septum hardware. Bag filling pumps and sets of critical orifice (as shown in Instrument 16-3c) are available.

SKC Inc. has 2-mil-thick Tedlar™ bags with septum fittings, two types of valve fittings in dual stainless steel or single inert polypropylene fittings, and hose/valve fittings and replaceable septum units (as shown in Instrument 16-3b).

16-4. Plastic Sampling Bag Systems

Alltech Associates, Inc.; Anspec Company, Inc.; BGI Incorporated; Calibrated Instruments, Inc.; Edlon Products Inc.; Plastic Film Enterprises; PMC, Inc.; SKC, Inc.; Supelco, Inc.

These bags are similar to those used and described in Section 16-3. The difference between the sections is that these bags are a part of a system used to obtain time-integrated air samples. Air collection systems are available from Calibrated Instruments, Inc., and consist of a five-layer, nonpermeable, high-density polyethylene bag (i.e., Cali- 5-Bond) available in sizes ranging from 100 ml to 200 L and a battery-operated air pump with an on-off cycle timer.

16-4-1. Drum Sampler

PMC, Inc.

Bags can be filled by displacing air or water from an airtight container that holds the sampling bag. As the fluids flow out, the bag is distended and the sample is drawn into the bag to fill the container. The advantage is that reactive analytes do not pass through a pump. A variety of bag materials are available depending on the reactivity of the analytes. In addition to the less reactive materials, bags that are black or opaque are available for the collection of light-sensitive gases and vapors.

16-4-2. Vacuum Bag Sampler

Supelco, Inc.

Vacuum bag samplers by Supelco include a 10 L bag sampler that allows capture of a discrete air sample in the bag without having the air sample pass through a vacuum pump first. This portable vacuum chamber-base sampler provides fast zero-cross-contamination sampling of refinery stack gases and volatile organic compound (VOC) vent gases. Powered by a 12-V DC battery, the bag sampler fills or empties a 10 L bag in 2 min. by applying pressure or a vacuum to the outside

of the sample bag. The automatic shut-off switch prevents the sample bag from overfilling. Pumping controllers meet the U.S. EPA requirement for taking a 10 L stack sample in 15 min. The sampler includes a battery, charger, flow controller, and pump timer (see Instrument 16-4-2). A 1 L size is also available, as well as an automatic, six 1 L-bag sampler that collects multiple air samples.

16-4-3. Vac-U-Chamber™

SKC, Inc.

The SKC Vac-U-Chamber™ is a rigid air sample box that allows the direct filling of an air sample bag using negative pressure provided by most personal air sampling pumps. When using the Vac-U-Chamber™, the air sample directly enters the bag without passing through the pump, thus eliminating the risk of contaminating the pump or the sample. Sample integrity is further assured by the stainless steel and Teflon construction of all surfaces the sample contacts. The large, vacuum-tight chamber accommodates bags up to 10 liters in size and features rigid walls that will not collapse under vacuum conditions. Use the Vac-U-Chamber for groundwater testing, soil gas sampling, stack sampling, ventilation studies, and Haz Mat testing. The Vac-U-Chamber is also available in a mini version with polypropylene fittings designed for use with 1-liter bags. Illustration 16-4-3a.

As an alternative, air sample bags are a convenient and accurate means of collecting airborne chemical hazards, particularly in areas where the concentration

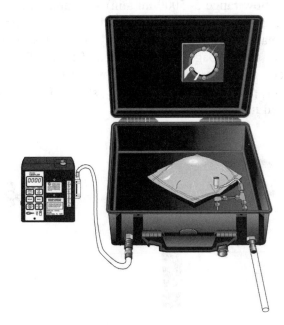

INSTRUMENT 16-4-3a. Vac-U-Chamber (SKC).

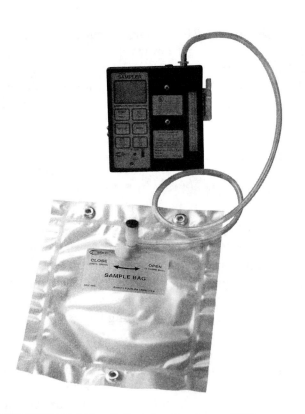

INSTRUMENT 16-4-3b. Air Sampling Bag System (SKC).

is above the detection limits of common analytical instruments. SKC's Air Sampling Bag System includes a 224-XR Series air sample pump with pressure port fitting (supplied with pump) and SKC air sample bags with fittings designed specifically for bag sampling. The combination of the low flow sampling flexibility of the battery-operated XR Series pump (flow range 5–5000 ml/min) with pressure port fitting and the quality-constructed SKC air sample bags make for an easy-to-use and efficient system. SKC air sample bags are constructed of Tedlar, Teflon, or black and clear layered Tedlar with fittings available in polypropylene, stainless steel, or Teflon. Pictured in Instrument 16-4-3b.

16-4-4. Vac-U-Tube™

SKC, Inc.

The new SKC Vac-U-Tube™ is an acrylic syringe with a removable faceplate that allows a specially designed 0.7-liter Tedlar sample bag to be placed inside it. For quick and easy bag samples, simply pull the plunger to fill the bag. Samples can be purged by pushing the plunger. The Vac-U-Tube™ requires no pump, takes less than 20 seconds to set up, is always available in an emergency, and is convenient for sampling and monitoring wells. Illustrated in Instrument 16-4-4.

16-5. Hypodermic Syringes

Cole-Parmer Instrument Co.; Fisher Scientific Company; General Supply Houses (see Table 16-I-6)

Syringes with volumes from 10 to 50 ml have been found satisfactory for air sampling[11]. Suitable syringes should be gas tight. They are available in glass and disposable plastic. Gas and vapor storage and decay curves for these devices must be determined. Advantages are their low cost, convenience, and ease of use. Large-volume acrylic syringes from 0.5 to 2.0 L, made by Hamilton, are available for special purposes from general supply houses.

16-6. Bubblers and Gas Washing Bottles

Ace Glass, Inc.; Cole-Parmer Instrument Co.; Corning Glass Works; Scientific Glass and Instruments Company; SKC, Inc.; Supelco, Inc.

The midget impinger is the most widely used in this group and is illustrated in Instrument 16-6a. It is designed for impacting particles at a flow rate of 2.8 L/min.; however, for industrial hygiene use as a gas and vapor bubbler, impingers are generally used with 10 to 20 ml of sorbing solution and flow rates of about 1.0 L/min. No more than 20 ml of sorbing solution should be added to the impinger flask. Air sampling is performed by connecting a personal pump or other source of suc-

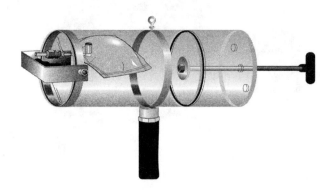

INSTRUMENT 16-4-4. Vac-U-Tube (SKC).

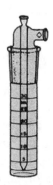

Model 9100

INSTRUMENT 16-6a. Midget Impingers (Ace Glass, Inc.).

tion to the outlet tube. The impinger is either hand-held or attached to the worker's clothing. Care must be taken that the impinger does not tilt, which could result in a loss of sorbing solution or reagent. Too much reagent solution or excessive flow rate will also lead to loss of sample. Spill-proof impingers have been designed to minimize this problem and are commercially available (e.g., SKC, Supelco). In-line traps are suggested to prevent carryover of liquids into the sample pumps.

SKC Midget Impingers (bubblers) collect airborne hazards into a liquid medium. Impinger can be worn either in a holster near the breathing zone for personal sampling, or mounted on an air sample pump for area sampling. SKC offers three types of impingers: the 'Standard Midget' with the tip placed precisely above the impingement surface; the 'Special Midget' with fritted tip to increase contact between the air sample and the liquid (many NIOSH and OSHA procedures specify fritted tip impingers); and the 'Spill-resistant Midget' with a special outlet to prevent spills if the impinger is tipped. Each SKC impinger contains a printed serial number on both sections to assist with sample identification and ensure impinger part matching. SKC recommends an impinger trap when using impingers to prevent impinger liquids from being drawn into the sample pump. Micro Impingers have been discontinued at SKC. See Instrument 16-6a.

Friedrichs and Milligan gas washing bottles are examples of spiral and helical sorbers (Instrument 16-6b). They may be used for collecting area samples of gaseous substances that are only moderately soluble or are slow reacting with reagents in the collection media. The spiral or helical structures provide for higher collection efficiency by allowing longer residence time of the contaminant within the tube. Slower reacting and less soluble substances are permitted more time to react with the sorbing solution.

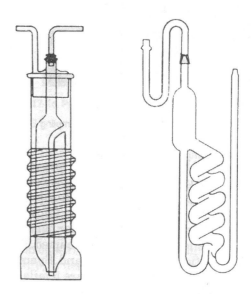

INSTRUMENT 16-6b. Spiral and Helical Absorbers.

Gases and vapors that are sparingly soluble in the collecting medium may be sampled in fritted bubblers (Instruments 16-6c and 16-6d). They contain sintered or fritted glass, or multiperforated plates at the inlet tube. Air streams drawn into these devices are broken into very small bubbles, and the heavy froth that develops increases the contact between gas and liquid.

Frits come in various sizes and grades, usually designated as fine, medium, coarse, and extra coarse. A coarse frit is usually best for gases and vapors that are appreciably soluble or reactive. A medium porosity frit may be used for gases and vapors that are difficult to collect, but the sampling rate must be adjusted to maintain a flow of discrete bubbles. For highly volatile gaseous substances that are extremely difficult to collect, a frit of fine porosity may be required to break the air into extremely small bubbles and ensure adequate collection efficiency. Airflow, however, must be controlled to avoid the formation of large bubbles by the coalescence of small bubbles.

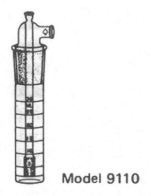

Model 9110

INSTRUMENT 16-6c. Midget Gas Bubbler (Corease Frit)
(Ace Glass, Inc.)

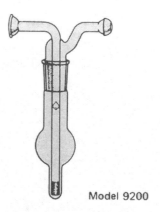

Model 9200

INSTRUMENT 16-6d. Nitrogen Dioxide Gas Bubbler
(Ace Glass, Inc.)

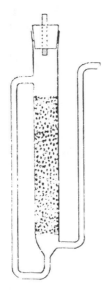

INSTRUMENT 16-7. Packed Glass-Bead Columns.

There is little value, for example, in using fine porosity frit while using increased airflow because a population of larger bubbles is produced. The finer the frit, however, the higher the pressure drop, which might require more powerful vacuum pumps. Selection of the proper frit should be made with all of these factors in mind. The collection efficiency of the sampling equipment must be determined for the specific contaminants involved. Generally, lower flow rates are used with fritted bubblers than with impingers of the same liquid capacity.

Supelco has a similar line of impingers and bubblers that are available with either conventional ground glass joints or convenient Teflon microconnectors.

16-7. Packed Glass-bead Columns

Cole-Parmer Instrument Co.; Fisher Scientific Company; General Supply Houses (See Table 16-I-6)

Packed glass-bead columns (Instrument 16-7) are used for special situations where a concentrated solution is needed. Glass pearl beads are wetted with the sorbing solution and provide a large surface area for the collection of the sample. It is of historical interest to note that the sorption of benzene and other aromatic hydrocarbon vapors in nitrating acid has been performed with this type of sorber. It is especially useful when a viscous sorbing liquid is required. The rate of sampling is necessarily low, 0.25 to 0.5 L/min. of air.

16-8. Cold Traps

Cole-Parmer Instrument Co.; Fisher Scientific Company; General Supply Houses (see Table 16-I-6)

Cold traps (Instrument 16-8) are generally component assemblies constructed on an as-needed basis. The cold trap usually consists of a U-shaped glass or copper section that is filled with the sorbent collection medium. The U-shaped section is immersed in liquid nitrogen, dry ice baths, or other cold mixtures to effect the trapping. The sorbent is chosen to optimize the collection of the types of contaminants (e.g., activated carbon is used for organics). Vacuum and reduced-pressure cold trapping utilize impinger-shaped traps that can be evacuated and the sample condensed on the inside of the outer wall. These traps can be used with or without sorbent, depending on the flow rate through the

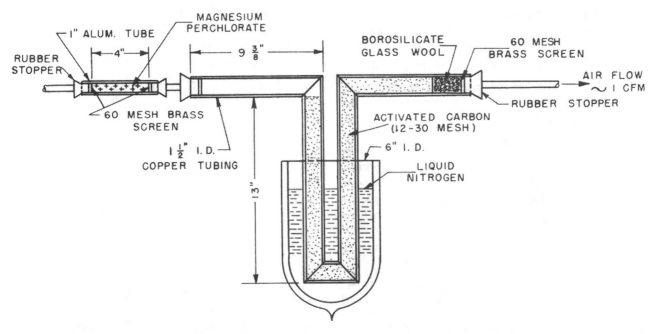

INSTRUMENT 16-8. Cold Trap.

trap. Faster flow or a higher sampling rate requires a more torturous path to efficiently trap the vapors. Please note that some makes of metal sampling canisters described in Instruments 16-1 can be used as cold traps with increased efficiency and capacity.

B. Active Samplers

16-9. Solid Sorbents, Active Sampling

Barneby Cheney Company; Columbia Scientific Industries; Fisher Scientific Company; Perkin-Elmer Corp., Pittsburgh; SKC, Inc.; Supelco, Inc.; Westvaco, Inc.; Witco

Sorption is used in the field of air quality monitoring as a means of collecting and concentrating airborne contaminants for analysis. The sorbent sample tubes collect airborne contaminants typically by sorption of the contaminant onto a solid sorbent surface and are the most widely used medium for collecting airborne gases and vapors. The large number of chemical vapors that may be encountered at varying concentrations ensures that no one sorbent material can effectively collect all contaminants under all conditions. The choice of the sorbent is designed to maximize collection efficiency while retaining low selectivity, and forms a critical part of an air sampling program. Several types of charcoal are commercially available for both active and passive sampling. Historically, the products used most frequently for air sampling are derived from coconut shells and lignite (Darco and Muchar). The mesh sizes employed vary considerably. NIOSH recommends 20/40 mesh coconut shell charcoal. Severs and Skory[21] found Pittsburgh PCB 12/30 mesh most suitable for sampling vinyl chloride, vinylidene chloride, and methyl chloride. The final

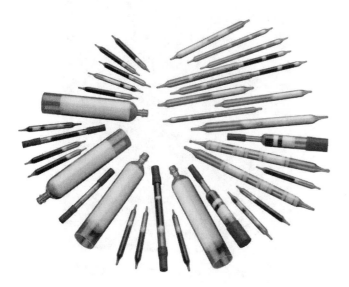

INSTRUMENT 16-9. Solid Sorbents, Active Sampling.

choice for a specific application should be made only after performance tests have been made. Other solid sorbents are available and are more desirable for polar and reactive analytes. Coated sorbents are also available for specific analytes as described in the NIOSH Manual of Analytical Methods, 4th edition.[17] An air sampling pump is used to draw air through the tube at rates from 0 to 1500 ml/min, although flow rates between 20 and 200 ml/min are more typical for long-term, low-flow sampling. Most tubes contain a primary sorbent section and a back-up bed that is used to indicate breakthrough.

The standard tube for low-flow, time-weighted average (TWA) sampling contains 150 mg of sorbent in 100- and 50-mg sections. Large tubes are also available that contain 600 mg of charcoal; 400 mg in the front section and 200 mg in the back section. Jumbo tubes contain 800 mg and 200 mg. The larger tubes allow higher flows for larger total capacities or higher air volumes for collecting very low concentrations. Other sizes of tubes can be ordered or prepared in the laboratory for special applications.

Sampling tubes need not always be made of glass. Many in use are constructed of stainless steel. One such unit, described by Severs and Skory,[21] measures 5.5 × 0.25 in. o.d. × 0.028 in. thick and is fitted with Swagelok™ caps.

SKC maintains a large selection of sorbent materials, including the Anasorb Series of sorbents. These and other sorbents are available in regular SKC sorbent devices. Currently, the most commonly used sorbents for sample tubes are coconut charcoal, petroleum charcoal, silica gel, XAD™-2 (a porous aromatic polymer), Tenax™, and Anasorb™ 747 (a beaded active carbon). In addition to the standard sorbents, there are more than 40 other sorbents that are used as well as sorbents that have been chemically coated or treated to enhance their collection efficiency. The amount and type of sorbent are specified by established air sampling methods published by OSHA, NIOSH[17], U.S. EPA, ASTM and some published proprietary methods for sampling individual chemicals or classes of chemicals in air. These reference methods specify the design and makeup of the tubes, the sampling procedure, and the details of chemical analysis. SKC stocks over 150 different sorbent tubes as standard items to collect over 1000 different airborne chemicals and will manufacture custom sorbent tubes for specific applications. See Table 16-I-3 for examples and Instrument 16-9.

16-9-1. OSHA Versatile Sampler

SKC, Inc.

Some airborne chemical hazards exist both in the gas and in the aerosol phase. A new tube design origi-

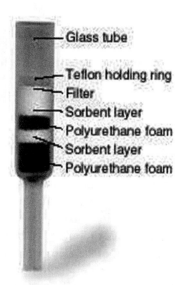

INSTRUMENT 16-9-1. OSHA Versatile Sampler (SKC).

nally developed by OSHA to overcome some of the inconveniences of a separate tube and pre-filter is often referred to as the OSHA Versatile Sampler (OVS). It is a specially-designed glass tube containing a filter to trap aerosols and a 2-section sorbent bed to adsorb vapors. Several OVS tubes are available from SKC including:

- OVS tube with XAD-2 and glass fiber filter for pesticides.
- OVS tube with XAD-2 and quartz filter for organophosphorus pesticides.
- OVS tube with Tenax and glass fiber filter for TNT and DNT explosives.
- OVS tube with XAD-7 and glass fiber filter for glycols.

See illustration in Instrument 16-9-1.

16-9-2. Solid Adsorbents, Coated, Active Sampling

SKC, Inc.

Some of the most important sorbent tube developments are the utilization of sorbents that have been washed, coated, or treated to allow the collection of specific chemicals that could not otherwise be sampled using standard sorbent tubes. For many chemicals, coated or treated tubes provide an alternative to wet chemistry methods using impingers. Some of the more notable sorbent tubes with special wash or treatment are:

- Silica gel, water washed for inorganic acids.
- XAD-2, coated with 2-HMP for formaldehyde.
- Zeolite molecular sieve, coated with TEA for NO/NO2.
- Carbon beads, treated with KOH for sulfur dioxide.
- Petroleum-based charcoal, coated with hydrobromic acid for ethylene oxide.

16-9-3. VOC Monitor SXC-20

Spectrex Corp.

An active, multi-gas, VOC Detector and quantitative sampler for solid adsorbent sampling with built in data logger and alarm has been developed by Spectrex. The SXC-20 VOC Monitor is a quantitative sampler/detector illustrated in Instrument 16-9-3. A miniature pump pulls an air sample across the surface of a heated, rapid, non-selective, semiconductor detector. The signal activates the following features: Color bar display of hazard levels—the VOC percentage is displayed on a color bar of green, orange, and red LEDs, and sounds an adjustable audible alarm at a preselected level; built-in data logger that supports both DOS and Windows, records VOC levels and tracks times of occurrences; at a preset gas level the output triggers a second miniature pump with an adjustable flow rate that pulls sample air through a charcoal or other sorbent tube that can be qualitatively and quantitatively analyzed on a gas chromatograph at the end of the sampling period. The system has been tested on over one hundred organic compounds down to OSHA, EPA and other regulatory-mandated levels.

16-9-4. Carbonyl Sampler

XonTech, Inc.

The Carbonyl Sampler, Model 925 is designed to collect ambient aldehydes, ketones, or their surrogates using Sep-pak adsorbent tubes. It meets all requirements of

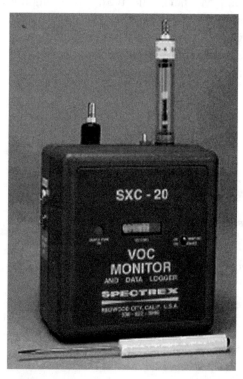

INSTRUMENT 16-9-3. VOC Monitor SXC-20 (SPECTREX).

USEPA Method TO-11 and PAMS (photochemical assessment monitoring stations). It consists of two modules: the tube module has ten sampling channels and a blank and the control module can operate up to four tube modules. The sampler is microprocessor controlled and can be remote controlled via modem. The instrument has a heated built-in ozone denuder to scrub the ozone. The software includes: leak test and scheduling days, start and end times, and flow check. It will also flag any power outages and continues sampling when power comes back on. At the end of each sample, a printer will automatically print the parameters of each sampling channel.

16-9-5. VOC Concentrator

XonTech, Inc.

The Model 930 Concentrator is used as a front-end accessory for unattended continuous and automatic collection and desorption of VOCs when used with a gas chromatograph. Dual sorption traps for C2 to C10 are alternately activated in order to provide continuous sampling. Sampling is performed at ambient temperature, while desorption and cleaning are accomplished by automated heating cycles. The concentrator is compatible with all GC systems with interfaced heated line. The Model 930 Concentrator is best used with the Model 940 Cryogenic Refocusing Trap.

16-9-6. Crygenic Refocusing Trap

XonTech, Inc.

The Model 940 Cryogenic Refocusing Trap is a liquid nitrogen-free refocusing trap using a split-Stirling Linear Drive cooler. The refocusing trap is used with the

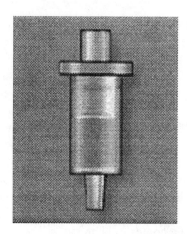

INSTRUMENT 16-9-7b. LpDNPH S10L.

Model 930 Concentrator and is automated with rapid cooling to –185 degrees C or other selectable temperatures. The trap also heats rapidly to +100 degrees C for desorption into a gas chromatograph.

16-9-7. Solid Adsorbents

Supelco, Inc.

Supelco specializes in solid sorbents and carries a wide range to fit diverse sampling needs: charcoal, graphitized carbon, carbon molecular sieve, silica gel, florisil, Hopcalite™, XAD®, Tenax™, Chromosorb™, and Porapak™. Many of the carbon adsorbents, Carbotrap™, Carbopack™, Carbosieve™, and Carboxen™ are proprietary and exhibit unique hydrophobic characteristics as well as being thermally stable. In addition to ORBO™ tubes, many of these carbons are used in multi-bed thermal desorption tubes as shown in Instrument 16-9-7a. Each manufacturer of thermal desorption

INSTRUMENT 16-9-7a. Thermal Desorption Tubes.

INSTRUMENT 16-9-7c. LpDNPH & ASSET.

equipment has a unique tube design, requiring the user to sample with the proper size tube for the instrument used for analysis.

Supelco manufactures some adsorbent tubes that are made of polypropylene and are in the configuration of a solid phase extraction (SPE) tube as shown in Instrument 16-9-7b. Samples are collected using similar flow rates to glass tubes and existing methodology, however, the sample prep is a solvent elution rather than 30-minute desorption which requires removal of the adsorbent from a glass tube. These single-bed tubes can be used in tandem to monitor breakthrough. The ASSET™ tube comes in charcoal, silica gel, and XAD-2. The LpDNPH tube is silica gel coated with 2,4-DNPH for sampling carbonyls and comes in several configurations and DNPH loading.

16-10. Automated Thermal Desorber Tubes/Desorbers

Perkin-Elmer Corp.; SKC, Inc.; Supelco, Inc., Alltech Associates

Thermal desorption techniques offer the advantage of greatly improved analytical sensitivity. Because a solvent is not used in this process, the collected sample is not diluted and, in most cases, analytical recovery is so close to 100% that desorption efficiency corrections are not required. To be suitable for thermal desorption, sorbents must meet exacting specifications that include low contaminant background, high thermal stability, and sufficient adsorptive strength to retain components of interest, but release them quickly when heat is applied.

16-10-1. Thermal Desorption Unit

Perkin-Elmer Corp.

The Perkin-Elmer Thermal Desorber is available with glass, stainless-steel, or glass-lined stainless tubes (empty or prepacked) for a variety of sampling methods, including EPA Method TO-14, indoor air, air toxics, ozone precursors, diffusive sampling, and vehicle emissions, as well as solid sample desorption of resins, films, and fragrances in ointments. A wide variety of volatiles can be sampled in the range from C_2 to C_{36} onto user-selectable single- or multibed sorbents. These samples are then loaded onto Perkin-Elmer's 50-sample ATD-400 autosampler for thermal desorption (see Instrument 16-10-1). No cryogen is used in the collection or analysis of samples because the ATD-400 utilizes electric (Peltier effect) cooling in conjunction with a sorbent-packed cold trap to concentrate the samples. Also available is a multitube automated sampler for indoor or outdoor environments.

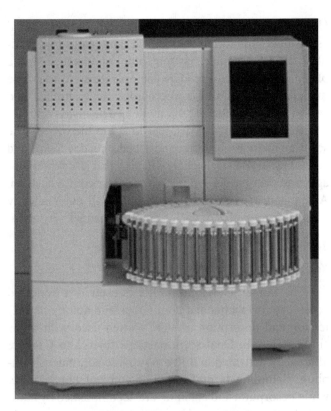

INSTRUMENT 16-10-1. Thermal Desorption Unit (Perkin-Elmer).

16-10-2. Unity™ Thermal Desorber

SKC, Inc.

SKC sorbent technology research has produced numerous materials for the collection of organic compounds for thermal desorption. Sorbents used in SKC Thermal Desorption Tubes include Anasorb 747, Anasorb CMS, Anasorb GCB1, Anasorb GCB2, Tenax TA, and Tenax GR. Thermal Desorption Tubes are available in stainless steel and glass in sizes to fit the SKC Unity™ Thermal Desorber and most commercially available thermal desorbers. The thermal tubes are illustrated in Instrument 16-10-2a.

INSTRUMENT 16-10-2a. Thermal Desorption Tubes.

INSTRUMENT 16-10-2b. Unity™ Thermal Desorber.

The Unity™ Thermal Desorber is a cost-effective, high-performance system for analysis of air samples, QA/QC of materials, method development, and special applications. Unity is compatible with Gas Chromatography/Mass Spectrometry (GC/MS). It features splitless operation even with high-resolution capillary columns offering unmatched detection limits for trace level analysis of organic compounds. Sample protection features include an inert flow path, automatic air and humidity purge, leak testing of each tube before analysis, and repeat analysis capability. Unity's electrically-cooled sorbent trap eliminates the need for liquid cryogen. Its single user-interchangeable analyzer assembly and Windows®-based operating software make Unity easy-to-use. Unity Thermal Desorber is compliant with key thermal desorption standards and methods including NIOSH 2449, US EPA TO-17, ASTM D 6196 and ISO CD 16017 (a diffusive method). See picture in Instrument 16-10-2b.

16-11. Polyurethane Foam, Active Sampling

16-11-1. Polyurethane Foam, Active Sampling Tubes

Supelco, Inc.

Several U.S. EPA and ASTM methods require a polyurethane foam (PUF) adsorbent cartridge for monitoring semivolatiles in stack, ambient, indoor, and work-place atmospheres. A low pressure drop across the cartridge facilitates high-volume sampling. Supelco provides precleaned high- and low-volume PUF cartridges, some in combination with XAD-2 to provide a wider collection range. Low volume, 1-5L/min, are ORBO-1000, 22mm OD × 7.6cm length PUF, and ORBO-1500 PUF/XAD-2/PUF, and are used for pesticides and PCBs. With a tapered stem, these cartridges can be connected to a standard personal sampling pump. An optional quartz filter can be attached to the inlet of the device to trap particulate matter. High volume, 20-225L/min, are the ORBO-2000, 6cm OD × 7.6cm length PUF, and ORBO-2500 PUF/XAD-2/PUF, and are used for pesticides, PCBs, PAHs, and dioxins. The high-volume cartridges require the use of a PS-1 type sampler. Instrument 16-11-1. The analytical method involves a solvent soxhlet extraction and concentration followed by gas chromatography. The PUF plug is reusable after appropriate solvent cleanup.

16-11-2. PUF Tubes

SKC, Inc.

Certain methods for sampling organochlorine pesticides, PCBs, or similar complex organic compounds specify the use of tubes containing a polyurethane foam (PUF) plug as the principal sampling medium. Precleaned and packaged to allow immediate use without further treatment, SKC PUF tubes meet or exceed method specifications for purity. SKC PUF tubes are suitable for ambient, indoor air, or personal sampling. Low-volume PUF tubes are for use with a personal sampling pump (1–5 L/min flow rate) for a sampling period of 4 to 24 hours. Large glass cylinder PUF tubes (65 mm OD) are used with a high-volume sampler in

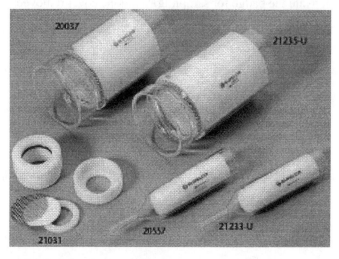

INSTRUMENT 16-11-1. PUF Cartridges for High- and Low-Volume Sampling.

environmental sampling. Custom PUF/sorbent combinations, repacking, and recertifying are all available from SKC. See Instrument 16-11-2.

C. Diffusive Samplers and Monitors

The various commercially available diffusive (or passive or dosimeter) monitoring systems and dosimeters are listed alphabetically by contaminant in Table 16-I-1. The table includes manufacturers and brand names. K & M distributes the TraceAir OVM passive badges. 3M diffusive monitors for ethylene oxide, carbon monoxide, formaldehyde, and mercury are listed in Table 16-I-2. The 'passive' badges supplied by SKC are found in Table 16-I-3. Addresses for manufacturers are found at the end of this chapter (Table 16-I-6). Note that the sampling rates for most diffusive monitors are controlled by the diffusion and/or permeation rates for each compound and by the specific geometry of the device (example sampling rates for different brands are shown in Table 16-I-4). Thus, each manufacturer must provide the sampling rates for the compounds of interest for their specific badges. Direct reading diffusive devices are more completely discussed and listed in Chapter 17. Descriptions of specific sampling systems are provided in the paragraphs that follow.

16-12-1. TraceAir System OVM Badges

K & M Environmental

The single-stage TraceAir organic vapor monitor (OVM) consists of a single strip of activated coconut charcoal sandwiched between two diffuser panels. Two covers are used to seal the unit. The charcoal strip contains 300 mg charcoal. The design of the badge allows the operator to remove either one or both of the covers to activate the badge for sampling. With one cover removed,

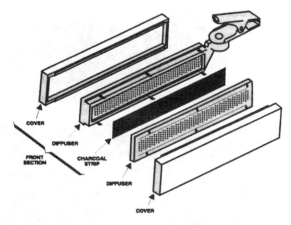

INSTRUMENT 16-12-1a. Single-Stage TraceAir Organic Vapor Monitor (OVM-1).

the unit samples at a rate of approximately 35 ml/min (for most organic solvent vapors). Removing both covers doubles the effective sampling rate to the 70 ml/min range. This higher range enables the monitor to monitor effectively for short periods of time for short-term exposure limits (STEL) sampling. The shortest sampling duration for this unit is approximately 0.2 ppm-hrs with both covers removed. The Trace Air OVM-2 contains a second charcoal strip that acts as a back-up. This second strip doubles the capacity of the OVM and facilitates sampling in those areas where potential exposures are expected to be higher than the capacity of the single-stage OVM-1. Only one cover can be removed from the OVM-2 because the back cover is used to seal the back-up section in the unit (diagrams of both units are shown in Instruments 16-12-1a and b). Thus, the shortest sampling duration for the OVM-2 is 0.4 ppm-hrs. The effective mass of sample components collected is based on the same equation given in the first part of the chapter. The TraceAir Technical Reference Guide provides instructions on the

INSTRUMENT 16-11-2. PUF Tubes (SKC).

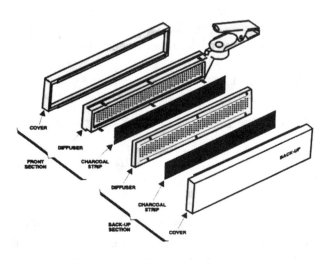

INSTRUMENT 16-12-1b. TraceAir OVM-2.

INSTRUMENT 16-12-2a. 3M Diffusive Samplers.

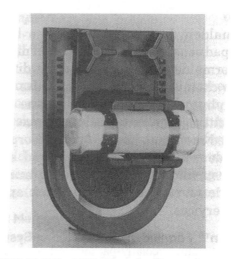

INSTRUMENT 16-12-3. ORSA 5 Sampler (Draeger).

use of the badges and the sampling rates for 132 organic compounds (see Table 16-I-4 for example rates).

16-12-2. *Organic Vapor Monitor #3500*

3M Company

The 3500 monitor contains a single charcoal sorbent pad separated from a diffusion membrane by spacers (Instruments 16-12-2a and b). The 3520 has a similar geometry, but incorporates a back-up section that collects gases and vapors when the capacity of the primary sorbent pad has been exceeded. The gas and vapor contaminants enter both types of monitors by molecular diffusion and are sorbed onto the charcoal pad(s). Sampling rates for gases and vapors are provided by 3M. At the end of a sampling period, the diffusion membrane is removed and a tight-fitting cap is snapped into place. If the 3520 back-up monitor is being used, the primary section is separated from the back-up section, and caps are snapped on both the primary and back-up sections. Design of the capped monitor allows *in situ* desorption of gases and vapors with carbon disulfide or other suitable solvents. Samples are desorbed for a minimum of 30 min. and an aliquot is removed for analysis by gas

INSTRUMENT 16-12-2b. Miscellaneous Diffusive Samplers.

chromatography. The organic vapor monitors are supplied with or without laboratory services.

Desorption efficiency values are known to vary with the amount of material on the charcoal and with the type and volume of desorbing solvent used. Therefore, actual desorption efficiencies should always be determined at the time of analysis. The recommended procedure for the 3M Organic Vapor Monitor #3500 is as follows:

The organic compound in the liquid state is introduced through the elutriation port onto a piece of filter paper placed between the elutriation cap and diffusion plate of the monitor. The port is closed and the organic compound is given sufficient time to vaporize and consequently be sorbed by the charcoal sorbent. The filter is removed and analyzed as a separate sample to determine if complete transfer of organic compound has occurred. Subsequently, the monitor is desorbed for an appropriate period and the aliquots for the gas chromatographic analysis are taken from the center port of the monitor.

The sampling rate should be the one supplied by the manufacturer and if possible verified by the user (see Table 16-I-4).

16-12-3. ORSA 5

National Draeger, Inc.

The ORSA is a diffusive, organic vapor sampler containing 400 mg of activated coconut shell charcoal. The measured contaminants flow automatically into the collecting tube during sampling, as a result of the concentration differential between the ambient air and the interior of the tube, where they are sorbed on the activated charcoal. For measurements in the TLV range, the sampling time may be as long as 8 hrs. Shorter or longer sampling times are possible for other ranges. The ORSA may be analyzed by the same laboratory methods used for active sampling charcoal tubes. Pannwitz[18, 19] had published a comparison of the ORSA 5 sampler with other methods, including detector tubes, pumped tubes, and liquid sorption methods. He concluded that there was no essential difference between the results obtained for the active and diffusive methods. See the illustration of the sampler in Instrument 16-12-3.

16-12-4. Passive Sampler Series 575
SKC, Inc.

A miniature sampling device for collecting airborne organic vapors by diffusion, SKC 575 Series Passive Samplers contain specially-packed sorbent in a holder with a diffusion barrier and are worn as a badge on the collar. Since not all organic vapors are effectively collected on charcoal, SKC samplers are available with three different sorbents: charcoal, Anasorb 727, and Anasorb 747, to better collect specific compounds. These samplers operate at known sampling rates that have been experimentally determined and validated. Currently, 85 organic chemicals have been validated and theoretical (calculated) rates for 167 other chemicals are available. The Passive Sampler operating instructions include sampling rates, minimum and maximum sampling times, validation levels, and desorption efficiencies for each compound. Full-shift 8-hour sampling can be performed for most compounds using just one sampler. In addition, because the sorbent layer is close to the diffusion barrier, contaminant molecules adsorb quickly enough to permit reliable short-term measurements. See Instrument 16-12-4.

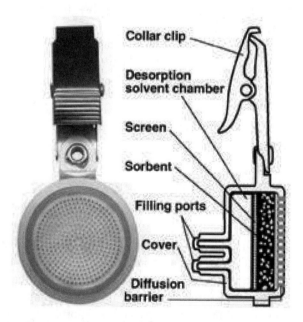

INSTRUMENT 16-12-4. Model 575 Organic Vapor Monitor (SKC).

16-12-5. ChemDisk™
Assay Technology

The thin ChemDisk™ Personal Sampler snaps into a monitor holder and is worn as a badge. A sampler, cap, return container, and mailing pouch protect the monitor for shipment to the laboratory. The monitor can be used with lab methods sanctioned by NIOSH, OSHA, and EPA for measuring air contaminants. The mechanically stable sampler with molded sampling ports generates accurate, reproducible sampling rates for OSHA compliance. Performance requirements (Specifications provided by the manufacturer are as follows):

- Operating Range (typical)
    1% of PEL – 1000% of PEL
- Optimal Accuracy Range
    10% of PEL – 200% of PEL
- Lowest Level Reliably Detected (typical)
    Aldehydes (Monitor 571)
    0.01 ppm (8-hr TWA)
    Organic Amines (Monitor 585)
    0.2 – 0.5 ppm (8-hr TWA)
- Organic Solvents (Monitor 541)
    0.05 – 0.5 ppm (8-hr TWA)
    (depends on analyte)

16-12-6. ChemChip™
Assay Technology

The ChemChip™ Personal Monitor is worn as a badge on pocket or lapel. After wearing, the test strips encased in the monitor are removed, developed, and inserted into a calibrated reflectance colorimeter (Electronic Reader) which provides on-site read-out of chemical exposure in parts per million (ppm). The monitor has been shown in two independent studies to comply with OSHA accuracy requirements. The ChemChip™ can be used for on-site monitoring of exposures to ethylene oxide, formaldehyde, or xylene. The manufacturers specifications follow:

- Operating Range (typical)
    10% of PEL – 500% of PEL
- Optimal Accuracy Range
    50% of PEL – 200% of PEL
- Lowest Level Reliably Detected (typical)
    8-hr TWA (Monitors 502, 510)
    0.1 ppm (8-hr TWA)
    15-min. STEL (Monitors 506, 511)
    1 ppm (8-hr TWA)

16-12-7. Radial Diffusive Sampler

Fondazione Salvatore Maugeri, Aquaria

The radial path diffusive sampler[76], radiello, is built by means of a sintered microporous polyethylene cylinder as the diffusive surface, and of a coaxial inner stainless steel net cylindrical adsorbing cartridge. The radial path in diffusion acts in such a way the sampling rate values are extraordinarily high and constant. As an example, using activated charcoal the sampling rate for benzene is 80 ml/min. and keeps constant in the 100-100,000,000 µg/m³ · min exposure, experimentally allowing accurate and precise measurements for exposure times as short as 1 hour and as long as three weeks.

Changing the granular sorbent, several volatile compounds can be sampled, such as aldehydes, anaesthetic gases and vapors, NO, NO2, SO2, O3, NH3, H2S, HCl, HF.

Furthermore, the cylindrical shape of the cartridge allows the thermal desorption recovery of VOCs by commercial devices, such as the Perkin-Elmer ATD 400.

All the radiello's parts are indefinitely reusable, except the adsorbing cartridge which is reusable by thermal recovery. Diagram is shown in Instrument 16-12-7.

16-12-8. Thermal Desorption Tubes for Diffusive Sampling

Some of the previously described thermal desorption tubes for active sampling in Section 16-10 can also be used for diffusive sampling.

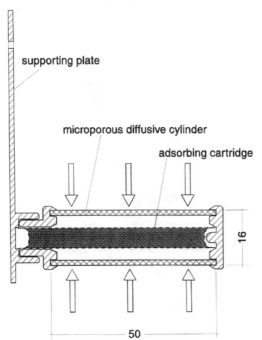

INSTRUMENT 16-12-7. Diffusion path in the radial symmetry passive sampler. Gaseous molecules reach the inner cylindrical adsorbing cartridge, diffusing along the radius of outermost microporous cylindrical diffusive surface (quotes are in mm).

16-13. Samplers for Formaldehyde

16-13-1. Formaldehyde Monitor

3M Company

The formaldehyde monitor contains a bisulfite-impregnated pad separated from a diffusion membrane by spacers. Formaldehyde is sampled by diffusion and collected by chemisorption on the bisulfite pad. At the end of sampling, the diffusion membrane is removed and a tight-fitting cap is snapped into place. The design of the capped monitor allows *in situ* desorption of the formaldehyde with deionized water and subsequent removal of an aliquot for colorimetric analysis. The formaldehyde monitor is supplied with or without laboratory services.

16-13-2. ACT Monitoring Card System™ for Formaldehyde

Envirometrics Products Co.

The ACT Electronic Reader is a sophisticated electronic device that is an integral part of the ACT Monitoring Card System™. Lightweight and portable, the reader contains a lead-acid battery that allows the analysis to be performed at the sampling site. One of the most unique features of the ACT reader is the ability to key in the sampling time and the temperature and humidity conditions. All diffusive monitoring results are affected by these variables. With this system, however, the data that are keyed in allow the reader to factor the conditions directly into the exposure calculations. Accurate results are available seconds after the monitoring period. In addition to formaldehyde, the system can be applied to glutaraldehyde, ammonia, hydrogen sulfide, chlorine, sulfur dioxide, carbon monoxide, nitrogen dioxide, ethylene oxide, and methyl ethyl ketone.

16-13-3. AirScan™ Formaldehyde Monitoring System

AirScan Environmental Technologies, Inc.

The AirScan™ Monitor is a diffusive monitor that employs a crystal growth and nucleation technology to measure concentration of toxic gases. During the sampling period, the gas diffuses through a sampling port onto a coated film inside the monitor. The film consists of microscopic (1 micrometer) sites where the analyte of interest reacts and forms crystal seeds. At the end of the sampling period, a supersaturated solution of the propietary reaction products is introduced onto the film and in 5 minutes, visible crystals grow on the sites where the analyte reaction has occurred. This results in a visible line whose length is proportional to the concentration of the measured compound. The user reads the length of

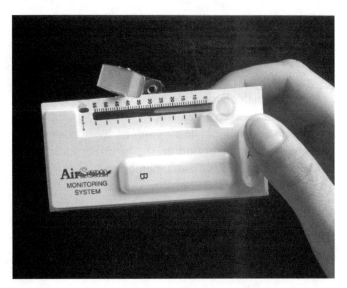

INSTRUMENT 16-13-3. AirScan Formaldehyde Monitoring System (AirScan).

this line in millimeters, and then refers to a lot-specific conversion chart to determine the exposure concentration. See Instrument 16-13-3.

AirScan™ Monitors exist for STEL (15-min), TWA (4- to 8-hr), and indoor air quality (IAQ) (24-hr) measurements, and are accurate to within 25% at the 95% confidence level. The monitor measures formaldehyde, but the crystal growth and nucleation technology is expandable to a large number of other compounds such as ethylene oxide and glutaraldehyde. Lower limits of detection are as follows: formaldehyde STEL: 0.30 ppm (ceiling value), A2, TWA: 0.03 ppm, IAQ: 0.03 ppm.

16-13-4a. Passive Bubbler 525 for Formaldehyde
SKC, Inc.

Small and lightweight, SKC's Passive Bubbler™ clips to a worker's shirt for personal sampling, or fits into a stand for area sampling. Airborne formaldehyde diffuses into an aqueous solution in the sampler through a diffusion disc that precisely controls the sample collection rate. A reagent is added to the sampling solution during analysis; formaldehyde concentration is then determined using a colorimeter. The complete Formaldehyde Monitoring Kit includes supplies to perform 100 tests and analyses. Validated by ASTM Method D 5014, SKC Passive Bubblers meet OSHA accuracy requirements for STEL and TWA measurements of airborne formaldehyde in both industrial and non-industrial indoor settings. Illustrated in Instrument 16-13-4a.

16-13-4b. Personal and Area Samplers for Formaldehyde
SKC, Inc.

Designed to accurately measure personal exposures to formaldehyde with accuracies of at least + or –25 %, SKC Personal and area formaldehyde diffusive samplers have been validated based on NIOSH Method 3500 and can be used for OSHA compliance monitoring. The PEL sampler collects formaldehyde at a low rate suitable for full-shift (8-hour) monitoring. The STEL sampler collects at a high rate suitable for 15-minute sampling. Use the SKC Formaldehyde Passive Sampler in the office or for industrial sampling. Samples can be sent to a laboratory for

INSTRUMENT 16-13-4a. Passive Bubbler for Formaldehyde *(SKC).*

INSTRUMENT 16-13-4b. SKC Formaldehyde Passive Sampler.

analysis where the chromotropic acid assay method is used. The SKC Formaldehyde 'Passive' Sampler is pictured Instrument 16-13-4b.

16-13-5. Indoor Air Formaldehyde Passive Sampler

SKC, Inc.

The Indoor Air Formaldehyde 'Passive' Sampler is the only device specifically designed for accurate indoor measurements of formaldehyde levels as low as the following: 0.01 ppm (+ or –30%), 0.02 ppm (+ or –15%) in the home, office, or industrial environment over a 5- to 7-day period. Small and unobtrusive, the sampler is simply hung from the ceiling in the area to be sampled. The sampler is inexpensive, accurate, and had been field validated by the Indoor Air Quality Program, Lawrence Berkeley Laboratories at the University of California at Berkeley. Samples can be sent to a laboratory for analysis where the chromotropic acid assay method is used. There are no known interference's from other substances.

16-14. Samplers for Oxides of Nitrogen and Ozone

16-14-1. Nitrox™

Landauer, Inc.

As a personal nitrous oxide monitor, Nitrox™, is a light weight, compact, diffusive monitor that provides an easy and cost-effective way to monitor and manage exposure to nitrous oxide. Nitrox accurately monitors exposure levels of 25 and 50 ppm as recommended by NIOSH, ACGIH and numerous professional societies. It has the ability to measure for less than a single hour and as long as 40 hours.

Nitrous oxide measurements can serve as a proxy for other waste anesthetic gases. Monitoring with Nitrox can help to ensure that scavenging devices and administrative

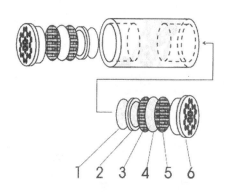

INSTRUMENT 16-14-2. Harvard Sampler for Ozone (Ogawa & Co.); 1 = Teflon disk, 2 = Teflon ring, 3 = stainless screen, 4 = coated collection filter, 5 = stainless screen, and 6 = diffuser end cap.

measures are working—important documentation should litigation ever occur. Laboratory analysis, rapid reporting and archival retention of exposure results are all part of Landauer's standard services in personnel monitoring. Illustrated in Instrument 16-14-1.

16-14-2. Harvard Passive Ozone Sampler

Ogawa & Co. USA, Inc.

The diffusive ozone sampler uses two multitube diffusion barriers with collection on coated glass fiber filters. The principle component of the coating is nitrite ion, which, in the presence of ozone, is oxidized to nitrate ion on the filter medium, $NO_2^- + O_3 = NO_3^- + O_2$. The filters are extracted with water and analyzed for nitrate ion by ion chromatography. The nominal lowest level is 200 ppb-hrs. Laboratory and field validation tests showed excellent agreement between the passive method and standard ozone monitoring techniques. Sampling is independent of temperature and relative humidity in typical ambient conditions. No significant interferences were found for other atmospheric pollutant gases. A protective cup is used as a wind screen for outdoor measurements and also acts as a rain cover. The samplers, coated filters, and protective cups are all commercially available (see diagram in Instrument 16-14-2).

16-14-3. Palmes Sampler

Gradko International Ltd.

The Palmes type diffusion samplers[16] are designed to enable monitoring of air pollutants such as nitrogen dioxide by the application of molecular gas diffusion principles. The sampler is constructed of an acrylic tube 7 cm. long and 1 cm. internal diameter, capped on both ends by polyethylene caps. One of the caps contains two stainless steel grids. Airborne nitrogen dioxide is collected on a

INSTRUMENT 16-14-1. Nitrox™ (Landauer).

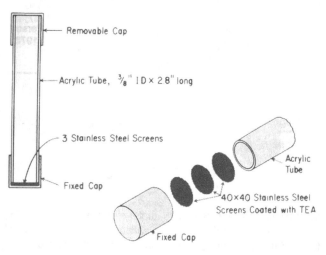

INSTRUMENT 16-14-3. Passive Palmes Sampler for NO₂.

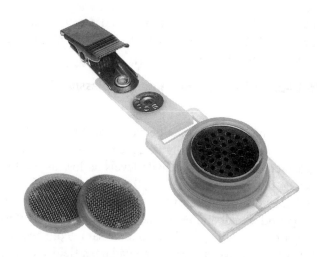

INSTRUMENT 16-15-3. Passive Mercury Sampler Badge, 520 Series (SKC).

chemical adsorbent, triethanolamine (TEA) which has been loaded onto the grids. The loaded tubes are exposed for predetermined periods: between 2-4 weeks, for example, after which the NO_2 concentration adsorbed by the TEA can be measured by spectrophotometric analysis at 540 nanometers, by a variation of the Saltzmann reaction described in NIOSH Method 6700.[17] The samplers are suitable for monitoring long-term trends, identifying "hot spots" of air pollution and cross checking other analysis equipment. Picture is seen in Instrument 16-14-3.

16-15. Mercury Samplers

16-15-1. Monitor #3600 Mercury Badge

3M Company

The 3M Company makes two badges for mercury monitoring. One corrects for chlorine interference, the other is used when chlorine is not present. The badge collects mercury on a gold film, facilitating the formation

of a mercury/gold amalgam. The badge is returned to the manufacturer for measurement of the change in electrical conductivity (see illustration in Instrument 16-15-1).

16-15-2. Mercury Badge

GMD Systems, Inc.

GMD's Mercury Badge weighs only 14 g and collects mercury via diffusion and sorption on Hydrar solid sorbent that is contained in a replaceable capsule element. The sorbent capsule is returned to the laboratory for chemical desorption and analysis by flameless atomic sorption.

16-15-3. Passive Mercury Sampler Badge, 520 Series

SKC, Inc.

Validated by OSHA Method ID 140, the SKC Inorganic Mercury Passive Samplers measure worker exposure levels as a TWA and permit positive analysis for mercury vapor at, above, or below the NIOSH/OSHA standard of 0.05 mg/m³. The miniature badges easily attach to a worker's collar or shirt pocket and operate on the principle of diffusion, eliminating the need for a sample pump. Designed for flameless atomic absorption analysis, there is no interference from chlorine or moisture; SKC Mercury Passive Samplers are highly accurate and reliable. The reusable capsule holder can be cleaned and reused, reducing sampling costs and eliminating false high readings caused by badge contamination. SKC Inorganic Mercury Passive Samplers may be used for long-term sampling up to 120 hours. Pictured in Instrument 16-15-3.

INSTRUMENT 16-15-1. Passive Sampler for Mercury (3M).

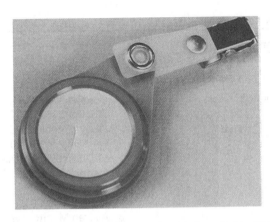

INSTRUMENT 16-16. Passive Monitor for Ethylene Oxide (3M).

16-16. Ethylene Oxide Monitor

3M Company

The ethylene oxide monitor contains a single, treated sorbent pad separated from a diffusion membrane by spacers. The ethylene oxide vapors are sampled by diffusion and sorbed on chemically treated activated charcoal, where they are converted to 2-bromoethanol. At the end of sampling, the diffusion cap is removed and a tight-fitting cap is snapped into place. When the sample is ready for chromatographic analysis, a 10% methylene chloride in methanol desorbing solution is added directly to the monitor. The ethylene oxide monitor is supplied with or without laboratory services and is illustrated in Instrument 16-16.

16-17. Other Diffusive Monitoring Systems

3M diffusive monitors for ethylene oxide, carbon monoxide, formaldehyde, and mercury are listed in Table 16-I-2. The types of badges supplied by SKC are found in Table 16-I-3. Table 16-I-1 lists various monitoring systems alphabetically, along with manufacturers and brand names. A field validation study compared three makes of badges for methylene chloride at a worksite and in a specially designed evaluation chamber[4]. An impregnated passive sampler for glutaraldehyde was compared with three other methods in a laboratory and hospital[22]. The ACT Monitoring Card System™ has been applied to TWA monitoring for the three gases of chlorine, chlorine dioxide and carbon monoxide in a bleached Kraft mill[1]. The Act Card System can also be applied to glutaraldehyde, ammonia, hydrogen sulfide, chlorine, sulfur dioxide, carbon monoxide, nitrogen dioxide, ethylene oxide, and methyl ethyl ketone. The Ogawa Company also markets a passive sampler that collects NO and NO_2 on coated filters which are analyzed by the Saltzmann procedure. The performance criteria for diffusive sampling was reviewed and discussed by Harper[8]. Addresses for manufacturers of diffusive devices are found in Table 16-I-6.

Additional Reading

Astrianakis, G.: Industrial Hygiene Aspects of a Sampling Survey at a Bleached-Kraft Pulp Mill in British Columbia. Am. Ind. Hyg. Assoc. J. 59: 694 (1998).

Brown, R.H.; Charlton, J.; Saunders, K.J.: The Development of an Improved Diffusive Sampler. Am. Ind. Hyg. Assoc. J. 42:865 (1981).

Calibrated Instruments, Inc.: Technical Bulletin A-5. Ardsley, NY.

Charron, K.A., Puskar, M.A. and Levine, S.P.: Field Validation of Passive Monitors for the Determination of Employee Exposures to Methylene Chloride in Pharmaceutical Production Facilities, Am. Ind. Hyg. Assoc. J. 59: 353 (1998).

Gisclard, J.B.; Robinson, D.B.; Kuezo, Jr., P.J.: A Rapid Empirical Procedure for the Determination of Acrylonitrile and Acrylic Esters in the Atmosphere. Am. Ind. Hyg. Assoc. J. 19:43 (1958).

Cocheo, V.; Boaretto, C.; Sacco, P.: High Uptake Rate Radial Diffusive Sampler Suitable for Both Solvent and Thermal Desorption. Am. Ind. Hyg. Assoc. J. 57, 897–904 (1996).

Guild, L.V.: Assessment of the Reliability of Backup Systems in Diffusive Sorbent Samples. Am. Ind. Hyg. Assoc. J. 52:198 (1991).

Harper, M.: Diffusive Sampling—Instrument Performance Criteria, Edited by M. L. Woebkenberg. Appl. Occup. Environ. Hyg. 13:(11) 759 (1998).

HSE/CAR Working Group 5. The Diffusive Monitor: Current News in the Field of Diffusive Sampling in the Workplace and in the Environment, Issue 10, Dec. 1998.

Hori, H.; Tanaka, I.: Response Characteristics of the Diffusive Sampler at Fluctuating Vapor Concentrations. Am. Ind. Hyg. Assoc. J. 54:95 (1993).

Lang, H.W.; Freedman, R.W.: The Use of Disposable Hypodermic Syringes for Collection of Mine Atmosphere Samples. Am. Ind. Hyg. Assoc. J. 30:523 (1969).

Levin, J.O.; Lindahl, R.: Diffusive Air Sampling of Reactive Compounds-A Review. Analyst, Vol.119, pp. 79–83 (1994).

Merino, M.: Passive Monitors. National Safety News 120:56 (1979).

Mine Safety Appliances Company: MSA Data Sheet 08-00-02. Pittsburgh, PA (1988).

NIOSH, Manual of Analytical Methods, 4th Edition, (1994).

Palmes, E.D.; et al.: Personal Sampler for Nitrogen Dioxide. Am. Ind. Hyg. Assoc J. 37:570 (1976).

Pannwitz, K.H.: Comparison of Active and Passive Sampling Devices. Drager Rev. 52:19 (1984).

Pannwitz, K.H.: ORSA 5, A New Sampling Device for Vapors of Organic Solvents. Drager Rev. 48:8 (1981).

Sampling and Analysis of Mine Atmosphere. Miners Circular No. 34 (Revised). U.S. Department of the Interior (1948).

Severs, L.W.; Skory, L.K.: Monitoring Personal Exposure to Vinyl Chloride, Vinylidene Chloride and Methyl Chloride in an Industrial Work Environment. Am. Ind. Hyg. Assoc. J. 36:669 (1975).

Wellons, S.L.; Trawick, E.G.; Stowers, M.F.; Jordan, S.L.P; Wass, T.L.: Laboratory and Hospital Evaluation of Four Personal Monitoring Methods for Glutaraldehyde in Ambient Air. Am. Ind. Hyg. Assoc. J. 59: 96 (1998).

West, P.W.; Reisner, K.D.: Field Tests of a Permeation-type Personal Monitor for Vinyl Chloride. Am. Ind. Hyg. Assoc J. 39:645 (1978).

TABLE 16-I-1. Manufacturers and Brand Names of Diffusion Monitors

System	Manufacturer	Brand Name
Acetone	SKC	SKC 575-002
Carbon Monoxide	3M	CO Monitor 3400
Carbon Monoxide	Wilson Safety Products	Dosimeter Badge
Ethylene oxide	3M	Monitor 3550
Formaldehyde	SKC	Passive Bubbler, 525-004
Formaldehyde	SKC	PEL Sampler, 526-201
Formaldehyde	SKC	STEL Sampler, 526-200
Formaldehyde	SKC	Indoor Air Kit, 526-100
Formaldehyde	3M	Monitor 3750
Formaldehyde	K & M	Chromair 380007
Formaldehyde	AirScan	AirScan Monitor System
Mercury	SKC	Mercury Vapor Badge
Mercury	3M	Monitor 3600, or 3600A
Methylene Chloride	SKC	Pass. Sampler 575-001MC
Nitrogen Dioxide	Ogawa, USA	Nitrogen Oxides Badge
Ozone	Ogawa, USA	Ozone Badge
Ozone	Cole-Parmer	
Nitrous Oxide	Landauer	Nitrox
Organic Vapor Monitor	Assay Technology	ChemChip & ChemDisk
Organic Vapor Monitor	K & M	TraceAir OVM Badge
Organic Vapor Monitor	National Draeger, Inc.	ORSA 5
Organic Vapor Monitor	SKC	Passive Sampler 575
Organic Vapor Monitor	3M	Org. Vapor Monitor 3500
Radial Diffusive Sampler	Aquaria	Radiello

TABLE 16-I-2. 3M Specific Passive Diffusion Monitoring Systems

System	Sampling Range	Interferences	Shelf Life	Brand Name
Organic vapors	Compound dependent	None	18 months	3500 (without analysis) 3500 (with analysis) 3520 (without analysis) 3530 (with analysis)
Ethylene oxide	0.2 to 600 ppm-hrs	None	18 months	3550 (with analysis) 3551 (with analysis)
Formaldehyde	0.8 to 40 ppm-hrs	Phenol, alcohols	18 months	3720 (with analysis) 3721 (without analysis)
Mercury	Up to 0.20 mg Hg/m3	Strong oxidizers such as halogen vapors. CO, O_3, NO_x, and SO_2 negligible. Organic vapors generally do not interfere.	12 months (for 3600) 6 months (for 3600A)	3600 (with 3M analysis) 3600A (with 3M analysis, for chlorine environments)

TABLE 16-I-3. SKC Monitoring Badges

System	Collection Medium	Analysis	Range
Organic Vapor Badges	Charcoal Anasorb 747 Anasorb 727	Solvent Desorption and GC Analysis	Varies with Compound
Passive Bubbler for Formaldehyde	0.05% Aqueous MBTH in Diffusion Cell	Colorimetric ASTM Method D5014	0.03 to 6 mg/m^3 (0.025 to 5 ppm)
Passive Badge PEL Formaldehyde STEL Formaldehyde Indoor Air	Dry (Proprietary)	Chromotropic Acid Assay	0.2 to 2 ppm 0.5 to 6 ppm 5 to 7 days
Passive Mercury Sampler	Anasorb C300	Acid digestion & cold vapor Atomic Absorption Spectrophotometry	0.061 to 0.020 mg/m^3

MBTH = 3-methyl-2benzothiazolone hydrazine hydrochloride

TABLE 16-I-4. Diffusive Monitoring Sampling Rates (cc/min.)

Compound	3M #3500 Badge Monitor	SKC 575-001	TraceAir K&M OVM-1
Acetone	40.1*	15.2*	41.5
Acrylonitrile	43.8	21.1	41.9
Allyl alcohol	40.4	18.8	40.4
n-Amyl acetate	26.0*	12.3	24.2
n-Amyl alcohol	31.2*	14.5	28.4
Benzene	35.5*	16.0*	36.5*
n-Butyl acetate	31.6	13.27	26.6
n-Butyl alcohol	34.3*	16.2	34.1*
Carbon tetrachloride	30.2*	14.1*	32.8
Chloroform	33.5	13.0*	35.2
Cyclohexane	32.4*	15.6*	28.5*
Dioxane	34.5	15.84	28.6*
Heptane	28.9*	13.9*	26.8*
Mesityl oxide	31.2*	13.4	22.1*
Methyl ethyl ketone	36.3*	17.1*	35.8*
Perchloro ethylene	28.3*	12.9*	31.6
Xylene, o-	27.3*	11.9	28.8
Xylene, m-	27.3*	12.5	27.2
Xylene, p-	27.3*	12.8	9.0*

* indicates sampling rate determined experimentally.

TABLE 16-I-5. Vendors of Gas and Vapor Collectors

Plastic Bags

ALT   Alltech Associates, Inc.
       2051 Waukegan Road
       Deerfield, IL 60015
       (847) 948-8600
       FAX (847) 948-1078

CA2   Calibrated Instruments, Inc.
       20 Saw Mill River Road
       Hawthorne, NY 10532
       (914) 741-5700 or (888) 779-2064
       FAX (914) 741-5711
       www.calibrated.com

EDL   Edlon-PSI
       117 State Rd.
       P.O. Box 667
       Avondale, PA 19311-0667
       (610) 268-3101 or (800) 753-3566
       FAX (610) 268-8898
       www.edlon-psi.com

MMM   3M Company
       3M Center
       Building 275-6W-01
       St. Paul, MN 55144-1000
       (612) 733-1110 or (800) 328-1667
       www.mmm.com

WSP   Wilson Safety Products
       P.O. Box 622
       Reading, PA 19603
       (215) 376-6161
       FAX (215) 371-7725

General Supply Houses

CPI   Cole-Parmer Instrument Co.
       625 East Bunker Court
       Vernon Hills, IL 60061
       (800) 323-4340
       FAX (847) 247-2929
       info@coleparmer.com
       www.cole-parmer.com

FSC   Fisher Scientific Company
       1801 Gateway Blvd., Suite 101
       Richardson, TX 75080-3750
       (800) 772-6733
       FAX (800) 772-7702
       www.fisherscientific.com

PFE   Plastic Film Enterprises
       1921 Bellaire
       Royal Oak, MI 48067
       (313) 399-0450 or (800) 336-3872
       FAX (313) 399-2534

SUP   Supelco, Inc.
       Supelco Park
       Bellefonte, PA 16823
       (814) 359-3441
       FAX (814) 359-5750
       supelco@sial.com

Bubblers and Gas Washers

AGI   Ace Glass, Inc.
       P.O. Box 688
       1430 North West Blvd.
       Vineland, NJ 08362-0688
       (800) 223-4524
       FAX (800) 543-6752
       www.aceglass.com

BGI   BGI Incorporated
       58 Cuinan Street
       Waltham, MA 02451
       (781) 891-9380
       FAX (781) 891-8151
       www.bgiusa.com

Adsorbents

BCC   Barneby-Cheney Company
       P.O. Box 2526
       Columbus, OH 43216
       (614) 258-9501

CSI   Columbia Scientific Industries
       Forney Corporation
       3405 Wiley Post Road
       Carrollton, TX 75006-5185
       (972) 458-6100
       FAX (972) 458-6650
       www.forneycorp.com

PIT   Pittsburgh
       Division of Calgon Corp.
       P.O. Box 1346
       Pittsburgh, PA 15230
       (412) 562-8301

CGW   Corning Glass Works
       P.O. Box 5000
       Corning, NY 14830
       (607) 974-4261

SGI   Scientific Glass & Instrument Co.
       P.O. Box 6
       Houston, TX 77001
       (713) 682-1481
       FAX (713) 682-3054

SUP   Supelco, Inc.
       Supelco Park
       Bellefonte, PA 16823
       (814) 359-3441
       FAX (814) 359-5750
       supelco@sial.com

Packaged Sampling Equipment

SKC   SKC Incorporated
       863 Valley View Road
       Eighty Four, PA 15330-1301
       (724) 941-9701 or (800) 752-8472
       FAX (724) 941-1369 or
       (800) 752-8476
       www.skcinc.com

SUP   Supelco, Inc.
       Supelco Park
       Bellefonte, PA 16823
       (814) 359-3441
       FAX (814) 359-5750
       supelco@sial.com

WES   Westvaco, Inc.
       P.O. Box 140
       Covington, VA 24426
       (540) 969-3700
       FAX (540) 965-0230
       www.westvaco.com

TABLE 16-I-6. Sources of Diffusive Monitors

ASE	AirScan Environ. Technologies, Inc. 197 Meister Avenue Branchburg, NJ 08876 (908) 725-1342 or (800) 639-2477 *www.airscaninc.com*	NDR	Draeger Safety, Inc. 101 Technology Drive Pittsburgh, PA 15275 (412) 787-8383 or (800) 922-5518 FAX (800) 922-5519 *www.draeger-usa.com*	SKC	SKC Incorporated 863 Valley View Road Eighty Four, PA 15330-1301 (724) 941-9701 or (800) 752-8472 FAX (724) 941-1369 or (800) 752-8476 www.skcinc.com
AST	Assay Technology 1252 Quarry Lane Pleasanton, CA 94566 (800) 833-1258 *askassay@assaytech.com* *www.assaytech.com*	NMS	National Mine Service Co. Safety Systems & Products U.S. Rt. 22 and 30 West Oakdale, PA 15071 (412) 429-0800	MMM	3M Company 3M Center Building 275-6W-01 St. Paul, MN 55144-1000 (612) 733-1110 or (800) 328-1667 *www.mmm.com*
DRW	Draegerwerk AG Moislinger Alle 53/55 Postfach 1339D-2400 Lubeck 1, Germany	OGA	Ogawa & Co., USA, Inc. 1230 S.E. 7th Avenue Pompano Beach, FL 33060 (305) 781-6223	WSP	Wilson Safety Products P.O. Box 622 Reading, PA 19603 (215) 376-6161 FAX (215) 371-7725
FSN	Fondazione Salvatore Maugeri-IRCCS Centro di Ricerche Ambientali Padova, Italy 0039 0498 064 511 FAX 0039 0498 064 555 *fsmpd@tin.it**	PEL	Perkin-Elmer Instruments 45 William Street Wellesley, MA 02481-4078 (781) 237-5100 FAX (781) 431-4145 *www.perkin-elmer.com*		
LAN	Landauer, Inc. 2 Science Road Glenwood, IL 60425-1586 (708) 755-7000 (708) 755-7016 *www.landaueriii.com*				

* The Radiello samplers are marketed by a company called AQUARIA in Milan, Italy, and represented in the U.S. by SEMPAIR ENVIRONMENTAL, P.O. Box 659, Chestertown, Maryland 21620. Phone 888-778-7829, fax 410-778-4470, Stephen W. Giesesr, President. sgieser!@juno.com.

Chapter 17

Detector Tubes, Direct-Reading Passive Badges, and Dosimeter Tubes

John Palassis, CIH, CSP, CHMM[A]; Jeff Bryant, MS, CIH[A]; and John N. Zey, MS, CIH[B]

[A]National Institute for Occupational Safety and Health, Cincinnati, Ohio;

[B]Central Missouri State University, Warrensburg, Missouri

CONTENTS

Detector Tubes

Development of Detector Tubes

Three types of direct-reading, colorimetric indicators have been in use for the determination of contaminant concentrations in air: liquid reagents; chemically treated papers; and glass detector tubes containing solid indicating chemicals. An early comprehensive bibliography in this area was prepared by Campbell and Miller.[1]

Liquid Reagents

Convenient laboratory procedures using liquid reagents have been simplified and packaged for field use. Reagents are supplied in sealed ampules or tubes, frequently in concentrated or even solid forms that are diluted or dissolved for use. Unstable mixtures may be freshly prepared when needed by breaking an ampule containing one ingredient inside a plastic tube or bottle containing the other. Commercial apparatus of this type is available for tetraethyl lead, tetramethyl lead, and TDI/MDI. Certain liquid reagents, such as those used for nitrogen dioxide sampling, produce a direct color upon exposure without requiring additional chemicals or manipulations. These permit simplified sampling equipment. Thus, relatively high concentrations of nitrogen dioxide may be determined directly by drawing an air sample into a 50- or 100-mL glass syringe containing a measured quantity of absorbing liquid reagent, capping, and shaking. Liquids containing indicators have been used for determining acid or alkaline gases by measuring the volume of air required to produce a color change. These liquid methods are somewhat inconvenient and bulky to transport and require a degree of skill to use. However, they are capable of good accuracy because measurement of color in liquids is inherently more reproducible and accurate than measurement of color on solids.

Chemically-Treated Papers

Chemically treated papers have been used to detect and determine gases because of their convenience and compactness. An early example of this detection method is the Gutzeit method in which arsine blackens a paper strip impregnated previously with mercuric bromide. Such papers may be freshly prepared and used wet, or stored and used in the dry state. Special chemical chalks or crayons have been used[2] to sensitize ordinary paper for phosgene, hydrogen cyanide, and other war gases. Semi-quantitative determinations may be made by hanging the paper in contaminated air. Inexpensive detector tabs are available commercially which darken upon exposure to carbon monoxide.[3] The accuracy of such procedures is limited by the fact that the volume of the air sample is rather indefinite and the degree of color change in the paper is influenced by air currents and temperature. More quantitative results may be obtained by using a sampling device capable of passing a measured volume of air over or through a definite area of paper at a controlled rate, as is done in commercial devices known as *tape samplers* which detect hydrogen sulfide, toluene diisocyanate, hydrogen fluoride and other compounds. Particulate matter contaminants such as chromic acid and lead may be determined similarly, usually by addition of liquid reagents to the sample on a filter paper. Visual evaluation of the stains on the paper may be made by comparison with color charts or by photoelectric instruments. Recording photoelectric instruments that use sensitized paper tapes operate in this manner; however, they are not described in this chapter. For more information, see the extensive discussions in the 7th Edition (1989) of this book, Chapter N entitled *"Sequential and Tape Samplers"* pages 291–303, and the 8th Edition, in Chapter 19, pages 445, 478–480.

Accuracy of these methods requires uniform sensitivity of the paper, stability of all chemicals used, and careful calibration. In the case of particulate matter analysis, it may be necessary to calibrate with the specific dust being sampled if the degree of a chemical's solubility is important.

Surface Wipe Sampling and Analysis

Recently, industrial hygienists have shown more interest in the detection of hazardous materials on surfaces, dust, soil, and metal. Consequently, several vendors developed test kits for the qualitative and semi-quantitative detection of toxic metals such as Pb, Ni, Cd, Hg, and others, as well as for various types of organic and inorganic compounds. Chemically-treated papers are often used. The colorimetric chemical reactions for such detection are the classic chemical reactions used for qualitative analysis in chemistry, and many reactions are similar to those listed on Table 17-1. Vendors for such products can be found at the end of this chapter, under the vendors' section on wipe tests.

Glass Detector Tubes

Glass detector tubes containing solid chemicals are another type of convenient and compact direct-reading device. The earliest detector tubes were made in 1920 to detect carbon monoxide in coal mines. A listing of early references was presented in a previous edition of this text.[4] There has been a great expansion in the development and use of detector tubes,[5–17] and more than 400 different types are now available commercially. Several manuals provide comprehensive descriptions and listings.[18–22] Because of the great popularity and wide use of glass detector tubes, the bulk of this discussion will deal with them, although much of the information will be applicable to the liquid and paper indicators as well.

Applications of Detector Tubes

There are many uses for detector tubes. They are convenient for qualitative[23] and semi-quantitative evaluation of toxic hazards in industrial atmospheres and for rapid evaluation of spills of hazardous materials.[24] They are also used for ambient air pollution studies, although in most situations, few available tubes have the required sensitivity. Detector tubes may be used for detection of explosive hazards as well as for process control of gas composition. Confirmation of carbon monoxide poisoning may be made by determining carbon monoxide in exhaled breath or in gas released from a sample of blood (after following an appropriate procedure to release the bound carbon monoxide). Determination of benzene in the exhaled air has also been reported.[25] Detector tubes have been used to evaluate and monitor permeation of chemicals through chemical-protective clothing.[26]

Detector tubes may be used for law enforcement purposes, such as determining ethyl alcohol in the

TABLE 17-1. Common Colorimetric Reactions in Gas Detector Tubes

1. Reduction of chromate or dichromate to chromous ion:

Draeger: Acetaldehyde 100/a; acetone 500/a-L; alcohol 100/a; aniline 0.5a; cyclohexane 100/a; diethyl ether 100/a; ethanol 500/a-L, 1000/a-D; ethyl acetate 200/a, 500/a-D; ethyl glycol acetate 50/a; ethylene oxide 25/a; n-hexane 100/a; hydrocarbons 100/a-L; methanol 50/a; n-pentane 100/a; o-toluidine 1/a.

Gastec: Acetone 151; aniline 181; amyl acetate 147; butane 104; butyl acetate 142, 142L; n-butanol 115; cyclo hexanol 118; ethanol 112, 112L; ethyl acetate 141, 141L; ethyl ether 161, 161L; ethylene oxide 163; gasoline 101, 101L; hexane 102H, 102L; hydrocarbons 103; isoamyl acetate 148; isoamyl alcohol 117; isobutyl acetate 144; isobutanol 116; isopropanol 113, 113L; isopropyl acetate 146; LP gas 100A; methanol 111, 111L, 111LL; methyl cyclohexanol 119; methyl ethyl ketone 152; methyl isobutyl ketone 153; methyl methacrylate 149; petroleum naphtha 106; propane 100B; propyl acetate 145; sulfur dioxide 5H; tetrahydrofuran 159; vinyl chloride 131.

Kitagawa: Acetone 102SA, 102SD; acrylonitrile 128SA, 128B; allyl alcohol 184S; butadiene 168SA, 168SC; butane 221SA; butanol 207U; butyl acetate 138SA, 138U; butyl acrylate 211U; butyl cellosolve 190U; cyclo hexane 115S; cyclohexanol 206U; cyclohexanone 197U; diacetone alcohol 195U; dichloroethane 235S, 230S; dimethyl ether 123; dioxane 154; ether 107SA, 107U; ethyl acetate 111SA, 111U; ethyl alcohol 104A; ethylene oxide 122; furan 161; gasoline 110S; general hydrocarbons 187S; n-hexane 113; isobutyl acetate 153; isoprene 241U; isopropanol 150; iso propyl acetate 149, 111U; methyl acetate 148; methyl acrylate 211U; methyl alcohol 119; methyl ethyl ketone 139B; methyl isobutyl ketone 155; naphthalene 226U; organic gases 186; propyl acetate 151; propylene oxide 163; sulfur dioxide 103; tetrahydrofuran 162; vinyl chloride 132.

MSA: Ethanol 804136; ethylene 804428; hexane 497664.

2. Reduction of iodine pentoxide plus fuming sulfuric acid to iodine:

Draeger: Benzene 2/a, 5/b, 20/a-L; carbon disulfide 5/a, 10/a-L; carbon monoxide 2/a, 5/c, 8/a, 10/a-L, 10/b, 50/a-L, 0.001%/a, 0.3%/b; ethyl benzene 30/a; hydrocarbon 0.1%/a; methylene chloride 100/a*, 50/a-I*, natural gas*; petroleum hydrocarbons 100/a; polytest; toluene 5/a, 200/a-L.

Gastec: Acetylene 171; benzene 121, 121L, 121S, 121SL; car bon monoxide 1HH, 1H, 1M, 1LK; chlorobenzene 126; o-dichlorobenzene 127; hydrocarbons 105; stoddard solvent 128; toluene 122, 122L; trichloroethylene 132HH; xylene 123.

Kitagawa: Benzene 118SB, SC; carbon monoxide 106, chlorobenzene 178S; chloropicrin 172; dichloroben zene 214S; ethylbenzene 179S; toluene 124SA, SB; xylene 134S.

MSA: Carbon disulfide 492514; carbon monoxide 803943, 487334, 804423, 487335, 488906(HP); dichloromethane 804416, gasoline 492870, per chloroethylene 804429, 487337; qualitest 497665; toluene 803947; trichloroethane 487343.

3. Reduction of chemical with ammonium molybdate plus palladium sulfate to molybdenum blue:

Draeger: Ethylene 0.1/a, 50/a; methyl acrylate 5/a; methyl methacrylate 50/a.

Gastec: Butadiene 174, 174L; ethylene 172, 172L.

Kitagawa: Acetylene 101S; Butadiene 168SB; carbon monoxide 106A, 106B, 106C*; ethylene 108B, 108SA; hydrogen sulfide and sulfur dioxide 120C.

MSA: Carbon monoxide 47134, ethyl mercaptan 804589

4. Reaction with potassium palladosulfite:

Draeger: Carbon monoxide 50/a-D.

Gastec: Carbon monoxide 1L, 1La, 1LL; hydrogen 30; hydro gen cyanide 12H

Kitagawa: Carbon monoxide 100, 106.

5. Color change of pH indicators (e.g. bromphenol blue, phenol red, thymol blue, methyl orange):

Draeger: Acetic acid 5/a, 5/a-L, 10/a-D; acetone 1000/a-D*; acid test; acrylonitrile 1/a*, 5/a*; amine; ammonia 2/a, 5/a, 10/a-L, 20/a-D; 0.5%/a; carbon dioxide 1000/a-L, 500/a-D, 1%/a-D; chlorobenzene 5/a*; cyanide 2/a*; cyclohexylamine 2/a; dimethylfor mamide 10/b*; formic acid 1/a; halogenated hydro carbons 100/a**; hydrazine 0.25/a; hydrochloric acid 2/a, 10/a-D, 10/a-L, 50/a; hydrocyanic acid 2/a*, 20/a-D*; hydrogen fluoride 2/a-L; nitric acid 1/a; per chloroethylene 50/a-L; phosphine 0.01/a*; phos phoric acid esters 0.05/a*; sulfur dioxide 0.1/a*, 2/a-L; triethylamine 5/a; vinyl chloride 0.5/a*.

Gastec: Acetaldehyde 92, 92M; acetic acid 81, 81L; acrolein 93; acrylonitrile 191, 191L; amines 180, 180L; ammo nia 3H, 3M, 3L, 3La; arsine 19LA; tert-butyl mercap tan 75, 75L; carbon dioxide 2HH, 2H, 2L, 2LL; carbon disulfide 13, 13M; carbonyl sulfide 21; chlorine 8HH; diborane 22; 1,2-dichloroethylene 139; dimethylac etamide 184; dimethylformamide 183; ethyl mercap tan 72L, formaldehyde 91M, 91L; hydrazine 185; hydrogen chloride 14M, 14L; hydrogen cyanide 12M, 12; hydrogen fluoride 17; mercaptans 70L; methacrylonitrile 192; nitric acid 15L; 2-penteneni trile 193; perchloroethylene 133HA, 133M, 133L, 133LL; phosphine 7, 7La; pyridine 182; sulfur dioxide 5M, 5L, 5La, 5Lb; trichloroethylene 132HA, 132HM, 132L, 132LL; vinyl chloride 131La, 131L; vinylidine chloride 130L.

Kitagawa: Acetaldehyde 133A, 133SB; acetic acid 216S; acetone 102SC; acrolein 136; acrylonitrile 128SC, 128SD; ammonia 105, 105B; arsine 121U; mercaptans 130U; carbon dioxide 126SA, 126SB, 126SC, 126SD, 126SH,

TABLE 17-1 (cont.). Common Colorimetric Reactions in Gas Detector Tubes

126UH; carbon disulfide 141SA, 141SB; carbonyl sulfide 239S; chloroprene; diborane 242S; dichloroether 223S; diethylamine 222S; formaldehyde 171SA; hydrazine 219S; hydrogen chloride 173SA; hydrogen cyanide 112B; nitric acid 233S; silane 240S.

MSA: Acetic acid 804138; ammonia 804405, 800300, 804406, carbon dioxide 497606, 487333, 804419, 488907(HP); formaldehyde 497649; hydrogen chloride 803948; hydrogen cyanide 803945, hydrogen fluoride 804142; sulfur dioxide 497661; triethylamine 804134.

6.   Reaction with o-tolidine:

Draeger: Chlorine 0.2/a, 0.3b, 1/a-L, 50/a; chloroform 2/a*; epichlorohydrin 5/b*; fluorine 0.1/a*; nitrogen dioxide 10/a-D; perchloroethylene 10/b, 200/a-D; trichloroethane 50/d*; trichloroethylene 2/a*, 10/a, 10/a-L*, 200/a-D; vinyl chloride 1/a*, 10/a-L*.

Gastec: Chlorine 8H, 8Ha; chloroform 137; methyl bromide 136H, 136L, 136La; methylene chloride 138; nitrogen dioxide 9L; nitrogen oxides 11HA, 11S, 11L; nitric oxide 10; 1,1,1-trichloroethane 135, 135L.

Kitagawa: Bromine 114; chlorine 109SA, 109SB; chlorine dioxide 116; nitrogen dioxide 117.

7.   Reaction with tetraphenylbenzidine:

MSA: Part 82399 for bromine, chlorine, chlorine dioxide. Part 83099 for nitrogen dioxide. Part 85833** for chlorobromomethane; 1,1-dichloroethane; dichloroethylene (cis-1,2 and trans-1,2); ethyl bromide; ethyl chloride; perchloroethylene (tetrachloroethylene); trichloroethylene; 1,2,3-trichloropropane; vinyl chloride (chloroethylene). Part 85834* for chlorobenzene (mono); 1,2-dibromoethane (ethylene dibromide); dichlorobenzene (ortho); 1,2-dichloroethane (ethylene dichloride); dichloroethyl ether; 1,1-dichloroethylene (vinylidine chloride); methyl bromide; methylene chloride (dichloromethane); propylene dichloride (1,2-dichloropropane); 1,1,2,2-tetrabromoethane; 1,1,2,2-tetrachloroethane; 1,1,3,3-tetrachloropropane; trichloroethane (beta 1,1,2); vinyl chloride (chloroethylene). Part 87042 for bromine; chlorine. Part 88536* for carbon tetrachloride; chlorobromomethane; 1-chloro-1,1-difluoroethane (Genetron 14213); chlorodifluoromethane (Freon 22); chloroform (trichloromethane); chloropentafluoroethane (Freon 115); chlorotrifluoromethane (Freon 13); 1,2-dibromoethane (ethylene dibromide); dichlorodifluoromethane (Freon 12); 1,1-dichloroethylene (vinylidine chloride); dichloroethylene (cis-1,2); dichlorotetrafluoroethane (Freon 114); fluorotrichloromethane (Freon 11); Freon 113; Freon 502; methyl bromide; methyl chloroform (1,1,1-trichloroethane); methylene chloride (dichloromethane); perchloro-

ethylene (tetrachloroethylene); trichloroethane (beta 1,1,2); trichloroethylene; 1,1,2-trichloro-1,2,2-trifluoroethane (Freon 113); trifluoromonobromomethane (Freon 13131). Part 91624** for acetonitrile; acrylonitrile; 1-chloro-l-nitropropane; cyanogen; 1,1-dichloro-l-nitroethane; dimethylacetamide; dimethylformamide; fumigants (Acritet, Insect-O-Fume, Fume-I-Gate, termi-Gas, Termi-Nate); methacrylonitrile; nitroethane; nitromethane, 1-nitropropane; 2-nitropropane; n-propyl nitrate; pyridine; vinyl chloride. Part 460225 for chlorine. Part 460424* for nitric oxide.

8.   Reaction of 2,4-dinitrophenylhydrazine forming a hydrazone:

MSA: Acetone 804141; methyl ethyl ketone 813334

9.   Oxidation by a iodate/sulfuric acid reagent:

MSA: Aromatic hydrocarbons 804132; benzene 807024, 804411.

10. Oxidation of an aromatic amine:

MSA: Chlorine 803944; chlorine dioxide 804133, nitrogen dioxide 487341, 804435; nitrous fumes 487336, 804425, 803946, 804426; trichloroethane 487342; vinyl chloride 803950.

11. Reaction with a lead or silver compound forming the sulfide or metal.

MSA: Hydrogen sulfide 487399, 487340, phosphine 497101, 485680, 489119

12. Reaction of chemical with copper iodide.

MSA: Mercury 497663.

13. Oxidation of indigotine to isatine.

MSA: Ozone 804140

14. Reaction of chemical with sulfuric acid:

MSA: Phenol 813778, styrene 804135.

15. Reaction with an aromatic aminoaldehyde and an aromatic amine.

MSA: Phosgene 803949.

16. Reaction of chemical with iodine, forming an iodide.

MSA: Sulfur dioxide 487338, 497662, SF_6 decomposition products 804433.

17. Precipitation of selenium from selenium/sulfuric acid.

MSA: Water vapor 488908.

* = Multiple reactions or multiple layer tube for improved specificity or preliminary reaction
** = Pyrolyzer required
HP = high pressure tube

breath, or to detect gasoline in soil in cases of suspected arson or leakage from underground tanks. Subsoil diffusion of volatile liquid contaminants can be inexpensively tracked.[27] One method of tracking subsoil diffusion involves hammering a Draeger-Stitz probe (a drill rod and probe tip) up to 6 meters deep into the soil at each sampling location. The rod is then withdrawn slightly to open the tip, and a detector tube in a capillary probe is lowered inside to the bottom to sample the vapor. It is connected through a capillary tube to a pump above ground. Alternatively, a charcoal tube or monitoring instrument probe may be inserted to sample the vapor. Minute quantities of ions in aqueous solutions also may be determined, such as sulfide in wastewater from pulp manufacturing, chromic acid in electrolytic plating wastewater, and nickel ion in wastewater of refineries.

Volatile contaminants in sewage or wastewater, such as ammonia, hydrogen cyanide, hydrogen sulfide, benzene, or chlorinated hydrocarbons, may be rapidly estimated with detector tubes. For this detection method, a 100-mL of sewage or wastewater sample is placed in a bubbler. One liter of air is then drawn through a train comprised of a charcoal tube, the bubbler, and a detector tube, which responds to the stripped contaminants.[28] Detector tubes were developed for the determination of aerosols such as oil, chromium (VI) oxide, cyanide, and sulfuric acid.[29] A special use of detector tubes for analyzing uranium hexafluoride hydrolysis products (uranyl fluoride and hydrofluoric acid) by utilizing hydrogen fluoride tubes has been reported.[30] Use of the detector tubes in the construction industry has been reported.[31]

Detector tubes have been widely advertised as being capable of use by unskilled personnel. Although it is true that the operating procedures are simple, rapid, and convenient, many limitations and potential errors are inherent in this method. The results may be dangerously misleading unless the sampling procedure is supervised and the findings are interpreted, preferably by an adequately trained occupational hygienist or by a professional with a science background. In the last few years, portable electronic instruments have been introduced by several manufacturers that can automatically do the air sampling and also the measurement, i.e., directly read the color changes, convert to ppb or ppm concentration and display it on an LCD screen, thus increasing accuracy and precision. Also, there are electronic units that can convert color changes on the passive badge samples and display their concentrations. Similarly, there are electronic colorimeter units that can convert color changes in passive bubblers and display them as concentration.[32]

Operating Procedures–Detector Tubes

Preparation for Sampling

The use of detector tubes is extremely simple. After its two sealed ends are broken open, place the glass tube in the manufacturer's holder which is fitted with a calibrated squeeze bellows or piston pump. Ascertain that the air flow direction is first through the tube to the pump. [Caution: use the same manufacturer's pump and tubes; do not mix different manufacturers' pumps and tubes; see further discussion section below under "Interchangeability of Detector Tube Brands."] Most detector tubes manufacturers have an arrow printed on the tube, indicating the direction that air should be entering the tube. The recommended air volume is then drawn through the tube by the operator. Adequate time must be allowed for each stroke. Even if a squeeze bellows is fully expanded, it may still be under a partial vacuum and may not have drawn its full volume of air. The manufacturer's sampling instructions must be followed closely.

Reading the Result

The observer then reads the concentration in the air by examining the exposed tube. Some of the earlier types of tubes were provided with charts of color tints to be matched with the solid chemical in the indicating portion of the tube. This visual judgment depended, of course, on the color vision of the observer and the lighting conditions. In an attempt to reduce the errors due to variations among observers, more recent types of tubes are based on a variable length of stain being produced on the indicator gel. There are a few tubes in which a variable volume of sample is collected until a standard length of stain is obtained; however, in most cases a fixed volume of sample is passed through the tube and the stain length is measured against a calibration scale. The scale may be printed either directly on the tube or on a provided chart.

The range in the interpretation of results by different observers is large because in many cases the end of a stain front is diffuse rather than sharp. Experience in sampling known concentrations is of great value in training an operator to know whether to measure the length at the beginning or end of the stain front or at some other portion of an irregularly shaped stain. In some cases, the stains change with time; thus, reading the stain length should not be unduly delayed.

Testing the Pump

Care must be taken to see that leak-proof pump valves and connections are maintained. A leakage test to ensure adequate performance may be made by inserting an unopened detector tube into the holder and squeezing the bulb; at the end of two minutes, any appreciable bulb expansion is evidence of a leak. If the apparatus is fitted with a calibrated piston pump, the handle is pulled back and locked. Two minutes later, it is released cautiously and the piston is allowed to pull back in; it should remain out no more than 5% of its original distance. Leakage indicates the need to replace check valves, tube connections, or the squeeze bellows or to grease the piston.

Flow Test

At periodic intervals, the flow rate of the apparatus should be checked and maintained within specifications for the tube calibrations (generally ± 10%). This may be done simply by timing the period of squeeze bellows expansion. A more accurate method is to place a used detector tube in the holder and draw an air sample through a calibrated rotameter. Alternatively, the air may be drawn from a burette in an inverted vertical position, which is sealed with a soap film, and the motion of the film past the graduations can be timed with a stop watch.[33] The latter method also provides a check on the total volume of the sample which is drawn. In some devices, the major resistance to the air flow is in the chemical packing of the tube; thus, each batch might require checking. An incorrect flow rate may indicate a partially clogged strainer or orifice that should be cleaned or replaced.

With most types of squeeze bellows and hand pumps, the sample air flow rate is variable, being high initially and low toward the end when the bulb or pump is almost filled. This variability has been claimed to be an advantage because the initially high rate gives a long stain and the final low rate sharpens the stain front. If the concentration is rapidly fluctuating during the sampling period, the variable flow rate will cause the reading to be an inaccurate mean value. The stain lengths may depend more on flow rate than on concentration. Colen[34, 35] found that flow patterns for six commonly used pumps were different. When five popular brands of carbon monoxide tubes were used with pumps other than their own, grossly erroneous results (with 268% variability) were observed, even with identical sample volumes. It should be noted that accuracy requires a close reproduction of the flow rate pattern for the calibrations to be correct.

Interchangeability of Detector Tube Brands

Interchanging brands can cause large errors that may not be apparent to the user. This practice could result in erroneous concentration levels of the toxic gas being measured. The detector tube and its pump form a system that is calibrated at the factory. A new calibration is conducted with each individual production lot of tubes.[36] The pump must provide the correct volume of air and provide a flow rate curve comparable to that used at the factory calibration in order for the calibration marks placed on that lot of tubes to remain valid.

Because all five major brands of detector tubes use a 100 cc pump, many people reason that the systems must be interchangeable. The fallacy of this reasoning lies in the fact that the rate of airflow also affects stain length. The target gas and the reagent in the analyzer layer of the tube must have adequate contact time to react with one another. If something is done to speed up the flow rate, the contact time is decreased and a smaller percentage of the reagent will react. The target gas will travel farther into the tube before it is consumed. The result is a stain that is paler in color but longer than the stain produced during factory calibration at that particular gas concentration. The typical user reads the length of the stain and not the color intensity; thus, the tube will be reading high, producing a loss in system accuracy.

Conversely, if the flow rate is slowed, the contact time is increased and a higher percentage of the reagent layer will react than was reacted at the factory calibration. This produces a shorter stain that is darker in color than the intended stain for that concentration. Now the tube is reading low, a potentially dangerous situation in applications like confined space entry testing, where a hazardous condition could be misconstrued as being safe.

Detector tube brands control flow rate in a variety of ways. Some systems use an orifice in the pump to slow the flow rate and produce the desired target gas/reagent-layer contact time. Other systems use densely packed detector tubes to provide the resistance. Interchanging systems that use these opposite flow restrictions can produce very wide accuracy swings. A nonrestricted tube with a non-restricted pump produces a flow rate that is much too fast and can produce high readings. A restricted pump mated with a restricted tube can slow the flow rate significantly below the tube's intended flow rate and produce low readings.

Even similarly restricted designs can produce faulty flow rates if interchanged. For example, piston pumps produce a much higher initial vacuum than bellows pumps. Even if two systems use tube restriction to control flow rate, the restrictions can be drastically different to accommodate the different pump vacuum levels intended for each brand of tube. Various agencies and organizations, both domestic and international, have been advocating the non-interchangeability of brands: the American Industrial Hygiene Association (AIHA),[22] National Institute for Occupational Safety and Health (NIOSH),[34] Occupational Safety and Health Administration (OSHA),[37] International Union of Pure and Applied Chemistry (IUPAC),[38] American National Standards Institute/Industrial Safety Equipment Association (ANSI/ISEA),[39] and Safety Equipment Institute (SEI).[40] The SEI certification on detector tubes is a system-certification that includes the pump and tube as a unit. Interchanging of components of different brands voids the certification. Therefore, <u>do not mix bellows pump/tube systems with piston pump tube systems</u>. However, there have been recent discussions and tests done for the interchangeability of the same piston-pump designs. A study conducted by a university laboratory for RAE Systems using their LP-1200 hand pump against Gastec/Sensidyne (GV/100) and Kitagawa (8014-400A) hand pumps indicated that these three manufacturers' gas detection tubes gave the same readings within 20% of the standard gas values (results can be viewed at *www.raesystems.com/tn129.htm*). The maximum deviation between any two pumps on a given tube was 10%. The conclusion was independent of the types of gases measured (10 different gases, total of 25 different gas concentrations in triplicate).

Special Air Sampling Techniques

Hard-to-Reach Areas. A number of special techniques may be used in appropriate cases. When sampling in inaccessible places, the indicator tube may be placed directly at the sampling point and the pump operated at some distance away. A rubber or Tygon tube extension of the same inside diameter as the indicator tube may be inserted between the pump and indicator tube. Such tubes are available commercially as accessories. Lengths as great as 60 ft have been successfully used without appreciable error, provided that more time is allowed between strokes of the pump to compensate for the reservoir effect and to obtain the full volume of sample. This method has the disadvantage that the detector tube cannot be observed during the sampling.

Hot-Temperature Areas. A second arrangement may be used when sampling hot gases such as from a furnace stack or engine exhaust. Cooling the sample is essential in these cases; otherwise, the calibration would be inaccurate and the volume of the gas sample uncertain. A probe of glass or metal, available commercially as an accessory, may be attached to the inlet end of the detector tube with a short piece of flexible tubing.[41] If this tube is cold initially, as little as 10 cm of tubing outside of the furnace is sufficient to cool the gas sample from 250°C to about 30°C. Such a probe has to be employed

with caution. In some cases, serious adsorption errors occur either on the tube or in condensed moisture. The dead volume of the probe should be negligible in comparison to the volume of sample taken. Solvent vapors should not be sampled by this method. When sampling air colder than 0°C, clasping the tube in the hand warms it sufficiently to eliminate any error.[41] Critical studies[42, 43] of applications for analysis of diesel exhaust showed serious errors for some tubes.

Other special techniques also may be employed. Combustion and decomposition gases from a fire may be rapidly evaluated with a set of five different tubes in a special holder that allows simultaneous sampling using one pump. These systems require special calibration. Some symmetrical tubes can be reversed in the holder and used for a second test. In certain special cases, tubes may be reused if a negative test was previously obtained; however, this should be done with caution and only after reading the manufacturer's recommendations for reuse of detector tubes. Two tubes also may be connected in series in special cases; e.g., first passing crude gas through a Kitagawa hydrogen sulfide tube and then through a phosgene tube to obtain two simultaneous determinations and remove interferences. These techniques may be used only after testing verifies that they do not impair the validity of the results.

High-Pressure Areas. Tubes also have been used for sampling in pressures as high as several atmospheres. This situation would exist, for example, in underwater stations. If both the tube and pump are in the chamber, the calibrations and sample volumes both may be altered. It has been reported[41, 44] that only the latter occurs for the following Draeger tubes: ammonia 5/a, arsine 0.05/a; carbon dioxide 0.1%/a; carbon monoxide 5/c,10/b; and hydrogen sulfide 1/c, 5/b. For these tubes, the corrected concentration is equal to the scale reading (ppm or vol %) divided by the ambient pressure (in atmospheres) at the pump. When tube tips are broken in a pressure chamber, the tube filling should be checked for possible displacement.

Specificity and Sensitivity of Detector Tubes
Multiple Uses

The specificity of the detector tubes is a major consideration for determining applicability and interpreting results. Most tubes are not specific. Chromate reduction is a common reaction used in tubes for detection of organic compounds. In the presence of mixtures, the uncritical acceptance of such readings can be grossly misleading. Comprehensive listings of reactions, as well as a discussion of other major aspects, are available.[18–21, 45] Seventeen common reactions and the associated tube types are listed in Table 17-1. The name of the compound listed on the tube often refers to its calibration scale rather than to a unique chemical reaction of its contents. [*Caution:* The detector tube may be useful for several compounds, but it is calibrated only for one, e.g., the aromatic hydrocarbons tube is calibrated for benzene.]

The lack of specificity of some tubes may be used to advantage for detection of substances other than those indicated by the manufacturer. In this respect, tubes using colorimetric reactions 1, 2, 6, and 7 (Table 17-1) are widely applicable. Thus, the Draeger Polytest screening tube (reaction 2) and ethyl acetate tube (reaction 1) may be used for qualitative indications of reducing and organic materials, respectively.[46] The Draeger trichloroethylene tube (reaction 6) is also applicable to chloroform, o-dichlorobenzene, dichloroethylene, ethylene chloride, methylene chloride, and perchloroethylene. The methyl bromide tube may be used for chlorobromomethane and methyl chloroform. The chlorine tube may be used for bromine and chlorine dioxide. The toluene tube may be used for xylene. Such use requires specific knowledge of the identity of the reagent and of the proper corrections to the calibration scales.

For some brands of indicator tubes, the units of the calibration scales are in milligrams per cubic meter. Although it has been said that this method of expression eliminates the necessity of making temperature and pressure corrections, such a claim is debatable because the scale calibrations themselves may be highly dependent on these variables. Units of parts per million or percent by volume are most common for industrial hygiene purposes and are used on most of the newer tubes. Conversions may be made from milligrams per cubic meter to parts per million by using the formula below (explained in Chapter 8, Calibration of Gas and Vapor Samples).

$$\text{ppm} = [(\text{mg/m}^3) \cdot 24.45] / \text{MW}$$

where MW = molecular weight

Low Levels of Detection

Although detector tubes are generally designed for detection of relatively high gas concentrations found in industrial workplaces, some have been applied to much lower outdoor air pollutant concentrations. Kitagawa[47] determined 0.01–2 ppm of nitrogen dioxide using two glass tubes in series, with the temperature controlled at 40°C. The first tube contained diatomaceous earth impregnated with a specific concentration of sulfuric acid to regulate the humidity of the air sample. The sec-

ond tube, 120 mm long × 2.4 mm inside diameter, contained white silica gel impregnated with o-tolidine (now available as Matheson-Kitagawa commercial tube no. 8014-117SB, has a range of 0.5–30 ppm, SEI-certified.) Air was drawn through the tubes for 30 min at 180 mL/min by an electric pump with a stainless steel orifice plate at its inlet. Grosskopf[48, 49] determined 0.007–0.5 ppm nitrogen dioxide by drawing air through a Draeger 0.5/a nitrous gas tube with a diaphragm pump for 10–40 min at the rate of 0.5 L/min. Readings were not affected by flow rates if the flow rates exceeded 0.5 L/min. No comments were given on the specificity, except that humidity from 30 L of air at 70% relative humidity did not impair the sensitivity. This tube responds to nitric oxide and to oxidants, both of which commonly may be present. Leichnitz[50] reported a tube (Draeger sulfur dioxide 0.1/a) capable of measuring 0.1–3 ppm of sulfur dioxide. This tube requires 100 strokes of a hand bellows pump (each taking 7–14 seconds) or use of the Draeger Quantimeter electric pump in which a motor-driven crank controlled by a microprocessor operates a bellows. This pump is described in the "Instrument Descriptions" section at the end of this chapter.

Less success was attained when carbon monoxide detector tubes were used for sampling periods of 4 hrs or longer with continuous pumps. It was found that at low concentrations, after an initial period, the stain lengths ceased to increase.[8] However, at higher concentrations, a new calibration could be made[51] (for 3- to 5-hr samples at 8 mL/min through a Kitagawa 100 tube in the range of 30–100 ppm of carbon monoxide). The latter investigator hypothesized that the oxygen in air bleached the black palladium stain and caused the front of stain produced by low concentrations to remain stationary after the first 20–30 min. Effects of water vapor and other contaminants must also be considered in this application. A new calibration is essential under the flow conditions to be used. Studies confirmed that secondary reactions, which bleached the indication and prevented long-term sampling, could be avoided with appropriate reagent systems.[52]

Long-Duration Sampling Tubes

Detector tubes have been developed[53, 54] for long duration sampling (4–8 hrs). These appear to be very similar to the tubes designed for short duration sampling and are effective within the same concentration ranges. They are calibrated for use with a continuous sampling pump, but they operate at lower flow rates. The application of these tubes is to provide time-weighted average concentrations rather than short-term (few minutes) values. To provide valid averages, the calibrations must be linear both with concentration and time and should display uniformly spaced markings for uniform increments of contaminants. The scales on these tubes usually are in terms of microliters of test gas (ppm × liters), rather than ppm, and the latter is calculated by dividing the scale reading by the liters of air sampled. Over 30 types of long-duration tubes are now available commercially. It should be noted that they must be used within the ranges of flow rate and total sampling time established during their calibration by the manufacturer, using the specified continuous sampling pump. A comparison of grab samples and long-term samples for ammonia in swine confinement buildings showed a consistent discrepancy; the long-term values were double the grab sample averages.[55] In other studies, low flow MDA Accuhaler pumps were used, with some loss of accuracy.[56] They generally are not suitable for analysis of concentrations in ranges lower than those of ordinary tubes designed for short duration sampling[57] because of the previously mentioned problems of water vapor, oxygen, and other contaminants.

Increasing Accuracy of Detector Tubes

Greater accuracy can be obtained when several detector tubes are used for replicate sampling. A simplified statistical approach based on an assumed normal distribution of values was recommended for taking 3–10 samples.[58] However, subsequent work indicated that most of the variations were due to the environmental fluctuations rather than to the relatively small analytical errors, and that a lognormal distribution was more appropriate. A step-by-step procedure was presented[59] which categorized the results into noncompliance (less than 5% chance of erroneously citing when actually compliance exists), no decision, and compliance (less than 5% chance of failing to cite when actually noncompliance exists).

Problems in the Manufacture of Detector Tubes

Color-Indicating Gel

The accuracy, limitations, and applications of detector tubes are highly dependent on the skill with which they were manufactured.[60] Generally, the supporting material is silica gel, alumina, ground glass, pumice, or resin. This material is impregnated with an indicator chemical that should be stable, specific, sensitive, and produce a color which strongly contrasts with the unexposed color and is non-fading for at least an hour. If the reaction with the test gas is relatively slow, a color is produced throughout the length of the tube because the gas is incompletely adsorbed and the concentration at the exiting end is an appreciable fraction of that at the

entrance. Such a color must be matched against a chart of standard tints. A rapidly reacting indicating chemical is much more desirable and yields a length-of-stain type of tube in which the test gas is completely adsorbed in the stained portion.

There is a very wide and unpredictable variation in the properties of different batches of indicating gel. The major portion of the chemical reaction probably occurs on the surface. Therefore, the number of active centers, which are highly sensitive to trace impurities, affect the reaction rate. These problems are well known in the preparation of various catalysts. Close controls must be kept on the purity and quality of the materials, the method of preparation, the cleanliness of the air in the factory or glove box in which the tubes are assembled, the inside diameter of the glass tubes, and even on the size analysis of the impregnated gel, which, in some cases, is important in controlling the flow rate. The manufacturer must also accurately calibrate each batch of indicating gel.

Some tube types are constructed with multiple layers of different impregnated gels with inert separators. Generally, the first layer is a precleansing chemical to remove interfering gases and improve the specificity of the indication. Thus, in the case of some carbon monoxide tubes, chemicals are provided to remove interfering hydrocarbons and nitrogen oxides. In carbon disulfide tubes, hydrogen sulfide is first removed. In hydrogen cyanide tubes, hydrogen chloride or sulfur dioxide are removed first. In other cases, the entrance layer provides a preliminary reaction essential to the indicating reaction. Thus, in some trichloroethylene tubes, the first oxidation layer liberates a halogen which is indicated in the subsequent layer. In some tubes for NO_x gases, a mixture of chromium trioxide and concentrated sulfuric acid is used to oxidize nitric oxide to nitrogen dioxide, which is the form to which the sensitive indicating layer responds. Although such multiple layer tubes are advantageous when properly constructed, they frequently have a shorter shelf life because of diffusion of chemicals between layers and consequent deterioration.

Shelf-Life Factors

A shelf-life of at least two years is highly desirable for practical purposes. A great deal of disappointment with various tube performances is no doubt due to inadequate shelf-life. Because some tubes have only been on the market for a short time, the manufacturer may have inadequate experience with the shelf-life of the product. Small variations in impurities, such as the moisture content, may have a large effect on the shelf-lives of different batches. The storage temperature, of course, greatly

TABLE 17-2. Shelf Life of Draeger Carbon Monoxide Detector Tubes

Temperature		
°C	°F	Shelf Life
25	77	>2 years
50	122	>1/2 year
80	176	weeks
100	212	1 week
125	257	3 days

[Note: above data may vary or may not be valid with other types of detector tubes or other brands. These data plot as an approximately straight line when the logarithm of the shelf-life time is plotted against a linear scale of the reciprocal of absolute temperature. Such a plot is usual for the reaction rate of a simple chemical reaction. In other cases, relationships may be more complex.]

affects the shelf-life, and it is highly desirable to store these tubes in a refrigerator or in a cool dark place. The accuracy of tubes stored on the back window shelf of a car may be rapidly destroyed by hot sunlight. In some cases, shelf-life has been estimated by accelerated tests at higher temperatures. Such a variation of shelf-life (length of time within which the calibration accuracy is maintained at ± 25%) is illustrated by the data listed in Table 17-2 received in a personal communication from Dr. Karl Grosskopf of the Draeger Company before 1965 and reconfirmed with Dr. Kurt Leichnitz of the Draeger Company in 1999, however, these data were not published.

Shipping

The shipping properties of tubes must also be controlled carefully. Loosely packed indicating gels may shift, causing an error in the zero point of scales printed directly on the detector tube, as well as an error in total stain length. When the size analysis indicates an appreciable range of fine particles, the fines may segregate during shipment to one side of the bore, causing different flow resistances and rates on each side of the tube. This may cause oval stain fronts that are not perpendicular to the tube bore. If the indicating gel is friable, the size analysis may change during shipping.

Obviously, satisfactory results can be obtained only if the manufacturers take great pains in the design, production, and calibration of tubes.

Theory of Calibration Scales

Up to now, calibration scales have been entirely empirical. The variables that can affect the length of stain are concentration of test gas, volume of air sample, sampling flow rate, temperature, and pressure, and a number of factors related to tube construction. There is a striking similarity in the fact that most of the length of stain calibration scales are logarithmic with respect to concentration in spite of the widely differing chemicals employed in different tube types. Although very few data are available for these relationships, a basic mathematical analysis was made by Saltzman.[61] The theoretical formulas discussed below will, of course, have to be modified as more data become available. The relationships were also studied by Grosskopf[49] and Leichnitz.[62]

In the usual case, although the test gas is adsorbed completely, equilibrium is not reached between the gas and the adsorbing indicator gel because the sampling period is relatively short and the flow rate is relatively high. The length of stain is determined by the kinetic rate at which the gas either reacts with the indicating chemical or is adsorbed on the silica gel. The theoretical analysis shows that the stain length is proportional to the logarithm of the product of gas concentration and sample volume:

$$L/H = ln\ (CV) + ln\ (K/H) \qquad (1)$$

where:
L = the stain length, cm
C = the gas concentration, ppm
V = the air sample volume, cm3
K = a constant for a given type of detector tube and analyte gas, cm^{-2}
H = a mass transfer proportionality factor having the dimension of centimeters, and known as the height of a mass transfer unit

The factor H varies with the sampling flow rate raised to an exponent of between 0.5 and 1.0, depending on the nature of the process that limits the kinetic rate of adsorption. This process may be diffusion of the test gas through a stagnant gas film surrounding the gel particles, the rate of surface chemical reaction, or diffusion in the solid gel particles. If the detector tube follows this mathematical model, a plot of stain length, L, on a linear scale, versus the logarithm of product CV (for a fixed constant flow rate) will be a straight line of slope H. It is important to control the flow rate because it may affect stain lengths more than gas concentrations due to its influence on the factor H.

If larger volume samples are taken at low concentrations and the value of L/H exceeds 4, the gel approaches equilibrium saturation at the inlet end, and calibration relationships are modified. The solution to the equations for this case has been presented graphically by Saltzman[61] in a generalized chart. However, there is little advantage to be gained in greatly increasing the sample size, because the stain front is greatly broadened and various errors are increased.

For some types of tubes such as hydrogen sulfide and ammonia and for long duration tubes, the reaction rate is fast enough that equilibrium can be attained between the indicating gel and the test gas. Under these conditions, there is a stoichiometric relationship between the volume of discolored indicating gel and the quantity of test gas adsorbed. In the simplest case, the stain length is proportional to the product of concentration and volume sampled:

$$L = K C V \qquad (2)$$

If adsorption is important, the exponent of concentration may differ from unity:

$$L = K C^{(1-n)} V \qquad (3)$$

The value of n is the same as that in the Freundlich isotherm equation for equilibrium adsorption, which states that the mass of gas adsorbed per unit mass of gel is proportional to the gas concentration raised to the power n. If the value of n is unity, which is not unusual, Equation 3 indicates that stain length is proportional to sample volume but is independent of concentration. The physical meaning of this is that all concentrations of gas are adsorbed completely by a fixed depth of gel. Such a tube is obviously of no practical value.

Equilibrium conditions may be assumed for a given type of indicator tube if stain lengths are directly proportioned to the volume of air sampled (at a fixed concentration) and are not affected by air sampling flow rate. A log-log plot then may be made of stain length versus concentration for a fixed volume. A straight line with a slope of unity indicates that Equation 2 applies; if another value of slope is obtained, Equation 3 applies.

In some of the narrower indicator tubes, manufacturing variation in tube diameters produces an appreciable percentage variation in tube cross-sectional areas. This results in an error in the calibration as high as 50% because the volume of sample per unit cross-sectional areas is different from that under standard test conditions. An additional complicating factor is the variation produced in flow rate per unit cross-sectional area. If an exactly equal quantity of indicating gel is put into each tube, variations in cross-sectional areas will be indicated by corresponding variations in the filled tube lengths. Correction charts are provided by one manufacturer on which the tube is positioned according to the filled

length and a scale is given for reading stain lengths. Although the corrections are rather complex, practically linear corrections are very close approximations that can reduce the errors to 10%. In most tubes, the tube diameters are controlled closely enough that no correction is necessary.

Temperature is another important variable for tube calibrations. The effect is different for different tubes. Because the color tint type of tube depends on the degree of reaction, it is most sensitive to temperature. For example, some old types of carbon monoxide tubes require correction by a factor of 2 for each deviation of 10°C from the standard calibration conditions.

Errors in judging stain lengths produce equal percentage errors in concentration derived from the calibration scale. Errors in measuring sample volume and in flow rate may also result in errors in the final value, although the exact relationships might vary according to the tubes.

Many other complications can be expected in calibration relationships. Thus, for nitrogen dioxide, the proportion of side reactions is changed at different flow rates. Changing sample volumes freely from calibration conditions is not recommended unless the tube is known to be thoroughly free from the effects of interfering gases and humidity in the air.

A crucial factor in the accuracy of the calibration is the apparatus used for preparing known low concentrations of the test gas (this subject is discussed more fully in "Calibration of Gas Vapor Samples" in Chapter 8). Some manufacturers have used static methods for calibration. However, experience has shown that losses of 50% or more by adsorption are not uncommon. Low concentrations of reactive gases and vapors are best prepared in a dynamic system. This has the further advantages of capabilities to generate extended volumes and to rapidly change concentrations as required. With either type of apparatus, it is highly desirable to check the concentrations using chemical methods of known accuracy. Some successful systems have been described.[63–69]

A simple and compact dynamic apparatus for accurately diluting tank gas (which may be either pure or a mixture) was developed by Saltzman[64, 65] and Avera.[66] The asbestos plug flowmeter measures and controls gas flows in the range of a few hundredths to a few milliliters per minute (note: because asbestos is no longer available, a similar inert packing material may be substituted). Air-vapor mixtures of volatile organic liquids may be prepared in a flow dilution apparatus using a motor-driven hypodermic syringe. High quality gears, bearings, and screws are needed in the motor drive to provide the uniform slow motion. Some commercial devices have been found unsatisfactory in this regard. Many types of permeation tubes now available have also been proven useful.

It is highly desirable for the user, as well as the manufacturer, to have facilities available for checking calibrations. Only in this manner may the user be confident that the tubes and corresponding technique are adequate for the intended purposes. Tubes may also be applied to gases other than those for which they have been calibrated by the manufacturer, in certain special cases, if the user can prepare a new calibration scale.

Performance Evaluation and Certification of Detector Tubes

Background

Evaluations by users of many types of tubes have been reported.[15, 72-95] Temperature and humidity were found to be significant factors in some cases.[96, 97] Accuracy was found to be highly variable. In some cases, the tubes were completely satisfactory; in others, completely unsatisfactory. Manufacturers, in their efforts to improve the range and sensitivity of their products, are rapidly changing the contents of their tubes, and these reports are frequently obsolete before they appear in print. Improved quality control, and perhaps greater self-policing of the industry, would greatly increase the value of the tubes, especially for the small consumer who is not in a position to check calibrations.

ACGIH Recommendations for Certification

After reviewing this need, a joint committee of the American Conference of Governmental Industrial Hygienists (ACGIH)—and AIHA in 1971 made the following recommendations:[98]

1. Manufacturers should supply a calibration chart (ppm) for each batch of tubes.
2. Length-of-stain tubes are preferable to those exhibiting change in hue or intensity of color.
3. Tests of calibrations should be made at 0.5, 1.0, 2.0, and 5.0 times the ACGIH Threshold Limit Value (TLV®).
4. The manufacturer should specify the methods of tests. Values should be checked by two independent methods.
5. Calibration at each test point should be accurate within ±25% (95% confidence limit).
6. Allowable ranges and corrections should be listed for temperature, pressure, and relative humidity.

7. Each batch of tubes should be labeled with a number and an expiration date. Instructions for proper storage should be given.
8. Tolerable concentrations of interferents should be listed.
9. Pumping volumes should be accurate within ±5%, and flow rates should be indicated.
10. Special calibrations should be provided for extended sampling for low concentrations, and flow rates should be specified.

Former NIOSH Certification Program

A performance evaluation program was initiated in the early 1970s by NIOSH. Known concentrations of test substances were generated in flow systems from sources such as cylinder mixtures, vapor pressure equilibration at known temperatures, or permeation tubes. Although few tubes achieved an accuracy of ±25%, many types showed accuracies in the range of ±25% to ±35%.[79-85]

A formal certification program [99, 100] was the next step. In addition to passing performance evaluation tests at the Morgantown, West Virginia, NIOSH laboratory, manufacturers were required to provide information on the contents of the tubes and to conduct a specified quality control program. Because of the dependence of the calibrations on the pumps used with the tubes, certifications were periodically updated and issued[101] for specified combinations of tubes and pumps. By 1981, tubes of four manufacturers for 23 contaminants had been certified. Unfortunately, the program was terminated in 1983 for lack of funding.[102]

The requirements for certification generally followed the recommendations of the joint ACGIH/AIHA committee. However, the accuracy requirement was modified to ±35% at 0.5 times the ACGIH TLV and ±25% at 1.0, 2.0, and 5.0 times the TLV, to be maintained until the expiration date, if the tubes were stored according to the manufacturer's instructions. At the TLV concentration, either the stain length had to be 15 mm or greater, or the relative standard deviation of the readings of the same tube by three or more independent tube readers had to be less than 10%. If the stain front was not exactly perpendicular to the tube axis (because of channeling of the air flow), the difference between the longest and shortest stain length measurements to the front had to be less than 20% of the mean length. Color intensity tubes had to have sufficient charts and sampling volume combinations to provide scale values including at least the following multiples of the TLV: 0.5, 0.75, 1.0, 2.0, 2.5, 3.0, 4.0, and 5.0; the relative standard deviation for readings of a tube by independent readers had to be <10%. Tests were to be conducted generally at 18.3°–29.5°C (65°–85°F) and at relative humidity of 50%, unless the humidity had to be reduced to avoid disturbing the test system. The manufacturer had to file a quality control plan and keep records of inspections of raw materials, finished tubes, and calibration and test equipment. Acceptable statistical quality levels for defects in finished tubes were as follows: critical 0% where tests were nondestructive; otherwise, 1.0%, major 2.5%, minor 4.0%, and accuracy 6.5%. Typical statistical calculations have been described.[103] Certification seals were affixed to approved devices. NIOSH reserved the right to withdraw certification for cause.

National and International Certification Programs

Because important legal and economic consequences depend on the accuracy of measurements of contaminant concentrations, enforcement agencies will most likely prefer certified equipment. Standards for detector tubes have been issued by more than 25 organizations,[104] including OSHA,[105] IUPAC,[106] Japanese Standards Association,[107] British Standards Institution,[108] ANSI,[109] the International Standards Organization (ISO),[110, 111] European Committee for Standardization (ECS/CEN),[112] France, the Soviet Union, the Council of Europe, and a variety of private organizations in the United States and Europe. Requirements are mostly similar to those cited above. A comparison study of the detector tube method versus European Standards has been reported. [113]

Certification by the Safety Equipment Institute

In 1986, the Safety Equipment Institute (SEI) announced a voluntary program for third-party certification of detector tubes.[114] Manufacturers submit tubes for testing as the schedule for each type is announced. Two AIHA-accredited laboratories were selected to evaluate the tubes according to the NIOSH protocol.[99] Another contractor makes onsite, quality assurance audits of manufacturing facilities every 6 months for three audits, and then annually. If the tubes meet all requirements, the manufacturer may apply the SEI certification mark. This program should provide a stimulus for greater acceptance and use and for further improvements in detector tube technology. Tubes will be retested every 3 years. Table 17-3 gives the current listing of certified tubes.[115] Types of tubes for more substances are currently in the testing process. For the latest information on SEI-certified detector tubes contact SEI at their Internet homepage at *http://www.SEInet.org*.

Passive Dosimeters (Stain Length)

Background

An important new advance has been the development of direct-reading, passive dosimeters. Passive dosimetry uses diffusion of the test gas and eliminates the need for a sampling pump and its calibration. These attractive devices are compact, convenient, and relatively inexpensive. (Description of lab-analyzed passive dosimeters is found in Chapter 16). In early work, detector tubes for toluene, ethanol, and isopropanol were cut open at the entrance of the chemical packing.[116] Later, glass adapters with a membrane (e.g., Millipore®, or silicone rubber) were used [117–119] to provide a draft shield, in some cases a pretreatment chemical layer, and a diffusion resistance. Simpler commercial devices merely provided a score mark which permitted breaking the tube at a controlled point.[120] Some allowed a controlled air space (e.g., 15 mm) upstream from the indicating gel to serve as the initial resistance to diffusion.[121] In some devices, rather than an indicating gel, a strip of chemically impregnated paper is inserted in the glass tube.

Theory

The theoretical calibration relationships for these devices rest upon Fick's First Law of Diffusion, which can be expressed as:

$$W = 10^{-6} \, C \, t \, D \, (A/X) \qquad (4)$$

where:
W = cm^3 of analyte gas collected
t = time, seconds
D = diffusion coefficient, cm^2/s
A = effective orifice cross-section area, cm^2
X = orifice length, cm

This equation assumes that the gas is completely adsorbed in the indicating gel and that there is no significant back diffusion pressure. A second common assumption is that the stain length is proportional to the amount adsorbed (analogous to Equation 2):

$$L = K \, W \qquad (5)$$

The test gas diffuses through a membrane or air space, then through the stained length of indicating gel, and is finally adsorbed at the stain front, which is assumed to be relatively narrow. It is convenient to express X in terms of L:

$$X = r + L \qquad (6)$$

TABLE 17-3. Certifications of Detector Tubes by the Safety Equipment Institute (SEI)[A] as of May 2000

Substance	Matheson/ Kitagawa	Mine Safety Appliances Co.	Draeger Safety, Inc.	Sensidyne/ Gastec	TLV/TWA (ppm)
Acetic acid	8014-216S	—	—	81	10
Acetone	8014-102SD	—	—	151L	750
Ammonia	8014-105SC	—	—	3La	25
Benzene	8014-118SC	—	—	121	10
Carbon dioxide	8014-126SA	—	—	2L	5000
Carbon disulfide	8014-141SB	—	—	13	10
Carbon monoxide	8014-106S	—	—	1La	50
Chlorine	8014-109SB	—	—	8La	1.0
Hydrogen chloride	8014-173SB	—	—	14L	5 [c]
Hydrogen cyanide	8014-112SB	—	2/a, CH25701	12L	10 [c]
Hydrogen sulfide	8014-120SD	487339	—	4LL	10
Methyl bromide	8014-157SB	—	—	136La	5
Nitric Oxide	8014-175U	—	—	10	5
Nitrogen dioxide	8014-117SB	—	0.5/c, CH30001	9L	3
Ozone	8014-182U	—	—	18L	0. 1 [c]
Phosphine	8014-121U	—	—	7La	0.3
Sulfur dioxide	8014-103SE	—	—	5Lb	2
Toluene	8014-124SA	—	—	122	100
Trichloroethylene	8014-134S	—	—	132M	50
Vinyl chloride	8014-132SC	—	—	131La	5
Xylene (pure)	8014-143S	—	—	—	100
(commercial grade)	—	—	—	123	—
Pump model[B]	8014-400A	—	—	—	—
Kwik draw	—	488543	—	—	—
Kwik draw deluxe	—	487500	—	—	—
Bellows pump, model 31	—	—	67 26065	—	—
Gastek 800 pump	—	—	—	GV-100	—
Gastek GV-100	—	—	—	GV-100S	—

[A] For up-to-date information, contact SEI on the web at *http://www.SEInet.org* or at (703) 442-5732

[B] Tubes are certified only when used with specified pump model of same manufacturer.

[c] = TLV Ceiling value

where: r = effective length, cm, corresponding to the diffusive resistance of the membrane or air space

Combining Equations 4–6 and rearranging yields:

$$r L + L^2 = (10^{-6} K D A) C t = K' C t \qquad (7)$$

where: K' = a constant equal to the bracketed expression, cm²/s

This equation has been shown to fit MSA tubes with a 15-mm air space.[121–123] When L was expressed in mm and t in hrs, r was taken as 15 mm, and K' was 0.59 mm²/hr for carbon monoxide, 11.0 for ammonia, 14.2 for nitrogen dioxide, 22.6 for hydrogen sulfide, 67.3 for sulfur dioxide, and 74.0 for carbon dioxide. For Draeger tubes, which do not use an air space, the equation applied with a zero value for r.[120] For membrane-type devices, the equation was modified by adding another constant: [117–119]

$$C t = a + b L + c L^2 \qquad (8)$$

where: a, b, c = empirical constants

These constants may differ for each individual membrane. The inapplicability of a general calibration is a disadvantage of this type of passive monitor.

A more complete mathematical analysis[124] showed that for rapidly changing concentrations the errors would be small. This was experimentally confirmed[125] for both passive dosimeters and for long-term tubes. Most of the published work on passive dosimeters has been studied by the staffs of manufacturers.[120–123, 125] Much larger errors were reported[126, 127] by users. Some of the stain boundaries were very diffuse and difficult to read, and some calibrations were inaccurate. Because these tubes are in an early state of development, the values should be checked as much as possible.

Colorimetric Badges

Background and Use

Another type of passive dosimeter is the direct-reading colorimetric badge. These dosimeters rely on the principle of diffusion. (Description of the theory of diffusive sampling can be found earlier in this chapter under the heading of "Passive Dosimeters", and also in Chapter 16). Colorimetric badges are extremely easy to use. Once the badge is exposed to a specific chemical gas or vapor, its reagent-coated film reacts and forms a color tint (stain) that is related to the product of time and concentration. The color is compared against known color standards. The ppm result is simply calculated by dividing the ppm • hours by the hours of sampling (exposure) time. Automated electronic colorimetric badge readers are available, see instrument descriptions in sections 17-1, 17-3, 17-11, and 17-33. Badges are utilized for short-term (STEL) monitoring or for 8-hour shift (TWA). Badges for low-level detection down to parts-per-billion are available for toluene diisocyanate (TDI), hydrazine, hydrides, and phosgene. Tables 17-5 through 17-9, 17-19, and 17-27 contain the instrument descriptions. Short- term sampling with low-level detection of ozone and formaldehyde are described in sections 17-16 through 17-18. Additional applications of colorimetric badges and instrument descriptions are found in sections 17-11, 17-16 through 17-19, 17-23 through 17-24, 17-27, and 17-33. All passive devices require a minimum air velocity at their entrances (0.008 m/s or 15 ft/min) to avoid "starvation" effects (depletion of the air concentration near the entrance).[70, 71]

Use Problems

When using a colorimetric badge for long-term sampling, such as for several days, it may result in an error of accumulation indicating higher than actual results. The solution to this problem is to change badges as often as possible. Some badges may "regenerate" after sampling if they are removed from the contaminated environment to a "clean" environment; they may fade over time and return to their original color. Re-use of an exposed badge should be discouraged because the level of sensitivity could not be guaranteed.

Conclusions

Use of direct-reading detector tubes and badges for analysis of toxic gas and vapor concentrations in air is a very rapid, convenient, and inexpensive technique that can be performed by semiskilled operators. These tubes and badges are in various stages of development, and highly variable results have been obtained. Accuracy depends on a high degree of skill in the manufacture of the tubes. At present, results may be regarded as only range-finding and approximate in nature. The best accuracy that can be expected from indicator tube systems of the best types is on the order of ±25%. Recent advances with electronic concentration readers have improved precision and accuracy. Because many of the tubes are far from specific, an accurate knowledge of the possible interfering gases present is very important. The quantitative effect of these interferences depends on the volume sampled in an irregular way. To avoid dangerously misleading results, the operation and interpretation should be under the supervision of a skilled occupational hygienist.

The manufacturers' descriptions for individual instruments are given in the pages which follow this discussion. It was not possible to check the accuracy of every detail of the description and claims made, and the responsibility for this material rests entirely with the individual manufacturers.

Acknowledgments

The authors wish to recognize the contributions of the original authors of this text in previous editions, Bernard E. Saltzman, Ph.D., and Paul E. Caplan.

References

1. Campbell, E.E.; Miller, H.E.: Chemical Detectors, A Bibliography for the Industrial Hygienist with Abstracts and Annotations. LAMS-2378. Los Alamos Scientific Laboratory, NM (Vol I., 1961; Vol. II, 1964).
2. U.S. Department of the Air Force: Individual Protective and Detection Equipment. In: U.S. Dept. of the Army Technical Manual, TM 3-290, pg. 56–80. Dept. of the Air Force Technical Order, TO 39C-10C-1 (September 1953).
3. McFee, D.R.; Lavine, R.E.; Sullivan, R.J.: Carbon Monoxide, A Prevalent Hazard Indicated by Detector Tabs. Am. Ind. Hyg. Assoc. J. 31:749 (1970).
4. Saltzman, B.E.; Caplan, P.E.: Detector Tubes, Direct Reading Passive Badges and Dosimeter Tubes. In: Air Sampling Instruments for Evaluation of Atmospheric Contaminants, 7th ed., Chapter T, pp. 449–476. S.V. Hering, Ed. American Conference of Governmental Industrial Hygienists, Cincinnati, OH (1989).
5. Ketcham, N.H.: Practical Air-Pollution Monitoring Devices. Am. Ind. Hyg. Assoc. J. 25:127 (1964).
6. Silverman, L.: Panel Discussion of Field Indicators in Industrial Hygiene. Am. Ind. Hyg. Assoc. J. 23:108 (1962).
7. Silverman, L.; Gardner, G.R.: Potassium Pallado Sulfite Method for Carbon Monoxide Detection. Am. Ind. Hyg. Assoc. J. 26:97 (1965).

8. Ingram, W.T.: Personal Air Pollution Monitoring Devices. Am. Ind. Hyg. Assoc. J. 25:298 (1964).

9. Linch, A.L.; Lord, Jr., S.S.; Kubitz, K.A.; Debrunner, M.R.: Phosgene in Air—Development of Improved Detection Procedures. Am. Ind. Hyg. Assoc. J. 26:465 (1965).

10. Linch, A.L.: Oxygen in Air Analyses—Evaluation of a Length of Stain Detector. Am. Ind. Hyg. Assoc. J. 26:645 (1965).

11. Leichnitz, K.: Determination of Arsine in Air in the Work Place (German). Die Berufsgenossenschaft (September 1967).

12. Leichnitz, K.: Cross-Sensitivity of Detector Tube Procedures for the Investigation of Air in the Work Place (German). Zentralblatt für arbeitsmedizin and Arbeitsschutz 18:97 (1968).

13. Linch, A.L.; Stalzer, R.F.; Lefferts, D.T.: Methyl and Ethyl Mercury Compounds—Recovery from Air and Analysis. Am. Ind. Hyg. Assoc. J. 29:79 (1968).

14. Peurifoy, P.V.; Woods, L.A.; Martin, G.A.: A Detector Tube for Determination of Aromatics in Gasoline. Anal. Chem. 40:1002 (1968).

15. Koljkowsky, P.: Indicator-tube Method for the Determination of Benzene in Air. Analyst 94:918 (1969).

16. Grubner, O.; Lynch, J.J.; Cares, J.W.; Burgess, W.A.: Collection of Nitrogen Dioxide by Porous Polymer Beads. Am. Ind. Hyg. Assoc. J. 33:201 (1972).

17. Neff, J.E.; Ketcham, N.H.: A Detector Tube for Analysis of Methyl Isocyanate in Air or Nitrogen Purge Gas. Am. Ind. Hyg. Assoc. J. 35:468 (1974).

18. National Draeger, Inc.: Draeger-Tube Handbook, 101 Technology Dr., Pittsburgh, PA 15275(1997).

19. Sensidyne/Gastec: Precision Gas Detector System Manual. Sensidyne, Inc., 12345 Starkey Road, Largo, FL 33543 (1985).

20. Detector Tube Handbook. Mine Safety Appliances (MSA) Co., 121 Gamma Drive, Pittsburgh, PA 15238 (1995).

21. Gas Detector Tube System Handbook. Matheson Instrument Group, 959 Route 46 East Parsippany, NJ 07054 (1995).

22. Direct Reading Colorimetric Indicator Tubes Manual, 2nd Ed., Peper, J.B.; Dawson, B.J. (Eds). American Industrial Hygiene Association, Fairfax, VA (1993).

23. Grote, A.A.; Kim, W.S.; Kupel, R.E.: Establishing a Protocol from Laboratory Studies to be Used in Field Sampling Operations. Am. Ind. Hyg. Assoc. J. 39:880 (1978).

24. Brown, V.R.: Gas and Vapor Detection During Spill Containment. In: Proceedings of the Haztech International Conference, August 11–16, 1986, Denver, CO, pp. 125B136. Colorado Ground Water Assoc. (1986).

25. Tarkan, N.: A Laboratory Method of Making Detector Tubes for the Determination of Benzene Concentration in the Exhaled Breath. J. Environ. Sci. Health A24 (2), 111–125 (1989).

26. Saner, S.F.; Henry, N.W.: The Use of Detector Tubes Following ASTM Method F-739-85 for Measuring Permeation Resistance of Clothing. Am. Ind. Hyg. Assoc. J. 50(6):298-302 (1989).

27. Loffelholz, R.: Investigation of Contaminated Areas by Means of Draeger Tubes. Draeger Review 63:2 (July 1989).

28. Sieben, O.: Investigation of Oil and Sludge Contaminated Industrial Waste Water by Means of the Draeger Air Extraction Method. Draeger Review 69:11 (May 1992).

29. Leichnitz, K.; Walton, J.: Determination of Aerosols by Means of Detector Tubes. Ann. Occup. Hyg. Vol 24, 1: 43–53 (1981).

30. Bostick, W.D.; Bostick, D.T.: Evaluation of Selected Detector Systems for Products Formed in the Atmospheric Hydrolysis of Uranium Hexafluoride. Report # ORNL/TM-10341. Oak Ridge National Laboratory, Oak Ridge, TN (December 1987).

31. Ruhl, R.; Knoll, M.: Detector Tubes in the Construction Industry. Proceedings of an International Symposium on "Clean Air at Work," Luxenburg, September 1991, Royal Society of Chemistry, London (1992).

32. Meneghelli, B.J.; Hodge, T.R; Robinson, L.J.; Lueck, D.E.: Development of an Automated Reader for the Analysis and Storage of Personnel Dosimeter Badge Data. The 1997 JAN-NAF Propellant Development and Characterization Subcommittee, Safety and Environmental Subcommittee, Joint Meeting, pp. 307-313, N 97-2000401-28, (March 1997).

33. Kusnetz, H.L.: Air Flow Calibration of Direct Reading Colorimetric Gas Detecting Devices. Am. Ind. Hyg. Assoc. J. 21:340 (1960).

34. Colen, F.H.: A Study of the Interchangeability of Gas Detector Tubes and Pumps. Report TR-71. National Institute for Occupational Safety and Health, Morgantown, WV (June 15, 1973).

35. Colen, F.H.: A Study of the Interchangeability of Gas Detector Tubes and Pumps. Am. Ind. Hyg. Assoc. J. 35:686 (1974).

36. Roberson, R.: Interchangeability of Detector Tube Brands. Sensidyne, Inc., 16333 Bay Vista Drive, Clearwater, FL 33760 (June 1997).

37. Occupational Safety and Health Administration (OSHA): Technical Manual (Appendix 1: I-1, Detector Tubes and Pumps), OSHA CD-ROM A97-1. OSHA, Washington, DC (February, 1997).

38. Collins, A.J.: International Union of Pure and Applied Chemistry (IUPAC): Performance Standard for Detector Tube Units Used to Monitor Gases and Vapors in Working Areas. Pure & Applied Chemistry, Volume 54, No. 9, pp. 1763–1767 (1982).

39. American National Standards Institute/Industrial Safety Equipment Association: A Standard for Detector Tube Performance, Appendix to ANSI/ISEA 102-1990 (1990).

40. Safety Equipment Institute: Certified Product List: Gas Detector Tube Units, April, 1998. SEI, 1307 Dolley Madison Blvd., Suite 3A, McLean, VA 22101 (1998).

41. Leichnitz, K.: Use of Detector Tubes Under Extreme Conditions (Humidity, Pressure, Temperature). Am. Ind. Hyg. Assoc. J. 38:707 (1977).

42. Carlson, D.H.; Osborne, M.D.; Johnson, J.H.: The Development and Application to Detector Tubes of a Laboratory Method to Assess Accuracy of Occupational Diesel Pollutant Concentration Measurements. Am. Ind. Hyg. Assoc. J. 43:275 (1982).

43. Douglas, K.E.; Beaulieu, H.J.: Field Validation Study of Nitrogen Dioxide Passive Samplers in a "Diesel" Haulage Underground Mine. Am. Ind. Hyg. Assoc. J. 44:774 (1983).

44. Leichnitz, K.: Effects of Pressure and Temperature on the Indication of Draeger Tubes. Draeger Rev. 31:1 (September 1973).

45. Linch, A.L.: Evaluation of Ambient Air Quality by Personnel Monitoring. CRC Press, Inc., Cleveland, OH (1974).

46. Leichnitz, K.: Qualitative Detection of Substances by Means of Draeger Detector Tube Polytest and Draeger Detector Tube Ethyl Acetate 200 A. Draeger Rev. 46:13 (December 1980).

47. Kitagawa, T.: Detector Tube Method for Rapid Determination of Minute Amounts of Nitrogen Dioxide in the Atmosphere. Yokohama National University, Yokohama, Japan (July 1965).

48. Drägerwerk AG: Information Sheet No. 44: 0.5a Nitrous Gas/Detector Tube, Drägerwerk, P.O. Box 1339, D-24 Lübeck 1, Federal Republic of Germany (November 1960).

49. Grosskopf, K.: A Tentative Systematic Description of Detector Tube Reactions (German). Chemiker Zeitung-Chemische Apparatus 87:270 (1963).

50. Leichnitz, K.: Determination of Low SO_2 Concentrations by Means of Detector Tubes. Draeger Rev. 30:1 (May 1973).

51. Linch, A.L.; Plaff, H.V.: Carbon Monoxide—Evaluation of Exposure Potential by Personnel Monitor Surveys. Am. ind. Hyg. Assoc. J. 32:745 (1971).

52. Leichnitz, K.: The Detector Tube Method and its Development Tendencies (German). Chemiker-Zeitung 97:638 (1973).

53. Leichnitz, K.: An Analysis by Means of Long-Term Detector Tubes. Draeger Rev. 40:9 (December 1977).

54. Leichnitz, K.: Some Information on the Long-Term Measuring System for Gases and Vapors. Draeger Rev. 43:6 (June 1979).

55. Donham, K.J.; Popendorf, W.I.: Ambient Levels of Selected Gases Inside Swine Confinement Buildings. Am. Ind. Hyg. Assoc. J. 46:658 (1985).

56. Heubener, D.J.: Evaluation of a Carbon Monoxide Dosimeter. Am. Ind. Hyg. Assoc. J. 41:590 (1980).

57. Dharmarajan, V.; Rando, R.J.: Clarification—re: A Recommendation for Modifying the Standard Analytical Method for Determination of Chlorine in Air. Am. Ind. Hyg. Assoc. J. 40:746 (1979).

58. National Institute for Occupational Safety and Health: Criteria for a Recommended Standard-Occupational Exposure to Carbon Monoxide. DHEW (NIOSH) Pub. No. HSM 73-11000. NIOSH, Rockville, MD (1972).

59. Leidel, N.A.; Busch, K.A.: Statistical Methods for Determination of Noncompliance with Occupational Health Standards. DHEW (NIOSH) Pub. No. 75-159. National Institute for Occupational Safety and Health, Cincinnati, OH (April 1975).

60. Leichnitz, K.: Detector Tubes. Proceedings of an International Symposium on "Clean Air at Work," Luxenburg, September 1991 Royal Society of Chemistry, London (1992).

61. Saltzman, B.E.: Basic Theory of Gas Indicator Tube Calibrations. Am. Ind. Hyg. Assoc. J. 23:112 (1962).

62. Leichnitz, K.: Attempt at Explanation of Calibration Curves of Detector Tubes (German). Chemiker-Ztg./Chem. Apparatus 91:141 (1967).

63. Scherberger, R.F.; Happ, G.P.; Miller, F.A.; Fassett, D.W.: A Dynamic Apparatus for Preparing Air-Vapor Mixtures of Known Concentrations. Am. Ind. Hyg. Assoc. J. 19:494 (1958).

64. Saltzman, B.E.: Preparation and Analysis of Calibrated Low Concentrations of Sixteen Toxic Gases. Anal. Chem. 33:1100 (1961).

65. Saltzman, B.E.: Preparation of Known Concentrations of Air Contaminants. In: The Industrial Environment—Its Evaluation and Control, Chap. 12, pp. 123–137. National Institute for Occupational Safety and Health, Contract HSM-99-71-45, Cincinnati, OH (1973).

66. Avera, Jr., C.B.: Simple Flow Regulator for Extremely Low Gas Flows. Rev. Sci. Instru. 32:985 (1961).

67. Cotabish, H.N.; McConnaughey, P.W.; Messer, H.C.: Making Known Concentrations for Instrument Calibration. Am. Ind. Hyg. Assoc. J. 22:392 (1961).

68. Hersch, P.A.: Controlled Addition of Experimental Pollutants to Air. J. Air Poll. Control Assoc. 19:164 (1969).

69. Hughes, E.E.; et al: Gas Generation Systems for the Evaluation of Gas Detecting Devices. NBSIR 73-292. National Bureau of Standards, Washington, DC (October 1973).

70. Zurlo, N.; Andreoletti, F.: Effect of Air Turbulence on Diffusive Sampling. In: Diffusive Sampling. An Alternative Approach to Workplace Air Monitoring. Proceedings of an International Symposium, Luxembourg, Sept. 22–26, 1986, A. Berlin, R.H. Brown, and K.J. Saunders, Eds., pp. 174–176. Royal Society of Chemistry, London (1987).

71. Pannwitz, K.H.: Influence of Air Currents on the Sampling of Organic Solvent Vapours with Diffusive Samplers. In: Diffusive Sampling. An Alternative Approach to Workplace Air Monitoring. Proceedings of an International Symposium, Luxembourg, Sept. 22–26, 1986. A. Berlin, R.H. Brown, and K.J. Saunders, Eds., pp. 157–160. Royal Society of Chemistry London, (1987).

72. Dittmar, P.; Stress, G.: The Suitability of Detection of Toxic Substances in the Air; I: Hydrogen Sulfide Detector Tubes (German). Arbeitsschutz 8:173 (1959).

73. Heseltine, H.K.: The Detection and Estimation of Low Concentrations of Methyl Bromide in Air. Pest Technology (England) (July/August 1959).

74. Kusnetz, H.L.; Saltzman, B.E.; LaNier, M.E.: Calibration and Eval-uation of Gas Detecting Tubes. Am. Ind. Hyg. Assoc. J. 21:361 (1960).

75. Banks, O.M.; Nelson, K.R.: Evaluation of Commercial Detector Tubes. Presented at the American Industrial Hygiene Conference, Detroit, MI (April 13, 1961).

76. LaNier, M.E.; Kusnetz, H.L.: Practices in the Field of Detector Tubes. Arch. Env. Health 6:418 (1963).

77. Hay, III, E.B.: Exposure to Aromatic Hydrocarbons in a Coke Oven By-Product Plant. Am. Ind. Hyg. Assoc. J. 25:386 (1964).

78. Larsen, L.B.; Hendricks, R.H.: An Evaluation of Certain Direct Reading Devices for the Determination of Ozone. Am. Ind. Hyg. Assoc. J. 30:620 (1969).

79. Morganstern, A.S.; Ash, R.M.; Lynch, J.R.: The Evaluation of Gas Detector Tube Systems; I: Carbon Monoxide. Am. Ind. Hyg. Assoc. J. 31:630 (1970).

80. Ash, R.M.; Lynch, J.R.: The Evaluation of Gas Detector Tube Systems: Benzene. Am. Ind. Hyg. Assoc. J. 32:410 (1971).

81. Ash, R.M.; Lynch, J.R.: The Evaluation of Detector Tube Systems: Sulfur Dioxide. Am. Ind. Hyg. Assoc. J. 32:490 (1971); also see Am. Ind. Hyg. Assoc. J. 33:11 (1972).

82. Ash, R.M.; Lynch, J.R.: The Evaluation of Detector Tube Systems: Carbon Tetrachloride. Am. Ind. Hyg. Assoc. J. 32:552 (1971).

83. Roper, C.P.: An Evaluation of Perchloroethylene Detector Tube. Am. Ind. Hyg. Assoc. J. 32:847 (1971).

84. Johnston, B.A.; Roper, C.P.: The Evaluation of Gas Detector Tube Systems: Chlorine. Am. Ind. Hyg. Assoc. J. 33:533 (1972).

85. Johnston, B.A.: The Evaluation of Gas Detector Tube Systems: Hydrogen Sulfide. Am. Ind. Hyg. Assoc. J. 33:811 (1972).

86. Jentzsch, D.; Fraser, D.A.: A Laboratory Evaluation of Long-term Detector Tubes: Benzene, Toluene, Trichloroethylene. Am. Ind. Hyg. Assoc. J. 42:810 (1981).

87. Septon, J.C.; Wilczek, Jr., T.: Evaluation of Hydrogen Sulfide Detector Tubes. Appl. Ind. Hyg. 1:196 (1986).

88. Leichnitz, K.: Survey of Draeger Long-Term Tubes with Special Consideration of the Long-Term Tubes Sulfur Dioxide 5/a-L. Draeger Review 48:16 (November 1981).

89. Leichnitz, K.: Draeger Long-Term Tubes Meet IUPAC Standard. Draeger Review 52:11 (January 1984).

90. Johansson, R.; Johnson, T.: Evaluation of Detector Tubes, Part XI, Phosgene (in Swedish). Arbetsmiljöinstitutet, Undersökningsrapport 1987:11 (1987).

91. Beck, S.W.; Stock, T.H.: An Evaluation of the Effects of Source and Concentration on Three Methods for the Measurement of Formaldehyde in Indoor Air. Am. Ind. Hyg. Assoc. J. 51:14 (1990).

92. Manninen, A.: Analysis of Airborne Ammonia: Comparison of Field Methods. Ann. Occ. Hyg. 32:399 (1988).

93. Droz, P.O.; Krebs, Y.; Nicole, C.; Guillemin, M.: A Direct Reading Method for Chlorinated Hydrocarbons in Breath. Am. Ind. Hyg. Assoc. J. 49:319 (1988).

94. Panova, N.; Velichkova, V.; Panev, T.: Industrial Evaluation of Hygitest Detector Tubes for Ethyl Acetate. Polish J. of Occup. Med. And Environ. Health. Vol. 6, No. 3 pp. 293-298 (1993).

95. Kirollos, K.S.; Mihaylov, G.M.; Nurney, B.: Validation Study of Direct-Read Formaldehyde Monitor. 1998 American Industrial Hygiene Assoc. Conference, Atlanta, GA, poster #344 by K&M Environmental (May 1998).

96. Stock, T.H.: The Use of Detector Tube Humidity Limits. Am. Ind. Hyg. Assoc. J. 47:241 (1986).

97. McCammon, Jr., C.S.; Crouse, W.E.; Carrol, Jr., H.B.: The Effect of Extreme Humidity and Temperature on Gas Detector Tube Performance. Am. Ind. Hyg. Assoc. J. 43:18 (1982).

98. Joint Comm. on Direct Reading Gas Detecting Systems, ACGIH–AIHA: Direct-Reading Gas Detecting Tube Systems. Am. Ind. Hyg. Assoc. J. 32:488 (1971).

99. National Institute for Occupational Safety and Health: Certification of Gas Detector Tube Units. Federal Register 38:11458 (May 8, 1973); also see 43 CFR 84.

100. Roper, C.P.: The NIOSH Detector Tube Certification Program. Am. Ind. Hyg. Assoc. J. 35:438 (1974).

101. National Institute for Occupational Safety and Health: NIOSH Certified Equipment List as of October 1, 1981. DHHS (NIOSH) Pub. No. 82-106. Cincinnati, OH (October 1981).

102. Centers for Disease Control, National Institute for Occupational Safety and Health: NIOSH Voluntary Testing and Certification Program. Fed. Reg. 48(191):44931 (September 30, 1983).

103. Leichnitz, K.: How Reliable are Detector Tubes? Draeger Rev. 43:21 (June 1979).

104. Leichnitz, K.: Comments of Official Organizations Regarding Suitability of Detector Tubes. Draeger Rev. 49:19 (May 1982).

105. U.S. Department of Labor: Directive 73-4. Use of Detector Tubes. USDOL, Washington, DC (March 1973).

106. International Union of Pure and Applied Chemistry (IUPAC): Performance Standards for Detector Tube Units Used to Monitor Gases and Vapours in Working Areas. Pure and Applied Chemistry 54:1763 (1982).

107. Japanese Industrial Standard: Detector Tube Type Gas Measuring Instruments. JIS K 0804-1985. Japanese Standards Assoc., Tokyo (1986).

108. British Standard: Gas Detector Tubes BS5343. Part 1. Specification for Short Term Gas Detector Tubes. British Standards Institution, 2 Park St., London W1A2BS, England (1986).

109. Industrial Safety Equipment Association: American National Standard for Detector Tube Units—Short Term Type for Toxic Gases and Vapors in Working Environments. ANSI/ISEA 102-1990. American National Standards Institute, New York (1990).

110. International Standards Organization, Technical Committee ISO/TC 146, Air Quality. Work-Place Air-Determination of Mass Concentration of Carbon Monoxide-Method Using Detector Tubes for Short-Term Sampling with Direct Indication. ISO 8760:1990. ISO, Geneva (1990).

111. International Standards Organization, Technical Committee ISO/TC 146, Air Quality. Work-Place Air—Determination of Mass Concentration of Nitrogen Dioxide—Method Using Detector Tubes for Short-Term Sampling with Direct Indication. ISO 8761:1989. ISO, Geneva (1989).

112. European Committee for Standardization (Comité Européen de Normalisation): Workplace Atmospheres—Requirements and Test Methods—Short-Term Detector Tube Measurement (EN 1231: 1996), Brussels (1996).

113. Leichnitz, K.: Comparison of Detector Tube Method with European Standards. Analyst, vol. 119, No. 1 (January 1994).

114. Wilcher, Jr., F.E.: SEI Gas Detector Tube Certification. Appl. Ind. Hyg. 3:R-7 (August 1988).

115. Safety Equipment Institute: Certified Products List, April 1998. SEI, 1307 Dolley Madison. Blvd., Suite 3A, McLean, VA 22101 (1998).

116. Hill, R.H.; Fraser, D.A.: Passive Dosimetry Using Detector Tubes. Am. Ind. Hyg. Assoc. J. 41:721 (1980).

117. Sefton, M.V.; Kostas, A.V.; Lombardi, C.: Stain Length Passive Dosimeters. Am. Ind. Hyg. Assoc. J. 43:820 (1982).

118. Gonzalez, L.A.; Sefton, M.V.: Stain Length Passive Dosimeter for Monitoring Carbon Monoxide. Am. Ind. Hyg. Assoc. J. 44:514 (1983).

119. Gonzalez, L.A.; Sefton, M.V.: Laboratory Evaluation of Stain Length Passive Dosimeters for Monitoring of Vinyl Chloride and Ethylene Oxide. Am. Ind. Hyg. Assoc. J. 46:591 (1985).

120. Pannwitz, K.H.: Direct-Reading Diffusion Tubes. Draeger Rev. 53:10 (June 1984).

121. McKee, E.S.; McConnaughey, P.W.: A Passive, Direct Reading, Length-of-Stain Dosimeter for Ammonia. Am. Ind. Hyg. Assoc. J. 46:407 (1985).

122. McConnaughey, P.W.; McKee, E.S.; Pretts, I.M.: Passive Colorimetric Dosimeter Tubes for Ammonia, Carbon Monoxide, Carbon Dioxide, Hydrogen Sulfide, Nitrogen Dioxide, and Sulfur Dioxide. Am. Ind. Hyg. Assoc. J. 46:357 (1985).

123. McKee, E.S.; McConnaughey, P.W.: Laboratory Validation of a Passive Length-of-Stain Dosimeter for Hydrogen Sulfide. Am. Ind. Hyg. Assoc. J. 47:475 (1986).

124. Bartley, D.L.: Diffusive Samplers Using Longitudinal Sorbent Strips. Am. Ind. Hyg. Assoc. J. 47:571 (1986).

125. Pannwitz, K.H.: The Direct-Reading Diffusion Tubes on the Test Bench. Draeger Rev. 57:2 (June 1986).

126. Cassinelli, M.E.; Hull, R.D.; Cuendet, P.A.: Performance of Sulfur Dioxide Passive Monitors. Am. Ind. Hyg. Assoc. J. 46:599 (1985).

127. Hossain, M.A.; Saltzman, B.E.: Laboratory Evaluation of Passive Colorimetric Dosimeter Tubes for Carbon Monoxide. Appl. Ind. Hyg. 4:119 (1989). Systems (EN 1231:1996), Brussels (1996).

Instrument Descriptions

Introduction

Detector tubes, direct-reading passive badges, and dosimeter tubes can be classified by certain general characteristics. For example, many detector tubes aspirate short-term air samples using a few strokes of a hand piston pump or rubber bulb. The long-term types use a continuous pump at a very low flow rate for periods as long as 8 hours to give time-weighted average (TWA) concentrations. No pump is required by passive dosimeter and badges that rely on diffusion of the analyte from air into the sensing absorbent. After each manufacturer's description of the instruments, a list of analytes with range of measurement and part number is included. The last table of this chapter lists the commercial sources and contact information of the instrument manufacturers described, as well their distributors.

17-1. ATI Air Monitoring System

Acculabs Technologies, Inc.

The ATI air monitoring system offers the user an accurate and cost effective method of determining personal and area exposure levels to hazardous chemicals. The lightweight monitoring cards are simply attached to the personnel or placed in the area to be monitored, and after the exposure period, a time-averaged hazard concentration can be read directly by means of a portable electronic reader. The ATI electronic reader automatically corrects for variables such as temperature and humidity, and converts a chemically-induced color change in the card to parts per million exposure. Results are available immediately and on-site without the pumps, tubes and the laboratory expense associated with conventional methods. All ATI monitoring cards are validated in accordance with the "NIOSH Protocol for the Evaluation of Passive Monitors." Sensitivity and range

INSTRUMENT 17-1. ATI Air Monitoring System.

are based on the Occupational Safety and Health Administration (OSHA) Permissible Exposure Limits (PELs) and Short-Term Exposure Limits (STELs). The precision and accuracy of the ATI system exceeds both OSHA and NIOSH requirements (see Instrument 17-1).

17-2. Personal Protection Indicators

American Gas & Chemical Company, Ltd.

Personal Protection Indicators (PPI) change color on exposure to low concentrations of various toxic gases or

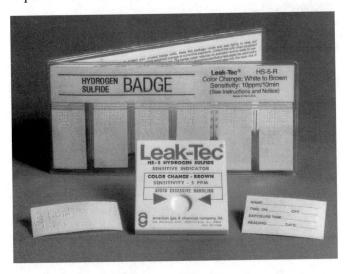

INSTRUMENT 17-2. LEAK-TEC Personal Protection Indicators.

vapors, allowing the user to evaluate ceiling or time weighted average (TWA) exposure. PPI are lightweight and easy to use. They do not require activation, chemical analysis or calibration. PPI change color as a result of a chemical reaction with the gas or vapor. As the time of exposure and/or the concentration of gas seen by the PPI increases, the color change becomes increasingly more pronounced. By using a color chart the user can estimate exposure.

Personal Protection Indicators (PPI) are available in three forms:

1. Disposable plastic badges with pocket clips
2. Reusable badge plaques, with inexpensive indicator refills
3. Area Contamination Monitors—adhesive-backed papers that can easily be attached to a wall to monitor the area or wrapped around a critical joint on a pipe to monitor fugitive leaks

PPIs are available for a number of gases including: ammonia, carbon monoxide, chlorine, hydrazine, hydrogen sulfide, nitrogen dioxide and ozone (see Table 17-I-2). PPIs are field-proven over 10 years and are used in a great variety of applications from home heaters and stoves, to aerospace fuels, to oil drilling and petrochemical production (see Instrument 17-2).

17-3. ChemChip™ Personal Monitoring System

Assay Technology

The ChemChip™ System includes a kit of Personal Monitoring Badges (Diffusive Samplers) and an Electronic Reader (Photometer) providing on-site measurement of time-weighted-average (TWA) and short-term exposure limit (STEL) exposures measured by the Badges in compliance with OSHA requirements (see Table 17-I-3).

Each Personal Monitor includes a chemically selective reagent system immobilized on a proprietary Monitor Strip encased within polymer membranes and incorporated in a lightweight (1/2 oz) Badge clipped to pocket or lapel. After the Badge is worn for a full shift (8-hr TWA) or short term (15-min STEL) exposure interval, the Monitor Strip is removed, developed, and inserted into a calibrated reflectance photometer (Electronic Reader) to obtain the result.

ChemChip Personal Monitors are available for monitoring ethylene oxide in 8-hr TWA and STEL formats. The ChemChip Electronic Reader is re-calibrated on-site to replicate a calibration determined at the factory for each manufacturing lot of Monitoring Badges. Lot-by-lot quality control and on-site calibration allow the

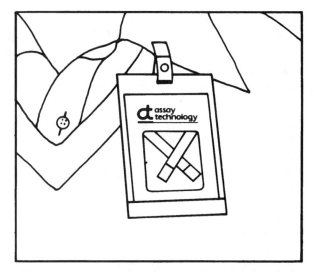

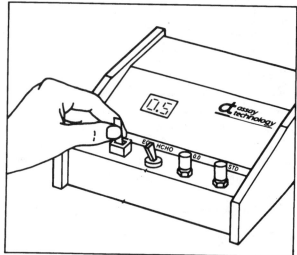

INSTRUMENT 17-3. ChemChip™ Personal Monitoring System.

optimization of accuracy in the range of 50% to 200% of the OSHA Permissible Exposure Limit (PEL) for each type of Monitor. ChemChip Monitors for Ethylene Oxide have been shown to comply with OSHA requirements for accuracy (±25% at PEL) in published third party investigations and in quality control tests performed on each manufactured lot (see Instrument 17-3).

17-4. Sure-Spot Dosimeter Badges
Bacharach, Inc.

Sections 17-5 through 17-9 describe the various types of Sure-Spot dosimeter badges (see Table 17-I-4).

17-5. Phosgene Dosimeter Badges
Bacharach, Inc.

The Bacharach Sure-Spot Phosgene Dosimeter Badges are available in three versions: the standard one-day badge; the one-day badge for use when chlorofor-

mates are present; the new two-day use badge. Each badge uses two separate windows. In the one-day badges, the first of these is the indicator window. It provides immediate color indication to the wearer when phosgene is present. The second window, the control window, is covered by a removable filter-barrier. This window extends the range to 300 ppm-min and the barrier protects against acid gases, sun bleaching, etc. This window is used for accurate dose measurement after peeling off the filter. The two-day badge uses the control window as an indicating window for a second exposure period. The dosimeter reads in ppm-min with the Phosgene Dose Estimator. It has two color wheels to allow rapid color matching of both windows in the dosimeter (see Instrument 17-4).

17-6. TDI Dosimeter Badges
Bacharach, Inc.

Exclusive isomer-independent paper tape detection technology provides an assessment of the total TDI dose in ppb-hours. The Bacharach Sure-Spot TDI Badge is a complete "passive" system which uses the principle of controlled diffusion to provide very accurate results. Its low-end threshold of sensitivity is also sufficient for approximating 15 minute short-term evaluations. In operation, a Sure-Spot TDI Badge card is activated and mounted in the reusable plastic holder. The unit is then clipped to the collar for sampling in the breathing zone. If TDI is present, a distinct stain develops in either the Control or Indicator windows or both, dependent on the concentration (see Instrument 17-4). At any time, the developed stain can be compared against known color standards and the dose in ppb-hours calculated. A similar badge is available for PPDI.

17-7. Hydrazine Dosimeter Badges
Bacharach, Inc.

The Bacharach Sure-Spot Hydrazine Dosimeter Badge meets the requirements for personal monitoring of exposure to both Hydrazine and Monomethyl

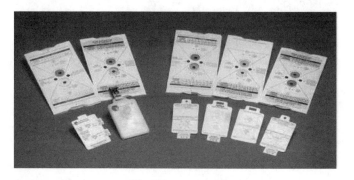

INSTRUMENT 17-4. Sure-Spot Dosimeter Badges.

Hydrazine (MMH). A unique design incorporates two separate paper tape chemistries in one badge housing. Two circular windows allow each tape to be exposed and observed. An automatic color change occurs in each tape in the presence of either substance. The reaction for Hydrazine is somewhat different from that for MMH and two Dose Estimators are required; one for each substance. The two tapes also have complimentary sensitivities, enabling their use in a wide variety of concentrations (see Instrument 17-4).

17-8. Hydrides Dosimeter Badges

Bacharach, Inc.

The Bacharach Sure-Spot Hydrides Dosimeter Badge is a lightweight, direct-reading dosimeter designed to be worn by individuals at risk of hydrides exposure (arsine, phosphine, diborane, silane) in their work environment. A highly specific color formation occurs in the presence of hydride gases. The color density produced is directly proportional to the concentration of hydrides and the total time of exposure. Dosage is expressed in ppm-hours. The highly accurate dosimeter works via diffusion with sensitivity well below the threshold limit values (TLV) of hydrides. The badge provides vivid visual warning of potentially harmful exposures. Any resulting stain is compared against color standards and the dose in ppm-hours determined directly. Personnel at risk will readily notice stain development on their own badge or on other's badges throughout the shift (see Instrument 17-4).

17-9. Bacharach Sure-Spot MDI/TDI/NDI (Isocyanates) Test Kit

Bacharach, Inc.

The Bacharach portable Sure-Spot Test Kit is a simple and fast means of testing for atmospheric concentrations of acutely toxic gases down to 1 ppb range. Operating from its internal rechargeable battery, it is ideal for quickly measuring vapor and aerosol levels of isocyanates during virtually any application involving MDI, TDI, HDI, or NDI. Based on well-proven colorimetric paper tape detection technology, the Sure-Spot Test Kit provides highly-accurate, dependable detection of toxic gas concentrations. In operation, a test card with the reactive paper tape is placed in a holder while a pre-calibrated pump pulls a measured air sample through it. The intensity of the resulting color stain is directly proportional to the concentration of gas present. The developed stain is visually matched against a concentration calculator, providing a readout of concentration in ppb. Each test card can serve as a record of test data, showing time, date, and location of the sample. The lightweight

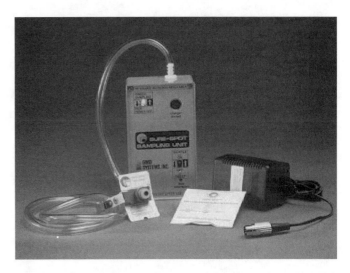

INSTRUMENT 17-9. Bacharach Sure-Spot MDI/TDI/NDI (Isocyanates) Test Kits.

Sure-Spot Test Kit can be located in areas where work involving acutely gases is conducted, or it can be carried by a worker using the supplied belt pouch. The Kit is designed to allow you to take spot samples in the breathing zone, as well as for remote, confined space sampling prior to entry of personnel. Everything needed for an entire series of spot tests is packaged in a rugged, fitted carrying case (see Instrument 17-9).

17-10. Carbon Monoxide Indicator

Bacharach, Inc.

The Bacharach Carbon Monoxide Indicator is a portable instrument for the detection of concentrations of carbon monoxide. Used by safety engineers and industrial hygienists, it finds applications in process industries, refineries, mines, tunnels, sewers, natural gas fields, and confined areas. Hazardous gas content, in ppm, is determined by measuring the length of the stain or bleach. Air is sampled with a hand-held sampling pump that has non-interchangeable scales and calibrated tubes. Measurements are read directly from the length of stain, and no color comparison charts or calibration curves are necessary. Kits for sampling carbon monoxide from hot flue-gases are available in 0–2,000 ppm and 0–5,000 ppm ranges (see Table 17-I-5).

17-11. CMS Chip Measurement System

Draeger Safety, Inc.

The Draeger CMS integrates established measurement principles with the power of new intelligent technology for determining gas and vapor concentrations. It integrates established measurement principles with the power of new intelligent technology. Draeger CMS

INSTRUMENT 17-11a. Draeger CMS Chip Measurement System.

analyzer utilizes chips to serve as chemical sensors that provide the system with the versatility to measure various gases and vapors. Alternating between different gases or vapors is as easy as inserting a new chip (see Instruments 17-11a and 17-11b). The basis of the CMS is the combination of electronics, optics, chemical reagent systems and a special pump system. Each chip contains 10 measurement capillaries (measurement channels), filled with a substance-specific reagent system. Low quantities of reagent ensure a quick and reproducible reaction with the contaminant gas or vapor. Each chip is calibrated during manufacturing and can be used for up to two years. The interaction between the various chips and the analyzer will be directed by the instructions in the bar code. This bar code is found on the chip and will be read by the analyzer optics and interpreted by the analyzer software. All fundamental information such as gas type, measurement range, and other necessary parameters for the measurement evaluation are included in the bar code. Temperature and humidity influences are accounted for during calibration. The entire operation of the measurement process is menu driven. The flow sensor is integrated with the pump system and functions according to a mass flow measurement principle. The smallest reaction can be detected by the optics and registered by the microprocessor. The established reaction effect is converted to a measurement value according to instructions from the software program. This measurement value is then displayed in the form of a concentration on the digital readout. In order to measure inaccessible areas, the analyzer can be equipped with a remote system which consists of an additional pump and extension hose. Maintenance of the analyzer is not necessary. The analyzer is intrinsically safe and certified according to UL (USA), CSA (Canada), and CENELEC (Europe). Additionally, the system is protected against dust and water spray according to the IP 54 standard, and it is resistant to electromagnetic radiation. The CMS analyzer now includes a data recorder and 28 chips for analyzing 17 different chemicals (see Table 17-I-6).

17-12. Short-Term Tubes and Accuro®, Quantimeter 1000 and Accuro 2000.

Draeger Safety, Inc.

The Draeger Safety Accuro bellows pump and short-term detector tubes form a portable sampling unit for measuring concentrations of various gases and vapors (Instrument 17-12a). Draeger Short-Term colorimetric detector tubes are currently available for determining and measuring more than 350 different gases, vapors, and aerosols. The Accuro pump (see Instrument 17-12b) delivers 100 mL of sampled air with each pump stroke. After a prescribed number of pump strokes, the stain

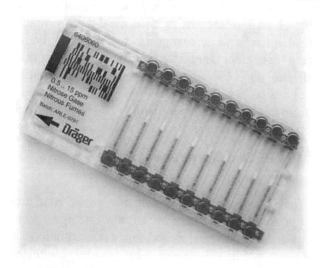

INSTRUMENT 17-11b. Draeger Chemical Measuring Sensor Chips.

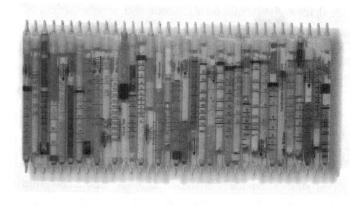

INSTRUMENT 17-12a. Draeger Short-Term Detector Tubes.

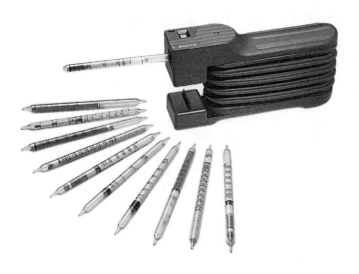

INSTRUMENT 17-12b. Draeger Accuro Bellows Pump.

INSTRUMENT 17-12d. Draeger Quantimeter 1000 Pump.

length or the discoloration of the tube gives a direct measure of the gas or vapor concentration. Calibration scales are printed directly on most types of tubes. The Accuro is a modular system. For large volume measurements, the Accuro slides into the electronically programmable Accuro 2000 (see Instrument 17-12c). The Quantimeter 1000 is a programmable, battery-operated bellows pump with the same flow characteristics as the hand-operated pump and it is intrinsically safe (see Instrument 17-12d). The complete Draeger Accuro deluxe pump kit with spare parts, tube opener, and screwdriver is contained in a vinyl carrying case and weighs approximately 1.5 kg (3.3 lbs). The detector tubes are essentially specific for particular gases or vapors. This specificity is achieved not only by the use of specific and stable reagents, but also by the use of precleansing layers placed in front of the actual reactive

layer to selectively adsorb interfering components that may be contained in the gas or vapor sample. *Aerosols*: Draeger also provides aerosol tubes for arsenic trioxide, cyanides, chromic acid, nickel, oil mist, and sulfuric acid. The reading deviations for many of the detector tubes are not more than ±25% from the true value. Table 17-I-7 indicates the use of Draeger short-term detector tubes with the Accuro® bellows pump.

17-13. Long-Duration Detector Tubes and Polymeter
Draeger Safety, Inc.

The Draeger long-term detector tubes and the Draeger Polymeter® measure the mean value of the contaminant concentration over periods of up to 8 hrs (see Instrument 17-13). The long-term detector tubes are calibrated in units of microliters and are designed for use

INSTRUMENT 17-12c. Draeger Accuro 2000 Pump.

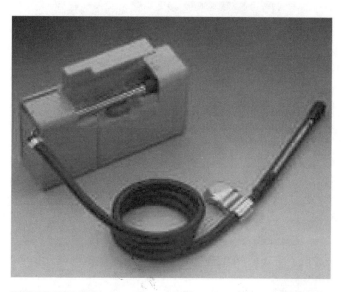

INSTRUMENT 17-13. Long-Duration Detector Tubes and Polymeter.

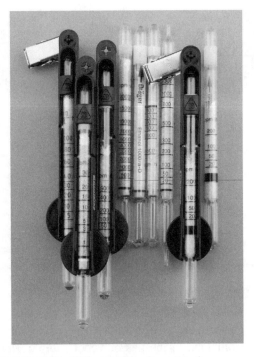

INSTRUMENT 17-14. Diffusion Tube (long-term).

over a flow rate range of 10 to 20 mL/min. The TWA concentration in ppm is calculated by dividing the detector tube indication by the sample volume in liters. The Polymeter is a battery-powered peristaltic pump that provides a continuous flow at approximately 15 mL/min. A counter on the pump records the number of revolutions so the volume drawn can be calculated. The unit is supplied in a leather carrying bag with a shoulder strap. An extension hose is available as an accessory. Table 17-I-8 indicates the measuring range and usage of Draeger long-term tubes.

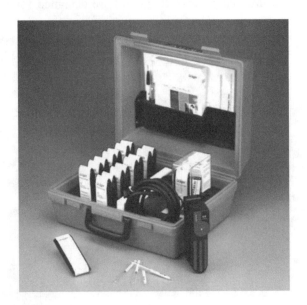

INSTRUMENT 17-15. Haz Mat Kit.

17-14. Diffusion Tubes (long-term)
Draeger Safety, Inc.

The direct-reading diffusion detector tubes from Draeger Safety, Inc. work on the principle of gaseous diffusion to give long-term, TWA measurements without a pump. The contaminant gas diffuses into the tube by means of the concentration gradient between the ambient atmosphere and the interior of the tube. The diffusion tubes have been calibrated in ppm × hrs, and/or volume % × hrs with the calibrated scale printed directly on the tube (see Instrument 17-14). This system consists of a tube holder and a diffusion tube that may be attached to a pocket or lapel. The range of measurement for various Draeger diffusion tubes is presented in Table 17-I-9.

17-15. Haz Mat Kit
Draeger Safety, Inc.

The Draeger Safety Haz Mat Kit is designed to aid in the initial assessment of potentially hazardous situations. The detector tubes included in the kit have been selected to utilize a systematic sampling matrix. Using the polytest tube as a starting point, the sampling matrix provides a systematic test sequence to obtain information about the chemical group to which an unknown substance may belong. The haz mat kit can also be used for providing quantitative measurement of specific gases and vapors. Evaluation of test results can be performed onsite. Housed in a durable, lightweight case (see Instrument 17-15), the haz mat kit includes the Draeger bellows pump with an automatic stroke counter, a 3-meter extension hose for testing inaccessible areas, and an air current kit for determining wind direction and velocity. Other contents include spare parts and tools for pump maintenance and 17 types of detector tubes. The components fit into a closed-cell foam insert which provides travel protection and organization. The detector tubes in the standard haz mat kit are listed in Table 17-I-10.

17-16. Bio-Check Ozone Badge
Draeger Safety, Inc.

The Draeger Bio-Check Ozone is a lightweight badge that changes color to indicate the ozone concentration at the immediate location. This easy-to-use badge (see Instrument 17-16) works indoors or outdoors, and provides accurate results as low as 30 $\mu g/m^3$ (20 minutes for indoor measurements, 10 minutes for outdoor). Ozone levels are determined by comparing a color change against the color standard provided with the kit. For personal exposure monitoring, the Bio-Check Ozone

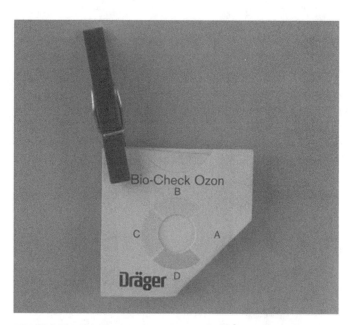

INSTRUMENT 17-16. Bio-Check Ozone Badge.

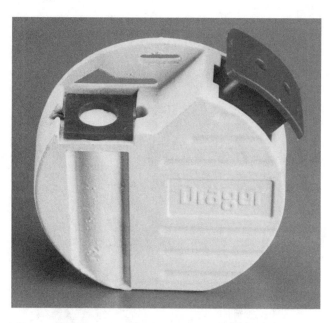

INSTRUMENT 17-18. Bio-Check F 0.05/a Formaldehyde Badge.

can be easily clipped onto clothing or for area monitoring, it can be placed on a flat surface. The measurement range is 30–240 μg/m³.

17-17. Bio-Check F Badge

Draeger Safety, Inc.

The Draeger Bio-Check F is a colorimetric badge specifically designed to measure *formaldehyde* in the air. The badge is so sensitive that within a 2-hour sampling period, concentrations from 0.02 to 0.7 ppm can be measured. The measurement results of the Draeger Bio-Check F are determined by comparing the discoloration of the badge with an enclosed set of color standards. The badge comes complete with a uniquely shaped holder that clips easily onto the user's clothing—or the device may be placed upright on a flat surface (see Instrument 17-17).

INSTRUMENT 17-17. Bio-Check F Badge.

17-18. Bio-Check F 0.05/a Badge

Draeger Safety, Inc.

The new Bio-Check F 0.05/a direct-reading badge is designed for measuring *indoor formaldehyde* levels at a workplace or home. Ideal for personal monitoring, this new badge provides an accurate reading in as few as 10 minutes, or as many as 120 minutes. Similar to the original Bio-Check F, the device measures formaldehyde concentrations in the range of 0.05–1.2 ppm and is simple to use (see Instrument 17-18).

17-19. Phosphine Badge

Draeger Safety, Inc.

The Draeger Phosphine Badge is a direct reading device utilizing color standards to determine the time-weighted average concentration down to 0.01 ppm. This badge is ideal for monitoring situations where background levels of phosphine are likely to exist (e.g. in the fumigation and semiconductor industries). The phosphine badges can be used for personal monitoring or area monitoring for up to 8 hours measuring phosphine in the range of 0.01–0.3 ppm. The range in absolute units is 0.1–2.4 ppm × hr (see Instrument 17-19).

17-20. Simultaneous Test Sets

Draeger Safety, Inc.

The Simultaneous Test system (see Instrument 17-20) offers three different sets which are used with an adapter (purchased separately) and a Draeger Pump for the semi-quantitative determination of inorganic gases and

INSTRUMENT 17-19. Phosphine Badge.

organic vapors. Each test set is able to test for five different gases simultaneously in less than one minute. The sets
provide on-site information to firefighters, Haz Mat
teams, and environmental agencies. Please note: Draeger
Tubes in the Simultaneous Test Sets are specifically calibrated to be used as a set. It is not recommended to mix
tubes from any of the sets or use standard tubes with the
Simultaneous Test Set Adapter. Table 17-I-11 provides a
list of the hazardous chemicals that can be analyzed.

17-21. Soil Gas Probe

Draeger Safety, Inc.

Hazardous contaminants in soil may be the result of
leaking underground storage tanks, previous use as
an industrial waste site, or as the result of a landfill

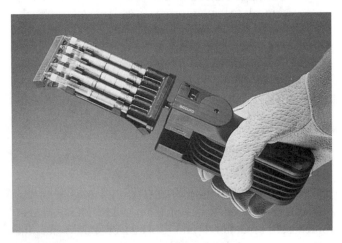

INSTRUMENT 17-20. Simultaneous Test Sets.

INSTRUMENT 17-21. Soil Gas Probe.

operation. A special method has been developed by
Draeger for measuring volatile subsurface contaminants
utilizing the Soil Gas Probe and Draeger Tubes. The
Draeger Soil Gas Probe is driven into the soil to a
desired depth. Next a Draeger Tube is inserted into the
sampling chamber and is then placed into the probe.
Using a short-term tube or sampling tube with a Draeger
pump provides an extremely simple, rapid, and economical method to conduct soil gas measurements (see Table
17-I-12). The Soil Gas Probe is suitable for centralizing
the concentration and tracing contaminant plumes. The
complete Soil Gas Probe Set for 1 meter depth includes
a drilling rod (25 × 1,000 mm), receiving chamber for
Draeger Tubes, 1.5M extension hose with socket,
grooved rod (25 × 1,000 mm), wrench set, brush set, and
transport box (see Instrument 17-21).

17-22. Draeger DLE Kit

Draeger Safety, Inc.

The Draeger Liquid Extraction Kit (DLE) is a contaminant detection method for analysis of liquid samples. The DLE Kit is the most economical screening
method available for fast, on-site determination of contaminants in industrial waste water, sewage treatment
and hazardous material spills. The measurement consists of two basic steps: 1) Extraction of the contami-

INSTRUMENT 17-22. Draeger DLE Kit.

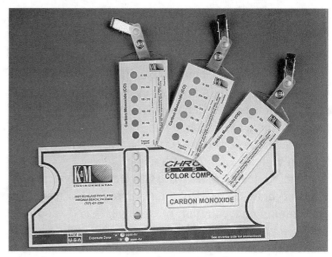

INSTRUMENT 17-23. ChromAir® Colormetric Badge System.

nant—during the extraction process, the contaminant in the liquid sample is transferred from the liquid phase into the gas phase. 2) Measurement of the contaminant—the extracted contaminant is measured from the headspace of the bottle by the Draeger Tube (see Table 17-I-13). The system for this special method consists of a Draeger Tube used with the Draeger Accuro Pump and a specially calibrated washing bottle. The method provides on-the-spot results at a fraction of the cost of other methods (see Instrument 17-22).

17-23. ChromAir® Colorimetric Badge System

K&M Environmental

The ChromAir System is a simple, reliable, economical monitoring system. It consists of a monitor and a color comparator. The ChromAir badge is a colorimetric direct-read monitor. It relies on the principle of diffusion. The ChromAir monitor provides the user with six exposure levels. Most ChromAir badges indicate from 1/10 to 2 times the time-weighted-average for an eight-hour work period. The scale printed on the badges and color comparators is based on exposure dose [parts per million times hour (ppm-hr)] (see Instrument 17-23). For higher resolution, the ChromAir badge may be used in conjunction with the ChromAir color comparator. ChromAir badges are available for different hazardous chemicals which are listed in Table 17-I-15.

17-24. SafeAir® Passive Monitoring Badges

K&M Environmental

The SafeAir system is a low-cost detection system which provides immediate visual indication of a chemical hazard. The SafeAir badge uses a coated technology

rather than impregnated filter paper, thus providing homogenous and stable color formation. A color change in the form of an "exclamation mark" warns the presence of the targeted hazard (see Instrument 17-24). The badge requires minimal training and no calibration, extra equipment, or laboratory analysis is required. For higher resolution and wider range, the SafeAir badge may be used in conjunction with the SafeAir color comparator. SafeAir badges are available for ammonia, aniline, arsine*, carbon dioxide*, carbon monoxide, chlorine, chlorine/chlorine dioxide, dimethyl amine, 1,1-dimethyl hydrazine*, formaldehyde, hydrazine*, hydrogen chloride*, hydrogen fluoride*, hydrogen sulfide, mercury, methyl chloroformate*, nitrogen dioxide, ozone, phosgene*, phosphine*, sulfur dioxide, and 2,4-toluene diisocyanate* (see Table 17-I-16). (* indicates SafeAir color comparators and validation reports are available).

INSTRUMENT 17-24. SafeAir® Passive Monitoring Badges.

17-25. Matheson-Kitagawa Toxic Gas Detector System

Matheson-Kitagawa

The Matheson-Kitagawa Toxic Gas Detector System is a complete sampling and analysis kit for on-the-spot readings. It is an excellent method for day-to-day checking, screening, QC in the lab, or plant spot testing. Nontechnical employees can operate the Matheson-Kitagawa System with a minimum of training.

The Model 8014KA Toxic Gas Detector System provides accurate, dependable, and reproducible results in determining concentration of toxic gases and vapors. It has been proven through extensive use by leading industrial companies and government agencies. One constant and reproducible sample volume reduces sampling and analysis errors—as opposed to other pump designs, there are no orifice changes or multiple strokes to keep track of. Calibrated detector tubes are available for many different gases and vapors, are shown in Table 17-I-17. The same basic sampling technique applies to all Matheson-Kitagawa Precision Detector Tubes (see Instrument 17-25). Only three easy steps are required to operate the detector: 1) break off the tips of a fresh detector tube, 2) insert the tube with arrow pointing toward the pump into the pump's sample inlet, 3) pull out the pump handle to automatically lock, drawing a 100-mL sample. A proprietary Sample Vue™, indicator shows when sampling is completed. Only one stroke is

needed for most analyses; no need for multiple volumes or stroke counters.

Matheson-Kitagawa precision detector tubes are formulated with high purity chemical reagents which absorb and react with the gas or vapor being measured. The reaction causes a colorimetric stain which varies in length to the concentration of the gas or vapor being measured. The length of stain is normally read directly off a scale printed on each tube. When used within their expiration date, the readings at 20°C (68°F) are designed to be within 5% to 10% of the true concentration. Temperature corrections for operating at other temperatures are normally unnecessary but are provided with those tubes requiring it.

17-26. Qualitative Analysis Tubes and Haz Mat Kit

Matheson-Kitagawa

Matheson's Qualitative Detector Tubes (Models 8070, 8075) provide fast on-the-spot identification of unknown gases and vapors. There is no need for cumbersome grab samples, time delayed laboratory analyses, expensive analytical instrumentation or complex decision-tree matrix approaches. And since calibration, electricity and battery charging are not necessary, the tubes are always ready for immediate use. Two types of tubes are available. Model 8014-186B identifies a broad range of organic compounds, such as gasoline, alcohols,

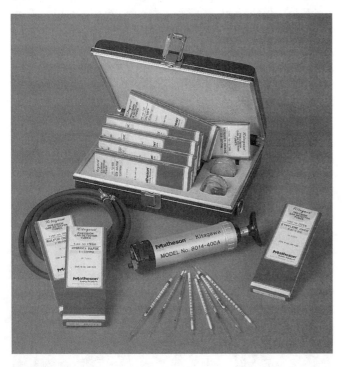

INSTRUMENT 17-25. Matheson-Kitagawa Toxic Gas Detector System.

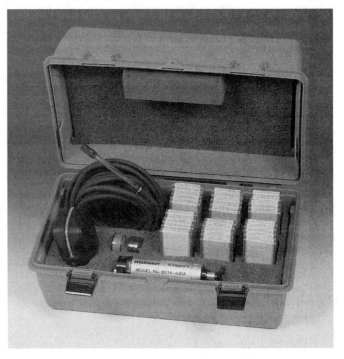

INSTRUMENT 17-26. Matheson-Kitagawa Qualitative Analysis Tubes and Haz Mat Kit.

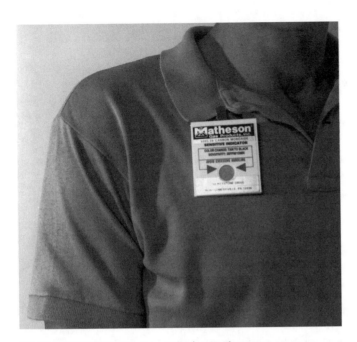

INSTRUMENT 17-27. Matheson Badge Dosimeter.

hydrocarbons, etc. Model 8014-131 identifies inorganic compounds, such as carbon monoxide, sulfur dioxide, chlorine, etc. Typically, both models are used in concert, to assure detection of both organic and inorganic compounds (see Instrument 17-26).

Principle of Operation: In operation, Matheson's Qualitative Tubes are used in the same way as conventional detector tubes. That is, the high precision Matheson-Kitagawa pump, Model 8014-400A, is used to draw the sample air through the tubes. However, unlike conventional tubes which are "length of stain" providing quantitative measurements, these tubes are comprised of several sections. Each section contains a unique, high purity blend of reagents that will adsorb and react with a particular gas or vapor, or family of gases and vapors. The resulting reaction causes a color to change. The unknown gas or vapor is determined by which section(s) changed color, and to what color they changed to. Model 8014-131 is used to detect inorganic compounds and consists of 5 sections, labeled "A" to "E." Only one tube is needed to provide a complete analysis for inorganics. Model 8014-186B is used to detect organic compounds and consists of 4 sections, labeled "A" to "D." Because of the extensive number of detectable organic compounds, two tubes are required for a complete analysis. One tube is used for "A" side sampling (section A closest to pump), and is followed by a second, fresh tube for "D" side sampling (section D closest to pump). The combined results are used to identify unknown substances. Table 17-I-18 lists the compounds that can be detected and their limit of detection.

17-27. Monitors & Badge Dosimeters

Matheson-Kitagawa

Matheson's Badge Dosimeters (Models 8005, 80069, 8007) provide an economical, easy-to-use method for monitoring personal exposure to seven toxic gases. Each badge is an instant reading detector and provides an immediate color change in the presence of specific toxic gases. Additionally a color comparison chart (not available for carbon monoxide or ozone) also allows the user to evaluate limited exposures as well as up to eight-hour time-weighted averages. These maintenance-free badges do not need activation and are ready for immediate use (see Table 17-I-19). No expensive analysis equipment or air samplers are required. Each badge clips easily to collar, shirt pocket or hat (see Instrument 17-27), or it can be used for area monitoring by affixing it to a wall or other support.

Each plaque weighs only 1/4 ounce and comes with a record form for recording employee name, date, time on, time off, elapsed exposure time and reading. Badges may be stored up to six months in unopened packs without deterioration. Usable life is one month after removing the badge from the package.

17-28. Matheson's Compressed Breathing Air Analysis Kit

Matheson-Kitagawa

The Matheson Model 8014BAK is an on-line analysis kit for ensuring the quality of compressed breathing

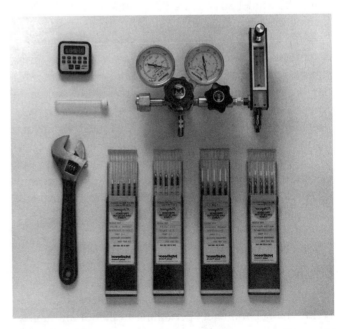

INSTRUMENT 17-28. Matheson Compressed Breathing Air Analysis Kit.

air. It quickly and easily measures the levels of carbon monoxide, carbon dioxide, oil mist, water vapor, and oxygen (see Table 17-I-20). Unlike other methods, there is no need to take a grab sample and analyze it off-line; the 8014BAK is designed to connect directly to the compressed breathing air source (see Instrument 17-28).

Principle of Operation: The Model 8014BAK system is essentially comprised of a pressure regulator, flowmeter, and a variety of detector tubes. In operation, measurements are made by passing the breathing air through each detector tube at a specified flow rate, pressure, and time interval. Each detector tube is formulated with a high purity reagent which adsorbs and reacts with the component being measured. This causes a colorimetric stain whose length is directly proportional to the amount of component in the breathing air. Its concentration can be read directly from the scale printed on each tube.

It is available with a choice of three connections- CGA 346, CGA 347, and 1/4" NPT Female. It is very important that the correct connection type be selected to match the application.

The Model 8014BAK-01 is fitted with a CGA 346 connection, and is rated for inlet pressures of 0-3000 psig. This model should be selected for analyzing compressed air in U.S. D.O.T.-approved cylinders with a Stamped Service Pressure in the range of 0–3000 psig. The Model 8014BAK-03 is fitted with a CGA 347 connection, and is rated for inlet pressures of 3001–5500 psig. This model should be selected for analyzing compressed air in U.S. D.O.T. approved cylinders with a Stamped Service Pressure in the range of 3001–5500 psig. The Model 8014 BAK-02 is fitted with a 1/4" NPT Female connection, and is rated for inlet pressures of 0–400 psig. This model should be selected for analyzing compressed air from non-cylinder sources having pressures no greater than 400 psig. [*Caution:* Adapters must not be used which connect a high pressure source to equipment rated at a lower pressure.] *Some Typical Applications:* The Model 8014BAK is ideal for anyone involved with the filling, generating, or usage of compressed breathing air. It has been proven through use in a variety of industries and applications such as: Emergency air packs/respirators, Fire departments/rescue squads, Scuba/diving, Hazardous waste cleanup.

17-29. Matheson's Indoor Air Quality Test Kit

Matheson-Kitagawa

The Matheson Model 8078 is a complete kit for analyzing many parameters pertaining to indoor air quality. All of the items included are also available as standalone products. The heart of the Model 8078 kit is the

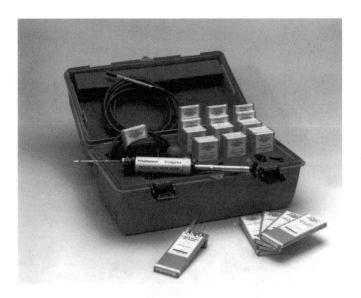

INSTRUMENT 17-29. Matheson Indoor Air Quality Test Kit.

Matheson-Kitagawa precision air sampling pump. It is used in conjunction with a variety of detector tubes. Included in the kit are tubes for measuring the concentration of formaldehyde, carbon monoxide, carbon dioxide and organic hydrocarbons. And although not included in the kit as standard, tubes are available for ammonia, ozone and a host of other gases and vapors. Qualitative tubes are also included for analysis of unknown materials. An air flow indicator kit (smoke tubes) is provided for determining ventilation patterns and efficiencies. And a 10-meter extension sampling hose is provided for remote sampling in hard-to-reach places. All of these products are packaged with relevant maintenance items in a convenient, extremely durable carrying case (see Instrument 17-29).

17-30. MSA Kwik-Draw Deluxe and Toximeter II Detector Tube Pumps

Mine Safety Appliances Company

The MSA Kwik-Draw Deluxe and Toximeter II pumps are designed for use with MSA's full selection of nearly 200 different detector tubes. They are partially listed in Table 17-I-21. Nearly all of MSA's detector tubes are printed with calibration scales that illustrate the concentration of the target contaminant. The Kwik-Draw Deluxe is a manual, bellows-type pump that delivers a 100 mL volume. The Kwik-Draw Deluxe features a patented end-of-stroke indicator and stroke counter. The Toximeter II is an automatic detector tube pump that requires no manual squeezing. It can be programmed for up to 250 pump strokes (see Instrument 17-30).

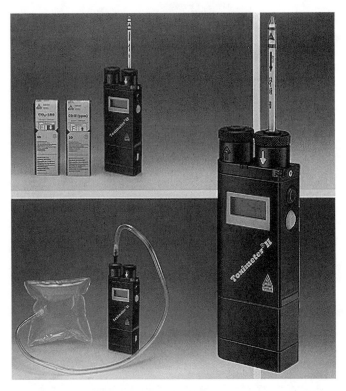

INSTRUMENT 17-30. MSA Kwik-Draw Deluxe and Toximeter II Detector Tube Pumps.

17-31. MSA Haz Mat Kit

Mine Safety Appliances Company

MSA offers a specialized Haz Mat Response Kit for use at Haz Mat spills. This kit features detector tubes for 12 different chemical classes which allows quick identification of hazards at spills (see Table 17-I-22). The MSA Haz Mat Response Kit offers a special manifold for testing four detector tubes at one time (see Instrument 17-31).

17-32. MSA Indoor Air Quality Kit

Mine Safety Appliances Company

The MSA Indoor Air Quality Kit checks for "sick building" syndrome, and can be used to identify worker complaints of headaches, dizziness, allergies, and nausea in the workplace. The kit includes tubes for carbon monoxide, carbon dioxide, formaldehyde, ozone and water vapor (see Table 17-I-23)—all of which are common components in office settings. The box of detector tubes includes a thermometer on the side to monitor temperature (see Instrument 17-32).

17-33. The PiezOptic Personal Dosimeter System

PiezOptic, Ltd.

Designed to overcome the shortcomings of most detector tube and passive diffusion direct-reading devices (poor sensitivity, accuracy and reproducibility), the PiezOptic system consists of a range of single-use passive badges and a generic reader used to quantify the results. The small ($4.0 \times 1.5 \times 0.5$ in), light-weight (<25g) badges are supplied foil-packed and ready-to-use. They can be clipped onto the clothing near the mouth to conveniently monitor the breathing zone. After the exposure period, which can be either short-term (15-30 min) or long-term (4-12h), the badge is placed into the reader and an accurate, quantitative value for exposure is obtained in a few seconds. The patented piezofilm sensing technology is extremely sensitive and precise allowing the detection of, for example, gluteraldehyde to ppb levels with a precision of around 5%. Bulky pumps and time-consuming pre-calibrations are not required and the result does not depend on the subjective evaluation of stain length or comparative color. Badges (long-term and short-term) are currently available for styrene, ozone, gluteraldehyde, formalde-

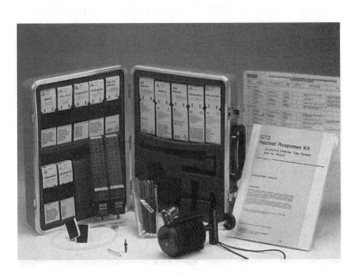

INSTRUMENT 17-31. MSA Haz Mat Kit.

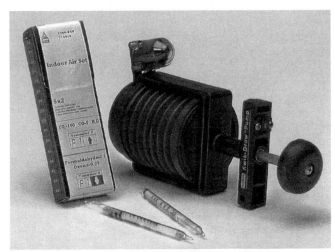

INSTRUMENT 17-32. MSA Indoor Air Quality Kit.

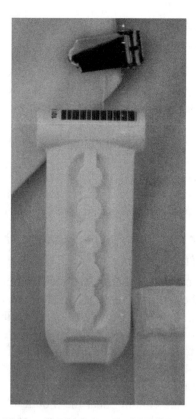

INSTRUMENT 17-33a. PiezOptic Personal Dosimeter Badge.

hyde, hydrazine, NO_2 and CO. Many others are becoming available (see Instruments 17-33a and 17-33b).

17-34. LP-1200, SampleRAE, RAE Tubes

RAE Systems, Inc.

RAE Systems is a manufacturer who started producing their own colorimetric detector tubes and pumps since 1997. As of 2000, they manufactured 46 different types of tubes for a total of 19 organic and inorganic gases and vapors (see Table 17-I-24). Their tubes were designed with a two-year shelf life (see Instrument 17-34a). Most of

INSTRUMENT 17-34a. RAE Systems Detector Tubes.

the tubes work with a one-pump stroke. RAE Systems offers a hand-held piston pump (see Instrument 17-34b) and an automatic sampling pump. The hand pump (model LP-1200) samples either 50 or 100 cc volumes. The automatic pump (model SampleRAE) (see Instrument 17-34c) can be used with detector tubes to sample fixed volume samples; or it can be used to sample high-volume samples (multiple pump strokes) from 50-950 cc in 50 cc increments. The 12 ounce intrinsically-safe unit is battery-powered (16-hours with 4 AA alkaline batteries), has a microprocessor, an LCD direct readout for the volume or flowrate, and comes with a calibration tube. SampleRAE can also be used with sorbent tubes and can collect a

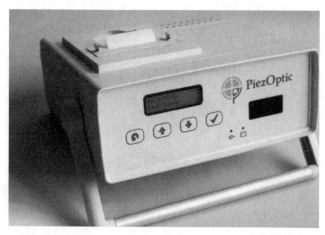

INSTRUMENT 17-33b. PiezOptic Personal Dosimeter Reader.

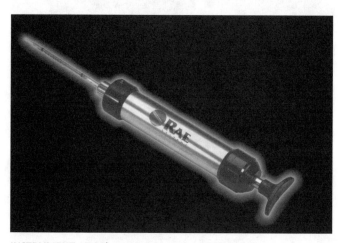

INSTRUMENT 17-34b. RAE Systems LP-1200 Pump.

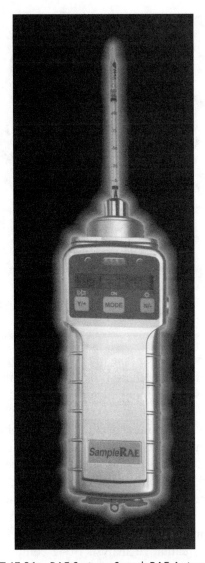

INSTRUMENT 17-34c. RAE Systems SampleRAE Automatic Pump.

fixed-volume of sample in a gas bag. RAE Systems funded an independent university lab study which resulted to a conclusion that the RAE tubes are interchangeable with Sensidyne and Kitigawa tubes; also, that their LP-1200 hand pump is interchangeable with the Gastek/Sensidyne GV/100 and Kitagawa 8014-400A hand pumps.

17-35. Sensidyne/Kitagawa Haz Mat Kits

Sensidyne, Inc.

The Sensidyne/Kitagawa Haz Mat Kit is a portable hazardous material detection kit requiring no electrical power or user calibration. The kit uses the Model AP-IS hand-held piston pump and extension cable and incorporates 15 different types of detector tubes for commonly encountered substances, as shown in Table 17-I-25. The kit includes two laminated sampling logic charts which allow the user to identify unknown compounds using the 15 tubes and incorporates all of these elements plus an air flow indicator, smoke tube kit, and 15 boxes of detector tubes in a hard-sided carrying case. This system is expandable and is ultimately capable of measuring over 200 gases (see Instrument 17-35).

17-36. Sensidyne/Kitagawa Precision Gas Detector System

Sensidyne, Inc.

Over 200 gases and vapors can be measured with the High-precision Gas Sampling System using the detector tubes which are partially listed in Table 17-I-26. The two major components are: 1) direct-reading detector tubes

INSTRUMENT 17-35. Sensidyne/Kitagawa Haz Mat Kit.

INSTRUMENT 17-36. Sensidyne/Kitagawa Precision Gas Detector System.

and 2) the high-precision, piston-type volumetric pump (see Instrument 17-36). Each detector tube contains a reagent that is specifically sensitive to a particular vapor or gas. These reagents are contained on fine-grain silica gel, activated alumina, or other adsorbing media (depending upon application requirements), inside a constant-inner-diameter, hermetically sealed glass tube. To sample, the operator snaps off both breakaway ends of a tube, inserts the tube into the hand-held pump, and pulls the pump handle out. A measured volume of ambient air is drawn inside the tube. The reagent changes color instantly and reacts quantitatively to provide a length-of-stain indication. The farther the color stain travels along the tube, the higher the concentration of gas. The calibration mark on the tube, at the point where the color stain stops, gives the concentration. Calibration scales for the detector tubes are printed on the basis of individual production lots. Calibration scales are in ppm, mg/L, or %, depending on the substance to be measured and the desired measuring range. Every tube and tube box carries the quality control number, chemical symbol, and the expiration date. The expandable measuring range permits measurement of concentrations above or below the printed scale simply by increasing or decreasing pump strokes.

17-37. Sensidyne/Kitagawa Qualitative Multistage Tubes

Sensidyne, Inc.

Provides qualitative detection of 12 inorganic gases simultaneously with simple single pump stroke operation and results in 20 seconds. Ideal for haz mat, confined space entry, and fire site re-entry testing. The Inorganic Qualitative Tube detector tube (Catalog No.

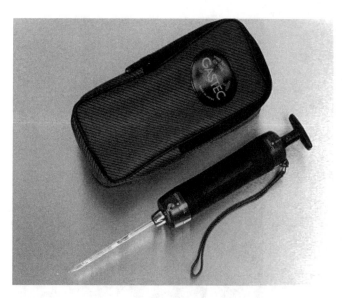

INSTRUMENT 17-37. Sensidyne/Kitagawa Qualitative Multistage Tubes.

131) incorporates a unique multi-layer design to qualitatively test for 12 common toxic gases simultaneously. These gases are ammonia, hydrogen chloride, hydrogen sulfide, chlorine, sulfur dioxide, nitrogen dioxide, carbon monoxide, amines, acetic acid, phosphine, acetylene, and methyl mercaptan. An operator simply breaks off the ends of the sealed glass tube and places it into the Model AP-IS piston pump, observing the directional arrow. After taking one pump stroke, the operator waits 20 seconds and then compares the tube to a color chart (see Instrument 17-37). A second tube (Catalog # 186B) for organic qualitative analysis of 41 compounds works in 30 seconds and utilizes the reverse side of the same color chart. The system is expandable with direct-reading quantitative detector tubes for over 200 substances.

TABLE 17-I-1. Index of Direct-Reading Instrument Types and Manufacturers

	Manufacturer	Detector Tubes		Passive Dosimeter Tubes	Passive Badges
		Short-Term	Long-Term		
1.	Acculab Technologies, Inc.				✓
2.	American Gas & Chemical Co., Ltd.				✓
3.	Assay Technology, Inc.				✓
4.	Bacharach, Inc.	✓			✓
5.	Draeger Safety, Inc.	✓	✓	✓	✓
6.	K & M Environmental				✓
7.	Matheson-Kitagawa	✓			✓
8.	Mine Safety Appliances Co.	✓	✓		
9.	PiezOptic, Ltd.				✓
10.	RAE Systems	✓			
11.	Sensidyne, Inc.	✓	✓		

TABLE 17-I-2. LEAK-TEC Personal Protection Indicators (Instrument 17-2)

Gas	Part #	Sensitivity	Color Change*
Ammonia	A-15	25 ppm/5 min	Yellow to blue
Carbon monoxide	CO-50	50 ppm	Tan to black
Chlorine	C-2	1 ppm/15 min	White to yellow
Hydrazine	H-5	0.1 ppm/15 min	White to yellow
Hydrogen sulfide	HS-5	10 ppm/10 min	White to brown
Nitrogen dioxide	N-1	5 ppm/15 min	White to yellow
Ozone	O-1	0.1 ppm/15 min	White to brown

* A color chart is available for all badges except carbon monoxide and ozone

TABLE 17-I-3. Assay Technology *ChemChip*™ Personal Monitoring Badges (Instrument 17-3)

Analyte	for Sampling	Range (ppm)	Item No.
Ethylene Oxide	8-hr TWA	0.1–6.0	502
Ethylene Oxide	15-min STEL	1–60	506

TABLE 17-I-4. Bacharach/GMD System (Instrument 17-4)

Part Number	Description
2755-0610	TDI Starter Kit
2756-0610	Hydrides Starter Kit
2750-1010	Phosgene Dosimeter Badge (std. model) – 1 day
1753-0610	Hydrazine Starter Kit
2780-0700	MDI Test Kit
2780-0500	TDI Test Kit
2780-0800	HDI Test Kit
2780-0900	NDI Test Kit

TABLE 17-I-5. Bacharach Carbon Monoxide Indicator Ranges

Complete Kit	Gas Type	Range
19-0240	CO	0–0.2%
19-0241	CO	0–0.5%
19-0244	CO	0–2,000 ppm
19-0245	CO	0–5,000 ppm

TABLE 17-I-6. Draeger CMS Chips (Instrument 17-11)

Chip	Measuring Range	Catalog No.
Ammonia	2–50 ppm	6406130
Ammonia	10–150 ppm	6406020
Benzene*	0.2–10 ppm	6406030
Benzene*	0.5–10 ppm	6406160
Benzene	10–250 ppm	6406280
Carbon Dioxide	200–3,000 ppm	6406190
Carbon Dioxide	1,000–25,000 ppm	6406070
Carbon Dioxide	1–20% vol	6406210
Carbon Monoxide	5–150 ppm	6406080
Chlorine	0.2–10 ppm	6406010
Hydrochloric Acid	1–25 ppm	6406090
Hydrochloric Acid	20–500 ppm	6406140
Hydrogen Cyanide	2–50 ppm	6406100
Hydrogen Sulfide	2–50 ppm	6406050
Hydrogen Sulfide	20–500 ppm	6406150
Hydrogen Sulfide	100–2,500 ppm	6406220
Mercaptan	0.25–6 ppm	6406360
Nitrogen Dioxide	0.5–25 ppm	6406120
Nitrous Fumes	0.5–15 ppm	6406060
Nitrous Fumes	10–200 ppm	6406240
Perchloroethylene	5–150 ppm	6406040
Petroleum Hydrocarb.	20–500 ppm	6406200
Petroleum Hydrocarb.	100–3,000 ppm	6406270
Phosgene	0.05–2 ppm	6406340
Phosphine	1–25 ppm	6406410
Phosphine	20–500 ppm	6406420
Phosphine	200–5,000 ppm	6406500
Propane	100–2,000 ppm	6406310
Sulfur Dioxide	0.4–10 ppm	6406110
Sulfur Dioxide	5–150 ppm	6406180
Toluene	10–300 ppm	6406250
Training Chip	—	6406290
Vinyl Chloride	0.3–10 ppm	6406170
Vinyl Chloride	10–250 ppm	6406230
Xylene	10–300 ppm	6406260

TABLE 17-I-7. Short-Term Draeger Tubes (Instrument 17-12)

Note: The list below is only a partial list of products. The full list can be found at *www.draeger-usa.com.*

Gases and Vapors	Draeger Tube	Measuring Range	Part No.
Acetaldehyde	Acetaldehyde 100/a	100–1,000 ppm	6726665
Acetic Acid	Acetic Acid 5/a	5–80 ppm	6722101
Acetic Anhydride	Formic Acid 1/a	Qualitative	6722701
Acetone	Acetone 100/b	100–12,000 ppm	CH22901
Acetylene	Petroleum Hydrocarbons100/a	100–2,500 ppm	6730201
Acid Compounds in air	Acid Test	Qualitative	8101121
Acrolein	Dimethyl Sulfide 1/a	0.1–10 ppm	6728451
Acrylonitrile	Acrylonitrile 0.5/a	0.5–20 ppm	6728591
Air Current	Smoke Tube	—	CH25301
Aliphatic Hydrocarbons	Hydrocarbon 2	2–23 mg/L	CH25401
(Boiling Range 50–200°C)			
2-Aminoethanol	Ammonia 0.25/a	0.5–6 ppm	8101711
2-Aminopropane	Cyclohexylamine 2/a	2–30 ppm	6728931
Ammonia 0.25/a	Ammonia	0.25–3 ppm	8101711
	Ammonia 2/a	2–30 ppm	6733231
	Ammonia 5/a	5–700 ppm	CH20501
	Ammonia 5/b	2.5–100 ppm	81019,41
	Ammonia 0.5%/a	0.05–10 Vol. %	CH31901
n-Amyl acetate	Ethyl Acetate 200/a	200–3,000 ppm	CH20201
Aniline	Aniline 0.5/a	0.5–10 ppm	6733171
	Aniline 5/a	1–20 ppm	CH20401
Antimony Hydride (Stibine)	Arsine 0.05/a	0.05–3 ppm	CH 25001
Arsenic Trioxide	Arsenic Trioxide 0.2/a	0.2 mg/m3	6728951
Arsine	Arsine 0.05/a	0.05–60 ppm	CH25001
Aziridine	Ammonia 0.25/a	0.25–3 ppm	8101711
Basic Cmpds. in air Amine Test	Qualitative	—	8101061

TABLE 17-I-8. Long-Term Draeger Tubes (Instrument 17-13)

Draeger Tube	Measuring Range	Maximum Sampling Time (hrs)	Part No.
Acetic Acid 5/a-L	1.25–40 ppm	4	6733041
Acetone 500/a-L	62.5–10,000 ppm	8	6728731
Ammonia 10/a-L	2.5–100 ppm	4	6728231
Benzene 20/a-L	10–200 ppm	2	6728221
Carbon Dioxide 1000/a-L	250–6,000 ppm	4	6728611
Carbon Disulfide 10/a-L	1.25–100 ppm	8	6728621
Carbon Monoxide 10/a-L	2.5–100 ppm	4	6728741
Carbon Monoxide 50/a-L	6.25–500 ppm	8	6728121
Chlorine 1/a-L	0.13–20 ppm	8	6728421
Ethanol 500/a-L	62.5–8,000 ppm	8	6728691
Hydrocarbons 100/a-L	25–3,000 ppm	4	6728571
Hydrochloric Acid 10/a-L	1.25–50 ppm	8	6728581
Hydrocyanic Acid 10/a-L	1.25–120 ppm	8	6728441
Hydrogen Sulfide 5/a-L	0.63–60 ppm	8	6728141
Methylene Chloride 50/a-L	12.5–800 ppm	4	6728881
Nitrogen Dioxide 10/a-L	1.25–100 ppm	8	6728281
Nitrous Fumes 5/a-L (NO + NO2)	1.25–50 ppm	4	6728911
Nitrous Fumes 50/a-L (NO + NO2)	12.5–350 ppm	4	6728191
Perchloroethylene 50/a-L	12.5–300 ppm	4	6728671
Sulfur Dioxide 2/a-L	0.5–20 ppm	4	6728921
Sulfur Dioxide 5/a-L	1.25–50 ppm	4	6728151
Toluene 200/a-L	25–4,000 ppm	8	6728271
Trichloroethylene 10/a-L	2.5–200 ppm	4	6728291
Vinyl Chloride 10/a-L	1–50 ppm	10	6728131

TABLE 17-I-9. Draeger Long-Term Diffusion Tubes (Instrument 17-14)

Draeger Tubes	Range in Absolute Units	Range of Measurement for Max. Period of Use (8 hrs)	Part No.
Acetic Acid 10/a-D	10–200 ppm × h	1.3–25 ppm	8101071
Ammonia 20/a-D	20–1,500 ppm × h	2.5–188 ppm	8101301
Butadiene 10/a-D	10–300 ppm × h	1.3–40 ppm	8101161
Carbon Dioxide 500/a-D	500–20,000 ppm × h	65–2,500 ppm	8101381
Carbon Dioxide 1%/a-D	1–30 Vol.% × h	0.13–3.8 Vol.%	8101051
Carbon Monoxide 50/a-D	50–600 ppm × h	6.3–75 ppm	6733191
Ethanol 1000/a-D	1,000–25,000 ppm × h	125–3,100 ppm	8101151
Ethyl Acetate 500/a-D	500–10,000 ppm × h	63–1,250 ppm	8101241
Hydrochloric Acid 10/a-D	10–200 ppm × h	1.3–25 ppm	6733111
Hydrocyanic Acid 20/a-D	20–200 ppm × h	2.5–25 ppm	6733221
Hydrogen Sulfide 10/a-D	10–300 ppm × h	1.3–38 ppm	6733091
Nitrogen Dioxide 10/a-D	10–200 ppm × h	1.3–25 ppm	8101111
Perchloroethylene 200/a-D	200–1,500 ppm × h	25–188 ppm	8101401
Sulfur Dioxide 5/a-D	5–150 ppm × h	0.6–19 ppm	8101091
Toluene 100/a-D	100–3,000 ppm × h	13–380 ppm	8101421
Trichloroethylene 200/a-D	200–1,000 ppm × h	25–125 ppm	8101441
Water Vapor 5/a-D	5–100 mg/liter × h	0.6–12.5 mg/liter	8101391
Diffusion Tube Holder	Package of 3		6733014

TABLE 17-I-10. Draeger Haz Mat Kit (Instrument 17-15)

Detector Tubes	Part Number
Polytest	CH28401
Ethyl acetate 200/a	CH20201
Methyl bromide 5/b	CH27301
Hydrazine 0.25/a	CH31801
Benzene 0.05	CH24801
Hydrocarbons 0.1%b	CH26101
Acetone 100/b	CH22901
Carbon monoxide 10/b	CH20601
Alcohol 100/a	CH29701
Carbon dioxide 0.1%/a	CH23501
Hydrocyanic acid 2/a	CH25701
Hydrogen sulfide 5/b	CH29801
Nitrous fumes 0.5/a	CH29401
Trichloroethylene 10/a	CH24401
Chlorine 0.2/a	CH24301
Oxygen 5%/B	6728081
Formic acid 1/a	6722701

TABLE 17-I-11. Draeger Simultaneous Test Sets (Instrument 17-20)

Description	Catalog No.
Simultaneous Test Set I, Inorganic fumes	8101735
Acid Gases, e.g., Hydrochloric Acid	
Basic Gases, e.g., Ammonia	
Carbon Monoxide	
Hydrocyanic Acid	
Nitrous Gases, e.g., Nitrogen Dioxide	
Simultaneous Test Set II, Inorganic fumes	8101736
Carbon Dioxide	
Chlorine	
Hydrogen Sulfide	
Phosgene	
Sulfur Dioxide	
Simultaneous Test Set III, Organic Vapors	8101770
Alcohols, e.g., Methanol	
Aliphatic Hydrocarbons, e.g., n-Hexane	
Aromatics, e.g., Toluene	
Chlorinated Hydrocarbons, e.g., Perchloroethylene	
Ketone, e.g., Acetone	
Accuro Pump, required	6400000
Adapter	6400090

TABLE 17-I-12. Draeger Soil Analysis with Soil Gas Probe (Instrument 17-21)

Substance	Draeger Tube	Measuring Range	Part No.
Inorganic / Organic Substances			
Acetic Acid	Acetic Acid 5/a	5–80 ppm	6722101
Acid Compounds	Acid Test	Qualitative	8101121
Ammonia	Ammonia 5/a	5–700 ppm	CH20501
Benzene	Benzene 2/a	2–60 ppm	8101231
BTX-Aromatics	Toluene 5/b	5–300 ppm	8101661
Carbon Dioxide	Carbon Dioxide 100/a	100–3,000 ppm	8101811
Carbon Dioxide	Carbon Dioxide 0.5%/a	0.5–10 Vol.%	CH31401
Carbon Dioxide	Carbon Dioxide 5%/A	5–60 Vol.%	CH20301
Carbon Tetrachloride	Carbon Tetrachloride 1/a	1–15 ppm	8101021
Chloroform	Chloroform 2/a	2–10 ppm	6728861
Dichloromethane	Methylene Chloride 100/a	100–2,000 ppm	6724601
Hydrocarbon Screening Test	Polytest	Qualitative	CH28401
Hydrocyanic Acid (Cyanide)	Hydrocyanic Acid 2/a	2–30 ppm	CH25701
Hydrogen Sulfide	Hydrogen Sulfide 1/d	1–200 ppm	8101831
Mercaptan	Mercaptan 0.5/a	0.5–5 ppm	6728981
Natural Gas (Methane)	Natural Gas Test	Qualitative	CH20001
n-Octane	Petroleum Hydrocarb.10/a	10–300 ppm	8101691
n-Octane	Petroleum Hydrocarb.100/a	100–2,500 ppm	6730201
Organic Arsenic Compounds and Arsine	Organic Arsenic Compounds and Arsine	Qualitative	CH26303
Organic Basic Nitrogen Compounds	Organic Basic Nitrogen Compounds	Qualitative	CH25903
Perchloroethylene	Perchloroethylene 0.1/a	0.1–4 ppm	8101551
Perchloroethylene	Perchloroethylene 2/a	2–300 ppm	8101501
Perchloroethylene	Perchloroethylene 10/b	10–500 ppm	CH30701
Phosgene	Phosgene 0.02/a	0.02–1 ppm	8101521
Phosgene	Phosgene 0.25/b	0.25–75 ppm	CH28301
1,1,1-Trichloroethane	Trichloroethane 50/d	50–600 ppm	CH21101
Thioether	Thioether	Qualitative	CH25803
Toluene	Toluene 50/a	50–400 ppm	8101701
Trichloroethylene	Trichloroethylene 2/a	2–200 ppm	6728541
Trichloroethylene	Trichloroethylene 10/a	50–500 ppm	CH24401
Vinyl Chloride	Vinyl Chloride 1/a	1–50 ppm	6728031
Water Vapor	Water Vapor 1/a	1–20 mg/L	8101081
o-Xylene	Xylene 10/a	10–400 ppm	6733161
Alternative to Direct Reading Tubes Listed Above			
Activated Charcoal Tube (Type B/G)			8101821
Activated Charcoal Tube (NIOSH size)			6728631

TABLE 17-I-13. Draeger Analysis with DLE Kit (Instrument 17-22)

Substance	Draeger Tube	Measuring Range	Part No.
Aliphatic Hydrocarbons			
Diesel fuels	Petroleum Hydrocarbons 10/a	0.5–5 mg/L	8101691
Diesel fuels (soil analysis)	Petroleum Hydrocarbons 10/a	Qualitative	8101691
Fuels	Petroleum Hydrocarbons 10/a	0.5–30 mg/L	8101691
Fuels (soil analysis)	Petroleum Hydrocarbons 10/a	Qualitative	8101691
Jet Fuels	Petroleum Hydrocarbons 10/a	0.5–5 mg/L	8101691
Jet Fuels (soil analysis)	Petroleum Hydrocarbons 10/a	Qualitative	8101691
n-Octane	Petroleum Hydrocarbons 10/a	0.1–2 mg/L	8101691
n-Octane	Petroleum Hydrocarbons 100/a	2–25 mg/L	6730201
Aromatic Hydrocarbons			
BTX-Aromatics	Toluene 5/b	0.2–5 mg/L	8101161
BTX-Aromatics (oil muds/emulsions)	Toluene 5/b	Qualitative	8101161
BTX-Aromatics (soil analysis)	Toluene 5/b	2–50 mg/Kg	8101161
Benzene	Benzene 2/a	0.5–5 mg/L	8101231
Toluene	Toluene 50/a	1–10 mg/L	8101701
Xylene (o, m, p)	Xylene 10/a	0.3–10 mg/L	6733161
Chlorinated Hydrocarbons (Volatile)			
Carbon Tetrachloride	Carbon Tetrachloride 5/c +		CH27401
	Activation tube	0.2–4 mg/L	8101141
Chloroform	Chloroform 2/a	0.05–0.75 mg/L	6728861
Dichloromethane	Methylene Chloride 100/a	5–100 mg/L	6724601
Multiphase system	Chloroform 2/a	Qualitative	6728861
Multiphase system	Methyl Bromide 0.5/a	Qualitative	8101671
Multiphase system	Perchloroethylene 0.1/a	Qualitative	8101551
Multiphase system	Perchloroethylene 2/a	Qualitative	8101501
Multiphase system	Trichloroethane 50/d	Qualitative	CH21101
Oil muds/emulsions	Chloroform 2/a	Qualitative	6728861
Oil muds/emulsions	Methyl Bromide 0.5/a	Qualitative	8101671
Oil muds/emulsions	Perchloroethylene 0.1/a	Qualitative	8101551
Oil muds/emulsions	Perchloroethylene 2/a	Qualitative	8101501
Oil muds/emulsions	Trichloroethane 50/d	Qualitative	CH21101
Perchloroethylene	Perchloroethylene 0.1/a	10–80 µg/L	8101551
Perchloroethylene	Perchloroethylene 2/a	0.1–4 mg/L	8101501
Soil analysis	Chloroform 2/a	Qualitative	6728861
Soil analysis	Perchloroethylene 0.1/a	Qualitative	8101551
Soil analysis	Perchloroethylene 2/a	Qualitative	8101501
1,1,1 -Trichloroethane	Trichloroethane 50/d	0.5–5 mg/L	CH21101
Trichloroethylene	Perchloroethylene 0.1/a	10–100 µg/L	8101551
Trichloroethylene	Perchloroethylene 2/a	0.1–1 mg/L	8101501
Trichloroethylene	Trichloroethylene 2/a	0.2–3 mg/L	6728541
Inorganic Substances			
Ammonia	Ammonia 0.25/a	1.5–10 mg/L	8101711
Ammonia	Ammonia 0.25/a	10–100 mg/L	8101711
Hydrocyanic Acid	Hydrocyanic Acid 2/a	0.5–10 mg/L	CH25701
Hydrogen Sulfide	Hydrogen Sulfide 0.2/a	50–500 µg/L	8101461
Hydrogen Sulfide	Hydrogen Sulfide 5/b	0.5–10 mg/L	CH29801
Organic acids			
Acetic Acid	Acetic Acid 5/a	0.5–20 g/L	6722101
Formic Acid	Acetic Acid 5/a	1–20 g/L	6722101
Organic Acids	Acetic Acid 5/a	0.5–15 g/L	6722101
Propionic Acid	Acetic Acid 5/a	0.3–10 g/L	6722101

TABLE 17-I-14. Draeger DLE Kit (Instrument 17-22)

Description	Catalog No.
Deluxe Kit includes instructions, tablet and items listed below	4052944
Model 31 Pump	6726065
Spare Parts Kit for Model 31 Pump	6727211
Screwdriver for Model 31 Pump	4039003
Wrench for Model 31 Pump	CH06754
Automatic Stroke Counter	6726124
Carbon Pre-tubes, 10	CH24101
Calibrated Gas Wash Bottle	6400016
Measuring Jug, 250 mL	6400029
Deluxe Tube Opener	6400010
Thermometer (–10 to 60°C)	6400028
Pocket Calculator	9099399
Deluxe Carrying Case, rigid	8711942

TABLE 17-I-15. ChromAir® Badges (Instrument 17-23)

Analyte	Threshold level (ppm-hr)	Minimum detectable conc. in 8 hours (ppm)	Part #
Acetone	20–24000	2.50	380020
Ammonia	4–300	0.50	380003
Carbon Disulfide[1]	0.5–30	0.06	380011
Carbon Monoxide	10–525	1.25	380008
Chlorine	0.4–13	0.05	380004
Chlorine Dioxide	0.1–1.4	0.013	380024
Ethanol	62–7360	7.75	380015*
Ethylene	5–800	0.63	380028
Formaldehyde	0.3–12	0.04	380007
Glutaraldehyde	STEL: (15min)	0.04–0.95	380017
Hydrazine	0.01–0.8	0.002	380012
Hydrogen Sulfide	1–240	0.13	380009
Mercury	0.125–1.6 mg/m³·hr	0.015 mg/m³	380018
Methanol	27–3200	3.38	380015
Methyl Ethyl Ketone	18–21600	2.25	380020*
Methyl Isobutyl Ketone	16–19200	2.0	380020*
Nitrogen Dioxide	0.5–13.0	0.06	380006
Ozone[2]	0.08–1.6	0.01	380010
Sulfur Dioxide	0.1–16	0.013	380005

[1] 3 ppm H_2S causes color development in cell 1; 10 ppm H_2S causes color development in cell 2.
[2] Ozone monitor is ten times more sensitive to ozone than to nitrogen dioxide.
* Coefficient must be applied to scale printed on badge.

TABLE 17-I-16. SafeAir® Badges (Instrument 17-24)

Analyte	Threshold level (ppm-hr)	MDC+ in 8 hours (ppm)	MRST* (hrs)	SRST** (mins)	Part #
Ammonia	4.0 ppm·hr	0.50 ppm	48	15	380020
Aniline	0.2 ppm·hr	0.025 ppm	48	5	382021
Arsine	18.0 ppb·hr	2.25 ppb	12	15	382030
Carbon Dioxide	8,000 ppm·hr	1,000 ppm	10	15	380003
Carbon Monoxide	7 ppm·hr	1 ppm	10	15	380011
Chlorine	0.18 ppm·Ahr	0.023 ppm	48	15	382009
Chlorine/Chlorine Dioxide	Cl₂: 0.18 ppm·hr ClO₂: 0.2 ppm·hr	Cl₂: 0.025 ppm ClO₂: 0.025 ppm	10	15	380004
Dimethyl Amine	5 ppm·hr	0.625 ppm	48	5	382019
1,1-Dimethyl Hydrazine	Front: 30 ppb·hr Back: 10 ppb·hr	Front: 3.75 ppb Back: 1.25 ppb	48	5	380015
Formaldehyde	0.4 ppm·hr	0.05 ppm	16	15	382011
Hydrazine	8.0 ppb·hr	1.0 ppb	48	5	382002
Hydrogen Chloride	2.0 ppm	STEL	15 min	15	382024
Hydrogen Fluoride	2.8 ppm	STEL	15 min	15	382029
Hydrogen Sulfide	2 ppm·hr	0.25 ppm	48	15	380017
Mercury/Dual Level	Front: 0.25 mg/m³·hr Back: 0.08 mg/m³·hr	Front: 0.031 mg/m³·hr Back: 0.01 mg/m³·hr	48	15	382005
Methyl Chloroformate	0.025 ppm·hr	0.0083 ppm (3 hrs)	3	10	382000
Nitrogen Dioxide	1 ppm·hr	0.125 ppm	10	15	380018
Ozone	0.05 ppm·hr	0.006 ppm	48	15	380015
Phosgene	0.9 ppm·min	0.015 ppm	3 days	1	382000
Phosphine	5.0 ppb·hr	0.625 ppb	12	15	382031
Sulfur Dioxide	0.2 ppm·hr	0.025 ppm	48	15	380020
2,4-Toluene diisocyanate (TDI)	5.0 ppb·hr	0.6 ppb	24	15	382001

* Maximum Recommended Sampling Time
** Shortest Recommended Sampling Time
+ Minimum Detectable Concentration

TABLE 17-I-17. Matheson-Kitagawa Precision Detector Tubes (Instrument 17-25)

Substance to be measured	Measuring Range (ppm)	Model No.
Acetaldehyde	5–140	8014-133sb
Acetic acid	1–50	8014-216s
Acetone	100–5000	8014-102sd
Acetylene	50–1000	8014-101s
Acrolein	0.005–1.8%	8014-136
Acrylonitrile	0.25–20	8014-128sd
Allyl alcohol	20–500	8014-184s
Ammonia	1–20 0.2–1	8014-105sd
Aniline	2–30 1–15	8014-181s
Arsine	0.05–2.0	8014-121u
Benzene in presence of gasoline and/or other aromatic Hydrocarbons	5–200	8014-118sb
Benzene	1–100	8014-118sc
Bromine	1–20	8014-114
Butadiene	2.5–100	8014-168sc
N-butane	0.05–0.6%	8014-221sa
1-butanol	5–100	8014-190u
Butyl acetate	15–400	8014-138u
Butyl acrylate	5–60	8014-211u
Butyl cellosolve	10–1000	8014-190u
Carbon dioxide	100–4000	8014-126sf
Carbon disulfide	2–50, 1–25	8014-141sb
Carbon monoxide	1–50	8014-106sc

TABLE 17-I-18. Matheson Qualitative Analysis Tubes and HazMat Kit (Instrument 17-26)

Compound	Detection Limit (ppm)	Compound	Detection Limit (ppm)
Inorganic Tube Model 8014-131		**Organic Tube Model 8014-186B (cont.)**	
Acetic Acid	15	Ethylamine	100
Acetylene	10	Ethyl Benzene	400
Amines	5	Ethyl Cellosolve	100
Ammonia	5	Ethylene	10
Carbon Monoxide	10	Ethylene Oxide	100
Chlorine	5	Formaldehyde	10
Hydrogen Chloride	20	Gasoline	0.1 mg/L
Hydrogen Sulfide	10	Heptane	10
Methyl Mercaptan	10	Hexane	10
Nitrogen Dioxide	5	Isopropyl Alcohol	500
Phosphine	2	Kerosene	0.1 mg/L
Sulfur Dioxide	10	Methyl Alcohol	100
		Methyl Ethyl Ketone	100
Organic Tube Model 8014-186B		Methyl Isobutyl Ketone	100
Acetaldehyde	100	Methyl Mercaptan	20
Acetone	500	Pentane	10
Acetylene	100	Phenol	10
Aniline	50	Propane	100
Benzene	100	Styrene	100
Butadiene	1000	Tetrachloroethylene	1100
Butane	10	Tetrahydrofuran	100
1-Butanol	100	Toluene	200
Butyl Acetate	100	Trichloroethane	1000
Carbon Disulfide	100	Trichloroethylene	1000
Cresol	20	Vinyl Chloride	10
Ethyl Acetate	500	Xylene	1000

TABLE 17-I-19. Matheson Badge Dosimeters (Models 8005, 8006, 8007) (Instrument 17-27)

Detected Gas	M#	Sensitivity	Color Change
Ammonia	17	25 ppm/5 min	Yellow to Blue
Carbon Monoxide	28	50 ppm/10 min	Tan to Black
Chlorine	71	1 ppm/15 min	White to Yellow
Hydrazine	32	0.1 ppm/15 min	White to Yellow
Hydrogen Sulfide	34	10 ppm/10 min	White to Brown
Nitrogen Dioxide	46	5 ppm/15 min	White to Yellow
Ozone	48	0.1 ppm/15 min	White to Brown

TABLE 17-I-20. Matheson Compressed Breathing Air Analysis Kit (Instrument 17-28)

Components	Measured Range	Sampling Time
Carbon Monoxide	5–100 ppm	2 min
Carbon Dioxide	100–3000 ppm	2 min
Oil Mist	0.3–5 mg/m^3	25 min
Water Vapor	20–160 mg/m^3	1 min
Oxygen	2–24%	1 min

TABLE 17-I-21. MSA Detector Tubes (Instrument 17-30)

Substance	Range	P/N
Acetic Acid	1–80	804138
Acetone	100–10,000	804141
Alcohols	100–6,000	804136
Amines	5–30	804134
Ammonia	20–1,000	800300
Benzene	1–25	807024
Bromine	0.2–30	803944
Bromoethane	30–720	487834
Butane	200–3800	804418
Carbon, Dioxide	100–2,000	488907
Carbon Monoxide	5–1,000	803943
Carbon Disulfide	2–300	492514
Chlorine	0.5 & 2.5	804412
Chlorine Dioxide	0.05–15	804133
Dichloroethane	30–720	804416
Formaldehyde	0.15–10	497649
Gasoline	30–6,000	492870
Hexane	100 & 1,000	804410
Hydrocarbons Halogen	20–170	487343
Hydrogen Chloride	1–30	803948
Hydrogen Cyanide	2–50	803945
Hydrogen Fluoride	50–630	804132
Hydrogen Sulfide	10–4,000	807340
Mercaptans	0.5–100	804597
Mercury	0.01–0.08	497663
Methylene Chloride	50–1,000	804416
Nitric Oxide	2–140	804425
Nitrogen Dioxide	0.05–50	487341
Oil Mist	1–3 mg/m^3	488909
Ozone	0.05–5	804140
Perchloroethylene	5–2,000	804429
Phosgene	0.1–20	803949
Phosphine	0.1–100	485680
Styrene	10–300	804135
Sulphur Dioxide	0.5–25	487338
Toluene	5–1,000	803947
Trichloroethylene	5–250	487342
Vinyl Chloride	1–70	803950
Water Vapor	5–160 mg/m^3	804438
Xylene	1.5–2,600	803947

TABLE 17-I-22. MSA Haz-Mat Response Kit (Instrument 17-31)

— P/N 807472 includes:

Kwik-Draw pump	P/N 487500
Ethanol tube, Box of 10	P/N 804136
Aromatic hydrocarbons tube, Box of 10	P/N 804132
Hexane tube, Box of 10	P/N 497664
Hydrogen cyanide tube, Box of 10	P/N 803945
Trichloroethane tube, Box of 10	P/N 487343
Nitrogen dioxide tube, Box of 10	PIN 487341
Carbon monoxide tube, Box of 10	P/N 803943
Triethylamine tube, Box of 10	P/N 804134
Carbon dioxide tube, Box of 10	P/N 487333
Hydrogen chloride tube, Box of 10	P/N 803948
Ethyl mercaptan tube, Box of 10	P/N 804589
Ethylene tube, Box of 10	P/N 804428

TABLE 17-I-23. MSA Indoor Air Quality Kit (Instrument 17-32)

—P/N 710981 includes:

P/N 487500 Kwik-Draw Deluxe pump
P/N 70918 Box of detector tubes, two each for:
    Carbon monoxide
    Carbon dioxide
    Formaldehyde
    Ozone
    Water vapor

TABLE 17-I-24. RAE Colorimetric Tubes

Analyte	Range (ppm)	Product No.
Acetone	0.1–2%	10-111-40-151
Ammonia	1–30	10-100-05-3L
Ammonia	25–500	10-100-15-3M
Ammonia	1–15%	10-100-40-3H
Ammonia	5–100	10-100-10-3La
Benzene	5–100	10-101-20-121
Benzene	0.5–40	10-101-05-121SL
Carbon Monoxide	5–100	10-102-20-1LL
Carbon Monoxide	20–500	10-102-30-1La
Carbon Monoxide	0.2–4%	10-102-45-1H
Carbon Dioxide	1–20%	10-104-50-2H
Carbon Dioxide	0.25–3%	10-104-45-2L
Carbon Dioxide	0.05–1%	10-104-40
Carbon Dioxide	300–5000	10-104-30-2LL
Chlorine	0.5–8	10-106-10-8La
Chlorine	5–100	10-106-20
Hydrocarbons	50–1000	10-110-30
Hydrogen Fluoride	0.5–20	10-105-10-17
Hydrogen Sulfide	0.2–3	10-103-05-120UP
Hydrogen Sulfide	2.5–60	10-103-10-4LL
Hydrogen Sulfide	100–2000	10-103-30-4H
Hydrogen Sulfide	50–800	10-103-20-4HM
Hydrogen Sulfide	25–250	10-103-18-4M
Hydrogen Sulfide	2–50	10-103-10-4LL
Hydrogen Sulfide	0.1–2%	10-103-40-4HH
Hydrogen Sulfide	10–120	10-103-15-4L
Hydrogen Sulfide	2–40%	10-103-50-4HT
Hydrogen Chloride	20–500	10-108-20-14M
Hydrogen Chloride	1–20	10-108-10-14L
MEK	0.02–0.6%	10-113-20-152
Mercaptans	5–120	10-129-20-70
Nitrogen Dioxide	0.5–30	10-117-10
Nitrogen Oxides*	1–50	10-109-20
Phosphine	5–50	10-116-10-7
Phosphine	25–50	10-116-20-7J
Phosphine	50–1000	10-116-25
Sulfur Dioxide	2–30	10-107-15-5La
Sulfur Dioxide	200–4000	10-107-30
Sulfur Dioxide	100–1800	10-107-25-5M
Sulfur Dioxide	5–100	10-107-20-5L
Sulfur Dioxide	0.2–5%	10-107-40-5H
Toluene	10–300	10-114-20-122
Water Vapor	0.03–0.16 mg/L	10-120-10-6LP
Water Vapor	0.1–0.64 mg/L	10-120-20-6LP
Xylene	10–200	10-112-20-123

* Box contains 5 colorimetric tubes and 5 pre-tubes with 1 connector for a total of 5 measurements

TABLE 17-I-25. Sensidyne/Kitagawa Deluxe HazMat Kit III and Tubes (Instrument 17-35)

Part Number 7013627 comes complete with the following equipment:

Item	Part Number
Model AP-IS pump kit	7013585
Hard Shell Case	7015574-1
5-m extension hose	7013596
Gas detection manual	7013584
Logic chart III	7013628
Color Chart for qualitative tubes	7013655
Air flow indicator kit	500
Detector tubes (one box each):	
Acetone	102SC
Carbon dioxide	126SA
Hydrogen sulfide	120SB
Trichloroethylene	134SB
Benzene	118SB
Methyl mercaptan	130U
Arsine	121U
Hydrogen fluoride	156S
Methyl bromide	157SB
Ethyl acetate	111U
Hydrogen	137U
Carbon tetrachloride	147S
Carbon disulfide	141SB
Organic qualitative test	186B
Inorganic qualitative test	131

TABLE 17-I-26. Sensidyne Detector Tubes (partial list of 300+ tubes, Instrument 17-36)

Note: **The list below is only a partial list of products. The full list can be found at *www.sensidyne.com*.**

Gas or Vapor	Range	Tube #
Acetaldehyde	0.004-1.0 %v	133A
	5-140 ppm	133SB
Acetic Acid	1-50 ppm	216S
Acetic Anhydride	1-15 ppm	216S
Acetone	0.1-5.0 %v	102SA
	0.01-4.0%	102SC
	40-5,000 ppm	102SD
Acetylene	50-1,000 ppm	101S
Acetylene & Ethylene (separate measure)	A: 20-300 ppm, E: 200-2,000 ppm	280S
Acetylene *Dichloride (see 1,2 Dichloroethylene)*		
Acrolein	0.005-1.8 %v	136
Acrylic Acid	1-50 ppm	216S
Acrylonitrile (Vinyl Cyanide)	0.1-3.5 %v	128SA
	1-120 ppm	128SC
	0.25- 20 ppm	128SD
Air (Breathing Air) Tubes (see Separate Listing)		
Allyl Alcohol (propenyl Alcohol)	20-500 ppm	184S
Allyl Chloride (3-Chloroprene) Amines (see specific Amines)	1-40 ppm	132S
Aminobenzene (see Aniline)		
2-Aminoethanol (see Monoethanol Amine)		
2-Aminopropane (see Isopropyl Amine)		
Ammonia	0.5-10 %v	105SA
	50-900 ppm	105SB
	5-260 ppm	105SC
	0.2-20 ppm	105SD
	0.5-30 %v	105SH
	0.1-1.0 %v	105SM
Amyl Acetate (see Pentyl Acetate)		
Aniline (Aminobenzene)	1-30 ppm	181S
Arsine	5-160 ppm	140SA
	0.05-2.0 ppm	121U
Benzene	1-100 ppm	118SC
Benzene (in presence of gasoline and/or other aromatic HCs)	5-200 ppm	115SB
Benzyl Chloride	1-16 ppm	132SC
Breathing Air Tubes (see Separate Listing)		
Bromine	1-20 ppm	114
Bromochloromethane (Chlorobromomethane)	5-400 ppm	157SB
Bromoform (Tribromomethane)	0.5-20 ppm	157SB
Bromomethane (see Methyl Bromide)		

TABLE 17-I-27. Commercial Sources of Colorimetric Indicators

Manufacturers

	Acculabs Technologies, Inc. 1018-E Morrisville Pkwy. Morrisville, NC 27560 (888) 853-5030 FAX (919) 468-0185 *E-mail: acculabs@mindspring.com*	NDR	Draeger Safety, Inc. 101 Technology Drive Pittsburgh, PA 15275 (412) 787-8383 or (800) 922-5518 FAX (800) 922-5519 *www.draeger-usa.com*		PiezOptic, Ltd. Viking House Ellingham Way Ashford Kent TN23 6NF United Kingdom 44(0)1233-641990 FAX +44(0)1233-645020 *www.itl.co.uk/www/piezoptic*
AGC	American Gas and Chemical Co. 220 Pegasus Ave. Northvale, NJ 07647 (800) 288-3647 or (201) 767-7300 FAX (201) 767-1741 *www.amgas.com*		K&M Environmental 2421 Bowland Pkwy., Suite 102 Virginia Beach, VA 23454 (757) 431-2260 or (800) 808-2234 FAX (757) 431-2255 *www.kandmenvironmental.com*	RAE	RAE Systems, Inc. 1339 Moffett Park Drive Sunnyvale, CA 94089 (408) 585-3523 or (888) 723-8823 FAX (408) 752-0724 *www.raesystems.com/home1.html*
	Assay Technology 1252 Quarry Lane Pleasanton, CA 94566 (800) 833-1258 or (925) 461-8880 FAX (925) 461-7149 *www.assaytec.com*	MGP	Matheson-Kitagawa Matheson Tri Gas 166 Keystone Drive Montgomeryville, PA 18936 (215) 641-2700 or (800) 416-2505 FAX (215) 641-2714 *www.matheson-trigas.com*	SEN	Sensidyne, Inc. 16333 Bay Vista Drive Clearwater, FL 33760 (727) 530-3602 or (800) 451-9444 FAX (727) 839-0550 *www.sensidyne.com*
BAC	Bacharach, Inc. 625 Alpha Drive Pittsburgh, PA 15238 (412) 963-2000 or (800) 736-4666 FAX (412) 963-2091 *help@bacharach-inc.com* *www.bacharach-inc.com*	MSA	Mine Safety Appliances Co. 121 Gamma Drive Pittsburgh, PA 15238 (800) 672-4678 FAX (412) 967-3451 *www.MSAnet.com*		

Distributors

AFC	AFC International, Inc. (distrib. Draeger Safety and K&M Environmental) P.O. Box 408 Cedar Lake, IN 46303 (219) 374-4623 or (800) 952-3293 FAX (219) 374-4625 *www.afcintl.com*		Environmental Specialty Products (distrib. PiezOptic, Ltd.) P.O. Box 365 Buford, GA 30515 U.S.A. (770) 995-6678 FAX (770) 995-6079	SMG	SMG (distrib. PiezOptic, Ltd.) 1631 Dorchester Suite 117 Plano, TX 75075 (972) 985-0883 FAX (972) 985-3252
	Brandt Instruments (distrib. PiezOptic, Ltd.) 750 East 1-10 Service Road Suite D Slidell, LA 70461 (504) 863-5597 FAX (504) 863-7112		O'Brien and Gere Companies 5555 East Genesse Street Fayettville, NY 13066 (888) 976-2477 FAX (315) 637-2015		
	Eirtech Instruments (distrib. PiezOptic, Ltd.) 1057 East Henrietta Road Rochester, NY 14623 U.S.A. (716) 424-2030 FAX (716) 424-2166		Safety First Environmental (distrib. K&M Environmental) 429 Hinsonton Rd. Meigs, GA 31765-3617 (912) 294-1926 (912) 294-0837 *www.safetyfirst.com*		

TABLE 17-I-27. (cont.). Commercial Sources of Colorimetric Indicators

Distributors, continued

ENM   ENMET Corp.   Safety & Hygiene Management Inc.
        P.O. Box 979   (distrib. of PiezOptic, Ltd.)
        Ann Arbor, MI 48106   430 Hazelwood Avenue
        (734) 761-1270   Waynesville, NC 28786 U.S.A.
        FAX (734) 761-3220   (828) 456-7798
        info@enmet.com   FAX (828) 456-7583
        www.enmet.com

Vendors for Wipe-Tests

BGI   BGI Incorporated   Gallard-Schlesinger Industries   HybriVet Systems, Inc.
        58 Cuinan Street   584 Mineola Avenue   P.O. Box 1210
        Waltham, MA 02451   Carle Place, NY 11514   Framingham, MA 01701
        (781) 891-9380   (516) 333-5600 or (800) 645-3044   (508) 651-7881 or (800) 262-5323
        FAX (781) 891-8151   FAX (516) 333-5628   FAX (508) 651-8837
        www.bgiusa.com   *www.gallard.com*   *www.leadcheck.com*

Chapter 18

Direct-Reading Gas and Vapor Instruments

Mary Lynn Woebkenberg, Ph.D.[A] and Charles S. McCammon, Ph.D.[B]

[A]U.S. Department of Health and Human Services, Public Health Service, Centers for Disease Control and Prevention, National Institute for Occupational Safety and Health, 4676 Columbia Parkway, MS: R-7, Cincinnati, Ohio;

[B]Tri-County Health Department, 4301 East 72nd Avenue, Commerce City, Colorado

CONTENTS

Introduction

This chapter presents useful information about direct-reading instruments for analyzing airborne gases and vapors. The instrumentation that will be discussed is that which provides an on-site indication, in useful units (e.g., ppm, mg/m³, etc.), of the presence of the contaminant(s) of interest. Frequently, these instruments are general, nonspecific detectors, but chemispecific detectors are also available. The instruments are commercially available.

Direct-reading instruments may be used for area, process, or personal monitoring, and it is convenient to describe three physical classifications for grouping these instruments: *personal* instruments are those instruments small enough to be worn by an individual; *portable* instruments are those easily carried by an individual; *transportable* instruments are those requiring a cart or other support for movement to or from the monitoring site. Ideally, these instruments operate from self-contained battery power, but many can also use, and some require, line current.

In this chapter, the reader will find information on operational, physical, and performance characteristics for each of the instruments described. The instruments are grouped into the following classifications: electrochemical instruments, spectrochemical instruments, thermochemical instruments, gas chromatographic instruments, mass spectrometers, paramagnetic instruments, and an aerosol formation and detection instrument. In each section, there is a general definition of the instrumentation to be described, an explanation of the principle of detection, and a brief discussion of conditions of application for the instruments, including capabilities, restrictions, and limitations. At the end of the chapter is a suggested reading list for the reader who requires more in-depth information about a particular technique.

Regardless of the instrument chosen for use and the capabilities of that instrument, there is no substitute for knowledge of the capabilities and limitations of the instrument as well as effects of the conditions in the proposed monitoring situation. Then, the most appropriate instrument can be chosen for a given application, meaningful data can be obtained, and, if necessary, effective solutions for contaminant control can be implemented.

Electrochemical Instruments

Electrochemical techniques involve the measurement of electrical signals associated with chemical systems.[1] These chemical systems are typically incorporated into electrochemical cells. Electrochemical techniques include instruments that operate on the principles of conductivity, potentiometry, coulometry, and ionization.

Conductivity (1.1–1.4)

Instruments that measure conductivity rely on the fact that charged species (ions) conduct electricity. Equally significant is the fact that at low concentrations, such as those concentrations typically found when these species are measured as workplace contaminants, conductivity is proportional to concentration. The fundamental equation for conductivity is given by

$$G = \frac{\Lambda C}{1000\,K} \tag{1}$$

where G = conductance in Siemens
          Λ = equivalent conductance in Siemens per centimeter-equivalent
          C = the concentration in equivalents per 1000 cm^3
          K = a geometric term describing the electrochemical cell

A conductivity measurement depends on the space between, and area (size) of, a pair of electrodes and also on the volume of solution between them. Because conductance is the reciprocal of resistance, that is,

$$G = \frac{1}{R} \tag{2}$$

where R is resistance in ohms, the latter is sometimes measured because it is a more fundamental property. It should be noted that species monitored by conductivity need not be in an ionic form in the vapor phase but may be gases or vapors that form electrolytes, by chemical reaction, in solution.

Conductivity measurements are temperature dependent, having a temperature coefficient that can be on the order of 2% per °C. Instruments that control temperature may use thermostatted cabinets; those that compensate for temperature effects do so electronically.

A special case of conductivity instrumentation is one wherein a gold film is used to amalgamate mercury (Hg). In the mercury conductivity detector, the change in resistance of the solid film is measured.

Conductivity is, typically, a nonspecific technique in that any species ionizable under the given conditions will affect the measurement. The specific conductance, λ, of each ionizable species is important, for only when the conductivity of interfering electrolytes is either constant and/or negligible can the conductivity of the species of interest be measured.

There also are several solid-state devices that exploit electronic conductivity charges induced in metal oxide semiconductors.[2] Their principle of operation is based on the change in surface conductivity of a semiconductor, such as SnO_2, as a result of gas adsorption. The adsorbed gas may either directly affect the conductivity, or interact with the surface oxygen coverage, which, in turn, affects the conductivity. These instruments are relatively inexpensive, are easy to use, and can be used in oxygen-depleted atmospheres. They are typically used in screening applications and for hazard warning.

Conductivity instruments are primarily used for detection of corrosive gases, e.g., ammonia (NH_3), hydrogen sulfide (H_2S), and sulfur dioxide (SO_2). The conductivity analyzers are numbers 1.1 through 1.4 in the "Instrument Descriptions" section. They are most effectively used in isothermal environments at or near room temperature. Environments with few potential interferences are preferred. Chemical prescrubbers can be helpful.

Potentiometry (2.1–2.45)

Instruments that use a change in electrochemical potential as their principle of detection are most commonly represented by the pH meter. Potentiometry is strictly defined as the measurement of the difference in potential between two electrodes in an electrochemical cell under the condition of zero current. Gases and vapors can react with reagents effecting an oxidation/reduction, the extent of which is proportional to the concentration of the reacting gas. The fundamental equation governing a potentiometric reaction is the Nernst equation:

$$E_{cell} = E_{cell}^o - \frac{RT}{nF} \ln \frac{[C]^c [D]^d}{[A]^a [B]^b} \quad\quad (3)$$

where: E_{cell} = cell potential
$E^o{}_{cell}$ = standard cell potential
R = gas constant
T = temperature
n = number of electrons involved in the electrode reaction
F = Faraday constant

Although the letters in brackets strictly represent the chemical activities of the reacting species, when considering dilute solutions, it is reasonable to approximate the activity using the concentration. This equation is simplified at nominal room temperature (25°C) by converting to the base ten logarithm and substituting for the constants: $R = 8.314$ Joules mol^{-1} T^{-1}, $T = 298$ K, $F = 96,485$ Coulombs/mol. This results in the following equation:

$$E_{cell} = E_{cell}^o - \frac{0.0591}{n} \frac{\log [C]^c [D]^d}{[A]^a [B]^b} \quad\quad (4)$$

The Nernst equation relates potential, E_{cell}, with temperature, the electronic state change of the species being oxidized or reduced, and the concentration of the species. When sampling with a potentiometer, the sampled analyte of interest would most likely be represented in this equation by one of the reactants, A or B.

Whereas potentiometry is basically a nonspecific technique, some degree of specificity may be obtained through the selection of the membrane through which the gaseous analyte must diffuse to enter the electrochemical cell, the selection of the reagent, the specific potential range, and the type of electrodes used.

Potentiometers are listed as numbers 2.1 through 2.45 in the "Instrument Descriptions" section. Some are diffusion monitors. They are used for the measurement of a variety of contaminants including carbon monoxide, chlorine, formaldehyde, hydrogen sulfide, oxides of nitrogen, oxides of sulfur, oxygen, and ozone. Preferable application is at, or near, room temperature for area samples, including confined space, and personal.

Coulometry (3.1–3.22)

Coulometric analyzers have as their principle of detection the determination of the quantity of electricity required to affect the complete electrolysis of the analyte of interest. The amount of electricity required is proportional to the amount of analyte present. This analyte may be the contaminant requiring monitoring, or it may be a chemical with which the contaminant quantitatively reacts. Regardless, the equation governing coulometry is Faraday's:

$$W = \frac{qM}{nF} \quad\quad (5)$$

where: W = mass of substance that is electrolyzed
q = charge, in Coulombs, required to completely electrolyze the substance
M = formula weight
n = number of electrons per molecule required for electrolysis
F = Faraday's constant: 96,485 Coulombs/mol

The quantity that an instrument must measure is q. This may be done either directly, by determining the integral (controlled-potential coulometry), or indirectly, by measuring the time required for electrolysis under conditions of constant current (constant-current coulometry). Both approaches work because of the following relationship:

$$q = \int i \, dt \quad\quad (6)$$

where: i = current in amperes
t = time

Coulometry is free of temperature dependencies. The technique, inherently very accurate, can be nonspecific. Judicious choice of filters, membranes, and electrolytes can be used to improve specificity. Coulometric analyzers are numbered 3.1 through 3.22 in the "Instrument Descriptions" section. The vast majority of these instruments are configured as oxygen or oxygen deficiency monitors, although coulometric analyzers are also available for carbon monoxide, chlorine, hydrogen cyanide, hydrogen sulfide, oxides of nitrogen, ozone, and sulfur dioxide. Coulometric detectors can be personal or area monitors, and pumped or diffusive samplers.

Ionization (4.1–4.15)

There are three types of ionization detectors: flame ionization (FID); photoionization (PID); and electron capture (ECD). All rely on the ability of their respective energy source (flame, lamp, or radioactivity) to ionize the species of interest.

Flame Ionization

In an FID, a gaseous sample is pyrolyzed in a hydrogen/air flame.[3] Pyrolysis produces ions and electrons that are carried through the plasma to an electrode gap, decreasing the gap resistance and allowing current to flow in the external circuit. Reactions in flame ionization include:

$$CH_3COCH_3 \rightarrow CH_3 + CH_3CO \text{ (cracking)}$$
$$H + C_3H_8 \rightarrow C_3H_7 + H_2 \text{ (stripping)}$$

There can be matrix reactions that give energetic intermediates exampled by:

$$H + HO_2 \rightarrow OH^* + O_2$$

and it is possible to ionize one species using excited components from the matrix as in:

$$CH_2 + OH^* \rightarrow CH_2OH^+ + e^-$$

Figure 18-1 shows a schematic of a typical flame ionization detector. The FID has a wide linear range, on the order of 10^6 to 10^7, and is a very sensitive detector able to detect on the order of nanogram quantities of organic compounds. As a result, this detector is excellent in trace analysis. Flame ionization is a nonspecific detection mechanism ideal for the detection of most organic compounds. The detector does not respond to, or responds very little to, common constituents of air, including water vapor. The user should be aware that electronegative compounds such as chlorine and sulfur (in the vapor phase) will depress the response.

Flame ionization detectors work well as portable survey instruments. Because the technique involves a flame, this must be considered when assessing the atmosphere wherein flame ionization would be used.

Photoionization

Photoionization is a flameless ionization technique wherein the contaminant gas or vapor is carried into an ionization chamber where an ultraviolet lamp of known constant voltage causes the ionization of any species having an ionization potential less than the energy emitted by the lamp.[4] That is, photoionization occurs when a molecule absorbs a photon of sufficient energy to cause the molecule to lose an electron and become a positively charged ion:

$$RH + hv \rightarrow RH^+ + e^- \tag{7}$$

where: RH = molecule to be ionized
 h = photon whose energy is greater than the ionization potential of RH
 RH^+ = ionized molecule
 e^- = electron lost in the process

The PID will have a high voltage positive bias electrode to repel the positively charged molecules accelerating them toward a negatively charged collector electrode. This, in turn, generates a signal at the collector which is proportional to the amount of ionized species. Figure 18-2 shows a schematic of a photoionization detector.

Photoionization is a nondestructive technique and somewhat selective through judicious selection of ultraviolet lamps of varying energies. Lamp energies are typically on the order of 10–11 eV, but others are available. PIDs are useful for detection of some permanent gases such as methane and ethane, but most light permanent gases (hydrogen, helium, nitrogen) have ionization energies higher than 10.6 eV and do not give a response. It is necessary to consider if water will interfere. PIDs have traditionally been area/survey instruments, but personal PIDs are now commercially available. While primarily

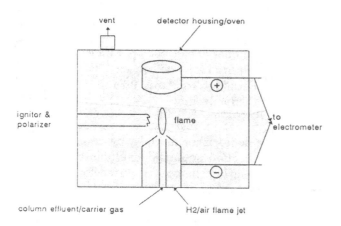

FIGURE 18-1. Schematic of Flame Ionization Detector.

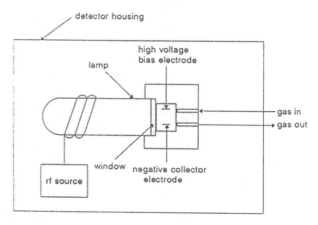

FIGURE 18-2. Schematic of a Photoionization Detector.

used for the detection of organic compounds, the PID has some utility for inorganic compounds such as nitric and sulfuric acids, hydrogen sulfide, arsine, and phosphine. Under optimum conditions, a PID can detect 5 pg of benzene and has a linear dynamic range on the order of 10^7.

Electron Capture

An ECD uses a radioactive source to generate the ions that are measured by this technique.[3] The radioactive source is usually H^3, Ni^{63}, or Kr^{85}. As the carrier gas, nitrogen, flows past the ion source, the nitrogen is ionized and slow electrons are formed that migrate to the anode, producing a steady current. Some molecules, said to have high electron affinity, have the ability to capture rapidly moving, free electrons from the radioactive source. When the molecules capture the electrons, they become stable, negative ions. This may happen by one of two mechanisms:

$$AB+e^- \rightarrow (AB)^- +energy \quad AB+e^- \rightarrow A^- +B\pm energy \quad \textbf{(8)}$$

where A and B = reactants

Figure 18-3 shows a schematic of an ECD. When samples with high electron affinity components are introduced into the chamber, the current flow, established through the ionization of the nitrogen, is reduced. Because the current reduction is a function of both the amount of sample present and its electron affinity, a calibration must be made separately for each sample component that is to be quantified.

An ECD is very selective, particularly for halogenated compounds, nitrates, conjugated carbonyls, and some organometallic compounds. It is useful for SF_6 and

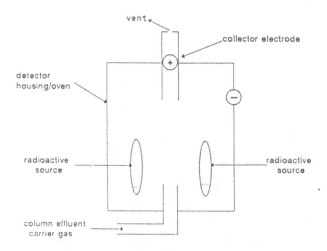

FIGURE 18-3. Schematic of an Electron Capture Detector.

pesticide identification. This detector is very sensitive (as low as 0.1 pg) for the compounds it will detect, but its linear range is very low, about 10^2–10^3.

Ionization detectors are numbers 4.1 through 4.15 in the "Instrument Descriptions" section. All three detectors are available in stand-alone instruments, as well as detectors for gas chromatographic systems, which will be discussed later in this chapter.

Spectrochemical Instruments

Instruments whose principle of detection is spectrochemical in nature include infrared analyzers, ultraviolet and visible light photometers, chemiluminescent detectors, and photometric analyzers.[5] Photometric analyzers include fluorescent and spectral intensity detectors. In general, spectrochemical analysis involves the use of a spectrum or some aspect of a spectrum to determine chemical species. A spectrum is a display of intensity of radiation that is emitted, absorbed, or scattered by a sample, versus wavelength. This radiation is related to photon energy via wavelength or frequency.

Infrared (5.1–5.16)

Infrared spectrometry (IR) involves the interaction of the infrared portion of the electromagnetic spectrum with matter. Specifically, that portion of the spectrum ranging in wavelength from 770 nm to 1000 μm, or 12,900 cm^{-1} to 10 cm^{-1} in wave number. The infrared portion of the spectrum is subdivided into three regions: the near-infrared (770 nm to 2.5 μm), the mid-infrared (2.5 to 50 μm), and the far-infrared (50 to 1000 μm). The terms "near," "mid," and "far" refer to proximity to the visible portion of the electromagnetic spectrum. Infrared radiation is not energetic enough to cause electronic transitions in molecules, but it does result in vibrational and rotational transitions. Nearly all molecules absorb infrared radiation, making the technique widely applicable. Because the IR spectrum of a given molecule is unique to that molecule, IR can be fairly specific and useful in compound identification. However, the possibility of overlapping peaks makes the use of any single wavelength IR measurement of an uncharacterized mixture risky.

Figure 18-4 shows a schematic of an infrared analyzer. These instruments consist primarily of six major sections: a source of infrared radiation; a wavelength selector; a sample cell; appropriate optics; a detector; and a signal processor/readout. Although Figure 18-4 shows the monochromator after the sample cell, wavelength selection can occur before the sample cell, after

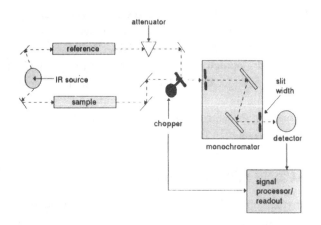

FIGURE 18-4. Schematic of Double Beam Infrared Analyzer.

the sample cell, or both. Infrared spectrometry may be either a nondispersive or a dispersive technique. A nondispersive IR is a filter photometer employing interference filters designed for the determination of a specific pollutant, whereas a dispersive IR uses prisms, gratings or interferometers to separate radiation into its component wavelengths to obtain a complete spectrum for qualitative identification.

Because it is an absorption technique, infrared spectrometry is governed by Beer's Law:

$$A = \varepsilon b c \qquad (9)$$

where: A = absorbance
ε = molar absorptivity
b = path length
c = concentration

This equation shows the relationships between the amount of energy absorbed and the length of the path through the sample, and between the absorbed energy and the concentration of the species of interest. The dependency of absorbance on path length is significant in discussing parameters of interest because the longer the path length of the instrument, the more sensitive the instrument should be. In introducing Beer's Law, it is significant to note that the absorbance, A, is log P_o/P, where P_o is the original incident radiation, and P is the energy remaining after some is absorbed by the sample. The linear range of IR is limited at any set path length.

An additional instrument parameter of interest is the slit width. The slit width defines the window of energy seen either by the sample or by the detector. Figure 18-4 shows the slit width at the detector end of the instrument. The width of this slit is inversely proportional to selectivity and peak resolution.

The direct-reading infrared instruments are given in the "Instrument Descriptions" section as numbers 5.1

through 5.16. Primarily area monitors, the instruments balance modest precision with selectivity and high throughput. Some instruments are designed as fixed wavelength monitors whereas others are capable of scanning the infrared spectrum. Some of these instruments are designed as general detectors for organics and subgroups such as hydrocarbons; others are more specific monitors for compounds such as methane, ethylene, ethane, propane, butane, vehicle emissions, carbon monoxide, carbon dioxide, and several freons. The user needs to be aware that certain ubiquitous compounds, like water, absorb very strongly in the infrared, and care must be exercised to avoid making measurements at or near these absorbances.

Ultraviolet and Visible Light Photometers (6.1–6.10)

Both ultraviolet (UV) and visible (VIS) light photometers operate on the principle of absorption of electromagnetic radiation. The UV is that portion of the electromagnetic spectrum having wavelengths from 10 to 350 nm. The actual spectral range for direct-reading UV instruments is closer to 180B350 nm, which is termed the "near UV," in deference to its proximity to the visible spectrum. The corresponding energy range for the UV is 3.6–7 eV for the near UV and 7–124 eV for the far, or vacuum, UV. The visible spectrum has longer wavelengths than the UV (350–770 nm) and correspondingly lower energies (1.6–3.6 eV). Like their infrared counterparts, the operational principle (energy absorption) of the UV-VIS instruments is governed by Beer's Law, and the techniques have the same relationships between absorption and concentration and between absorption and path length. Although the relationship between absorbance and concentration is linear, the value typically measured in spectrophotometry is transmittance, T, whose relationship with absorbance is given by:

$$A = 2 - log\ \%T \qquad (10)$$

Transmittance is the ratio of the amount of energy passing through the sample (not absorbed) to the amount of incident energy.

Figure 18-5 shows a schematic of a typical UV-VIS photometer. The instruments operating on the principle of energy absorption in the UV-VIS region are given in the "Instrument Descriptions" section as numbers 6.1 through 6.10. Most of these instruments are designed to analyze gaseous samples such as ammonia, mercury vapor (which absorbs very strongly at 253.7 nm), oxides of nitrogen, ozone, and sulfur dioxide.

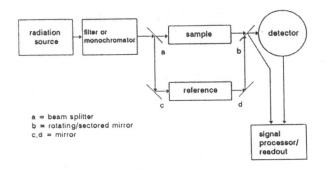

a = beam splitter
b = rotating/sectored mirror
c,d = mirror

FIGURE 18-5. Schematic of a UV-VIS Spectrophotometer.

A special case of visible spectroscopy is colorimetry, wherein the sample is mixed with a reagent selected to react with the contaminant of interest, forming a colored product. The ability of this colored, liquid product to absorb light in the visible region is exploited. This type of instrument, governed by the same chemical principles, can be used as a continuous monitor for a variety of compounds. The UV-VIS instruments, primarily area monitors, are capable of detecting contaminants in the ppm range.

Chemiluminescence (7.1–7.6)

Chemiluminescence is a form of emission spectroscopy wherein spectral information is obtained from nonradiational activation processes.[6] In this case, the emitted energy results from species that are excited by chemical reactions and are returning to the lower energy state by emission of a photon. Chemiluminescence is based on the fact that in some chemical reactions, a significant fraction of the intermediates or products are produced in excited electronic states. The emission of photons from these excited electronic states is measured and, if the reaction conditions are arranged appropriately, is proportional to the concentration of the contaminant of interest. Two common chemiluminescence mechanisms are:

$$A + B \rightarrow I + I^* \qquad A + B \rightarrow P + P^* \qquad \textbf{(11)}$$
$$\downarrow \qquad + \qquad \downarrow$$
$$I + h\nu \rightarrow P \qquad P \rightarrow h\nu$$

where: A and B = reactants
I = intermediate
P = product
$*$ = excited state
$h\nu$ = emitted energy

Three conditions must be met in order to have chemiluminescence take place. First, there needs to be enough energy to produce the excited state; second, there must be a favorable reaction pathway to produce the excited state; and, third, photon emission must be a favorable deactivation process.

The direct-reading chemiluminescent detectors are numbered 7.1 through 7.6 in the "Instrument Descriptions" section. They analyze gas phase samples and have been developed primarily for oxides of nitrogen and ozone. Because of the chemical reactions involved, the instruments have a high degree of specificity and have typical limits of detection on the order of 10 ppb.

Photometric Analyzers

This category includes fluorescence analyzers, flame photometric detectors, spectral intensity analyzers, and photometers, primarily reflectance. The first three techniques are all examples of emission spectroscopy wherein the excitation process is radiative in nature; the last category includes automated media advance samplers, branched sequential samplers, and paper tape stain development, all of which utilize photometric analysis.

Fluorescence (8.2, 8.3, 8.5)

Fluorescence is the emission of photons from molecules in excited states when the excited states are the result of the absorption of energy from some source of radiation. For most molecules, electrons are paired in the lowest energy or ground state. If a molecule absorbs energy from a sufficiently powerful radiation source, such as a mercury or xenon arc lamp, the molecule will become "excited," moving an electron to a higher energy state. When the electron returns to the lower, more stable energy condition, it releases the absorbed energy in photons. A significant characteristic of fluorescence is that the emitted radiation is of a longer wavelength (lower energy) than the exciting radiation. Figure 18-6 shows a block diagram of the components of a fluorescence instrument. An excitation wavelength selector is used to limit the energy to that which will cause fluorescence of the sample while excluding energy wavelengths that may interfere with the detection. The emission wavelength selector isolates the fluorescence peak. Detection is at right angles to allow measurement of the longer wavelength light emitted from the sample while avoiding detection of light from the source, which could cause large errors in measurement. A narrow band of excitation and emission wavelengths can make the instrument very selective and often specific.

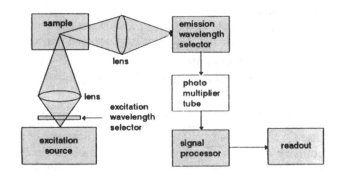

FIGURE 18-6. Schematic of a Fluorescence Spectrometer.

Fluorescence instruments are available for carbon monoxide and sulfur dioxide. They are numbers 8.2, 8.3, and 8.5 in the "Instrument Descriptions" section. Typical limits of detection are in the 5–10 ppb range.

Flame Photometric (8.4, 8.6, 8.14)

Flame photometric detectors can be adjusted to obtain selectivity for nanogram quantities of sulfur or phosphorous compounds. The detector works by measuring the emission of light from a hydrogen flame. Light from the flame impinges upon a mirror and is reflected to an optical filter that allows only light of either 526 μm (for phosphorous) or 394 μm (for sulfur) to pass through to the photomultiplier tube. Calibration with a flame photometric detector is critical because this detector exhibits little or no linearity. From the "Instrument Descriptions" section, numbers 8.4, 8.6, and 8.14 are flame photometric detectors. They have limits of detection in the low ppb range.

Spectral Intensity (8.8, 8.13)

Spectral intensity analyzers measure the radiant power of emission from an analyte due to nonradiational excitation. Two such instruments are available (numbers 8.8 and 8.13 in the "Instrument Descriptions" section). Both instruments are used for halide detection by measuring the increased spectral intensity of an AC arc (or spark) in the presence of halogenated hydrocarbons. The increased intensity can be related to the concentration of the halogenated compound by using a calibration curve based on the specific compound of interest, as each response curve for each halogenated compound will be different. These instruments have limits of detection in the tens of ppm range and have limited selectivity, i.e., they can differentiate halogenated compounds from non-halogenated compounds, but cannot differentiate between halogenated compounds.

Photometers [Other](8.1, 8.7, 8.9, 8.12)

The remaining instruments in this category are simply referred to as photometers. The instruments, numbers 8.1, 8.7, and 8.9 through 8.12 in the "Instrument Descriptions" section, have unique sampling characteristics and detection principles relative to the other instruments in this category (spectrochemical techniques), but they operate on spectrochemical principles nonetheless. The majority of these instruments allow for unattended sampling through the use of automated sampling media advance (i.e., tape samplers, rotating drum samplers, rotating disc samplers, and turntable samplers measuring reflectance) or branched sequential sampling trains. These samplers typically involve a color change of the sampling medium and the analytic finish is measurement of the light reflected from the sampling medium. These instruments are useful for such toxic species as toluene diisocyanate, ammonia, phosgene, arsine, and hydrogen cyanide. The reflectance instruments can be quite specific through judicious selection of the chemistry for the sampler, and the ability to change the chemistry makes these instruments potentially useful for a wide variety of compounds.

Other instruments in this category include one designed for the determination of CO. This instrument actually measures mercury that is generated via reduction of solid-state mercury oxide. The amount of mercury generated is equal to the quantity of carbon monoxide oxidized in the sample. The mercury is measured using a UV filter photometer. The other three instruments rely on the development of a color stain, wherein the intensity change or the development of the intensity change is measured via a photoelectric cell. These last three instruments are useful primarily for hydrogen sulfide, although one will determine other analytes as a function of the chemically impregnated paper used for color development. All the photometers have limits of detection in the low ppm range and are very specific for the contaminant(s) of interest.

Thermochemical Instruments

Gases and vapors have certain thermal properties that can be exploited in their analysis.[7] Of the instruments available for industrial hygiene applications, one of two thermal properties, conductivity or heat of combustion, is measured.

Thermal Conductivity (9.1, 9.3)

Thermal conductivity detectors are relatively simple devices that operate on the principle that a hot body will lose heat at a rate that depends on the composition of the surrounding gas. That is, the ability of the surrounding gas to conduct heat away from the hot body can be used as a measure of the composition of the gas. In actual practice, a thermal conductivity detector consists of an electrically heated element, or sensing device, whose temperature at constant electrical power depends on the thermal conductivity of the surrounding gas. The resistance of the sensing device is used as a measure of its temperature. Thermal conductivity detectors are universal detectors, responding to all compounds. They have large linear dynamic ranges, on the order of 10^5, and limits of detection on the order of 10^{-8} gram of solute per mL of carrier gas (10–100 ppm for most analytes). Thermal conductivity detectors require good temperature and flow control. They are numbered 9.1 through 9.3 in the "Instrument Descriptions" section.

Heat of Combustion (10.1–10.35)

Heat of combustion detectors, comprising the largest single class of direct-reading instruments for analyzing airborne gases and vapors, measure the heat released during combustion or reaction of the contaminant gas of interest. The released heat is a particular characteristic of combustible gases and may be used for quantitative detection. There are two main mechanisms for the operation of heat of combustion detectors. The first relies on heated filaments. Upon introduction of the contaminated air into the sample cell, the contaminant comes into contact with a heated source, igniting the contaminant. The resulting heat changes the resistance of the filament. The measured change in filament resistance is related to the gas concentration through the use of calibration standards.

The second mechanism used in heat of combustion instruments employs the use of catalysts via catalytically heated filaments or oxidation catalysts. This second mechanism may use one of two methods of detection: a measured resistance change, or temperature changes measured via thermocouples or thermistors.

Like thermal conductivity detectors, heat of combustion detectors are nonspecific, universal detectors. Some specificity can be introduced by manipulation of the temperature; that is, the combustion temperature may be controlled so that it is insufficient to combust interfering gases. From the second mechanism, some specificity may be introduced by careful selection of the oxidation catalyst.

As the category name implies, heat of combustion detectors are available as generic detectors for combustible gases. Some more specific heat of combustion detectors are available for carbon monoxide, ethylene

oxide, hydrogen sulfide, methane, and oxygen deficiency. Most of these monitors read out in terms of percent of the lower explosive limit (LEL) or hundreds of ppm and the limits of detection are a function of the analyte of interest. These instruments are numbered 10.1 through 10.35 in the "Instrument Descriptions" section.

Gas Chromatographs (11.1–11.7)

In terms of detection of airborne gases and vapors, the detectors used in gas chromatographic analyzers have, for the most part, been discussed earlier in this chapter.[3,4] The most frequently used detectors in GCs designed for industrial hygiene applications are the FID and the PID. The reason gas chromatographs are being discussed separately is fourfold: there are several direct-reading gas chromatographs commercially available; they represent a distinct family of instruments in that they very specifically address the issue of separation (specificity), as well as detection, in industrial hygiene monitoring; they represent one area where a great deal of research and development is ongoing; and they most closely approximate the transfer of laboratory analytical techniques into the field.

Figure 18-7 shows a schematic of a GC. The sample is either injected into the GC using a gas-tight syringe or the instrument may be capable of obtaining its own sample via a built-in sampling pump. If the sample is a liquid, the instrument must be capable of vaporizing the sample (e.g., using a heated injection port).

The actual separation of the sample into its component parts takes place on the GC column.[8] Columns are typically long tubes made of metal, glass, Teflon®, or fused silica. Columns in portable, direct-reading GCs are of two kinds: packed and wall-coated. A packed column contains a granular material used as a solid support which is coated with a chemical chosen for its ability to interact with the components of the sample. This chosen chemical is referred to as the stationary phase. Packed columns are generally from 4 or 5 cm to 1 meter or more in length and have external diameters on the order of

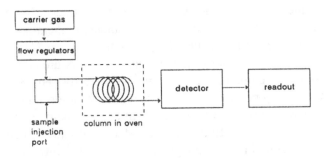

FIGURE 18-7. Schematic of a Gas Chromatograph.

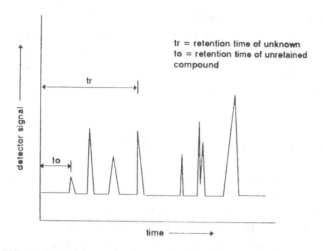

FIGURE 18-8. Schematic of a Typical Gas Chromatogram.

0.3 cm (1/8 in). A wall-coated column tends to be longer (5 cm to 3 m or more) and narrower (i.d. from 0.1 to 1 mm) than packed columns. In a wall-coated column, there is no granular solid support for the stationary phase. It is, as the name implies, coated directly on the inner walls of the column. The long, thinner columns (i.d.'s < 0.5 mm) are sometimes referred to as capillary columns.

The sample is carried through the column by an inert (relative to the sample) carrier gas, which, depending on the direct-reading GC, may be helium, hydrogen, nitrogen, argon, carbon dioxide, or air. The separation is governed by the degree of interaction of the sample with the stationary phase and the properties of the carrier gas. All components of a mixture spend the same amount of time in the carrier gas, so their different elution times is a function of the time partitioning between the stationary phase and the gas phase. The elapsed time from injection until the detector sees a component of a mixture is that component's retention time. The retention time is a function of the physical properties of a component in a sample, whereas the size of the peak is a function of the amount. Figure 18-8 shows the component parts of a typical chromatogram.

The degree of separation of two components, as well as their relative retention times, depends, in part, on the temperature at which the system operates; the higher the temperature the shorter the retention times. Some portable GCs operate only at ambient temperatures; others are capable of heating the column.

As each of the component parts of a mixture elutes from the column, they go into the detector. Portable GC detectors include flame ionization, photoionization, electron capture, ultraviolet, flame photometric, and thermal conductivity (which have already been addressed in this chapter), as well as nitrogen-phosphorous and argon ionization.

Because of their separation capabilities, GCs offer excellent selectivity combined with low limits of detection. The limits of detection are primarily a property of the individual detectors and are given in the detector discussions, but portable GCs generally have limits of detection at sub-ppm levels. Some limitations associated with portable GCs include the need for more user knowledge of the technique, the size, and cost. The portable GCs are numbers 11.1 through 11.7 in the "Instrument Descriptions" section.

Mass Spectrometers (12.1–12.2)

As the name implies, mass spectrometers determine mass of molecular fragments. Specifically, a mass spectrometer will determine the masses of individual fragments that have been converted into ions. A mass spectrometer determines mass by measuring the mass-to-charge ratio of ions formed from the molecule(s). After the ions are formed, they are separated in the mass analyzer according to their mass-to-charge ratio and collected by a detector wherein the ion flux is converted into an electrical signal proportional to the ion flux. [9] The components of a typical mass spectrometer are shown schematically in Figure 18.8.

Separation by, for example, gas chromatography of the components in a mixture prior to mass spectral analysis provides for unambiguous identification of mixture components. Mass spectrometry is the only technique currently available that will provide for such identification of compounds in the field.

Mass spectrometers are currently used only for area samples because of their size and power requirements and are primarily used for volatile organic compounds.

Summary

Many instruments are available for direct-reading analysis of gases and vapors. They operate on a variety of principles of detection and vary in performance characteristics such as linear range, specificity, and limits of detection. Direct-reading instruments represent a powerful tool in developing sampling strategies. That is, direct-reading instruments, when correctly used, can determine, in real or near-real time, those areas of high concentration, those workers at highest risk, and those processes with the highest emissions. Such information is useful in solving a variety of gas and vapor exposure problems. This information can guide the hygienist or safety professional in obtaining other more informative and useful samples requiring laboratory analyses. Used properly, direct-reading instruments can conserve resources, eliminating samples with results of "none detected."

References

1. Strobel, H.A.; Heineman W.R.: Chemical Instrumentation: A Systematic Approach, 3rd ed. John Wiley and Sons, New York (1989).
2. Gentry, S.J.: Instrument Performance and Standards. Appl. Occup. Environ. Hyg. 8(4), 260–266 (1993).
3. David, D.J.: Gas Chromatographic Detectors. Wiley-Interscience, New York (1974).
4. HNU Systems: Gas Chromatograph, Model 301, Instruction Manual. Newton Highlands, MA (1986).
5. Ingle, J.D.; Crouch, S.R.: Spectrochemical Analysis. Prentice-Hall, Inc., New Jersey (1988).
6. Hodgeson, J.A.: A Review of Chemiluminescent Techniques for Air Pollution Monitoring. Toxicol. Environ. Chem. Rev. 11:81 (1974).
7. Skoog, D.A.; West, D.M.; Holler, F.J.: Fundamentals of Analytical Chemistry, 5th ed. Saunders College Publishing, New York (1988).
8. McNair, H.M.; Miller, J.M.: Basic Gas Chromatography. John Wiley and Sons, New York (1997).
9. What is Mass Spectrometry?, 3rd Edition, American Society for Mass Spectrometry, Santa Fe, NM (1998).

Additional Reading

American Conference of Governmental Industrial Hygienists: Volume 1, Dosimetry for Chemical and Physical Agents. William D. Kelley, Ed. ACGIH, Cincinnati, OH (1981).
Cohen, B.S.: Air Sampling Instrument Performance. Appl. Occup. Environ. Hyg. 8(4) 227–229 (1993).
Cralley, L.J.; Cralley, L.V.: Patty's Industrial Hygiene and Toxicology, Vol. III, Theory and Rationale of Industrial Hygiene Practice. Wiley-Interscience, New York (1981).
Environmental Instrumentation Group, Lawrence Berkeley Laboratory: Instrumentation for Environment Monitoring Air. LBL-1. Technical Information Division, Lawrence Berkeley Laboratory, Berkeley, CA (1973).
Hosey, A.D.: History of the Development of Industrial Hygiene Sampling Instruments and Techniques. American Conference of Governmental Industrial Hygienists, Cincinnati, OH (1981).
Instrumentation for Monitoring Air Quality. R.C. Barras, Symposium Chairman. ASTM Special Publication 555 (74-76066). ASTM, Philadelphia, PA (1974).
Nader, J.S.: Source Monitoring. In: Air Pollution, 3rd ed., Vol. 3., Ch. 15, pp. 589–645. A.C. Stem, Ed. Academic Press, Inc., New York (1976).
Stevens, R.K.; Herget, W.F.: Analytical Methods Applied to Air Pollution Measurements. Ann Arbor Science Publ., Inc., Ann Arbor, MI (1974).
Christian, G.D.: Analytical Chemistry, 5th Edition, John Wiley and Sons, New York (1994).

Instrument Descriptions

This section contains tables and short descriptions of the commercially available direct-reading instruments for gases and vapors. Not all instruments from each manufacturer are listed. Rather, representative instruments are included. The tables are designed to provide an overview of the instrument features, sizes, and capabilities, whereas the descriptions give more detailed infor-mation and photographs. Each description is numbered and is cross-referenced in the tables that appear at the end of the chapter. The descriptions are grouped by the operating principle upon which the measurement is based. The following instrument tables are included:

Table 18-I-1. Electrical Conductivity Analyzers
Table 18-I-2. Potentiometric Analyzers
Table 18-I-3. Coulometric Analyzers
Table 18-I-4. Ionization Detectors
Table 18-I-5. Infrared Photometers
Table 18-I-6. Ultraviolet and Visible Light Photometers
Table 18-I-7. Chemiluminescent Detectors
Table 18-I-8. Photometric Analyzers
Table 18-I-9. Thermal Conductivity Detectors
Table 18-I-10. Heat of Combustion Detectors
Table 18-I-11. Gas Chromatograph Analyzers
Table 18-I-12. Multi-Gas Monitors
Table 18-I-13. Portable Mass Spectrometers

These tables reference instrument manufacturers by code letters; complete names and addresses are given in Table 18-I-14.

18-1. Electrical Conductivity Analyzers

18-1-1. Jerome 431-X Mercury Vapor Analyzer
Arizona Instrument Corporation

The Model 431-X Mercury Vapor Analyzer is a portable instrument designed for mercury surveys in workplace environments. The Model 431-X uses a patented Gold Film microsensor as the basis of detection. The sensor absorbs and integrates the mercury present in the sample, registering this as a proportional change in electrical resistance. The sensor's selectivity to mercury eliminates many interferences common to atomic absorption, such as water vapor, SO_2, aromatic hydrocarbons, and particulates. The Model 431-X incorporates an internal pump and digital display with microprocessor control. Activating either the 10-second sample or the 1-second survey mode starts the pump that draws a precise volume of air over the Gold Film sensor. Mercury in the sample is adsorbed and integrated by the sensor. The microprocessor computes the concentration of mercury in mg/m^3 and displays the results on the digital meter until the next sample cycle is activated. Response time: sample mode, 13 seconds; survey mode, 4 seconds. Meter: LCD display. Construction: aluminum alloy. Flow rate: 0.75 L/min. Can be attached to a datalogger or PC for automatic sampling.

18-1-2. Jerome 631-X Hydrogen Sulfide Analyzer
Arizona Instrument Corp.

The new Jerome 631-X Hydrogen Sulfide Analyzer utilizes the same gold film technology in its proven mercury monitor to detect H_2S in the range of 0.001 to 50 ppm with the push of one button. The instrument works the same as for mercury where the H_2S is adsorbed onto the gold film. Potential interferences from SO_2, water vapor, CO, and CO_2 are eliminated. In the survey mode, response times are as quick as 3 s. The instrument can be connected to a PC for unattended operation. The unit is designed for use in leak detection, odor and corrosion control, and safety in industries such as wastewater treatment, oil and gas, and pulp and paper.

18-1-3. UltraGas-U3S Sulfur Dioxide Analyzers
Calibrated Instruments, Inc.

The UltraGas-U3S is a sampling and analysis device for measuring the concentration of SO_2 in air by the conductivity method. Existing interference components can be eliminated in most cases through suitable absorption traps so that measurement is selective. In the instrument, a constant and continuous stream of air and reagent mix in a reaction chamber. The conductivity of the solution changes in proportion to the concentration of SO_2. The conductivity change is determined in the detector by two electrode sections. The conductivity of the reagent is measured first in one section, and after reaction with SO_2, the conductivity is measured in the second section. The difference in the two alternating currents flowing through the two electrode sections is selected electronically by the recorder. A temperature-dependent resistance compensates for temperature changes.

18-1-4. Mikrogas® Series Gas Analyzers
Calibrated Instruments, Inc.

Mikrogas® instruments continuously measure concentrations of SO_2, HCl, H_2S, NH_3, Cl_2, $COCl_2$, COS, CS_2, HCN, and other gases in ambient air, industrial process streams, standing tanks, waste treatment facilities, and stack and incineration emissions using the conductimetric measuring principle. Extremely accurate and precise streams of sample gas and a liquid reagent of measured conductivity are volumetrically forwarded to a wet sampling head where they are combined. Thoroughly mixed in a reaction line, the sample gas and chemically changed reagent are then again separated. The reacted conductivity level of the reagent solution is then monitored as it passes a temperature-compensated continuous measuring electrode. An elec-

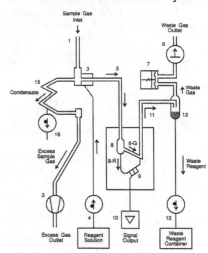

INSTRUMENT 18-1-4. Mikrogas® Series Analyzers.

tronic circuit determines the change in conductivity of the reagent solution. This change in conductivity is proportional to the concentration of gas being sampled. Operating temperature: 2 to 40°C.

18-1-5. Ultragas® Series Gas Analyzers
Calibrated Instruments, Inc.

Utilizing the principle of conductivity measurement, Ultragas® instruments provide continuous or batch analysis for laboratory and closed chamber research, and environmental, industrial process, ambient air, and stack applications involving one or more of the following gases: CO, CO_2, CH_4, NH_3, H_2S, SO_2, HCl, $COCl_2$, COS, CS_2, HCN, and other hydrocarbons. High precision pumps continuously forward a liquid reagent and sample gas stream to a temperature-compensated reaction line where both are combined in a constant volumetric ratio and thoroughly mixed. The conductivity of the reagent changes in proportion to the concentration of the gas being sampled and is measured by an electrode. Operating temperature: 2 to 35°C.

18-2. Potentiometric Analyzers (see 18-12, Multi-Gas Monitors)

18-2-1. Canary® Single Gas Monitor
Bacharach, Inc.

The Canary is a compact, personal monitor which can be fitted to monitor for 10 different gases (O_2, CH_4, H_2S, CO, SO_2, Cl_2, NO_2, HCl, HCN, and NH_3). The range of applications is vast, including steel manufacturing, pulp and paper mills, oil and gas refineries, waste and wastewater treatment, chemical plants, and offshore drilling. The unit weighs only 10 oz and includes visual and audible alarms, direct LED readout of gas concentration, a confidence light, single-button operation. Easy zeroing, calibration and sensor replacement. Powered by 4 AA alkaline or Ni-Cd batteries. Operating temperature: $-10°$ to $50°$ C.

18-2-2. CO Sniffer® Multi-Purpose Detector

Bacharach, Inc.

Hand-held monitor continuously measures CO between 0 and 2000 ppm in less than one minute. Applications are for CO surveys and for breath analysis (with special module). Concentration is displayed via a highly visible LCD with backlight. The unit has an internal pump, a rugged design with case to match, weighs only 21 oz, and is operated by 4 C-size alkaline batteries.

18-2-3. Toxi Series of Single Gas Detectors

Biosystems, Inc.

The Toxi Series of gas detectors are small, rugged personal monitors for the detection of oxygen and toxic gases in applications where low cost, ease of use, and durability are prime considerations. Sensors are available for O_2, CO, H_2S, SO_2, Cl_2, NH_3, NO, NO_2, or the new dual-purpose "CO Plus" sensor for the simultaneous detection of both CO and H_2S. The Toxi units feature a microprocessor controller which allows choice of alarms (including optional vibrator alarm) for ceiling values, STELs, or TWAs; true one-button operation; choice of batteries; automatic calibration adjustment, and a variety of accessories. The unit also allows the downloading of data to a PC. All Toxi monitors are shipped complete with sensor, calibration adapter, belt clip, a lanyard, and an alligator clip. Toxi monitors are intrinsically safe for use in hazardous locations Class I, Groups A, B, C, and D.

18-2-4. TOXYCLIP AND TOXYCLIP2 Personal Gas Alarm

BW Technologies

The TOXYCLIP and now the TOXYCLIP2 offer a miniature personal gas alarm for carbon monoxide, oxygen and hydrogen sulfide. The TOXYCLIP2 offers 2 years of protection without any need for factory renewal (CO and H_2S). The units feature easy one-button operation, life remaining of sensor, self-test of batteries and sensor, LCD readout, two levels of visual and loud audible alarms, and impact-resistant case with RF shielding. Weighing just 2 ounces, the TOXYCLIP2 is one of smallest personal alarms available. Also available is the TOXYCLIP Test Station for easy calibration.

18-2-5. MINIMAX Personal Gas Monitor

BW Technologies

The MINIMAX offers a wide variety of options in a personal gas monitor. The Plug-in sensors are available for CO, H_2S, SO_2, Cl_2, HCN and NO. The unit features five year battery life, automatic self-test, auto zero calibration, intelligent alpha-numeric display, two user selectable alarm levels for TWA and instantaneous high level, hi-output visual and audible alarms, microprocessor controller, fast response time (3–9 s), and rugged and is light weight (5.8 ounces). Each time the unit is turned on, it will automatically execute a self-test and advise the current alarm setpoints. Calibration requires only one step and the zero is set automatically.

18-2-6. Gasman and Gas Baron Personal Gas Detectors

CEA Instruments, Inc.

The Gasman is a series of single gas, shirt-pocket-sized personal gas monitors for toxic gases, combustible gases, or O_2. All models have a large, front-mounted digital display with built-in back light. Visual and audible alarms are available for toxic gases to provide instantaneous and TWA warnings. The monitors are rugged and water resistant. They are radio frequency (RF) shielded and are powered by four AA or rechargeable batteries. Models are available for O_2, H_2S, CO, SO_2, Cl_2, NO_2, NO, HCl, HCN, NH_3, and combustibles. The Gas Baron is a palm-sized personal monitor for O_2, H_2S, CO, Cl_2, ClO_2, and HCN. It features field replaceable sensors, 6–12 month battery life, vibrating alarm, test button, UL and CSA intrinsic approvals.

18-2-7. Series U Toxic Gas Detectors

CEA Instruments, Inc.

The Series U instruments are dedicated gas detectors in portable, wall-mounted, or multipoint configurations for a variety of contaminants. All instruments in this series use electrochemical-type or catalytic sensors and are available for H_2S, CO, SO_2, NH_3, hydrocarbons, H_2, combustibles, freons, EtO, alcohols, and diesel/gasoline vapors. The diffusion-type sensors are guaranteed for 2 years, provide rapid response, are solid-state, and are UL approved. Other features include low battery warning lights; built-in battery charger; high poison resistance to sulfur, lead, silicon, and halogenated compounds; and rugged, compact, leather carrying case. Operating temperature: -20 to $+65°C$.

18-2-8. TG-KA Series Portable Toxic Gas Analyzers

CEA Instruments, Inc.

The TG-KA Series Analyzers are available for formaldehyde, phosgene, ozone, hydrogen fluoride, and hydrides. The portable units can be used with AC/DC operation for continuous monitoring. Other features include a rechargeable battery for 30 hrs of operation, remote refillable sensor for spot checking, digital readout, recorder output, user adjustable audible and visual alarms, light weight (10 oz).

18-2-9. Gasman II Personal Gas Monitor

Crowcon Detection Instruments

The Gasman II is a lightweight personal monitor housed in a rugged orange NEMA 4X rated housing. The unit has a large LCD display with a back lighting button, audible and visual alarms, up to 4 month operation from one set of AA batteries, and operates over a wide range of temperatures (−10 to 50°C) and humidity (0–95% non-condensing). Sensors are available for combustibles, CH_4, CO, H_2S, O_2, Cl_2, HCl, HCN, NH_3, O_3, SO_2, and NO. Other features include RFI shielding, intrinsically safe (UL Class I, Division I, Groups A, B, C, and D), and an operation light.

18-2-10. MONOGARD and dynaMite Personal Monitors

Dynamation, Inc.

The MONOGARD and dynaMite Series of pocket-sized instruments combine digital LCD and diffusion chemical cell sensing for CO, H_2S, O_2, SO_2, and NO. The units feature an audible, pulsating alarm and a visual flashing light when unsafe atmospheres are encountered. Each unit has a low battery alarm, test switch, and illuminated display switch for reading in dark areas. All alarm points are factory set and customer adjustable. MONOGARDs are enclosed in rugged aluminum cases with leatherette carrying cases. The dynaMite gives more than 250 hrs of continuous operation from its replaceable lithium battery. Operating temperature for the monitors ranges from 0 to 41°C or to 52°C. Response time is 90% of full reading in 30 seconds. Monitors warm up in less than 10 seconds. The expected sensor life is 1.5 years (6-month warranty).

18-2-11. SPECTRUM Series Pocket-Size Gas Detectors

ENMET Corp.

The new SPECTRUM Series are inexpensive, high performance, microprocessor-based personal gas monitors. These unique gas detectors feature simple operation,

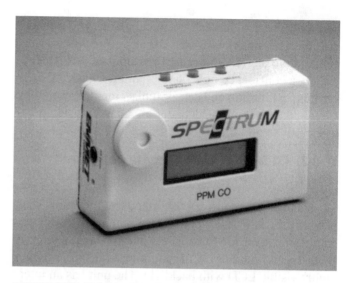

INSTRUMENT 18-2-11. Spectrum Series Gas Detector.

large easy-to-read continuous digital display, and a new generation of electrochemical toxic sensors in a small, durable package. Instruments are available to monitor CO, H_2S, Cl_2, HCN, SO_2, NH_3, or O_2. These instruments have been designed for use by firefighters and rescue personnel as well as for industrial applications. The instruments also feature maximum/minimum value tracking, "Intelli-Cal" calibration procedure, dual-level audio and visual alarms, password protected maintenance functions, low battery alarm and temperature display.

18-2-12. Smart Logger Series

ENMET Corp.

The SMART LOGGER series personal pocket-size gas detector features interchangeable, precalibrated

INSTRUMENT 18-2-12. Smart Logger Series.

SMART BLOCK sensors. This modular design enables the user to simply and easily convert one unit for the detection of many different gases. SMART BLOCK modules are available for : H_2, CO, Cl_2, HCN, H_2S, HCl, NO, NO_2, NH_3, O_3, ClO_2, ETO, and O_2. As your requirements for monitoring a specific gas change all you do is to plug in a different precalibrated SMART BLOCK module that meets your current needs. Instrument provides continuous datalogging for industrial hygiene applications. It stores up to 120 hours of exposure and event information which can be downloaded to a serial printer or an IBM-compatible personal computer. Instrument also features on-board datalogging, TWA and STEL alarms, and continuous digital readout.

18-2-13. Formaldehyde, Glutaraldehyde, and Ethylene Oxide Monitors

Environmental Sensors Co.

The ESC line of electrochemical sensors include the MVN-100A & B for ethylene oxide, the MVN-200D for glutaraldehyde, and the MVN-300 for formaldehyde. These units read ppm concentrations for these specialty gases. Applications include in hospitals, clinics, laboratories, sterilization facilities, chemical processing plants, and in the construction industry. The units are compact and lightweight, have audible and visual alarms, continuous LCD display, instant data output, long life sensors, and are easy to operate and affordable. Datalogging is optional.

18-2-14. GT and GTD Series Multi-Gas Monitors

GasTech, Inc.

The GT Series multi-gas monitor is ideal for pretesting of confined spaces in refineries, wastewater treatment plants, fire departments, off shore oil wells, fish processing, utilities, tanneries and other applications. This unit can be configured to monitor up to four gases in several different combinations for the detection of combustibles in the LEL and ppm range, oxygen content and a choice of one or two toxics. A built-in internal sampling pump will draw gas samples from up to 100 feet. The GT Series features include built-in datalogger, TWA/STEL alarms, and exclusive LIP (Liquid Inhibiting Probe). This unit is UL classified and CSA certified with a one year warranty.

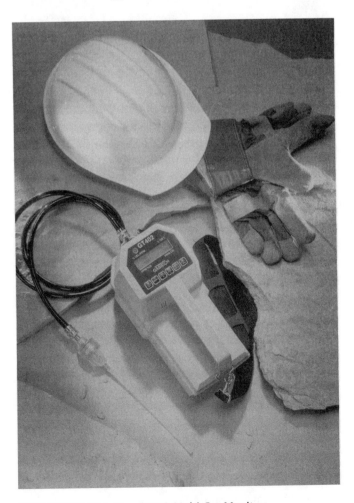

INSTRUMENT 18-2-14a. GT402 Multi-Gas Monitor.

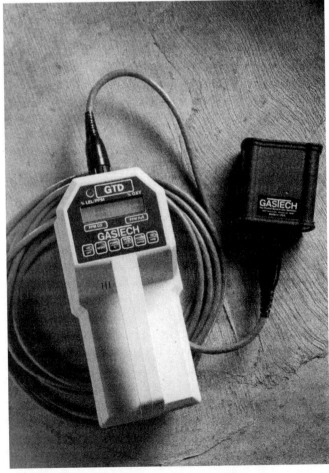

INSTRUMENT 18-2-14b. GTD Multi-Gas Monitor.

The GTD Series diffusion version of the GT multi-gas monitor is ideal for pre-testing of confined spaces and use in applications such as railcars, tanneries, refineries, off shore oil wells, and utilities areas where extreme conditions such as moisture, dust or contamination exist. This unit can be configured to monitor up to four gases in several different combinations for the detection of combustibles in the LEL and ppm range, oxygen content and a choice of one or two toxics. The GTD provides real-time response to gas by lowering the sensor module in confined space by the optional 20- to 50-ft extender cable. Features built-in datalogger, and LCD display.

18-2-15. *GT-2400 Multi-Gas Monitor*

GasTech, Inc.

The GT-2400 **low cost multi-gas monitor** is a complete monitoring instrument for multiple applications, such as municipalities, electric and gas utilities, breweries, petrochemical, and a host of other applications. This personal portable instrument can detect up to four gases; combustible, oxygen content and a choice of one or two toxics. The standard sampling method is 'diffusion' or use the optional sample-draw pump. The GT-2400 features include visual and audible alarms, simple operation and calibration, and peak hold mode. This unit is UL classified and CSA certified with a one year warranty.

18-2-16. *STM 2100 Multi-Gas Monitor*

GasTech, Inc.

The STM 2100 **multi-gas monitor** is designed to protect workers from hazardous gases in confined spaces and other industrial work sites. This monitor can detect up to four gases, combustibles in the % LEL

INSTRUMENT 18-2-16. STM 2100 Multi-Gas Monitor.

range (ppm and %VOL range when using the sample-draw pump), oxygen content and a choice of one or two toxics. The standard sampling method is 'diffusion' or use the optional sample-draw pump. The STM 2100 features built-in datalogger, TWA/STEL alarms, simple two button operation, toxic sensor recognition and adjustable calibration reminder. This unit is UL classified and CSA certified with a one year warranty.

18-2-17. *95 Series Single Gas Monitor*

GasTech, Inc.

The 95 Series **single gas monitor** is ideal for personal protection from hazardous gases. These low cost microprocessor controlled instruments are available in models that detect carbon monoxide, hydrogen sulfide or oxygen content. A top-mounted LCD (liquid crystal display) shows readings of the gas being measured, a built-in audible and visual alarm light to warn users of hazardous conditions. The 95 Series instruments can operate up to 3000 hours of operation, and an integrated pocket/belt clip provides hands-free monitoring. This unit is UL classified and CSA certified and comes with a one year warranty.

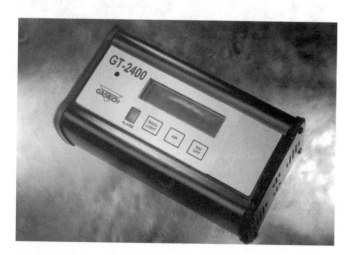

INSTRUMENT 18-2-15. GT-2400 Multi-Gas Monitor.

INSTRUMENT 18-2-17. 95 Series Single Gas Monitor.

INSTRUMENT 18-2-21. STX Single Gas Monitor.

18-2-18. Portable, Personal Monitors for Toxic Gases
GC Industries

GCI offers a complete range of small, portable monitors/alarms for the detection of O_2, CO, H_2S, SO_2, NO, and NO_x. Monitors are designed as personal monitors for field use, providing continuous monitoring and alarm of a particular gas. Pocket-sized monitors feature LCD display, audible and visual alarms, 9 V power supply, and a patented electrochemical sensor. Units have a 90% response time of 20 seconds, sensor life of 1 year, easy field replacement of sensors, and operates at temperatures of 0–40° C.

18-2-19. G3000 MicroTox® Series
GfG dynamation

The G3000 Series toxic monitors are hand-held, lightweight monitors available for CO (Microco®) and H_2S (Microtox®). Both the Microco and Microtox utilize diffusion input electrochemical cells. The cells are designed to last 1–2 years with little maintenance. A steel mesh diffusion screen and a Teflon membrane protect the unit from dust and splash water. The rechargeable, sintered metal Ni-Cd battery pack powers the unit for over 100 hrs of continuous operation on one charge. Both units use a three-chamber, 8-mm, high digital display. Operating temperature for both units is 0 to 53°C; response time is 15 seconds (T_{90}).

18-2-20. G111 Toxitector®-CO
GfG dynamation

The Toxitector® is a very small detector used to continuously monitor for CO concentrations releasing audible and visual alarms. The units provide three alarm thresholds: 30, 60 and 300 ppm. Applications include in the steel industry, power stations, chemical industry, mining indus-

try and various authorities. The battery operated units run for 150 hours between charges. The very small units are rugged and feature true pocket size, a stainless steel body, and a >2 years sensor life. The units operate in temperatures of –15 to +50° C and humidities of 5–95% RH.

18-2-21. STX Single Gas Monitor
Industrial Scientific Corp.

The STX70 continuously monitors oxygen or any of eight toxic gases. Available sensors include NH_3, CO, Cl_2, HCN, H_2S, NO, NO_2, and SO_2. The instrument is available in three distinct configurations. A Non-display version operates in "alarm-only" mode, while the Display version provides digital readout of toxic gases in ppm and oxygen in percent of volume. Also available is a Datalogging/Industrial Hygiene version (display or non-display) that provides 60 hours of data storage logged at one minute intervals. Additional features include RFI protection, audible and visual alarms and an easy to read top mounted LCD display.

18-2-22. T80 Single Gas Monitor
Industrial Scientific Corp.

The T80 monitors any one of eight gases with electrochemical smart sensors including CO, Cl_2, HCN, H_2S, NO_2, O_2 and SO_2. The T80 recognizes each installed sensor and an LCD display provides continuous gas concentration readings. The audible alarm is rated at 90 dB at three feet, and an optional internal vibrating alarm is available. Features include one-button auto calibration, built-in STEL and TWA readings and peak/hold memory. The high impact, RFI resistant case adds durability for almost any environment. A 9-volt alkaline battery provides continuous operation for up to 2600 hours or 4400 hours with a 9-volt lithium battery.

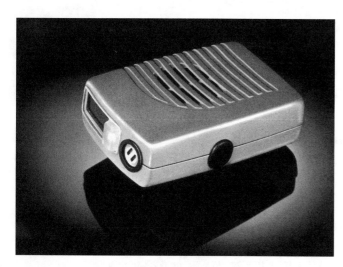

INSTRUMENT 18-2-22. T80 Single Gas Monitor.

18-2-23. GasBadge Single Gas Monitor

Industrial Scientific Corp.

GasBadge delivers continuous non-display monitoring for potentially hazardous levels of CO, H_2S, or O_2 with audible and visual alarms. After the unit is activated it will run continuously for one full year. At the conclusion of each year, the unit shuts down and can then be cost-effectively renewed for another year of usage. With the exception of daily battery testing and periodic gas bump testing, GasBadge requires no additional maintenance. Weighing less than three ounces, the instrument is extremely durable and offers outstanding RFI protection. The instrument does not require any calibration, though an optional automatic calibrator is available for users requiring periodic calibration.

18-2-24. Portable Gas Analyzers, Series 1000/4000/7000

Interscan Corp.

Interscan's field proven portable analyzers are available for CO, Cl_2, ClO_2, ethylene oxide, formaldehyde, hydrazine, HCl, HCN, H_2S, NO, NO_2, and SO_2. The 1000 Series is the original workhorse unit while the 4000 is a smaller version providing most of the same features. The 7000 in addition includes an internal datalogging capability. The units are $178 \times 102 \times 225$ mm, and weigh 2 kg. Accuracy is rated 2% of full scale, while zero and span drift are less than 2% of full scale (24 hour). All the units are calibrated against a known standard.

18-2-25. LD Series Continuous Monitoring Systems

Interscan Corp.

Interscan's LD Series Continuous Monitoring Systems are available for CO, Cl_2, ClO_2, ethylene

oxide, formaldehyde, hydrazine, HCl, HCN, H_2S, NO, NO_2, and SO_2. Industrially hardened, the systems can be provided with a variety of alarm and packaging features. Integral datalogging is also available. Dimensions are $356 \times 508 \times 222$ mm, exclusive of alarm strobe light. Weight is 10.4 kg. Various multipoint versions of the LD Series provided with full SCADA features are available.

18-2-26. RM Series Continuous Monitoring Systems

Interscan Corp.

Interscan's RM Series rack mountable analyzers are available for CO, Cl_2, ClO_2, ethylene oxide, formaldehyde, hydrazine, HCl, HCN, H_2S, NO, NO_2, and SO_2. They are intended to either be used in laboratory applications, or to be installed as part of a larger instrument system. Continuously adjustable alarm relays, and 4–20 mA analog output are provided. The dimensions are $178 \times 483 \times 305$ mm, and the weight is 5.2 kg, when provided with a pump and rotameter.

INSTRUMENT 18-2-23. GasBadge Single Gas Monitor.

18-2-27. ToxiBEE and GasBUG Series Personal Monitors

Luminor Safety Products

The ToxiBEE and GasBUG Series are small, light-weight personal warning monitors for toxic gases. The ToxiBEE weighs 2 oz, is less than an inch thick, and is available for CO and H_2S. The units have loud audible and bright visual alarms and are intrinsically safe (UL Class I, Division I, Groups A, B, C, and D). The units feature quick response (3–5 sec), easy calibration, 2-yr warranty, and low battery indicators.

18-2-28. UniMAX Personal Single Gas Detector

Lumidor Safety Products

The new UniMAX is a UL classified compact, pocket-sized, microprocessor controlled single gas monitor. The UniMAX continuously detects and displays gas concentration and instrument status on a large, easy to read digital display. The UniMAX can be configured to monitor O_2, CO, NH_3, PH_3, SO_2, H_2S, NO_2 Cl_2, and ClO_2. The UniMAX utilizes state of the art electrochemical interchangeable plug-in sensors, with smart sensor technology, offering easy adaptability and versatility for monitoring a wide variety of areas and to ensure both worker safety and simple operation. The UniMAX is housed in a rugged, water-resistant, ergonomic case with a strong, durable belt clip. It is suitable for potentially hazardous environments in pulp/paper mills, utilities, waste water treatment facilities, chemical, construction, manufacturing, mining, and petroleum industries. The unit has multiple alarms to alert the user to a dangerous situation which include a horn, bright flashing LEDs and an optional internal vibrator for noisy environments.

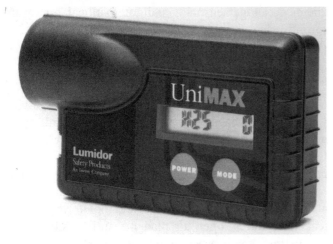

INSTRUMENT 18-2-28. UniMAX Personal Single Gas Detector.

18-2-29. Series PM-7700 Personal Toxic Gas Monitors

Metrosonics, Inc.

The PM-7700 uses interchangeable electrochemical sensors to detect CO, H_2S, SO_2, Cl_2, NO_2, NO, O_2, ETO, O_3, HCN, NH_3, H_2, and HCl for personal protection. The sensors utilize capillary diffusion barrier technology, which results in a direct response to volume concentration. The sensors are also cable mounted and can be clipped to clothing for readings taken in the breathing zone. Visual and ear piece audible alarms can be set for two user-selected gas concentrations. The PM-7700 can operate 720 hrs on a 9V alkaline battery. Operating temperature: –5 to 40°C. Response to 90% of final reading ranges from 30 to 90 seconds, depending on gas.

18-2-30. MiniCO® Responder Carbon Monoxide Indicators

Mine Safety Appliances Company

The MSA MiniCO® Responder provides fast, accurate sampling of CO concentrations from 0–99 ppm in a pocket-sized device. They operate on the principle of an electrochemical sensor cell. The MiniCO Responder features simple three-button control for easy operation, even with heavy gloves. With the press of a button, the instrument's back-lit display provides readings in low-light conditions for 30 s. The rugged case provides impact protection and RF shielding. A Goretex® filter over the inlet provides exceptional water resistance. This unit was designed for use by firefighters. An adapter with aspirator bulb, using standard MSA sampling lines, is available for remote sampling. The units are battery powered. The alarm set point is adjustable over the range of 25 to 500 ppm. All models have ±2% precision and accuracy, 90% response time in 30 seconds, a span drift less than 2% full scale/day, and zero drift less than 1% full scale/day. MiniCO indicators can be field-calibrated using the MSA Calibration Check Kit, Model R. Common interferents include SO_2, H_2S, NO_2, ethyl alcohol, and H_2.

18-2-31. Cricket® Personal Alarms

Mine Safety Appliances Co.

The Cricket® Personal Alarms are very lightweight, weighing less than 3 oz and can be worn clipped to a pocket, belt, lapel, or hard hat for hands-free operation. The instrument is only $1.25 \times 3 \times 2$ in. in size. The Cricket series of personal alarms are miniature, battery-powered instruments designed to provide users with an inexpensive yet dependable way to monitor for O_2 deficiency, CO, or H_2S. The instruments operate continuously and sound an

INSTRUMENT 18-2-32. MSTox 8600 Personal Toxic Gas Monitor.

alarm if levels in the environment exceed preset levels (for CO and H_2S) or fall below a preset level (for O_2).

18-2-32. MSTox 8600 Personal Toxic Gas Monitor

MST Measurement Systems, Inc.

The MSTox is an intrinsically safe personal toxic gas monitor. The unit can detect a wide range of gases including exotic gases used in semiconductor manufacturing. The MSTox is compact and lightweight. It has a rugged, high impact exterior that stands up to demanding use. The MSTox can be used in a variety of applications. Features include field settable, dual level concentration alarms, easy to read digital display, earphone jack and audio and visual alarm indications. CSA approved for Class 1, Division 1 and 2, Groups A, B, C and D.

18-2-33. Satellite

MST Measurement Systems, Inc.

The Satellite is a microprocessor based monitoring system for the detection of toxic, combustible and pyrophoric gases. It has a local, graphical LCD display that shows gas type, concentration and alarm status.

INSTRUMENT 18-2-33. Satellite.

Features include continuous sensor self test, user settable dual alarm and the ability to change gas types by simply plugging in a new sensor.

The Satellite is offered in both analog and digital versions. The analog version utilizes a 4–20 mA signal designed to interface with a controller, PLC or DCS system. The Digital version utilizes LonWorks™ technology, addressable input/output modules and MST's Wonderware based visualization software. UL listed for Class 1, Division 2, Groups A, B, C, and D.

18-2-34. Pac III Personal Monitor

National Draeger, Inc.

The Pac III is a single gas monitor which features Draeger's intelligent electrochemical sensors, loud (95 dBA at 1 ft) and bright alarms, large backlit displays, RFI shielding, and a rugged ABS/polycarbonate housing. The sensors come equipped with their own electronics which allows for interchangeability of sensors, increased temperature range stability, minimum cross sensitivities to other gases, and constant sensor validation of status and performance. Sensors are available for CO, O_2, CO_2, Cl_2, HCN, NH_3, H_2S, NO, NO_2, hydrides (PH_3, AsH_3, and B_2H_6) and SO_2. The unit operates from a 9V battery and over a temperature range of –20 to 50°C.

18-2-35. MicroPac Personal Gas Instrument

National Draeger, Inc.

The MicroPac is a compact gas warning instrument available for CO, H_2S and O_2. Uses include in the ambient air in industry, during the transport of dangerous goods and at other workplaces exposed to gas hazards. The instrument features very long operating periods, pocket size, infrared interface for data transmission, wide temperature and humidity range, small robust housing, different display variants, optimized electrochemical Drägersensors®, and international approvals for intrinsic safety.

18-2-36. Neotox® XL Pocket-Size Single Gas Monitors

Neotronics

The Neotox® XL monitors offer individual, lightweight, pocket-sized protection against the hazards of O_2 deficiency and enrichment, flammable gas (volume or LEL),CO, H_2S, Cl_2, SO_2, and NO_2. The monitor incorporates a top mounted backlit LCD display, one button operation, three levels of alarm, and audible and visual alarms. Applications include oil field monitoring, chemical laboratories, open air plan measuring, silos, food processing, steel processing, residential monitoring, underground parking garages, and handling bulk chlorine. Other features include small lightweight design, watchdog beep for unit function, "Lock On" and real time alarms, color coded, peak reading, optional use of rechargeable or alkaline batteries, auto-zero, optional datalogging, and a rubber boot for protection. The Firetox® CO unit is designed specifically to assist fire-fighters in locating CO leaks.

18-2-37. Solotox® Disposable Monitor

Neotronics

The Solotox is a disposable, personal, portable single gas detector. This unit requires zero maintenance (no battery replacement or calibration). This device is an inexpensive way to provide personal safety when hazardous levels of either hydrogen sulfide or carbon monoxide are present. The Solotox is sealed for life and has a preset alarm which will sound for the length of time the gas concentration exceeds the set point or until the battery is exhausted. The unit features a simple go/no go design, two year maintenance free life, visual and audible alarms, low life battery indicator, UL certified intrinsically safe, and a compact rugged design.

18-2-38. Sulfur Dioxide Analyzer/Recorder

Process Analyzers, Inc.

The Titrilog II is an automatic instrument for the determination of oxidizable sulfur compounds such as H_2S, SO_2, mercaptans, thiophene, and organic sulfides and disulfides. This instrument can be used for measurement in the atmosphere, in gas streams, and in stack gases. The measurement cell consists of an electrolyte containing potassium bromide from which free bromine is being generated electrolytically. In addition to the generating electrodes, there is a set of electrodes sensitive to free bromine. The potential of these electrodes varies with the concentration of free bromine in the solution. To distinguish between some of the different sulfur compounds, liquid absorptive filters are furnished as an accessory. These filters absorb one or more of the compounds of interest, enabling their concentration to be determined by difference. A programming system will route the sample through either of the filters, bypass the filters, and establish a zero level on an automatic repetitive cycle.

18-2-39. ToxiRAE Personal Gas Monitors

RAE Systems, Inc.

The ToxiRAE Series of Personal Gas Monitors features small, lightweight pocket-sized units available for a variety of toxic gases including O_2, combustibles, VOCs, CO, H_2S, SO_2, NO, NO_2, Cl_2, HCN, NH_3, and PH_3. The

units feature a rugged, weatherproof composite case, a large digital display with backlight, loud audio and flashing visual alarms, a vibrating alarm, RF protection, plug-in interchangeable sensors, datalogging (PGM-30D), and STEL, TWA and Peak information.

18-2-40. Series 94 Personal Gas Monitor

RKI Instruments, Inc.

These pocket sized, personal monitors are available for the detection of oxygen deficiency, carbon monoxide, and hydrogen sulfide. These units are intrinsically safe, compact, accurate, and have microprocessor controlled functions in a slim and light package. They can easily fit in your shirt pocket. The units feature digital readout, visual and audible alarms (buzzer) for high concentrations and low battery, peak value function, TWA and STEL values for toxics, and touch operated controls.

18-2-41. Model SC-90 Portable Toxic Gas Monitor

RKI Instruments, Inc.

The Model SC-90 is a survey electrochemical monitor which will react to over 17 different gases in the ppm range. This monitor is useful as a personal safety monitor for detecting dangerous levels of toxic gases in semiconductor and other industries. The unit continuously reads gas concentrations until a preset alarm level is triggered, and audible and visual alarms are activated. The unit averages gas readings over time and has a peak reading function. The unit uses replaceable alkaline batteries which will run the monitor for up to 20 hours. Other features include a dot matrix self-illuminated display, microprocessor controller, 30 sec response times, and various carrying options.

18-2-42. S100 Series Portable Gas Indicators

Scott Aviation

The S100 Series are intrinsically safe, portable instruments that can be applied to area monitoring, confined space entry, or personal monitoring. Instruments are available that can monitor for one, two, or three of the following: combustible gas, O_2, H_2S, and CO. Features of the S100 Series include dual low-battery alarms, liquid crystal display that illuminates for low ambient light conditions, and audible alarms for each measured variable. The S100 Series has memory capability to store the highest combustible gas concentration, peak CO or H_2S concentration, or the lowest O_2 measurement. Response time to 63% change: 10 seconds to LEL, 20 seconds for O_2, 45 seconds for H_2S, and 25 seconds for CO. Operating temperature ranges are −10 to 60°C for LEL and 0 to 40°C for the others.

18-2-43. Portable Gas Monitors
Sensidyne, Inc.

Sensidyne markets a wide range of pocket-sized personal monitors (Mini Monitors), portable survey monitors (Series SS2000 and SS4000 for semiconductor gases), and a variety of fixed gas detection systems. The Mini Monitors and Series SS2000 monitors utilize diffusion electrochemical cells specifically designed for each gas to be detected. The lightweight (7 oz), pocket-sized Mini Monitors feature a continuous LED light-illuminated digital display, dual alarm set points, intrinsically safe design, replaceable batteries, RFI/EMI protection, and easy calibration. Additional features on the hand-held Series SS2000 include long-life sensors (3 years expected life), rechargeable batteries, optional continuous operation from AC power, triple alarm system, and ability to withstand temperature extremes. Response time: <20 seconds for Mini Monitors, 10–15 seconds for SS2000, <30 seconds for SS4000. Battery life: over 100 hrs for Mini Monitors, 20 hrs for SS2000, 35 hrs for SS4000. Humidity range: 5% to 95% for Mini Monitors and SS2000; 20% to 90% for SS4000. Temperature range: 0 to 40°C for all monitors.

18-2-44. Portable Flue Gas Analyzer
Teledyne Analytical Instruments

The Model 990 is a completely portable, battery-powered flue gas analyzer designed to rapidly monitor the O_2 and CO content of a combustion process. When these two measurements are combined for the purpose of maximizing fuel-burning efficiencies, boilers and heaters can be fine-tuned for optimum air/fuel ratios.

The CO trace measurement is accomplished by an electrochemical sensor (6-month warranty). The sensor output is directly proportional to the CO concentration. Zero and span drifts are less than 2% in 24 hrs. A 90% of full-scale response is attained in 30 seconds or less. Operating temperature: 0 to 50°C. O_2 analysis is accomplished with Teledyne's Micro-Fuel Cell (1-year warranty), which produces an electrical signal that is directly proportional and specific to the O_2 concentration in the flue gas. A 90% of full-scale response is attained in 13 seconds or less.

18-2-45. MDA Scientific Lifeline Gas Monitors
Zellweger Analytics

Designed to eliminate many of the shortcomings of traditional cell-based technologies, the new MDA Scientific LIFELINE product line features quick installation, minimal maintenance, and "smart" sensors which provide early warning diagnostics of a sensor's condition and effective life cycle. Four versions of LIFELINE monitors cover virtually all semiconductor applications, as well as many other industrial applications requiring a rugged toxic gas detection system. A passive, diffusion-type system and a remote system capable of locating sensors up to 50 feet from the transmitter feature an intrinsically safe design. Extractive and pyrolyzing models are capable of regulating samples and drawing air from streams up to 50 feet away. Each version has a compact footprint and features easy maintenance. Gases monitored include AsH_3, PH_3, B_2H_6, SiH_4, HCl, HBr, HF, NF_3, CO, TEOS, NH_3, H_2, Cl_2 and others. All units use smart, digital sensors that monitor cell performance and advise of recalibration requirements. Electro-chemical cell-based sensors are easily exchanged at six-month intervals through a sensor exchange program. Complete diagnostics are reported locally on unit's LCD display and transmitted via 4–20 mA outputs. There is also a LonWorks compatible version available.

18-3: Coulometric Analyzers (see 18-12, Multi-Gas Monitors)

18-3-1. Model OX630 Oxygen Analyzer
Engineering Systems and Designs

The OX630 is a portable unit that utilizes a maintenance-free galvanic electrode to measure atmospheric oxygen levels from 0% to 100%. The electrode has an expected life of 3–5 years at 25°C, 1 atmosphere pressure, and a concentration of 20.9% O_2. The OX630 is sold in a kit containing the meter, an electrode on a 5-ft cable, screwdriver for calibration, and a carrying case. Electrodes up to 100 ft can be manufactured upon request. Operating temperature range: 0 to 40°C. Response time: 95% of final reading in 30 seconds. Power: 9V battery. Calibration: 20.9% in air.

18-3-2. Oxytector® Oxygen Warning Device
GfG dynamation

The Oxytector® is an ultra-compact, easy to use, battery powered instrument for continuous oxygen monitoring. The main application is monitoring oxygen content in air when going into manholes, canals, tank vessels and other narrow areas with the danger of reduced oxygen concentration. Applications include water supplies, municipal service, fire brigades, civil engineering power stations, mining industry, and various authorities. The unit features variable warning optical and acoustic alarms, 150 hours of operation between charges, a battery function check, a stainless steel body, and 3 alarm thresholds (17, 19, 22%).

18-3-3. Models 8060/8061 Oxygen Deficiency Monitors

Matheson Gas Products

Models 8060/8061 Oxygen Deficiency Monitors are portable personal monitors or guarding against oxygen deficient exposure. Using an electrochemical sensor, the instruments each provide a large digital concentration readout, and an audible alarm when the level drops below 19.5%. The model 8060 uses a short 3 foot coiled cable connecting the sensor to the instrument. The 8061 uses a 16-ft cable on a spool making it ideal for lowering into confined spaces.

18-3-4. NGA 2000 Trace Oxygen Analyzer Module (TO2)

Rosemount Analytical, Inc.

The TO2 is designed for applications like analyzing trace impurities in pure gases, controlling inert atmospheres in heat treat application, and monitoring inerting operations where the presence of oxygen is undesirable. At the heart of the TO2 Analyzer Module is a one-depleting electrode sensor which provides accurate oxygen measurements in the ppm oxygen range. With sensitivity to less than 10 ppb, the sensor is truly designed to provide precision trace oxygen measurement with superior performance. The TO2 module is a self contained unit complete with detector and microprocessor-based electronics. It is ideal for air separation plants, heat treat applications and inerting operations.

18-3-5. Model 55 Oxygen Deficiency Monitor

Sierra Monitor Corp.

The Model 55 is a hand-held oxygen monitor with digital display to measure oxygen depletion in confined spaces. This monitor is accurate, rugged and simple to use and includes a digital display and audible alarm that sounds at 19.5% oxygen. An external senor with a 20-foot extension is optional.

18-4: Flame Ionization Detectors

18-4-1. Total Hydrocarbon Analyzer

Columbia Scientific Corporation, a division of Forney Corporation

The Model HC500-2D performs real time and continuous dry analysis of hydrocarbon gases utilizing a flame ionization detector (FID). Emphasis is focused in stable and reliable performance without a requirement for clean hydrocarbon-free combustion air. The temper-

ature of the sample air, hydrogen, and exhaust gas is controlled within ± 1% over 10 to 40 C. The HC500-2D closely approximates ppm hydrocarbon molecules rather than approximate methane equivalents as provided by FIDs operating in the gas chromatograph (GC) mode. The Model HC500-2C is available to differentiate methane and non-methane hydrocarbons. Outputs are available as 0–100 mV, 0–1 V (other outputs optional). Ranges: 0 to 10, 50, 100, 500, 1000 ppm. Minimum Detectable Sensitivity: 0.1 ppm CH_4. Noise: ±0.05 ppm CH_4. Lag time: <15 seconds. Rise Time/Fall Time to 90%: <30 seconds. Zero Drift/Span Drift: ±2 % Full Scale (FS)/day; ±3.0 % FS/3 days. Linearity: ±1.0 % FS. Selectable time constant: 1 second or 10 seconds. Operational specifications: unattended operation (no adjustment of flow or electrical systems), 7 days. Sample flow rate: 200 ml/min; hydrogen flow rate: 140 ml/min.

18-4-2. Century TVA-1000 Toxic Vapor Analyzer

Foxboro Company

The TVA-1000 is an over-the-shoulder portable vapor analyzer which offers both flame ionization and photoionization. An optional Enhanced Probe is available which allows single hand operation. The power and fuel supplies the unit for eight hours of operation in a fully certified intrinsically safe package. Other features include on-board datalogging, multipoint calibration, multiple calibration, menu driven operation, easy-to-read LCD displays, large keypad, datalogging thumb switch on hand-held probe, automatic autoranging, internal diagnostics, and an optional flame re-ignite function.

18-4-3. Models PI-101, IS-101, DL-101 Portable Photoionizers

HNU Systems, Inc.

The PI-101 is the original photoionization analyzer developed in 1974. The units are portable analyzers used for the measurement of gases in industrial atmospheres. The IS-101 in an intrinsically safe unit for explosive conditions. The new DL-101 is a microprocessor-controlled unit equipped with datalogging and four operating modes. The basic sensor consists of a sealed UV light source that emits photons which are energetic enough to ionize many trace species (particularly organics) but do not ionize the major components of air such as O_2, N_2, CO, CO_2, or H_2O. The field created on an electrode drives any ions formed by adsorption of the UV light to the collector electrode where the current (proportional to concentration) is measured.

This instrument consists of two separate units: a sensor and a readout, connected by a 3-ft, shielded, multiconductor cable with electrical connector. The case for the readout module is constructed of drawn aluminum. The sensor's outer body is of aluminum and engineering thermoplastic. The DL-101-2 has two automatic modes: survey and hazardous waste. The DL-101-4 has two additional modes of operation which include industrial hygiene and leak detection.

18-4-4. Passport® PID II Monitor
Mine Safety Appliances Co.

The Passport PID II Organic Vapor Monitor is a convenient, portable instrument for detecting low concentrations of VOCs. Applications include soil remediation in connection with hazardous material spills and underground tank leakage, arson investigation, industrial hygiene applications, emissions monitoring and general leak detection. The unit is ideal for Method 21 analysis since it can directly measure 0.1–10,000 ppm VOCs. Features include humidity insensitivity, autoranging capability, 69 pre-programmed plus 10 user-defined response factors, three buttons for simplified operation, graphic and numeric display, peak, STEL, and TWA measurements, simple fresh air setup, and labeling features. The unit has certification for various Class I, II, and III instrinsically safe operations.

18-4-5. Model 2020 Miniature Photoionization Monitor
PE Photovac

The PE Photovac Model 2020 is a hand-held Photoionization Detector System for use in the monitoring of ambient air for the presence of organic vapors. Virtually any non-methane organic volatile can be observed and measured to levels as low as 0.5 ppm. The 2020 is easily used in the field for hours using its rechargeable battery. Internal datalogging allows the operator to record hundreds of sampling points that can be downloaded to a computer for recordkeeping. The 2020 can also be pre-set to sample at intervals for tracking of VOC levels over time, or for calculation of 8 hour exposure levels. The 10.6 ev lamp is guaranteed for one year, and is easily cleaned and replaced by the user. An optional 11.7 ev lamp can be used where the ionization potential of the compound(s) of interest is (are) higher than optimal with the standard lamp. Applications included confined space pre-entry, fugitive emissions, plume characterization, leak detection, workplace monitoring, storage vessels, site characterization, and SVE system monitoring.

18-4-6. MicroFID™ Flame Ionization Detector
PE Photovac

The MicroFID is a battery-powered, hand-held flame ionization detector (FID). It is designed to detect and measure total hydrocarbons, or total VOCs in the range of 0.1 to 50,000 ppm. The MicroFID is completely self-contained and is intrinsically safe for use in potentially hazardous environments. The MicroFID is the world's lightest weight direct-reading FID. Its ergonomic design includes a built-in handle, and a rubberized keyboard that can be used while wearing protective equipment. Up to 750 sample points can be recorded with date, time and instrument status at selectable intervals or in Method 21 Format. The data can be downloaded to any PC for data management. PE Photovac Star 21 Software is designed to address the requirements of Method 21 data reporting and sample scheduling. Using Star 21, monitoring schedules can be uploaded to the MicroFID datalogger. The MicroFID has a wide linear range, responds to almost all organic compounds, is quite stable, and is virtually immune to possible interferences such as water vapor. The sample air serves as the source of oxygen to support the flame. A glow plug starts the flame automatically. After the sample passes through the detector, it is vented through a flame arrestor, preventing the ignition of any flammable gases surround the sampling location. The rugged design of the MicroFID combines with small size, light weight, and ease of operation, to make it an ideal choice for use in the field.

18-4-7. MiniRAE 2000, MiniRAE PLUS, and UltraRAE Portable VOC Monitors
RAE Systems, Inc.

The MiniRAE 2000 is the newest member of the RAE family of portable photo-ionization monitors. It is advertised as the smallest handheld VOC monitor available. The MiniRAE PLUS is a hand-held PID monitor useful for leak testing, fugitive emissions testing, etc. The UltraRAE is a quick, spot check Monitor used with a colorimetric tube to get readings for total hydrocarbons or classes of hydrocarbons for pre-screening during tank entry, hazardous material response, and refinery downstream monitoring. The MiniRAE 2000 features the new RAE 3-D sensor, provides easy access to the lamp and sensor, reduces moisture interference, has improved linearity and sensitivity, and has an extended range of 0–10,000 ppm. Other features include a built-in sample pump, external NMH batteries, automotive charger available, large keys operable with gloves, back lit display which is alarm- or darkness-activated, and preset alarms for STEL, TWA, low, and high peak values. There are many other features for all models.

18-4-8. NGA 2000 McFID
Rosemount Analytical, Inc.

The NGA 200 McFID is an advanced, high sensitivity Flame Ionization Detection system. The McFID utilizes state of the art "variable pressure" chromatographic separation of methane and non-methane hydrocarbons to provide a fast and accurate analysis of these constituents. Four standard measured variables corresponding to CH_4, non-CH_4, Total Hydrocarbons and real-time chromatogram plot. Exclusive split electronic-pneumatic bench design provides intrinsically safe operation, no continuous dilution purge required. Meets or exceeds U.S. EPA requirements CFR 40 pt. 60, App. A Method 25 (process) and CFR 40 pt. 86, (light duty automotive) hydrocarbon monitoring requirements.

18-4-9. Sensidyne Intrinsically Safe Portable FID
Sensidyne, Inc.

Sensidyne's Portable Flame Ionization Detector (FID) allows one to measure low levels of volatile organic compounds (VOCs) in virtually any hazardous environment since it is intrinsically safe (Class I, Division I, Groups A, B, C, and D). The FID is rugged, portable (3.9 kg), and has quick responses (<10 sec). It can measure VOCs over the range of 0.1 to 10,000 ppm (0.05 ppm minimum limit of detection for methane) and has up to 29 hrs of continuous operation. An optional datalogger allows for storage of measurements with easy downloading to a serial printer or IBM PC. Temperature range: 0 to 40°C. Humidity Range: 10-95% continuous, non-condensing. Other features include a multi-purpose sampling probe, a gas chromatograph version for GC separation of VOCs, a choice of hydrogen bottles, audible and visual alarms, back lit display, FID range extenders (up to 10x), Integraph software to interpret GC results, and low battery indicator.

18-4-10. Models 580EZ Portable PID and Models 55C Direct Methane, Non-Methane, 51 Total Hydrocarbon FID Analyzers
Thermo Environmental Instruments

The 580EZ is a portable PID monitor for VOC analysis. It features datalogging, microprocessor control, graphic LCD display, three button operation, weatherproof cases and keys, backlit display, and Smart Battery indication of battery status. Models 55C and 51 use FID for Total Hydrocarbon Analysis. Model 55C allows differentiation of methane, non-methane hydrocarbons while Model 51 is a heated FID for determining total hydrocarbons (THC). The units each have individual features and specifications for their specific applications.

18-5: Infrared Photometers

18-5-1. Model 1301 Gas Analyzer
Bruel and Kjaer Instruments, Inc.

The Model 1301 Gas Analyzer is a fully self-contained, transportable Fourier Transform Infrared (FTIR) spectometer that utilizes photoacoustic detection and is designed for field use. The unit can be used as both an analyzer to determine what gases are present and as a monitor for concentration measurements. Any gas or vapor that has an infrared absorbance between 4000/cm and 650/cm can be detected. Detection limits are typically in the range from 0.1 to 10 ppm. The dynamic range is 4 orders of magnitude. The unit has extensive internal data handling and data storage capabilities, along with a built-in disk drive and graphics screen. Serial and parallel interfaces allow for the transfer of data to various peripherals and computers. Zero drift: detection limit over 3 months. Span drift: 5% of reading over 3 months.

18-5-2. Toxic Gas Monitor Type 1302
Bruel & Kjaer Instruments, Inc.

The Toxic Gas Monitor Type 1302 is designed for the continuous measurement of various toxic gases. Typical applications are area monitoring for process emissions and perimeter monitoring for accidental releases. The monitor can operate unattended for months at a time. The Multi-gas Monitor 1302 is a portable unit that has typical applications for occupational exposure, tracer gas analysis, and indoor air quality assessment. The measurement technique used in both instruments is based on infrared photoacoustic spectroscopy. This method is based on the fact that when a gas absorbs modulated light, it emits sound proportional to the concentration of the gas. During operation, air is pumped into the measurement chamber. The chamber is sealed and irradiated with modulated, narrow band, infrared light. If the toxic gas of interest is in the air sample, sound is emitted and measured with a microphone. The signal is processed and the result is transmitted to the controlling computer. Selectivity is controlled by fitting the monitor with the appropriate optical filter for the gas of interest. A wide range of filters is available, covering the useful region of the infrared spectrum.

The Toxic Gas Monitor is remotely controlled from a personal computer that can be positioned a considerable distance from the monitor. The monitoring system can incorporate from 1 to 254 monitors connected to one computer. The Model 1302 has 32 KB of memory and an

80-character display. It has a measurement time of 30 seconds for one gas and up to 100 seconds for five gases. Span drift: 2.5% of reading in 3 months. Zero drift: detection threshold concentration in 3 months.

18-5-3. Models CEA 266 & 104 Wall Mounted CO₂; CEA 105, GD-444 and GD-344 Portable and Personal CO₂ Monitors

CEA Instruments, Inc.

Models CEA 266 and 144 are wall mounted analyzers for CO_2. The units feature digital display, internal pumps, linear 4–20 mA outputs, low maintenance and water resistant housing. These units are designed for continuous area monitoring of CO_2 concentrations. Models CEA 105 and GD-444 are portable CO_2 analyzers for continuous or spot check analysis of CO_2. The CEA-105 is a hand-held unit with audible and visual alarms, AC/DC operation, 0–2000 ppm display, and has optional temperature and RH probes. The GD-444 is pocket-sized, autoranging unit with a built-in pump, optional datalogger, large backlit display, and push button operation. The GD-344 is a lightweight (9 oz) personal CO_2 monitor which monitors 400–15,000 ppm and is easy to operate, weather and dust resistant, has a long-life battery, and digital display with audio and visual alarms.

18-5-4. Fourier Transform Infrared Analyzer (FTIR)

Columbia Scientific Corporation, a division of Forney Corporation

The Quantum 7000 FTIR Gas Analyzer is a transportable gas analyzer with the latest monitoring hardware and software technology to provide a complete, high-speed gas analysis system. The Quantum 7000 uses a patented interferometer with cube corner reflectors. The advantage of this design is that it is significantly less sensitive to temperature variations and to vibration. The Quantum 7000 utilizes a Fourier Transform Infrared (FTIR) spectrometer to analyze the entire IR spectrum of the gas sample and allows for detection, identification, and quantification of up to 32 different components of a gas sample. Sample cell: various configurations available. Data System: internal 486 computer, DOS operating system, CALCWARE analysis software, RS-232 Centronics port. Analyzer Performance: up to 12 measurements a second. Lowest detection limits: typically 10–100 ppb.

18-5-5. Miran SapphIRe Gas Analyzers

Foxboro Environmental Company

The Miran SapphIRe is the new line of NDIR gas analyzers from Foxboro which replaces the old 101, 1A, 1B, 1B2, etc. line of portable IR analyzers. The SapphIRe

units are lightweight and portable, ergonomically designed, has an eight-line 40-character display that prompts user through each step, has interference compensation which helps eliminate the additive effects of other gases, and are easily upgraded. The SapphIRe-1 is dedicated to specific gases requested by the user, which can be tuned to virtually any wavelength or pathlength needed. The SapphIRe-5 is dedicated to measure a multicomponent application defined by the user. Up to 5 components can be measured and displayed, plus up to 5 components can be defined that are to be factored out of the analysis. The SapphIRe-30 and 30E are calibrated for 30 single gas applications. In addition, the unit is calibrated for one multi-component application per user specifications. The SapphIRe-100 and 100E analyzers employ factory calibrated multicomponent interference capabilities for the user's specific application. The instrument has the ability to "dial up" any of over 100 factory calibrated single gas applications from the fixed library.

18-5-6. APBA-250E Indoor Air CO₂ Monitor

Horiba Instrument Corp.

The APBA-250E Indoor CO_2 Monitor is designed for convenient monitoring and control of atmospheric carbon dioxide content and is ideal for greenhouses, food storage facilities and confined indoor spaces. This wall mounted unit uses NDIR to monitor CO_2 over the ranges of 0–3000 ppm, 0–1%, and 0–5%. The unit features a visual alarm, recorder output, and LED readout.

18-5-7. APMA-360 Ambient CO Monitor

Horiba Instrument Corp.

The APMA-360 is designed for continuous ambient air monitoring of CO using cross flow modulation, NDIR. This instrument features excellent long-term stability with low zero drift. The unit features ranges from 0 to 10 up to 100 ppm CO, various alarms, four language on-screen messages, variable input/output ranges, and excellent performance.

18-5-8. CDU440/MDU420 Infrared Single Gas Monitors

Industrial Scientific Corp.

Both the CDU440 and MDU420 feature Industrial Scientific's patented infrared sensor technology and internal sampling pumps. The CDU440 provides continuous auto-ranging detection of carbon dioxide levels as low as 10 ppm to 6% by volume (60,000 ppm). The MD420 detects both low and high levels of methane gas, from 0 to 100% LEL and auto ranging to 100% of volume. Both instruments feature a five-year infrared

INSTRUMENT 18-5-8. CDU440/MDU420 Infrared Single Gas Monitor.

sensor warranty, peak hold memory and interchangeable Ni-Cd or lithium battery packs.

18-5-9. Aq-511 CO_2 Monitor

Metrosonics, Inc.

The Model Aq-511 is designed to monitor and display CO_2 concentrations in indoor environments. It features a built-in pump for continuous monitoring. Other models (Aq-501, 502, 512, 513) will also monitor temperature, humidity, air velocity, light, VOCs, particulates, noise and other gas levels (see 18-12-27).

18-5-10. Lira® Nondispersive Infrared Analyzer

Mine Safety Appliances Company

The Lira® NDIR Analyzer is designed for fixed station use in the detection of a single gas or vapor in chemical process streams. Lira Analyzers are commonly used to measure CO, CO_2, SO_2, ammonia and virtually all hydrocarbons and chlorinated hydrocarbons. The Model 202X Lira Analyzer monitors many hazardous gases in the ppm range and is suited to CO in the range of 0–50 ppm and organic compounds at their TLV levels. The Model 3000 Lira offers somewhat lower sensitivity and is available in both digital and analog versions. All models provide signals for activating alarms, records, and process control or automatic shutdown equipment.

18-5-11. ML®9830B Carbon Monoxide Analyzer

Monitor Labs, Inc.

The ML®9830B CO Analyzer uses a combination of non-dispersive infrared and gas filter correlation (GFC) techniques. The GFC minimizes the sensitivity to potentially interfering gases such as H_2O and CO_2. IR radiation

is emitted from a long life electrically heated element and projected through a rotating filter wheel which is heated and contains one glass cell of CO and another filled with nitrogen. As the wheel rotates, the reference beam is detected when the cell with the CO is in the light beam, and a measurement of the unknown gas is detected when the cell with nitrogen is in the light beam. The unit features a heated measurement cell to minimize maintenance, automatic zero background correction for minimum zero drift, temperature stabilized filter and gas filter wheel to minimize span drift, and automatic zero/span check with internal zero source and external span gas.

18-5-12. Riken RI-411A Portable Infrared CO_2 Monitor

RKI Instruments, Inc.

The Riken RI-411A is a lightweight CO_2 infrared gas monitor with digital readout and audible alarm. The unit is applicable to food-related industries, brewers, mushroom growers, greenhouse horticulture, welding, office ventilation systems, cooling systems, hazardous environments, laboratory and research projects, etc. The Riken RI-411A utilizes NDIR absorption to measure CO_2 in air. The unit is Ni-Cd battery operated and microprocessor controlled. The readings of CO_2 concentrations can be continuous or averaged over 1, 3, or 15 min. Averaged readings are held on the display until needed by the user. The RI-411A has a solid-state detector, an illuminated dot-matrix digital display, and a recorder output and can operate on AC using an optional DC power supply. Audible alarms: high CO_2, 5000 ppm (short pulse, optional 25%), averaging period (long tone), and low battery (continuous tone). Response time: 10 seconds to 90% indication. Calibration: zero, calibration using nitrogen or air cylinder (zero gas); span, calibration using cylinder of CO_2 in air. Ambient temperature range: –10 to 40°C. Ambient humidity range: 10% to 90% RH. Recorder output: 0- to 10-mv DC (linear). Auxiliary charger available for charging or continuous operation on 115-VAC adaptor. Operating hrs: about 6 hrs continuous.

18-5-13. Riken RI-413A Portable Halocarbon Indicator

RKI Instruments, Inc.

The Riken Infrared Gas Analyzer Model RI-413A is an NDIR analyzer coupled with a microprocessor to allow the measurement of halogenated carbon gases, specifically, R-11, R-12, R-22, R-113, R-114, R502, and R-134a. The unit has an internal pump that pulls a sample through a probe and into the infrared source. The source emits a broad band of energy which is focused on a solid state detector through a narrow band filter selected to transmit only a certain range of frequencies which are selectively absorbed by halogenated carbon gas. The unit features three options for battery use (alkaline, Ni-Cd, carbon-zinc), continuous readout and averaging concentration display, audible alarm for continuous display or averaging display plus low battery indicator, and a recorder output.

18-5-14. Sieger Searchline Excel Open Path Infrared Gas Detector

Sieger Instruments

The Searchline Excel is a state-of-the-art open path infrared gas detection system. It would be mounted in a potentially hazardous environment to detect the presence and build up of potentially explosive concentrations of hydrocarbon gas clouds. The receiver unit produces an analogue or digital output signal proportional to actual gas concentrations within the volume of air located between a transmitter and receiver. The signal indicates the level of potential hazard in units of LEL. Applications range from offshore platforms and

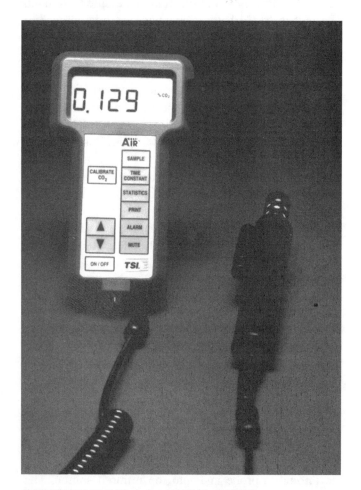

INSTRUMENT 18-5-15. InspectAir™ Model 8560.

vessels, downstream chemical processing plants, gas transport and pipelines. Features: Totally solar immune using a uniquely modulated high-intensity lamp source (Eye-safe), smart heated optical windows, patented fog filter, early warning dirty optics, instrument assisted alignment, T90 response time of less that 3 seconds, operating distances range from 15 to 650 feet, operating temperature range between –40 and +150 EF.

18-5-15. InspectAir™ Model 8560

TSI, Inc.

The Model 8560 Carbon Dioxide Monitor is a portable, hand-held instrument for measuring high levels of carbon dioxide. The NDIR sensor has very fast response to changes making it an ideal survey type instrument. It also has two alarm levels that enable it to be used as an area monitor. Statistics for a measurement period are also available. It is typically battery powered but can be run extended times with an AC adapter. The instrument is designed for concentrations of carbon dioxide up to 5% in air. This same CO_2 sensor is available in the Q-TRAK™, which also measures temperature, % RH and CO.

18-6: Ultraviolet (UV) and Visible Light Photometers

18-6-1. Model TGM555 Portable Toxic Gas Monitor

CEA Instruments, Inc.

The TGM555 is a portable, ambient air monitor that can be used for continuous colorimetric analysis of numerous compounds. The TGM555 contains a rechargeable DC power source and a constant-volume adjustable air pump. An air sample is continuously drawn into the unit and scrubbed with an absorbing reagent that removes a trace pollutant from the air stream and transfers it into the liquid reagent system. The subsequent color formation is read by a colorimeter and displayed on a built-in meter or on the optional digital readout. A recorder output is also provided.

Operating period: 20 hrs, fully charged internal batteries. Signal output: 0–1.0 V at 0–2.0 mA. Calibration: <1% drift/72 hrs. Sensitivity: 1% of full scale. Nonlinearity: <2%. Zero and span drive: <2%/72 hrs. Air flow drift: <1%/72 hours. Noise: 0.75% of full scale. Lag time: 4 min. Rise time to 90%: 4 min. Fall time 90%: 2.5 min. Temperature range: 4.5 to 49°C. Temperature drift: at laboratory conditions ±3°C, ±1%; from 15 to 30°C, ±2%; from 30 to 50°C, ±4%; from 14 to 50°C, ±8%. Relative humidity range: 5% to 95%.

Reagent requirements: SO_2, 3.4 L/week modified West and Gaeke; 3.4 L/week demineralized water; NO_2, 3.4 L/week modified Saltzman (Lyshkow).

18-6-2. *Ultraviolet Photometric Ozone Analyzer*

Columbia Scientific Corporation, a division of Forney Corporation

The Photomet™ 3100 Analysis method, which measures ozone directly, is based on the absorption of UV radiation by ozone at the 254 nm band using the Beer-Lambert Law to arrive at the exact concentration. Ozone concentration is displayed digitally in ppm or mg/m^3 and can be corrected automatically for temperature and pressure. Ranges of 0.5 and 1.0 ppm are the EPA designated ranges for this equivalent method. Potential instrument drift is compensated for by continuous alternating measurements of zero air and ozone air. Ranges: .05 to 1.0 ppm. Lower Detectable Limit: 2.0 ppb. Noise: $<\pm$ 1 ppb. Lag Time: < 20 seconds. Rise Time/Fall Time: 15 seconds. Zero/Span Drift: < 2 ppb/24 hours. Precision: $\pm$ 2 ppb. Linearity: $\pm$ 0.2%. Cycle time: Updated results every 10 seconds. Operational Specifications: Sample flow rate: 2.5 $\pm$ 0.5 lpm. Outputs: display, 0–1 V (other voltages available), RS-232 (optional). Optional internal ozone generator for span checks requires 5 lpm of clean dry air.

18-6-3. *Model 1003 Ozone Monitor*

Dasibi Environmental Corporation

The Model 1003 Ozone Monitor continuously monitors the concentration of ozone in the air in ppm. An analog output is available for continuous strip-chart recording, and a binary-coded decimal (BCD) output enables direct interfacing with a computer or a printer. Ozone concentration is measured by detecting the absorption level of UV light within a sample volume of air. Accuracy: ±3%. Scale factor: adjustable to any standard. Drift: <0.001 ppm/week noncumulative. Zero span: ±0.4%/°C, corresponding to much less than 0.001 ppm. Interval: 8 or 30 seconds. Flow rate: 7 L/min at 8-second intervals; 1.0 L/min at 30-second intervals. Zero return: 1 interval from 1.0 ppm. Temperature: 0 to 49°C. Meets vibration and shock constraints typically encountered in shipping, aircraft, and mobile vans; maintenance, 1000-hr mean time between maintenance under typical conditions.

18-6-4. *ML®9810B Ozone Analyzer*

Monitor Labs, Inc.

The ML®9810B utilizes ultraviolet photometric measurement to quantify ozone concentrations. The unit is designed to be a cost-efficient solution for ambient air quality monitoring systems (AQMS) and continuous emission monitoring systems (CEMS). The units include sample pumps and particulate filters. UV light is generated by a temperature controlled mercury arc lamp and is filtered to select the wavelength of interest, collimated, and directed through a single glass measurement cell to the detector. The unit features easy to maintain single-cell construction, classical narrow band UV measurement, programmable automatic zero/span calibration checks, optional internal UV driven ozone generator for automatic span checks, and exceptional stability.

18-6-5. *ML®9850B (MB) Sulfur Dioxide Analyzer*

Monitor Labs, Inc.

The ML®9850B and ML®9850MB are based on detection of the UV excited fluorescence of the SO_2 molecules. When the unknown gas sample is exposed to specific wavelengths of UV light, the SO_2 molecules are excited and emit a photon when they relax to their normal energy state. The excitation intensity is sensed by a reference detector so that the measurement can be processed using a ratiometric technique. The units are designed for AQMS and CEMS monitoring. The units feature rugged reliable measurements, internal scrubbers to remove potential interfering hydrocarbons, programmable automatic zero/span, long life UV lamp, and no continuously moving parts.

18-6-6. *Instantaneous Vapor Detector*

Sunshine Scientific Instruments

The Instantaneous Vapor Detector is intended primarily for the detection of mercury vapor but can be used for the detection of other vapors in specified ranges of concentration. Applications include the manufacture of electrical apparatus, instruments, bulbs, glassware, fur, and salt; use in the chemical, metal mining, and smelting industries; and use by insurance companies and laboratories. Operation of the detector is based on UV light absorption by mercury vapor. This same principle is also used for the detection of certain other vapors that have selective absorption characteristics for UV radiation. For this reason, the identity of the vapor under test must be known and the vapor must be free from other substances which will absorb or obstruct UV light. In addition, the vapor should be relatively uncontaminated by extraneous substances such as fog, dust, or smoke. Features: warm-up time <15 minutes; <1% change in reading for 10% line voltage variation. Low power consumption permits operation from a battery-powered inverter for complete portability. Special options include explosion-resistant Model 38E, recorder output, single-

or dual-set point meter (Model 38F), panel or rack mounting, audible/visible alarms, and systems for monitoring multiple locations.

18-7: Chemiluminescence

18-7-1. Chemiluminescent Nitrogen Oxides Analyzer

Columbia Scientific Corporation, a division of Forney Corporation

The Model 5600 NO_X Analyzer performs specific, real-time and continuous dry analysis of ambient level nitric oxide and nitrogen dioxide in gas mixtures by detecting chemiluminescence resulting from the nitric oxide/ozone reaction. The reaction chamber and photomultiplier tube are thermally stabilized to ensure accurate readings over widely varying ambient temperatures without the need for moving parts such as a chopper. An exclusive photon counting technique filters photomultiplier noise far more effectively then conventional techniques. The logical menu-driven operation is extremely user-friendly and features a liquid crystal display. Range: 0 to 50,000 ppb. Noise: 0.05 ppb. Linearity (0 to 50,000 ppb): NO and NO_X 1% Full Scale (FS), NO_2 1.5 % FS. Lower Detection Limit: 0.1 ppb. Zero Drift: < 1ppb/day. Span Drift: < ± 1 %/day. Precision: ± 3 ppb. Rise Time: 120 seconds. Lag Time: 8 seconds. Interference Equivalent: < 20 ppb. Operational Specifications: Sample flow rate: 500 mL/min. Ozone generator flow rate: 125 mL/min. Temperature Range: 10 to 35 C. Humidity Range: 0 to 95% RH non-condensing. Output: Display, RS-232 (other outputs available).

18-7-2. Chemiluminescent Ozone Analyzer

Columbia Scientific Corporation, a division of Forney Corporation

EPA-designated Model OA325-2R/OA350-2R detects chemiluminescence from the reaction between excess ethylene and ozone in ambient air. Unlike the UV photometric, its chemiluminescent measurement is free from interferences of other gases. The Model OA325-2R provides cost-effective ozone monitoring while the Model OA350-2R incorporates internal calibration capability. Temperature control of the photomultiplier tube and flow controllers ensure repeatable results. Range: 0 to 0.01, 0.1, 0.5, 1.0, 5.0, and 10.0 ppm. Noise: 0.0005 ppm at 20% URL: 0.002 ppm at 80% URL. Lower Detectable Limit: 0.001. Zero Drift : ± 0.002 ppm/day. Span Drift: ± 1.5 % at 20% URL, ± 2.5 % at 80 % URL. Lag Time: 1.0 min. Rise Time/Fall Time: 0.5 min. Precision: 0.001 at 20% URL, 80% URL. Linearity: + 1% Full Scale.

18-7-3. NGA 2000 Heated NO/NOx Analyzer Module (WCLD)

Rosemount Analytical, Inc.

The NGA 2000, Wet Chemiluminescence Detection Analyzer Module (WCLD) is the industry's first modular analyzer designed to meet environmental regulations on NO_x emissions. It's heated sample handling module for hot/wet gas samples eliminates loss of NO_2 in sample condensate. A built-in Peltier Cooler removes moisture after NO_2 conversion. Applications such as internal combustion engine emissions (ICEE), continuous emissions monitoring systems (CEMS), and NO_x scrubber efficiency monitoring can all benefit from increased accuracy and simplified sample handling requirements associated with the WCLD.

18-7-4. NGA 2000 CLD NO/NOx Analyzer Module

Rosemount Analytical, Inc.

The NGA 2000 Chemiluminescence Detector (CLD) NO/NO_x Analyzer module provides quick and accurate measurement of oxides of nitrogen (NO/NO_x) over a wide dynamic range from 0 to 10 ppm through 0 to 10,000 ppm. It's thermoelectrically cooled solid-state detector insures high stability. Designed for such varied applications as continuous emissions monitoring systems (CEMS), process gas analysis, and internal combustion engine emissions (ICEE), the CLD may be a stand alone analysis instrument with the addition of the Input/Output (I/O) Module and Platform or it can be integrated into a sophisticated multi-component analysis network. It is an efficient, interferent free vitreous carbon NO_2 Converter.

18-8: Photometric Analyzers

18-8-1. AutoStep Plus Portable Toxic Gas Monitor

Bacharach, Inc.

The AutoStep Plus is a microprocessor-controlled portable paper tape toxic gas detector. The AutoStep Plus utilizes a removable, gas-specific ranging or gas-type module to monitor for TDI, MDI, hydrazines, Phosgene "A," Phosgene "B," or acid gas. Multi-gas modules are available to monitor for TDI, MDI, and HDI or chlorine, hydrides, and acid gas. The detection principle is colorimetric paper tape and reflected light, level measurement controlled by the microprocessor. Other features of the Autostep Plus include 2,000-point datalogging capability, built-in audible alarm, and external battery and data connectors. The precision depends on gas and model: 15% of reading or 1 ppb/0.1 ppm, whichever is greater.

The AutoStep Plus can operate from −10 to 40°C and in 5% to 95% relative humidity, non-condensing.

18-8-2. Sure-Spot Test Kit, GMD Systems

Bacharach, Inc.

The Sure-Spot Test Kit is a portable, colorimetric paper tape test system for a variety of toxic gases, such as MDI, TDI, HDI or NDI. Air is drawn through a test card with a pre-calibrated pump and a colored stain is developed based on the reaction with the toxic material of interest. The intensity of the color stain is directly proportional to the concentration of gas present. Can detect isocyanates down to the 1 ppb with an accuracy of +/− 25%. The unit is easily carried by a worker and can operate at temperatures of 10° to 40° C, and 20–85% RH.

18-8-3. Remote Intelligent Sensor (RIS)

Bacharach, Inc.

The Remote Intelligent Sensor (RIS) is housed in an environmentally sealed (NEMA 4), tamper resistant enclosure and operates from an external 12 VDC power supply. The unit utilizes colorimetric paper tape with reflected light level measurement as the detection principle, in a datalogging, remote housing. Paper tape units are available for TDI, MDI, HDI, PPDI, hydrazine, MM hydrazine, phosgene (A and B), chlorine, arsine, acid gases, and HCl. The data can be sent to a printer on a regular basis or will store data up to 8–10 hrs. Large LCD readout display. Unit detects gases in the ppb ranges with accuracy of +/− 15% (+/− 1 ppb). Operates over the temperature range of −40° to 80°C, 40–90% RH (non-condensing).

18-8-4. Flame Photometric Total Sulfur Detector

Columbia Scientific Corporation, a division of Forney Corporation

The continuous, dry optical principle of the Flame Photometric Detector (FPD) is used in the Model SA285 to measure total sulfur. The Model SA285 incorporates continuous performance reliability, high sensitivity, speed of response, and level of precision needed for area monitoring and for basic research. Model SA285E and Model SA285H are available for specific applications to measure sulfur dioxide and hydrogen sulfide respectively. Four linear ranges : 0 to 50 ppb, 100 ppb, 500 ppb, 1 ppm. Noise at 80% URL: ± 2.5 ppb. Lower Detectable Limit: 1.0 ppb. Zero Drift: ± 2 ppb. Span Drift: ± 10 % at 20%URL, ± 3 % at 80% URL. Lag Time: 5 seconds. Rise Time/Fall Time to 95%: 0.5 to 5 ppb. Precision: 1.0 at 20% URL, 4.0 at 80% URL. Linearity: ± 1.0 %.

Temperature Range: 10 to 40°C. Unattended Operation: 14 to 28 days. The Model SA260 is also available for flame photometric detection of sulfur up to 10 ppm.

18-8-5. Ultraviolet Fluorescence Sulfur Dioxide Analyzer

Columbia Scientific Corporation, a division of Forney Corporation

The Model 5700 SO_2 Analyzer performs specific, real-time and continuous dry analysis of ambient level sulfur dioxide in gas mixtures by detecting fluorescence resulting from illumination with ultraviolet light. The reaction chamber and photomultiplier tube are thermally stabilized to ensure accurate readings over widely varying ambient temperatures without the need for moving parts such as a chopper. An exclusive photon counting technique filters photomultiplier noise far more effectively then conventional techniques. Range: 0 to 20,000 ppb. Noise: 0.2 ppb. Minimum Detectable Sensitivity: 0.4 ppb. Interference Equivalent: < 20 ppb. Zero Drift (24 hour): ± 1ppb. Span drift (24 hour): ± 1 %. Precision: 1 ppb at 200 ppb, < 1% above 1000 ppb. Lag Time: 20 seconds. Rise Time: 140 seconds. Linearity: ± 1 %. Temperature Range: 0 to 35°C. Humidity Range: 0 to 95% RH non-condensing.

18-8-6. Flame Photometric Phosphorous Gas Analyzer

Columbia Scientific Corporation, a division of Forney Corporation

The operating principle of the Model PA260 utilizes the photometric detection of the 526 millimicron band emitted by compounds which contain phosphorous in a hydrogen rich air flame. The specificity of the measurement results from the geometric arrangement which optically shields the photomultiplier tube from the primary flame and from the employment of a narrow bandpass interference filter. Selectable current ranges on a log scale for concentrations of 0.005 to 1 ppm may be used with good quantitative results to a maximum of 10 ppm. Minimum Detectable Sensitivity: 0.5 ppb. Noise: ± 0.5 ppb. with 1 to 10 sec time constant. Lag Time: < 15 seconds. Rise Time/Fall Time: < 30 seconds. Precision: + 1.0 ppb. Zero Drift: + 0.5 ppb/ day. Span drift: ± 2% of reading per day. Operational specifications: (7) days unattended operation. Sample flow rate is approximately 240 ml/min. Hydrogen flow rate is approximately 125 ml/min. Relative Humidity: 0 to 99% non-condensing. Ambient temperature range: 10 to 40°C. Model SA260 is also available for flame photometric detection of sulfur up to 10 ppm.

18-8-7. Model 722R H₂S Gas Analyzer

Houston Atlas, Inc.

The 722R can be used either as a stand alone hydrogen sulfide analyzer or coupled with any of the range extension, sample conditioning or hydrogenation systems to measure ppb, ppm or percent levels. The Model 722R operates continuously and automatically with no interferences. Conforms with ASTM methods D 4084, D 4323, and D 4468. Applications include environmental monitoring, trace H_2S measurements, reformer cycle gas, sweeting plants, personnel protection, amine and caustic treaters, stack monitoring, and fuel gas boilers.

18-8-8. Dräger CMS Portable Gas Monitor

National Draeger, Inc.

The Drager Chip Measurement System (CMS) utilizes a chip containing 10 capillaries filled with specific reagents which react with the gas of interest, producing a color reaction. The color change is read optically and is displayed in a digital readout. The monitor itself is a hand-held unit with push button operation and any of 15 different chips can be inserted for different gases. Currently available chips include benzene, NH_3, CO_2, CO, Cl_2, HCl, HCN, H_2S, NO_x, NO_2, SO_2, and perchloroethylene. Features include a constant mass flow unaffected by atmospheric fluctuations, predetermined temperature and humidity effects, and is intrinsically safe (UL Class I, Division I, Group A, B, C, and D).

18-8-9. Rotorod H₂S Gas Sampler, Model 721

VICI Metronics, Inc.

The Model 721 H_2S Gas Sampler provides a fast, simple and economical way to measure low concentrations of H_2S. Detector discs contain a chemically treated pad that develops a dark color in the presence of H_2S. The spinning disc samples a high volume of air to give a sensitivity in the ppb range in less than 3 minutes. Color reference standards printed on each disc allow instant dosage evaluation. The detector is also used to monitor ambient levels of H_2S by mounting the tag in a simple tubular shelter.

18-8-10. MDA Scientific Model CM4 Continuous Gas Monitor

Zellweger Analytics, Inc.

The MDA Scientific Model CM4 continuously monitors ambient air and ventilation systems for toxic, corrosive, and pyrophoric gas levels. The instrument

INSTRUMENT 18-8-9. Rotorod H₂S Gas Sampler, Model 721.

simultaneously analyzes up to four locations, 300 feet (90 meters) or more from the monitor. The Model CM4 uses the Chemcassette® colorimetric detection method, measuring down to ppb levels with virtually no interference from other common gases. Chemcassette technology provides physical evidence of the presence of the target gas. Model CM4 provides hazardous gas detection and monitoring protection for the semiconductor, pharmaceutical, and chemical industries. Several communications configurations are available, including LonWorks®, Wonderware In Touch®, Intellutions Fix 32®, and Profibus® to link multiple instruments and provide hundreds of monitoring points. Calibrations are available for more than 25 toxic and corrosive gases,

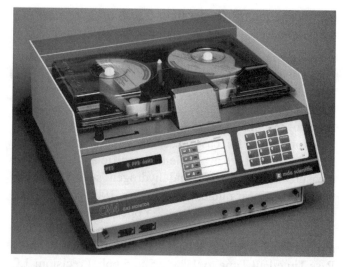

INSTRUMENT 18-8-10. MDA Scientific Model CM4 Continuous Gas Monitor.

including ammonia, oxidizers, hydrides, mineral acids, hydrogen cyanide, hydrogen sulfide, PFCs, phosgene, and others. Nitrogen trifluoride (NF_3) can be detected by the Model CM4-P, a pyrolyzer version, which converts NF_3 to hydrogen fluoride via pyrolysis.

18-8-11. MDA Scientific ChemKey TLD Gas Monitor

Zellweger Analytics, Inc.

The MDA Scientific ChemKey TLD gas monitor is a portable, direct-reading instrument capable of detecting and measuring over 40 different toxic, corrosive or pyrophoric gases. The unit is ideal for use by emergency response teams, industrial hygienists, or anyone who needs to monitor for different hazardous gases. The instrument features a unique ChemKey Gas Selection System, a programmed key which, when installed, allows operators to switch monitoring modes from one gas to another. All that is required is a simple change of key and a Chemcassette colorimetric tape. The exclusive Chemcasstte tape technology provides physical evidence of the presence of the target gas. Gases detected by the ChemKey TLD include ammonia, arsine, chlorine, diborane, diisocyanates, hydrazines, hydrogen fluoride, hydrogen chloride, hydrogen cyanide, hydrogen sulfide, phosgene, phosphine, silane, and many other hazardous gases. It features built-in audio/visual alarms, 4–20 mA output, and mechanical alarm relays.

18-8-12. MDA Scientific Model EGM Exhaust Gas Monitor

Zellweger Analytics, Inc.

The MDA Scientific Model EGM Exhaust Gas Monitor is a compact, simple to operate, low-

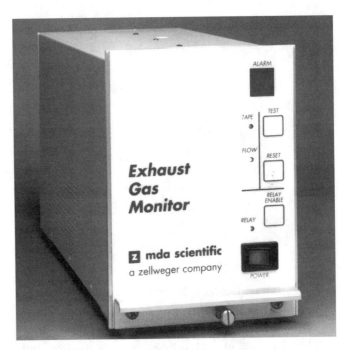

INSTRUMENT 18-8-12. MDA Scientific Model EGM Exhaust Gas Monitor.

maintenance gas monitor designed to continuously monitor for toxic gases in gas cabinet ventilation, laboratory vent hoods, and other pre- and post-process exhaust streams. The unit is considered ideal for meeting environmental regulations, process control and safety concerns. Model EGM offers low level gas detection in the parts-per-billion and higher levels using the exclusive Chemcassette colorimetric detection system. The instrument features low maintenance, ease of use, and a compact, panel mountable enclosure. Gases monitored include aliphatic amines, chlorine/oxidizers, hydrides, hydrogen cyanide, hydrogen peroxide, hydrogen sulfide, mineral acids, ozone, phosgene, and sulfur dioxide.

18-8-13. MDA Scientific IsoLogger® Personal Gas Detection System

Zellweger Analytics, Inc.

The MDA Scientific Isologger is a compact personal gas monitoring system for datalogging and reporting exposures to isocyanate gases such as those found in urethanes, chemical, and petrochemical production. The unit may also be used as an area monitor for survey and temporary monitoring. The system uses the well-proven Chemcassette colorimetric method of detection for part-per-billion sensitivity, provides excellent selectivity without interference and requires no gas calibration. User-selectable audio and visual alarms are provided as well as a real-time display. A Windows-based datalogging software package is provided for tabular and

INSTRUMENT 18-8-11. MDA Scientific ChemKey TLD Gas Monitor.

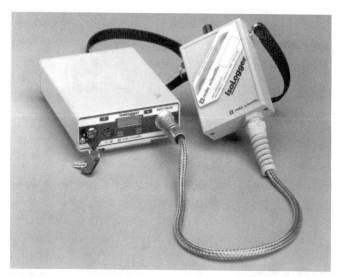

INSTRUMENT 18-8-13. MDA Scientific IsoLogger® Personal Gas Detection System.

INSTRUMENT 18-8-14. MDA Scientific Model SPM Gas Monitor.

graphic reporting of data. Detectable gases include HDI, IPDI, MDI, TDI, AsH$_3$, HCl, COCl$_2$.

18-8-14. MDA Scientific Model SPM Gas Monitor
Zellweger Analytics, Inc.

MDA Scientific Model SPM single point monitor provides a rugged, portable, fast response gas detector. The unit can be used as a fixed point monitor or a portable instrument for surveying, and is designed for harsh industrial applications, including chemical, petrochemical, and urethanes production environments. Model SPM detects specific target gases with sensitivity at ppb levels, and utilizes exclusive Chemcassette colorimetric technology to provide physical evidence of gas detection. Over 50 toxic, corrosive, or pyrophoric gases calibrations are available. The list of gases and gas families includes: amines, ammonia, bromine, chlorine, chlorine dioxide, diisocyanates, hydrazines, hydrides, hydrogen cyanide, hydrogen peroxide, hydrogen sulfide, mineral acids, nitrogen dioxide, ozone, phosgene, and sulfur dioxide.

18-8-15. MDA Scientific Series 7100 Gas Monitor
Zellweger Analytics, Inc.

The MDA Scientific Series 7100 continuous gas monitor responds instantly to hazardous gas leaks and pinpoints concentrations as low as 1 ppb with full documentation and self-diagnostics. It is considered ideal for applications where a quick starting, fast response, user-friendly gas detection and monitoring instrument is needed. These applications include monitoring gas storage areas, gas cylinder changes, bulk chemical transfers, equipment repairs, survey work,

spill clean-ups, and emergency response situations. The instrument ignores non-target gases which can trigger false alarms in other detection systems. Exclusive Chemcassette colorimetric technology provides physical evidence of the target gas, even at ppb levels. The Chemcassette tape is continually advanced during monitoring which prevents the tape sensor media from "falling asleep" or becoming permanently poisoned. Series 7100 instruments detect and monitor over 50 toxic and corrosive gases, including amines, diisocyanates, hydrazines, hydrides, mineral acids, and oxidizers.

18-8-16. MDA Scientific Series 16 Gas Monitor
Zellweger Analytics, Inc.

The MDA Scientific System 16 is a modular toxic gas monitoring system designed for simple in-field expansion and is capable detecting and monitoring one or two different toxic, corrosive, or pyrophoric gases or gas families at 4 to 16 individual locations (points). A System 16 modular multipoint monitor provides extremely cost-effective surveillance of work areas and chemical storage areas in semiconductor manufacturing, chemical and pharmaceutical production.

INSTRUMENT 18-8-15. MDA Scientific Series 7100 Gas Monitor.

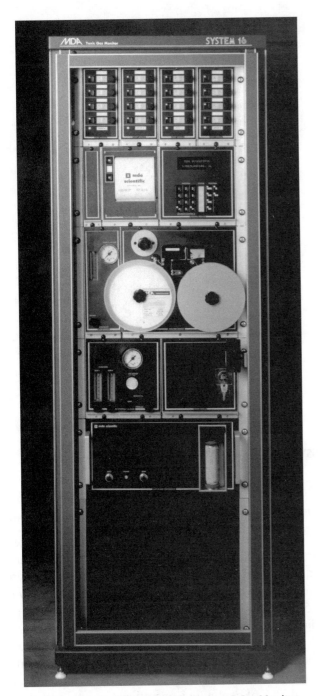

INSTRUMENT 18-8-16. MDA Scientific Series 16 Gas Monitor.

System 16 uses the exclusive Chemcassette colorimetric detection system for specific, low-level gas detection. The instrument can be expanded by simply plugging the appropriate sampling or detection module into the chassis. Additional monitoring points are brought under surveillance by adding a sample module; the number of gases monitored is increased by adding a detection/analyzer module. Up to three 4-point sampling modules and one additional analyzer module may be added to the each system. The instrument also features a built-in printer, local digital

display, alarm relays and complete self-diagnostics. System 16 offers calibrations for more than 25 toxic gases including ammonia, oxidizers, hydrides, mineral acids, hydrogen cyanide, hydrogen sulfide, PFCs, and phosgene. System 16 is available for the detection and measurement of dozens of gases including ammonia, chlorine, arsine, diborane, phosphine, silane, HCN, HCl, HF, phosgene and nitrogen trifluoride.

18-9: Thermal Conductivity Detectors (also see 18-12 Multi-Gas Monitors)

18-9-1. Leak Hunter Plus Model 8066
Matheson Gas Products

The Model 8066 Leak Hunter *Plus* is a portable handheld leak detector used for pinpointing gas leaks in gas handling systems. Using a thermal conductivity sensor and ambient air as its reference, the Leak Hunter *Plus* can be used for detecting helium, hydrogen, carbon dioxide, argon, sulfur hexafluoride and halocarbons. With a response time of less than one second, the unit provides both audible and visual readouts indicating positive leak location and leak size. The instrument's probe attaches to the body for single handed operation, or can be detached allowing access to tighter spaces. Memory resident calibration data, user selectable measurement units and a peak hold function facilitate the measurement of leak size.

18-10: Heat of Combustion Detectors (see also 18-12, Multi-Gas Monitors)

18-10-1. Leakator™ 10
Bacharach, Inc.

The Leakator™ is a battery operated, hand-held, combustible and toxic gas leak detector. Ten bright LEDs show level of gas presence while 3 separate LEDs indicate sensor status, power levels and power on. The instrument is inexpensive, has a longer probe for small area testing, earphone, and 20 hours of battery operation.

18-10-2. TLV SNIFFER®
Bacharach, Inc.

This instrument operates on the principle of catalytic combustion (a process of oxidizing a combustible gas/air mixture on the surface of a heated catalytic bead element). Designed for low level combustible gas detection (0–100 ppm scale) for use as a leak detector, assessment of contaminant conditions, or arson investigations. Eight-hour continuous operation is possible with six size D, Ni-Cd batteries or approximately 3 hrs with six size D, carbon-zinc batteries. Sampling rate: 1.65 L/min, nominal. Readout mode: meter, audible alarm, earphone output, and recorder output. Response time: initial response within 12 seconds of exposure. Its stability is in keeping with instruments of similar sensitivity and construction.

18-10-3. Sniffer® 300 Series

Bacharach, Inc.

The Sniffer Series 300 gas indicators are designed for intermittent measurement of combustible gases and oxygen deficiency. The internal pump allows sampling up to 100 feet away. The rugged instruments are in ergonomically designed, rugged, weatherproof cases with a built-in handle with locking trigger-style switch and a shoulder harness. Applications include general industry and government, plant and welding surveys, leak detection and confined space entry. Units are powered by four size D batteries (alkaline, C-Zn, or Ni-Cd); operate in temperature ranges from –20 to 50°C (limited by battery specifications), and have non-latching visual and audible alarms.

18-10-4. ABL-50, 741 and 4021 Series Respiratory Air Monitors

GfG dynamation

The Model ABL-50 is a CO monitor/alarm specifically designed for respiratory airline breathing applications. It will continuously indicate the level of CO in ppm on its built-in meter and activate external alarms if the concentration exceeds the preset alarm threshold. The Models ABL-741 and 4021 also monitor for oxygen levels and organic vapors. The Model ABL-50 is connected to a tee fitting in the airline that bleeds off a small, continuous sample of air flowing between the compressor and the user. This sample is filtered for particulate matter, has the oil mist removed, and is regulated to 10 psig before passing over the solid-state, catalytic semiconductor sensor. Enclosure: polyester fiberglass NEMA 4 with cover latch. Controls: calibration and alarm threshold internal. Meter size: 2.5 in. Response: 90% of maximum reading within 2 min with 20 ppm CO concentration; faster at higher concentrations. Alarm adjustment range: 2–50 ppm CO. Recorder output: 0–1 mA. Interferences: other types of organic vapors will be detected if present in high concentrations or at their TLV. Sensor purge period: 1 min nominal. Sensor stabilization period: 10 min nominal.

INSTRUMENT 18-10-5. Model EX-10 Personal Gas Detector.

18-10-5. Model EX-10 Personal Gas Detector

ENMET Corp.

The EX-10 is a pocket-sized combustible gas detector designed for personal protection of workers in hazardous areas or for confined space entry. The catalytic sensor in this instrument has a typical life expectancy of 3–5 years. Alarms are audible and visual, and are adjustable over the full scale. Response time: less than 10 seconds when exposed to 50% LEL methane. Operating temperature range: –10 to 55°C for intermittent exposures. Rechargeable 3.6-V Ni-Cd batteries provide 45 hrs of continuous operation.

18-10-6. Macurco Gas Detectors

Macurco, Inc.

Macurco offers a wide range of stationary monitors for CO, combustible gases, ammonia refrigerants, various toxics, and hydrogen. These units cover a wide range of applications including use in homes, offices, warehouses, commercial buildings, maintenance facilities, parking garages, bathrooms, conference rooms, chillers, coolers, poultry operations, etc. The units may be plugged into 120 VAC for continuous use or operated on 12 VDC and use low maintenance, long-life (7–10 years) solid-state semiconductor sensors. Dozens of different units are available depending on the applications, size, power requirements, etc. A variety of alarm and output options are available.

18-10-7. Model 8057 Hazardous Gas Leak Detector
Matheson Gas Products

The Model 8057 General Purpose Leak Detector is a portable handheld leak detector ideal for pinpointing leaks of a variety of hazardous gases including ammonia, chlorine, hydrogen sulfide, sulfur dioxide, and hydrides. Its general purpose sensor enables detection for a wide variety of gases and vapor without requiring individual sensors, making it an extremely versatile leak detector.

18-10-8. D-Series Combustible Gas Systems
Scott Aviation

The D-Series line of portable instruments detects most combustible gases or vapors in air measured in ppm or LFL. The unit is used to safeguard lives and property. Applications include in petroleum refineries, petrochemical plants, steel mills, on gas transmission and distribution pipe lines, off shore, in the fire and police service, occupational hazard surveys, in the maritime industry, the military, telephone and radio communications, or mines. Scott portables detect hazardous conditions in virtually any industry, commercial or even residential environment.

18-10-9. Pocket Ozone Detector
Spectrex Corporation

The Spectrex Model A-20ZX Ozone Detector utilizes a semiconductor sensor to reliably and simply check ozone concentrations. Application include around water purifiers, ozone generators, air purifiers, laser printers and copiers, pulp bleaching, power generators and specialty manufacturing. The unit can detect down to 0.01 ppm of ozone over the range of 0-1 ppm. Weighs only 142 gm, and fits easily into a shirt pocket. Other features include a carrying case and rechargeable batteries. The Model C-30Z is a continuous, fixed monitor for ozone that displays concentration with a series of color bars. The unit comes with various warning displays and audible alarms.

18-10-10. Model SXC-20 VOC Monitor
Spectrex Corporation

The Model SXC-20 is a portable VOC monitor that utilizes a semiconductor sensor. The unit can detect over 100 VOCs and the response is indicated in a color bar display. The unit will run for up to 8 hours and has a built-in datalogger. The SXC-20 features an internal sampling pump to provide active sampling and faster response and the ability to connect a VOC sampling tube which automatically turns the sample on when a preset VOC level is detected.

18-11: Gas Chromatographic (GC) Analyzers

18-11-1. Crowcon ISFID-ISFIDGC Portable Gas Chromatograph
Crowcon Detection Instruments

The Crowcon ISFID unit can be used as a direct-reading flame ionization detector, or the ISFIDGC allow for chromatographic separation. The ISFID has a sampling pump that draws air through a hand held probe and passes it to the FID. The unit weighs 8 pounds, 11 for the GC, and provides a visual and audible output within 2 seconds. The magnitude of the signal is linear and proportional to the concentration of the organic sample. The unit has flame arrestors on every entry and is FM approved for use in Class I, Division I, Groups, A, B, C, and D. The unit has automatic ignition and warms up to use in just 2 minutes. An LED status light assures flow conditions are constantly maintained. The meter is backlit for use in low light applications, and the cylinder will last up to 50 hours. Battery life is 45 hours based on 6 alkaline cells. The GC offers a large selection of interchangeable columns and the unit has datalogging capabilities.

18-11-2. CENTURY TVA-1000 Toxic Vapor Analyzer (see Instrument 18-4-2)
Foxboro Company

18-11-3. Model 311D Portable Gas Chromatograph
HNU Systems, Inc.

The 311D gas chromatograph offer a range of compact versatility for environmental analysis of organic compounds. Four different detectors are available, including two detectors operational at once. Packed or capillary columns can be used, as well as a a wide range of isothermal or programmed temperatures, including temperature ramping. The units feature all the extras of a laboratory gas chromatograph, but in a rugged, compact package. Other features include Peakworks® software for data handling and report generation, and a concentrator to go down to parts per trillion levels.

18-11-4. Voyager™ Portable Gas Chromatograph
PE Photovac

Weighing just 15 pounds, the PE Voyager gas chromatograph (GC) is a truly portable, battery operated, point-and-press instrument ergonomically designed for Environmental Site Characterization and Occupational Safety and Health monitoring. The Voyager uses gas chromatography to isolate and measure the compounds of interest. For users who need to monitor for specific VOCs in the presence of other compounds, this is the best way to eliminate interferences. Typical examples are BTEX in gasoline environments and 1,3-butadiene in C4 isomeric hydrocarbon backgrounds. With its on-board carrier gas and rechargeable battery, the Voyager can operate independently on-site for up to 8 hours. Users can view chromatograms and/or tabular results using the built-in backlit LCD. If desired, results can be uploaded to a computer in the field, or back at the lab or office. The unique Voyager Analytical Engine includes a smart selection of three columns, preinstalled, and selectable just by pressing a key. The three columns offer overlapping selectivity and confirmational analysis of "light," "medium," and "heavy" compounds of interest within complex environments. There is no need to worry about changing columns. Each Voyager is delivered with the user's choice of one or more Applications Assays. Each Assay automates the setup of all the instrument operating conditions and all of the data reduction parameters pertinent to a predefined set of compounds. Running the application is as simple as press and go.

18-11-5. Scentograph "Plus II" and Scentoscreen Portable Gas Chromatographs

Sentex Systems, Inc.

The Scentograph is a portable gas chromatograph designed to provide onsite field analysis with laboratory gas chromatographic quality. Five detector options are available, along with a choice of commercially available columns, either capillary or packed. The units can be heated isothermally or ramped to temperatures up to 180°C. The unit has an internal battery and gas supply for total portability or can be connected to AC power for prolonged use. A detachable laptop PC with applicable software controls the system and conducts the sample analysis, storing the results on disk for future recall.

18-12: Multi-Gas Monitors

These instruments contain detectors utilizing more than one operating principle. The most common example of this instrument is the confined space monitor which has a sensor for oxygen (coulometric analyzers), combustible gases (heat of combustion detectors), and

various toxic gases, commonly CO and H_2S (potentiometric analyzers). Similar instruments for Indoor Air Quality applications are listed in the ACGIH publication, *Air Sampling Instrument Selection Guide: Indoor Air Quality*.

18-12-1. Series 200 and 300 Gas Detectors

AIM USA

The Series 200 and 300 Gas Detectors are designed to detect combustible gases, oxygen, and toxic gases. Applications include: confined space entry survey work, industrial safety and hygiene, fugitive emissions and leak detection. Instruments in the 200 and 300 Series are available to detect one, two, or three separate gases. Three datalogging formats are available: alarm incident, time-interval sampling, and location-survey testing. Sensors are chemical-specific electrochemical for toxic gases and O_2, and nonspecific metal oxide for combustible gases. The metal oxide sensor comes standard with the one sensor version and is optional in the two or three sensor versions.

18-12-2. Series 500 Gas Detectors

AIM USA

The Series 500 Gas Detectors are designed to monitor for combustibles, O_2, H_2S, and CO in confined space entry applications. These instruments use metal oxide or Pellister sensors to detect combustibles and electrochemical sensors to detect O_2, H_2S, and CO.

Features of the 500 Series include audible and visual alarms, shock resistance case, and datalogging capabilities for 1300 preset time intervals. Accessories included with the instruments include: power supply, manual, quick start card, confined space booklet, calibration hood, carrying case, QC sheet, and tool kit. Operating temperature range: –20 to 50°C. Operating relative humidity range: 5% to 99%, noncondensing.

18-12-3. Bodyguard® Personal Gas Monitor

Bacharach, Inc.

A personal monitor designed as a one-, two- or three-gas monitor (oxygen, combustibles, and toxics). A simple, rugged instrument design is perfect for daily field use in confined space entry, oil and gas refining, waste water treatment, coal mines and general industry. Uses alkaline or Ni-Cd batteries, has a water-resistant case, and visual and audible alarms. Other features include, one-step zeroing, alternating tone alarms, built-in diagnostics, a confidence light (to show it is on), automatic calibration and easy sensor replacement.

18-12-4. GasPointer® II Combustible Gas Detectors

Bacharach, Inc.

The GasPointer® II detectors are intrinsically safe, battery-powered portable instruments designed to measure concentrations of methane (or propane) and carbon monoxide and/or oxygen in ambient air and flue gas. The instrument is particularly designed for locating natural gas or propane leaks around pipes, fittings and appliances; for measuring excessive levels of carbon monoxide in heating and ventilation systems; and sampling for confined space entry. Available as a one-, two- or three-gas detector, it has a high visibility LCD display which automatically ranges from % LEL to % gas, or % gas by volume (oxygen), or 1–2000 ppm CO. Options include audible and visual alarms, three types of charges, probe assembly with holster, flexible SS probe and calibration equipment and gases. The GasPointer® II can operate in a temperature range of –15 to 50°C and a relative humidity range of 5% to 99%, non-condensing.

18-12-5. Sentinel® 44T Personal Multi-Gas Monitor

Bacharach, Inc.

The Sentinel® 44T is designed to measure O_2, combustible gases, CO, and H_2S in confined-space entry applications. The Sentinel® is equipped with audible and visual alarms that are activated when preset instantaneous levels, short-term exposure limits (STELs), or time-weighted averages (TWAs) are exceeded. Other features include: multi-gas LCD display, radio frequency interference (RFI) protection, datalogging capabilities, optional 44Talk software to download to a computer or printer, and removable battery packs. Accessories include: hand aspirator or motorized pump, 10-in. or 36-in. probes, calibration kits and gases, remote earphone or pocket alarms. Rechargeable lead acid gel cells can operate the instrument for 8–10 hrs per charge. The Sentinel® can operate in a temperature range of –20 to 50°C and a relative humidity range of 05% to 99%, non-condensing.

18-12-6. SNIFFER® 500 Series Portable Area Monitors

Bacharach, Inc.

The SNIFFER® 500 Series Portable Area Monitors are instruments designed to alert personnel to the hazards of O_2 deficiency and the presence of dangerous concentrations of combustible gases, CO, or H_2S. The SNIFFER 500 Series combines sensors for two or three different contaminants. The sensors include a heated catalytic bead for combustible gases and electrochemical cells for O_2, H_2S, and CO. Any combination of these contaminants, up to three, is available. Various visual (steady or pulsing LEDs) and audible alarms (using steady, alternating, or pulsed tones) are used for different instruments. In addition to the various alarm options, the 500 Series includes an integral sampling pump, a variety of concentration ranges for combustibles, analog displays, low flow and battery alarms, and use in hazardous areas. Operating temperature: –20 to 50°C. Response time: variable from 5 seconds to 60 seconds (90% response). Operating time: 10 hrs.

18-12-7. PhD Series and Cannonball® Multi-Gas Monitors

Biosytems, Inc.

The Biosystems PhD5 monitors up to five atmospheric hazards simultaneously: oxygen, combustible gas, and up to three channels of toxic gas detection. Choose from a wide variety of toxic sensors; including Biosystems' innovative new "Duo-Tox" two-channel CO/H_2S sensor to measure both carbon monoxide and hydrogen sulfide when both hazards are present without cross interference.

Biosystems' PhD Lite monitors up to four atmospheric hazards simultaneously: oxygen, combustible gas, carbon monoxide, and hydrogen sulfide. The PhD Lite allows you to choose specific sensors for the detection of CO or H_2S when you have only one toxic hazard to deal with; or choose Biosystems' innovative new dual channel CO/H_2S sensor to measure both carbon monoxide and hydrogen sulfide when both hazards are simultaneously present.

The Biosystems PhD Plus is an economical, confined space gas detector which measures oxygen, combustibles, and up to two additional toxic gases. All normal procedures automatic calibration adjustment are controlled through the single On/Off Mode button. The Biosystems PhD Plus is housed in a rugged, compact case with choice of snap-in NiCad or alkaline battery packs.

The Biosystems PhD Ultra datalogging confined space gas detector measures oxygen, combustibles, and up to two toxic gases. All normal procedures including automatic calibration adjustment are controlled with a single button. The instrument automatically logs gas readings, user and location ID and other important information. The PhD Ultra is housed in a rugged, compact case with choice of snap-in NiCad or alkaline battery packs.

The Biosystems Cannonball2 portable gas detector is also designed to measure oxygen, combustible gas and up to two toxic gases. It offers the ease of use and dependability of our other gas detection products with the additional advantage of being housed in a gasketed, waterproof case designed to take on the roughest industrial environments and survive.

The Biosystems PhD2 gas detector measures oxygen, combustibles, and up to two additional toxic gases. The PhD2 offers the widest range of toxic sensor choices including CO, H_2S, SO_2, Cl_2, NH_3, NO, NO_2, HCl and HCN. The instrument automatically logs gas readings and other important information.

18-12-8. DEFENDER and MULTIMAX Multi-Gas Detector

BW Technologies

The DEFENDER is designed as a confined space monitor for the detection of combustibles, oxygen, carbon monoxide, and hydrogen sulfide. The unit features a compact and lightweight design, nonintrusive battery/sensor replacement and calibration, two alarm levels per sensor, auto zero and auto span functions, automatic self-test, one button operation, password security, use of Black and Decker VersaPak Ni-Cd batteries, and a rugged ergonomic design. The MULTIMAX offers 2, 3 or 4 gas detectors. In addition to O_2 and combustibles, detectors are available for CO, H_2S, SO_2, Cl_2, ClO_2, HCN, and NO_2. It features an optional sampling pump, data hold, radio linked remote alarm, multiple audible and visual alarms, one switch operation, up to 160 hr of operation on one set of alkaline batteries, intrinsically safe approval and all in a rugged case.

18-12-9. LMS-40 and Personal Surveyor Multi-Gas Monitors

CEA Instruments, Inc.

The LMS-40 is equipped with methane and carbon dioxide infrared measurements, oxygen cell measurements, and temperature and pressure measurements. It is also available for numerous other toxic gases. The LMS-40 has a built-in sample pump with long battery life, an internal datalogger that stores all measurements, time, date, locations, and is user adaptable, has built in 'help' function keys and built-in status checks to avoid erroneous readings. The unit has excellent accuracy and stability in a water resistant lightweight housing plus a serial interface to a computer or directly to a printer.

The Personal Surveyor simultaneously monitors and displays: O_2, Combustibles (%LEL), CO, and H_2S in any one to four gas configurations. The unit is: extremely rugged, has easy one button operation, long life rechargeable and/or alkaline battery packs that easily slide on and

off, push button calibration and electronic zeroing, UL, CSA and other intrinsic safety approvals, an internal datalogger with built-in memory back up battery, optional miniature air pump, water resistant and RF shielded, and with a high contrast display with bright backlight.

18-12-10. Custodian, Triple Plus, and Detective Multi-Gas Monitors

Crowcon Detection Instruments

The Triple Plus is a 2-, 3-, or 4-gas monitor, the Custodian and Detective are 3- or 4-gas units. All units feature rugged design in NEMA 4X rated housings and are UL approved for use in hazardous atmospheres. All contain audible and visual alarms, simple controls, datalogging, downloading to printers or personal computers, backlit digital displays, temperature ranges of –10 to 50°C, humidity of 0–95% non-condensing (Triple Plus will float if dropped in water), wide range of sensors, and RFI shielding. Units for combustibles can be made to auto-range which utilizes both a pellistor and a thermal conductivity sensor. In addition to the %LEL and % by volume combustible and O_2 sensors, toxic sensors are available for CO, H_2S, SO_2, Cl_2, HCN, and O_3. The Detective is a transportable multi-gas monitor which can be set up in a hazardous location within minutes and provide area warnings of gas clouds. The unit is mounted on a tripod frame and features an omni-directional sounder and 4 red LED clusters on top. Multiple units can be connected via 30-ft cables to provide a broader warning area.

18-12-11. EXPLORER™ Multi-Gas Monitor

CSE Corp.

The EXPLORER™ is a one-button, three gas, handheld gas monitor. The unit toggles to read each individual gas when the button is pushed. Audible and visual alarms sound if one or more of the gases exceeds a preset limit. The unit can be configured to monitor any of 12 combinations of gases including combustibles, O_2, CO, CH_4, or H_2S. Features include EasyCal 100 calibration unit, backlit LCD display, fast rechargeable Ni-metal hydride batteries, remote sampling pump and probe, and is intrinsically safe.

18-12-12. Multiwarn II

Draeger Safety

The Multiwarn II is a versatile measuring instrument for: 1 to 4 Gases (without infrared sensor) or 1 to 5 gases (with infrared sensor). It is available in two models: basic easy-to-use, or extended with advanced operating software. Sensors available include one or two infrared

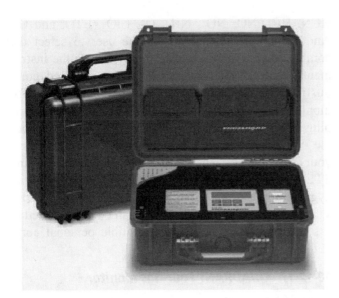

INSTRUMENT 18-12-13. BLAQ Box 1300 Series Indoor Air Quality System.

INSTRUMENT 18-12-14. Telaire 7001 Carbon Dioxide and Temperature Measurement Monitor.

sensors, a plug-in catalytic Ex-sensor (for all explosive gases and vapors), or choose from a selection of 13 plug-in, intelligent, interchangeable electrochemical sensors for the measurement of more than 35 different gases. Equipment options include an integrated pump and 50-hour datalogger and versatile operating software.

18-12-13. BLAQ Box 1300 Series Indoor Air Quality System

Engelhard Sensor Technologies

The BLAQ Box is designed for monitoring indoor air quality and ventilation rates over extended periods of time. The unit is enclosed in a rugged, theft and tamper resistant carrying case and includes a cable and padlock. The streamlined push-button interface panel and LCD display give real time readings of carbon dioxide, temperature, relative humidity and carbon monoxide (optional). Data is recorded and can be graphed on a PC or laptop using Windows-based VG Graphing software (included). The CO_2 sensor uses a non-dispersive infrared detector and has a range of 0-5,000 ppm. The electrochemical CO sensor has a range of 0–300 ppm. RH measurements range is 20–95%. Temperature measurement is 0–40°C or 32–112°F. The built-in datalogger has a storage capacity of 14,000 data points and has user selectable sampling intervals.

18-12-14. Telaire 7001 Carbon Dioxide and Temperature Measurement Monitor

Engelhard Sensor Technologies

This hand-held CO_2 and Temperature monitor was designed to check ventilation rates in commercial or res-

idential applications. The 7001 is equipped with Engelhard's patented dual beam Absorption Infrared technology and has a range of 0–10,000 ppm. The 7001 has the ability to display CO_2 in less than 30 seconds, and is an ideal tool for identifying energy savings opportunities in over-ventilated spaces, determining if air quality complaints are due to insufficient ventilation, or locating the presence of combustion fumes generated from vehicles and appliances. This unit is also equipped with an output for recording data via accessory cable and datalogger. The 7001 comes with an AC adapter or runs on 4 AA batteries for 70 hours.

18-12-15. CGS-90R Portable Gas Detectors; Omni-4000 and Quadrant Multi-Gas Monitors

ENMET Corp.

The CGS-90R portable gas detector for confined space entry features "broad range" toxic and com-

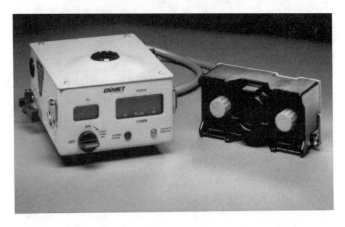

INSTRUMENT 18-12-15. CGS-90R Portable Gas Detectors.

INSTRUMENT 18-12-16. Omni-4000 Portable Programmable Gas Detector.

bustible sensors as referenced in the May 19, 1994 Amendment to the Federal OSHA CFR 1910.146 confined space rule. The instrument also includes a detachable sensor head with 20 foot sensor cable for remote pre-entry testing, separate toxic and combustible bar graphs, and separate digital oxygen display.

The Quadrant is set up to monitor four gases: O_2, combustibles, CO, and H_2S. The Omni 4000 can monitor multi-gases including NH_3, CO, Cl_2, HCN, HCl, H_2S, NO, NO_2, and SO_2, ClO_2, ETO, O_2, and combustibles.

18-12-16. Omni-4000 Portable Programmable Gas Detector

ENMET Corporation

The OMNI-4000 features interchangeable, precalibrated SMART BLOCK sensors, and monitors four gases simultaneously. This unique modular design enables the user to simply and easily convert the instrument for the detection of many different gases. SMART BLOCK modules are available for H_2, CO, H_2S, Cl_2,

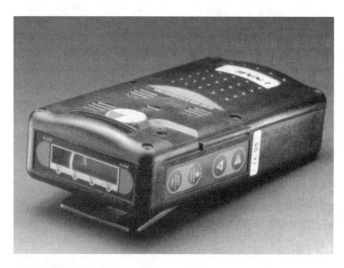

INSTRUMENT 18-12-17. Quadrant Four-Gas Monitor.

HCN, HCl, NO, NO_2, NH_3, O_3, ClO_2, ETO, and O_2. Another unique feature allows the user to select the combustible gas they wish to monitor from the instrument's internal memory of 25 preprogrammed combustible gases/vapors. As your requirements for monitoring a specific gas change, all you have to do is plug in a different precalibrated SMART BLOCK module or choose a different combustible gas that meets your current needs. The OMNI-4000 provides TWA and STEL alarms, and continuous datalogging for industrial hygiene applications. It stores up to 48 hours of exposure and event information which can be downloaded to a serial printer or an IBM compatible personal computer.

18-12-17. QUADRANT Four-Gas Monitor

ENMET Corporation

The QUADRANT portable gas detector is a very small, light-weight four gas instrument. Compact design with surface-mount electronics provides a truly pocket-size, four-function gas detector. Flexibility in sensor combination (CO, H_2S, O_2, and a programmable combustible sensor) enables the user to customize the instrument to meet their needs. This compact unit can be used with a chest/shoulder harness or belt clip. QUADRANT can be used as a simple warning device or as a sophisticated instrument providing TWA/STEL alarms and datalogging. It was specifically designed with the budget customer in mind.

18-12-18. STM2100 Portable Gas Monitor

GasTech, Inc.

GasTech's STM2100 personal portable gas monitor is designed to protect workers from hazardous gases in confined spaces and other industrial sites. This portable unit can detect up to four gases: oxygen content, combustibles (%Vol, LEL or ppm), and up to two toxics (NH_3, CO, Cl_2, HCN, H_2S, NO, NO_2, or SO_2). The unit features, simultaneous display of all four gases simultaneously, one button operation, two UL listed NiMH battery packs or alkaline batteries, programmable ID locations, fail detection sensor circuitry, high stability and low ppm sensitivity for hydrocarbon monitoring, built-in datalogger with XPRESS CAL software, back-lit LCD display, and UL classified.

18-12-19. G750 Polytector II

GfG dynamation

The G750 offers the choice of up to six "smart" sensors at one time, including IR detection of CO_2. Sensors available include O_2, LEL, percent volume and

INSTRUMENT 18-12-21. Multi-Gas Monitors.

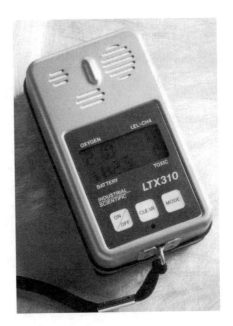

INSTRUMENT 18-12-22. LTX310 Multi-Gas Monitor.

infrared CO_2, and two toxics (CO, H_2S, SO_2). The G750 automatically sets sensor parameters, allowing the flexibility to interchange sensors in the field. Applications include confined space or industrial hygiene monitoring. Features include AutoCal easy calibration; datalogging with software; visual, vibrating, and audible alarms; large graphic numeric display; and a sturdy polyamide case.

18-12-20. CGM II 900 AutoCal® Series Multi-Gas Monitor

GfG dynamation

The CGM combines smart sensor technology to provide four gas monitors plus a ToxAlert™ broad range sensor to detect hundreds of unknown toxic gases. The unit has the AutoCal features which provides simplified calibration. Sensors are available for combustibles (LEL), O_2, CO, H_2S, and ToxAlert. The unit features a power take off port for datalogging, an optional minipump, remote alarms, visual and audible (buzzer) alarms, and RF resistant casing.

18-12-21. ATX612 and 620 Multi-Gas Monitors

Industrial Scientific Corp.

The ATX612 and ATX620 can be configured for up to four gas monitoring, allowing for continuous detection of combustible gases (%LEL, methane 0–5% of volume and ppm hydrocarbons), O_2 and two toxic gases (including CO, Cl_2, H_2S, NO_2, and SO_2). The internal constant flow pump will draw samples from up to 100 feet in the ATX612. The ATX612 and 620 feature one

button auto calibration, an ultra-bright light bar and 90 dB audible alarm. The instruments are available with optional datalogging for up to 100 hours of data storage and industrial hygiene functions. The ATX612 and 620 are available with Ni-Cd batteries and built-in charger or replaceable alkaline batteries.

18-12-22. LTX310 Multi-Gas Monitor

Industrial Scientific Corp.

The LTX simultaneously monitors up to three gases: combustible gases (% LEL or methane % by volume), O_2 and any one of eight toxic gases (NH_3, CO, Cl_2, HCN, H_2S, NO, NO_2, and SO_2). The LTX310 automatically recognizes and identifies the installed sensors and adjusts alarm and calibration parameters accordingly. Additional features include one-button auto calibration, combustible sensor over-range protection, plug-in sensors, RFI protection, audible and visual alarms and peak hold readings. User selectable override features include latching alarm, alarm set points and calibration settings.

18-12-23. TMX-412 Multi-Gas Monitor

Industrial Scientific Corp.

The TMX-412 simultaneously monitors up to four gases: combustible gases (% LEL or methane % of volume), O_2, and one or two toxic gases (CO, Cl_2, H_2S, NO_2, or SO_2). The instrument automatically recognizes all installed sensors and displays instantaneous readings on the LCD display. The TMX412 is available with an optional datalogging board with real-time clock that

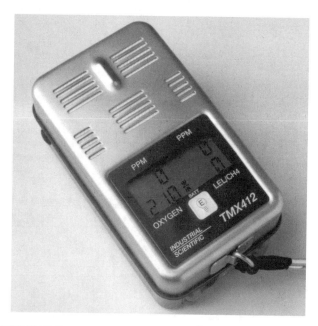

INSTRUMENT 18-12-23. TMX-412 Multi-Gas Monitor.

calculates and records STEL and TWA readings (up to 110 hours of data stored). Additional features include one-button auto calibration, combustible sensor over-range protection, plug in sensors, RFI protection, audible and visual alarms and peak hold readings.

18-12-24. MicroMAX Multi-Gas Monitor

Lumidor Safety Products

The MicroMAX series is a UL classified and CSA certified small, lightweight, microprocessor controlled multi-gas monitor (1-5) with an internal sampling pump to provide very short response times. Weighing under one-pound, the MicroMAX simultaneously monitors concentrations of O_2, combustibles and up to three toxic gases (NH_3, CO, Cl_2, HCN, ClO_2, H_2S, SO_2, PH_3). The MicroMAX is compact, fast responding, and easy to use with one button operation. The MicroMAX features

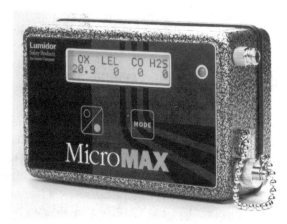

INSTRUMENT 18-12-24. MicroMax Multi-Gas Monitor.

interchangeable sensors, datalogging port, interchangeable batteries (Ni-Cd or alkaline), automatic backlit display, RFI resistance, and easy calibration. The MicroMAX is ideal for use in and around electrical, water or gas utility manholes, water and wastewater treatment plants, landfill operations, silos, tunnels, oil fields, gas pipelines, petrochemical facilities, pulp and paper mills and in confined areas in virtually any municipal or industrial environment. The unit has an optional datalogging feature which allows downloading of data to a PC, providing the ability to record readings to comply with record keeping and analysis requirements.

18-12-25. Model IQ 1000 MEGA-Channel Gas Detector

Matheson Gas Products

The Model IQ 1000 MEGA-Channel Gas Detector System, equipped with the MEGA-Gas sensor, can detect over 100 different gases and vapors. The IQ 1000 is ideal for Haz Mat or emergency response operations, where the substance to be detected is an unknown, or if the substance to be detected can be different each time out.

18-12-26. Gasport®, Passport®, Passport® Personal, and Watchman® Multi-Gas Monitors

Mine Safety Appliances Co.

The Gasport Gas Tester is a handheld instrument that can be used by gas utility workers to simultaneously test for the presence of combustible methane (CH_4), carbon monoxide (CO), hydrogen sulfide (H_2S), and Oxygen (O_2) enrichment or deficiency. Depending on the user's needs, gases can be monitored across seven different ranges, including: 0–5000 ppm CH_4; 0–100% LEL (Lower Explosive Limit) CH_4 or 0–5% by volume CH_4; 0–100% by volume CH_4, 0–25% O2; 0–50 ppm (parts per million) H_2S; and 0–1000 ppm CO.

The Passport FiveStar Alarm is compact, weighing less than 17.3 ounces (500 grams). Designed with MSA's Plug-and-Play Sensor Memory System, the Passport FiveStar Alarm allows you to install the best combinations of sensors to match the task at hand. As many as five sensors can be used at one time. If you require data downloading, the instrument will soon be available with an infrared IrDA-compatible link for easy data communications.

The Passport Personal Alarm is a very compact instrument used to monitor combustible gases, oxygen and toxic gases in workplace atmospheres, especially in confined spaces such as manholes, storage tanks, tank cars, vaults, mines and sewers. The face of the instrument simultaneously displays the readings of all gases monitored. The instrument's metal-filled polycarbonate case provides protection from electromagnetic interference (EMI) and radio frequency interference (RFI).

The Watchman Multi-Gas Monitor is a hand-held instrument used to detect and monitor combustible gases, oxygen and toxic gases in workplace atmospheres, especially in confined spaces such as manholes, storage tanks, tank cars, vaults, mines and sewers. Designed for rugged handling, the monitor incorporates the state-of-the-art technology of the Passport Personal Alarm in the strong, aluminum housing of the Portable Indicator and Alarm, Model 360. The Watchman Monitor also is easy to operate. The face of the instrument simultaneously displays the readings of all gases monitored. Microprocessor-based, the instrument is menu-driven for ease of operation. The user can "page through" various menu activities for calibration and display options.

18-12-27. Model pm-7400 and pm-7440 Multi-Gas Monitors

Metrosonics, Inc.

The pm-7400 monitors up to four gases simultaneously: O_2, LEL, NH_3, CO, Cl_2, HCN, ClO_2, H_2S, SO_2, PH_3, NO, NO_2, ETO. The hand-held unit provides alarms, 1 button operation and calibrates up to 12 sensors with unique calibration setup. The pm-7440 is a desktop unit that monitors up to 4 gases simultaneously (same gases as above plus O_3, H_2, and HCl) and prints instantaneous gas levels and reports. Both units can use available Metrosonics Software for Data Analysis.

18-12-28. Dräger MiniWarn Multi-Gas Personal Monitor

National Draeger, Inc.

The Dräger MiniWarn is a hand-held personal monitor for 4 gases (choice of 20 sensors including combustibles, CO, O_2, and H_2S). A simple 3-button keypad makes for ease of operation with datalogging capabilities. The unit features the new DrägerSensors® XS intelligent sensors with up to a 3-yr warranty (for selected gases). The unit features a loud audible and very bright visual alarms and an attachable pump for remote sampling.

18-12-29. MultiRAE PLUS and VRAE Hand-Held Multi-Gas Monitors

RAE Systems, Inc.

Both units from RAE are 1–5 gas monitoring units with electrochemical sensors and also features the very small RAE PID sensor for VOCs (MultiRAE PLUS only). The gases available include oxygen, combustibles, VOCs, CO, H_2S, SO_2, NO, NO_2, Cl_2, HCN, NH_3, and

PH_3. Other features include large backlit LCD display, multiple audio and visual alarms, 10–12 hours of run time, 16,000 data point datalogging download to PC, built-in pump, runs on a variety of battery packs, and has a rigid sampling probe. Applications include refineries and petro chemical plants (confined space, hot work permits), utilities (cable vaults and transformer stations), waste water treatment plants, marine and off shore wells, landfill operations, and food processing. The units both have fast response times, are intrinsically safe, and work over large temperature and humidity extremes.

18-12-30. RKI Multiple Gas Monitors, Models GX-82, 86, 86A, and 94

RKI Instruments, Inc.

RKI offers a wide range of three or four gas personal monitors for use in confined spaces.

The units offer detection of oxygen, combustibles, and one or two toxics (CO, H_2S, and SO_2). Applications include sewage treatment plants, utility manholes, chemical plants, hazardous waste sites, sulfur plants, nuclear plants, logging operations, mines, refineries, tunnels, paper mills, drilling rigs, storage tanks, and pipelines. The units have many features including datalogging, microprocessor control, continuous monitor and readout, variable audible and visual alarms with multiple set points, intrinsically safe, touch control panel, rugged dot matrix LCD readout with back lighting, user replaceable batteries, and many carrying options.

18-12-31. RKI EAGLE® Portable Gas Detector

RKI Instruments

The RKI EAGLE offers an ergonomically designed package to monitor up to 4 different gases with a powerful internal pump. Sensors available include hydrocarbons (in LEL or ppm), O_2, CO, H_2S, NH_3, AsH_3, Cl_2, F_2, HF, HCl, NO_2, O_3, PH_3, SlH_4, and SO_2. Other features include the ability to draw a sample through 75 feet of tubing, large LCD display with backlighting, 2 alarms per channel plus TWA and STEL, audible and visual alarms, autocalibration and demand zero, 30 hr of operation on one set of batteries, intrinsically safe approval, internal dust and hydrophobic filters, methane elimination for hydrocarbon analysis, status indicator, and compliance with EPA Method 21 protocol for fugative emissions testing.

18-12-32. GMP 1000M Monitor

Rosemount Analytical, Inc.

The GPM 1000M is a packaged approach to measuring up to five gases plus opacity. The measurement options include: CO, CO_2, SO_2, NO, NO_x, O_2, THC and Opacity. The GPM 1000M is manufactured under ISO 9001 certified quality standards. Temperature controllers for both probe and heated sample line are included. Non-freon thermoelectric sample conditioner containing integral pre-cooler, sample pump, condensate removal system, and water intrusion monitor for sample pump shutdown in the unlikely event of a conditioner failure; provides unsurpassed analyzer protection. Local/remote calibration capability for diagnostic testing of analyzers, and compliance with the EPA's 40 CFR 60, Appendix F, and 40 CFR 75.

18-12-33. Scott SA2000 Portable Gas Detector

Scott Aviation

The SA2000 four-gas portable instrument detects the following gases: CO, H_2S, LEL, O_2. Scott's SA200 is available in one, two, three-gas options, in addition to the standard four-gas detection system. For users needing one gas monitoring, SA2000 Single Series offers any choice of the four gases listed above. The Dual Series SA2000 offers O_2 and combustible gas detection. For more complex monitoring needs, choose the SA2000 Triple Series. The Triple Series is upgradable to the standard four-gas SA2000 model allowing maximum flexibility in an expanding gas monitoring program. All SA2000 Series instruments are available in either alkaline or nicad battery configurations.

18-12-34. ProtectAir™ Model 8570

TSI, Inc.

Model 8570, Personal Multi-gas monitor is a small, portable and intrinsically safe instrument primarily designed for confined space entry applications. Housed in a rugged stainless steel case, it has exceptional durability and RFI protection. The standard instrument has two sensors, oxygen (O_2) and combustibles (CH_4). The measurement can be expanded to include two additional toxic gases. It will accept either a rechargeable (NiMH) battery pack or six AA alkaline batteries to provide long duty cycles. The instrument will log data for 40+ hours and is compatible with TRAKPRO software. The instrument has many options to customize a package for unique situations. Among the popular options are a hand-aspirated pump, a powered pump, a miniature remote alarm, calibration kits, and protective carrying cases.

18-12-35. Q-Trak® Model 8551

TSI, Inc.

Models 8550 and 8551, Indoor Air Quality Monitors are available in two versions. Both versions

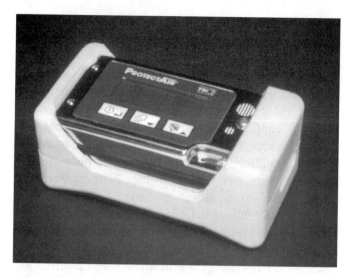

INSTRUMENT 18-12-34. ProtectAir™ Model 8570.

measure and record 3 parameters: CO_2, temperature and humidity. Model 8551 adds carbon monoxide as a 4th measurement parameter. All sensors, including the NDIR CO_2 sensor, are located in the detachable probe. Both models will log all parameters for 3+ weeks at a 1-minute interval, longer at extended intervals. The instrument can download data to the TRAKPRO data analysis software. They are battery operated for ease of portability or can operate on AC for extended monitoring

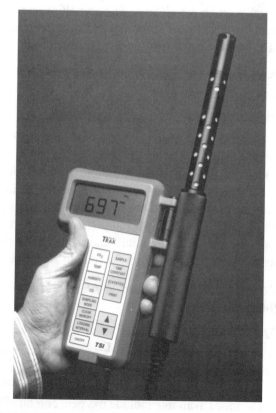

INSTRUMENT 18-12-35. Q-Trak® Model 8551.

553

requirements. The new IAQ-CALC™ will monitor the same parameters, but new software allows calculation of % outside air.

18-12-36. Minigas-XL
Zellweger Analytics, Inc.

The MiniGas-XL can be configured to monitor and datalog up to four gases—oxygen, flammable, and two toxics (CO, H_2S, Cl_2, and SO_2). The pocket-sized unit helps ensure safe work conditions required for confined space and hazardous area working. Applications include water and sewage treatment, telecommunications, construction, oil and gas, chemical, transport, process and mining. Features include a tough die-cast metal case; protective rubber boot; instantaneous, STEL and TWA dual tone audible alarms and visual red alarms; either dry cell or NiMH batteries, EasyCal calibration and calibration reminder; field settable alarms; optional sample pump or aspiration pump; earphones; and comprehensive datalogging software.

18-13: Portable Mass Spectrometers

There are many brands of laboratory-housed mass spectrometers which we will not duplicate here. Of interest to our readers is the class of portable analytical instruments (see also gas chromatographs, 18-11).

18-13-1. MG2100 Portable Mass Spectrometers
Industrial Scientific Corp.

The MG2100 Portable Mass Spectrometer delivers full mass spectrometry in a package that weighs less than 35 pounds. The unit operates over the range of 10-200 amu (higher levels are available) and is a completely self-contained system, housing an electron impact (EI) ionizer, mass analyzer, detector, 386X microprocessor, analytical software, vacuum pump, controller electronics, a handle for easy transportation, sampling inlets and a digital readout. The unit can be interfaced through an RS232 port to a PC for expanded capabilities.

TABLE 18-I-1. Electrical Conductivity Analyzers

Instrument No.	Mfg./ Supp.*	Model	Analytes	Range (ppm)	Detection Limit (ppm)	Precision (±)	Dimensions (cm) H	W	L	Weight (kg)	Power	Alarms Aud.	Vis.	Comments
18-1-1	AIC	431-X	Hg	0.001–2	0.001	5%	5.1	2.4	1.6	2.3	110 VAC	–	–	A
18-1-2	AIC	631-X	H_2S	0–50	–	–	5.1	2.4	1.6	2.3	115 VAC	–	–	A
18-1-3	CAL	U3S	SO_2	–	0.005	–	6.3	5.5	8.3	–	115 VAC	–	–	B, C
18-1-4	CAL	Mikrogas®	SO_2, H_2S, HCl, Cl_2, $COCl_2$, NH_3, COS, CS_2, HCN	variable	variable	–	–	–	–	–	115 VAC	–	–	D
18-1-5	CAL	Ultragas®	same as above	variable	variable	–	–	–	–	–	115 VAC	–	–	D

* Manufacturer codes given in Table 18-I-14.
A. Collects a 1- or 10-second sample on a gold film sensor.
B. Absorbs sample in acidified H_2O_2.
C. Converts SO_2 to H_2SO_4.
D. For process operation.

TABLE 18-I-2. Potentiometric Analyzers

Instrument No.	Mfg./Supp.*	Model	Analytes	Range (ppm)	Detection Limit (ppm)	Precision (±)	Dimensions (cm) H	W	L	Weight (kg)	Power	Alarms Aud.	Vis.	Comments
18-2-1	BAC	Canary®	O_2, CO, H_2S, SO_2, NH_3, NO_2, Cl_2, HCN, HCl, CH_4	variable	variable	variable	6.4	3.8	11.4	0.3	4-AA	X	X	A, B, C
18-2-2	BAC	CO Sniffer®	CO in air or in breath	0–2000	–	5%	22	9	7.4	0.6	4-C cells	–	–	A
18-2-3	BI2	Toxi	O_2, CO, H_2S, SO_2, NH_3, NO, NO_2, Cl_2	variable	variable	–	1.1	5.2	2.8	0.13-1.1	3-AAA cells	X	X	A, B, C
18-2-4	BWT	Toxiclip	O_2, CO, H_2S	0-100, 300	1-CO 0.3-H_2S	–	7.2	3.8	2	0.056	–	X	X	A, B, C
18-2-5	BWT	MiniMax	SO_2, H_2S, Cl_2, CO, HCN, NO	variable	0.1-1.0	–	8.7	7	2.7	0.16	5-yr battery	X	X	A, B, C
18-2-6	CEA	Gasman	CO, CO_2, CH_4, NH_3, H_2S, SO_2, HCl, $COCl_2$, COS, CS_2, HCN	variable	0.1	3% FS	–	–	–	–	4-AA cells	X	X	A, B, C
18-2-7	CEA	Series U	O_2, H_2S, CO, SO_2, Cl_2, NO_2, NO, HCl, HCN, NH_3, %LEL	0-250, 500, 100 or %LEL	variable	1% FS	11.5	6.6	3.9	0.3	4-AA cells	X	X	A, D
18-2-8	CEA	TG-KA Series	HCHO, phosgene, ozone, HF, hydrides	variable	variable	–	1.7	2.9	1.1	0.09	AC/DC	X	X	A
18-2-9	CDI	Gasman II	O_2, H_2S, CO, SO_2, Cl_2, NO_2, NO, HCl, HCN, NH_3, O_2, CH_4, NO	variable	variable	–	11	6.6	4	0.27	4-AA	X	X	A, B, C
18-2-10	DYN	Monogard	CO, H_2S, O_2, SO_2, NO	–	–	–	–	–	–	–	Lithium battery	X	X	B
18-2-11	ENM	Spectrum	CO, H_2S, Cl_2, HCN, SO_2, NH_3, O_2	–	–	–	–	–	–	–	–	X	X	B

TABLE 18-I-2. (cont.). Potentiometric Analyzers

Instrument No.	Mfg./Supp.*	Model	Analytes	Range (ppm)	Detection Limit (ppm)	Precision (±)	Dimensions (cm) H	W	L	Weight (kg)	Power	Alarms Aud.	Vis.	Comments
18-2-12	ENM	Smart Logger	H_2, CO, Cl_2, HCN, H_2S, HCl, NO, NO_2, NH_3, O_3, ClO_2, ETO, O_2	varies by analyte	–	1%	1.3	6.3	2.8	0.2	9V	X	X	A, B, E
18-2-13	ENS	MVN-100A	ETO, formaldehyde, glutaraldehyde	0.5-100 ETO	1.0	–	11	6	2.5	0.11	9V	X	–	A, B
18-2-14	GAT	GT, GTD	CH_4, O_2, CO, H_2S	variable	variable	–	25.4	15	13	2.25	4-D cells	X	X	A, E
18-2-15	GAT	GT-2400	CO, NH_3, Cl_2, HCN, H_2S, NO, NO_2, SO_2, O_2, CH_4	variable	variable	0.2-10	15	9	5.6	0.7	2-C cells	X	X	A, C
18-2-16	GAT	STM2100	NH_3, CO, Cl_2, HCN, H_2S, NO, NO_2, SO_2, O_2, CH_4	variable	variable	0.2-10	–	–	–	0.7	alkaline and NiCd	X	X	A, B, C, E
18-2-17	GAT	95 Series	CO, H_2S, O_2	0-500 0-100 0-30%	variable	±10%	11.4	6.4	2.5	0.2	9V	X	X	A, B, C
18-2-18	GCI	701	H_2S, CO, SO_2, NO_x, NO	0-100 variable	3	±3ppm	13.3	6.3	3.2	0.2	9V	X	X	A, B
18-2-19	DYN	G 3000	CO, H_2S	0-700/50	–	2-3%	1.5	0.9	0.6	0.2	NiCd	X	X	A, B
18-2-20	DYN	Toxitector®	CO	0-1000	30	–	9.2	3.4	–	0.15	NiCd	X	X	A, B
18-2-21	ISC	STX	CO, NH_3, Cl_2, HCN, H_2S, NO, NO_2, SO_2	variable	variable	–	10.2	6.4	3.0	0.2	Lithium	X	X	A, B
18-2-22	ISC	T80	CO, Cl_2, HCN, H_2S, NO_2, O_2, SO_2	variable	variable	–	10.2	6.8	3.4	0.2	9V	X	X	A, B, C, E
18-2-23	ISC	GasBadge	CO, H_2S, O_2	variable	variable	–	7.6	4.3	3.2	0.08	6V Li	X	X	A, B, C
18-2-24	ITS	Series 1000, 4000, 7000	CO, SO_2, H_2S, Cl_2, NO, NO_2, hydrazines, ClO_2, HCN, ETO, HCHO	0.1-10 times TLV	2% FS	1%FS	2.9	2.4	4.5	3.6	NiCd	X	X	A, C, E

TABLE 18-I-2. (cont.). Potentiometric Analyzers

Instrument No.	Mfg./ Supp.*	Model	Analytes	Range (ppm)	Detection Limit (ppm)	Precision (±)	Dimensions (cm) H	W	L	Weight (kg)	Power	Alarms Aud.	Vis.	Comments
18-2-25	ITS	LD Series	same as above	variable	variable	–	35.6	50.8	22	10.4	110 VAC	X	X	A, E, F
18-2-26	ITS	RM Series	CO, Cl₂, ClO₂, ETO, HCHO, hydrazines, HCl, HCN, H₂S, NO, NO₂, SO₂	variable	variable	±2%FS	17.8	48	30.5	5.2	110 VAC	X	X	A, F
18-2-27	LSP	ToxiBEE GasBUG	CO, H₂S, O₂	35, 10 .95%	variable	±5%	7.6	6.4	4	0.085	–	X	X	A, B, C
18-2-28	LSP	UniMAX	O₂, CO, NH₃, PH₃, SO₂, H₂S, NO₂, Cl₂, ClO₂	variable	variable	±2%	10.4	6.4	2.8	0.14	3-AAA NiMH	X	X	A, B, C
18-2-29	MET	PM-7700	CO, H₂S, SO₂, Cl₂, NO₂, NO, O₂, ETO, HCN, NH₃, H₂, HCl	variable	variable	1%	7.6	10.2	2.3	0.3	9V	X	X	A, E, F
18-2-30	MSA	MiniCo® Responder	CO	0–999	2	2%	–	–	–	–	9V	X	X	A, B
18-2-31	MSA	Cricket®	CO, H₂S, O₂	variable	variable	–	–	–	–	–	3V Li	X	X	A, B
18-2-32	MST	MSTox 8600	18+ gases	variable	variable	–	9.3	4.7	2.1	0.09	Recharge or Disposable	X	X	A, B, C
18-2-33	MST	Satellite	CO, H₂S, O₂	variable	variable	–	11.4	4.4	3.2	0.08	Li	X	X	A, B, C
18-2-34	NDR	Pac III	CO, NH₃, CO₂, Cl₂, ETO, PH₃, HCN, H₂S, NO₂, NO, O₂, LEL	variable	variable	–	6.7	11.6	3.2	0.2	Alk, Li, or NiCd	X	X	A, B, C, E
18-2-35	NDR	MicroPac	CO, H₂S, O₂	0–400 0–100 0–25%	–	–	5.4	8.4	3.2	0.1	–	X	X	A, B, C
18-2-36	NEO	Neotox® XL	CO, H₂S, Cl₂, SO₂, NO₂, O₂, LEL	0–100 0–500 0–35%	0.1–1	1–5%	4.9	6.5	11.3	0.203	3-AA NiCd	X	X	A, B, C
18-2-37	NEO	Solotox®	CO, H₂S	0–500 0–100	1.8 7.5	–	7.5	6.4	4.1	0.08	2 yr. disposable	X	X	A, B, C

TABLE 18-I-2. (cont.). Potentiometric Analyzers

Instrument No.	Mfg./Supp.*	Model	Analytes	Range (ppm)	Detection Limit (ppm)	Precision (±)	Dimensions (cm) H	W	L	Weight (kg)	Power	Alarms Aud.	Vis.	Comments
18-2-38	PRA	Titrilog II	H_2S, SO_2	–	0.01–0.02	–	5.6	5.6	8.3	30	115 VAC	–	–	G, H
18-2-39	RAE	ToxiRAE	O_2, LEL, VOCS, CO, H_2S, SO_2, NO, NO_2, Cl_2, HCN, NH_3, PH_3	variable	0.1–1	–	15.2	4.4	2.5	0.13	2-AAA	X	X	A, B, C
18-2-40	RKI	Series 94	O_2, CO, H_2S	0–40% 0–500 0–99	–	5–10%FS	14.5	4.4	2.5	0.12	2-AAA	X	X	A, B, C
18-2-41	RKI	SC-90	17+ gases	variable	variable	10%FS	20	8	14	1.9	Alkaline or NiCd	X	X	A
18-2-42	SCA	S100	O_2, H_2S, CO	variable	variable	–	3	2.5	3	1.8	battery	–	X	A
18-2-43	SEN	SS2000 SS4000	Br_2, Cl_2, CO_2, F_2, H_2, HCl, HCN, HF, NH_3, SO_2, PH_3, AsH_3	variable	variable	10%	21.6	11	17	1.5	NiCd	X	X	A, C
18-2-44	TEL	990	O_2, CO in combustion products	0–5, 10% 0–100, 500	2%	5%	4.8	5.1	2.7	5	NiCd	X	X	A, I, J
18-2-45	ZEL	Lifeline	AsH_3, PH_3, Br_2H_6, SiH_4, HCl, HBr, HF, CO, NH_3	variable	variable	–	12.5	8	5.7	1.4	NiCd	X	X	A, B, C

* Manufacturer codes given in Table 18-I-14.

A.	Electrochemical sensor	F.	Available in a variety of fixed units.
B.	Diffusion sampling.	G.	Cell reagent is KBr, where Br_2 is generated.
C.	Intrinsically safe.	H.	Liquid prefilters are required for some analytes.
D.	Explosion proof units available.	I.	Designed for combustion process measurements.
E.	Datalogger capabilities.	J.	Separate sensors for CO and O_2.

TABLE 18-I-3. Coulometric Analyzers

Instrument No.	Mfg./Supp.*	Model	Analytes	Range (%)	Detection Limit (ppm)	Precision (±)	Dimensions (cm) H	W	L	Weight (kg)	Power	Alarms Aud.	Vis.	Comments
18-3-1	ESD	OX630	O$_2$	0–100	–	1%FS	15.3	8.9	3.8	–	9V	–	–	A, B
18-3-2	GFG	Oxytector®	O$_2$	0–25	0.1	0.5%	3.4	9.2		0.15	NiCd	X	X	A, B
18-3-3	MGP	8060, 8061	O$_2$	0–40	–	3%	12	6.6	2.9	0.3	2-AA NiCd	X	–	A
18-3-4	RAI	NGA 2000 TO2	O$_2$	0–100	10 ppb	–	rack mounted			–	110 VAC	X	X	C
18-3-5	SMC	55	O$_2$	0–25	–	–	hand-held			–	–	X	–	A, B

* Manufacturer codes given in Table 18-I-14.
A. Electrolytic cell for oxygen.
B. Diffusion sensor.
C. Electrode sensor.

TABLE 18-I-4. Ionization Detectors

Instrument No.	Mfg./Supp.*	Model	Analytes	Range (ppm)	Detection Limit (ppm)	Precision (±)	Dimensions (cm) H	W	L	Weight (kg)	Power	Alarms Aud.	Vis.	Comments
18-4-1	FRC	HC500-2D	hydrocarbons	0–10 0–1000	0.1 CH$_4$	0.1ppm CH$_4$	4.8	7.5	7.9	18.2	110 VAC	–	–	A, C
18-4-2	FOX	TVA-1000	organic and toxic compounds	0–2000PID 0–50000FID	0.1 benzene 0.3 hexane	1%	34.3	26.3	8.1	5.6	NiCd	X	X	A or B D, E, F
18-4-3	HNU	PI-101, IS-101, DL-101	organic and toxic compounds	0–20 0–2000	0.2 benzene	1%FS	10.9	5.3	8.1	4.1	NiCd 12V DC	X	X	B, D, F
18-4-4	MSA	Passport® PID II	organic and toxic compounds	0–10000	0.1	–	–	–	–	–	NiCd	–	–	B, D, F
18-4-5	PEC	2020	organic compounds	0.5–2000	0.5	–	hand-held			0.79	–	–	–	B, D, E, F
18-4-6	PEC	MicroFID®	organic compounds	0.1–50000	0.1	–	hand-held			3.7	–	–	–	A, D, E, F
18-4-7	RAE	MiniRAE 2000, Plus	VOCs	0–10000	0.1	10%	21.8	7.6	5.8	0.55	NiCd, NiMH Alk 4-AA	X	X	B, D, E, F
18-4-8	RAI	NGA 2000 McFID	hydrocarbons	–	–	–	rack mounted			–	110 VAC	–	–	A, F
18-4-9	SEN	Portable FID	VOCs	0.1–10000	0.05 CH$_4$	10%	–	–	–	3.9	NiCd Alk	X	–	A, D, E, F
18-4-10	THE	580EZ, 55C, 51	VOCs	0.1–2000	0.1	10%	17.1	14.6	25	2.7	NiCd	X	–	B, D, E

* Manufacturer codes given in Table 18-I-14.
A. Flame Ionization Detector.
B. Photoionization Detector.
C. Microprocessor controlled.
D. Portable, self-contained instrument.
E. Datalogging capabilities.
F. Intrinsically safe.

TABLE 18-I-5. Infrared Photometers

Instrument No.	Mfg./Supp.*	Model	Analytes	Range (ppm)	Detection Limit (ppm)	Precision (±)	Dimensions (cm) H	W	L	Weight (kg)	Power	Alarms Aud.	Vis.	Comments
18-5-1	BKJ	1301	IR absorbing gases	4 orders of magnitude	0.01–1	1%FS	70.5	43	15	18	115 VAC	–	–	A
18-5-2	BKJ	1302	IR absorbing gases	5 orders of magnitude	0.01–1	1%FS	17.5	39.5	30	9	VAC or 12V	X	X	A, B
18-5-3	CEA	GD 444	CO_2	0–10,000	20	1%	13	8	4.4	0.43	MHz	X	X	D
18-5-4	CSC	Quantum 7000	IR absorbing gases	–	10–100ppb	–	–	–	–	–	–	–	–	–
18-5-5	FOX	Sapphire	>100 IR absorbing gases	varies by gas	tp 50 ppb	varies	36.6	55	19.3	10	120 VAC	X	X	D, C
18-5-6	HOR	APBA-250	CO_2	0–3000, 1%, 5%	–	1.5%	26	22	8.5	2.7	VAC	X	–	–
18-5-7	HOR	APMA-360	Ambient CO	0–10,000	0.05	1%FS	43	22	55	20	115 VAC	X	X	F
18-5-8	ISC	CDU 440	CO_2, CH_4	0–60,000	100	–	12	7	5	0.85	NiCd	–	–	C, D, G
18-5-9	MET	Aq-511	CO_2	0–2,000, 20,000	100	3%	9.1	18	3.3	0.83	NiCd Alk	–	–	D, G
18-5-10	MSA	Liva®	CO, CO_2, SO_2 or other hydrocarbons	0–50	1	1%	–	–	–	–	VAC	X	X	–
18-5-11	MLI	ML® 9830B	CO, SO_2, NO_x	0–200	0.01	0.1	43.2	17.8	64.8	20.9	VAC	–	–	H
18-5-12	RKI	RI-411A	CO_2	0–5,000	25	10%	25	19	11.3	2.4	Alk, CZn, NiCd	X	X	D
18-5-13		RI-413A		0–10,000 0–20%										
18-5-14	ZEL	Searchline Excel	hydrocarbons	variable	variable	variable	rack mounted			–	VAC	–	–	–
18-5-15	TSI	InspectAir™ 8560	CO_2	0–50,000	10	3%	9.9	16.8	3.8	0.6	4-AA	X	–	D

* Manufacturer codes given in Table 18-I-14.
A. Utilizes FTIR photo-acoustic spectroscopy.
B. Measures up to 5 gases simultaneously.
C. Intrinsically safe.
D. Portable, self-contained instrument.

E. Uses FTIR.
F. Utilizes cross flow modulation NDIR.
G. Datalogger.
H. Utilizes gas filter correlation NDIR.

TABLE 18-I-6. Ultraviolet and Visible Light Photometers

Instrument No.	Mfg./Supp.*	Model	Analytes	Range (ppm)	Detection Limit (ppm)	Precision (±)	Dimensions (cm) H	W	L	Weight (kg)	Power	Alarms Aud.	Vis.	Comments
18-6-1	CEA	TGM555	SO_2, NO_2, NO_x, NH_3, Cl_2, TDI, HCHO, HCN, halides	variable	0.025%	1%	4.7	7.9	2.2	11.4	12VDC	–	–	A, C
18-6-2	CSI	Photomet™ 3100	O_3	0–1.0	0.002	2 ppb	14	48	56	16	VAC	–	–	B
18-6-3	DEZ	1003	O_3	0.01–9.99	0.01	2%	5	15	18.5	20.5	VAC	–	–	B
18-6-4	MLI	ML® 9810B	O_3	0–20	0.001	1%	43.2	17.8	64.8	16	VAC	–	–	B
18-6-5	MLI	ML® 9850B	SO_2	0–20	0.001	1%	–	–	–	25	VAC	–	–	D
18-6-6	SSI	Instantaneous vapor detector	O_3	0–1.0	0.002	2 ppb	14	48	56	16	VAC	–	–	B

* Manufacturer codes given in Table 18-I-14.
A. Liquid reagents required.
B. UV absorption.
C. Visible light absorption.
D. UV fluorescence.
E. Dual Beam.

TABLE 18-I-7. Chemiluminescent Detectors

Instrument No.	Mfg./ Supp.*	Model	Analytes	Range (ppm)	Detection Limit (ppm)	Precision (±)	Dimensions (cm) H	W	L	Weight (kg)	Power	Alarms Aud.	Vis.	Comments
18-7-1	CSI	5600 NO$_x$	NO, NO$_2$, NO$_x$	0–50	0.0001	0.002 ppm	43	22	57	18.6	VAC	–	–	A, C, D
18-7-2	CSI	OA325-2R OA350-2R	O$_3$	0–10	0.001	0.001 ppm	31	43	51	18.2	VAC	–	–	A, B
18-7-3	RAI	NGA 2000 WCLD	NO, NO$_x$	0–250	–	–	–	–	–	–	VAC	–	–	A, C, D, E
18-7-4	RAI	NGA 2000 CLD	NO, NO$_x$	0–250	–	–	–	–	–	–	VAC	–	–	A, C, D

* Manufacturer codes given in Table 18-I-14.
A. Intended for unattended operation.
B. Uses chemiluminescent reactions of O$_3$ with ethylene.
C. Uses chemiluminescent reaction of NO with ozone as basis for detection.
D. NO$_2$ converted to NO for analysis.
E. Handles wet samples.

TABLE 18-I-8. Photometric Analyzers

Instrument No.	Mfg./Supp.*	Model	Analytes	Range (ppm)	Detection Limit (ppm)	Precision (±)	Dimensions (cm) H	W	L	Weight (kg)	Power	Alarms Aud.	Vis.	Comments
18-8-1	BAC	AutoStep	TDI, MDI, hydrazines, phosgene acid gas	0–0.2 0–5 0–1	0.001–0.01	15%	24.4	21.7	9.8	2.2	Pb-acid	X	–	A, B, C
18-8-2	BAC	SureSpot	MDI, TDI, HDI, NDI	0–10	0.001	25%	16.2	9.5	6	0.8	Pb-acid	–	–	A, E
18-8-3	BAC	RIS	TDI, MDI, IPDI, HDI, hydrazine, phosgene, acid gas	0–0.2, 1.0, 5.0	0.001–0.01	15%	30	20	17.8	5.2	VAC 12VDC	X	X	A, D
18-8-4	CSI	SA 285	SO_2, H_2S, total sulphur	0–1000 ppb	0.0004	1%FS	12	17	20	22.7	VAC	X	X	D
18-8-5	CSI	5700	SO_2	0–20	0.0004	0.001	43	22	57	24	VAC	–	–	D, F
18-8-6	CSI	PA260	phosphorous gas	0–10	0.001	19	7.5	4.8	7.9	18.2	VAC	–	–	C, D
18-8-7	HAI	722R	H_2S	0–100	1	3%	8.3	5.1	5.1	27.3	VAC	X	–	A
18-8-8	NDR	CMS	NH_3, C_6H_6, CO_2, CO, Cl_2, HCl, HCN, H_2S, NO_2, NO_x SO_2	variable	0.2–10	–	20.5	9.2	4.5	0.73	4-AA	–	–	B, C
18-8-9	VIC	721	H_2S	–	ppb	25%	–	–	–	–	battery	–	–	A, B, E
18-8-10	ZEL	CM4	25 toxic gases, NH_3, HCN, H_2S, phosgene	variable	ppb	15%	rack mounted			25	VAC	X	X	A, D
18-8-11	ZEL	ChemKey TLD	>40 gases, TDI, MDI, NH_3, AsH_3, Cl_2, H_2S, etc.	variable	ppb	15%	16.5	21.2	17.7	4.1	Pb-acid VAC	X	X	A, B, C, D
18-8-12	ZEL	EGM	exhaust gases, e.g., aliphatic amines, HCN, H_2S, O_3, SO_2	variable	ppb	15%	10.8	14.4	27	3.4	VAC	X	X	A, D
18-8-13	ZEL	IsoLogger®	HDI, TDI, MDI, IPDI, AsH_3, HCL, $COCl_2$	variable	ppb	15%	7	9.5	2.9	0.2	NiCd	X	X	A, B

TABLE 18-I-8. Photometric Analyzers (Continued)

Instrument No.	Mfg./Supp.*	Model	Analytes	Range (ppm)	Detection Limit (ppm)	Precision (±)	Dimensions (cm) H	W	L	Weight (kg)	Power	Alarms Aud.	Vis.	Comments
18-8-14	ZEL	SPM	>50 gases, iso-cyanates and gases	variable	ppb	15%	30.5	30.5	17.3	6.6	VAC battery operated	X	X	A, B, D
18-8-15	ZEL	7100	>50 gases (see above)	variable	1 ppb	5% at TLV	16.5	43.2	46	204	VAC	X	X	A, D
18-8-16	ZEL	Series 16	>50 gases (see above)	1/10–10 x TLV	1 ppb	5% at TLV	wall rack			204	VAC	X	X	A, D

* Manufacturer codes given in Table 18-I-14.

A. Tape sampler.
B. Portable, self-contained instrument.
C. Intrinsically safe.
D. Intended for unattended operation.
E. Grab sample.
F. UV Fluorescence.

TABLE 18-I-9. Thermal Conductivity Detectors

Instrument No.	Mfg./Supp.*	Model	Analytes	Range (ppm)	Detection Limit (ppm)	Precision (±)	Dimensions (cm) H	W	L	Weight (kg)	Power	Alarms Aud.	Vis.	Comments
18-9-1	MGP	Leak Hunter 8066	nonflammable gases	–	CO_2:3.5×1, O^{-5}; Freon 12:1.2×10^{-5}; cc/sec leak	–	1.4	3.9	5.5	2.3	9V	X	X	–

* Manufacturer codes given in Table 18-I-14.

TABLE 18-I-10. Heat of Combustion Detectors

Instrument No.	Mfg./Supp.*	Model	Analytes	Range (ppm)	Detection Limit (ppm)	Precision (±)	Dimensions (cm) H	W	L	Weight (kg)	Power	Alarms Aud.	Vis.	Comments
18-10-1	BAC	Leakator® 10	combustible gases	50-50,000 CH$_4$	50	–	21.6	5.7	4.4	0.5	5-C cells	–	–	B, D, E
18-10-2	BAC	TLV Sniffer®	combustible gases	0-100 0-100%LEL	3	5%FS	22.8	9.5	16.8	2.3	NiCd 6-D cells	X	–	A, D
18-10-3	BAC	Sniffer® 300 Series	combustible gases, O$_2$ def.	0-2000 0-100%LEL 0-25% O$_2$	–	5%FS	18	8.1	26.2	1.8	NiCd 4-D cells	X	X	A, D
18-10-4	GFG	ABL-50	CO in airlines	2-50	2	10%FS	5.1	5.5	2.2	7.3	VAC 12VDC	X	X	B, F
18-10-5	ENM	Ex-10	combustible gases	0-100%LEL	–	5%FS	4	6	18	0.5	NiCd VAC	X	X	A
18-10-6	MAC	GD-1	combustible gases	0-20%	–	–	11.4	12.7	3.8	0.45	VAC	X	X	A
18-10-7	MGP	8057	Cl$_2$, AsH$_3$, H$_2$, H$_2$S, PH$_3$, etc.	–	variable	–	1.1	2.4	0.5	0.4	NiCd 4-AA	X	X	B, E
18-10-8	SCA	D Series	combustible gases	%LEL % gas	–	–	5.75	3.25	5.75	2.3	8-D cells	X	X	A
18-10-9	SPE	A-21ZX	O$_3$	0-0.3 0-10	0.01	20%	5	10	2.5	0.14	NiMH	X	X	B, C
18-10-10	SPE	SXC-20	VOCs	–	3-5	–	10.7	11.7	6.1	0.9	battery or VAC	–	X	B, G

* Manufacturer codes given in Table 18-I-14.

A. Heated catalytic combustion sensor.
B. Metal oxide semiconductor sensor.
C. Diffusion sampler.
D. Intrinsically safe.
E. Designed as leak detector.
F. Airline monitor.
G. Data logging.

TABLE 18-I-11. Gas Chromotograph Analyzers

Instrument No.	Mfg./ Supp.*	Model	Analytes	Range (ppm)	Detection Limit (ppm)	Dectectors	Dimensions (cm) H	W	L	Weight (kg)	Power	Alarms Aud.	Vis.	Comments
18-11-1	CDI	ISFID ISFIDGC	organics	0–100, 1000, 10000	0.05	FID	–	–	–	5	6-C cells	X	X	A, B
18-11-2	FOX	TVA1000	See instr. 18-4-2	–	–	–	–	–	–	–	–	–	–	–
18-11-3	HNU	311D	organics	1, 10, 100	–	PID, FID, ECD, TCD, FUVAD	26.7	35	28	11.3	VAC	–	–	A, C
18-11-4	PEE	Voyager®	organics	wide	0.01	PID, ECD	39	27	15	6.8	10–18 VDC	X	–	A, B, C
18-11-5	SST	Scento- graph Plus II	organics	0.1–2000	0.1	PID, AID, MAID, ECD, TCD	15.2	52	50.8	24	12VDC VAC	X	–	A, C

* Manufacturer codes given in Table 18-I-14.
A. Intrinsically safe.
B. Data logging.
C. Designed for portable operation.

FID = Flame Ionization Detector
PID = Photoionization Detector
ECD = Electron Capture Detector

TCD = Thermal Conductivity Detector
AID = Argon Ionization Detector
MAID = Micro Argon Ionization Detector

TABLE 18-I-12. Multi-Gas Monitors

Instrument No.	Mfg./Supp.*	Model	Analytes	Range (ppm)	Detection Limit (ppm)	Precision (±)	Dimensions (cm) H	W	L	Weight (kg)	Power	Alarms Aud.	Vis.	Comments
18-12-1	AIM	Series 200 Series 300	%LEL, O_2, CO, H_2S, SO_2, NO, NO_2, Cl_2, H_2, HCN, HCl	variable	variable	–	39.9	64 dia		1.0	battery	X	X	A, B, D
18-12-2	AIM	Series 500	%LEL, O_2, CO, H_2S	0–100%LEL 0–25% O_2 0–200 H_2S 0–500 CO	variable	2.5%FS	19	10.2	6.4	1.4	Pb-acid	X	X	B, C, D
18-12-3	BAC	Body-guard®	%LEL, O_2, CO, H_2S	0–100%LEL 0–30% O_2 0–999 CO 0–500 H_2S	variable	–	7.3	4.1	12	0.4	2-AA	X	X	A, B
18-12-4	BAC	Gas Pointer II®	%LEL, O_2, CO	0–100%LEL 0–25% O_2 0–2000 CO	variable	–	20.5	9.4	5.6	0.9	NiCd	X	X	A, B
18-12-5	BAC	Sentinel® 44T	%LEL, O_2, CO, H_2S, SO_2, HCN, Cl_2, NO_2	0v100%LEL 0–25% O_2 variable	variable	–	11.4	19.7	5.1	1.13	Pb-acid	X	X	A, B, D
18-12-6	BAC	Sniffer® 500	%LEL, O_2, CO, H_2S	0–100%LEL 0–25% O_2 0–500 CO 0–100 H_2S	variable	–	25.4	19.4	15.9	4.3	Pb-acid	X	X	A, B
18-12-7	BI2	PhD Series Ultra	%LEL, O_2, CO, H_2S, Cl_2, ClO_2, HCN, NO_2, NO	variable	variable 0.1	–	17	9.5	5	0.6	NiCd or 3-AA	X	X	A, B, C, D
18-12-8	BWT	Defender	%LEL, O_2, CO, H_2S, SO_2, Cl_2, ClO_2, HCN, NO_2	variable	variable	–	11.8	10.3	2.9	0.4	NiCd	X	X	A, B, C, D
18-12-9	CEA	LMS-40	CH_4, CO_2, O_2, temp., press.	variable	variable	–	–	–	–	–	battery	X	X	C
18-12-10	CDI	Custodian	%LEL, O_2, 2 toxics	variable	variable	–	15.5	10.2	5.2	0.5	NiCd	X	X	A, B, C, D
18-12-11	CSE	EXPLORER®	3 gases, % LEL, O_2, CO, H_2S	variable	variable	–	7.6	15.6	4.2	0.54	NiMH	X	X	A, B, D

TABLE 18-I-12. (cont.). Multi-Gas Monitors

Instrument No.	Mfg./Supp.*	Model	Analytes	Range (ppm)	Detection Limit (ppm)	Precision (±)	Dimensions (cm) H	W	L	Weight (kg)	Power	Alarms Aud.	Vis.	Comments
18-12-12	NDR	Multiwarn II	1–5 gases, % LEL, O_2, 35 toxic gases, IR sensor available	variable	variable	–	–	–	–	1	NiCd	X	X	A, B, C
18-12-13	EST	BLAQ Box 1300	CO_2, CO, temp., %RH	0–5000 CO_2 0–500 CO	100 CO_2	5%	34	30	15.2	7.7	VAC	–	–	C
18-12-14	EST	Telaire 7001	CO_2, temp.	0–10,000	1	5%	–	–	–	–	4-AA	–	–	D
18-12-15	ENM	CGS-90R	%LEL, O_2, CO, + toxics	0–100%LEL 0–30% O_2 0–100 toxics	variable	–	19.2	11.1	6.1	1.77	9.6VDC	X	X	A, B, C
18-12-16	ENM	Omni 4000	CO, H_2S, NH_3, SO_2, Cl_2, HCN, HCl, NO, NO_2, %LEL, O_2, H_2	variable	variable	–	19.4	11.9	5.8	1	NiCd	X	X	A, B, C
18-12-17	ENM	Quadrant 4 gas	%LEL, O_2, CO, H_2S	0–100%LEL 0–25% O_2 0–500 CO 0–100 H_2S	variable	–	15	8	4	0.45	NiCd	X	X	A, B, C, D
18-12-18	GAT	STM 2100 4 gas	%LEL, O_2, CO, NH_3, Cl_2, HCN, H_2S, NO, NO_2, SO_2	variable	variable	5–10%	–	–	–	0.68	NiMH or alk	X	X	A, B, C, D
18-12-19	GFG	G750 Polytector II	6 gases: IR-CO_2, O_2, %LEL, CO, H_2S, SO_2	0–100%LEL 0–200 CO 0–25% O_2	variable	2 ppm CO .2% O_2	3.3	1.4	0.9	0.9	–	–	–	B, D
18-12-20	GFG	CGM II 900	%LEL, O_2, CO, H_2S, Tox Alert	0–100%LEL 0v25% O_2 0–1000 CO	± 1%LEL ± 1 ppm	1%LEL 1ppm CO	19	11.2	5.7	1.4	NiCd or alk	X	X	A, B, D
18-12-21	ISC	ATX612	4 gases: %LEL, O_2, CO, Cl_2, H_2S, NO_2, SO_2	0–100%LEL 0–30% O_2 0–999 toxic	1%	–	20.8	9.4	8.1	1.5	NiCd or alk	X	X	A, B, D
18-12-22	ISC	LTX310	3 gases: %LEL, O_2, CO, Cl_2, HCN, H_2S, NO, NO_2, SO_2	same as above	1%	– 1	12.1	7.0	4.2	0.6	NiCd or alk	X	X	A, B, D

TABLE 18-I-12. (cont.). Multi Gas Monitors

Instrument No.	Mfg./Supp.*	Model	Analytes	Range (ppm)	Detection Limit (ppm)	Precision (±)	Dimensions (cm) H	W	L	Weight (kg)	Power	Alarms Aud.	Vis.	Comments
18-12-23	ISC	TMX-412	4 gases: %LEL, O$_2$, CO, Cl$_2$, H$_2$S, NO$_2$, SO$_2$	0–100%LEL, 0–5, 30% O$_2$, 0–999 variable	0.1%, 0.1–1.0	–	12.1	7.0	5.0	0.7	NiCd or alk	X	X	B, C, D
18-12-24	LSP	Micro-MAX	5 gases: %LEL, O$_2$, CO, H$_2$S, SO$_2$,HCN, PH$_3$, NH$_3$, Cl$_2$, ClO$_2$	0–100%LEL, 0–25% O$_2$, toxics vary	variable	2%	12.1	7.6	4.6	0.4	NiCd or alk	X	X	A, B, C* D
18-12-25	MGP	IQ 1000	Up to 100 gases	variable	variable	–	–	–	–	–		–	–	–
18-12-26	MSA	Passport®	5 gases: %LEL, O$_2$, CO, H$_2$S, SO$_2$, NO, NO$_2$	variable	variable	–	–	–	–	0.5	NiCd	X	X	A, B, C, D
18-12-27	MET	pm-7400	4 gases: %LEL, O$_2$, CO, NH$_3$, Cl$_2$ + 7 others	variable	variable	–	–	–	–	–	NiCd	X	X	A, B, C, D
18-12-28	NDR	MiniWarn	1–5 gases: %LEL, O$_2$, VOCs, + 9 toxics	0–100% variable	variable	–	7.8	14.3	5.8	0.45	NiCd alk NiMH	X	X	A, B, C, D
18-12-29	RAE	MultiRAE	1–5 gases: %LEL, O$_2$, VOCs, + 9 toxics	0–100%LEL, 0–30% O$_2$, 0–5, 10, 50, 100, 500	0.1%, 0/1-1	–	11.8	7.6	4.8	0.45	NiCd or alk	X	X	A, B, C, D
18-12-30	RKI	GX-Series GX-94	3–4 gases: % LEL, O$_2$, CO, H$_2$S, SO$_2$	0–100%LEL, 0–40% O$_2$, 0–100 H$_2$S, 0–200 CO	variable	0.5–10%FS	9	6.4	16	0.7	NiCd 2-C cells	X	X	A, B, C, D
18-12-31	RKI	Eagle®	4 gases: %LEL, O$_2$, CO, H$_2$S, + 11 super toxics variable	0–100%LEL, (0–50,000 ppm) 0–40% O$_2$	variable	5%FS	26.7	15	17.8	2.3	NiCd 4-D cells	X	X	A, B, C*
18-12-32	RA2	GMP 1000	5 gases + opacity: CO, CO$_2$, SO$_2$, NO, NO$_x$, O$_2$, THC	variable	variable	0.1%	rack mounted			–	VAC	–	–	–

TABLE 18-I-12. (cont.) Multi Gas Monitors

Instrument No.	Mfg./ Supp.*	Model	Analytes	Range (ppm)	Detection Limit (ppm)	Precision (±)	Dimensions (cm) H	W	L	Weight (kg)	Power	Alarms Aud.	Vis.	Comments
18-12-33	SCA	SA2000	4 gases: %LEL, O_2, CO, H_2S	0–100% 0–25% 0–200 0–1000	–	3%–5 ppm	7.5	3.75	1.75	0.68	NiCd	X	X	A, B, C
18-12-34	TSI	8570	2–4 gases, % LEL, O_2, CO, H_2S, SO_2, NO, NO_2	0–100%LEL 0–30% 0–20, 400	variable	–	14.6	7.6	5.1	0.64	NiMH or alk	X	X	A, B, C, D
18-12-35	TSI	Q-Trak® 8551	CO_2, CO, temp., %RH	0–5000 CO_2 0–500 CO 0–50° C 5–95%RH	1 1 0.1° C 0.1%RH	3 %	10.7	18.3	3.8	0.6	4-AA VAC	–	–	C, D
18-12-36	ZEL	MiniGas XL	4 gases: % LEL, O_2, CO, H_2S, Cl_2, SO_2	0–100% 0–25% 0–100, 500, 1000	0.1% 0.1% 0.1–1	0.3% 3% .5–5ppm	5.5	7.2	18	0.86	NiMH NiCd alk	X	X	A, B, C, D

* Manufacturer codes given in Table 18-I-14.
A. Intrinsically safe.
B. Good for confined space entry.
C. Data logging capabilities.
D. Diffusion sampling.

TABLE 18-I-13. Portable Mass Spectrometers

Instrument No.	Mfg./ Supp.*	Model	Analytes	Range (ppm)	Detection Limit (ppm)	Precision (±)	Dimensions (cm) H	W	L	Weight (kg)	Power	Alarms Aud.	Vis.	Comments
18-13-1	ISC	MG 2100	VOCs	10–200amu	–	–	23	33	58.4	15.5	VAC	–	–	–

* Manufacturer codes given in Table 18-I-14.

TABLE 18-I-14. List of Instrument Manufacturers

Code	Manufacturer	Code	Manufacturer	Code	Manufacturer
AI2	AIM USA 8403 Cross Park Drive, #1C Austin, TX 78754 (512) 832-5665 or (800) 275-4246 FAX (512) 832-2188	BKJ	Bruel & Kjaer Instruments, Inc. 185 Forest Street Marlborough, MA 01752-3029 (508) 481-7000 FAX (508) 485-0519 www.bkhome.com	FRC	Forney Corporation 3405 Wiley Post Road Carrolton, TX 75006 (972) 458-6100 or (800) 356-7740 FAX (972) 458-6455 www.forneycorp.com
AIC	Arizona Instrument Corp. 1912 W. 4th Street Tempe, AZ 85281 (602) 470-1414 or (800) 528-7411 FAX (602) 804-0656 www.azic.com	BWT	BW Technologies #242, 3030-3 Ave. NE Calgary AB Canada T2A 6T7 (403) 248-9226 or (800) 663-4164 FAX (403) 273-3708 info@bwtnet.com www.bwtnet.com	CDI	Crowcon Detection Instruments 2001 Ford Circle, Suite F Milford, OH 45150 (513) 831-3877 or (800) 527-6926 FAX (513) 831-4263 crowcon@aol.com www.crowcon.com
BAC	Bacharach, Inc. 625 Alpha Drive Pittsburgh, PA 15238 (412) 963-2000 or (800) 736-4666 (412) 963-2091 help@bacharach-inc.com www.bacharach-inc.com	CA2	Calibrated Instruments, Inc. 20 Saw Mill River Road Hawthorne, NY 10532 (914) 741-5700 or (888) 779-2064 FAX (914) 741-5711 www.calibrated.com	CSE	CSE Corporation 600 Seco Road Monroeville, PA 15146 (412) 856-9200 or (800) 245-2224 FAX (412) 856-9203
BI2	Biosystems, Inc. 651 S. Main Street Middletown, CT 06457 (860) 344-1079 or (800) 711-6776 FAX (860) 344-1068 www.biosystems.com	CEA	CEA Instruments, Inc. 16 Chestnut Street Emerson, NJ 07630 (201) 967-5660 or (888) 893-9640 FAX (201) 967-8450 ceainstr@aol.com www.CEAinstr.com	DEZ	Dasibi Environmental Corporation 506 Paula Avenue Glendale, CA 91201 (818) 247-7601 FAX (818) 247-7614 www.dasibi.com
DYM	GFG/Dynamation Gas Monitors 3784 Plaza Drive Ann Arbor, MI 48108 (734) 769-0573 or (800) 959-0329 (734) 769-1888 gfg@dynamationinc.com www.dynamationinc.com	ENS	Environmental Sensors Co. 3201 North Dixie Highway Boca Raton, FL 33431 (561) 338-3116 or (888) 338-4230 FAX (561) 338-5737 www.environmentalsensors.com	HNU	HNU Systems 25 Walpole Park So. Drive Walpole, MA 02081 (508) 660-5001 FAX (508) 660-5040 sales@hnu.com www.hnu.com
ESD	Engineering Systems & Design 17 W. Jefferson Street, Suite 5 Rockville, MD 19713 (302) 456-0446 or (800) 328-0516 FAX (302) 456-0441	FOX	Foxboro Environmental 600 North Bedford Street East Bridgewater, MA 02333 (508) 378-5556 or (888) 369-2676 FAX (508) 378-5505 www.foxboro.com	HOR	Horiba Instrument Corp. 17671 Armstrong Avenue Irvine, CA 92614 (949) 250-4811 or (800) 446-7422 FAX (949) 250-0924 labinfo@horiba.com www.neptune.net/horibal

TABLE 18-I-14. (cont.). List of Instrument Manufacturers

Code	Manufacturer	Code	Manufacturer	Code	Manufacturer
ENM	Enmet Corporation P.O. Box 979 Ann Arbor, MI 48106 (734) 761-1270 FAX (734) 761-3220 info@enmet.com www.enmet.com	GAT	GasTech, Inc. 8407 Central Avenue Newark, CA 94560 (510) 745-8700 or (877) 427-8324 FAX (510) 794-6201 sales@gastech.com www.gastech-inc.com	HAI	Houston Atlas, Inc. 22001 North Park Drive Kingwood, TX 77339 (281) 348-1700 FAX (281) 348-1286 www.houstonatlas.com
EST	Engelhard Sensor Technologies 6489 Calle Real Goleta, CA 93117 (805) 964-1699 or (800) 472-6075 FAX (805) 964-3680	GCI	G.C. Industries, Inc. 5696 Stewart Avenue Fremont, CA 94538 (510) 226-1329 FAX (510) 226-1112	ISC	Industrial Scientific Corporation 1001 Oakdale Road Oakdale, PA 15071 (412) 788-4353 or (800) 338-3287 FAX (412) 788-8353 www.indsci.com
ITS	Interscan Corp. P.O. Box 2496 Chatsworth, CA 91313 (818) 882-2331 or (800) 458-6153 FAX (818) 341-0642 www.gasdetection.com	MST	MST Measurement Systems, Inc. 975 Deerfield Parkway Buffalo Grove, IL 60089 (847) 808-2500 or (800) 547-2900 FAX (847) 808-9976 info@mst-us.com www.mst-us.com	NDR	National Draeger, Inc. 101 Technology Drive P.O. Box 120 Pittsburgh, PA 15230 (412) 787-8383 or (800) 922-518 www.draeger-usa.com
LSP	Lumidor Safety Products 11221 Interchange Circle S. Miramar, FL 33025 (954) 433-7000 or (800) 433-7220 FAX (954) 433-7730 www.lumidor.com	MET	Metrosonics, Inc. P.O. Box 23075 Rochester, NY 14692 (716) 334-7300 FAX (716) 334-2635 www.metrosonics.com	NEO	Neotronics 4331 Thurmond Tanner Road P.O. Box 2100 Flowery Branch, GA 30542 (770) 967-2196 or (800) 535-0606 FAX (770) 967-1854 www.zelana.com/neotron/
MAC	Macurco, Inc. 3946 S. Mariposa Street Englewood, CO 80110 (303) 781-4062 FAX (303) 761-6640 info@macurco.com www.macurco.com	MSA	Mine Safety Appliances Company P.O. Box 427 Pittsburgh, PA 15230 (724) 776-8600 or (800) 672-4678 FAX (724) 776-3280 www.msanet.com	PEC	PE Photovac 2851 Brighton Road Oakville, Ontario Canada L6H 6C9 (905) 829-0030 FAX (905) 829-4701 www.can-am.net/suppliers/photovac.htm
MGP	Matheson Gas Products 166 Keystone Drive Montgomeryville, PA 18936 (215) 641-2700 or (800) 416-2505 FAX (215) 641-2714 www.mathesontrigas.com	MLI	Monitor Labs, Inc. 76 Inverness Drive East Englewood, CO 80112-5189 (303) 792-3300 or (800) 422-1499 FAX (303) 799-4853 www.monitorlabs.com	PRA	Process Analyzers, Inc. 231 Lower Morrisville Road Fallsington, PA 19058 (215) 736-2596 FAX (215) 736-8194

TABLE 18-I-14. (cont.). List of Instrument Manufacturers

RAE	RAE Systems, Inc. 1339 Moffett Park Drive Sunnyvale, CA 94089 (877) 723-2878 or (408) 752-0723 FAX (408) 752-0724 www.raesystems.com/home1.html	SCA	Scott Aviation 309 West Crowell Street Monroe, NC 28112 (704) 282-8400 FAX (704) 282-8424 www.scottaviation.com	SPE	Spectrex Corporation 3580 Haven Avenue Redwood City, CA 94063 (650) 365-6567 or (800) 822-3940 FAX (650) 365-5845 www.spectrex.com
RAI	Rosemount Analytical, Inc. 4125 La Palma Avenue Anaheim, CA 90631 (330) 682-9010 or (800) 433-6076 FAX (330) 684-4434 www.frco.com/proanalytic	SEN	Sensidyne, Inc. 16333 Bay Vista Drive Clearwater, FL 33760 (727) 530-3602 or (800) 451-9444 FAX (727) 839-0550 www.sensidyne.com	SSI	Sunshine Scientific Instruments 1810 Grant Avenue Philadelphia, PA 19115 (215) 673-5600 or (800) 343-1199 FAX (215) 673-5609
RKI	RKI Instruments, Inc. (510) 441-5656 or (800) 754-5165 FAX (510) 441-5650 www.rkiinstruments.com	SST	Sentex Sensing Technology 553 Broad Avenue Ridgefield, NJ 07657 (201) 945-3694 or (800) 736-8394 FAX (201) 941-6064	TEL	Teledyne Analytical Instruments 16830 Chestnut Street Industry, CA 91749 (626) 934-1500 or (888) 789-8168 FAX (626) 961-2538 www.teledyne.com
SMC	Sierra Monitor Corp. 1991 Tarob Court Milpitas, CA 95035 (408) 262-6611 or (800) 727-4377 FAX (408) 262-9042 www.sierramonitor.com sierra@sierramonitor.com	THE	Thermo Environmental Instruments 8 West Forge Parkway Franklin, MA 02038 (508) 520-0430 FAX (508) 520-1460 www.thermoei.com	TSI	TSI, Inc. P.O. Box 64394 St. Paul, MN 55164 (651) 483-0900 FAX (651) 490-2748 info@tsi.com www.tsi.com
VCI	VICI Metronics, Inc. 2991 Corvine Drive Santa Clara, CA 95051 (408) 737-0550 FAX (408) 737-0346 metronics@vici.com www.vicimetronics.com	ZEL	Zellweger Analytics 405 Barclay Boulevard Lincolnshire, IL 60069 (847) 634-2800 or (800) 323-2000 FAX (847) 634-1371 www.zelena.com		

Chapter 19

Denuder Systems and Diffusion Batteries

Yung-Sung Cheng, Ph.D.

Lovelace Respiratory Research Institute, P.O. Box 5890, Albuquerque, NM 87185

CONTENTS

Introduction

The diffusion technique is used to collect ultrafine particles and vapors and to determine the size distribution of ultrafine particles (<0.2 μm). The technique was first conceived following the observation that losses of atmospheric nuclei in tubes were related to their diffusion coefficients.[1] Mathematical equations for diffusion losses in rectangular or circular tubes were subsequently derived.[2, 3] This enabled accurate determination of diffusion coefficients and submicrometer particle sizes from measurement of particle penetration through these tubes.

Diffusion samplers are devices that separate particles or vapors by differential diffusion mobilities. Two types of diffusion samplers are often used in air sampling: a diffusion battery can be used to measure the size distribution of submicrometer particles, and a diffusion denuder is designed to separate and collect gases or vapors from airborne particles. These devices are useful for particles smaller than 0.1 μm, including condensation nuclei, radon progeny, and gas or vapor molecules.

The diffusion coefficient of a particle or molecule is inversely proportional to the particle size. The diffusion battery was initially designed to measure the particle size of condensation nuclei in the atmosphere, whereas the diffusion denuder was first designed to determine the diffusion coefficient of gas and vapor molecules. Current applications of diffusion denuders have been extended to atmospheric sampling of SO_2, NH_3, and NO_x, or as a scrubber in an acidic aerosol sampling system to remove certain types of gases to avoid absorption of these gases on particles collected in the filter. Diffusion batteries are frequently used in the laboratory to measure size distributions of ultrafine particles and in indoor environments to determine activity size distributions of radon and thoron progeny.

Diffusion samplers include tubes of different shapes and stacks of fine mesh screens of well-defined characteristics. Principles and operations of diffusion denuders and diffusion batteries have been reviewed.[4, 5] A diffusion battery is often used with a condensation nucleus counter in a sampling train to determine the concentration and particle size distribution of an aerosol. However, radioactive aerosols can be sampled by a diffusion battery and the substrates counted directly for radioactivity. This chapter describes the operating principle, theory, design, applications, and data analyses of diffusion batteries and denuders.

Theories of the Diffusion Technique

The mathematical expressions relating collection or penetration of vapors and particles through cylindrical and rectangular tubes and screens have been derived. These expressions can be used to calculate diffusion coefficients or particle sizes from experimental measurements through diffusion samplers.

These mathematical expressions were derived from the convective diffusion equation describing the concentration profile (c) in various geometries and flow profiles:

$$\frac{D}{r}\frac{\partial}{\partial r}\left(r\frac{\partial c}{\partial r}\right) = u(r)\frac{\partial c}{\partial z} \tag{1}$$

where: D = diffusion coefficient
r = radial direction
z = axial direction
$u(r)$ = velocity profile in the axial direction

Several assumptions were made in the derivation of Equation 1: (1) the concentration is in a steady-state condition; (2) the flow field in the device is a function of radial position, r, only; (3) the effect of diffusion in the direction of flow is neglected; (4) no production or reaction of the gas or aerosol occurs in the device; and (5) the sticking coefficient of the gas or particle is 100% on the collection surface (walls or screens). Diffusion devices can be classified as tube (channel)-type or screen-type with different flow profiles. Solutions to Equation 1 for different types of diffusion samplers are summarized in the following section.

Tubes and Channels

Penetration (P) of particles or gases due to the diffusional mechanism has been derived for channels of different geometries, including cylindrical, rectangular, disk, and annular shapes. The general solution of Equation 1 can be expressed as a series of exponential functions:

$$P = \sum_{n=1}^{\infty} A_n \exp(-\beta_n \mu) \tag{2}$$

where: μ = dimensionless argument relating the diffusion coefficient, channel length, and flow rate.

The right hand side of Equation 2 is an infinite series, however, only a finite number of terms is needed to compute or converge to the accurate solution. Convergence of Equation 2 depends on the magnitude of μ. For larger values of μ (low penetration), fewer terms are needed for convergence; whereas, at small values of μ (high penetration), many terms are required. For high penetration, alternative equations have been derived.

Solutions of the diffusion equation for tubes or channels depend strongly on the flow field in the tubes, $u(r)$. The flow profile can be laminar or turbulent depending on the flow rate (or Reynolds number), and transient or fully developed depending on the length of the tube and the Reynolds number. One always designs a diffusion sampler to satisfy the fully-developed laminar flow condition. Discussions for the effects of flow profile are reported by Ingham.[6] Specific solutions for fully-developed laminar flows in tubes or channels of different shapes are described below.

Cylindrical Tubes

Penetration through a circular tube (Figure 19-1) at a flow rate, Q, for particles with a diffusion coefficient of D has been derived by several investigators as a function of the parameter μ defined as $\pi DL/Q$.[3, 6–10] The numerical solution obtained by Bowen et al.[11] for μ between 1×10^{-7} and 1 is most accurate. Results obtained by Davis and Parkins,[8] Tan and Hsu,[9] and Sideman et al.,[7] and Lekhtmakher[10] agree substantially with those of Bowen et al.[11] By comparison of various expressions, the following analytical solutions have the accuracy of four significant numbers as compared to Bowen's result in the entire range of μ:[12]

$$P = 0.81905 \exp(-3.6568\mu) + 0.09753 \exp(-22.305\mu) + 0.0325 \exp(-56.961\mu) + 0.01544 \exp(-107.62\mu) \tag{3}$$

for $\mu > 0.02$ and

$$P = 1.0 - 2.5638\,\mu^{2/3} + 1.2\mu + 0.1767\mu^{4/3} \quad for\ \mu \leq 0.02 \tag{4}$$

The formula for small values of μ is taken from Gormley and Kennedy,[3] Newman,[13] and Ingham.[6]

Rectangular Channels and Parallel Circular Plates

Particle penetration through a parallel narrow rectangular tube (Figure 19-1) of width W and separation H, where $H \ll W$, has been derived as a function of μ

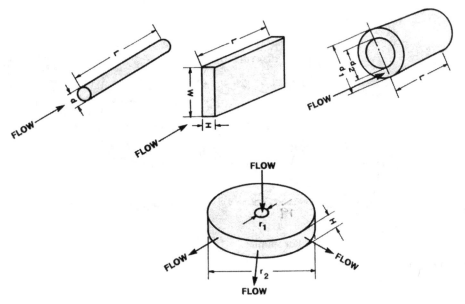

FIGURE 19-1. Schematics of different shapes of tubes.

defined as $8DLW/3QH$ (Gormley cited by Nolan and Nolan).[2, 7, 11, 14–16] The same equation can be used to calculate penetration for inward flow between parallel circular plates (Figure 19-2), where the diffusion parameter μ is defined as $8\pi D(r_2^2 - r_1^2)/3QH$, where r_2 and r_1 are outer and inner radii of the disks.[15] The most accurate solution was given by Tan and Thomas[16] and Bowen *et al.*;[11] other investigators agreed substantially with their results. The most accurate analytical formulae (as to the accuracy of four significant numbers) for the entire range of μ are:[12]

$$P = 0.9104 \exp(-2.8278\mu) + 0.0531 \exp(-32.147\mu) + \tag{5}$$
$$0.01528 \exp(-93.475\mu) + 0.00681 \exp(-186.805\mu)$$
for $\mu > 0.05$ and

$$P = 1 - 1.526\,\mu^{2/3} + 0.15\mu + 0.0342\mu^{4/3} \; for\; \mu \le 0.05 \tag{6}$$

The formula for small values of μ is given by Ingham.[17] Kennedy (quoted by Nolan and Kennedy[18]) derived a similar formula with different coefficients, but the results vary by only 1%.

Annular Tubes

Theoretical calculation of diffusional losses through an annular tube (Figure 19-2) has been derived by solving Equation 1 using a fully developed flow in the annular tube.[19] The penetration is then a function of the inner to outer radius of the denuder, k, and the diffusional parameter, $\mu = \pi DL(d_2 + d_1)/2Q(d_2 - d_1)$]:

$$P = A_0\,(k)\exp(-\beta_0^2(k)\mu) + A_1\,(k)\exp(-\beta_1^2(k)\mu) + \tag{7}$$
$$A_2\,(k)\exp(-\beta_2^2(k)\mu) + A_3\,(k)\exp(-\beta_3^2(k)\mu) +$$

where d_2 and d_1 are outer and inner diameters, respectively. Values of $A_n(k)$ and $\beta_n(k)$ for selected values of k are given in Table 19-1. For $k = 0$ (cylindrical tube) and $k = 1$ (parallel plate), Equations 3 and 5 are the asymptotic solutions. A numerical solution of the diffusional equation in an annular tube is obtained by solving the velocity profile instead of using the fully developed flow has been obtained.[20] From the numerical solution, a fitted equation is obtained:

$$P = 1 - \exp[-0.03711(1 - k)^{1.317} Pe^{0.678}] \tag{8}$$

where Pe $(= u_m\,d_2^2/4DL)$ is the Peclet number and u_m is the mean velocity. These equations are in general agreement with experimental data.[21]

Wire Screens

Aerosol penetration through a stack of fine mesh screens with circular fibers of uniform diameter and arrangement has been derived.[22–24] A stack of fine mesh screens simulates a fan model filter[25, 26] in terms of flow resistance and aerosol deposition characteristics.[27] The theoretical penetration was derived based on the aerosol filtration in the fan model filter:

$$P = \exp[-Bn(2.7Pe^{-2/3} + \frac{1}{k}R^2 + \tag{9}$$
$$\frac{1.24}{k^{1/2}}Pe^{-1/2}R^{2/3})]$$

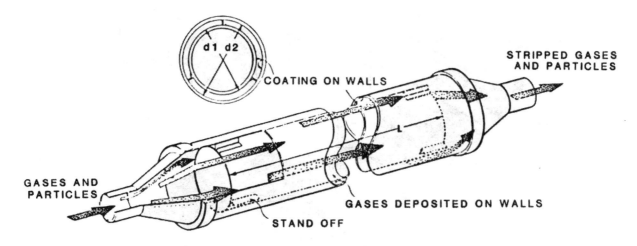

FIGURE 19-2. Schematic of an annular denuder.

where: $B = 4\alpha h/\pi(1-\alpha)d_f$
n = number of screens
d_f = fiber diameter
h = thickness of a single screen
α = solid volume fraction of the screen
k = hydrodynamic factor of the screen
$= -0.5\ ln(2\alpha/\pi) + (2\alpha/\pi) - 0.75 - 0.25(2\alpha/\pi)^2$
$R = d_p/d_f$, the interception parameter
$Pe = Ud_f/D$ is the Peclet number
U = superficial velocity

Equation 9 includes the diffusional and interceptional losses of aerosol on screens and is valid for particles up to 1 µm in size.[27] For particles larger than 1 µm, inertial impaction becomes an important mechanism, and Equation 9 may not be adequate. For smaller particles ($d_p < 0.01$ µm), diffusional deposition is the dominant mechanism, and Equation 9 is simplified to:

$$P = exp(-2.7BnPe^{-2/3})\qquad(10)$$

Diffusion Denuders

Gas or vapor molecules diffuse rapidly to the wall of a diffusion sampler and adsorb onto the wall coated with material suitable for collecting the gas. Diffusion tubes have been used to measure diffusion coefficients of several gases in the air.[28–30] Since 1980, diffusion denuders followed by a filter pack have been developed to sample atmospheric nitric acid vapors and nitrate particulate aerosols. Using this sampling technique, called the denuder difference method, one can separate gaseous species, such as HNO_3 and NH_3, from particulate nitrates and thus minimize sampling artifacts due to the presence of these gases.[31–37] Diffusion denuders are also used to monitor vapors, such as formaldehyde, chlorinated organic compounds, and tetra alkyl lead, in the ambient air or work environments.[38–40] Some personal samplers have also been developed for industrial hygiene use.[41, 42]

TABLE 19-1. A_n and β_n for Selected Values of k

k	0.05	0.1	0.25	0.5	0.75	0.85	0.95
β_0	3.493	3.610	3.765	3.851	3.878	3.882	3.884
β_1	7.906	8.086	8.315	8.436	8.473	8.479	8.481
β_2	12.31	12.56	12.87	13.03	13.08	13.09	13.09
β_3	16.71	17.02	17.42	17.63	17.70	17.71	17.71
A_0	0.879	0.889	0.900	0.908	0.910	0.911	0.911
A_1	0.036	0.026	0.012	0.003	0.001	0.000	0.000
A_2	0.044	0.046	0.05	0.052	0.053	0.053	0.053
A_3	0.006	0.005	0.002	0.001	0.000	0.000	0.000

Description of Diffusion Denuders

Two types of diffusion denuders have been designed: the cylindrical tube and annular tube.

Cylindrical Denuders

In cylindrical denuders, a single cylindrical glass or Teflon tube is often used for collecting gases or vapors. The diameter and length of the tube and the sampling flow rate are designed to have greater than 99% collection efficiency. For example, a glass tube of 3 mm i.d. and 35 cm long would have over 99% efficiency for ammonia ($D = 2.47 \times 10^{-5}\ m^2\ s^{-1}$) at 3 L min^{-1}.[43] For higher sampling flow rates, parallel tube assemblies have been designed,[31, 35] consisting of 16 glass tubes 5 mm i.d. and 30 cm long. The sampling flow rate was 50 L min^{-1}, and the collection efficiency for ammonia was over 99%.[31] A 2-stage diffusion denuder consisting of 212 glass honeycomb tubes, each with a height of 2.54 cm and an inside diameter of 0.2 cm has been designed to removes gases for collection of ambient particles.[32]

Penetration through the tube-type denuders can be estimated by taking the first term of Equation 3 only:

$$P = 0.819\ exp(-3.66\ \mu) \qquad (11)$$

This simplified equation is accurate at higher values of μ (> 0.4) and at lower penetrations ($P < 0.190$). The error of the estimated penetration from Equation 3 increases with the decreasing value of μ (−0.25% error for $\mu = 0.2$ and $P = 0.395$, and −1.8% for $\mu = 0.1$ and $P = 0.579$). Equation 11 is applicable for the fully developed laminar flow region in the tube. The flow Reynolds number in the tube should be less than 2300 for laminar flow:

$$Re = \frac{4\rho Q}{\eta \pi D} < 2300 \qquad (12)$$

where: Q = volumetric flow rate
D = diameter of the tube
ρ = gas density
η = gas viscosity

In the entrance of the tube, the flow is in a transition region from plug flow to developed flow. The length of entrance, L_e, is defined by the following equation and should be minimized:

$$L_e = 0.035\ d\ Re \qquad (13)$$

Annular Denuders

Higher sampling flow rates are desirable, especially for sampling trains consisting of denuders and filters or dichotomous samplers.[34] An annular tube denuder designed for this purpose consists of two coaxial cylinders with the inner one sealed at both ends, so that air is forced to pass through the annular space (Figure 19-3).[21] The collection efficiency of the annular tube can be estimated from Equation 7 or 8 for lower Reynolds numbers ($Re < 2300$) defined as:

$$Re = \frac{4\rho Q}{\eta \pi (d_1 + d_2)} \qquad (14)$$

Comparing the performance of the cylindrical and annular denuders in removing a gas from an air stream, a typical annular denuder ($d_2 = 3.3$ cm and $d_1 = 3.0$ cm) is possible by equating Equations 7 and 11. It can be shown that:

$$\left.\frac{Q}{L}\right|_{annular} = 31.5 \left.\frac{Q}{L}\right|_{cylinder} \qquad (15)$$

This relationship shows that for a given tube length, the annular denuder can operate at 30 times the flow rate of the cylindrical denuder and still have the same removal efficiency. In addition, the Reynolds number would still indicate laminar flow conditions for the annular tube system. A multi-channel annular diffusion denuder has been tested and used in ambient air sampling.[38]

Compact Coil Denuder

A compact coil denuder consisting of a 1.0 cm i.d. and 95 cm long (L) glass tube bent into a three-turn helical coil with a 10 cm diameter (Figure 19-3) has been designed by Pui et al.[44] The heat and mass transfer rates to the tube wall in a curved tube are much higher than those in a straight tube operated at the same conditions.[45, 46] This denuder is operated at 10 L min^{-1} (Q) with a Reynolds number of 1400. The penetration through the denuder can be expressed as:[44]

$$P = 0.82 \exp\left(-\frac{\pi LD}{Q}\ Sh\right) \qquad (16)$$

$$Sh = \frac{0.864}{\delta}\ D_e^{1/2}(1 + 2.35\ D_e^{-1/2}) \qquad (17)$$

where: Sh = Sherwood number
D_e = Dean number, the flow Reynolds number divided by the square root of the radius of curvature
δ = thickness ratio of the concentration

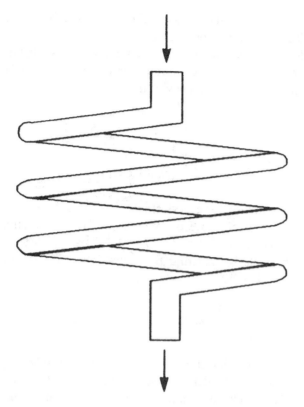

FIGURE 19-3. Schematic of a compact coil denuder.

$$P = exp(-\frac{2\pi\alpha L}{Q}) \qquad (18)$$

where: L = length of the active surface
        Q = flow rate
        α = rD/δ is a function of the diffusion coefficient D
        r = radius of the tube
        δ = boundary thickness, which is a function of the flow Reynolds number

The penetration must be determined empirically. Operating the denuder at 16.1 L min^{-1} (Re = 2500), Durham et al.[47] obtained a retention of 0.911 for HNO$_3$.

Scrubber-Type Diffusion Denuders

Most diffusion denuders have a solid coating on the wall to collect gaseous species, and the coating substrates are washed after sampling for analysis. For continuous analysis of gas species, diffusion scrubbers are used, where the absorbent or solvent in the liquid form is continuously flowing along the tube wall, and the analyte can be analyzed in real-time.[48, 49] A tubular scrubber is made by inserting a membrane tube into the glass tube to form a jacket between the glass wall and the membrane (Figure 19-4). Porous membranes, such as PTFE and polypropylene, allow gases but not particles to permeate and dissolve in solution which flows continuously through the jacket. The collection characteristics of this diffusion scrubber should be similar to the tube diffusion denuder. Another type of diffusion denuder consists of a tube with a membrane tube at the center in which air flows through the annular space, while the solvent passes through the membrane tube.

Coating Substrates

Absorbent material can be coated onto the tube wall of a denuder to collect the gas of interest from the air stream. Table 19-2 lists substrates for removal of some gases as reported in the literature. Some materials absorb more than one gas. For example, sodium carbonate can absorb acidic gases found in the ambient air, including HCl, HNO$_2$, HNO$_3$ and SO$_2$. The method of application of material to the tube wall depends largely on the nature of the material. Most materials are first dissolved and then applied to the tube wall. Solvents are allowed to evaporate, leaving the absorbent on the glass tube wall. In some cases, the glass denuder wall has been etched by sandblasting the surface to increase the capacity of walls

boundary to the momentum boundary layer

This unit has a 99.3% collection efficiency for SO$_2$ with less than 6% loss for particles between 0.015 to 2.5 μm in diameter. It is compact and easy to operate.

Transition-Flow Denuder

Both cylindrical and annular denuders are operated in laminar flow conditions, and they are designed to remove all gases of interest from the air stream. Particle evaporation may increase the concentration of some gases in passing through such denuders, especially in the case of the decomposition of NH$_4$NO$_3$ into HNO$_3$ and NH$_3$ gases. To avoid biases due to evaporation of particles, one approach to sampling such gases is to collect only a known fraction of gases in the denuder and then calculate the gas concentration.

A transition-flow denuder was designed by Durham et al.[47] to permit higher sampling flow. The cylindrical denuder has an inside diameter of 0.95 cm with a 6 cm distance of the first active surface to allow for development of a stable flow profile. The denuder section is lined with a 3.2 cm long nylon sheet. By assuming complete mixing in the active section, the penetration can be expressed as:

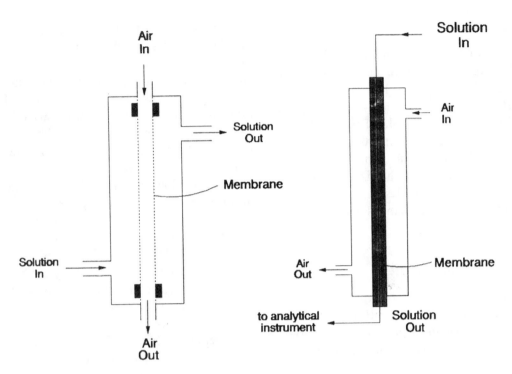

FIGURE 19-4. Schematics of diffusion scrubbers.

to support the denuding chemical substrate.[21] Absorbent paper impregnated with liquid or solution substrate, such as oleic acid, has been used to line the inside of the denuder wall.[28] A nylon sheet has also been used as a liner.[47] Anodized aluminum surfaces have recently been found to be a good absorbing surface for nitric acid. Annular denuders made of anodized aluminum do not need coating.[54] Tenax or silica gel powder is more difficult to apply; however, these materials adhere to the glass wall coated with silicon grease.[38, 42]

Sampling Trains

When sampling ambient or working atmospheres, it is sometimes necessary to collect gases and particulate materials separately. In this case, a sampling train consisting of diffusion denuders and a filter pack has been used. A more complex system as shown in Figure 19-5, including a cyclone precutter, two Na_2CO_3-coated annular denuders, and a filter pack with a Teflon and a nylon filter, has been used to collect acidic gases (HNO_3, HNO_2, SO_2 and HCl) separately from nitrate and sulfate particles.[37] The first denuder removes gases quantitatively, whereas the second accounts for the interference from particulate material deposited on the wall under the assumption that particle deposition on each denuder is the same.[40] The denuders are placed vertically to avoid particle deposition on the walls by sedimentation. A dif-

fusion scrubber can be connected to an ion chromatograph or other analytical instruments for real-time analysis of gases.[49, 55] Further discussion can be found in Chapter 4.

TABLE 19-2. Materials for Absorbing Gases in the Diffusion Denuder

Coating Material	Gas Absorbed	Reference
Oxalic acid	NH_3, aniline	41, 43
Oleic acid	SO_3	28
H_3PO_3	NH_3	31
K_2CO_3	SO_2, H_2S	50
Na_2CO_3	SO_2, HCl, HNO_3, HNO_2	35
$CuSO_4$	NH_3	28
PbO_2	SO_2, H_2S	50
WO_3	NH_3, HNO_3	51
MgO	HNO_3	31
NaF	HNO_3	52
NaOH and guaiacol	NO_2	53
Bisulfite-triethanolamine	formaldehyde	39
Nylon sheet	SO_2, HNO_3	47
Tenax powder	chlorinated organics	38
Silica gel	aniline	42
ICl	tetra alkyl lead	40

Diffusion Batteries

Diffusion batteries were originally developed to measure the diffusion coefficient of particles less than 0.1μm in diameter. They have since been used for determination of particle size distributions by converting the diffusion coefficient to the particle size. Diffusion batteries are one type of only a few instruments that are applicable in measuring ultrafine particles between 0.1 μm and about 1 nm, corresponding to the size of molecular clusters. In this section, various designs of the instrument, detection of particles, and methods of data analysis will be discussed.

Description of Diffusion Batteries

Several types of diffusion batteries have been designed. Those based on rectangular channels and parallel circular plates are single-stage diffusion batteries. Cylindrical-tube and screen-type diffusion batteries usually have several stages.

Rectangular Channel

Rectangular channel diffusion batteries usually consist of many rectangular plates forming parallel channels

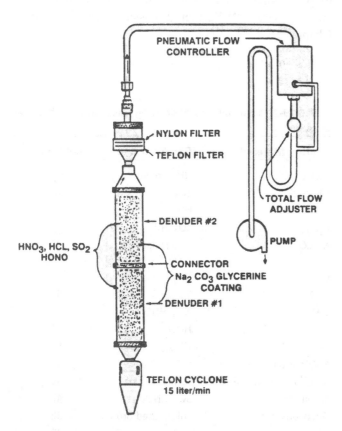

FIGURE 19-5. An ambient acidic aerosol sampler consisting of a precutter, two annular denuders, and a filter pack.[37]

of equal width. These plates are separated by spacers and glued to a container with an airtight seal. For example, a diffusion battery consisting of 20 parallel channels (0.01 cm wide, 12.7 cm high, and 47.3 cm long) made from graphite plates has been designed for a 1 L min⁻¹ sampling flow rate.[28] Other instruments have been made from aluminum or glass plates with similar construction.[2, 56–59] Each channel should be parallel and have the same width. Deviation of channel width results in a nonuniform flow rate through each channel, which in turn causes the deviation of penetration from the theoretical prediction of Equations 5 and 6. A diffusion battery of 10 single channels, each separately housed in a box, has been designed by Pollak and Metnieks.[60]

A single-stage diffusion battery can be used to measure the diffusion coefficient of monodisperse aerosols at one flow rate. When it is used to measure polydisperse aerosols, such as those found in ambient air, several measurements taken at different flow rates are necessary to determine the distribution of diffusion coefficients.

Parallel Disks

A diffusion sampler has been designed by Kotrappa *et al.*[61] It is based on the diffusional losses of particles from a fluid flowing radially inward between two coaxial, parallel, or circular plates as originally proposed by Mercer and Mercer.[15] Stainless steel plates (3.77 cm diameter) with a central hole of 0.2 cm diameter in the upper plate are the collecting substrate. Separation between the plates is 0.225 cm. An absolute filter is used to collect material penetrating the device. This sampler has been used to determine the diffusion coefficient of radon decay products, which have diffusion coefficients of the order of 0.05 cm² sec⁻¹. The amount of radioactivity collected at the plates and absolute filter was determined, and the diffusion coefficient calculated from Equation 5, simplified to contain only the first term:

$$P = 0.9104 \exp(-2.8278 \frac{8\pi D(r_2^2 - r_1^2)}{3QH}) \quad (19)$$

Cylindrical Tubes

Tube-type diffusion batteries made of cylindrical tubes usually consist of a cluster of thin-walled tubes with diameters less than 0.1 cm i.d. Large equivalent length (actual length × number of tubes) is required for measurement of particle size because particles have much smaller diffusion coefficients than do gas molecules. Several cluster tube diffusion batteries have been designed.[62–64] Figure 19-6 shows the schematic of a tube-type diffusion battery as reported by Scheibel and

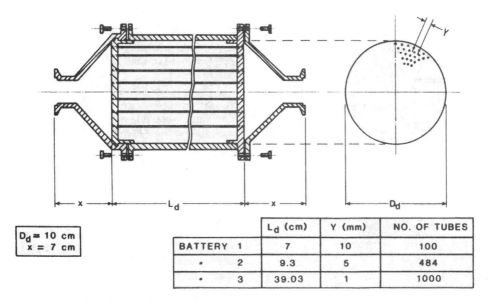

	L_d (cm)	Y (mm)	NO. OF TUBES
BATTERY 1	7	10	100
" 2	9.3	5	484
" 3	39.03	1	1000

D_d = 10 cm
x = 7 cm

FIGURE 19-6. Schematic of a cluster-tube diffusion battery[64].

Porstendörfer.[64] Three diffusion batteries with 100, 484, and 1000 single tubes were used with lengths of 5.0, 9.3, and 39.03 cm, respectively. Tube-type diffusion batteries use materials that are commercially available and are easier to construct than the parallel-plate diffusion battery. A lightweight material such as aluminum is often used; however, this type of diffusion battery is still heavy, bulky, and expensive. Most cluster tube diffusion batteries consist of one to three stages,[63, 64] although an eight-stage diffusion battery has been constructed.[62]

Compact diffusion batteries with many stages have been designed by using collimated hole or honeycomb structures (CHS). The CHS are discs containing a large number of near circular holes. Figure 19-7 shows a 1 3/4-inch-diameter CHS disc made from stainless steel containing 14,500 holes of 0.009-inch-diameter (Brunswick, CO, Chicago, IL). With a thickness of 1/8 to 1 inch, the equivalent length ranged from 46 to 369 meters. A portable 11-stage diffusion battery has been designed with CHS elements.[65] The total length is 60 cm, and the equivalent length is 5094 m. Figure 19-8 shows the schematic of a five-stage diffusion battery made from CHS elements. A multiple-stage diffusion battery is required to measure the size distribution of a polydisperse aerosol. The development of a multiple-stage CHS diffusion battery opens the possibility of routine measurements of submicrometer aerosols. However, the commercial sources for the CHS are not currently available and therefore, only a few diffusion batteries are still being used in some laboratories. Other CHS discs made from glass capillary tubes of 25 or 50 mm in diameter and a thickness of 0.5 to 2.0 mm are

commercially available (Galileo Electro Optical Corp., Struburg, MA). A six-stage CHS diffusion battery made from glass has been designed.[66]

Wire Screens

Diffusion batteries using stacks of filters as the cell material have been used by Sinclair and Hinchliffe[67] and Twomey and Zalabsky.[68] This filter material is lightweight and inexpensive to build. However, commercial fiber or membrane filters are not ideal materials because

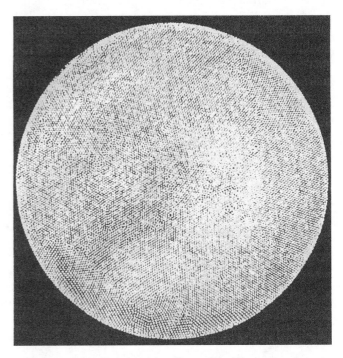

FIGURE 19-7. A stainless steel, collimated hole structure disc.

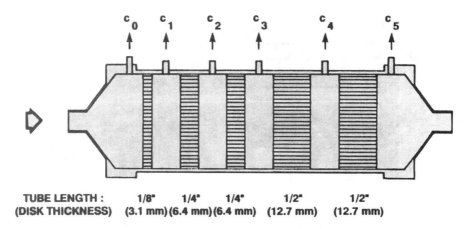

FIGURE 19-8. Schematic of a five-stage diffusion battery consisting of a stainless steel collimated hole structure.

of the nonuniformity in the fiber diameter and porosity. Aerosol penetration through the filter may not be consistent and could not be predicted accurately by filtration theory. Sinclair and Hoopes[69] designed a 10-stage unit using stainless steel mesh 635 screens of uniform diameter, opening, and thickness (Figure 19-9). The designed flow rate ranged from 4 to 6 L min^{-1}. Stacks of these well-defined screens simulate a fan model both in geometry and in flow resistance.[27] Penetration through screens can be predicted by the fan model filtration theory (Equation 8).[22, 23] Subsequently, this unit has become commercially available (Model 3040, TSI Inc., St. Paul, MN). Other types of screens have also been tested and found useful.[24, 27] Table 19-3 lists characteristics of the different screens as shown in Figure 19-10. Screen-type diffusion batteries are compact in size and simple to construct. Screens can be cleaned and replaced easily when they are contaminated or worn out.

Most multi-stage diffusion batteries described here are arranged in a series so that the aerosol concentration decreases continuously through the cells. Aerosol penetration is usually detected by a condensation nucleus counter (CNC). Based on parallel flow and mass collection principles, a parallel flow diffusion battery (PFDB)

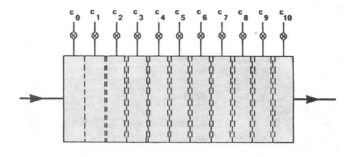

FIGURE 19-9. Schematic of a 10-stage screen-type diffusion battery.

has been designed by Cheng et al.[70] This unit measures the penetration by mass or radioactivity without a particle detecting unit. It is also more useful to detect unstable aerosols with fluctuating size and concentration. A schematic diagram of the PFDB is shown in Figure 19-11. The PFDB consists of a conical cap and a collection section containing seven cells. Each diffusion cell contains a different number of stainless steel 200-mesh screens followed by a 25 mm Zefluor filter (Gelman, Ann Arbor, MI). The seven cells typically contain 0 through 35 screens. Critical orifices provide a 2 L min^{-1} flow rate through each cell resulting in a total flow rate of 14 L min^{-1}. Gravimetric determination of collected filter samples from each cell provides the direct mass penetration as a function of screen number for the determination of aerosol size distribution, thereby eliminating the sometimes inaccurate conversion of number to mass.

Screen diffusion batteries have been used routinely to determine the activity size distribution of radon progeny. A single screen and a filter have been used to estimate the unattached fraction. The screen and flow rate were chosen such that 4 to 10 nm particles can penetrate the screen with 50% efficiency. The radioactivity collected on the screen and the filter are counted, and the amount of activity collected on the screen is assumed to be the unattached radon progeny. The activity size distribution of the radon progeny can be measured with either a graded screen diffusion battery[71] or a PFDB.[72, 73] A graded diffusion battery consists of several stages, each with a different type of screen. The screens with a lower mesh number are used to collect particles in the nm size range, whereas screens of a larger mesh number are used to collect larger particles. Figure 19-12 shows a schematic of a GDB including five stages of screens and a backup filter.[74] After a

TABLE 19-3. Characteristic Dimensions and Constants for Various Types of Screens in Screen-Type Diffusion Batteries

Screen Mesh	145	200	400	400	635
Weave	Square	Square	Square	Twill	Twill
Screen diameter (μm)	55.9	40.6	25.4	25.4	20
Screen thickness (μm)	122	96.3	57.1	63.5	50
Solid volume fraction	0.244	0.230	0.292	0.313	0.345
k*	0.330	0.352	0.269	0.246	0.216

* hydrodynamic factor of the screen (Equation 9)

brief sampling time (5 to 10 min), the screens and filters are counted directly for radioactivity. Because of the geometry of the screen, the efficiency of the alpha counting is not 100%, and corrections should be made to allow for a lower counting efficiency on screens.[75] Parallel flow diffusion batteries with much higher flow rates (over 30 L min⁻¹) are used to collect indoor radon progeny, which often have very low concentrations. Usually, only the backup filters for each stage of the PFDB are counted for radioactivity. Thus, this method eliminates possible errors due to the attenuation of alpha rays by the screen.

Use and Data Analysis

Aerosol penetration through a diffusion battery provides data for the determination of particle size distribution. Aerosol penetration through a diffusion cell is obtained by measuring the number, mass, or activity concentrations at the inlet and outlet of each cell. A CNC is used to measure the number concentration. Figure 19-13 shows a schematic diagram of a system including a diffusion battery, automatic switching valve, and a CNC. With the automatic sampling system, it takes 3 min to complete an 11-channel measurement.

For radioactive aerosols, penetration based on activity can be obtained by collecting samples at the diffusion cell and a backup filter at the end of the diffusion battery. The single-stage parallel disk diffusion sampler[61] and screen diffusion batteries have been used for this purpose. Screens can be counted directly for radioactivity.[72] Penetration based on the mass can be obtained by using a PFDB.

FIGURE 19-10. Photomicrograph of stainless mesh screens.

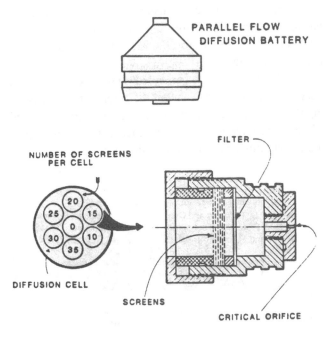

FIGURE 19-11. Schematic of a parallel flow diffusion battery.

Data Analysis

Monodisperse Aerosol

Particle size distributions are calculated from penetration data obtained from the measurements by the diffusion battery. For a monodisperse aerosol, the diffusion coefficient, D, can be calculated directly from the corresponding Equations 3–9. The particle size, d_p, is then calculated from the following relationship:

$$D = \frac{kTC(d_p)}{3\pi\eta d_p} \tag{20}$$

$$C(d_p) = 1 + \frac{2\lambda}{d_p}\left[1.142 + 0.558\exp(-0.999\frac{d_p}{2\lambda})\right] \tag{21}$$

where: k = Boltzmann constant (1.38×10^{-16} erg K^{-1})
   T = absolute temperature (K)
   C = "Cunningham" slip correction factor
   λ = mean free path of air (0.0673 μm at 23°C and 760 mm Hg)

With monodisperse particles, measurements from a single-stage diffusion battery are sufficient, and measurements from multiple-stage devices should improve the accuracy.

Polydisperse Aerosols

Most aerosols in ambient environments and work places have polydisperse size distributions, and the method described in the previous section does not apply. Three penetration data points are the minimum required, but more will improve the accuracy of the size determination. Both graphical and numerical inversion methods have been developed for the size determination from penetration data.

Fuchs et al.[76] have generated a family of penetration curves for rectangular channel diffusion batteries assuming the aerosol size distribution is lognormally distributed. Mercer and Greene[77] have provided curves representing the penetration of aerosols in both cylindrical and rectan-

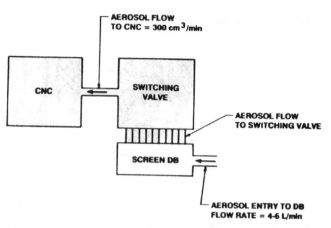

FIGURE 19-13. Schematic of a diffusion battery, automatic switch valve, and condensation nuclei counter.

gular channels as functions of the diffusion parameter, μ, and the geometrical standard deviation from 1 to 5. Once the data are properly aligned with one of the curves, this method gives a rough estimate of the mean and geometric standard deviation of the diffusion coefficient. Similar curves have been derived for screen-type diffusion batteries.[78] This method does not apply to aerosols, which do not follow lognormal size distributions.

Sinclair[62] used a graphical "stripping" method to estimate the particle size distribution from penetration data through a multi-stage, cylindrical-type diffusion battery. A family of penetration curves has been calculated for monodisperse particles over a range of equivalent lengths. The experimental penetration data are plotted on a different paper of the same scale. The experimental curve is matched against the theoretical curves, and the one having the best fit at the right end of experimental curves (i.e., where penetration is least) is subtracted, leaving a new experimental curve. The process is repeated until the original experimental curve is entirely eliminated. Particle size and fractions of each size in the original aerosol are indicated by the matched theoretical curves and their intercepts with the ordinate of the graph. A similar method has been applied to the screen-type diffusion battery.[79] This method does not assume a certain size distribution and thus is more useful. However, results for both graphical methods depend on judgement in matching curves.

More consistent results can be obtained by using numerical inversion methods. In a diffusion battery, aerosol penetration through stage i, P_i can be expressed mathematically as the integral of the aerosol penetration equation for monodisperse aerosol $P_i(x)$:

$$P_i = \int_0^\infty P_i(x)\,f(x)\,dx \tag{22}$$

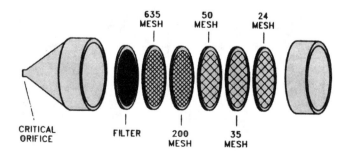

FIGURE 19-12. Schematic of a graded diffusion battery.

where: $f(x)$ = size distribution
$P_i(x)$ = aerosol penetration of size x in stage i

Each observed penetration for stage $i = 1$ through n can be expressed in the form of Equation 22. Several numerical inversion methods have been developed to obtain the aerosol size distribution, $f(x)$. Raabe[80] has developed a nonlinear least square regression to solve Equation 22 under the assumption of a lognormal distribution for $f(x)$. A similar method is used by Soderholm[12] for diffusion battery data analysis. A nonlinear iterative method proposed by Twomey[81] was applied to diffusion batteries by Knutson and Sinclair.[82] Modification of Twomey's method is used for data analysis of screen-type diffusion batteries.[83, 84] An expectation-maximization algorithm has been developed for the screen-type diffusion battery and appears to work as well as or better than the least square regression and Twomey's method.[85] Further recommendations for calculating particle size distribution from data obtained with a diffusion battery can be found in Chapter 6.

Summary and Conclusions

Diffusion denuders are instruments used to separate and collect gas/vapor from particles, and diffusion batteries are used to determine the particle size distribution of ultrafine particles. These instruments are based on similar sampling principles and mathematical formulas. These techniques are important tools in studying ultrafine particles, gases/vapors, and molecular clusters such as radon progeny.

Acknowledgments

The author is grateful for his many colleagues at the Inhalation Toxicology Research Institute who reviewed this chapter. This research was supported by the Office of Health and Environmental Research of the U. S. Department of Energy under Contract No. DE-AC04-76EV01013.

References

1. Nolan, J. J.; Guerrini, V. H.: The Diffusion Coefficient of Condensation Nuclei and Velocity of Fall in Air of Atmospheric Nuclei. Proc. R. Iri. Acad. 43:5–24 (1935).
2. Nolan, J. J.; Nolan, P. J.: Diffusion and Fall of Atmospheric Condensation Nuclei. Proc. R. Iri. Acad. A45:47–63 (1938).
3. Gormley, P. G.; Kennedy, M.: Diffusion from a Stream Flowing Through a Cylindrical Tube. Proc. Roy. Irish Academy A52:163–169 (1949).
4. Cheng, Y. S.: Diffusion Batteries and Denuders. In: Air Sampling Instruments for Evaluation of Atmospheric Contaminants. S.V. Herring, Ed., pp. 406–419. ACGIH, Cincinnati, OH (1989).
5. Cheng, Y. S.: Diffusion and Condensation Techniques. In: Air Measurement, Principles, Techniques and Applications. K. Willeke, and P. A. Baron, Eds., pp. 427–451. Van Nostrand Reinhold, New York (1993).
6. Ingham, D. B.: Diffusion of Aerosols from a Stream Flowing through a Cylindrical Tube. J. Aerosol Sci. 6:125–132 (1975).
7. Sideman, S.; Luss, D.; Peck, R. E.: Heat Transfer in Laminar Flow in Circular and Flat Conduits with (Constant) Surface Resistance. Appl. Sci. Res. A14:157–171 (1965).
8. Davis, H. R.; Parkins, G. V.: Mass Transfer from Small Capillaries with Wall Resistance in the Laminar Flow Regime. Appl. Sci. Res. 22:20–30 (1970).
9. Tan, C. W.; Hsu, C. J.: Diffusion of Aerosols in Laminar Flow in a Cylindrical Tube. J. Aerosol Sci. 2:117–124 (1971).
10. Lekhtmakher, S. O.: Effect of Peclet Number on the Precipitation of Particles from a Laminar Flow. J. Eng. Physics 20:400–402 (1971).
11. Bowen, B. D.; Levine, S.; Epstein, N.: Fine Particle Deposition in Laminar Flow Through Parallel-plate and Cylindrical Channels. J. Colloid Interface Sci. 54:375–390 (1976).
12. Soderholm, S. C.: Analysis of Diffusion Battery Data. J. Aerosol Sci. 10:163–175 (1979).
13. Newman, J.: Extension of the Leveque Solution. J. Heat Transfer 91:177–178 (1969).
14. DeMarcus, W.; Thomas, J. W.: Theory of a Diffusion Battery. Oak Ridge National Laboratory Report ORNL-1413. Oak Ridge, TN (1952).
15. Mercer, T. T.; Mercer, R. L.: Diffusional Deposition from a Fluid Flowing Radially Between Concentric, Parallel, Circular Plates. J. Aerosol Sci. 1:279–285 (1970).
16. Tan, C. W.; Thomas, J. W.: Aerosol Penetration Through a Parallel-plate Diffusion Battery. J. Aerosol Sci. 3:39–43 (1972).
17. Ingham, D. B.: Simultaneous Diffusion and Sedimentation of Aerosol Particles in Rectangular Tubes. J. Aerosol Sci. 7:373–380 (1976).
18. Nolan, P. J.; Kennedy, P. J.: Anomalous Loss of Condensation Nuclei in Rubber Tubing. J. Atmos. Terrestrial Phys. 3:181–185 (1953).
19. Winiwarter, W.: A Calculation Procedure for the Determination of the Efficiency in Annular Denuders. Atmos. Environ. 23:1997–2002 (1989).
20. Fan, B. J.; Cheng, Y. S.; Yeh, H. C.: Gas Collection Efficiency and Entrance Flow Effect of an Annular Diffusion Denuder. Aerosol Sci.Technol. 25:113–120 (1996).
21. Possanzini, M.; Febo, A.; Aliberti, A.: New Design of a High-performance Denuder for the Sampling of Atmospheric Pollutants. Atmos. Environ. 17:2605–2610 (1983).
22. Cheng, Y. S.; Yeh, H. C.: Theory of a Screen-type Diffusion Battery. J. Aerosol Sci. 11:313–320 (1980).
23. Cheng, Y. S.; Keating, J. A.; Kanapilly, G. M.: Theory and Calibration of a Screen-type Diffusion Battery. J. Aerosol Sci. 11:549–556 (1980).
24. Yeh, H. C.; Cheng, Y. S.; Orman, M. M.: Evaluation of Various Types of Wire Screens as Diffusion Battery Cells. J. Colloid Interface Sci. 86:12–16 (1982).
25. Kirsch, A. A.; Fuchs, N. A.: Studies of Fibrous Aerosol Filters-iii. Diffusional Deposition of Aerosols in Fibrous Filters. Ann. Occup. Hyg. 11:299–304 (1968).
26. Kirsch, A. A.; Stechkina, I. B.: The Theory of Aerosol Filtration with Fibrous Filter. In: Fundamentals of Aerosol Science, D. T. Shaw, Ed., pp. 165–256. John Wiley, New York (1978).
27. Cheng, Y. S.; Yeh, H. C.; Brinsko, K. J.: Use of Wire Screens as a Fan Model Filter. Aerosol Sci. Technol. 4:165–174 (1985).
28. Thomas, J. W.: The Diffusion Battery Method for Aerosol Particle Size Determination. J. Colloid Sci. 10:246–255 (1955).
29. Fish, B. R.; Durham, J. L.: Diffusion Coefficient of SO_2 in Air. Environ. Lett. 2:13–21 (1971).

30. Durham, J. L.; Spiller, L. L.; Ellestad, T. G.: Nitric Acid-nitrate Aerosol Measurements by a Diffusion Denuder: A Performance Evaluation. Atmos. Environ. 21:589–598 (1987).

31. Stevens, R. K.; Dzubay, T. G.; Russwurm, G.; Rickel, D.: Sampling and Analysis of Atmospheric Sulfates and Related Species. Atmos. Environ. 12:55–68 (1978).

32. Sioutas, C.; Koutrakis, P.; Wolfson, J.M.: Particle Losses in Glass Honeycomb Denuder Sampler. Aerosol Sci.Technol. 21:137–148 (1994).

33. Appel, B. R.; Tokiwa, Y.; Haik, M.: Sampling of Nitrates in Ambient Air. Atmos. Environ. 15:283–289 (1981).

34. Shaw, R. W.; Stevens, R. K.; Bowermaster, J. et al.: Measurements of Atmospheric Nitrate and Nitric Acid; the Denuder Difference Experiment. Atmos. Environ. 16:845–853 (1982).

35. Forrest, J.; Spandau, D. J.; Tanner, R. L.; Newman, L.: Determination of Atmospheric Nitrate and Nitric Acid Employing a Diffusion Denuder with a Filter Pack. Atmos. Environ. 16:1473–1485 (1982).

36. Ferm, M.: A Na_2CO_3-coated Denuder and Filter for Determination of Gaseous HNO_3 and Particulate NO in the Atmosphere. Atmos. Environ. 20:1193–1201 (1986).

37. Stevens, R. K.: Modern Methods to Measure Air Pollutants. In: Aerosols: Research, Risk Assessment and Control Strategies, S. D. Lee, Ed., pp. 69–95. Lewis Publishers, Chelsea, MI (1986).

38. Johnson, N. D.; Barton, S. C.; Thomas, G. H. S.; et al.: Development of Gas/Particle Fractionating Sampler of Chlorinated Organics. Presented at the 78th Annual Meeting of Air Pollution Control Assocation, Detroit, MI (1985).

39. Cecchini, F.; Febo, A.; Possanzini, M.: High Efficiency Annular Denuder for Formaldehyde Monitoring. Anal. Lett. 18:681–693 (1985).

40. Febo, A.; DiPalo, V.; Possanzini, M.: The Determination of Tetraalkyl Lead Air by a Denuder Diffusion Technique. Sci. Total Environ. 48:187–194 (1986).

41. DeSantis, F.; Perrino, C.: Personal Sampling of Aniline in Working Site by Using High Efficiency Annular Denuders. Ann. Chimica 76:355–364 (1986).

42. Gunderson, E. C.; Anderson, C. C.: Collection Device for Separating Airborne Vapor and Particulates. Am. Ind. Hyg. Assoc. J. 48:634–638 (1987).

43. Ferm, M.: Method for Determination of Atmospheric Ammonia. Atmos. Environ. 13:1385–1393 (1979).

44. Pui, D. Y. H.; Lewis, C. W.; Tsai, C. J.; Liu, B. Y. H.: A Compact Coiled Denuder for Atmospheric Sampling. Environ. Sci. Technol. 24:307–312 (1990).

45. Mori, Y.; Nakayama, W.: Study on Forced Convective Heat Transfer in Curved Pipes. Int. J. Heat Mass Transfer 10:37–59 (1967).

46. Mori, Y.; Nakayama, W.: Study on Forced Convective Heat Transfer in Curved Pipes. Int. J. Heat Mass Transfer 10:681–695 (1967).

47. Durham, J. L.; Ellestad, T. G.; Stockburger, L. et al.: A Transition-flow Reactor Tube for Measuring Trace Gas Concentrations. J. Air Pollut. Control Assoc. 36:1228–1232 (1986).

48. Dasgupta, P. K.: A Diffusion Scrubber for the Collection of Atmospheric Gases. Atmos. Environ. 18:1593–1599 (1984).

49. Dasgupta, P. K.; Dong, S.; Hwang, H. et al.: Continuous Liquid-phase Fluorometry Coupled to a Diffusion Scrubber for the Real-time Determination of Atmospheric Formaldehyde Hydrogen Peroxide and Sulfur Dioxide. Atmos. Environ. 22:946–963 (1988).

50. Durham, J. L.; Wilson, W. E.; Bailey, E. B.: Application of an SO_2 Denuder for Continuous Measurement of Sulfur in Submicrometric Aerosols. Atmos. Environ. 12:883–886 (1978).

51. Braman, R. S.; Shelley, T.; McClenny, W. A.: Tungstic Acid for Preconcentration and Determination of Gaseous and Particulate Ammonia and Nitric Acid in Ambient Air. Anal. Chem. 54:358–364 (1982).

52. Slanina, J.; Lamoen-Doornebal, L. V.; Lingerak, W. A.; Meilof, W.: Application of a Thermo-denuder Analyzer to the Determination of H_2SO_4, HNO_3 and NH_3 in Air. Int. J. Environ. Anal. Chem. 9:59–70 (1981).

53. Buttini, P.; DiPalo, V.; Possanzini, M.: Coupling of Denuder and Ion Chromatographic Techniques for NO_2 Trace Level Determination in Air. Sci. of Total Environ. 61:59–72 (1987).

54. John, W.; Wall, S.M.; Ondo, J.L.: A New Method for Nitric Acid and Nitrate Aerosol Measurement Using the Dichotomous Sampler. Atmos. Environ. 22:1627–1635 (1988).

55. Lindgren, P. F.; Dasgupta, P. K.: Measurement of Atmospheric Sulfur Dioxide by Diffusion Scrubber Coupled Ion Chromatography. Anal. Chem. 61:19–24 (1988).

56. Nolan, P. J.; Doherty, D. J.: Size and Charge Distribution of Atmospheric Condensation Nuclei. Proc. R. Ir. Acad. 53A:163–179 (1950).

57. Pollak, L. W.; O'Conner, T. C.; Metnieks, A. L.: On the Determination of the Diffusion Coefficient of Condensation Nuclei Using the Static and Dynamic Methods. Geofis. Pura Applicata 34:177–194 (1956).

58. Megaw, W. J.; Wiffen, R. D.: Measurement of the Diffusion Coefficient of Homogeneous and Other Nuclei. J. Rech. Atmos. 1:113–125 (1963).

59. Rich, T. A.: Apparatus and Method for Measuring the Size of Aerosols. J. Rech. Atmos. 2:79–85 (1966).

60. Pollak, L. M.; Metnieks, A. I.: New Calibration of Photoelectric Nucleus Counters. Geofis. Pura Appl. 43:285–301 (1959).

61. Kotrappa, K.; Bhanti, D. P.; Dhandayutham, R.: Diffusion Sampler Useful for Measuring Diffusion Coefficients and Unattached Fractions of Radon and Thoron Decay Products. Health Phys. 29:155–162 (1975).

62. Sinclair, D.: Measurement and Production of Submicron Aerosols. In: Proceedings of the 7th International Conference on Condensation and Ice Nuclei, pp. 132–137, Prague, Vienna (1969).

63. Breslin, A. J.; Guggenheim, S. F.; George, A. C.: Compact High Efficiency Diffusion Batteries. Staub-Rein. Luft 31(8):1–5 (1971).

64. Scheibel, H. G.;. Porstendörfer, J.: Penetration Measurements for Tube and Screen-type Diffusion Batteries in the Ultrafine Particle Size Range. J. Aerosol Sci. 15:673–682 (1984).

65. Sinclair, D.: A Portable Diffusion Battery. Am. Ind. Hyg. Assoc. J. 33:729–735 (1972).

66. Brown, K. E.; Beyer, J.; Gentry, J. W.: Calibration and Design of Diffusion Batteries for Ultrafine Aerosols. J. Aerosol Sci. 15:133–145 (1984).

67. Sinclair, D.; Hinchliffe, L.: Production and Measurement of Submicron Aerosols. In: Assessment of Airborne Particles, T. T. Mercer et al., Eds., pp. 182–199. C. C. Thomas, Springfield, IL (1972).

68. Twomey, S. A.; Zalabsky, R. A.: Multifilter Technique for Examination of the Size Distribution of the Natural Aerosol in the Submicrometer Size Range. Environ. Sci. Technol. 15:177–184 (1981).

69. Sinclair, D.; Hoopes, G. S.: A Novel Form of Diffusion Battery. Am. Ind. Hyg. Assoc. J. 36:39–42 (1975).

70. Cheng, Y. S.; Yeh, H. C.; Mauderly, J. L.; Mokler, B. V.: Characterization of Diesel Exhaust in a Chronic Inhalation Study. Am. Ind. Hyg. Assoc. J. 45:547–555 (1984).

71. Holub, R. F.; Knutson, E. O.; Solomon, S.: Tests of the Graded Wire Screen Technique for Measuring the Amount and Size Distribution of Unattached Radon Progeny. Radiat. Protect. Dosim. 24:265–268 (1988).

72. Reineking, A.; Porstendörfer, J.: High-volume Screen Diffusion Batteries and α-spectroscopy for Measurement of the Radon Daughter Activity Size Distributions in the Environment. J. Aerosol Sci. 17:873–880 (1986).

73. Ramamurthi, M.; Hopke, P. K.: An Automated, Semicontinuous System for Measuring Indoor Radon Progeny Activity-weighted

Size Distributions, dp: 0.5-500 nm. Aerosol Sci. Technol. 14:82–92 (1991).

74. Cheng, Y. S.; Su, Y. F.; Newton, G. J.; Yeh, H. C.: Use of a Graded Diffusion Battery in Measuring the Radon Activity Size Distribution. J. Aerosol Sci. 23:361–372 (1992).

75. Solomon, S.; Ren, T.: Counting Efficiencies for Alpha Particles Emitted from Wire Screens. Aerosol Sci. Technol. 17:69–83 (1992).

76. Fuchs, N. A.; Stechkina, I. B.; Starosselskii, V. I.: On the Determination of Particle Size Distribution in Polydisperse Aerosols by the Diffusion Method. Br. J. Appl. Phy. 13:280–281 (1962).

77. Mercer, T. T.; Greene, T. D.: Interpretation of Diffusion Battery Data. J. Aerosol Sci. 5:251–255 (1974).

78. Lee, K. W.; Connick, P. A.; Gieske, J. A.: Extension of the Screen-type Diffusion Battery. J. Aerosol Sci. 12:385–386 (1981).

79. Sinclair, D.; Countess, R. J.; Liu, B. Y. H.; Pui, D. Y. H.: Automatic Analysis of Submicron Aerosols. In: Aerosol Measurement, W. E. Clark and M. D. Durham, Eds., pp. 544–563. University Press of Florida, Gainesville, FL (1979).

80. Raabe, O. G.: A General Method for Fitting Size Distributions to Multi-component Aerosol Data Using Weighted Least-squares. Environ. Sci. Technol. 12:1162–1167 (1978).

81. Twomey, S.: Comparison of Constrained Linear Inversion and an Alternative Nonlinear Algorithm Applied to the Indirect Estimation of Particle Size Distribution. J. Comput. Phys. 18:188–200 (1975).

82. Knutson, E. O.; Sinclair, D.: Experience in Sampling Urban Aerosols with the Sinclair Diffusion Battery and Nucleus Counter. In: Proc. Advances in Particle Sampling and Measurement, W. B. Smith, Ed., pp. 98–120. EPA 600/7-79-065. US EPA, Washington DC (1979).

83. Kapadia, A.: Data Reduction Techniques for Aerosol Size Distribution Measurement Instruments. Ph.D. Thesis, University of Minnesota (1980).

84. Cheng, Y. S.; Yeh, H. C.: Analysis of Screen Diffusion Battery Data. Am. Ind. Hyg. Assoc. J. 45:556–561 (1984).

85. Maher, E. F.; Laird, N. M.: EM Algorithm Reconstruction of Particle Size Distributions from Diffusion Battery Data. J. Aerosol Sci. 16:557–570 (1985).

Instrument Descriptions

19-1. Screen Diffusion Battery, Model 3040

TSI Incorporated

The TSI diffusion battery has 10 stages with stainless steel 635 mesh screens. The screen holder is made of either stainless steel or aluminum. The flow rate through the diffusion battery is between 4 and 6 L min^{-1}.

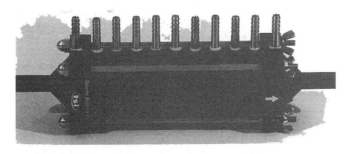

INSTRUMENT 19-1. Screen diffusion battery, TSI Model 3040.

INSTRUMENT 19-2. Parallel flow diffusing battery.

The useful particle size range is from 0.003 to 0.5 μm. The aerosol concentration in each stage is usually measured by a continuous flow condensation nucleus counter (Model 3022 or 3025), and an automatic switch valve (Model 3042) is used to measure the concentration in successive stages automatically. Dimensions: 25 cm × 6.3 cm × 9 cm.

19.2 Parallel Flow Diffusion Battery

In-Tox Products

The parallel flow diffusion battery utilizes the principle of screen diffusion battery and parallel flow. The unit is made of aluminum. It consists of a conical cap and a collection section containing seven cells. Each diffusion cell contains a different number of stainless steel 200 mesh screens followed by a 25-mm Zefluor filter (Gelman, Ann Arbor, MI). The seven cells typically contain 0 through 35 screens. Critical orifices provide a 2 L min^{-1} flow rate through each cell resulting in a total flow rate of 14 L min^{-1}. Gravimetric determination of collected filter samples from each cell provides the direct mass penetration as a function of screen number for the determination of aerosol size distribution. Dimensions: 23 cm × 23 cm × 23 cm.

19-3. Annular Denuders

URG Incorporated

The company has several types of annular denuders designed to be used alone for collecting gases or to be used in a series with a filter, impactor, or cyclone for collecting ambient and indoor aerosols as well as acidic

INSTRUMENT 19-3. Annular denuders from URG showing single-channel and three-channel glass denuders.

TABLE 19-I-1. Commercial Sources

Symbol	Source
TSI	500 Cardigan Rd. St. Paul, MN 55164 (800) 677-2708/(651) 483-0900 FAX: (651) 490-2748 EMAIL: *info@tsi.com* *http://www.tsi.com*
INT	In-Tox Products P.O. Box 2070 Moriarty, NM 87035 (505) 832-5107 FAX: (505) 832-5092 EMAIL: *intoxpd@aol.com* *http://www.intoxproducts.com*
URG	URG Corporation 116 S. Merritt Mill Rd. Chapel Hill, NC 27516 (919) 942-3522

gases. Model URG-2000-30B is a glass denuder with a length of 242 mm and a diameter of 30 mm. It has 1 mm annular space. Inside glass surfaces are etched to provide greater surface area for coating.

URG's Multichannel Annular Denuders are composed of concentric glass tubes. The inner surface of these tubes is etched to provide greater surface area for coating. They are sealed into a Teflon-coated stainless steel sheath. Air flow is laminar through the annular denuders, thus ensuring that the gases separate from the fine particles. The annular denuder is coated with a chemical sink that allows for capture of the species of interest. Acidic gases, basic gases and organic gases are denuded from the air stream and the subsequent rinse can be analyzed. URG offers a variety of multichannel glass and stainless steel annular denuders available in different lengths.

The URG-2000-25A is a personal monitor for ambient, indoor air and personal sampling. The 25A consists of a size-selective impactor inlet, which separates coarse particles from fine particles by impaction, a filter pack, and PUF cartridge. The unit is housed in rugged, high-density polyethylene. This personal monitor operates at 4 Lpm with a cut point of 1(m, 2.5(m, or 10(m. Particles, on appropriate filter media, are analyzed for organic and elemental carbon, anions, cations, and trace metals. The PUF is analyzed for PAH's, pesticides, SVOC's and PCB's.

An optional URG-2000-25F consists of a size-selective inlet (1μm, 2.5μm or 10μm) followed by a filter pack in a rugged, high-density polyethylene housing. Both units can be used with Tissuquartz or Teflon filter media. The units require a 4 Lpm portable sampling pump to operate.

Chapter 20

Sampling from Ducts and Stacks

Dale A. Lundgren, Ph.D., P.E.[A] , Charles E. Billings, Ph.D.[B]

[A]Environmental Engineering Sciences, University of Florida, 408 Black Hall, Gainesville, Florida 32611;

[B]Massachusetts Institute of Technology, 77 Massachusetts Avenue, Room 20C-204, Cambridge, MA 02139

CONTENTS

Introduction

Sampling of a flowing gas stream has a number of important applications in industrial hygiene and environmental health. The purposes of this chapter are 1) to develop the general principles and procedures for those with little previous experience (basic considerations) and 2) to review current techniques used primarily for air pollution control.

The objectives of gas stream sampling are 1) to obtain a representative sample (or specimen) from a flowing gas stream and 2) to determine flow characteristics, fluid composition, or properties of constituents.

The objective of the sampling activity must be clearly identified and correctly stated because it will constrain selection and application of sampling methods. Typical purposes include:

1. The measurement of air pollution source emission rates for specific constituents (e.g., gases or particulate matter).

2. Evaluation of total collection efficiency and pressure drop of a gas cleaning device.

3. Measurement of source emission rate by species and spectra (e.g., particle size distribution, or an individual substance such as lead).

4. Evaluation of the particle size-efficiency of a control device or system.
5. Research and development on processes, apparatus, or methods.

One common application of gas stream sampling is to determine stack discharge (emission) rates or concentrations from an industrial air pollution source. Data to be obtained in this case include gas composition, temperature, pressure, gas velocity, volumetric flow rate, nature and concentration of contaminants, and process information related to emissions. The purpose is to provide valid information to management with respect to legally allowable contaminant discharge concentration or rate; data are also provided for evaluating the performance of control devices.

This procedure is typically called air pollutant source emission sampling (or source sampling, stack sampling, stack testing, emission testing, or source evaluation). During the past 30 years, and especially since the formation of the U.S. Environmental Protection Agency (U.S. EPA) in 1971, highly specific methods of manual gas stream sampling have been developed and promulgated in regulations for individual industrial source emission categories. These methods will be discussed in more detail below.

Principles

Essential Aspects

Selection of the general method to be followed (i.e., the apparatus and procedures) includes consideration of the system, flow, and contaminants that interact to affect selection of a particular sampling method and where it is best applied to the flowing gas system to obtain valid (i.e., statistically, physically, and chemically representative) samples of components. Information on the general process and specific operation under consideration must be obtained. Such information includes process flow sheets; nature, composition, and quantities of materials in the operation and their phases; flow rates; possible plans and specifications; principal dimensions; whether the process is continuous or steady-state (e.g., enclosed or automated) or cyclic, intermittent, or otherwise non-continuous (e.g., batch, open, or manual); and details on the control system design (plans and specifications on hoods, ductwork, fans, collectors, and stack design). This background information is used to plan the sampling activity.

For many air pollutant sources, the U.S. EPA now requires continuous sampling and analysis of emissions,

and this may be done by *in-situ* procedures. The usual method for industrial hygiene and air pollution control sampling (and for process sampling) involves manual extraction of a sample of the fluid (gas) and separation of the components of interest for subsequent analysis either directly at the site or in an analytical laboratory. The following discussion refers specifically to extractive sampling for components and contaminants of the fluid (gas) stream.

Nature of Flow and Selection of Sampling Location

To select an appropriate location for introduction of extractive sampling apparatus, consideration must be given to the effects on velocity and concentration profiles caused by the gas flow system (i.e., ductwork, elbows, tees or junctions, fans, collectors, expansions, contractions, valves, gates, dampers, and flowmeters) and by the velocity profile of the flow stream. Each flow system component produces specific effects on the gas flow pattern transverse to the flow at each location in the stream, and the velocity and concentration profiles are modified by each of the components in the ductwork, at points immediately downstream.

Illustrations of these effects have been presented from field studies in breechings of coal-fired boilers.[1] Figure 20-1 shows an example of velocity profiles and particle concentration after a junction and elbow. Velocities and particle concentrations (actually flux or mass/area-time) are shown here as the ratio of the value determined at a point (an equal area traverse point, as will be discussed below) to the numerical average value, across the whole flue, such that the concentration or velocity at each point is expressed as its ratio to the overall average. Hawksley et al., in discussing effects illustrated in Figure 20-1, state:[1]

Figure [20-1] gives data at sampling positions after [a] 90° bend. Solids are centrifuged to the outside of the bend but the uniformity is gradually restored by turbulent mixing, as shown, although in this instance the pattern is complicated by an unequal supply of solids from the two i.d. fans. The degree of uniformity obtained at two to six diameters from a bend is shown. About three to five diameters is sufficient to establish a tolerable uniformity, although this may not be the case when the quantity of grit is high. When there is little grit, one to three diameters is adequate. Even at the bend or within one diameter or so of it [Figure 20-1] the local mass flows do not vary greatly. The direction of variation lies in the plane of the bend and the effect of previous bends is not

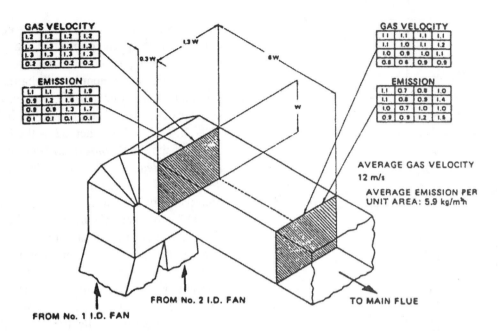

FIGURE 20-1. Measured velocity and concentration profiles after a bend.[1] Average gas velocity, 39 ft/sec (11.9 m/s); average emission per unit area = 1.2 lb/ft²-hr (1.6 g/m² -sec); equivalent round diameter, D_e = 2LW/(L+W) = 6.8 ft (2.1 mm); distance between planes 1 and 2 = 36/6.8 = 5.4D.

usually apparent . . . Gentler bends tend to give less centrifuging of solids. . . .

If the velocity is high and the bend sharp, the gas flow may separate from the inner wall of the bend and not become reattached until one or two flue diameters downstream from the bend. The fluid in the dead space will tend to circulate in a large eddy, the direction of rotation being forwards near the main stream and backwards near the wall. The circulation velocity is not high but may be sufficient for the reversed flow to be detected by means of a Pitot tube. The solids flow is contained in the main stream and the emission can be measured in the presence of a reversed flow if it is possible to define the effective cross-sectional area of the main stream. Separation may occur if a flue diverges too rapidly; it arises also, with the shedding of free eddies or vortices, after obstacles such as the blades of dampers. A converging flue tends to produce a more uniform distribution of gas and solids flow.

There is usually a steep gradient of solids flow immediately after a fan. . . and the solids may be displaced also to the side of the flue opposite to the inlet of the fan. But depending on the design features of the fan, the distribution may not be markedly non-uniform. . . .

One widely used experimental procedure for laboratory and pilot plant studies to produce uniform (flat)

velocity and concentration profiles at 5 duct diameters downstream is the annular orifice or Stairmand disc (D_{disc} = 0.707D_{duct}). This device works best at duct velocities of the order of 2000–4000 fpm (10–20 m/s), at the cost of an unrecoverable pressure drop of about one velocity head. It can be mounted on a single axial shaft and rotated out of the stream when not in use. With suitable static taps (one duct radius upstream and downstream of the obstruction), it can be used as a conventional total flowmeter. Because of the pressure loss penalty, it has not found wide use in process, fuel, or waste gas streams of high volume flow rate.

Selection of a suitable location for sampling thus involves judgment based on experience with flow phenomena. As a general rule, sampling should be done at least 5–8 diameters downstream from a disturbance and at least 2 diameters upstream from one. A selection guideline from the U.S. EPA standard method is discussed below when these criteria cannot be met.

Selection of Number and Arrangement of Traverse Points

To obtain a representative sample of a fluid property, substance, or characteristic that varies across the duct (flue, etc.), a series of samples is taken at a number of points in an equal area traverse in a plane transverse to flow. The total cross-sectional area of the duct is divided into a number of equal areas, and samples are taken at locations which best represent the center of the smaller

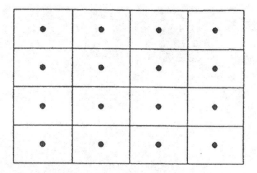

Example showing rectangular stack cross section divided into 16 equal areas, with a traverse point at centroid of each area.

Number of Traverse Points	Matrix Layout
9	3 × 3
12	4 × 3
16	4 × 4
20	5 × 4
25	5 × 5
30	6 × 5
36	6 × 6
42	7 × 6
49	7 × 7

FIGURE 20-2a. U.S. EPA cross-section layout for rectangular stacks.

areas (centroid). In the case of square or rectangular ducts, the equal areas will be squares or rectangles, and samples are taken from the center of each of them. For round ducts, the smaller areas are concentric circles, and samples are commonly taken on two perpendicular diameters, although segments may be used. The procedure required by the U.S. EPA for dividing a rectangular or round duct into smaller areas is illustrated in Figure 20-2a and b.[2] Selection of the appropriate number of points depends upon the size of the duct and an estimate of degree of disturbance of the flow and expected concentration profiles (i.e., distance to upstream and downstream disturbances).

The U.S. EPA uses a selection guide that adjusts the minimum number of sampling points in a traverse according to distance of the sampling plane with respect to upstream or downstream disturbance, increasing the number above 12 when the traverse or sampling plane is closer than 8 diameters downstream from the disturbance or 2 diameters upstream, as shown in Figure 20-3.[2] The U.S. EPA selection rule requires a smaller number of sampling points for velocity and nonparticulate sampling traverses as contrasted to the number required for particulate sampling traverses. Smaller numbers of traverse points are also required in small ducts (< 24 in. in diameter [< 610 mm]) as shown by the lower curves in Figure 20-3a and 20-3b. Use of the U.S.

EPA selection rules may lead to a conservatively high number of sampling points for many situations if the flow is reasonably uniform and constituents are well mixed.[3] The equivalent round diameter (D_e) of square and rectangular ducts used to determine the required number of sampling points is given in Figure 20-1 (viz., $D_e = 2LW/[L + W]$; where: L = the height and W = the width of the duct). Industrial Ventilation—A Manual of Recommended Practice[4] provides tables (9-2, 9-3, and 9-4) that give actual dimensions for 6, 10, and 20 velocity traverse points for duct diameters from 3 to 80 in. (80 to 2000 mm), as summarized in Table 20-1.

After choosing the sampling plane or site and determining the number of sampling points, openings are made in the duct wall to permit the apparatus to be inserted. For lighter gauge ventilation ductwork under slight negative pressure, a hole-saw is used to cut a suitable round hole (typically 1.5 to 3 in. diameter [38 to 76 mm]). For heavier ductwork, (>16 gauge; flues, breeching, stacks, etc.), a 3- or 4-in.-diameter (76- or 101-mm-diameter) coupling (or short nipple) is welded onto the wall, and a round hole is flame-cut out of the wall. Access to the sampling ports, platforms, ladders, rails, toeboards, jib booms, and other fixtures and utilities at the sampling location are arranged as required (e.g., temporary scaffolding and extension cords versus fabricated platforms and ladders welded in place). (See the Occupational Safety and Health Administration [OSHA] regulations [29 CFR 1910 and 1926] for proper designs.)

Measurement of Velocity Profile and Gas Flow Rate

Composition, temperature, and pressure of the gas are determined by methods appropriate to the particular industry and gas stream. Standard guides for these determinations are provided in the U.S. EPA methods for specific industrial stack emission sources,[2] discussed below.

Temperature and static pressure are measured in the duct at the sampling point with a thermometer and static pressure tap, respectively.[4] If the operation (or process) involves combustion, flue gas analysis is performed for oxygen, carbon dioxide, and, if suspected, carbon monoxide, by standard Orsat Analysis or Fyrite® analyzers.[3, 5] For process gases, gas analysis, temperature, and pressure are determined by individual industry or company practices. Methods for specific industrial air pollution source emissions to the atmosphere are also presented below.[2] Water vapor will be present in combustion gases to the extent of 5% to 10%, depending on the amount of excess air, hydrogen, and water in fuel (e.g., gas versus coal or oil). Water scrubber outlets contain water vapor up to

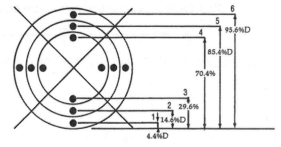

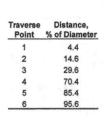

Traverse Point	Distance, % of Diameter
1	4.4
2	14.6
3	29.6
4	70.4
5	85.4
6	95.6

Example showing circular stack cross section divided into 12 equal areas, with location of traverse points indicated (6 points on a diameter).

Traverse Point No. on a Diameter	Number of Traverse Points on a Diameter											
	2	4	6	8	10	12	14	16	18	20	22	24
1	14.6	6.7	4.4	3.2	2.6	2.1	1.8	1.6	1.4	1.3	1.1	1.1
2	85.4	25.0	14.6	10.5	8.2	6.7	5.7	4.9	4.4	3.9	3.5	3.2
3	—	75.0	29.6	19.4	14.6	11.8	9.9	8.5	7.5	6.7	6.0	5.5
4	—	93.3	70.4	32.3	22.6	17.7	14.6	12.5	10.9	9.7	8.7	7.9
5	—	—	85.4	67.7	34.2	25.0	20.1	16.9	14.6	12.9	11.6	10.5
6	—	—	95.6	80.6	65.8	35.6	26.9	22.0	18.8	16.5	14.6	13.2
7	—	—	—	89.5	77.4	64.4	36.6	28.3	23.6	20.4	18.0	16.1
8	—	—	—	96.8	85.4	75.0	63.4	37.5	29.6	25.0	21.8	19.4
9	—	—	—	—	91.8	82.3	73.1	62.5	38.2	30.6	26.2	23.0
10	—	—	—	—	97.4	88.2	79.9	71.7	61.8	38.8	31.5	27.2
11	—	—	—	—	—	93.3	85.4	78.0	70.4	61.2	39.3	32.3
12	—	—	—	—	—	97.9	90.1	83.1	76.4	69.4	60.7	39.8
13	—	—	—	—	—	—	94.3	87.5	81.2	75.0	68.5	60.2
14	—	—	—	—	—	—	98.2	91.5	85.4	79.6	73.8	67.7
15	—	—	—	—	—	—	—	95.1	89.1	83.5	78.2	72.8
16	—	—	—	—	—	—	—	98.4	92.5	87.1	82.0	77.0
17	—	—	—	—	—	—	—	—	95.6	90.3	85.4	80.6
18	—	—	—	—	—	—	—	—	98.6	93.3	88.4	83.9
19	—	—	—	—	—	—	—	—	—	96.1	91.3	86.8
20	—	—	—	—	—	—	—	—	—	98.7	94.0	89.5
21	—	—	—	—	—	—	—	—	—	—	96.5	92.1
22	—	—	—	—	—	—	—	—	—	—	98.9	94.5
23	—	—	—	—	—	—	—	—	—	—	—	96.8
24	—	—	—	—	—	—	—	—	—	—	—	98.9

FIGURE 20-2b. U.S. EPA location of traverse points in circular stacks (percent of stack diameter from inside wall to traverse point).[2]

saturation at the temperature of the outlet gas and may equal from 15% (outlet temperature of 130°F [55°C]) to 70% (outlet temperature of 195°F [90°C]). Kilns, dryers, and refuse incinerators may have outlet gas moisture contents from 10% to 30% depending on fuel, material, moisture content, excess air, etc. Initially, estimation of moisture content may be based on experience and judgment; then, moisture content can be measured during sampling by condensation in a cooled impinger followed by silica gel (the water volume is measured and the silica gel is weighed). Conventional psychrometric wet bulb and dry bulb thermometers are used under certain conditions, i.e., relatively clean air at moderate temperatures.[6] Knowing the molecular composition and the moisture content of the gas, density is calculated at the stack temperature and pressure using the fractional composition and gas laws. Details of this procedure can be found in the literature.[2, 3]

Velocity pressure (h_v) at each traverse point is measured using a Pitot tube of standard design for relatively particulate-free gases or using reverse-impact-type tubes for dusty gas (Figures 20-4a through 4c).[2, 4] Each leg of the Pitot tube is connected by rubber tubing to a pressure measurement device. Effects of flow characteristics (e.g., turbulence, yaw, or vorticity), wall proximity, viscosity, and related problems affect Pitot tube measurements.[2, 7] The standard pitot-static tube design has a calibration coefficient near 1 for most typical air flows, but it is not exactly constant.[7] In general, the reverse-impact tube has a calibration coefficient of $Kp \approx 0.84$, from the induced negative pressure due to the wake downstream of the reverse-static opening (i.e., it will produce a slightly higher h_v, about 20%). This device should be calibrated in about the same conditions in which it will be used and used in the same orientation as it was calibrated.[2] Calculations for velocity are indicated elsewhere.

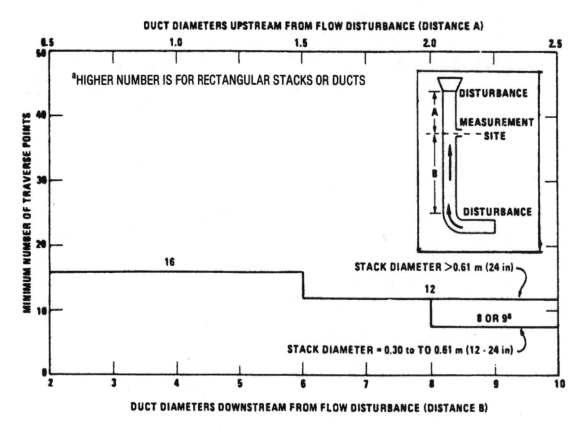

FIGURE 20-3a. Minimum number of traverse points for velocity (nonparticulate) traverses.[2]

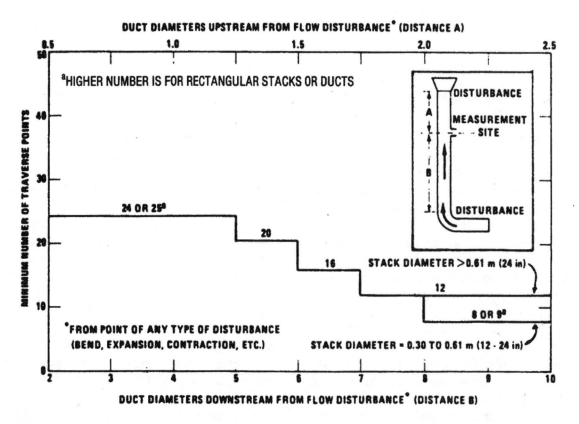

FIGURE 20-3b. Minimum number of traverse points for particulate traverses.[2]

TABLE 20-1. Selection of Number of Velocity Traverse Points [4]

Diameter			Cross-sectional Area		Number of Test Points
in	ft	mm	ft²	m²	
3–6	0.25–0.5	76–152	0.05–0.2	0.005–0.0186	6
4–48	0.3–4	102–1219	0.09–12.6	0.008–1.17	10
40–80	3.3–6.7	1016–2032	8.7–34.9	0.81–32.4	20

For practical purposes in the field, with normal fluctuations in flow, the standard Pitot tube or the reverse-impact-type tube can be used with an inclined manometer down to about 0.1 in. (2 mm) of water (i.e., 1.0 in. [25 mm] of displacement on a 10:1 inclined manometer) and can measure velocities down to about 1000 fpm ($\approx$5 m/s) with a small error.[7] To measure lower velocities, a variety of more sensitive devices are available.[7] Further details on measurement of air velocity and instruments are presented in Chapter 9 of *Industrial Ventilation—A Manual of Recommended Practice*.[4] Velocities at all traverse points are summed and averaged to determine the average velocity at the cross section.[2, 4, 6] The total volumetric flow rate is the product of the average velocity and the cross-sectional area at the plane of measured velocity.

Special instruments or procedures are required for measurement of flow characteristics and properties in the general regime of gas dynamics, which includes high Reynolds number flow ($Re > 10^6$), non-negligible Mach number ($M > 0.1$), or where sufficient energy is present in the flow to cause important interactions with probes. These situations include high velocities (>25,000 fpm [>127 m/s]), compressible flow, shock flows, and high temperatures and pressures.[7] Other flow phenomena requiring special procedures include measurement of turbulence structure and intensity. A recommended method to detect the presence of a rotary component of flow (vortical, spiralling flow) is described in the U.S. EPA Method 1.[2] Cyclic or random pulsations in flow also require special consideration.

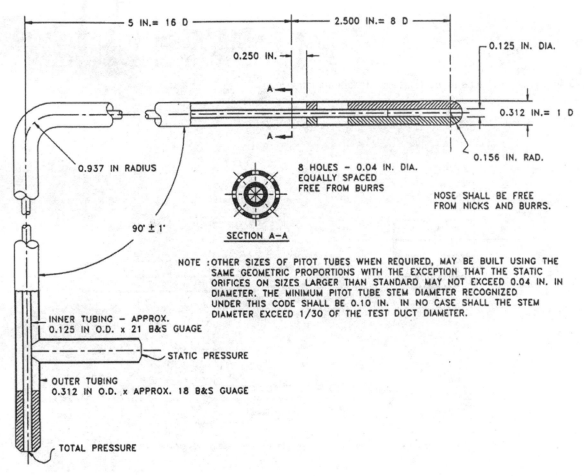

FIGURE 20-4a. Standard Pitot tube for duct gas velocity determination.[4]

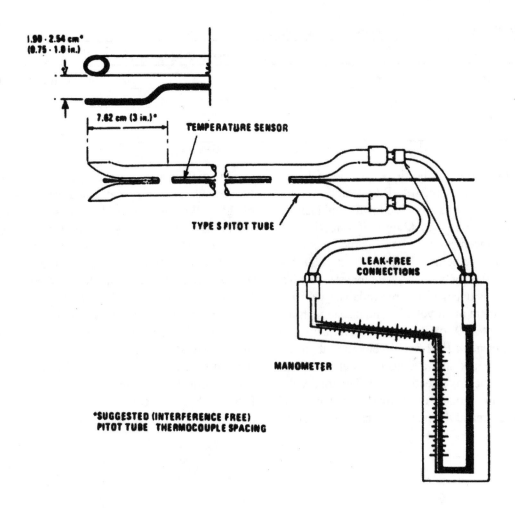

FIGURE 20-4b. Type S Pitot tube for duct gas velocity determination.[2]

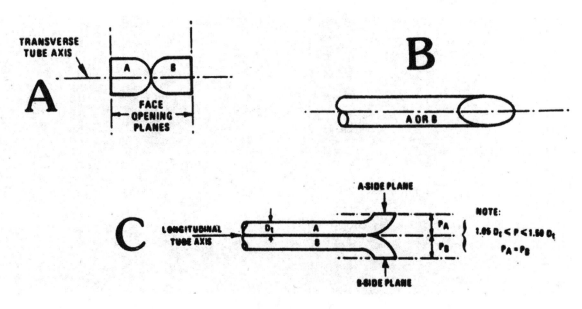

FIGURE 20-4c. Type S Pitot tube. A: end view; B: side view; C: top view.[2]

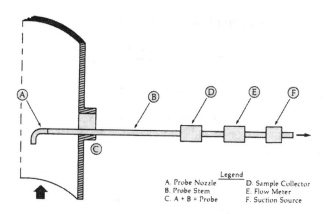

FIGURE 20-5a. Gas stream sampling train schematic: out-stack collector.

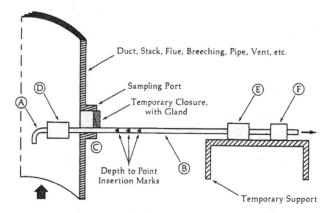

FIGURE 20-5b. Gas stream sampling train schematic: in-stack collector.

Sampling and Analysis

General Considerations

The most general form of an extractive gas sampling system is shown in Figure 20-5a. A sampling probe is connected to a sample collection device followed by a flowmeter and a source of suction. If only well-mixed gases are to be extracted, a straight pipe may be inserted and used as the sampling probe. Sampling of gases from process streams flowing under pressure may be accomplished from a stub pipe or short nipple with a gate valve connected to the main line. When the valve is opened, a sample is obtained in any suitable container or collector (e.g., evacuated flask or Mylar® bag).

When sampling particulate material from the stream, a curved tip is added to point into the oncoming stream to sample correctly as shown at (A) in Figure 20-5a. Particulate matter will deposit on the interior surfaces of the probe nozzle (A) and probe stem (B). This material must be cleaned out and included in the total amount caught in the collector (D) to obtain a reliable estimate of gas stream concentration. In the case of dry granular materials, particulate matter may be brushed or washed out of the nozzle and stem easily. In other situations where condensation or reaction occurs, such as with oil smokes, asphalt fumes, or other tarry or sticky materials, cleaning the probe becomes more of a chore and may require washing with suitable solvents.

To overcome particulate deposition problems and to eliminate condensation effects on the collected sample in hot moist gases, the collector may be mounted on the end of the probe stem and inserted into the stream, as shown in Figure 20-5b. A nozzle is added to face into the stream to properly sample for particulate matter. The order of the major components is nozzle, collector, probe, flowmeter, and suction source. Other auxiliary

apparatus shown in Figure 20-5b include a closure with gland to seal the sampling port (necessary if the stream pressure is positive to the outside, to reduce exposures of sampling personnel or protect the process stream) and a temporary support to hold the sampling train steady while a sample is obtained at each traverse point in the stack, duct, flue, breeching, pipe, etc. Other auxiliary apparatus not shown may include provision for a jacketed probe to be heated (or cooled) to maintain the sample temperature to the collector, moisture or other condensible collectors, other flowmeters, stack Pitot-static tube, stack thermocouple, thermometers in sample stream at collector and at flowmeter, pressure gauges and manometers, and flow volume totalizer. These apparatus are illustrated below (e.g., see "U.S. EPA Appendix Methods 1–5 and 17," Figure 20-11). Advantages and disadvantages associated with collector location are given in Table 20-2. Typical stack sampling collector, flowmeter, and pump alternatives are illustrated schematically in Figure 20-6. The "Instrument Descriptions" section at the end of this chapter describes these major components and indicates suppliers.

Isokinetic Sampling for Particulate Matter

Because aerosol particles have an inertial behavior different than the gas in which they are suspended, a representative sample must be extracted from a flowing gas stream at the stream velocity. That is, the nozzle tip opening area A_n, (ft^2 or mm^2) and sample volumetric flow rate Q_m, (cfm or L/min at stack *temperature and pressure* conditions) must be adjusted to obtain a velocity $V_n = Q_m/A_n$ equal to the gas stream velocity, V_s, at the point of sampling. The sampling constraint $V_n = V_s$ is called "isokinetic" or "equal-velocity" sampling. Because V_s varies across the transverse section at the sampling location (and has been determined by the Pitot traverse above), the sampling volume flow rate, Q_m, is

TABLE 20-2. Advantages and Disadvantages of In-stack versus Out-stack Collector Location

Advantages

In-stack:
1. Immediate collection in or close to gas stream.
2. No deposition in probe.
3. No condensation in collector.
4. Smaller equipment generally required.

Out-stack:
1. Large sample volume may be taken (long time).
2. Large holding capacity in collector.
3. Smaller and simpler probe design (small stack hole).
4. Easier to change collector or remove sample.
5. Sample can be cooled before collection.
6. Less likelihood of sample loss.
7. More flexibility in choice of collector.
8. Optimum collector velocity can be used.
9. Sample volume may be metered before collector, if desired.

Disadvantages

In-stack:
1. Choice of collector limited.
2. Sampling volume limited.
3. Holding capacity smaller.
4. Larger stack hole.
5. Stack suction may cause some sample loss.
6. Optimum collector velocity may be exceeded.
7. Sample may be more difficult to remove.

Out-stack:
1. Deposition of material occurs in probe.
2. Condensation may occur in probe or collector.
3. Probe cleanout required between samples.
4. Larger equipment may be required, such as heated sampling filter box, heated probe, etc.

varied as the sampling probe nozzle tip is sequentially located at each of the sampling points in the traverse. A sampling probe nozzle diameter D_n, (in. or mm) is selected to yield the appropriate velocities with a knowledge of the sampling pump volume flow rate capability. Typically, 1.0 ft³/min (28.3 L/min) sampling volume flow rate is used because this rate is within the capability of most portable vane-type vacuum pumps with a 1/4- or 1/3-HP electric motor.

Typical calculations for determination of the isokinetic sampling nozzle tip size are derived from the requirement $V_n = V_s$. Assume that the total mass concentration is to be determined, that the total flow is steady, the temperature and pressure are near ambient, and the process that generates the particulate matter is continuous. When the velocity at one traverse point in the duct has been measured to be 3000 fpm (15.24 m/s), and it is desired to sample at 1.0 cfm (28.3 L/min), the appropriate nozzle tip size and sampling volume flow rate is determined as follows:

$$V_n A_n = (1.0 \text{ cfm})(144 \text{ in.}^2 / \text{ft}^2), \text{ or} \qquad (1)$$

$$= 28.3 \text{L/min}\left(16.67 \frac{\text{m/s}}{\text{L/m}^3}\right)(\text{metric units})$$

$$V_n = V_s = 3000 \text{ ft/min, or} \qquad (2)$$

$$15.24 \text{ m/s (metric units), then}$$

$$A_n = 1 \times \frac{144}{3000} = 0.048 \text{ in.}^2, \text{ or} \qquad (3)$$

$$= 28.3 \times \frac{16.67}{15.24} = 31.7 \text{ mm}^2 \text{ (metric units)}$$

The diameter of the nozzle tip is therefore:

$$D^2 = 4\frac{A_n}{\pi} = 0.061 \text{ in.}^2, \text{ or} \qquad (4)$$

$$= 4\frac{A_n}{\pi} = 40.3 \text{ mm}^2 \text{ (metric units)}$$

$$D_n = 0.25 \text{ in., or} \qquad (5)$$

$$= 6.35 \text{ mm (metric units)}$$

There are various practical precautions in the gas stream sampling literature regarding minimum size of probe tip, particularly for larger particles at higher concentrations (e.g., stoker-fired coal fly ash grits), but as a general rule, 0.25 in. (6 mm) or larger is a reasonable size for most field situations.

If the temperature (or pressure) in the gas stream is substantially different from the temperature and pressure of the sample flowmeter and pump, temperature and pressure corrections to gas volume must be made in accordance with the perfect gas laws (Boyle's Law, Charles' Law). Temperature and pressure changes cause a change in volume (a reduction as temperature drops) as the sample is drawn out of the stack and passes through the collector. The temperature and pressure at the flowmeter is measured, and the volume corrected back to stream conditions, to achieve isokinetic conditions. In addition, the orifice flowmeter has a characteristic performance equation:

$$Q_n = K\left(\frac{\Delta h}{\rho}\right)^{1/2} \qquad (6)$$

where: Q_n = sample volume rate (cfm)
  K = dimensional constant containing area, coefficients, etc.
  Δh = orifice pressure drop (inches of water)
  ρ = gas density

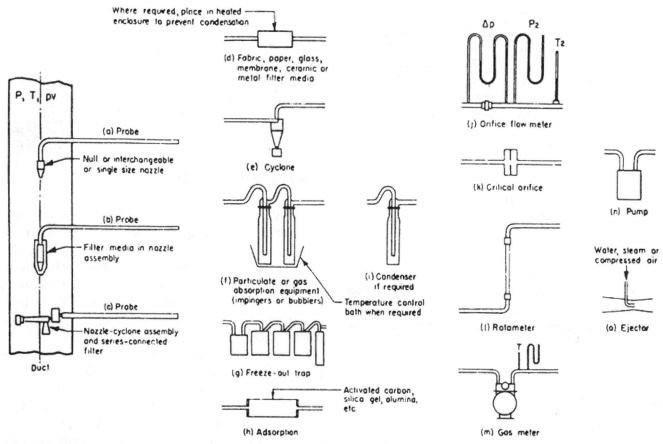

FIGURE 20-6. Sampling system components. *(Courtesy of Academic Press, New York)*

Gas density varies with absolute temperature (T_i) and pressure (P_i) as follows:

$$\frac{\rho_i}{\rho_o} = \left(\frac{P_i}{P_o}\right)\left(\frac{T_o}{T_i}\right) \qquad (7)$$

so that if the gas passing through the flowmeter is not at the same temperature and pressure when calibrated, then these corrections must be performed as well. All of the above calculations are usually combined in a standard meter rate equation in typical sampling methods. Effects on gas volume of removal of condensibles (e.g., moisture) in the collection train prior to the flowmeter must be included in the meter rate equation as well.[2, 6]

Sampling at the probe nozzle tip with a velocity that is substantially different from the stream velocity is termed nonisokinetic sampling. This procedure may cause a size-selective segregation of particle matter entering the proble tip. Broadly, for larger, heavier particles (i.e., for $\rho_p d^2_p \gg 1$, $\mu m^2 - g/cm^3$), oversampling (nozzle velocity > stack velocity, $V_n > V_s$) gives underestimation of the mass concentration because of the inability of larger particles to turn with the gas flow into the nozzle tip. The actual effect on measured concentration due to nonisokinetic sampling is shown elsewhere.

Undersampling ($V_n < V_s$) results in overestimation of concentration, with parallel arguments to those given above, which lead to inclusion of greater numbers of larger, heavier particles in the sample. Isokinetic and null-balance sampling probes are discussed below in "Special Apparatus and Applications."

Effects of Flow on Representative Sampling

The situation for isokinetic sampling assumes that the probe walls are of ideally thin material, of negligible thickness, far from all other flow disturbances. The actual situation in practice is rather more complicated, as shown in Figures 20-7a through 7c. First, the probe nozzle is typically made of thick-walled tubing tapered to a fine, sharp edge at the inlet opening (Figures 20-7a and 7b). Second, within an inch or two (25–50mm) from the opening, other apparatus distort the flow field upstream of the nozzle (Figures 20-7c and 7d). Third, the flow over the outside of the nozzle is accelerating and a laminar boundary layer is forming, possibly with some flow separation at the junction of the nozzle taper and the main tubing which effects the external flow field upstream of the nozzle. Finally, flow in any real gas stream is turbulent approaching the probe nozzle opening with an unknown (i.e., unmeasured) amount of transverse

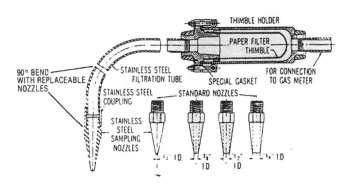

FIGURE 20-7a. Paper filter thimble holder with replaceable nozzles. The filtration tube has 90° bend.

motion of the fluid and the particles, not necessarily in phase (because the larger particle motions will lag the fluid motion due to greater particle inertia). These latter effects are illustrated schematically in Figure 20-8. Rouillard and Hicks[8] have made measurements of the velocity field in the upstream vicinity of common sampling probes and tips.

Applications

Introduction

This section contains details on specific methods of manual gas stream sampling (MGSS) that may be applied to individual industrial processes and operations. A general guideline for planning a survey is given, followed by a presentation of MGSS methods promulgated and proposed as regulations by the U.S. EPA[2] and a summary review of methods developed, recommended, or required by other groups. Certain special applications of gas stream

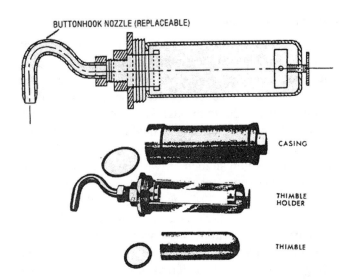

FIGURE 20-7b. Alundum® filter thimble holder with replaceable buttonhook nozzle. (Alundum is a registered trademark of Norton Company, Worcester, MA.)

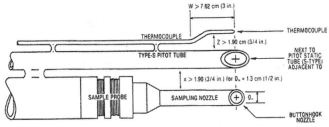

FIGURE 20-7c. Proper thermocouple placement to prevent interference on the U.S. EPA probe.[2]

sampling for issues of current interest are reviewed. Accuracy of methods, qualifications of technical personnel, and other selected issues are considered.

There are three kinds of methods that may be considered in any gas sampling situation: 1) individual methods developed as above, using apparatus selected for the project, 2) standard methods (apparatus and procedures) to be discussed below, and 3) modified standard methods.

Some useful information can be obtained from a simple sample of a dusty gas taken with an out-of-stack filter holder (closed face) or an impinger, attached to a piece of bent soft copper refrigeration tubing for a probe and connected through a critical orifice to a vacuum pump. These types of inexpensive data can be used in many situations to assess process wastestream loadings, to estimate collector efficiency, to define a need for collectors on emission streams, etc. Standard methods (described below) require more elaborate equipment and procedure time but yield more valid information. When planning a gas stream sampling project, judgment must be used to select a method consistent with project objectives.

There are also two levels of sampling difficulty to be considered:

1. Routine sampling, in which the procedure has been done before, where there are data available on previ-

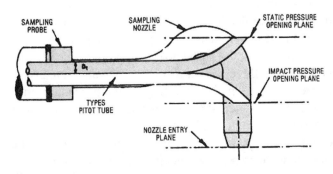

FIGURE 20-7d. Side view of probe.[2] To prevent Pitot tube from interfering with gas flow streamlines approaching the nozzle, the impact pressure opening plane of the Pitot tube shall be even with or above the nozzle entry plane.

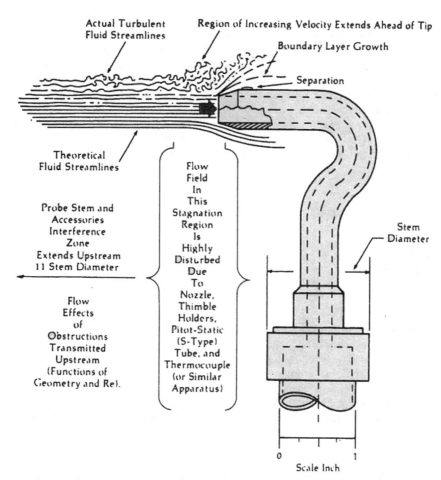

FIGURE 20-8. Actual approach zone flow phenomena in gas stream sampling.

ous tests, and the project can be redone by one or two senior technicians in a relatively straightforward way.
2. Nonroutine sampling, in which the procedure has not been done before, or new types of data are required. Examples of nonroutine sampling include data on particle-size and/or composition parameters,[9] data on a new process or one substantially changed, data on a high temperature-high pressure gas line, or data on newer sampling methods with limited field evaluation experience.[10, 11]

In all cases, there is need for information on the nature, characteristics, and quantities of materials to be expected in the gas stream. Each is process-dependent, and a general rule is to obtain information on the process initially. Expected emission concentrations can be calculated from air pollutant emission factors, with data on process parameters (i.e., size of process), flue gas flow rates, fuel rate, operation rate, etc.[12]

After determining the general process parameters and expected concentrations of substances of interest contained in the gas stream, apparatus and procedures are selected or developed to obtain required data. The following sections consider these topics.

Guideline for Planning and Implementing a Gas Stream Sampling Project

Table 20-3 has been prepared to identify necessary steps in the process of preparing for and conducting, assisting with, or observing a stack sampling project. It is a reasonably complete outline of necessary steps in the process, based on extensive experience. Each item can be developed in greater detail to prepare a checklist for conduct of each part of the project.

U.S. EPA Reference Methods for Stationary Source Air Pollutant Emission Sampling

Gas stream sampling and the underlying principles are used by the U.S. EPA to measure emissions from air pollution sources for 1) determination of quantity and composition of emission species; 2) development of emission inventories; 3) development of emission factors; 4) source emission surveillance, reporting, and assessment; 5) monitoring status of compliance with emission standards and enforcement activities; 6) permit and application support for a variety of regulatory purposes; 7) validation of source continuous emission

TABLE 20-3. Guidelines for Planning and Implementing a Gas Stream Sampling Project*

1. Identify purpose of tests: pollutant emission compliance, control equipment tests, process evaluation, other.
2. Identify data to be obtained; process, operation rates, etc.; gas composition and properties; flow character.
3. Obtain plant and process information; operational plans and specifications, general and specific site details, materials, flows, etc.
4. Select and specify test methods; individual design, standard method, modified method; obtain written agreement on method(s).
5. Site visit; discuss plans, prepare preliminary schedule of activities assess and evaluate status of equipment.*
6. Select site(s): location(s) for tests and need for supporting facilities, such as arrangements for space, laboratories, etc.*
7. If to be contractor implemented (or for budget estimating purposes): prepare work statement, identify bidders select evaluation criteria, send request for proposal, get back proposals, apply evaluation criteria, evaluate, select contractor, meet, negotiate award contract or purchase order.
8. Select equipment.*
9. Assign crew.*
10. Plan for access arrangements: ladders, scaffolds, platforms, jib booms and tackle, shelters, utilities space, etc.
11. Schedule activities (final) (budget final).
12. Assemble equipment.
13. Prepare necessary supplies, reagents, weighed filters, etc.
14. Calibrate flowmeters, lead checks, etc.
15. Pack for shipment.
16. Ship or take.
17. Travel to site.
18. Meet with plant personnel, test observers, inspect, advise, etc.*
19. Unpack equipment, set up, lead check, calibration checks.
20. Run preliminary test, check for cyclonic flow, unusual moisture, etc.
21. Analyze preliminary data, prepare preliminary result.
22. Discuss preliminary result with the plant personnel and observers, modify test procedures.*
23. Conduct test(s): three repetitions for U.S. EPA methods may require 3 separate days in sequence.*
24. Analyze preliminary samples and data; calculate % isokinetic (%I) for particulate tests; identify problems and need to modify, change, or abort test plans.*
25. Evaluate and discuss preliminary results with plant personnel/observers.
26. Remove equipment, clean up site, secure ports, etc.
27. Repack equipment, prepare samples for transport to analytical laboratory.
28. Ship or take back.
29. Travel back.
30. Unpack equipment, clean up, repair, recalibrate or recheck leaks.
31. Discuss analysis of samples for desired data with analyst.
32. Analyze samples.
33. Prepare results and calculations.
34. Prepare report:
    Report contents (minimum required for legally valid reporting): title page, letter of transmittal, table of contents, list of figures, list of tables, summary or abstract page, 1) introduction, 2) background, 3) apparatus and procedure, 4) test results, 5) discussion of results, 6) conclusion, 7) recommendations 8) appendices including copies of all field data, process flow chart, sample and analytical results (raw data), fuel analysis, data pages and tables filled out in the field, field observations, chain-of-custody affirmations, calibrations, preliminary filing data, test crew names and other identifiers (e.g., SSN), etc., and all other appropriate field test and backup data, e.g., copies of standard test methods or other pertinent regulations, stack opacity observations, typical calculations performed, etc.
35. Present report results, decide on course of action.

*State and federal air pollution control agency personnel may require pretest review and approval and onsite supervision of tests for regulatory compliance.

monitoring systems (CEMS); 8) assessment of best control technology in use; 9) validating performance of control devices; 10) development of better, novel, or new control technology; 11) development of better or new measurement methods, devices, etc.; and 12) developing or validating air dispersion models, etc.

One major provision of the Clean Air Act Amendments of 1970 was the establishment of uniform national standards or new source performance standards (NSPS).[2] These are specific allowable emissions for individual new (or substantially modified) stationary source categories and facilities. Each NSPS, when issued as a regulation, applies to new construction. These are also generally used by state air pollution control agencies as a guide to good practice for existing sources. Principal contents of the regulation include

TABLE 20-4. Sample of the Standards of Performance Table—40 CFR Part 60[2]

Source Category	Affected Facility	Pollutant	Emission Level	Monitoring Requirement[A]
Subpart D: Fossil-fuel-fired steam generators for which construction commenced after August 17, 1971 (>250 million Btu/hr)	Fossil-fuel-fired boilers[B]	Particulate Opacity SO_2 NO_x	0.10 lbs/million Btu 20% (27% for 6 min/hr) 1.20 lbs/million Btu 0.70 lbs/million Btu	No requirement Continuous Continuous Continuous

[A]Continuous monitors are used to determine excess emissions only, unless noted as "continuous compliance."
[B]Includes boilers firing solid or liquid fuel and wood residue mixtures.

emission standards for each substance and methods for determination of compliance. Emission standards are typically expressed either in mass of material per unit of material processed, or per unit of energy input or product output rate, or in terms of outlet concentration directly. Table 20-4 lists an example of the promulgated NSPS for one source and the facilities affected (e.g., unit operation or process equipment).[13] The remainder can be obtained directly from a U.S. EPA publication. Table 20-4 also lists the allowed emission level from the facility and whether a CEMS is required on the stack discharge for compliance. Each NSPS also lists the reference test methods required to demonstrate compliance and for recordkeeping and reporting purposes (surveillance procedures).

Standards of Performance for New Stationary Sources are compiled and updated annually.[2] A semi-annual update service is available from the U.S. EPA. Most practicing professionals accumulate the *Federal Register* announcements on a daily review basis and also use current awareness reporting systems.[14]

As indicated above, the U.S. EPA has developed and published (and occasionally revised) standard reference methods for MGSS for about 70 source/facility categories. Table 20-5 lists the appendix test method number and its title for methods used in the NSPS and 16 methods used in National Emissions Standards for Hazardous Air Pollutants (NESHAPS). With a few exceptions (e.g., Methods 9, 22, 24, and 28), these are all manual methods involving specific apparatus and procedural instructions. Applications of each method to individual NSPS are indicated in the Code of Federal Regulations.[2] For example, Subpart D (steam generators) requires use of Methods 1 through 5 for determination of particulate emissions, Methods 6 and 7 for SO_2 and NOx, and Method 9 for opacity (visual). From the U.S. EPA publication[2] and updates, method details can be obtained for other applications.

U.S. EPA Appendix Methods 1–5 and 17

U.S. EPA Appendix Methods 1–5 or 17 are used in more than half of the NSPS. These methods are described briefly here. The full text describing apparatus and procedures may be obtained from Reference 2.

Method 1 describes the selection of a sampling site and determination of the appropriate number of sampling traverse points, as described previously in conjunction with Figure 20-3.

Method 2 describes the conduct of a velocity traverse. Upon determining the traverse points, the traverse is conducted by the principles discussed above. Actual details of procedures are given in Method 2.[2] Training and apprenticeship experience are most valuable for valid implementation of any of these procedures.

Method 3 contains details for determination of flue gas composition in conjunction with moisture content determination by Method 4. Figure 20-9 indicates apparatus for withdrawing a grab sample to a Fyrite™ (or Orsat) analyzer. A more representative sample can be obtained by pumping an integrated sample into a plastic bag held in a box, as shown in Figure 20-9. A sample rate of 0.5–1.0 L/min and a bag volume of the order of 50 L are recommended (30 L final sample volume). The integrated sample can be made more representative of the total gas stream by traversing across the flow field during sampling (see Reference 2 for details). Analysis for carbon dioxide and oxygen (dry molecular weight determination) can be made with a Fyrite™ analyzer. For more accurate data, an Orsat analysis is required. Procedures for leak testing, calibration, use, and calculations are also given in Reference 2.

Method 4 describes apparatus and procedures for determination of moisture in stack gases from combustion sources, pyrometallurgical processes, or in the discharge from wet scrubbers. As shown in Figure 20-10, a heated probe conducts a gas stream sample through a

TABLE 20-5. U.S. EPA Manual Gas Stream Sampling (MGSS) Reference Methods for New Source Performance Standards (NSPS) and National Emissions Standards for Hazardous Air Pollutants (NESHAPS) (7/92)*

Appendix A Test Method No.	Test
1	Sample and velocity traverses for stationary sources
1A	Sample and velocity traverses for stationary sources with small stacks or ducts
2	Determination of stack gas velocity and volumetric flow rate (Type S Pitot tube)
2A	Direct measurement of gas volume through pipes and small ducts
2B	Determination of exhaust gas volume flow rate from gasoline vapor incinerators
2C	Determination of stack gas velocity and volumetric flow rate in small stacks or ducts (standard Pitot tube)
2D	Measurement of gas volumetric flow rates in small pipes and ducts
3	Gas analysis for carbon dioxide, oxygen, excess air, and dry molecular weight
3A	Determination of oxygen and carbon dioxide concentrations in emissions from stationary sources (instrumental analyzer procedure)
4	Determination of moisture content in stack gases
5	Determination of particulate emissions from stationary sources
5A	Determination of particulate emissions from the asphalt processing and asphalt roofing industry
5B	Determination of nonsulfuric acid particulate matter from stationary sources
5C	[Reserved]
5D	Determination of particulate emissions from positive-pressure fabric filters
5E	Determination of particulate emissions from the wool fiberglass insulation manufacturing industry
5F	Determination of nonsulfate particulate matter from stationary sources
5G	Determination of particulate emissions from wood heaters from a dilution tunnel sampling location
5H	Determination of particulate emissions from wood heaters from a stack location
6	Determination of sulfur dioxide emissions from stationary sources
6A	Determination of sulfur dioxide, moisture, and carbon dioxide emissions from fossil fuel combustion sources
6B	Determination of sulfur dioxide and carbon dioxide daily average emissions from fossil fuel combustion sources
6C	Determination of sulfur dioxide emissions from stationary sources (instrumental analyzer procedure)
7	Determination of nitrogen oxide emissions from stationary sources
7A	Determination of nitrogen oxide emissions from stationary sources—ion chromatographic method
7B	Determination of nitrogen oxide emissions from stationary sources (ultraviolet spectrophotometry)
7C	Determination of nitrogen oxide emissions from stationary sources—alkaline-permanganate/colorimetric method
7D	Determination of nitrogen oxide emissions from stationary sources—alkaline-permanganate/ion chromatographic method
7E	Determination of nitrogen oxides emissions from stationary sources (instrumental analyzer procedure)
8	Determination of sulfuric acid mist and sulfur dioxide emissions from stationary sources
9	Visual determination of the opacity of emissions from stationary sources
9A	Determination of the opacity of emissions from stationary sources remotely by lidar method 1
10	Determination of carbon monoxide emissions from stationary sources
10A	Determination of carbon monoxide emissions in certifying continuous emission monitoring systems at petroleum refineries
10B	Determination of carbon monoxide emissions from stationary sources
11	Determination of hydrogen sulfide content of fuel gas streams in petroleum refineries
12	Determination of inorganic lead emissions from stationary sources
13A	Determination of total fluoride emissions from stationary sources—SPADNS zirconium lake method
13B	Determination of total fluoride emissions from stationary sources—specific ion electrode method
14	Determination of fluoride emissions from potroom roof monitors for primary aluminum plants
15	Determination of hydrogen sulfide, carbonyl sulfide, and carbon disulfide emissions from stationary sources
15A	Determination of total reduced sulfur emissions from sulfur recovery plants in petroleum refineries
16	Semicontinuous determination of sulfur emissions from stationary sources
16A	Determination of total reduced sulfur emissions from stationary sources (impinger technique)
16B	Determination of total reduced sulfur emissions from stationary sources
17	Determination of particulate emissions from stationary sources (in-stack filtration method)
18	Measurement of gaseous organic compound emissions by gas chromatography
19	Determination of sulfur dioxide removal efficiency and particulate, sulfur dioxide, and nitrogen oxides emission rates
20	Determination of nitrogen oxides, sulfur dioxide, and diluent emissions from stationary gas turbines
21	Determination of volatile organic compound leaks
22	Visual determination of fugitive emissions from material sources and smoke emissions from flares

TABLE 20-5 (cont.). U.S. EPA Manual Gas Stream Sampling (MGSS) Reference Methods for New Source Performance Standards (NSPS) and National Emissions Standards for Hazardous Air Pollutants (NESHAPS) (7/92)*

Appendix A Test Method No.	Test
23	Determination of polychlorinated dibenzo-p-dioxins and polychlorinated dibenzofurans from stationary sources
24	Determination of volatile matter content, water content, density, volume solids, and weight solids of surface coatings
24A	Determination of volatile matter content and density of printing inks and related coatings
25	Determination of total gaseous nonmethane organic emissions as carbon
25A	Determination of total gaseous organic concentration using a flame ionization analyzer
25B	Determination of total gaseous organic concentration using a nondispersive infrared analyzer
26	Determination of hydrogen chloride emissions from stationary sources
27	Determination of vapor tightness of gasoline delivery tank using pressure-vacuum test
28	Certification and auditing of wood heaters
28A	Measurement of air to fuel ratio and minimum achievable burn rates for wood-fired appliances
101	Determination of particulate and gaseous mercury emissions from chlor-alkali plants—air streams
101A	Determination of particulate and gaseous mercury emissions from sewage sludge incinerators
102	Determination of particulate and gaseous mercury emissions from stationary sources (hydrogen streams)
103	Beryllium screening method
104	Determination of beryllium emissions from stationary sources
105	Determination of mercury in wastewater treatment plant sewage sludges
106	Determination of vinyl chloride from stationary sources
107	Determination of vinyl chloride of inprocess wastewater samples, and vinyl chloride content of polyvinyl chloride resin, slurry, wet cake, and latex samples
107A	Determination of vinyl chloride content of solvents, resin-solvent solution, polyvinyl chloride resin, resin slurry, wet resin, and latex samples
108	Determination of particulate and gaseous arsenic emissions
108A	Determination of arsenic content in ore samples from nonferrous smelters
108B	Determination of arsenic content in ore samples from nonferrous smelters
108C	Determination of arsenic content in ore samples from nonferrous smelters
111	Determination of ^{210}Po emissions from stationary sources
114	Test methods for measuring radionuclide emissions from stationary sources
115	Monitoring for ^{222}Rn Emissions
201	Determination of PM_{10} emissions (exhaust gas recycle procedure)
202	Determination of condensible particulate emissions from stationary sources

*Source: Reference 2.

heated filter to an ice-water-cooled condenser. Moisture condenses and its volume is measured. The condenser consists of four Greenburg–Smith impingers or a coil of tubing, immersed in an ice bucket. In the first arrangement, it is recommended that the fourth impinger be filled with silica gel. In the second case, silica gel or other dessicant should be included after the tubing coil to dry the gas completely and to protect downstream components from acid attack. For example, condensibles and (sulfur) acids in the gas passing through a dry gas meter rapidly rust, corrode, and freeze up the motion, rendering the meter inoperable.

Method 5 is used to sample for particulate matter, using an out-of-stack filter contained in a heated box. Standard components, considered in series, are: 1) a glass-lined heated probe with a button-hook nozzle (outside taper); 2) an attached thermocouple; 3) an attached reverse-impact (Type S) Pitot-static tube (these three items comprise the pitobe); 4) a heated, fibrous filter holder and chamber; 5) four Greenburg-Smith impingers in a series; 6) a leak-free vacuum pump; 7) a dry gas meter; and 8) an outlet orifice. Pressure gauges, temperature probes, flowmeter manometer and Pitot tube manometer, flow control and shutoff valving, and electrical switching, fuses, etc., are arranged at the point of use in the system or are transmitted to gauges in a central meter box. Construction details for standard models are contained in a U.S. EPA technical publication.[15] Methods for maintenance, calibration, and operation of the equipment are described in a companion U.S. EPA publication.[16] The apparatus and procedures were developed during the early 1960s by the U.S. Bureau of Mines and the National Air Pollution Administration. [17, 18]

Procedures for calibration, leak checks, preparation of consumable supplies, use in the field, sample recov-

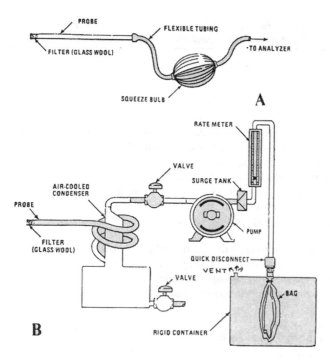

FIGURE 20-9. U.S. EPA Reference Method 3 gas sampling train.[2]
A. Grab Sampling train. B. Integrated gas sampling train.

ery, recording of data and observations, and calculations required are contained in Reference 2. Each of the several manufacturers listed in Table 20-I-1 (located in the "Instrument Descriptions" section at the end of this chapter) provides an operating instruction manual for their specific design. To use the system, the sampling probe is mounted on the sample box and the sample box is connected to the meter box by means of the umbilical cord. The sampling probe or its support is marked with glass cloth tape at traverse cross-sectional depth of insertion points. After leak-checking, the assembly is then mounted on a suitable framework for sliding in and out of the stack. Typically, operation is maintained at isokinetic flow conditions at about 1.0 cfm (28.3 L/min) at each traverse point by adjusting sampling volume with a fixed probe nozzle tip size. Three major methods of adjusting the flow rate are commonly used: 1) a nomograph that solves the isokinetic equation graphically, as described by Rom;[16] 2) a meter rate equation that contains all the factors and is reduced for simple calculation in the field on a hand-held scientific calculator; or 3) a small, programmable, hand-held computer may be used.

Sampling with the Method 5 train is somewhat more complex than with others used in the past. It is especially necessary to obtain some training and experience before using Method 5.

To reduce some of the operational complexity from the Method 5 train used for sampling from gas streams where the particulate concentration is independent of the temperature, the U.S. EPA has developed Method 17, as shown in Figure 20-11.[2] The heated filter holder in the sample box has been removed as a requirement, and a flat (or thimble) filter holder is attached to the stack end of the probe for direct insertion into the gas stream. This in-stack filter method collects particulate matter at stack temperature and removes the need to clean out the probe liner and filter front half with acetone or other solvents at the end of each test. A simple, unheated steel (instead of glass) probe is attached to the moisture condenser by a flexible hose (Teflon®-lined). Moving this simple filter-probe assembly to each point in the traverse is easier and cleanup is quicker at the end. Both Method 5 and Method 17 usually require two persons to operate them in the field. Prior training and supervised experience are necessary in order to be aware of and compensate for problems with leaks, breakage, equipment, etc.

Safety Precautions

Safety and health considerations for gas stream sampling teams include hazards associated with climbing and working at heights, electrical shock, confined space entry and work, and exposure to chemicals from pressurized openings in ducts.

Electrical or fire hazards can occur occasionally when a ground fault develops. The glass-lined Pitot tube is electrically heated and the outer sheath can reach 110 volts, unless it is grounded. None of the standard, commercially available Method 5 equipment is intrinsically safe for Class I Group D explosive atmospheres. Exposed resistance heaters exist in older designs. One of the major difficulties with the Method 5 train is obtaining a satisfactory leak check (<0.02 cfm [<0.57 L/min] at 0.5 atm;

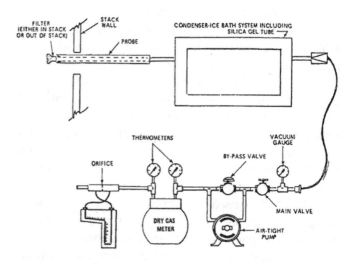

FIGURE 20-10. U.S. EPA Reference Method 4 moisture sampling train.[2]

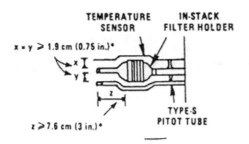

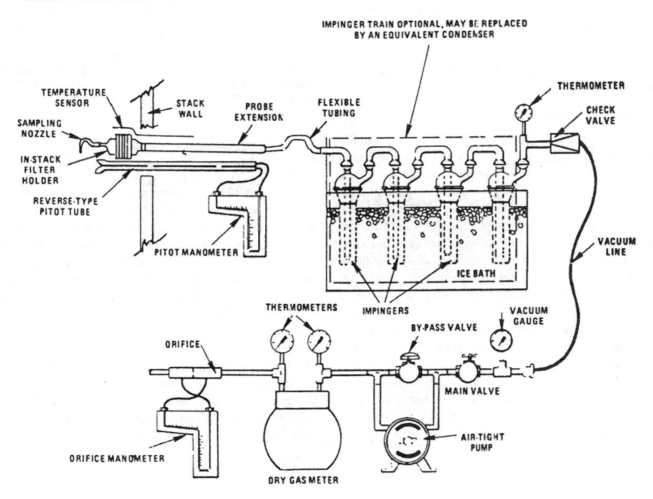

FIGURE 20-11. U.S. EPA Reference Method 17 particulate sampling train, equipped with in-stack filter.[2]

Method 5 section 4.1.4[2]). There are about 100 joints, junctions, connections, etc. (permanent and temporary), and any one or several may leak on any given test setup. In addition, the glass-lined probe has been known to break during setup. There are no specific universal guidelines for leak location; each instance is unique when it occurs. Experience and judgment are essential.

Other MGSS Guidelines, Codes, and Standards

Many other organizations and agencies have developed stack sampling methods for various operations or emissions. Methods are also discussed in the Intersociety Committee Manual, Methods of Air Sampling and Analysis.[19] There are also companion developments of stack sampling methods in other industrialized countries such as the design of the British Coal Utilization Research Association.[1]

Special Apparatus and Applications

This section deals briefly with four major topical areas alluded to in previous sections: 1) isokinetic sampling probes; 2) high-volume stack samplers; 3) high

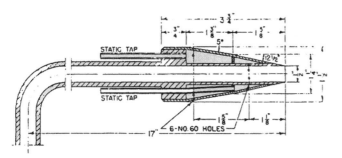

FIGURE 20-12a. Typical null-type isokinetic sampling nozzle, probe A. *(Courtesy of Western Precipitation Corp.)*

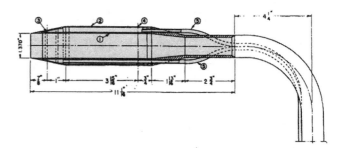

FIGURE 20-12b. Typical null-type isokinetic sampling nozzle, probe B: 1.5-in. (38.1-mm) o.d. tube, No. 16 gauge, 1.370-in. (34.8-mm) i.d. (1); 2. 125-in. (54-mm) o.d. tube, No. 18 gauge, 2.037-in. (51.7-mm) i.d. (2); inside static 8.125-in.- (206.4-mm) diameter holes equally spaced (3); 12.125-in.- (308-mm-) diameter holes equally spaced (4); static tubes, 3/16-in. (4.8-mm) o.d., 0.177-in. (3-mm) i.d. (5). *(Courtesy Bethlehem Steel Company)*

temperature-high pressure (HiT-HiP) sampling systems; and 4) selected systems.

Isokinetic Sampling Nozzles

Null-type isokinetic sampling nozzles are usually three-chamber designs that operate by measuring static pressure on the outside of the probe nozzle body and static pressure inside the inlet opening of the nozzle, as illustrated in Figures 20-12a and 12b.[20] The null-static pressure balance is achieved when zero static pressure differential (null) is developed between the inside and outside static pressure taps. This is assumed to indicate that isokinetic velocity is being obtained.

Dennis *et al.*[20] found that calibration of these probes was required because errors in isokinetic velocity of –28% to +8% occurred at null static pressure balance over a velocity range from 1000 to 7000 fpm (5.08–35.56 m/s). The general conclusions from this extensive investigation are that 1) null balance static pressure probes must be calibrated in a duct of the same general size and configuration as the field installation and at the velocity, temperature, etc. at which they will

be used; 2) isokinetic velocity actually will be achieved at some positive (or negative) value of the static pressure difference; and 3) the calibration will change with duct velocity and size/configuration. Null-type sampling probes cannot theoretically achieve isokinetic velocities except possibly in a very limited range because the location of static tap holes, probe and nozzle size and shape, inlet configuration, etc., all interact with boundary layer growth and consequent static pressure distribution over the outside of the probe nozzle and along the inside of the nozzle inlet tube (in the entrance length region). Other designs of isokinetic null-balance probe nozzles are illustrated in Figures 20-13a through 13c.[21–23]

Shrouded aerosol sampling probes offer an advantage in non-ideal airflow conditions. Studies of the airflow and particle sampling characteristics have been reported by Rodgers *et al.*[24] and Cain, Ram, and Woodward.[25]

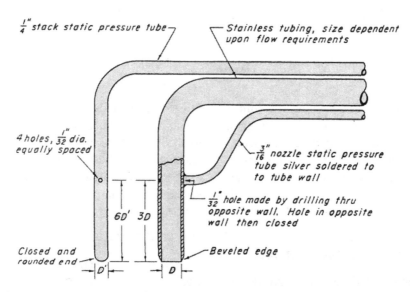

FIGURE 20-13a. Industrial Hygiene Foundation simplified "null" nozzle design.[21]

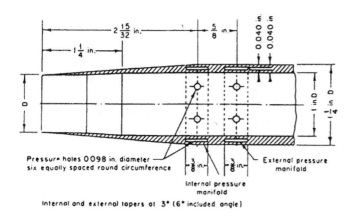

FIGURE 20-13b. Sampling probe (pressure leads from manifold not shown).[22]

High-Volume Stack Samplers

Normal in- and out-stack sampling apparatus are designed to operate usually with a small, easily portable vane-type vacuum pump having a flow capability of about 1.0 cfm (28.3 L/min). There are a number of applications in which a higher volume flow rate is desirable

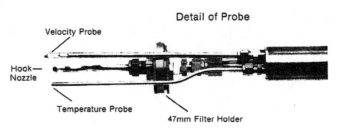

FIGURE 20-13c. Automatic isokinetic Method 5 sampling attachment.[23]

and a suitable collector and vacuum source combination must be obtained, e.g., for sampling of the outlet concentration of a fabric filter operating at 99.99% plus efficiency or for fuel and waste gas streams containing particulate concentrations below 1.0 mg/m^3 (<0.5 grains/1000 ft^3).

A 50- to 75-cfm (1400—2100 L/min) model, developed as the Boubel-CS3 hi-volume sampler, is shown in Figure 20-14.[26] A cast aluminum, gasketed filter holder contains a flat 8-in. × 10-in. (203 × 245mm) rectangle of all-glass, high-efficiency fiber filter paper (MSA-1106B

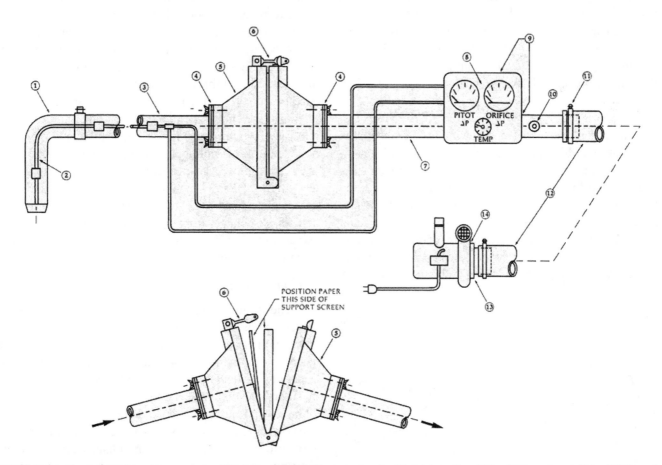

FIGURE 20-14. Boubel-CS3 hi-volume sampler;[26] inlet nozzle (1); Pitot-static tube (2); inlet section, 30 in. (762 mm) and 48 in. (1219 mm) (3); Neoprene gaskets (4); filter housing assembly (5); housing fastener (6); control section (7); test panel (8); flow control orifice (9); control valve (10); hose clamps (11); flex hose (12); blower hose adaptor (13); suction blower (Cadillac) (14).

or HV-70). Flow is conducted through (1) sampling nozzle tip (1 7/8-in. [47.6-mm] i.d. for low velocity, < 2000 fpm [10 m/s]; 1 3/8-in. [34.9-mm] i.d. for midrange velocities, 2000–4000 fpm [10–20 m/s]; and 15/16-in. [23.8-mm] i.d., high velocity, >4000 fpm [> 20 m/s]), then through (3) aluminum extension tubing (3 or 4 ft [760–1020 mm]), through the filter and housing (5), through an orifice meter whose upstream and downstream static pressure taps are connected to a 0- to 4-in. water gauge (w.g.) (0–100 mm w.g.) Magnehelic® gauge, through a butterfly control valve, and through flexible tubing to a conventional Cadillac® Blower. Flow is adjusted with valve (10). The technical instructions contain nomographs to adjust flow rate using the Pitot tube (2) readings shown on the second 0- to 2-in. w.g. (0–50 mm w.g.) Magnehelic gauge and the temperature at the measuring orifice (8).

A high-volume stack sampler (HVSS) was developed by the Aerotherm Division of Acurex Corporation for the U.S. EPA. It is a U.S. EPA Method 5 train operating at 4 cfm (113 L/min) with a stainless steel probe liner and General Electric's Lexan® polycarbonate impingers. It operates at 6 cfm (170 L/min) with a heated cyclone in place and 7.5 cfm (212 L/min) when the cyclone is removed. The pump is a 10 cfm, oil-less vane type, custom modified and rated at 10 cfm (283 L/min) at 0 mm Hg vacuum. Stated weight is 38 lbs (17 kg) for the sample box and 50 lbs (33 kg) for the control unit.

High Temperature–High Pressure Gas Stream Sampling

Sampling from a confined flowing gas stream at high temperature or high pressure may require special consideration for insertion of apparatus to reduce ambient leakage outward for personnel safety or process reasons and to preserve sampled material. These are common problems in the pyrometallurgical, natural and producer fuel-gas, chemical process, petroleum, and related industries. Typical devices developed for leak-limiting insertion of MGSS systems are shown in Figure 20-15.[27, 28]

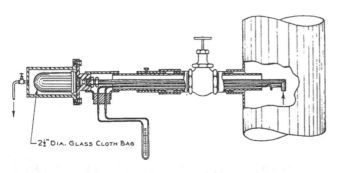

FIGURE 20-15a. Sampling high pressure blast furnace gas (25 mm Hg, 540°C).[27]

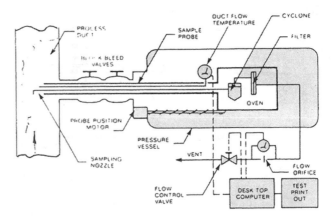

FIGURE 20-15b. Sampling high temperature and pressure fluid-bed combustion gas (12 atm, 1100°C).[28]

The out-stack design for a few millimeters of mercury pressure indicates a typical method of insertion of the probe through a gate valve and sliding gland arrangement. It has a large capacity fabric filter bag, plus provision for determining local velocity (Figure 20-15a).[27] The Accurex sampling system, shown in Figure 20-15b, is designed to determine the efficiency of particulate removal before fluidized-bed coal combustion gas enters a gas turbine. A remotely located computer automatically positions the water-cooled sampling probe at a given traverse point and maintains isokinetic sampling rate. As shown in Figure 20-15b, the sampling probe enters the pressurized duct through block and bleed valves. The gas stream velocity and temperature are measured by instruments at the end of the probe. A computer controls sampling by opening a gas flow control valve and maintains the isokinetic sampling rate by adjusting it. Large particles are removed in a cyclone, and fine particles are caught by a filter. Gas passes out through an orifice used to calculate sampling flow rate. The entire sampling system is suspended from a hanger to allow for thermal expansion. The computer controls probe movement to each sample point.[28]

Special Systems

The source assessment sampling system (SASS), developed by Accurex for the U.S. EPA, represents a special, commercially available system designed for research purposes. It extracts a 4-cfm (113-L/min) sample and separates particulate matter by size fractions (cyclones). It filters the undersize fraction (filter holder), removes organics (porous sorbent), and then removes condensibles (in impingers). "SASS is an integrated sampling system, capable of measuring particulate loading and size distribution, determining trace element concentration, and trapping organic substances. Stainless . . . or water-cooled, quartz-lined probes can be supplied for high temperature sampling."[29] It contains three cyclones plus a filter to separate particulate matter by aerodynamic diameter. Organics are collected on an adsorber and trace elements are collected in impingers.[10]

Determination of Particle Size and Chemical Composition

Introduction

This section discusses special requirements in gas stream sampling methods used to determine size and composition of particulate materials transported by the stream. Objectives for determination of size and composition include evaluation of properties or effects that vary with size or composition, or evaluation of process yield or equipment performance in terms of size or composition.

Particulate Air Pollutant Source Emissions

The traditional approach to the control of particulate air pollutants from stationary sources has been accomplished by measurement of the emission concentration of the total amount of particulate matter conveyed by the gas stream. Continuing developments in the understanding of the complexity of particulate air pollution have, over the past 25 years or so, led to the design, commercial production, and use of a variety of size-discriminating instruments and apparatus with application to particulate air pollution source emissions. Devices used for source emission sampling are broadly divided into 1) inertial classifiers (mechanical collectors) and 2) others such as light scatter, electrical mobility, or diffusion battery analyzers. In general, the devices in the first group include cascade impactors and cyclones. Their use for source evaluation is a technically sophisticated procedure. Both of these

---

TABLE 20-6. Summary of Recommendations for In-stack Impactor Use[30]

Selection of Collection Surfaces
1. Grease (spray silicone or Apiezon H)
   a. Use at a temperature less than 200°C.
   b. Apply a thickness equal or greater than the size particles to be impacted.
   c. Precondition the surface for 1 hour at a temperature about 25°C above the expected sampling temperature.

2. Glass fiber
   a. Use at temperatures less than approximately 500°C.
   b. Precondition to avoid SO_2 uptake.
   c. Precondition for 1 hour at 25°C above the temperature of sampling.
   d. Note that the measured aerosol mass median diameter could be 30% greater than actual.
   e. If possible, avoid sampling "hard" aerosols because of increased particle bounce.

3. Uncoated metal
   a. Use at temperatures up to 500°C.
   b. Precondition for 1 hour at or above the temperature of sampling.
   c. Avoid sampling oil or "hard" aerosols because of bounce and unstable collection characteristics.

General Selection of Flow Rate
1. Maintain jet velocities less than 75 m/sec when sampling either oil or hygroscopic-type aerosols.
2. Maintain jet velocities less than 50 m/sec when sampling "hard" aerosols.
3. Choose a flow rate that will provide sizing information over the range of the expected mass median diameter, within the above limits.

Stage Loadings
1. Hygroscopic-type aerosols: 5–7 mg maximum per stage.
2. Oil aerosols: 15 mg maximum per stage.
3. Stage loading checks:
   a. Observe back side of nozzle for increased deposition.
   b. Observe primary deposits for uniformity.
4. General minimum stage loading:
   a. Collect 10 times the stage weighing sensitivity.

Treatment of Interstage Losses
Exclude losses from calculations of particle size distribution due to errors involved in trying to recover these losses. Include losses as collected mass if calculating total aerosol mass concentration.

Treatment of Sizing Data
Consider that the error associated with a mass measurement is inversely proportional to the mass collected; therefore, when constructing the distribution, give greater weight to the data points representing the majority of the collected mass.

TABLE 20-7. Impactor Decision-making[31]

Item	Basis of Decision	Criteria
Impactor	Loading and size estimate	a. If concentration of particles smaller than 5 µm is less than 0.46 g/am³ (0.2 grain/acf), use high flow rate impactor (0.5 acfm [14.a L/min]). b. If concentration of particles smaller than 5 µm is greater than 0.46 g/am³ (0.2 grain/acf), use low flow rate impactor (< 0.05 acfm [1.41 L/min]).
Sampling rate	Loading and gas velocity	a. Fixed, near isokinetic. b. Limit so that last jet velocity does not exceed: -60 m/sec greased -35 m/sec without grease.
Nozzle	Gas velocity	a. Near isokinetic, ± 10%. b. Sharp edged; minimum 1.4-mm i.d.
Precutter	Size and loading	If precutter loading is comparable to first stage loading, use precutter.
Sampling time	Loading and flow rate	a. Refer to Reference 59, Section 5.5. b. No stage loading greater than 10 mg.
Collection substrates	Temperature and gas composition	a. Use metallic foil or fiber substrates whenever possible. b. Use adhesive coatings whenever possible.
Number of sample points	Velocity distribution and duct configuration	a. At least two points per station. b. At least two samples per point.
Orientation of impactor	Dust size, port configuration, and size	Vertical impactor axis whenever possible.
Heating	Temperature and presence of condensible vapor	a. If flue is above 177°C, sample at process temperature. b. If flue is below 177°C, sample at 11°C above process temperature at impactor exit external heaters.
Probe	Port not accessible using normal techniques	a. Only if absolutely necessary. b. Precutter on end in duct. c. Minimum length and bends possible.

device categories have particle removal characteristic curves that are functions of particle properties (size, shape, density, etc.); operating flow characteristics (flow rate or velocity, temperature, gas composition, and content of moisture or other condensibles, etc.); and the amount of material presented to and contained (deposited) within.

Comparisons of results of particle size analysis obtained with four different cascade impactors were found to vary among the various available designs and upon circumstances of use.[30] Recommendations for use of in-stack impactors are summarized in Table 20-6.[30] Decision-making guidelines are presented in Table 20-7.[31] Also see the operating instruction manual for each device to obtain information on stage collection performance calibrations, recommended collection substrates, adhesives versus operating temperature, interstage losses, tolerable deposit per stage versus analytical sensitivity required and its relation to expected particulate concentration in the gas stream, interferences between substrate or adhesive and collected particulate matter or its analysis, etc.[32]

Use of impactors for higher temperature sampling in process gas streams introduces additional limits on adhesives, volatilization (weight loss) and reaction with process or flue gases (weight gain). The adhesive selected should also be considered with respect to possible analytical interferences, i.e., organic adhesives might provide too high a background for subsequent analysis for organic components. In some circumstances, manufacturers may furnish properly cut-out pieces of fibrous substrate (e.g., all glass fiber filter material) to collect impacted particles and reduce bounce-off or to reduce tare weight of substrate. For additional information on the use of impactors for ambient air sampling, the reader should check the materials and references in Chapter 14.

Comparative interstage wall losses have been reported, as shown in Figure 20-16.[32] Wall losses are a function of impactor design, substrate treatment, particle size, type of aerosol material (solid versus liquid), and jet velocity and flow rate. In practice, operating flow rate, collector plate adhesive, and the amount of material

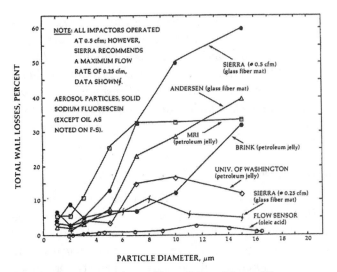

FIGURE 20-16. Comparison of impactor wall losses.[32]

collected on a stage will interact to affect wall loss and deposit bounce-off. The amount of material collected on a single stage is recommended not to exceed 5–15 mg (see Table 20-6). Operating time thus depends on the concentration of particulate material in the gas stream as well as expected particle size. Impactors have significant limitations for *in-situ* separation of solid particulate matter into size fractions based on aerodynamic equivalent diameter. Upper stages may be overloaded, causing material to bounce off and to be deposited on a lower stage, thus distorting apparent size. They overload rapidly at high concentrations (e.g., on the inlet to a collector). To overcome some of these overload limits, the U.S. EPA and manufacturers have developed precutter or scalping cyclone inlets for use as a preseparator ahead of the impactor.

A series cyclone design has been developed and calibrated by SoRI, as shown in Figure 20-17.[33, 34] These instruments may be used to evaluate source emission of thoracic or inhalable particles. A formal sampling method for PM_{10} emissions has also been developed as U.S. EPA Methods 201 and 201A.[2]

Particle Size-Efficiency of Collection Systems

There has been an increasing interest in the particle size-specific collection performance of control devices during the past several years. A number of field studies have been undertaken on inlet and outlet particle size distributions for a variety of sources and collection systems.[35–37] A review of the coal fly ash emission program undertaken by the Electric Power Research Institute (EPRI) for electrostatic precipitators has been reported.[38] Principal test apparatus used in these studies include in-stack cascade impactors for upstream or downstream concentrations. Some data on submicrometer fractions have been obtained in various studies using

an electric mobility analyzer, light-scattering photometers, condensation nuclei counters, and diffusion batteries; in at least one instance, electron microscopic analysis was used as well. The low-pressure cascade impactor described by Hering *et al.*[39] has been adapted for gas stream sampling by Pilat.[40]

Method Validation, Accuracy, Precision, and Sampling Statistics

Validation Between Methods

The foregoing discussions have indicated the diversity of methods that have been used within the United States to determine characteristics or components in flowing gases. Prior to 1972, intermethod comparisons were not generally undertaken. The original use of U.S. EPA Method 5 (commencing in the late 1960s) to determine particulate emissions required the inclusion of the dried impinger residue as part of the total emission concentration, in addition to material collected in the probe and cyclone and on the filter. These two quantities of particulate material are referred to as front-half (i.e., material brushed or washed from nozzle, probe liner, and cyclone plus filter particulate matter) and back-half (dried residue from three impingers).

For many years, combustion engineers and related specialists have used apparatus suggested in the American Society of Mechanical Engineers (ASME) Power Test Code used for performance tests on steam generators, incinerators, fly ash collectors, etc. When the U.S. EPA began to include the back-half catch as a surrogate for materials that form particulate matter later in the plume downwind, concentrations from the sources were determined to be greater than had been customarily obtained with the front-half of Method 5 or with an in-stack filter thimble alone.[41] Increased particulate matter in out-stack configurations has been attributed to sulfur dioxide conversion to sulfates.

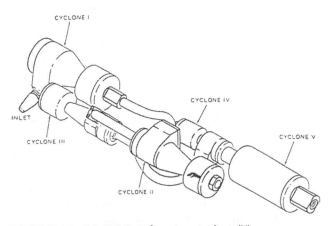

FIGURE 20-17. U.S. EPA/SoRI five-stage cyclone.[34]

There have been a number of reports on intermethod results from pulverized coal-fired boilers, oil-fired boilers, electric furnace fume, incinerators, etc., under the sponsorship of the U.S. EPA, the American Society for Testing and Materials' (ASTM) Project Threshold, ASME, and individual industrial organizations.[42] Typical issues and early data are contained in the U.S. EPA reports and other references.[43, 44] Subsequent studies have determined that careful field practices and maintenance of U.S. EPA Method 5 probe and filter heated to stack temperature yield largely equivalent mass loadings for both methods (ASME/WP-50 and U.S. EPA Method 5) at least for steam generator fly ash, if condensation is avoided.

Education, Training, Certification, and Accreditation for Gas Stream Sampling

In order to obtain a reasonably representative and valid sample of a flowing gas stream, one needs to have some education, training, and experience. Source evaluation sampling has become fairly complex, costly, and more widespread in the past 25 years. Typical instruction may be obtained on U.S. EPA Methods 1–8 from U.S. EPA course number 450, Source Sampling (Air Pollution Training Institute, Research Triangle Park, NC). Courses are also provided by manufacturers of the sampling equipment.

There is also a national association of source sampling principals (Source Evaluation Society [SES], P.O. Box 12124, Research Triangle Park, NC), with about 300 members. The SES has a biennial (about every 1.5 years) meeting under the auspices of the Engineering Foundations at a Conference on Stack Sampling and Stationary Source Evaluation. It also meets *ad hoc* as part of the annual meeting of the Air and Waste Management Association (formerly the Air Pollution Control Association).

The SES has prepared to establish programs for the accreditation of organizations involved with source emissions testing and analysis, and for the certification of individuals who conduct or direct emissions testing or analysis. Current recommendations for laboratory accreditation are summarized below.

- *Education, Training, Experience—one individual onsite (e.g., team leader) to be certified by test in the method that will be used.*
- *Proficiency Demonstration—review of test reports and analysis of audit samples.*
- *Equipment Specifications—possess up to 80% of major listed equipment specified in a method including calibration, field, and laboratory.*

- *Recordkeeping/Reporting—retain records for 10 years, including quality assurance (QA), calibration and test data, and sample retention for reanalysis.*
- *Quality Assurance—formal plan including calibration, sample identification and custody, QA responsibility chart, implementation procedures, and onsite audits.*[14]

Dedication

This chapter is respectfully dedicated to the memory of Bernard D. Bloomfield (1922–1971). He was the original author of this chapter which appeared in the fourth and fifth editions of the *Air Sampling Instruments* manual. He lectured widely and wrote chapters for several texts on this topic from his broad experience with the Michigan Department of Public Health.

References

1. Hawksley, P.G.W.; Badzioch, S.; Blackett, J.H.: Measurement of Solids in Flue Gases, 2nd ed., pp. 114–116. The Institute of Fuel, London (1977).
2. U.S. Environmental Protection Agency: Standards of Performance for New Stationary Sources, pp. 195–1142. 40 CFR 60 (Rev. July 1, 1998). U.S. Government Printing Office, Washington, DC (and U.S. Government Bookstores in 22 U.S. cities) (1998).
3. American Society of Mechanical Engineers: Flue and Exhaust Gas Analyses, Part 10, Instruments and Apparatus, Supplement to ASME Performance Test Codes, PTC 19.10-1981, p. 9. ASME, United Engineering Ctr., 345 E. 47th St., New York, NY (1980).
4. American Conference of Governmental Industrial Hygienists: Industrial Ventilation—A Manual of Recommended Practice, 23rd ed., Chap 9. ACGIH, Cincinnati, OH (1998).
5. Air Test Kit Bacharach 5220. Bacharach Instruments Co., Div. of AMBAC Industries, Inc., 625 Alpha Dr., Pittsburgh, PA.
6. American Society of Mechanical Engineers: Performance Test Code PTC 38-1980. In: Determining the Concentration of Particulate Matter in a Gas Stream, p. 79. ASME, United Engineering Ctr., 345 E. 47th St., New York, NY (1980).
7. Ower, E.; Pankhurst, R.C.: The Measurement of Air Flow, 5th ed., Chap. III. Pergamon Press, Inc., Elmsford, NY (1977).
8. Rouillard, E.E. A.; Hicks, R.E.: Flow Patterns Upstream of Isokinetic Dust Sampling Probes. J. Air Pollut. Control Assoc. 20(6):599 (1978).
9. Moseman, R.F.; Bath, D.B.; McReynolds, J.R.; *et al.*: Field Evaluation of Methodology for Measurement of Cadmium in Stationary Source Stack Gases. EPA/600/S4-86/048. NTIS No. PB 87-145 355/AS. National Technical Information Service, 5285 Port Royal Road, Springfield, VA (April 1987).
10. Schlickenrieder, L.M.; Adams, J.W.; Thrun, K.E.: Modified Methods and Source Assessment Sampling System Operator's Manual. EPA/600/S8-85/003. NTIS No. PB 85-169 878/AS. National Technical Information Service, 5285 Port Royal Road, Springfield, VA (April 1985).
11. Farthing, W.E.; Williamson, A.D.; Dawer, S.S.; *et al.*: Investigation of Source Emission PM$_{10}$ Particulate Matter: Field Studies of Candidate Methods. EPA/600/S4-86/042. NTIS No. PB 87-132 841/AS. National Technical Information Service, 5285 Port Royal Road, Springfield, VA (March 1987).

12. U.S. Environmental Protection Agency: Compilation of Emission Factors, AP-42, 3rd ed., and all supplements to current year (1998). National Technical Information Service, 5285 Port Royal Road, Springfield, VA, and U.S. Government Printing Office, Washington, DC (and U.S. Government Bookstores in 22 U.S. cities).

13. Pahl, D.: EPA's Program for Establishing Standards of Performance for New Stationary Sources of Air Pollution. J. Air Pollut. Control. Assoc. 33(5):486 (1983).

14. Source Evaluation Society Newsletter, P.O. Box 12124, Research Triangle Park, NC.

15. Martin, R.M.: Construction Details of Isokinetic Source-Sampling Equipment. EPA Report No. APTD-0581. U.S. EPA, Research Triangle Park, NC (1971).

16. Rom, J.J.: Maintenance, Calibration, and Operation of Isokinetic Source-Sampling Equipment. EPA Report No. APTD-0576. U.S. EPA, Research Triangle Park, NC (1972).

17. Gerstle, R.W.; Cuffe, S.T.; Orning, A.A.; Schwartz, C.H.: Air Pollution Emissions from Coal-Fired Power Plants, Report No. 2. J. Air Pollut. Control Assoc. 15(2):353 (1964).

18. Gerstle, R.W.; Cuffe, S.T.; Orning, A.A.; Schwartz, C.H.: Air Pollution Emissions from Coal-Fired Power Plants, Report No. 2. J. Air Pollut. Control Assoc. 15(2):59 (1965).

19. Intersociety Committee: Methods of Air Sampling and Analysis, 3rd ed., Lewis Publishers, Inc., Chelsea, MI (1989).

20. Dennis, R.; Samples, W.R.; Anderson, D.M.; Silverman, L.: Isokinetic Sampling Probes. Ind. Eng. Chem. 49(2):294 (1957).

21. Haines, G.R.; Hemeon, W.C.L.: Measurement of Dust Emission in Stack Gases. Information Circular No. 5. to the American Iron and Steel Institute. Industrial Hygiene Foundation of America, Inc., Pittsburgh, PA (1953); test results appear in Air Repair. J. Air Pollut. Control Assoc. 4:159 (1954).

22. Toynbee, P.A.; Parks, W.J.S.: Isokinetic Sampling Probes. Int. J. Air Water Pollut. 6:13 (1962).

23. Kurz Instruments, Inc., 24H Garden Road, Monterey, CA.

24. Rodgers, J.C.; Fairchild, C.I.; Wood, G.D.; et al.: Single Point Aerosol Sampling: Evaluation of Mixing and Probe Performance in a Nuclear Stack. Health Phys. 70:25–35 (1996).

25. Cain, S.A.; Ram, M.; Woodward, S.: Qualitative and Quantitative Wind Tunnel Measurements of the Airflow Through a Shrouded Airborne Aerosol Sampling Probe. J. Aerosol Sci. 29:1157–1169 (1998).

26. Boubel, R.W.: A High Volume Stack Sampler. J. Air Pollut. Control Assoc. 21(12):783 (1971).

27. Arbogst, A.H.: The Quantitative Determination of Dust in Gas. Iron and Steel Engineer, pp. 1–8 (October 1948).

28. Accurex Corp.: Aerotherm Accurex High Temperature-High Pressure Sampling System Product Literature. Accurex Corp., Mountain View, CA (1981).

29. U.S. Environmental Protection Agency: Modified Method 5 Train and Source Assessment Sampling Systems Operations Manual. EPA Report 600/8-85; NTIS No. PB85-1G9578. National Technical Information Service, 5285 Port Royal Road, Springfield, VA (1985).

30. Lundgren, D.A.; Balfour, W.D.: Size Classification of Industrial Aerosols Using In-stack Cascade Impactors. J. Aerosol Sci. 13:181 (1982).

31. Harris, D.B.: Procedures for Cascade Impactor Calibration and Operation in Process Streams. U.S. EPA Report No. EPA-600/2-77-004. U.S. EPA, IERL/ORD, Research Triangle Park, NC (1977).

32. Cushing, K.M.; McCain, J.D.; Smith, W.B.: Experimental Determination of Sizing Parameters and Wall Losses of Five Commercially Available Cascade Impactors. Paper No. 76-37.4, presented at the APCA Annual Meeting, Portland, OR. APCA, Pittsburgh, PA (1976); also see Environ. Sci. Technol. 13(6):726 (1979).

33. Smith, W.B.; Cushing, K.M.; Wilson, R.R.: Cyclone Samplers for Measuring the Concentration of Inhalable Particles in Process Streams. J. Aerosol Sci. 13(3):259 (1982).

34. Smith, W.B.; Wilson, Jr., R.R.; Harris, D.B.: A Five-Stage Cyclone System for in-situ Sampling. Environ. Sci. Technol. 13(11):1387 (1979).

35. Bradway, R.M.; Cass, R.W.: Fractional Efficiency of a Utility Boiler Baghouse: Nucla Generating Plant. Report No. EPA-600/2-75-013a to U.S. EPA. GCA Corp., Bedford, MA (August 1975).

36. Cass, R.W.; Bradway, R.M.: Fractional Efficiency of a Utility Boiler Baghouse: Sunbury Steam-Electric Station. Report No. EPA-600/2-76-077a to U.S. EPA. GCA Corp., Bedford, MA (March 1976).

37. Cass, R.W.; Langley, J.E.: Fractional Efficiency of an Electric Arc Furnace Baghouse. Report No. EPA-600/7-77-023 to U.S. EPA. GCA Corp., Bedford, MA (March 1977).

38. McElroy, M.W.; Carr, R.C.; Ensor, D.S.; Markowski, G.R.: Size Distribution of Fine Particles from Coal Combustion. Science 215(4528):13 (1982).

39. Hering, S.V.; Friedlander, S.K.; Collins, J.J.; Richards, L.W.: Design and Evaluation of a New Low-pressure Impactor. Environ. Sci. Technol. 13(2):184 (1979).

40. Lundgren, D.A.; Lippmann, M.; Harris, F.S.; et al., Eds.: Aerosol Measurement. University Presses of Florida, Gainesville, FL (1979).

41. Hemeon, W.C.L.; Black, A.W.: Stack Dust Sampling: In-stack Filter or EPA Train. J. Air Pollut. Control Assoc. 22(7):516 (1972).

42. Selle, S.J.; Gronhovd, G.H.: Some Comparisons of Simultaneous Gas Particulate Determinations Using the ASME and EPA Methods. ASME Reprint 72-WA/APC-4. ASME, United Engineering Ctr., 345 E. 47th St., New York, NY.

43. Govan, F.A.; Terracciano, L.A.: Source Testing of Utility Boilers for Particulate and Gaseous Emissions. Paper No. 72-72, presented at APCA Annual Meeting, Miami Beach, FL. Air Pollution Control Association, Pittsburgh, PA (1972).

44. Crandall, W.A.: Determining Particulates in Stack Gases. Mech. Eng. 14 (December 1972).

Instrument Descriptions

U.S. EPA Method 5 type stack gas sampling equipment is available from the manufactures listed in Table 20-I-1. Fundamentally, all equipment is similar and will meet the U.S. EPA requirements for isokinetically withdrawing a gas sample from a stack and collecting out the particulate matter on a glass fiber filter maintained at a prescribed temperature. Most of the units described below consist of the following components:

Sampling probe: Available in various standard effective lengths (e.g. 3, 5, and 10 ft [0.9, 1.5, and 3 m]) with standard probe liners of stainless steel or borosilicate glass (some manufacturers also have inconel and quartz liners). Probes are equipped with a Type S Pitot tube and thermocouple. Normally, the probe liner is covered by a resistance wire heater and probe thermocouple.

Sampling unit: A two-piece, modular unit consisting of a heated filter box (capable of maintaining a set temperature around the filter holder) and a condenser unit. The condenser is normally an ice-bath compartment capable of holding about six glass impinger assemblies (minimum of four required), with a quick-connect fitting for the umbilical cord.

Umbilical cord: A flexible length of tubing that connects the sampling unit outlet to the meter box (or sample control box) inlet. Incorporated with the sampling line tubing are electrical lines and thermocouple cable that transmit temperature readings to the control unit. Standard 25, 50, 75, and 100 ft (7.5, 15, and 30 m) lengths are available from most manufacturers. Some umbilical cords can be connected to form longer lengths.

Control unit (or meter box): Contains the system's vacuum pump, gas volume meter, gas flow rate meter, two manometers (or other pressure gauges), temperature gauges, operating values, and electrical connections. The control unit is used to adjust and monitor the sample flow rate to achieve isokinetic sampling conditions.

20-1. Universal Stack Sampler

Andersen Instruments, Inc.

This is a typical U.S. EPA Method-5-type stack sampling system. The control unit (meter box) contains a 4 cfm rotary-vane pump, dry gas meter, double-column manometer, LED readout digital temperature meter, operating values, and all electrical connections (with circuit breakers). Control unit weight is 73 lbs (33 kg). The sampling unit (heated filter housing and condenser) is a modular, two-piece case of stainless steel construction. Umbilical cords of 25–300 ft (7.5–90 m) lengths can be connected for longer lengths. Sampling probes have detachable Pitot tube tips and stack and probe liner thermocouples.

20-2. Emission Parameter Analyzer

Andersen Instruments, Inc.

This is another U.S. EPA Method 5-type sampling system. This equipment was formerly manufactured by the Western Precipitation Division of Joy Manufacturing.

20-3. U.S. EPA Method 5 Source Sampling Equipment

CAE Express

CAE Express manufactures complete Method 5 sampling trains, as well as equipment items or systems to be used with several of the other U.S. EPA methods. CAE Instrumental Rental also rents Method 5 and other

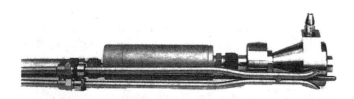

INSTRUMENT 20-1. Universal Stack Sampler.

equipment. Many of the components are interchangeable between manufacturers.

20-4. Isokinetic Source Sampler

Apex Instruments

Apex Instruments manufactures a complete line of source sampling instruments, including meter box consoles, sample cases, probe assemblies, and many glassware items. This equipment is of standard design and operation and meets the normal U.S. EPA criteria for source sampling.

20-5. AP 5500-E-S Stack Sampling System

Scientific Glass and Instruments, Inc.

SGI has a complete line of Method 5 (particle), Method 6 (SO_2), Method 7 (NO_x), Method 8 (H_2SO_4), and other sampling equipment. This equipment is similar to the Method 5 equipment previously described.

20-6. Model 2010 Method 5 Stack Sampling System

Andersen Instruments, Inc.

This Method 5 stack sampling equipment is constructed to meet the U.S. EPA specifications for isokinetic sampling. The Model 2010 system includes a control console, sample case, stainless steel probe, umbilical cord, and glassware. Nutech also manufactures the volatile organic sampling train (VOST); the semi-volatile organic sampling train, which is now the U.S. EPA Method 23 (previously called Modified Method 5); and many other stack sampling equipment items.

20-7. Stack Sampling Systems

NAPP, Inc.

Since 1969, NAPP Inc., has been manufacturing a line of stack sampling equipment. They now produce a very complete line of equipment for most of the U.S. EPA sampling method requirements. In addition to the general Method 5 equipment, NAPP provides a volatile organic sampling system, in-stack thimble filter holder assemblies, and PM_{10} stack sampling equipment, and recently has listed a computerized Model 31 stack sampler.

20-8. Automated Method 5 (AST)

Andersen Instruments, Inc.

A new automated Method 5 sampler is available from Andersen Instruments, Inc. This system is programmed to solve the U.S. EPA Method 5 sampling equations, calculate sampling traverses, record temperatures and pressures, and automatically adjust the isokinetic sample rate. All measurements are displayed

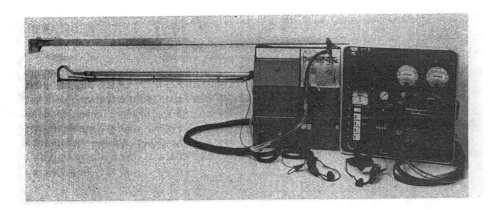

INSTRUMENT 20-7. Manual stack sampling equipment.

on a single LED screen and automatically recorded for a final printout. The manufacturer has stated that this system conforms to 40 CFR 60 requirements.

20-9. PM₁₀ Source Sampling System

Andersen Instruments, Inc.

The PM_{10} source sampler is designed to collect the PM_{10} size particulate emissions from an industrial source. Procedures for measurement of size-specific particulate emissions are more complex, but similar to, particulate sampling by the U.S. EPA Method 5. Size-specific source emission measurements are obtained using an inertial size-separation device such as a cyclone or cascade impactor. A proper size sampling nozzle is used to allow isokinetic sampling at a calculated flowrate to obtain the 10 μm cut size for the cyclone.

Incorporation of emission gas recycling (EGR) allows a variable fraction of conditioned filtered exhaust gas from the sampler to be added back to the sample nozzle and the inertial classifier. This action maintains a constant flowrate through the inertial classifier while the gas flowrate into the sample nozzle is adjusted to remain isokinetic while traversing the stack.

20-10. Stacksamplr LCO™

Andersen Instruments, Inc.

The Stacksamplr™ is a portable, stack gas sampler used for isokinetically collecting solids, mists, and gaseous pollutants from most chemical and combustion processes. The probe is a combination probe and Pitot tube assembly. Several interchangeable inlet nozzles are provided. Probes are available in various effective lengths. The sampling case contains a filter holder for particle removal and impingers for water and gas removal. The control case contains flowmeter, draft gauge, temperature controls, valves, timer, switches, and all the necessary components for control of the isokinetic sampling. A separate pump is used. An umbilical cord connects the sampling case and the control case.

20-11. Automatic High-Volume Stack Sampler

CS-3, Inc.

The automatic high-volume stack sampler is designed for high-volume stack sampling. The integrated sampling system consists of a probe with Pitot tube, large filter holder, and air mover with orifice flowmeter. Isokinetic setting is controlled by a microprocessor. Flow rates are from 10 to 60 cfm (300–1700 L/min). The system is also available in a manual configuration. Probe assembly is constructed of aluminum with three nozzles available for velocities of 800–1200 fpm (4–6 m/s). The filter housing is designed for an 8- × 10- in. (2- × 25-cm) glass fiber filter. The balance of the sampler consists of a control system; a flexible sampling hose; a two-speed, heavy-duty suction-blower; and a microprocessor.

20-12. Dust and Fume Determination Assembly, Models D-1000 and D-1027

Andersen Instruments, Inc.

This equipment was designed to measure aerosol concentrations in a gas as it passes through a flue. The measurement involves withdrawing a controlled flow rate of gas from the flue and separating the particulate matter from the gas. The type of particle, the temperature, and the moisture content of the carrier gas will determine the method to be used in making the separation. The paper thimbles (Model D-1000) may be used up to 120°C (250°F) and suction pressures of up to 4 in. (10 cm) Hg. Alundum thimbles (Model D-1027) are used where it is important to have high wet strength, chemical resistance, or high temperature resistance.

The complete dust and fume sampling equipment employing the Alundum thimble method (Model D-1027) consists of an aspirating eductor, thermometer, vacuum gauge, dry gas meter, condenser, heavy-duty rubber hose, Alundum thimble holder, and various nozzles. The paper thimble equipment differs from the Alundum arrangement only in that different sampling nozzles are used and the thimble holder is made of aluminum for placement outside the flue. Filter thimble holders are described in greater detail elsewhere.

20-13. Stack Sampling Nozzles and Thimble Holders
BGI Incorporated

These are components of sampling trains for in-stack particulate sampling. BGI thimble holders have been designed for use with the U.S. EPA-type probes for Methods 16 and 17. They can also be used for Method 5 and other sampling applications. They are constructed of polished 316 stainless steel. The 30 × 100 mm unit can also be used with Alundum thimbles. The BGI button-hook nozzle is designed to fit the U.S. EPA-type sampling trains that accept nozzles terminating in 5/8-in. (1.59-cm) o.d. tubing. Nozzles are available in nominal inside diameters from 1/8 to 1 in. (0.32–2.5 cm).

20-14. Isokinetic Sampling Systems
Kurz Instruments, Inc.

The Kurz systems employ thermal mass flow sensors to sense both stack and sample gas velocities to control sampling rate automatically to achieve isokinetic sampling. The Series 1275 systems are singlepoint systems in which the stack and sample sensors can be operated in a differential mode to provide automatic sample flow control and isokinetic sampling at a single point. The Series 4200 systems are multipoint systems in which the average stack velocity is measured by sensors at several points, and the average sample velocity is measured by a single sensor in the combined sample from all the sample points. Although this ensures overall average isokinetic conditions, sampling at individual points may be anisokinetic.

Series 1275 components include probe assembly consisting of isokinetic sampling head, probe support, and filter box (or other collecting device at user's option); system enclosure housing flow sensor electronics, electronic sample valve controller, and sample valve; and pump. Series 4200 components include multipoint stack velocity sensor probe; single or dual sampling nozzles and manifold; and system electronics, including flow control valve, pump, and sample collection device at user option.

20-15. Stack Sampling Equipment
Environmental Supply Company, Inc.

Environmental Supply Company, Inc. of Durham, NC manufactures a line of stack sampling equipment suitable for most of the U.S. EPA sampling method requirements. Systems are available for Methods 2, 3, 5, 6, 17, 23, 26, 29, 201 and VOST.

TABLE 20-I-1. List of Manufacturers

	Apex Instruments 125 Quantum St. P.O. Box 727 Holly Springs, NC 27540 (919) 557-7300 or (800) 882-3214 FAX (919) 557-7110 info@apexinst.com www.apexinst.com	CSS	CS-3, Inc. P.O. Box 5186 Bend, OR 97708 (541) 388-4729 or (800) 910-9398 FAX (541) 382-1807 www.envirometrics.com/cs3	KRZ	Kurz Instruments, Incorporated 2411 Garden Road Monterey, CA 93940 (800) 424-7356 FAX (831) 646-8901 www.kurz-instruments.com
BGI	BGI Incorporated 58 Guinan Street Waltham, MA 02451 (781) 891-9380 FAX (781) 891-8151 www.bgiusa.com		Environmental Supply Company 2142 East Geer Street Durham, NC 27704 (800) 782-2575 or (919) 956-9688 FAX (919) 682-0333 esc@environsupply.com www.environsupply.com		NAPP, Inc. 2104 Kramer Lane Austin, TX 78763 (512) 479-7509 FAX (512) 837-4532
	CAE Express 500 W. Wood Street Palatine, IL 60067 (800) 223-3977 FAX 847 991-3385 cae@cleanair.com www.cleanair.com	GRA	Andersen Instruments, Inc. 500 Technology Court Smyrna, GA 30082-5211 (404) 319-9999 or (800) 241-6898 FAX (770) 319-0336 www.graseby.com	SGI	Scientific Glass & Instruments, Inc. P.O. Box 6 Houston, TX 77001 (281) 682-1481 FAX (281) 682-3054

Chapter 21

Sampling Airborne Radioactivity

Beverly S. Cohen, Ph.D., Maire S.A. Heikkinen, Ph.D.
Nelson Institute of Environmental Medicine, New York University School of Medicine, Tuxedo, NY

CONTENTS

Introduction

Radioactivity is the spontaneous transformation of the nucleus of an atom by the emission of corpuscular or electromagnetic radiation. Radioactive contaminants have historically been considered apart from chemical contaminants because it is their radiological properties that determine their biological and environmental impact. Additionally, they have been regulated by special government agencies concerned with radiological protection. Prior to the 1940s, there was essentially no concern about airborne radioactivity. The role of the short-lived decay products of radon in the etiology of lung cancer in underground miners was not yet appreciated. Small amounts of naturally occurring radionuclides were released to air from burning of fossil fuels, but there was almost no potential for other contaminant air-

borne radionuclides. Protection from significant exposure to ionizing radiation was required for only a limited number of scientists and physicians. This was provided by adherence to guidelines recommended by groups such as the International Commission on Radiological Protection (ICRP) and the National Council on Radiation Protection and Measurements (NCRP).

Radioactive contaminants are also distinguished by the specialized and very sensitive methods available for the detection of radioactivity. Measurements of concentrations of a few thousand atoms per liter are not uncommon. Average indoor air concentrations of radon (^{222}Rn), for example, are less than 2×10^4 atoms per liter, or about 7×10^{-13} ppm. Concentrations of the short-lived decay products of radon normally total fewer than 30 atoms per liter. The sensitivity with which radioactivity can be detected results from the ionization produced in matter by the radiation. This ionization also produces responses in biological tissue at very low levels of irradiation, so that in a sense, the measurement capabilities are commensurate with the significance of the quantities measured. Yet, complex questions result from the ability to measure very small quantities of radiation, e.g., "What is the significance of a radiation dose to tissue that is a small fraction of the dose from natural background radiation?" and "How low is 'as low as reasonably achievable'(ALARA),[1] when implementing radiation protection guidelines?"

Special constraints on sampling methods that derive from the specific radiological properties of a contaminant must be integrated with good basic air sampling practices. Guidance may be obtained from other parts of this text for activities ranging from design of appropriate sampling strategies through design and calibration of the entire sampling train; including consideration of inlet bias, isokinetic sampling, efficiency of the collection substrate, sample loss and stability, and air flow calibration. Special consideration must be given to the radiometric properties of the particular nuclide to evaluate the need for sample processing. Chemical separations are frequently unnecessary because of the ease with which radioactive materials can be detected. Source preparation and the radiation detection system must be suited to the type of radiation emitted, and rapid decay of the sample is sometimes a significant problem.

This chapter will discuss some special aspects of sampling that result from the radioactivity of the airborne material. Very good information on radiation is available through the Internet homepage of the University of Michigan (http://www.umich.edu/~radinfo).

Units

Background

Three separate physical entities must be considered: 1) the source of the radiation, 2) the radiation, and 3) the absorber. It is important to recognize the separateness of these items. Sources of ionizing radiation include the sun and other extraterrestrial objects, radioactive isotopes, and particle accelerators (including common X-ray machines). The only airborne sources are radioactive isotopes. The radiation travels outward from the source carrying away energy; it can continue indefinitely with essentially undiminished energy if traversing a vacuum. The absorber is the material in which the radiation will deposit energy by ionization and excitation of the atoms. In some cases, source and absorber are inextricably meshed, but they are nonetheless inherently separate

entities with different physical properties which are not transferable from one to the other.

Convenient measurement units such as the curie, roentgen, and rad (see below) were developed over the years by scientists working with ionizing radiation. As knowledge and measurement processes improved, these historical units were occasionally reevaluated and standardized. As a result of international agreement, a new set of units consistent with the System Internationale (SI) was adopted in 1975.[2] These new units have been generally accepted since 1985.[3] A few important units are given below. These units apply to 1) sources, 2) the radiation, and 3) the absorber. Both historical and SI units are listed. A complete list of units with conversion factors is presented in Table 21-1.

Definitions

Sources

The quantity of a radioactive source is defined by its "activity" or the rate of spontaneous nuclear transformation (see Equation 1). The unit of activity is the becquerel (Bq):

$$1 \text{ Bq} = 1 \text{ } s^{-1}$$

Thus, 1 Bq represents one transformation, or disintegration, per second.

The historical unit of activity is the curie (Ci):

$$1 \text{ Ci} = 3.7 \times 10^{10} \text{ } s^{-1} \text{ (exactly)}$$

Radiation

Exposure is a measure of the quantity of X or gamma radiation. It is defined by the quantity of electric charge the radiation produces as it traverses an air mass. Exposure does not have a special unit in the SI system but combines the basic units of charge in coulombs (C) and mass in kilograms (kg). The units of exposure are $C \text{ kg}^{-1}$.

The conventional unit of exposure is the roentgen (R):

$$1 \text{ R} = 2.58 \times 10^{-4} \text{ C kg}^{-1} \text{ (exactly)}$$

Thus, 2.58×10^{-4} C is the charge of the ions of one sign produced in one kg of air by one roentgen of X or gamma radiation.

Energy: Corpuscular radiation is generally described by stating the particle identity and its kinetic energy. The SI unit of energy is the joule, but conventional units in multiples of the electron volt (eV) are used almost exclusively. Common multiples are keV (10^3 eV) and MeV (10^6 eV). One eV is the kinetic energy acquired by an electron accelerated through a potential difference of 1 volt.

TABLE 21-1. Conversion Between SI and Conventional Units*

Quantity	Symbol for Quantity	Expression in SI Units	Expression in Symbols for SI Units	Special Name for SI Units	Symbols Using Special Name	Conventional Unit	Symbol for Conventional Unit	Value of Conventional Unit in SI Units
Activity	A	1 per second	s^{-1}	becquerel	Bq	curie	Ci	3.7×10^{10} Bq
Absorbed dose	D	joule per kilogram	Jkg^{-1}	gray	Gy	rad	rad	0.01 Gy
Absorbed dose rate	$\dot{D}$	joule per kilogram second	$Jkg^{-1}\,s^{-1}$		Gy/s	rad	$rads^{-1}$	0.01 Gy/s
Average energy per ion pair	W	joule	J			electron volt	eV	1.602×10^{-19} J
Equivalent dose	H	joule per kilogram	Jkg^{-1}	sievert	Sv	rem	rem	0.01 Sv
Equivalent dose rate	$\dot{H}$	joule per kilogram second	$Jkg^{-1}\,s^{-1}$		Svs^{-1}	rem per second	$rems^{-1}$	0.01 Svs^{-1}
Electric current	I	ampere	A			ampere	A	1.0 A
Electric potential difference	U, V	watts per ampere	WA^{-1}	volt	V	volt	V	1.0 W/A
Exposure	X	coulomb per kilogram	Ckg^{-1}			roentgen	R	2.58×10^{-4} Ckg^{-1}
Exposure rate	$\dot{X}$	coulomb per kilogram second	$Ckg^{-1}\,s^{-1}$			roentgen per second	Rs^{-1}	2.58×10^{-4} $Ckg^{-1}s^{-1}$
Fluence	F	1 per meter squared	m^{-2}			1 per centimeter squared	$1cm^2$	$1.0 \times 10^4 m^{-2}$
Fluence rate	$\dot{F}$	1 per meter squared second	$m^{-2}\,s^{-1}$			1 per centimeter squared second	$1cm^{-2}s^{-1}$	$1.0 \times 10^4 m^{-2}s^{-1}$
Kerma	K	joule per kilogram	Jkg^{-1}	gray	Gy	rad	rad	0.01 Gy
Kerma rate	$\dot{K}$	joule per kilogram second	$Jkg^{-1}\,s^{-1}$		Gys^{-1}	rad per second	$rads^{-1}$	0.01 Gys^{-1}
Lineal energy	y	joule per meter	Jm^{-1}			kiloelectronvolt per micrometer	$keV\mu m^{-1}$	1.602×10^{-10} Jm^{-1}
Linear energy transfer	L	joule per meter	Jm^{-1}			kiloelectronvolt per micrometer	$keV\mu m^{-1}$	1.602×10^{-10} Jm^{-1}
Mass attenuation coefficient	$\mu\rho^{-1}$	meter squared per kilogram	m^2kg^{-1}			centimeter squared per gram	cm^2g^{-1}	0.1 m^2kg^{-1}
Mass energy transfer coefficient	$\mu_{tr}\rho^{-1}$	meter squared per kilogram	m^2kg^{-1}			centimeter squared per gram	cm^2g^{-1}	0.1 m^2kg^{-1}
Mass energy absorption coefficient	$\mu_{en}\rho^{-1}$	meter squared per kilogram	m^2kg^{-1}			centimeter squared per gram	cm^2g^{-1}	0.1 m^2kg^{-1}
Mass stopping power	$S\rho^{-1}$	joule meter squared per kilogram	Jm^2kg^{-1}			million electron volts centimeter squared per gram	$meV\,cm^2g^{-1}$	1.602×10^{-14} Jm^2kg^{-1}
Power	P	joule per second	Js^{-1}	watt	W	watt	W	1.0W
Pressure	P	newton per meter squared	Nm^{-2}	pascal	Pa	torr	torr	(101325/760)Pa
Radiation chemical yield	G	mole per joule	$molJ^{-1}$			molecules per 100 electron volts	molecules/100 eV	1.04×10^{-7} $molJ^{-1}$
Specific energy	z	joule per kilogram	Jkg^{-1}	gray	Gy	rad	rad	0.01 Gy

*Adapted from NCRP Report No. 82.[3]

$$1 \text{ eV} = 1.602 \times 10^{-19} \text{ J}$$

$$= 1.602 \times 10^{-12} \text{ ergs}$$

Dose

Absorbed Dose: Dose is the quantity of energy transferred to the absorber by the ionizing radiation. The SI unit of absorbed dose (D) has been given a special name, the gray (Gy):

$$1 \text{ Gy} = 1 \text{ Jkg}^{-1}$$

The historical unit of absorbed dose is the rad, which is equal to 100 ergs per gram of absorber.

$$1 \text{ rad} = 10^2 \text{ Gy}$$

Equivalent Dose: A special unit used in radiation protection is the equivalent dose ($H_{T,R}$). It is the product of the average absorbed dose in a specified organ or tissue ($D_{T,R}$) and a radiation weighting factor (W_R) that accounts for biological effectiveness of the ionizing radiation producing the dose. Thus, equivalent dose is:

$$H_{T,R} = W_R D_{T,R}$$

where: $D_{T,R}$ = the absorbed dose in a specified tissue
W_R = the radiation weighting factor

The subscript T refers to the specific tissue; subscript R refers to the specific radiation. Values of W_R are given in Table 21-2. When several types or energies of radiation are present, the average equivalent doses must be summed. The unit of equivalent dose is the sievert (Sv):

$$1 \text{ Sv} = 1 \text{ Jkg}^{-1}$$

The historical unit is the rem:

$$1 \text{ rem} = 10^{-2} \text{Jkg}^{-1}$$

TABLE 21-2. Radiation Weighting Factor, w_R[A,B]

Type and energy range		w_R
x and γ rays, electrons, positrons and muons		1
Neutrons, energy	<10 keV	5
	10 keV to 100 keV	10
	>100 keV to 2 MeV	20
	>2 MeV to 20 MeV	10
	>20 MeV	5
Protons, other than recoil protons and energy >2 MeV		2
Alpha particles, fission fragments, nonrelativistic heavy nuclei		20

[A]Adapted from NCRP No. 116.[4]
[B]For detailed explanations and constraints on usage, see ICRP No. 60[5] or NCRP no. 116.[4]

More information on units and terminology is available through the U.S. Nuclear Regulatory Commission (NRC) Internet homepage (http://www.nrc.gov).

Fundamentals of Radioactivity

Radioactive Decay

The transformation, or decay, of a nucleus is a random process so that if there is a large number (N) of identical radioactive atoms, the rate at which they decay (dN/dt) in a given time period will be a constant fraction of N.

$$\frac{dN}{dt} = -\lambda N \tag{1}$$

where: dN = the number of unstable nuclei which transform in a time interval dt.
λ = the proportionality constant, or the fraction which decay per unit time.

λ is known as the decay constant and is characteristic of a given nuclide or atomic species. dN/dt is the "activity" of a source. Decay is a stochastic or random process; thus, Equation 1 only applies to sufficiently large samples of a nuclide.

Integration of Equation 1 from time $t = 0$ to t yields the number of nuclei which survive to time t:

$$N = N_o e^{-\lambda t} \tag{2}$$

where: N_o = the number of nuclei at $t = 0$
N = the number present at time t.

The time (T) at which half the nuclei will have transformed or decayed ($t = T$ when $N/N_o = 1/2$) is then:

$$T = \frac{0.693}{\lambda} \tag{3}$$

where: T = the half life of the species; a characteristic time that is always the same for a particular nuclide.

Radiation Properties

The physical properties of the emitted radiation determine both the biological significance of the radiation and various requirements for sampling and detection. The most common corpuscular radiations are alpha or beta particles. Electromagnetic radiation is emitted in the form of high energy photons called gamma rays.

Alpha Particles

Alpha particles are helium nuclei. They are emitted mainly from nuclei with high atomic mass leaving behind an atom with atomic number reduced by 2 and mass reduced by 4 mass units. Their energies range from about 2 to 11 MeV. Alpha particles emitted from a given nuclear species are monochromatic; that is, they all have the same kinetic energy. Alpha particles are massive enough so that they are not easily deflected as they traverse matter and, typically, their paths are straight lines. The double charge and relatively high mass causes dense ionization along the track. A 5.0-MeV alpha particle, for example, will cause several thousand ion pairs per micrometer (μm) of water or tissue, transferring about 100 keV of energy per μm to the molecules of the absorber. The rate at which energy is transferred per unit path length of an absorber is called the linear energy transfer (LET). Alphas are classified as high LET particles. They can only traverse a few centimeters of air or a few micrometers of tissue before losing all of their initial kinetic energy. This very limited range prevents alpha particles from penetrating the skin. Unless an alpha particle source (i.e., a radioactive alpha-emitting particle) is inhaled or ingested, significant irradiation of internal tissue cannot occur. Any absorber in the path of an alpha particle will significantly reduce its energy. Self-absorption by the source can be substantial. The efficiency with which alpha particles may be detected when particulate material is collected on a filter is highest if samples are very thin. The detection efficiency for alpha particles on a dust-laden filter will be reduced significantly because of self-absorption by the dust.

Beta Particles

Beta particles are positive or negative electrons. When an atom decays by beta emission, the atomic number changes by ± 1, but the atomic mass does not change if an electron (e^-) is emitted because an orbital electron will replace the lost mass to balance the extra positive charge gained by the nucleus. If a positron (e^+) is emitted, the atomic mass is reduced by twice the mass of an electron. When a nucleus decays by beta emission, a neutrino or antineutrino is also emitted and the energy loss is shared between the particles. Thus, betas from a given species are emitted with a range of energies up to a maximum that is specific to the nuclear transition. The average share of the energy carried off by the beta particle (from a collection of the same atoms) is about one-third of the total energy of the nuclear transition. Typical energies range from 10 keV to 4.0 MeV. Beta particles are easily deflected by interactions with orbital electrons because they have the same mass, so they travel erratic paths causing ionization and excitation of atoms as they pass until all of their initial kinetic energy has been transferred to the absorber. The trail of ion pairs left behind will be much less dense than that of an alpha particle. Beta particles will typically lose energy to the absorber at a few keV per micrometer and are thus low LET radiation. Positrons will ultimately interact with an electron causing both to annihilate with the emission of two 0.511-MeV gamma rays. Beta particles, depending on energy, may travel from a few centimeters to 10 or 15 m in air, or from a few micrometers to about 2.0 cm in tissue.

Gamma Rays

Gamma rays are photons and exhibit both wave and particle properties. The energy (E) is proportional to the frequency (f) of the radiation; $E = hf$, where h is Planck's constant. Photons from a particular nuclear transition are monochromatic, but some nuclear decays result in emission of several different photons. Typical energies range from a few keV to a few MeV. The manner in which high energy photons, or gamma rays, interact with matter to ionize atoms in the absorber varies with energy and the specifics of the absorbing material. The energy of a beam of gamma radiation will be attenuated exponentially because interactions between the gamma rays and the atoms of the absorber are stochastic. Gamma rays do not exhibit a finite range but the mean free path, i.e., the average distance a photon will travel before having a collision, gives a measure of the penetration. The mean free path is also known as the relaxation length. The mean free path in air for a 1.0-MeV gamma ray is about 120 m; it is about 14 cm in water or tissue.

Other Emissions

A variety of particles other than alpha particles, beta particles, and gamma rays are emitted less commonly in nuclear transformations. These include protons, neutrons, conversion electrons, Auger electrons, and X-rays. Further information may be found in NCRP Report No. 58[6] and Knoll.[7] A comprehensive listing of detailed decay schemes is presented in Firestone *et al.*[8]

Table 21-3 presents a list of major radiations of some isotopes used in medicine and industry, identified in materials or air around accelerators, or found in reactor coolant and corrosion products.

Radiation Detectors

Radiation detectors in common use are gas-filled chambers, scintillation detectors, semiconductor detectors, thermoluminescent dosimeters, and etched-

TABLE 21-3. Half Life and Major Radiations of Selected Isotopes[A]

Nuclide	Half life[B]	Major Radiations	Approximate Energies (MeV) and Intensities[B]
$^{3}_{1}$H	12.33y	β^-	0.0186 max
$^{7}_{4}$Be	53.3d	γ	0.478 (10.3%)
$^{14}_{6}$C	5730y	β^-	0.156 max
$^{13}_{7}$N	9.96m	β^+ γ	1.19 max 0.511 (200%, annihilation radiation)
$^{15}_{8}$O	122.s	β^+ γ	1.723 max 0.511 (200%, annihilation radiation)
$^{22}_{11}$Na	2.602y	β^+ γ	0.545 max (90.57%) 1.275 (100%)
$^{24}_{11}$Na	15.02h	β^- γ	1.389 max 1.369 (100%), 2.754 (100%)
$^{32}_{15}$P	14.28d	β^-	1.711 max
$^{35}_{16}$S	87.4d	β^-	0.167 max
$^{41}_{18}$Ar	1.837h	β^- γ	2.49 max, 1.198 max 1.293(99%)
$^{42}_{19}$K	12.36h	β^- γ	3.519 max 1.524 (18.8%), 0.312 (0.3%)
$^{47}_{20}$Ca	4.536d	β^- γ	1.988 max (16%), 0.684 max (83.9%) 1.297 (77%), 0.807 (7%), 0.49 (7%)
$^{51}_{24}$Cr	27.70d	 γ	V X-rays 0.320 (10.2%)
$^{54}_{25}$Mn	312d	 γ	Cr X-rays 0.835 (100%)
$^{55}_{26}$Fe	2.7y		Mn X-rays
$^{59}_{26}$Fe	44.6d	β^- γ	0.273 max (48.5%), 0.475 max (51.2%), 1.573 max (0.3%) 0.143 (1.02%), 0.192 (3.08%), 1.099 (56.5%), 1.292 (43.2%)
$^{57}_{27}$Co	271d	γ	0.122 (86%), 0.136 (11%), 0.014 (9%), Fe X-rays
$^{60}_{27}$Co	5.271y	β^- γ	0.318 max (99.88%) 1.173 (99.90%), 1.332 (99.98%)
$^{85}_{36}$Kr	10.7y	β^- γ	0.672 max 0.514 (0.43%)
$^{89}_{38}$Sr	50.5d	β^-	1.488 max (99.99%)
$^{90}_{38}$Sr	28.8y	β^-	0.546 max
$^{90}_{39}$Y	64.1h	β^-	2.288 max (99.98%)

TABLE 21-3 (cont.). Half Life and Major Radiations of Selected Isotopes[A]

Nuclide	Half life[B]	Major Radiations	Approximate Energies (MeV) and Intensities[B]
$^{99m}_{43}$Tc	6.02h	γ	Tc X-rays 0.141(89%)
$^{125}_{53}$I	60.2d	γ	0.035 (6.7%) Te X-rays
$^{131}_{53}$I	8.04d	β^- γ	0.336 max (13%), 0.606 max (86%), 0.81 max (0.6%) Xe X-rays 0.284 (6.04%), 0.0802 (2.61%), 0.364 (81%), 0.637 (7.21%), 0.723 (1.79%)
$^{138}_{54}$Xe	14.1m	β^- γ	2.720 max, 2.460 max 0.605 (32%), 0.434 (20%), 1.768 (17%), 2.015 (12%), 0.396 (6%)
$^{137}_{55}$Cs	30.17y	β^- γ	0.5116 max (94.6%), 1.176 max (6%) 0.662(85%)
$^{192}_{77}$Ir	74.2d	β^- γ	0.672 max (47%), 0.536 max (41%), 0.256 max (6%) 0.316 (83%), 0.468 (48%), 0.308 (30%), 0.296 (28.7%), 0.588 (4.6%), 0.604 (8.3%) Os X-rays, Pt X-rays
$^{198}_{79}$Au	2.696d	β^- γ	0.961 max, 0.290 max 0.4118 (96%), 0.676 (1%), 1.088 (2.5%)
$^{210}_{82}$Pb	22.26y	β^- γ	0.063 max (18%), 0.016 max (82%) 0.0465 (4%) Bi L X-rays
$^{222}_{86}$Rn	3.8235d	α	5.489 (100%)
$^{224}_{88}$Ra	3.66d	α γ	5.686 (95%), 5.449 (5%) 0.241(4%) Rn X-rays
$^{226}_{88}$Ra	1600y	α γ	4.784 (94%), 4.602 (6%) 0.186 (3%) Rn X-rays
$^{241}_{95}$Am	433y	α γ	5.486 (86%), 5.443 (13%), 5.387 (1.3%) 0.060(36%) Np L X-rays

[A]After Schleien and Terpilak.[9]
[B]Common time units: y (years); d (days); m (minutes); s (seconds). Data from *Table of Isotopes*.[8]

track detectors. Ionization chambers, proportional counters, and Geiger–Mueller counters are gas-filled chambers. The incident radiation interacts with the gas to form ion pairs. An electric field is established across the gas volume by collecting electrodes. The electrons are collected at the anode and the positive ions at the cathode. Semiconductor detectors similarly collect the ion pairs produced in a small volume of a semiconducting solid. Scintillation counting is based on the detection of visible light that is emitted by certain materials when they are irradiated. Recent technical developments have increased the use of etched-track detectors and thermoluminescent dosimeters. Other less used methods include photographic film, calorimetric measurements, and chemical reaction vessels. These latter methods are not normally used for air sampling and will not be discussed further. Additional information may be obtained from NCRP,[6] Knoll,[7] and Eichholz and Poston.[10]

Gas-Filled Detectors

Ionization Chambers

In an ionization chamber, the ions produced in the gas by radiation are collected as a result of the applied electric field, the electrons moving to the anode and the positive ions to the cathode. With sufficient voltage across the electrodes, all ions will be collected before recombination can occur. The current produced is measured by a microammeter or a sensitive current integrating device. Either the total amount of charge or the rate at which charge is collected is a measure of the intensity of the radiation. Small portable ionization chambers are available for use as survey meters. If they are to be used for alpha or beta particle detection, there must be a very thin "window" that the particles can penetrate to reach the detection volume. For photons, penetration is not a problem, but few ion pairs will be produced in a small gas volume, resulting in very low detection efficiency. The number of ion pairs formed in the gas depends on the gas density; thus, increased sensitivity may be obtained by increasing the gas pressure. Pressure ionization chambers containing argon, which operate at about 20 atmospheres, can be used to measure environmental gamma ray fields.

Ionization chambers may be used for detecting individual pulses rather than current flow. If an ionizing particle produces a number of ion pairs in the chamber, a current pulse will result, and the rate at which pulses are registered is a measure of the radiation intensity. The number of events in a measured time period may also be used, with calibration and geometric corrections, to determine source activity. If all of the energy of the original ionizing particle is absorbed in the gas volume, the size of the pulse will be proportional to the initial energy of the particle. Suitable electronics must be used to shape the pulse and provide time resolution. Ionization chambers are particularly useful for radiation with high linear energy transfer, such as alpha particles, which produce many ion pairs within the detection volume.

Proportional Counters

As the voltage across a chamber increases, the initial electron from each ion pair gains sufficient kinetic energy as it moves toward the anode to ionize some of the gas molecules. The resultant secondary ion pairs will amplify the pulse. The higher the applied voltage, the more energetic the initial electrons will become, and the more secondary ion pairs will be produced. The pulse size thus increases with voltage. A chamber operating in this region of amplification is a proportional counter.

Proportional counters generally utilize a cylindrical configuration, with a central high-voltage electrode as the anode and an outer conducting surface as the cathode (Figure 21-1). This configuration produces a high gradient field around the central electrode. If the voltage is carefully maintained, the pulse size will be proportional to the original quantity of ionization. The pulses of current may then be sorted and recorded electronically according to size by a multichannel analyzer, or specific sizes may be selected for counting using discriminators to remove smaller and larger pulses. The size of the pulse represents the amount of energy absorbed in the gas volume, and with proper calibration, the energy resolution can be used to identify specific nuclides. The presence of a particular nuclear emission will be indicated by a peak occurring at a given energy which can be separated from the general spectrum of background radiation. Gas-filled detectors for energy analysis have largely been replaced by crystalline and solid-state detectors, which are described below.

Gas-filled counters may be operated in the proportional region at atmospheric pressure with one end open so that the source can be placed directly into the counting volume. This is valuable for very low energy radiations which cannot penetrate the window of a counting chamber.

Geiger–Mueller Counters

As the voltage across a gas volume increases further, a region will be reached where a single ionization within the chamber will result in secondary ionization of all

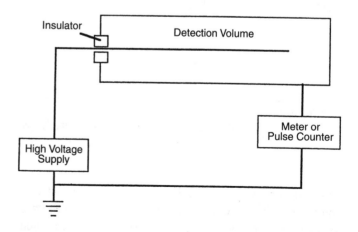

FIGURE 21-1. Block diagram of a gas-filled radiation detector system. A pulse height analyzer may be added when the detector is used as a proportional counter.

of the gas molecules in the volume. This is the Geiger–Mueller (G–M) operating region. The response of the chamber will be nearly constant over a considerable voltage range (the plateau) until the voltage becomes so high that the applied electric field will pull electrons from the gas molecules and the chamber will enter a self-discharge region. Along the G–M plateau, any ionization will result in a pulse of the same magnitude, and the chamber is used to simply count the number of ionizing events which take place within. If a chamber is operated in the G–M region, quenching gases or electronic quenching must be used to stop the electrical discharge after each pulse. The external circuitry provides pulse shaping for time resolution, but the counter will not be able to respond to a second ionizing event during the discharge, and measured count rates need to be corrected for dead time. In regions of very high gamma-ray flux, such instruments may be unable to respond. They should be designed to then give maximum readout; otherwise, a false zero may be indicated.

G–M tubes are useful for detecting gamma radiation that may cause only a single ionizing event in a gas volume; however, the detection efficiency is low. They can be built with thick walls and are relatively sturdy. Many radiation survey instruments are comprised of a small portable power supply and meter to which a G–M tube "probe" is attached by a flexible cable. G–M tubes with thin end-windows can be used to scan surfaces for beta or alpha particle contamination or to count small sources.

Scintillation Detectors

Many substances emit visible light when exposed to ionizing radiation. These include phosphors such as zinc sulfide crystals, sodium iodide and cesium iodide crystals, and various organic materials. Liquid scintillators to detect low energy beta particles are frequently used in biological studies but are rarely used with air samples. NCRP[6] provides references to information sources on the subject.

Detector crystals are made with specific impurities to improve their scintillation properties. NaI and CsI crystals activated with thallium are commonly used for photon detection. They are much more efficient absorbers of photons than gas-filled chambers. If the photon is completely absorbed in the scintillator, the quantity of light emitted will be proportional to the energy of the incident photon. Because the amount of energy absorbed increases with the volume of the absorber, large crystals are desirable. To be useful, the scintillator must be transparent to the emitted light; therefore, single crystals are needed for large detectors.

The emitted light is converted to an electrical pulse by a photomultiplier tube. The signals from the photomultiplier tube are amplified and electronically counted (Figure 21-2). Pulses greater than or less than a certain size may be counted or the pulses may be accumulated by size in a multichannel analyzer, as described above for a proportional counter. Relatively good energy resolution for gamma rays may be obtained with a crystal scintillation detector.

An example of the use of scintillation crystals to quantify airborne radioactivity is the measurement of radon gas adsorbed on charcoal. Charcoal-containing canisters are deployed for periods of 4–7 days. Radon gas will diffuse into the container and be adsorbed onto the charcoal. The radon gas will decay through a series of short-lived nuclides, several of which emit photons (Table 21-4). The photons specific to radon decay can be selectively counted by a scintillation detection system.

ZnS activated with silver is the most commonly used phosphor for alpha particle detection. It is available coated on Mylar® sheets or discs which may be placed directly onto an alpha particle source for high detection efficiency. An extremely low background arrangement for alpha-particle counting uses phosphor-coated Mylar discs.[11] It is particularly effective for very low activity samples such as filters used to collect environmental levels of the short-lived decay products of ^{222}Rn. Each filter is placed on a small plastic mount, covered with the phosphor disc, and then wrapped with Mylar. Alternatively, phosphor-coated material may be incorporated into a fixed detection system and the sample filter placed in a source holder near the phosphor. ZnS phosphor is also used to coat the interior surface of grab samplers known as "Lucas flasks" for detection of alpha radioactivity in air (Figure 21-3).[12]

Semiconductor Detectors

Detectors fabricated from solid semiconducting materials, primarily silicon or germanium, are essentially solid-state ionization chambers. Ionization takes place within the detector volume, producing pairs of charge carriers consisting of an electron and a hole. The charge is collected by a voltage placed across the detection volume. The electrons can be collected from the solid, at least for very thin solids (maximum thickness about 1.5–2.0 cm), because they have been raised to conduction bands by the excitation; the holes in the valence band move toward the opposite electrode. As with gas ionization chambers, the collected charge is proportional to the amount of energy deposited in the sensitive volume. Various methods are used to create as large a

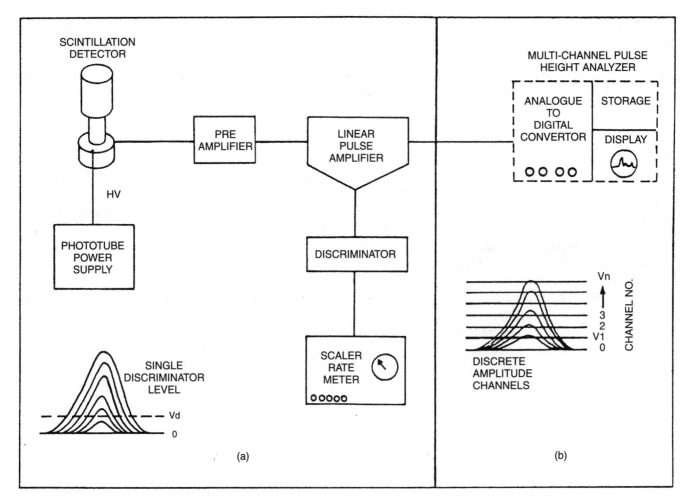

FIGURE 21-2. Block diagram of typical scintillation-counter systems: (a) for integral count-rate measurements for photon energies above those corresponding to a discriminator voltage V_d; (b) for "pulse-height-spectrometer" measurements at different photon energies corresponding to discriminator voltage intervals V_1, V_2...V_n.[6] *Reprinted with permission of National Council on Radiation Protection and Measurements.*

sensitive region in the solid as possible. This requires that holes and electrons be balanced properly when a collecting voltage is applied. One of the methods involves drifting lithium ions through the detector. It is then necessary to keep the detector at cryogenic temperatures to maintain the lithium gradient. Other methods result in detectors that can be maintained at room temperature, e.g., silicon surface barrier detectors. These detectors have very thin detection regions that are useful for alpha or beta particle spectrometry but not for gamma ray photons. High-purity germanium (HPGe) detectors, which are not lithium drifted, must be operated at cryogenic temperatures because the electrons can be raised to the conduction band by transfer of thermal energy at room temperatures, thus adding unwanted background noise to the system. Semiconductor detectors, because they are solids, have much higher detection efficiency for gamma rays than do gas-filled detectors.

Ge, with a higher density and atomic number than Si, is preferred for gamma detection. The energy resolution is superior to that of a NaI crystal (Figure 21-4).

Etched Track Detectors

Solid-state nuclear track detectors consist of a large group of inorganic and organic dielectrics that register tracks when traversed by heavy charged particles. The first observation of charged particle tracks in a crystal was reported in 1958.[13] The track is more vulnerable than the bulk material to dissolution by etching agents, which makes possible enlargement of the tracks to a size that can be observed optically. There is a threshold in the amount of linear energy transfer by an ionizing particle to the detector that must be exceeded for tracks to register. Only a few materials have been identified which will respond to alpha particles. These materials include

TABLE 21-4. Uranium Series from ^{222}Rn to ^{210}Pb

Nuclide	Historical Name	Half life*	Major Radiation Energies (MeV) and Intensities [7,9]		
			α	β	γ
$^{222}_{86}$Rn	Emanation radon (Rn)	3.823 d	5.49 (99.92%)	—	—
$^{218}_{84}$Po	Radium A	3.05 m	6.00 (99.98%)	—	—
$^{214}_{82}$Pb	Radium B	26.8 m	—	0.178 (2.4%) 0.665 (46.1%) 0.722 (40.8%) 1.02 (9.5%)	0.295 (18.4%) 0.352 (35.4%) 0.768 (1.04%)
$^{216}_{85}$At	Astatine	~2 s	6.65 (6.4%) 6.99 (90%) 6.76 (3.6%)	—	—
$^{214}_{83}$Bi	Radium C	19.9 m	—	1.06 (5.56%) 1.15 (4.25%) 1.41 (8.15%) 1.50 (16.9%) 1.54 (17.5%) 1.89 (7.56%) 3.27 (19.8%)	0.609 (44.8%) 0.768 (4.76%) 1.12 (14.8%) 1.24 (5.83%) 1.76 (15.3%) 2.20 (4.98%)
$^{214}_{84}$Po	Radium C'	164 μs	7.69 (100%)	—	—
$^{210}_{81}$Tl	Radium C"	1.30 m	—	1.86 (24%) 2.02 (10%) 2.41 (10%) 4.20 (30%) 4.38 (20%)	0.298 (79.1%) 0.800 (99.0%) 1.07 (12%) 1.21 (17%) 1.36 (21%)
$^{210}_{82}$Pb	Radium D	22.3 y	—	0.0165 (87%) 0.063 (18%)	0.046 (4.18%)

Branching: $^{218}_{84}$Po → 99.98% to $^{214}_{82}$Pb; 0.02% to $^{216}_{85}$At. $^{214}_{83}$Bi → 99.98% to $^{214}_{84}$Po; 0.02% to $^{210}_{81}$Tl.

*Common time units: y (years); d (days); m (minutes); s (seconds).

cellulose nitrate and polycarbonates. Most detectors do not respond to light, beta particles, or gamma ray photons and thus provide a very low background system for the measurement of extremely low levels of alpha radioactivity. However, the ambient temperature and relative humidity have been shown to have an effect on the detectors.[14] No power source or electronic equipment is required. Detectors are exposed, collected, and returned to the laboratory for chemical and/or electrochemical etching and the counting of tracks. Extreme care must be taken in the handling and calibration of these detectors in order to obtain reproducible and reliable results. Allyl diglycol carbonate and cellulose nitrate detectors are currently in use for long-term integrated sampling of environmental radon.[15]

Thermoluminescent Dosimeters

Thermoluminescent dosimeters (TLDs) are crystalline materials in which electrons displaced by an interaction with ionizing radiation become trapped at an elevated energy level and emit visible light when released from that energy level. The number of trapped electrons is related to the radiation exposure. The trapped electrons are released by heating the TLDs, and the amount of light emitted during the heating process (the glow curve) is related to the exposure via calibration. The TLDs must be annealed prior to an exposure measurement. The crystals used most commonly are CaF_2 and LiF, with volumes of a few cubic millimeters. As with etched-track detectors, no power or electronic equipment is needed at a measurement site. A

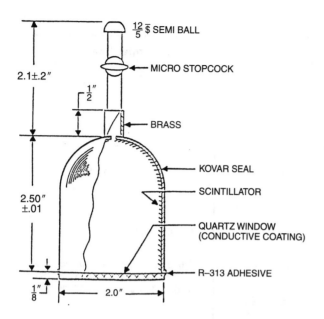

FIGURE 21-3. Alpha particles produced in the gas volume during decay of radon and short-lived progeny interact at the walls to produce scintillations. The light is transmitted through the circular quartz window at the base to a photomultiplier tube. The signal may then be amplifed and counted.[12] *Reprinted with permission of Atomic Industry Forum.*

laboratory-based readout unit is required. TLDs respond to alpha, beta, and gamma radiation. These detectors are useful for long-term environmental monitoring and have been incorporated into integrating radon detection systems.[16, 17]

Detector Calibration

The proper calibration of most radiation detection systems requires knowledge of the properties of both source and detector. Careful investigation of calibration methods must be made for specific cases. An extensive discussion on the preparation of calibration sources can be found in NCRP Report Nos. 58 and 97.[6, 18] Standard sources can be obtained from the National Institute of Standards and Technology (NIST), formerly the National Bureau of Standards (NBS), and some government laboratories. The use of such sources, while an essential ingredient in quality control, does not itself ensure measurement accuracy. Use of a standard source with a given geometry will determine the counting efficiency for a specific setup, but any change in source characteristics (e.g., source substrate, thickness) can introduce significant differences. One or two specialized laboratories maintain chambers with well-

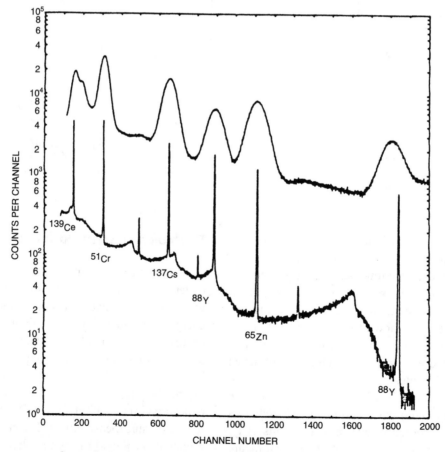

FIGURE 21-4. Gamma-ray spectra of a 5-ml mixed-radionuclide-solution source taken with the source within a 5-in. NaI(Tl) well crystal (upper curve) and at the face of a 60-cm³ Ge(Li) detector (lower curve). The counting time in each case was 2000 s (from measurements made at the National Bureau of Standards).[6] *Reprinted with permission of National Council on Radiation Protection and Measurements.*

characterized atmospheres of radon that are available for calibration of radon detectors.

The energy of the radiation to be measured may have important effects on the detector response. It is therefore prudent to calibrate for the specific radiation to be measured. Etched-track detectors, for example, will respond to ionizing particles only when the linear energy transfer is within a specific range. Calibration, as a function of gamma-ray energy, is important for survey instruments with gas-filled detectors (for guidance see NCRP Report No. 112).[19] Response may decrease rapidly for low energy gamma rays because of absorption in the chamber walls. Energy calibrations are essential in the case of spectrometric analysis (e.g., with a scintillation crystal or a semiconductor detector) or when electronic discrimination is used.

Background Reduction

Because background radiation is ubiquitous, it is always desirable and frequently essential to reduce the background count rate in order to count samples with very low activity. Lead shielding is commonly used around photon detectors to reduce terrestrial and cosmic gamma ray background. Very heavy lead shielding is required for sensitive crystalline photon detectors. Little shielding is needed for thin, alpha-detection phosphors or surface-barrier detectors. Methods other than shielding include counting only simultaneous beta-gamma emissions, energy spectrometry, or very sophisticated double crystal (CsI/NaI combination) scintillation counting.

Statistical Considerations

Counting Statistics

Counting data belong to a population where events are discrete and a relatively small number of events occur in the time that is available. This type of population is best described by the Poisson distribution. For this distribution, the variance is equal to the number of counts, and the best estimate of the standard deviation is the square root of the number of counts. If a total of N sample counts is acquired in time t, the standard deviation (SD) of N is $\sqrt{N}$. The probability that the true mean count lies within the interval $N \pm \sqrt{N}$ is 0.675, within $N \pm 2\sqrt{N}$ it is 0.95, and within $N \pm 3\sqrt{N}$ it is 0.997. The uncertainty (often called the "error") in the count may be represented by the standard deviation. The count rate, $R = N/t$; the standard deviation of the count rate, $SD_R = \sqrt{N}/t$; and the coefficient of variation of the count rate, $CV_R = SD_R/R = 1/\sqrt{N}$.

When establishing sampling and counting protocols, sampling times should be balanced against counting time for the desired level of precision. If a source with a count rate of 100 counts per minute (cpm) is counted for 1, 10, 100, or 1000 min, the CVs of the count rate will be 0.10, 0.03, 0.01, and 0.003. An increase in counting time from 1 to 10 min reduces the uncertainty from 10% to 3%. This is desirable if other sampling errors are in the range of a few percent. An increase in counting time from 10 min to 1.67 hrs to reduce the uncertainty to 1%, or to 16.7 hrs to reduce it to 0.3%, may not be warranted. Similar considerations apply to increasing sampling duration in order to increase the count rate of the sample. Clearly, for radionuclides with half lives that are the same order of magnitude as the sampling duration, the loss of sample by decay must be considered.

Background counts are always detected because of the presence of natural terrestrial and cosmic radiation. Then:

$$R_n = R - R_b \tag{4}$$

where: R_n = the net sample count rate
R = the total sample count rate
R_b = the background count rate.

The errors in the background (SD_b) and sample count rates (SD_R) are independent and are therefore propagated by summing the variances. The SD of the net count rate, SD_n, is then estimated as the square root of the total variance (Chapter 10).

$$SD_n = \left[(SD_R)^2 + (SD_b)^2 \right]^{1/2}$$
$$= \left[\frac{N}{t_s^2} + \frac{N_b}{t_s^2} \right]^{1/2} \tag{5}$$

where: N_b = the number of background counts
N = the total number of counts
t_s = the sample counting time
t_b = the background counting time.

Lower Limits of Detection

When the amount of radioactivity contained in a sample will result in a count rate that is very close to background, it is not always clear whether activity is present and, if so, how well the quantity of activity can be measured. There are a number of ways in which detection limits are defined for counting data.

Three lower limits of activity have been distinguished.[20, 21] The first is a limit at which it is decided whether activity is present, the second is the amount of activity that may be detected with a given level of reliability, and the third is the quantity of activity that may be measured with given precision. Detailed discussions and derivations of these limits are found in Currie,[20] Altshuler and Pasternack,[22] Pasternack and Harley,[23] and ANSI 13.30.[24] Which of these limits should be chosen depends on the specifics of the measurement, but the limit reported should be defined clearly.

A convenient measure is given by Pasternack and Harley.[23] They define the "lower limit of detection" (LLD) of a radioactivity counter as "the smallest amount of sample activity that will yield a net-count sufficiently large so as to imply its presence."

The LLD is approximated as:

$$LLD = \gamma\,(k_\alpha + k_\beta)\left(\frac{N}{t_s^2} + \frac{N_b}{t_b^2}\right)^{1/2} \qquad (6)$$

where:

k_α and k_β represent the value corresponding to the preselected risk for concluding falsely that activity is present (α) and the pre-determined degree of confidence for detecting its presence ($1 - \beta$). For $\alpha = \beta = 0.05$, $k_\alpha = k_\beta = 1.645$

γ =   a calibration constant to convert counts into activity

N =   the measured sample plus background counting time t

N_b =   the measured background counting time t_b

For $\alpha = \beta = 0.05$, this can be written as:

$$LLD = 3.29\,\gamma\,SD_n \qquad (7)$$

LLDs may be reduced by repeated measurements to obtain better estimates for the variance of the background count rate (Equation 5).

When very low levels of activity are sampled proper determination of the LLD is critical. The background count rate and its standard deviation should be determined from radiometric determinations on procedure blanks. Then blanks should have every characteristic of the sample except for the presence of the radionuclide to be measured.[25]

Sampling Methods

Sampling methods must be designed specifically for particular nuclides to incorporate an appropriate radiation detection system. Detectors must be fitted to both the type and energy of the radiation. Half lives, if short, may limit the procedure, but a simple measurement of half life may permit identification and quantification, even in the presence of interferences. Specific air sampling methods for certain nuclides have been published. Some are contained in NCRP publications.[21, 26, 27] The third edition of *Methods of Air Sampling and Analysis*,[28] a publication of the American Public Health Association, gives methods for measuring atmospheric ^{131}I, ^{222}Rn, elemental tritium, and tritium present as water vapor. For further discussion of radioactive aerosols in general, and specific information on radon and its short-lived decay products, References 29 and 30 are recommended, as well as the U.S. Department of Energy, Environmental Measurements Laboratory (EML) Procedures Manual, which is also available on their Internet homepage (http://www.eml.doe.gov). A review of tritium sampling and measurement can be found in Reference 31.

Recognition of the magnitude of the radiation dose to the population from naturally occurring levels of the short-lived decay products of radon[32–36] has resulted in the development and improvement of a substantial number of measurement techniques. Information on recent publications on radon can be found on the U.S. EPA Internet homepage (http://www.epa.gov/iaq/radon). The decay series from ^{222}Rn through ^{210}Pb is shown in Table 21-4. The terms short-lived "decay products" or "daughters" or "progeny" refer to the series from ^{218}Po through ^{214}Po. As seen from the varying decay rates and emissions, either alpha, beta, or gamma ray detectors may be used and energy resolution or series decay times employed to separate the various decay products. It is difficult to quantify each decay product, or "daughter," in most environments because of the extraordinarily low levels of activity. However, the concentration of each of the short-lived progeny must be known for a complete determination of the radiation dose to the respiratory tract tissue. If significant concentrations of ^{220}Rn (commonly known as thoron) and its decay products are present, these too must be quantified.

Either radon (gas) or progeny (particles) concentrations may be measured. Concentrations of radon as low as 3.7 Bqm^{-3} (0.1 pCi·l^{-1}) can be measured in grab samples, and much lower concentrations can be measured with integrating samplers. When the progeny are measured, the concentration is frequently reported in "working level" (WL). The WL is a measure of the potential alpha energy concentration in air (PAEC). PAEC is defined as the number of decay product atoms in air volume multiplied by the alpha particle energy that will be released as each atom decays to ^{210}Pb. The SI unit of PAEC is Jm^{-3}. The WL is any combination of short-lived decay products in one liter of air that will result in the emission of 1.3×10^5 MeV potential alpha energy and is equal to 2.08×10^{-5} Jm^{-3}. This is equivalent to 3700 Bqm^{-3} (100 pCi·l^{-1} units) of ^{222}Rn in equilibrium with its short-lived progeny. If each decay product formed in the series remained airborne and in the space, the concentrations would be in equilibrium, but removal by ventilation and deposition to walls and other surfaces disturbs the equilibrium. The WL is a historical unit that avoids the problem of equilibrium.

Very low background alpha particle detection systems permit detection of radon progeny concentrations as low as 0.0005 WL for a 5-min filter sample. The air is filtered for exactly 5 min onto a 0.8-μm pore size membrane filter. A ZnS (Ag)-coated phosphor is placed over the filter, and the light flashes are counted by a photomultiplier tube with appropriate electronics. Counts are taken for three or four specific time intervals. The activity concentration in air of each of the short-lived nuclides and the WL level can then be calculated by taking into account the decay time of each nuclide, the efficiency of counting, and the air volume sampled.[37–42] A method based on beta counting the filter has also been developed.[43] Table 21-5, adapted from the U.S. EPA National Radon Proficiency Program (RPP) listing (http://www.epa.gov/radonpro/methods), summarizes many of the instruments and methods for measuring radon and its short-lived decay products in air. The EPA RPP was terminated September 30, 1998, and a replacement program was started by the National Environmental Health Association (NEHA) in October 1998 (http://www.neha.org.radonpage). An extensive review of the instrumentation can also be found in George,[44] and a report detailing radon measurement methods may be obtained from NCRP.[18] The U.S. EPA has published a set of recommended protocols for measurement of radon in homes.[45] The purposes of these protocols are to ensure quality control and to obtain measurements under stable and standardized conditions. The latter will facilitate comparisons but may not represent average concentrations in the home. The working level month (WLM) is a historical unit of exposure and is an exposure rate of 1 WL for a working month of 170 hrs (1 WLM = 0.0035 Jhm^{-3}). Remedial action is recommended by NCRP if an annual exposure exceeds 2 WLM (7×10^{-3} Jhm^{-3}) (Table 21-6). This translates roughly to an air concentration of 300 Bqm^{-3} (8 pCi·l^{-1}) of ^{222}Rn,

but will vary because the radiation dose from radon daughter inhalation depends on occupancy factors and other variables. U.S. EPA recommends that indoor air concentrations of radon be reduced to below 148 Bqm^{-3} (4 pCi·l^{-1}).

Sampling Strategy

Radioactive gases or particles in air may be sampled by grab sampling, continuous monitoring, or integrated sampling methods. Grab samples will give the concentration of a contaminant at a particular location at the instant the sample was collected. The equipment is usually quite simple, and this sampling method is useful for screening on a small scale. Periodic grab sampling may be used to assess average concentrations. For average ambient concentrations, grab sampling should span several seasons. Samples may be counted immediately, as is necessary for short-lived isotopes (e.g., the overall half life of the short-lived radon progeny is about 30 min), but it is frequently possible to return the sample to the laboratory for counting, thereby avoiding the difficulties associated with transporting electronic counting equipment to the measurement site.

Continuous air monitoring is required for an in-depth assessment of airborne radioactivity because of spatial and temporal variations. With continuous sampling it is possible to observe variability and concentration peaks. Sources can be identified and effects of ventilation or weather patterns on ambient concentrations may be observed. Continuous monitoring is normally required for protective surveillance at nuclear reactors and processing facilities. Several regulatory guides issued by the U.S. Nuclear Regulatory Commission (NRC) contain guidance on air sampling (Regulatory Guides 8.21, 8.23, 8.24, 8.25, and 8.30); the most detailed is "Air Sampling in the Workplace," Regulatory Guide 8.25[46] (homepage: http://www.nrc.gov).

Integrated sampling over extended time periods will result in a single average concentration value. Detection methods are often simpler and less expensive than either continuous or grab sampling methods. For the case of radon and its short-lived decay products, both passive and active samplers are available for integrated monitoring of environmental indoor air concentrations.

Gas Phase Sampling

Radioactive gases or vapors may be sampled directly into a detector volume. The Lucas flask (Figure 21-3) for ^{222}Rn is an example. The bottom of the container is made of optically clear glass. The remaining interior surfaces

TABLE 21-5A. Methods for Measuring Radon in Air

Method	Principle of Operation	Analysis	Exposure Time
Activated Charcoal Adsorption	Radon adsorbed on charcoal in a canister	Gamma counting of canister	2–7 days
Alpha Track Detection; Filtered	Detector films in a container with an inlet filter, radon diffusion	Films etched and alpha tracks counted	3–12 months
Alpha Track Detection; Unfiltered	Detector films in a container without an inlet filter; all alpha emitters will be registered	Films etched and alpha tracks counted	3–12 months ($0.35 < F_{eq} < 0.6$)
Charcoal Liquid Scintillation	Radon adsorbed on charcoal	Charcoal treated with scintillation fluid, which is then counted in a scintillation counter	2–7 days
Continuous Monitors	Real-time, continuous measurement devices	Scintillation cell or ionization chambers	—
Electret Ion Chamber	Electrostatically charged disk detector in a container with an inlet filter	Drop in the voltage of the electret measured	1–2 months; long term 2–7 days; short term
Grab sample/Activated Charcoal	Air drawn through activated charcoal	Gamma counting or liquid scintillation counting	15 min–1 hour
Grab sample/Scintillation Cell	Filtered air drawn through a scintillation cell with a pump or by opening a valve of an evacuated cell	Cell scintillation counted	—
Grab sample/Bag	Sample bag filled with air	Sample transferred into a scintillation cell, which is then counted	—
3-Day Evacuated Scintillation Cell	Filtered air drawn through a restrictor valve of an evacuated scintillation cell or other container	Cell scintillation counted or sample transferred into a scintillation cell, which is then counted	3 days
1-Day Bag	Air drawn intermittently in small amounts into a sample bag	Sample transferred into a scintillation cell, which is then counted	1 day

Adapted from the U.S.EPA National Radon Proficiency Program listing (EPA homepage).

TABLE 21-5B. Methods for Measuring Radon Progeny in Air

Method	Principle of Operation	Analysis	Exposure Time
Continuous Working Level Monitoring	Air drawn continuously through a filter	Alpha particles emitted from the filter counted	1 day
Grab Working Level Measurement	Air drawn through a filter for 5 min	Filter alpha counted during predetermined time intervals*	5 min
Integrated Radon Progeny Measurement	Air drawn continuously through a filter	Alpha particles registered by TLD, track etch film, or an electret	> 3 days

Adapted from the U.S.EPA National Radon Proficiency Program listing (EPA homepage).
* see References [37–42]

are lined with ZnS (Ag) scintillator. Air is drawn into the evacuated flask by opening the valve on top. The dimensions of the flask ensure that most alpha particles emitted in the flask will reach the walls and produce scintillations. The scintillations are detected and quantified by placing the bottom of the flask into contact with a photomultiplier tube coupled to a counting device. Integrated air samples may be metered into an impermeable, nonreactive sampling bag or tank, and later trans-ferred to an appropriately designed detection volume. Gases may be collected on charcoal actively or by passive diffusion[47] and either de-emanated into a counting volume or gamma-counted directly if the emitted radiation is sufficiently energetic to penetrate the container. The entire charcoal container may be placed on a NaI crystal or a lithium-drifted germanium detector. In the latter case, geometric relationships have a significant effect on counting efficiency, and the counting efficiency

TABLE 21-6. Summary of Recommendations on Limits for Exposure to Ionizing Radiation[A,B,C,D]

A. Occupational exposures[D]	
1. Effective dose limits	
a) Annual	50 mSv
b) Cumulative	10 mSv x age
2. Equivalent dose annual limits for tissues and organs	
a) Lens of eye	150 mSv
b) Skin, hands, and feet	500 mSv
B. Guidance for emergency occupational exposure[D]	(see Section 14 of Reference 4)
C. Public exposures (annual)	
1. Effective dose limit, continuous or frequent exposure[D]	1mSv
2. Effective dose limit, infrequent exposure[D]	5 mSv
3. Equivalent dose limits for tissues and organs[D]	
a) Lens of eye	15 mSv
b) Skin, hands and feet	50 mSv
4. Remedial action for natural sources:	
a) Effective dose (excluding radon)	> 5 mSv
b) Exposure to radon decay products	$> 7 \times 10^{-3}$ Jh/M³
D. Education and training exposures (annual)[D]	
1. Effective dose limit	1 mSv
2. Equivalent dose limit for tissues and organs	
a) Lens of eye	15 mSv
b) Skin, hands, and feet	50 mSv
E. Embryo-fetus exposures[D] (monthly)	
1. Equivalent dose limit	0.5 mSv
F. Negligible individual dose (annual)[D]	0.01 mSv

[A]Adapted from NCRP Report No. 116[(4)]
[B]Excluding medical exposures.
[C]See Tables 21-2 and 21-7 for recommendations on w_R and w_T, respectively.
[D]Sum of external and internal exposures but excluding doses from natural sources.

for the distributed source must be determined. As an example, Tries et al.[(48)] provide a detailed discussion on preparation of carbon cartridge standards for ^{125}I. Effects of interferences, such as water vapor, on collection efficiency must be evaluated. Canisters containing 100 g of activated charcoal are commonly used to detect environmental radon. Exposure is for 7 days, followed by counting with a NaI crystal detector system.[(49, 50)]

A passive monitor for tritiated water vapor (HTO) consists of a standard scintillation vial containing water, or other sorbent, that has a precisely defined orifice in the lid. The vapor diffuses into the sampler and is collected by the sorbent. Vial and contents are then prepared for liquid scintillation counting.[(31)] This sampler has recently been adapted for sampling $^{14}CO_2$ in air.[(51)] It has also been evaluated for sampling other noble gases, e.g., ^{85}Kr, ^{41}Ar, ^{133}Xe.[(52)]

Internal ionization chambers are also used for radioactive gases. A known volume of sample is admit-

ted into an evacuated ionization chamber and the current is measured. The current may be compared with that of an identical chamber containing pure, aged air.

Particle Sampling

The concentration of radioactive particles in air is most frequently determined by collecting all the particles in a known volume of air onto a filter and counting the activity on the filter. The counting efficiency of the system must be calibrated for the specific source, filter, and detector geometry. Continuous air monitors frequently operate in a semicontinuous manner by filtering airborne particles onto a portion of continuous tape for a specified time period. The sample is then counted by a detector just above or beneath the tape after which the tape moves to provide a clean substrate for the next sample. For alpha particles or very low energy beta particles, substantial absorption may occur in the filter or even in the air gap between source and detector. Calibration specific for radiation energy is required for each geometry unless it has been determined previously that the response is not energy dependent.

The size distribution of the airborne radioactive particles may introduce sample bias. Overall sampling efficiency for aerosols depends strongly on aspiration efficiency, entry efficiency, and transport efficiency of the collecting probe[53-55] (Chapter 11). These are all particle-size-dependent processes. Sampling of large particles ($d > 10$ μm) is particularly susceptible to bias as a result of these factors. Large radioactive airborne particles may result directly from accidental releases, or indirectly from deposition of fallout onto soil particles that are subsequently resuspended. The number concentration may be very low so that detection depends on sampling large volumes of air. A review of these problems, which includes analysis of equipment designed for large particle sampling in this context, is presented in Reference 56. If some specific fraction of the ambient aerosol is desired (e.g., only the inspirable, thoracic, or respirable mass fraction),[57] such separation may be incorporated into the sampling train (62, Chapter 5). Other appropriate sampling instruments, such as the cascade impactor,[58] may also be used. Counting of the individual stages will determine an activity-weighted particle size distribution (Chapter 14).

The radon progeny size distributions are very often bimodal; the smaller particle diameter mode, so-called "unattached" fraction, is generally <10 nm and the larger mode, "attached" fraction, varies from 50 nm to 300 nm. These modes are often measured separately. The particle size of the unattached fraction is determined with a single wire screen or graded screen array and the size distribution of the attached progeny is measured with a cascade impactor or a diffusion battery. Chapter 19 includes the descriptions of these devices. New methods are continuously being developed for the measurement of the radon/thoron progeny size distributions both for dosimetric purposes and tracer studies.[59-62] Radon and thoron and their progeny are used as tracers, e.g., in earthquake prediction, volcanic surveillance, mineral exploration, geothermal surveys, and atmospheric studies.

Radiation Safety Sampling Programs

An operational radiation safety program requires continuous air monitoring systems coupled to alarm systems.[63, 64] Surveillance for airborne contaminants is most commonly done by continuous air monitors (CAMs). Particles are collected on a filter and counted with a conventional detector, which is usually a thin window G–M counter, scintillation detector, or a solid-state detector. Energy discrimination may be incorporated in the detection system when monitoring for a known emitter. CAM performance for continuous alpha air monitoring is an area of ongoing investigation.[65] Fiber-supported membrane filters such as Fluoropore FSLW (Millipore Corp.) are recommended for monitoring alpha emitters[66, 67] (see Chapter 13). Selecting a specific pulse height to be counted will significantly reduce interferences from background such as radon daughters. Where plutonium and other alpha emitters are of concern, alpha spectrometry will provide specificity in the detection-alarm system.

A two count method for separating the contribution of the short lived component from the air sample has been developed by Allen.[68] The American National Standards Institute (ANSI) has issued a number of guides for the performance of instrumentation used to monitor airborne radioactivity (e.g., ANSI N13.1[69] and ANSI N42.17B).[70] When using monitor-alarm systems, a check source should be used routinely to ensure that the system is operating properly. Also, appropriate placement of the CAMS is critical to detection of releases of alpha emitters. It has been reported that CAMS may alarm less than 30% of the time in some facilities depending on how released aerosol is transported to the CAM. [71]

Radiation Protection Criteria

Evaluation of airborne radioactive contaminant concentrations for radiation protection differs from that of chemical contaminants because protection criteria for ionizing radiation are based on the radiation dose ultimately delivered to an individual. Evaluation of airborne contamination must be based on the complex relationships between exposure to a given concentration and dose. For protection purposes, the dose, or energy delivered to tissue, is modified to include the concept of biological *equivalence* for different types of radiation, as well as *effectiveness*, which normalizes for organ sensitivity.

Responsibility for recommending limits for exposure to ionizing radiation for both the occupational and nonoccupational (general public) exposures in the United States has been delegated by Congress to the NCRP. Regulations are promulgated by various agencies including the U.S. EPA, the U.S. DOE, the NRC, the Occupational Safety and Health Administration (OSHA), and others; e.g., 10CFR Part 835 (DOE) and 10CFR Part 20 (NRC). A guide to the many radiation standards and guidances can be found in the *Handbook of Health Physics and Radiological Health*,[72] but it is necessary to check for current published standards. Information on air standards is available through the Internet homepages of these organizations and also through the U.S. EPA homepage (http://www.epa.gov/radiation). The dose limits recommended by NCRP[4] are given in Table 21-6. These guidelines conform with, but extend, the recommendations of the ICRP,[5] which are in use in most other countries.

"The goal of radiation protection is to prevent the occurrence of serious radiation induced conditions (acute and chronic deterministic effects) in exposed persons and to reduce stochastic effects in exposed persons to a degree that is acceptable in relation to the benefits to the individual and to society from the activities that generate such exposure."[4] Radiation protection guidance for occupational exposure in the United States is thus based on risk and benefit considerations. There are three basic principles: 1) any activity involving radiation exposure should be justified as useful enough to society to warrant the exposure, 2) exposures that result from carrying out such activities should be kept as low as reasonably achievable, and 3) the maximum annual dose to an individual worker should be limited to specified numerical values. The numerical value specified is an upper limit of acceptability rather than a design criterion. Exposure of any individual to the maximum dose for any substantial portion of a lifetime is discouraged. Limits for exposure of the general public are based on these same principles.

The limiting numerical values for assessed dose are based on risk of fatal cancer. They are specified as "effective dose" (*E*) to an individual. *E* is defined as:

$$E = \sum_T w_T H_T \tag{8}$$

where: $w_T =$ a tissue weighting factor (Table 21-7)
$H_T =$ the equivalent dose received by tissue T

The limits are established based on radiation risk for individual organs. The factors (w_T) provide for weighting if more than one organ is exposed in order to limit the total risk for an individual.

Dose limits apply to the sum of external and internal exposures. External exposures are assessed via effective dose. For internal exposures, "committed effective dose" must be calculated. This type of dose takes into account the continuing irradiation of organs and tissues that occurs after intake of a radionuclide.[5] Because of this continuing radiation, ICRP established the "Annual Limit on Intake" (ALI) (ICRP 61). The ALI limits the committed effective dose from an intake in a single year to 20 mSv. NCRP recommends that ALI be used as reference levels and adopts the ICRP values as "Annual Reference Levels of Intake" (ARLI). Air concentrations must be controlled so that individuals will not be exposed to more than the ARLI.

The maximum air concentration to which a worker may be exposed, in compliance with these limits, is called the "Derived Reference Air Concentration"

TABLE 21-7. Tissue Weighting Factor (w_T) for Different Tissues and Organs[A,B]

0.01	0.05	0.12	0.20
Bone surface	Bladder	Bone marrow	Gonads
Skin	Breast	Colon	
	Liver	Lung	
	Esophagus	Stomach	
	Thyroid		
	Remainder[C,D]		

[A]Adapted from ICRP Report No. 60[5] and NCRP Report No. 116.[4]

[B]The values have been developed for a reference population of equal numbers of both sexes and a wide range of ages. In the definition of effective dose, they apply to workers, to the whole population, and to either sex. These w_T values are based on rounded values of the organ's contribution to the total detriment.

[C]For purposes of calculation, the remainder comprises the following additional tissues and organs: adrenals, brain, small intestine, large intestine, kidney, muscle, pancreas, spleen, thymus, and uterus. The list includes organs that are likely to be selectively irradiated. Some organs in the list are known to be susceptible to cancer induction. If other tissues and organs subsequently become identified as having a significant risk of induced cancer, they will then be included either with a specific w_T or in this additional list constituting the remainder. The remainder may also include other tissues or organs selectively irradiated.

[D]In those exceptional cases in which one of the remainder tissues or organs receives an equivalent dose in excess of the highest dose in any of the 12 organs for which a weighting factor is specified, a weighting factor of 0.025 should be applied to that tissue or organ, and a weighting factor of 0.025 should be applied to the average dose in the other remainder tissues or organs (see ICRP Report No. 60[5]).

(DRAC). The DRAC for each radioactive nuclide is derived from the ARLI. A series of calculations relates the air concentration to organ and tissue concentrations via inhalation and metabolic processes, along with dosimetric calculations based on the emitted radiation. The calculation requires knowledge of the physical properties of the inhaled nuclide, lung deposition efficiency, solubility of the particle in the lung, transfer coefficients between body compartments, retention times, organ and tissue geometric factors, and so forth. The publications *Reference Man*[73] and *Annual Limits for Intake of Radionuclides by Workers*[74] provide numerical values and models for these calculations. Both NCRP and ICRP have published the lung models used for the dosimetry of inhaled nuclides.[75, 76]

In addition to annual dose limits, NCRP recommends for occupational exposure that the cumulative effective dose not exceed the age of the individual in years × 10 mSv. Exposure to individuals under 18 is discouraged, but occasional exposure for educational and training purposes is acceptable within strict guidelines.[4]

National radiation protection standards for the public are also shown in Table 21-6. The limits for public exposures do not include background or medical exposure. Remedial action is recommended for natural exposures beyond the limits specified in Table 21-6. The numerical values are considered an upper limit, and all exposures should be kept as low as practicable. There is, however, a level of risk considered to be so low as to be negligible and to require no attention or action. This "Negligible Individual Dose" (NID) is set at an effective dose of 0.01 mSv (1 mrem) per year.

Summary

Airborne radioactive contaminants must be sampled by methods appropriate to the type and energy of the radiation emitted. Sampling and detection equipment must be selected and calibrated for specific nuclides. Very sensitive detection methods are currently available and extremely low levels of contamination may be quantitated with properly selected equipment. Natural background radiation will limit the level of radioactivity that can be measured because of the statistical nature of the decay process so that efforts to reduce the detection of background radiation are often needed. Sampling of airborne radioactivity in the workplace and in the environment must ensure that recommended dose limits for both workers and the public are not exceeded and that all exposures remain as low as reasonably achievable.

References

1. Code of Federal Regulations: Radiation Protection Guidance to Federal Agencies for Occupational Exposure. Fed. Reg. 52:2822 (1987).
2. International Commission on Radiological Protection: Fundamental Quantities and Units for Ionizing Radiation. ICRU Report 60. ICRP, Washington, DC 20014 (1998).
3. National Council on Radiation Protection and Measurements: SI Units in Radiation Protection and Measurements. NCRP Report No. 82. NCRP, Bethesda, MD (1985).
4. National Council on Radiation Protection and Measurements: Limitation of Exposure to Ionizing Radiation. NCRP Report No. 116. NCRP, Bethesda, MD (1993).
5. International Commission on Radiological Protection: Radiation Protection: 1990 Recommendations of the International Commission on Radiological Protection. ICRP Publication 60. Annals of the ICRP, 21(1–3). Pergamon Press, Elmsford, NY (1991).
6. National Council on Radiation Protection and Measurements: A Handbook of Radioactivity Measurement Procedures, 2nd ed. NCRP Report No. 58. NCRP, Bethesda, MD (1985).
7. Knoll, G.F.: Radiation Detection and Measurement, 2nd ed. John Wiley and Sons, Inc., New York (1989).
8. Firestone, R.B.; Shirley, V.S.; Baglin, C.M.; et al., Eds.: Table of Isotopes, John Wiley and Sons, Inc., New York (1996).
9. Schleien, B.; Terpilak, M.S.: The Health Physics and Radiological Health Handbook, Supplement 1. Nucleon Lectern Associates, Inc., Silver Springs, MD (1986).
10. Eichholz, G.G.; Poston, J.W.: Principles of Nuclear Radiation Detection. Ann Arbor Science Publishers, Ann Arbor, MI (1979).
11. Hallden, N.A.; Harley, J.H.: An Improved Alpha-counting Technique. Anal. Chem. 32:1861 (1960).
12. Lucas, Sr., H.F.: Alpha Scintillation Radon Counting in Workshop on Methods for Measurement of Radiation In and Around Uranium Mills. Atomic Indust. Forum, Vol. 3, No. 9 (1977).
13. Young, D.A.: Etching of Radiation Damage in Lithium Fluoride. Nature 182:375 (1958).
14. Homer, J.B.; Miles, J.C.H.: The Effects of Heat and Humidity Before, During, and After Exposure on the Response of PADC (CR-39) to Alpha Particles. Nucl. Tracks 12(1-6): 133–136(1986).
15. Durrani, S.A.; Ilid, R.; Eds.: Radon Measurements by Etched Track Detectors, Applications in Radiation Protection, Earth Sciences and the Environment. World Scientific, London (1997).
16. Schiager, K.J.: Integrating Radon Progeny Air Sampler. Am. Ind. Hyg. Assoc. J. 35:165 (1974).
17. Maiello, M.L.; Harley, N.H.: EGARD: An Environmental X-ray and ^{222}Rn Detector. Health Phys. 53:301 (1987).
18. National Council on Radiation Protection and Measurements: Measurement of Radon and Radon Daughters in Air. NCRP Report No. 97. NCRP, Bethesda, MD (1988).
19. National Council on Radiation Protection and Measurements: Calibration of Survey Instruments Used in Radiation Protection for the Assessment of Ionizing Radiation Fields and Radioactive Surface Contamination. NCRP Report No. 112. NCRP, Bethesda, MD (1991).
20. Currie, L.A.: Limits for Qualitative Detection and Quantitative Determination. Anal. Chem. 40:586 (1968).
21. National Council on Radiation Protection and Measurements: Tritium Measurement Techniques. NCRP Report No. 47. NCRP, Bethesda, MD (1976).
22. Altshuler, B.; Pasternack, B.: Statistical Measures of the Lower Limit of Detection of a Radioactivity Counter. Health Phys. 9:293 (1963).
23. Pasternack, B.S.; Harley, N.H.: Detection Limits for Radionuclides in the Analysis of Multi-component Gamma Ray Spectrometer Data. Nucl. Instrum. Methods 91:533 (1971).

643

24. ANSI: American National Standard on Performance Criteria for Radiobioassay. ANSI N13.30-1996. American National Standards Institute, New York (1996).

25. Rodgers, J.C.; Kenney, J.W.: Issues in Establishing an Aerosol Radiological Baseline for the Waste Isolation Pilot Plant near Carlsbad, New Mexico. Health Phys. 72:300–308 (1997).

26. National Council on Radiation Protection and Measurements: Environmental Radiation Measurements. NCRP Report No. 50. NCRP, Bethesda, MD (1976).

27. National Council on Radiation Protection and Measurements: Carbon-14 in the Environment. NCRP Report No. 81. NCRP, Bethesda, MD (1985).

28. Lodge, Jr., J.P., Ed.: Methods of Air Sampling and Analysis, 3rd ed. Intersociety Committee. Lewis Publishers, Inc., Chelsea, MI (1989).

29. Hoover, M.D.; Newton, G.J.: Radioactive Aerosols. In: Aerosol Measurement: Principles, Techniques, and Applications, pp. 768–798. K. Willeke and P.A. Baron, Eds. Van Nostrand Reinhold, New York (1993).

30. Cohen, B.S.: Radon and Its Short-Lived Decay Product Aerosols. In: Aerosol Measurement: Principles, Techniques, and Applications, pp. 799–815. K. Willeke and P.A. Baron, Eds. Van Nostrand Reinhold, New York (1993).

31. Wood, M.J.; McElroy, R.G.C.; Surette, R.A.; Brown, R.M.: Tritium Sampling and Measurement. Health Phys. 65:610–627 (1993).

32. National Council on Radiation Protection and Measurements: Exposures from the Uranium Series with Emphasis on Radon and Its Daughters. NCRP Report No. 77. NCRP, Bethesda, MD (1984).

33. National Council on Radiation Protection and Measurements: Evaluation of Occupational and Environmental Exposures to Radon and Radon Daughters in the United States. NCRP Report No. 78. NCRP, Bethesda, MD (1984).

34. National Council on Radiation Protection and Measurements: Ionizing Radiation Exposure to the Population of the United States. NCRP Report No. 93. NCRP, Bethesda, MD (1987).

35. National Research Council: Biological Effects of Ionizing Radiation: Health Risks of Radon and Other Internally Deposited Alpha-Emitters. BEIR IV. National Academy Press, Washington, D.C. (1988).

36. National Research Council: Biological Effects of Ionizing Radiation: Health Effects of Exposure to Indoor Radon. BEIR VI. National Academy Press, Washington, D.C. (1997). (Also available on the Internet, e.g., via EPA homepage http://www.epa.gov/iaq/radon)

37. Kusnetz, H.L.: Radon Daughters in Mine Atmospheres—A Field Method for Determining Concentrations. AIHA Quart. 17:85–88(1956).

38. Raabe, O.G.; Wrenn, M.E.: Analysis of the Activity of Radon Daughter Samples by Weighted Least Squares. Health Phys. 17:593–605(1969).

39. Rolle, R.: Rapid Working Level Monitoring. Health Phys. 22:233–238(1972).

40. Thomas, J.W.: Measurement of Radon Daughters in Air. Health Phys. 23:783 (1972).

41. Scott, A.G.: A Field Method for Measurement of Radon Daughters in Air. Health Phys. 41:403–405(1981).

42. Nazaroff, W.W.: Optimizing the Total-Alpha Three-Count Technique for Measuring Concentrations of Radon Progeny in Residences. Health Phys. 46(2):395–405(1984).

43. Papp, Z.; Daróczy, S.: Measurement of Radon Decay Products and Thoron Decay Products in Air by Beta Counting Using End-Window Geiger–Müller Counter. Health Phys. 72:601–610 (1997).

44. George, A.C.: State-of-the-Art Instruments for Measuring Radon/Thoron and Their Progeny in Dwellings, Health Phys. 70:451–463 (1996).

45. U.S. Environmental Protection Agency: Interim Indoor Radon and Radon Decay Product Measurement Protocols. EPA 520/1-86-04. U.S. EPA, Office of Radiation Programs, Washington, DC (1986).

46. U.S. Nuclear Regulatory Commission. Air Sampling in the Workplace, Regulatory Guide 8.25. Superintendent of Documents, U.S. Government Printing Office, Washington, DC (1992).

47. Underhill, D.W.: Basic Theory for the Diffusive Sampling of Radon. Health Physics 65:17–24 (1993).

48. Tries, M.A.; Ring, J.P.; Chabot, G.E.: Carbon Cartridge Standards for ^{125}I and Suggested Applications. Health Phys. 73:502–511 (1997).

49. Cohen, B.L.; Nason, R.: A Diffusion Barrier Charcoal Absorption Collector for Measuring Rn Concentrations in Indoor Air. Health Phys. 50:457 (1986).

50. George, A.C.; Weber, T.: An Improved Passive Activated C Collector for Measuring Environmental ^{222}Rn in Indoor Air. Health Phys. 58:583–589 (1990).

51. Wood, M.J.; Surette, R.A.; Mohindra, J.K.; Patterson, J.G.: $^{14}CO_2$ in Air Sampling with Passive Diffusion Samplers. Health Phys. 74:253–258 (1998).

52. Wood, M.J.; Hong, A.; Cross, W.G.; et al.: Calibration of a Portable Tritium-in- air Monitor for Various Radioactive Gases. Health Phys. 72:423–430 (1997).

53. Vincent, J.H.: Aerosol Sampling: Science and Practice. John Wiley and Sons, New York (1989).

54. Willeke, K.; Baron, P.; Eds.: Aerosol Measurement: Principles, Techniques, and Applications. Van Nostrand Reinhold, New York (1993).

55. Parulian, A.; Rodgers, J.C.; McFarland, A.R.: A Constant Flow Filter Air Samplers for Work Place Environments. Health Phys. 71:870–878 (1996).

56. Garland, J.A.; Nicholson, D.W.: A Review of Methods for Sampling Large Airborne Particles and Associated Radioactivity. J. Aerosol Sci. 22:479–499 (1991).

57. American Conference of Governmental Industrial Hygienists: Particle Size-selective Sampling for Health-related Aerosols, J. Vincent, Ed. ACGIH, Cincinnati, OH (1999).

58. Lodge, J.P.; Chan, T.L.; Eds.: The Cascade Impactor. American Industrial Hygiene Association, Akron, OH (1986).

59. Heikkinen, M.S.A.: A Portable, Integrating Diffusion Battery for the Measurement of Radon Progeny Particle Size Distribution in Indoor Air. Ph.D. Thesis, New York University, NY (1997).

60. Tokonami, S.; Takahashi, F.; Iimoto, T.; Kurosawa, R.: A new Device to Measure the Activity Size Distribution of Radon Progeny in a Low Level Environment. Health Phys. 73:494–497 (1997).

61. Tymen, G.; Kerouanton, D.; Huet, C.; Boulaud, D.: An Annular Diffusion Channel Equipped with a Track Detector Film for Long-Term Measurements of Activity Concentration and Size Distribution of Nanometer ^{218}Po Particles. J. Aerosol Sci. 30:205–216 (1998).

62. Yu, K.N.; Guan, Z.J.: A Portable Bronchial Dosimeter For Radon Progenies. Health Phys. 75:147–152 (1998).

63. National Council on Radiation Protection and Measurements: Operational Radiation Safety Program. NCRP Report No. 59. NCRP, Bethesda, MD (1978).

64. National Council on Radiation Protection and Measurements: Radiation Alarms and Access Control Systems. NCRP Report No. 88. NCRP, Bethesda, MD (1986).

65. Hoover, M.D.; Newton, J.G.: Statistical Limitations in the Sensitivity of Continuous Air Monitors for Alpha-emitting Radionuclides. In: Inhalation Toxicology Research Institute Annual Report 1991–1992, pp. 1–4. Lovelace Biomedical Environmental Research Institute, Albuquerque, NM (1992).

66. Moore, M.E.; McFarland, A.R.; Rodgers, J.C.: Factors that Affect Alpha Particle Detection in Continuous Air Monitor Applications. Health Physics 65:69–81 (1993).

67. Hoover, M.D.; Newton, J.G.: Update on Selection and Use of Filter Media in Continuous Air Monitors for Alpha-emitting Radionuclides. In: Inhalation Toxicology Research Institute Annual Report 1991–1992, pp. 5–7. Lovelace Biomedical Environmental Research Institute, Albuquerque, NM (1992).

68. Allen, D.E.: Determination of MDA for a Two Count Method for Stripping Short- Lived Activity Out of an Air Sample. Health Phys. 73:512–517 (1997).

69. ANSI: American National Standard for Sampling and Monitoring Releases of Airborne Radioactive Substances from the Ducts and Stacks of Nuclear Facilities. ANSI N13.1-1999. American National Standards Institute, New York (1999).

70. ANSI: American National Standard on Performance Specifications for Health Physics Instrumentation—Occupa-tional Airborne Radioactivity Monitoring Instrumentation. ANSI N42.17B-1989. American National Standards Institute, New York (1989).

71. Whicker, J.; Rodgers, J.C.; Fairchild, C.I.; et al.: Evaluation of Continuous Air Monitor Placement in a Plutonium Facility. Health Phys. 72:734–743 (1997).

72. Shleien, B.; Birky, B.; Slaback, Jr., L.A.: Handbook of Health Physics and Radiological Health, 3rd ed. Lippincott Williams & Wilkins, Baltimore, MD (1998).

73. International Commission on Radiological Protection: Reference Man: Anatomical, Physiological, and Metabolic Char-acteristics. ICRP Publication 23. Pergamon Press, Oxford (1975).

74. International Commission on Radiological Protection: Annual Limits on Intake of Radionuclides by Workers Based on the 1990 Recommendations. ICRP Publication 61. Annals of the ICRP 21(4). Pergamon Press, Elmsford, NY (1991).

75. National Council on Radiation Protection and Measurements: Deposition, Retention and Dosimetry of Inhaled Radioactive Substances. NCRP Report No. 125. NCRP, Bethesda, MD (1997).

76. International Commission on Radiological Protection: Human Respiratory Tract Deposition Model for Radiological Protection. ICRP Publication 66. Annals of the ICRP 24(1-3). Pergamon Press, Oxford (1994).

Instrument Descriptions

Airborne radiation is frequently collected onto an adsorbent or on a filter which is subsequently assayed to quantify the activity. Real-time monitors are available in which the radiation detectors are physically located very close to the collection media so that the radiation that has been collected will interact with the detector to produce quantitative results after proper calibration. Detectors can also be deployed with which the radiation interacts to leave tracks which are later counted. Some of these are listed in Tables 21-I-1 and 21-I-2. A few are described in detail in the instrument section following the tables.

A complete list of instrument manufacturers is provided in Table 21-I-3, located at the end of this section.

21-1. Electret-Passive Environmental Radon Monitoring (E-PERM®) System

Rad Elec Inc.

The E-PERM system employs electret ion chamber technology to provide an accurate, passive measurement of airborne radon and thoron. The system consists of an

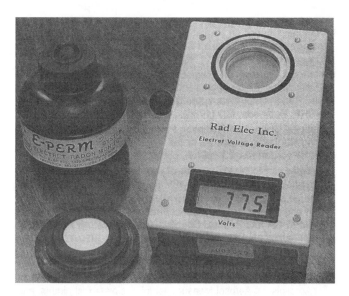

INSTRUMENT 21-1. Electret-passive Environmental Radon Monitoring (E-PERM®) System.

electret (permanently charged Teflon® piece) set into an electrically conducting plastic chamber. Radon diffuses passively into the chamber through a filtered inlet and causes ionization. The positively charged electret collects negative ions, thereby depleting the positive charge on the electret. A portable electret voltage reader is used to measure the change in surface voltage of the electret, which is related to the radon concentration and the exposure duration. Suitable calibration algorithms are used for calculating radon/thoron concentrations. Different combination of electrets and chambers are used to cover a wide range of radon concentrations (2 to 100,000 pCi·l⁻¹-days).

21-2. RA-DOME Electret Radon Monitor

RTCA

For the long term measurements of environmental radon concentrations, (3 months to a year), RTCA developed the passive Ra-dome monitor. It consists of two

INSTRUMENT 21-2. RA-DOME Electret Radon Monitor.

chamber systems each having an ionization volume of 0.68 liters. At the bottom of each hemispherical chamber a charged thin electret is placed to sense the ionization inside the sensitive volume due to radon and other ionizing radiations. The drop in voltage on the surface of the electret is measured with a voltmeter and is proportional to the concentration of radon. The second chamber, which is impervious to radon, is used to measure the contribution of the background radiation and is subtracted from the voltage drop on the passive radon monitor. The lower limit of detection is about 7 Bq m^{-3} or 0.2 pCi·l^{-1}.

21-3. Radon Measurement System, Models AB-4 and AB-4A

Pylon Electronics Inc.

The AB-4 and AB-4A are portable measurement systems designed to measure radon gas concentrations using photomultiplier tube and Lucas type scintillation cell technology. The minimum detectable level is 18 Bq m^{-3} (0.3 pCi·l^{-1}) at a 95% confidence level and a sensitivity of 0.034 cpm/Bq m^{-3} (1.25 cpm/pCi·l^{-1}) nominal. The AB-4A includes a sensor package which provides simultaneous measurements of ambient temperature, humidity and pressure as well as an internal printer. Applications include waste site monitoring, compliance testing, environmental audits, continuous monitoring, and home testing.

21-4. Portable Radiation Monitor, Models AB-5 and AB-5R

Pylon Electronics Inc.

The AB-5 and AB-5R are portable monitors offering a built-in photomultiplier tube to detect and measure radon gas concentrations using a wide variety of detectors (e.g. Lucas type scintillation cells) and accessories. The AB-5 is a laboratory grade instrument and the AB-5R is a ruggedized version of the AB-5 that is specifically designed for use in the field. Using the Model TEL detector, the minimum detectable level is 0.93 Bq m^{-3} (0.025 pCi·l^{-1}) at a 95% confidence level and a sensitivity of 0.62 cpm/Bq m^{-3} (23 cpm/pCi·l^{-1}) nominal. Applications with the appropriate detectors and accessories include continuous monitoring, grab sampling, working level measurements, environmental studies, field surveys, surface contamination, airborne particulates and more.

21-5. Active Continuous Radiation Monitor, Model CRM-1

Pylon Electronics Inc.

The CRM-1 is a non-portable measurement system using photomultiplier tube and Lucas type scintillation cell technology and designed to provide unattended continuous measurements of radon gas concentrations in industrial process and exhaust gases. A pump is used to draw and return gases from the monitored process eliminating the danger of exposure to personnel via the measurement system. The system provides laboratory grade measurements and is packaged in a lockable stainless steel cabinet for use in severe environments. The minimum detectable level is 18 Bq m^{-3} (0.3 pCi·l^{-1}) at a 95% confidence level and a sensitivity of 0.034 cpm/Bq m^{-3} (1.25 cpm/pCi·l^{-1}) nominal. Applications include stack emission monitoring, autonomous continuous monitoring, process monitoring, and mining/ore processing.

21-6. Passive Continuous Radiation Monitor, Model CRM-2

Pylon Electronics Inc.

The CRM-2 is a non-portable measurement system using photomultiplier tube and Lucas type scintillation cell technology and designed to provide unattended continuous measurements of radon gas concentrations for area monitoring. The local atmosphere diffuses into the remotely located cell allowing for real time measurements of the area radon concentration. The system provides laboratory grade measurements and is packaged in a lockable stainless steel cabinet for use in severe environments. The minimum detectable level is 18 Bq m^{-3} (0.3 pCi·l^{-1}) at a 95% confidence level and a sensitivity of 0.034 cpm/Bq m^{-3} (1.25 cpm/pCi·l^{-1}) nominal. Applications include stack waste site monitoring, radioactive site clean-ups, auto-nomous continuous monitoring, mining/ore processing, industrial site monitoring, and perimeter monitoring.

21-7. RAD7 Electronic Radon Detector, Model 711

DURRIDGE Co.

Air is sampled into a sensing volume which contains a passivated ion implanted planar silicon detector. An electrostatic field drives the radon progeny onto the detector surface. A real time spectral analysis is made of the alpha energies of the subsequent decays. Thus, thoron and radon may be independently measured. Background is less than 0.01 pCi·l^{-1}. Cycle time ranges from 2 to 24 hours, for up to 99 cycles. A 2 day protocol, for example, may be 48 one-hour cycles. Outputs include LCD display and IR printer. Output format may include a bar chart of individual cycle radon levels, a cumulative distribution, and an alpha particle energy spectrum. Internal, rechargeable batteries power the instrument for three days. Internal pump permits remote sampling. Remote control software is available.

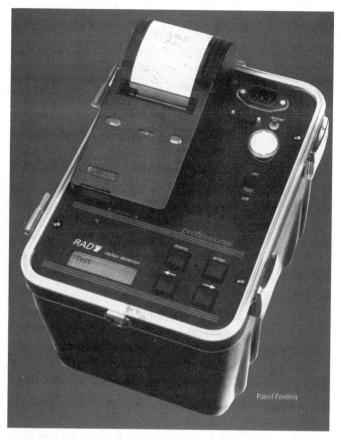

INSTRUMENT 21-7. RAD7 Electronic Radon Detector, Model 711.

There is an optional radon in water accessory that gives a reading within one hour of taking a sample.

21-8. Alpha-II Radon Gas Monitor

Diversified Research Inc.

The Alpha-II radon monitor is designed for professional use in continuous radon measurements according to U.S. EPA protocols. Radon diffuses into a chamber and subsequent progeny are precipitated by an electric field onto a solid state detector. This microprocessor controlled solid state, digital instrument will measure radon concentrations of 0.1 to 1000 pCi·l⁻¹ in intervals of 1 to 999 hours continuously repeating the cycle and storing the data for display on the LCD or printing a hard copy. Operates in temperatures of −40°F to 185°F at all humidities. For mitigation application, a +5 V output is provided when radon concentrations exceed a set threshold in pCi·l⁻¹. Measurement precision is ±2.1% for standard error at 10 pCi·l⁻¹ concentrations with total error of less than ±3%. Tampering is detected and stored for printing in report. The Alpha-II operates on AC power or 3 C-cells with battery life over three days and is unaffected by power interruptions or outages. The monitor is 6.25 × 6.25 × 8.5 inches, weighing 3.5 lb.

21-9. Airborne Alpha Radiation Detector, CRM-510

femto-TECH Inc.

The femto-TECH, Inc. Model CRM-510 is a precision airborne alpha radiation detection instrument based on passive air diffusion and pulsed ion chamber technology. Due to a unique electrometer and open grid probe design, the Model CRM-510 is suited for a wide range of radon measurement application. Due to the low current requirement of the electrometer detector and onboard computer, the Model CRM-510 is a portable self-contained continuous radon monitor that can read and store test data for four days of stand-alone operation. In the "PASSIVE" mode of operation the Model CRM-510 is suited for screening and follow-up type testing. Barometric pressure, temperature and relative humidity data are also collected.

INSTRUMENT 21-8. Alpha-II Radon Gas Monitor.

INSTRUMENT 21-9. Airborne Alpha Radiation Detector, CRM-510.

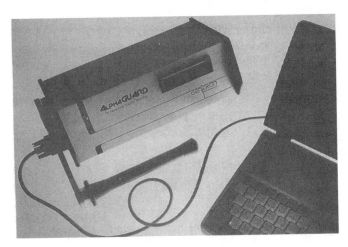

INSTRUMENT 21-10. Radon Monitoring System.

21-10. Radon Monitoring System

Genitron Instruments GmbH, Frankfurt, and Rad Elec, Inc., Frederick, MD

AlphaGUARD family of radon monitors provide a lightweight and reliable monitoring system for the continuous measurement of radon in air, soil, water, building materials and for the determination of radon progeny. AlphaGUARD incorporates a pulse-counting ionization chamber and records relevant climatic parameters. It provides its high detection efficiency (1.7 cpm per $pCi \cdot l^{-1}$), wide measuring range (<0.05 to 54,000 $pCi \cdot l^{-1}$), and fast response time (30 min for a change from ambient to 30 $pCi \cdot l^{-1}$). 1-, 10- and 60-min. measuring cycles are available and data can be stored up to >6 months. Battery capacity is ≥10 days. It is sensitive to both high humidity and vibrations. The software package AlphaEXPERT provides a sophisticated tool

for routine graphical work up, presentation and evaluation of data.

21-11. Radon Gas Monitor

Eberline Instrument Corp.

The Model RGM-3 is a microcomputer-based portable system for continuously measuring radon gas. The detector assembly consists of a metal chamber coated on the inside with zinc sulfide doped with silver powder for alpha sensitivity. Prefiltered air is drawn through the chamber by a vacuum pump. The RGM-3 supports two modes of operation. In the continuous mode, the pump runs while the onboard counting computer maintains a 1-hr file of count versus time. In the grab sampling mode, a sample is pulled into the chamber, monitored, and then flushed. Background compensation is automatically calculated based on a calibration flush with zero activity air and an estimate of previous plate out of the short-lived progeny. Sensitivity is 5 $cpm/pCi \cdot l^{-1}$, pump flow rate is 10 $l^{-1}min$, and battery life is 2–4 hrs for continuous sampling. The monitor is 41 × 25 × 30 cm and weighs 13.6 kg.

21-12. Rapid Radon Monitor

Technical Associates

The Model FR-5R-FS (not pictured) provides rapid testing; first appraisal results in just 6 minutes. Radon is collected onto charcoal in a canister at a high flow rate. Sensitivity is 0.3 $pCi \cdot l^{-1}$ at the 90% confidence level. Easy-to-operate digital scaler/analyzer. High-volume pump pulls over 200 $l^{-1}min$. Optional shielding permits in-house testing for standard 2 1/2-in. and 2 7/8-in. canisters.

21-13. Radon Gas Monitor

Sun Nuclear Corporation

The Model 1027 radon gas monitor is an update of the models 1023 and 1026. It is a microcomputer-based continuous radon monitor. Radon gas is allowed to diffuse into a chamber. Decay products are blocked from entry by a filter. In the chamber, the decay products plate out on a diffused-junction photodiode detector by means of an electrostatic field. The system is completely solid-state with software calibration. Overall sensitivity is 2.5 $cph/pCi \cdot l^{-1}$. On demand, it displays the longer-term (since last reset) and short-term (12 hr) average radon gas concentration on a 3-digit LED display in $pCi \cdot l^{-1}$. It provides a hard-copy printout of the long-term memory data. It is line operated through the use of an AC to DC power adapter. It contains non-recharging batteries which provide up to 7 hours of

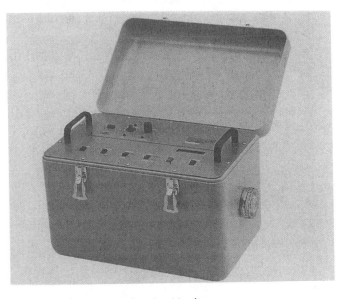

INSTRUMENT 21-11. Radon Gas Monitor.

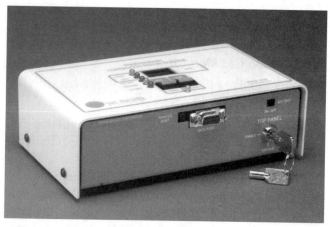

INSTRUMENT 21-13. Radon Gas Monitor.

INSTRUMENT 21-15. Portable Radon/Radon Daughter Detector, RDA-200.

uninterrupted power. The monitor contains mercury switches that place an indication, on the printout, of any movement of the instrument. The system is cylindrical in shape, approximately 5 in. in diameter and 7 in. high; it weighs 2 lbs.

21-14. Passive Radon Monitor, UltraTrack

USInspect

UltraTrack is a passive ^{222}Rn monitor (diameter 7.5 cm, height 3 cm) suitable for measurement of indoor or outdoor environmental concentrations (2.2 to 900 kBq m^{-3} h, 2.5 to 1000 pCi·l^{-1} day). It has a background protection guard that prevents ^{222}Rn exposure until the desired measurement is initiated. The monitor utilizes triplicate 9 mm × 9mm CR-39 alpha track detection film. The lower limit of detection is 2.2 kBq m^{-3} h (2.5 pCi·l^{-1} day). The alpha track efficiency for each film is 2.6 tracks per kBq m^{-3} h (2.3 tracks per pCi·l^{-1} day). The background on each pristine film is 5 to 10 tracks. The monitor is fabricated from conducting plastic, and the alpha track film is covered with thin aluminized Mylar during measurement. No charge artifacts disturb the calibration. The alpha tracks on the entire area of all 3 films (3 × 81 mm^2) are counted to determine the ^{222}Rn concentration. Positive controls and background alpha track films are processed with each batch of field samples.

21-15. Portable Radon/Radon Daughter Detector, RDA-200

Scintrex Ltd.

The RDA-200 detector system consists of a Detection Console and accessories and it can be customized to measure radon gas, radon daughters, radon/thoron soil gas, and radon from water samples using a degassing unit. This unit utilizes various detector systems such as a filter collector tray with a ZnS(Ag)

phosphor coated Mylar or a flow through ZnS(Ag) phosphor coated cell. The counting unit is battery operated for use in the field. Sensitivity depends on the counting sequence selected; for radon it is less than 1 pCi·l^{-1}.

21-16. Working Level Monitor, WLM-30

Scintrex Ltd

The WLM-30 measures the indoor air alpha activity in a variety of locations and conditions. It contains a servo-controller pump assembly which operates at constant 1 l min^{-1} flow rate. The air is drawn through a 25 mm, 0.8 μm membrane filter which is seated in a removable housing at a fixed distance from a solid state detector. The instrument has a 256-channel counter to indicate the alpha energy distribution of the sample. It can be programmed by an IBM compatible PC computer for several parameters, e.g., sampling interval (from 1 to 999 minutes), channel windows, start and stop times. The software package ENVILOG 111 is provided by the manufacturer. The instrument records the collective total

INSTRUMENT 21-16. Working Level Monitor, WLM-30.

INSTRUMENTS 21-17 & 18. Radon WL Meter, Model TN-WL-02 and Instant Radon Progeny Meter.

count from radon and thoron progeny for each sampling period or for each sampling interval in cpm or mWL, counts from individual daughters in a given sampling period and for each sampling interval in cpm. Sensitivity is 1 mWL/7.5 Bq m^{-3}.

21-17. Radon WL Meter, Model TN-WL-02

Thomson & Nielson Electronics Ltd.

The Model TN-WL-02 Radon WL Meter is an active monitor designed to measure the radon progeny inside buildings. The lightweight, versatile instrument is easy to operate and very reliable. The Radon WL Meter samples air and collects airborne radionuclides on a filter. The alpha activity from a sample is counted using a semiconductor detection system. The count data can be converted to the appropriate units by manual calculation or via a TN Data Recorder Accessory. The Radon WL Meter is calibrated against U.S. and other internationally recognized standards. Options include a Data Recorder Package (Model TN-DU-01) that is IBM PC compatible with software included. Graph or raw data printout are available. It is also modem compatible, has a rechargeable battery with 24 hour capacity. A thorium-230 check source allows field validation of the detector.

21-18. Instant Radon Progeny Meter

Thomson & Nielson Electronics Ltd.

The Model TN- IR-21 employs an integral regulated 8 lmin^{-1} pump to collect radon progeny on a glass microfiber filter that is positioned over a semiconductor alpha detector. The pump, filter, detector, display, rechargeable battery, and controlling electronics are all contained in a single 21 × 20 × 18 cm package weighing

6 kg. The instrument provides an estimate of radon progeny concentration in only 5 minutes and calculates more accurate (× 10% at 20 mWL) results in 22.5 minutes.

21-19. Working Level Measurement System

Pylon Electronics Inc.

The Pylon Model WLx is a sophisticated, portable instrument designed to measure Working Level radon and thoron gas concentrations in air using a solid-state detector. It is a laboratory grade instrument that is also well suited for use in the field. The detection range is 0.6 mWL to 50 WL. The servo controlled pump allows unattended area level monitoring for extended periods in adverse environments. Applications include simultaneous radon and thoron progeny measurements, area and building monitoring (with alarm outputs), radiological protection of personnel, and health physics studies.

21-20. Xenogard

Nuclear Associates

With the Xenogard Monitor, concentrations of Xenon-133 in room air and gas trap effluent can be quantitatively monitored continuously and accurately. To continuously monitor and integrate room air concentration, room air is drawn into the fully-shielded counting chamber and counted by a thin window GM tube while the air is exchanged more than 3 times per minute. An analog meter continuously displays MPC units while two digital registers display integrated MPC-Hours and total hours (running time), respectively. If Xe-133 room air concentration exceeds full scale, the digital register flashes on and off as a warning to personnel. In addition, an audible alarm can be activated. Setting the Xenogard analog meter multiplier to ×100 or ×1000 displays 10^{-3} μCiml^{-1} or 10^{-2} μCiml^{-1} full scale. Concentrations approaching the latter level at the trap's exhaust port can result in a xenon room air concentration approaching

INSTRUMENT 21-20. Xenogard.

INSTRUMENT 21-21. Iodine Air Monitor.

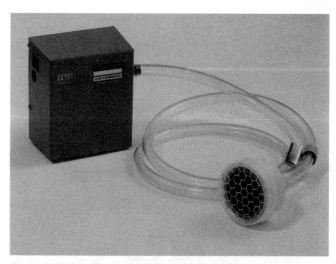

INSTRUMENT 21-22. Persair, Personal Air Sampler.

1 MPC. Therefore, the Xenogard simplifies the monitoring of effluent air from any xenon trap and can be used periodically to verify trap performance.

21-21. Iodine Air Monitor

Biodex Medical Systems Inc.

The Iodine Air Monitor provides integrated exposure information and can provide hard copy printouts or send results to a connected computer. Sensitive and versatile, the monitor meets new lower EPA requirements, I-131, 2.1×10^{-3} μCicm^{-3} and I-125, 1.2×10^{-13} μCicm^{-3} and is also used to monitor hoods or workplace air. Using a standard charcoal filter cartridge in its specially designed no-leak trap, the Iodine Air Monitor will trap any airborne radioiodine. The filter is under constant surveillance with a scintillation detector. Air is drawn by a regulated pump through a three-foot inlet hose using a flow meter and then exhausted through another three-foot hose. These hoses can easily be extended to accommodate hoods, glove boxes, stacks, etc. Time is recorded by an odometer type timer wired into the pump circuit. Completely self-contained and portable, the monitor includes 10 filter cartridges.

21-22. Persair, Personal Air Sampler

Nuclear Associates

The Persair is a personal sampler to be used for sampling atmospheric contamination in the user's breathing zone. Samples are collected on standard filter paper or a special, highly efficient charcoal filter cartridge designed for iodine nuclides. The unit features a high flow rate, about 5 to 7 lmin^{-1}. A 5-digit register indicates the total air volume sampled. A 3-cell rechargeable battery pack allows at least 8 hours of sampling time

between charges. The system is useful for sampling ^{131}I and ^{125}I inhaled during radio-iodination procedures.

21-23. Iodine Air Monitor

Technical Associates

The Model FM-5-ABNI air monitor samples airborne radioiodine contamination and has alarm record capability. It also provides integrated exposure information and provides hard copy. The contaminant is collected in a standard charcoal filter cartridge. The filter is under constant surveillance via a scintillation detector. Limit of sensitivity (as ^{131}I) is 10^{-9} μCi ml^{-1} min^{-1} exposure.

21-24. Particulate, Iodine, and Noble Gas Air Monitoring System

Eberline Instrument Corp.

The Model PING-1A is shielded with 3 in. of lead in a 4π geometry to reduce background. Particulate is

INSTRUMENT 21-23. Iodine Air Monitor.

INSTRUMENT 21-24. Particulate, Iodine, and Noble Gas Air Monitoring System.

collected onto a 47 mm diameter filter and monitored by a 2 in. diameter × 0.010 in. thick plastic beta scintillation detector and a solid-state alpha particle detector for radon background subtraction. Iodine is adsorbed on TEDA-impregnated charcoal and counted with a 2 × 2 in. NaI(Tl) detector and pulse height analyzer. The system is mounted on a cart.

21-25. Noble Gas Monitoring Carts and Equipment
Qualprotech

Qualprotech designs and builds transfer carts for noble gas and stack monitoring. Two different carts comprise a set. The larger cart carries a multi-channel analyzer and the electronics equipment, the other the cryogenic equipment and a Germanium detector for insertion into a lead lined Marinelli chamber. The carts are connected on site. A fine screw adjustment raises the detector into the Marinelli chamber. The liquid nitrogen cylinder is raised and lowered with the detector. The detector head must be maintained at cryogenic temperatures. Indication lights check the level of nitrogen. The carts are fitted with a lighting system for easy viewing of the Marinelli chamber during insertion of the detector head. Electrical power sources are provided and Velcro straps are used for nitrogen bottles.

21-26. Particulate, Iodine and Tritium Samplers
Qualprotech

Qualprotech custom manufactures PIT samplers in 6 × 2 × 2 ft standing cabinets or in mobile carts that can be moved to the measurement site. The stack effluent sampler provides continuous compliance monitoring of sampled air taken from the exhaust ducting of nuclear power stations. The instrument collects samples of particulates, iodine, and tritium. Particulate

INSTRUMENT 21-25. Noble Gas Monitoring Carts and Equipment.

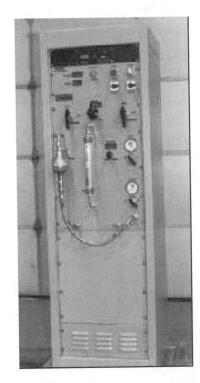

INSTRUMENT 21-26. Particulate, Iodine and Tritium Samplers.

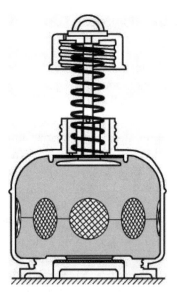

INSTRUMENT 21-27. E-PERM® Passive Tritium Air Monitor.

matter is collected on a 47 mm filter paper, iodine vapour is collected on a sodium zeolite cartridge, and the tritiated vapour is captured in a dessicant cell. The units are all easily removable for weekly replacement. Flows through the units can be measured by rotameters or by mass flow meters or controllers. The instrument can be used as a compliance monitor if used in conjunction with a calibrated velocity probe in the ducting, and a totalizer to record the total flow and the combined sample flow which has passed through the collection bottles. Power requirement is 120 VAC at 60 Hz, and 15 A.

21-27. E-PERM® Passive Tritium Air Monitor

Rad Elec Inc.

The E-PERM system employs electret ion chamber technology to provide an accurate, passive measurement of airborne tritium. The system consists of an electret (permanently charged Teflon® piece) set into an electrically conducting plastic chamber. Tritium gas and triturated water vapor diffuse passively into the chamber through a large area filter and cause ionization. The positively charged electret collects negative ions, thereby depleting the positive charge on the electret. A portable electret voltage reader is used to measure the change in surface voltage of the electret, which is related to the tritium concentration and the exposure duration. Suitable calibration algorithms are used for calculating the concentration of airborne tritium. Different combination of electrets and chambers are used to cover a wide range of radon concentrations (0.8 to 100 pCi m^{-3} day).

21-28. Tritium in Air Monitor to Public Release Levels

Technical Associates

The Model STG-5ATL is a sensitive, rugged, down-to-public-release-level monitor (10^{-7} μCicm³). It measures tritium as HTO in the presence of radioactive noble gases and varying external radiation background. The air being monitored is filtered, deionized, and divided into two streams. In one of the streams, HTO is removed. The two streams are then passed through two balanced detectors operated in subtractive mode. The net reading is presented directly with outputs for alarm (included) and printer and computer interface. Alarm and hard copy are standard features. Provision is made for interface with external computer or control system.

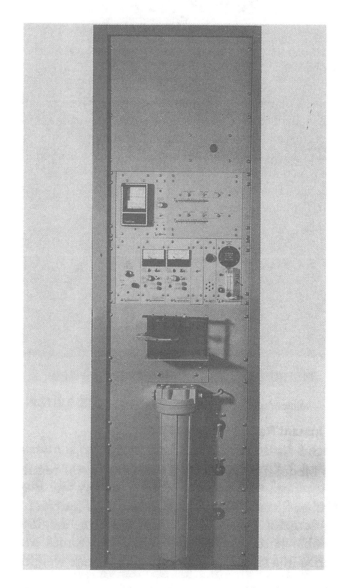

INSTRUMENT 21-28. Tritium in Air Monitor to Public Release Levels.

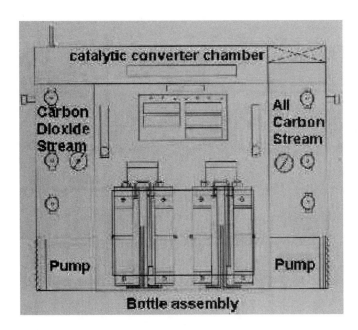

INSTRUMENT 21-29. Carbon-14 Monitoring System.

21-29. Carbon-14 Monitoring System

Qualprotech

Qualprotech manufactures carbon-14 collection/sampling equipment for sampling exhaust gases from the outlet ventilation ducting in nuclear power stations. The sampling flow rate is 400 ml min^{-1}, divided into two separate 200 ml min^{-1} streams. Carbon-14 is collected as carbon dioxide in one stream and as total carbon in the other. The total carbon flow differs from the carbon dioxide flow in that it is directed through a catalytic converter before the collection of the carbon in a sodium hydroxide solution. Sample bottles containing the entrapped carbon-14 are removed and replaced weekly for measurement in a scintillation counter. The flow rates of the sampling gases in each stream are controlled, measured and totalized through mass flow controllers. The quantities of chemicals supplied permit continuous flow for a week without replenishment. Carbon-14 cabinets are wall mounted and measure 3 ft (w) × 2 ft (h) × 9 in (d). Operating valves are surface mounted together with a mimic depiction of the circuit. The equipment requires 120 V, 60 Hz, and 15 A power supply. The equipment is controlled through a single circuit board containing all of the control data and most of the instrumentation. Fail safe interlocks are provided in the event of malfunction, and indicator lights show that everything is working.

21-30. Alpha Air Monitor

Eberline Instrument Corp.

Model Alpha-6 is a continuous alpha monitor employing a solid-state detector with a 256-channel analyzer to separate energies in order to identify specific isotopes and minimize interference from radon-thoron alpha emitters. The air flow is adjustable from 10 to 100 lmin^{-1}. The instrument provides immediate readout and connection to computer storage. An adjustable alarm is provided. The detector head comes in a variety of configurations to support in line and ambient monitoring of the air. The instrument measures approximately 35 × 32 × 38 cm and weighs about 6.6–7 kg, depending on the head. A CAM (ACS-l) system based on the Alpha-6 monitor is also available.

21-31. Beta Air Monitor

Ludlum Measurements Inc.

The Model 333-2 is a beta particulate air monitor utilizing 2 pancake type Geiger–Mueller tubes configured back to back with one tube facing a filter paper and

INSTRUMENT 21-30. Alpha Air Monitor.

INSTRUMENT 21-31. Beta Air Monitor.

the other for gamma background subtraction. Air is pulled through a filter by the 333-1P giving an air flow capability from 20 to 70 lmin⁻¹. Filters and detectors are enclosed in a lead shielded stainless steel chamber to reduce spurious background counts. Continuous counts are recorded by a striking strip chart recorder, and alert and alarm points can be set from 0 to 100,000 cpm. An alarm is indicated by a red strobe on the top of the instrument as well as an audible indicator, which is rated at greater than 92 dB at 10 feet. Typical (4pi) efficiencies for the unit are 5% – ^{14}C; 22% – ^{90}Sr/^{90}Y; 19% – ^{99}Tc; 32% – ^{32}P. Power is provided by any standard 110 VAC receptacle, however the instrument can be configured to operate at 220 VAC as well.

21-32. Beta Particle Air Monitor

Eberline Instrument Corp.

The AMS-4 collects and monitors beta radiation on particles. The AMS-4 provides direct concentration read-out that is continuously compared to alarm setpoints. The user can specify the DAC to concentration unit conversion factor. The detector head assembly may be placed up to 1000 feet from the instrument. The head supports two 2 in. diameter, gas-proportional sensor detectors; one for monitoring the beta activity of the particulate filter and a second for real-time gamma background subtraction. An optional pump module supplies a nominal 60 lmin⁻¹ flow

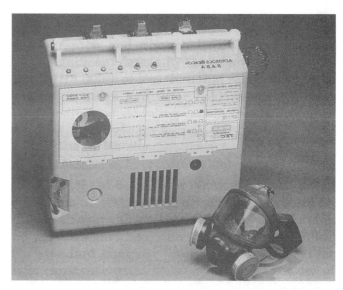

INSTRUMENT 21-33. Portable Continuous Airborne Monitor, BAB.

rate. A hot wire anemometer monitors actual flow rate. All count data are stored in the instrument's memory for later retrieval to a computer or printer. The radial sampling head and optional pump weigh 31 lbs.

21-33. Portable Continuous Airborne Monitor, BAB

Novelec, North America Inc.

The BAB is a family of real-time portable airborne alpha- and beta-emitting particulate monitors. Particles are collected on filter paper at a continuously monitored air flow rate. Using solid-state detectors and digital electronics, it reports ambient airborne concen- trations directly in μCicm⁻³ over a wide range of concentrations. Detection sensitivity depends on background and counting time. This instrument operates in typical nuclear power plant background gamma radiation and radon levels. Gamma background compensation is accomplished with a guard detector; radon background via several spectrometry channels. Its portability allows it to be used at the work site to provide continuous coverage of the job. It provides preset alarm thresholds and built-in self testing. Data capture options allow for real-time control point monitoring of remote job sites by radio or RS-232 link to a PC and/or historical data capture on a RAM card for reading and storing in a PC.

21-34. Air Monitor-Gas, Gross Beta-Gamma Particulate, and Iodine (or Gross Alpha)

Technical Associates

The Model BAM-3H three-channel air monitor simultaneously measures gross beta-gamma particulates, gaseous radioactivity, and iodine (gross alpha in BAM-3HC). It is a continuous duty, high capacity, rugged skid or

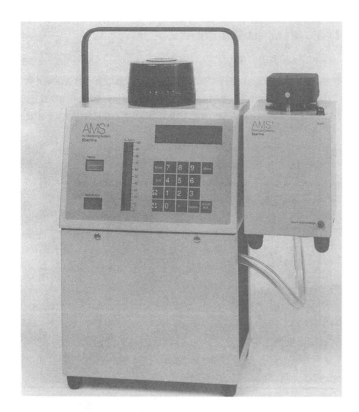

INSTRUMENT 21-32. Beta Particle Air Monitor.

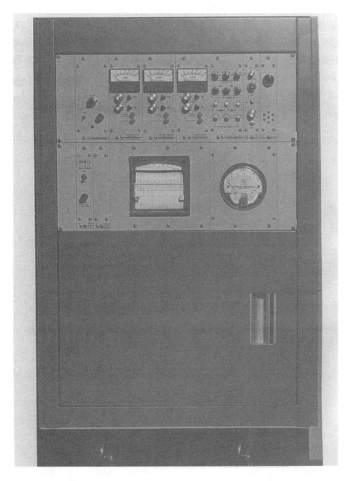

INSTRUMENT 21-34. Air Monitor-gas, Gross Beta-gamma Particulate, and Iodine (or gross alpha).

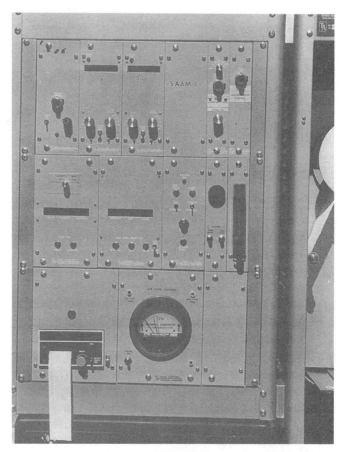

INSTRUMENT 21-35. Moving Filter Alpha Plus Beta-gamma System.

caster-mounted system. Electronics are plug-in modules allowing change or addition of function and rapid repair by substitution of modules in the field. Shields are 3 in., void-free, and lead encased in welded steel; they have stainless steel liners for easy decontamination. Filters are changed via a quick disconnect, o-ring, sealed filter holder. Air flow is factory set at 3 ft³min⁻¹, but can be set at flows up to 10 ft³min⁻¹ without equipment change.

21-35. Moving Filter Alpha Plus Beta-Gamma System

Technical Associates

Model SAAM-l filter tape A/BG air monitor system is a line-operated, continuous-duty, long period air monitor that automatically integrates over a preset period and prints results. It incorporates an alarm as well as hard copy. Standard configuration reads gross alpha and gross beta-gamma. It has plug-in capacity for one or more channels of specific radiation energies (e.g., ^{239}Pu, ^{238}U). This system draws air through a section of filter tape and measures the activity on the tape after a predetermined time; usually 20 min. It has a dwell time of 20 min, and sensitivity on the order of 1% of the most restrictive alpha or beta-gamma contamination standard.

TABLE 21-I-1. Collectors for Measurement of Radon in Air

Name	Collector	Exposure Duration	Notes	Company	Figure No.
ST 100 Charcoal Liquid Scintillation Detector	Passive diffusion onto charcoal. Desorbed into liquid scintillator for beta counting.	1–4 days	Wt. 0.4 oz. Near 100% counting efficiency. Bar coded ID.	REM	21-I-1
AT-100 Alpha Track Detector	Diffusion based. CR-39 track detector. Housing filters out dust and radon progeny.	LLD 0.8 pCi·l⁻¹ month. For a 3 month 4 pCi·l⁻¹ exposure, uncertainty is 12%.	Electrochemical etching; counting with computer aided image analysis.	REM	21-I-2
Activated Carbon Collectors Type I	3 (or 4) inch diameter can containing 50 g (90g) activated charcoal. Diffusion barrier covers.	Up to 7 days	Gamma counting Rn decay products of ^{214}Bi and ^{214}Pb. LLD 3.7-7.5 Bq m⁻³ (0.1 to 0.2 pCi·l⁻¹) Rn for 4 days. Regenerated by heating at 125°C overnight.	RTC	21-I-3
Liquid Scintillation Vials Type 2	20 ml plastic vial with diffusion barrier cover, contains 2g activated carbon.	Up to 7 days	Analysis is liquid scintillation to alpha count ^{220}Rn, ^{218}Po and ^{214}Po. LLD 7 Bq m⁻³ (0.2 pCi·l⁻¹) Rn for 4 day measurement interval. Calibrated for different RH conditions.	RTC	—
Radtrak™	Alpha-track process, Track-Etch.	3 mos. to 1 yr.	Return to company for etching and counting analysis.	RAD	21-I-4
Historical Reconstruction Detector Model 3.0	Two parts: 1) A conventional alpha track detector to measure current radon and 2) A dual chip strip attached to a glass surface for deposited and implanted.	—	Developed to reconstruct past exposures from a glass surface in a room, based on a room specific atmospheric model.	MRD	—
ARMS 1 DBCA Charcoal Canister	Diffusion barrier, 25g activated coconut charcoal, contains moisture desiccant.	2–7 days	Gamma spectroscopy with NaI scintillation detector.	ARM	—
Charcoal Canister	Charcoal canisters for the collection of radon in various sampling systems.	—	2 ¾ & 4" diameter canister available in 4 mesh sizes; open face and diffusion barrier models available. Analysis by gamma spectrometry (not provided by FJS)	FJS	—

TABLE 21-I-2. Collectors for Airborne Activity

Name	Collector	Exposure Duration	Notes	Company	Figure No.
Teda Impregnated Charcoal Cartridges	Radioiodine Collection Filter Cartridges; Coconut shell charcoal impregnated with 5% triethylene diamine.	—	Metal and plastic cartridges available with various standard dimensions to fit into sampling heads in air sampling systems; 4 mesh sizes for required pressure drop. Collection efficiency vs. flow rate data available. Quality control measurements and compliance data provided.	FJS; HIQ	—
Silver Zeolite	Silver impregnated zeolite (molecular sieve) cartridges contain a highly efficient inorganic adsorbent for the collection of elemental and organic forms of radioactive iodine. Noble gases not retained to any significant degree.	—	Cartridge may be preceded by a particle filter. Fits into sample head in an air monitoring system. Used in post accident standby systems.	FJS; HIQ	21-I-5

FIGURE 21-I-2. AT-100 Alpha Track Detector.

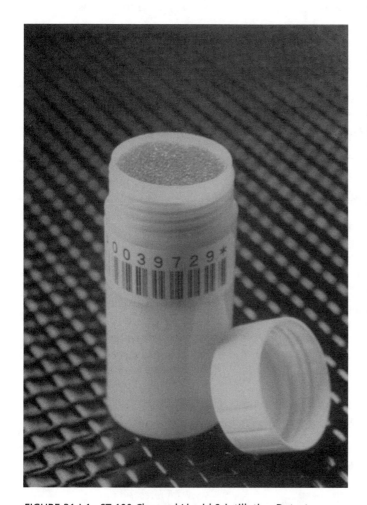

FIGURE 21-I-1. ST 100 Charcoal Liquid Scintillation Detector.

FIGURE 21-I-3. Activated Carbon Collectors Type I.

FIGURE 21-I-4. Radtrak.

FIGURE 21-I-5. Silver Zeolite.

TABLE 21-I-3. List of Instrument Manufacturers

ARM	Advanced Radiation Monitoring Service, Inc. 48705 Hickory Lane Mattawan, MI 49071 (616) 668-5246 or (888) 447-2366 FAX (616) 668-5353	FEM	Femto-TECH, Inc. 325 Industry Drive P.O. Box 8257 Carlisle, OH 45005 (513) 746-4427 FAX (513) 746-9134	PYL	Pylon Electronics Inc. 147 Colonade Road Nepean, Ontario, Canada K2E 7L9 (613) 226-7920 or (800) 896-4439 FAX (613) 226-8195 www.pylonelectronics.com
BMS	Biodex Medical Systems 20 Ramsay Road Box 702 Shirley, NY 11967-0702 (631) 924-9000 or (800) 224-6339 FAX (631) 924-9338 www.biodex.com	GEN	Genitron Instruments GMBH Heerstrasse 149 D-60488 Frankfurt a.M., Germany +49-69-976 514-0 FAX +49-69-765 327 sales@genitron.de www.genitron.de	QUA	Qualprotech 2333 Wyecroft Road, Unit #9 Oakville, Ontario, Canada L6L 6L4 (905) 825-0697 FAX (905) 825-0716 www.qualprotech.com
DRI	Diversified Research, Inc. 1155 Redwood Road Merritt Island, FL 32952 (407) 453-6496	LUD	Ludlum Measurements, Inc. P.O. Box 810 501 Oak Street Sweetwater, TX 79556 (915) 235-5494 or (800) 622-0828 FAX (915) 235-4672 ludlum@ludlums.com www.ludlums.com	RAD	Rad Elec Inc. 5714-C Industry Lane Frederick, MD 21704 (301) 694-0011 or (800) 526-5482 FAX (301) 694-0013 www.radelec.com
DUR	Durridge Co. 7 Railroad Ave., Suite D P.O. Box 71 Bedford, MA 01730 (781) 687-9556 sales@durridge.com www.durridge.com	NOV	Novelec, North America, Inc. 113 W. Outer Drive P.O. Box 6621 Oak Ridge, TN 37831 (615) 482-9287 FAX (615) 483-030	RTC	RTCA 2 Hayes Street Elmsford, NY 10523 (914) 345-3380 or (800) 457-2366 FAX (914) 345-8546 www.rtca.com
EBE	Eberline Instruments 504 Airport Road Santa Fe, NM 87505 (505) 471-3232 or (800) 274-4212 FAX (505) 473-9221 www.eberlineinst.com	NUC	Nuclear Associates 100 Voice Road P.O. Box 349 Carle Place, NY 11514-0349 (516) 741-6360 or (888) 466-8257 FAX (516) 741-5414 sales@nucl.com www.nucl.com	SCI	Scintrex, Ltd. 222 Snidercroft Road Concord, Ontario, Canada L4K 1B5 (905) 669-2280 FAX (905) 669-6403 scintrex@idsdetection.com www.idsdetection.com

TABLE 21-I-3 (cont.). List of Instrument Manufacturers

SNC	Sun Nuclear Corporation 425-A Pineda Court Melbourne, FL 32940-7508 (321) 259-6862 FAX (321) 259-7979 *www.sunnuclear.com*		TNL	Thomson & Nielson Electronics Ltd. 1050 Baxter Road Nepean, Ontario, Canada K2C 3P1 (613) 596-4563 FAX (613) 596-5243
TEA	Technical Associates 7051 Eton Avenue Canoga Park, CA 91303 (818) 883-7043 FAX (818) 883-6103		USI	USInspect 22560 Glenn Drive Sterling, VA 20164 (703) 444-6229 FAX (703) 444-6471

Chapter 22

Sampling Biological Aerosols

Janet M. Macher, Sc.D., M.P.H.[A] and Harriet A. Burge, Ph.D.[B]

[A]Environmental Health Laboratory, California Department of Health Services, Berkeley, California;

[B]Department of Environmental Health, Harvard School of Public Health, Boston, Massachusetts

CONTENTS

Introduction

Environmental scientists rely on air samples to measure exposures to gases, vapors, and airborne particles from natural and human activities in indoor and outdoor air. Among these air contaminants are materials from living or once-living sources that may cause disease in humans, animals, or plants. Interest in aerobiology and the sampling of bioaerosols has increased as investigators recognize the potential for serious morbidity—even mortality—that may result from inhalation of particles of biological origin.

The term "bioaerosol" refers to airborne particles containing: 1) intact living or dead microorganisms (as single units, in homogeneous or heterogeneous groups, or attached to other particles); 2) microbial spores (resistant reproductive structures produced by many fungi and some bacteria); and 3) fragments of microor-

ganisms and larger organisms (excreta and body parts from arthropods, skin scales from mammals, and pollen and plant debris). The term "microorganism" refers to fungi, bacteria, viruses, and protozoa; however, this chapter does not specifically address air sampling for the latter two categories of agents. The agents of primary interest from the first two categories are: 1) saprophytic (or saprobic) fungi and bacteria (those that live on dead organic matter and that may grow indoors on building materials and furnishings or be transported indoors from outdoor sources); and 2) facultatively and obligately parasitic fungi and bacteria (those that live on or in other organisms and may be found in humans, animals, or the environment). These microorganisms may cause allergic or toxic reactions or infectious diseases. Indoor "biological contamination" has been defined as the presence of: 1) biologically derived aerosols of a kind and concentration likely to cause disease or predispose people to disease; 2) inappropriate concentrations of outdoor bioaerosols, especially in buildings designed to prevent their entry; or 3) indoor microbial growth and remnants of biological growth that may become aerosolized and to which people may be exposed.[1]

Gases of biological origin may be of interest because the compounds are themselves hazardous or because they serve as markers for the presence of particular biological agents. Examples of this category of air contaminants are carbon dioxide from exhaled breath, ammonia from animal wastes, and the microbial volatile organic compounds (MVOCs) that many bacteria and fungi produce.[2–4] Instruments for collecting and analyzing gases and vapors of biological origin are identical to the methods discussed in Chapters 16, 17, and 18, even though the immediate origin of such compounds is a microorganism, plant, or animal rather than an industrial chemical or other product used in a work environment.

It is neither possible nor desirable in this chapter to review the extensive literature on the occurrence, health effects, and sampling of bioaerosols. Several publications cover the broad field of bioaerosol effects and significance.[5–15] A number of review articles have described bioaerosol samplers and discussed their uses and limitations.[16–27] For a historical view of bioaerosol samplers, readers should consult the references from previous editions of this chapter. Many important, earlier publications were omitted from this edition to allow space for more recent and accessible material. This chapter discusses the diversity of bioaerosols and their importance as health hazards. We also suggest how to select the most suitable equipment for collecting various biological agents in different sampling situations. Other ACGIH publications address related and complementary aspects of bioaerosol sampling.[1, 15, 25, 28, 29]

Researchers from many fields collect air samples to study the concentrations and distributions of airborne microorganisms and other bioaerosols. For example, a variety of bioaerosol samplers are used by allergists and immunologists;[5, 10, 19, 30, 31] infection control professionals;[32, 33] animal and plant pathologists;[34, 35] atmospheric researchers;[36, 37] and quality control specialists in the biotechnology, food processing, and drug manufacturing industries.[20, 38–44] This chapter focuses on investigations of problems related to human health and well-being in non-manufacturing workplaces such as offices, but this information may also apply to exposures in residential, commercial, and recreational environments. In these settings, bioaerosols are of concern for their potential infectivity, allergenicity, or toxicity.[3, 7, 9, 11, 35, 45–62]

Air sampling for many chemical and physical agents is conducted to determine compliance with ambient and workplace exposure limits. ACGIH has developed threshold limit values (TLVs®) for certain substances of biological origin, including cellulose; some wood, cotton, and grain dusts; nicotine; pyrethrum; starch; subtilisins (proteolytic enzymes); sucrose; and vegetable oil mist. However, there are no health-based exposure limits against which to compare environmental air concentrations of most materials of biological origin to which persons may be exposed in non-manufacturing workplaces.[20, 63–65] A statement in the ACGIH TLV®/BEI® book explains why it has not yet been possible to establish such limits for various categories of biological agents.[1]

Selection of a Bioaerosol Sampler

Bioaerosol samplers typically use one of three basic methods of particle collection, that is, filtration, impaction, or impingement into a liquid, the latter being inertial impaction followed by capture in a collection liquid.[22, 27, 66, 67] Chapters 13 and 14 explain these particle collection mechanisms. Table 22-1 describes several commercially available bioaerosol samplers. The column "Application" identifies the general categories of bioaerosol analysis compatible with individual samplers (e.g., microorganism culture, immunoassay, or molecular detection methods). The "Instrument Descriptions" section of this chapter provides illustrations and detailed information on the samplers listed in Table 22-1 however, samplers are often modified and current configurations may differ from those listed. Table 22-2 outlines the primary methods of sample collection and analysis for major groups of bioaerosols along with the information that each assay can provide about the biological agents (e.g., number or mass concentration in air or taxonomic identification). Further information about the health effects and environmental reservoirs of these agents as well as methods for their detection, identification, and quantification is available elsewhere.[8, 12–15, 68]

Users have attempted to list features that an ideal air sampler would possess;[29, 69] however, no single, currently available bioaerosol sampler is optimal for all applications. To select an appropriate device, users must clearly outline: 1) their reasons for collecting air samples and the intended use of the data; 2) the biological agents they wish to study and the information they need about them: 3) how samples will be assayed and the detection limits of the methods; 4) the estimated or previously measured bioaerosol concentration at the test site; 5) the aerodynamic diameters of the particles to be collected; 6) the velocity of the sample air stream; and 7) environmental conditions (e.g., air temperature and moisture

TABLE 22-1. Widely Used, Commercially Available Samplers for Collecting Bioaerosols

Sampler[A]	Principle of Operation	Sampling Rate (L/min)	Manufacturer/Supplier[B]	Commercial Name	Application[C]
SLIT AGAR IMPACTORS					
1. Rotating slit or slit-to agar impactors (vp,sc)	Impaction onto agar in 9- or 15-cm plates on rotating surfaces	28 175, 350, 525, 700	BAR CAS	1a. Mattson-Garvin Air Sampler 1b. Casella Airborne Bacteria Sampler	C C
		15-55 15-30	NBS NBS	New Brunswick Slit-to-Agar Air Samplers: 1c. STA-203 1d. Slit-to-Agar Air Sampler: STA-204	C C
MULTIPLE-HOLE IMPACTORS					
2. One-stage, 100-hole impactor (sc)	Impaction onto agar in 9-cm plates	10, 20	BMC	2. Burkard Portable Air Sampler for Agar Plates	C
3. One-stage, 400-hole impactors (vp, sc)	Impaction onto agar in 9-cm plates	28	AND	3a. Andersen Single Stage Viable Particle Sampler; N6 Single Stage Viable Impactor	C
		28	AER	3b. Aerotech 6 Bioaerosol Sampler	C
4. One-stage, 12-hole impactors (vp, sc)	Impaction onto agar in 9-cm plates	28-71 28, 142	VAI VAI	Sterilizable Microbiological Atrium (SMA): 4a. SMA Micro Sampler 4b. SMA Micro Portable Viable Air Sampler	C C
5. One-stage, 219- or 487-hole impactor (contact plates) or 400-hole impactor (9-cm plates) (sc)	Impaction onto agar in 5.5- or 8.4-cm contact plates or 9-cm plates	100	PBI, BSI	Surface-Air-Sampler (SAS): 5a. SAS Super 100 Sampler	C
One-stage, 219-hole impactor (sc)	Impaction onto agar in 5.5-cm contact plates	100	PBI, BSI	5b. SAS HiVac Impact	C
One-stage, 400-hole impactor (sc)	Impaction onto agar in 9-cm plates	100	PBI, BSI	5c. SAS HiVAC Petri	C
One-stage, 220-hole impactor (sc)	Impaction onto agar in 5.5-cm contact plates	100	PAR	5d. MicroBio Air Samplers: MB1, MB2	C
One-stage, 106- or 318-hole impactor (sc)	Impaction onto agar in 9-cm plates	100, 200	MBI	5e. Sampl' Air Air Sampler	C
One-stage, 400-hole impactor (sc)	Impaction onto agar in 9-cm plates	100	MER/VWR	5f. Merck Air Sampler MAS 100	C
6. One-stage, 967-hole impactor (sc)	Impaction onto agar in 7.5-cm cassette	140, 180	MIL	6a. M Air T, Millipore Air Tester	C
One-stage, 219- or 487-hole impactors (contact plates) or 400-hole impactor (9-cm plates) (sc)	Impaction onto agar in 5.5- or 8.4-cm contact plates or 9-cm plates	180	PBI, BSI	6b. SAS Super 180 Sampler	C

TABLE 22-1 (cont.). Widely Used, Commercially Available Samplers for Collecting Bioaerosols

Sampler[A]	Principle of Operation	Sampling Rate (L/min)	Manufacturer/ Supplier[B]	Commercial Name	Application[C]
7. Two-stage, 200-hole impactor (vp)	Impaction onto agar in 9-cm plates	28	AND	7. Andersen Two Stage Viable Sampler/Cascade Impactor	C
8. Six-stage, 400-hole impactor (vp)	Impaction onto agar in 9-cm plates	28	AND	8. Andersen Six Stage Viable Sampler/Cascade Impactor	C
9. Eight-stage, personal impactor (vp)	Impaction onto 3.4-cm substrates	2	AND	9. Andersen Personal Cascade Impactor, Series 290 Marple Personal Cascade Impactor	O
FILTERS					
10. Cassette filters (vp)	Filtration, generally 25-, 37-, or 45-mm filters in pre-loaded, disposable cassettes or reusable filter holders	1–5	see Chapter 13		H, M, O
CENTRIFUGAL SAMPLERS					
11. Centrifugal agar impactors (sc)	Impaction onto agar in plastic strips			Reuter Centrifugal Samplers (RCS):	
		40	BDC	11a. Standard RCS	C
		50	BDC	11b. RCS Plus	C
		100	BDC	11c. RCS High Flow Microbial Air Sampler	C
12. Wetted cyclone samplers (vp, sc)	Tangential impingement into thin liquid layer	50–55, 167, 500	HAM	12a. AEA Technology PLC Aerojet Cyclones	C, M, O
		167	PAR	12b. MicroBio MB3 Portable Cyclone	C, M, O
		700–1000	LRI	12c. Aerojet-General Liquid-Scrubber	C, M, O
		100–800	MRI	12d. SpinCon High-Volume Cyclonic Liquid Sampler	C, M, O
		≤300	ITI	12e. Mini-Cyclone Aerosol Collector	C, M, O
13. Dry cyclone sampler (sc)	Reverse flow cyclone	≤20	BMC	13. Burkard Cyclone Sampler	H, M, O
14. Three-jet, tangential sampler (vp)	Tangential impaction onto glass or filter surface or impingement into liquid	12.5	SKC	14. BioSampler	C, M, O
LIQUID IMPINGERS					
15. All-glass impingers (vp)	Impingement into liquid	12.5	AGI, HAM, MIL	15. All-Glass Impingers (AGI): AGI-4, AGI-30	C, M, O
16. Three-stage impingers (vp)	Impingement into liquid	10, 20, 50	BMC	16a. Burkard Multiple-Stage Liquid Impinger	C, M, O
		10, 20, 50	HAM	16b. Hampshire Glass Three-Stage Impinger	C, M, O

TABLE 22-1 (cont.). Widely Used, Commercially Available Samplers for Collecting Bioaerosols

Sampler[A]	Principle of Operation	Sampling Rate (L/min)	Manufacturer/ Supplier[B]	Commercial Name	Application[C]
POLLEN, SPORE, AND PARTICLE IMPACTORS					
17. One- to seven-day tape/slide impactors (sc)	Impaction onto rotating drum with tape strip or glass slide	10	BMC	17a. Burkard Recording Volumetric Spore Trap	M
		10	LAN	17b. Lanzoni Volumetric Pollen and Particle Sampler	M
		25	GRM	17c. Kramer-Collins Suction Trap	M
18. Moving slide impactors (sc)	Impaction onto moving glass slides	15	ALL, MCC	18a. Allergenco Air Sampler (MK-3)	M
		10	BMC	18b. Burkard Continuous Recording Air Sampler	M
		10	LAN	18c. Lanzoni Volumetric Pollen and Particle Sampler	M
19. Stationary slide impactor (sc)	Impaction onto stationary glass slide	10	BMC	19. Burkard Personal Volumetric Air Sampler	M
20. Cassette slide impactor (vp)	Impaction onto stationary glass slide	15	ZAA, AER, MCC, SKC	20. Air-O-Cell Sampling Cassette	M
21. Cassette tape sampler	Deposition onto stationary tape	10	MET	21. Partrap FA52	M
22. Rotating rod impactor (sc)	Impaction onto rotating rods	48	STI	22. Rotorod	M
23. Rotating arm impactor	Impaction onto rotating arm with liquid rinse	125	MST	23. BioCapture Air Sampler	C, H, M, O

[A] Letters in parentheses:
　vp = Requires a vacuum pump and flow control device, which the sampler manufacturer/supplier may provide.
　sc = Self-Contained with built-in air mover.
[B] See Table 22-I-1.
[C] C = Culture of sensitive and hardy microorganisms (e.g., vegetative bacterial and fungal cells and spores).
　H = Culture of hardy microorganisms only (e.g., spore-forming bacteria and fungi).
　M = Microscopic examination of collected particles.
　O = Other assay (e.g., immunoassays, bioassays, chemical assays, or molecular detection methods).

TABLE 22-2. Collection and Analysis of Air Samples for Biological Agents

Biological Agent	Sampling Method	Sample Analysis	Data Obtained
AMEBAE	Impingers, agar or slide impactors	Culture Microscopy: cell morphology Immunosassay—labeled antibody stains	Concentration: number/m³ Isolate identification Confirmation of presence of specific ameba
BACTERIA	Agar or slide impactors impingers, wetted cyclones, filters	Direct microscopy/total counts Immunoassay—labeled antibody stains Nucleic acid probe (molecular hybridization), nucleic acid amplification (PCR)	Concentration: cells/m³ Confirmation of presence of specific bacterium Confirmation of presence of specific bacterium
		Culture: colony morphology	Concentration: CFU/m³
		Microscopy following culture: cell morphology, Gram-stain characteristics	Isolate identification (general)
		Biochemical assay following culture: substrate utilization assay	Isolate identification (specific)
Bacterial cell-wall components:			
Endotoxin (Gram-negative bacteria)	Filters, impingers	LAL assay	Concentration (endotoxin): endotoxin units/m³
		Chemical assay: GC-MS, HPLC	Concentration (lipopolysaccharide): µg/m³
Other (total bacteria)	Filters	Chemical assay: GC-MS, HPLC	Concentration (muramic acid diaminopimelic acid): µg/m³
Bacterial whole-cell lipids; phospholipids	Filters, impingers	Chemical assay: GC FAME, GC-MS	Community profile
FUNGI	Slide or agar impactors, filters, impingers, wetted cyclones	Direct microscopy/total counts Culture: Colony morphology Microscopy following culture spore and hyphal morphology	Concentration: spores/m³ Concentration: CFU/m³ Isolate identification
Yeasts (whole-cell lipids, fatty acids)	Agar impactors, impingers, wetted cyclones	Chemical assay: GC/FAME Biochemical assay: substrate utilization	Isolate identification
Fungal cell-wall components	Filters	LAL, immunoassay	Concentration (glucan): units/m³ or µg/m³
		Chemical assay: GC-MS, HPLC	Concentration (glucan or ergosterol): units/m³ or µg/m³

Fungal toxins (see also myco-toxins)	Slide or agar impactors filters, impingers wetted cylones	For toxigenic fungi: (see above for fungi—Direct microscopy, Culture, and Microscopy following culture)
		For toxins:
		    Chemical assay (TLC, HPLC, GC-MS)
		    Immunoassay
		    Cytotoxicity assay
			Confirmation of toxin presence
			Concentration (toxin): ng/m³
			Confirmation of toxin presence
			Detection of toxic activity without toxin identification

TABLE 22-2 (cont.). Collection and Analysis of Air Samples for Biological Agents

Biological Agent	Sampling Method	Sample Analysis	Data Obtained
POLLEN	Slide or tape impactors	Microscopy: pollen count	Concentration (pollen grains): number/m^3
		Microscopy: pollen morphology	Pollen identification
VIRUSES	Impingers, wetted cyclones, agar or slide impactors, filters	Cell culture	Concentration (cytopathic units): number/m^3; isolate identification
		Immunoassay—labeled antibody stain	Confirmation of presence of specific virus
		Electron microscopy	Isolate identification
		Molecular assay: nucleic acid probes (molecular hybridization), nucleic acid amplification (PCR)	Confirmation of presence of specific virus

CFU	=	colony-forming unit	LAL	=	limulus amebocyte lysate
FAME	=	fatty acid methyl ester	MS	=	mass spectrometry
GC	=	gas chromotography	PCR	=	polymerase chain reaction
HPLC	=	high performance liquid chromotography	TLC	=	thin-layer chromotography

content at the sampling site) that may restrict the choice of sampling method. Investigators must also consider the cost and availability of samplers and related supplies as well as the levels of technical skills required of field and laboratory personnel.

Sampling Strategy

Every air sampling program should be based on a well-considered strategy that begins with collection of general health and environmental data intended to lead investigators to the formulation of hypotheses to explain their observations.[70, 71] After investigators develop hypotheses, environmental sampling may help them test the validity of their theories about bioaerosol exposures. When collecting bioaerosol samples, investigators must consider: 1) possible sampling locations; 2) the number of samples to collect; 3) bioaerosol concentration as well as seasonal and temporal fluctuations in concentration; 4) the volume of air to collect; 5) the sampling time required to collect a given volume of air; and 6) the limitations of available assay systems.[72] Investigators should not overlook the value of information they can gain by means other than air sampling. For example, visual inspections for sources of biological agents and moisture can be valuable as can collection of source samples from surfaces, waters, and bulk materials to determine if they harbor biological agents that may become airborne.[71–74]

Concentration Considerations

Form of Recovered Particles

When studying airborne microorganisms, investigators may only need to determine the air concentration of an agent without concern for whether collected particles contain one or more separate units (e.g., individual bacterial or fungal cells). Agar impactors are widely used to obtain information on the number of particles containing culturable bacteria or fungi. Such counts are reported as the number of "Colony-Forming Units" (CFUs) because it is not known if the colonies arose from: 1) single particles carrying single microorganisms; 2) single particles composed of clusters of identical or different microorganisms; or 3) the coincidental impaction of multiple culturable particles. In other situations (e.g., to estimate inhaled dose), it may be desirable to know the air concentration of separate units (e.g., the total number of potentially infectious cells per cubic meter of air) or to measure the mass concentration of a particular agent (e.g., nanograms of endotoxin per cubic meter). Liquid impingers and wetted cyclones may separate aggregated particles into smaller units (possibly single cells),[75] and data from these samplers may better approximate the total number of culturable or countable airborne cells. Slide or tape impactors and filter devices generate samples that can be analyzed microscopically. By visually examining particles, analysts can obtain information on the number of particle clusters from which to estimate the total number of individual units. For example for fungal spores, an analyst can count the number of spore chains of different lengths and calculate the total number of spores in a sample. Samplers that collect bioaerosols in a form from which biological agents can be eluted or extracted may provide data on the total mass of material collected (e.g., nanograms of a specific allergenic protein) or the potency of a biological agent (e.g., the cytotoxicity of a mycotoxin as measured in a bioassay). Some samplers have been adapted to allow investigators to choose among collection surfaces or substrates, depending on the needs of a study. For example, a slot sampler was designed to collect particles onto agar or an adhesive surface or into a liquid,[76, 77] and Instrument 22-14 can be used dry or with a collection liquid.[78]

Air Flow Rate and Sampling Time

Investigators must anticipate bioaerosol concentrations at study sites to determine what volume of air to collect for a particular biological agent and assay. Identification of a desired sample volume allows investigators to select samplers with air flow rates that permit reasonable sampling times. Estimates of bioaerosol concentration may come from preliminary sampling results or published reports of studies in similar settings. The amount of material in an infectious, allergenic, or toxic dose of a bioaerosol may be very small compared to the amount of airborne dust or the number of fibers needed to produce diseases due to nonbiological agents. Theoretically, a single, infectious virus, bacterium, or fungus may enter the body and multiply to sufficient numbers to cause disease. In sensitized persons, allergens appear to exert their effects at ng/m^3 levels, and some biological toxins may cause effects at $\mu g/m^3$ or ng/m^3 levels. Microbiologists frequently use the ability of microorganisms to multiply as well as synthetic amplification of selected nucleic acid sequences to achieve very sensitive detection and very specific identification of biological agents. Nevertheless, to collect bioaerosols present in low concentrations, large-volume samples may be required; for example, short-term samples collected at high air flow rates or long-term samples collected at low air flow rates. In the former case, investigators should consider how well short-term samples represent actual exposures. In the latter case, investigators should evaluate the potential effects of prolonged sampling on the integrity of the biological agents being collected.

Sampling when bioaerosol concentration is high may present problems of rapid sample overload on direct agar or slide impactors (Figures 22-1 and 22-2). Investigators often decrease air flow rates or shorten collection times to avoid overloading impactors when sam-

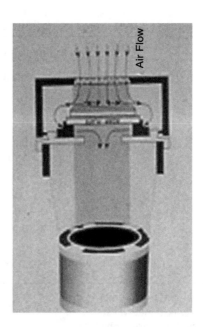

FIGURE 22-1. Direct agar impactor (Instrument 22-5a).

pling in heavily contaminated environments. Before making such adjustments, investigators should consider what effect these changes may have on particle collection efficiency and the representativeness of the samples collected. For example, decreasing sampling rate will increase the d_{50} cutpoint of an impactor, lowering the efficiency with which smaller particles are collected. Likewise, if collection time is only a few seconds, separation of particles in a multiple-stage sampler will be incorrect if the air does not reach a steady flow rate through all stages of the device.

The air flow rates for the bioaerosol samplers in Table 22-1 range from 1 to 1000 L/min, allowing investigators to identify devices that may be suitable for sampling in various indoor and outdoor environments. Equation 1 gives a formula to calculate sampling time, t, when the desired surface density, δ, deposit area, A, average air concentration, C_a, and airflow rate, Q, are known[79, 80]

$$t = \frac{\delta A}{C_a Q}$$

Sampling time is often measured in minutes and surface density in CFU or number of particles per unit of collection surface area. Typical target densities are 1–5 CFU/cm^2 on an agar surface and 10^4–10^5 spores or pollen grains/cm^2 on a slide to be examined under a microscope. Deposit area may be measured in cm^2, air concentration in CFU/m^3 or particles/m^3, and airflow rate in L/min or m^3/min. Possible interactive effects of air flow rate and sampling time have been explored for several bioaerosol samplers.[81–84]

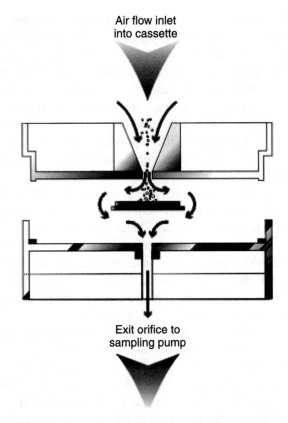

Air flow inlet
into cassette

Exit orifice to
sampling pump

FIGURE 22-2. Slide impactor (Instrument 22-20).

Equation 2 gives a formula to calculate sampling time, t, when the desired concentration of a liquid suspension, C_l, and collection, wash, or extraction liquid volume, V_l, are known:

$$t = \frac{C_l V_l}{C_a Q}$$

Suspension concentrations are typically measured as particle count or amount of material per unit volume of liquid (e.g., CFU/ml or endotoxin units/ml), and liquid volumes are measured in milliliters or liters. Investigators may choose to collect several air volumes, when studying previously untested environments, so that at least one sample yields an acceptable result (i.e., one within the upper and lower detection limits of the collection and analytical methods).[72]

Colony Overgrowth and Particle Masking

The density of colonies on a culture plate or particles on a surface affects the reliability of the information that can be obtained from a sample. Too few or many colonies or particles are difficult to count and may lead to inaccurate estimates of bioaerosol concentration. Investigators should consult a knowledgeable laboratory analyst to learn the ideal particle surface density for the intended analysis. Investigators should

realize that in conjunction with air flow rate and sample collection time, these restrictions determine a sampling method's detection limits (i.e., the minimum and maximum bioaerosol concentrations a method can measure reliably).[72]

With multiple-hole impactors, there is the likelihood that more than one particle will enter a hole through which another particle has traveled. The likelihood of this occurring increases with the number of particles collected. Closely spaced colonies may not be distinguishable, leading to underestimation of the bioaerosol concentration. Tables are available that provide corrected colony counts and standard deviations on these counts for multiple-hole impactors to adjust plate counts for the probability that more than one particle impacted at each collection site.[27, 85–87] Colony-count corrections may also be calculated by various convenient methods.[80, 88, 89] Table 22-3 lists approximate correction factors for multiple-hole impactors.[80] For example, if 120 of 400 holes were filled (i.e., colonies grew at this number of total impaction sites), the "filled fraction" would be 0.30 [120/400 = 0.30], the correction factor would be 1.189 (from Table 22-3), and the corrected count would be 143 [120 × 1.189 = 143]. For intermediate fill fractions, f, Equation 3 can be used to calculate the corrected colony count, n_c.

$$n_c = n_f \left(\frac{1.075}{1.052 - f} \right)^{0.483} \quad \text{for } f \langle 0.95$$

where n_f is the number of CFUs or filled impaction sites. For example, if the plate count was 130 of 400 holes, f would be 0.325 [130/400 = 0.325] and n_c would be 157 [130 × (1.075/1.052 – 0.325)$^{0.483}$ = 157].

Macroscopic colonies (i.e., those that can readily be seen by eye) are most often counted. In general, 25 to 250 bacterial colonies and 10 to 50 fungal colonies are considered optimal for accurate counting and identification of CFUs on standard, 9-cm plates.[72] Target colony counts are adjusted proportionally for smaller and larger plates. Therefore, another rule of thumb microbiologists have used is a maximum surface density of approximately 1 colony/cm². For multiple-hole impactors, investigators should also consider the greater variability associated with higher plate counts when using a positive-hole correction.

Colony overlap or particle masking may also occur when bioaerosols are collected with slit impactors onto a stationary or moving collection surface (Instruments 22-1, 22-17–22-20; Figures 22-2 and 22-3). However, calculating the probability of multiple impactions in slit samples is not simple because of the long narrow inlet

TABLE 22-3. Correction Factors for Particle Collection with Multiple-Hole Impactors

Filled Fraction[A]	Correction Factor[B]
0.05	1.026
0.10	1.054
0.15	1.084
0.20	1.116
0.25	1.151
0.30	1.189
0.35	1.231
0.40	1.277
0.45	1.329
0.50	1.386
0.55	1.452
0.60	1.527
0.65	1.615
0.70	1.720
0.75	1.848
0.80	2.012
0.85	2.232
0.90	2.559
0.95	3.154
1.00	>5.878

[A] Fraction of impaction sites with CFUs.
[B] Total number of viable particles collected equals the number of filled sites times the correction factor.

and the possibly advancing, rather than fixed, collection surface. Therefore, investigators may consider that each colony on a slit-to-agar sample resulted from impaction of a single culturable particle and that each cluster of particles on a tape sample was deposited as an aggregate. Bacterial and fungal colonies may also be counted with the aid of a microscope before they are readily visible. A CFU correction based on colony diameter, colony density, and impaction area has been proposed.[90–92]

Bioaerosols in highly contaminated environments have been collected with membrane filters of various

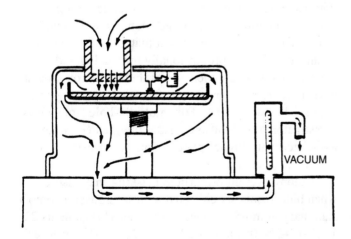

FIGURE 22-3. Rotating slit agar impactor showing flowmeter and adjustable stage height (Instrument 22-1).

kinds, the type of filter depending on the requirements of the subsequent sample analysis. Particles with distinctive morphological features may be examined directly with a microscope or following contrast staining if particle density is in an appropriate range (i.e., neither too sparse for efficient counting or so dense that masking occurs).[62, 75, 93–99] Unlike counting of asbestos and other fibers on cleared cellulose ester filters, polycarbonate filters typically are used to count stained microbial cells that are often examined under fluorescent lighting.

Provided colony density is within an acceptable range, bacteria and fungi may be allowed to grow into CFUs directly on a cellulose ester filter that was placed face-up on agar-based culture medium or a support pad wetted with liquid culture medium.[33] Bacterial or fungal cells may also be washed from filters or other collection surfaces for transfer to an assay system, such as growth on agar-based culture media, staining with labeled antibodies, measurement of enzymatic activity, or molecular detection methods.[81, 84, 97, 100–105]

Membrane filters composed of compressed gelatin foam were designed to reduce desiccation of vegetative microorganisms relative to other types of filter media. Particles collected on a gelatin filter may be assayed by placing it directly on nutrient agar before incubation or by dissolving the filter in liquid and determining the microbial content of the suspension by suitable assay.[81, 84, 102, 106, 107]

Problems of too high or low particle concentrations may be avoided when bioaerosols are collected into a liquid or collected particles can be transferred to a liquid. High-concentration samples can be diluted prior to processing and it may be possible to centrifuge or filter low-concentration samples.[19, 108] Therefore, an eluted membrane filter, wetted cyclone, wetted tangential sampler, or impinger is useful when it is difficult to anticipate bioaerosol concentration or when the concentration may vary greatly (Instruments 22-10, 22-12, 22-14–22-16, or 22-23).

The final volume of collection or wash fluid must be measured and taken into account when calculating sample concentration. For example, the original collection volume in an all-glass impinger (AGI) may have been 20 ml, but the final volume will be somewhat less due to water evaporation and transfer losses. At 47% relative humidity, evaporation of water from the AGI-4 and AGI-30 was found to be approximately 0.2 ml/min or 1% of the initial volume of sampling liquid.[109] The use of a nonevaporating collection fluid (e.g., mineral oil) in a tangential impactor (Instrument 22-14) has been proposed to reduce problems related to evaporation of water-based collection media as well as particle bounce and reaerosolization that may occur in impingers and wetted cyclones.[78] Hydrophobic biological particles (e.g., some fungal spores) may be collected poorly in water-based sampling liquids because such particles may not be wetted sufficiently to be retained in the collection fluid.[77, 110] Reaerosolization of particles collected in impinger samplers has also been observed.[77, 109] To capture such particles, a downstream membrane filter may be added.[103]

Sampling Efficiency

The physics of removing particles from air and the general principles of accurate sample collection apply to the sampling of any airborne material, whether of biological or other origin. Therefore, many of the basic principles of particle collection discussed in Chapters 13 and 14 can be adapted to bioaerosol sampling. The overall efficiency of a bioaerosol sampler can be divided into four components, that is, inlet, particle removal, biological recovery, and assay efficiencies.[27, 44] *Inlet sampling efficiency* measures the ability of a sampler to entrain particles from the ambient environment without bias as to particle size, shape, or aerodynamic behavior. Grinshpun *et al.*[111] modeled the inlet characteristics of several bioaerosol samplers for collecting particles of different aerodynamic diameters at typical indoor and outdoor wind speeds for varying sampler orientations. Griffiths and Stewart[44] also examined the effect of inlet orientation on bioaerosol sampler performance. *Particle removal efficiency* measures a device's ability to separate particles from the sampled air stream and to deposit them on or in a collection medium. *Biological recovery efficiency* measures a sampler's ability to deliver the collected particles to an assay system without altering the viability, biological activity, physical integrity, or other essential characteristic of the biological agents. *Assay efficiency* refers to the accuracy with which collected particles are counted and correctly identified.[44, 112, 113] To ensure that samples are analyzed properly and bioaerosols identified correctly, testing laboratories should participate in appropriate programs for performance evaluation or proficiency testing. If no such programs are available, laboratories should consider conducting inter-laboratory evaluations in collaboration with other researchers and commercial testing groups.[114]

In filtration, particles are collected primarily by interception, and sampling efficiency depends on particle and filter pore diameters. The collection efficiency of inertial impactors is determined primarily by particle characteristics (such as density, diameter, and surface

features) and air velocity. In part, jet velocity determines an impactor's cutpoint (d_{50}). The distance between an impactor nozzle or jet and a collection surface is also critical to particle collection efficiency. Slit-to-agar impactors may have adjustable stage heights to ensure the proper slit-to-agar distance (Figure 22-3). For other impactors that collect particles on an agar surface, the distance varies with the type of culture plate used and the amount of agar added. Most impactors are designed for a jet-to-plate distance approximately equal to nozzle width (e.g., 0.25 to 1 mm). For size-separating samplers, agar volume should comply closely to a manufacturer's specification to ensure collection of particles near an impactor's d_{50} cutpoint. Agar depth is less critical for general bioaerosol sampling as long as: 1) the agar surface is fairly level and smooth; 2) all plates have approximately the same amount of agar; 3) the jet-to-plate distance in "overfilled" plates is not less that half the nozzle width; and 4) agar in "under-filled" plates is not so thin that it dries out during sample collection or incubation.

Typical surfaces for collecting bioaerosols are agar-based media and coated tapes or glass slides. Surface properties affect particle bounce, which may complicate the efficient collection of bioaerosols with inertial impactors. A biological particle striking moist agar generally will stick, but a "bouncy" particle hitting a previously deposited particle may rebound and be carried away in the air stream. Particles may also bounce from the impaction surfaces in impingers.[77] Glass microscope slides and transparent tapes are usually coated to improve particle retention (Instruments 22-17–22-22). Commonly used coatings include glycerine jelly, silicone grease, petroleum jelly (sometimes with added liquid paraffin), and polyvinyl alcohol.[115, 116] Double-sided, self-adhesive, acrylic tape (Scotch 3M 9425, St. Paul, MN) was found to have higher capture efficiency than a glycerine/gelatin adhesive.[117] A transparent "acrylic" substrate is used in a cassette slide impactor (Instrument 22-20). In addition to good particle retention properties, slides, tapes, and coatings must have good optical qualities and be compatible with any stains or mounting media the analytical laboratory will use on the specimens.

A potential source of confusion with impactors is satellite colonies or particles that are occasionally seen at the perimeter of an impaction zone. Particles that have bounced or blown off eventually may deposit beyond an impaction area. Air motion downstream of an impaction site may also become turbulent, and previously unimpacted particles in this turbulent air may move sufficiently close to a collection surface to deposit at some distance from the primary impaction point.[27] A build-up of deposited particles, deformation of a collection surface at an impaction site (e.g., a dimple formed in drying agar), or interference between adjacent jets may disturb smooth air flow. Collected particles may also migrate from an impaction area if water or excess adhesive spreads during or after sample collection.

Changes in critical features of samplers can alter the efficiency of particle collection and biological recovery for culturable microorganisms. For example, the AGI (Instrument 22-15) comes in two versions, the AGI-4 and the AGI-30 with corresponding jet-to-plate distances of 4 and 30 mm. For the same sampling time and air flow rate, an AGI-30 may collect fewer small particles than an AGI-4 (because of the increased jet-to-plate distance); however, more of the collected cells may remain viable because they suffer less damage. One group found that damage to microbial cells during impaction in liquid impingers could be greater than subsequent loss of cell viability due to agitation during continued sampling.[75] Viability loss has also been seen in agar impactors when operated at varying impaction velocities.[118]

In addition to the factors outlined above, investigators may need to consider the efficiency with which collected particles are transferred to an assay system. To illustrate, consider a 0.4-μm-pore membrane filter, which has very high particle collection efficiency. Transfer and detection efficiencies may be good if culturable particles are allowed to grow in place on the filter or the filter surface is examined directly with a microscope. However, there may be transfer losses if collected particles must be washed from the filter. Isolation of culturable microorganisms may be lower from filter than impactor or impinger samples as a consequence of cell damage due to particle dehydration and rehydration that occur, respectively, during sample collection and assay. It is a combination of such considerations that determines which device is most suitable for a given bioaerosol sampling situation. It may be necessary to try a sampler at several flow rates and sample collection times as well as with different collection media to optimize efficiency for both particle collection and biological preservation. It may also be necessary to evaluate variations in an analytical method to optimize retrieval and detection of biological agents from air samples.

Size-Selective Sampling

Individual viral particles range from approximately 0.01 to 0.25 μm, bacterial cells from 0.1 to 10 μm, fungal spores and fragments from 1 to 100 μm, and pollen grains from 10 to 100 μm. Dimensions, as quoted in biol-

ogy texts, typically are made from microscopic examinations of prepared specimens and may not accurately predict bioaerosol aerodynamic diameters. These measurements generally are of hydrated cells whereas airborne organisms may be desiccated. Some fungal spores are released singly or separate into single units after release. However, airborne bacteria and viruses are seldom found as individual units (unless they were generated as fine sprays of dilute water suspensions). From this observation, investigators have concluded that these microorganisms were aerosolized from liquids containing other materials or were reaerosolized along with other particles. The other material may be respiratory or oral secretions, skin flakes, soil particles, or plant fragments.

Large biological particles of less than unit density as well as nonspherical particles may behave aerodynamically like smaller, unit-density, spherical particles. The density of pollen grains and fungal spores may be less than that of water, and many bacteria, fungal spores, and mycelial fragments are not spherical. Allergens and microbial toxins are non-living components of source organisms and may be airborne on intact cells as well as on fragments of cells or other materials. Some such particles are larger than 10 μm (e.g., mite fecal pellets, large fungal spores, and pollen grains), but other biological agents are found on particles less than 1 μm (e.g., ragweed allergen and bacterial endotoxin have been detected in submicrometer particles).

The category "d_{50}" in the "Instrument Description" section lists measured or calculated cutpoints for samplers if this information is available. Knowledge of approximate particle size is important when selecting a suitable sampler and choosing between sampling air or potential sources to detect biological agents.[119] Large particles may not remain airborne very long and, thus, may be missed if only air samples are collected. Allergens from dust mites and cockroaches are associated with particles larger than 5 μm diameter and become airborne only when sources are disturbed whereas cat allergen is present on particles less than 2.5 μm.[119–121] Thus, while exposure to cat allergen may be assessed by collecting air samples, the presence of dust mite and cockroach allergens typically is measured in settled dust.[14, 122]

Many accepted methods for determining the size distribution of airborne dusts and mineral particles are also suitable for sizing culturable bioaerosols. Several groups have found that particles of biological origin can be collected in air samplers and captured in air cleaners similarly to nonbiological particles of the same aerodynamic diameter. Therefore, substitution of aerodynamically equivalent nonbiological particles for bioaerosols should be acceptable in some types of research (e.g., for the evaluation of sampler performance and respirator efficiency).[123–128] It also may be safer and more convenient to work with nonbiological particles. Researchers often use many of the same bacteria and fungi to evaluate bioaerosol samplers and efforts are underway to identify representative test organisms.[44, 129, 130]

The best method of bioaerosol size separation is that which is simplest, provides the required size information, and is compatible with the chosen assay system. A widely used, multiple-stage impactor collects particles in six size fractions (Instrument 22-8, Figure 22-4). A similar, two-stage impactor separates particles larger and smaller than 7 μm (Instrument 22-7). The curved neck of an AGI traps particles larger than 8 μm, and these particles can be recovered by rinsing the inlet tube (Instrument 22-15).[103] A three-stage impinger approximates particle collection in the nasopharyngeal, tracheobronchial, and alveolar regions of the lungs and allows size separation for bioaerosols that are best collected into liquid (Instrument 22-16).[131] Several bioaerosol samplers have been designed to match various sampling conventions (see Sampler Performance).

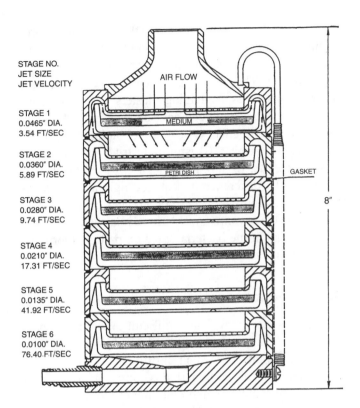

FIGURE 22-4. Six-stage cascade impactor (Instrument 22-8).

Rapid Bioaerosol Detection

Direct-reading samplers to measure particle size distribution as well as number or mass concentration are widely available (Chapter 15). However, the concentration of airborne microorganisms usually cannot be predicted from the concentration of total suspended or size-fractionated particles even when an agent's aerodynamic diameter is known. The number of microorganisms and other biological agents in ambient air is usually small relative to the total particle load, except where other airborne particles have been removed intentionally and a bioaerosol is known to dominate.

Currently, the most rapid method for detecting many bioaerosols is direct microscopic examination immediately after particle collection. This method requires that the particles of interest be morphologically distinctive and, therefore, is restricted to the detection of fungal spores, pollen grains, and stained bacterial or fungal cells.[36, 37, 62, 67, 75, 84, 93–99, 102, 103, 132–137] With microscopic sample analysis, results are available in minutes to hours after sample collection as compared to days for culture-based assays. Methods for rapid identification and enumeration of microorganisms in water have been adapted to the study of airborne bacteria and fungi collected in impingers[138–140] and to the measurement of enzymes or cell metabolites collected with bubblers or wetted cyclones.[19, 107, 141–143] While rapid identification systems

are available for culturable bacteria and yeasts, these systems are used primarily to identify microorganisms after they have been isolated in culture.[144, 145] Before bacteria or yeasts can be identified, they must be grown in pure culture, typically for at least 18 hrs. At present, even rapid detection methods based on gene amplification as well as biological and chemical assays require several hours of laboratory processing.

Research is underway to develop direct-reading or real-time instruments for rapid detection of airborne biological agents.[142, 146–153] These devices have been based on particle light scattering, electrical mobility, or inertia in an accelerated air flow.[154] An airborne biosensor that can be operated remotely used a cyclone air sampler to collect particles and immunoassay to detect targeted microbial agents.[142] One commercially available, direct-reading device for rapid bioaerosol detection is the ultraviolet, aerodynamic, particle sizer spectrometer (UV-APS, Model 3312 (Instrument 15-7-2); TSI, St. Paul, Minn; *http://www.tsi.com*).[150] This instrument uses particle time of flight to calculate aerodynamic size and to predict when a particle arrives in the UV detection area. The UV laser is triggered to irradiate the particle at an excitation wavelength of 349/355 nm. Resulting fluorescence between 400 and 580 nm is detected with a photomultiplier tube. This fluorescence indicates the presence of "bio-molecular, life-indicating" material, such as riboflavin or NADH (the reduced form of nicotinamide adenine dinucleotide). The intensity of the fluorescence is read by a high-speed analog-to-digital converter. This device is expensive, best suited for detecting particles containing clusters of microorganisms rather than single-cell aerosols, and not yet widely used for investigating exposures to biological agents. The BIRAL Aspect (Bristol Industrial and Research Associates, Ltd, Bristol, England; *http://www.biral.com*) measures the difference in scattered light at three locations; the magnitude of these differences being a measure of particle asymmetry.[151] This device has been used to identify airborne bacteria based on particle morphology (size and shape).

Sampler Performance

Investigators may find it difficult to learn how performance compares among available bioaerosol samplers and to decide which instrument is best for a particular application. In part, this confusion arises from inconsistent evaluation and reporting of bioaerosol sampler performance.[28, 29, 130] To date, researchers have devised evaluation systems independently and with particular applications in mind. The lack of standard test methods has led to a wide array of study designs with arbitrary selection of samplers, reference methods, challenge aerosols, analytical procedures, sample collection times, sample air volumes, and laboratory or field conditions. Although regrettable, this varied approach is inevitable, given the diversity of biological agents, the environments in which sampling is conducted, and the methods of bioaerosol collection and analysis.

To evaluate air sampler performance, instruments may be compared with another sampler (i.e., a reference device such as Instruments 22-1, 22-3a, 22-8, or 22-15) or with a sampling convention. The latter approach is preferred to establish correlations between exposure measurements and health outcomes because aerodynamic diameter determines where a particle deposits in the respiratory tract and many diseases are associated with deposition of particles in particular respiratory regions.[155, 156] Increasingly, investigators are evaluating bioaerosol samplers by comparing sampler collection efficiency to fractional deposition in the respiratory tract as a function of particle size.[94, 100, 131, 156–162] The inhalable convention refers to materials hazardous when deposited anywhere in the respiratory tract, while the thoracic convention refers to the lung airways and gas-exchange region. These conventions may be appropriate to assess the performance of samplers for collecting airborne allergens carried on large particles (e.g., pollen grains, arthropod feces and body parts, and certain fungal spores). Respirable particles are small enough to reach the gas-exchange region of the lungs. Some bio-aerosols are known to be hazardous primarily in the alveolar region of the respiratory system (e.g., certain infectious agents, endotoxin, and antigens involved in hypersensitivity pnenmonitis). Therefore, investigators should use samplers that are efficient for collecting smaller particles to measure exposures to respirable particles.

Minimum criteria for bioaerosol sampler evaluations and the uniform reporting of test results have been proposed.[29] Adoption of common protocols for testing bioaerosol sampler performance along with standardization of procedures for collecting and analyzing bioaerosol samples would help investigators by providing a more consistent basis for comparing findings from different research laboratories and test environments.[28, 29, 113, 145, 163] While striving for the most correct and representative results, investigators should bear in mind that "good" air measurements often may suffice and "excellent" measurements may be beyond their means or needs.[164] Not all bioaerosol sampling requires highly accurate, precise, or efficient sample collection and particle-size discrimination. For example, it may be sufficient to detect or confirm the presence of a specific microorganism, such as an infectious agent, without accurately measuring its air concentration. Likewise, comparisons that an investigator makes with a less-than-ideal sampler may still be valid if the device performs consistently in all test environments and in the hands of all users.[29]

Collection of Bioaerosol Samples

Sampling from Moving and Still Air

Investigators occasionally need to collect samples from moving air streams (e.g., within ventilation ducts or at outdoor air intakes or supply air diffusers). Ambient air velocity is one of several parameters on which particle collection efficiency depends.[27] The efficiency with which particles enter the orifice of a sampler is also a function of aerodynamic particle size, inlet velocity, and the relative directions of ambient and inlet air flows. Tests conducted in laminar flow clean benches and biological safety cabinets are other examples of bioaerosol sampling in moving air streams. Sampling may be performed to measure containment of bioaerosols generated within an enclosure and exclusion of bioaerosols from outside a unit. Not surprisingly, the operation of an air sampler itself can disturb air flow patterns in such systems, which may confound sampling results.[38, 165]

Outdoor air velocities can range from 0 to 15 m/s, with rapid and radical changes of speed and direction. Therefore, measures should be taken to ensure that representative air samples are collected when sampling outdoors (see Chapter 3). Exposed culture plates or glass slides onto which airborne particles can settle by gravity are occasionally used outdoors and indoors to estimate bioaerosol concentrations.[34, 166] Such samples may be informative when measuring indoor or outdoor contamination of surfaces by bioaerosol deposition. However, particle settling is strongly affected by changes in wind direction and air velocity, and gravitational sampling is not suitable for bioaerosol collection outdoors. Even in relatively still indoor air, large particles are over sampled with gravity collection. A glycerol-soaked, cellulose-acetate and nitrate membrane filter has been tested as a "passive" sampler (i.e., particle collection by natural convection or diffusion without use of an air mover such as a vacuum pump).[108] The passive filters demonstrated good correlation with closed-face samplers for total spore counts; however, storage beyond one hour reduced the culturability of collected fungal spores on the glycerol-soaked filters.

Low ambient air speeds can usually be expected indoors, typically below 0.5 m/s. Therefore, a sampler's inlet may be oriented in any convenient position relative to gravity and ambient airflow direction without compromising the representativeness of the particles collected. Some bioaerosol samplers are offered with inlets designed for "remote" sampling, that is, collection of air samples from difficult-to-reach locations through the use of a sampling wand or probe or extension of the inlet via flexible tubing. While convenient, lengthy inlet extensions and sharp bends should be avoided because particle losses can be considerable. Devices that operate at fairly high flow rates but are unobtrusive are recommended for sampling in residences, offices, laboratories, and health care settings where the bioaerosol concentrations may be low and disturbances that would alter normal activity patterns would be undesirable.

Sampler Calibration and Maintenance

Air Flow Calibration

Power supply for vacuum pumps in bioaerosol samplers may be electrical line current or replaceable or rechargeable batteries. Some bioaerosol samplers with constant or non-adjustable flow rates have only an indicator light to show that the sampler is operating. On other instruments, sampling rate may be controlled manually or a critical orifice may be used to maintain constant flow. For example, the capillary nozzle in an AGI may be operated as a critical orifice. However, the diameter of these nozzles as well as the jet-to-impaction surface distance varies from unit to unit and this variability has been found to affect particle collection efficiency.[109] For individual air samplers, the availability of flow control or measurement devices is noted in the "Instrument Description" section.

Air flow should be measured with all sampling elements included; for example, filters, culture plates, slides, or collection fluids. Samplers with high flow rates (e.g., >10 L/min) can be calibrated with wet test or dry gas meters, precision rotameters, rotating vane anemometers, critical flow orifices, and commercial airflow calibrators.[27] However, such devices may create air pressure artifacts in the calibration and sampling train that can affect calibration accuracy. Therefore, line or chamber pressure should be monitored near a sampler's inlet to identify pressure-related problems. If a pressure drop is noted, air can be supplied to the front of a calibration and sampling train to equalize downstream air pressure. Readers should consult Chapter 7 for further directions on air flow calibration. For samplers that cannot be calibrated using methods that add resistance, other approaches must be used (e.g., large bubble tubes, evacuation of measured volumes of air from inflated bags, or average measurements of air flow using hot-wire anemometers).[167, 168] Some samplers can be placed in sealable, airtight chambers for calibration provided the chamber does not create a pressure drop and the inside pressure remains equal to ambient pressure.[27] Users should confirm air flow rate periodically with an independent flow measuring device and verify the accuracy of built-in flow meters and critical orifices.

Avoiding Contamination

Several types of contamination can occur during collection of bioaerosol samples, which may lead to various kinds of errors. Obstruction of sampling inlets may change air flow characteristics, particle collection efficiency, and particle separation by size. In addition, corrections for multiple-hole impactors would be inaccurate if a significant number of the sampling holes were obstructed. Unique to sampling viable organisms is a requirement for aseptic equipment and specimen handling. Microorganisms are present almost everywhere, for example, in air and water, on inanimate surfaces, and on the human body. In addition, microorganisms can begin multiplying shortly after collection if moisture and temperature conditions allow them to do so. Failure on an investigator's part to recognize the ubiquity and nature of microorganisms as living, reproducing entities may lead to unrepresentative findings. It is essential that sample specimens be protected from contamination as well as temperature and humidity extremes during collection and while being transported to and from a laboratory.[169]

The category "Materials" in the "Instrument Description" section identifies if the particle collection portion of a bioaerosol sampler can be autoclaved or sterilized by dry heat or ethylene oxide prior to use. If a sampling device is not cleaned between uses when sampling in high-concentration environments, contaminants may be carried from one sampling session to another (e.g., when sampling outdoors, during remediation processes, or in agricultural environments where hardy bacterial and fungal spores may be abundant). Sanitization procedures commonly used in the field are dipping sampler inlets in or wiping them with 70% ethanol and air drying.[170]

Special considerations for avoiding contamination also apply to collection of biological agents other than culturable microorganisms. For example, endotoxin is ubiquitous and sampling devices and laboratory equipment that will come into contact with samples for this material must be processed to ensure that they are free of endotoxin and other compounds that may interfere with sample assays.[171] Investigators should confer with laboratory personnel on the proper preparation of sampling equipment to ensure that contamination does not occur prior to or during sample collection and to maintain the integrity of bioaerosol samples after they are collected. Inclusion of positive and negative control samples and other quality assurance practices should be routine in the field and in the laboratory.[14, 172]

Detection of Biological Agents

Critical to any bioaerosol evaluation are identification of the biological agents of interest and the assay methods that will be used. Investigators must make these determinations before selecting an air sampler for a study. Saprophytic bacteria and fungi have been and continue to be the focus of much of the attention on bioaerosols. Although encompassing other biological agents, the following discussion reflects this focus. The most commonly used detection and identification methods for bioaerosols have been based on culture for viable microorganisms (i.e., determining colony counts and identifying microorganisms grown in laboratory culture) and microscopic examination of pollen grains, fungal spores, and other distinctive biological particles. Therefore, the largest bodies of data available for com-

paring air sampling results are for culture-based and microscope-based methods. Additional analytical procedures (e.g., immunological, biological, chemical, and molecular assays) are standard for measuring air concentrations of certain biological agents (e.g., allergens, infectious agents, endotoxin, peptidoglycan, mycotoxins, glucans, and ergosterol).[16, 24, 32, 68, 99, 101, 105, 134, 145, 147, 170, 173–186] For the near future, culture and microscopy likely will remain the primary tools for identifying and quantifying the bioaerosols for which these techniques are suitable. Alternative assays may become more widely available for bioaerosol sampling as other methods are adapted for use on environmental air samples and more commercial laboratories become equipped to offer these services.[29] Analytical details are best left to laboratory personnel experienced, as required, in microbiology, molecular biology, immunology, entomology, biochemistry, or chemistry. Even so, persons collecting bioaerosol samples must know enough about assay methods to handle samples appropriately. A few general points follow, but readers should consult other texts for more complete information on the analysis of bioaerosol samples.[14, 24, 68, 114, 187]

Scope of Analytical Methods

Analytical methods (Table 22-2) can generally be categorized as one of three approaches, which have been referred to as broad, indicator, and focused methods.[114] Often, saprophytic fungi or bacteria are the biological agents of primary interest but an investigation has not yet targeted any specific genera or species of microorganisms. Investigators may seek information on the overall kinds and relative concentrations of microorganisms that are present to determine if they appear to be typical or unusual for a particular environment. In these cases, investigators choose collection and analytical methods that provide the broadest possible information. At present, culture-based methods are used most commonly in broad studies because they are widely available, simple, sensitive, and specific.[114, 163, 188] Direct microscopic examination of sample material is also widely applicable for the identification of many fungi and for enumeration of total fungi and bacteria after staining.[26, 62, 67, 75, 93–96, 99, 104, 132–135, 137, 140, 189, 190]

Besides the culture and direct microscopic methods mentioned above, investigators have measured quantitative indicators of the presence of large groups of microorganisms. Examples of this approach are the identification of *Escherichia coli* (an indicator of contamination with raw sewage), the analysis of the glucan or ergosterol content of samples (measures of fungal biomass), the monitoring of protein in air (a marker of indoor biological air contamination), or the detection of guanine (an indicator of dust mite presence). The term "indicator" is used in a similar context when referring to microorganisms or chemical markers whose detection may reflect the simultaneous occurrence or presence of the actual biological agents responsible for adverse health effects (e.g., allergens, toxins, or other cell products). Some fungi have also been suggested as indicators of the presence of excessive moisture, for example, *Aspergillus fumigatus*, species of *Trichoderma, Exophiala, Stachybotrys, Phialophora, Fusarium, Ulocladium,* and yeasts *(Rhodotorula)*.[191] Other fungi have been cited as indicators of health hazards, for example, *A. fumigatus, Aspergillus flavus,* and other species that cause aspergillosis as well as *Fusarium moniliforme, Histoplasma capsulatum,* and *Cryptococcus neoformans*.[163, 192]

Focused analyses are used to document the presence of specific biological agents associated with particular health effects. Methods that focus on single organisms or agents include culture, nucleic acid probes, chemical assays, and immunoassays. The decision to use a focused analysis requires that investigators hypothesize that specific biological agents may be present. Such hypotheses may be based on observed health effects (associated with specific agents) or environmental conditions (that lead investigators to suspect the potential presence of particular agents).

Limitations of Analytical Methods

Analysis and interpretation of air sampling data are critical parts of bioaerosol investigations and covered in other texts.[14, 172, 191–196] To interpret bioaerosol sampling results correctly, investigators must keep in mind the limitations of analytical methods. Broad methods, such as culture and microscopy, may not detect some biological agents and may allow only limited identification of those that are detected. Although many fungi and bacteria will grow on general culture media, many others are not culturable under any conditions or grow poorly in the laboratory. Consequently, these microorganisms are not detectable by analytical methods that rely on cell growth in culture. Given that microorganism viability is not a factor in allergic diseases, inhalation fevers, or organic dust toxic syndrome, investigators should not rely solely on culture-based bioaerosol analysis to measure exposure to organic dusts that may include microorganisms.[98, 101, 183, 187] Similarly, microscopic examination of collected particles is only suitable for relatively large, morphologi-

cally distinctive bioaerosols and those that can be stained or labelled to make them detectable and perhaps identifiable. Often, multiple sampling and analytical approaches are needed to obtain representative information on the range of biological agents that may be present and their relative concentrations.

Uses of Bioaerosol Samplers

People are exposed to bioaerosols inside buildings as well as in the open air. Consequently, bioaerosol sampling is conducted in a variety of indoor and outdoor settings to measure human exposure and study the physical, biological, and chemical behavior of airborne biological agents.[197] Researchers often evaluate bioaerosol samplers in environmental chambers before they test sampler performance in typical use situations. Thus, indoor and outdoor environments and environmental chambers are three general categories in which bioaerosol samplers are used. A review of the bioaerosol samplers others have chosen for various settings can help an investigator identify instruments suitable for a similar application.

The Indoor Environment

In the indoor environment, bioaerosols may originate from human or animal occupants, indoor microbiological growth, or outdoor air that enters via ventilation systems and other means. All humans release bacteria and viruses from the upper and lower respiratory tract and bacteria from the skin and scalp. Animals housed indoors (e.g., birds and mammals) are potential sources of allergens and infectious agents. Many bacteria and fungi as well as arthropods such as cockroaches and dust mites live indoors in water reservoirs, on damp items, and in other suitable habitats. Sources of biological particles are abundant outdoors and, not surprisingly, outdoor air contains many suspended particles of biological origin. Thus, outdoor air used for dilution ventilation and that which infiltrates buildings via incidental openings may introduce a variety of outdoor bioaerosols to the indoor environment.

Bioaerosol samplers have been used to study biological agents in a variety of indoor workplaces (e.g., office buildings, residences, health care centers, microbiology laboratories, biotechnology industries, pharmaceutical preparation and packaging settings, food production and processing sites, animal confinement buildings, agricultural industries, buildings undergoing remediation for fungal contamination, and various manufacturing workplaces).[2, 9, 20, 24, 32, 40, 41, 93, 96, 98, 101, 104, 106, 131, 134, 139, 169, 176, 183, 190, 198–207] Bioaerosol sampling is also conducted indoors to evaluate the performance of air samplers and air cleaners under realistic field conditions.[34, 62, 67, 106, 167, 168, 208–211]

Personal sampling for assessment of bioaerosol exposure has not received a great deal of attention, although the importance of such measurements for understanding exposure–response relationships is clear. To assess exposure to fungal propagules, the participants in a 1994 workshop on the *Health Implications of Fungi in Indoor Environments* recommended that personal air samples be collected in addition to environmental (i.e., area or static) samples.[191] Similar recommendations have been made for monitoring indoor air[64] and exposures in biotechnology industries.[20] Several bioaerosol samplers may be worn conveniently in the breathing zone, for example, cassette filters and impactors (Instruments 22-10 and 22-20).[29–32, 42, 81, 84, 98–100, 103, 105, 108, 176, 178–180, 183, 212, 213] These methods of bioaerosol collection are often paired with assays not based on microorganism culture (e.g., microscopic particle examination, immunoassays, bioassays, and chemical or molecular assays).

The Outdoor Environment

Outdoor air may contain infectious agents and biological allergens as well as toxins from animals, vegetation, soil, and water.[36, 197, 214] When outdoors, people are exposed directly to these airborne biological agents and, when indoors, people may be exposed indirectly if outdoor bioaerosols enter buildings. There are a number of differences between indoor and outdoor environments that determine what bioaerosols may be present. Further, the physical characteristics of the ambient environment may affect efficient particle collection and the ability of biological agents to multiply or retain other properties associated with adverse health effects. For example, as compared to indoor settings, bioaerosols in the outdoor environment: 1) may be more diverse because they are generated from a wider variety of sources; 2) often are suspended in moving rather than virtually still air; 3) must be collected under a range of weather conditions; and 4) may have been subjected to greater stress than usually occurs indoors. Outdoor factors that may inactivate sensitive microorganisms or change the chemical or physical structure of particles and associated allergens or toxins include exposure to ultraviolet radiation, air pollutants, and extreme temperatures or relative humidities.[137, 214–219]

Outdoor air sampling stations are widely used to measure ambient concentrations of pollen grains and fungal spores to aid allergists in managing patients with known sensitivities.[19, 220–222] Plant pathologists employ

bioaerosol sampling to understand the mechanisms by which agents of plant diseases are released and to monitor the migration of bacteria and fungi that can seriously damage crops, timber, and other vegetation.[217, 223] Outdoor sources of biologically derived airborne contaminants related to human activities include emissions from wastewater treatment plants, solid waste handling facilities, construction and demolition projects, and farming operations.[224, 225] Particles of biological origin also play an important role in cloud physics, for example, by accumulating water and acting as ice nuclei.[36, 37] Samplers used outdoors must be robust enough to withstand adverse weather conditions and the inlets must be protected so that rain, snow, frost, direct sunlight, and other factors do not compromise samples.[223, 226]

Experimental Chamber Studies

In addition to investigations of naturally occurring or manmade bioaerosols in indoor and outdoor environments, researchers have learned a great deal from studying test aerosols in chambers of varying size and design. Investigators have used bioaerosol samplers: 1) to assess the survival of airborne viruses and bacteria to better understand the transmission of agents of respiratory diseases;[42, 43, 174, 178, 179, 181, 182, 224, 227–232] 2) to explore the effects of environmental factors and particle size on organism survival in air;[233] and 3) to learn how aerosolization, relative humidity, and temperature affect bioaerosol size and survival.[44, 95, 103, 24, 129, 233–239] Test chambers are also used to evaluate the performance of particulate respirators,[123–125, 127, 128, 238, 240, 241] to test air filters and other air cleaners,[126, 242, 243] and to study release or emission of waterborne particles.[138, 244] Many air sampling instruments and analytical methods are evaluated in chambers[42, 44, 75–77, 82–84, 102, 108, 118, 132, 135, 139, 160, 174, 178, 179, 237, 245–248] and in wind tunnels to determine the effects of wind speed and other parameters on sampling efficiency for various particle sizes.[94, 157, 226, 249, 250]

Summary

Bioaerosol sampling is a multifaceted challenge that requires familiarity with the diverse origin, composition, and aerodynamic behavior of biologically derived airborne contaminants. Of equal importance is knowledge of the prevalence of airborne biological agents in indoor and outdoor environments and the potential health effects associated with inhalation exposure. Many sampling and analytical methods are available to collect and assay air samples for biological agents. These methods vary in their suitability for the study of particular biological agents.

Successful interpretation of sampling results also depends on the appropriateness of the study design that the investigators followed. Decisions on what agents to study as well as where and when to collect samples are as critical as bioaerosol sampler selection.[71]

Acknowledgments

Both authors are members of the ACGIH Bioaerosols Committee and have incorporated in this chapter text from the committee's publication *Bioaerosols: Assessment and Control*.[15] Janet Macher is also a member of the ACGIH Air Sampling Instruments Committee and has incorporated text from the committee's publication *Air Sampling Instrument Selection Guide: Indoor Air Quality*[25] and a column on bioaerosol sampling from a series on the performance of air sampling instruments.[28] The authors thank the sampler manufacturers and suppliers as well as Sergey Grinshpun, Tiina Reponen, and Klaus Willeke of the University of Cincinnati, Cincinnati, Ohio, for their help compiling the information in the "Instrument Description" section of this chapter.

References

1. American Conference of Governmental Industrial Hygienists: Biologically Derived Airborne Contaminants. In: 2000 TLVs® and BEIs®. Threshold Limit Values for Chemical Substances and Physical Agents. Biological Exposure Indices, pp. 11–14, ACGIH, Cincinnati, OH (2000).

2. Batterman, S.A.: Sampling and Analysis of Biologic Volatile Organic Compounds: In: Bioaerosols, pp. 249–268. H.A. Burge, Ed., Lewis Publishers, Boca Raton, FL (1995).

3. Yang, C.S.; Johanning, E.: Airborne Fungi and Mycotoxins. In: Manual of Environmental Microbiology. pp. 651-660. C.J. Hurst, G.R. Knudsen, M.J. McInerney, et al. Eds. American Society for Microbiology, Washington, DC (1997).

4. Ammann, H.M.: Microbial Volatile Organic Compounds. In: Bioaerosols: Assessment and Control. pp. 26-1-26-17. J.M. Macher, H.M. Ammann, H.A. Burge, et al., Eds. American Conference of Governmental Industrial Hygienists, Cincinnati, OH (1999).

5. Institute of Medicine: Indoor Allergens: Assessing and Controlling Adverse Health Effects. A.M. Pope, R. Patterson, and H. Burge, Eds. National Academy Press, Washington, DC (1993).

6. Flannigan, B.; Miller, J.D.: Health Implications of Fungi in Indoor Environments—An Overview. In: Health Implications of Fungi in Indoor Environments, pp. 3–28, R.A. Samson, B. Flannigan, M.E. Flannigan, et al., Eds. Elsevier, New York, NY (1994).

7. Lacey, J.; Dutkiewicz, J.: Bioaerosols and Occupational Lung Disease. J. Aerosol Sci. 25:1371–1404 (1994).

8. Lighthart, B.; Mohr, A.J., Eds.: Atmospheric Microbial Aerosols: Theory and Application. Chapman and Hall, New York, NY (1994).

9. Rylander, R.; Jacobs, R.R.; Eds.: Organic Dusts: Exposure, Effects, and Prevention. Lewis Publishers, Boca Raton, FL (1994).

10. Trudeau, W.L.; Fernández-Caldes, E.: Identifying and Measuring Indoor Biologic Agents. J. Allergy Clin. Immunol. 94:393–400 (1994).

11. Wald, P.H.; Stave, G.M.; Eds.: Physical and Biological Hazards of the Workplace. Van Nostrand Reinhold, New York, NY (1994).

12. Burge, H.A., Ed.: Bioaerosols. Lewis Publishers, Boca Raton, FL (1995).

13. Cox, C.S.; Wathes, C.M.; Eds.: Bioaerosols Handbook. Lewis Publishers, Boca Raton, FL (1995).

14. American Industrial Hygiene Association: Field Guide for the Determination of Biological Contaminants in Environmental Samples. AIHA, Fairfax, VA (1996).

15. American Conference of Governmental Industrial Hygienists: Bioaerosols: Assessment and Control, J.M. Macher, H.M. Ammann, H.A. Burge, et al., Eds. ACGIH, Cincinnati, OH (1999).

16. Griffiths, W.D.; DeCosemo, G.A.L.: The Assessment of Bioaerosols: A Critical Review. J. Aerosol Sci. 25:1425–1458 (1994).

17. Henningson, E.W.; Ahlberg, M.S.: Evaluation of Microbiological Aerosol Samplers: A Review. J. Aerosol Sci. 25:1459–1492 (1994).

18. Jensen, P.A.; Lighthart, B.; Mohr, A.J.; et al.: Instrumentation Used With Microbial Bioaerosols. In: Atmospheric Microbial Aerosols: Theory and Application. pp. 226–284. B. Lighthart and A.J. Mohr, Eds. Chapman and Hall, New York, NY (1994).

19. Cage, B.R.; Schreiber, K.; Barnes, C.; et al.: Evaluation of Four Bioaerosol Samplers in the Outdoor Environment. Ann. Allergy Immunol. 77:401–406 (1996).

20. Crook, B.: Review: Methods of Monitoring for Process Microorganisms in Biotechnology. Ann. Occup. Hyg. 40:245–260 (1996).

21. Eduard, W.: Measurement Methods and Strategies for Non-infectious Microbial Components in Bioaerosols at the Workplace. Analyst. 121:1197–1201 (1996).

22. Buttner, M.P.; Willeke, K.; Grinshpun, S.A.: Sampling and Analysis of Airborne Microorganisms. In: Manual of Environmental Microbiology. pp. 629–640. C.J. Hurst, G.R. Knudsen, M.J. McInerney, et al., Eds. American Society for Microbiology, Washington, DC (1997).

23. Crook, B.; Sherwood-Higham, J.L.: Sampling and Assay of Bioaerosols in the Work Environment. J. Aerosol Sci. 28:417–426 (1997).

24. Flannigan, B.: Air Sampling for Fungi in Indoor Environments. J. Aerosol Sci. 28:381–392 (1997).

25. American Conference of Governmental Industrial Hygienists: Air Sampling Instrument Selection Guide: Indoor Air Quality. ACGIH, Cincinnati, OH (1998).

26. Horner, W.E.; Lehrer, S.B.; Salvaggio, J.E.: Aerobiology. In: Allergens and Allergen Immunotherapy. 2nd Ed., pp. 53–72, R.F. Lockey and S.C. Bukantz, Eds. Marcel Dekker, New York, NY (1999).

27. Willeke, K.; Macher, J.M.: Air Sampling. In: Bioaerosols: Assessment and Control. pp. 11-1-11-25. J.M. Macher, H.M. Ammann, H.A. Burge, et al., Eds. American Conference of Governmental Industrial Hygienists, Cincinnati, OH (1999).

28. Macher, J.M.: Instrument Performance Criteria: Bioaerosol Samplers. Appl. Occup. Environ. Hyg. 12:723–729 (1997).

29. Macher, J.M.: Evaluation of Bioaerosol Sampler Performance. Appl. Occup. Environ. Hyg. 12:730–736 (1997).

30. Fiorina, A.: A Personal Sampler to Monitor Airborne Particles of Biological Origin. Aerobiologia. 14:299–301 (1998).

31. Fiorina, A.; Scordamaglia, A.; Mincarini, M.; et al.: Aerobiologic Particle Sampling by a New Personal Collector (Partrap FA51) in Comparison to the Hirst (Burkard) Sampler. Allergy 52:1026–1030 (1997).

32. Sawyer, M.H.; Chamberlin, C.J.; Wu, Y.N.; et al.: Detection of Varicella-Zoster Virus DNA in Air Samples from Hospital Rooms. J. Infect. Dis. 169:91–94 (1994).

33. Rath, P.M.; Ansorg, R.: Value of Environmental Sampling and Molecular Typing of Aspergilli to Assess Nosocomial Sources of Aspergillosis. J. Hosp. Infect. 37:47–53 (1997).

34. Dykstra, M.J.; Loomis, M.; Reininger, K.; et al.: A Comparison of Sampling Methods for Airborne Fungal Spores During an Outbreak of Aspergillosis in the Forest Aviary of the North Carolina Zoological Park. J. Zoo Wildlife Med. 28:454–463 (1997).

35. Leveti, E.: Aerobiology of Agricultural Pathogens. In: Manual of Environmental Microbiology. pp. 693–702. C.J. Hurst, G.R. Knudsen, M.J. McInerney, et al., Eds. American Society for Microbiology, Washington, DC (1997).

36. Matthias-Maser, S.; Jaenicke, R.: Examination of Atmospheric Bioaerosol Particles with Radii > 0.2μm. J. Aerosol Sci. 25:1605–1613 (1994).

37. Matthias-Maser, S.; Jaenicke, R.: The Size Distribution of Primary Biological Aerosol Particles with Radii > 0.2μm in an Urban/rural Influenced Region. Atmos. Res. 39:279–286 (1995).

38. Ljungqvist, B.; Reinmüller, B.: The Biotest RCS Air Samplers in Unidirectional Flow. J. Pharm. Sci. Technol. 48:41–44 (1994).

39. Sayre, P.; Burckle, J.; Macek, G.; et al.: Regulatory Issues for Bioaerosols. In: Atmospheric Microbial Aerosols: Theory and Application, pp. 331–364, B. Lighthart and A.J. Mohr, Eds. Chapman and Hall, New York, NY (1994).

40. Harding, L.; Fleming, D.O.; Macher, J.M.: Biological Hazards. Chapter 14. In: Fundamentals of Industrial Hygiene. 4h ed. pp. 403–449, B.A. Plog, J. Niland, P.J. Quinlan, Eds. National Safety Council, Itasca, IL (1996).

41. Speight, S.E.; Hallis, B.A.; Bennett, A.M.; et al.: Enzyme-linked Immunosorbent Assay for the Detection of Airborne Microorganisms Used in Biotechnology. J: Aerosol Sci. 28:483–492 (1997).

42. Nugent, P.G.; Cornett, I.; Stewart, I.W.; et al.: Personal Monitoring of Exposure to Genetically Modified Microorganisms in Bioaerosols: Rapid and Sensitive Detection Using PCR. J. Aerosol Sci. 28:525-538 (1997).

43. Stärk. K.D.C.; Nicolet, J.; Frey, J.: Detection of Mycoplasma hyopneumoniae by Air Sampling with a Nested PCR Assay. Appl. Environ. Microbiol. 64:543–548 (1998).

44. Griffiths, W.D.; Stewart, I.W.: Performance of Bioaerosol Samplers Used by the UK Biotechnology Industry. J. Aerosol Sci. 30:1029-1040 (1999).

45. Gerberding, I.L.; Holmes, K.K.: Microbial Agents and Infectious Diseases. In: Textbook of Clinical Occupational and Environmental Medicine, pp. 699–716, L. Rosenstock and M.R. Cullen, Eds. W.B. Saunders Co., Philadelphia, PA (1994).

46. Hodgson, M.I.: Exposures in Indoor Air. In: Textbook of Clinical Occupational and Environmental Medicine pp. 866–875, L. Rosenstock and M.R. Cullen, Eds. W.B. Saunders Co., Philadelphia, PA (1994).

47. Merchant, I.A.: Plant and Vegetable Exposures. In: Textbook of Clinical Occupational and Environmental Medicine pp. 693–699, L. Rosenstock and M.R. Cullen, Eds. W.B. Saunders Co., Philadelphia, PA (1994).

48. Merchant, I.A.; Thorne, P.S.; Reynolds, S.I.: Animal Exposures. In: Textbook of Clinical Occupational and Environmental Medicine, pp. 688–699, L. Rosenstock and M.R. Cullen, Eds. W.B. Saunders Co., Philadelphia, PA (1994).

49. Rose, C.S.: Fungi. In: Physical and Biological Hazards of the Workplace. pp. 392–416. P.H. Wald and G.M. Stave, Eds. Van Nostrand Reinhold, New York, NY (1994).

50. Rose, C.S.: Hypersensitivity Pneumonitis. In: Textbook of Clinical Occupational and Environmental Medicine, pp. 242–248 L. Rosenstock and M.R. Cullen, Eds. W.B. Saunders Co., Philadelphia, PA (1994)

51. Cookingham, C.E.; Solomon, W.R.: Bioaerosol-induced Hypersensitivity Diseases. In: Bioaerosols. pp. 205–233. H.A. Burge, Ed., Lewis Publishers, Boca Raton, FL (1995).

52. Burge, H.A.: Airborne Contagious Disease. In: Bioaerosols, pp. 25–47m H.A. Burge, Ed., Lewis Publishers, Boca Raton, FL (1995).

53. Collins, C.H.; Aw, T.C.; Grange, I.M.: Microbial Diseases of Occupations, Sports and Recreations. Butterworth-Heinemann, Boston, MA (1997).

54. Fields, B.S.: Legionellae and Legionnaires' Disease. In: Manual of Environmental Microbiology, pp. 666–675, C.J. Hurst, G.R. Knudsen, M.I. McInerney, et al., Eds. American Society for Microbiology, Washington, DC (1997).

55. Olenchock, S.A.: Airborne Endotoxin. In: Manual of Environmental Microbiology, pp. 661–665, C.J. Hurst, G.R. Knudsen, M.I. McInerney, et al., Eds. American Society for Microbiology, Washington, DC (1997).

56. Sattar, S.A.; Ijaz, M.K.: Airborne Viruses. In: Manual of Environmental Microbiology, pp. 682–692, C.J. Hurst, G.R. Knudsen, M.I. McInerney, et al., Eds. American Society for Microbiology, Washington, DC (1997).

57. Macher, J.M.; Rosenberg, I.: Evaluation and Management of Exposure to Infectious Agents. In: Handbook of Occupational Safety and Health, pp. 287–371, L. DiBernardinis, Ed. Wiley and Sons, New York, NY (1999).

58. American Conference of Governmental Industrial Hygienists: Health Effects of Bioaerosols. In: Bioaerosols: Assessment and Control, pp. 3-1-3-12. J.M. Macher, H.M. Ammann, H.A. Burge, et al., Eds. American Conference of Governmental Industrial Hygienists, Cincinnati, OH (1999).

59. Krake, A.M.; Worthington, K.A.; Wallingford, K.M.; et al.: Evaluation of Microbiological Contamination in a Museum. Appl. Occup. Environ. Hyg. 14:499-509 (1999).

60. Nardell, E.A.; Macher, J.M.: Respiratory Infections—Transmission and Environmental Control. In: Bioaerosols: Assessment and Control, pp. 9-1-9-13, I.M. Macher, H.M. Ammann, H.A. Burge, et al., Eds. American Conference of Government Industrial Hygienists, Cincinnati, OH (1999).

61. Rose, C.S.; Kreiss, K.; Milton, D.K.; et al.: Medical Roles and Recommendations. In: Bioaerosols: Assessment and Control; pp. 8-1-8-9. J.M. Macher, H.M. Ammann, H.A. Burge, D.K. Milton, and P.R. Morey. Eds. American Conference of Governmental Industrial Hygienists, Cincinnati, OH (1999).

62. Stern, M.A.; Allitt, U.; Corden, J.; et al.: The Investigation of Fungal Spores in Intramural Air Using a Burkard Continuous Recording Air Sampler. Indoor Built Environ. 8:40–8 (1999).

63. Maroni, M.; Axelrad, R.; Bacaloni, A.: NATO's Efforts to Set Indoor Air Quality Guidelines and Standards. Am. Ind. Hyg. Assoc. J. 56:499–508 (1995).

64. Rao, C.Y.; Burge, H.A.; Chang, I.C.S.: Review of Quantitative Standards and Guidelines for Fungi in Indoor Air. J. Air Waste Manage. Assoc. 46:899–908 (1996).

65. ACGIH: Introduction. In: Bioaerosols: Assessment and Control, pp. 1-1-1-5. J.M. Macher, H.M. Ammann, H.A. Burge, et al., Eds. American Conference of Governmental Industrial Hygienists, Cincinnati, OH (1999).

66. Crook, B.: Non-inertial Samplers: Biological Perspectives. In: Bioaerosols Handbook, pp. 269–283 C.S. Cox and C.M. Wathes, Eds., Lewis Publishers, Boca Raton, FL (1995).

67. Moschandreas, D.J.; Cha, D.K.; Qian, J.: Measurement of Indoor Bioaerosol Levels by a Direct Count Method. J. Environ. Engin.122:374-378 (1996).

68. American Society for Microbiology: Manual of Environmental Microbiology. C.J. Hurst, G.R. Knudsen, M.J. McInerney, et al. Eds. ASM, Washington, DC (1997).

69. Crook, B. : Non-inertial Samplers: Biological Perspectives. In: Bioaerosols Handbook; pp. 269–283, C.S. Cox and C.M. Wathes, Eds. Lewis Publishers, Boca Raton, FL (1995).

70. Burge, H.A.: Bioaerosol Investigations. In: Bioaerosols, pp. 1–23, H.A. Burge, Ed. Lewis Publishers, Boca Raton, FL (1995).

71. ACGIH: Developing an Investigation Strategy. In: Bioaerosols: Assessment and Control. pp. 11-1-11.25. J.M. Macher, H.M. Ammann, H.A. Burge, et al., Eds. American Conference of

Governmental Industrial Hygienists, Cincinnati, OH pp. 2-1–2-10 (1999).

72. ACGIH: Developing a Sampling Plan. In: Bioaerosols: Assessment and Control, pp. 5-1–5-13. J.M. Macher, H.M. Ammann, H.A. Burge, *et al.*, Eds. American Conference of Governmental Industrial Hygienists, Cincinnati, OH (1999).

73. ACGIH: The Building Walkthrough. In: Bioaerosols: Assessment and Control, pp. 4-1–4-11. J.M. Macher, H.M. Ammann, H.A. Burge, *et al.*, Eds. American Conference of Governmental Industrial Hygienists, Cincinnati, OH (1999).

74. Martyny, J.W.; Martinez, K.F.; Morey, P.R.: Source Sampling. In: Bioaerosols: Assessment and Control. pp. 12-1–12-8. J.M. Macher, H.M. Ammann, H.A. Burge, *et al.*, Eds. American Conference of Governmental Industrial Hygienists, Cincinnati, OH (1999).

75. Terzieva, S.; Donnelly, J.; Ulevicius, V.; *et al.*: Comparison of Methods for Detection and Enumeration of Airborne Microorganisms Collected by Liquid Impingement. Appl. Environ. Microbiol. 62:2264–2272 (1996).

76. Juozaitis, A.; Willeke, K.; Grinshpun, S.; *et al.*: Impaction onto a Glass Slide or Agar versus Impingement into a Liquid for the Collection and Recovery of Airborne Microorganisms. Appl. Environ. Microbiol. 60:861–870 (1994).

77. Grinshpun, S.A.; Willeke, K.; Ulevicius, V.; *et al.*: Effect of Impaction, Bounce and Reaerosolization on the Collection Efficiency of Impingers. Aerosol Sci. Technol. 26:326-342 (1997).

78. Willeke, K.; Lin, X.; Grinshpun, S.A.: Improved Aerosol Collection by Combined Impaction and Centrifugal Motion. Aerosol Sci. Technol. 28:439-456 (1998).

79. Nevalainen, A.; Pastuszka J.; Liebhaber, F.; *et al.*: Performance of Bioaerosol Samplers: Collection Characteristics and Sampler Design Considerations. Atmos. Environ. 26A:531–540 (1992).

80. Hinds, W.C.: Aerosol Technologsy. pp. 394–401. John Wiley & Sons, New York, NY (1999).

81. Li, C.-S.; Hao, M.-L.; Lin, W.-H.; *et al.*: Evaluation of Microbial Samplers for Bacterial Microorganisms. Aerosol Sci. Technol. 30:100–108 (1999).

82. Li, C.-S.; Lin, W.-H.: Sampling Performance of Impactors for Bacterial Bioaerosols. Aerosol Sci. Technol. 30:280–287 (1999).

83. Li, C.-S.; Lin. Y.-C.: Sampling Performance of Impactors for Fungal Spores and Yeast Cells. Aerosol Sci. Technol. 31:226–230 (1999).

84. Lin, W.-H.; Li, C.-S.: Evaluation of Impingement and Filtration Methods for Yeast Bioaerosol Sampling. Aerosol Sci. Technol. 30:119–126 (1999)

85. Andersen, A.A.: New Sampler for the Collection, Sizing and Enumeration of Viable Airborne Particles. J. Bacteriol. 76:471–484 (1958).

86. Peto, S.; Powell, E.O.: The Assessment of Aerosol Concentration by Means of the Andersen Sampler. J. Appl. Bacteriol. 33:582–598 (1970).

87. Macher, J.M.: Positive-Hole Correction of Multiple-Jet Impactors for Collecting Viable Microorganisms. Am. Ind. Hyg. Assoc. 1: 50:561–568 (1989).

88. Leopold, S.S.: "Positive Hole" Statistical Adjustment for a Two-Stage, 200-Hole-per-Stage Andersen Air Sampler. Am. Ind. Hyg. Assoc.1: 49: A88–A90 (1988).

89. Somerville, M.C.; Rivers, J.C.: An Alternative Approach for the Correction of Bioaerosol Data Collected with Multiple Jet Impactors. Am. Ind. Hyg. Assoc.1: 55:127–131 (1994).

90. Chang, C.-W.; Hwang, Yo-H.; Grinshpun, S.A.; *et al.*: Evaluation of Counting Error Due to Colony Masking in Bioaerosol Sampling. Appl. Environ. Microbiol. 60:3732–3738 (1994).

91. Chang, C.-W.; Grinshpun, S.A.; Willeke, K.; *et al.*: Factors Affecting Microbiological Colony Count Accuracy for Bioaerosol Sampling and Analysis. Am. Ind. Hyg. Assoc. J. 56:979–986 (1995).

92. Chen, C.C.; Yu, T.S.; Chang, J.Y.; *et al.*: A Computer Simulation Study on Bioaerosol Colony Counting Error Due to Masking Effect. Ann. Occup. Hyg. 42:501–510 (1998).

93. Jones, W.G.; Dennis, J. W.; May, J.J.; *et al.*: Dust Control During Bedding Chopping. Appl. Occup. Environ. Hyg. 10:467–475 (1995).

94. Gao, P.; Dillon, H.K.; Farthing, W.E.: Development and Evaluation of an Inhalable Bioaerosol Manifold Sampler. Am. Ind. Hyg. Assoc. J. 58: 196–206 (1997).

95. Heidelberg, J.F.; Shahamat, M.; Levin, M.; *et al.*: Effect of Aerosolization on Culturability and Viability of Gram-negative Bacteria. Appl. Environ. Microbiol. 63:3585–3588 (1997).

96. McCammon. C.S.; Martinez, K.F.; Bullock, D.K.; *et al.*: An Evaluation of Biological and Other Exposures During an Indoor Stock Show. Appl. Occup. Environ. Hyg. 12:315–322 (1997).

97. Gazenko, S. V.; Reponen, T.A.; Grinshpun, S.A.; *et al.*: Analysis of Airborne Actinomycete Spores with Fluorogenic Substrates. Appl. Environ. Microbiol. 64:4410–4415 (1998).

98. Kullman, G.J.; Thorne, P.S.; Waldron, P.F.; *et al.*: Organic Dust Exposures from Work in Dairy Barns. Am. Ind. Hyg. Assoc. J. 59:403–413 (1998).

99. Mahar, S.; Reynolds, S.J.; Thorne, P.S.: Worker Exposure to Particulates, Endotoxins, and Bioaerosols in Two Refuse-Derived Fuel Plants. Am. Ind. Hyg. Assoc. J. 60:679–683 (1999).

100. Kenny, L.C.; Stancliffe, J.D., Crook, B.; *et al.*: The Adaptation of Existing Personal Inhalable Aerosol Samplers for Bioaerosol Sampling. Am. Ind. Hyg. Assoc. J. 59:831–841 (1998).

101. Krahmer, M.; Fox, K.; Fox, A.; *et al.*: Total and Viable Airborne Bacterial Load in Two Different Agricultural Environments Using Gas Chromatography-Tandem Mass Spectrometry and Culture: A Prototype Study. Am. Ind. Hyg. Assoc. J. 59:524–531 (1998).

102. Lin, W.-H.; Li, C.-S.: The Effect of Sampling Time and Flow Rate on the Bioefficiency of Three Fungal Spore Sampling Methods. Aerosol Sci. Technol. 28:511–522 (1998).

103. Lin, W.-H.; Li, C.-S.: Collection Efficiency and Culturability of Impingement into a Liquid for Bioaerosols of Fungal Spores and Yeast Cells. Aerosol Sci. Technol. J. 30:109–118 (1999).

104. Rautiala, S.; Reponen, T.; Nevalainen, A.; *et al.*: Control of Exposure to Airborne Microorganisms During Remediation of Moldy Buildings; Report of Three Case Studies. Am. Ind. Hyg. Assoc. J. 59:455–460 (1998).

105. Alwis, K.U.; Mandryk, J.; Hocking, A.D.: Exposure to Biohazards in Wood Dust: Bacteria, Fungi, Endotoxins, and (1→3)-â-D-Glucans. Appl. Occup. Environ. Hyg.14:598–608 (1999).

106. Parks, S.R; Bennett, A.M.; Speight, S.E.; *et al.*: An Assessment of the Sartorius MD8 Microbiological Air Sampler. J. Appl. Bacteriol. 80:529–534 (1996).

107. Stewart, I.W.; Leaver, G.; Futter, S.J.: The Enumeration of Aerosolized *Saccharomyces cerevisiae* Using Bioluminescent Assay of Total Adenylates. J. Aerosol Sci. 28:511–523 (1997).

108. Nasman, A.; Blomquist, G.; Levin, J.-O.: Air Sampling of Fungal Spores on Filters: An Investigation on Passive Sampling and Viability. J. Environ. Monit. 1:361–365 (1999).

109. Lin, X.; Willeke, K.; Ulevicius, V.; *et al.*: Effect of Sampling Time on the Collection Efficiency of All-glass Impingers. Am. Ind. Hyg. Assoc. J. 58:480–488 (1997).

110. Muilenberg, M.L.: Aeroallergens assessment by microscope and culture. Immunol. Allergy Clinics North Amer. 9:245–268 (1989).

111. Grinshpun, S.A.; Chang, C.W.; Nevalainen, A.; *et al.*: Inlet Characteristics of Bioaerosol Samplers. J. Aerosol Sci. 25:1503–1522 (1994).

112. Frenz, D.A.; Scamehom, R.T.; Hokanson, J.M.; *et al.*: A Brief Method for Analyzing Rotorod® Samples for Pollen Content. Aerobiologia. 12:51–54 (1996).

113. Di-Giovanni, R: A Review of the Sampling Efficiency of Rotating-arm Impactors Used in Aerobiological Studies. Grana. 37:164–171 (1998).

114. ACGIH: Sample Analysis. In: Bioaerosols: Assessment and Control, pp. 6-1–6-13. J.M. Macher, H.M. Ammann, H.A. Burge, *et*

al., Eds. American Conference of Governmental Industrial Hygienists, Cincinnati, OH (1999).

115. The British Aerobiology Federation. Airborne Pollens and Spores: A Guide to Trapping and Counting. European Aeroallegen Network (UK). University of Worcester, United Kingdom (1995).

116. Madelin, T.M.; Madelin, M.F.: Biological Analysis of Fungi and Associated Molds. In: Bioaerosols Handbook. pp. 361–386, C.S. Cox and C.M. Wathes, Eds., Lewis Publishers, Boca Raton, FL (1995).

117. Alcazar, P.; Comtois, P.: A New Adhesive for Airborne Pollen Sampling. Aerobiologia. 15:105–108 (1999).

118. Stewart, S.L.; Grinshpun, S.A.; Willeke, K.; et al.: Effect of Impact Stress on Microbial Recovery on an Agar Surface. App. Env. Microbiol. 61:1232–1239 (1995).

119. Luczynska, C.M.: Sampling and Assay of Indoor Allergens. J. Aerosol Sci. 28:393–399 (1997).

120. Montoya, L.D.; Hildemann, L.M.: Size-resolved Quantification of Nonviable Indoor Aeroallergens: A Review. J. Environ. Eng. 123:965–973 (1997).)

121. Platts-Mills, T.A.; Carter, M.C.: Asthma and Indoor Exposure to Allergens. New Engl. J. Med. 336:1382–1384 (1997).

122. Rose, C.S.: Antigens. In: Bioaerosols: Assessment and Control. pp. 25-1–25-11. J.M. Macher, H.M. Ammann, H.A. Burge, et al., Eds. American Conference of Government Industrial Hygienists, Cincinnati, OH (1999).

123. Brosseau, L.M.; Chen, S.-K.; Vesley, D.; et al.: System Design and Test Method for Measuring Respirator Filter Efficiency Using Mycobacterium Aerosols. J. Aerosol Sci. 25:1567–1577 (1994).

124. Brosseau, L.M.; McCullough, N.V.; Vesley, D.: Mycobacterial Aerosol Collection Efficiency of Respirator and Surgical Mask Filters Under Varying Conditions of Flow and Humidity. Appl. Occup. Environ. Hyg. 12:435–445 (1997).

125. Willeke, K.; Qian, Y.; Donnelly, J.; et al.: Penetration of Airborne Microorganisms Through a Surgical Mask and a Dust/mist Respirator. Am. Ind. Hyg. Assoc. J. 57:348–355 (1996).

126. Maus, R.; Urnhauer, H.: Collection Efficiencies of Coarse and Fine Dust Filter Media for Airborne Biological Particles. J. Aerosol Sci. 28:401–415 (1997).

127. Wake, D.; Bowry, A.C.; Crook, B.; et al.: Performance of Respirator Filters and Surgical Masks Against Bacterial Aerosols. J. Aerosol Sci. 28: 1311–1329 (1997).

128. Qian. Y.; Willeke, K.; Grinshpun, S.A.; et al.: Performance of N95 Respirators: Filtration Efficiency for Airborne Microbial and Inert Particles. Am. Ind. Hyg. Assoc. J. 59: 128–132 (1998).

129. Griffiths, W.D.; Stewart, I. W.; Reading, A.R.; et al.: Effect of Aerosolization, Growth Phase and Residence Time in Spray and Collection Fluids on the Culturability of Cells and Spores. J. Aerosol Sci. 27:803–820 (1996).

130. Miller, S.L.; Cheng, Y.S.; Macher, J.M.: Guest Editorial. Aerosol Sci Technol. 30:93–99 (1999).

131. Asking, L. ; Olsson, B.: Calibration at Different Flow Rates of a Multistage Liquid Impinger. Aerosol Sci. Technol. 27:39–49 (1997).

132. Muilenberg, M.L.; Burge, H.A.: Filter Cassette Sampling for Bacterial and Fungal Aerosols. In: Health Implications of Fungi in Indoor Environment, pp. 75–89, R.A. Samson, B. Flannigan, M.E. Flannigan, et al., Eds. Elsevier, New York, NY (1994).

133. Morris, K.J.: Modern Microscopic Methods of Bioaerosol Analysis. In: Bioaerosols Handbook. pp. 285–316. Lewis Publishers, Boca Raton, FL (1995).

134. Johanning, E.; Biagini, R.; Hull, D.; et al.: Health and Immunology Study Following Exposure to Toxigenic Fungi (Stachybotrys chartarum) in a Water-damaged Office Environment. Int. Arch. Occup. Environ. Health 68:207–218 (1996).

135. Kildesø, J.; Nielsen, B.H.: Exposure Assessment of Airborne Microorganisms by Fluorescence Microscopy and Image Processing. Ann. Occup. Hyg. 41:201–216 (1997).

136. Schafer, A.; Harms, H.; Zehnder, A.J.B.: Bacterial Accumulation at the Air-water Interface. Environ. Int. 32:3704–3712 (1998).

137. Tong, Y.; Lighthart, B.: Diurnal Distribution of Total and Culturable Atmospheric Bacteria at a Rural Site. Aerosol Sci. Technol. 30:246–54 (1999).

138. Henningson, E.W.; Lundquist, M.; Larsson, E.; et al.: A Comparative Study of Different Methods to Determine the Total Number and the Survival Ratio of Bacteria in Aerobiological Samples. J. Aerosol Sci. 28:459–469 (1997).

139. Lange, J.L.; Thorne, P.S.; Lynch, N.: Application of Flow Cytometry and Fluorescent in situ Hybridization for Assessment of Exposures to Airborne Bacteria. Appl. Environ. Microbiol. 63: 1557–1563 (1997).

140. Hernandez, M.; Miller, S.L.; Landfear, D. W., et al.: A Combined Fluorochrome Method for Quantitation of Metabolically Active and Inactive Airborne Bacteria. Aerosol Sci. Technol. 30: 145–160 (:1999).

141. Nitescu, I.; Cumming, R.H.; Rowell, F.J.; et al.: Evaluation of a Model for the Detection of Aerosol-containing Protease in Real-time Using Chromogenic Substrates in the Sampling Fluid of a Cyclone and a Bubbler. J. Aerosol Sci. 28:501–510 (1997).

142. Ligler, F.S.; Anderson, G.P.; Davidson, P.T.; et al.: Remote Sensing Using an Airborne Biosensor. Environ. Sci. Technol. 32:2461–2466 (1998).

143. Pendergrass, S.M.; Jensen, P.A.: Application of the Gas Chromatography-fatty Acid Methyl Ester System for the Identification of Environmental and Clinical Isolates of the Family Micrococcaceae. Appl. Occup. Environ. Hyg. 12:543–546 (1997).

144. NIOSH: Sampling and Characterization of Bioaerosols. In: NIOSH Manual of Analytical Methods. pp. 82–112. P.A. Jensen and M.P. Schafer; Bioaerosol Sampling (Indoor Air). NIOSH Analytical Method 0800. M.K. Lonon; Aerobic Bacteria by GC-FAME. NIOSH Analytical Method 0801. S.M. Pendergrass; Mycobacterium tuberculosis, Airborne. NIOSH Analytical Method 0900. Millie P. Schafer (1998).

145. National Institute for Occupational Safety and Health: Sampling and Characterization of Bioaerosols. In: NIOSH Manual of Analytical Methods. pp. 82–112. P.A. Jensen and M.P. Schafer; Bioaerosol Sampling (Indoor Air). NIOSH Analytical Method 0800. M.K. Lonon; Aerobic Bacteria by GC-F AME. NIOSH Analytical Method 0801. S.M. Pendergrass; Mycobacterium tuberculosis, Airborne. NIOSH Analytical Method 0900. Millie P. Schafer (1998).

146. Evans, B.T.N.; Yee, E.; Roy, G.; et al.: Remote Detection and Mapping of Bioaerosols. J. Aerosol Sci. 25:1549–1566 (1994).

147. Pinnick, R.G.; Hill, S.C.; Nachman, P.; et al.: Fluorescence Particle Counter for Detecting Airborne Bacteria and Other Biological Particles. Aerosol Sci. Technol. 23:653–664 (1995).

148. Pinnick, R.G.; Hill, S.C.; Nachman, P.; et al.: Aerosol Fluorescence Spectrum Analyzer for Rapid Measurement of Single Micrometer-sized Airborne Biological Particles. Aerosol Sci. Technol. 28:95–104 (1998).

149. Gieray, R.A.; Reilly, P.T.A.; Yang, M.; et al.: Real-time Detection of Individual Airborne Bacteria. J. Microbiol. Methods. 29: 191–199 (1997).

150. Hairston, P.P.; Ho, J.; Quant, F.R.: Design of an Instrument for Real-time Detection of Bioaerosols Using Simultaneous Measurement of Particle Aerodynamic Size and Intrinsic Fluorescence. J. Aerosol Sci. 28:471–482 (1997).

151. Inman, C.M.; Lewis, D.J.: Characterising Aerosol Particles Using Spatial Light Scattering. J. Aerosol Sci. 29:S407–S408 (1998).

152. Cheng, Y.S.; Barr, E.B.; Fan, B.J.; et al.: Detection of Bioaerosols Using Multiwavelength UV Fluorescence Spectroscopy. Aerosol Sci. Technol. 30:186–201 (1999).

153. Seaver, M.; Eversole, J.D.; Hardgrove, J.J.; et al.: Size and Fluorescence Measurements for Field Detection of Biological Aerosols. Aerosol Sci. Technol. 30:174–185 (1999).

154. Lacey, J.: Guest Editorial. Proceedings: Sampling and Rapid

Assay of Bioaerosols, June 14, 1995, IARC-Rothamsted, England. J. Aerosol Sci. 28:345–538 (1997).

155. Liden, G.: Performance Parameters for Assessing the Acceptability of Aerosol Sampling Equipment. Analyst. 119:27 –33 (1994).

156. American Conference of Governmental Industrial Hygienists: Appendix D: Particle Size-Selective Sampling Criteria for Airborne Particulate Matter. In: 2000 TLVs® and BEIs®. Threshold Limit Values for Chemical Substances and Physical Agents. Biological Exposure Indices, pp. 83–86. ACGIH, Cincinnati, OH (2000).

157. Upton. S.L.; Mark, D.; Douglass, E.J.; et al.: A Wind Tunnel Evaluation of the Physical Sampling Efficiencies of Three Bioaerosol Samplers. J. Aerosol Sci. 25: 1493–1501 (1994).

158. Comité Européen de Normalisation: Draft Eurpoean Standard. Workplace Atmospheres: Assessment of Performance of Instruments for Measurement of airborne Particle Concentrations. CEN/TC137/WG3/N192 (1997).

159. American Society for Testing and Materials: Standard Practice for Evaluating the Performance of Respirable Aerosol Samplers. ASTM D 6061–6096. ASTM, West Conshohocken, PA (1996).

160. American Society for Testing and Materials: Standard Guide for Personal Samplers of Health-Related Aerosol Fractions [Metric]. ASTM D 6062M–96. ASTM, West Conshohocken, PA (1996).

161. Griffiths, W.D.; Boysan, F.: Computational Fluid Dynamics (CFD) and Empirical Modelling of the Performance of a Number of Cyclone Samplers. J. Aerosol Sci. 27:281–304 (1996).

162. Gao, P.; Dillon, H.K.; Baker, J.; et al.: Numerical Prediction of the Performance of a Manifold Sampler with a Circular Slit Inlet in Turbulent Flow. J. Aerosol Sci. 30:299–312 (1999).

163. American Industrial Hygiene Association: Viable Fungi and Bacteria in Air Bulk, and Surface Samples. In: Field Guide for the Determination of Biological Contaminants in Environmental Samples, pp. 37–74. AIHA, Fairfax, VA (1996).

164. Cohen, B.S.: Preface—Proceedings of the International Symposium on Air Sampling Instrument Performance. Appl. Occup. Environ. Hyg. 8:225-226 (1993).

165. Ljungqvist, B.; Reinmüller, B.: Interaction Between Air Movements and the Dispersion of Contaminants: Clean Zones with Unidirectional Air Flow. J. Parenter. Sci. Technol. 47:60–69 (1993).

166. Whyte, W.: In Support of Settle Plates. J. Pharm. Sci. Technol. 50:201-2-4 (1996).

167. Jensen, P.A.: Evaluation of Standard and Modified Sampling Heads for the International PBI Surface Air System Bioaerosol Sampler. Am. Ind. Hyg. Assoc. J. 56:272-279 (1995).

168. Mehta, S.K.; Mishra, S.K.; Pierson, D.L.: Evaluation of Three Portable Samplers for Monitoring Airborne Fungi. Appl. Environ. Microbiol. 62:1835–1838 (1996).

169. Thorne, P.S.; Lane, J.L.; Bloebaum, P.; et al.: Bioaerosol Sampling in Field Studies: Can Samples be Express Mailed? Am. Ind. Hyg. Assoc. J. 55:1072–1079 (1994).

170. American Industrial Hygiene Association: HA: Total Fungi and Substance Derived from Fungi. Field Guide for the Determination of Biological Contaminants in Environmental Samples. pp. 119–130. (1996c).

171. Milton, D.K.: Endotoxin and Other Bacterial Cell-Wall Components. In: Bioaerosols: Assessment and Control, pp. 23-1–23-14, J.M. Macher, H.M. Ammann, H.A. Burge, et al.; Eds. American Conference of Governmental Industrial Hygienists, Cincinnati, OH (1999).

172. Macher, J.M.: Data Analysis. In: Bioaerosols: Assessment and Control, pp. 13-1–13-6. J.M. Macher, H.M. Ammann, H.A. Burge, et al., Eds. American Conference of Governmental Industrial Hygienists, Cincinnati, OH (1999).

173. Alvarez, A.J., A.J.; Buttner, M.P.; Toranzos, G.A.; et al.: Use of Solid-phase PCR for Enhanced Detection of Airborne Microorganisms. Appl. Environ. Microbiol. 60:374–376 (1994).

174. Mukoda, T.J.; Todd, L.A.; Sobsey, M.D.: PCR and Gene Probes for Detecting Bioaerosols. J. Aerosol Sci. 25:1523–1532 (1994).

175. MacNeil, L.; Kauri, T.; Robertson, W.: Molecular Techniques and their Potential Application in Monitoring the Microbiological Quality of Indoor Air. Can. J. Microbiol. 41:657–665 (1995).

176. Milton, D.K.; Walters, M.D.; Hammond, K.; et al.: Worker Exposure to Endotoxin, Phenolic Compounds, and Formaldehyde in a Fiberglass Insulation Manufacturing Plant, Am. Ind. Hyg. Assoc. J. 57:889–896 (1996).

177. Poruthoor, S.K.; Dasgupta, P.K.; Genfa, Z.: Indoor Air Pollution and Sick Building Syndrome. Monitoring Aerosol Protein as a Measure of Bioaerosol. Environ. Sci. Technol. 32:1147–1152 (1998).

178. Schafer, M.P.; Fernback, J.E.; Jensen, P.A.: Sampling and Analytical Method Development for Qualitative Assessment of Airborne Mycobacterial Species of the Mycobacterium tuberculosis Complex. Am. Ind. Hyg. Assoc. J. 59:540–546 (1998).

179. Schafer, M.P.; Fernback, J.E.; Ernst, M.K.: Detection and Characterization of Airborne Mycobacterium tuberculosis H37Ra Particles, a Surrogate for Airborne Pathogenic M. tuberculosis. Aerosol Sci. Technol. 30:161–173 (1999).

180. Selim, M.I.; Juchems, A.M.; Popendorf, W.: Assessing Airborne Aflatoxin B1 During On-farm Grain Handling Activities. Am. Ind. Hyg. Assoc. J. 59:252–256 (1998).

181. Mastorides, S.M.; Oehler, R.L.; Greene, J.N.; et al.: Detection of Airborne Mycobacterium tuberculosis by Air Filtration and Polymerase Chain Reaction [letter; comment]. Clin. Infectious Diseases. 25:756–757 (1997).

182. Mastorides, S.M.; Oehler, R.L.; Greene, J.N.; et al.: The Detection of Airborne Mycobacterium tuberculosis Using Micropore Membrane Air Sampling and Polymerase Chain Reaction. Chest. 115:19–25 (1999).

183. Simpson, J.C.G.; Niven, R.McL.; Pickering, C.A.; et al.: Comparative Personal Exposures to Organic Dusts and Endotoxin. Ann. Occup. Hyg. 43:107–115 (1999).

184. Burge, H.A.: Fungal Toxins and ß-(1-3)-D-Glucans. In: Bioaerosols: Assessment and Control, pp. 24-1–24-13. J.M. Macher, H.M. Ammann, H.A. Burge, et al., Eds. American Conference of Governmental Industrial Hygienists, Cincinnati, OH (1999).

185. Chapman, M.D.: Analytical Methods: Immunoassays. In: Bioaerosols. pp. 235–248. H.A. Burge, Ed., Lewis Publishers, Boca Raton, FL (1995).

186. Miller, J.D.; Young, J.C.: The Use of Ergosterol to Measure Exposure to Fungal Propagules in Indoor Air. Am. Ind. Hyg. Assoc. 1: 58:39–42 (1997).

187. Eduard, W.; Heederik, D.: Methods for Quantitative Assessments of Airborne Levels of Noninfectious Microorganisms in Highly Contaminated Work Environments. Am. Ind. Hyg. Assoc. J. 59:113–127 (1998).

188. Dillon, H.K.; Miller, J.D.; Sorenson, W.G.; et al.: Review of Methods Applicable to the Assessment of Mold Exposure to Children. Environ. Health Perspec. 107(suppl 3):473–480 (1999).

189. Thorne, P.S.; DeKoster, J.A.; Subramanian, P.: Environmental Assessment of Aerosols, Bioaerosols, and Airborne Endotoxins in a Machining Plant. Am. Ind. Hyg. Assoc. J. 57:1163–1167 (1996).

190. Rautiala, S.; Reponen, T.; Hyvarinen, A.; et al.: Exposure to Airborne Microbes During the Repair of Moldy Buildings. Am. Ind. Hyg. Assoc J. 57:279–284 (1996).

191. Samson, R.A.; Flannigan, B.; Flannigan, M.E.; et al., Eds.: Recommendations and Media. In: Health Implications of Fungi in Indoor Environments, pp. 589–592. Elsevier, New York, NY (1994).

192. Health Canada: Fungal Contamination in Public Buildings: A Guide to Recognition and Management. Federal-Provincial Committee on Environmental and Occupational Health. Environmental Health Directorate, Ottawa, Ontario (1995).

193. Health Canada: Indoor Air Quality in Office Buildings: A Technical Guide. Federal-Provincial Advisory Committee on Environmental and Occupational Health. Environmental Health Directorate, Ottawa, Ontario (1993).

194. International Society of Indoor Air Quality and Climate: Control of Moisture Problems Affecting Biological Indoor Air Quality. TFI-1996. ISIAQ, Ottawa, Canada (1996).

195. ACGIH: Data Interpretation. In: Bioaerosols: Assessment and Control, pp. 7-1–7-0. J.M. Macher, H.M. Ammann, H.A. Burge, *et al.*, Eds. American Conference of Governmental Industrial Hygienists, Cincinnati, OH. (1999).

196. Burge, H.A.; Macher, J.M.; Milton, D.K.; *et al.*: Data Evaluation. In Bioaerosols: Assessment and Control, pp. 14-1–14-11. J.M. Macher, H.M. Ammann, H.A. Burge, *et al.*, Eds. American Conference of Governmental Industrial Hygienists, Cincinnati, OH (1999).

197. Lighthart, B.; Stetzenbach, L.D.: Distribution of Microbial Bioaerosols. In: Atmospheric Microbial Aerosols: Theory and Application. pp. 693–702. B. Lighthart and A.J. Mohr, Eds. Chapman and Hall, New York, NY (1994).

198. Burge, H.A.: Bioaerosols in the Residential Environment. In: Bioaerosols Handbook. pp. 579–597. C.S. Cox and C.M. Wathes, Eds., Lewis Publishers, Boca Raton, FL (1995).

199. Crook, B.; Olenchock, S.A.: Industrial Workplaces. In: Bioaerosols Handbook. pp. 531–545. C.S. Cox and C.M. Wathes, Eds. Lewis Publishers, Boca Raton, FL (1995).

200. DeKoster, J.A.; Thorne, P.S.: Bioaerosol Concentrations in Noncomplaint, Complaint, and Intervention Homes in the Midwest. Am. Ind. Hyg. Assoc. J. 56:573–580 (1995).

201. Wathes, S.M.: Bioaerosols in Animal Houses. In: Bioaerosols Handbook. pp. 547–557. C.S. Cox and C.M. Wathes, Eds., Lewis Publishers, Boca Raton, FL (1995).

202. Godish, D.; Godish, T.; Hooper, B.; *et al.*: Airborne Mould Levels and Related Environmental Factors in Australian Houses. Indoor Built Environ. 5:148–154 (1996).

203. Robertson, L.D.: The Identification of Baseline Indoor Air Quality Parameters for a Renovated Building Prior to Occupancy. Am. Ind. Hyg. Assoc. J. 57:1058–1061 (1996).

204. Sigler, L.; Abbott, S.P.; Gauvreau, H.: Assessment of Worker Exposure to Airborne Molds in Honeybee Overwintering Facilities. Am. Ind. Hyg. Assoc. J. 57:484–490 (1996).

205. Woskie, S.R; Virji, M.A.; Kriebel, D.; *et al.*: Exposure Assessment for a Field Investigation of the Acute Respiratory Effects of Metalworking Fluids. I. Summary of Findings. Am. Ind. Hyg. Assoc. J. 57:1154–1162 (1996).

206. Levy, J.I.; Nishioka, Y.; Gilbert, K.; *et al.*: Variabilities in Aerosolizing Activities and Airborne Fungal Concentrations in a Bakery. Am. Ind. Hyg. Assoc. J. 60;317–325 (1999).

207. Zacharisen, M.C.; Kadambi, A.; Schlueter, D.P.; *et al.*: The Spectrum of Respiratory Disease Associated With Exposure to Metal Working Fluids. J. Occup. Environ. Med. 40:640–647 (1998).

208. Straja, S. ; Leonard, R.T. : Statistical Analysis of Indoor Bacterial Air Concentration and Comparison of Four RCS Biotest Samplers. Environ. Int. 22:389–404 (1996).

209. Cheng, Y.S.; Lu, J.C.; Chen, T.R : Efficiency of a Portable Indoor Air Cleaner in Removing Pollens and Fungal Spores. J. Aerosol Sci. 29:92–101 (1998).

210. Foarde, K.K.; Hanley, J.T.; Ensor, D.S.; *et al.*: Development of a Method for Measuring Single-Pass Bioaerosol Removal Efficiencies of a Room Air Cleaner. Aerosol Sci. Technol. 30:223–234 (1999).

211. Foarde, K.K.; Myers, E.A.; Hanley, J.T.; *et al.*: Methodology to Perform Clean Air Delivery Rate Type Determinations with Micro-biological Aerosols. Aerosol Sci. Technol. 30:235–245 (1999).

212. Kalatoor, S.; Grinshpun, S.A.; Willeke, K.: New Aerosol Sampler with Low Wind Sensitivity and Good Filter Collection Uniformity. Atmos. Environ. 29:1105–1112 (1995).

213. Aizenberg, V.; Bidinger, E.; Grinshpun, S.A.; *et al.*: Airflow and Particle Velocities Near a Personal Aerosol Sampler with a Curved, Porous Aerosol Sampling Surface. Aerosol Sci. Technol. 18:247–258 (1998).

214. Muilenberg, M.L.: The Outdoor Aerosol. In: Bioaerosols,. 163–204, H.A. Burge, Ed., Lewis Publishers, Boca Raton, FL (1995).

215. Kim, J.: Atmospheric Environment of Bioaerosols. In: Atmospheric Microbial Aerosols: Theory and Application, pp. 28–67, B. Lighthart and A.J. Mohr, Eds. Chapman and Hall, New York, NY (1994).

216. Cox, C.S.: Stability of Airborne Microbes and Allergens. In: Bioaerosols Handbook, pp. 77–99, C.S. Cox and C.M. Wathes, Eds. Lewis Publishers, Boca Raton, FL (1995).

217. Lacey, J.; Venette, J.: Outdoor Air Sampling Techniques. In: Bioaerosols Handbook, pp. 407–471, C.S. Cox and C.M. Wathes, Eds., Lewis Publishers, Boca Raton, FL (1995).

218. Lin, W.-H.; Li, C.-S.: Size Characteristics of Fungus Allergens in the Subtropical Climate. Aerosol Sci. Technol. 25:93–100 (1996).

219. Mohr, A.J.: Fate and Transport of Microorganisms in Air. In: Manual of Environmental Biology, pp. 641–650, C.J. Hurst, G.R Knudsen, M.J. McInerney, *et al.*, Eds. American Society for Microbiology, Washington, DC (1997).

220. Rantio-Lehtimaki, A.: Aerobiology of Pollen and Pollen Antigens. In: Bioaerosols Handbook, pp. 387–406, C.S. Cox and C.M. Wathes, Eds., Lewis Publishers, Boca Raton, FL (1995).

221. Mullins, J.; Emberlin, J.: Sampling Pollens. J. Aerosol Sci. 28:365–370 (1997).

222. American Academy of Asthma, Allergy, and Immunology: Pollen and Spore Report. AAAAI, Milwaukee, WI (1997).

223. McCartney, H.A.; Fitt, B.D.; Schmechel, D.: Sampling Bioaerosols in Plant Pathology. J. Aerosol Sci. 28:349–364 (1997).

224. Palmer, C.J.; Bonilla, G.F.; Roll, B.; Paszko-Kolva, C.; *et al.*: Detection of *Legionella* Species in Reclaimed Water and Air with the EnviroAmp Legionella PCR Kit and Direct Fluorescent Antibody Staining. *Appl. Environ. Microbiol.* 61:407–412 (1995).

225. Rosas, I.; Salinas, E.; Yela, A.; *et al.*: *Escherichia coli* in Settled-Dust and Air Samples Collected in Residential Environments in Mexico City. Appl. Environ. Microbiol. 63:4093–4095 (1997).

226. Marshall, W.A. : Laboratory Evaluation of a New Aerobiological Sampler for Use in the Antarctic. J. Aerosol Sci. 28:371–380 (1997).

227. Bartlett, M.S.; Lee, C.H.; Lu, J.J.; *et al.*: *Pneumocystis carinii* Detected in Air. J. Eukaryotic Microbio, 41:75S (1994).

228. Wakefield, A.E.: Detection of DNA Sequences Identical to *Pneumocystis carinii* in Samples of Ambient Air. J. Eukaryotic Microbiol. 41: 116S (1994).

229. McCluskey, R; Sandin, R.; Greene, J.: Detection of Airborne Cytomegalovirus in Hospital Rooms of Immunocompromised Patients. J. Virological Methods. 56:115–118 (1996).

230. Olsson M.; Sukura A.; Lindberg LA.; *et al.*: Detection of *Pneumocystis carinii* DNA by Filtration of Air. Scandinavian J. Infect. Dis. 28:279–282 (1996).

231. Latouche, S.; Olsson, M.; Polack, B.; *et al.*: Detection of *Pneumocystis carinii* f. sp. in Air Samples Collected in Animal Rooms. J. Eukaryotic Microbiol. 44:46S–47S (1997).

232. Ko, G.; Burge, H.A.; Muilenberg, M.; *et al.*: Survival of Mycobacteria on HEPA Filter Material. J. Am. Biolog. Safety Assoc. 3:65–78 (1998).

233. Lighthart, B.; Shaffer, B.T.: Increased Airborne Bacterial Survival as a Function of Particle Content and Size. Aerosol Sci. Technol. 27:439–446 (1997).

234. Qian, Y.; Willeke, K.; Ulevicius, V.; *et al.*: Dynamic Size Spectrometry of Airborne Microorganisms: Laboratory Evaluation and Calibration. Atmos. Environ. 29: 1123–1129 (1995).

235. Reponen, T.; Willeke, K.; Ulevicius, V.; *et al.*: Effect of Relative Humidity on the Aerodynamic Diameter and Respiratory Deposition of Fungal Spores. Atmos. Environ. 30:3967–3974 (1996).

236. Reponen, T.; Willeke, K.; Ulevicius, V.; et al.: Techniques for Dispersion of Microorganisms into Air. Aerosol Sci. Technol. 27:405–421 (1997).

237. Reponen, T.A.; Gazenko, S.V.; Grinshpun, S.A.; *et al.*: Characteristics of Airborne Actinomycete Spores. Appl. Environ. Microbiol. 64:3807–3812 (1998).

238. McCullough, N.F.; Brosseau, L.M.; Vesley, D.; *et al.*: Improved Methods for Generation, Sampling, and Recovery of Biological Aerosols in Filter Challenge Tests. Am. Ind. Hyg. Assoc. J. 59:234–241 (1998).

239. Johnson, D.L.; Pearce, T.A.; Esmen, N.A.: The Effect of Phosphate Buffer on Aerosol Size Distribution of Nebulized *Bacillus subtilis* and *Pseudomonas fluorescens* Bacteria. Aerosol Sci. Technol. 30:202–210 (1999).

240. Johnson, B.; Martin, D.D.; Resnick, I.G.: Efficacy of Selected Respiratory Protective Equipment Challenged with *Bacillus subtilis* subsp. *niger*. Appl. Environ. Microbiol. 60:2184–2186 (1994).

241. Qian, Y.; Willeke, K.; Ulevicius, V.; *et al.*: Particle Reentrainment from Fibrous Filters. Aerosol Sci. Technol. 27:394–404 (1997).

242. Maus, R.; Umhauer, H.: Determination of the Fractional Efficiencies of Fibrous Filter Media by Optical in situ Measurements. Aerosol Sci. Technol. 24:161–173 (1996).

243. Miller-Leiden, S.; Lobascio, C.; Macher, J.M.; *et al.*: Effectiveness of In-room Air Filtration for Tuberculosis Infection Control in Health-care Settings. J. Air Waste Management Assoc. 46:869–882 (1996).

244. Tyndall, R.L.; Lehman, E.S.; Bowman, E.K.; *et al.*: Home Humidifiers as a Potential Source of Exposure to Microbial Pathogens, Endotoxins, and Allergens. Indoor Air. 5:171–178 (1995).

245. Neef, A.; Amann, R.; Schleifer, K.-H.: Detection of Microbial Cells in Aerosols Using Nucleic Acid Probes. System. Appl. Microbiol. 18:113–122 (1995).

246. Fogelmark, B.; Rylander, R: (1-3)-ß-D-Glucans in Some Indoor Air Fungi. Indoor Built Environ. 6:291–294 (1997).

247. Macnaughton, S.J.; Jenkins, T.L.; Alugupalli, S.; *et al.*: Quantitative Sampling of Indoor Air Biomass by Signature Lipid Biomarker Analysis: Feasibility Study in a Model System. Am. Ind. Hyg. Assoc. J. 58:270–277 (1997).

248. Mainelis, S.A.; Grinshpun, S.A.; Willeke, K.; *et al.*: Collection of Airborne Microorganisms by Electrostatic Precipitation. Aerosol Sci. Technol. 30:127–144 (1999).

249. Banks, L.; DiGiovanni, P.: A Wind Tunnel Comparison of the Rotorod and Samplair Pollen Samplers. Aerobiologia. 10:141–145 (1994).

250. Griffiths, W.D.; Stewart, I.W.; Futter, S.J.; *et al.*: The Development of Sampling Methods for the Assessment of Indoor Bioaerosols. J. Aerosol Sci. 28:437–457 (1997).

251. Jensen, P.A.; Todd, W.F.; Davis, G.N.; Scarpino, P.V.: Evaluation of Eight Bioaerosol Samplers Challenged with Aerosols of Free Bacteria. Am. Ind. Hyg. Assoc. J. 53:660–667 (1992).

252. Lach, V.: Performance of the Surface Air System Air Samplers. J. Hosp. Infection. 6:102–107 (1985).

253. Rubow, K.L.; Marple, V.A.; Olin, J.; McCawley, M.A.: A Personal Cascade Impactor: Design, Evaluation and Calibration. Am. Ind. Hyg. Assoc. J. 48:532–538 (1997).

254. Clark, S.; Lach, V.; Lidwell, O.M.: The Performance of the Biotest RCS Centrifugal Air Sampler. J. Hosp. Inf. 2:181–186 (1981).

255. Macher, J.M.; First, M.W.: Personal Air Samplers for Measuring Occupational Exposures to Biological Hazards. Am. Ind. Hyg. Assoc. J. 45:76–83 (1984).

256. Fannin, K.F.; Vana, S.C.: Development and Evaluation of an Ambient Viable Microbial Air Sampler. EPA Report No. 600/1-81-069, Washington, DC: U.S. Environmental Protection Agency (1981).

257. Cown, W.B.; Kethley, T.W.; Fincher, E.L.: The Critical-orifice Liquid Impinger as a Sampler for Bacterial Aerosols. Appl. Microbiol. 5:119–124 (1957).

258. May, K.R.: Multistage Liquid Impinger. Bacteriol. Rev. 30:559–570 (1966).

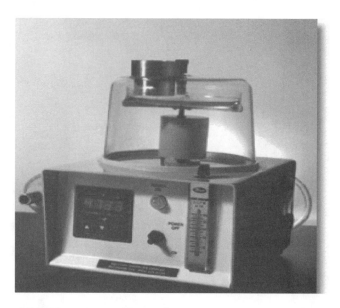

INSTRUMENT 22-1a. Mattson-Garvin Air Sampler, Model 220, a rotating slit agar impactor; shown with HEPA filter and probe assembly.

Instrument Descriptions

Rotating Slit or Slit-to-Agar Impactors

22-1a. *Mattson-Garvin Air Sampler (BAR)*

Air is drawn at 28 L/min through a fixed 0.15- × 41-mm slit and impacted directly onto semisolid medium on a rotating turntable. Model M/G 220 designed to sample room air and M/G P-320 to test compressed gases. Model 220 has a built-in vacuum pump; Model P-320 has a constant pressure reducing valve (maximum 860 kPa, minimum 33 kPa) (645 cm Hg, minimum 25 cm Hg). Both models have air flow gauges and adjustment valves. Samplers available with drive motors for rotational speeds of 5, 15, 30, or 60 min per revolution. Adjustable electric timer shuts off sampler at cycle end. Gauge to adjust slit-to-agar distance to 2–3 mm. High efficiency particulate air (HEPA) filter on air exhausts (Models 220 and P-320).

d_{50} cutpoint: 0.5 μm;[251] 0.53 μm (calculated).[18, 22]

Collection substrate: 40–50 ml semisolid medium in 15- × 1.5-cm disposable culture plates.

Power: 120 V 60 Hz or 240 V 50 Hz.

Materials. Dome: Lexan plastic; base: aluminum; nozzle: aluminum.

Dimensions: 30 cm high × 25 cm wide × 30 cm front to back.

Weight: Model 220: 7 kg; Model P-320: 5 kg.

22-1b. *Casella Airborne Bacteria Sampler (CAS)*

Air is drawn at 175, 350, 525, or 700 L/min through four 1- × 44.5-mm slits and impacted directly onto

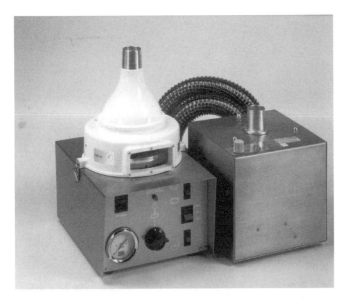

INSTRUMENT 22-1b. Casella Airborne Bacteria Sampler, Model MK-II, a rotating slit agar impactor.

semisolid medium on a rotating turntable. Airflow gauge and adjustment valve included. Rotational speed variable at 0.5, 2, or 5 min per revolution with automatic shutoff at cycle end. Turntable height adjusts for slit-to-agar distance of 2 mm. Template available for reading individual segments of impaction plates. Carrying handle attached.

Collection substrate: 40–50 ml semisolid medium in 15-cm disposable culture plates.

Power: 120 V, 6 amp, 60 Hz.

Auxiliary air mover: Approximately 0.5-hp vacuum pump able to move up to 700 L/min air flow at 4-kPa (3.0-cm-Hg) pressure drop.

INSTRUMENT 22-1c, d. New Brunswick Slit-to-Agar Air Sampler, a rotating slit agar impactor.

Materials: Aluminum and stainless steel.

Dimensions: Sampler: 32 × 25 × 37 cm; pump: 22 × 22 × 26 cm.

Weight: Sampler: 10 kg; pump: 6 kg.

22-1c, d. New Brunswick Slit-to-Agar Air Samplers (NBS)

Air is drawn through a fixed 25-mm-long × 0.1- to 0.2-mm-wide slit and impacted directly onto semisolid medium on a rotating turntable. Model STA-203 samples at 15–55 L/min; Model STA-204 samples at 15–30 L/min. The latter has a built-in vacuum pump. Thermal mass flow controller to measure air flow rate. Microprocessor controlled with LED display and visual and audible alarms for power interrupt, variation in flow rate, or end of sample. Rotational speed adjustable from 2–99.9 min per revolution with automatic shutoff at cycle end. Turntable height adjustable. Template available for reading individual segments of impaction plates.

Collection substrate: 40–50 ml semisolid medium in 15-× 1.5-cm disposable culture plates.

Auxiliary air mover: Model STA-203: Vacuum source required.

Power: Model STA-204: 120/100 or 220/240 V, 50/60 Hz.

Materials: Bonnet: plastic composite; base: metal and plastic; nozzle: stainless steel.

Dimensions: Model STA-203: 29 cm high × 27 cm wide × 28 cm front to back; Model STA-204: 46 × 27 × 28 cm.

Weight: Model STA-203: 5 kg; Model 204: 11.4 kg.

Multiple-Hole Impactors

22-2. Burkard Portable Air Sampler for Agar Plates (BMC Sampler 2B) (BMC)

Air is drawn at 10 or 20 L/min through 100 1-mm holes and impacted directly onto semisolid medium. Sampler can be run continuously or using a built-in 1- to 9-min timer. Two covers available: open or with cone.

d_{50} cutpoint: 4.18 μm at 10 L/min (calculated),[18, 22] 2.56 μm at ≥20 L/min (calculated).[168]

Collection substrate: 27 ml semisolid medium in 9-cm disposable culture plates.

Power: Rechargeable NiMH battery or AC adaptor (110 V/60 Hz or 240 V/50 Hz). LED indicator light and audible signal.

Materials: Aluminum inlet (autoclavable); plastic housing.

INSTRUMENT 22-2. Burkard Portable Air Sampler for Agar Plates, a single stage, 100-hole agar impactor.

Dimensions: 15 cm high × 11 cm wide.
Weight: 0.7 kg.

22-3a. Andersen Single Stage Viable Particle Sampler; N6 Single Stage Viable Cascade Impactor (AND)

Air is drawn at 28 L/min through 400 0.25-mm holes and impacted directly onto semisolid medium. Model 10-880 includes sampler only; Model 10-890 includes sampler, 115-V/60-Hz vacuum pump, and carrying case. Field calibrator and tripod available.

d_{50} cutpoint: 0.65 μm,[85] 0.57 μm (calculated),[79] 0.58 μm (calculated),[18, 22]

Collection substrate: 40–50 ml semisolid medium in 10- × 1.5-cm disposable culture plates.

Auxiliary air mover: Rotary-vane or diaphragm vacuum pump able to move 28 L/min at 2.7-kPa (2-cm-Hg) pressure drop.

Materials: Aluminum (autoclavable).

Dimensions: 7.3 cm high × 10.8 cm wide.

Weight: 0.6 kg.

22-3b. Aerotech 6 Bioaerosol (AER)

Air is drawn at 28 L/min through 400 0.25-mm holes and impacted directly onto semisolid medium. Sampler pre-drilled for tripod attachment. High-volume vacuum pump and in-line calibrator available. Sampler is modeled after Instrument 23-3a.

22-4a, b. Sterilizable Microbiological Atrium (SMA) Micro Sampler, SMA Portable Micro Sampler (VAI)

Air is drawn through twelve 2.4-, 6.4-, or 12.7-mm holes and impacted directly onto semisolid medium. Independent units may be operated at 28–71 L/min; portable unit operates at 28 or 142 L/min. SMA Remote Atrium may be used to sample compressed gases.

d_{50} cutpoint: Portable unit: 95% efficient for 0.2-μm particles (manufacturer).

Collection substrate: 25 or 32 ml semisolid medium in 9- × 1.5-cm disposable culture plates.

Power: May be used with a 1- to 10-port control system with vacuum pump, central vacuum system, or vacuum pump able to maintain a pressure difference of 61 kPa (18 in Hg) required to operate critical orifice. Portable unit has rechargeable battery, LCD indicator light, and timer.

Materials: Aluminum or stainless steel (autoclavable, dry heat or ethylene oxide sterilizable).

INSTRUMENT 22-3a. Andersen Single Stage Viable Particle Sampler, a single-stage, 400-hole agar impactor.

INSTRUMENT 22-4a. SMA Micro Sampler, a single-stage, 12-hole agar impactor.

INSTRUMENT 22-4b. SMA Portable Micro Sampler, a single-stage, 12-hole agar impactor.

Dimensions: Independent unit: 3.8 cm high × 10.8 cm wide; portable unit: 18 cm high × 15 cm wide × 15 cm front to back.

Weight: Portable unit: 5.4 kg.

22-5a, b, c, 22-6b. Surface-Air-Samplers (SAS) Super 100 and Super 180; High Volume Samplers HiVAC Impact and HiVAC Petri (PBI, BSI)

Air is drawn through multiple holes and impacted directly onto semisolid medium. SAS Super 100, Super 100 CR (for clean rooms), HiVAC Impact, and HiVAC Petri sample at 100 L/min; SAS Super 180 samples at 180 L/min. Heads for Super 100 and Super 180 available with 219 1-mm holes (5.5-cm contact plate), 400 0.25-mm holes (9-cm plate), or 487 1-mm holes (8.4-cm contact plate). HiVAC Impact head has 219 1-mm holes (5.5-cm contact plate); HiVAC Petri has 400 0.25-mm holes (9-cm plate). Pitot tube or vane anemometer available for air flow validation. Carrying cases included ("Bio-Safe Carrier" case for Super 100 CR). GLP-GMP data transfer; RS232 output for PC printer connection. Tripods available.

d_{50} cutpoint: (219/220 1-mm holes) 180 L/min: 1.9 µm,[252] 1.39 µm,[167] 1.45 µm (calculated),[79] 1.35 µm (calculated).[167]

Collection substrate: Semisolid medium in disposable culture plates: 15–18 ml in 5.5-cm contact plates; 20–25 ml in 8.4-cm contact plates; 27 ml in 9-cm plates.

Power: Battery pack or AC adapter, indicator lights and timers. HiVAC Impact and HiVAC Petri: built-in batteries, indicator lights, and timers.

Materials: Sampler heads: autoclavable aluminum or stainless steel. Super 100 and 180: polyurethane single-body resin housing. High Volume Sampler: enameled aluminum housing. HiVAC Impact and HiVAC Petri: poliammide (molded high-impact plastic). "Bio-Safe Carrier" case: autoclavable polycarbonate.

Dimensions: Super 100 and 180: 29 cm high × 11 cm wide × 10.5 cm front to back, (battery pack: 18 × 11 × 16 cm). HiVAC Impact and HiVAC Petri: 28.5 × 15.5 × 11 cm.

Weight: Aluminum sampling head: 0.115 kg; stainless steel sampling head: 0.32 kg. Super 100 and 180 housings: 1.75 kg (battery pack: 5.0 kg). HiVAC Impact: 1.4 kg. HiVAC Petri: 1.5 kg.

22-5d. MicroBio MB1 and MB2 Air Samplers (PAR)

Air is drawn at 100 L/min through 220 1-mm holes and impacted directly onto semisolid medium. Flowrate factory calibrated. Model MB2 is microprocessor controlled with time-delay feature that allows independent operation. Carrying case available for Model MB1; carrying case included with Model MB2.

d_{50} cutpoint: 1.8 µm (calculated).

INSTRUMENT 22-5a. SAS Super 100, a single-stage, multiple-hole agar impactor.

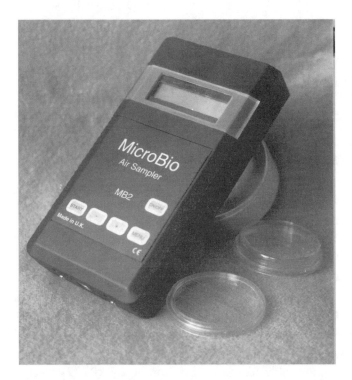

INSTRUMENT 22-5d. MicroBio MB2 Air Sampler, a single-stage, 220-hole agar impactor.

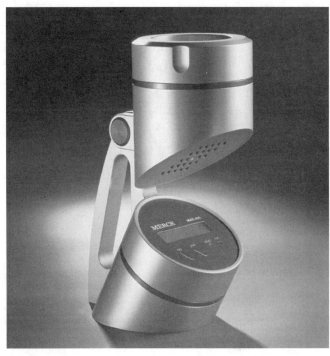

INSTRUMENT 22-5f. Merck Air Sampler MAS 100, a single-stage, 400-hole agar impactor.

Collection substrate: 12–14 ml semisolid medium in 5.5-cm disposable culture (contact) plates.

Power: Rechargeable NiMH battery. MB1: LED display; MB2: LCD display. Indicator lights and timers. Remote control operation optional.

Materials: Sampler heads: autoclavable aluminum; sampler housing: high-impact polystyrene.

Dimensions: MB1: (sampler head) 3.5 cm high × 9.5 cm wide, (case) 18.8 cm high × 11 cm wide × 6 cm front to back; MB2: (sampler head and support) 7 cm high × 9.5 cm wide, (housing) 19.6 cm high × 10 cm wide × 4 cm front to back.

Weight: (Sampler and sampling head) MB1: 0.76 kg; MB2: 0.86 kg.

22-5e. Sampl'Air Multiple-Head Air Sampler (MBI)

Air is drawn at 100 or 200 L/min through 318 or 106 3-mm holes, standard and high speed grids respectively, and impacted directly onto semisolid medium. Unit can be operated with one or two sampling heads simultaneously.

d_{50} cutpoint: High speed grid; low flow: 1.5 μm (manufacturer).

Standard grid; low flow: 5 μm.

Standard grid; high flow: 3 μm.

Collection substrate: 16–17 ml semisolid medium in 9-cm disposable culture plates.

Power: Rechargeable battery with LED display.

Materials: Sampler heads: aluminum or stainless steel; control unit: high-impact plastic.

Dimensions: 16.5 cm high × 13.5 cm wide × 25 cm front to back.

Weight: 5.4 kg.

INSTRUMENT 22-5e. Sampl'Air Air Sampler, a single-stage, 106- or 318-hole agar impactor.

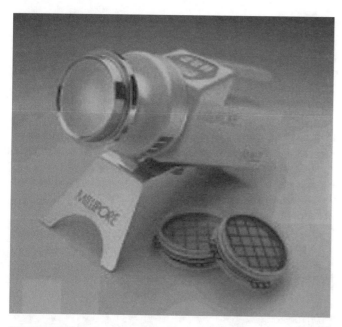

INSTRUMENT 22-6a. M Air T™ Millipore Air Tester (MIL), a single-stage, 967-hole agar impactor.

22-5f. Merck Air Sampler MAS 100 (MER, VWR)

Air is drawn at 100 L/min through 400 0.75- or 0.34-mm holes and impacted directly onto semisolid medium. Explosion-proof models available. Calibration unit available.

d_{50} cutpoint:. (0.75-mm holes) 1.72 μm.[82]

Collection substrate: 20–25 ml semisolid medium in 9-cm disposable culture plates.

Power: Rechargeable battery or AC adaptor (110 V/60 Hz or 240 V/50 Hz). LED indicator light.

Materials: Aluminum housing; lid sterilizable.

Dimensions: 29 cm high × 11 cm wide.

Weight: 2.2 kg.

22-6a. M Air T™ Millipore Air Tester (MIL)

Air is drawn at 140 L/min (first 500 L) and 180 L/min (second 500 L) through 967 0.5–0.6-mm holes and impacted directly onto semisolid medium in a 75-cm-diameter cassette. Sampler comes with battery charger, tripod, and carrying case. Calibration unit available.

d_{50} cutpoint: 3.5 μm (manufacturer).

Collection substrate: Semisolid medium (34 ml) in pre-filled disposable cassettes with gridded base.

Power: Rechargeable (type) battery. LED display. Indicator lights and timers, including delay timer.

Materials: Multiple-hole inlet: stainless steel (auto-clavable); sampler head: aluminium; sampler housing: aluminum.

Dimensions: 25.5 cm high × 12.5 cm wide × 10.5 cm deep.

Weight: 1.9 kg.

22-6b. Surface-Air-Samplers (SAS) Super 180 (see 23-5a–23-5c)

22-7. Andersen Two Stage Viable Sampler; Two Stage Viable Cascade Impactor (AND)

Air is drawn at 28 L/min through two stages, with 200 1.5- or 0.4-mm holes, and impacted directly onto semisolid medium. Model 10-860 includes sampler only; Model 10-850 includes sampler, 115-V/60-Hz vacuum pump, and carrying case. Field calibrator and tripod available.

d_{50} cutpoint: Stages 0 and 1: 8.0 and 0.95 μm (calculated),[18, 251] 6.28 and 0.83 μm (calculated).[18, 22]

Collection substrate: 20–25 ml semisolid medium in 10- × 1.5-cm disposable culture plates.

Auxiliary air mover: Rotary-vane or diaphragm vacuum pump able to move 28 L/min at 2.7-kPa (2-cm-Hg) pressure drop.

Materials: Aluminum (autoclavable).

Dimensions: 12.7 cm high × 12 cm wide.

Weight: 1.3 kg.

22-8. Andersen Six Stage Viable Sampler; Six Stage Viable Particle Cascade Impactor (AND)

Air is drawn at 28 L/min through six stages, each with 400 holes, and impacted directly onto semisolid medium. Hole diameters from Stage 1 to 6: 1.18, 0.91,

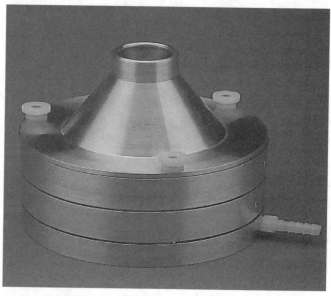

INSTRUMENT 22-7. Andersen Two Stage Viable Sampler, a two-stage, 200-hole agar impactor.

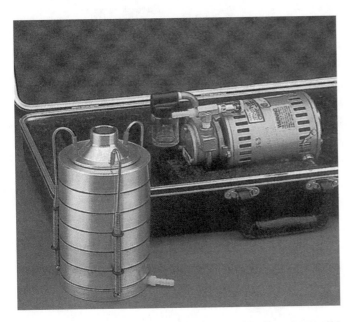

INSTRUMENT 22-8. Andersen Six Stage Viable Sampler, a six-stage, 400-hole agar impactor.

0.71, 0.53, 0.34, and 0.25 mm. Model 10-830 includes sampler only; Model 10-800 includes sampler, 115-V/60-Hz vacuum pump, and carrying case. Field calibrator and tripod available.

d_{50} cutpoint: Stages 1 to 6: 7.0, 4.7, 3.3, 2.1, 1.1, and 0.65 μm;[85] 6.24, 4.21, 2.86, 1.84, 0.94, 0.58 (calculated);[18, 22] Stages 1 and 6, respectively (calculated): 6.61 and 0.57.[79]

Collection substrate: 27 ml in manufacturer's glass plates.

Auxiliary air mover: Rotary-vane or diaphragm vacuum pump able to move 28 L/min at 2.7-kPa (2-cm-Hg) pressure drop.

Materials: Aluminum (autoclavable).

Dimensions: 20 cm high × 11.4 cm wide.

Weight: 1.25 kg.

22-9. Andersen Personal Cascade Impactor; Series 290 Marple Personal Cascade Impactor (AND)

Air is drawn at 1–5 L/min through one to eight stages, each with six tapered radial slots or pairs of holes, and impacted onto removable substrates. The slot or hole diameters from Stage 1 to 8 are 2.6 × 9.5, 1.4 × 9.5, 0.81 × 9.5, 0.43 × 9.5, 0.25 × 9.5, 0.17 × 4.8, 0.46, and 0.32 mm. Inlet shroud available. Respectively, Models SE292, SE294, SE296, and SE298 include stages 1–2, 1–4, 1–6, and 1–8. Models in -K series also include a Model SE190-P personal sampling pump with battery charger (115 V/60 Hz or 220 V/50 Hz), interconnecting tubing, and carrying case.

d_{50} cutpoint: (2 L/min). Stages 1 to 8: 21.3, 14.8, 9.8, 6.0, 3.5, 1.55, 0.93 and 0.52 μm, plus a back-up filter.[253]

Collection substrate: 34-mm removable Mylar or stainless steel substrates; 34-mm, 5-μm pore, PVC back-up filter.

Auxiliary air mover: Personal vacuum pump able to move 0.5–5 L/min at 2.0 kPa (1.5-cm-Hg) pressure drop.

Materials: Aluminum with nickel-plated impactor stages.

Dimensions; Depending on number of stages used, 5.6–8.6 cm high × 5.6 cm wide × 4.1 cm front to back.

Weight: Depending on number of stages used, 0.17–0.20 kg.

Filters

22-10. Filters (see Chapter 13)

Centrifugal Samplers

22-11a, b, c. Standard RCS and RCS Plus Centrifugal Air Samplers and RCS High Flow Microbial Air Sampler (BDC)

Air is drawn by a rotating impeller and impacted onto semisolid medium. The RCS Standard sampler

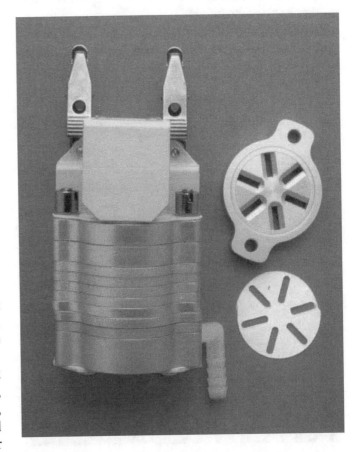

INSTRUMENT 22-9. Andersen Personal Cascade Impactor, an eight-stage impactor.

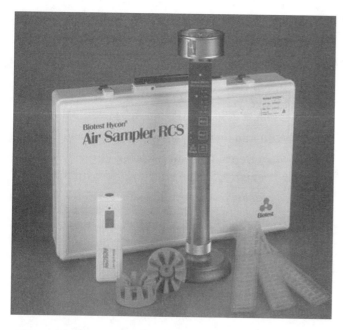

INSTRUMENT 22-11a. Standard RCS Centrifugal Air Sampler, an agar impactor.

operates at a nominal rate of 40 L/min, the RCS Plus sampler at 50 L/min, and the RCS High Flow at 100 L/min. Air enters and exits through the same opening in the RCS Standard and through separate ports in the RCS Plus and RCS High Flow. An air direction ring on the RCS High Flow reduces turbulence in unidirectional flow areas.Explosion-proof models available. Calibration mold available for RCS Standard; anemometer available for RCS Plus and RCS High Flow calibration. Samplers available with carrying cases. Tripods available. Programmable by sample volume.

d_{50} cutpoint: Standard RCS: ~5 μm (estimated),[254] ~3–4 μm (estimated),[22, 251, 255] 7.5 μm (calculated)[22]

RCS Plus: 80% efficient for 2-μm particles and ≥98% efficient above 4 μm (manufacturer), 6 μm (calculated),[168] 0.82 μm (reported).[22]

Collection substrate: Semisolid medium in disposable culture strips. Strips contain 34 sections, 1 cm² × 0.5 cm.

Power: Standard RCS: operates on four "D" cell batteries or AC adapter, LED display with indicator light and timer, infrared remote control available.

RCS Plus: rechargeable 7.2-V NiCad battery or AC adaptor, indicator display, audible signal, infrared remote control available.

RCS High Flow: rechargeable 9.6-V NiCad battery, indicator display, audible signal, delay timer, infrared remote control available.

Materials: Standard RCS: housing and impeller drum—chrome-plated brass, keypad—polyester, impeller blades—aluminum, protection cap—chrome-plated brass, drum, blades, and cap autoclavable

RCS Plus: housing—polycarbonate, rotor—aluminum, protection cap—stainless steel, rotor and cap autoclavable.

RCS High Flow: housing—polycarbonate, rotor—aluminum, protection cap—stainless steel, air direction ring—polycarbonate, rotor, cap, and ring autoclavable.

Dimensions: Standard RCS: 6-cm head diameter, 34 cm high; RCS Plus: 13 cm high × 11 cm wide × 30 cm long; RCS High Flow: 11 cm high × 13 cm wide × 33 cm long.

Weight: Standard RCS: 1.3 kg; RCS Plus: 1.5 kg; RCS High Flow: 1.5 kg, all weights with batteries.

INSTRUMENT 22-11b. RCS Plus Centrifugal Air Sampler, an agar impactor.

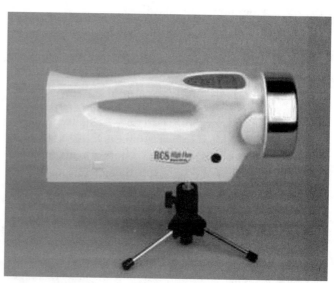

INSTRUMENT 22-11c. RCS High Flow Microbial Air Sampler, an agar impactor.

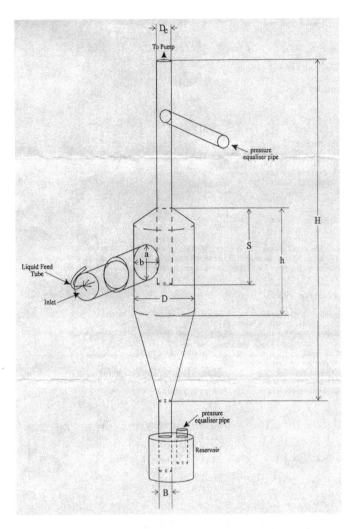

INSTRUMENT 22-12a. AEA Technology PLC Aerojet Cyclone, a wetted cyclone sampler.

22-12a. *AEA Technology PLC Aerojet Cyclone (HAM)*

Air is drawn at 167 or 500 L/min through, respectively, a 1.24- or 2.15-cm orifice connected to the body of a glass cyclone by elliptical pipework;[250] 50–55 L/min version also available. Particles are collected in liquid, which can be recirculated or continuously delivered and collected.

d_{50} cutpoint: 167 L/min: 0.8 µm; 500 L/min: 1.5 µm.[250]

Collection substrate: Collection fluid delivery rate: 1.5–3 L/min.

Auxiliary air mover: Vacuum pump.

Materials: Glass (autoclavable, dry heat sterilizable), stainless steel, electrically conductive plastic.

Dimensions: 167 L/min: 11.4 cm high × 3.75 cm wide; 500 L/min: 19.8 cm high × 6.5 cm wide.

Weight: (glass samplers) 167 L/min: 0.12 kg; 500 L/min: 0.35 kg.

22-12b. *MicroBio MB3 Portable Cyclone Sampler for Bioaerosols (PAR)*

Air is drawn at 167 L/min through a 1.0-cm orifice connected to the body of a cyclone by elliptical pipework.[250] Particles are collected in liquid, which can be recirculated or continuously delivered and collected.

d_{50} cutpoint: 167 L/min: 0.8 µm.[250]

Collection substrate: Collection fluid delivery rate: 1.5-3 L/min.

Power: Battery powered (6v NiMH rechargeable).

Materials: Stainless steel (autoclavable, dry heat sterilizable).

Dimensions: Cyclone: 4.0 cm wide × 20 cm high; control unit: 20 cm high × 20 cm wide × 25 cm front to back.

Weight: 2 kg.

22-12c. *Aerojet-General Liquid-Scrubber Air Sampler (LRI)*

Air is drawn at 700–1000 L/min through a 2.1-cm orifice into a glass cyclone. Particles are collected in liquid, which can be recirculated or continuously delivered and collected.

d_{50} cutpoint: 52% (geometric mean) efficient for *Bacillus subtilis* spores relative to the AGI-30.[256]

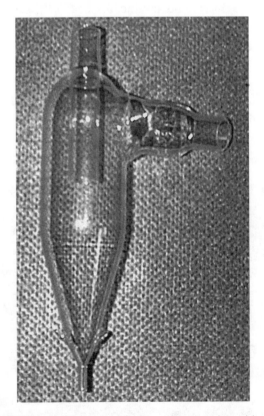

INSTRUMENT 22-12c. Aerojet-General Liquid-Scrubber Air Sampler, a wetted cyclone sampler.

INSTRUMENT 22-12d. SpinCon High-Volume Cyclonic Liquid Sampler, a wetted cyclone sampler.

INSTRUMENT 22-13. Burkard Cyclone Sampler, a dry cyclone sampler.

Collection substrate: Collection fluid aspirated by sampled air stream or pump-delivered at a rate of 1–4 ml/min.

Auxiliary air mover: Vacuum pump.

Materials: Glass (autoclavable, dry heat sterilizable).

Dimensions: 25 cm high × 6.5 cm wide × 16 cm front to back (including inlet).

Weight: 0.26 kg.

22-12d. SpinCon High-Volume Cyclonic Liquid Sampler (MRI)

Air is drawn at 100–800 L/min through a tangential, 20-cm, vertical slit. Particles are collected in 5–10 ml of

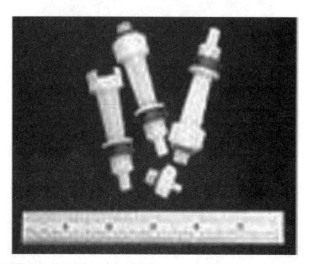

INSTRUMENT 22-12e. Mini-Cyclone Aerosol Collector.

liquid, which is recirculated and replenished.

Collection substrate: Collection fluid pump-delivered at a rate of 1.5 ml/min.

Power: 120 V, 60 Hz, 320 W max.

Materials: Cyclone: clear acrylic, polycarbonate plastic, or glass (autoclavable, dry heat sterilizable); Body: aluminum.

Dimensions: Cyclone: 20 cm high × 5–15 cm wide; Unit: 46 cm high × 32 cm wide × 20 cm front to back.

Weight: ~77 kg.

22-12e. MiniCyclone Aerosol Collector (ITI)

Contact manufacturer for information.

22-13. Burkard Cyclone Sampler (BMC)

Air is drawn at 10 L/min through a 0.9- × 0.3-cm vertical orifice into a reverse-flow cyclone. Flowmeter available for calibration.

d_{50} cutpoint: 1.2 μm (data available from manufacturer).

Collection substrate: Particles collected dry in catchpot or in liquid medium.

Power: Rechargeable NiMH battery or AC adaptor (110 V/60 Hz or 240 V/50 Hz). Indicator light and timer.

Materials: Collecting vessel: 1.5-ml eppendorf vial.

Dimensions: 33 cm high × 17 cm wide × 22 cm front to back.

Weight: 4.6 kg.

INSTRUMENT 22-14. BioSampler, a three-jet, tangential impactor.

INSTRUMENT 22-15. All-Glass Impinger (AGI-4), a single-stage liquid impinger.

22-14. BioSampler (SKC)

Air is drawn at 12.5 L/min through three 0.75-mm nozzles. The nozzles are directed at an angle to the inner sampler wall, which induces a swirling air motion.

d_{50} cutpoint: For 0.5-, 1-, and 2-μm particles, respectively, water as collection fluid: 89%, 95%, and 98%; viscous collection fluid or surface coated with glycerol or mineral oil: 82%, 94%, and 97%; no collection fluid or coating: 60%, 85%, and 98% efficient.[78]

Collection substrate: 20–30 ml water-based or non-evaporating (e.g., mineral oil) liquid in 50-ml capacity container. Sampler can also be used dry or with the inner surface coated with an adhesive material (e.g., mineral oil or glycerol, respectively, for culture-based or non-culture-based analysis) or lined with filter paper treated with mineral oil.

Auxiliary air mover: Vacuum pump able to maintain a pressure difference of at least 51 kPa (0.5 atm) required to operate critical orifices.

Materials: Glass (autoclavable, dry heat sterilizable).

Dimensions: 22.9 cm high × 5.1 cm wide × 6.6 cm front to back (including vacuum attachment).
Weight: 0.16 kg.

Liquid Impingers

22-15. All-Glass Impingers (AGI-4 and AGI-30) (AGI, HAM, MIL)

Air is drawn at 12.5 L/min through a 1-mm capillary (critical orifice) jet and impacted against a wetted surface 4 or 30 mm from the jet (respectively, Models AGI-4 and AGI-30). Samplers come in two sections with ground-glass connections.

d_{50} cutpoint: AGI-30: 0.3 μm (calculated),[22, 251, 257] 0.31 μm (calculated).[79]

Collection substrate: 10–20 ml liquid in a 125-ml capacity bottle.

Auxiliary air mover: Vacuum pump able to maintain a pressure difference of 26.6 kPa (20 cm Hg) required to operate critical orifice.

Materials: Glass (autoclavable, dry heat sterilizable).

Dimensions: 28 cm high × 4 cm wide × 10 cm front to back (including vacuum attachment).

Weight: 0.15 kg.

22-16a. Burkard Multiple-Stage Liquid Impinger (BMC Sampler 4D) (BMC)

Air is drawn at 20 L/min through three stages and impacted against wetted surfaces. Entrance has a stagnation-point shield to improve aspiration efficiency in moving air. A critical orifice is incorporated into the sampler.

d_{50} cutpoint: Stages 1 to 3: ≥10, 4 to 10, and <4 μm.[258]

Collection substrate. 6 ml liquid in each stage.

Auxiliary air mover: Vacuum pump able to maintain a pressure difference of 24 kPa (18 cm Hg) required to operate critical orifice.

Materials: Aluminum alloy or stainless steel (heat sterilizable).

Dimensions: 14.5 cm high × 10 cm wide × 11.5 cm front to back.

Weight: 0.7 kg.

INSTRUMENT 22-16a. Burkard Multiple-Stage Liquid Impinger, a three-stage liquid impinger.

22-16b. Hampshire Glass Three-Stage Impinger (HAM)

Air is drawn at 10, 20, or 55 L/min through three stages and impacted against wetted surfaces. Jet diameters for stages 1 to 3, 10-L/min model: 0.8, 0.5, and 0.14 cm; 20-L/min model: 1.5, 0.75, and 0.2 cm; 55-L/min model: 1.5, 1.0, and 0.33 cm. A critical orifice may be incorporated into the sampler.

d_{50} cutpoint: Stages 1 to 3, 10-L/min model: ≥ 7, ≥ 3, and ≥ 1 m; 20-L/min model: 10 μm, 4 μm, and lower respiratory tract equivalent; 55-L/min model: same as 10-L/min model.[258]

Collection substrate: 10-L/min model: 2 ml liquid each stage; 20-L/min model: 4 ml liquid each stage; 55-L/min model: 7 to 10 ml liquid in stages 1 and 2, 10 ml liquid in stage 3.

Auxiliary air mover: Vacuum pump able to maintain a pressure difference of 1.3–17.3 kPa (10–13 cm Hg) required to operate critical orifice.

Materials: Hand-blown glass (autoclavable, dry heat sterilizable).

Dimensions: Outer diameters, 10-L/min model: 4.5 cm; 20-L/min model: 6.0 cm; 55-L/min model: 7.5 cm.

Pollen, Spore, and Particle Impactors

22-17a. Burkard 1- to 7-Day Recording Volumetric Spore Trap (BMC Sampler 9) (BMC)

Air is drawn at 10 L/min through a 2- × 14-mm slot (0.5- × 14-mm orifice also available) and impacted onto a tape supported on a drum rotating at 0.2 cm/hr (7-day collection) or onto a glass slide (24-hr collection). A wind vane maintains sampler orientation. Flowmeter available for calibration.

d_{50} cutpoint: 2- × 14-mm slot, 10 L/min: 3.70 μm (calculated).[18]

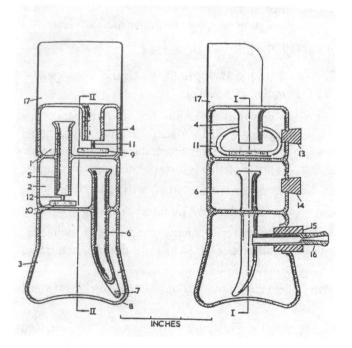

INSTRUMENT 22-16b. Hampshire Glass Three-Stage Liquid Impinger.

INSTRUMENT 22-17a. Burkard Recording Volumetric Spore Trap, a 1- to 7-day impactor.

Collection substrate: Adhesive-coated transparent plastic tape or 75- × 25-mm glass slide.

Power: Rechargeable NiMH battery or AC adaptor (110 V/60 Hz or 240 V/50 Hz).

Materials: Parts treated to prevent corrosion.

Dimensions: 94 cm high, 53-cm vane radius.

Weight. 16 kg.

22-17b. Lanzoni 1- to 7-Day Volumetric Pollen and Particle Sampler (L Sampler VPPS 2000) (LAN)

Air is drawn at 10 L/min (adjustable from 0.5 to 11.0 L/min) through a 2- × 14-mm slot and impacted onto a tape supported on a drum rotating at 0.2 cm/hr (7-day collection) or onto a glass slide (24-hr collection). A wind vane maintains sampler orientation. Flowmeter available for calibration.

d_{50} cutpoint: 10 μm; 70% efficient for 30-μm particles (manufacturer).

Collection substrate: Adhesive-coated transparent plastic tape or 75- × 25-mm glass slide.

Power: 220 V, 50 Hz, 115 V, 60 Hz, or 12-V DC with or without solar panels.

Materials: Stainless steel, brass and aluminum alloy.

Dimensions: Without legs: 77 cm high, 47-cm vane radius; legs: 35 or 85 cm long.

Weight: 16 kg.

22-17c. Kramer-Collins Suction Trap with Rotating Drum (GRM)

Air is drawn at 25 L/min through a 2- × 14-mm slot and impacted onto a tape supported on a drum rotating at an adjustable rate for collection times between 1 h and 32 days. A wind vane maintains sampler orientation.

Collection substrate: Adhesive-coated, double-sided, transparent plastic tape.

Auxiliary air mover: Vacuum pump able to move 15–28 L/min air flow.

Materials: Plastic.

Dimensions: 21 cm high, 46 wide.

Weight: 1.6 kg (without vacuum pump).

22-18a. Allergenco Air Sampler - MK-3 (ALL, MCC)

Air is drawn at 15 L/min for a preset sampling time of 10 min through a 1- × 14-mm slot and impacted in discrete deposits on a glass slide. Samples can be collected at intervals ranging from 1 min to 24 hrs.

d_{50} cutpoint: 2.0 μm (calculated).

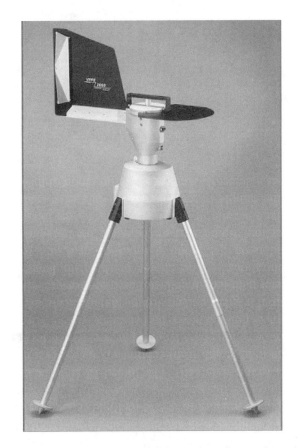

INSTRUMENT 22-17b. Lanzoni Volumetric Pollen and Particle Sampler, a 1- to 7-day impactor.

Collection substrate: Adhesive-coated 75- × 25-mm glass slide.

Power: 110/120-V AC power or 12-V DC adapter. Battery rechargeable by AC adapter. LCD display, indicator lights.

Materials: Rigid metal case; brass slide-carrier tray.

Dimensions: 13 cm high × 9.5 cm wide × 16 cm front to back.

Weight: 2 kg.

INSTRUMENT 22-18a. Allergenco Air Sampler (MK-3), a moving slide impactor.

INSTRUMENT 22-18b. Burkard Continuous Recording Air Sampler, a moving slide impactor.

INSTRUMENT 22-18c. Lanzoni Volumetric Pollen and Particle Sampler, a moving slide impactor.

22-18b. Burkard Continuous Recording Air Sampler (BMC)

Air is drawn at 10 L/min through a 1- × 14-mm slot and impacted onto a glass slide moving beneath the slot at a rate of 0.2 cm/hr for 24 hrs, 0.4 cm/hr for 12 hrs, or 0.8 cm/hr for 6 hrs (total travel distance: 4.8 cm). Flowmeter available for calibration.

d_{50} cutpoint: 1- × 14-mm slot, 10 L/min: 2.52 μm (calculated).[79]

Collection substrate: Adhesive-coated 75- × 25-mm glass slide.

Power: Electrical connection required. Indicator light and timer.

Materials: Rigid metal case; stainless steel moving parts.

Dimensions: 11.5 cm high × 12 cm wide × 21.5 cm front to back.

Weight: 2.6 kg.

22-18c. Lanzoni Volumetric Pollen and Particle Sampler (L Sampler VPPS 1000) (LAN)

Air is drawn at 10 L/min through a 2- × 14-mm slot and impacted onto a glass slide moving behind the slot at a rate of 0.2 cm/hr for 24 hrs. Cover protects unit from rain and snow. Flowmeter available for calibration. VPPS 1000 Wood is a smaller and lighter version.

d_{50} cutpoint: 10 μm; 70% efficient for 30-μm particles (manufacturer).

Collection substrate: Adhesive-coated 75- × 25-mm glass slide.

Power: Operates off electrical connection, 12-V DC external battery or rechargeable internal battery.

Materials: External housing: stainless steel, brass, and aluminum alloy or wood.

Dimensions: Metal unit: 23 cm high × 21 cm wide × 50 cm front to back; wood unit: 18 cm high × 17 cm wide × 29 cm front to back.

Weight: Metal unit: 6 kg; wood unit: 3.3 kg.

22-19. Burkard Personal Volumetric Air Sampler (BMC)

Air is drawn at 10 L/min through a 2- × 14-mm slot and impacted onto a glass slide. Can be run continuously or using a built-in 1- to 9-min timer. Flowmeter available for calibration.

d_{50} cutpoint: 2- × 14-mm slot, 10 L/min: 5.2 μm (calculated); 1- × 14-mm slot, 10 L/min: 2.52 μm (calculated).[22, 79]

INSTRUMENT 22-19 Burkard Personal Volumetric Air Sampler, a stationary slide impactor.

INSTRUMENT 22-20. Air-O-Cell Sampling Cassette, a stationary slide impactor.

Collection substrate: Adhesive-coated 75- × 25-mm glass slide.

Power: Rechargeable NiMH battery or AC adaptor (110 V/ 60 Hz or 240 V/50 Hz). LED indicator light and audible signal.

Materials: Chrome-plated orifice; plastic housing.

Dimensions: 11.5 cm high × 9 cm wide.

Weight: 0.64 kg.

22-20. Air-O-Cell Sampling Cassette (ZAA, AER, MCC, SKC)

Air is drawn at 15 L/min through a 1.05- × 14.4-mm slot and impacted onto a removable glass slide included with sampler. Mini-pump kit available with adjustable flow rate, battery pack, charger, timer, rotameter, and carrying case. Desktop stand and tripod available.

d_{50} cutpoint: Respectively 3.2, 2.6, 2.2, 2.0, or 1.8 μm at 10, 15, 20, 25, or 30 L/min (manufacturer).

Collection substrate: Transparent adhesive on glass slide.

Auxiliary air mover: Vacuum pump able to move 15 L/min at 0.12-kPa (0.5-in water) pressure drop. Mini-pump: rechargeable battery or AC adaptor, indicator light.

Materials: Clear polystyrene.

Dimensions: 3.2 cm high × 3.8 cm wide.

Weight: 0.02 kg.

22-21. Personal sampler Partrap FA52 (MET)

Air is drawn at 10 L/min into a disposable sampling chamber with 12 cm² of collection tape. Battery charger and carrying case included. Timer and tripod available.

Collection substrate: Adhesive-coated transparent plastic tape.

Power: Rechargeable battery, LED indicator light.

Dimensions: 18 × 9 × 4 cm.

Weight: 0.48 kg.

22-22. Rotorod Samplers (STI)

Two vertical rods rotate at a nominal rate of 2400 rpm impacting particles on the leading edges of the rods. Air sampling rate is 21.7 L/min for each rod when 22 mm of rod length is examined (total air flow rate: 43.4 L/min). Two models are available: Model 20—stationary sampler and Model 40—stationary area sampler. Model 20 accepts either fixed or retracting heads; Model 40 accepts retracting head. For intermittent operation, Model 20 can be controlled by an external timing device; Model 40 has a programmable internal timer.

INSTRUMENT 22-21. Partrap FA52, a cassette tape sampler.

INSTRUMENT 22-22. Rotorod, a rotating rod impactor.

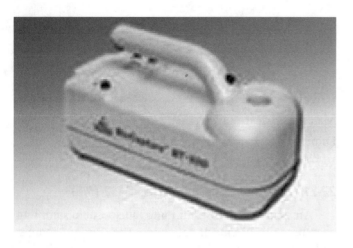

INSTRUMENT 22-23.   BioCapture Air Sampler, a rotating arm impactor.

d_{50} cutpoint: Collection rods approximately 100% efficient for 20-µm particles (manufacturer).

Collection substrate: Adhesive-coated 1.52- × 1.52- × 32-mm collection rods.

Power: Model 20: 12-V DC or AC adaptor; Model 40: 12-V DC or AC adaptor.

Materials: Protective cases: molded polycarbonate; collection rods: clear polystyrene.

Dimensions: Model 20: 12.5 × 9.0 × 7.5 cm; Model 40: 25.5 × 19 × 15 cm.

Weight: Model 20: 0.5 kg; Model 40: 1.5 kg.

22-23. *BioCapture Air Sampler (MST)*

Contact manufacturer for information.

TABLE 22-I-1. Manufacturers and Suppliers of Samplers to Collect Bioaerosols

AER	Aerotech Laboratories, Inc. 2020 W. Lone Cactus Drive Phoenix, AZ 85027-2640 602-780-4800 or 800-651-4802 FAX: 602-780-7695 *www.aerotechlabs.com*	BAR	Barramundi Corporation P.O. Drawer 4259 Homosassa Springs, FL 34447-4259 352-628-0200 FAX: 352-628-0203 *barra@citrus.infi.net* *www.mattson-garvin.com*	CAS	Casella Limited Regent House, Wolseley Road Kempston, Bedford MK42 7JY England 44(0)1234-841441 or 44(0) 1234-841468 FAX: 44(0)1234-841490 *www.casella.co.uk*
AGI	Ace Glass Inc. P.O. Box 688 1430 Northwest Boulevard Vineland, NJ 08362-0688 856-692-3333 or 800-223-4524 FAX: 800-543-6752 *www.aceglass.com*	BDC	Biotest Diagnostics Corporation 66 Ford Road, Suite 131 Denville, NJ 07834-1300 973-625-1300 or 800-522-0090 FAX: 973-625-9454 *www.biotest.com*	GRM	G-R Manufacturing Company 1317 Collins Lane Manhattan, KS 66502-9577 785-537-7276 FAX: 785-537-4462
ALL	Allergen LLC dba Allergenco/Blewstone Press P.O. Box 8571 Wainwright Station San Antonio, TX 78208-0571 210-822-4116 FAX: 210-822-4116 *51, 210-805-8518 *www.txdirect.net/corp/allergen* *allergen@txdirect.net*	BMC	Burkard Manufacturing Company, Limited Woodcock Hill Industrial Estate Rickmansworth, Hertfordshire WD3 1PJ England 44(0)1923-773134 FAX: 44(0)1923-774790 *sales@burkard.co.uk* *www.burkard.co.uk*	HAM	Hampshire Glassware 77-79 Dukes Road, Hampshire Southampton SO14 0ST England 44(0)1703-553755 FAX: 44(0)1703-553020
AND	Andersen Instruments 500 Technology Court Smyrna, GA 30082-5211 770-319-9999 or 800-241-6898 FAX: 770-319-0336 *www.anderseninstruments.com*	BSI	Bioscience International 11607 Magruder Lane Rockville, MD 20852-4365 301-230-0072 FAX: 301-230-1418 *www.biosci-intl.com*	ITI	InnovaTek Incorporated 350 Hills Street Suite 104 Richland, WA 99352-5511 509-375-1093 FAX: 509-375-5183 *www.tekkie.com*

TABLE 22-I-1. (cont.) Manufacturers and Suppliers of Samplers to Collect Bioaerosols

LAN	Lanzoni, S.R.L. Via Michelino 93/B 40127 Bologna Italy 39(0)51-504810-39(0)51-501334 FAX: 39(0)51-6331892 *www.lanzoni.it*	MIL	Millipore Corporation 80 Ashby Road Bedford, MA 01730-2271 781-533-6000 or 800-645-5476 FAX: 781-533-3110 *www.millipore.com*	SKC	SKC Incorporated 863 Valley View Road Eighty Four, PA 15330-1301 724-941-9701 or 800-752-8472 FAX: 724-941-1369 or 800-752-8476 *www.skcinc.com*
LRI	Life's Resources, Incorporated 114 E. Main Street P.O. Box 260 Addison, MI 49220-0260 517-547-7494 or (800) 553-8880 FAX: 517-547-5444 *www.lifes-resources.com*	MRI	MidWest Research Institute 425 Volker Boulevard Kansas City, MO 64110-2299 816-753-7600 FAX: 816-753-8420 *www.mriresearch.org*	STI	Sampling Technologies Inc. 10801 Wayzata Boulevard Suite 340 Minnetonka, MN 55305-1533 612-544-1588 or 800-264-1338 FAX: 612-544-1977 or 800-880-8040 *rotorod@rotorod.com* *www.rotorod.com*
MBI	Microbiology International 97H Monocacy Boulevard Frederick, MD 21701-5778 301-662-6835 or 800-396-4276 FAX: 301-662-8096 *info@microbiology-intl.com* *www.microbiology-intl.com*	MST	MesoSystems Technology Incorporated 3200 George Washington Way Richland, WA 99352-1626 509-375-1111 FAX: 509-375-0115 *www.mesosystem.com*	VAI	Veltek Associates, Inc. Environmental Control Monitoring Division 1039 West Bridge Street Phoenixville, PA 19460-4218 610-983-4949 FAX: 610-983-9494
MCC	McCrone Microscopes and Accessories 850 Pasquinelli Drive Westmont, IL 60559-5539 630-887-7100 or 800-622-8122 FAX: 630-887-7764 *www.mccrone.com*	NBS	New Brunswick Scientific Company, Inc. P.O. Box 4005 44 Talmadge Road Edison, NJ 08818-4005 800-631-5417 or 732-287-1200 FAX: 732-287-4222 *bioinfo@nbsc.com* *www.nbsc.com*	VWR	VWR Scientific Products Corporation Goshen Corporate Park West 1310 Goshen Parkway West Chester, PA 19380-5985 610-431-1700 FAX: 610-436-1760 *www.vwrsp.com*
MER	Merck KGaA 64271 Darmstadt, Germany +49-6151-720 FAX +49-6151-722000 *www.merck.de/english*	PAR	F.W. Parrett Limited 65 Riefield Road London SE9 2RA England 44(0)1181-8504226 [UK 0181-8593254] FAX: 44(0)1181-8504226 *fparrett@aol.com*	ZAA	Zefon Analytical Associates 2860 23rd Avenue North St. Petersburg, FL 33713-4211 727-327-5449 or 800-282-0073 FAX: 727-323-6965 *www.zefon.com*
MET	*Metha*pharm Inc. 131 Clarence Street Brantford, Ontario N3T 2V6 Canada 800-287-7686 or 519-751-3602 FAX: 519-751-9149 *methapharm@sympatico.ca* *www.methapharm.com*	PBI	International PBI Via Novara, 89 20153 Milan Italy 39 2-40-090-010 FAX: 39 2-40-353695 *www.wheatsonsci.com*		

Chapter 23

Direct-Reading Fourier Transform Infrared Spectroscopy for Occupational and Environmental Air Monitoring

Lori A. Todd, Ph.D.
Department of Environmental Sciences and Engineering, University of North Carolina at Chapel Hill, Chapel Hill, NC 27599

CONTENTS

Introduction

Direct-reading instruments are among the most important tools available to industrial hygienists and environmental scientists for detecting and quantifying airborne concentrations of gases and vapors. Direct-reading instruments have the temporal resolution to measure concentrations of acutely toxic chemicals, measure short-term exposures to chemicals, measure fugitive emissions, evaluate equipment or process leaks, and help optimize industrial processes. However, many direct-reading instruments are limited in the number of chemicals they can monitor at one time and do not have the sensitivity and selectivity needed for identifying and quantifying chemicals in mixtures. Some instruments can simultaneously monitor multiple chemicals; however, they are usually survey instruments that cannot differentiate among the chemicals.

The ability to identify and quantify individual chemicals in mixtures, in real-time, and at low limits of detection (below regulatory exposure limits) has become very important in both environmental and industrial hygiene applications. For example, in the environmental arena, the establishment of the 1990 Clean Air Act Amendments requires maximum achievable control technology (MACT) to comply with emission standards for hazardous air pollutants (HAPs).[1] To comply with these regulations, real-time techniques that measure mixtures greatly assist with process modifications.

In industrial hygiene, the ability to measure low levels of mixtures in indoor and outdoor air is important for evaluating adverse health effects in the community and in non-industrial and industrial indoor environments.[2] Indoor air is composed of a complex mixture of pollutants and it is possible that many of the adverse health effects are due to the impact of exposures to mixtures of air pollutants at trace concentrations. Currently, to measure exposures to mixtures of chemicals in air at part per billion (ppb) levels, time-weighted average samples are obtained over many hours. In addition, the range of possible air contaminants requires collection on a variety of collection media. The long sampling times required to obtain the low limits of detection makes it difficult to effectively evaluate human exposures and characterize source emissions. The time averaged sampling smoothes fluctuations in concentrations; these peaks may be important for locating and characterizing source emissions and for estimating exposures. In addition, the diverse media required for collection may limit the number of measurements that can reasonably be obtained over a large area at one time. Real-time measurement of mixtures is vital for linking symptoms to exposures and generating dose-response curves. This is important in understanding causes of adverse health effects and in identifying, controlling, and preventing the hazards.

Fourier transform infrared (FTIR) spectroscopy has the potential to provide the flexibility, sensitivity (low limits of detection), and specificity for making real-time field measurements of known and unknown components in mixtures. FTIR spectroscopy has been used for many years as a laboratory technique; however, advances in computer technology have stimulated the development of FTIR spectroscopy for field applications and bringing it to the forefront of monitoring technology. The ability of FTIR spectrometers to simultaneously measure multiple chemicals in real-time, down to ppb concentrations, has attracted industrial hygienists, air pollution scientists, and regulatory agencies.

In theory, direct-reading instruments that use infrared (IR) absorption to detect chemicals in air can be used to identify and quantify hundreds of chemicals. IR absorption techniques use the principle that compounds selectively absorb energy in the IR region of the electromagnetic spectrum. When molecules absorb incident electromagnetic radiation at specific wavelengths, the energy can boost electrons from their ground state to an excited state, or cause the molecules to stretch, bend, or rotate.[3] To absorb in the infrared region, the movement of the atoms in the molecule must cause a rhythmic change in the dipole moment of the compound; molecules with strong dipoles will usually have strong IR absorptions.[3, 4] With the exception of elements, inorganic salts, and diatomic molecules (oxygen, nitrogen, chlorine), most substances absorb in the IR region. Noble gases, such as helium, cannot be detected because they exist as individual molecules. Diatomic molecules do not have infrared spectra because of their symmetry.

In the mid-infrared region of the electromagnetic spectrum (4000 to 400 cm⁻¹), absorption usually occurs at several different wavelengths of IR light, which results in patterns or bands. This pattern of energy absorption, called the absorption spectrum, creates a unique fingerprint for each molecule and can be used to identify the chemical. Small differences in chemical structures can result in large differences in absorption spectra. This can be seen in Figure 23-1 that shows spectra for ortho-, meta-, and para-xylene. The mid-IR region can be divided into regions with absorption bands from certain functional groups (4000 – 1300 cm⁻¹), the fingerprint region (1600 – 1000 cm⁻¹), and the aromatic region (1000 – 400 cm⁻¹); see Table 23-1. The fingerprint region is the most complex region, with a large number of bands, and is the best region to distinguish whether two samples are different. Functional groups tend to absorb IR radiation in the same wavenumber region regardless of the structure of the rest of the molecule; therefore, unknown molecules can be identified from the molecule's IR spectrum.

To identify a compound, the pattern of absorption bands (number, location, and shape) in a sample's absorption spectrum is compared to a reference absorption spectrum for the chemical of interest. A reference spectrum is created using known concentrations of the pollutant of interest under controlled conditions of temperature and pressure. While the pattern of absorption is used for identification of compounds, the intensity of the spectral bands is used for quantification. Within constraints, there is a linear relationship between the intensity of the spectral bands and the concentration of the compound.[3]

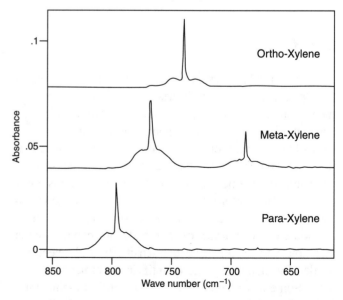

FIGURE 23-1. Absorption spectra for ortho-, meta-, and para-xylene. The x-axis is wave number and the y-axis is absorbance.

TABLE 23-1. Characteristic Infrared Absorptions for Types of Bonds and Functional Groups

Bond	Compound Type	Wave Number Range cm⁻¹
C-H	Alkanes	2850-2960
C-H	Alkanes	1350-1470
C≡C	Alkynes	2100-2260
C-N	Amines	1180-1360
N-H	Amines	3300-3500
C-O	Alcohols, ethers, esters, carboxylic acids	1180-1300
C=C	Alkenes	1640-1680
C=O	Aldehydes, ketones, carboxylic acids, esters	1690-1760
C≡N	Nitriles	2210-2260
C-H	Aromatic rings	3000-3100
C=C	Aromatic rings	< 1000
	Mono-substituted	710-770
	Ring bend	680-700
	Di-substituted (meta)	735-770
	Ring bend	680-700
	Di-substituted (ortho)	750-810
	Di-substituted (para)	790-860
O-H	Monomeric alcohols, phenols	3610-3640
O-H	Hydrogen bonded alcohols, phenols	3200-3600

The optical device used in FTIR spectrometers, to simultaneously and rapidly (seconds) scan the entire mid-IR region, is an advance over traditional IR optics and is called an interferometer.[4] An interferometer is much more efficient at collecting and analyzing the radiation than traditional IR optics, and it can achieve higher spectral resolution (the ability to distinguish between absorption bands of compounds), greater specificity, a higher signal-to-noise ratio, and lower limits of detection (ppb).[5–7]

With an FTIR spectrometer, the entire IR spectra is collected simultaneously and is saved with each measurement; thus, overlapping peaks can be resolved, and unknown as well as known analytes in mixtures can be identified and quantified. Most commonly used field IR spectrometers differ significantly from FTIR spectrometers in the optics used to select wavelengths of IR light. Most field IR spectrometers use non-dispersive (NDIR) optics, which means some method is used to pre-select a limited wavelength region in the IR. When a limited wavelength region is used, only a single absorption peak for the target analyte of interest can be quantified. Therefore, if there are unknowns in the sample that have peaks which overlap with the target analyte peak, quan-

tification will be inaccurate. When measurements are obtained for only a small portion of the IR region, it is difficult to evaluate the accuracy of the concentration quantified for the target analyte. If multiple chemicals are present which absorb in different regions of the IR spectra, the operator must then select the next wavelength of interest to monitor. Some commonly used field IR instruments only allow a few regions to be selected; this limits the number of chemicals that can be identified. Some field IR spectrometers can sequentially scan through all the wavelengths in the mid-IR region; however, sequential scanning of IR wavelengths is slow and results in a low signal-to-noise ratio.

FTIR spectrometer systems use an interferometer, transfer optics, an enclosed or open gas cell (White cell), an IR source, and a detector. FTIR spectrometers use IR sources that emit broad-band electromagnetic radiation and operate at a color temperature from 1200° to 1500°K. The spectral content of the thermal emission depends upon the temperature of the heated source. The two most common types of IR sources are the Nernst glower and the globar. IR light is transmitted through the sample at a known optical path length (meters); the concentrations of chemicals measured are integrated over the entire length of the optical path and are reported as the product of concentration (ppm) and path length (meters). The ppm-meter concentration is then divided by the path length to obtain a path-averaged concentration (ppm).

FTIR systems that use a closed cell are called extractive spectrometers and those using an un-enclosed cell are called open-path spectrometers. Extractive FTIR spectrometers are point sampling devices that pump or flow air into a closed cell for analyses. In open-path FTIR spectrometers, the IR light beam is completely open to the atmosphere and can be hundreds of meters long; therefore, large areas can be monitored non-invasively in real-time. Both types of spectrometers use the same software to analyze target compounds in air and both require training for proper operation and data analysis.

Extractive and open-path FTIR spectrometers have been used for a variety of environmental applications, to complement traditional point sampling methods, including stack sampling for HAPs to comply with emission standards, process modifications to reduce mass emission rates, and fence line monitoring.[8–18] Extractive FTIR spectroscopy is of interest as a test method for environmental compliance monitoring because of its speed and effectiveness in quantitative analyses of gas mixtures. It has the ability to reduce costs and improve data quality. For example, extractive FTIR spectrometers have been used to measure multi-component mixtures at a coal fired burner and volatile organic emissions from a solvent recovery unit at a manufacturer of magnetic data storage tapes.[12, 19] The U.S. Environmental Protection Agency (U.S. EPA) has proposed test methods for measuring vapor phase organic and inorganic emissions using extractive FTIR spectroscopy. [20, 21]

Open-path FTIR spectrometers have been used along the perimeters of industrial facilities and on a centrally located platform to scan retroreflectors placed along the edges of the facility to monitor fugitive emissions.[22–24] They have also been employed for ambient air monitoring at Superfund site.[25] Several methods using open-path FTIR spectrometers have been developed and applied in the past to estimate emission rates from fugitive sources, such as landfills, coal mines, and water treatment plants.[10, 19, 21, 26–29] Recently, open-path FTIR spectrometers have been used to measure emission rates of greenhouse gases from a wastewater treatment system using tracer gases.[30] A method for determining emission rates from non-homogeneous sources has been suggested that combines concentrations measured using multiple open-path FTIR spectrometers with plume dispersion modelling.[31] U.S. EPA has developed a guidance document and a compendium method for using open-path FTIR spectrometers in the field.[25, 27]

In industrial hygiene, extractive and open-path FTIR technology have been used in a variety of workplaces including the semi-conductor industry, hazardous waste sites, hospitals, dry cleaners, and in research studies.[27, 28, 32–40] The use of extractive FTIR is increasing, and there is now a National Institute for Occupational Safety and Health (NIOSH) Analytical method for using extractive FTIR for organic and inorganic gases.[41] Future applications using FTIR spectroscopy include evaluating non-industrial indoor air quality problems and community exposures. Indoor air is composed of a complex mixture of pollutants and FTIR technology could provide real-time simultaneous measurements of the contaminants at low limits of detection (ppb). When evaluating community exposures, open-path FTIR spectrometers could provide average concentrations over long path lengths (meters) in a neighborhood and these concentrations may provide a more representative picture of chemicals than a single point sample in one or several isolated locations.

In addition to using a single open-path FTIR spectrometer to measure an average concentration over a long distance, a new method is being developed which uses multiple open-path FTIR spectrometers whose beams overlap to create a network of measurements. This network of concentration measurements is combined with mathematical techniques used in computer-assisted tomography (CAT) to map concentrations of chemicals in near real-time for an entire workplace or outdoor area;[42–45] see Figure 23-2. This "environmental CAT scanning" technique creates spatially and temporally resolved maps in real-time that can be used for exposure assessment, source monitoring, and ventilation evaluation. Each map represents a snapshot in time; when the maps are linked together a video is generated of contaminant movement and concentration.

In the industrial hygiene field, this method would monitor multiple locations throughout a room simultaneously. The industrial hygienist does not have to first determine the few important hot spots in order to place point sampling devices in a room. An environmental CAT scanning system may enable real-time evaluations of short-term or chronic peak exposures, at any location in a measurement space, and provide a tool to determine ventilation efficiency and pollutant transport.[45] The concentration maps generated with this CAT system provide a powerful way to track the rapid movement of chemicals in the room (see Figure 23-3). When maps are combined with ventilation measurements, tracer gas releases, information on process changes and workload, or information on the movement and location of workers in a room, these data can be used to quantify

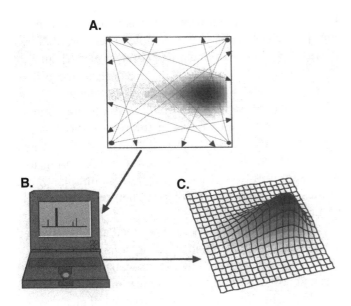

FIGURE 23-2. Sequence of events for creating concentration maps using the environmental CAT scanning system. A. Four open-path FTIR spectrometers scanning an area. B. Computer using a tomographic algorithm to reconstruct concentration maps from open-path measurements. C. Reconstructed concentration map. The peak height represents the concentration and the square grid of cells represents an area. *(Reprinted with the permission of The British Occupational Hygiene Society.)*

chemical emission rates, evaluate the effectiveness of ventilation systems, and track human exposures.

A new mobile instrument is being developed for industrial and non-industrial indoor air applications which uses features from both extractive and open-path FTIR spectrometers. This instrument has a long, folded, open-path (up to 120 meters) which can be enclosed for calibration.[46] While the instrument is designed with an extended open-path, the folded geometry is housed within a relatively small unenclosed sample volume of 75 liters; see Figure 23-4. The folded optical path passively measures concentration; therefore, pumps are not required to obtain a sample, and trace contaminants are not lost due to adsorption on sampling lines or bags, during transportation, or during analysis. The open cell is approximately three feet high, a height that can measure concentrations in the vicinity of the breathing zone. This spectrometer is particularly unique because it will potentially quantify mixtures, in real-time, in the vicinity of the breathing zone, at very low limits of quantification (ppb). Therefore, patterns of exposure to indoor air contaminants could be evaluated.

This chapter describes instrument design, theory, and future directions of using FTIR spectroscopy for environmental and occupational air sampling.

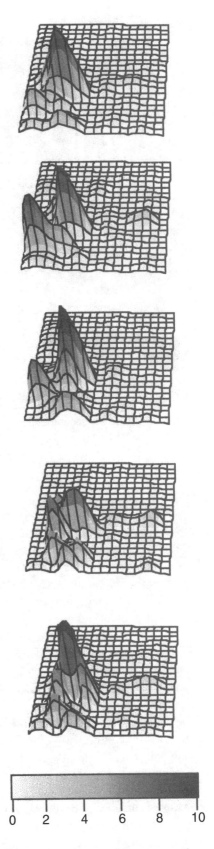

FIGURE 23-3. A series of maps of five consecutive maps created every seven minutes in an exposure chamber. Each map represents a 12 by 14 foot chamber and the height of the peaks represents concentration in ppm. *(Reprinted with the permission of The British Occupational Hygiene Society.)*

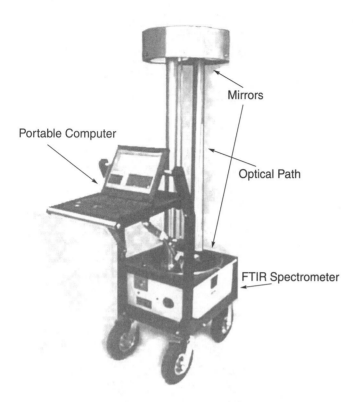

FIGURE 23-4. Schematic of the new generation, mobile, localized open-path spectrometer.

Instrument Design

Extractive FTIR Spectrometers

An extractive FTIR spectrometer system uses an interferometer, transfer optics, a closed gas cell, a detector, and an IR source (see Figure 23-5). The overall setup is similar to traditional portable IR spectrometers. Using extractive FTIR spectrometers, broad-band infrared radiation is directed into an interferometer that modulates the beam and sends it into one or more enclosed sample cells. The gas cells may be made of glass or stainless steel and can be single-pass cells or can incorporate mirrors to create multiple reflections of IR light through the sample to increase the optical path length (multi-pass cell). The folded IR light beam can vary in total length from centimeters up to several kilometers. Very long path lengths have been used for studying trace pollutants in the ambient air.[47] For most industrial hygiene and environmental applications, 10 or 20 meter path length cells are used to measure < 1 ppm concentrations.[32] The choice of the path length will depend upon the required limits of detection for the application. Some extractive systems have several fixed path length cells in one instrument. After interacting with the sample in the cell, the IR light exists and illuminates a detector. Accurate determination of the effec-

tive optical absorption path length is vital for determining the concentration of the sample. Therefore, before, during and after field operation of the system, quantifying standards, such as ethylene, are used in the cell at known concentrations to calculate the pathlength.

Extractive spectrometer systems usually have methods to control and/or measure the temperature of the sampling probe, gas delivery system, and closed gas cell. Stack sampling or other process monitoring use long sampling probes and heated sampling lines; probes are equipped with filters at their inlets to remove particulates.

U.S. EPA and NIOSH have outlined an extensive set of quality assurance and quality control procedures when using extractive FTIR spectrometers for pre-, during, and post sampling as well as quantitative analysis.[20, 41] When using an extractive spectrometer, a clean background is obtained in the field by first evacuating the cell to ≤ 5 mm Hg and then filling it with high purity nitrogen to ambient pressure, or by purging the cell with ten volumes of nitrogen. After obtaining the background, additional spectra can be obtained of known or suspected interferences and a calibration transfer standard. Analytical interferences could include water, carbon dioxide, and other analytes with spectral features that overlap with the compounds of interest. Calibration transfer standards refer to gases that can be used to verify the system's path length and performance. In the field, the cell can be filled with certified sample gases as described above for the background.

Extractive systems can be used for point sampling at a variety of locations or continuous environmental monitoring. Extractive systems can range in size from mobile models placed on a cart to large fixed-laboratory sized models.

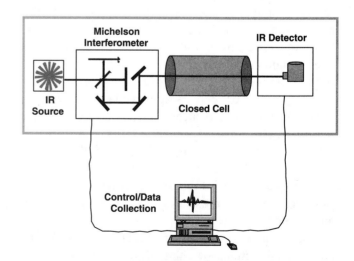

FIGURE 23-5. A schematic of an extractive FTIR spectrometer.

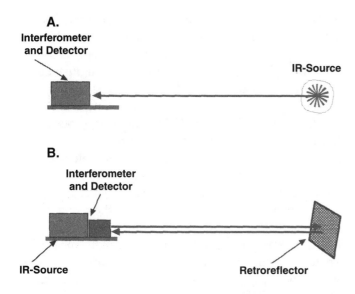

FIGURE 23-6.   A. Schematic of a bistatic open-path FTIR spectrometer.   B.   Schematic of a monostatic open-path FTIR spectrometer.

Open-path FTIR Spectrometers (Remote Sensing)

In the open-path remote sensing design, the closed gas cell is removed and the IR light is transmitted across a long open-path. The open-path FTIR spectrometer system can be operated in two different configurations: bistatic and monostatic; see Figure 23-6 (A and B). These configurations differ in the transfer optics and the number of times that the IR beam passes through the air. In the bistatic mode, the IR source is placed at the opposite end of the optical path from the detector (can be greater than 500 meters apart), and the IR beam passes through the air once. In the monostatic mode, the IR source and detector are located together at one end of the path, and the beam is sent out to a reflector placed at the other end of the path. In the monostatic mode, the reflector (flat mirror, corner cube or cat's eye retroreflector) sends the light directly back to the detector, and the IR light passes through the air twice. Telescopes can be used to collimate, send, and receive the IR beam to and from the interferometer to the retroreflector (monostatic systems) or to receive the IR light at the detector (bistatic systems). Telescopes may not be necessary when short path lengths are used for indoor workplace or exposure chamber applications.

In the bistatic mode, the view of the telescope is larger than the IR source; therefore, energy is recorded from surrounding areas as well as the IR source. Thermal radiation coming from the environment (usually ~ 300 °K) as well as the energy from the transmitted IR beam (1500 °K) passes through the interferometer to the detector. Given that the background terrestrial

energy could represent up to 20% of the energy of the source, the background energy needs to be measured and subtracted from the field spectra before quantitative analysis. Thermal background energy is measured in the field by taking a black body spectrum; the IR light source is turned off (or blocked) and a measurement is obtained with the open-path FTIR spectrometer. If there are large swings in temperature during the day, multiple black body spectrums must be obtained.

Monostatic systems do not require a black body spectrum to be obtained because they modulate the light; however, these systems may generate internal scattered light within the instrument which must be subtracted from the spectra before quantification. This scattered, or stray light, is usually small and is not path length dependent. Stray light can be obtained by either turning the spectrometer away from the retroreflector or by placing an opaque non-reflective material over the front of the telescope and acquiring spectra. Stray light must be measured at the beginning of the operation and any time that an optical component changes.

One challenge facing users of open-path FTIR spectrometers is the acquisition of a clean background spectrum, which does not contain any target analyte(s), and is ideally taken under the same experimental conditions as those taken for the sample spectrum.[25, 26] Using an extractive FTIR spectrometer, the beam is enclosed in a gas cell; the cell can be purged with clean air or nitrogen to obtain contaminant-free background data and then filled with known concentrations of contaminants to determine accuracy and verify the path length. In contrast, the optical path of an open-path FTIR spectrometer cannot be evacuated to obtain background readings, nor can it be filled with known concentrations of chemicals.

In most industrial and environmental applications, it is difficult to find air that is free of target analytes. Even if a clean air background can be obtained at the start of the day, instrument fluctuations and changing environmental conditions over the course of the day (varying partial pressure of water vapor), may require obtaining additional backgrounds for greater accuracy. If possible, indoors or outdoors, backgrounds are taken when the target analyte concentration drops to zero. In the workplace, it may be impossible to obtain a clean background when processes occur 24 hours a day. Outdoors, it may not be feasible to obtain a clean background when measuring chemicals that are always present in ambient air. Although it is recognized that multiple backgrounds should be obtained during a day or sampling period, there is little guidance available as to the frequency for obtaining backgrounds.

Three main methods are used to obtain backgrounds outdoors when contaminants are consistently generated by the source: upwind, cross-wind, and synthetic. Upwind backgrounds are taken perpendicular to the wind field at least two times a day; this method can be labor intensive, especially for a large site. When using these backgrounds, it is assumed that the air is free of target analytes or that analytes present in the air are not originating at the site being monitored. Cross-wind backgrounds are taken with the optical path along the side of the site and with the wind parallel to the path. If the wind speed is low, contaminants may diffuse from the site into the path. Synthetic backgrounds are generated by taking field spectra and subtracting all the spectral features in the region of quantification. Software is then used to generate a best-fit curve for the region using a high-order polynomial. It can be difficult to select points to generate a baseline when large wave number regions are used in the analyses or when there is a curvature of the baseline.

Validation of the path length and instrument operation is difficult using open-path FTIR spectrometers because there is no way to introduce a chemical of known concentration into the open-path. This is sometimes overcome by placing a short closed cell (centimeters) in the beam within the instrument, or a longer cell (meter) in front of the instrument in the beam path.[48] Measurements are obtained through the cell filled with several known concentrations of a gas, at the same range of ppm-meter concentrations that will be measured in the field.

Open-path FTIR technology is attractive because these instruments can simultaneously measure a wide variety of compounds, non-invasively, in real-time and over large areas that would normally require the use of large numbers of point sampling devices. With open-path FTIR spectrometers, there is no contact of the instrument with the sample, and there are no sampling lines, bags, or canisters, which can adsorb compounds and result in unpredictable contaminant losses. In addition, samples do not require shipment to a laboratory for analysis.

Open-path spectrometers can be used to monitor large distances, inaccessible areas, hazardous waste sites, and outside industries for fugitive emissions (spills or leaks) of highly toxic chemicals. They can be used along with dispersion models or tracer gases to calculate emissions from open areas such as chemical and agricultural waste lagoons. In the occupational arena, open-path FTIR spectrometers have the potential to provide the temporal resolution and sensitivity required for measuring short-term exposures to acutely toxic chemicals and measuring chemicals in complex mixtures, over large areas. In the community, long path lengths can be used through neighborhoods, to measure fugitive emissions of chemicals from industrial sites or from operations such as agricultural pesticide spraying.

A number of outdoor field studies have been performed to evaluate the accuracy and precision of open-path FTIR spectrometers; chemicals were released into the atmosphere, and open-path FTIR spectrometer measurements were compared to point sample measurements taken along the path.[49–52] In these studies, qualitative and quantitative results varied both by the open-path FTIR system used and by the contaminant measured. There are inherent limitations to performing outdoor studies because meteorological conditions are unstable and uncontrollable. Changing wind speed and direction can alter the path of the plume and the released gas may intersect the IR beam where there are no point samples, generating discrepancies unrelated to measurement accuracy. A limited number of indoor open-path FTIR spectrometer studies have been successfully performed using a controlled room-size ventilation chamber, an outdoor exposure chamber, and a calibration cell.[36, 38, 47] In indoor and outdoor chamber studies, the open-path FTIR measurements agreed with point samples within 15% and 36%, respectively. In the calibration cell studies, there was no significant difference found between the concentrations generated in the calibration cell and open-path FTIR measured concentrations. In the chamber studies, accuracy was related to the spectral library used for quantification, the method of quantification (classical least squares, integration, or subtraction), and the clean background used in the analyses.

Mobile, Localized Open-Path Infrared Spectrometer

A new generation, mobile open-path FTIR spectrometer is being developed for the industrial hygiene field to simultaneously monitor multiple contaminants in air, in real-time, at very low limits of detection (ppb). This instrument was designed at 3M Corporation for industrial and non-industrial indoor air applications and has been constructed by MIDAC Corporation (Irvine, CA).[46] This mobile instrument could fill an important gap in methods available to industrial hygienists and researchers for understanding and evaluating health-related problems due to indoor air pollution, measuring mixtures of known and unknown chemicals in workplace air at ppb concentrations, and improving identification of pollutant sources. The system would be wheeled around a workplace to obtain area measurements in the vicinity of a worker's breathing zone.

The prototype instrument uses technology that is a cross between extractive and open-path FTIR spectroscopy. The open-path design is achieved through the use of a folded optical path created using mirrors to reflect the light. This open-path cell is capable of achieving a large absorption path length up to 120 meters. The long path length allows for very low limits of detection, but because of the folded geometry, there is a relatively small unenclosed sample volume of 75 liters. The folded optical path is open to the atmosphere and passively measures concentration; therefore, pumps are not required to obtain a sample, and trace contaminants are not lost due to adsorption on sampling lines or bags, during transportation, or during analysis. The open cell is located at a height that can measure concentrations in the vicinity of the breathing zone.

The extractive feature is achieved through the use of removable covers that completely enclose and seal the sample volume; this enables the introduction of calibration standard gases immediately before the analysis of sample gases. This clearly distinguishes this instrument from standard open-path FTIR spectrometers in which the sample volume is orders of magnitude larger, for a similar path length, and cannot be isolated for calibration. The entire unit is housed on a mobile cart with a portable computer and battery power; therefore, it can easily be transported through a room.

Preliminary tests were performed to evaluate this instrument using a 40-meter path length: nitrous oxide, trichloroethylene, and methyl ethyl ketone, were generated in a chamber and were accurately measured at 0.1–1.0 ppm.[53] These concentrations are orders of magnitude lower than current occupational exposure limits. In the field, this instrument has been evaluated against analytical methods approved by NIOSH in hospital recovery rooms, dental and histology laboratories, and a medical examiner's office. More extensive laboratory and field tests, at longer path lengths, must be performed to fully evaluate this instrument for measuring mixtures of chemicals in air, in real-time.

Instrument Hardware

Interferometer

At the heart of the FTIR spectrometer is the interferometer, which enables all wavelengths of IR light to be measured simultaneously. One type of interferometer

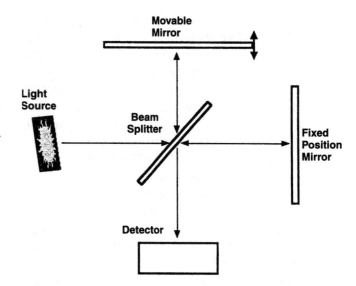

FIGURE 23-7. Schematic of a Michelson interferometer.

is called the Michelson interferometer, which was introduced by A. A. Michelson in 1881; see Figure 23-7 for a schematic of an interferometer. The interferometer consists of a beam splitter and a stationary and moveable mirror which are orthogonal to each other. A Nernst glower or globar is used to generate IR light that travels to the beam splitter. The beam splitter divides the light into two beams of equal intensity; one-half of the light is sent to the moveable mirror and one-half is sent to the fixed mirror. After hitting the mirrors, the beams recombine at the beam splitter and exit to the detector. When the moving mirror and the fixed mirror are at the same distance from the beamsplitter, the IR light beams reflected off the mirrors are in phase with each other, and all the wavelengths of light constructively interfere in the recombined beam. This results in the highest intensity of light for all possible mirror positions; this mirror position is called the zero path difference (ZPD). When the moving mirror moves along the optic axis, the half of the optical beam that hits the moving mirror travels a different (shorter) distance than the optical beam that hits the fixed mirror. Therefore, when the light beams recombine, the light beams may be out of phase and destructively interfere; this results in decreased light intensity. At any mirror position, the intensity of light is governed by constructive and destructive interferences in the recombined beam.

When the light recombines, it hits the detector and a signal is recorded; the pattern of constructive and destructive interference is called the interferogram, and it depends upon the wavelengths of light present in the beam. The interferogram is a plot of light intensity versus optical path difference (mirror position); see

Figure 23-8. Every data point in the interferogram contains intensity information about every infrared wavelength transmitted from the source to the detector. The zero position on the x-axis, the ZPD, is quite large and is called a centerburst. Using an IR broad-band radiation source, the incident IR radiation contains many wavelengths; therefore, at each position of the moving mirror, the signal recorded by the detector is an integral of the intensities of all the wavelengths of light. The interferogram is used in all subsequent analyses.

FTIR spectrometers have greater sensitivity (lower limits of detection) than their dispersive counterparts. Detector sensitivity increases as the intensity of the IR light reaching the detector increases. With FTIR spectrometers there are no slits to restrict the wavenumber range and reduce the intensity of IR radiation that strikes the detector. In an FTIR instrument, all wavelengths of the light reach the detector at the same time, as compared to a dispersive instrument where only a few wavelengths reach a detector at any one time.

Important measures of data quality have to do with the shapes of the various stages of the spectra. Interferograms should always have the same basic shape, with an intense peak at the center, and two sidearms. The intensity of the centerburst should not change drastically over time. A drop in intensity could indicate a misalignment of the optics or clouding of optical windows or the beamsplitter. If there is too much IR radiation striking the detector, the detector could be saturated. The IR intensity reaching the detector can be decreased by placing commercially available fiberglass or metal mesh screens between the IR source and the interferometer. The use of screens to eliminate non-linearity results in a decrease in intensity, which can result in a decreased signal-to-noise ratio and increased minimum detection limits.

Infrared Detectors

After the IR light passes through the air sample and interferometer, the optical beam is focused onto a detector filament. The sensitivity of the FTIR spectrometer depends on the type of detector. There are three main detectors used for IR spectroscopy: mercury, cadmium, and telluride (MCT); deuterated triglycerine sulfite (DTGS); and indium antinomide (InSb). Most FTIR spectrometers use the MCT detector; when cryogenically cooled, it is very sensitive to IR radiation. MCT and DTGS detectors usually respond in the 500 to 5000 cm^{-1} regions; MCT detectors are up to ten times more sensitive than DTGS detectors. DTGS detectors are sensitive to mechanical and acoustical vibration and require special mounting techniques. InSb detectors respond above 1800 cm^{-1} and are used to detect compounds such as hydrogen fluoride and hydrogen chloride. MCT detectors saturate much more easily than DTGS detectors. For an MCT detector, below 700 cm^{-1}, the single beam spectrum should be flat; if it is above the baseline, the detector could be saturated.

For proper operation, an MCT or InSb detector must be cooled by placing it in a Dewar which is filled with liquid nitrogen or by using a Stirling engine cooler. Liquid nitrogen usually needs to be replenished every 7 hours, whereas the Stirling engine cooler usually can work for at least 6 months; however, the Stirling engine cooler is considerably more expensive.

In addition to knowing the spectral response regions, detectors are defined by their noise equivalent power (NEP) and sensitivity ($D*$). The NEP is a measure of the inherent noise in the detector; see Equation 1. $D*$ is a ratio of the square root of the area of the detector to the NEP; see Equations 2–4. The operator can do little about these numbers once the detector is purchased. Detectors used in FTIR spectrometers usually have NEP values of about 5×10^{-2}; smaller numbers are better than larger numbers.

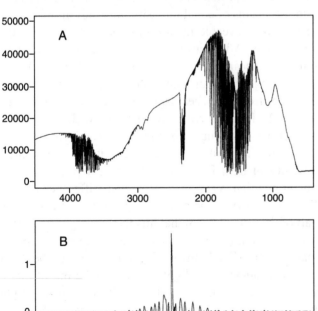

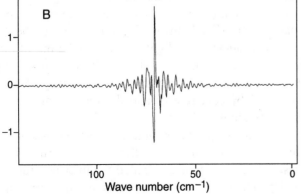

FIGURE 23-8. A. Single beam spectrum. B. An interferogram.

$$NEP = W/(Hz^{1/2}) \qquad \qquad \textbf{(1)}$$

$$D^* = cm \times (Hz^{1/2})/W \qquad \qquad \textbf{(2)}$$

$$cm = (\text{area of detector})1/2 \qquad \qquad \textbf{(3)}$$

$$D^* = (\text{area of detector})1/2 \, / \, (NEP), \qquad \textbf{(4)}$$

where: NEP = noise equivalent power;
   W = watts;
   Hz = Hertz;
   D^* = detector sensitivity.

The most commonly used MCT detectors are photo-conductive detectors; when the light hits the detector, IR photons are absorbed, and the electrons in the detector filament move from the valence band to the conduction band. A voltage is applied across the detector filament, and the electrons in the conduction band generate an electrical current. This current is proportional to the number of photons striking the filament (the intensity of the IR light). The signal produced by the detector is then digitized using an analog-to-digital converter (ADC). The dynamic range of the ADC is the difference between the highest and lowest signals it can handle; the lowest signal is reserved for recording system noise. For a 20-bit ADC, the highest signal that can be recorded is 2^{20} above noise.

In FTIR spectrometry, the size of the centerburst in the interferogram is the signal strength that determines how much of the ADC dynamic range is used. If the centerburst is too large, the interferogram is cut off or clipped; this results in inaccurate quantification. To adjust the size of the interferogram centerburst, hardware and software gain settings are adjusted by the user.

Spectral Resolution

The resolution of an FTIR spectrometer refers to the minimum separation that two absorption features can have and still be distinguished from one another. One criterion used for determining resolution is called the Raleigh criterion; this states that two features are resolved when the maximum intensity of one feature falls at the minimum intensity of the other feature. Another criteria which is commonly used states that the minimum separation in wave numbers (cm^{-1}) of two spectral features is the reciprocal of the maximum optical path difference in centimeters (the difference in the distance the two beams travel) of the two mirrors in the Michelson interferometer. Therefore, a 0.5 cm^{-1} wave number resolution spectrometer means the mirror travels a maximum of 2 cm, and a 0.25 cm^{-1} wave number resolution spec-

trometer means the mirror travels a maximum of 4 cm. This translates into an increase in the scan time as the resolution increases (the number describing resolution decreases). In addition, as the resolution increases, the size of the data file generated increases as well.

The appropriate resolution to use is not always a simple choice and depends upon the required signal-to-noise ratio (SNR), required scan time, and the spectral characteristics (band widths of the absorption lines in the spectra) of the target compound(s) and interfering species. The quantitative relationship between the SNR ratio, resolution, and measurement time is referred to as "trading rules."[8] In general, the SNR is directly proportional to the resolution and to the square root of the measurement time. For example, going from a 1.0 cm^{-1} to 0.5 cm^{-1} wave number resolution (doubling the resolution) would double the noise. Therefore, to obtain the same SNR when using a higher resolution, the measurement time would need to be quadrupled. Thus, the operator must balance an SNR with adequate sampling time.

In general, measurements should be made at the lowest possible resolution that adequately resolves spectral features of chemicals in the sample. As resolution increases, the degree of overlap between spectral features decreases; however, the SNR increases. For resolving most compounds in mixtures, resolutions of greater than 4 cm^{-1} would usually be necessary. If water vapor and CO_2 must be characterized to accurately quantify compounds in the sample, a high resolution instrument (0.125 cm^{-1}) would be required because some absorption bands are as narrow as 0.1 cm^{-1}.

Most FTIR spectrometers used for occupational or environmental field applications use resolutions of 1.0 or 0.5 cm^{-1}; most NDIR field instruments have much lower resolutions of 7–13 cm^{-1}. Therefore, FTIR spectrometers are capable of resolving spectral features that in many cases would be difficult using NDIR instruments.

FTIR spectrometers usually have a better SNR than NDIR field IR spectrometers because scans can be coadded, each wave number is sampled longer, and all the wavelengths are measured simultaneously. When scans taken at constant resolution are coadded, random fluctuations in the signal cancel each other out, which decreases the SNR. Because the SNR is proportional to the square root of the time spent measuring a wave number, the SNR is higher using FTIR instruments because they spend much more time observing each wavelength during each scan than NDIR instruments. Finally, FTIR instruments measure all wavelengths simultaneously; therefore, noise is shared among all of the wavelengths. In contrast, in conventional NDIR instruments, only one wavelength is sampled at a time; to monitor multiple wavelengths, NDIR instruments must sequentially scan each wavelength. Therefore, for a specific measurement time period, with FTIR spectrometers all wave numbers are measured for the full time period, whereas for dispersive spectrometers, each wavenumber is measured for only a small fraction of the time period.[4]

Data Analysis

Identification and quantification of pollutants is performed on absorption spectra using mathematical and statistical techniques based on Beer's Law. Beer's Law states that at a constant path length, the intensity of the IR energy diminishes exponentially with the concentration of chemicals in the path; see Equation 5. To measure concentration, the intensity of the IR light with contaminants (I) and without contaminants (I_o) must be obtained and converted to an absorption spectra; see Equations 5–8. Absorption (Equation 8) is proportional to the concentration of the chemical and the distance (path length) that the light travels through the chemical. Therefore, the longer the path length that the IR light travels through the air containing the compound of interest, before it reaches the detector, the greater the sensitivity and the lower the limits of detection.

Deviation from Beer's Law can be caused by spectroscopic factors at high absorbances, such as > 0.7 – 1.0, where very little light is transmitted through the sample, and at very low absorbances, where the signal from a small sample peak becomes lost in the noise < 0.004.

$$I = I_o \, exp \, -(aCL) \qquad (5)$$

$$Transmission = I/I_o \qquad (6)$$

$$Absorption = - log \, (I/I_o) \qquad (7)$$

$$Absorption = aCL \qquad (8)$$

where: I = the intensity of IR light with absorbing chemical present;
I_o = is the intensity of IR light without absorbing chemical present;
a = the molecular absorption coefficient of the chemical;
C = the concentration of the chemical;
L = the path length of the IR radiation though the chemical.

As can be seen in Equation 5, to generate an absorption spectra, both a sample spectra (I) and a clean background spectra (I_o) must be obtained. For accurate quantification, the pathlength (L) must be determined. Background spectra are generated experimentally using

air that does not contain target compounds, or they are generated synthetically. In extractive systems, it is fairly straightforward to generate a clean background (see extractive spectrometers); however, it is more difficult using open-path FTIR spectrometers (see open-path spectrometers). To create an absorption spectra, first the interferogram is converted to a single beam spectrum by performing a fast Fourier transform, using appropriate software; see Figure 23-8. Basically, intensity information is recovered as a function of wavelength. Two single beam spectra (sample and background) are then ratioed to get a transmission spectra; see Equation 6. The absorption spectra is generated by taking the negative logarithm of the transmission spectra; see Equation 7. Black body and stray light measurements are subtracted before generating the absorption spectra.

The path length (L) of a cell is determined using a calibration standard (CTS) gas. Usually, ethylene in nitrogen is recommended with 2% accuracy; however, almost any stable compound can be used. One or more spectra are obtained of a CTS gas at the gas temperature and pressure of the reference library spectrum. The path length is then determined by quantifying the CTS gas and determining the path length of the instrument based upon the known concentration-path length of the CTS gas.

Absorption spectra are used for identification and quantification; see Figure 23-1. Data analysis is performed by selecting a wave number region in the absorption spectrum of the gas that ideally is free from interferences from other contaminants. The wave number region is selected by examining the appropriate reference spectrum for the contaminant of interest. The wave number region selected is usually the region covered by the entire peak and is outside a strong absorbance band of water and carbon dioxide. Wave number regions can be rejected whose relative absorbance is less than 1% of the peak.

Quantitative analysis can be performed using a variety of methods including peak height, peak area, spectral subtraction, partial least squares, iterative least squares and classical least squares (CLS). CLS is most commonly used to match the measured spectra to a library of calibration reference spectra for the pollutants of interest.[54, 55] CLS analysis has resulted in improved precision and accuracy in multi-component spectral analysis. This has been successful even in those cases where there is strong overlap of infrared spectral features.

The reference spectra are obtained under carefully controlled conditions of temperature and pressure for known concentrations of pure gases (NIST-traceable standards 2% accuracy). Ideally, the reference library can be prepared on the field instrument; however, this is usually impractical. Usually a short cell, with a known path length, is filled with a high concentration of pure gas; the spectrum is then taken at the appropriate wave number resolution. Once a reference spectrum is obtained, it usually can be used for all subsequent analyses. Reference spectra can be created by the user and are available commercially from manufacturers for a range of chemicals, from the U.S. EPA for the HAPs, and from the National Institute of Standards and Technology (NIST).[56] Reference spectra can be developed over a range of concentrations to establish linearity for the concentrations encountered in the specific application. At this time, there are no universally agreed upon procedures for creating reference spectra nor is there a universally accepted library of spectra.

One of the greatest challenges in quantifying chemicals is the impact of water vapor on analysis. Water vapor is present throughout the entire mid-IR wavelength region and can overlap with spectral bands in the target analytes and can affect quantification. In addition, changes in water vapor concentration throughout the sampling period can result in a curvature of the baseline in the absorption spectra and can affect quantification. Therefore, water vapor spectra should be used as an interferent in the analyses. Water vapor spectra can be prepared from field spectra, when using an open-path FTIR spectrometer, or by generating different concentrations of water vapor in the cell of an extractive FTIR spectrometer.

Quality Assurance

The quality of FTIR spectrometer measurements can be influenced by the system (detector sensitivity, IR source output, quality of reference spectra, software package used for analyses) and by the sample (spectral interferents). A variety of procedures have been suggested for quality assurance and quality control for extractive and open-path FTIR spectrometers. Two of them are discussed here: root mean square (RMS) noise and minimum detection limit (MDL). RMS noise is an important parameter for tracking instrument performance over time and should be plotted on a quality control chart. It is a linear least squares fit of selected data points in the spectra to the baseline and is calculated using spectra that have been acquired using the same operating parameters as the field spectra. Two back-to-back single beam spectra (no time lag between them) are used to measure the RMS noise in three wave number regions ($968-1008$ cm^{-1}, $2480-2520$ cm^{-1}, $4380-4420$ cm^{-1}). Other regions that cover the target analyte of interest can be used as well.

The MDL is the minimum concentration of a compound that can be detected by an instrument with a given statistical probability. Usually the detection limit is given as three times the standard deviation of the noise. Detection limits will change depending upon interfering species and the variability of water vapor and other atmospheric conditions during sampling. One method for calculating the MDL is to obtain 16 back-to-back spectra, and then create absorption spectra from them (the first spectra is the background for the second, and the second is the background for the third, etc.). These absorption spectra are analyzed in exactly the same way as the field spectra and should result in a set of numbers that are close to zero. Three times the standard deviation of this calculated set of concentrations is considered the MDL.

In addition to RMS noise and the MDL, some of the other parameters that should be evaluated include: measurement of return beam intensity; precision; accuracy, resolution; and nonlinear instrument response.

Future Directions

Environmental CAT Scanning: Mapping Chemicals in Real-Time

A new method, called "environmental computer-assisted tomography (CAT) scanning" is being developed for measuring and visualizing the concentration of multiple chemicals in air; it combines the chemical detection technology of open-path FTIR spectroscopy, with the mapping capabilities of computed tomography. [40, 42–44, 57–62] Computed tomography has been used extensively in fields as diverse as radio astronomy, electron microscopy, and holography; however, its greatest achievement and progress has been in medicine where it is used in medical CAT scans to reconstruct organs in the body.[63–66] For the application of CAT scanning to the occupational and environmental field, maps of chemical plumes are reconstructed from multiple FTIR absorption measurements through the section of the air of interest.

While a single open-path spectrometer can detect low concentrations of multiple chemicals simultaneously, in near real-time, it provides only an average concentration across the space that the open-path beam traverses. Thus, it cannot be used to pinpoint the location of a peak; see Figure 23-6. When multiple, intersecting open-path measurements are obtained for a given plane, tomographic reconstruction algorithms can be used to create a two-dimensional map of the near real-time concentration profile; see Figure 23-2. In theory, one or more scanning open-path spectrometers can be placed in an area, along with optical sources, mirrors, and/or retroreflectors, to create a network of open-path infrared beams. Maps generated from the path averaged measurements show the concentration as well as the location of contaminant plumes. These spatially resolved maps can be obtained using far fewer measurements than would be required using traditional point sampling measurements.

Each map provides a snapshot, over a short period of time (minutes), of the concentration and location of multiple chemicals in air. As measurements are obtained over the course of a day, the reconstructed concentration maps are linked together to visualize the flow of contaminants over space and time. Figure 23-3 shows an example of maps reconstructing a chemical at five different time periods.

Concentration maps can be used to improve our understanding of contaminant dispersion, monitor waste emissions, measure emission rates, develop and validate better contaminant transport models, evaluate ventilation in a workplace, and estimate human exposure to pollutants. This method is a major departure from currently used methods that use point samples placed at specific locations to monitor a limited number of chemicals in air.

The degree of spatial and temporal resolution that can be obtained with this method depends upon the number and spatial placement of the open-path spectrometer(s), retroreflectors, and mirrors in an area; the open-path spectrometer scan time, the time required to completely sample an area; and the reconstruction algorithm.

Research on this method has included theoretical studies evaluating tomographic algorithms and open-path FTIR spectrometer numbers and orientations, laboratory chamber studies, and a limited number of field tests. Smaller and more rugged open-path FTIR spectrometers, cheaper retroreflectors, and new methods for obtaining background spectra will aid implementation of this system.

To evaluate this system outdoors, a small scale environmental CAT scanning system was tested outdoors using a simulated volume source that released trace gases at known emission rates.[67, 68] The concentration maps generated with the system were compared with plumes generated using a plume dispersion model. For many of the reconstructed time periods, the plume shapes, directions and locations reconstructed in the tomographic maps compared fairly well with the model predictions. However, the model generated maps underpredicted the tomographic concentration maps for almost all of the time periods. A large-scale study to measure ammonia emissions was performed with the CAT scanning system over a 6-acre swine waste lagoon at an intensive swine confinement facility.[69] Two open-path FTIR spectrometers and twelve retroreflectors were positioned around a lagoon and the system scanned the entire lagoon every 2 minutes (see Figure 23-9). Tracer gases were combined with the tomographic maps to determine emissions. With this system, the entire surface of the lagoon was monitored in real-time with good spatial resolution. Figure 23-10 shows pictures of reconstructed maps of ammonia for three different time periods.

Conclusions

Extractive and open-path FTIR spectrometers are an important addition to available field instrumentation for both occupational and environmental applications. This technology has the potential to identify and quantify known and unknown chemicals in mixtures, in near real-time, at low limits of detection (ppb). In the environmental arena, extractive FTIR spectrometers are being used by some industries to comply with MACT standards and perform process modifications; U.S. EPA has proposed test methods using this technology. Open-path FTIR spectrometers are used less frequently than extractive FTIR spectrometers; however, the ability to non-invasively measure compounds over long distances is important for fenceline, community, hazardous waste site, and chemical and agricultural waste lagoon sampling. It would be difficult to perform the same type of sampling using point sampling devices.

The occupational arena has been slower at adopting this technology than the environmental arena. However, with the addition of a NIOSH method, this technology should gain wider acceptance and use. It fills an important gap and provides the flexibility, sensitivity (low limits of detection), and specificity for real-time field measurements of multi-component air samples. Most workers are exposed to multiple chemicals in air; the industrial hygienist usually must select a few chemicals to sample, from a long list, to minimize sampling time and expense. FTIR technology enables an industrial hygienist to simultaneously monitor, in near-real time,

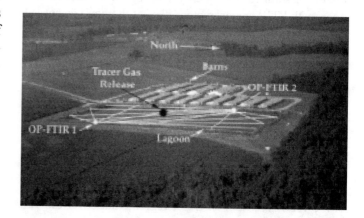

FIGURE 23-9. Aerial photo of the swine waste lagoon, showing the location of the OP-FTIR spectrometers, configuration of the optical rays, and the release point of the tracer gas.

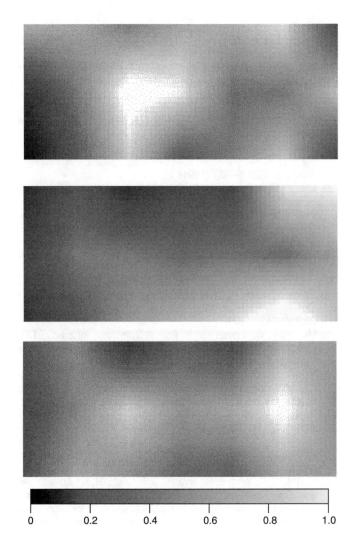

FIGURE 23-10. Three two-minute maps of ammonia concentration in ppm over the swine waste lagoon.

many more chemicals than currently feasible. In addition, having a hard copy record of absorbances over the entire IR spectra would allow the industrial hygienist to evaluate exposures to chemicals that they might not have previously selected as part of their analyses. Occupational exposure limits usually decrease in magnitude rather than increase over time; instruments are needed that can measure low concentrations.

The number of workers who are affected in non-industrial environments by adverse health effects due to indoor air exposures has increased dramatically over the years. Multiple pollutants are usually present at very low concentrations in air, orders of magnitude below typically documented adverse health effects and occupational exposure limits. Instrumentation that could measure low ppm concentrations of multiple chemicals in real-time is important for understanding and evaluating health-related problems due to indoor air pollution.

Using this technology to spatially and temporally map concentrations in indoor and outdoor air would provide visualization of contaminant flow that is unavailable at this time. This visualization will provide us with a tool for understanding contaminant generation and dispersion, determining chemical emission rates, developing effective controls, and evaluating human exposure.

The advantages of this technology are balanced by the fact that effective use of FTIR spectroscopy requires significant training. These instruments are not black boxes, and operators must understand IR spectroscopy, especially when unknown environments are being evaluated. In addition, there are still unresolved issues, which need to be fully addressed, related to quality of the reference spectra and instrument quality assurance and quality control, especially in regards to open-path FTIR spectroscopy. While the instruments are not inexpensive, the type of data obtained can provide information that may be invaluable for improving our environment and our workplaces.

Acknowledgments

The author acknowledges Dr. Kathleen Mottus for her technical assistance on this chapter and the Presidential Faculty Fellow Award of the National Science Foundation (94-53433) for supporting this work.

References

1. Clean Air Act Amendments. Pub. L. No. 101-549, 104 Stat. 2399 codified at 42 U.S.C. 7400 *et seq.* (1990).
2. Berglund, B.; Brunekreef, B.; Knoppel, H.; *et al.*: Effects of Indoor Air Pollution on Human Health. Indoor Air 2:2–25. (1992).
3. Silverstein, R. M.; Bassler, G. C.; Morrill, T. C.: Spectrometric Identification of Organic Compounds, 5th ed. John Wiley and Sons, Inc., New York (1991).
4. Smith, B. C.: Fundamentals of FTIR Spectroscopy. CRC Press, Boca Raton, FL (1996).
5. Horlick, G.: Introduction to Fourier Transform Spectroscopy. Appl. Spectroscopy 22:617–634. (1968).
6. Bell, R. J.: Introductory Fourier Transform Spectroscopy. Academic Press, New York (1972).
7. Griffiths, P. R.; de Haseth, J. A.: Fourier Transform Infrared Spectrometry. John Wiley and Sons, Inc., New York (1986).
8. Spellicy, R. L.; Crow, W. L.; Draves, J. A.; *et al.*: Spectroscopic Remote Sensing-addressing Requirements of the Clean Air Act. Spectroscopy 8:24–34. (1991).
9. Grant, W. B.; Kagan, R. H.; McClenny, W. A.: Optical Remote Measurement of Toxic Gases. J. Air Waste Manage. Assoc. 42:18–30 (1992).
10. Reagen, W. K.; DePuydt, M. M.; Wright, B. D.: Comprehensive VOC Source Emissions Assessment. A Combined Approach of EPA Method TO-14, EPA Method TO-11, Extractive FTIR. Presentation at the 88th Annual Meeting & Exhibition of the Air & Waste Management Association. 95-TA32.02 (1995).
11. Tuazon, E. C.; Winer, A. M.; Pitts, J. R.: Trace Pollutant Concentrations in a Multiday Smog Episode in the California

South Coast Air Basin by Long Path Length Fourier Transform Infrared Spectroscopy. Environ. Sci. Technol. 15:1232–1237. (1981).

12. —Field Validation Test Using Fourier Transform Infrared (FTIR) Spectrometry to Measure Formaldehyde, Phenol and Methanol at a Wool Fiberglass Facility. Draft Report. U.S. Environmental Protection Agency, Entropy Environmentalists, Inc. EPA Contract No. 68D20163, Research Triangle Park, NC.(September 1994).

13. —Fourier Transform Infrared (FTIR) Method Validation at a Coal-fired Boiler. U.S. Environmental Protections Agency Entropy Environmentalists, Inc., Document 454R95004, EPA Contract 68D20163. Research Triangle Park, NC (July 1993).

14. Simonds, M.; Xiao, H.; Levine, S. P.: Optical Remote Sensing for Air Pollutants–Review. Am. Ind. Hyg. Assoc. J. 55:953–965 (1994).

15. Eldridge, J. S.; Stock, J. W.; Reagen, W. K.; Osborne, J. M.: Extractive FTIR: Manufacturing Process Optimization Study. Presentation at the 88th Annual Meeting & Exhibition of the Air & Waste Management Association. 95-TA32.06 (1995).

16. Hall, M. J.; Lucas, M.; Koshl, C. P.: Measuring Chlorinated Hydrocarbons in Combustion by Use of Fourier Transform Infrared Spectroscopy. Environ. Sci. Technol. 25:250–267 (1991).

17. Herget, W. W.; Brasher, J. D.: Remote Measurement of Gaseous Pollutant Concentrations Using a Mobile Fourier Transform Interferometer System. Appl. Optics 18: 3404–3420 (1979).

18. Herget, W. F.: Analysis of Gaseous Air Pollutants Using a Mobile FTIR System. Am. Lab. 4:72B78. (1982).

19. Reagen, W. K.; Wright, B. D.; Kreueger, D. J.; Plummer, G. M.: Fourier Transform Infrared Method at a Carbon Bed Solvent Recover Unit for Four Gaseous Hydrocarbons. Environ. Sci. Technol. 33:1752–1759 (1999).

20. U.S. Environmental Protection Agency Addendum to Test Method 320. Protocol for the use of Extractive Fourier Transform Infrared (FTIR) Spectrometry for the analyses of gaseous emissions from stationary sources. Office of Air Quality Planning and Standards (OAQPS) Emissions Measurement Center (EMC).

21. U.S. Environmental Protection Agency Proposed Test Method 320. Measurement of Vapor Phase Organic and Inorganic Emissions by Extractive Fourier Transform Infrared (FTIR) Spectroscopy. Title 40. CFR, Part 63, Appendix A.

22. Kump, R. L.; Hommrich, D. N.: Fenceline and Ambient Air Monitoring with Open Path FTIR Optical Remote Sensing. In: Proceedings of the SP-81: Optical Remote Sensing Applications to Environmental and Industrial Safety Problems, Houston, TX, pp. 269–272. Air and Waste Management Association, Pittsburgh, PA (1992).

23. Spellicy, R. L.; Draves, J. A.; Crow, W. L.; et al.: A Demonstration of Optical Remote Sensing in a Petrichemical Environment. In: Proceedings of the SP-81: Optical Remote Sensing Applications to Environmental and Industrial Safety Problems, Houston, TX, pp. 273–285. Air and Waste Management Association, Pittsburgh, PA (1992).

24. Minnich, T. R.; Scotto, R. L.; Leo, M. R.; et al.: a Practical Methodology Using Open-path FTIR Spectroscopy to Generate Gaseous Fugitive-source Emission Factors at Industrial Facilities, In: Proceedings of the SP-81: Optical Remote Sensing, Application to Environmental and Industrial Safety Problems, Houston, TX, pp. 273–285. Air and Waste Management Association, Pittsburgh, PA (1992).

25. Mickunas, D. B.; Zarus, G. M.; Turpin, R. D.; Campagna, P. R.: Remote Optical Sensing Instrument Monitoring to Demonstrate Compliance with Short-term Exposure Action Limits During Cleanup Operations at Uncontrolled Hazardous Waste Sites. J. Hazard. Mater. 43:55–65 (1995).

26. Russwurm, G. M.; Childers, J. W.: FTIR Open-Path Monitoring

Guidance Document. 2nd Ed. EPA/600/R-96/040. ManTech Environmental Technology, Inc. Research Triangle Park, NC (1996).

27. U.S. Environmental Protection Agency: Compendium of Methods for the Determination of Toxic Organic Compounds in Ambient Air. 2nd Ed. Compendium Method TO-16. Long Path Open-Path Fourier Transform Infrared Monitoring of Atmospheric Gases. ORD, USEPA. EPA/625/R-96/010B. U.S. EPA, Research Triangle Park, NC (1997).

28. Xiao, H. K.; Levine, S. P.; Herget, W. F.; et al.: A Transportable Remote Sensing Infrared Air-Monitoring System. Am. Ind. Hyg. Assoc. J. 52:449–457 (1991).

29. Levine, S. P.; Puskar, M. A.; Geraci, C.; et al.: Fourier Transform Infrared Spectroscopy Applied to Hazardous Waste: I. Preliminary Test of Material Analysis for Improvement of Personal Protection Strategies. Am. Ind. Hyg. Assoc. J. 46:181–186 (1985).

30. Eklund, B.: Comparison of Line- and Point-source Releases of Tracer Gases. Atmos. Environ. 33:1065–1071 (1999).

31. Hashmonay, R. A.; Yost, M. G.; Memane, Y.; Benayahu, Y.: Emission Rate Apportionment from Fugitive Sources Using Open-path FTIR and Mathematical Inversion. Atmos. Environ. 33:735–743. (1999).

32. Herget, F. W.; Levine, S. P.: Fourier Transform Infrared (FTIR) Spectroscopy for Monitoring Semiconductor Process Gas Emissions. App. Ind. Hyg. 1:110–112. (1986).

33. Ying, L. S.; Levine, S. P.; Strang, C. R.; Herget, W. F.: Fourier Transform Infrared (FTIR) Spectroscopy For Monitoring Airborne Gases & Vapors of Industrial Hygiene Concern. Am. Ind. Hyg. Assoc. J. 50:354–359 (1989).

34. Strang, C. R.; Levine, S. P.: The Limits of Detection for the Monitoring of Semiconductor Manufacturing Gas Vapor Emissions by Fourier Transform Infrared (FTIR) Spectroscopy. Am. Ind. Hyg. Asso. J. 50:78–84 (1989).

35. Levine, S. P.; Ying, L. S.; Strang, C. R.; Xiao, H. K.: Advantages and Disadvantages in the Use of Fourier Transform Infrared (FTIR) and Filter Infrared (FIR) Spectrometers for Monitoring Airborne Gases and Vapors of Industrial Hygiene Concern. Appl. Ind. Hyg. 4:180–187 (1989).

36. Todd, L. A.: Direct-Reading Instrumental Methods for Gases, Vapors, and Aerosols. In: The Occupational Environment—Its Evaluation and Control. S. DiNardi, Ed. pp. 176–208. AIHA Press, Fairfax, VA (1997).

37. Yost, M. G.; Xiao, H. K.; Spear, R. C.; Levine, S. P.: Comparative Testing of an FTIR Remote Optical Sensor with Area Samplers in a Controlled Ventilation Chamber. Am. Ind. Hyg. Assoc. J. 53: 611–616 (1992).

38. Ying, L. S.; Levine, S. P.: Fourier Transform Infrared Spectroscopy for Monitoring Airborne Solvent Vapors in Workplace Air. Am. Ind. Hyg. Assoc. J. 50:360–365 (1989).

39. Todd, L. A.: Evaluation of an Open-path Fourier Transform Infrared Spectrophotometer Using an Exposure Chamber. Appl. Occup. Environ. Hyg. 11:1327–1334 (1996).

40. Chaffin, C. T.; Marshall, T. L.; Jaakkola, P. T.; et al.: The Assessment of Indoor Air Quality Using Extractive Fourier Transform Infrared (FTIR) Measurements. In: Proceedings of the SPIE International Symposium on Optical Sensing for Environmental and Process Monitoring. pp. 140–150. Air and Waste Management Association, Pittsburgh, PA (1995).

41. Organic and Inorganic Gases by Extractive FTIR Spectrometry. NIOSH Manual of Analytical Methods (NMAM) Fourth Edition. Issue 1. Interim, May 2000. Method 3800.

42. Samanta, A.; Todd, L.: Mapping Air Contaminants Indoors using a Prototype Computed Tomography System. Ann. Occup. Hyg. 40:675–691 (1996).

43. Todd, L.; Leith, D.: Remote Sensing Computed Tomography in Industrial Hygiene. Am. Ind. Hyg. Assoc. J. 51: 224–233 (1990).

44. Todd, L. A.: Computed Tomography in Industrial Hygiene, In: Patty's Industrial Hygiene, 5th Ed., Volume 1. R. Harris, Ed. 411–446 John Wiley & Sons, Inc. New York (2000).

45. Yost, M. G.; Gagdil, A. J.; Drescher, A. C.; *et al.*: Imaging Indoor Tracer Gas Concentration with Computed Tomography: Experimental Results with a Remote Sensing FTIR System. Am. Ind. Hyg. Assoc. J. 55:395–402 (1994).

46. –US Patent Application; Serial # 08/723,433; filed September 30th, 1996.

47. Hanst, P. L.; Wong, N. W.; Bragin, J.: A Long-Path Infrared Study of Los Angeles Smog. Atmos. Environ. 16:969–981 (1982).

48. Cone, A. L.; Farhat, S. K; Todd, L.: Development of QA/QC Performance Standards for Field Use of Open Path FTIR Spectrometers In: Proceedings of the SPIE International Symposium on Optical Sensing for Environmental and Process Monitoring. 334–338. Air and Waste Management Association, Pittsburgh, PA (1995).

49. Spartz, M. L.; Witkowski, M. R.; Fateley, J. H.; *et al.*: Evaluation of a Mobile FTIR System for Rapid VOC Determination. Part 1: Preliminary Qualitative/Quantitative Calibration Results. Am. Environ. Lab. :15–30 (1989).

50. Carter, R. E.; Thomas, M. J.; Marotz, G. A.; Lane, D. D.; Hudson, J.L.: Compound Detection Concentration Estimation by Open-Path Fourier Transform Infrared Spectroscopy Canisters Under Controlled Field Conditions. Environ. Sci. Technol. 26:2175–2181 (1992).

51. Spartz, M. L.; Witkowski, M. R.; Fateley, J. H.; *et al.*: Comparison of Long Path FTIR Data to Whole Air Canister Data from a Controlled Upwind Point Source. In: Proceedings, EPA/AWMA International Symposium on the Measurement of Toxic Related Air Pollutants. pp. 685–692. Air and Waste Management Association, Pittsburgh, PA (1990).

52. Hudson, J. L.; Thomas, M. J.; Arello, J.; *et al.*: An Overview Assessment of the Intercomparibility Performance of Multiple FTIR Systems as Applied to the Measurement of Air Toxics In: Proceedings of the SP-81: Optical Remote Sensing Applications to Environmental and Industrial Safety Problems, Houston, TX, pp. 112–122 Air and Waste Management Association, Pittsburgh, PA (1992).

53. Dodson, A.; Todd, L. A.: Chamber Validation of a Localized Open-Path FTIR Spectrometer. Presentation at the ACGIH 1998 Applied Workshop on Occupational and Environmental Exposure Assessment, Chapel Hill, NC (1998).

54. Haaland, D. M.; Easterling, R. G.: Application of New Least Squares Methods for the Quantitative Infrared Analysis of Multicomponent Samples Appl. Spectroscopy 36:665–673 (1982).

55. Haaland, D. M.; Easterling, R. G.: Improved Sensitivity of Infrared Spectroscopy by the Application of Least Squares Methods. Appl. Spectroscopy 34:539–548 (1980).

56. National Institiue of Standards and Technology: Standard Reference Database 79 v1.00. Quantitative Infrared Database. NIST, Gaithersburg, MD (1998).

57. Todd, L. A.; Ramachandran, G.: Evaluation of Algorithms for Tomographic Reconstruction of Chemicals in Indoor Air. Am. Ind. Hyg. Assoc. J. 55:403–417 (1994).

58. Todd, L. A.; Bhattacharyya, R.: Tomographic Reconstruction of Air Pollutants: Evaluation of Measurement Geometries. Appl. Optics 36:7678–7688 (1997).

59. Bhattacharyya, R.; Todd, L. A.: Spatial Temporal Visualization of Gases & Vapors in Air Using Computed Tomography: Numerical Studies. Ann. Occup. Hyg. 41:105–122 (1997).

60. Drescher, A. C.; Gadgil, A. J.; Price, P. N.; Nazaroff, W. W.: Novel Approach for Tomographic Reconstruction of Gas Concentration Distributions in Air: Use of Smooth Basis Functions and Simulated Annealing. Atmos. Environ. 30:929–940 (1996).

61. Todd, L. A.; Ramachandran, G.: Evaluation of Optical Source-Detector Configurations for Tomographic Reconstruction of

Concentrations in Indoor Air. Amer. Indus. Hyg. Assoc. J. 55:1133–1143 (1994).

62. Todd, L. A.; Yost, M. G.; Hashmonay, R. A : Trends and Future Applications of Optical Remote Sensing and Computed Tomography to Map Air Contaminants. Proceedings of SPIE Environmental Monitoring and Remediation Technologies Conference, Boston. Volume 3534. 399–404 (1998)

63. Hounsfield, G. N.: Computerized Transverse Axial Scanning (Tomography): Part I. Description of a System. Br. J. Radiol. 46:1016–1022 (1973).

64. Cormack, A. M.: Representation of a Function by Its Line Integrals, with Some Radiological Applications. II. J. Appl. Phys. 35:2908–2913 (1964).

65. DeRosier, D. J.; Klug, A.: Reconstruction of Three-Dimensional Structures from Electron Micrographs. Nature 217:130–134 (1968).

66. Wolfe, D. C.; Byer, R.: Model Studies of Laser Absorption Computed Tomography for Remote Air Pollution Measurement. Appl. Optics 21:1165–1177 (1982).

67. Piper, A. R.; Todd, L. A.; Mottus, K.: A Field Study Using Open-path FTIR Spectroscopy to Measure and Map Air Emissions from Volume Sources. Field Anal. Chem. Technol. 3:69–79 (1999).

68. Todd, L. A.: Mapping the Air in Real-Time to Visualize the Flow of Gases and Vapors: Occupational and Environmental Applications. Appl. Occup. Environ. Hyg. Vol 15(1) 106-113 Jan. 2000.

69. Todd, L. A.; Ramanathan, M.; Mottus, K.: Measuring Chemical Emissions Using Environmental CAT Scanning. Presentation at the Workshop on Atmospheric Nitrogen Compounds II: Emissions, Transport, Transformation, Deposition and Assessment. Chapel Hill, NC (1999).

Instrument Descriptions

The instruments described here include: field-ready extractive FTIR spectrometers, compact extractive FTIR spectrometers that might be suitable for the field, and open-path FTIR spectrometers. A number of laboratory bench-top FTIR spectrometer models are mentioned by name.

1. Field-ready Extractive FTIR Spectrometers

23-1-1. IFS 120M

Bruker Analytik GmbH

The IFS 120M has been specifically designed for stratospheric analysis, atmospheric emission studies, and remote monitoring of atmospheric pollutants. The 120M requires power and liquid nitrogen. It uses Bruker's OPUS™ software which is a multi-tasking environment; it allows manipulation and plotting of previously acquired data while the spectrometer simultaneously acquires new spectra. Interchangeable scanners are available for resolutions of 0.008 cm⁻¹ or 0.0035 cm⁻¹. Simultaneous data collection from MCT and InSb detectors are available. The instrument is telescope-ready and sun trackers or other instrument mounts are available.

FIGURE 23-1-1. IFS 120M.

FIGURE 23-2-1. The MIDAC Illuminator.

23-1-2. M Series Spectrometers
MIDAC Corporation

M Series spectrometers provide a range of resolutions, from 2 cm^{-1} for routine analysis to 0.5 cm^{-1} for high resolution (all models are step selectable down to 32 cm^{-1}). Available detectors include: air-cooled DTGS and liquid nitrogen-cooled MCT or InSb. Switching detectors is done by turning a key on dual detector systems.

23-1-3. The Prospect-IRTM
MIDAC Corporation

The Prospect-IRTM is designed for routine sampling, dedicated analysis, and educational applications. It is designed to be rugged, with a cast aluminum enclosure. It is designed to be resistant to low frequency vibration and can be operated in any orientation. It provides a range of resolutions, from 2 cm^{-1} for routine analysis to 0.5 cm^{-1} for high resolution (all models are step selectable down to 32 cm^{-1}). Available detectors include: air cooled DTGS and liquid nitrogen cooled MCT or InSb.

2. Open-Path FTIR Spectrometers

23-2-1. The MIDAC Illuminator
MIDAC Corporation

The MIDAC Illuminator is a rugged, compact, and high performance FTIR module that that can interface with many configurations (bistatic, monostatic, telescope). The illuminator has a high energy, air-cooled IR source and the patented MIDAC interferometer. This instrument is resistant to low frequency vibration interference and is able to operate reliably in virtually any orientation. The Illuminator has a resolution of 0.5 cm^{-1}.

23-2-2. Model 100 FTIR Spectrometers
BLOCK Engineering

Block Engineering Model 100 FTIR Spectrometer provides remote passive data collection on stationary or moving air and ground platforms, such as helicopters or reconnaissance aircraft. These sensors are small, compact, rugged instruments. The Model 100 can be optimized to collect data in either the 1.5 to 5.5 μm or the 7 to 14 μm spectral band. MCT or InSb detectors are available cooled to 80K by either liquid nitrogen or a Stirling cooler. Spectral resolutions are operator selectable from 4 cm^{-1} to 16 cm^{-1}, with scan rates ranging from 5 to over 40 scans per second.

23-2-3. Model 195-P
BLOCK Engineering

The Block Engineering Model 195-P FTIR is a rugged, portable, battery-powered spectrometer that is capable of hand-held, tripod-mounted, or vehicle-mounted high-resolution measurements in the 2–5 μm band. An integrated TV shows, and synchronously documents the target whose spectrum is measured. Voice annotation and housekeeping display features provide

FIGURE 23-1-2. M Series Spectrometers.

situational data. All collected information is recorded on standard VHS format for subsequent playback and data reduction. The instrument has a spectral resolution of 4 cm^{-1} and can operate for more than 2 hours of continuous spectra/image collection with each battery charge.

23-2-4. Model 160

BLOCK Engineering

The Block Engineering Airborne Pod-Mounted FTIR Spectrometer Systems are designed for airborne reconnaissance pods on high-speed aircraft. The systems can obtain spectral data in the 0.2 to 12 μm region from a variety of targets. The pod-mounted instrumentation records the IR emission of the target and the background. IR data analysis systems are available for airborne and ground-based data reduction and processing. This instrument is available in shortwave infrared, longwave infrared, and broadband configurations with a selectable field of view, spectral rate, and spectral resolution.

23-2-5. Model 500

BLOCK Engineering

Block Engineering's Model 500 FTIR spectrometers provide for remote passive data collection. These sensors are man-portable self-contained instruments including on-board data collection and data storage. The Model 500 can simultaneously collect FTIR, FLIR, video, GPS, and meteorological information. Power must be provided from a remote power supply or an optional battery pack. The Model 500 can be optimized to collect spectroradiometric data in either the 1.5 to 5.5 μm or the 7 to 14 μm spectral band. MCT or InSb detectors are available cooled to 80K by either liquid nitrogen or an integral Stirling cooler. Spectral resolutions are selectable from 2 cm^{-1} to 16 cm^{-1} with scan rates ranging from 2 to 50 scans per second.

3. Mobile, Localized Open-Path FTIR Spectrometers

23-3-1. Aries Roaming Air Monitoring System

MIDAC Corporation

This instrument uses the Illumintor in a system that combines an FTIR with a very long path gas cell (100 meters). The cell has removable covers so the gas cell can be opened to monitor ambient air conditions at breathing height. The Illuminator has a resolution of 0.5 cm^{-1}.

FIGURE 23-3-1. Aries Roaming Air Monitoring System.

4. Benchtop FTIR Spectrometer Models

23-4-1. Avatar 360 E.S.P.

Nicolet

23-4-2. Vector 22

Bruker Analytik

23-4-3. Spectrum RX FT-IR Spectrometer

Perkin-Elmer

23-4-4. Satellite Series FTIR

Mattson

FIGURE 23-4-1. Avatar 360 E.S.P.

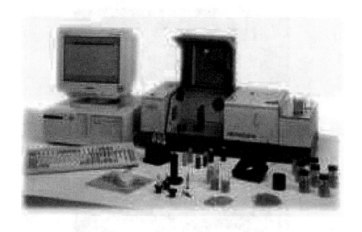

FIGURE 23-4-2. Vector 22.

FIGURE 23-4-4. Satellite Series FTIR.

TABLE 23-I-1.   List of Manufacturers

BLE	Block Engineering 164 Locke Drive Marlborough, MA 01752-1178 (508) 480-9643 FAX: (508) 480-9226 *www.blockeng.com*	MAT	Mattson Instruments 5225 Verona Road Madison, WI 53711-4495 (800) 423-6641 or (608) 276-6300 FAX: 608-273-6818 *info@mattsonir.com* *www.mattsonir.com*	NIC	Nicolet Instrument Corporation 5225 Verona Road Madison, WI 53711 (800) 642-6538 FAX: (608) 273-5046 *nicinfo@nicolet.com* *www.nicolet.com*
BRA	Bruker Analytik GmbH Division IX, EPR/minispec Silbersteifen D-76287 Rheinstetten/Karlsruhe Germany ++49-721-5161-141 FAX: ++49-721-5161-237 *epr@bruker.de* *www.bruker.de/analytic/* *analytic.htm*	MID	MIDAC Corporation 17911 Fitch Avenue Irvine, CA 92614 (949) 660-8558 FAX: (949) 660-9334 *info@midac.com* *www.midac.com*	PEI	PerkinElmer Instruments 761 Main Avenue Norwalk, CT 06859-0001 (203) 762-4003 or (800) 762-4003 FAX: (203) 762-4054 *info@perkin-elmer.com* *www.perkin-elmer.com*

Index

Note: This index was compiled by the authors for their individual chapters.